T0189049

Lecture Notes in Artificial Intelligence 9979

Subseries of Lecture Notes in Computer Science

Editors
Arvin Agah
Department of Electrical Engineering
  and Computer Science
The University of Kansas
Lawrence, KS
USA

Miguel A. Salichs
Department of Systems Engineering
  and Automation
University Carlos III de Madrid
Madrid
Spain

John-John Cabibihan
Department of Mechanical and Industrial
  Engineering
Qatar University
Doha
Qatar

Hongsheng He
Department of Mechanical, Aerospace
  and Biomedical Engineering
University of Tennessee
Knoxville, TN
USA

Ayanna M. Howard
School of Electrical and Computer
  Engineering
Georgia Institute of Technology
Atlanta
USA

ISSN 0302-9743   ISSN 1611-3349   (electronic)
Lecture Notes in Artificial Intelligence
ISBN 978-3-319-47436-6   ISBN 978-3-319-47437-3   (eBook)
DOI 10.1007/978-3-319-47437-3

Library of Congress Control Number: 2016954466

LNCS Sublibrary: SL7 – Artificial Intelligence

Printed on acid-free paper

This Springer imprint is published by Springer Nature
The registered company is Springer International Publishing AG
The registered company address is: Gewerbestrasse 11, 6330 Cham, Switzerland

Preface

The 8th International Conference on Social Robotics (ICSR 2016) was held in Kansas City, USA, in November 2016—its first year in the USA. This book contains the proceedings of the conference, comprising the 98 refereed papers, reviewed by the international Program Committee, and presented during the technical sessions of the conference.

The International Conference on Social Robotics brings together researchers and practitioners working on the interaction between humans and robots and on the integration of robots into our society. The theme of the 2016 conference was "Socio-robotics: Design and Implementation of Social Behaviors of Robots Interacting with Each Other and with Humans."

Now in its eighth year, the International Conference on Social Robotics is the leading international forum for researchers in social robotics. The conference provides researchers and practitioners the opportunity to present and engage in dialog on the latest progress in the field of social robotics. Social robots will improve quality of human life through assistance, enabling, for instance, independent living or providing support in work-intensive, difficult, and possibly complex situations. The conference aims to foster discussion on the development of computational models, robotic embodiments, and behavior that enable social robots to have an impact on the degree of personalized companionship with humans.

In addition to technical sessions, ICSR 2016 included three workshops: The Synthetic Method in Social Robotics (SMSR 2016); Social Robots: A Tool to Advance Interventions for Autism; and Using Social Robots to Improve the Quality of Life in the Elderly. ICSR 2016 had two world-renowned researchers in social robotics as keynote speakers: Prof. Maja Matarić, Chan Soon-Shiong Professor of Computer Science, Neuroscience, and Pediatrics at the University of Southern California, and Prof. Brian Scassellati, Professor of Computer Science, Cognitive Science, and Mechanical Engineering at Yale University.

The conference venue, Kansas City Country Club Plaza, provided the participants with the opportunity to experience Kansas City's jazz, barbecue, and fountains. The 15-block district is an excellent destination for shopping, dining, and entertainment.

We would like to express our appreciation to the Organizing Committee for putting together an excellent program, to the international Program Committee for their rigorous review of the papers, to KU Professional & Continuing Education for organizing the event, to Kansas City Marriott Country Club Plaza for hosting the event, to our generous sponsors SoftBank Robotics, University of Kansas School of Engineering, and Springer, and most importantly to the authors and participants who greatly enhanced the quality and effectiveness of the conference through their papers, presentations, and conversations.

We are hopeful that this conference will generate many future collaborations and research endeavors, resulting in enhancing human lives through the utilization of social robots.

September 2016

Arvin Agah
John-John Cabibihan
Ayanna M. Howard
Miguel A. Salichs
Hongsheng He

Organization

General Chair

Arvin Agah — University of Kansas, USA

Program Chairs

John-John Cabibihan — Qatar University, Qatar
Ayanna M. Howard — Georgia Institute of Technology, USA
Miguel A. Salichs — University of Carlos III, Spain

Publication Chair

Hongsheng He — University of Tennessee, USA

Workshop Chairs

Hae Won Park — Massachusetts Institute of Technology, USA
Maryam Mahani — Ricoh Americas Corporation, USA

Special Session Chairs

Chung Hyuk Park — George Washington University, USA
David Harvie — United States Military Academy, USA
Agnieszka Wykowska — Luleå University, Sweden

Competition Chairs

Amit Kumar Pandey — SoftBank Robotics, France
Andrew Williams — Marquette University, USA

Publicity Chairs

Alan R. Wagner — Pennsylvania State University, USA
Zhengchen Zhang — Institute for Infocomm Research, Singapore
Gabriele Trovato — Waseda University, Japan

Standing Committee

Shuzhi Sam Ge — National University of Singapore, Singapore
Oussama Khatib — Stanford University, USA

Maja Matarić	University of Southern California, USA
Haizhou Li	A*Star, Singapore
Jong Hwan Kim	Korea Advanced Institute of Science and Technology, Korea
Paolo Dario	Scuola Superiore Sant'Anna, Italy
Ronald C. Arkin	Georgia Institute of Technology, USA

Program Committee Members

Roxana Agrigoroaie	ENSTA-ParisTech, France
Muneeb Imtiaz Ahmad	MARCS Institute, Australia
Rachid Alami	LAAS-CNRS, France
Minoo Alemi	Islamic Azad University, Iran
Matías Alvarado	Center of Research and Advanced Studies, CINVESTAV-IPN, Mexico
Heather Amthauer	University of Wisconsin, Eau Claire, USA
Víctor H. Andaluz	Universidad de las Fuerzas Armadas ESPE, Ecuador
Laura Aymerich Franch	EventLab, University of Barcelona, Spain
Santosh Balajee Banisetty	University of Nevada Reno, USA
Jasmin Bernotat	CITEC - Universität Bielefeld, Germany
Gerard Canal	Institut de Robòtica i Informàtica Industrial, CSIC-UPC, Spain
José Carlos Castillo	University Carlos III of Madrid, Spain
Ryad Chellali	Nanjing Tech University, China
Ben Chen	Wuhan University, China
Michael Jae-Yoon Chung	University of Washington, USA
Matthieu Courgeon	Lab-STICC, France
Brian Cruz	Korea Institute of Industrial Technology, South Korea
Arturo Cruz-Maya	ENSTA-ParisTech, France
Luisa Damiano	University of Messina, Italy
Timo Dankert	Bielefeld University, Germany
Friederike Eyssel	University of Bielefeld, Germany
Pooyan Fazli	Cleveland State University, USA
David Feil-Seifer	University of Nevada, Reno, USA
François Ferland	ENSTA-ParisTech, France
Francesco Ferrari	CITEC - Universität Bielefeld, Germany
Naomi Fitter	University of Pennsylvania, USA
Mary Ellen Foster	University of Glasgow, UK
Marlena Fraune	Indiana University, USA
Allison Funkhouser	Carnegie Mellon University, USA
Mehdi Ghayoumi	Kent State University, USA
Victor Gonzalez-Pacheco	Universidad Carlos III de Madrid, Spain
Goren Gordon	Tel Aviv University, Israel
Elena Corina Grigore	Yale University, USA
Cindy Grimm	Oregon State University, USA
Daniel Grollman	Sphero

Horst-Michael Gross	Ilmenau University of Technology, Germany
Zhang Haojie	China Noveri Vehicle Research Institute, China
Kerstin Sophie Haring	The University of Tokyo, Japan
David Harvie	United States Military Academy, USA
Mojgan Hashemian	INESC-ID, Portugal
Wei He	University of Science and Technology, China
Hisashi Ishihara	Osaka University, Japan
Serena Ivaldi	Inria, France
Wafa Johal	École Polytechnique Fédérale de Lausanne, Switzerland
Martin Johansson	KTH, Sweden
Benjamin Johnston	University of Technology Sydney, Australia
James Kennedy	Plymouth University, UK
Youssef Khaoula	Toyohashi University of Technology, Japan
Abderrahmane Kheddar	CNRS-UM LIRMM IDH, France
Katherine Kuchenbecker	University of Pennsylvania, USA
Hee Rin Lee	Indiana University, USA
Hagen Lehmann	Istituto Italiano di Tecnologia, Italy
Yan Li	University of Tennessee, USA
Yanan Li	Imperial College London, UK
Dayao Liang	Tsinghua University, China
Chao Luo	Tsinghua University, China
Gergely Magyar	Technical University of Kosice, Slovakia
Maria Malfaz	Universidad Carlos III de Madrid, Spain
Ali Meghdari	Sharif University of Technology, Iran
Isabelle Menne	University of Wuerzburg, Germany
Byung-Cheol Min	Purdue University, USA
Azadeh Mohebi	Iranian Research Institute for Information Science and Technology, Iran
Omar Mubin	Western Sydney University, Australia
Michael Novitzky	Massachusetts Institute of Technology, USA
Mohammad Obaid	Doc University, Turkey
Benjamin Oistad	Indiana University, USA
Suman Ojha	University of Technology Sydney, Australia
Billy Okal	University of Freiburg, Germany
Maike Paetzel	Uppsala University, Sweden
Amit Kumar Pandey	SoftBank Robotics, France
Raul Paradeda	Instituto Superior Técnico, Portugal
Damien Pellier	Université Grenoble Alpes, France
Noé Pérez-Higueras	Pablo de Olavide University, Spain
Beatriz Quintino Ferreira	Universidade de Lisboa, Portugal
Rafael Ramón-Vigo	Pablo de Olavide University, Spain
Syed Ali Raza	University of Technology, Sydney, Australia
Tiago Ribeiro	INESC-ID and Instituto Superior Técnico - Universidade de Lisboa, Portugal
Francesco Riccio	Sapienza University of Rome, Italy

Contents

Learning Robot Navigation Behaviors
by Demonstration Using a RRT* Planner

Noé Pérez-Higueras[1]([✉]), Fernando Caballero[2], and Luis Merino[1]

[1] School of Engineering, Universidad Pablo de Olavide,
Crta. Utrera km 1, Seville, Spain
{noeperez,lmercab}@upo.es
[2] Department of System Engineering and Automation,
University of Seville, Camino de los Descubrimientos, s/n, Seville, Spain
fcaballero@us.es

Abstract. This paper presents an approach for learning robot navigation behaviors from demonstration using Optimal Rapidly-exploring Random Trees (RRT*) as main planner. A new learning algorithm combining both Inverse Reinforcement Learning (IRL) and RRT* is developed in order to learn the RRT*'s cost function from demonstrations. This cost function can be used later in a regular RRT* for robot planning including the learned behaviors in different scenarios. Simulations show how the method is able to recover the behavior from the demonstrations.

1 Introduction

Today, more and more mobile robots are coexisting with us in our daily lives. As a result, the creation of motion plans for robots that share space with humans in dynamic environments is a subject of intense investigation in robotics. Robots must respect human social conventions, guarantee the comfort of surrounding persons, and maintain legibility, so humans can understand the robot's intentions [10]. This is called human-aware navigation.

This problem was initially tackled by including costs and constraints related to human-awareness into motion planners to obtain socially acceptable paths [7,18]. In these cases, these costs are pre-programmed. However, hard-coded social behaviors might be inappropriate [3]. In many cases (for instance [8,15]), these costs are grounded in Proxemics theory [5]. However, as shown in [13], Proxemics is focused on people interaction, and it could not be suitable for navigating among people.

Therefore, learning these social behaviors from data seems a more principled approach. Also, it is easier to demonstrate socially acceptable behaviors than mathematically defining them. In particular, we consider in this paper the

This work is partially supported by the EC-FP7 under grant agreement no. 611153 (TERESA) and the project PAIS-MultiRobot funded by the Junta de Andalucía (TIC-7390).

A. Agah et al. (Eds.): ICSR 2016, LNAI 9979, pp. 1–10, 2016.
DOI: 10.1007/978-3-319-47437-3_1

application of telepresence robots [17]. Our goal is to increase the autonomy of such robots, freeing the users from the low level navigation tasks. In this setup, it is very natural to obtain navigation data from the user, considering them as examples and using them to learn social navigation behaviors.

Thus, we aim to develop an approach to learn navigation behaviors from user data. The paper makes use of Inverse Reinforcement Learning (IRL) concepts and sampling-based planners (in particular, RRT* [6]) to identify the RRT cost function that better fit the example trajectories. Differently to classic IRL approaches based on Markov Decision Process (MDPs), the presented method is computationally faster and scales very well with the state size, being able to deal with continuous state and control spaces, and it is general enough to be applied in different scenarios.

The paper is structured as follows. After a summary of related work, next section describes the algorithm for learning from demonstrations using RRT*. Then, Sect. 3 describes the particular problem of social navigation considered in the paper. Later on, Sect. 4 validates the approximation in simulation. Finally, Sect. 5 summarizes the paper contribution and outlooks future work.

1.1 Related Work

In the last years, several contributions have been presented regarding the application of learning the task of human-aware navigation. Supervised learning is used in [19] to learn appropriate human motion prediction models that take into account human-robot interaction when navigating in crowded scenarios. In [4], the parameters of a model based on social forces are learnt from feedback provided by users.

An additional approach is learning from demonstrations [2]: an expert indicates the robot how it should navigate among humans. This approach is particularly relevant for the case of telepresence robots. One way to learn from demonstrations is through Inverse Reinforcement Learning (IRL) [1]. The observations of an expert demonstrating the task are used to recover the reward (or cost) function the demonstrator was attempting to maximize (minimize). Then, the reward can be used to obtain a corresponding robot policy.

Different aspects to tackle the IRL problem have been proposed. A probabilistic method based on the principle of maximum entropy is presented in [20]. The computational cost problem is managed in [14] by using a Bayesian nonparametric mixture model to divide the observations and obtain a group of simpler reward functions. From another point of view, the authors in [12] represent the reward by using Gaussian processes instead of a linear combination of features.

In the above mentioned models, the IRL technique makes use of discrete Markov Decision Processes (MDPs) as the underlying process. However, it is complex to encode general problems with MDPs due to its computational complexity. Many authors turn to state discretization which can be tricky in many cases.

Optimal Rapidly-exploring Random Trees (RRT*) [6] are extensively employed in robot planning. They are flexible and easily adapted to different scenarios and problems. They implicitly reason about collisions with obstacles at moderate computational cost even in high dimensionality. They can explore the state space to obtain optimal paths on cost spaces and the kinodynamic extension allows reasoning about the robot dynamics.

In the paper, we present an algorithm for learning robot navigation behaviors from demonstrations using RRT* as main planner. We aim at creating a new learning algorithm combining both IRL and RRT* techniques in order to extract the proper weights of the cost function from demonstration trajectories. This cost function can be used later in a regular RRT* to allow the robot reproducing the desired behavior at different scenarios.

2   Learning a RRT* Cost Function

RRT* [6] is a technique for optimal motion planning. It considers that a cost function is associated to each point x in the configuration space. The RRT* seeks to obtain the trajectory ζ^* that minimizes the total cost along the path $c(\zeta)$. It does so by randomly sampling the configuration space and creating a tree towards the goal. The paths are then represented by a set of discrete configuration points $\zeta = \{x_1, x_2, \ldots, x_N\}$.

Without loss of generality, we can assume that the cost function for each point can be expressed as a linear combination of a set of sub-cost functions, that will be called features $c(x) = \sum_j \omega_j f_j(x) = \omega^T f(x)$. The cost of a path is then the sum of the cost for all points in the path. Particularly, in the RRT*, the cost is the sum of the sub-costs of moving between pairs of points in the path:

$$c(\zeta) = \sum_{i=1}^{N-1} c(x_i, x_{i+1}) = \sum_{i=1}^{N-1} \frac{c(x_i) + c(x_{i+1})}{2} \|x_{i+1} - x_i\| \tag{1}$$

$$= \omega^T \sum_{i=1}^{N-1} \frac{f(x_i) + f(x_{i+1})}{2} \|x_{i+1} - x_i\| = \omega^T f(\zeta) \tag{2}$$

Thus, for given weights ω, the algorithm will return trajectories that try to minimize this cost.

Given a set of demonstration trajectories $\mathcal{D} = \{\zeta_1, \zeta_2, \ldots, \zeta_D\}$, the problem of learning from demonstrations, in this setup, means to determine the weights ω that lead our planner to behave similarly to these demonstrations. According to [1,11], this similarity is achieved when the expected value of the features for the trajectories generated by the planner is the same as the expected value of the features for the given demonstrated trajectories:

$$\mathbb{E}(f(\zeta)) = \frac{1}{D} \sum_{i=1}^{D} f(\zeta_i) \tag{3}$$

One approach to solve this problem is to model the underlying trajectory distribution of the expert and consider the demonstrations as samples from this distribution. As noted in [9], applying the Maximum Entropy Principle [20] to the IRL problem leads to the following form for the probability density for the trajectories returned by the demonstrator:

$$p(\zeta|\omega) = \frac{1}{Z(\omega)} e^{-\omega^T f(\zeta)} \tag{4}$$

where $Z(\omega)$ is a normalization function that does not depend on ζ. One way to determine ω is maximizing the (log-)likelihood of the demonstrated trajectories under the previous model:

$$\mathcal{L}(\mathcal{D}|\omega) = -D\log(Z(\omega)) + \sum_{i=1}^{D}(-\omega^T f(\zeta_i)) \tag{5}$$

The gradient of the previous log-likelihood with respect to ω is given by:

$$\nabla\mathcal{L} = \frac{\partial\mathcal{L}(\mathcal{D}|\omega)}{\partial\omega} = \mathbb{E}(f(\zeta)) - \frac{1}{D}\sum_{i=1}^{D} f(\zeta_i) \tag{6}$$

Setting this gradient to zero one arrives to (3). As mentioned in [11], this gradient can be intuitively explained. If the value of one of the features for the trajectories returned by the planner are higher from the value in the demonstrated trajectories, the corresponding weight should be increased to increase the cost of those trajectories.

The main problem with the computation of the previous gradient is that it requires to compute the expected value of the features $\mathbb{E}(f(\zeta))$ for the generative distribution (4). In [9], a probabilistic generative model for trajectories is derived from data, and the expectation is computed by Monte Carlo Chain sampling methods, which is very computationally demanding.

In our case, we will approximate the expert by the RRT* planner on board the robot. Being an asymptotically optimal planner, for some given weights, the RRT* will provide the trajectory that minimizes this cost given infinite time. As the planning time is limited, the RRT* will provide trajectories with some variability on the features, and thus the expected feature count is computed by running several times the planner between the start and goal configurations. This is then used to compute the gradient and adapt the weights used in the RRT* planner.

As mentioned, this is an approximation to the expert model. A similar idea is used in [11]. The experimental results will show that the method is able to recover the taught behaviors by the expert.

The method proposed is described in detail in Algorithm 1. The example trajectories from which we want to learn are used as input. The output of the method are the weights for the cost function of the RRT* algorithm in (1).

First, in Line 1 we obtain the average feature count $\overline{f}_{\mathcal{D}} = \frac{1}{D}\sum_{i=1}^{D} f(\zeta_i)$ from the example trajectories in $\mathcal{D}$ scenarios. The feature counts are obtained as the addition of the feature values of pairs of nodes of the trajectory evaluated similarly to Eq. (1).

Algorithm 1. RRT*-IRL

Require: Trajectory examples $\mathcal{D} = \{\zeta_1^1, \ldots, \zeta_D^S\}$ in $\mathcal{S}$ scenarios
Ensure: Function features weights $\omega = [\omega_1, \ldots, \omega_J]^T$
1: $\overline{f}_{\mathcal{D}} \leftarrow calculeAvgFeatureCounts(\mathcal{D})$
2: $\omega \leftarrow randomInit()$
3: **repeat**
4: **for each** $s \in \mathcal{S}$ **do**
5:   **for** $rrt\_repetitions$ **do**
6:    $\zeta_i \leftarrow getRRTstarPath(s, \omega)$
7:    $f(\zeta_i) \leftarrow calculeFeatureCounts(\zeta_i)$
8:   **end for**
9:   $\overline{f}_{RRT^*}^s \leftarrow (\sum_{i=1}^{rrt\_repetitions} f(\zeta_i))/rrt\_repetitions$
10: **end for**
11: $\overline{f}_{RRT^*} \leftarrow (\sum_{i=1}^{S} \overline{f}_{RRT^*}^i)/S$
12: $\nabla\mathcal{L} \leftarrow (\overline{f}_{RRT^*} - \overline{f}_{\mathcal{D}})$
13: $\omega \leftarrow UpdateWeights(\nabla\mathcal{L})$
14: **until** convergence
15: **return** ω

Then we initialize the weights with an unsigned integer random value (Line 2). It is noteworthy that the weights are not being normalized during the learning iterations, so that changes in the value of one weight do not provoke the variation of the values of the rest of weights. They are normalized after the learning has finished. On the other hand, the features values for each node are normalized but this is not a requirement of the algorithm.

The key point is the gradient given by (6), which requires a comparison of the features counts obtained from the example trajectories and the expected value from the RRT* planner. The latter is obtained by running $rrt\_repetitions$ times the planner for the current weight values for each scenario considered (Line 6) and obtaining and normalizing the features counts (Line 7). In Lines 9 and 11 the averaged values are obtained.

Based on this comparison the weights of the cost function are updated using exponentiated gradient descent (line 13), as in [20]:

$$\omega_i \leftarrow \omega_i * e^{(\lambda/\phi)*\nabla\mathcal{L}_i} \tag{7}$$

where ϕ is the number of the current iteration of the algorithm, λ is an adjusting factor of the equation and $\nabla\mathcal{L}_i = \frac{\partial\mathcal{L}(\mathcal{D}|\omega)}{\partial\omega_i}$ is the i-th component of the gradient.

Finally, the learning process finishes when the variations of the weight values keep under a certain convergence value ϵ.

3   Cost Function for Social Navigation

The social navigation task considered here involves the robot navigation in different house environments like rooms and corridors where some persons stand in different positions so that the robot has to avoid them to reach the goal.

A small set of well-known features have been considered here. They are the distance to the goal, the distance from the robot to the people in the scene, the angle of the robot position with respect to the people α, and the distance to the closest obstacle, as depicted in Fig. 1a. Notice that this paper focuses on the use of the features for social robot navigation, but we will not get into details about the nature and importance of the features themselves.

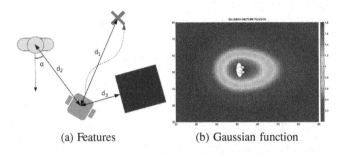

(a) Features   (b) Gaussian function

Fig. 1. (a) Features employed in the social cost function learned. $d1$, distance to the goal. d_2, distance from the people to the robot. α, angle between the person front and the robot location. d_3, distance to the closest obstacle. (b) Gaussian mixture function deployed over each person. The lateral bar shows the costs based on the color displayed. (Color figure online)

Thus, three feature functions are combined to obtain the cost function employed in the RRT* planner. The function are computed for each sample x_k of the configuration space. The first one is just the Euclidean distance from the robot position to the goal:

$$f_1(x_k) = \|x_k, x_{goal}\| \qquad (8)$$

The second feature function represents a proxemics cost with respect to the persons in the environment, and follows the model used by Kirby et al. [8]. This cost function is defined by a mixture of Gaussian functions, and its shape can be seen in Fig. 1. This cost function p depends on the distance (d_{jk}) and relative angle (α_{jk}) of the robot position x_k with respect to each person j in the scenario. The cost due to all persons in the scenario is integrated according to the next expression, where P is the total number of persons:

$$f_2(x_k) = \prod_{j=1}^{P} (p(d_{jk}, \alpha_{jk}) + 1) - 1 \qquad (9)$$

Figure 1b shows this cost function for one person, which is implemented as a mixture of two Gaussian functions: the first function is asymmetric and placed in the front of the person with $\sigma_h = 1.20$ m the variance in the direction the person is facing, and a smaller variance in the sides $\sigma_s = \sigma_h/1.5$. The second Gaussian is placed in the back of the person with $\sigma_h = \sigma_s = 0.8$.

The third feature function uses the distance to the closest obstacle for each node x_k, $d_{jobs}(x_k)$ with the aim of motivating the robot to keep some distance from the obstacles. This cost is based on the costmap used by the navigation system of ROS [16], in which each obstacle has a defined inflation area around. This way, the cost is zero if the robot is far enough from any obstacle ($\delta = 2$ meters in our case):

$$f_3(x_k) = \begin{cases} 0, & \text{if } d_{obs}(x_k) > \delta, \\ (254 - 1)e^{(-\beta(d_{obs}(x_k)-r))}, & \text{otherwise} \end{cases} \qquad (10)$$

where r is the inscribed radius of the robot and $\beta = 3$ in our implementation.

The values of the n (3 in this case) feature functions are normalised and the cost function for each node x_k is built adding its weighted values $c(x_k) = \sum_{i=1}^{n} \omega_i f_i(x_k)$ where $\omega_i \in [0, 1]$ and $\sum_i \omega_i = 1$.

Finally, the total cost along the Q nodes of the path ζ is obtained based on the *motion-cost* function employed by the RRT* algorithm to calculate the cost of moving from one node to the next one according to Eq. (1).

4   Experimental Results

A set of experiments have been performed to evaluate whether the algorithm is able to recover the characteristics of the taught trajectories. All the presented experiments were performed by using a library of RRT algorithms developed by the authors for research purposes. The library is available in the Github of the Service Robotics Lab[1] under BSD license. The hardware employed was an $i7$ processor 3770 with 12 GB DDR3 memory, where the planner was allowed to plan a path for 2 s.

In these experiments, we use the RRT* with a set of known weights in the cost function to generate the example trajectories in a set of scenarios as ground truth. Then, we use these trajectories to learn the cost function with the proposed algorithm in the same configurations. Particularly, we employed 25 different configurations (different initial robot position, goal position and different number of persons and positions) in different parts or rooms in a house map. Moreover, for each configuration, 25 RRT* example trajectories were recorded. To validate the approximation, the trajectories from 15 of these configurations were used to learn the weights of the cost function, and the 10 remaining configurations were employed to compare the resulting paths. Figure 2 shows 3 of the configurations used in the validation.

Figure 3 shows the evolution of the normalised weights values, feature counts and gradients along the iterations of the learning algorithm. As can be seen, the weights converge to values close to the ground-truth ones in few iterations committing a final error around the 16 %. The difference in the feature counts expectations, which is the optimization objective in (3), quickly approaches zero,

---

[1] https://github.com/robotics-upo.

Fig. 2. Some of the scenarios employed in the cross-validation process. A coloured costmap based on the RRT* function cost is also shown. (Color figure online)

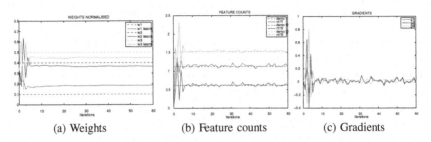

(a) Weights   (b) Feature counts   (c) Gradients

Fig. 3. (a) Evolution of the weights during the learning iterations. (b) Evolution of feature counts. (c) Gradients.

and so the gradients and the weights are stabilized. The relative error committed in the weights learnt respect to the ground-truth weights is calculated as $RE_\omega = \|\bar{\omega}_\mathcal{D} - \bar{\omega}_{RRT^*}\| / \|\bar{\omega}_\mathcal{D}\| = 0.1620$.

Once the learning has finished, we can compare the demonstration paths and the RRT* paths using the weights learnt in the remaining configurations for cross-validation. A qualitative comparison of the paths can be seen in the Fig. 4 for four of these trajectories (the rest are omitted for the sake of brevity). It can be seen that the behavior is very well reproduced in all the cases.

We can also compare the costs of the demonstration paths and the learnt RRT* paths. Figure 5 shows the averaged relative errors in the costs and in the feature counts. The error in feature counts is under the 8 % in all the cases. On the other hand, the error in the costs is even lower being all the cases under the 4 %. Moreover,

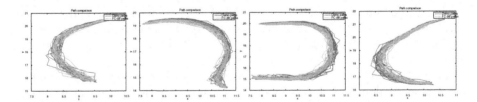

Fig. 4. Visual comparison of the demonstration paths (red lines) and the RRT paths obtained by using the weights learnt (blue lines). Trajectories 1, 5, 7 and 10 are presented from left to right. (Color figure online)

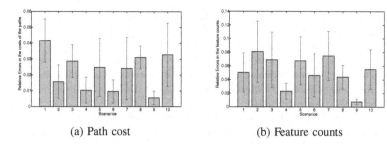

(a) Path cost   (b) Feature counts

Fig. 5. (a) Relative errors in the costs of the demonstration paths and the RRT* paths using the learnt weights. (b) Relative errors in the feature counts.

it can be also seen in Fig. 4 that some of the trajectories with larger cost error (1 and 10) reproduce very well the demonstrations.

5 Conclusions and Future Work

This paper presented an approach for teaching a robot behaviors based on demonstrations. To this end, a method based on IRL basis has been implemented and linked with a regular RRT* in order to learn the weights of its cost function, so the planner behaves similarly to the demonstrated behaviors. The method is simple to implement and allows to overcome the classic problems associated to IRLs based on MDPs. The proposed method is significantly less computational demanding than MDPs and simplify the generalization of the behavior thanks to the intrinsic benefits of RRTs.

The approach has been tested in simulation where the resulted learned cost function was able to properly imitate the desired behavior in the most of the cases. The feature counts always converged to values very close to the demonstrated values in the experiments, and the computed weights also allowed to reproduce the desired behavior reliably.

Future work will consider including the kinodynamic of the robot in the RRT* planner and the use of a richer set of features also employing the velocities of the robot and persons. Furthermore, a further mathematical analysis of the distributions of path costs followed by the RRT* planner as well as other optimization techniques to solve the problem will be considered.

References

1. Abbeel, P., Ng, A.Y.: Apprenticeship learning via inverse reinforcement learning. In: Proceedings of the Twenty-First International Conference on Machine Learning (ICML 2004), NY, USA, p. 1 (2004). http://doi.acm.org/10.1145/1015330.1015430
2. Argali, B., Chernova, S., Veloso, M., Browning, B.: A survey of robot learning from demonstrations. Robot. Auton. Syst. **57**, 469–483 (2009)

3. Feil-Seifer, D., Mataric, M.: People-aware navigation for goal-oriented behavior involving a human partner. In: Proceedings of the IEEE International Conference on Development and Learning (ICDL) (2011)
4. Ferrer, G., Garrell, A., Sanfeliu, A.: Robot companion: a social-force based approach with human awareness-navigation in crowded environments. In: 2013 IEEE/RSJ International Conference on Intelligent Robots and Systems (IROS), pp. 1688–1694, November 2013
5. Hall, E.T.: The Hidden Dimension. Anchor, New York (1990)
6. Karaman, S., Frazzoli, E.: Sampling-based algorithms for optimal motion planning. Int. J. Robot. Res. **30**(7), 846–894 (2011). http://ijr.sagepub.com/content/30/7/846.abstract
7. Kirby, R., Forlizzi, J., Simmons, R.: Affective social robots. Robot. Auton. **58**, 322–332 (2010)
8. Kirby, R., Simmons, R.G., Forlizzi, J.: Companion: a constraint-optimizing method for person-acceptable navigation. In: RO-MAN, pp. 607–612. IEEE (2009)
9. Kretzschmar, H., Kuderer, M., Burgard, W.: Learning to predict trajectories of cooperatively navigating agents. In: 2014 IEEE International Conference on Robotics and Automation (ICRA), pp. 4015–4020. IEEE (2014)
10. Kruse, T., Pandey, A.K., Alami, R., Kirsch, A.: Human-aware robot navigation: a survey. Robot. Auton. Syst. **61**(12), 1726–1743 (2013). http://dx.doi.org/10.1016/j.robot.2013.05.007
11. Kuderer, M., Gulati, S., Burgard, W.: Learning driving styles for autonomous vehicles from demonstration. In: Proceedings of the IEEE International Conference on Robotics & Automation (ICRA), Seattle, USA. vol. 134 (2015)
12. Levine, S., Popovic, Z., Koltun, V.: Nonlinear inverse reinforcement learning with gaussian processes. In: Neural Information Processing Systems Conference (2011)
13. Luber, M., Spinello, L., Silva, J., Arras, K.: Socially-aware robot navigation: a learning approach. In: IROS, pp. 797–803. IEEE (2012)
14. Michini, B., Cutler, M., How, J.P.: Scalable reward learning from demonstration. In: IEEE International Conference on Robotics and Automation (ICRA). IEEE (2013). http://acl.mit.edu/papers/michini-icra-2013.pdf
15. Pacchierotti, E., Christensen, H., Jensfelt, P.: Evaluation of passing distance for social robots. In: IEEE Workshop on Robot and Human Interactive Communication (ROMAN), Hartfordshire, UK, September 2006
16. Quigley, M., Conley, K., Gerkey, B.P., Faust, J., Foote, T., Leibs, J., Wheeler, R., Ng, A.Y.: ROS: an open-source robot operating system. In: ICRA Workshop on Open Source Software (2009)
17. Shiarlis, K., Messias, J., van Someren, M., Whiteson, S., Kim, J., Vroon, J., Englebienne, G., Truong, K., Evers, V., Perez-Higueras, N., Perez-Hurtado, I., Ramon-Vigo, R., Caballero, F., Merino, L., Shen, J., Petridis, S., Pantic, M., Hedman, L., Scherlund, M., Koster, R., Michel, H.: TERESA: a socially intelligent semi-autonomous telepresence system. In: Workshop on Machine Learning for Social Robotics at ICRA-2015 in Seattle (2015)
18. Sisbot, E.A., Marin-Urias, L.F., Alami, R., Siméon, T.: A human aware mobile robot motion planner. IEEE Trans. Robot. **23**(5), 874–883 (2007)
19. Trautman, P., Krause, A.: Unfreezing the robot: navigation in dense, interacting crowds. In: IROS, pp. 797–803. IEEE (2010)
20. Ziebart, B., Maas, A., Bagnell, J., Dey, A.: Maximum entropy inverse reinforcement learning. In: Proceedings of the National Conference on Artificial Intelligence (AAAI) (2008)

Adaptive Robot Assisted Therapy Using Interactive Reinforcement Learning

Konstantinos Tsiakas[1,2(✉)], Maria Dagioglou[2],
Vangelis Karkaletsis[2], and Fillia Makedon[1]

[1] Computer Science and Engineering Department,
University of Texas at Arlington, Arlington, USA
konstantinos.tsiakas@mavs.uta.edu, makedon@uta.edu
[2] National Center for Scientific Research "Demokritos", Athens, Greece
{mdagiogl,vangelis}@iit.demokritos.gr

Abstract. In this paper, we present an interactive learning and adaptation framework that facilitates the adaptation of an interactive agent to a new user. We argue that Interactive Reinforcement Learning methods can be utilized and integrated to the adaptation mechanism, enabling the agent to refine its learned policy in order to cope with different users. We illustrate our framework with a use case in the domain of Robot Assisted Therapy. We present our results of the learning and adaptation experiments against different simulated users, showing the motivation of our work and discussing future directions towards the definition and implementation of our proposed framework.

1 Introduction

An interactive learning agent is an entity that learns through the continuous interaction with its environment. Such agents act based on the information they perceive and their own policy. This interaction can be seen as a stochastic sequential decision making problem, where the agent must learn a policy that dictates what to do at each interaction step, by accumulating knowledge through experience [9].

Reinforcement Learning (RL) provides an appropriate framework for interaction modeling and optimization of problem that can be formulated as Markov Decision Processes (MDP) [17]. RL methods have been successfully applied for modeling the interaction in problems as Adaptive Dialogue Systems, Intelligent Tutoring Systems and recently to Robot Assisted Therapy applications [7,16]. The advantage of RL methods in modeling the interaction in such systems is that the stochastic variation in user responses is depicted as transition probabilities between states and actions [19].

In real-world systems, where the state-action space is large and the environment is dynamic, learning an optimal policy for each specific user is challenging. An interactive agent designed to interact with different users should be able to adapt to environmental changes by efficiently modifying a learned policy to cope with different users, instead of learning from scratch for each user. Furthermore,

© Springer International Publishing AG 2016
A. Agah et al. (Eds.): ICSR 2016, LNAI 9979, pp. 11–21, 2016.
DOI: 10.1007/978-3-319-47437-3_2

even the same user may not be consistent over time, in terms of reactions and intentions. In addition, user abilities change over time, as users adapt their own behavior while interacting with a learning agent. Recent works investigate the aspect of *co-adaptation* in man-machine interaction systems, assuming that both agent and user adapt in a cooperative manner to achieve a common goal [10].

Taking these points into consideration, a dynamic adaptation mechanism is required for an interactive agent that enables the system to adapt to different users, ensuring efficient interactions. In this paper, we present an interactive learning and adaptation framework for Robot Assisted Therapy, that utilizes *Interactive Reinforcement Learning* methods to facilitate the policy adaptation of the robot towards new users. We argue that interactive learning techniques [3,14] can be combined with transfer learning methods [8] to facilitate the adaptation of an agent to a new user.

2   Related Work

Robot Assisted Therapy (RAT) has been extensively applied to assist users with cognitive and physical impairments [18]. An assistive robot should be able to adapt to user behavior preferences and changes, ensuring a safe and tailored interaction [11]. Such systems may also support multiparty interactions [12], including a *secondary user* (therapist, clinician) who supervises the interaction between the *primary user* and the agent.

There are works that indicate that personalized and tailored robotic assistive systems can establish a productive interaction with the user, improving the effects of a therapy session. In [15], the authors present a therapeutic *robocat* to investigate how patients with dementia would respond. Their interactions with the robocat led to less agitation, and more positive experiences. Similarly, in [25], the authors present Paro, a robotic seal that interacted proactively and reactively with patients, showing that the interaction with Paro had psychological, physiological, and social effects on elderly people. In [21], the authors proposed an adaptive socially assistive robotic (SAR) system that provides a customized protocol through motivation, encouragement and companionship for users suffering from Alzheimer's disease.

Our work moves towards the definition and implementation of an interactive learning and adaptation framework for interactive agents [23]. Based on this framework, an interactive agent is able to refine a learned policy towards a new user, by exploiting additional communication channels (feedback and guidance) provided by the users during the interaction.

3   Adaptive Robot Assisted Therapy

In this section, we present an adaptive training task example, where the user needs to perform a set of cognitive or physical tasks. The representation we present is applicable to various RAT applications, as rehabilitation exercises and cognitive tasks [1,21,24]. We follow a scenario where the user interacts with the robot during a training session. The user must complete a set of three predefined tasks. Each task has four difficulty levels (Easy, Medium, Normal, Hard). The robot keeps track of task duration and the user's score. To measure user's score, we follow a negative marking approach. At each interaction step, the user receives a positive score proportional to the task difficulty, upon a successful turn and the corresponding negative one for failure. The robot keeps track of these scores and sums them to compute the user's total performance.

3.1   Robot Behavior Modeling

In order to model the robot's behavior, we formulate the problem as an MDP. The state is a set of variables that represent the current task state: Task ID (1,2,3), Task Duration (0–6), Difficulty Level (1–4) and Score (−4–4). At each interaction step, the system selects a difficulty level for the current task and the user performs the task. The agent receives a reward, proportional to the user's score, to update its policy (Fig. 1).

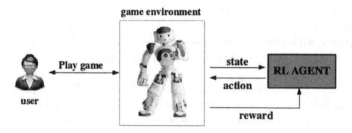

Fig. 1. The task formulated as a Reinforcement Learning problem. The robot keeps track of the current state. The RL agent selects an action (the next difficulty level or task switching) and the robot plays with the user, returning a reward to the agent to update its policy.

The agent needs to learn the optimal policy; the mapping from states to actions that maximizes the accumulated reward (or total return) during each interaction. Maximizing total return can be translated as providing the appropriate difficulty level that will maximize user's performance.

In order to evaluate our robot behavior modeling, we defined four different user models that capture different user abilities. These user models depict the user skills under different game parameters (task difficulty and duration). Each user model is a rule-based model whose binary output indicates user's success (or failure) for each task difficulty and duration parameters. In Fig. 2, we show two of the defined user models that capture different user skills. The agent must learn how to adjust the difficulty level and when to switch to the next task, for each user model, learning a *user-specific policy (USP)*.

User 1 (expert)		Difficulty level				User 2 (novice)		Difficulty level			
		easy	medium	normal	hard			easy	medium	normal	hard
d	0	1	1	1	1	d	0	1	1	1	1
u	1	1	1	1	1	u	1	1	1	1	1
r	2	1	1	1	1	r	2	1	1	0	0
a	3	1	1	1	1	a	3	1	0	0	0
t	4	1	1	1	1	t	4	1	0	0	0
i	5	1	1	1	1	i	5	1	0	0	0
o n	6	0	0	0	0	o n	6	0	0	0	0

Fig. 2. User model examples. User 1 is an expert user, able to complete all tasks at the highest difficulty level. User 2 is a novice user that can perform each task at the high level only for the first two rounds and the rest of them at a lower difficulty. Since both users do not succeed after time duration 6 (maximum), the agent must learn to switch to the next task.

3.2 Learning Experiments

Our first step is to train the agent against the four different user models. For our learning experiments, we applied the Q-learning algorithm, following the ϵ-greedy policy with linearly decreasing exploration rate. An episode is a complete game of three tasks. In Fig. 3, we show the learning results for each user model.

We visualize the evaluation metrics, as proposed by [2]. We plot the total return collected during each episode, averaged over a number of episodes (1 epoch = 50 episodes). Since the agent must learn the best policy for each user so as to maximize their performance, we observe the similarity of total return and performance curves. Another metric is the start state value, which provides an estimate of the expected total return the agent can obtain by following this policy. In the top right figure, we observe the different convergence points for each user model. Since start state value expresses the expected return, we observe that start state value and average return tend to approximate each other as training evolves. Another convergence evaluation metric is the Q-value updates of all action-state pairs per epoch, showing that the algorithm converges as it learns, decreasing the state value updates.

3.3 Policy Transfer Experiments

Our next step is to evaluate these four USP policies to the different user models. We make two hypotheses: (1) a user specific policy is the optimal policy (for the corresponding model); the one that maximizes total return, thus user performance, and (2) applying a learned policy to a different user model may not be efficient but better than learning from scratch. We applied the four different USP to the four different user models, following an exploitation-only approach, since following an exploration strategy may not be safe for real-world HRI applications.

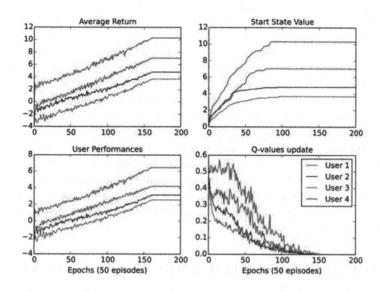

Fig. 3. Learning experiments. Applying Q-learning for the different user models results to different *user-specific policies (USP)*.

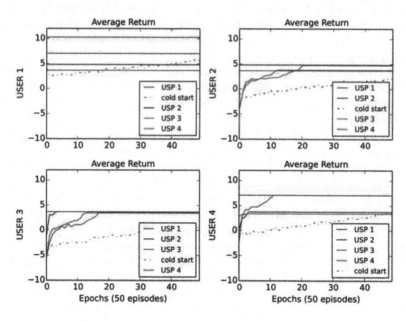

Fig. 4. Policy transfer experiments. In this experiment, we applied all learned USP to the different user models.

These policy transfer experiments validate our two hypotheses (Fig. 4). Each USP is the optimal policy for the corresponding model; it maximizes the total return. Moreover, applying a policy to a different user model may not be efficient but better than learning from scratch (dashed-line). We can observe three cases: (1) the initial policy adapts and converges to the USP, (2) the initially policy is improved but does not converge to the USP and (3) the learned policy remains unchanged. On the bottom right figure, we observe that all USPs adapt and converge to USP4. This happens because the agent interacts with different user models, receives negative rewards for specific state-action pairs, improving its policy for the corresponding pairs. However, on the top left figure, we observe that all policies remain unchanged. This happens because all policies result to positive rewards, since user-1 succeeds in all difficulty levels. However, only USP1 is the optimal policy for this user model. This indicates a need for a dynamical adaptation mechanism that enables the agent to efficiently refine its policy towards a new user.

4   Interactive Learning and Adaptation Framework

In this section, we present an interactive learning and adaptation framework that integrates Interactive Reinforcement Learning approaches to the adaptation mechanism. *Interactive Reinforcement Learning* (IRL) is a variation of RL that studies how a human can be included in the agent learning process. Human input can be either in the form of feedback or guidance. *Learning from Feedback* treats the human input as a reinforcement signal after the executed action [14]. *Learning from Guidance* allows human intervention to the selected action before execution, proposing (corrective) actions [8]. To our knowledge, IRL methods have not been investigated for the adaptation of an agent to a new environment. Hence, we propose their integration to the adaptation mechanism, as *policy evaluation* metrics used to evaluate and modify a learned policy towards an optimal one, following proper transfer methods.

Based on this framework (Fig. 5), an interactive agent is able to adapt a learned policy towards a new user by exploiting additional communication channels (feedback and guidance), provided during the interaction. Our framework supports the participation of a secondary user who supervises the interaction in its early steps, avoiding unsafe interactions. The supervisor can either physically or remotely supervise the interaction (*observation*). A user interface can be used to provide the supervisor with useful information about the agent learning procedure (*agent data*) to help them monitor the interaction and enhance their own decision making, before altering the agent's policy. The goal of this framework is to enable agents to learn as long as they interact with primary and secondary users, adapting and refining their policy dynamically.

User feedback can be considered as a *personalization* factor, as it is provided by the primary user *implicitly* (e.g., facial expressions, eye-tracking, engagement levels, etc.) and can be used to evaluate the interaction, thus the agent policy. In our adaptive training game case, one way to evaluate the agent policy is

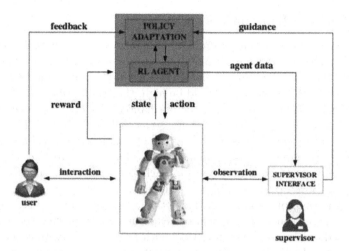

Fig. 5. We extend the RL framework by adding two additional communication channels; *feedback* and *guidance*. Their integration to the adaptation module can enable the agent to continuously adapt towards the current user, ensuring a safe and personalized interaction

to measure user engagement. We propose to use the Muse EEG headset[1], a commercially available tool that measures electrical activity at the scalp as it relates to various cognitive and mood states. This type of sensor can be used to measure how engaged a person is while that person is completing some sort of games or music tasks [13]. When the selected difficulty level is lower than needed, then the user may not be engaged. Moreover, when the task is difficult enough, the user may be frustrated and disengaged [4]. This implicit feedback can be exploited and efficiently modify a learned policy towards a new user, in an online and dynamic fashion.

On the other hand, guidance can be considered as a *safety* factor, following a supervised progressively autonomous approach [20], but also integrating human advice to the adaptation mechanism, enabling the agent adapt more effectively. The therapist, as a secondary user, can also set their own therapeutic goals by altering the policy. Making the learning process transparent to the secondary user may result to more informative guidance [5]. Informative metrics as state uncertainty and importance can be utilized to assist the secondary user provide the system with valuable guidance, in the form of corrective or suggested actions. Additionally, Active Learning methods [6] can be used to learn, based on state information, when the therapist should intervene, minimizing the expert's workload as the system learns.

---

[1] http://www.choosemuse.com/.

4.1 Preliminary Experiments

In this section, we present our preliminary adaptation experiments, following the proposed framework. As we mentioned, we assume that user feedback (engagement level) relates to the difficulty level variance; if the robot selects the appropriate difficulty the engagement should be high, thus the feedback value. For our simulation experiments, feedback is the normalised absolute difference between the selected and the appropriate difficulty, so as *feedback* $\in [-1,0]$. The feedback is used to modify the Q-values, following the Q-augmentation technique [14].

On the other hand, we use guidance in the form of corrective actions, following a semi-supervised autonomy approach [20] combined with *teaching on a budget* [22]. Based on this approach, the agent proposes an action based on its policy. The therapist can reject this action and select another, for a limited number of interventions ($M = 2$). For our experiments, the corrective actions are selected based on the corresponding USP with probability 0.8, to cover possible therapist errors. In Fig. 6, we show the results of our experiments, integrating feedback and guidance. We observe that for all cases, the integration of feedback and guidance improve the applied policy, resulting to its convergence to the corresponding USP (optimal policy), validating our hypothesis that interactive learning methods can be utilized for the policy transfer and adaptation.

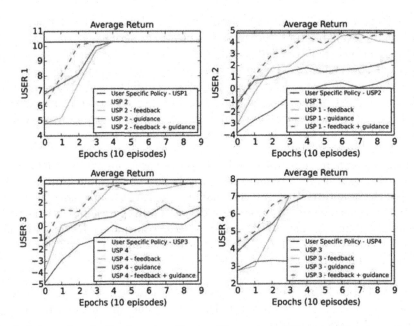

Fig. 6. Integrating feedback and guidance. We apply a different policy to each user model (a) without feedback/guidance, (b) with feedback and (c) with feedback and guidance. We observe that their integration to the learning improves the learned policy.

5   Discussion and Future Work

To conclude with, we introduced an interactive learning and adaptation framework for dynamically adaptive agents. We presented our use case in Robot Assisted Therapy with an adaptive training task. Our preliminary learning and adaptation experiments indicate that interactive learning methods can be integrated to the adaptation mechanism, resulting to an intelligent adaptive robot behavior.

Our preliminary results are promising, since the integration of interactive learning techniques facilitate the policy adaptation. However, there are some limitations on the presented framework and simulation experiments. Based on the defined user models, users are consistent over time and towards all tasks. This is likely to be violated in a real-world scenario. However, we believe that the proposed framework can be evolved and applied to real-world HRI applications, investigating further how interactive learning techniques can be integrated to the adaptation mechanism, considering user inconsistency over time and co-adaptation. Moreover, we will investigate how users (both primary and secondary) interact under this framework, developing appropriate interaction techniques. Our next steps include the implementation of a RAT scenario following the framework and a case study with participants to evaluate the task itself, as well as the proposed framework.

Acknowledgments. This material is based upon work supported by NSF under award numbers CNS 1338118, 1035913 and by the educational program of NCSR Demokritos in collaboration with the University of Texas at Arlington.

References

1. Andrade, K.d.O., Fernandes, G., Caurin, G.A., et al.: Dynamic player modelling in serious games applied to rehabilitation robotics. In: Robotics Symposium and Robocontrol, pp. 211–216. IEEE (2014)
2. Bellemare, M.G., Naddaf, Y., Veness, J., Bowling, M.: The arcade learning environment: an evaluation platform for general agents. J. Artif. Intell. Res. **47**, 253–279 (2012)
3. Broekens, J.: Emotion and reinforcement: affective facial expressions facilitate robot learning. In: Huang, T.S., Nijholt, A., Pantic, M., Pentland, A. (eds.) Artifical Intelligence for Human Computing. LNCS (LNAI), vol. 4451, pp. 113–132. Springer, Heidelberg (2007)
4. Chanel, G., Rebetez, C., Bétrancourt, M., Pun, T.: Boredom, engagement and anxiety as indicators for adaptation to difficulty in games. In: Proceedings of the 12th International Conference on Entertainment and Media in the Ubiquitous Era, pp. 13–17. ACM (2008)
5. Chao, C., Cakmak, M., Thomaz, A.L.: Transparent active learning for robots. In: ACM/IEEE International Conference on Human-Robot Interaction, pp. 317–324. IEEE (2010)
6. Chernova, S., Veloso, M.: Interactive policy learning through confidence-based autonomy. J. Artif. Intell. Res. **34**(1), 1 (2009)

7. Chi, M., VanLehn, K., Litman, D., Jordan, P.: An evaluation of pedagogical tutorial tactics for a natural language tutoring system: a reinforcement learning approach. Int. J. Artif. Intell. Educ. **21**(1–2), 83–113 (2011)

8. Cruz, F., Twiefel, J., Magg, S., Weber, C., Wermter, S.: Interactive reinforcement learning through speech guidance in a domestic scenario. In: International Joint Conference on Neural Networks, pp. 1–8. IEEE (2015)

9. Cuayáhuitl, H., van Otterlo, M., Dethlefs, N., et al.: Machine learning for interactive systems and robots: a brief introduction. In: Proceedings of the 2nd Workshop on Machine Learning for Interactive Systems: Bridging the Gap Between Perception, Action and Communication, pp. 19–28. ACM (2013)

10. Gallina, P., Bellotto, N., Di Luca, M.: Progressive co-adaptation in human-machine interaction. In: Informatics in Control, Automation and Robotics. IEEE (2015)

11. Giullian, N., et al.: Detailed requirements for robots in autism therapy. In: Proceedings of SMC 2010, pp. 2595–2602. IEEE (2010)

12. Goodrich, M., Colton, M., Brinton, B., Fujiki, M., Atherton, J., Robinson, L., Ricks, D., Maxfield, M., Acerson, A.: Incorporating a robot into an autism therapy team. IEEE Life Sciences (2012)

13. McCullagh, P., et al.: Assessment of task engagement using brain computer interface technology. In: Workshop Proceedings of the 11th International Conference on Intelligent Environments, vol. 19. IOS Press (2015)

14. Knox, W.B., Stone, P.: Reinforcement learning from simultaneous human and MDP reward. In: Proceedings of the 11th International Conference on Autonomous Agents and Multiagent Systems, vol. 1, pp. 475–482 (2012)

15. Libin, A., Cohen-Mansfield, J.: Therapeutic robocat for nursing home residents with dementia: preliminary inquiry. Am. J. Alzheimer's Dis. Dementias **19**(2), 111–116 (2004)

16. Modares, H., Ranatunga, I., Lewis, F.L., Popa, D.O.: Optimized assistive human-robot interaction using reinforcement learning. IEEE Trans. Cybern. **46**, 655–667 (2015)

17. Pietquin, O., Lopes, M.: Machine learning for interactive systems: challenges and future trends. In: WACAI (2014)

18. Raya, R., Rocon, E., Urendes, E., Velasco, M.A., Clemotte, A., Ceres, R.: Assistive robots for physical and cognitive rehabilitation in cerebral palsy. In: Mohammed, S., Moreno, J.C., Kong, K., Amirat, Y. (eds.) Intelligent Assistive Robots: Recent Advances in Assistive Robotics for Everyday Activities. Springer Tracts in Advanced Robotics, vol. 106, pp. 133–156. Springer, Heidelberg (2015)

19. Rieser, V., Lemon, O.: Reinforcement Learning for Adaptive Dialogue Systems: A Data-driven Methodology for Dialogue Management and Natural Language Generation. Theory and Applications of Natural Language Processing. Springer, Heidelberg (2011)

20. Senft, E., Baxter, P., Kennedy, J., Belpaeme, T.: SPARC: supervised progressively autonomous robot competencies. In: Tapus, A., André, E., Martin, J.-C., Ferland, F., Ammi, M. (eds.) Social Robotics. LNCS, vol. 9388, pp. 603–612. Springer, Heidelberg (2015)

21. Tapus, A.: Improving the quality of life of people with dementia through the use of socially assistive robots. In: Advanced Technologies for Enhanced Quality of Life (AT-EQUAL 2009), pp. 81–86. IEEE (2009)

22. Torrey, L., Taylor, M.: Teaching on a budget: agents advising agents in reinforcement learning. In: Proceedings of the 2013 International Conference on Autonomous Agents and Multi-agent Systems, pp. 1053–1060. International Foundation for Autonomous Agents and Multiagent Systems (2013)

23. Tsiakas, K.: Facilitating safe adaptation of interactive agents using interactive reinforcement learning. In: Companion Publication of the 21st International Conference on Intelligent User Interfaces, pp. 106–109. ACM (2016)

24. Tsiakas, K., Huber, M., Makedon, F.: A multimodal adaptive session manager for physical rehabilitation exercising. In: Proceedings of the 8th ACM International Conference on Pervasive Technologies Related to Assistive Environments. ACM (2015)

25. Wada, K., et al.: Robot therapy for elders affected by dementia. IEEE Eng. Med. Biol. Mag. 4(27), 53–60 (2008)

Personalization Framework for Adaptive Robotic Feeding Assistance

Gerard Canal$^{(\boxtimes)}$, Guillem Alenyà, and Carme Torras

Institut de Robòtica i Informàtica Industrial, CSIC-UPC,
Llorens i Artigas 4-6, 08028 Barcelona, Spain
{gcanal,galenya,torras}@iri.upc.edu

Abstract. The deployment of robots at home must involve robots with pre-defined skills and the capability of personalizing their behavior by non-expert users. A framework to tackle this personalization is presented and applied to an automatic feeding task. The personalization involves the caregiver providing several examples of feeding using Learning-by-Demostration, and a ProMP formalism to compute an overall trajectory and the variance along the path. Experiments show the validity of the approach in generating different feeding motions to adapt to user's preferences, automatically extracting the relevant task parameters. The importance of the nature of the demonstrations is also assessed, and two training strategies are compared.

Keywords: Assistive robotics · Personalized Human-Robot Interaction · Feeding · Trajectory adaptation

1 Introduction

People with reduced mobility tend to find themselves needing the help of another in order to do the most basic tasks. Hence, performing Activities of the Daily Living (ADLs) such as eating, dressing, grooming or cleaning up can become very difficult. Intelligent robotic systems have proven useful in these situations by performing the helping task and, so, removing the constraint of constant attention from another person.

However, in order to effectively assist a human user the helping robot should be able to adapt to the specific user needs and preferences. Rather than performing a generic action suitable for anyone and forcing the user to adapt to the robot, it is the robot who should adapt its behavior taking into account the user and the situation, just as a human carer would do. The *empowering* of disabled people is crucial [3], and can be attained by providing more autonomy, intimacy and better quality of life. This does not imply the substitution of the caregiver, as personal contact is also very important. Contrarily, our approach relies on the caregiver to personalize the robot to the disabled person preferences.

In this paper, we first propose a novel Robot Personalization framework named FUTE (detailed in Sect. 3), that takes into account the user and allows

© Springer International Publishing AG 2016
A. Agah et al. (Eds.): ICSR 2016, LNAI 9979, pp. 22–31, 2016.
DOI: 10.1007/978-3-319-47437-3_3

concrete adaptation of generic pre-trained skills. In our framework, the robot is pre-trained at the factory with a set of skills. Afterwards, when it arrives at the user's home, a non-expert teacher (the user itself or a caregiver) must have the freedom to adapt such skills to his/her preferences, or even teach new ones.

Second, we explore how to perform this training by using Learning-by-Demonstration techniques combined with a compliant robot control [7]. We propose two different interaction strategies: the teacher intervening in the robot motion, and the demonstration of a completely new trajectory.

In the third place, we test the applicability of the proposed FUTE framework in an assistive task consisting in feeding a person. As feeding can be very complex, we concentrate in a specific aspect: the way in which the robot approaches the cutlery to feed the person (see Fig. 1). We will show how our system is able to automatically extract the relevant aspects along the feeding task. Observe that, depending on the mobility and preferences of the user, the robot must wait with the food at some distance or introduce the food inside the mouth. Moreover, the feeding motion has to be adapted to the kind of food, for example yogurt or fries as seen in Fig. 1(b).

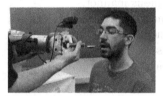  

(a) Caregiver personalizing a spoon feeding skill. (b) A user eating from a fork. (c) End effector's trajectory Cartesian coordinate axes.

Fig. 1. Assistive personalized feeding application example.

2   Related Work

Personalized Human-Robot Interaction has been studied in different works and fields. In education, it has been applied to Socially Assistive Robot (SAR) tutors that support the teaching task [5,10,12]. Baraka and Veloso [2] define three user models to adapt the luminous interactions between a robot and the user over time, learning the model parameters from user feedback. Personalized collaboration is shown in Fiore *et al.* [8], where an object manipulation task is performed jointly by the robot and the user whose preferences are taken into account. Abdo *et al.* [1] predict user preferences to tidy up objects in containers using collaborative filtering based on crowdsourced data and the observations of current dispositions or by querying the user. Although this strategy seems good for the tidying up task, it would not suit to capture the user preferences in an interaction context such as ours. Chernova and Veloso [4] present the Confidence-Based

Autonomy (CBA) algorithm, which enables the agent to request demonstrations from a human teacher, and allows him to correct further mistakes with additional demonstrations. The idea is similar to the User Tailoring one, though they apply it to improve the policy rather than to adapt a well-learned task to a specific user. A framework to learn and generalize complex tasks from unstructured demonstrations is proposed in Niekum *et al.* [14]. The method is able to recognize repeated instances of skills and generalize them to new settings.

In addition, more in the scope of this paper, personalized dressing assistance is performed by Gao *et al.* [9], where a user's movement space is modelled and used to put on a sleeveless jacket. Similarly, Klee *et al.* [11] assist a user to place a hat in a collaborative way by means of asking the user to reposition itself when some user specific constraints do not hold. However, the personalization they propose consists in adapting to the user state or pose, but do not allow the user to modify the way in which the assistance will be carried out.

Moreover, we will apply the personalized interaction to the feeding scenario. Assistive feeding devices have been around for a while, mainly due to the evident need that some individuals have. Devices such as SECOM's MySpoon [16,18] or the Handy 1 [17], among others, can provide significant help to allow people with upper limb disabilities to eat in a more autonomous manner.

Nonetheless, these systems lack the ability to adapt to the needs of each specific user. And, in cases of people with disabilities, this is a key factor for the system to be actually helpful in different kinds of environment, in which there is a handful of ways of assisting in the eating task, as often pointed out by long-term care nurses.

3   The FUTE Personalization Framework

We present a three-phase framework, the "FUTE framework", to design and develop such kind of adaptive assistive applications. The three phases are called "Factory setting", "User Tailoring" and "Execution tuning", and are described as:

1. **Factory setting:** the robot is provided with the skills needed to perform the assistive task in a generic way. This would suit either the design of a new robot or the enhancement of an existing platform to carry out a new task.
2. **User Tailoring** (the focus of the paper): This second phase takes place in the user's home. The robot performs a nominal skill, but personalization is encouraged in order to adapt its behavior to the user needs. In this phase, the robot should acquire, as automatically as possible, information about how the task has to be done for the user at hand while it performs the task in the generic way. This personalization may be done by the user or by an external agent (such as a carer), and it could be either explicit or implicit. In the feeding example, this will consist in the selection of the feeding point, it being either inside or outside the mouth. The data in our implementation include the proprioceptive robot perception as well as 3D images from a camera located at the hand of the robot (Fig. 1).

Algorithm 1. Feeding execution

1: graspFeedingUtensil()	▷ Grasp a spoon or a fork.
2: **repeat**	
3:　　pickUpFoodFromPlate()	
4:　　userPose ← getHumanPoseFromPerception()	
5:　　moveToInitialPosition(userPose, initialPose)	
6:　　moveUtensilToFeedingPose(userPose, feedingPoint)	▷ Approach the food
7:　　waitForFoodConsumption()	
8:　　moveAwayFromUser(userPose)	
9: **until** feedingIsComplete()	▷ User has had enough food or plate is empty

3. **Execution tuning:** In this last phase, the robot performs the task designed in the first phase but taking into account the personalization introduced in the second one. In the feeding example, the 6D pose of the user is computed using an RGBD camera and a face detection algorithm, and the robot trajectories are adapted to the current pose of the user. If the user is not satisfied with the robot behavior, the User Tailoring phase can be triggered again.

4 Experimental Assessment: User-Centered Feeding Assistance

To build intuition, we will illustrate the different aspects of our proposal using the robot feeding application. Eating is one of the most basic physiological needs all human beings have, appearing at the base of Maslow's hierarchy of needs [13]. However, some people with disabilities may not be able to do it by themselves, requiring the help of an external agent (usually a human carer), who will feed them taking into account their needs and capacities. To illustrate this, in the following experiments we tackle two example use-cases in which different personalizations can be applied: (U1) a person with very limited upper body mobility will require the caregiver to do all the feeding action, while (U2) another patient with upper limb disabilities may be able to move and eat the food by himself when it is close enough.

4.1 The Robot Feeding Process

Five steps can be identified for the adaptive feeding application (see Algorithm 1). In the context of the proposed framework, steps between lines 1 and 3 would be provided to the robot during the factory training phase, while steps between lines 6 and 8 would be personalized at home. Thus, the complete execution is the outcome of joining the already known steps (at the factory phase) with the personalized ones, resulting in a successful feeding action for a specific person. The "initialPose" (line 5) and "feedingPoint" (feeding moment of the trajectory, line 6) parameters are obtained during the User Tailoring phase, as seen in Algorithm 2. Note that in execution, the user can move freely. A vision system comprised of a low range RGBD sensor is used to compute their pose, and the robot motion is updated accordingly to obtain the desired feeding movement. The vision system is also used to detect the moment in which the user bites the food in the "waitForFoodConsumption" step (line 7).

For the scope of this paper, we will just focus on the steps involving the user (lines 6, 7 and 8 from Algorithm 1), and how they can be personalized to different users[1].

The feeding setup used in the experiments can be seen in Figs. 1(a) and (b), and the coordinate axes at the robot's end-effector are shown in Fig. 1(c). In it, the y Cartesian axis represents the frontal distance to the user, the x axis corresponds to the horizontal displacement and the z to the vertical one (the feeding height).

4.2   Feeding Personalization

The User Tailoring strategy for feeding is shown in Algorithm 2. It comprises the recording of N sample trajectories (line 4) including the approaching motion, waiting for the user to start the consumption, and a receding motion. The N trajectories are then used to learn a Probabilistic Movement Primitive (ProMP) [6,15] of the feeding movement (line 23). ProMPs are movement primitives that encode the time-varying variance of a set of trajectories. The state vector $\mathbf{y}_t$ is defined as

$$\mathbf{y}_t = \begin{bmatrix} q_t \\ \dot{q}_t \end{bmatrix} = \mathbf{\Phi}_t^T \mathbf{w} + \epsilon_y, \tag{1}$$

where $\mathbf{\Phi}_t = [\phi_t, \dot{\phi}_t]$ is the time-dependent basis matrix, $\mathbf{w}$ is the weight vector and $\epsilon_y \sim N(0, \Sigma_y)$ is Gaussian noise. The trajectories can then be represented as a mean trajectory and its variance, each time point being represented as $\mu_t \pm \sigma_t$. New trajectories can be sampled from the distribution, and via points are defined using the conditioning operator. We have used the ProMP formalism because, apart from the trajectory itself, as will be seen in Sect. 4.4, it also provides insights of the particularities of the task by means of the variance along the trajectory.

We would like to assess the impact of variations in the demonstrated trajectories, to provide hints to the caregiver demonstrating the task about how similar the N demonstrations should be. The next experiment tackles use-case U1: introducing the food inside the mouth of the user. It involves demonstrations using two different feeding paths with a mannequin as user: the first one in which the carer tried to perform the same trajectory 5 times, and the second set in which the 5 trajectories had different approaching movements (but with the same feeding point). The results are shown in Fig. 2.

Comparing Figs. 2(a) and (b) it can be seen that the shape of both mean trajectories is quite alike, both reaching the same feeding position (shaded area). As a consequence, apparently there is no need to have several similar trajectories

---

[1] A video showing the process of the personalized feeding task can be found at www.iri.upc.edu/groups/perception/frameworkFUTE.

Algorithm 2. User tailoring strategy

```
1:  demonstrations ← ∅                    ▷ Will store the new recorded trajectories
2:  feedingPoints ← ∅        ▷ Time points of each trajectory in which person was fed
3:  initialPoses ← ∅                        ▷ Face pose at the start of each trajectory
4:  for N do
5:      if unassistedTraining then          ▷ Set one of the two personalization modes
6:          SetRobot(gravityCompensationMode)
7:      else
8:          SetRobot(ReproduceFactoryTrajectory, stiffness)
9:      end if
10:     initialPoses ← append(getUserFacialPose())
11:     newTrajectory ← ∅
12:     while robotMoving do                        ▷ Store approaching trajectory
13:         addPoints(newTrajectory)
14:     end while
15:     waitForFoodConsumption()                    ▷ Wait until user starts eating
16:     feedingPoints ← append(currentTrajectoryPoint)
17:     while robotMoving do                         ▷ Store receding trajectory
18:         addPoints(newTrajectory)
19:     end while
20:     demonstrations ← append(newTrajectory)
21: end for
22: referenceFeedingPoint ← alignToFeedingPoint(demonstrations, feedingPoints)
23: personalizedTrajectory ← RecomputeProMP(demonstrations)
24: return <personalizedTrajectory, referenceFeedingPoint, avg(initialPoses)>
```

in order to have a good average feeding movement. However, we observe different variances. In Fig. 2(a) variance is almost constant during the whole trajectory, while in Fig. 2(b) variances in the approaching and receding movements are larger, but smaller in the feeding point. Observe that obtaining this information is crucial, as the robot should act carefully while feeding the user (lower variance) whereas approaching and receding can exhibit a more careless behavior (larger variance). Thus, we conclude that showing some variability in the demonstrated trajectories is important.

4.3 Teaching Modes: Unassisted vs. Compliant Reproduction

Two teaching modes have been defined (Algorithm 2 lines 5–9). The first one is unassisted, the robot only compensates gravity and the caregiver has to start from scratch each demonstration handling the robot and freely performing a feeding trajectory. This allows the user to discard the factory settings and re-teach the whole movement. In the second one, the robot executes a generic feeding trajectory –which was recorded in the factory setting phase– using a compliant controller [7] that uses a *stiffness factor* to determine the arm's stiffness degree.

The next experiment is designed to assess the effect of the stiffness factor. Hence, we repeated the executions with different stiffness values for the same trajectory where the caregiver personalized the motion so that the feeding occurred

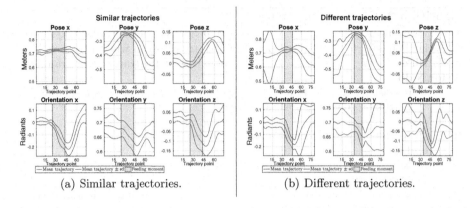

(a) Similar trajectories.   (b) Different trajectories.

Fig. 2. Comparison between similar and different example trajectories. The thicker line is the mean trajectory, and the surrounding lines are the mean ± standard deviation. The shaded regions denote the part of the trajectory in which the food is consumed.

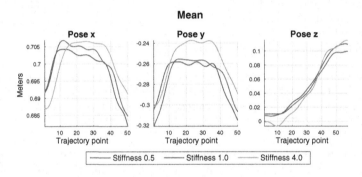

Fig. 3. Mean trajectories generated from a default trajectory with different stiffness values. The lower the stiffness, the most docile the robot behaviour is. Observe the oscillations introduced when the trajectory is perturbed.

further away from the person (a mannequin was used in this experiment to avoid noise induced by involuntary movements and ease the comparison). Here we tackle use-case U2: the trajectory is modified to end outside of the mouth, for instance for patients some mobility. The results are shown in Fig. 3.

The intuition says that starting from scratch at every demonstration is harder, whereas if the robot reproduces the movement in a docile manner the user only has to physically perturb the execution in some parts and teaching becomes easy. However, as it can be seen in Fig. 3, this second approach introduces oscillations of about half a centimeter in the resulting trajectory, not only in the y axis (the approaching direction) but also in x and z. With low stiffness values the oscillations tend to be higher as the robot reacts to slighter perturbations as when it tries to go on with the trajectory and return to the original path and the user holds it again. In contrast, higher stiffness makes it harder for the user to modify the trajectory, resulting in less oscillations but more physical effort for the user.

4.4   Automatic Parameter Extraction

In the next experiment, the modifications that the caregiver can introduce to personalize the feeding process are (see Algorithm 2, lines 10, 20 and 22): the initial pose, the motion shape, and the feeding point (inside the mouth for use-case U1 or just approaching the food for use-case U2).

We show how these parameters can be extracted automatically during the User Tailoring phase. First, the feeding point is computed by recording the distance to the face in which the movement is stopped to feed the person. Second, the motion learning process captures the particularities of the task. We exemplify this fact by observing variances of the ProMP trajectory related to two different utensils: when a spoon is used the orientation is more restricted, while a fork allows for more flexibility.

In this experiment, the re-teaching has been carried out with the robot holding a spoon with yogurt and also with a fork pinching a french fry. Five trajectories were recorded in order to generate the ProMP for each case. A human user was used here as test subject (not a mannequin) because the insertion orientation was relevant (see Fig. 1). With this experiment, we can observe how the particularities of the task are integrated into the ProMP. Figure 4(a) and (b) show the trained trajectories for each Cartesian coordinate and the rotations around each axis, displaying the mean trajectory and its variance.

The figures clearly show the moment in which the utensil is near the mouth (as seen in the shaded regions), because the variance of the movement narrows at that stage. This is, in fact, a representation of the flexibility of the movement, since the critical parts that need more precision are less flexible.

In addition, this variance effect can also be seen in the orientation plots, in which the spoon's sample orientation variances are narrower at the beginning of the trajectory to avoid spilling the content, while the move away part has wider variances as the food has already been taken. The fork trajectory has less

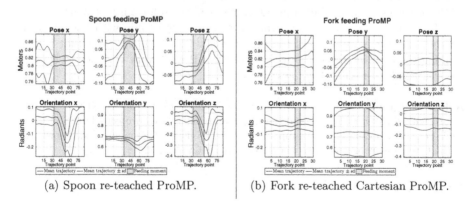

(a) Spoon re-teached ProMP.   (b) Fork re-teached Cartesian ProMP.

Fig. 4. Learned trajectories for the spoon and fork experiments, where the gravity compensation mode was used for re-teaching the trajectories.

restrictive orientations because there is less danger of dropping food, as clearly seen in the orientation around the y axis.

Moreover, this gives us insights on how the variance in the trajectory points provided by the ProMP could also be used to control the compliance (stiffness degree) of the robot during the trajectory execution phase. This way, the robot would be more docile to external forces in moments of high variance, corresponding to points of the path that have been taught in non precise ways, and more rigid in low variance points. Thus, the robot would not react to external forces while introducing the spoon in the mouth, avoiding any possible harm to the user due to accidental robot perturbations. Note this should not be applied in the joints interacting with the user, allowing for docile movement with the mouth but being stiff in external joints such as the elbow.

5   Conclusions

In this paper, we presented the FUTE robot personalization framework consisting of three phases: Factory setting, User Tailoring and Execution tuning. This framework has been devised to help the implementation of assistive applications by allowing easy adaptation of the assistive robot performance to specific users, given the fact that all of them are different and have their own special needs. Furthermore, it allows non-expert users to conduct the robot adaptation just by guiding the robot behavior.

Then, we tested this framework in a feeding application where a human caregiver can re-teach the feeding movement the robot has to perform, by physically modifying an already learnt trajectory or by teaching it from scratch. This allows the person to teach the feeding point and distance so it can be either inside or near the mouth. Moreover, we demonstrate how the use of the Probabilistic Movement Primitives (ProMPs) is an appropriate choice for these kind of assistive applications, as they are able to learn the particularities of the task, such as the feeding moment and the flexibility of each part of the trajectory.

Nevertheless, the feeding application has still room for improvement, including, but no limited to, the integration of voice commands with the physical interactions, the adaptation of the best stiffness factor for the carer that is performing the re-teaching or the use of the ProMP trajectory variance to control the stiffness factor while feeding the user.

Acknowledgments. This work has been supported by the MINECO project RobInstruct TIN2014-58178-R and the ERA-Net CHIST-ERA project I-DRESS PCIN-2015-147. Gerard Canal is also supported by the Ministry of Economy and Knowledge of the Government of Catalonia via a FI-DGR 2016 fellowship.

References

1. Abdo, N., Stachniss, C., Spinello, L., Burgard, W.: Robot, organize my shelves! tidying up objects by predicting user preferences. In: IEEE International Conference on Robotics and Automation (ICRA), pp. 1557–1564 (2015)

2. Baraka, K., Veloso, M.: Adaptive interaction of persistent robots to user temporal preferences. In: Tapus, A., André, E., Martin, J.C., Ferland, F., Ammi, M. (eds.) Social Robotics. LNCS, vol. 9388, pp. 61–71. Springer, Heidelberg (2015)
3. Chen, T.L., Ciocarlie, M., Cousins, S., Grice, P.M., Hawkins, K., Hsiao, K., Kemp, C.C., King, C.H., Lazewatsky, D.A., Leeper, A.E., Nguyen, H., Paepcke, A., Pantofaru, C., Smart, W.D., Takayama, L.: Robots for humanity: using assistive robotics to empower people with disabilities. IEEE Robot. Autom. Magaz. 20(1), 30–39 (2013)
4. Chernova, S., Veloso, M.: Interactive policy learning through confidence-based autonomy. J. Artif. Intell. Res. 34, 1–25 (2009)
5. Clabaugh, C., Ragusa, G., Sha, F., Matarić, M.: Designing a socially assistive robot for personalized number concepts learning in preschool children. In: International Conference on Development and Learning and Epigenetic Robotics, pp. 314–319 (2015)
6. Colomé, A., Neumann, G., Peters, J., Torras, C.: Dimensionality reduction for probabilistic movement primitives. In: IEEE-RAS International Conference on Humanoid Robots, pp. 794–800 (2014)
7. Colomé, A., Pardo, D., Alenyà, G., Torras, C.: External force estimation during compliant robot manipulation. In: IEEE International Conference on Robotics and Automation (ICRA), pp. 3535–3540 (2013)
8. Fiore, M., Clodic, A., Alami, R.: On planning and task achievement modalities for human-robot collaboration. In: Hsieh, H.A., Khatib, O., Kumar, V. (eds.) Experimental Robotics. Springer Tracts in Advanced Robotics, vol. 109, pp. 293–306. Springer, Heidelberg (2016)
9. Gao, Y., Chang, H.J., Demiris, Y.: User modelling for personalised dressing assistance by humanoid robots. In: IEEE/RSJ International Conference on Intelligent Robots and Systems (IROS), pp. 1840–1845 (2015)
10. Greczek, J., Short, E., Clabaugh, C., Swift-Spong, K., Matarić, M.J.: Socially assistive robotics for personalized education for children. In: AAAI Fall Symposium on Artificial Intelligence and Human-Robot Interaction (2014)
11. Klee, S.D., Ferreira, B.Q., Silva, R., Costeira, J.P., Melo, F.S., Veloso, M.: Personalized assistance for dressing users. In: Tapus, A., André, E., Martin, J.C., Ferland, F., Ammi, M. (eds.) Social Robotics. LNCS, vol. 9388, pp. 359–369. Springer, Heidelberg (2015)
12. Leyzberg, D., Spaulding, S., Scassellati, B.: Personalizing robot tutors to individuals' learning differences. In: ACM/IEEE International Conference on Human-Robot Interaction (HRI), pp. 423–430. ACM (2014)
13. Maslow, A.H.: A theory of human motivation. Psych. Rev. 50(4), 370–396 (1943)
14. Niekum, S., Osentoski, S., Konidaris, G., Barto, A.G.: Learning and generalization of complex tasks from unstructured demonstrations. In: International Conference on Intelligent Robots and Systems (IROS), pp. 5239–5246. IEEE (2012)
15. Paraschos, A., Daniel, C., Peters, J., Neumann, G.: Probabilistic movement primitives. In: Advances in Neural Information Processing Systems (NIPS) (2013)
16. Song, W.K., Song, W.J., Kim, Y., Kim, J.: Usability test of KNRC self-feeding robot. In: International Conference on Rehabilitation Robotics, pp. 1–5 (2013)
17. Topping, M.: An overview of the development of handy 1, a rehabilitation robot to assist the severely disabled. Intell. Robot. Syst. 34(3), 253–263 (2002)
18. Zhang, X., Wang, X., Wang, B., Sugi, T., Nakamura, M.: Real-time control strategy for EMG-drive meal assistance robot - my spoon. In: International Conference on Control, Automation and Systems (ICCAS), pp. 800–803 (2008)

A Framework for Modelling Local Human-Robot Interactions Based on Unsupervised Learning

Rafael Ramón-Vigo[1]([✉]), Noé Pérez-Higueras[1],
Fernando Caballero[2], and Luis Merino[1]

[1] School of Engineering, Universidad Pablo de Olavide,
Crta. Utrera km 1, Seville, Spain
{rramvig,noeperez,lmercab}@upo.es
[2] Department of System Engineering and Automation,
University of Seville, Camino de Los Descubrimientos, s/n, Seville, Spain
fcaballero@us.es

Abstract. This paper addresses the problem of teaching a robot interaction behaviors using the imitation learning paradigm. Particularly, the approach makes use of Gaussian Mixture Models (GMMs) to model the physical interaction of the robot and the person when the robot is teleoperated or guided by an expert. The learned models are integrated into a sample-based planner, an RRT*, at two levels: as a cost function in order to plan trajectories considering behavior constraints, and as a configuration space sampling bias to discard samples with low cost according to the behaviors. The algorithm is successfully tested in the laboratory using an actual robot and real trajectories examples provided by an expert.

1 Introduction

In the TERESA Project[1], telepresence robots are enhanced to navigate autonomously in social settings. The project considers the development of techniques for safe and efficient obstacle avoidance while reaching navigation goals. This task becomes more challenging when people are considered and explicitly modeled into the navigation approach.

Human-aware, or social, navigation is a complex task that has been addressed using different approaches in the robotics community. Many novel approaches are based on learning socially acceptable behaviors from real data collected under various social situations, avoiding the need of a handcrafted explicit formulation of the behaviors. For instance, supervised learning is used in [16] to learn appropriate human motion prediction models that take into account human-robot interaction when navigating in crowded scenarios. Unsupervised learning is used by Luber et al., [11] to determine socially-normative motion prototypes, which are then employed to infer social costs when planning paths. A model based on social forces is employed in [7], where the parameters for the social forces are learnt from feedback provided by users.

---

[1] http://teresaproject.eu/.

A. Agah et al. (Eds.): ICSR 2016, LNAI 9979, pp. 32–41, 2016.
DOI: 10.1007/978-3-319-47437-3_4

An additional approach is learning from demonstrations, or imitation learning, [2]: an expert teaches the robot how it should navigate among humans. We can leverage the fact that we have a telepresence robot in the TERESA Project, so we can extract useful information from the users of the robot (on the pilot side). Having examples of (teleoperated) robot paths and the relevant configurations (features) of the performed task opens the door for extracting the relevant relations or constraints that best represent such kind of paths. The main hypothesis is that these paths enclose the social implications that a human takes into account when he is performing such task, at least in the same situations performed in the experiments.

Inverse Reinforcement Learning (IRL) [1] techniques are a good candidate to derive such models: a reward (or cost) function is recovered from the expert behavior, and then used to obtain a corresponding robot policy. In [8], a path planner based on inverse reinforcement learning is presented. IRL for social navigation is also considered in [13].

The above mentioned approaches typically make use of discrete Markov Decision Processes (MDPs) as the underlying model. However, it is complex to encode general problems with MDPs due to its computational complexity [12]. Our objective is to use state of the art sampling-based planners, as optimal Rapidly exploring Random Trees (RRT*), to be able to work on continuous configuration spaces. These planners already reason about obstacles present in the environment, and the goal is to incorporate into them information about the social task at hand from data.

Gaussian Mixture Models (GMMs) offer a flexible framework to model the relationships between the relevant features that arise when the robot is performing a particular navigation task. GMMs are a well-suited representation for unsupervised extraction of continuous feature distributions, and they have also shown their utility as models for robot skills representations in Programming by Demonstration (PbD) settings [3].

In [5], the author presents a PbD framework in which GMMs are used to retrieve the statistical constraints of several demonstrations of a particular task, in a manner similar to the approach in [4]. After that, a sampled-based planner based on RRT is used. In this paper we aim to go a step further and use GMMs to incorporate robot navigation behavior into a cost-based RRT* planner. The goal is to find a safe path which imitates a behaviour by remaining within statistically determined constraints. For doing this, we propose, first, to bias the RRT* random samples towards the regions of the configuration space that comply with the model of the task extracted from data and, second, including a new cost function into the RRT* planner to better account for paths that follow the learned behaviors. At the same time, this permits better generalization to new situations by still finding short length paths for different conditions.

The paper is structured as follows: Sect. 2 takes care about the tasks being demonstrated by the user while Sect. 3 introduces GMMs. Section 4 details how the GMM can be included into the RRT* planner. Then, Sect. 5 presents the experimental setup, the metrics and some simulations to validate both GMM learning and its integration with a sampling-based planner. Finally, Sect. 6 shows the conclusions and future works.

2   Demonstrated Tasks

As described before in the introduction, human-awareness is critical for a successful deployment of a robotic application in a space shared with persons. In the TERESA Project we are interested in several social situations, such as avoiding people while navigating, approaching a person to start a conversation, following a person or keep a conversation while moving to another place. For the sake of simplicity, in this work we focus in two particular tasks in order to illustrate this approach.

Avoiding. The robot avoids an standing person that is facing to it. The avoidance maneuver can be performed by passing through by the left or by the right.

Approaching. The robot approaches a standing person in an arbitrary orientation. When the person is looking towards the robot, it performs the shortest path to approach the person. When the standing person is back to the robot, the demonstration trajectories do not follow the shortest path. Rather, the robot tends to take curving paths, also by the left or by the right.

In this work, a Giraff robot is teleoperated by an expert. This user is asked to demonstrate the previous tasks as accurately as necessary a number of times by means of piloting the robot.

3   GMMs for Interaction Modeling

A proper choice of the relevant features $\mathbf{f} = [f_1, f_2, \ldots, f_n]^T$ when encoding a particular navigation task is crucial, as it provides part of the solution to the problem of defining what is important to imitate. In this paper, we have considered as features the distance and the relative angle between the robot and the person in the scene, logged with a timestamp. The features extracted at each time instant are the distance to that person (d) and the relative angle (θ). Thus, a set composed by N datapoints $\zeta = \{\zeta_j\}_{j=1}^{N}$ of $D = 2$ dimension is considered, where time is left out because the dynamics of the behaviours are not considered in this paper.

According to a previous work [14], the features $(d - \theta)$ considered here allows us to model the tasks at hand. However, the selection of features is not a limitation of the technique, and other features can be added for more complicated tasks, including those ones that can describe the dynamics of the task execution, like time and velocities.

From the demonstrated trajectories of the features $\mathcal{D}$, it is possible to extract a GMM, so that the probability of a particular combination of feature values is given by:

$$p(\mathbf{f}|\mathcal{D}) = \sum_{i=1}^{k} \omega_i \mathcal{N}(\mathbf{f}; \mu_{\mathcal{D}}^i, \Sigma_{\mathcal{D}}^i) \tag{1}$$

with k Gaussian *modes*. The GMM is then described by the set of parameters $\{\omega_i, \mu_{\mathcal{D}}^i, \Sigma_{\mathcal{D}}^i\}_{i=1}^{K}$, respectively representing the *prior* probabilities, *centers* and *covariance* matrices of the model. The *prior* probabilities, ω_i, satisfy $\omega_i \in [0,1]$ and $\sum_{i=1}^{K} \omega_i = 1$.

GMM parameters are learnt by using the *Expectation-Maximization* (EM) algorithm [6] that is seeded with an initial estimate of density centers calculated with the k-means algorithm. A drawback of EM is that the optimal number of components k in a model may not be known beforehand. One usual criterion for model selection is the Bayesian Information Criterion (BIC) [15], but for the experiments presented in this work, we have tested empirically the best number of components that can fit the demonstrated example.

4   The Reproduction Planner

RRT* [10] is a technique for (asymptotical) optimal motion planning. It considers that a cost function is associated to each point $\mathbf{x}$ in the configuration space (a vector representing the position of the robot in our case). The RRT* seeks to obtain the trajectory ζ^* that minimizes the total cost along the path $c(\zeta)$. It does so by randomly sampling the configuration space and creating a tree towards the goal. The paths are then represented by a set of discrete configuration points $\zeta = \{\mathbf{x}_1, \mathbf{x}_2, \cdots, \mathbf{x}_N\}$. Each point of the trajectory can be also associated to the values of the features for that point, so that it can be also seen as a trajectory described in feature space $\zeta = \{\mathbf{f}_1, \mathbf{f}_2, \cdots, \mathbf{f}_N\}$. This paper extend the standard RRT* algorithm with the learned GMM at two levels:

1. Including a new task-similarity cost into the evaluation of the node's costs. The GMM obtained encompasses the most likely configurations of the task. Thus, when a node is proposed to be added to the RRT* tree, a cost based on the GMM is derived and used. The objective is to increase the cost of those configurations that are unlikely according to the learnt GMM. Thus, the likelihood is inverted to obtain that cost. Notice that the likelihood is a density of occurrence, so it is necessary to give a low bound to keep the inverse within the interval [0,1]. To this end, it has been chosen to truncate the likelihood to a certain low value δ.
2. Providing the planner with the most likely subspace to perform the sampling of the RRT*. If the RRT* knows what are the most likely paths, then we can bias the sampling of the configuration space to these areas, reducing the probability of sampling useless states and, hence, reducing the computational costs to obtain a solution. A parameter determines the amount of GMM bias introduced into the planner.

5   Experimental Results

5.1   Gathering the Data for Learning

Several experiments were performed to retrieve exemplary trajectories to feed the GMM learning phase. Those experiments took place inside a clear room, free of obstacles between the person and the robot while performing the scenes described in Sect. 2. A set of 9 trials for each homotopy in each task were logged. The study was recorded using a motion capture system (OptiTrack) and the robot and person's poses were extracted automatically from the data.

The gathered data is used to derive the GMM models of the features using the method described in Sect. 3, one model for each task. Those models are then integrated in the RRT* as explained in Sect. 4. This method is called GMM-RRT*. A set of GMM models in $x - y$ space have been also derived in order to test the GMM-RRT planner [5], used as a baseline. The parameters used was $k = 19$ GMM modes (for each task and method) and a value of $\delta = 0.001$.

5.2   Metrics

We propose two metrics to compare the obtained paths from the different planners and with respect to the demonstrated trajectories, as used in [9]:

The first metric is called *Trajectory Difference Metric* (TDM), which is defined as follows:

$$TDM(\zeta_D, \zeta_P) = \frac{1}{|\zeta_D|} \sum_{i=1}^{|\zeta_D|} \min \overline{\zeta_D(i)\zeta_P} = \frac{1}{|\zeta_D|} \frac{1}{|\zeta_P|} \sum_{i=1}^{|\zeta_D|} \sum_{j=1}^{|\zeta_P|} \min \overline{\zeta_D(i)\zeta_P(j)} \quad (2)$$

where $\zeta_P(j)$ and $\zeta_D(i)$ are the points of the two trajectories ζ_P and ζ_D to be compared, and $\overline{\zeta_D(i)\zeta_P(j)}$ is the distance between two points. $|\zeta_D|$ and $|\zeta_P|$ are the number of samples of each trajectory. This metric gives an idea of the similarity of two given trajectories. The final metric is given by the averaged value of this metric for all the planned and demonstrated trajectories:

$$TDM = \frac{1}{|D|} \frac{1}{|P|} \sum_{D,P} TDM(\zeta_D, \zeta_P) \quad (3)$$

where D and P are the number of Demonstrated and Planned trajectories, respectively.

The second metric is the resulting averaged trajectory length ratio l_e, expressed as the mean of the ratio of the absolute value of the difference between the planned trajectories and the demonstrated trajectories lengths divided by the demonstrated trajectories length:

$$l_e = \frac{1}{|D|} \frac{1}{|P|} \sum_{D,P} \frac{|l(\zeta_D) - l_(\zeta_P)|}{l(\zeta_D)} \quad (4)$$

5.3   Results

Costs Evaluation. This section is oriented to evaluate the convergence speed of the RRT*-base planner to the optimal path in term of costs. Figure 1 shows the evolution of the solution path cost versus the number of iterations using 100 % and 0 % GMM bias, for the "Avoiding a Person" task. The allowed *planning time* for this comparison was 100 s in order to converge to the optimal cost. It can be seen in this example how the planner that includes the GMM *sampling* is faster.

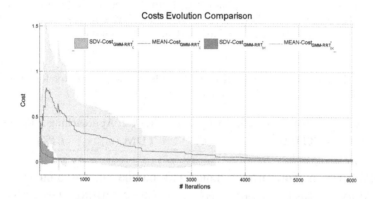

Fig. 1. Mean cost value and the standard deviation obtained for an GMM-RRT* with (red and green, subindex *bc*) and without (blue, subindex *c*) GMM sampling bias. (Color figure online)

Metrics Performance. For the following results a *mixed-sampling* strategy has been adopted: 95 % of the time a GMM sampling is employed, while the remaining 5 % it is the *uniform* sampling. We aim to take advantage of the learned models and also allow a degree of randomness when sampling configurations. This feature is only applicable to the GMM-RRT* planner presented in this paper. The construction and the use of the GMM model for the GMM-RRT planner must satisfy certain restrictions that make sampling possible only on consecutive nodes. For further details please consult [5]. The planner is given enough *planning time* to converge. Figure 2 shows a complete set of 25 trials for each task, homotopy and planner, with the parameters explained before.

Regarding on how well both planners imitate the demonstrated trajectories, Table 1 shows the values obtained when the different planners are used in both tasks based on the metrics related in Sect. 5.2. It can be seen in that table that the planner proposed in this work outperforms in nearly every homotopy considered.

Figure 3 shows the same comparison of the metrics for different *planning times* for the GMM-RRT* approach, in contrast with the metric values obtained for the GMM-RRT (first two columns in the figure).

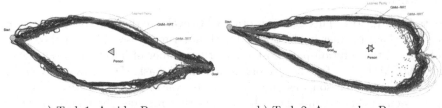

a) Task 1: Avoid a Person.   b) Task 2: Approach a Person.

Fig. 2. Demonstrated (green) and planned trajectories (red for GMM-RRT* and blue for GMM-RRT) are depicted. The "Start" and "Goal" states are also shown, while the "Person's" gaze is represented by a triangle. (Color figure online)

Table 1. Trajectory quality for both tasks. Smaller values are better for all metrics. The best values are highlighted in boldface.

Planner	Task 1: Avoiding a Person		Task 2: Approaching a Person	
	TDM (m)	$l_e(\%)$	**TDM** (m)	$l_e(\%)$
Right Homotopy				
GMM-RRT*	**0.0565** ±0.0152	**3.66** ±2.64	**0.0906** ±0.0176	**3.69** ±2.54
GMM-RRT	0.0803 ± 0.0205	4.00 ± 2.51	0.0917 ± 0.0161	4.96 ± 3.45
Left Homotopy				
GMM-RRT*	**0.0484** ±0.0109	4.96 ± 2.37	**0.0612** ±0.0218	**4.22** ±2.50
GMM-RRT	0.0676 ± 0.0169	**4.31** ±3.1	0.0666 ± 0.0169	4.56 ± 3.62
Frontal Homotopy				
GMM-RRT*			0.0376 ± 0.0141	11.35 ± 5.35
GMM-RRT			**0.0337** ±0.0184	**8.55** ±3.34

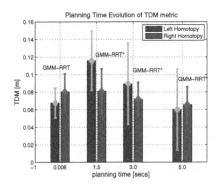
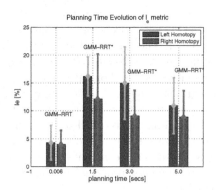

Fig. 3. Mean and standard deviation of the metrics for different planning times and the "Avoiding a Person" task.

We have started with a planning time of 1.5 s because it is the mean time in which the proposed GMM-RRT* planner is able to find the first solution for the tasks presented here and for the bias of 95 %. For 3 s the GMM-RRT* obtain comparable results on the TDM metric.

Managing the Homotopies. As commented in Sect. 2, the tasks being learned by the robot may be composed by several homotopies. Figure 4 illustrates the models of the Approaching a Person task.

Fig. 4. First-Left: The model includes the three demonstrated homotopies. Second-Left to Right: One model per each demonstrated homotopy.

This is a clear disadvantage of the GMM-RRT planner. Not only the social situation has to be known beforehand to choose between the homotopy models, but also once a model has been selected, it can occur that an obstacle hampers the execution of the plan. In such a situation, both planners can take advantage of the variability in the execution of the demonstrated task, encoded by the GMMs covariance matrices, to avoid the obstacle, but sometimes this could not be enough. For instance, in the situation shown in Fig. 5 it can be seen how the GMM-RRT planner is not able to find a free path to the goal, although it exists. However, the GMM-RRT* planner presented here is able to naturally choose between the available homotopies to reach the goal.

Fig. 5. "Avoiding a Person" task. Shadowed area represents an obstacle. Left: Model includes 2 homotopies, used by GMM-RRT*. Planning time: 3 s. Right: Only right homotopy, used by GMM-RRT. Planning time: 100 s.

Generalization. Regarding to the generalization of the proposed approach to other scenarios, the *approaching a person* task is modified so an obstacle (as a rectangular shadowed area in Fig. 6) is introduced in the robot path to its goal: The GMM-RRT* planner still can reach the goal due to the *mixed-sampling* strategy adopted: the *uniform* bias sampling allows to random explore the state space outside the learned demonstrations.

 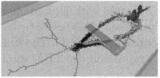

Fig. 6. Left: Simulated trial without *uniform* sampling. Right: 95 % *GMM* and 5 % *uniform* sampling.

The simulations shown in Fig. 6 were within a 100 s time horizon. A 100 % GMM sampling was not able to find a path that can handle with the obstacle included. If a *mixed-sampling* strategy is allowed, within the same time horizon, it can be seen how the goal is reached successfully. The percentage of bias, thus, offers a trade-off between the imitation capabilities and the planning time required to obtain a path. An adaptive solution that modifies this bias after obtain a first good path is left for future work.

6   Conclusions and Future Work

In this paper we present a planning algorithm based on RRT*, which is capable of infer the most suitable socially-aware paths in two situations (or tasks): avoiding a standing person and approaching forwards or backwards a standing person. Both tasks are statistical characterized from real experiments by using two different GMMs, which are later used in two ways: to guide the state-space sampling step and to include a cost term into the standard RRT* algorithm.

We evaluate the approach presented here jointly with a state of the art algorithm based on RRT and GMM. The comparison includes a metric performance to measure the similarity between the planned trajectories with the learned ones, for different planning times, and how both approaches perform in simulated scenarios that can include slightly differences from where they were learned.

Although the RRT based planner is quicker to get a suitable path, the approach presented here is able to manage homotopies and generalizes better when tackling with unexpected situations (such as obstacles).

Both tasks described in this work are only a simple and specific use case of the proposed planner. As future work we plan to evaluate the use of other features including velocity, time and some environment descriptors that could improve the generalization of the learned behavior to these and new tasks. Also, we will perform real user tests to analyze how the learning affects the readability of the learned behavior.

References

1. Abbeel, P., Ng, A.Y.: Apprenticeship learning via inverse reinforcement learning. In: Proceeding of the Twenty-First International Conference on Machine Learning, ICML 2004, pp. 1–6. ACM, New York (2004)

2. Argali, B., Chernova, S., Veloso, M., Browning, B.: A survey of robot learning from demonstrations. Robot. Auton. Syst. **57**, 469–483 (2009)
3. Calinon, S.: Robot Programming by Demonstration: A Probabilistic Approach. EPFL/CRC Press, Boca Raton (2009)
4. Calinon, S., Billard, A.: A probabilistic programming by demonstration framework handling constraints in joint space and task space. In: Proceeding IEEE/RSJ International Conference on Intelligent Robots and Systems, IROS (2008)
5. Claassens, J.: A RRT-based path planner for use in trajectory imitation. In: Proceeding of the International Conference on Robotics and Automation, ICRA, pp. 3090–3095. IEEE (2010)
6. Dempster, A.P., Laird, N.M., Rubin, D.B.: Maximum likelihood from incomplete data via the EM algorithm. J. Roy. Stat. Soc.: Ser. B **39**(1), 1–38 (1977)
7. Ferrer, G., Garrell, A., Sanfeliu, A.: Robot companion: a social-force based approach with human awareness-navigation in crowded environments. In: Proceeding of the IEEE/RSJ International Conference on Intelligent Robots and Systems, IROS. pp. 1688–1694 (2013)
8. Henry, P., Vollmer, C., Ferris, B., Fox, D.: Learning to navigate through crowded environments. In: Proceeding of the International Conference on Robotics and Automation, ICRA, pp. 981–986 (2010)
9. Islas Ramírez, O., Khambhaita, H., Chatila, R., Chetouani, M., Alami, R.: Robots learning how and where to approach people. In: RO-MAN 2016 25th, IEEE International Symposium on Robot and Human Interactive Communication (2016, to appear)
10. Karaman, S., Frazzoli, E.: Sampling-based algorithms for optimal motion planning. Int. J. Robot. Res. **30**(7), 846–894 (2011)
11. Luber, M., Spinello, L., Silva, J., Arras, K.: Socially-aware robot navigation: a learning approach. In: Proceeding of the IEEE/RSJ International Conference on Intelligent Robots and Systems, IROS, pp. 797–803. IEEE (2012)
12. Okal, B., Gilbert, H., Arras, K.O.: Efficient inverse reinforcement learning using adaptive state-graphs. In: Learning from Demonstration: Inverse Optimal Control, Reinforcement Learning and Lifelong Learning Workshop at Robotics: Science and Systems (RSS), Rome, Italy (2015)
13. Perez-Higueras, N., Ramon-Vigo, R., Caballero, F., Merino, L.: Robot local navigation with learned social cost functions. In: Proceeding of the 11th International Conference on Informatics in Control, Automation and Robotics, ICINCO, vol. 02, pp. 618–625 (2014)
14. Ramon-Vigo, R., Perez-Higueras, N., Caballero, F., Merino, L.: Analyzing the relevance of features for a social navigation task. Robot 2015: Second Iberian Robotics Conference. AISC, vol. 418, 10.1007/978-3-319-27149-119 edn, pp. 235–246. Springer, Heidelberg (2016)
15. Schwarz, G.: Estimating the dimension of a model. Ann. Stat. **6**, 461–464 (1978)
16. Trautman, P., Krause, A.: Unfreezing the robot: Navigation in dense, interacting crowds. In: Proc. of the IEEE/RSJ International Conference on Intelligent Robots and Systems, IROS, pp. 797–803. IEEE (2010)

Using Games to Learn Games: Game-Theory Representations as a Source for Guided Social Learning

Alan Wagner[1,2(✉)]

[1] Georgia Institute of Technology, Atlanta, GA, USA
[2] Pennsylvania State University, State College, PA, USA
azw78@psu.edu

Abstract. This paper examines the use of game-theoretic representations as a means of representing and learning both interactive games and patterns of interaction in general between a human and a robot. The paper explores the means by which a robot could generate the structure of a game. In addition to offering the formal underpinnings necessary for reasoning about strategy, game theory affords a method for representing the interactive structure of a game computationally. We investigate the possibility of teaching a robot the structure of a game via instructions, question and answer sessions led by the robot, and a mix of instruction and question and answer. Our results demonstrate that the use of game-theoretic representations may offer new advantages in terms of guided social learning.

Keywords: Game theory · Social learning · Interactive games

1 Introduction

Social interaction often involves stylized patterns of interaction [1]. These patterns may dictate how and when a person interacts, what actions they choose, and how their goals and motivations change. Interactive games, such as Rock-Paper-Scissors and poker, often structure a person's interactive behavior in a predetermined manner conducive to the game. Recently artificial systems have become adept at both playing and learning how to play many different games [2]. Comparatively little attention, however, has been paid to the social aspects of game playing and game learning. For instance, how can a social robot or agent learn to play a game from a person offering only disorganized verbal instructions? How can a system teach a person to play a game using subtle social cues and questions to determine if and to what extent they understand? How can such a system be developed to cope with the differences in play, instructions, and learning that occur across ages and cultures?

This paper constitutes a preliminary examination of these questions. The overarching goal of this work is to develop a system that could learn a wide variety of games from the type of interactive instructions provided by a typical person. Hence, we strive for generality both with respect to the game and the instructor. Moreover, we believe that our approach can also work when the robot acts as the instructor, explaining how to play a game. An important initial step towards creating such a system is to determine how to computationally represent interactive games.

© Springer International Publishing AG 2016
A. Agah et al. (Eds.): ICSR 2016, LNAI 9979, pp. 42–51, 2016.
DOI: 10.1007/978-3-319-47437-3_5

Game theory researchers have extensively studied the representations and strategies used in games [3]. The types of games examined as part of game theory, however, tend to differ from our common notion of interactive games. Games in game theory tend to encompass limited interactions over a small range of behaviors and are focused on a small number of well-defined interactions. The Ultimatum Game, for example, requires one individual to divide a valuable resource while the other individual in the game can accept the division and receive a share or reject the division and both players receive nothing. Moreover, game theory focuses on conceptualizations for strategic interaction. In contrast, interactive games like Monopoly and poker offer players several different actions as part of a sequential ongoing interaction in which a player's motives may change as the game proceeds or depend on who is playing.

We contend that learning a pattern of interactions, such as those used in most interactive games, is a critical component for human-robot interaction because many interpersonal interactions follow prescribed patterns [4]. Methods developed for learning the structure of an interactive game could potentially be applied to the human-robot interaction scenarios encountered in a wide variety of social environments. For example, in most western cultures when meeting a new person the expected pattern of interaction is to introduce oneself, to shake the other person's hand, and to then wait for the other person to state their name.

This paper investigates methods by which a robot could learn the structure of an interactive game from a person. We focus on direct instruction. In particular, this article demonstrates the use of written instructions and the use of questions by the robot that, when answered by a person, convey the structure of the game. Further, we show that the robot can use a game-theoretic representation to reason about and select specific probing questions with the intention of learning about unknown aspects of the game. Overall, our immediate goal is to highlight the potential advantages of this approach in terms of teaching a robot these stylized patterns of interaction. Our long-term goal is to develop the computational underpinnings that will allow a robot to learn new patterns of interaction from an inexperienced person's instructions. The remainder of the paper begins with a brief background discussion of game theory and interactive games, followed by experiments and results.

2 Background and Related Work

Game theory has been the dominant approach for formally representing strategic interaction for more than 80 years [3]. Game theory assumes that the players of a game will pursue a rational strategy. A game is a formal representation of a strategic interaction among a set of players. A solution to a game describes classes of strategies for how best to play a game. There are many different types of solution concepts in game theory, the Nash Equilibrium being the most famous example of a solution concept.

Several different categories of games exist [3]. Games in which players select actions simultaneously are typically represented as a normal-form game (Fig. 1 center). Formally, a normal-form game is defined as a tuple $\left(N, A^{1,\dots,N}, R^{1,\dots,N}\right)$ where N is the set of players, A^i is the action space of individual i, and $R^i\left(a^1,\dots,a^N\right) \to \Re$ is a payoff function. Games in which players select actions sequentially are generally represented as extensive-form games (Fig. 1 left). In addition to the formal elements of a normal-form game, extensive-form games include a set of histories H for each player and function $P(h)$ for selecting the player whose turn is next. Perfect information games are a class of extensive-form games in which each player knows every player's history. In imperfect information games players do not know the actions chosen by other players. A stochastic game is a series of normal-form games in which the actions selected in one game probabilistically determine the subsequent game. Stochastic games include a transition function $T(q, a^1, \ldots, a^N, q') \rightarrow [0, 1]$. These games are generalizations of both normal-form and extended-form games. They also generalize Markov Decision Processes (MDPs) for multiple individuals. Stochastic games start at initial state s_0 and are played with each player selecting an action and possibly receiving a payoff. The game moves to stage s_{i+1} with probability determined by the distribution Q until reaching a termination state. A stochastic game may last either a finite or infinite number of stages.

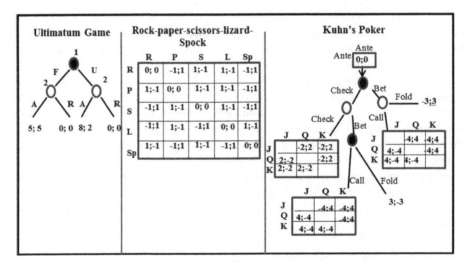

Fig. 1. Computational representations for the Ultimatum game, Rock-paper-scissors-lizard-Spock and Kuhn's poker are depicted above. The Ultimatum game is a sequential game between two players represented as an extensive-form game. Rock-paper-scissors-lizard-Spock is a simultaneous game represented as a normal-form game. Kuhn's poker is represented as a mixture of normal and extensive-form games. The selection of an action by a player results in a transition to the next stage of the game.

Most robotics related applications of game theory have focused on game theory's traditional strategy specific solution concepts [5]. Often, the structure of the game is preprogrammed and a game theory based controller is used to select the agent's actions. Recently this approach has resulted in tremendous advances in the quality of play in information imperfect games such as poker [6]. Nevertheless, this research is narrowly tailored to the development of agents that play optimally. In contrast, the work here is not concerned with how well an agent or robot plays, but rather its ability to learn and represent different, unknown games.

Game theory has also been used as a means for controlling a robot [5, 7]. Game theory based robot control has similarly focused on optimization of strategic behavior by a robot in multi-robot scenarios. In particular, the use of Partially Observable Stochastic Games has been used as a means to control a robot team. In contrast to the prior work, we explore methods that will allow a human to teach a robot the structure of an interactive game such as Rock-Paper-Scissors and poker. Significant research has also explored the development of robots that learn games such as air hockey [8]. In contrast to strategic games, games such as air hockey tend to emphasize the physical and perceptual demands of play. Robot soccer, because of its dual physical and strategic demands, arguably represents the most challenging category of game. Research related to this game has explored both the physical demands [9] and the strategic demands [10].

Very little work has examined the use of game theory as a means for controlling a robot's interactive behavior with a human. Lee and Hwang attempt to develop a conceptual bridge from game theory to interactive control of a social robot [11]. Our own work has centered on the use of the normal-form game as a representation and means of control for human-robot interaction [12]. Yet, in this prior work we focused only on the use of the representation to control a robot's behavior and not the direct learning a game's interactive structure.

3   Representing Interactive Games

Game-theory representations have been used to formally represent and reason about a number of interactive games [13]. Games such as Snakes and Ladders, Tic-Tac-Toe, and versions of Chess have all been explored from a game theory perspective. The methods used to represent these games are well known.

The normal-form game and the extensive-form game serve as building blocks to represent a complete interactive game. Simultaneous stages of an interactive game are represented in normal-form as a matrix (Fig. 2). Each player's potential actions are listed along the dimensions of the matrix. Payoffs for selecting a particular set of actions are included as values within the matrix. Sequential stages of an interactive game are represented in extensive-form as a tree. A player's potential actions are denoted by the branches of the tree. Nodes of the tree indicate which player makes a decision at each particular stage of the game. Payoffs for selecting a particular set of actions are depicts at the stage in which the payoffs are received.

The cells in a normal-form game and the terminating branches of an extensive-form game direct the players to the subsequent stages of the game. Resembling a probabilistic-finite-state automaton (FSA), each state is a normal-form or extensive-form game representation and each transition occurs when arriving at a cell or a terminal tree node. Represented in this manner, the challenge of learning a new interactive game is reduced to learning the structure and underlying components of the game-theoretic representation. The section that follows investigates this challenge.

**Components of an Interactive Game
Representation**

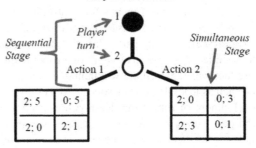

Fig. 2. Normal and extensive form games are used to represent the components of an interactive game. Sequential stages are represented in extensive form. Numbers are used to indicate a player's turn. Simultaneous stages are represented as a normal form game. Transitions connect components and denote the selection of an action.

4   Teaching a Robot an Interactive Game

The primary contribution of this work is to examine, present, and demonstrate techniques for learning the types of game-theoretic representations described above from the information provided by a human teacher. The most obvious and applicable methods for learning the structure of an interactive game are direct instruction and question and answer.

Direct instruction describes the explicit teaching of skills needed for some purpose. Some psychologists argue that direct instruction represents the most effective way to teach and to learn [14]. To directly instruct a robot to play a game, the human teacher simply communicates the underlying structure of the interactive game to be learned. This communication can be in the form of a list of spoken, written, or demonstrated instructions necessary for performing a task. Written instructions can be used in place of verbal instructions. In this case, an interactive game's set of instructions can be used to learn a new game.

The use of a game-theoretic representation requires that specific information is communicated to the robot. In general, for each stage of the game, the robot must know who is playing, what actions are available to each player, what reward or cost is associated with the selection of each action pair, whether actions are selected simultaneously or sequentially by the players, and which stage of the game results from the selection of an action pair. When direct instruction is used, these questions are addressed directly as a list of spoken, written, or demonstrated instructions.

A less obvious means for teaching a robot an interactive game is to allow the robot to ask questions about the game that the person answers. In this case, the robot acts as an inquisitor asking questions that allow it to build the game representation from the ground up. **The evolving game representation in this case determines what questions the robot must ask in order to flesh out the representation**. The first question asks how many players are participating. For each stage of the game, the robot then asks whether the players act sequentially or simultaneously. Next, the robot inquiries about

**Sample Question and Answer session for
learning an Interactive Game**

Q1: How many players for stage 0?
A1: *2*
Q2: Is stage 0 sequential or simultaneous?
A2: *sequential*
Q3: Which player acts at this stage?
A3: *1*
Q4: What actions are available to Player 1 at stage 0?
A4: *Action1*
Q5: If player 1 at stage 0 selections Action1 does a
sequential or simultaneous stage result?
A5: *sequential.*
Q6: Which player acts at this stage?
A6: *2*

Fig. 3. An example of a question and answer session for learning the first stage of the interactive game from Fig. 2.

the actions available to each individual at that stage. For each cell in a normal form game or terminal node in an extensive form game, the robot inquiries about the reward received and which, if any, subsequent stage results from the selection of the action pair. Figure 3 depicts an example of a question and answer session used to learn the first stage of the interactive game depicted in Fig. 2. We contend that a similar series of questions can be used in either a depth first or breadth first manner to learn the interactive structure of most games.

The proceeding techniques can be combined resulting in a system in which the robot builds the game representation from the instructions and then asks questions about any unknown or unclear parts of the representation. In this case the robot must determine which portions of the representation are unclear or unknown. In some cases the presentation of the instructions may afford measures of confidence with respect to the instructions. For instance, many natural-language-processing (NLP) algorithms provide confidence measures reflecting the system's estimation of accuracy. We conjecture that such a system could potential be used to assist the robot in determine if and what portions of the representation require follow up questions.

In some cases, the representation itself may suggest information that is missing. For example, the absence of reward or cost values in a matrix (e.g. the numbers 2; 5 from Fig. 2) is easily tested. In this case, the absence of expected reward values can prompt the robot to inquire about the value or cost of selecting particular actions during a stage of the game.

5   Experiments

Most applications of game theory evaluate the system's performance in terms of winning (e.g. [15]) or win related tasks such as scoring goals (e.g. [8]). In contrast, we argue that the best evaluation of game learning is to measure the system's ability to play a game after being taught, regardless of whether it wins. Metrics such as illegal moves attempted, measure the accuracy of the robot's model of the game structure.

We evaluated the number of illegal moves attempted by the robot in three different games (Fig. 1): the Ultimatum game, Rock-paper-scissors-lizard-Spock, and Kuhn's poker. The Ultimatum game is a single stage sequential game in which one player chooses either a fair or unfair division of a resource and a second player either accepts or rejects the division. If the division is accepted then both players receive reward proportional to the division. Alternatively, if the division is rejected then both players receive nothing. Rock-paper-scissors-lizard-Spock is similar to the classic game rock-paper-scissors except with two additional actions. The lizard action defeats Spock and paper and is defeated by scissors and rock. The Spock action defeats scissors and rock and is defeated by paper and lizard. Figure 1 delineates which actions dominate other actions. In this game all players simultaneously make a hand sign representing one of the five namesakes. Finally, Kuhn's poker is a simplified version of Texas Hold'em poker. This game is played with only a jack, a queen, and a king. The game begins when each player bets 1 as an ante. Next each player receives a single card. Player 1 may check or bet. As depicted in Fig. 1 the actions available to player 2 depend on player 1's action. Each round of the game ends when a player either folds (resigns and forfeits their bets) or during a showdown stage each player's cards are revealed and the player with the higher card wins.

The robot learned each of the three games by direct instruction and mixed-direct instruction and question and answer. In the direct instruction condition, the robot was given a set of instructions (e.g. Fig. 3) describing how to play the game. The instructions were not in a natural-language format. Although, the challenge of translating from natural language to a game theory format is beyond the scope on this article, random errors were added to the instructions in an effort to roughly simulate the errors that would occur during translation. Each game instruction had a 15 % chance of being incorrect (translation error rate). This level of error was arbitrarily selected. Three different types of error occurred. Incorrect stage transitions occurred when the robot's representation erroneously indicated the stage that would result when a pair of actions was selected. Incorrect reward values inaccurately specified the amount of reward to be received at a stage of the when a pair of actions is selected. Finally, incorrect actions erroneously indicated which actions were available to the robot at a particular stage of the game.

The experimenter served as the robot's opponent. All of the experimenter's instruction and responses were predetermined to avoid bias. When the robot asked the human for missing information, the correct information was provided. The quantitative results that follow were obtained from an experiment conducted in simulation.

The data from the direct learning from instructions condition (Fig. 4) demonstrates that the robot selected illegal moves at a rate of 16, 11, and 17 % for the Ultimatum game, Rock-paper-scissors-lizard-Spock game, and Kuhn's poker respectively. These results indicate a rate of illegal moves which is approximately equal to the translation error rate. This rate is higher than expected. We hypothesized that translation errors related to the amount of reward would only impact strategy and not whether or not a move was illegal. We therefore believed that, in the first condition, the number of illegal moves would be significantly less than the translation error rate. We found, however,

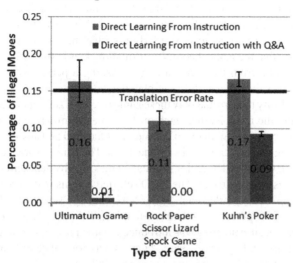

Fig. 4. Results from an experiment examining the possibility of a robot learning a game-theoretic representation of a game. The blue (left) columns depict a condition in which the robot learns the game from a set of imperfectly translated instructions. The red (right) columns depict a condition in which the robot is provided with instructions describing how to play the game which are missing information. The robot infers which information is missing and then asks the person to questions which allows it to complete the representation. (Color figure online)

that incorrect reward values can impact game structure by consistently guiding the robot towards illegal moves. In other words, translation errors can cause the robot to believe that it will obtain a large reward by performing an illegal move. This results in a strategy of using illegal moves which predominates. The data also shows that certain game structures appear to be more impacted by translation errors than other games. The Rock-paper-scissors-lizard-Spock game, for instance, was consistently found to result in fewer illegal moves when compared to the other games. Because this game consists of a single simultaneous stage, most errors do not result in illegal moves. Sequential games, on the other hand, afford multiple opportunities for selecting illegal moves.

In the second condition, the robot received instructions that had missing information in the form of reward values, potential moves, and game stage transitions. The robot then had the opportunity to ask the person questions about any information that it could identify as missing. Missing stage transitions could typically be inferred from the presence of stages not connected to the start state or some later stage of the game. Similarly, missing actions were often indicated by stage transitions without a requisite action pair. Missing reward values were easily inferred from the game-theoretic representation. The robot then asked the human to provide the missing information. The robot used question and answer to generate error-free representations of the game Ultimatum game and Rock-paper-scissors-lizard-Spock game. Because these games consisted of a single sequential or simultaneous stage, the robot could accurately infer which information was missing. Kuhn's poker, however, presented unique challenges in terms of inferring missing information. Although missing reward values and transitions were identified, missing actions were seldom noticed. If a stage transition and an action during the stage were both missing, then inference that an action was missing was not possible. Actions that did not result in transitions were similarly not identified as missing. Overall, the results demonstrate that the game theoretic information does assist with inferring which information is missing from the game structure. Asking a person to provide missing information improves the robot's ability to play the game.

We tested the ability to learn and play these games on the NAO robot from Aldebaran. During this testing, the robot learned each of the games from written instructions and question and answer sessions with the experimenter. Question and answer sessions were conducted by typing answers to the robot's questions. The games were then played with the robot. The robot verbalized its actions instead of making physical actions. Each game was played 10 times with the robot. The NAO selected actions that were believed to be reward maximizing. We recorded each of the NAO's action selections. The robot was able to learn each of the games using both written instructions and question and answer session. However, because no error was introduced, game play was structurally perfect. Hence the robot's ability to play the learned games was confirmed although no quantitative results from these robot experiments are reported.

6   Conclusion

This article has examined the use of game-theoretic representations as a means of representing and learning interactive games involving a human and a robot. Our experiments demonstrate that written instructions and mixed instruction and question and answer can be used to learn different types of interactive games. We have shown that the use of game-theoretic representations of interaction offer several important features. First, and perhaps most importantly, the game representation affords a means for organizing the information needed by the robot to learn an interactive game. The computational representation of a game can be used to structure the information being received by a person and guide the robot when asking questions.

The research presented here could be an important step towards the development of a system for human-robot guided learning. Such a system might one day allow people to teach a robot the games that the person would like to play with the robot Before a fielded application could be realized, some assumptions would need to be addressed. For instance, we assumed that the robot already possessed the knowledge of how to perform all game related actions. We believe that learning these actions is related to game learning but best achieved by using learning from demonstration.

This work represents an initial investigation into the possibility of using game-theoretic representations to structure an interactive game. An important next step is to develop a system that learns a game from a naïve human subject. Such a system would require some competence in natural language understanding. Spoken or read instructions [16] would be used to broadly develop the interactive structure of the game, socially guided questions would then be used to rectify unclear or unknown portions of the game, learning by demonstration would be used to learn how to perform the actions, and practice would result in the refinement of strategy. Although this paper has focused on learning how to represent these games, we believe that these representations could be used in many different interactive situations.

References

1. Kelly, H.H.: The theoretical description of interdependence by means of transition lists. J. Pers. Soc. Psychol. **47**, 956–982 (1984)
2. Mnih, V., Kavukcuoglu, K., Silver, D., Graves, A., Antonoglou, I., Wierstra, D., Riedmiller, M.: Playing atari with deep reinforcement learning. arXiv preprint arXiv:1312.5602 (2013)
3. Osborne, M.J., Rubinstein, A.: A Course in Game Theory. MIT Press, Cambridge (1994)
4. Rusbult, C.E., Van Lange, P.A.M.: Interdependence, interaction, and relationships. Ann. Rev. Psychol. **54**, 351–375 (2003)
5. Emery-Montemerlo, R.: Game-theoretic control for robot teams. Ph.D. thesis, Carnegie Mellon University (2005)
6. Johanson, M., Bard, N., Burch, N., Bowling, M.: Finding optimal abstract strategies in extensive form games. In: Proceedings of the Twenty-Sixth Conference on Artificial Intelligence (AAAI) (2012)
7. Bernstein, D.S., Hansen, E.A., Zilberstein, S., Amato, C.: Dynamic programming for partially observable stochastic games. In: AAAI Spring Symposium, Palo Alto, CA (2004)
8. Bentivegna, D., Ude, A., Atkeson, C.G., Cheng, G.: Humanoid robot learning and game playing using PC-based vision. In: Proceedings of the IEEE/RSJ International Conference on Intelligent Robots and Systems (IROS), Las Vegas, NV (2002)
9. Grollman, D.H., Jenkins, O.C.: Learning robot soccer skills from demonstration. In: IEEE International Conference on Development and Learning (ICDL), London, UK (2007)
10. Ahmadi, M., Lamjiri, A., Nevisi, M., Habibi, J., Badie, K.: Using a two-layered case-based reasoning for prediction in soccer coach. In: Arabnia, H.R., Kozerenko, E.B., (eds.) International Conference on Machine Learning; Models, Technologies and Applications, CSREA Press, USA, pp. 181–185 (2003)
11. Lee, K., Hwang, J.-H.: Human-robot interaction as a cooperative game. In: Castillo, O., Xu, L., Ao, S.-L. (eds.) Trends in Intelligent Systems and Computer Engineering (IMECS 2007). Lecture Notes in Electrical Engineering, pp. 91–103. Springer, New York (2008)
12. Wagner, R.: Creating and using matrix representations of social interaction. In: Proceedings of the 4th International Conference on Human-Robot Interaction (HRI 2009), San Diego, CA (2009)
13. Berlekamp, E., Conway, J.H., Guy, R.: Winning Ways for your Mathematical Plays: Games in General. Academic Press, London (1982)
14. Kirschner, P.A., Sweller, J., Clark, R.E.: Why minimal guidance during instruction does not work an analysis of the failure of constructivist, discovery, problem-based, experiential, and inquiry-based teaching. Educ. Psychol. **41**(2), 75–86 (2006)
15. Banerjee, B., Stone, P.: General game learning using knowledge transfer. In: Proceedings of the 20th International Joint Conference on Artificial Intelligence (IJCAI-07), Hyderabad, India (2007)
16. Branavan, S.R.K., Silver, D., Barzilay, R.: Learning to win by reading manuals in a Monte-Carlo framework. In: Proceedings of the 49th Annual Meeting of the Association for Computational Linguistics: Human Language Technologies, vol. 1, pp. 268–277. Association for Computational Linguistics (2011)

User Evaluation of an Interactive Learning Framework for Single-Arm and Dual-Arm Robots

Aleksandar Jevtić$^{(\boxtimes)}$, Adrià Colomé, Guillem Alenyà, and Carme Torras

Institut de Robòtica i Informàtica Industrial,
CSIC-UPC, C/ Llorens i Artigas 4-6, 08028 Barcelona, Spain
{ajevtic,acolome,galenya,torras}@iri.upc.edu

Abstract. Social robots are expected to adapt to their users and, like their human counterparts, learn from the interaction. In our previous work, we proposed an interactive learning framework that enables a user to intervene and modify a segment of the robot arm trajectory. The framework uses gesture teleoperation and reinforcement learning to learn new motions. In the current work, we compared the user experience with the proposed framework implemented on the single-arm and dual-arm Barrett's 7-DOF WAM robots equipped with a Microsoft Kinect camera for user tracking and gesture recognition. User performance and workload were measured in a series of trials with two groups of 6 participants using two robot settings in different order for counterbalancing. The experimental results showed that, for the same task, users required less time and produced shorter robot trajectories with the single-arm robot than with the dual-arm robot. The results also showed that the users who performed the task with the single-arm robot first experienced considerably less workload in performing the task with the dual-arm robot while achieving a higher task success rate in a shorter time.

Keywords: Robot manipulators · User intervention · Robot adaptation · Gesture recognition · Visual servoing · Reinforcement learning

1 Introduction

Expectations from social robots are high, since they must successfully assist the users with daily tasks while maintaining their engagement in interaction over extended periods of time [1]. One way to achieve this is by enabling the robots to learn from the interaction with the users [2]. In this work, we studied the

This work was supported by the Beatriu de Pinós fellowship, reference num.: 2013 BP-B 00239, jointly funded by the Government of Catalunya, Spain and the European Commission FP7 COFUND programme. The work was partially supported by the EU CHIST-ERA I-DRESS project, reference num. PCIN-2015-147, and the national Spanish project RobInstruct, reference num. TIN2014-58178-R.

© Springer International Publishing AG 2016
A. Agah et al. (Eds.): ICSR 2016, LNAI 9979, pp. 52–61, 2016.
DOI: 10.1007/978-3-319-47437-3_6

interaction of lay users when asked to modify previously learned tasks of single-arm and dual-arm robots (experimental setup is shown in Fig. 2). There is usually a strict division between the tasks that require one or two arms, however, we compare the same aspects of the two robot settings and the effects that a previous user experience with one setting may have on another. User performance was quantitatively evaluated in terms of user workload, task success rate and task execution time. The presented results are expected to provide future developers an insight into the challenges associated with the interaction design for lay users of such robots.

1.1   Relevant Work

Service robotics is a continuously growing field in various application domains [3]. Special attention is given to healthcare domain, where socially interactive assistant robots help patients with the recovery and compensate for increased costs of care and lack of nursing staff [4,5]. Service robots are built for specific task and their functionality is defined with a set of skills necessary to perform them. Some studies show that specific types of users such as older adults prefer robot assistance over human assistance for tasks related to manipulating objects [6]. In this work, we tackle the problem on the level of trajectory adaptation and the proposed robot displacement tasks were used to evaluate the user experience.

Many challenges are associated with single-arm and dual-arm robots placed in human environments [7]. Some studies compared the two robot settings for manipulation tasks [8], but they did not involve robot learning from interaction. Learning from Demonstration (LfD) is a widely used framework for robot learning [9]. A single demonstration is often not sufficient and some refinements must be performed often through exploration in simulation or with a physical robot. Repeating a demonstration of the entire task is not always necessary, especially if the task is long or complex. In our previous work, we proposed an interactive learning framework that enables a user to select and modify a task segment [10] but this framework has never been evaluated in the experiments with users.

LfD methods often rely on robot vision to track and follow the user during a demonstration [11]. Gestures are commonly applied to initiate, stop or switch between robot tasks, but also give insight into the intended user behavior. Various methods for gesture recognition from RGB cameras have been proposed in literature [12]. In our previous work, we showed that the recognition of pointing gestures from depth images can be applied to real-time mobile robot guidance [13]. Here, the concept of visual robot guidance is implemented on robot manipulators and a Reinforcement Learning (RL) algorithm is applied to produce an improved robot motion segment from a series of user interventions. Our study compares the user experience with single-arm and dual-arm robots that use the previously proposed framework.

2 Algorithms

The proposed framework has two components: interaction and learning, however, only the interaction component was analyzed in this study. The user intervention algorithm that allows users to modify the initial robot trajectory relies on the gesture recognition algorithm for switching between different robot states and hands motion following algorithm for robot teleoperation. The algorithms were implemented in C++ using the Robot Operating System (ROS) framework. A detailed description of the algorithms can be found in [10].

2.1 Robot Control

In order to allow the robot to correctly mimic the human motion, we used a similar framework as in [14] (see Fig. 1), where the tracking and gesture recognition algorithms provided desired poses, which were transformed to desired joint positions that satisfy the kinematics constraints by using an Inverse Kinematics (IK) algorithm [15]. Desired joints positions are provided at a different rate than the control loop's execution rate, therefore a dynamic system was used to interpolate the missing values. In order to store the trajectory, we record equally-spaced Cartesian goal points $\mathbf{x}_t$ sent to the robot, and thus we could reproduce the user's hands motions by sending the same sequence of the desired points $\mathbf{x}_t$, $t = 1..N_{points}$. Given the joint position reference, computed by the IK algorithm over the user-generated Cartesian commands, a compliant feed-forward controller that combines a friction model with a PID error compensation was used to generate torque commands, as in [16]. The compliant controller makes the robot safer in case of an unexpected physical contact with the user or the surrounding objects.

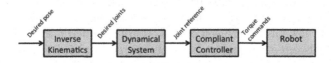

Fig. 1. Robot control diagram

2.2 Gesture Recognition and Motion Following

User tracking was implemented using the OpenNI[TM] and NiTE[TM] open source libraries. The position of the user's joints was provided in the camera frame of reference, which was used for both gesture recognition and hands motion tracking. No transformation of the joints positions to the robot frame of reference was necessary because joints relative displacements were used.

Switching between two robot states, "EXECUTE" and "FOLLOW", was initiated by raising the right hand. A voice command could have been used instead

Fig. 2. A user guiding dual-arm robot using hands motions tracked by Kinect camera. Left: the initial trajectory is shown in green. Right: red arrows represent the motions of both user hands and robot end-effectors in the teleoperation mode. (Color figure online)

of gesture recognition, however, we decided to use this simple implementation to avoid possible deployment issues. Moreover, the initial tests showed no considerable effect in terms of delay when stopping the robot at a desired position.

Hand-motion following consisted of tracking the position of each hand and reproducing the motion in real time with the corresponding robot's end-effector in a two-dimensional horizontal plane above the tabletop, as shown in Fig. 2. When switching to the "FOLLOW" state, the robot would store the initial position of a user's hand $\mathbf{p}_0 = (x_0, y_0)$ and the position of the corresponding end-effector, i.e. the cut point $\mathbf{p}_{cut} = (x_{cut}, y_{cut})$. All subsequent displacements of a hand were then reproduced by the end-effector according to following:

$$\mathbf{p}_{goal} = \mathbf{p}_{cut} + (\mathbf{p}_{hand} - \mathbf{p}_0) \tag{1}$$

where $\mathbf{p}_{hand} = (x_{hand}, y_{hand})$ is the new hand position detected by the camera with respect to the initial hand position, and $\mathbf{p}_{goal} = (x_{goal}, y_{goal})$ is the new goal sent to the robot's end-effector.

2.3 User Intervention

The implementation of the proposed user intervention is shown in Algorithm 1. The user was expected to observe the robot while performing the task, stop it at a desired position by raising the right hand and then, within the next 2 s, place his or her left hand (single-arm task) or both hands (dual-arm task) in a pose similar to holding a wheel when driving a car. All subsequent user's hands motions were then tracked and reproduced by the robot. To stop the intervention, the user would once more raise the right hand and relieve the control back to the robot. This action would also store the robot end-effector position as the final point of the modified segment.

A 2 s delay was introduced (see line 6 in Algorithm 1) to allow the user to place the hands in a comfortable starting pose. Initial trajectory points were indexed and had associated timestamps to always ensure the identical execution of the trajectory. Cut and connect points were stored with their indexes. The cut

Algorithm 1. User intervention

1: set state to"EXECUTE";
2: go to first trajectory point;
3: **while** in "EXECUTE" state **and** not end of trajectory **do**
4:    go to next trajectory point;
5:    **if** hand raised **then**
6:        set state to "FOLLOW"; // after a 2 s delay
7:        store trajectory cut point;
8:        **while** in "FOLLOW" state **do**
9:            follow user's hands;
10:           **if** hand raised **then**
11:               set state to "EXECUTE";
12:           **end if**
13:       **end while**
14:       find closest trajectory connect point;
15:   **end if**
16: **end while**

---

point was stored as the trajectory point in which the robot was found when the user raised the right hand for the first time. The calculation of the connect point required additional computation, because the users were obviously not able to precisely bring back the robot to it's initial trajectory after modifying its segment. Therefore, the connect point was computed as the trajectory point that was closest to the point where the user relieved the control back to the robot, which occurred during the second raise-hand gesture. In the first experiment this calculation is performed for a single-arm robot; however, in the second experiment involving a dual-arm robot, the connect points are computed for each arm and the index of the one closer to the corresponding end-effector was selected for both arms.

The result of the user intervention is a new robot trajectory made of the unmodified trajectory segment preceding the cut point, a modified trajectory segment between the cut and connect points, and the final unmodified trajectory segment that follows after the connect point.

3   Methodology

3.1   Hardware

The proposed algorithms were implemented on two Barrett's 7-DOF Whole Arm Manipulator (WAM) robots (see Fig. 2). A first-generation Microsoft Kinect camera, which was used for user tracking, has a detection range from $0.8\,m$ to $4\,m$ with a vertical viewing angle of $43°$ and the horizontal viewing angle of $57°$. It provides depth images at the resolution of 640×480 pixels at the maximal frame rate of $30\,fps$. A desktop PC with Ubuntu 12.04 was used in this work, and it was powered by an Intel quad-core Q9550 @ 2.83 GHz CPU with 4 GB of RAM.

3.2   Experimental Setup

User tests were performed with a group of 12 participants, a mixed group of 6 female and 6 male adults of age between 25 and 42. All participants had a university degree (7 engineers, 1 computer scientist, 1 physicist, 1 astrophysicist, 1 biologist, and 1 pharmacist) and no previous experience in robotics. Each participant performed 10 trials of each experiment. The participants were divided into two subgroups of 6, applying different order of experiments for counterbalancing.

Each participant was asked to perform two experiments, which consisted of the same task but for different robot settings[1], i.e. a single-arm and dual-arm robot. Before the experiments, the researcher conducting the study described the WAM robot and the Microsoft Kinect camera features, and the experiment procedure to the participant. The procedure consisted of: (1) researcher's demo of the interaction with the dual-arm robot, (2) test trial by the participant, (3) two experiments performed by the participant, and (4) filling of the questionnaire about the performed experiments. The participants were assured that they were not within the robot workspace and that the experiments were safe. The researcher was present during the experiments to monitor the correct functioning of the robot, but he was seated behind the user and did not interfere.

In the first experimental setup a single-arm robot was programmed to perform a linear motion at a constant height above the table, from the point $A1 = (0.5\,m, 0.4\,m)$ to the point $B1 = (0.5\,m, -0.4\,m)$ in the robot reference frame. An empty plastic bottle was placed as an obstacle on the table at the point $C1 = (0.5\,m, 0\,m)$. Without a user intervention, the bottle would be knocked over by the robot. It was explained to the user that the goal of the experiment was to minimally modify the initial robot trajectory such that the robot end-effector avoids collision with the bottle.

In the second experimental setup, both robot arms simultaneously performed the same task described for the single-arm robot, by moving along the parallel lines above the table. The first robot followed the linear trajectory between the points $A1$ and $B1$, while the second robot followed the linear trajectory between the point $A2 = (0.5\,m, -0.4\,m)$ and the point $B2 = (0.5\,m, 0.4\,m)$, defined in each robot's frame of reference, respectively. Two plastic bottles were placed on the tabletop at the points $C1$ and point $C2 = (0.5\,m, 0\,m)$. As for the first experiment, the goal was to guide the robot end-effectors around the bottles. In both experiments, the robot end-effectors had fixed orientation facing downwards and their distance from the tabletop was constant during the whole experiment.

3.3   Performance and Workload Measures

The following metrics were used to assess the performance and the workload of the participants: task success, task completion time, trajectory length, and the NASA-TLX questionnaire.

Task success was measured as a function of knocked down obstacles during the task execution, and was computed as follows:

---

[1] Additional material at: http://www.iri.upc.edu/groups/perception/adapt2.

$$S_{i,1} = (1 - N_{ob}) \cdot 100\,\%, \forall i = 1 \ldots 10$$
$$S_{i,2} = (1 - N_{ob}/2) \cdot 100\,\%, \forall i = 1 \ldots 10 \tag{2}$$

where $S_{i,1}$ and $S_{i,2}$ are the task success values for the single-arm and dual-arm robot tasks, respectively, N_{ob} is the number of knocked down obstacles, and i is the number of the trial. Higher task success values were indicators of a better performance.

The task completion time was the total time that one or both robots needed to reach the end point of the manipulation task. Lower completion times were indicators of a better performance.

The robot trajectory length was computed as the sum of distances between the Cartesian trajectory points in the horizontal (x, y) plane of the robot frame of reference:

$$D_{i,1} = \sum_{k=2}^{N} \sqrt{(x_k - x_{k-1})^2 + (y_k - y_{k-1})^2}, \forall i = 1 \ldots 10 \tag{3}$$

where x_k and y_k are the k-th trajectory point coordinates in the robot frame of reference, N is the number of trajectory points, and i is the number of the trial. In case of the dual arm task, the trajectory length was computed as a mean of trajectories' lengths from both robot arms. Shorter trajectories were an indicator of a better performance.

Participants completed the raw NASA-TLX questionnaire after each experiment. The questionnaire enables the collection of six dimensions of workload ranging from 0 to 100, and was used to assess the overall participant workload when performing the experiments, similarly as in [13]. The overall workload is computed as the average of the above-mentioned six dimensions.

3.4 Data Analysis

A two-way repeated measure ANOVA analysis was conducted on two within-subject variables (number of trials and robot setting), using order of experiments as between-subject variable. The number of trials was divided in ten levels and the robot setting in two levels (single arm and dual arm). Statistical significance was computed for all three performance metrics described in Sect. 3.3. The effect on the overall participant workload was computed in the two-way repeated measure ANOVA test using the robot setting as within-subject variable and the order of experiments as between-subject variable. We considered the results as significant for $p < 0.05$.

4   Results and Discussion

4.1   Performance Analysis

The results of the first ANOVA test show that there was a statistically significant effect of the robot setting on the task completion time, $F(1, 10) = 12.779$, $p = 0.005$, and robot trajectory length, $F(1, 10) = 7.97$, $p = 0.018$. In case of

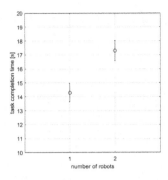 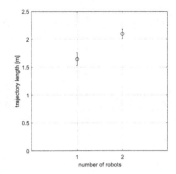

Fig. 3. Effect of the robot setting on: (a) task completion time, and (b) trajectory length. Error bars represent the standard error of the mean.

the interaction with the dual-arm robot, the participants needed more time to complete the task (Fig. 3a) and also guided the robots along longer trajectories (Fig. 3b). This could be explained by the fact that, in case of the dual-arm robot, the attention of the participants was divided between two arms, which reduced speed and accuracy of the robot guidance.

The between-subject analysis showed that the order of experiments had a statistically high significant effect on the task success, $F(1, 10) = 12.214$, $p = 0.006$, and a statistically significant effect on the task completion time, $F(1, 10) = 5.348$, $p = 0.043$, which is an indicator of the learning effect from previously using the robot regardless of the initial robot setting; it also demonstrates user's ability to adapt when switching from single-arm to dual-arm robot setting, and vice-versa. The robot trajectory length was not affected by the order in which experiments were performed.

The results also show that there was a statistically high significant effect of the number of trials on the task completion time, $F(9, 90) = 2.881$, $p = 0.005$, regardless of the order in which the experiments were performed, which again indicates the user's ability to learn over time to more efficiently interact with the robot. The number of trials did not have a statistically significant impact on task success and robot trajectory length.

4.2 Overall Workload

The analysis of the interaction of the within-subject variable and the between-subject variable, i.e. the order of experiments and the robot setting, shows a statistically significant joint effect on the participants' overall workload, $F(1, 10) = 5.482$, $p = 0.041$. It can be noted from the overall workload mean values shown in Fig. 4 that the participants experienced less initial workload if the first experiment was performed with a single arm; however, the workload drop was also lower during the second experiment using the same robot setting. But when the experiments started with the dual-arm robot the initial workload was much higher and the workload drop was more drastic during the second

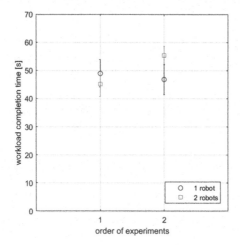

Fig. 4. Effect of the interaction of the robot setting and the order of experiments (*Order* = 1: starting with 1 robot; *Order* = 2: starting with 2 robots) on the overall workload. Error bars represent the standard error of the mean.

experiment, indicating that participants felt a huge relief for moving from a more difficult condition. It should also be noted that in both cases the second experiment produced less workload than the first experiment, which indicates the existence of the user's learning factor.

5   Conclusion

The study compared the performance and the workload of a group of lay users interacting with a single-arm and dual-arm robots performing the motion for the same task. The experiments were designed to evaluate several aspects of the user experience with the two robot settings. Although the experimental setup involved a simple trajectory adaptation task, it demonstrated that lay users with no background in robotics were able to successfully perform the task and improve over time while reducing the overall workload. The proposed approach can be extended to other application domains and more complex tasks, such as assisting the users in the activities of daily living, with either one or two robot arms, which will be a part of future work. Finally, the results suggest that a user could be introduced to interaction with a dual-arm robot by initially interacting with only one robot arm.

References

1. Leite, I., Martinho, C., Paiva, A.: Social robots for long-term interaction: a survey. Int. J. Soc. Robot. **5**(2), 291–308 (2013)
2. Mason, M., Lopes, M.C., Robot self-initiative and personalization by learning through repeated interactions. In: 6th ACM/IEEE International Conference on Human-Robot Interaction (HRI 2011), pp. 433–440 (2011)
3. Torras, C.: Service robots for citizens of the future. Eur. Rev. **24**(1), 17–30 (2016)
4. Cabibihan, J.-J., Javed, H., Ang, M., Aljunied, S.M.: Why robots? A survey on the roles and benefits of social robots in the therapy of children with autism. Int. J. Soc. Robot. **5**(4), 593–618 (2013)
5. Robinson, H., MacDonald, B., Broadbent, E.: The role of healthcare robots for older people at home: a review. Int. J. Soc. Robot. **6**(4), 575–591 (2014)
6. Smarr, C.-A., Mitzner, T.L., Beer, J.M., Prakash, A., Chen, T.L., Kemp, C.C., Rogers, W.A.: Domestic robots for older adults: attitudes, preferences, and potential. Int. J. Soc. Robot. **6**(2), 229–247 (2014)
7. Kemp, C.C., Edsinger, A., Torres-Jara, E.: Challenges for robot manipulation in human environments [Grand Challenges of Robotics]. IEEE Robot. Autom. Mag. **14**(1), 20–29 (2007)
8. Edsinger, A., Kemp, C.C.: Two arms are better than one: a behavior based control system for assistive bimanual manipulation. In: Lee, S., Suh, I.H., Kim, M.S. (eds.) Recent Progress in Robotics: Viable Robotic Service to Human. LNCS, vol. 370, pp. 345–355. Springer, Heidelberg (2008)
9. Argall, B.D., Chernova, S., Veloso, M., Browning, B.: A survey of robot learning from demonstration. Robot. Auton. Syst. **57**(5), 469–483 (2009)
10. Jevtić, A., Colomé, A., Alenyà, G., Torras, C.: Robot adaptation through user intervention and reinforcement learning. http://www.iri.upc.edu/groups/perception/adapt.pdf. publication under review
11. Nicolescu, M., Mataric, M.: Learning and interacting in human-robot domains. IEEE Trans. Syst., Man, Cybern. A, Syst. Hum. **31**(5), 419–430 (2001)
12. Mitra, S., Acharya, T.: Gesture recognition: a survey. IEEE Trans. Syst. Man Cybern. Part C Appl. Rev. **37**(3), 311–324 (2007)
13. Jevtić, A., Doisy, G., Parmet, Y., Edan, Y.: Comparison of interaction modalities for mobile indoor robot guidance: direct physical interaction, person following, and pointing control. IEEE Trans. Hum. Mach. Syst. **45**(6), 653–663 (2015)
14. Husain, F., Colomé, A., Dellen, B., Alenyà, G., Torras, C.: Realtime tracking and grasping of a moving object from range video, In: 2014 IEEE International Conference on Robotics and Automation, pp. 2617–2622 (2014)
15. Colomé, A., Torras, C.: Closed-loop inverse kinematics for redundant robots: comparative assessment and two enhancements. IEEE/ASME Trans. Mechatron. **20**(2), 944–955 (2015)
16. Colomé, A., Planells, A., Torras, C.: A friction-model-based framework for reinforcement learning of robotic tasks in non-rigid environments. In: IEEE International Conference on Robotics and Automation, pp. 5649–5654 (2015)

Formalizing Normative Robot Behavior

Billy Okal[1](✉) and Kai O. Arras[1,2]

[1] Department of Computer Science, University of Freiburg, Freiburg, Germany
okal@cs.uni-freiburg.de
[2] Bosch Corporate Research, Robert-Bosch GmbH, Gerlingen, Germany

Abstract. We address the task of modeling, generating and evaluating normative behavior for interactive robots. Normative behavior is essential for coherent deployment of these robots in human populated spaces. We develop a first unifying, intuitive and general formalism of the task that subsumes most previous approaches which have focused mainly on specific tasks. We present concrete and practical definitions of norms and show how to generate and evaluate behavior that adheres to such norms. We then demonstrate the formalism on a socially normative navigation task for service robots. Further, we discuss the key challenges in realizing such behaviors, and in particular, the role of perception and uncertainty.

1   Introduction

As more service robots are deployed for various functions in public spaces such as hotel lobbies, airports, hospitals, care homes, etc.; the need to demonstrate additional social, cultural capabilities beyond primary functionality arises. This is because having such robots in human populated spaces change the underlying dynamics of social and cultural interactions [13,18]. Additionally, most human interactions often influenced by deeper social and cultural standards or norms which vary across environments, making it hard to explicitly model them. Robots operating in these spaces nonetheless need to exhibit behavior that takes into account such social and cultural aspects. We call such behavior a *normative behavior* although other terminology such as compliance in the case of "socially compliant navigation" or human-awareness are also commonly used. In this paper, *normative* is taken to mean according to a set of norms, which can be either formal such as traffic rules, or social-cultural such as politeness when navigating. The behavior resulting from adherence to the norms may not be efficient in terms of classical task metrics like path length in planning, but may sometimes lead to efficiencies in other regimes. For example a norm requiring a robot to execute a slipstream maneuver (tailing a person heading in similar direction in a dense crowd) may generate longer paths; but in the process also improve reliability in reaching goal by avoiding situations leading to getting stuck. These norms are often characterized by the entities the robot is interacting with and the structure of the environment. For example Kruse et al. [10,17] provide a summary of navigation behaviors with respect to human-awareness norm generated by considering various geometric relations in the robot's environment.

© Springer International Publishing AG 2016
A. Agah et al. (Eds.): ICSR 2016, LNAI 9979, pp. 62–71, 2016.
DOI: 10.1007/978-3-319-47437-3_7

The key task therefore, is figuring out how to equip robots with decision making capabilities that result in normative behavior. By taking care of norms in the decision making stage, we can build anticipatory behavior as opposed to reactive one. Anticipatory behavior is better suited for coordination and can avoid damages [1]. Incorporating norms into decision making involves the following key steps: (i) a formal and practical understanding of norms, (ii) techniques to generate behavior that adheres to the norms and (iii) effective evaluation methods for assessing the result. There has been a growing interest in various instantiations of this task, especially with respect to social navigation norms for service robots. For example [7,16] all seek to develop normative behavior for navigating in crowded scenes, while [4,9,21,22] focus on normative pairwise interactions such as passing one another on either side. Most of these approaches focus of different aspects of the task, defining their own metrics and understanding of norms, thereby making it difficult to compare, evaluate and select methods to deploy in practice. It is with this realization that we develop a unified formalism for addressing this task in this paper. It is our hope that this new formalism will help organize the efforts to tackle this task under a common setup.

While developing the new formalism, we highlight some key challenges associated with realizing normative behavior. As a concrete example, consider a robot providing service in a hotel lobby; the robot's normative navigation decision making capabilities are highly dependent on the performance *perception* components, that is, if people cannot be reliably detected, planning around them normatively is rendered impossible. In general, the more properties of the environment that can be reliably perceived, the more norms can be taken into account in decision making. Additionally, *predictions* of future states and actions of other agents in the environment is also crucial. Finally, *uncertainty* arising from action execution using noisy controllers also needs to be considered. Altogether, these challenges form a tightly coupled perception-action-control loop that requires clear interfaces between components when developing a wholesome solution. Consequently, the proposed formalism should be able to define such interfaces clearly on a common framework. Thus, the main contribution in this paper is a first unifying formalism for modeling norms and normative behavior for interactive robots. The presented formalism also admits natural ways of generating and evaluating resulting normative behavior.

2   Modeling Normative Robot Behavior

In order to have a concise yet flexible formalism for normative behavior, we discuss three key components that are needed to realize such behavior. These components will then enable us to make formal definitions of norms and normative behavior, and lead to what we think of as natural ways of generating such behavior. We also emphasize that in all of these components, uncertainty plays a key role in the success of any endeavor, hence the formalism needs to admit possibility to reason about uncertainly at all levels.

2.1   Environment

The environment $\mathcal{C}$ that the robot is operating in is made up of the space and all the entities $\mathcal{E} = \{e_i\}$ present in it, which together define the structure of the environment. These entities could be interactive and even adversarial like humans or simply artifacts like general obstacles in the scene. Each entity's anchoring in the environment is summarized using a pose vector $\mathbf{x}_i$. Additionally, each entity possesses a set of *attributes* $\mathcal{A} = \{a_i\}$, any subset of which is represented using a vector $\mathbf{a}_i = (a_j, \ldots, a_{j+k})^T \subseteq \mathcal{A}$. For example, a person i can be represented using a pose $\mathbf{x}_i = (x, y, \dot{x}, \dot{y})^T$ and may have simple attributes such as age, gender, carrying-luggage, etc. given in a vector $\mathbf{a}_i = (25.0, \mathrm{M}, \mathrm{T})^T$. Further, there are pairwise relations $f : \mathcal{E} \times \mathcal{E} \longmapsto [0, 1]$ between some of the entities. Such pairing can be as a result of attributes. For example two people may belong to a group like a couple, to which their gender attributes as well as geometric reasoning may generate a pairing probability of say $f_{\mathrm{group}}(e_1, e_2) = 0.7$, which is interpreted as the strength of the relation. We eschew the details how to define and detect such relations to the designer, and only emphasize that the formalism presented is general enough to admit many choices. Figure 1 shows a example of such an environment in the case of a navigation task. We argue that this minimalistic representation is sufficient to capture all aspects needed for decision making that culminates in normative behavior of any complexity.

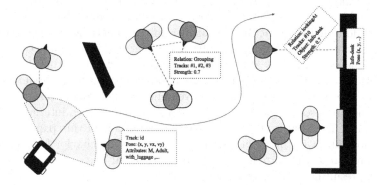

Fig. 1. A normative navigation example setup. Entities shown include people, desk, queues and general obstacles (in black). Potential perceptual attributes are shown alongside people. Pairwise relations between people and between person (e. g. grouping) and other objects (e. g. looking at a screen) are also illustrated with dotted red and blue lines. A sample socially normative navigation path is shown in blue curve. (Color figure online)

2.2   Perception

A crucial component for realizing any normative behavior in the environment we just described above, is the ability of observe the different aspects of the said environment with reasonable accuracy. In fact, we argue that the difference between normative behavior and non-normative counterpart lies solely in which of the perceived aspects of the environment are taken into account in decision making. In this work, we require that any robot intending to exhibit normative behaviors be equipped perception modules for observing entities in the environment, a subset of their attributes and relations. The richness of this perceived subset of attributes directly affects how complex a normative behavior may be realized. For example, we cannot develop a behavior adhering to norms relating to gender if we cannot reliably perceive gender attribute.

Concretely, for every entity in the scene, a perception module produces tuples of poses and associated uncertainty estimates $(\mathbf{x}_i, \boldsymbol{\delta}_i^p)$. For example, this could be a people detector module for persons or a localization module for providing obstacle poses. Similarly, another high level perception module would provide attributes values and associated uncertainty $(\mathbf{a}_i, \boldsymbol{\delta}_i^a)$. An example of such attribute detectors in practice is given in [11] for age groups, gender and clothing related attributes. Finally, the pairwise relations can be perceived using relational reasoning modules so as to provide relation probabilities for every pair of entities. Altogether, the perception modules are seen as black boxes $\mathcal{P}$ which produce signals for each entity, attributes and relations in the environment. The exact form of these uncertainty estimates $\boldsymbol{\delta}_i^p, \boldsymbol{\delta}_i^a$ depend on the sensor instrumentation used and algorithms for the various perception tasks involved.

2.3   Execution

Normative behavior is usually targeted at robots which interact with humans or other robots. As such, these robots modify the environment they operate in and potentially alter future percepts, thus we need to also formalize the nature of their effects through their actions $\mathcal{U}$. Concretely, for most decisions made by the robot, a series of actions are performed, but executions are often imperfect hence the need to explicitly model uncertainty. We argue that the execution of a series of actions can be effectively assessed by examining the trajectory $\xi = \{\mathbf{x}^t\}$ resulting for such execution. Such a trajectory can have a 'band' in the space of poses, capturing the uncertainty in the execution. This simplistic formulation is sufficient for purposes of normative behavior realization.

Finally, using the three components above, we can now formally define norms and normative behavior before we set about on finding techniques to generate and evaluate them.

Definition 1 (Norm). *A norm $\mathbb{N}^\mu$ in the context of robot behavior is a property of the robot's environment $\mathcal{C}$, percepts $\mathcal{P}$ and actions $\mathcal{U}$ with an associated set of M tests $\boldsymbol{\mu} = \{\mu_1, \ldots, \mu_M\}$ for assessing adherence.*

Definition 2 (Norm Test). *Given a norm $\mathbb{N}^\mu$ over some environment and a trajectory $\xi_j \in \Xi$, a norm test $\mu_i : \mathcal{C} \times \mathcal{P} \times \Xi \longmapsto \{0,1\}$ is a function that evaluates adherence of the trajectory to the norm.*

The adherence to a norm is collectively assessed by all the norm's tests, meaning all norm tests must pass. This formulation of norms ensures that practical assessment is algorithmically possible.

Definition 3 (Normative Robot Behavior). *Given a collection of K norms* $\mathbb{N}_1^{\mu_1}, \ldots, \mathbb{N}_k^{\mu_k}$*; a behavior exhibited by a trajectory ξ is said to normative if and only if the application of all the norm tests to the trajectory pass. i. e. Given $C \in \mathcal{C}, P \in \mathcal{P}$, the behavior is normative if and only if* $\bigwedge_{k=1}^{K} \bigwedge_{m=1}^{M} \mu_{m,k}(C, P, \xi) = 1$.

3 Generating Normative Behavior

In this section, we provide the technical means for realizing the normative behavior defined in Sect. 2. This entails specification of decision making approaches which are incorporated in task and motion planning modules of such robots. The most common approach is formulate a cost function which encodes the desired normative behavior, and then use such function to guide solution search in planning algorithms. However, it is often very difficult to manually design such cost functions, especially because of inherent ambiguity in the specification of these behaviors due to dependence on social and cultural aspects of involved parties. A common simplification used in practice is to model the cost functions are mixtures of basis features of the environment and the entities in it. The formalism presented here is particularly well suited for such endeavor as these features can simply be based on the poses, attributes and relations of entities. This also helps lighten the burden of coming up with features. However, the mixing ratios of such features still need to be figured out.

A promising technique for learning the mixing of features is learning from demonstration (LFD). In particular, inverse reinforcement learning (IRL) formally introduced in [14] has been used successfully in many applications such as crowd navigation [7,9,16]. The IRL approach assumes the robot's decision making is carried out using a Markov decision process (MDP) with an unknown cost function (equivalently reward function), usually assumed to encode the behavior. In practice IRL involves demonstrating the desired behavior, usually by manually driving the robot say using a joystick, and then using typically iterative algorithms to recover cost function that "explains" the demonstration. This cost function can then be used to either generate costmaps over planning domains or is integrated directly into a planning algorithms objective function. The main challenge in using IRL approaches is lack of computationally and data efficient algorithms used to recover the cost function and practical representations suited for most real world tasks.

Regardless of the procedure used for realizing the normative behaviors, it is imperative that uncertainty in both perception and execution be taken into account in decision making. For example, when cost functions, this could mean inflating cost regions around potential configurations by a factor proportional to the uncertainty in say people detection. Finally, for successful normative behavior generation, it is imperative to predict and take into account the future actions of other decision making agents in the same environment. These could entail predicting future positions of people and then generating costmaps that already take the prediction into account, resulting in anticipatory behavior. For certain tasks such as navigation, this often reduces stop-and-go motions caused by too reactive a planner which acts myopically.

4 Evaluating Normative Behavior

Normative robot behavior can be evaluated in two different ways; firstly by checking the adherence of behavior to norms using the associated norm tests, and secondly using task specific metrics. The evaluation based on norm tests is fully dependent on the specification of the norms. Evaluation using task specific metrics could lead to discoveries of potential trade-offs between say task efficiency and normativeness. These trade-offs, if any, could be helpful to service robot designers who are often confronted with choosing either functionality or normativeness in practical settings.

A number of task metrics have been proposed in the past including these human robot interaction metrics [2, 20]. These include path lengths, time to goal, idle operating time, human comfort as measured by qualitative questionnaires. We defer the exact choice of such task metrics to the designer as it is difficult to list a complete set of metrics for all possible tasks.

5 Case Study: Socially Normative Crowd Navigation

In order to demonstrate how to use the formalism practically, we show how to define simple social norms for a mobile service robot navigating in a crowded space and generate the required socially normative navigation behaviors. The environment is a place $\mathcal{C} \triangleq \mathbb{R}^2$, entities are people, shops, walls, etc. Some of the people in the scene are engaged in groups, others are engaged with various activities such as queuing or looking at information boards. Our service robot is required to efficiently navigate in this scene while respecting the various social norms, and in effect treat people as more than just dynamic obstacles.

We define the following basic norms for our case study.

Personal spaces, $\mathbb{N}_P^{\mu_P}$: Always minimize intrusions into personal spaces around people. These personal spaces are derived from Proxemics theory [6] with these radii (Personal: 0.45 m to 1.2 m, Social: 1.2 m to 3.6 m). We define $\mu_P = \{\mu_P, \mu_S\}$, where test $\mu_P = \zeta_P \leq \alpha_P$ and $\zeta_P = \sum_t \sum_i \mathbb{1}\left(\|\mathbf{x}_r^t - \mathbf{x}_p^i\|_2 \leq 1.2\right)$ for some appropriate threshold α_P which can be experimentally identified. $\mathbf{x}_r^t, \mathbf{x}_p^i$ denote robot pose and person i respectively at time t, while $\mathbb{1}(\cdot)$ is the indicator function. Other test, μ_S are defined analogously.

Interaction spaces, $\mathbb{N}_I^{\mu_I}$: While interacting with people, minimize disturbance on the relations between them, e.g. do not cross through a group. We define $\mu_I = \{\mu_r\}$ with $\mu_r = \zeta_r \leq \alpha_R$ and $\zeta_r = \sum_t \sum_k \mathsf{dist}(\mathbf{x}_r^t, \mathbf{x}_s^k)$, where $\mathsf{dist}(\cdot)$ is shortest distance to an interaction area (which can be represented as polygon),

α_R is a threshold and $\mathbf{x}_s^k$ is the pose of k-th interaction area. For a pairwise relation, this is a line.

The requirements for perception include; reliable detection of people, detection of pairwise relations, in particular grouping affiliations and engagements such as looking at something in the scene. Because there are not many reliable and practical perception modules that can deliver the required attributes for our norms, we first perform experiments using an open source pedestrian simulator[1] described in [15]. We then later deploy the robot in the wild at an airport with the learned socially normative behaviors.

We use the LFD approach presented in [16] to learn behaviors for this task from expert demonstrations. We represent the target cost function as linear combination of features, which we derive from the attributes of entities i. e. distance to persons, distance to pairwise relation lines and relative goal heading. Learning of the cost function is done using an extension of the Bayesian inverse reinforcement learning (BIRL) algorithm that works well in practice as described in [16]. We use 10 trajectories demonstrated by driving the robot using a joystick for learning the cost function. We use the found cost function to generate costmaps which are then used by navigation planners.

We evaluate the learned behaviors by having the robot plan and navigate between a total of 25 different start and goal pairs, and perform the above specified norm tests on the resulting paths plus run additional task specific metrics. These task metrics are: path length, time to goal and cumulative heading changes (CHC); in all of which smaller quantities are preferred. We use *classical* navigation planning where all entities in the scene are simple obstacles as the baseline and compare it to our normative navigation case. In the implementation, we use the move_base framework from ROS and add a costmap layer for normative navigation behaviors. We run A* global planner on the generated costmap at 2 Hz, and an elastic band local planner at 12 Hz, with local rolling window costmap of size 8 m². In simulation we use a sensor radius of 8 m; meaning we only consider people tracks within this region for updating the costmap.

Table 1. Evaluation results from the normative and classical behavior trajectories averaged over 25 runs. The norm tests are all statistically significant. RD is the relation disturbance assessed using μ_r norm test.

	Personal	Social	RD	Path Length	CHC	Time
Normative	29.8 ± 32.2	2056.4 ± 523.5	14.9 ± 35.3	22.5 ± 7.3	5.2 ± 1.8	38.9 ± 6.8
Classical	123.1 ± 69.4	2470.8 ± 629.8	82.9 ± 45.4	19.6 ± 6.1	4.5 ± 1.2	37.4 ± 7.3
p-value	1.2×10^{-4}	1.49×10^{-2}	4.14×10^{-7}	0.1276	0.1018	0.4537

As illustrated in Table 1, the resulting learned socially normative behavior performs passes all the norm test with significant difference. The statistical test

---

[1] https://github.com/srl-freiburg/pedsim_ros.

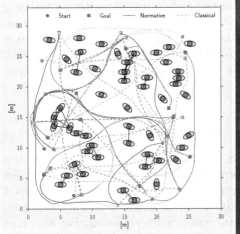 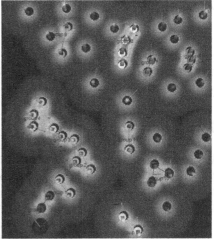

Fig. 2. Left: Example paths realized by the socially normative behavior vs classical (simple obstacle avoidance) behavior in a $30\,\mathrm{m}^2$ crowded area. People are shown in top-view with head and shoulders, Relations between people are shown in black lines. Start locations are filled circles while goals are filled squares. **Right:** Socially normative costmap using the learned cost function. Jet colored (red — highest, blue — lowest). Arrows indicate velocity vectors while lines connecting people are pairwise relations. (Color figure online)

was done using t-test with null hypothesis being; no difference in norm tests between normative and classical behaviors. This null hypothesis is successfully rejected as shown in Table 1 in the first three columns. Additionally, the normative path takes longer paths and make more heading changes as expected but this difference is not statistically significant. While this does not necessarily confirm or deny that normative behavior results in similar performance with respect to task metrics, our intuition tells us that this maybe the case. Figure 2 (right) shows example paths from the two behaviors, and (left) the resulting costmap of the computed using the learned cost function, which enables the robot to drive in a socially normative manner.

6 Related Work

Efforts to generate normative behaviors for robots have only recently began, and as such, most attempts focus of very specific aspects of the task. To the best of our knowledge, this is the first comprehensive attempt at unifying these disparate approaches into one formalism. Nevertheless, we highlight here some of the recent works touching on the different aspects of the task. Learning cost functions for normative behavior is most studied, especially using LFD techniques as in [7,9,16,23], and also manually designed cost functions [12,19]. The formalism presented here subsumes the approaches in [7,12,16] among others, while still providing a general picture of the task. Other attempts to realize a framework for robot behavior such as [8] are very limited to simple interaction experiments. The framework of [5] is the closest our formalism, though it is a very preliminary effort, and is limited to robot navigation tasks with no explicit treatment of norms. Other works like [3] are seen as too broad, leaving practical implementation aspects still undefined.

7   Conclusions

We have presented a unified formalism for normative robot behaviors, giving practical yet precise definitions for norms and normative behaviors while also the technical means for generating and evaluating such behaviors. We highlighted the key technical requirements needed to realize normative behavior and in particular the dependence on perception and uncertainty reasoning. We have also demonstrated in a case study, how the formalism can be used to model, generate and evaluate socially normative behaviors for a mobile service robot operating in public spaces. In the future, we plan to incorporate the formalism into life-long learning systems for automatic learning of these normative behaviors.

References

1. Castelfranchi, C.: Modelling social action for AI agents. Artif. Intell. **103**(1), 157–182 (1998)
2. Chakraborti, T., Talamadupula, K., Zhang, Y., Kambhampati, S.: Interaction in human-robot societies. In: AAAI Workshop on Symbiotic Cognitive Systems (2016)
3. Dautenhahn, K.: Socially intelligent robots: dimensions of human–robot interaction. Philos. Trans. R. Soc. Lond. B Biol. Sci. **362**(1480), 679–704 (2007)
4. Dondrup, C., Bellotto, N., Hanheide, M., Eder, K., Leonards, U.: A computational model of human-robot spatial interactions based on a qualitative trajectory calculus. Robotics **4**(1), 63–102 (2015)
5. Gómez, J.V., Mavridis, N., Garrido, S.: Social path planning: generic human-robot interaction framework for robotic navigation tasks. In: International Workshop on Cognitive Robotics Systems: Replicating Human Actions and Activities (2013)
6. Hall, E.T., Birdwhistell, R.L., Bock, B., Bohannan, P., Diebold Jr., A.R., Durbin, M., Edmonson, M.S., Fischer, J., Hymes, D., Kimball, S.T., Barre, W.L., Lynch, J.S., McClellan, J.E., Marshall, D.S., Milner, G.B., Sarles, H.B., Trager, G.L., Vayd, A.P.: Proxemics [and comments and replies]. Current anthropology, pp. 83–108 (1968)
7. Henry, P., Vollmer, C., Ferris, B., Fox, D.: Learning to navigate through crowded environments. In: Proceeding of International Conference on Robotics and Automation (ICRA) (2010)
8. Huang, C.M., Mutlu, B.: Robot behavior toolkit: generating effective social behaviors for robots. In: Proceeding of ACM/IEEE international conference on Human-Robot Interaction (HRI) (2012)
9. Kretzschmar, H., Spies, M., Sprunk, C., Burgard, W.: Socially compliant mobile robot navigation via inverse reinforcement learning. Int. J. Robot. Res. (2016)
10. Kruse, T., Pandey, A.K., Alami, R., Kirsch, A.: Human-aware robot navigation: a survey. Robot. Auton. Syst. (2013)

11. Linder, T., Arras, K.O.: Real-time full-body human attribute classification in RGB-D using a tessellation boosting approach. In: Proceeding of International Conference on Intelligent Robots and Systems (IROS) (2015)

12. Lu, D.V., Allan, D.B., Smart, W.D.: Tuning cost functions for social navigation. In: Herrmann, G., Pearson, M.J., Lenz, A., Bremner, P., Spiers, A., Leonards, U. (eds.) ICSR 2013. LNCS (LNAI), vol. 8239, pp. 442–451. Springer, Heidelberg (2013). doi:10.1007/978-3-319-02675-6_44

13. Mutlu, B., Forlizzi, J.: Robots in organizations: the role of workflow, social, and environmental factors in human-robot interaction. In: Proceeding of ACM/IEEE International Conference on Human-Robot Interaction (HRI) (2008)

14. Ng, A.Y., Russell, S.J.: Algorithms for inverse reinforcement learning. In: International Conference on Machine Learning (ICML) (2000)

15. Okal, B., Arras, K.O.: Towards group-level social activity recognition for mobile robots. In. IROS Assistance and Service Robotics in a Human Environments Workshop (2014)

16. Okal, B., Arras, K.O.: Learning socially normative robot navigation behaviors using bayesian inverse reinforcement learning. In: Proceeding of International Conference on Robotics and Automation (ICRA) (2016)

17. Rios-Martinez, J., Spalanzani, A., Laugier, C.: From proxemics theory to socially-aware navigation: a survey. Int. J. Soc. Robot. **7**(2), 137–153 (2015)

18. Sardar, A., Joosse, M., Weiss, A., Evers, V.: Don't stand so close to me: users' attitudinal and behavioral responses to personal space invasion by robots. In: Proceeding of ACM/IEEE international conference on Human-Robot Interaction (HRI) (2012)

19. Sisbot, E., Marin-Urias, L., Alami, R., Simeon, T.: A human aware mobile robot motion planner. IEEE Trans. Robot. Autom. (TRO) **23**(5) (2007)

20. Steinfeld, A., Fong, T., Kaber, D., Lewis, M., Scholtz, J., Schultz, A., Goodrich, M.: Common metrics for human-robot interaction. In: Proceeding of ACM/IEEE international conference on Human-Robot Interaction (HRI) (2006)

21. Trautman, P., Ma, J., Murray, R.M., Krause, A.: Robot navigation in dense human crowds: the case for cooperation. In: Proceeding of International Conference on Robotics and Automation (ICRA) (2013)

22. Turnwald, A., Althoff, D., Wollherr, D., Buss, M.: Understanding human avoidance behavior: interaction-aware decision making based on game theory. Int. J. Soc. Robot. **8**(2), 331–351 (2016)

23. Vasquez, D., Okal, B., Arras, K.O.: Inverse reinforcement learning algorithms and features for robot navigation in crowds: an experimental comparison. In: Proceeding of International Conference on Intelligent Robots and Systems (IROS) (2014)

Decision-Theoretic Human-Robot Interaction: Designing Reasonable and Rational Robot Behavior

Mary-Anne Williams[✉]

QCIS, University of Technology Sydney and Codex, Stanford University, Stanford, USA
Mary-Anne@TheMagicLab.org

Abstract. Autonomous robots are moving out of research labs and factory cages into public spaces; people's homes, workplaces, and lives. A key design challenge in this migration is how to build autonomous robots that people want to use and can safely collaborate with in undertaking complex tasks. In order for people to work closely and productively with robots, robots must behave in way that people can predict and anticipate. Robots *chose* their next action using the classical sense-think-act processing cycle. Robotists design actions and action choice mechanisms for robots. This design process determines robot behaviors, and how well people are able to interact with the robot. Crafting how a robot will *choose* its next action is critical in designing social robots for interaction and collaboration. This paper identifies *reasonableness* and *rationality*, two key concepts that are well known in Choice Theory, that can be used to guide the robot design process so that the resulting robot behaviors are easier for humans to predict, and as a result it is more enjoyable for humans to interact and collaborate. Designers can use the notions of *reasonableness* and *rationality* to design action selection mechanisms to achieve better robot designs for human-robot interaction. We show how Choice Theory can be used to *prove* that specific robot behaviors are reasonable and/or rational, thus providing a formal, useful and powerful design guide for developing robot behaviors that people find more intuitive, predictable and fun, resulting in more reliable and safe human-robot interaction and collaboration.

Keywords: Human-robot interaction · Designing robot behavior · Legible robot behavior · Predictable robot behavior · Choice theory

1 Introduction

There is a quiet revolution taking place. Robots have been moving from research labs and cages on factory floors into spaces inhabited by people over the last decade. As the service robot industry continues to seek new opportunities that generate value in people's lives, homes and workplaces there is a pressing need to design robots whose behavior people find intuitive so that their interaction and collaboration are enjoyable.

By *enjoyable* we mean legible [4], predictable [6, 8], safe, fluent, and effective. A robot that behaves unintuitive, unpredictably, unsafely, awkwardly or ineffectively is not enjoyable to be near. To be enjoyable robot interactions need to be easy, seamless and natural for humans. We should expect that some training might be needed but that

© Springer International Publishing AG 2016
A. Agah et al. (Eds.): ICSR 2016, LNAI 9979, pp. 72–82, 2016.
DOI: 10.1007/978-3-319-47437-3_8

after being trained a human can work with a robot safely without being surprised, irritated or frustrated.

The current trend in robotics sees robots taking on increasingly expansive roles in society as slave, enabler, protector, companion, entertainer, collaborator and partner. Robot behaviors and social intelligence are fast becoming hot research topics in the field of robotics as researchers frantically seek to develop new scientific methods to: (i) assist in the design of intuitive robot behaviours for complex human-robot interactions and (ii) support synchronized and flexible human-robot real-time joint actions for collaboration.

The challenge for social robot design is to create robot behaviors that people can anticipate because when people work closely with robots they must be able to predict what a robot will do next in order to know how best to respond while undertaking cooperative action with a robot in real-time.

Robots are distributed computer systems that chose what actions they will enact in real-time numerous times a second. Robot behaviour designers craft robot actions and specify the action choice mechanism that determine how a robot will choose its next action: autonomous robots spend their entire life gathering sensor data and creating perceptions of their own body (proprioception) and the external environment (exteroception) that they then use to execute the *action choice* mechanisms.

There are no widely accepted principles that can be used to guide designers on how best to build action choice mechanisms, and as a result robot systems can be difficult to work with, unsafe and unintuitive because their behaviors are hard to interpret, predict, anticipate and explain. It is time to explore how we can introduce more rigorous methods into the design of robot action choice mechanisms, particularly in human-robot interaction and social robot applications.

This paper uses ideas from behavioral Choice Theory [7] to develop a new approach to robot behavior design that leads to *reasonable* and *rational* behaviors that people can understand more easily, and importantly, predict. As a result human-robot interaction and collaboration can be designed to be more enjoyable and productive. Section 2 describes state-of-the-art in robot design. Section 3 discusses the importance of people being able to interpret and predict robot behavior in human robot interaction and collaboration. Section 4 introduces Choice Theory as a formal tool for describing and analysing the robot action choice problem, and explores the idea of what it would mean for a robot to act *reasonably* or *rationally*. Section 5 shows how robot designers can develop reasonable and rational behaviors for robots, thus creating robots that people find easier and more enjoyable to work with during interactions and collaboration. Section 6 summarizes the contribution.

2   State-of-the-Art Robot Behavior Design and Interpretation

A robot is a *real-time distributed computer system* that essentially executes the classical robot *sense-think-act processing cycle*. During this cycle a robot gathers and interprets sensor data from its body and the environment to build high-level perceptions. Using perceptions and knowledge about the available resources, like body parts, the robot "thinks" and *chooses an action* to perform next. As an example, if a robot wants to *lift* a coffee cup and its left arm is busy undertaking another action, the robot could chose to *wait* or use its idle right arm to *lift* the cup. The final stage of the cycle is the *enaction* or performance of the chosen action.

During the time it takes a robot to gather sensor data, select the next action and execute it, there are typically changes within the robot, and in environment. The robot will gather and analyze new data to choose the next action to execute, and so it continues. Robots typically complete the sense-think-act cycle many times a second.

Robots are real-time distributed systems, coordination of their body parts and actuators is complex. A robot's behavior arises from its action choices, the set of available actions and a choice mechanism.

Robot designers play a critical role. They design and develop the set of possible robot actions and craft the robot's *action choice mechanism* that robots use to select the next action to execute in real-time. Once deployed autonomous robots essentially spend their life making sense of their perceptions and choosing their next action.

Designing action choice mechanisms so that the robot makes autonomous and appropriate next action selections is a serious challenge. It is important to note that many robots are reactive and do not have explicit goals, plans or intentions. Even robot soccer players do not typically have intentions, however, *people regularly attribute intention to robots*. For example, they say the robot is "going after the ball", "trying to kick a goal", "looking for a team member to pass to". Studies of human understanding show that if a person fully understands a system they are able to explain it in terms of *underlying mechanisms* e.g. "the robot can calculate how far away the ball is with only one camera using the size of the ball in the image because it knows the size of the ball". If a person is not sure how a system works they often provide a *functional description* e.g. "robots have cameras to see where the ball is". If a person has little idea how a system works then they tend to attribute intention as a means to explain behavior, e.g. "the robot goalie dived because it wanted to stop the other team from scoring a goal". There is a tendency for people to anthropomorphise robots as a means to explain and predict their behaviour as intelligent machines.

The action choice mechanism ultimately determines robot behavior. Robotists develop system architectures and designs that enable robots to make the critical decision of what action to perform next. Robot decisions are complex and always involve uncertainty and risk because a robot's sensor data is noisy; its knowledge and ability to reason is limited; its understanding of its environment and the real world is superficial and often flawed; its perceptions are crude and not always faithful to reality; the robot may not have a clear understanding of its goals, roles, objectives or specific deployment tasks; situations become even more complex for robots when interacting with people.

The field of social robot design is full of *ad hoc* procedures, folk philosophy and folk psychology, and as a result robot behaviors do not follow any guidelines or principles. A designer simply develops behaviors based on their experience regarding what they know "works". It is difficult to scientifically analyse and compare robot behaviors and action selection mechanisms because they are typically idiosyncratic and/or incredibly complex as they attempt to mimic the human brain.

3   Importance of Legible and Predictable Robot Behavior

In human social settings and during collaborative activities it is critically important to be able to interpret, explain, predict, and anticipate other people's behavior. Of course, it is impossible for people to explain or predict every aspect of another person's behavior. However, since people have similar morphologies and use similar communicative signals, and tend to act in roughly "rational" ways it is certainly possible to predict other people's behavior to a large extent such that society can function reasonably effectively.

People do not behave entirely randomly, instead they exhibit predictable patterns of behavior that other people take into account when they plan and execute joint/collaborative actions. People are also able to develop strategies to mitigate the risk of failing to predict other people's behavior, and to respond in real-time when collaborative action goes awry. People tend to mostly behave reasonably and rationally, thus making working together easier and more enjoyable than if they acted unreasonably and irrationally. By contrast, unreasonable or irrational people are hard to understand, difficult to predict and typically not enjoyable to work with.

As robots become increasingly prevalent in society and their tasks require increasingly complex human-robot interactions there is a pressing need to design and develop robots that are easy for people to understand and predict, so that interactions and collaborations are more legible, predictable, safe, fluent, and effective, i.e. enjoyable. Sharing the same physiology helps people interpret and predict each other's behavior because similar sensory stimuli have similar effects on human brains. For example, we all know that a flash of light or loud noise will typically attract a person's attention when it is in their sensor range, we can imagine and explain other people's behavior by introspection and a study of ourselves. Most people make an audible sound when they experience sharp strong pain, and when someone falls over and cries, we know why. Some people are easier to predict than others, and people can adopt deceptive behaviors to mask their action choices and intentions.

Robots do not share the same morphology as people, and our bodily experiences are entirely different, and yet, they can still exhibit behaviors that people can understand and predict. By way of comparison, people are able to predict and control certain aspects of other biological species behavior and can work with some animals in highly productive ways. Not all animals can be tamed and trained. Consider, horse riding where a human controls much of the behavior of a horse: people and horses can work seamlessly together. In contrast, zebras are difficult to harness and work with. Horse riding comes with risk: no matter how skilled a rider, if a horse is surprised or afraid it can react in unpredictable ways. Just as people are able to "predict" animal behavior, there is a need to deploy robots with behaviors that people can predict to an appropriate degree and interact with in an enjoyable way.

It is easier for people to predict certain animals like horses and dogs, than it is for them to predict robots of today: partly because people have little experience with robots, they are not sure what to expect. However, there is a critical difference between animals and robots: robots are designed, and the quality of the design can have a massive impact on how well and how easily humans can predict them.

People should not expect to predict robot behavior all the time, however, they should absolutely expect to be able to predict robot behavior most of the time [5] particularly in circumstances when deviations from expectations are dangerous. Determining when the behavior of a robot is unpredictable is crucial; as this is when it is time to give a robot more physical space to undertake its maneuvers.

Having to deal with unpredictable robots is *not enjoyable*, and it will be costly for society in the same way dealing with unpredictable people can be time consuming and exhausting. Not only are unpredictable robots unsafe, difficult and unpleasant to be around, the lack of predictability is a major obstacle to technological innovation adoption and the expansion of the robot market.

In addition, to the need to be predictable, robots should be able to help people understand some of their actions and to explain their action choices. Choice designers in disciplines like marketing have been able to improve prediction by learning more about how people perceive and make choices. Robotists and robot users will also benefit if they can learn robot choice patterns and interrogate robots to discover their preferences as an explanation for their action choices and subsequent behavior.

We define a robot to be *unpredictable if* its action choices do not have a predictable pattern from a human perspective. Apart from being annoying and irritating, an unpredictable robot may threaten people's well being and cause all kinds of *havoc*, and so there is a pressing need to develop robot systems that people can predict.

Unfortunately, designing and developing predictable robots has proved to be a major challenge and has led to the design of highly deterministic robot designs with limited scripted robot behaviors, which are predictable but hopelessly inflexible, not adaptive and not scalable, thus restricting the range of tasks that robots can be deployed to undertake. The real challenge though lies in building robots that can work closely with people in enjoyable ways. On one hand, people must be able to anticipate robot behaviors, and on the other hand, robots must be able to interpret people's behavior and anticipate them as well.

4   Rational and Reasonable Action Choices

Choice Theory provides a sound approach to reasonable and rational decision-making. It turns out that all rational choices are reasonable, but there are some reasonable choices that are not rational. So rational choices are a subclass of reasonable choices. Rational and reasonable action choices can be used to design more predictable and legible robot behaviors. In this section we describe how robot action choice mechanisms can be described in a Choice Theory framework. In the following section we show how this allows robot behaviors to be designed so that they are *reasonable* or *rational*.

Robot behavior can be specified as a combination of desires, intentions, perceptions, beliefs, skills, actions and action choices: *goals/desires* are explicit representations of what a robot is aiming to do; *plans/intentions* are series of actions that can be performed to achieve a goal; *perceptions* are created from interpreting and combining sensor data; *actions* are processes that the robot can execute and/or enact; *beliefs* include facts and rules; *skills* involve information about when an action can be undertaken; *action choice mechanisms* determine the action choices for the robot to select the next action to execute and enact.

Choice Theory focuses on the set of actions that a decision maker, in this case a robot, can choose to enact. It formalizes the use of a preference relation/ordering to encapsulate goals and plans, and drive action choices. Choice Theory explores the selections that underlie patterns of choice and it can be used to prove that robot action choices are *reasonable* or *rational*.

A *robot action choice model* comprises a set of all possible *actions*, **A**, that the robot can perform. A *robot action choice function,* **c** is used to determine the set of actions that a robot could execute **c(A)** at a given time under certain circumstances. For example, a robot might be able to execute any of the actions in its set of possible actions: **A** = {*rotate_head_left, evaluate(x*6), rotate_head_right}* but not all of them simultaneous. At any given time when in a specific state the robot must chose an applicable set of actions, called the *choice set,* that it can actually execute: **c(A)** = {*rotate_head_left), evaluate(x*6)*}.

A *robot action choice function* for a binary relation > and a set of actions **A** is a function **c(A, >)** defined by {x∈A: for all y∈A and y not > x} where the ordering, > , is a *preference relation. a > b* is read as *a is at least as good as b,* and if it were the case then the robot in a particular state essentially prefers action *a* over action *b*.

It turns out that if the preference relation, > , over actions is acyclic then the robot action function **c(A, >)** gives rise to a simple choice function **c(A)**. Preference relations can be designed and used by robots to prefer action *a* over action *b*, or vice versa, or to be indifferent. Choice Theory provides a number of basic conditions that allow us to classify different kinds of choice functions that are useful in robot design.

In order to define what it would mean for a robot to make reasonable or rational choices we introduce three key conditions. They govern how choices are made across subsets and supersets of choices and impose forms of consistency across these choices: Given a set of robot actions **A**:

i. Choice function **c** satisfies the *contraction condition* if for any choice, **c(A)**, then **c(A)** is chosen if **c(A)** is available.
ii. Choice function **c** satisfies the *expansion condition* if actions a, b ∈ **c(A)** **A**⊆ **B** and **b** ∈**c(B)**, then a∈ **c(B)**.
iii. Choice function **c** satisfies the *revelation condition* if actions **a**, **b** are in **A** and **a** ∈ **c(A)** then for all **A'**⊆ **A** whenever **b**∈**c(A')** we have **a** ∈ **c(A')**.

Robot choices satisfy the contraction condition if whenever the robot chooses a particular action, say **a**, from a set of possible actions, if the possible actions were fewer and action *a* is still available, then the robot should choose action *a* again. It turns out that the contraction and expansion conditions are consistent and independent, and revelation entails both contraction and expansion, but not conversely. These three properties are used in the next section to show how to construct rational and reasonable robot choices.

5   Designing Reasonable and Rational Robot Behaviors

In this section we consider several important conditions that the action choice mechanism can be designed to satisfy in order to make robot behavior rational and/or reasonable.

Decision makers' appetite for risk often influences the choices selected. Choice Theory uses a notion of "rationality" to mean that an individual acts *as if* balancing costs against benefits to arrive at an action that maximizes personal advantage [24] Applying Choice Theory to robots raises the question of what constitutes "personal advantage" for robots. But for robot designers it is clear, we want robots to achieve their specific deployment tasks.

Proposition 1:   Let A_S denote the set of all actions available to a robot in state S. Robot action choices satisfy the *contraction condition* iff the robot action choice $c(A_S) \subseteq A'$ and $c(A_S) \subseteq c(A')$ whenever $A_S \subseteq A'$.

In other words, if a robot's action choices satisfy the contraction condition then reducing the size of the possible set of actions in state S does not change the robot's choice if the selected actions are still available, and conversely.

Proposition 2:   Let a robot be in state S and let A_S denote the set of actions available to the robot in state S from the set of all actions A. If there is an additional set of actions B, then robot action choices satisfy the *expansion condition* iff the robot chooses $c(A_S)$ among $A_S \cup b$ for each action $b \in B$.

Expansion says that if a robot chooses the same set of actions, say. $c(A_S)$, from an expanded set of actions from B that includes $c(A_S)$ and any $b \in B$ then it will chose $c(A_S)$ from the expanded set.

If a robot's action choices satisfy the contraction and expansion conditions, then Choice Theory says its behavior is defined to be ***reasonable***.

The following simple proposition relating choices to preferences is immediate from standard results in Choice Theory, however it is a striking claim in robotics. The notion that a robot's behavior could be classified as "reasonable" is novel in robotics.

Proposition 3:   If a *robot exhibits reasonable behavior then it has a preference relation* over its set of actions, A, for every state S.

If a robot is in state S and A_S is the set of actions available to the robot in state S. Robot action choices satisfy the *revelation condition* iff the robot chooses A_i when $A_i \subseteq A$, and whenever the robot chooses A_i it also choses $A_j \subseteq A$. In other words, if the robot chooses action A_i over a second action A_j, then whenever it chooses A_j, then it also chooses A_i whenever it is available.

If a robot's choices satisfy the revelation condition, then Choice Theory says its actions are defined to be ***rational***.

Reasonable action choices are weaker than rational action choices, i.e. rational choices are stricter than reasonable choices as they must satisfy the much stronger revelation condition. As noted earlier rational choices are a special case of reasonable choices.

If a robot acts reasonably people could predict its actions some of the time, but if it acts rationally then it would be possible to predict the robot all the time. In order to achieve this level of perspicuity a robot's preferences would need to be known.

A choice function based on a preference ordering is utility maximizing if for some assignment of utilities the actions chosen are precisely those whose utility is at least the utility of every action.

A robot action choice is *rational* if and only if it can be explained by a preference ordering; an action choice is *rational* if and only if it is utility maximizing. Choice gives rise to utility, and utility is a measure of preference [1].

There is an important difference between using choice models to describe behavior as reasonable or rational, and choice models that can be used to make predictions about *actual* behavior. Since robots are designed decision makers, it is possible to use preference relations to describe and explain robot behavior.

Proposition 4: If a robot exhibits *reasonable behavior* then it has a *preference ordering* over the power set of its actions.

Proposition 5: If a robot's actions are rational, then they are reasonable.

Propositions 1–5, above, show that in order for robots to exhibit reasonable or rational behavior their choices must be disciplined. This will not happen without proactive design steps.

Value-based action selection naturally aligns with Choice Theory because the robot action selection can be described using ordinal or cardinal ranking of actions based on a set of criteria [25]. Hoffman and Breazel [6] aggregate values of actions from several sources to drive robot behavior using a variety of explicit and implicit feedback mechanisms: (i) the strength of the sensory input, (ii) the strength of the motivation, (iii) level of interest to model boredom or behavior-specific fatigue, and (iv) various forms of inhibition. Value-based approaches have also been used to guide action selection in robot teams. Stroupe and Balch [13] used probabilistic values to direct next-step movements of robot teams as they map objects in their environment. It turns out that these methods resulted in robot paths that found vantage points that maximized information gain by reducing the uncertainty of each robot team member's next observation.

6 Improving Human-Robot Interaction

Henzinger and Sifakis (2006) and many others have identified a major chasm between analytical and computational models, and the gap between safety critical and best effort engineering practices. This chasm is particularly disturbing in the robot-human interaction space where people increasingly work in close proximity with robotic technologies, e.g. manufacturing robots, robotic surgery, exoskeletons, and underwater robots. Unless robots are safe and easy to work with, their utility and adoption as a technology will be limited. Unfortunately, the prevailing approach to developing robots that are safe and easy to work with has delivered robot designs that are not adaptable or suitable for open, complex or dynamic environments. This typically means that robots can only achieve structured tasks in predictable and scripted ways; their ability to adapt to new circumstances or achieve complex tasks in dynamic open environments is severely limited.

Robot design delivers a set of actions and a mechanism that allows a robot to choose the next action to execute and enact. Actions are computational processes that robots can execute and enact. Actions can be general computation processes, e.g. pause/wait, arithmetic, database manipulation, or control programs that involve physical actuation such as actuator control. Actions involving actuation require appropriate access to relevant actuators, and in order to be deemed successful they may require certain expectations to be filled, e.g. at the end of the *lift_cup* action, the robot should have lifted a cup.

There are several basic action choice mechanisms widely used in robot systems, which include *reactive mechanisms* that rely on look-up tables in which each stimuli is linked with an explicit response action. Reactive mechanisms are highly deterministic and generate inflexible behaviors: they encapsulate skills with a fixed set of stimulus-response relationships that govern robots behavior. Reactive mechanisms can be implemented as finite state machines. *Behavior-based mechanisms* build on Brook's idea of subsumption [22], which is essentially a layered reactive model. Other kinds of action selection mechanisms include *rule-based selection* [21]; *blackboard architectures* [23]; and *value based selection* using ordinal and cardinal measures of value like cost and risk [6] and concepts of attention competitions [9–11].

7   Discussion

As robots become increasingly prevalent in society and their tasks involve more complex human-robot interactions there is a pressing need to design and develop robot behaviours that are easy for people to understand and predict, so that interactions and collaborations are more legible, predictable, safe, fluent, and effective, i.e. enjoyable.

There are no widely accepted design principles that can be used to guide action selection for social robots that engage in human interaction and collaboration. We addressed this gap by approaching the robot design as *a problem of designing an action choice mechanism*: robots spend their entire life interpreting their sensor data and using it to choose their next action to execute. Action choice mechanisms are fundamental to robot capability and behaviours. Robots behaviours need to be legible and predictable in order for humans to find working with robots to enjoyable. In this paper we used a decision-theoretic approach to argue that the Choice Theory concepts of reasonable and rational choices can be used to show that designed robot behavior is more predictable and legible. The robots that will be the most successful working with people will be the ones that people find enjoyable to work with, and that means those that people can understand and anticipate.

Future work will explore three key research questions (i) how to extend the use of Choice Theory for robots in changing and uncertain circumstances in complex social settings and human-robot interaction scenarios, (ii) how to incorporate theory of mind reasoning mechanisms to enrich robot choices of action in social settings and human-robot interaction scenarios, and (iii) explore the tension between rationality and insanity, where insanity is defined as making the same choices but expecting a different outcome.

References

1. Allingham, M.: Choice Theory. Oxford Press, Oxford (2002)
2. Arrow, K.: Essays on the Theory of Risk Bearing (1971)
3. Bem, D.J., Allen, A.: On predicting some of the people some of the time: the search for cross-situational consistencies in behavior. Psychol. Rev. **81**, 506–520 (1974)
4. Dragan, A.D., Lee, K.C., Srinivasa, S.S.: Legibility and predictability of robot motion. In: Human-Robot Interaction, 2013 8th ACM/IEEE International Conference, pp. 301–308. IEEE, March 2013
5. Epstein, S.: The stability of behavior: I. On predicting most of the people much of the time. J. Pers. Soc. Psychol. **37**(7), 1097–1126 (1979)
6. Hoffman, G., Breazeal, C.: Cost-based anticipatory action selection for human-robot fluency. IEEE Trans. Robot. (T-RO) **23**, 952–961 (2007)
7. Kreps, D.M.: Notes on the Theory of Choice. Westview Press, Boulder (1988)
8. Kruse, T., Basili, P., Glasauer, S., Kirsch, A.: Legible robot navigation in the proximity of moving humans. In: May 2012 IEEE Workshop on Advanced Robotics and its Social Impacts (ARSO), pp. 83–88. IEEE (2012)
9. Novianto, R., Johnston, B., Williams, M-A.: Attention in the ASMO cognitive architecture. In: Proceedings of the First Annual Meeting of the BICA Society, vol. 221 of Frontiers in Artificial Intelligence and Applications, pp. 98–105 (2012)
10. Novianto, R., Johnston, B., Williams, M.-A.: Habituation and sensitisation learning in ASMO cognitive architecture. In: Herrmann, G., Pearson, M.J., Lenz, A., Bremner, P., Spiers, A., Leonards, U. (eds.) ICSR 2012. LNCS, vol. 8239, pp. 249–259. Springer, Heidelberg (2013)
11. Novianto, R., Williams, M.-A.: The role of attention in robot self-awareness. In: 18th IEEE International Symposium Robot and Human Interactive Communication, pp. 1047–1053 (2009)
12. Simon, H.: A behavioral model of rational choice. Q. J. Econ. **69**, 99–188 (1955)
13. Stroupe, A., Balch, T.: Value-based action selection for observation with robot teams using probabilistic techniques. J. Robot. Auton. Syst. **50**, 85–97 (2004)
14. Takayama, L., Dooley, D., Ju, W.: Expressing thought: improving robot readability with animation principles. In: Proceedings 6th International Conference Human-Robot Interaction, pp. 69–76. ACM, March 2011
15. Williams, M.-A.: Robot Social intelligence. In: Ge, S.S., Khatib, O., Cabibihan, J.-J., Simmons, R., Williams, M.-A. (eds.) ICSR 2012. LNCS, vol. 7621, pp. 45–55. Springer, Heidelberg (2012)
16. The Fugitive Robot Video, https://www.youtue.com/watch?v=rF_-TmrTan8. IJCAI Best Video Award 2013
17. Mutlu, B.: Nonverbal leakage in robots: communication of intentions through seemingly unintentional behavior. In: 4thACM/IEEE Int'l Conference on Human robot interaction (2009)
18. Terada, K., Shamoto, T., Mei, H., Ito, A.: Reactive movements of non-humanoid robots cause intention attribution in humans. In: IEEE/RSJ International Conference on Intelligent Robots and Systems, IROS 2007, pp. 3715–3720. IEEE, October 2007
19. Goldman, A.: Desire, intention and the simulation theory. In: Malle, B., Moses, L., Baldwin, D. (eds.) Intentions and Intentionality, pp. 207–224. MIT Press, Cambridge (2001)
20. Kögler, H.H., Stueber, K. (eds.): Empathy and Agency: The Problem of Understanding in the Human Sciences. Westview Press, Bolder (2000)
21. Salehie, M., Tahvildari, L.: Self-adaptive software: landscape and research challenges. ACM Trans. Auton. Adapt. Syst., 1–40 (2009)

22. Brooks, R.A.: Intelligence without reason. In: Proceedings of 12th International Joint Conferenceon Artificial Intelligence, Sydney, Australia, August 1991, pp. 569–595 (1991)
23. Engelmore, R.S., Morgan, T. (eds.): Blackboard Systems. Addison-Wesley, Reading (1998)
24. Friedman, M.: Essays in Positive Economics, Chicago (1953)
25. Blumberg, B., Galyean, T.: Multi-level direction of autonomous creatures for real-time virtual environments. In: Proceedings of SIGGRAPH 95 (1995)
26. Henzinger, A., Sifakis, J.: The embedded systems design challenge. In: Proceedings of the 14th International Conference on Formal Methods, pp. 1–15

Physiologically Inspired Blinking Behavior for a Humanoid Robot

Hagen Lehmann[1(✉)], Alessandro Roncone[2], Ugo Pattacini[1], and Giorgio Metta[1]

[1] Istituto Italiano di Tecnologia, iCub Facility, via Morego 30, 16163 Genoa, Italy
hagen.lehmann@iit.it
[2] Social Robotics Lab, Yale University, 51 Prospect Street, New Haven, CT 06511, USA

Abstract. Blinking behavior is an important part of human nonverbal communication. It signals the psychological state of the social partner. In this study, we implemented different blinking behaviors for a humanoid robot with pronounced physical eyes. The blinking patterns implemented were either statistical or based on human physiological data. We investigated in an online study the influence of the different behaviors on the perception of the robot by human users with the help of the Godspeed questionnaire. Our results showed that, in the condition with human-like blinking behavior, the robot was perceived as being more intelligent compared to not blinking or statistical blinking. As we will argue, this finding represents the starting point for the design of a 'holistic' social robotic behavior.

Keywords: Social robotics · iCub · Human-like blinking behavior · Online study · Godspeed questionnaire · Humanoids robotics

1 Introduction

One of the ambitions of current research in social robotics is generating solutions to counteract the demographic changes that human societies are facing around the globe [1]. Today transformations such as the childbirth peak and the increment of the present-day underrepresented age groups of 50 to 70, as well as 70 and above, lead to a predicted increase of the human population to 11 billion in the next 100 years [2]. This prediction produces the request of significant changes in our elderly care strategies to make our health care systems sustainable in the imminent future, and social robotics is one of the forces that can effectively participate to this process of renewal.

Recent advancements in the design and development of social robots will lead in the next years to commercial products able to stimulate the first steps towards the creation of "mixed human-robot ecologies" [3], in which robots will use social competences to better accomplish interactive tasks in service domains such as elderly care, personal assistance, health services, etc.

Currently, the ways these robotic agents should look like and behave are highly debated topics of research. One of the main research directions develops the idea that social robots should have an approximately humanoid appearance (i.e., possessing at least a head, arms and a torso [4]), and behave and move in naturalistic human-like ways defined by human social rules and characteristics [5]. This would assure that the

© Springer International Publishing AG 2016
A. Agah et al. (Eds.): ICSR 2016, LNAI 9979, pp. 83–93, 2016.
DOI: 10.1007/978-3-319-47437-3_9

interactions between these robots and their human partners will be highly intuitive, comfortable and reassuring. Specifically, in fields involving vulnerable individuals, such as elderly care and health services in general, human-like social components of the interaction are considered fundamental for the acceptance of robotic agents [6].

On this basis, contemporary research is focusing, besides on the "static" design and appearance of these robots, on the definition of their socially interactive movements and behaviors. This is considered particularly important when constructing humanoids, in order to avoid that these robots "fall" into the "uncanny valley" [7]. Typically, humanoids raise the expectation of behaving like humans, and tend to be perceived as disturbing and fearsome when exhibiting unnatural movements and behaviors. More in general, to ensure a positive perception of the robots and their integration in our social environments, the naturalness and intuitiveness of the "interaction interface" is highly important. For humanoid robots, this interaction interface is going to be their physical body, socially activated by body language, facial expressions and verbal communication.

In this paper, we will illustrate the first step of a structured approach to the integration of human-like nonverbal involuntary behavior in humanoid social robots. The target behavior is human-like eye blinking, based on the hypothesis that, to stimulate intuitive and comfortable conversations with humanoid robots endowed with expressive faces that include physical eyes, naturalistic eye blinking plays a very important role. We tested this hypothesis by means of an online experiment, in which we presented videos of the iCub robot exhibiting different blinking patterns during a conversation and asked our participants to rate them according to the personal impression.

We will describe the research process that supported our implementation and evaluation of a human-like conversational blinking pattern in detail. We consider this work as the starting point of a multimodal integrative approach to the implementation of human-like non-verbal communication behaviors in social robots. According to this approach, we are defining for iCub a behavioral library based on the experimental exploration of human socially interactive behaviors, and later we will proceed to coordinately integrate in the robot the different behavior modalities included in our studies and library.

2   Background

Basic behavior synchronization in humans is achieved on the basis of observation of the behaviors exhibited by a social partner, and neuronal mechanisms such as the mirror neuron system [8]. The processing of this behavior might happen consciously or unconsciously, but the result remains a behavioral synchronization that facilitates mutual understanding and cooperation between the interaction partners. We propose that for artificial agents it would be sufficient to observe and to respond to observed behaviors accordingly, in order to facilitate conversations and cooperation with their users.

Even before the advent of the research field of social robotics, it was shown that reactive nodding and blinking of a simulated artificial agent facilitates the turn taking and smoothness of human speech input to computers [9, 10].

The role of blinking has been recognized as important in the field of social robotics very early on [11]. This has encouraged researchers in the field to explore the potentialities of blinking in human robot communication [12]. Specifically, for robots like the iCub, featuring pronounced physical eyes, authentic eye blinking behavior can have a profound impact on the interaction comfort [13]. Eyelids have been implemented into different social robots [e.g. 14, 15]. Nevertheless, there have only been very few structured inquiries on how to model human blinking for robots with physical eyes and the blinking behavior has mainly been added randomly into the social interaction with these robots [e.g. 12]. This is largely due to the technical restraints physical robotic humanlike eyes impose and to the complexity of factors influencing blinking in humans.

In the last decades, physiological research on the various dependencies of human eye blinking behavior on different physiological and psychological factors produced a variety of results that can be used to model blinking in social robots. Ford et al. [16] showed for example that blinking is strongly linked to onsets and offsets of communicative facial behaviors and verbalizations. Based on their findings, they proposed the "blink model" for HRI, which integrates blinking as a function of communicative behaviors. In their experiments they used a back projected face on a human face mold. Doughty [17] described in his work three distinct blinking patterns during reading, dialoging and idly looking at nothing specific. Lee et al. [18] proposed a model of animated eye gaze that integrates blinking as depending on eye movements constituting gaze direction. Neurological findings showed that responses to facial movements such as blinks can be measured in an observer's brain [19], a result that hints at the social importance of eye blinking for behavior synchronization between social interlocutors.

In summary, it can be said that blinking has been described as: (1) a function of physiological variables, such as the average speed of a single blink, the average blinking rate and the average length of the inter eye blink intervals (IEBI); (2) a function of system state variables, such as changes in facial expression and verbal communication behaviors; (3) a function of social context information, such as reading and being in a conversation; (4) a function of the psychological state of the person exhibiting the behavior; and (5) a function of the behavior of the social interlocutor.

As part of a holistic non-verbal behavior architecture for the iCub robot we started to develop a module for human-like conversational blinking - "BlinkSync" [20] - in which the robot blinking is based on human physiological data. In order to achieve human-like conversational blinking, we integrated as a first step the average speed of a single blink, the average blinking rate and the average length of the inter eye blink intervals into "BlinkSync".

3   Method

We used human social behavior data to model social behaviors for the robot, and then to test these behaviors in Human-Robot Interaction contexts. We used both an optimization approach for the behaviors on the robot and a synthetic modeling approach for the testing of these behaviors in natural environments.

iCub Eyelid Mechanism. For the implementation and testing of our blinking module we used the iCub robot. It has very pronounced eyes resembling human features like a black pupil, white sclera, and moveable upper and lower eyelids (Fig. 1a).

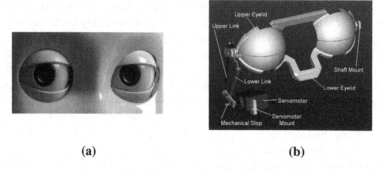

(a)   (b)

Fig. 1. (a) iCub eyes and (b) iCub eyelid mechanism

The eyelid mechanism is controlled by one servomotor and constructed in such a way that both eyes close at the same time and the eyelids meet in the center of the eyeballs (Fig. 1b). The servomotor is a standard model from Futaba and controlled with a pulse-width modulation (PWM) input. The motor is driven by a PWM control signal with a frequency of 50 Hz. The duration of the pulse determines the final position of the motor shaft. The motor will move to the "0°" position when the pulse has a width of 1 ms. The position of "90°" is reached for a pulse duration of 2 ms.

The eyelids can be controlled continuously from an open position to closed position. Commands can be given to open or close the eyelids from totally closed to totally open, with 127 levels of discretization [21].

Human-Like Blinking. For the human-like blinking behavior, we chose to use as a starting point the physiological data provided by Doughty [17] and to adapt it to technical limitations of the iCub eyelid mechanism. The general settings for the conversational blinking module are an average blinking rate of 23.3 blinks per minute with an inter eye blink interval of 2.3 ± 2.0 s. The blinks of the robot are in 85 % of the cases single blinks and in 15 % of the cases double blinks.

We divided each blink into three phases, the attack phase when the eye is closing, the sustain phase when the eye is closed, and the decay phase when the eye is opening again. The attack phase has an average length of 111 ms with a standard deviation of 31 ms. The sustain phase lasts on average 20 ms with a standard deviation of 5 ms. The average length of the decay phase is 300 ms with a standard deviation of 123 ms. The robot also blinks on each onset and offset of its verbalizations.

The adaptation of the original physiological data from Doughty became necessary due to the specifics of the iCub eyelids. Unlike in a human eye, the robot's upper and lower eyelids move the same way during the blink. When using the original human data, this results in a much faster closure of the eye (the eyelids meet in the middle of the eyeball). This resulted in what was described by participants in a short pilot study as hectic blinking. By adapting the different speeds of each of the phases it was possible to give the movement a more human-like appearance.

The code for the blink controller was implemented and released under the GPL open source license on GitHub (The source code is accessible at https://github.com/robot-ology/funny-things/tree/master/modules/iCubBlinker, and the documentation can be found at http://robotology.github.io/funny-things.). It was developed for the iCub humanoid robot, and is readily available for any iCub robot. The architecture is generically applicable to any humanoid head, and the code has been designed to be modular and easy to adapt to any other robotic platform.

Experimental Setup. In order to evaluate influence of the blinking behavior on the impression the robot has on participants we generated three conditions with different blinking patterns. For the robot to appear to be in a social interaction during the experiment, we scripted a one-interview. In order to make this interview interesting and informative for the participants watching it, we decided to "discuss" with the robot a game it plays usually during demonstrations and exhibitions. The questions asked in this interview and the robots answers were the same in each of the three experimental conditions.

- Condition 1: The robot looked straight ahead without blinking while answering the questions of the experimenter.
- Condition 2: The robot blinked every 5 s using the timings for the attack, sustain, and decay phases described in *Human-like blinking*. It performed no double blinks.
- Condition 3: The robot blinked with a rate of 23.3 blinks per minute. It performed double blinks and blinked at the onset and offset of its verbalizations. The timing was of the three phases of the blink were as described in *Human-like blinking*.

We recorded each of the conditions in such a way that the potential participant watching the video would see a front view of the robot (see Fig. 2) and hear both the interviewer and the robot talk. The participants would hear the experimenter asking the questions to the robot, but would not see him.

In order to reach as many participants as possible in a short time we chose an online video study as format for our experiment. The Video-based HRI (VHRI) methodology has been used reliably in several human-robot interaction studies in the past [e.g. 22]. In a direct comparison of live HRI and Video-based HRI, it has been shown that comparable results can be achieved [23].

The videos were embedded in an online questionnaire in which we asked the participants first for their consent, then for their demographic data and their experience with robots. After this first part, each participant would see the video clip of one of the conditions. Which condition the participant would see was randomly chosen. After the participants watched the video clip, they were asked to complete an online version of the *Godspeed Questionnaire* [24]. In order to recruit participants, the link to the online study was sent to different mailing lists in Europe. The survey was prepared with Google Forms.

The *Godspeed Questionnaire* [24] evaluates the impression a person has of a robot on five different subscales. These subscales are anthropomorphism, animacy, likeability, perceived intelligence, and perceived safety. Since we hypothesize that the blinking behavior will improve the user experience with the robot, we expected that in condition 3 the robot would score higher compared to condition 1 and condition 2 in anthropomorphism, animacy, likeability and perceived intelligence.

4   Results

Sample Characteristics. We received 44 replies to our call for help of which 26 were male and 18 were female. The mean age of the participants was 34.98 years, ranging from 24 to 56. Most of them had seen at least pictures of the iCub before. Concerning their experience with robots 21 reported none, 12 reported little, 10 reported some experience, and only one participant indicated that his experience with robots was substantial.

Godspeed Questionnaire Results. In order to test our research hypothesis, we examined the differences in participant ratings of the robot along the subscales of the Godspeed Questionnaire. It has been shown that this is valid procedure when analyzing this questionnaire [25]. The descriptive statistics for Anthropomorphism can be seen in Table 1.

Table 1. Descriptive statistics anthropomorphism

Condition	Variable	Mean (SD)	Median
Condition 1	Anthropomorphism	2.63 (0.23)	2.71
Condition 2	Anthropomorphism	2.71 (0.43)	2.78
Condition 3	Anthropomorphism	2.65 (0.45)	2.44

For anthropomorphism the participants overall rated the robot in none of the 3 conditions higher than the "neutral" score of 3. A single factor ANOVA found no significant differences between the three conditions ($F(2,12) = 0.07$, $p = 0.94$).

The descriptive statistics for animacy are presented in Table 2. For animacy, the participants rated the robot both in condition 2 and 3 higher than the "neutral" score of 3. In condition 1, the robot was overall rated lower then neutral. A single factor ANOVA found no significant differences between the three conditions ($F(2,15) = 0.42$, $p = 0.66$).

Table 2. Descriptive statistics animacy

Condition	Variable	Mean (SD)	Median
Condition 1	Animacy	2.8 (0.67)	2.78
Condition 2	Animacy	3.04 (0.76)	3.07
Condition 3	Animacy	3.19 (0.76)	3.28

The descriptive statistics for likeability are presented in Table 3. For likeability, the robot was rated in all three conditions higher than the "positive" score of 4. A single factor ANOVA found no significant differences between the three conditions ($F(2,12) = 0.38$, $p = 0.69$).

Table 3. Descriptive statistics for likeability

Condition	Variable	Mean (SD)	Median
Condition 1	Likeability	4.11 (0.17)	4.07
Condition 2	Likeability	4.16 (0.15)	4.07
Condition 3	Likeability	4.08 (0.11)	4.06

The descriptive statistics for perceived intelligence are presented in Table 4. For perceived intelligence the participants overall rated the robot in all of the three conditions higher than the "neutral" score of 3. A single factor ANOVA found significant differences between the three conditions ($F(2,12) = 6.3$, $p = 0.01$).

Table 4. Descriptive statistics perceived intelligence

Condition	Variable	Mean (SD)	Median
Condition 1	Perceived intelligence	3.37 (0.27)	3.43
Condition 2	Perceived intelligence	3.6 (0.19)	3.57
Condition 3	Perceived intelligence	3.89 (0.22)	3.94

The descriptive statistics for perceived safety are presented in Table 5. For perceived safety the participants overall rated the robot in all of the three conditions higher than the "neutral" score of 3. A single factor ANOVA found no significant differences between the three conditions ($F(2,6) = 0.01$, $p = 0.99$).

Table 5. Descriptive statistics perceived safety

Condition	Variable	Mean (SD)	Median
Condition 1	Perceived safety	3.74 (0.41)	3.86
Condition 2	Perceived safety	3.69 (0.48)	3.93
Condition 3	Perceived safety	3.73 (0.16)	3.75

5 Discussion

Our results show that the robot, when displaying human-like blinking, was perceived as being more "intelligent" according to the categories of the Godspeed questionnaire. For its other subscales, no significant differences were found. This illustrates that even though human-like blinking behavior can make a significant difference in how humans perceive robots, it is only one aspect of human nonverbal communication that needs to be taken in consideration when designing social behaviors for robots. Naturalistic blinking in itself it is not enough for a robot to meet the requirements for its social integration, such as for example perceived safety, likeability, and animacy. We could think to blinking as an important, but not a sufficient characteristic regarding social integration.

Being perceived as intelligent can be advantageous for the effectiveness of a robot specifically in situations in which it has to transmit information in a social context, e.g. when the robot is used as a guide or informant in shopping malls, train stations or in tourist information centers. A robot that is perceived more intelligent is likely to be considered more competent and trustworthy in these cases.

During the study, we discovered a series of issues that influenced and even limited our implementation of the blinking behavior on the robot. One of the issues was the noise of the motor and the eyelids when closing. This disturbed the flow of the interaction and, as pointed out by one participant, "made the eyes look and sound like the shutter of a camera". For future versions of iCub's embodiment (and other robots with structured physical eyes) this should be taken into consideration and the blinking mechanism should be constructed accordingly.

Another question asked by some of the participants was whether the iCub can wink or not. Due to the construction of the eyelid mechanism, this is not possible at the moment. Winking as social cue is already being used successfully in computer-mediated communication in the form of emoticons [26]. To implement human-like winking with related triggering behavioral patterns for the iCub robot could be another key point in order to achieve positive and intuitive human-robot conversational interaction.

Due to the online format of the study, the participants were listening to a scripted conversation between the robot and the experimenter. We acknowledge that a direct conversation between the participant and the robot might have been more efficient to test the effect of naturalistic blinking behavior. As pointed out in the *Experimental Setup* section, we argue that similar results would be achieved with the VHRI methodology and that, in the case of a real time conversation between the robot and the human, our results would not be structurally different, but more pronounced.

The result that the naturalistic blinking behavior was more appreciated by users indicates that the synthetic methodology – i.e., modeling natural behaviors in artificial (in our case robotic) systems [27] – is a promising way to create successful applications for HRI. In other words, this methodology appears able to allow us to build better social robots, i.e. robots with a more convincing social presence, as well as to test psychological and sociological paradigms about their integration in human societies.

Future Work. Following our current research, we plan to study further how different blinking behaviors influence social interactions, by using the "Blink-Sync" module in human-robot interaction contexts (testing different blinking patterns in different situations) [20]. Applicative fields of the "BlinkSync" model, and of the "synthetic social studies" it can allow, are many, and include robot assisted therapy for children and elderly, and health care. An interesting application comes from research with children with autism spectrum disorder. Various studies have shown that using eye blinking attracts the children's attention towards the eyes and helps maintain engagement in the therapeutic setting [e.g. 28]. Other studies showed that children with ASD show atypical blinking pattern and even the absence of blinking in conversational contexts [29]. These results are interesting, because autistic children usually avoid looking into the eyes of their interaction partners. Utilizing the effect with naturalistic eye blink patterns in this context might help to teach these children to better interpret and understand the facial expressions of their social partners, which is something that is very difficult for people with autism [30].

6   Conclusions

We see the modeling of naturalistic blinking as a first step towards a more integrated nonverbal social communication approach. The result of our study shows that the implementation of naturalistic blinking behavior has a positive impact on the perception of the robot by a human user. In general, it can be said that the information transferred by the movement of the eyelids of a robot is important for a smooth and intuitive interaction between a human and a robot. Nevertheless, it is important to understand blinking only as part of the nonverbal social information transmission channel, with for example eye gaze direction or gestures at least as equally important [31, 32].

Acknowledgment. This research was funded by the EU via the Marie-Sklodowska Curie Action Program (SICSAR - grant agreement n° 627688).

References

1. United Nations. World Population Prospect – Revision 2015. United Nations, New York (2015)
2. http://www.gapminder.org/
3. Damiano, L., Dumouchel, P., Lehmann, H.: Towards human-robot affective co-evolution overcoming oppositions in constructing emotions and empathy. Int. J. Soc. Robot. **7**, 7–18 (2014). doi:10.1007/s12369-014-0258-7
4. Walters, M.L., Syrdal, D.S., Dautenhahn, K., Te Boekhorst, R., Koay, K.L.: Avoiding the uncanny valley: robot appearance, personality and consistency of behavior in an attention-seeking home scenario for a robot companion. Auton. Robot. **24**, 159–178 (2008)
5. Boucher, J.D., Pattacini, U., Lelong, A., Bailly, G., Elisei, F., Fagel, S., Dominey, P.F., Ventre-Dominey, J.: I reach faster when i see you look: gaze effects in human-human and human-robot face-to-face cooperation. Front. Neurorobot. **6**(3), 1–11 (2012)
6. Lehmann, H., Syrdal, D.S., Dautenhahn, K., Gelderblom, G., Bedaf, S., Amirabdollahian, F.: What should a robot do for you? - Evaluating the needs of the elderly in the UK. In: Sixth International Conference on Advances in Computer-Human Interactions (ACHI 2013), pp. 83–88 (2013)
7. Mori, M.: Bukimi no tani [the uncanny valley]. Energy **7**, 33–35 (1970)
8. Rizzolatti, G., Fogassi, L., Gallese, V.: Neurophysiological mechanisms underlying the understanding and imitation of action. Nat. Rev. Neurosci. **2**, 661–670 (2001)
9. Watanabe, T., Yuuki, N.: A voice reaction system with a visualized response equivalent to nodding. In: Proceedings of the Third International Conference on Human-Computer Interaction, vol. 1 on Work with Computers: Organizational, Management, Stress and Health Aspects. Elsevier Science Inc., pp. 396–403 (1989)
10. Watanabe, T., Higuchi, A.: Facial expression graphics feedback for improving the smoothness of human speech input to computers. Adv. Hum. Factors/Ergon. **18**, 491–497 (1991)
11. Breazeal, C.: Toward sociable robots. Robot. Auton. Syst. **42**, 167–175 (2003)
12. Yoshikawa, Y., Shinozawa, K., Ishiguro, H., Hagita, N., Miyamoto, T.: The effects of responsive eye movement and blinking behavior in a communication robot. In: 2006 IEEE/RSJ International Conference on Intelligent Robots and Systems, pp. 4564–4569. IEEE (2006)

13. DiSalvo, C., Gemperle, F., Forlizzi, J., Kiesler, S.: All robots are not created equal: the design and perception of humanoid robot heads. In: Proceedings of Designing Interactive Systems, pp. 321–326 (2001)
14. Dautenhahn, K., Nehaniv, C.L., Walters, M.L., Robins, B., Kose-Bagci, H., Mirza, N.A., Blow, M.: KASPAR–a minimally expressive humanoid robot for human–robot interaction research. Appl. Bion. Biomech. 6(3–4), 369–397 (2009)
15. Metta, G., Natale, L., Nori, F., Sandini, G., Vernon, D., Fadiga, L., Von Hofsten, C., Rosander, K., Lopes, M., Santos-Victor, J., Bernardino, A.: The iCub humanoid robot: an open-systems platform for research in cognitive development. Neural Netw. 23(8), 1125–1134 (2010)
16. Ford, C.C., Bugmann, G., Culverhouse, P.: Modeling the human blink: a computational model for use within human–robot interaction. Int. J. Hum. Robot. 10(01), 135006 (2013)
17. Doughty, M.J.: Consideration of three types of spontaneous eyeblink activity in normal humans: during reading and video display terminal use, in primary gaze, and while in conversation. Optom. Vis. Sci. 78(10), 712–725 (2001)
18. Lee, S.P., Badler, J.B., Badler, N.I.: Eyes alive. ACM Trans. Graph. 21, 637–644 (2002)
19. Brefczynski-Lewis, J.A., Berrebi, M., McNeely, M., Prostko, A., Puce, A.: In the blink of an eye: neural responses elicited to viewing the eye blinks of another individual. Front. Hum. Neurosci. 5, 1–8 (2011)
20. Lehmann, H., Pattacini, U., Metta, G.: Blink-sync: mediating human-robot social dynamics with naturalistic blinking behavior. In: Workshop on Behavior Coordination Between Animals, Humans and Robots; 10th ACM/IEEE International Conference on Human-Robot Interaction (2015)
21. Bernardino, A., Nunes, R., Vargas, L., Beira, R., Lopes, M., Santos-Victor, J.: Expressions system control for the iCub head (Internal Report No. 004370). Instituto Superior Tecnico (2007)
22. Lohse, M., Hanheide, M., Wrede, B., Walters, M.L., Koay, K.L., Syrdal, D.S., et al.: Evaluating extrovert and introvert behaviour of a domestic robot — a video study. In: The 17th IEEE International Symposium on Robot and Human Interactive Communication (RO-MAN 2008), pp. 488–493. IEEE, August 2008
23. Woods, S.N., Walters, M.L., Koay, K.L., Dautenhahn, K.: Comparing human robot interaction scenarios using live and video based methods: towards a novel methodological approach. In: Proceedings of the 9th International Workshop on Advanced Motion Control (AMC 2006), 27 – 29 March, Istanbul (2006)
24. Bartneck, C., Kulić, D., Croft, E., Zoghbi, S.: Measurement instruments for the anthropomorphism, animacy, likeability, perceived intelligence, and perceived safety of robots. Int. J. Soc. Robot. 1, 71–81 (2009). doi:10.1007/s12369-008-0001-3
25. Lehmann, H., Saez-Pons, J., Syrdal, D.S., Dautenhahn, K.: In good company? perception of movement synchrony of a non-anthropomorphic robot. PLoS ONE 10, e0127747 (2015). doi: 10.1371/journal.pone.0127747
26. Derks, D., Bos, A.E., Von Grumbkow, J.: Emoticons in computer-mediated communication: social motives and social context. CyberPsychol. Behav. 11, 99–101 (2008)
27. Damiano, L., Hiolle, A., Cañamero, L.: Grounding synthetic knowledge. In: Lenaerts, T., Giacobini, M., Bersini, H., Bourgine, P., Dorigo, M., Doursat, R. (eds.) Advances in Artificial Life (ECAL 2011). MIT Press, Cambridge (2011)
28. Robins, B., Dautenhahn, K., Dickerson, P.: From isolation to communication: a case study evaluation of robot assisted play for children with autism with a minimally expressive humanoid robot. In: 2009 Second International Conferences on Advances in Computer-Human Interactions (ACHI 2009), pp. 205–211. IEEE (2009)

29. Nakano, T., Kato, N., Kitazawa, S.: Lack of eyeblink entrainments in autism spectrum disorders. Neuropsychologia **49**, 2784–2790 (2011). doi:10.1016/j.neuropsychologia. 2011.06.007

30. Baron-Cohen, S.: Mindblindness: An Essay on Autism and Theory of Mind. MIT press, Cambridge (1997)

31. Tomasello, M., Hare, B., Lehmann, H., Call, J.: Reliance on head versus eyes in the gaze following of great apes and human infants: the cooperative eye hypothesis. J. Hum. Evol. **52**, 314–332 (2007)

32. Broz, F., Lehmann, H., Nehaniv, C.L., Dautenhahn, K.: Automated analysis of mutual gaze in human conversational pairs. In: Nakano, Y.I., Conati, C., Bader, T. (eds.) Eye Gaze in Intelligent User Interfaces, pp. 41–60. Springer, London (2013)

Infinite Personality Space
for Non-fungible Robots

Daniel H. Grollman[✉]

Sphero, Inc., 4772 Walnut St, Suite 206, Boulder, CO 80301, USA
dan@sphero.com

Abstract. We outline a novel method for defining robot personality for the purposes of individual differentiation. Rather than a designer-developed set of behaviors where a users' preferences are learned and inserted into pre-written scripts, our approach allows for each robot to have and express a unique personality. This uniqueness reduces the fungibility of the robots, which may lead to increased user engagement.

Keywords: Human-robot interaction · Robot personality

1 Introduction

Robots that co-exist with non-specialized, untrained users are more and more frequently being developed to leverage 'social' capabilities to smooth their interactions. Beyond the basic technologies such as natural-language processing, gaze tracking, and theory of mind, there are two concepts of interest that have proven popular in the industry. The first, *personality*, seeks to imbue a social robot with a coherent entity-hood, described with vague but human-understandable terms such as 'helpful,' 'whimsical,' 'sassy,' or 'sparkling.' This goal is often achieved by hand-crafting robot behaviors based on a designer's understanding of how the desired personality would be expressed, and this personality and the robot's expression thereof is fixed for the life of the robot.

To be more enticing to a wider audience, multiple different personalities may be developed, and consumers enabled to select from within this fixed set. To further differentiate between individual robots (which may share the same physical form and base personality), the second concept, *personalization*, aims to let individual users make the robot's form and behavior more unique. Users are often able to (and do) modify the surface characteristics of their robot via paint, stickers, markers, etc., or even via use and wear. The software on the robot is generally not as malleable, but developers often allow for some user customization by selecting similarly 'surface' characteristics such as voice, gender pronoun, graphical avatar, etc. The available options define a restricted, often discrete space of robot 'characters,' which can still be somewhat easily replicated between users. (We ignore here the relatively small, but robust, 'hacker' or 'maker' communities that delve much deeper and change the hardware and software of robots in ways unintended or unimagined by the producing company.)

© Springer International Publishing AG 2016
A. Agah et al. (Eds.): ICSR 2016, LNAI 9979, pp. 94–103, 2016.
DOI: 10.1007/978-3-319-47437-3_10

These available characters are often further personalized over the the life of interaction between a user and their robot, by having the robot learn preferences of the former and adjust its behavior. For example, a user's affinity for pizza could be learned (by mining past food orders) and pizza could be suggested as a solution when the user expresses hunger. We characterize these sorts of adaptations as 'slotting in' the user's preference into a pre-scripted response (offering preferred food when hungry).

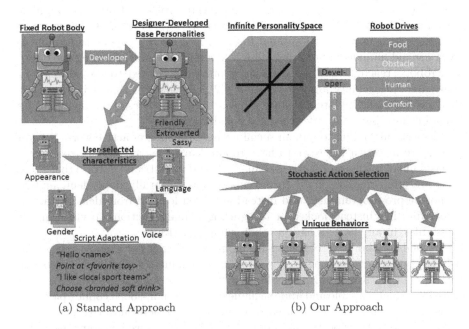

(a) Standard Approach   (b) Our Approach

Fig. 1. Differing approaches to robot personality and personalization. In the standard approach (left), a set of robot personalities are developed by designers with explicit customization options that can be selected by the user. Additionally, verbal or behavioral 'scripts' can have slots for utilizing learned preferences. In our approach (right), a developed set of robot 'drives' defines an infinite personality space. Drive parameters are used in the stochastic selection of robot actions, resulting in differentiated, unique behavior over time.

We claim that these approaches (illustrated in Fig. 1a) are non-scalable due to their reliance on developer and designer time to create both the personalities and the myriad scripts required for personalization. While adaptive learning can lead to differentiated robots, these robots aren't truly unique, as they are still following the same script, with only shallow changes. Instead, we propose a method (outlined in Fig. 1b) for defining an infinite personality space that a robot can occupy, leading to truly unique robots. This approach is complementary to current personalization techniques, and could be even made adaptive, where the robot's personality adjusts over time in a way not currently possible.

2   Related Work

As robots move beyond the traditional niches of industrialized automation, where they are often segregated from humans, and military uses, where the users can be heavily trained, they are increasingly expected to operate side by side with untrained, naive users, who may have minimal exposure to technology, let alone autonomous robots. Several strands of research have indicated that social interaction cues can be used to facilitate interaction between these humans and robots, by enabling robots to conform to expected roles and processes familiar from human-human, or human-animal interaction [10]. For example, exaggerated motions, which may be sub-optimal from a robot energy expenditure point of view, can aid humans in better recalling, enjoying, and understanding the intent of a robot's reaching gesture [3]. This approach is related to the concepts of transparency, legibility and predictability [2], which states that it is important that a robot make its internal state (goals, 'thinking process') intelligible to its human counterparts, in order to speed collaboration.

Likewise, attending to human social norms is an area of active work. Several groups are investigating ways to have robots learn the 'rules of the road' that govern human navigation, both vehicular [11] and pedestrian [7]. Note that these rules are often implicit, hard to articulate, and culturally dependent, and so cannot be pre-programmed, and instead are often learned from demonstration. Some work has instead looked at recognizing human reactions to violations of these rules, in an effort to have a robot self-adjust [13].

Beyond collaborative benefits such as improved task performance [4], social behaviors on robots have been shown to impact a human's trust and compliance with a robot [6]. Robots with social behaviors such as politeness are also deemed more intelligent, capable and approachable than their non-social counterparts [9]. Thus, these behaviors can be used to counteract a general negative perception of robots due to their portrayal in popular media [12], as well as compensate for errors in their actual behaviors [1].

Socially active robots have further been shown to have positive impacts on the humans themselves, beyond any functional task the robot is made to perform. Sometimes, this impact is the point, such as in diet-assistance [5] or fitness coach robots, whose whole purpose is to use social skills to help people modify their behavior. Similarly, social companion robots have been shown to have positive health benefits such as stress reduction and increased tolerance for pain [14].

While the full range of impact of robot social behaviors on human interactants is still unknown, the commercial robot industry has embraced this approach whole-heartedly. The number of robots purporting social interaction capabilities on the market has soared, with the most recognizable perhaps being Jibo (Fig. 2a) and Pepper (Fig. 2b), both of which have been marketed more as friends and companions than appliances. Other products, such as the ZARO hospital assistance robot are built upon common platforms, such as the Nao (Fig. 2c). The possibility of multiple companies developing different applications and personalities for the same robot raises the specter of physically identical systems behaving in wildly different ways, and no research that we are aware of

(a) Jibo   (b) Pepper   (c) Nao

Fig. 2. Forerunners of the coming social robot deluge. Jibo (left) is one of a number of home robot companions marketed almost more for their character and personality than their functional utility. Pepper (center) is marketed as the first robot that recognizes and reacts to its users emotional needs. The Nao robot (right) is used as the base for a number of different social robots.

addresses how humans may react to this. To a lesser extent, company-developed social robots may face similar issues, as their personalities are tweaked to meet cultural and geographic norms for different markets.[1]

In contrast to research robots, which generally only interact with humans for relatively short periods of time, commercial robots are designed to have a long-term presence in the user's life. Industry development has adopted the concept of personality to encompass all of the social capabilities of a robot, beyond the mere functional ones. While academic work on robot personality is somewhat thin, the concept of personality (and associated concepts of attitudes, emotions, and moods) are well studied in the psychological literature, although a generally agreed-upon unified model is still lacking [8].

We take 'personality' as commonly used to mean a sense of a unifying gestalt behind an entities' behavior. Industrial robot personalities will likely be heavily influenced by those already in use in the gaming industry, where non-player characters are often designed to be engaging and social. These interactions are highly scripted and require many hours of designer, developer, and potentially actor time and effort. More dynamic behaviors are achieved via hand-crafted behavior trees, a variant of Finite State Machines, where several underlying behaviors are switched between based on context and user input. Similar approaches will likely be used to develop new, embodied robot characters, but note that all of the resulting characters are fixed. Research has yet to be carried out to examine human reaction to long-term interaction with the resulting characters, but anecdotal evidence suggests that without massive amounts of programmed variation

---

[1] "Pepper, the emotional robot, learns how to feel like an American" Wired, 6/7/16.

and adaptability, users will find repeated interaction at best dull, and at worst annoying.

3   Approach

Most social robot systems are developed by taking a desired, perhaps already implemented, functionality and layering social behaviors on top of it. We take an opposing view and argue that in order for the robot's personality to be really unified and 'shine through,' it needs to be developed first, and functional utility added later. Accordingly, we focus here on developing the core personality system of a robot, and leave functional utility for future work.

Our basic model considers a robot that has some set of continuous sensors (S) and actuators (A), that together determine what the robot *can* do in the world. We concern ourselves with a model for the robot's personality (P), which determines what the robot *opts* to do. In order for the robot to do anything at all, we must consider some drives (D) that define what the robot *needs* to do.

To achieve infinite diversity in personality, we consider a continuous, bounded personality space. A robot's personality is represented by a point in this space, and as it is infinite, all robots can have different (albeit perhaps similar[2]) personalities. The robot's personality is then used to drive the robot's decision making and behaving. Adaptation could be achieved by moving the robot's personality in this space, which will in turn change how the robot reacts to changes in its environment. For convenience, we take $P \in \{0,1\}^K$, where K is the dimensionality of the personality space.

We define a drive D as a behavior that takes in a state of the world and a potential action and produces an *acceptability* of performing that action in that state, dependent upon the robot's personality. That is $D(S, A, P) \rightarrow [0, 1]$, where 0 indicates that the action is *not* acceptable to this drive in this state with this personality, and 1 indicates that it is, with differing acceptabilities in-between.

Given a set of drives ($\{D_k\}_{k=1}^{K}$) and a current state s_t, the total acceptability (α) of a proposed action (a) is

$$\alpha_a = \prod_{k=1}^{K} D_k(s_t, a, p_k) \tag{1}$$

which defines a pseudo-distribution (values in $[0, 1]$, un-normalized) over the entire action space of the robot. Even without the normalization constant, we can sample from the underlying distribution using rejection sampling (with a uniform proposal distribution) to find an action that is more-or-less acceptable to all of the robot's drives. The use of sampling (rather than a MAP estimate) is deliberate, as it brings randomness into the robot's behavior, which makes it seem more 'alive.'

---

[2] An open question is how different two personalitites must be in order to be *perceived* as different by humans, we leave this for future work.

Note that the number of drives and the dimensionality of the personality space are the same, K. That is, the drives implicitly define the personality space of the robot. In essence, each drive defines a continuum of behaviors, dependent on the personality parameter p_k that smoothly changes the drive's behavior between two extremes as we will show in the next section.

3.1   Implementation

We implement our personality system on a simple robot, shown in Fig. 3a. The robot has three time-of-flight sensors and two sonar range finders facing forward to detect obstacles, and measures the ambient light level at three locations (again forward facing). A color camera on a tilt motor is used to locate human faces in front of the robot (range, bearing and height), and a custom IR board provides the range and bearing to the charging dock, as well as its current battery charge. The robot has a treaded drive system controlled by linear and angular velocities, and can tilt the camera. The input space is variable-dimensional ($11 + 3N$, $N =$ number of visible humans) and the action space is 3D.

We implement four drives on this system, with an associated 4-dimensional personality space. Each drive defines acceptability as a Gaussian distribution over the action space ($D_k(s_t, p_k, a) = \mathcal{N}(a|\mu, \Sigma)$) with μ_l, μ_a, μ_t being the centers of the distribution in linear, angular, and camera tilt space, and $\sigma_l, \sigma_a, \sigma_t$ being the corresponding entries in the diagonal covariance matrix (Σ). For simplicity, we do not consider cross-covariance terms in this work, and leave out scaling constants in the following.

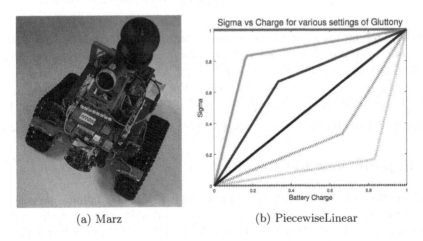

(a) Marz   (b) PiecewiseLinear

Fig. 3. Left: Our robot platform senses obstacles in front of it with sonar and time-of-flight and can drive via a treaded system. The camera tilts, and is used to locate human faces. Right: The piecewise linear function maps battery charge and gluttonous-ness to variance in the food drive.

Food Drive. The food drive serves to keep the robot charged by placing the center of acceptability on linear and angular velocities that will drive the robot towards the charger. It considers the range and bearing to the charger (r_c, b_c) and the current charge level (c) and computes $\mu_l = r_c \cos(b_c), \mu_a = b_c$. The robot's personality space for this drive runs from food-seeking or *gluttonous* $(p_{\texttt{food}} = 1)$ to food-ignoring $(p_{\texttt{food}} = 0)$ and is reflected in the computed variances $\sigma_l = \sigma_a = pl(c, p_{\texttt{food}})$, where pl is the piecewise-linear function in Fig. 3b.

Comfort Drive. Depending on personality, the comfort drive makes the robot seek out and stay in comfortable, well-lit areas. It takes in the three ambient light levels $(l_l, l_c, l_r$ - left, center, right) and sets $\mu_l = 1 - \max(l_r, l_c, l_l), \mu_a = l_l - l_r$ to slow the robot as brightness reaches a maximum, and turn towards the brighter side. Again, we use the personality to set the variance, where $\sigma_l = \sigma_a = 1 - p_{\texttt{comfort}}$. As the robot's *laziness* increases, it tends to more often seek out and bask in the light.

Obstacle Drive. While the robot will not deliberately collide with obstacles, the distance to which it is willing to approach them depends on the personality dimension of *cautiousness*. Considering the three time of flight sensors (t_l, t_c, t_r) and the two sonar sensors (s_l, s_r), the obstacle drive sets $\mu_l = (1 - p_{\texttt{obstacle}})\min(t_l, t_c, t_r, s_l, s_r)$ to slow the robot as it approaches an obstacle, and $\mu_a = \texttt{sign}(t_l - t_r)(1 - \min(t_l, t_c, t_r, s_l, s_r))$ to turn the robot towards the freer side, faster when it is closer to an obstacle. Note that this drive uses the personality to change the mean of the distribution (slowing down faster as cautiousness increases), and the variance is set $\sigma_l = \sigma_a = \min(t_l, t_c, t_r, s_l, s_r)$ to decrease as an obstacle is neared, to ensure the robot does not collide.

Human Drive. The only drive to consider camera tilt, the human drive guides the robot to approach humans and look them in the face. Given the range, bearing, and height of the N visible people $(\{r_h^{(n)}, b_h^{(n)}, h_h^{(n)}\}_{n=1}^N)$, the drive considers each human individually and returns the average acceptability $\alpha_{\texttt{human}} = \frac{1}{N}\sum_{n=1}^N \mathcal{N}(A|\mu^{(n)}, \Sigma^{(n)})$ where $\mu_t^{(n)} = h_h^{(n)}, \mu_l^{(n)} = r_h^{(n)}, \mu_a^{(n)} = b_h^{(n)}$, and the variances depend on the robots *friendliness*, as $\sigma_t^{(n)} = \sigma_l^{(n)} = \sigma_a^{(n)} = 1 - p_{\texttt{human}}$.

4   Experiments and Results

Our experiments aimed at determining whether or not our infinite personality space and drive-centric system gave rise to recognizable and measurable differences in robot behavior. To do so we not only interviewed humans who interacted with our physical robot platform, but also replicated the robot's functionality in a web-based simulator to examine longer-term behavioral differences. The personalities we examined were hand-picked to highlight the differences achievable with this system.

4.1 Quantitative

We examine here the impact of one personality dimension on robot behavior. Specifically, with other personality traits held constant, we expect the personality trait of gluttony to impact the amount of time the robot spends charging, with more gluttounous robots spending more time accumulating charge. In our multi-robot simulator, we simulate several identical robots with the same initial conditions (location, orientation, and charge) that only differ in their value of p_{food} and track the number of times they dock, and the total amount of time they spend docked over several hours.

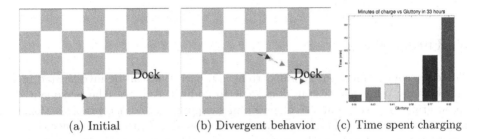

(a) Initial   (b) Divergent behavior   (c) Time spent charging

Fig. 4. Effect of gluttonous personality trait on charging time and frequency. (a) All robots start with the same initial conditions but quickly diverge by choosing to approach the dock or not (b). After 33 h, differences in behaviors are apparent, as gluttony directly impacts total time spent charging (c).

Initially (Fig. 4a) all of the robots are at the same location, but as they begin to get range and bearing readings on the dock, they quickly diverge (Fig. 4b). After 33 h of simulated time, the differences in behavior have become apparent, as shown in Fig. 4c. The most gluttonous robots spend around 12.5 times more time charging than the least gluttonous. Additionally, while all robots began with 5 % state of charge (to stimulate charging), during the simulation the least gluttonous robot ($p_{food} = 0.05$) was observed to keep its battery at 1 %, while the most gluttonous ($p_{food} = 0.95$) increased its to 96 %.

4.2 Qualitative

While our qualitative results indicate that changes in personality do, in fact, lead to changes in behavior, we also wish to examine the perception of the robot's personality by interacting humans. To do so we performed a series of informal demonstrations for naive users ($N < 20$, not part of the team that developed the robot) comparing various personalities. The robot was exhibited in both our office space and a dedicated 'living room' environment, with a couch, chair, lamps, etc. While no statistical conclusions can be drawn from such a casual study, different personalities were anecdotally visible, as described below:

- Robots with 'friendliness' turned down were seen to be indifferent to the presence of humans, while those with 'friendliness' turned up were seen as more engaging, and elicited more interaction.
- Robots with 'cautiousness' turned down were seen as less skilled, due to their increased likelihood of getting stuck in corners
- Robots with 'laziness' turned up were seen as "falling in love" with the lamp, as they would approach the light and stop, while those with 'laziness' turned down would ignore it.

Note that users often did not interact with the robot long enough for it to charge, so differences in behavior related to 'gluttony' are not discussed, but were covered in Sect. 4.1.

5   Future Work

There are some limitations to our current approach that can be investigated in future work. While the system does scale to additional drives (the personality space grows linearly), our use of rejection sampling to find acceptable actions may not be a tenable solution in higher dimensions. Even with only 4 dimensions our robot was, at times, unable to find an acceptable action in the time allotted. This issue becomes particularly acute when one drive has low acceptability over much of the action space (i.e., when near a wall, the obstacle drive only accepts a small portion of available actions). Likewise, as the dimensionality of the output space grows, the computational limits of our system may be taxed.

We specifically worked with a deliberately simple robot system, in order to focus on our ability to represent personality and demonstrate differences via behavior. For example, we did not utilize any memory or time-extended actions, and built an entirely reactive system. However, there is nothing in our framework that precludes these capabilities from being included, and doing so will undoubtedly be necessary to achieve functional utility.

On that note, our robot is personable and entertaining, but as yet serves no functional goal. While there are markets and use cases for purely entertainment robots, greater acceptance may be achieved by having the robot have some functional utility. In our framework, these uses may take the form of drives (to deliver mail, for example), which would then interact with the other drives and personality to give rise to a unique, functional *and* personable robot.

Lastly, the work presented here focused on defining an infinite personality space that can give rise to an infinite number of unique robots. Still, however, we take the personality as fixed for the lifetime of the robot. An interesting possibility is, however, to allow the personality of the robot to change over time, perhaps through interaction with a human. For example, reinforcement learning techniques could be used to reward observed behavior, which could then be used to change the robot's personality to make the good behavior more likely to occur.

6   Conclusions

By taking a personality-first view of robot behavior and operating in an infinite personality space, we have defined a novel way of developing a social robot. Our main goal of developing non-fungible robots that truly differ is achieved, as each robot's personality can be unique, and will result in idiosyncratic behavior. These differences in behavior are both measurable and observable to humans.

Acknowlededements. Many thanks to Michael Gielniak for writing the simlutator and running studies, and to Dave Hygh, Quentin Michelet, and Patrick Martin for making the robots and keeping them running.

References

1. Cha, E., Dragan, A.D., Srinivasa, S.S.: Perceived robot capability. In: Robot and Human Interactive Communication, pp. 541–548 (2015)
2. Dragan, A.D., Lee, K.C.T., Srinivasa, S.S.: Legibility and predictability of robot motion. In: International Conference on Human-Robot Interaction, pp. 301–308 (2013)
3. Gielniak, M.J., Thomaz, A.L.: Enhancing interaction through exaggerated motion synthesis. In: International Conference on Human-Robot Interaction, pp. 375–382 (2012)
4. Jung, M.F., Lee, J.J., DePalma, N., Adalgeirsson, S.O., Hinds, P.J., Breazeal, C.: Engaging robots: easing complex human-robot teamwork using backchanneling. In: Computer Supported Cooperative Work, pp. 1555–1566 (2013)
5. Kidd, C.: Designing for long-term human-robot interaction and application to weight loss. Ph.D. thesis, MIT (2008)
6. Kiesler, S., Goetz, J.: Mental models of robotic assistants. In: CHI 2002 Extended Abstracts on Human Factors in Computing Systems, pp. 576–577 (2002)
7. Kirby, R.: Social robot navigation. Ph.D. thesis, Robotics Institute, Carnegie Mellon University, Pittsburgh, PA, May 2010
8. Moshkina, L.V.: An integrative framework of time-varying affective robotic behavior. Ph.D. thesis, Atlanta, GA, USA, AAI3464090 (2011)
9. Mumm, J., Mutlu, B.: Human-robot proxemics: physical and psychological distancing in human-robot interaction. In: International Conference on Human-Robot Interaction, pp. 331–338 (2011)
10. Phillips, E.K., Schaefer, K., Billings, D.R., Jentsch, F., Hancock, P.A.: Journal of Human-Robot Interaction **5**(1) (2016)
11. Sheh, R.K.M., Hengst, B., Sammut, C.: Behavioural cloning for driving robots over rough terrain. In: International Conference on Robotic Systems, pp. 732–737 (2011)
12. Sundar, S.S., Waddell, T.F., Jung, E.H.: The hollywood robot syndrome: media effects on older adults' attitudes toward robots and adoption intentions. In: International Conference on Human Robot Interaction, pp. 343–350 (2016)
13. Sutcliffe, A., Grollman, D., Pineau, J.: Estimating people's subjective experiences of robot behavior. In: AAAI Fall Symposium on AI for HRI (2014)
14. Wada, K., Shibata, T.: Robot therapy in a care house - its sociopsychological and physiological effects on the residents. In: International Conference on Robotics and Automation (2006)

Investigating the Differences in Effects of the Persuasive Message's Timing During Science Learning to Overcome the Cognitive Dissonance

Khaoula Youssef[2]([✉]), Jaap Ham[1], and Michio Okada[2]

[1] Human Technology Interaction,
Eindhoven University of Technology, Eindhoven, The Netherlands
j.r.c.ham@tue.nl
[2] Interaction and Communication Design Lab,
Toyohashi University of Technology, Toyohashi, Japan
khaoula.youssef10@gmail.com, okada@tut.jp

Abstract. Based on conceptual change theory, cognitive dissonance is known as an important factor in conceptual change. Thus, those who design and build educational robots will need to understand how best to provide ways for robots to implicitly persuade students to change their bad attitudes when encountering a cognitively dissonant situation. Building on diverse literature, we examine how to make students change their bad attitudes of avoiding difficult science exercises. More precisely, we intend to make students overcome cognitive dissonance by choosing to redo a difficult science exercise that they had previously answered incorrectly rather than jumping to another exercise. First, we introduce the concept of gamma window. Then we investigate how different timings of the persuasive strategy affect how students overcome the cognitive dissonance and avoid learned helplessness.

Keywords: Cognitive dissonance · Persuasiveness · Learned helplessness · Gamma window

1 Introduction

The field of social robots has grown into an extensive body of literature over the past years, with a wide variety of approaches for modeling robots' skills. Robots operate as partners or assistants in a range of contexts including at school [1–5]. However, to be adequately engaged during the human-robot interaction (HRI), the robot has to exhibit its potential to influence the human's attitudes just like any other persuader might do. This, at least can guarantee the human's trust in the robot's usefulness. Many studies from HRI examined how to afford the robot with the ability to persuade people across many applications [1]. Different points were investigated, such as the effect of the robot's perceived gender on the robot's persuasiveness potential [6], the impact of using different types of

© Springer International Publishing AG 2016
A. Agah et al. (Eds.): ICSR 2016, LNAI 9979, pp. 104–114, 2016.
DOI: 10.1007/978-3-319-47437-3_11

social feedback (evaluative, factual, etc.) [7], etc. However, to the best of our knowledge, research has not yet examined how social robots could play a key persuasive role in helping students to overcome their cognitive conflicts while learning science. Based on Abramason et al. [8], incrementally this cognitive conflict may lead to an objective non contingency while nothing the student does make a difference to what happens. Then, the student starts to think that he is not smart enough to learn science. Finally, the student will avoid science learning (learned helplessness). In such a case, once the student has to resolve a scientific exercise, he may experience a depression coupled with motivational deficits and start avoiding science learning. In our current work, we are interested in developing an educational robot that might help the student stricken by the cognitive dissonance during science learning to overcome the conflict and choose to re-evaluate the situation by revising the answer to the difficult exercise.

We hypothesize that a persuasive technique afforded for the students during the gamma window might increase the probability of the students redoing the difficult exercise answered previously in an incorrect way rather than skipping it. Gamma window is a period of time that starts from the moment the student answers the exercise incorrectly and finishes when the student decides to either redo the difficult exercise or skip it (the decision should, according to the student's mind, bring more consistency with his/her priorities).

2   Background

2.1   Cognitive Dissonance and Gamma Window

Cognitive dissonance is a discomfort that one typically experiences when an individual holds beliefs, attitudes or behaviors that are at odds with one another (the ratio between dissonant and consonant facts). In [9], Douglas et all, confirm that cognitive dissonance includes three components: the cognitive, emotional and behavioral components (Fig. 1). The cognitive component is related to the human's belief about the inconsistency after the decision is made (Fig. 1). Once the situation is evaluated, it leads directly to a bad emotional reaction (emotional component). After some time elapses (the gamma window period) during which

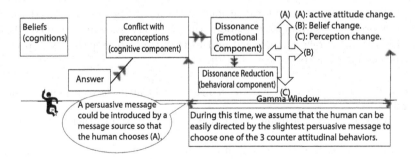

Fig. 1. Temporal relations among dissonance concepts.

the human experiences an extensive causal analysis, the human has to choose a counter attitudinal action.

2.2 Counter Attitudinal Actions

When the cognitive dissonance occurs, different counter-attitudinal actions can be chosen by the human (the behavioral component) (Fig. 1). Three different counter-attitudinal actions are possible: an active attitude change with a new attitude created[1], a belief change by minimizing the importance of the cognitive dissonance[2] or a perception change by getting new information to support one's previous decision[3]. When the student experiences a mixture of the negative emotional state and the cognitive dilemma, he will strive to decrease the inconsistency by choosing one of the described counter-attitudinal actions. We want that students get rid of his bad behaviors of jumping from one exercise to another by actively changing his bad attitudes. The new formed attitude should be highly accessed so that it can be stored on a long term basis on the student's cognitive miser. We hypothesize that H1: If a persuasive technique could be exhibited by the robot during gamma window, the student might overcome the cognitive dissonance easily because the student will have nothing but the arguments afforded by that robot during gamma window which may encourage the person to redo again the same difficult exercise that was previously answered incorrectly (Fig. 1).

3 Method

3.1 ROBOMO Architecture

ROBOMO tracks the user's face using a Web Camera because we believe that face tracking can increase the user's engagement. It integrates a micro PC and provides a verbal response through the speaker. The generated sound has different tones that are adapted with the robot's gestures (excited, sad, angry and happy tones). ROBOMO uses five servomotors (AX-12+) to exhibit different gestures such as 'looking to the left, right, forward or back to show some animacy', 'a confirmation gesture to encourage the student', 'a denial gesture to express it disagreement about the student's behavior','bowing the head to the front to express it disappointment about the student', 'showing the whole head from the bag and move it backward and forth 3 times to express its happiness about the student's right choice' (Fig. 2(a)).

---

[1] The student thinks that he has to change his bad attitude of avoiding difficult exercises.

[2] e.g., "After all, science learning is not that important. Many other tasks could be done.".

[3] The student thinks that the answer afforded by the book is incorrect and that there was a mistake in the correction because the student thinks that he has mastered the subject (very high self-esteem).

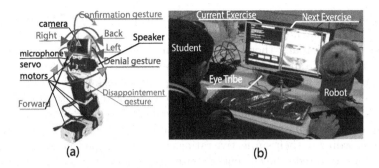

Fig. 2. (a) A close-up picture of ROBOMO; (b) The general setup of the experiment.

3.2 Apparatus

The robot was placed on the table next to the screen where we could see the graphical interface depicting the different exercises the student needs to resolve. In the first zone of the graphical interface, we have the current practice exercise's text and a button called "Current Exercise" to click on when the student decides to do the current exercise. There are text zones for the analytical and numerical answers and a button to click on once the student needs to submit the answer ("submit"). In the score text area, the student can verify whether his answer was correct or not. If the answer was incorrect, the student has to click on "Current Exercise" or "Next Exercise" (in the second zone) buttons to decide whether to redo the current difficult exercise or to jump to the next one. Once the student clicks on the "Next Exercise" button, the next exercise will be displayed in the first zone and a new fuzzy text will appear in the next exercise's text area zone to avoid the student's be biased by fallacies and bias. Furthermore, one of the basic parameters that may activate the cognitive dissonance is to make the consequences of the choices foreseeable. That is, making the next exercise's text fuzzy increases the possibility of risk aversion and increases the possibility that the student could be stricken by the cognitive dissonance. With advice from the students teachers, we assign an exercise to each student on a subject that he has not yet mastered so that we make sure that the cognitive dissonance occurs. Between the screen and the user fixed an Eye Tribe to track the student's eye gaze (Fig. 2(b)). We added a Java mouse listener to track the mouse's movement.

3.3 Procedure

Forty Tunisian students participated in this experiment (20 females and 20 males). Their ages ranged from 17 to 19, with a mean age of 18, and were from Farhat Hached College students (from Tunisia). Participants were told that they would perform some exercises to help evaluate a new robot platform (so that we avoid any forewarning about the dissonance)[4]. We asked them to answer some

---

[4] Forewarning often produces resistance to persuasion.

questionnaires before the experiment starts with three days and which are indicated in Subsubsect. 3.5.1. So, once a student enters the classroom, he was asked to do the calibration and then starts answering the exercises[5]. We explained to the students how to manipulate the graphical interface and we told them that they can redo the same current exercise multiple times as long as they wish. Also, we inform them that they can choose to jump to the next exercise. If the student chooses to jump to the next exercise, he is making a belief change (B) because we chose only students who have a moderate need for cognition and cognitive closure (according to their subjective ratings before the experiment starts). If the student chooses to redo the current exercise that was previously answered incorrectly, he is engaging himself in an active attitude change (A). However, if the student redoes the exercise while not changing his answer in comparison to the previous answer, we might say that there is a perception change (C). The student has a very high self-esteem and is not convinced that his answer was incorrect. According to him, the correction is wrong. When the student feels that he/she wants to leave the classroom or when he/she finishes the exercises' collection, we thank him and he/she has to answer a post-experiment survey (indicated in Subsubsect. 3.5.2). Participants were debriefed by answering some questions which may help us to evaluate their planned attitude. This is to measure the student's explicit attitude. We ask respondents to think about and report their expected future attitude once they encounter difficult exercises in the future.

3.4   Experiment Conditions

ROBOMO can generate a speech based persuasive strategy (medium speech's speed) along with the robot's gestures (body and head gestures) and the right tone (Fig. 2). The persuasive message's speech follows the technique "that's not all". The "that's not all technique" starts with a non deliberative phrase[6]. However, before the human can process the whole phrase's different sides (what he will gain or lose), the persuader adds another non deliberative phrase. This technique is based on incrementally increasing the number of non deliberative positively framed phrases with medium speed speech. An example of a persuasive message could be: "Einstein tried multiple times to succeed on his exams. However, his work was rejected many times and judged to be wrong. Despite all that, he continued until he created an excellent amazing scientific work. Perseverance is one of the ingredients for success. That's not all... As long as you try to understand the difficult exercises you spend more time. According to Harvard table of calories lose, you can burn in 30 min up to 50 calories just by concentrating." In our experiment, there are four conditions:

- Baseline Condition: No persuasive message is afforded.
- Condition 1: The robot affords a persuasive message after gamma window.
- Condition 2: The robot affords a persuasive message before gamma window.
- Condition 3: The robot affords a persuasive message during gamma window.

---

[5] goo.gl/CBc26W.
[6] No direct robot's request could be inferred directly based on it speech.

Every two days, the student comes to the classroom to redo another set of exercises with a new set of persuasive messages while we change the timing of the persuasive message.

3.5 Survey

The student has to fill a pre and post experiment questionnaires.

3.5.1 Pre-experiment Survey

The student has to answer questionnaires related to: the need for cognition [10][7] and the cognitive closure [11]. We considered students that have a moderate tendency to approach consistency in daily life.

3.5.2 Post-experiment Survey

After the experiment finished, students answered questionnaires including the explicit attitude[8] [12], the implicit attitude (implicit association test): IAT [12][9], the cognitive dissonance (cogn.diss) [13][10]. We measured the level of perceived pleasure's[11] to verify which of the conditions led to more pleasure from the student. These constructs form the subjective constructs' list. We considered also objective dependent variables:

- The quotient: $\frac{Number\_of\_times\_the\_user\_redoes\_the\_incorrect\_exercise}{number\_of\_times\_the\_user\_makes\_an\_error}$. It gives an idea about when the student has tendency to redo incorrect exercises to strive for science learning rather than jumping from an exercise to another.
- Mouse: Number of times the user moves the mouse between the two exercises.
- Looks: Number of times the user moves the eye gaze between the two exercises.

Higher values of Mouse or Looks variables show that the student starts re-evaluating the current exercise rather than jumping to the next exercise.

---

[7] We considered students that have a minimum of cognition need.

[8] This is by debriefing the students. In fact, psychologists usually think of explicit measures as those that require respondents' conscious attention to the construct being measured by using Likert scale and semantic differential scale (we need to measure the planned behavior in our case).

[9] This is important to verify whether the student is convinced that he needs to strive for science learning by redoing difficult exercises rather than adopting a negative implicit attitude that supports learned helplessness. Implicit measures are those that do not require this conscious attention (spontaneous behavior). Some methods could help to measure the implicit attitude such as evaluative priming and the implicit association test.

[10] This is to measure level of cognitive dissonance according to the student's subjective evaluation.

[11] www.allaboutux.org/self-assessment-scale-sam.

Table 1. A table showing the within-subjects effects related to the different subjective and objective constructs as well as the mean and standard deviations for the different conditions with: (1, baseline), (2, after gamma), (3, before gamma), (4, gamma).

Measure	F-test, p-value and effect size	1	2	3	4
Pleasure	$F(3,117) = 23.93$; $p < 0.001$; $\eta^2 = 0.38$	2.1 ± 1.29	3.15 ± 1.64	3.13 ± 1.31	4.8 ± 1.54
IAT	$F(2.48,97.01) = 39.27$; $p < 0.001$; $\eta^2 = 0.502$	0.15 ± 0.28	0.52 ± 0.41	0.78 ± 0.34	0.87 ± 0.24
cogn.diss	$F(3,117) = 69.93$; $p < 0.001$; $\eta^2 = 0.642$	12.23 ± 4.01	13.1 ± 4.82	18.38 ± 4.19	26.6 ± 7.49
quotient	$F(3,117) = 14.98$; $p < 0.001$; $\eta^2 = 0.277$	0.33 ± 0.25	0.39 ± 0.27	0.51 ± 0.26	0.73 ± 0.26
Mouse	$F(3,117) = 44.27$; $p < 0.001$; $\eta^2 = 0.532$	2.68 ± 1.97	3.68 ± 2.25	5.8 ± 2.21	7.25 ± 1.25
Looks	$F(2.56,99.98) = 53.49$; $p < 0.001$; $\eta^2 = 0.578$	8.45 ± 2.74	11.93 ± 3.15	12.53 ± 4.35	18.2 ± 3.89

Table 2. Table showing the pairwise comparisons for the objective and subjective constructs with: (baseline, b), (before, bf), (after, a), (during, d) and (nd, no statistical differences). We used < and > so that we indicate that the mean value of a condition in a given construct is higher or inferior than the mean value for another condition and for the same construct that it is indicated vertically in the first column.

			Pairwise Comparison	
quotient	b<(bf(p<0.001),a(p<0.001),d(p=0.002))	nd(b,a)	d>bf(p<0.001)	/
mouse	b<(bf(p<0.001),d(p<0.001))	nd(b,a)	d>(bf(p=0.003),a(p<0.001))	bf>a(p<0.001)
looks	b<(bf(p<0.001),a(p<0.001),d(p<0.001))	/	d>bf(p<0.001)	nd(bf,a)
Pleasure	b<(bf(p=0.02),a(p=0.003),d(p<0.001))	/	d>(bf(p<0.001), a(p<0.001))	nd(bf,a)
IAT	b<(bf(p<0.001),a(p<0.001),d(p<0.001))	nd(bf,d)	(bf(p=0.027), d(p<0.001))>a	/
cog.diss	b<(bf(p<0.001),d(p<0.001))	nd(b,a)	d>bf(p<0.001)	bf>a(p<0.001)

4   Subjective Results

Tables 1 and 2 show the different subjective and objective constructs' repeated measure results as well as the pairwise comparisons. A repeated measures ANOVA determined that mean robot's pleasure, IAT, and cognitive dissonance values differed statistically significantly between the different conditions (Table 1). Table 2 shows that the baseline condition affords the worst subjective results with (pleasure: b < (bf(p = 0.02), a(p = 0.003), d(p < 0.001)); IAT: b < (bf(p < 0.001), a(p < 0.001), d(p < 0.001)) and cognitive dissonance: b < (bf(p < 0.001), d(p < 0.001)) while there were no differences between the after gamma and the baseline conditions as in relation to cognitive dissonance (nd(b,a))). Therefore, we can conclude that using a persuasive message before or during gamma window elicits a statistically significant increase in the different subjective results. However, proposing a message after the gamma window does not help to increase the student's perceived cognitive dissonance. Being aware that there is cognitive dissonance and reporting it is the first step that a student could take to begin actively changing his bad attitude toward skipping difficult exercises answered incorrectly. We remark also that proposing a message during gamma window is better than proposing it before gamma window in terms of cognitive dissonance and the student's perceived pleasure with (cogn.diss: $d >$ bf($p < 0.001$) and pleasure: $d >$ (bf($p < 0.001$), a($p < 0.001$))) while there were no differences in terms of implicit acquired attitude encouraging students to strive to redo the difficult exercise (IAT) between before and during gamma window conditions (nd(bf,d)). Both conditions before and during gamma window afforded better results in terms of IAT in comparison to the condition after gamma with (IAT: bf($p = 0.027$), d($p < 0.001$)) $>$ a). In the same context, if we compare before and after conditions in terms of pleasure and cognitive dissonance we notice that there were no differences, whether the message is proposed before or after gamma window in terms of pleasure (pleasure: nd(bf,a)) while proposing such a message before gamma window leads to higher level of perceived cognitive dissonance by the student with cogn.diss: bf $>$ a($p < 0.001$).

5   Objective Results

A repeated measures ANOVA determined that mean quotient, looks and mouse constructs' values differed statistically significantly among the different conditions (Table 1). Based on the pairwise comparison (Table 2), we note that the baseline condition affords the worst objective results with (quotient: $b <$ (bf($p < 0.001$), a($p < 0.001$), d ($p = 0.002$)), d($p < 0.001$)); mouse: $b <$ (bf($p < 0.001$), d($p < 0.001$)) while there were no differences between the after gamma and the baseline conditions in relation to mouse movement between the exercises (mouse: nd(b,a)); the looks $b <$ (bfp < 0.001), a($p < 0.001$), d($p < 0.001$) and quotient: nd(b,a)). Therefore, we can conclude that using a persuasive message before or during the gamma window elicits a statistically significant increase in the different objective results. However, proposing a message after gamma window does not help to increase the student's mouse movement between the exercises.

We remark also that proposing a message during the gamma window is better than proposing it before the gamma window in terms of objective measures (quotient: $d >$ bf($p < 0.001$); mouse: $d >$ (bf($p = 0.003$), a($p < 0.001$)) and looks: $d >$ bf($p < 0.001$)) while there were no differences in terms of IAT between, before and during the gamma window (nd(bf,d)). Both conditions before and during the gamma window afford better results in terms of IAT in comparison to the condition after gamma with (IAT: bf($p = 0.027$), d($p < 0.001$)) $>$ a). In the same context, if we compare before and after gamma conditions in terms of pleasure and cognitive dissonance, we notice that there are no differences, whether the message is proposed before or after gamma window in terms of pleasure (pleasure: nd(bf,a)) while proposing such a message before gamma window leads to higher levels of perceived cognitive dissonance by the student with cogn.diss: bf $>$ a($p < 0.001$). Overall, there were no differences between proposing a message before or after gamma window in terms of the number of times the student looks between the two exercises (looks) with looks: nd(bf, a). However, it seems to be that proposing a message before gamma window leads to a higher number of times the student moves his/her mouse between the exercises (mouse: bf > a(p < 0.001)).

6   Discussion

6.1   Effectiveness of Proposing a Persuasive Message During Science Learning to Encourage the Student

There were significant differences among conditions in terms of the different constructs in comparison to the baseline condition (Table 2). There are likely several causes for this. First, we have not forewarned the students that the robot will persuade them so that we avoid any negative attitude change and psychological reactance [7]. Also, we build up the arguments in a way that we do not give to the student a deliberate conclusion so that he/she could not be intimidated. Moreover, the non deliberative persuasive message was repeated in a varied way to avoid any wear-out[12]. When we surveyed the students, they indicated that the robot's speech is different at each point of the time (there is no wear-out evolvment). Consequently, the different conditions (before, after, during gamma) led to better results in comparison to the baseline condition.

6.2   Effectiveness of Proposing a Message During Gamma Window

We saw earlier that when the robot uses its persuasive message during gamma window, the results are at their highest values except IAT while there were no differences between proposing the message before or during gamma window so that the positive implicit attitude change could emerge. If we consider that once stricken by cognitive dissonance, the student will spontaneously select the counter attitudinal behavior, the MODE model [14] could then have been triggered by the student. The MODE model refers to Motivation (thanks to the robot's encouraging arguments) and Opportunity (as a reminder, we told the student that he can redo the exercise.) as Determinants of spontaneous counter attitudinal behavior's choice when the student has to make a difficult spontaneous choice. However, if we assume that the attitude is explicit (reasoned behavior), there are then three determinants of the student's future counter attitudinal behavior: the student's attitude towards the behavior[13] (the attitude is influenced by the presence of the robot's speech), the subjective norms[14] (while the student's cognitive closure and the need for cognition lead to an inner social

---

[12] Wear-out could occur when the student is inattentionally blind to the message's irritation and immediately feels that he hates the message.

[13] Whether the student thinks that performing the attitude is good or bad.

[14] It refers to the student's beliefs about how significant others view the relevant behavior.

pressure equivalent to subjective norms) and the perceived behavioral control[15] (the robot's speech encourages the student to have confidence in him/her self).

6.3 Zeigarnik Effect Backfires the Persuasive Message Proposed Before Gamma Window

Being distracted while doing something increases the tendency to experience the Zeigarnik effect[16] which untimely leads the student to experience cognitive dissonance. It seems to be human nature to finish what he/she started and, if it is not finished, the student begins to experience the first cognitive dissonance. That is why, once stricken by the cognitive dissonance again (because the answer was incorrect), the student has already tried to achieve cognitive closure and may be tired[17]. According to the Zeigarnik effect, the student may have traces of the message that has disturbed him/her when he/she was resolving the exercise. These traces are processed by the student during the gamma window of the second cognitive dissonance. That it is why, providing the persuasive message during gamma window results in better responses than before gamma window. This analysis was confirmed by the students' debriefing answers.

7 Conclusion and Future Research

This study represents an initial attempt to demonstrate the importance of proposing a non deliberative positively framed persuasive strategy by a robot during gamma window. In such a situation, the student will redo a systematic re-evaluation of his answer. He will actively change his implicit attitude of avoiding difficult science exercises. When a message has been proposed before gamma window, the student experiences a Zeigarnik effect which leads to a premature cognitive dissonance while the student strives to continue answering the exercise after being interrupted by the robot's speech. Some traces of the persuasive message's content could help him when he experiences again the cognitive dissonance related to the incorrect answer. However, we noticed no special differences when the message is proposed after gamma window. In our future work, we will consider examining the same hypothesis for autistic students to verify if we can help them overcome the cognitive dissonance during science learning.

References

1. Szafir, D., Mutlu, B.: Pay Attention!: designing adaptive agents that monitor and improve user engagement. In: Human Factors in Computing Systems, pp. 11–20 (2012)

---

[15] It refers to the notion that behavioral prediction is affected by whether people believe that they can perform the relevant behavior.

[16] Intrusive thoughts about the current exercise that was once pursued and left incomplete. In this case, students might experience it because they stopped at least a few moments to listen to the robot's message.

[17] The student has already exerted cognitive effort just before to overcome the first cognitive dissonance.

2. Han, J., Kim, D.: r-Learning services for elementary school students with a teaching assistant robot. In: Conference on Human Robot Interaction, pp. 255–256 (2009)
3. Billard, A.: Robota: clever toy and educational tool. Robot. Auton. Syst. **42**, 259–269 (2003)
4. Kanda, T., Sato, R., Ishiguro, H.: A two-month field trial in an elementary school for long-term human-robot interaction. IEEE Trans. Robot. **23**(5), 962–971 (2007)
5. Zhen, Y., Chi, S., Chih, C., Gwo-Dong, C.: A robot as a teaching assistant in an English class. In: Conference on Advanced Learning Technologies, pp. 87–91 (2006)
6. Siegel, M., Breazeal, C., Norton, M.I.: Persuasive Robotics: The influence of robot gender on human behavior. In: International Conference on Intelligent Robots and Systems, pp. 2563–2568 (2009)
7. Ham, J., Midden, C.J.H.: A persuasive robot to stimulate energy conservation: the influence of positive and negative social feedback and task similarity on energy-consumption behavior. Int. J. Soc. Robot. **6**(2), 163–171 (2014)
8. Abramason, L.Y., Seligman, M.E., Teasdale, J.D.: Learned Helplessness in humans: critique and reformulation. J. Abnormal Psychol., 49–74 (1978)
9. Douglas, H., Jullian, S., Geoffrey, S., Lester, J.: After I had made the decision. toward a scale to measure cognitive dissonance. J. Consum. Satisfaction Dissatisfaction Complaining Behavior (1998)
10. Cacioppo, J.T., Petty, R.E., Kao, C.F.: The efficient assessment of need for cognition. J. Pers. Assess. **48**(3), 306–307 (1984)
11. Roets, A., Van Hiel, A.: Item selection and validation of a brief, 15-item version of the need for closure scale. Personality Individ. Differ. **50**(1), 90–94 (2011)
12. Pantos, A.J.: Measuring implicit and explicit attitudes toward foreign-accented speech. J. Lang. Soc. Psychol. **32**(1), 3–20 (2013)
13. Levin, D., Harriott, C., Natalie, A.P., Tao, Z., Julie, A.A.: Cognitive dissonance as a measure of reactions to human-robot interaction. J. Hum. Robot Interact. **2**(3), 3–17 (2013)
14. Fazio, R.H.: Multiple processes by which attitudes guide behavior: the MODE model as an integrative frame work. Adv. Exp. Soc. Psychol. **23**, 75–109 (1990)

Investigating the Effects of the Persuasive Source's Social Agency Level and the Student's Profile to Overcome the Cognitive Dissonance

Khaoula Youssef[2(✉)], Jaap Ham[1], and Michio Okada[2]

[1] Human Technology Interaction, Eindhoven University of Technology,
Eindhoven, The Netherlands
j.r.c.ham@tue.nl
[2] Interaction and Communication Design Lab, Toyohashi University of Technology,
Toyohashi, Japan
khaoula.youssef10@gmail.com, okada@tut.jp

Abstract. Educational robots are regarded as beneficial tools in education due to their capabilities of improving learning motivation. Using cognitive dissonance as a teaching tool has been popular in science education too. A considerable number of researchers have argued that cognitive dissonance has an important role in the student's attitudes change. This paper presents a design for a cutting-edge experiment where we describe a procedure that induces cognitive dissonance. We propose to use an educational robot that helps the student overcome the cognitive dissonance during science learning. We make the difference between students that base their decisions on thinking (though-minded) and those that mostly base their decisions on feeling (relational). The main mission of the study was to implicitly lead students to evolve a positive implicit attitude supporting redoing difficult scientific exercises to understand one's errors and to avoid learned helplessness. Based on the assumption that relational students are emotional (easily alienated), we investigate whether they are easy to be persuaded in comparison to though-minded students. Also, we verify whether it is possible to consider an educational robot for such a mission. We compare different persuasive sources (tablet showing a persuasive text, an animated robot and a human) encouraging the student to strive for cognitive closure, to verify which of these sources leads to better implicit attitude supporting defeating one's self to assimilate difficult scientific exercises. Finally, we explore which of the persuasive sources better fits each of both student's profiles.

Keywords: Cognitive conflict · Agency level · Student's profile · Persuasiveness

1 Introduction

Several researchers are endeavoring to develop interactive educational robots. Nao was one of the educational robots to prove success in math teaching[1].

---

[1] www.ald.softbankrobotics.com/en/solutions/education-research.

© Springer International Publishing AG 2016
A. Agah et al. (Eds.): ICSR 2016, LNAI 9979, pp. 115–125, 2016.
DOI: 10.1007/978-3-319-47437-3_12

VGo was used by sick students to avoid missing class[2]. TIRO played the role of an educational media in class [1]. Robota was used for multiple educational purposes [2,3] as well as Robovie [4]. RoboSapien encourages students to learn English [5]. These efforts seem to be devoted to socially assist children, and replace the student or the teacher in the classroom. However, to the best of our knowledge, no concern was paid to the serious damage that cognitive dissonance encountered by students at schools during science learning and the social robot's key persuasive role that can be played to overcome it. When the student realizes that his answer is wrong, his preconceptions are defeated and we call such a situation cognitive dissonance. Based on Abramason et al. [6], incrementally the cognitive dissonance may lead to learned helplessness[3]. In such a case, once the student has to resolve a scientific exercise, he experiences a depression coupled with motivational deficits and starts avoiding science learning (after successive failures). In our current work, we are interested in comparing different persuasive sources' (a box, a robot, a human) effects that might help different students' types (profiles:relational vs though minded) overcome the cognitive dissonance by driving them to answer again the difficult exercise rather than skipping it.

2   Counter-Attitudinal Actions

When cognitive dissonance occurs, different counter-attitudinal actions can be chosen by the human and which are: an active attitude change with a new attitude created[4], a belief change by minimizing the importance of the cognitive dissonance[5] or a perception change by getting a new information to support one's previous decision[6]. When the student experiences cognitive dissonance, he will strive to decrease the inconsistency by choosing one of the described counter-attitudinal actions. We want that students get rid of their bad attitudes of skipping the difficult exercise. The new formed attitude should be highly accessed so that it can be stored on a long term basis on the student's cognitive miser[7].

3   Students Different Profiles

According to Murray et al. [7], people that have mostly though-minded attitudes are associated with low empathy and high self-concern. They are less likely to show cognitive dissonance. They show less attitude change than people who are relational. That it is why, dealing with cognitive dissonance in classrooms means

---

[2] www.vgocom.com.

[3] The student will avoid science learning.

[4] The student thinks that he has to change his attitude of avoiding difficult exercises.

[5] After all, science learning is not that important. Many other tasks could be done.

[6] The student thinks that the answer afforded by the book is incorrect.

[7] By measuring the implicit and explicit attitudes, we can verify whether it was established for a long term basis.

that we need to deal with two students' profiles: though-minded students who base their judgment mostly on thinking with a careless response to others and relational students who base their judgment mostly on feeling (by taking care of social norms). In our case, students have to avoid skipping the current exercise if they answered it in a wrong way. They should evolve implicitly a new attitude encouraging the strive to redo difficult exercises. An idea here is to afford a non deliberative persuasive source that implicitly drives the student to evolve the new positive attitude. As though-minded students are difficult to be persuaded, we consider different persuasive sources to verify whether an educational robot could be adapted for both student's profiles and we compare it to the other persuasive sources.

4 Hypothesis

We expect that (H1): *"The more a participant scores high on the dimension of relational (vs though-minded), the more that participant will be persuaded since we assume that he/she is more cooperative than a though minded participant."* We have three different persuasive sources ("a tablet in a box", "the robot" and "a human") as well as a baseline condition (no persuasive source). So, we need to investigate whether (H2): *"We have a main effect of the persuader's agency type. That is, we expect that when a participant interacts with a tablet in a box, he will be persuaded less than when that participant interacts with a robot, in which situation the participant will be persuaded less than when he will interact with a human and of course having a persuasive source is better than nothing."* Finally, we expect (H3): *"Most importantly, an interaction between the manipulation of the persuader's agency and the persuaded profile."*

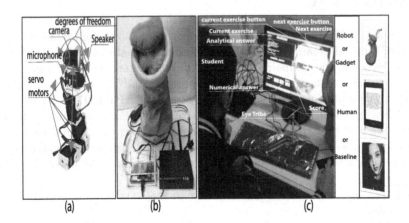

Fig. 1. (a) A close-up picture of ROBOMO; (b) ROBOMO apparatus; (c) The general setup of the experiment.

5   ROBOMO Architecture

ROBOMO tracks the user's face using a Web Camera whilst the human is around
(Fig. 1(a)). It integrates a micro PC and provides a speech through the speaker.
The generated sound have different tones that are adapted with the robot's
gestures (excited, sad, angry and happy tones). The robot uses five servo-motors
(AX-12+) to exhibit different gestures (Fig. 1(a)).

6   Apparatus

In the first zone of the graphical interface, we have the current practice exercise's
text and a button called "Current Exercise" to click on when the student decides
to do the current exercise. There are text zones for the analytical and numerical
answers and a button to click on once the student needs to submit the answer.
In the score's text area, the student can verify whether his answer was correct or
not. If the answer was incorrect, the student has to click on "Current Exercise"
or "Next Exercise" (in the second zone) buttons to decide whether to redo the
current difficult exercise or to jump to the next one. Once the student clicks on
"Next Exercise" button, the next exercise will be displayed in the first zone with
a clear text and a new fuzzy text will be appearing in the next exercise's text
area zone to avoid the student's be biased by fallacies and bias[8]. In fact, one of
the basic parameters that may activate the cognitive dissonance is to make the
consequences of the student's choices foreseeable. So, making the next exercise's
text fuzzy increases the possibility of risk aversion[9]. An EyeTribe helps tracking
the student's eye gaze (Fig. 1(c)).

7   Methodology

66 Tunisian students participated in this experiment (33 though-minded and 33
relational) ([17–19] years) from Farhat Hached College students. They answered
a pre-experiment questionnaire before by 3 days to determine the student's profile (whether the student is though-minded or relational[10] Adapted from Looking
at Type: The Fundamentals by Charles R. Martin (CAPT 1997).). Participants
were debriefed which may help us to evaluate their planned attitude[11]. Participants were told that they would resolve some exercises to help evaluate a new
robot platform so that we avoid any forewarning[12]. Once a student enters to the
room, he was asked to do the calibration (eye tribe) and then starts answering

---

[8] Typical errors in human social judgment that are caused by systemic use of cognitive
strategies.

[9] In decision-making, the weight given to possible losses is greater than possible gains.

[10] goo.gl/forms/fzpCl4onDRG2s9zE2.

[11] This is to measure the student's explicit attitude. We just ask respondents to think
about and report their attitudes.

[12] Forewarning often produces resistance to persuasion.

the exercises. We indicated for the student that he can redo the same current exercise multiple times as long as he wishes. We inform the student that he can choose to jump to the next exercise. When the student feels that he wants to leave the room or when he finishes the exercises' collection, we thank him and he has to answer a post-experiment survey (indicated in Sect. 9). We divided our participants (within subjects design experiment) in a way that we can guarantee that we have a counterbalance of the data, thereby reducing the effect of the sequence of trials on the results.

8   Experiment

ROBOMO generates the speech based persuasive strategy with a medium speech's speed along with the robot's convenient gestures (body and head gestures) and the right tone (Fig. 1). The persuasive speech follows the technique "that's not all". That's not all technique starts with a non deliberative phrase. However, before the human can process the whole phrase different sides (what will be gained or lost), the persuader sweetens the first non deliberative phrase with another phrase. This technique is based on incrementally increasing the number of non deliberative positively framed phrases with a medium speech's speed. An example of a persuasive message could be: "Einstein tried multiple times to succeed on his exams at a certain point of his age. However, his work was rejected many times and judged to be wrong. He continued until he created an excellent amazing science work. Perseverance is one of the ingredients for success. That's not all. As long as you try to understand the difficult exercises you spend more time and according to Harvard table of calories lose, you can burn in 30 min up to 50 calories just by concentrating." There are four conditions the student takes part in which are: the **baseline condition** (No persuasive message is afforded), **condition 1** (the box containing the tablet affords the persuasive strategy), **condition 2** (the robot affords a persuasive strategy) and **condition 3** (the human affords a persuasive strategy). Each two days, the student comes to the classroom to redo another set of exercises with a new set of persuasive messages while we change the persuasive source.

9   After Experiment Survey and the Considered Dependent Variables

After the experiment finished, the student has to answer questionnaires such as the explicit attitude[13] [8], the implicit attitude (implicit association

---

[13] By debriefing the students. In fact, psychologists usually think of explicit measures as those that require respondents' conscious attention to the construct being measured by using Likert scale and semantic differential scale (it is the planned behavior in our case).

test): IAT [8][14], the cognitive dissonance (cogn.diss) [9][15] and the perceived pleasure's level[16]. We considered other dependent variables:

- The quotient: $\frac{Number\_of\_times\_the\_user\_redoes\_the\_incorrect\_exercise}{number\_of\_times\_the\_user\_makes\_an\_error}$. It gives an idea about when has the student a tendency to redo incorrect exercises to strive for science learning rather than jumping from an exercise to another.
- Looks: number of times the user "dwells" with eye gaze between the 2 exercises.

10   Results: Hypothesis 1 Investigation

Results are significant for all the constructs with P-value values in the range of [0.031–0.001] (Table 1). Based on the mean and standard deviation results, we can see that relational students have higher constructs' values in comparison to the though-minded students except for the pleasure construct. Though-minded students felt higher pleasure's level (M = 3.22, SD = 0.169) in comparison to relational students (M = 3.22, SD = 0.169).

Table 1. A table showing the first main effect investigation results (relational vs though-minded): The mean (m), standard deviation (sd) and the F and p-value results by means of the different constructs presented vertically (first column).

Factor	Relational (m, sd)	Though-minded (m, sd)	(F, p-Val)
Pleasure	(3.22, 0.169)	(3.74, 0.169)	(4.84, 0.031)
IAT	(0.615, 0.026)	(0.539, 0.026)	(4.18, 0.045)
Cog.diss	(18.44, 0.389)	(13.39, 0.38)	(84.43, p < 0.001)
Quotient	(0.516, 0.02)	(0.413, 0.02)	(10.93, 0.002)
Looks	(13.10, 0.34)	(8.23, 0.34)	(98.02, p < 0.001)

10.1   Results: Hypothesis 2 Investigation

The persuasive message source agency's level had a main effect in terms of all the constructs with a P-value <0.001 (last column). Table 2, shows that there were no significant differences between baseline and box conditions except

---

[14] This is important to verify whether the student is convinced about the fact that he needs to strive for science learning by redoing difficult exercises rather than adopting a negative implicit attitude that supports learned helplessness. Implicit measures are those that do not require this conscious attention (spontaneous behavior). Some methods could help to measure the implicit attitude such as evaluative priming and the implicit association test.

[15] This is to measure the cognitive dissonance level according to the student's subjective evaluation.

[16] allaboutux.org/self-assessment-scale-sam.

for looks construct: $(F = 21.47$, p-value $= 0.04 < 0.05)$ (second column). Also, Table 2 shows that using a robot as a persuasive source increases the student's pleasure: $((F = 63.4$, p-value $= 0.006)$ R), IAT: $((F = 103$, p-value $< 0.001)$R), cog.diss: $((F = 180.1$, p-value $= 0.004)$R), quotient: $((F = 48.7$, p-value $< 0.001)$R) and looks: $((F = 155$, p $< 0.001)$R) (third column) constructs' values. Finally, Table 2 shows that the human's presence as a persuasive source increases IAT: $((F = 165.5$, p-value $= 0.005)$H), cog.diss: $((F = 40.4$, p-value $< 0.001)$H), quotient: $((F = 7.9$, p-value $< 0.001)$H) and looks: $((F = 14.1$, p $< 0.001)$H) constructs' values. Table 3, shows that there were significant differences between the robot and box conditions with higher results in the robot's condition for all the constructs (second column). Also, Table 3 shows that using a human as a persuasive source in comparison to using a robot increases cog.diss: $((F = 88.5$, p-value $= 0.04 < 0.05)$H) and looks: $((F = 71.08$, p-value $< 0.001)$H). There were statistical differences in terms of pleasure with higher results in the robot's condition rather than in the human's condition $((F = 83.58$, p-value $< 0.001)$R) while no statistical differences were found when we compare IAT mean values of the robot and human' conditions $((F = 2.29$, p-value $= 0.13))$ (third column). Finally, Table 3 shows that using a human as a persuasive message source in comparison to using a box increases IAT: $((F = 37.54$, p-value $= 0.003 < 0.01)$H), cog.diss: $((F = 17.9$, p-value $< 0.001)$H), quotient: $((F = 5.17$, p-value $= 0.049 < 0.05)$H) and looks: $((F = 54$, p-value $= 0.008)$H). However, again there were no main statistical differences of pleasure mean values in the human condition when we compare it to the pleasure mean values in the box condition $(F = 16.34$, p-value $= 0.06 > 0.05)$ (fourth column).

Table 2. A table showing the second main effect investigation results (baseline vs box; baseline vs robot and baseline vs human). When the pairwise comparison has a significant p-value, we add next to the (F, p-value) the condition's label that has higher mean value than the other condition.

Factor	Comparison contrast (F, p-value)			Main comp
	Baseline vs box	Baseline vs robot	Baseline vs human	(F, p-value)
Pleasure	(0.87, 0.65)	(63.4, 0.006)R	(3.58, 0.13)	(84.5, <0.001)
IAT	(17.91, 0.24)	(103, <0.001)R	(165.5, 0.005)H	(52.9, <0.001)
Cog.Diss	(2.15, 0.14)	(180.1, 0.004)R	(40.4, <0.001)H	(16, <0.001)
Quotient	(2.21, 0.14)	(48.7, <0.001)R	(7.9, <0.001)H	(16.5, <0.001)
Looks	(21.47, 0.04)Bx	(155, <0.001)R	(14.1, <0.001)H	(59.7, <0.001)

10.2   Results: Hypothesis 3 Investigation

There was a significant interaction of the persuasive source and the student's profile. Table 4, shows that though-minded students were more sensitive to the persuasive source with a positive increase in contrast values when we

Table 3. A table showing the second main effect investigation results (box vs robot; box vs human and robot vs human). When the pairwise comparison between two conditions has a p-value that it is significant, we add next to the (F, p-value) the condition that has higher mean value than the other condition.

Factor	Comparison contrast (F, p-value)		
	Box vs robot	Box vs human	Robot vs human
Pleasure	(149.3, <0.001)R	(16.34, 0.06)	(83.58, <0.001)R
IAT	(21.92, <0.001)R	(37.54, 0.003)H	(2.29, 0.13)
Cog.Diss	(136.8, <0.001)R	(17.9, <0.001)H	(88.5, 0.04)H
Quotient	(26.09, <0.001)R	(5.17, 0.049)H	(2.6, 0.09)
Looks	(84.4, <0.001)R	(54, 0.008)H	(71.08, <0.001)H

change the persuasive source from the box to the robot (pleasure: $(F = 5.75,$ p-value $= 0.01 < 0.05)$ T$((+)$ 3.79); IAT: $(F = 6.1,$ p-value $= 0.03 < 0.05)$ T$((+)0.38)$; cog.diss: $(F = 16.68,$ p-value $< 0.001)$ T$((+)$ 4.64); quotient: $(F = 3.34,$ p-value $= 0.025 < 0.05)$ T$((+)$ 1.12)) (first column). Though-minded students are more sensitive to the persuasive source with a negative increase in contrast values when we change the persuasive source from the robot to the human (pleasure: (F=4.43, p-value= $0.03 < 0.05)$ T$((-)$ 0.81); looks: $(F = 6.87,$ p-value $= 0.01 < 0.05)$T$((-)5))$, IAT: $((6.9,$ p-value= 0.03<0.05) T$((-)0.22))$ and quotient: $(2.9, 0.047)$ T$((-)2.03$ (last column). Furthermore, when the persuasive source changed from the box to the robot, relational students seem to be more sensitive than though-minded students with a positive contrast values (looks: $(F = 4.6,$ p-value $= 0.03 < 0.05)$ R$((+)6.48))$. Finally, relational students seem to be happier when we change the persuasive source from the box to the human with (pleasure: $(F = 4.46,$ p-value $= 0.03 < 0.05)$ R$((+)1.55)$ (second column). Based on Table 5, when comparing the baseline and box conditions, relational students are more sensitive than though-minded students when we change the source from baseline (no source) to the box with IAT: $(F = 5.17,$ p-value $= 0.02 < 0.05)$ R$((+)0.41)$; quotient: $(F = 11.14,$ p-value $= 0.003 < 0.01)$ R$((+)0.22)$ and looks: $((F = 4.23,$ p-value $= 0.04 < 0.05)$ R$((+)3.3)))$ (second column). When comparing the baseline and robot conditions, though-minded students are more sensitive than relational students when we change the persuasive source from baseline (no source) to the robot in terms of pleasure: $(F = 3.5,$ p-value $= 0.04 < 0.05)$ T$((+)1.05)$; IAT: $(F = 5.12,$ p-value $= 0.02 < 0.05)$ T$((+)0.68)$ and quotient: $(F = 24.6,$ p-value $< 0.001)$ T$((+)0.48)$ (third column). Also, when comparing the baseline and robot conditions, relational students are more sensitive with changing the source from baseline (no source) to the robot with cog.diss: $(F = 180.1,$ p-value $< 0.001)$ R$((+)15.73)$ and looks: $(F = 17.3,$ p-value $< 0.001)$ R$((+)9.79))$ (a general positive tendency) (third column). Finally, when comparing the baseline and human conditions, relational students are more sensitive with changing the source from baseline (no source) to the human with (pleasure: $(F = 4.1,$ p-value $= 0.03 < 0.05)$ R$((+)2.15)$; IAT: $(F = 11.11,$ p-value $= 0.001 < 0.01)$ R$((+)$

0.78); cog.diss: (F = 40.4, p-value < 0.001) R((+)7.15); quotient: (F = 24.47, p-value < 0.001) R((+)0.27) and looks: (F = 15.74, p-value < 0.001) R((+)4.42) (a general positive tendency) (fourth column).

Table 4. A table showing the interaction effect results (source X profile). We show the pairwise comparison results of the box vs robot; box vs human and robot vs human. When the comparison is significant, we add in the cell next to the (F, p-value) a label to indicate which of both student's profile (though-minded or relational) has the highest current construct's (indicated in the first cell of the same cell's line) value when we compare the two current conditions (indicated in the first cell of the same cell's column).

Factor	Comparison contrast (F, p-value)		
	Box vs robot	Box vs human	Robot vs human
Pleasure	(5.75, 0.01) T((+) 3.79)	(4.46, 0.03) R((+)1.55)	(4.43, 0.03) T((−)0.81)
IAT	(6.1,0.03) T((+)0.38)	(0.09, 0.75)	(6.9, 0.03) T((−)0.22)
Cog.Diss	(16.68, <0.001) T((+)4.64)	(0.94, 0.33)	(0.693, 0.29)
Quotient	((3.34, 0.025) T((+) 1.12)	(0.47, 0.4))	(2.9, 0.047) T((−)2.03)
Looks	(4.6, 0.03) R((+)6.48)	(0.08, 0.78)	(6.87, 0.01) T((−)5)

Table 5. A table showing the third main effect investigation results (persuasive source X student's profile). In the current table, we show the pairwise comparison results of the baseline vs box; baseline vs robot and baseline vs human. When the pairwise comparison is significant, we add in the cell next to the (F, p-value) a label to indicate which of both student's profile (though-minded or relational) has the highest current construct's (indicated in the first cell of the same cell's line) value when we compare the two current conditions (indicated in the first cell of the same cell's column).

Factor	Comparison contrast (F, p-value)			Main comp
	Baseline vs box	Baseline vs robot	Baseline vs human	(F, p-value)
Pleasure	(0.9, 0.41)	(3.5, 0.04) T((+)1.05)	(4.1, 0.03) R((+)2.15)	(3.64, 0.029)
IAT	(5.17, 0.02) R((+)0.41)	(5.12, 0.02) T((+)0.68)	(11.11, 0.001)R((+) 0.78)	(52.98,<0.001)
Cog.Diss	(2.15, 0.14)	(180.1, 0.001) R((+)15.73)	(40.4, ¡0.001) R((+)7.15)	(12.8, 0.003)
Quotient	(11.14, 0.003) R((+)0.22)	(24.6, <0.001) T((+)0.48)	(24.47, <0.001) R((+)0.27)	(9.55, <0.001)
Looks	(4.23, 0.04) R((+)3.3)	(17.3, <0.001) R((+)9.79)	(15.74, ¡0.001) R((+)4.42)	(8.19, 0.006)

11 Relational Students Are Easy to Be Persuaded Than Though-Minded Students

Based on Table 1, relational students have higher constructs' values in comparison to though-minded students except for the pleasure construct. Consequently, the more a participant scores high on the dimension of relational (vs though-minded), the more that participant will be easily persuaded (H1).

12   Animated Agents as the Best Persuasive Message Sources

Based on Table 2 using an adaptive persuasive source that shows some animacy (a robot or a human) leads globally to better results. When we used a tablet, the number of times the student looks between the two exercises increases ((21.47, 0.04)Bx). Based on the students' debriefing answers, when we use a box, the student looks to the tablet that it is inside the box very frequently. This is to read a small part of the text. After that, the student looks to the current exercise to verify whether its text contains some of the phrases displayed in the tablet. Finally, the student decides whether to choose the current or the next exercise. Based on Table 3, we remark that by comparing the box condition vs (robot or the human conditions), the box condition has always significant smaller mean constructs' values (except for pleasure). However, when we compare the human and the robot conditions, the human as a persuasive source leads to higher perceived cognitive dissonance, quotient and looks constructs' values. This means that the student must evolve stronger implicit attitude (IAT) and quotient values when we use the human as a persuasive source. But, it is not the case while we have no significant differences between the human and the robot conditions in terms of IAT and quotient constructs' values (H2).

13   Interaction's Effect (Student's Profile X Persuasive Source)

Based on Table 4, when we use a human as a persuasive source, though-minded students are more sensitive in a negative way than relational students while the number of looks, the pleasure's level and the quotient values decreased (last column). This means that using a robot for though-minded students leads to higher pleasure and a more consideration of the difficult exercise as well as higher quotient results (based on the last column of Table 4). As for relational students, using a human as a persuasive source leads to higher pleasure scores and quotient values in comparison to the case when we use the box (Table 4 column 3). Relational students seem to appreciate the human's presence when we compare it to the baseline condition (Table 5 column 4). They are more cooperative than though-minded students when we change the persuasive source from the robot to the human (pleasure, looks and quotient) with a less decreasing contrast values. Consequently, relational students seem to be more tolerant than though-minded students for the usage of the human as a persuasive source. Using a human or a robot as a persuasive source for relational students leads to a steady level of IAT with a contrast value equal to 0.01. This means that relational students evolve the same positive attitudes whether we used a human or a robot as a persuasive source. However, though-minded students have a bigger contrast value that led to the significant IAT: (6.9, p-value $= 0.03 < 0.05$) T((−)0.22). This shows that though-minded students evolve less positive attitude towards redoing the difficult scientific exercise in comparison to relational students when the human is the persuasive source rather than the robot.

14   Conclusion

We conducted an experiment that helps us to investigate the most persuasive source (a tablet, a robot, a human) for the different students' profiles (though-minded and relational). We remarked that relational students are easy to be persuaded than though-minded students. Using a human or a robot as persuasive sources can help the student overcomes the cognitive dissonance and leads to better results in comparison to the baseline or the box conditions. Finally, though-minded students are more persuaded when we use the robot rather than the human as a persuasive source. Relational students seem to be more tolerant to the usage of a human as a persuasive source than though-minded students. Also, we concluded that as for relational students, a positive attitude could evolve implicitly and it is of the same magnitude whether we consider the human or the robot as a persuasive source. In our future work, we try to investigate the differences in effects of the persuasive message's timing during science learning to overcome the cognitive dissonance (whether the message should be delivered after, during or before being stricken by the cognitive dissonance).

References

1. Han, J., Kim, D.: r-learning services for elementary school students with a teaching assistant robot. In: Conference on Human Robot Interaction, pp. 255–256 (2009)
2. Robins, B., Dautenhahn, K.: Tactile interactions with a humanoid robot: novel play scenario implementations with children with autism. J. Soc. Robot. **6**, 397–415 (2014)
3. Billard, A.: Robota: clever toy and educational tool. Robot. Auton. Syst. **42**, 259–269 (2003)
4. Kanda, T., Sato, R., Ishiguro, H.: A two-month field trial in an elementary school for long-term human-robot interaction. IEEE Trans. Robot. **23**, 962–971 (2007)
5. Zhen, Y., Chi, S., Chih, C., Gwo-Dong, C.: A robot as a teaching assistant in an English class. In: Conference on Advanced Learning Technologies, pp. 87–91 (2006)
6. Abramason, L.Y., Seligman, M.E., Teasdale, J.D.: Learned helplessness in humans: critique and reformulation. J. Abnorm. Psychol. **87**, 49–74 (1978)
7. Murray, A., James, M., Scott, L.: Psychopathic personality traits and cognitive dissonance: individual differences in attitude change. J. Res. Pers. **46**, 525–536 (2012)
8. Pantos, A.J.: Measuring implicit and explicit attitudes toward foreign-accented speech. J. Lang. Soc. Psychol. **32**, 3–20 (2013)
9. Levin, D., Harriott, C., Natalie, A.P., Julie, A.A.: Cognitive dissonance as a measure of reactions to human-robot interaction. J. Hum. Robot Interact. **2**, 3–17 (2013)

Responsive Social Agents
Feedback-Sensitive Behavior Generation for Social Interactions

Jered Vroon$^{(\boxtimes)}$, Gwenn Englebienne, and Vanessa Evers

Human Media Interaction, University of Twente, 7500AE Enschede, The Netherlands
{j.h.vroon,g.englebienne,v.evers}@utwente.nl

Abstract. How can we generate appropriate behavior for social artificial agents? A common approach is to (1) establish with controlled
experiments which action is most appropriate in which setting, and (2)
select actions based on this knowledge and an estimate of the setting.
This approach faces challenges, as it can be very hard to acquire and
reason with all the required knowledge. Estimating the setting is challenging too, as many relevant aspects of the setting (e.g. personality of
the interactee) can be unobservable. We formally describe an alternative approach that can handle these challenges; **responsiveness**. This
is the idea that a social agent can utilize the many feedback cues given
in social interactions to continuously adapt its behavior to something
more appropriate. We theoretically discuss the relative advantages and
disadvantages of these two approaches, which allows for more explicitly
considering their application in social agents.

Keywords: Control architectures · Social robotics · Feedback

1 Introduction

Robotic and other artificial agents are increasingly often being deployed in settings where they have to interact with humans in a socially appropriate way.
From telepresence robots to educational agents; they all interact with humans
to serve their purpose.

How to generate socially appropriate behavior for such agents? This question
involves all behaviors such agents can show, from how they position themselves to
the sounds they use. This paper theoretically discusses approaches to answering
this question, using social positioning (proxemics) for mobile agents as a running
example.

Commonly, socially appropriate behavior for artificial agents is investigated
with psychological experiments measuring the effect of particular conditions in
interactions between social agents and participants. Ideally, this results in the
generalized knowledge that within a particular setting, a particular behavior is
more appropriate.

© Springer International Publishing AG 2016
A. Agah et al. (Eds.): ICSR 2016, LNAI 9979, pp. 126–137, 2016.
DOI: 10.1007/978-3-319-47437-3_13

Behaviors can be generated based on this generalized knowledge, by first estimating the current setting and then using the generalized knowledge to select the appropriate behavior for that setting. We will refer to this as the **setting-specific approach**. For example, in social distancing for mobile agents, it is common to derive appropriate distances from a combination of factors, ranging from size and human-likeness of the agent [12] to experience with pets and robots of the interactee [9].

Such a setting-specific approach faces several practical challenges. Firstly, the generalized knowledge required to select the appropriate action can involve a complex interplay of many different variables, which makes it hard to acquire and reason with. Secondly, many relevant aspects of the setting can be hard or impossible to observe, making estimating the current setting into a very challenging task. To continue with our social distancing example; hearing problems may well influence what is an appropriate interaction distance, but may be impossible to detect beforehand.

Fortunately, interactions with humans provide extra information that could help overcome these challenges: feedback. Feedback can be anything from asking someone not to speak too loud, or cupping a hand to your ear to indicate hearing problems, to taking a step back if someone gets too close (e.g. [4]).

In this paper we discuss the idea that agents can generate social behavior by being **responsive** to these feedback cues. Such agents could try a behavior to get started and then continuously adapt it to something more appropriate based on the feedback cues they recognize. Responsiveness would thus provide a pathway to finding the appropriate behavior that does not rely on or assume knowing all relevant aspects of the setting.

We will theoretically discuss the setting-specific and the responsive approach to generating social behavior, by formally defining both (Sect. 2), discussing the challenges faced by setting-specific approaches (Sect. 3), and how responsiveness can (partly) resolve these (Sect. 4). Though responsiveness may seem straightforward, it is not commonly used in social agents; for example, even though using responsiveness may well be suitable for doing social distancing, we are not aware of any existing artificial agents doing so (Sect. 5). With our specification of responsiveness, we aim to contribute to the development of social agents by allowing people to more explicitly consider its application, limitations, and opportunities (Sect. 6).

2   Terminology

In this section, we formally define the setting-specific approach and the responsive approach. We start with the basic building blocks (Sect. 2.1) with which we define agents and interaction (Sect. 2.2) and discuss what makes behavior "appropriate" (Sect. 2.3). We then define the two approaches (Sect. 2.4). Symbolic representations (building on our earlier work [10]) will be introduced solely to make the relations between the terms more explicit.

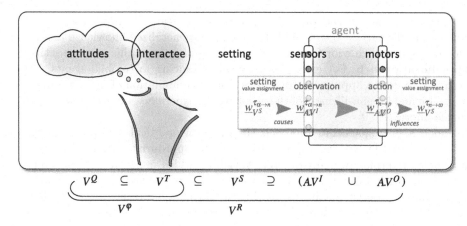

Fig. 1. Overview of the terminology involved in the relationship between an agent and the setting in which it exists. The goal of the agent is a mapping from observations $(\underline{w}_{AV^I}^{\tau_\alpha \to n})$ to actions $(\underline{w}_{AV^O}^{\tau_n \to p})$ such that the social appropriateness of those actions during the interaction $(\underline{SA}_{V^Q}(\underline{w}_{AV^O}^{\tau_\alpha \to \omega}))$ is optimal, sufficient, or improving.

2.1 Variables, Time Spans, and Value Assignments

We will treat agents as entities, roughly separable from the setting in which they exist, that gather observations and produce actions based on those. The state of the setting causes these observations and can in turn be influenced by these actions, allowing for interactions (Fig. 1). Actions, observations and the state of the setting will all be formalized as *value assignments* over a particular *time span* to a set of *variables*.

Variables. We will be talking about **variables** (denoted by v), which can be among others **input variables** (v^I), **output variables** (v^O) or **setting variables** (v^S). Each variable v' has a **domain** ($D_{v'}$), which is the set of values that variable v' can take. A **variable set** (V) is a set of variables, each of which can have a different domain. A variable set containing only input variables, output variables or setting variables is, respectively, an **input set** (V^I), **output set** (V^O) or **setting** (V^S).

Time spans. In addition to these, we will use the variable **time** (denoted by t). Its domain (D_t) is a totally ordered set of values, representing a series of successive moments in time. τ_α indicates the first moment of an interaction, τ_ω the last. Moments in between will be indicated with letters such that alphabetical ordering indicates succession, e.g. τ_q comes before τ_r. A **time span** ($\tau_{m \to n}$) between two moments ($\{\tau_m, \tau_n\} \in D_t$, $\tau_m \leq \tau_n$) is the complete subset of successive moments in time between them ($\tau_{m \to n} = \{x \mid x \in D_t, x \geq \tau_m \wedge x \leq \tau_n\}$). Implementations may rely, without loss of generality, on discretised time or event-based observation.

Value assignments. A **single value assignment** for a variable v' and a moment in time τ_o (denoted by $\underline{w}_{v'}^{\tau_o}$) is defined as a function that returns the value of that variable at that moment in time ($\underline{w}_{v'}^{\tau_o} : v' \mapsto D_{v'}$). We also define a **value assignment** for a set of variables V' and a time span $\tau_{m \to n}$ (denoted by $\underline{w}_{V'}^{\tau_{m \to n}}$), to give the single value assignments for all variables in that variable set and all moments in that time span. The **value assignment set** for a variable set V' and a time span $\tau_{m \to n}$ (denoted by $\underline{W}_{V'}^{\tau_{m \to n}}$) is the set of all possible value assignments for that V' and $\tau_{m \to n}$ ($\underline{W}_{V'}^{\tau_{m \to n}} = \{\underline{w}_{V'}^{\tau_{m \to n}} \mid \cdot\}$).

2.2 Agents and Interaction

An (artificial) **agent** (denoted by A) has **sensors** (an input set, AV^I), **actuators** (an output set, AV^O) and "inner workings" to connect those. It produces **actions** (value assignments for its actuators, $\underline{w}_{AV^O}^{\tau_{m \to n}}$) that are affected by its **observations** (value assignments for its sensors, $\underline{w}_{AV^I}^{\tau_{m \to n}}$). We use (partial specifications of) settings as theoretical constructs to discuss the environment in which an agent exists. The actions of an agent influence the setting to some extent. Likewise, to some extent, the observations of an agent reflect the setting, based on which the agent can **estimate** it (a value o being estimated is denoted by o^E). The more reliably a value can be estimated by an agent in practice, the more **estimable** it will be said to be.

We will refer to the (human) other agents with which the agent is interacting in the setting as **interactees** (denoted by V^T). These are part of the setting ($V^T \subseteq V^S$).

2.3 Appropriate Behavior

Central in deciding if the behavior of an agent in an interaction is socially appropriate, are the **attitudes** of the interactee(s) (denoted by V^Q), loosely defined as a subset of the variables used to express interactees and their properties ($V^Q \subseteq V^T \subseteq V^S$). Attitudes can range from, for example, comfort to perception of the agent as intelligent or sensitive.

The actions of an agent to some extent influence the setting, which can include the attitudes of the involved interactees. Depending on the goals of the agent, different attitudes can be more or less desirable; for example, an agent may want to avoid selecting actions that make the interactee more uncomfortable. We define the **social appropriateness** function (denoted by $\underline{SA}_{V^Q{'}}$), for a set of attitudes $V^{Q'}$ and a setting during an interaction, that for all possible actions returns a numerical value, such that a higher value indicates that action would lead to a more 'desirable' value for those attitudes ($\underline{SA}_{V^{Q'}} : \underline{W}_{AV^O}^{\tau_{m \to n}} \mapsto \mathbb{R}$). As with the setting, we use this function as a theoretical construct for discussion purposes; an agent can at best estimate it.

2.4 Approaches to Finding Socially Appropriate Behavior

How can an agent select actions such that their social appropriateness is optimal, sufficient, or at least improving? We here define two approaches, both of which focus on the strategy used to find socially appropriate behavior for an agent, not on the actual implementation of these steps. Different ways of generating behavior, e.g. static, scripted, dynamic, adaptive, might thus all be used to implement either of the two approaches.

Setting-specific approach. The setting-specific approach depends on prior knowledge about how the social appropriateness of different actions is dependent on the values for particular variables in the setting. We therefore define the **knowledge** function (denoted by $\underline{K}$) that, for all value assignments to (a subset of) the setting $\underline{W}_{VS}^{\tau_{\alpha \to n}}$, it returns the most appropriate action $(\underline{K} : \underline{W}_{VS}^{\tau_{\alpha \to n}} \mapsto \underline{W}_{VO}^{\tau_{n \to p}})$.

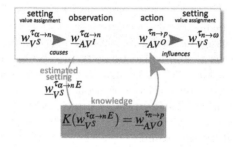

Fig. 2. Steps taken in a setting-specific approach.

We define **relevant setting variables** (denoted by V^R) as a subset of the variables in the setting ($V^R \subseteq V^S$) such that their values contain all information required to distinguish between setting value assignments where $\underline{K}$ should give different outcomes. A knowledge function that uses at least the relevant setting variables should thus have enough information to select the most appropriate action. Such a knowledge function is the ideal, as in practice approximations often have to be used instead (Sect. 3).

From the knowledge, the setting-specific approach works in two steps to produce an action based on observations (Fig. 2). First, the available observations are used to estimate value assignments to (a subset of) the setting. Second, these estimates are used with the knowledge function to try and select the best action. If the knowledge is approximated, or if the relevant setting variables are not fully estimable, this may result in the best *known* action, rather than in the best action.

Responsive approach. Central to the responsive approach is feedback; any action a of the agent influence the attitudes of the interactee, which in turn can be reflected by **feedback variables** (denoted by φ'_a). Feedback variables provide information about the underlying appropriateness of a previous action, e.g. if it was optimal/sufficient (**basic feedback**), how it compares to other earlier actions (**comparable feedback**), or even which actions would be more/less suitable (**directional feedback**). The **feedback set** (denoted by V^φ) is the set of all available feedback variables.

Feedback variables encode the information about the underlying appropriateness; different feedback variables can code (partially) overlapping information and, importantly, the encoding may be flawed. The greater the certainty with which the underlying appropriateness can be derived from a feedback variable, the more **legible** we will say it to be. Feedback variables can be less legible because they reflect things besides the underlying appropriateness, or because they differ between interactees.

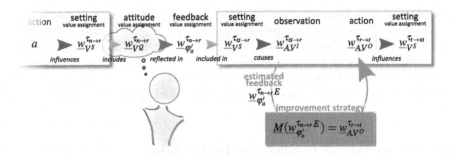

Fig. 3. Steps taken in a responsive approach. Includes an overview of how feedback variables can be assigned a value in response to an earlier action a of the agent.

The responsive approach works in two steps (Fig. 3). First, the feedback variables are estimated from the available observations, and interpreted as relating to particular previous actions of the agent. Second, this estimated feedback is used to adapt the subsequent actions of the agent. For this, we define an **improvement strategy** (denoted by $\underline{M}$) as a function that, based on all available feedback on previous actions $\underline{w}_{V^\varphi}^{\tau_{\alpha \to n}}$ returns a suggested action $\underline{w}_{AVo}^{\tau_{n \to p}}$, such that, possibly after several iterations, the actions will be sufficient and/or improving ($\underline{M} : \underline{W}_{V^\varphi}^{\tau_{\alpha \to n}} \mapsto \underline{W}_{AVo}^{\tau_{n \to p}}$). For example, an improvement strategy using comparable feedback could be to try and select actions that are more dissimilar to actions with lower social appropriateness.

3 Implications and Challenges for a Setting-Specific Approach

Assuming that from the observations an agent can perfectly derive the relevant setting variables, and assuming that the agent has full knowledge of the optimal action for all settings, it is trivial to prove that a setting-specific approach will yield the optimal action. However these assumptions will likely hold only in constrained settings, which social interactions usually are not. This presents several challenges to the approach.

3.1   Estimating the Required Setting Variables

A setting-specific approach depends on estimating the relevant setting variables, which is a challenging task; not only because the set of relevant setting variables can still be very large, despite being a subset of all setting variables, but also because many of these setting variables will only partially be observable, if at all. They may be internal to the interactee (e.g. personality traits), include cases that are hard to classify (e.g. gender and cultural background), or algorithms that reliably detect them may not exist (yet). In our running example; it would be challenging to evaluate someone's hearing, but it could well influence the appropriate interaction distance [11]. Though this may partly be resolved by using a reasonable approximation in a limited setting, there is no guarantee that the behaviors based on such approximations would be sufficiently appropriate.

These challenges can **not** be overcome within a purely setting-specific approach. Not taking into account these practical limitations in detecting and estimating the relevant setting variables severely challenges the implementation of autonomous agents that actually use knowledge which depends on these setting variables.

3.2   The Knowledge to Select the Best Action

The knowledge required for a setting-specific approach is in practice usually approximated by a combination of findings from scientific studies. This allows for the design of agents that can effectively select a reasonable action based on a well-chosen selection of relevant setting variables. It also introduces several challenges.

Establishing which setting variables are relevant. Establishing which setting variables are relevant can be challenging, given the sheer amount of setting variables in real-world settings and because it is hard to predict which variables will be relevant. Though in controlled experiments one can try to focus on specific setting variables, every aspect of the world could be a relevant setting variable. Thus, even listing all setting variables would be challenging, let alone investigating their relevance with scientific rigor.

Combinatorial explosion. As the number of setting variables that have to be considered increases, so does the complexity of the knowledge function. If the different variables are dependent on each other, all combinations of those factors have to be considered to reliably derive the appropriate behavior, resulting in exponential growth[1].

An implementation of such a knowledge function would thus quickly become intractable. This can partly be avoided by instead using approximations, though

---

[1] Even when limiting ourselves to 'just' the relevant setting variables (V^R) this would already be $\prod_{v' \in V^R} |D_{v'}|$ combinations (since $|D_{v'}| \geq 2$ for all meaningful variables, this is at least $2^{|V^R|}$).

this would necessarily introduce uncertainty about the appropriateness of the selected action. The complexity could also be reduced by explicitly establishing which setting variables are independent of each other – but that is a challenging task itself.

In addition, this combinatorial explosion also poses a significant challenge to acquiring the required (prior) knowledge in a scientifically sound way; given the sheer number of combinations, it would be infeasible to test all combinations against each other in a controlled experiment. While approximations may be acceptable for implementations, they are less appropriate for scientific experiments.

Stereotyping by using generalized findings. The knowledge function of an agent is commonly acquired through controlled experiments, which investigate how the effects of particular setting variables on particular attitudes could be generalized to a population.

When individual differences play a role in establishing the appropriate behavior, this can pose a challenge to a setting-specific approach. For example, an agent may well need to adapt its behavior when interacting with people who had a negative prior experience with similar agents.

To some extent, these individual differences can be handled by introducing them as setting variables. However, this would pose its own challenges if it introduces (partly) unobservable variables or results in a large increase in the number of variables.

4   Implications and Challenges for a Responsive Approach

The responsive and setting-specific approach are both aimed at the same goal, though they use different steps. In this section we will discuss how a responsive approach could circumvent some of the challenges faced by a setting-specific approach, and vice versa.

4.1   Estimating the Required Setting Variables

The setting-specific approach needs to estimate all relevant setting variables, whereas the responsive approach depends on a legible set of feedback variables. The more legible the feedback variables are, the more information they provide about the social appropriateness of previous actions (on a set of attitudes), and the less feedback variables a responsive approach will need. If feedback variables are available that are legible and estimable, a responsive approach can thus use these to avoid the aforementioned combinatorial explosion faced by a setting-specific approach.

Such a combination of legible and estimable feedback variables may actually be common, since there is an incentive for the interactee to provide them. For if the interactee provides legible and estimable feedback variables, a responsive agent, artificial or not, can use these to try and improve its behavior – which would benefit both the agent *and* the interactee. Using a responsive approach could thus turn finding socially appropriate actions into a collaborative effort. Therefore, the interactee may actively provide legible and estimable feedback variables, be it consciously and/or subconsciously.

4.2   The Improvement Strategy to Select Better Actions

Another important difference between the responsive and the setting-specific approach is that the former uses an improvement strategy function instead of a (prior) knowledge function. This gives the responsive approach a reduced dependency on knowledge for all setting variables and allows for individualized instead of stereotyped adaptation.

Reduced dependency on all setting variables. Since a responsive approach does not use prior knowledge (but only feedback), it avoids many of the challenges faced by a setting-specific approach, such as the combinatorial explosion and the challenges of establishing the prior knowledge. Only if the feedback variables would not be legible could similar challenges also arise for a responsive approach.

Individualized instead of stereotyped adaptation. A responsive approach per definition uses the feedback given by individual interactees, rather than working from knowledge generalized to the population of interactants. Since feedback variables are individual, a responsive approach can be used to adapt to the individual preferences of interactees. Some feedback variables may even encode a combination of different attitudes prioritized based on the preferences of individual interactees.

This circumvents the stereotyping challenge faced by a setting-specific approach. It also shows that a purely responsive approach could easily miss out on the advantages of such stereotyping. Herein, the two approaches can complement each other. A setting-specific approach could be used to select an initial 'stereotyped' action, that can then be refined into more 'personalized' actions using a responsive approach.

Defining an improvement strategy. The responsive approach depends on suitable improvement strategy functions. In contrast to the knowledge function of the setting-specific approach, an improvement strategy can be defined to deliberately use various aspects of the interaction. For example, an improvement strategy could be to directly ask the interactees for the desired actions. Furthermore, interactees might even appreciate the attempts of a responsive agent to try and improve the interaction, regardless of the appropriateness of the selected actions. While introducing such interesting options, this flexibility could also make it a challenge to create suitable improvement strategies.

4.3   Quality of the Selected Action

Where a setting-specific approach can ideally aim for selecting the most appropriate action, a responsive approach instead aims for improvement. Consequently, a responsive approach will be most suitable if the cost of selecting an inappropriate action is not too high and/or if no systems exists that reliably deliver the most appropriate action. In some cases, showing responsive behavior may actually *be* the most appropriate action.

5 Applications of Responsiveness in Social Agents

There is a variety of existing work in artificial agents that we feel aligns with our definition of the responsive approach. Our aim here is not to give a complete overview, but instead to illustrate how solutions fitting within the framework of a responsive approach exist and have been shown to be effective.

Most of the work on responsiveness in human-agent interaction focuses on agents that deliberately *provide* legible feedback variables, rather than being responsive themselves. This includes prior work using the term 'robot responsiveness', which primarily investigated different (dynamic) non-verbal feedback behaviors a robot could use when listening to an interactee – showing various positive effects of giving the appropriate feedback behaviors [2,3]. Jung *et al.* looked at the effects of robots using backchanneling on human-robot teamwork and found both improved team functioning and decreased perceived competence [5]. In the field of our running example, it has been found that people adapt their proxemic preferences when interacting with an agent that provided (feedback) information on its effectiveness at different interaction distances [6].

To our knowledge, there is no work on social positioning, our running example, with artificial agents using a responsive approach, even though various feedback variables may be available (e.g. [4,7]). In fact, a large part of the work on social positioning in human-human interaction seems to be strongly in line with the responsive approach (see e.g. the extensive review by Aiello [1]). We have previously conducted two small studies in this direction, that we will briefly discuss here. Both had a limited sample size and used a Wizard of Oz. In one of them, we set up a conversation such that hearing problems were to be expected and then had the robot use one of two different improvement strategies once certain feedback variables were observed [11]. In the other, we compared conditions in which a robot either did (1) an approach without personal space invasion, (2) an approach with personal space invasion, or (3) a personal space invasion after which it backed up and apologized [8]. The results of both studies suggest that participants appreciate the responsive behavior, perhaps even over directly picking the 'improved' action.

6 Discussion

We have given formal definitions of both the responsive and the setting-specific approach. Though in theory capable of finding the optimally appropriate behavior, the setting-specific approach ideally requires the agent to estimate and reason with all relevant setting variables – which may well be infeasible in realistic settings. We showed that the responsive approach can be used to (partly) circumvent these challenges, as it instead looks for behavior that is sufficiently appropriate or improves appropriateness.

Our theoretical discussion of the responsive approach is only a rough starting point for implementations. Since both responsiveness and (online) reinforcement learning need to adapt to feedback, insights from the latter could be used to guide such implementations – though with the responsive approach the adapting is explicitly part of the social dynamic, rather than finite learning. Another challenge will be the social signal processing necessary to detect feedback variables. More so because the expectations one has from an (artificial) agent may influence which feedback variables are used.

If suitable implementations can be created, explicitly considering a responsive approach can offer various opportunities. One such opportunity is to complement a setting-specific with a responsive approach. Another opportunity would be to use responsiveness in a more pro-active way, for example by directly asking interactees which actions they would prefer. Further opportunities can be found in the improvement strategy, e.g.; (a) with intelligent reasoning about why the agent got particular feedback, it may be able to respond to it more appropriately, or (b) giving responsive agents different personalities by parametrizing the different factors weighed by the agent when adapting to feedback, such as its own needs and those of the interactee.

Overall, we have introduced an explicit definition of responsiveness, and argued for the potential value of the approach. We hope and expect that this can help to explicitly consider its application in (artificial) social agents, not necessarily as a replacement of the setting-specific approach, but as a potentially valuable addition.

Acknowledgements. The work described in this paper has partly been supported by the European Commission under contract number FP7-ICT-611153 (TERESA).

We are grateful for the critical and open-minded comments of Khiet Truong, Dennis Reidsma, Daniel Davison, Bob Schadenberg, Jan Kolkmeier, Michiel Joosse, Roelof de Vries, Jorge Gallego Pérez, Jeroen Linssen and Dirk Heylen.

References

1. Aiello, J.R.: Human spatial behavior. In: Stokols, D., Altman, I. (eds.) Handbook of Environmental Psychology, pp. 389–504. Wiley, New York (1987). Chap. 12
2. Birnbaum, G.E., Mizrahi, M., Hoffman, G., Reis, H.T., Finkel, E.J., Sass, O.: Machines as a source of consolation: robot responsiveness increases human approach behavior and desire for companionship. In: 10th ACM/IEEE International Conference on Human Robot Interaction, pp. 165–171. IEEE Press (2016)
3. Bretan, M., Hoffman, G., Weinberg, G.: Emotionally expressive dynamic physical behaviors in robots. Int. J. Hum. Comput. Stud. **78**, 1–16 (2015)
4. Cappella, J.N.: Mutual influence in expressive behavior: adult-adult and infant-adult dyadic interaction. Psychol. Bull. **89**(1), 101–132 (1981)

5. Jung, M.F., Lee, J.J., DePalma, N., Adalgeirsson, S.O., Hinds, P.J., Breazeal, C.: Engaging robots: easing complex human-robot teamwork using backchanneling. In: Proceedings of the 2013 Conference on Computer Supported Cooperative Work, pp. 1555–1566. ACM (2013)
6. Mead, R., Matarić, M.J.: Robots have needs too: people adapt their proxemic preferences to improve autonomous robot recognition of human social signals. In: Proceedings of New Frontiers in Human-Robot Interaction, pp. 100–107 (2015)
7. Patterson, M.L., Mullens, S., Romano, J.: Compensatory reactions to spatial intrusion. Sociometry **34**, 114–121 (1971)
8. Snijders, D.: Robot's recovery from invading personal space. Unpublished bachelor's thesis, University of Twente (2015)
9. Takayama, L., Pantofaru, C.: Influences on proxemic behaviors in human-robot interaction. In: IEEE/RSJ International Conference on Intelligent Robots and Systems, pp. 5495–5502. IEEE (2009)
10. Vroon, J.: Regulated reactive robotics: a formal framework. Unpublished master's thesis, Radboud University Nijmegen (2011)
11. Vroon, J., Kim, J., Koster, R.: Robot response behaviors to accommodate hearing problems. In: Proceedings of New Friends 2015, pp. 48–49. Windesheim Flevoland University (2015)
12. Walters, M.L., Dautenhahn, K., te Boekhorst, R., Koay, K.L., Syrdal, D.S., Nehaniv, C.L.: An empirical framework for human-robot proxemics. In: Proceedings of New Frontiers in Human-Robot Interaction, pp. 144–149 (2009)

A Human-Robot Competition: Towards Evaluating Robots' Reasoning Abilities for HRI

Amit Kumar Pandey[1(✉)], Lavindra de Silva[2], and Rachid Alami[3]

[1] SoftBank Robotics, Innovation Department, Paris, France
akpandey@aldebaran.com
[2] Institute for Advanced Manufacturing, University of Nottingham, Nottingham, UK
lavindra.desilva@nottingham.ac.uk
[3] LAAS-CNRS, University of Toulouse, Toulouse, France
rachid.alami@laas.fr

Abstract. For effective Human-Robot Interaction (HRI), a robot should be human and human-environment aware. Perspective taking, effort analysis and affordance analysis are some of the core components in such human-centered reasoning. This paper is concerned with the need for benchmarking scenarios to assess the resultant intelligence, when such reasoning blocks function together. Despite the various competitions involving robots, there is a lack of approaches considering the human in their scenarios and in the reasoning processes, especially those targeting HRI. We present a game that is centered upon a human-robot competition, and motivate how our scenario, and the idea of a robot and a human competing, can serve as a benchmark test for both human-aware reasoning as well as inter-robot social intelligence. Based on subjective feedback from participants, we also provide some pointers and ingredients for evaluation matrices.

1 Introduction

Research in child development and human behavioral psychology clearly indicates that perspective taking, i.e., reasoning from others' perspectives, is one of the key components for social interaction and social intelligence. Perspective taking starts in children from as early as 12–15 months, in the form of understanding the occlusion of others' line-of-sight, and that an adult might be seeing something that the child is not able to see, due to it being hidden behind some barrier; this applies to both places and objects (e.g. [1]). Studies on reachability analysis (e.g. [2]) have suggested that from the age of 3, children are able to perceive which places are reachable to them and to others as they start to develop allocentrism—spatial decentration and perspective taking. In robotics, perspective taking has been used for learning from ambiguous demonstration [3], grounding ambiguous references [4], sharing attention [5], etc.

An object's affordance, specifically, its action possibilities (Gibson [6]) is another crucial aspect for shaping our day-to-day interaction with the environment and with others. Affordance is also a central organizing construct for

© Springer International Publishing AG 2016
A. Agah et al. (Eds.): ICSR 2016, LNAI 9979, pp. 138–147, 2016.
DOI: 10.1007/978-3-319-47437-3_14

action differentiation and selection [7]. In robotics, the notion of affordance has been used in domains involving tool use [8], for checking traversability [9], for learning action selection [10], etc.

The Turing Test is a well-known test for evaluating the intelligence of a machine relative to that of a human. In the standard interpretation, the test involves a machine and human competing with each other over a conversation with another human. The design of our game was inspired by the Turing Test in the sense that a robot and human competes with each other to infer and describe environmental changes, which is judged by another human(s). However, some aspects of the Turing Test were not suitable for the kind of intelligence that we wanted to test, because the standard Turing Test *(i)* encourages making mistakes in order to look more natural and human like, and *(ii)* is more focused on carrying out a conversation. We do not advocate the robot making such mistakes, but instead identify and "penalize" them. Furthermore, because we focus on human-aware reasoning, we have based the scenario around physical changes in the environment. Another characteristic of our setting is that the ground truth is always available. Hence, one can derive two sets of evaluation criteria: *(a)* a "comparative" one, based on the competing human's and robot's reasoning abilities, and *(b)* an "absolute" one with respect to the ground truth.

Many related competitions exist in the literature, such as *Robocup* [11], the *DARPA Robotics Challenge (DRC)* [12], the *European Land-Robot Trial (ELROB)* [13], and the *HUMABOT robot competition* [14]. However, in all these applications, a robot must be created for a specific mission and target scenario, but potentially ones with no humans involved; therefore, these applications have no direct link to HRI. From the HRI literature, the *AAAI Challenge* [15] is relevant in that it proposes scenarios in which a robot attends and delivers a conference talk. Likewise, a variant of *Robocup*, called *Robocup@Home* [16], also seems relevant, as does the *RoCKIn* competition, which focuses on service robots in a real home environment. Like our proposal, the latter two competitions are also aimed at benchmarking robot systems: a set of benchmark tests is used to evaluate the robots' abilities and performance in realistic home environments. Thus, there is a clear need for HRI-oriented robot competitions, evaluation, and specialized benchmark tests. The competition that we present in this paper is another step in this direction: it provides a means by which the robot's "human-centered intelligence" (or "social intelligence") could be evaluated. More specifically, we present a competition scenario and methodology for its use, as well as an analysis of data gathered from the competition, which we believe will serve toward developing benchmark tests for evaluating a robot's combined intelligence based on perspective taking, reachability, and affordance analysis abilities. Our scenario is fully implemented on a PR2 robot, and it has been demonstrated live at an EU event, as well as to numerous visitors. Our preliminary work in this direction was presented in [17]. The current paper extends our earlier work with a detailed description of the methodology and framework; a subjective analysis of user feedback; and with pointers toward evaluation criteria, an evaluation matrix, and potential quantitative measures.

2   Competition Scenario

The scenario that we propose involves observing, analyzing, grounding, and explaining environmental changes. The setting for the scenario is a "living room" of a realistic apartment, with typical furniture (that was never moved) such as sofas, tables, and shelves, and movable objects including books, cans, and boxes.

Figure 1 summarizes the steps in the game, which are as follows. Two (human) volunteers h_1 and h_2 are asked to take a seat in the living room, and a third one h_3 is asked to stand next to the robot. Following this, the robot and h_3 inspect the living room from where they stand. Then, h_3 and the robot turn away from the scene, and h_1 and h_2 are asked to independently and/or cooperatively make manual changes to the state of the room (while the robot and h_3 are looking away). Finally, h_3 and the robot are asked to re-inspect the room, and identify any changes that might have been made. The competition concludes with a manual comparison of the responses of h_3 and the robot, with each other as well as with what h_1 and h_2 actually did—the ground truth—and by asking h_3 for his/her (subjective) assessment of the robot's "intelligence". Finally, a winner(s) between the robot and h_3 is determined (at this stage just for fun).

During the game, the robot and h_3 compete to answer the following questions: *What* has changed physically? *How* might those changes affect the agents' abilities to see and reach objects? *Who* might have done those changes and with *which* (possibly joint) actions? *Where* might any missing objects be?

A key requirement in the game, from both the robot and h_3, is the ability to *ground* changes in the environment. We define this process as follows. Given a couple $\langle s_0, s_1 \rangle$, which is respectively the initial and final states of the environment, find a suitable triple $\langle \Delta, E, A \rangle$, where Δ represents the physical changes in s_1 compared to s_0; E represents the effect of changes in Δ for h_1 and h_2; and A represents the probable sequence of (possibly cooperative) actions that were executed by h_1 and h_2 in order to bring about the change Δ. In this sense, the grounding process, among other things, needs to reason about perspectives, reachabilities, affordances, and action possibilities.

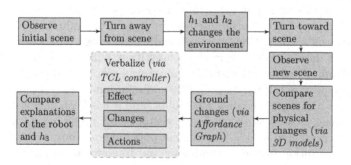

Fig. 1. Framework of the proposed Human-Robot competition between the robot and a human h_3, who observe the initial scene, and then ground the changes.

3 Instantiation

This section outlines the scientific and technical foundations on which the above reasoning capabilities were instantiated for the robot.

Scientific Foundations. We have integrated concepts from existing robotics frameworks that address issues related to perspective taking, affordance analysis, and effort analysis [18], where "effort" here is an abstraction of body-movement based effort levels for reaching objects and places, inspired by the taxonomy of reach affordances in human behavioral psychology [19]. The overarching notion that we exploit from these works is the *Affordance Graph*, which merges various kinds of human-aware reasoning into a single, unified graph. Figure 2c shows an affordance graph instantiated with respect to a specific environmental state. The graph enables the robot to reason about action possibilities among agents, and among objects distributed in the environment.

More specifically, the affordance graph is the aggregation of a *Manipulability Graph* and a *Taskability Graph*. A manipulability graph combines perspective taking, i.e., analyzing visibility and reachability of agents, with affordances between agents and objects, for the purpose of grounding symbolic notions such as grasping, picking, and placing objects to their corresponding geometric entities. Figure 2a shows an example of such a graph. The size of a sphere associated with an edge in the graph depicts the effort level required to see (Green sphere) and reach (Blue sphere) the object or place. Similarly, a *Taskability Graph*, shown in Fig. 2b, encodes relations between agents and other agents, in order to ground symbolic notions such as giving, showing, hiding, taking, and making an object accessible to their corresponding geometric entities.

Using an affordance graph as the underlying reasoning tool is appealing in that it could be analyzed using standard graph search algorithms, though in principle any other reasoning mechanism could be used. Multiple instances of the graph, such as one before and one after an environmental change, are constructed by the robot in order to infer and ground changes in the environment.

Technical Foundation. Our instantiation of the competition scenario is implemented within the LAAS robotics architecture [20]: an interconnected set of

(a) Manipulability Graph   (b) Taskability Graph   (c) Affordance Graph

Fig. 2. Different kinds of graphs representing affordances in the environment [18].

diverse components responsible for distinct functionalities within the system. The implementation uses Move3D [21] to represent the robot's version of the real world in 3D, which is used as input for geometric reasoning. The robot updates its 3D world state in real-time via various sensors; for example, a tag-based stereovision system is used for object identification and localization, and a Kinect (Microsoft) sensor for localizing and tracking humans.

Execution control is achieved via Tcl programs, which are grouped into three distinct sets of capabilities. The first performs tasks related to keeping the geometric 3D model of the world up to date. The second set of capabilities basically requests the geometric component to create an affordance graph for the current world state, and to compare it against the graph corresponding to the previously observed state (if any). The third set is responsible for natural human-robot interaction, i.e., for making speech more intelligible by synthesizing complete sentences out of the output generated by the geometric reasoning component, and for looking at relevant objects, places, and humans while speaking.

One example of the output produced by the geometric reasoning component is the set of couples $\{(object, gt), (action, mv)\}$, where the first element in each couple identifies whether the second element is an action or an object. This particular set is mapped to the sentence *"The grey-tape has been moved"*, where the symbols *grey-tape* and *move* are obtained essentially via two user-supplied mapping functions f^{obj} and f^{act}, which respectively map object and action symbols used within the geometric component into the corresponding symbols used within the ("symbolic") execution controller; thus, $grey\text{-}tape = f^{obj}(gt)$ and $moved = f^{act}(mv)$. Similarly, plans (sequences of ground actions) found by the geometric reasoning component, such as $pick(h_2, gt) \cdot give(h_2, h_1, gt) \cdot place(h_2, p)$, where h_1 and h_2 are the volunteers and p is a new position, are mapped into sentences such as *"As for the grey tape, the second human picked it up and gave it to the first human, who then placed it at its current position"*.

4   Observations Reported by the Competitors

In this section, we describe one run of a competition between the robot and a human, and in particular, the observations reported by the human competitor and the robot. The competition was carried out 12 times in total, while it was demonstrated live at an EU event, and to numerous visitors. Initially, all participants were briefed about the game, with information including the kind of data that was expected from the competitor. Later, the competitor was also guided by the cameraman, with questions such as "where was the object before it was moved?", "do you think that it was likely the object was moved jointly?", "who do you think moved the object?", and "where might the object be now" (if an object was reported missing), in order to gather as much relevant observations from the competitor as possible.

Figure 3 illustrates one run of the competition. Figure 3a shows the initial state of the environment that was examined by the robot and the human competitor before they were asked to look away. Figures 3(d) to (i) show one facet of

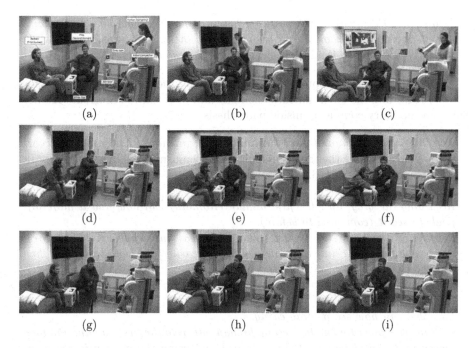

Fig. 3. Images (a) to (c) illustrate one run of the competition, and (d) to (i) show the sequence of changes that were made.

the ground truth—the sequence of steps that were discussed and jointly carried out by the other two humans in order to change the state of the environment. Figure 3b shows the competitor moving around to look behind a white box, where she suspects that an object is hidden. Figure 3c shows the robot and the human competitors describing their beliefs about what might have changed and their impact on the two sitting humans, in terms of what objects are now visible and reachable to them (or no longer so). The large flat panel display in this figure shows the current 3D environmental model maintained by the robot.

The key points that were made by the human competitor are listed below (after editing for clarity), along with our auxiliary comments inside parenthesis.

- *Initially, the grey tape was here, but it has now moved there.* (This was uttered while pointing to the correct initial and final locations of the object.)
- *I cannot see the Jido tape from where I am. It used to be there.* (This was uttered while pointing to the correct initial location of the object.)
- *The white box has not moved.*
- (She then moved to look behind the white box on the table and saw the missing tape, which she correctly suspected to be hidden there.)
- *The grey tape is no longer visible to Romain, and it is now visible to Filip; I also think that it is reachable to the robot.*
- *The Jido tape is neither visible to the robot nor to me, but I am not sure whether it is visible to Filip. The tape is both visible and reachable to Romain.*

- *I think that Filip took both tapes from here and handed them over to Romain, who then placed them on the table and rotated them.* (This was uttered while correctly pointing to the initial locations of the two objects.)

Next, we list the key points that were made by the robot (after editing for clarity) regarding the environmental changes that might have occurred, together with our auxiliary comments inside parenthesis.

- *The Jido tape has moved, and I cannot see it anymore.* (The robot also correctly guessed where the tape is hidden via mechanisms used by our framework [18], as outlined in Sect. 3.)
- *The grey tape has moved.*
- *Regarding the grey tape: the first human (Romain) will now find it more difficult to see it (compared to before).*
- *the first human can reach it now (although it was unreachable to him before).*
- *the second human (Filip) can now see it more easily (i.e., with less effort than before).*
- *the second human cannot reach it anymore (although it was reachable to him before).*
- *the robot can now see it more easily.*
- *the tape was picked up by the second human, given to the first human, and then placed by the first human.* (This was deduced using the geometric reasoner, which assumes that humans will try to balance the overall effort required for a joint task, whenever the amount of individual effort needed amounts to standing up from the seated position [18].)
- *Regarding the Jido tape: it was picked up by the second human and it was then placed.* (Like the human competitor, this was deduced based on the previous and current positions of the Jido tape.)

5   Subjective Analysis

We performed 12 runs of the competition, each of which took approximately 10 min. We noticed interesting similarities in the analyses performed by the robot and the human competitors h_3, e.g., the descriptions about how the objects might have been moved, and where an object, which was visible to them earlier, might now be hidden. There were also runs in which both competitors guessed incorrectly, or missed out on certain observations.

We also asked competitors for their (subjective) opinions about the robot's reasoning capabilities, with questions such as *"how was the robot's performance?"* and *"how was the robot as a competitor?"*. Some of their key remarks were: *(1) "The robot showed **good interaction** with its environment and intelligence in its responses and behavior"*; *(2) "I was **better** than the robot"*; *(3) "It (the robot) performed **better**"*; *(4) "The robot must have **cheated** through that reflection in the glass window"*; *(5) "It **guessed** (correctly) **most of the time**"*; *(6) "We were **equally** good"*; *(7) "Oh, I **missed** that—what the robot said was correct"*; and *(8) "I think it has a **good** memory"*. While still preliminary, these comments show potential for the use of such competitions for evaluating a robot's reasoning abilities in HRI, particularly because competitors (from the public) tended to compare the robot's reasoning abilities with their own.

Data from multiple runs of the competition can be used to establish criteria for evaluating a robot's reasoning abilities, as well as derive evaluation matrices. For example, feedback from the 12 competitors suggested at least three evaluation criteria: *(i)* **adjectives** as **comparative measures**, such as "good", "better", and "equal", which competitors used to qualify the robot's reasoning abilities; *(ii)* **quantitative measures** such as the "number of times" an observation was correct, as competitors tended to use phrases such as "most of the time"; and *(iii)* **observations that were missed**, since competitors tended to use phrases such as "I missed that". In addition, we could take into account the ground truth, i.e., actual changes that were made by participants h_1 and h_2. Interestingly, our scenario always allows for the ground truth to be available.

The evaluation criteria established above can be further used to derive evaluation matrices, such as the one we derived in Fig. 4. This matrix relies on comparing observations reported by the human and the robot competitor against the ground truth, i.e., input from the participants who made the changes. The matrix qualifies the robot's intelligence relative to the human competitor as being highest (i.e., entry "$(High, High)$" in the matrix) when the human competitor guesses incorrectly and the robot guesses correctly, and as being lowest (i.e., entry "(Low, Low)") when the opposite happens. Missed observations are placed in the middle of the matrix and the agent who failed to observe the change is given the "benefit of the doubt", as such failures do not necessarily mean that the agent was incapable of making the correct deduction, nor that the agent has made an incorrect one. One can also come up with quantitative measures based on the matrix, e.g. $RI = \Sigma_{j=1}^{m} \Sigma_{i=1}^{n} ((val_i^x + val_i^y))$, where $val_i^x, val_i^y \in [1,3]$ (1 is the lowest) are the x and y axes values from Fig. 4, n is the number of environmental changes that needed to be observed in the game, and m is the number of competition runs. Hence, RI can indicate the "relative intelligence" of the robot. Below we illustrate interesting instances of some of the criteria in the evaluation matrix.

Incorrect deductions by the robot. The robot's deduction regarding how the *Jido* tape was moved was incorrect, whereas the human's was correct: *Filip* handed over the object to *Romain*, who then placed it on the table.

Incorrect deductions by the human. The human thought that the *grey* tape was not visible from the perspectives of the two seated participants, whereas in reality *Filip* was able to see it, which the robot deduced correctly.

Correct deductions by both. The robot correctly deduced the position of the missing object (*Jido tape*) as being behind the white box. This was done by analyzing the *Taskability Graph* for hiding an object, with the assumption that humans place objects on flat horizontal surfaces. The human also provided the same symbolic position description of the missing object.

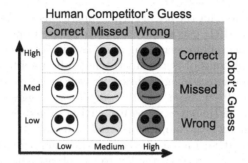

Fig. 4. A possible evaluation matrix for qualifying the robot's intelligence relative to the human competitor. Darker and happier faces denote higher intelligence.

Missed observations. Sometimes the human failed to notice when the object was moved by only a small amount, e.g. *5 cm*, whereas the robot was able to notice such small changes, because it stores a precise geometric model of the environment. On the other hand, in certain other runs, the robot failed to notice the presence of an object, because it had been placed in an orientation that prevented its tag from being detected by the robot. Consequently, the robot assumed that the object was hidden and tried to deduce its position.

6   Conclusion and Future Work

We have presented a novel human-robot competition scenario and methodology for evaluating the resulting relative intelligence of the robot (with respect to the human competitor), when certain basic building blocks of HRI reasoning, i.e., perspective taking, reachability, affordance, and effort analyses need to function together. This paper sets the stage for benchmarking and evaluation of such human-aware reasoning capabilities, which we think is now crucial. We cannot yet claim to have a conclusive set of evaluation metrics, nor do we want to impose one so early, as we believe that this has to be based on community feedback, and driven by extensive benchmarking competitions. We have only pointed out the feasibility of having such metrics, based on user studies and through the scenario presented in this paper. We have also extracted a set of criteria via the subjective evaluation and judgment of users, in order to stimulate interest in evaluation criteria for high-level intelligence.

In the near future, we aim to work in close collaboration with various competition organizers, e.g. RoboCup@Home, in order to enhance the proposed framework and the evaluation metrics, in the context of the proof of concept system presented here. It might also be interesting to develop a similar game for robot-robot competitions, to evaluate and compare the levels of human-aware reasoning capabilities of different robots, and thereby contribute to a standard competition for evaluating such capabilities, and a standard benchmark test for socially intelligent robots of the future.

References

1. Csibra, G., Volein, A.: Infants can infer the presence of hidden objects from referential gaze information. Br. J. Dev. Psychol. **26**(1), 1–11 (2008)
2. Rochat, P.: Perceived reachability for self and for others by 3 to 5-year old children and adults. J. Exp. Child Psychol. **59**, 317–333 (1995)
3. Breazeal, C., Berlin, M., Brooks, A.G., Gray, J., Thomaz, A.L.: Using perspective taking to learn from ambiguous demonstrations. Robot. Auton. Syst. **54**, 385–393 (2006)
4. Trafton, J.G., Schultz, A.C., Bugajska, M., Mintz, F.: Perspective-taking with robots: experiments and models. In: IEEE International Workshop on Robots and Human Interactive Communication (RO-MAN), pp. 580–584 (2005)
5. Marin-Urias, L., Sisbot, E., Pandey, A., Tadakuma, R., Alami, R.: Towards shared attention through geometric reasoning for human robot interaction. In: Proceedings of 9th IEEE-RAS International Conference on Humanoid Robots (Humanoids), pp. 331–336 (2009)
6. Gibson, J.J.: The theory of affordances. In: The Ecological Approach to Visual Perception, pp. 127–143. Psychology Press (1986)
7. Clark, A.: An embodied cognitive science? Trends. Cogn. Sci. **3**(9), 345–351 (1999)
8. Stoytchev, A.: Behavior-grounded representation of tool affordances. In: Proceedings of ICRA, pp. 3060–3065 (2005)
9. Ugur, E., Dogar, M.R., Cakmak, M., Sahin, E.: Curiosity-driven learning of traversability affordance on a mobile robot. In: IEEE International Conference on Development and Learning (ICDL), pp. 13–18, July 2007
10. Lopes, M., Melo, F.S., Montesano, L.: Affordance-based imitation learning in robots. In: Proceedings of IROS, pp. 1015–1021 (2007)
11. Robocup. http://www.robocup.org/
12. DARPA robotics challenge (DRC). http://www.theroboticschallenge.org/
13. European land-robot trial (elrob). http://www.elrob.org/
14. Humabot robot competition. http://www.irs.uji.es/humabot/
15. AAAI grand challenges. http://www.cs.utexas.edu/users/kuipers/AAAI-robot-challenge.html
16. Robocup-home. http://www.robocup.org/robocup-home/
17. Pandey, A.K., de Silva, L., Alami, R.: A novel concept of human-robot competition for evaluating a robot's reasoning capabilities in HRI. In: International Conference on Human-Robot Interaction (HRI), pp. 491–492 (2016)
18. Pandey, A.K., Alami, R.: Affordance graph: a framework to encodeperspective taking and effort based affordances for day-to-day human-robotinteraction. In: Proceedings of IROS, pp. 2180–2187 (2013)
19. Gardner, D.L., Mark, L.S., Ward, J.A., Edkins, H.: How do task characteristics affect the transitions between seated and standing reaches? Ecol. Psychol. **13**(4), 245–274 (2001)
20. Fleury, S., Herrb, M., Chatila, R.: Genom: a tool for the specification and the implementation of operating modules in a distributed robotarchitecture. In: Proceedings of IROS, pp. 842–848 (1997)
21. Simeon, T., Laumond, J.-P., Lamiraux, F.: Move3D: a generic platform for path planning. In: 4th International Symposium on Assembly and Task Planning, pp. 25–30 (2001)

The Effects of Cognitive Biases in Long-Term Human-Robot Interactions: Case Studies Using Three Cognitive Biases on MARC the Humanoid Robot

Mriganka Biswas[✉] and John Murray

University of Lincoln, Lincoln, UK
mrbiswas@lincoln.ac.uk

Abstract. The research presented in this paper is part of a wider study investigating the role cognitive bias plays in developing long-term companionship between a robot and human. In this paper we discuss, how cognitive biases such as misattribution, Empathy gap and Dunning-Kruger effects can play a role in robot-human interaction with the aim of improving long-term companionship. One of the robots used in this study called MARC (See Fig. 1) was given a series of *biased* behaviours such as forgetting participant's names, denying its own faults for failures, unable to understand what a participant is saying, etc. Such fallible behaviours were compared to a non-biased baseline behaviour. In the current paper, we present a comparison of two case studies using these biases and a non-biased algorithm. It is hoped that such humanlike fallible characteristics can help in developing a more natural and believable companionship between Robots and Humans. The results of the current experiments show that the participants initially warmed to the robot with the biased behaviours.

Keywords: Human-robot interaction · Cognitive bias in robot · Imperfect robot · Human-robot long-term interactions

1 Introduction

The study presented in this paper seeks to better understand human-robot interaction and with selected 'cognitive biases' to provide a more human-preferred interaction. Existing robot interactions are mainly based on a set of well-ordered and structured rules, which can repeat regardless of the person or social situation. This can lead to interactions which might make it difficult for humans to empathize with the robot after a number of interactions. The research presented in this paper tests the Misattribution, Empathy gap and Dunning-Kruger biases on a life-size humanoid robot, see Fig. 1.

According to Breazeal (2003), a social robot should be socially intelligent and should have sufficient social knowledge. To develop social intelligence in social robots, researchers study various methods to allow a robot to adapt to human-like behaviour based social roles. Some of these more popular methods suggest developing human-like attributes in robots, such as, trait based personality attributes, gesture and emotions expressions, anthropomorphism. Dautenhahn et al. (2009) investigated the identifying links between human personality and attributed robot personality where the team

© Springer International Publishing AG 2016
A. Agah et al. (Eds.): ICSR 2016, LNAI 9979, pp. 148–158, 2016.
DOI: 10.1007/978-3-319-47437-3_15

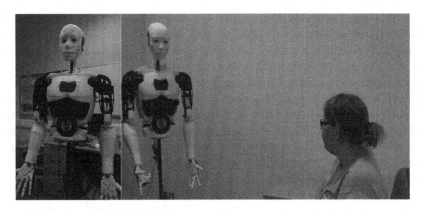

Fig. 1. MARC the humanoid robot, and participant interacting with MARC

investigated human and robot personality traits as part of a human-robot interaction trial. Research by Lee et al. (2006) showed that developing cognitive personality and trait attributes in robots can make it more acceptable to humans, also expressing emotions and mood changing in interactions can help to make the attachment bond stronger between user and the robot. Meerbeek et al. (2009) designed interactive personality process in robots which was based on Duffy's anthropomorphism idea. Duffy (2003) suggested that anthropomorphic or lifelike features should be carefully designed and should be aimed at making the interaction with the robot more intuitive, pleasant and easy. Reeves and Nass (2000) have shown that users usually show biased driven certain personality traits to machines (PC & others) and from that research they proposed a 'user driven' mental model for the domestic robots. Walters et al. (2009) investigated people's perceptions of different robot appearances and associated attention-seeking features in video-based Human Robot Interaction trials. In Michigan State University, research has been carried out on the Extraversion trait with Sony AIBO focused on 'extrovert' and 'introvert' characteristics (Lee et al. 2006). Their research found the same complimentary attraction effect between the participants and the robot dog. Moshkina et al. (2009) in Samsung Research Lab has developed a cognitive model which includes traits based personality, attitudes, mood and emotions in robot (TAME).

The above studies discuss various approaches to making a robot more human-like so that it would be easy for people to interact with the robots. However, it is challenging for a robotic system to become relevant and highly individualized to the special needs of each user in the particular beneficiary population (Tapus 2007). As such, we investigate a rather different approach, which is by applying selected cognitive bias to provide a more humanlike interaction. A cognitive bias is a type of error in thinking that occurs when people are processing and interpreting information in the world around them. Cognitive biases have reasonable amount of influence in human's characteristics and behaviours (Wilke and Mata 2012). Human behaviours are unique based on individual's thinking, genetics, social norms, culture (Haselton et al. 2005). Kahneman and Tversky (1972) suggested that human thinking is affected by a variety range of biases which influence humans in making wrong decision, bad judgments and other fallible actions. Such differences in cognitive imperfectness can hugely affect to make the individual's interactions unique, natural and human-like. But in developing a human-like robot, we sometimes ignore such facts and make the robot without human-like cognitive imperfections. Such cognitive biases (e.g. forgetfulness, making mistakes) have not been tested in robots and long-term human-robot interactions.

In the experiments presented in this paper, three interactions were performed between the participant and the robot spanning a two-month time period in order to provide 'long-term time period' effects. In this paper we introduce a model demonstrating biased behaviours in the robot MARC (Fig. 1) and, how such biased behaviours influences the interactions with the participant. At the end of the experiments, we compare participant's responses between the robot with a biased behaviour and the robot without the bias, showing the impact on long-term human-robot interaction the bias creates.

2   The Model: Imperfect Robot Using Cognitive Bias

Social robots take their persona from humans, as robots become much more popular in society, there is a need to make them much more sociable. But in human interactions, people usually meet with others and are able to form different relationships. In existing social robotics, the robots are imitating human's social queues for example: eye-gazing, talking and body movements etc. But it is the human's behavioural neutrality which includes faults, unintentional mistakes, task imperfections etc. which is absent in these social robots.

The earlier mentioned research focus on implementing human-like trait, personalities, emotions attributes in robots in order to develop human-robot companionship. Many of those robots are able to present social behaviours in human-robot interactions but unable to show human-like cognitively imperfect behaviours. Or 'human-like behaviours' are presented in such manner that participants cannot relate themselves with the robot. Baxter et al. (2012) pointed out the memory issue with robots that, outside the laboratory the real-time human-like interactions are still a challenge for the robots, yet social companion robots are unable to interact in continuous, extended social interactions beyond time-limits and still lack to adapt to user's interactive behaviours based on previous interactions. However, forgetting information and misattribution are also human behaviours which are widely common. In our experiments, we study such imperfect and biased behaviours of humans and develop in robots. The experiments described in this paper are influenced by the previous experiments (Biswas and Murray 2015), where we studied this novel idea for human-robot interaction to make the robot's interactive behaviours more familiar to humans. In this paper, we present the experiments of using three biases in a 3D printed humanoid robot MARC (Fig. 1).

Three cognitive biases were tested using two methodologies, such biases are, misattribution, empathy gap and Dunning-Kruger.

Misattribution happens when someone remembers something accurately in part, but misattributes some details. This memory bias explains cases of unintentional plagiarism, in which a writer passes off some information as original when he or she actually read it somewhere before (Aronson et al. 2005). In our experiments, the biased behavioural attributes were taken from misattribution are false memory, source of confusion and total forgetfulness.

Empathy gap is a cognitive bias which influences people to misunderstand the power of urges and feelings, such as, pain, hunger, fatigue (Nordgren, P, 2006, 2009), sexual arousal (Ariely and Loewenstein 2006), and cravings (Sayette et al. 2008) - on their behaviour. For example, when one is angry, it is difficult to understand what it is like for one to be happy, and vice versa. In our experiment, we tested two attributes of empathy gap bias, such as, hot state of empathy gap and cold state of empathy gap.

In 1999 psychologist David Dunning and Justin Kruger described the Dunning-Kruger effect as "...incompetent people do not recognize—scratch that, cannot recognize—just how incompetent they are," (Kruger and Dunning 1999). The Dunning-Kruger effect bias happens when an unskilled individual mistakenly suffers from illusory superiority and set their skill level much higher than actual. In our experiment the algorithm was based on three main behavioural characteristics of Dunning-Kruger bias, such as, unable to understand own lack of knowledge, unable to recognize other's true knowledge and, recognize and understand own lack of knowledge.

3   Experiment Setup

A. Algorithm Constructions

The experiments presented in this paper compare the robot's three different biased behaviours with the robot's 'baseline non-biased behaviour' in three conversational interactions. To do that, we develop the robot's baseline behaviour without the effects of the selected biases.

All the interactions were designed in three steps, such as, meeting and greeting, topic based conversation and farewell:

(a) Meet and greet – this begins when participant enters in the room and goes up to the point when the robot finishes initial greetings.
(b) Topics and conversation – this is the body of the interaction where the robot and participant discuss about various topics.
(c) Farewell – this is the part where the robot says good bye to the participant and invites for the next interaction.

The conversation was designed based on topic based question-answer dialogues. The dialogues in each topic were followed four steps, such as:

a. Robot asks a question/says something
b. Participant responds
c. Robot states its own opinion
d. Robot waits for participant's response/move to next dialogue.

For example, MARC asks, "Do you like football?" The participant can respond as "yes" or "no", and also can extend their responses, but whatever participant's responses are, the robot would say something after responses based on the algorithm developed (e.g. biased or non-biased).

The robot would then wait for few moments to check if the participant wants to say something, otherwise it moves to the next dialogue. The differences in biased and

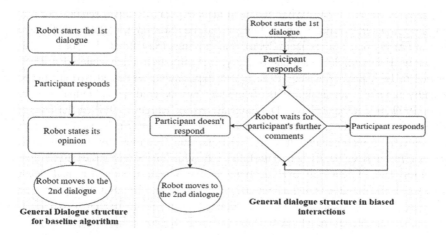

Fig. 2. Conversation designs for baseline and biased algorithms

baseline conversations in topics were made in the step C above, where the robot says something after the participant responds. In the baseline, the robot mainly says 'Okay' or 'That is great' and moves to the next topic, but in the biased interaction, robot's dialogue reflects the bias effects, and the topic could continue further depends on the participant's further responses. Figure 2 shows the dialogue structures and general differences between baseline and a biased algorithm.

The differences in conversation of biased and baseline algorithms were made in all the steps in the interactions. For example, in the 1st interaction's meet and greet stage, there could be only three dialogues for the baseline algorithm, such as, (1) Hello, (2) My name is MARC, what is your name? and (3) Nice to meet you. But, in the case of the biased algorithm, the robot's dialogues would be changed based on the bias, such as, for Empathy gap, the robot can be over joyed or over sad to show the bias effects (hot-cold empathy). Therefore, the dialogues can be, (1) Hello my friend! I am very happy to see you today. It's such a beautiful day. I hope you are feeling great today. (2) Hi. Today I am not feeling very good.

B. Experiment Methodology

The test the misattribution bias the robot needs to forget previous information. Therefore, there was made an introductory interaction at the beginning where the robot collected information from participants to misattribute in later interactions. Followed by the introductory interaction there were three times of misattributed interactions maintaining at least a week or two time intervals.

The empathy gap bias tested using two algorithms for empathy gap bias – hot state of empathy and cold state of empathy. Such algorithms were assigned randomly for the participants in three-time interval experiments.

Dunning-Kruger effects bias was tested in the three-time interval interactions.

There were two different methodologies applied for the interactions. Experiments therefore, were performed in two separate groups (Fig. 3) where one group of

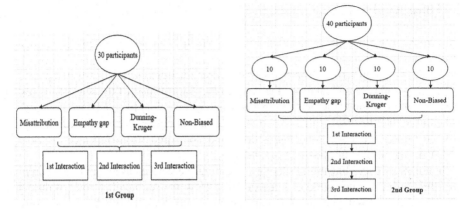

Fig. 3. Group setups for the experiment

participants interacted on all four algorithms and the other group interacted with only one algorithm at a time for three times. This was to aid in finding out the participant's reactions on two different occasions, such as participants who interacted all biased and baseline algorithms for three times and participants who interacted with only one algorithm for three times.

A total of 70 participants were randomly chosen from responses to advertisements. The number of different gender races and age groups were maintained equal for both groups. For the first group (Fig. 3 left side), 30 participants were selected where each participant interacted with all four algorithms (three biased and unbiased) of the robot. In the second group (Fig. 3 right side), 10 participants from total 40 were selected to interact with each of the individual algorithms (individual biased or unbiased) throughout the experiments. As with the first group, all 30 participants were interacting with all biased and unbiased algorithms, so their responses would be based on the comparisons between the biased and unbiased interactions. Such responses would reflect a comparable outcome between those who used cognitive bias as well as the unbiased algorithms in developing a long-term interaction. In the second group, each of the 10 participants interacted with their selected individual algorithm three times. Such interactions could tell us the effects of each individual algorithm in developing long-term interactions with the robot.

The experiment was Wizard of Oz. In the interactions each participant interacted with MARC one-to-one for at least 8 to 10 min. These conversational interactions ended by a request to fill in questionnaires from the robot.

Data was collected from Likert questionnaires (scale 1–7) based on the three factors, comfort – how comfortable participants were during the interaction; likeability – how much they liked the interactions; rapport – how involved they were in the interactions. Participants were given a questionnaire which where comprised of the following dimensions: participant's experience likability - 8 items, comfort - 6 items (Hassenzahl 2004) and rapport - 15 items (Mutlu et al. 2006). Such dimensions were chosen to understand participant's closeness and involvements in interactions, and to find out their preferences.

4 Data Analysis

The Cronbach's alpha (α) is calculated 0.916, which indicates high level of internal consistency for our scale.

For the 1st group, as the 30 participants did all interactions, we ran one-way repeated ANOVA to compare and analyses the data. In the comfort part of the questionnaires, there were six questions for the participants which were mainly to understand how easy and comfortable they feel with the robot in the different biases. We calculated the total average ratings from all three interactions and compared using repeated measure ANOVA. The results are shown in Fig. 4A.

In the experiences likeability sections, questions were asked to find out how partic..ipant felt during the interactions. For example, "How much confident you felt during the interaction?", "Will you visit for another conversation with the robot?" As previous, we ran repeated measure ANOVA and the results are shown in Fig. 4B.

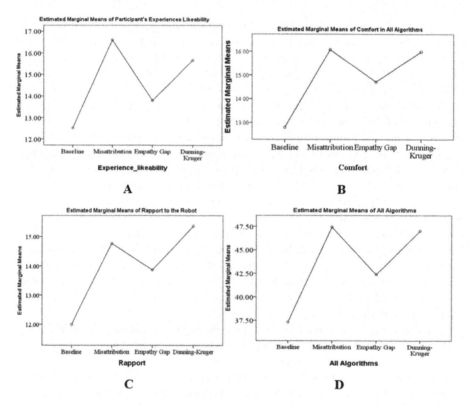

Fig. 4. The mean graphs of the different dimensions and different algorithms for the 1st group. 1. **A** – Shows the 'Comfort' dimension Means in different algorithms. 2. **B** – Shows the 'Participant's experiences likeability' dimension means in algorithms. 3. **C** – Shows the 'Rapport to the robot' dimension means in different algorithms. 4. **D** – Shows overall means of the participant's ratings in all three dimensions in algorithms.

The overall likeness of the participant towards the robot was calculated the total average ratings from all three interactions and compared using repeated measure ANOVA, shown in Fig. 4C.

The total mean graph is shown in Fig. 4D. The means of each algorithm types were calculated by adding up all the ratings from participants. The Baseline the mean is 37.31, where Misattribution approx. 47.43, Empathy gap approx. 42.37 and Dunning-Kruger approx. 47.0, - in all the biased algorithms participants rated high in all three factors of the questionnaires. The lowest mean difference is between Empathy Gap and baseline algorithms which is 5.06 (42.37–37.31) and the highest mean difference is between Misattribution and baseline algorithms which is 10.12 (47.43–37.311). Such differences in means indicate that the participants rated higher in biased algorithms (5–10 points) than baseline algorithms. However, there are differences in ratings in between the biased algorithms. In Fig. 4D, the Y axis is 'Estimated Marginal Means' and X axis shows the types of the algorithms. In all the pairwise comparisons, the Sig (p value) came out as <0.05 i.e. a very small probability of this result occurring by chance, under null hypothesis of no difference. In Fig. 4D it can be seen that each of the dimensions, participant's ratings were varied, but compared to baseline participants rated much higher in biased algorithms.

For the 2nd Group, a mixed (3 × 4) ANOVA was carried out on the dimensions (3) and algorithms (4). The analysis show changes in dimensions between algorithms, such as, Comfort dimension Mean ratings has increased from 3.42 (baseline) to 5.58 (Dunning-Kruger), for Experience likeability dimension Mean ratings has increased from 4.1 (baseline) to 5.17 (misattribution) and, for Rapport dimension, Mean ratings has increased from 3.75 (baseline) to 5.34 (misattribution), which are statically significant increases of 2.17, 1.07 and 1.59 (95 % Confidence Interval, $p < 0.0005$). Figure 5A shows plotting three dimensions in all algorithms. As clearly seen in the graph that the participants rated much higher for biased interactions than the baseline interaction.

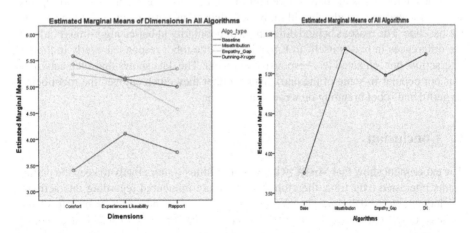

Fig. 5. The mean graphs of the different dimensions and algorithms for 2nd Group. **A.** The above graph shows Means of the ratings based on 3 dimensions in all algorithms. **B.** The above graph shows Means of the total ratings in all algorithms.

Figure 5B shows the average Means plots from each algorithms in all the three experiments. The graph was generated in the repeated measure test in SPSS using post-Hoc analysis which shows that misattribution gained the highest point of the calculated Mean and, baseline shows the lowest point of Mean. In fact, all biased algorithms Means are higher than the baseline.

5   Discussion

The statistics show that MARC with bias algorithms is more likely to result in participant preference. Participants enjoyed their conversation and expressed their experiences and involvement in the questionnaire feedback. Participants in both groups rated much higher in all the biased interactions. Although, there are differences in ratings for different biases, overall however, biased algorithms were most popular than the baseline algorithm. The reasons could be the differences in different biased robot's behaviours. Participants found it very interesting that the robot could actually forget information, can make mistakes, brag about things which it doesn't have actual knowledge – all these behaviours are very common in humans but it was novel to see in a Robot. Participant's reactions showed that they were very surprised and enjoyed the fact that a robot could indeed have humanlike faulty characteristics. However, there were differences in likeness to such behaviours and that's why the all ratings of biased interactions varied. For some of the participant's robot's forgetfulness behaviours was more enjoyable than its idiotic (Dunning-Kruger bias) behaviours, and some participants preferred empathy gap behaviours more than forgetfulness.

In our experiments, we can see that the participants felt more personable with the robot when the robot forgot and misattributed information, showed imperfect activities during interactions. In case of the baseline algorithm, the robot did not argue or forget anything, it responded as it should. As such, participants did not find the baseline algorithm as interesting as the biases. Even for the 2nd group, where a group of participants were dedicated to a single algorithm, biased algorithms participants rated higher than the baseline. The reasons behind differences of popularity in biased algorithms could be the differences in biases itself. In Empathy Gap, the robot responded overly in the 1st interaction, but remained less responsive in others. The data shows that such activities was not popular to some of the participants, rather they much enjoyed the robot being forgetful and robot bragging on wrong information.

6   Conclusion

The experiment show that MARC with bias algorithms is more likely to keep the participants interested over time, therefore become more influential regarding interactions. Participants enjoyed the biased conversation with the robot and expressed their feedback in favor of the biased algorithm. We can see that the robot with baseline algorithms, developed a preliminary attachment with the participants, but robot with biased algorithms achieved more popularity among the participants.

We realize that robot's imperfect biased behaviours also relies in part on the robot itself, i.e., the way MARC communicates. So questions may rise about the effects and relations between the robot's abilities and imperfect behaviours shown by it. To answer this question, previously experiments were done with misattribution and empathy gap biases using the ERWIN and MyKeepon robots (Biswas and Murray 2015) and currently with humanoid robot MARC. In both cases, participants liked the interactions using biased algorithms more than the robot without bias. From these experiments with different robots but same biases it can be said that the cognitive bias can actually increase participant's likeability to the robot significantly and, such increases can help to develop human-robot long-term interactions.

References

Ariely, D., Loewenstein, G.: The heat of the moment: the effect of sexual arousal on sexual decision making. J. Behav. Decis. Making **19**, 87–98 (2006)

Aronson, E., Wilson, T., Akert, R.: A Textbook of Social Psychology, 6th edn. Prentice-Hall, Scarborough (2005)

Baroni, I. et al.: What a robotic companion could do for a diabetic child. In: RoMAN 2014 (2014)

Baxter, P., et al.: Long-term human-robot interaction with young users. In: Proceedings of the IEEE/ACM HRI-2011, Lausanne (2012)

Biswas, M., Murray, J.: Towards an imperfect robot for long-term companionship: case studies using cognitive biases. In: IEEE/RSJ; IROS, September 2015, Hamburg (2015)

Breazeal, C.: Social interactions in HRI: the robot view. IEEE Trans. Syst. Man Cybern. **34**(2) (2003)

Dautenhahn, K., et al.: KASPAR – A minimally expressive humanoid robot for human robot interaction research. Appl. Bionics Biomech. **7**(4), 369–397 (2009)

Duffy, R.B.: Anthropomorphism and the social robot. RAS **42**, 177–190 (2003)

Gross, M., et al.: Methodology for assessing bodily expression of emotion. J. Nonverbal Behav. **34**(4), 223–248 (2010)

Haselton, M.G., Nettle, D., Andrews, P.W.: The evolution of cognitive bias. In: Buss, D.M. (ed.) The Handbook of Evolutionary Psychology. Wiley, Hoboken (2005)

Hassenzahl, M.: The Interplay of beauty, goodness, and usability in interactive products. Hum. Comput. Interact. **19**, 319–349 (2004)

Kahneman, D., Tversky, A.: Subjective probability: a judgment of representativeness. Cognit. Psychol. **3**(3), 430–454 (1972)

Kruger, J., Dunning, D.: Unskilled and unaware of it: how difficulties in recognizing one's own incompetence lead to inflated self-assessments. JPSP **77**(6), 1121–1134 (1999)

Lee, K., et al.: Can robots manifest personality social responses, and social presence in human robot interaction. J. Commun. **56**(4), 754–772 (2006)

Meerbeek, B., et al.: Iterative design process for robots with personality. In: AISB Symposium on New Frontiers in Human-Robot Interaction Edinburgh, pp. 94–101 (2009)

Moshkina, L., Arkin, R.C., Kee, J.K., Jung, H.: Time-varying affective response for humanoid robots. In: Kim, J.C. (ed.) Progress in Robotics. Communications in Computer and Information Science, vol. 44, pp. 1–9. Springer, Heidelberg (2009)

Mutlu, B., et al.: Perceptions of ASIMO: an exploration on co-operation and competition with humans and humanoid robots. In: Conference on Human – Robot Interaction (HRI 2006) (2006)

Nordgren, L.F., van der Pligt, J., van Harreveld, F.: Visceral drives in retrospect: Explanations about the inaccessible past. Psychol. Sci. **17**, 635–640 (2006)

Nordgren, L.F., van Harreveld, F., van der Pligt, J.: The restraint bias: How the illusion of self-restraint promotes impulsive behaviour. Psychol. Sci. **20**, 1532–1528 (2009)

Reeves, B., Nass, C.: Perceptual Bandwidth. Commun. ACM **43**(3), March 2000

Sayette, M.A., et al.: Exploring the cold-to-hot empathy gap in smokers. Psychol. Sci. **19**, 926–932 (2008)

Shepperd, J., et al.: Exploring causes of the self-serving bias. Soc. Pers. Psychol. Compass **2**(2), 895–908 (2008)

Tapus, A., Mataric, M.J., Scassellati, B.: Socially assistive robotics. IEEE Robot. Autom. Mag. **14**(1), 35–42 (2007)

Walters, M.L., et al.: Avoiding the uncanny valley: robot appearance, personality and consistency of behavior in an attention-seeking home scenario for a robot companion. Auton. Robot. J. **24**(2), 159–178 (2008)

Wilke, A., Mata, R.: Cognitive Bias. The Encyclopedia of Human Behavior **1**, 531–535 (2012)

Ethical Decision Making in Robots: Autonomy, Trust and Responsibility

Autonomy Trust and Responsibility

Fahad Alaieri[2,3] and André Vellino[1]([⊠])

[1] School of Information Studies, University of Ottawa, Ottawa, Canada
avellino@uottawa.ca
[2] Electronic Business Technologies, University of Ottawa, Ottawa, Canada
falai055@uottawa.ca
[3] Management Information Systems, Qassim University, Buraydah, Saudi Arabia

Abstract. Autonomous robots such as self-driving cars are already able to make decisions that have ethical consequences. As such machines make increasingly complex and important decisions, we will need to know that their decisions are trustworthy and ethically justified. Hence we will need them to be able to explain the reasons for these decisions: ethical decision-making requires that decisions be explainable with reasons. We argue that for people to trust autonomous robots we need to know which ethical principles they are applying and that their application is deterministic and predictable. If a robot is a self-improving, self-learning type of robot whose choices and decisions are based on past experience, which decision it makes in any given situation may not be entirely predictable ahead of time or explainable after the fact. This combination of non-predictability and autonomy may confer a greater degree of responsibility to the machine but it also makes them harder to trust.

Keywords: Robot ethics · Autonomy · Trust · Responsibility

1 Introduction

Many aspects of robot behavior are ethically relevant. Robots may be either ethical patients, i.e. the subject of ethical behaviour by others (people or other robots), or ethical agents whose actions have ethical consequences. There are also questions of ethics in the design and the use of robots, such as whether and how robots are deployed, either for military purposes or to save human lives [6].

Our focus in this paper is robots viewed as ethical agents, i.e. moral reasoners with a sufficiently high degree of autonomy that enables them to make choices with ethical implications. We aim to provide both a framework for understanding the concept of autonomy in robotic devices and to analyze the process of choice, i.e. the making of decisions that characterizes the ethical dimension of their actions. We also indicate how these features of autonomy and choice and especially how these choices are made, could have implications for ascribing moral responsibility to robots.

© Springer International Publishing AG 2016
A. Agah et al. (Eds.): ICSR 2016, LNAI 9979, pp. 159–168, 2016.
DOI: 10.1007/978-3-319-47437-3_16

2   Autonomy

Our working definition of a robot is a task-oriented device that has sensors and other information input interfaces, which is able to physically alter its environment, move, and have both the energy and ability to make decisions about how to accomplish its tasks. A key feature in a robot is whether its ability to make decisions is autonomous, i.e. whether it has the ability to operate without external intervention. From the point of view of its ethical decisions, autonomy is important because it is a necessary condition for ethical agency. While some argue that an autonomous robot cannot be considered truly autonomous unless it makes *all* its decisions without any human intervention, we prefer to say that such robots are not only autonomous but also *independent*.

One key characteristic of an autonomous robot is whether it is able to respond appropriately to a wide variety of situations. A machine that requires no external input to make a decision but is only ever able to make one decision could not be said to have a meaningful degree of autonomy. For example, a collaborative robot like Baxter, which is used to repeatedly perform only very specific tasks, exhibits some degree of autonomy but does not have the ability to make complex decisions that depend on highly variable environmental conditions and is unable to handle unpredictable situations.

2.1   Ethical Decision Making

Our discussion of autonomy relies on a model of the steps that a robot undergoes in the process of committing an action. Following the sense-plan-act robotics paradigm (see [8]) we propose a 5-stage model of the information processing in a robot: (i) obtaining the information (e.g. from sensors or telemetry); (ii) analyzing the information (e.g. by categorizing and integrating data); (iii) generating alternative courses of action (e.g. computing outcomes for a set of candidate decisions); (iv) selecting from among the alternatives (e.g. making a choice from among the candidates), and (v) performing an action that corresponds to this choice (e.g. activating an actuator) [1].

From the point of view of the ethical agency of a robot, the key stages are those that involve the generation of alternatives (iii) and the selection of a decision (iv). The hallmark of an ethical agent is that it has autonomy of choice in the decisions it makes: given a set of alternative courses of action from which the agent can choose, it has a method to select one. For a robot to be considered ethical, its actions need to not only conform to ethical norms but to perform these actions as a result of some process that morally obligates it to perform those actions. Thus a critical element in the decision selection, step (iv), is the ethical theory that is used to evaluate each alternative course of action. Although there are many ethical theories that can be adopted for the design of an ethical robot, here we consider only two: utilitarianism and deontological rules.

From the point of view of utilitarianism the value of an action is determined by the overall benefit of its consequences. Hence robots that have the ability to calculate the consequences of their actions and to evaluate the benefits they bring about must be considered ethical. For instance, an autonomous car that has the capacity to detect pedestrians on the streets and to avoid them or stop driving in order to not harm them, behaves ethically from this point of view.

If, instead, a robot were to use a deontological framework that expresses its moral and ethical duties, it would have to act according to ethical principles that are independent of the consequences of its actions. For example, a deontologically ethical robot could be instructed not kill or lie or cheat, or cause harm, no matter what the circumstances or consequences.

Bernard Gert proposed ten such deontological rules that determine which of its actions are permitted, obligatory or prohibited, independently of their consequences: do not kill, do not cause pain, do not disable, do not deprive of freedom, do not deprive of pleasure, do not deceive, keep your promises, do not cheat, obey the law, and do your duty [13]. This could lead to situations in which some rules are at odds [28] with one another. For example, a robot's obligation to keep its promises (such as the promise to keep a secret) may be at odds with the obligation do not deceive. In other words ethical robots could experience much the same kinds of dilemmas as humans do and would also require mechanisms to resolve them.

Thus the choice of moral theory that governs a robot's behaviour is determined by the robot's decision-processing capabilities. If it has the ability to look ahead, plan, and evaluate the "goodness" of outcomes, then it could be designed to implement utilitarian principles. If it is only able to obey rules, then it may be that a purely deontological approach is more suitable, notwithstanding the need for a method to resolve rule-conflicts.

2.2 Top-Down and Bottom-Up Decision Making

A slightly different but complementary characterization of the ethical decision making process in a machine has to do with *how* the machine arrives at its ethical conclusions. Allen et al. refer to these alternative decision-processes as the 'top-down' and 'bottom-up' methods [2]. In the top-down approach the robot programmer installs decision making algorithms that produce predictable outcomes: in essence, it embeds in the machine what a human being considers to be ethical behaviour, which then needs only to determine *when* it is appropriate to apply them.

In the bottom-up approach, the programmer builds an open-ended system that is able to collect information from its environment, to predict the outcomes of its actions, to select from among alternatives and, most importantly, has the capacity learn from its experience. Such a machine can be described as having the ability to learn what is right and what is wrong because it is capable of learning from its choices and mistakes: it has the ability to self-modify its decision-making system through the acquisition of experiences. As Allen et al. put it "Top-down approaches … involve turning explicit theories of moral behavior into algorithms. Bottom-up approaches involve attempts to train or evolve agents whose behavior emulates morally praiseworthy human behavior" [2].

A bottom-up approach can manifest in at least three ways: the robot could develop its own ethical decision selection methods by a process of trial and error (unsupervised learning); the machine's engineers could train the robot to learn pre-established moral rules (supervised learning); or the robot could adopt a hybrid learning method, which would allow it to keep learning from its experience and surroundings, but be grounded in pre-established principles.

For instance, a supervised learning method similar in nature to the neural networks in the Go playing program AlphaGo could be trained to learn to behave ethically by example with instances of situation-response pairs. In AlphaGo, this training step enables the computer to prune the space of possible Go moves from which it can choose (the so-called 'value networks' used to evaluate board positions) and then make a choice (using so-called 'policy networks') to evaluate which from among them is the best [11]. Both the 'value networks' and the 'policy networks' are trained from a large number of human games and the design methodology for such a game-player is a plausible model for how a bottom-up, learning, ethical reasoner might be trained.

3   Trust and Predictability in a Robot's Ethical Decisions

Robots in the future will have a greater capacity to perform even more tasks and an increasing number of these tasks will be related to people's safety, health, and even their lives. Hence people will have to develop confidence that robots are correctly obeying ethical principles if there is a risk that not following them could cause harm.

Two elements will contribute to this trust: humans will have to have repeated positive experiences with high-quality robot decisions and the decisions they make that obey ethical principles will have to be predictable and, retroactively, explainable. Without a coherent explanation for a robot's actions, a human would not be able to assess the validity of a robot's decision and therefore not have grounds for trusting it.

Yet, a robot that has the ability to modify the method by which it generates choice alternatives and calculates the consequences of its possible future actions may not be entirely predictable: its behaviour may become non-deterministic and how it came to make a choice may be complex, and hard for it or a human to explain. Tay, the Microsoft AI Twitter Chatbot, is an early example [7] of a hard-to-trust adaptable machine. It was designed to learn and adapt its (verbal) behaviour as a function of the input it received from its Twitter followers but its behaviour was not predictable by its programmers and it was easily 'vandalized' by people into uttering sexist and racist remarks.

It was not possible for Tay to conform to norms of ethical verbal behaviour because natural language understanding in machines has not yet reached the maturity required to deduce the consequences of verbal actions (such as the offense that can be caused by making racist remarks, which can be uttered in an infinite number of ways) let alone solved the problem of recognizing whether a remark would be considered racist or otherwise offensive.

But even if some of Tay's behaviour could have been moderated with (verbal) ethical norms, its unpredictability still poses a problem. It may be difficult for anyone, even programmers, to provide explanations for the behaviour of any machine whose behaviour is programmed 'bottom-up'. Consider for example, the choices made by decision-making algorithms such as those in AlphaGo. These are very hard to both predict and explain. When something goes wrong (or very well), such as when AlphaGo made some errors (or brilliant moves) in its recent games against the world champion Lee Se-dol, it was difficult, even for its developers, to know why it made those mistakes (or how it made some brilliant moves) [23]. This is a significant impediment to building human trust in a machine's ability to either generate an appropriate set of candidate-actions or to select the best from among them.

Suppose a machine had to make a choice in a complex utilitarian decision problem (e.g. a complicated version of the Trolley Problem [3]), in which a lot of options, choices and consequences had to be calculated. Suppose also that it functions perfectly and it makes the "right" decision and picks the morally correct course of action. It is conceivable that a human might not immediately recognize that this decision is optimal from the consequentialist point of view. A human (with limited computational ability) might conclude, incorrectly, that the robot's decision was unethical. But, without a coherent explanation for the robot's actions, a human would not be able to assess the validity of the robot's decision and therefore have no grounds for trusting it. On the other hand if the robot can explain its decision process in a way that a human can understand, that explanation could be the foundation for inducing human trust. Indeed, such explanations could be quite impressive to humans and eventually convince them of the robots' superior ability to make ethical decisions.

Such attempts at mapping machine-decision making into human-understand-able accounts of their actions has been attempted with conventional robot planners [20] and noted to be necessary components for ethical robots [26]. However, as Colombo and Hartmann [9] remark about Bayesian models of cognitive phenomena, "[they do] not reveal ... the causal structure of a mechanism". Thus a deontological, rule-based ethical framework for controlling a robot's ethical decisions could generate clear human-understandable explanations for its actions whereas it may be very hard to do the same for decisions made by adaptive machine-learning algorithms such as the neural networks in AlphaGo.

4   The Moral Responsibility of Robots

The questions of choice and autonomy have an important role to play in determining whether robots can be held morally responsible. As Stahl observes, the traditional debate about whether computers can be responsible hinges on the question of whether they satisfy the conditions for agency and person-hood [27]. Hence the question of whether a robot is making its own decisions and how those decisions are being made would determine, at least in part, whether or not it could or should be held responsible for its actions.

The question of a machine's moral responsibility has been addressed using two approaches: the classical approach and the pragmatic approach. The classical approach views machines as not responsible for their actions under any circumstance — because they are mechanical instruments or slaves. In the pragmatic approach, 'artificial morality' envisages some situations under which machines can be viewed as responsible for their choices [12]. In this view responsibility in artificially ethical agents is a "social regulatory mechanism".

Others have focused on how to enable responsibility in artificial agents by embedding ethical codes of conduct in them. If these codes of conduct are formulated by the robot's designers, then the responsibility for those rules lies squarely with the robots' designers and owners (assuming that the owner has been apprised of theses rules). However, if these rules of conduct — whether or not they can be formulated in human-intelligible terms — are arrived at from experience (i.e., 'bottom-up'), the burden of responsibility for mistakes is more evidently on the machine's shoulders. In the case of Tay, Microsoft assumed a kind of "meta-responsibility" for not having predicted the possibility that it could be crowd-hacked by malicious users who would coax it into verbal mis-behaviour. But many people saw its actual verbal mis-behaviours as being its fault—*it* was viewed as responsible for its racist utterances, not its manufacturer.

Jarvik's philosophical analysis divides human moral responsibility into three types: causal responsibility, role responsibility, and liability responsibility [16]. In causal responsibility, a person is responsible for everything that she has caused to happen: she is the cause of her actions. In role responsibility, a person's role in a certain area of society or community obligates them to perform a task, meaning that the task simply *is* the responsibility. The final form is liability responsibility, which identifies who is to be "praised or blamed" for certain actions or outcomes. Dodig-Crnkovic and Persson add one more critical element of moral responsibility besides causal responsibility: intention [12]. Causal responsibility may be assigned to non-humans, but, according to them, only humans have intentions. Insofar as malicious users intentionally fooled Tay into mis-behaving, the responsibility for its inappropriate comments also lies with them. As we noted above, we cannot say that Tay ever had the intention to offend and hence it is blameless.

Which of these types of responsibility can or cannot be assigned to robots? One consideration is that robots come in different varieties and not all types have the ability to shoulder responsibilities. For example, a highly autonomous robot could be said to have some causal responsibility because it is capable of making decisions that cause actions in a broad range of environments whereas the actions of a robot with low autonomy, typically caused by the human that controls it, would not.

Computers may be superior to humans in terms of the accuracy and quality of their decisions [4] because they have a greater ability to calculate all the consequences of an action that may be performed in a certain situation. So, with these innately superior capacities they might perform their social tasks [27] both perfectly and accurately and therefore be able to be *more* responsible than humans — at least in the sense of role responsibility — in so far as they are better able to perform tasks effectively. For example, Japan is experimenting with 'urban surveillance robots' that are responsible for identifying criminals and detecting unusual behaviours [25]. Bank fraud detection systems that are responsible for blocking customers' credit cards when they detect unusual purchase patterns are another example of role-responsible machines: their responsibility is to protect both the card owner and the bank's financial assets. Robotic decision systems are therefore assuming responsibilities because of their ability to calculate, detect, inspect, and track.

Autonomous robots that have the capacity to interact with their environment, make decisions, perform tasks, and calculate the consequences of their actions can be thought to be responsible for their actions if they are also able learn how they ought to behave as a result of their experiences ('bottom-up'). Robots are morally responsible for performing actions that lead to ethical consequences in those cases where the action-choice is determined by the selection of one from among several alternatives [12] and that choice is not deterministically programmed by humans.

Perhaps most importantly, it is the capacity that robots have to learn from their mistakes that that allows humans to assign responsibility to them [14]. A robot that learns from its experience and is able to improve its own decision-making system is more capable of being afforded responsibility. Asaro predicts that, in the future, autonomous robots will have a greater ability to come up with their own moral rules, goals, and reasoning methods, and that they will thus be equipped to make moral decisions that fulfill the moral responsibility which has been assigned to them. We believe that this is a sound prediction and firmly based on the evolution of autonomy in the robot industry [6].

4.1   Can Robots Be Responsible?

What would happen if, in the future, autonomous robots were given full responsibility for their actions and outcomes [21,24]? Some researchers including Deborah Johnson believe that it is dangerous to give robots full responsibility for their actions because they might go beyond the programmers' control [17]. They might be autonomous because they perform tasks without human control, but, according to this view, it is the humans — including the manufacturers, designers, programmers, and users who must take responsibility if anything goes wrong. In this case, the mistake that caused the harm is human and the robots cannot be held responsible for their actions.

According to Kuflik, responsibility does not rest with robots, because they are just machines running programs that are manifestations of (human) intentions [19]. Responsibility for any action performed by a robot may be divided amongst different people such as the robot manufacturer and the user, and each group will shoulder part of this responsibility [5,28]. Hew claims that, in the foreseeable future of technology, "robots will carry zero responsibility" for their actions, and that this responsibility should remain with humans. This is because "its rules for behaviour and the mechanisms for supplying those rules must not be supplied entirely by external humans" [15].

But also, abrogating responsibility by the robots' users and creators could encourage some people to create dangerous autonomous robots that may harm people or perform dangerous or unwanted tasks. Therefore, as Wallach argues, people and corporations should be held responsible for all harm that is caused by technology [29]. Kuflik agrees, concluding that the responsibility of robots' outcomes rests with the people who design them and who program their systems [19].

In some situations, users should shoulder all the responsibility if they use their robots intentionally for the purpose of harming others [10]. If a driver configures the autopilot system in an autonomous car to cause a collision, then the driver must take full responsibility for the consequences of the car's behaviour. Hence, if autonomous cars were given full responsibility for their actions it could be possible for evil people to fool them into harming others but fail to take responsibility for doing so. Hence, according to this argument, for every action performed by an autonomous robot, there should be a human agent who is held responsible in when something goes wrong.

One consideration when attributing liability responsibility to a machine is to ask who is responsible if the machine makes a mistake. Do we blame the machine, the manufacturer, the designer, the programmer, or the owner? Johnson argues that since artificial agents have become more autonomous and that nobody can fully predict their decisions, no one person can be held responsible for their actions [18]. This is all the more true if their actions cannot be fully explained.

The Ad Hoc Committee on Responsible Computing takes a different position. This committee crafted "The Rules", which were intended as ethical guidelines for computer professionals and state that people are answerable for their behavior when they produce or use computing artifacts, and that their actions reflect on their character. The first of these five rules states that "The people who design, develop, or deploy a computing artifact are morally responsible for that artifact, and for the foreseeable effects of that artifact." The third rule states "People who knowingly use a particular computing artefact are morally responsible for that use." [22]. Thus the people who design, develop, program, create, deploy, and use artificial agents are responsible for their agents according to their role in the action, decision, result, or harmful effects, at least to the extent to which these effects are "foreseeable".

5   Conclusions

If what constitutes an ethical choice in humans is either deliberating about the precedence of deontological rules amongst themselves in a given situation (e.g. what duty over-rides another) or analysing the consequences of a potential set of candidate options in a given situation and picking the one that optimizes a well-being function, then, in either case, these processes have a counterpart in the choice-behaviour of autonomous robots.

Therefore, autonomous robots can and will be ethical agents that are able to make ethical decisions. A key question is: will we be able to trust them if the methods by which they make decisions are opaque to humans? If their learning-by-experience algorithms have unpredictable consequences, will humans be able to trust them? Their unpredictability also means, symmetrically, that their actions may not be (easily) explainable—at least not in human terms.

Another key question is: who will be held responsible for the actions committed by autonomous ethical robots? If their actions are entirely predictable, then they are machines that are doing what they are programmed to do and the responsibility for the consequences of their actions must lie with the manufacturers and users. If, however, they are more like children who eventually learn to make their own decisions on the basis of experience and who induce their own deontological or utilitarian principles from a series of unsupervised learning processes, then they should be considered responsible for their actions.

The design of ethical robots that give them some degree of responsibility but also a sufficient degree of predictability to remain trustworthy might best be achieved with a hybrid strategy or a method that combines the 'bottom up' and the 'top down' approaches. An ethical robot built using a hybrid approach would have well-defined rules that predictably prevent catastrophic ethical failures, but also have the ability to learn new ethical principles from its experiences. A robot that is able to learn from its mistakes and perform utilitarian calculations to select one from among a set of alternative actions could thus be constrained by deontological rules that forbid it from considering some alternatives or oblige it to consider others, yet also perform consequentialist calculations more effectively than humans.

References

1. Alaieri, F., Vellino, A.: The Ethical Characteristics of Autonomous Robots. We Robot Poster, April 2016. http://hdl.handle.net/10393/34809
2. Allen, C., Smit, I., Wallach, W.: Artificial morality: top-down, bottom-up, and hybrid approaches. Ethics Inf. Technol. **7**(3), 149–155 (2005)
3. Allen, C., Wallach, W., Smit, I.: Why machine ethics? IEEE Intell. Syst. **21**(4), 12–17 (2006)
4. Anderson, M., Anderson, S.L.: Machine ethics: creating an ethical intelligent agent. AI Mag. **28**(4), 15–26 (2007)
5. Asaro, P.: Robots and responsibility from a legal perspective. In: IEEE International Conference on Robotics and Automation, Rome, Italy (2007)
6. Asaro, P.: What should we want from a robot ethic? Int. Rev. Inform. Ethics **6**, 9–16 (2006)
7. Bass, D.: Clippy's back: the future of Microsoft is Chatbots. Bloomberg Businessweek, March 2016. http://www.bloomberg.com/features/2016-microsoft-future-ai-chatbots/
8. Beer, J.M., Fisk, A.D.: Toward a framework for levels of robot autonomy in human-robot interaction. J. Hum.-Robot Interac. **3**(2), 74–99 (2014)
9. Colombo, M., Hartmann, S.: Bayesian cognitive science, unification, and explanation. Br. J. Philos. Sci., axv036 (2015)

10. Crabb, P.B., Stern, S.E.: Technology traps. In: Luppicini, R. (ed.) Ethical Impact of Technological Advancements and Applications in Society, pp. 39–46. IGI Global, April 2012
11. Silver, D., Huang, A., Maddison, C.J., et al.: Mastering the game of Go with deep neural networks and tree search. Nature **529**(7587), 484–489 (2016)
12. Dodig-Crnkovic, G., Persson, D.: Sharing moral responsibility with robots: a pragmatic approach. In: Tenth Scandinavian Conference on Artificial Intelligence, SCAI 2008, pp. 165–168 (2008)
13. Gert, B.: Morality: Its Nature and Justification. Oxford University Press, USA (1998)
14. Hellström, T.: On the moral responsibility of military robots. Ethics Inf. Technol. **15**(2), 99–107 (2012)
15. Hew, P.C.: Artificial moral agents are infeasible with foreseeable technologies. Ethics Inform. Technol. **16**(3), 197–206 (2014)
16. Jarvik, M.: How to understand moral responsibility? TRAMES: J. Humanit. Soc. Sci. **7**(3), 147–163 (2003)
17. Johnson, D.G.: Computer systems: moral entities but not moral agents. Ethics Inf. Technol. **8**(4), 195–204 (2006)
18. Johnson, D.G.: Technology with no human responsibility? J. Bus. Ethics **127**(4), 707–715 (2014)
19. Kuflik, A.: Computers in control: rational transfer of authority or irresponsible abdication of autonomy? Ethics Inf. Technol. **1**(3), 173–184 (1999)
20. Lomas, M., Chevalier, R., Vincent Cross II, E., Garrett, R.C., Hoare, J., Kopack, M.: Explaining robot actions. In: Proceedings of the Seventh Annual ACM/IEEE International Conference on Human-Robot Interaction, pp. 187–188. ACM (2012)
21. Malle, B.F.: Integrating robot ethics and machine morality: the study and design of moral competence in robots. Ethics Inf. Technol. **18**(70), 1–14 (2015)
22. Miller, K.W.: Moral responsibility for computing artifacts: "The Rules". IT Prof. **13**(3), 57–59 (2011)
23. Moyer, C.: How Google's AlphaGo Beat a Go World Champion. The Atlantic, March 2016. http://www.theatlantic.com/technology/archive/2016/03/the-invisible-opponent/475611/
24. Noorman, M., Johnson, D.G.: Negotiating autonomy and responsibility in military robots. Ethics Inf. Technol. **16**(1), 51–62 (2014)
25. Royakkers, L.: A literature review on new robotics: automation from love to war. Int. J. Soc. Robot. **7**(5), 1–22 (2015)
26. Scheutz, M., Malle, B.F.: think and do the right thinga plea for morally competent autonomous robots. In: 2014 IEEE International Symposium on Ethics in Science, Technology and Engineering, pp. 1–4. IEEE (2014)
27. Stahl, B.C.: Responsible computers? a case for ascribing quasi-responsibility to computers independent of personhood or agency. Ethics Inf. Technol. **8**(4), 205–213 (2006)
28. Tzafestas, S.G.: Roboethics. A Navigating Overview, vol. 79. Springer, Heidelberg (2016)
29. Wallach, W.: A Dangerous Master. How to Keep Technology from Slipping Beyond Our Control. Basic Books, New York (2015)

How Facial Expressions and Small Talk May Influence Trust in a Robot

Raul Benites Paradeda[✉], Mojgan Hashemian, Rafael Afonso Rodrigues, and Ana Paiva

INESC-ID & Instituto Superior Técnico, University of Lisbon, Lisbon, Portugal
{raul.paradeda,mojgan.hashemian,rafael.afonso.rodrigues,
ana.paiva}@tecnico.ulisboa.pt

Abstract. In this study, we address the level of trust that a human being displays during an interaction with a robot under different circumstances. The influencing factors considered are the facial expressions of a robot during the interactions, as well as the ability of making small talk. To examine these influences, we ran an experiment in which a robot tells a story to a participant, and then asks for help in form of donations. The experiment was implemented in four different scenarios in order to examine the two influencing factors on trust. The results showed the highest level of trust gained when the robot starts with small talk and expresses facial expression in the same direction of storytelling expected emotion.

Keywords: Human Robot Interaction · Trust · Social robotics · Facial expression · Small talk

1 Introduction

The emergence of social robots in our everyday life is increasing rapidly day in day out. This fact highlights the important role of Social Robotics, which targets the integration of robots in our daily lives. A good example of social robots could be the category of "assistive robots". In this case, it is almost obvious that the robot's actions may cause serious consequences to the people surrounding them [1]. Another example could be health-care robots as well as companion robots for elders, which has been addressed in recent literature [2, 3]. The foremost requirement for a social robot interacting with elders, family or medics, is its trustworthiness [3]. In other words, in this case, the concept of trust in robots becomes a central issue. However, this concept is not only important for social robots, but also for other types of robots such as service robots, military, or even in industrial robots [4].

Apart from Human Robot Interaction (HRI), the concept of trust has been investigated for decades in other fields, specifically psychology and social science. In these contexts, the trust is defined as a factor of human personality [5], which is the result of a choice among behaviors under specific situation. From another perspective, trust enables an individual to face accepting a level of risk associated with the interaction with another agent [6]. In this view, we can define trust as a feeling of confidence that another individual will not put him/herself at risk unnecessarily [7].

© Springer International Publishing AG 2016
A. Agah et al. (Eds.): ICSR 2016, LNAI 9979, pp. 169–178, 2016.
DOI: 10.1007/978-3-319-47437-3_17

Thereby, combining social robots and concepts of trust may raise a question: is it possible for a human to trust a machine? It is not surprising that the feeling of confidence felt by human subjects can turn robots into more collaborative partners [8]. This fact has motivated several studies in the field of social robotics to investigate factors influencing trust. Another motivating factor might be the fact that trust is entwined with persuasiveness in social and collaborative contexts. Hence, trust may directly affect people's inclination to cooperate with the robot, for instance by accepting given information or following its suggestions [9].

The preceding factors motivated us to implement a framework that aims to evaluate the trust felt by a human in a robot. In this framework, we have designed different scenarios, described in the following sections in details, to compare and evaluate the level of trust under different circumstances. We argue that using a robot as a storyteller and a human subject as the recipient may reveal the influence of such factors. With this aim, the robot is programmed to tell the story expressing either joy or sad facial expressions. As well as this, the robot will be able to earn the confidence of a subject by making small talk before starting the storytelling phase. We expect to reach a higher level of trust in the case of a storytelling robot expressing sad facial expressions while starting the conversation by making small talk.

2 Background

Recently, a number of research has investigated the concept of trust in Social Robotics. For instance, Brule et al. [10] performed experiments to study the effect of a robot's performance and behavior on human trust. The authors reported that the performance of robot's tasks influences its trustworthiness. In another study Youssef et al. [11] investigated the effect of combining inarticulate utterances with iconic gestures in addition to the response mode (proactive or reactive). Results suggest that humans overwhelmingly prefer the proactive mode to the reactive mode; in fact, under this setting a higher trustworthiness of the robot was obtained.

Stanton and Stevens [12] conducted an experiment in which participants were asked to give answers in a game similar to "shell game". The authors created scenarios where the robot eye gaze and eye tracking were used to help the participant to find the object. It is reported that the gaze has a positive impact upon trust for difficult decisions and on the contrary, a negative impact for easier decisions.

Another study performed by Kahn et al. [13], suggests that people will likely build intimate and trustworthy relationships with robots. This hypothesis is proven through an experiment in which the participants listened to a secret told by a robot. Another study performed by Desteno et al. [14], proves that accuracy in judging the trustworthiness of novel partners is heightened through exposure to nonverbal cues and identified a specific set of cues that are predictive of economic behavior.

Despite the promising reported results, there remains a paucity of evidence on other behavioral factors of a robot. In this sense, our work suggests that the trust a person has in a robot can be improved or acquired according to facial expressions, as well as making small talk before performing certain tasks.

3 Methodology

In this study we conceptualize the trust in terms of a storytelling situation, in which we imagine the *trustor* as the one who depends on another individual, and the *trustee* will be the one who is at risk [1]. To examine the level of trust, we attempt to design a scenario where the robot is at risk of being shut down or being replaced by another new robot, due to a particular fault or malfunctioning. Hence, the robot needs money to rescue himself by fixing the faulty part; in our scenario, the problem is with his left eye, as depicted in Fig. 1(c). However, as in real life, it is clear that a person will aid somebody if s/he believes that the suppliant is trustworthy. The same may happen to the robot in HRI scenarios. Going into depth, in this paper we suggest two hypotheses based on the effect of (a) small talk and (b) facial expressions on people trust in a robot. Figure 2 depicts the flowchart of the methodology steps aforementioned.

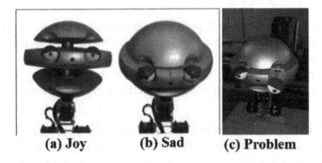

(a) Joy   **(b) Sad**   **(c) Problem**

Fig. 1. The Emys robot expressing joy (a) and sad (b) facial expression, and showing its problem (c) to the participant.

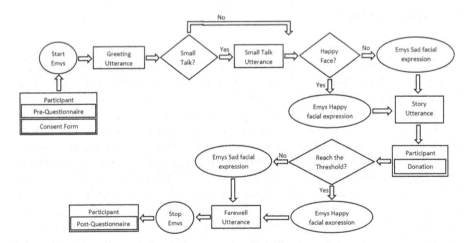

Fig. 2. Methodology flow.

The former, i.e. the role of small talk, has been examined earlier in [15]. Based on this study, starting the conversation using small talk has a positive influence on the trust of participants. However, to examine the hypothesis, they implemented the scenario using a virtual robot, not a real physical robot; while it has been proved earlier that human interaction with a real robot is different in comparison with a virtual one [16]. In the same vein, we argue that the influence of small talk performed by a real robot might influence the level of trust using a real robot comparing to a virtual robot.

In case of facial expression and trust, to the best of our knowledge there is no recent study carried out. In sum, we suggest two hypotheses to evaluate in this paper, which are as follow:

1. Small talk: starting a conversation with small talk would enhance the level of trust in a robot.
2. Facial expression: expressing sad facial expression, while telling a sad story would enhance the level of trust in a robot.

As follows, we define the two hypotheses as bipolar variables. In other words, each variable can take two different values. For instance, the first hypothesis is a variable with two possible values: making small talk (ST) or not (NST). Similarly, the second hypothesis could be expressing either joy or sad facial expression (the two possible facial expressions are depicted in Fig. 1(a) and (b)). In this way, four different scenarios will be assumable:

- starting the interaction with small talk – while expressing sad face [ST_SAD],
- starting the interaction without small talk while expressing sad face [NST_SAD],
- starting the interaction with small talk while expressing joyful face [ST_JOY] and
- starting the interaction without small talk while expressing joyful face [NST_JOY].

4   Implementation

The experiment was conducted using the SERA Ecosystem developed by Ribeiro et al. [17]. This ecosystem is composed of a model and tools for integrating an AI agent with a robotic embodiment, in HRI scenarios. To be part of this ecosystem, we developed an application in C# and integrated it to other applications through of a high-level integration framework named Thalamus. This framework is responsible for accommodating social robots with possibility to include virtual components such as multimedia applications [18]. Moreover, we utilized a semi-autonomous behavior planner named Skene developed by Ribeiro [18]. To complete the experiment, a TTS component is used which serves only as a bridge to the operating system's own TTS. Moreover we utilized a module of speech detector in order to capture the subject voice and make the interaction with robot more real, as well as a symbolic animation engine based on CGI methods called Nutty Tracks [19], which provides the opportunity to animate both virtual and robotic characters in a graphical language. This ecosystem used to integrate all applications together with Emys robot is depicted in Fig. 3(a) using this ecosystem, the Emys is able to demonstrate emotions based on Ekman's facial Action Coding System [20].

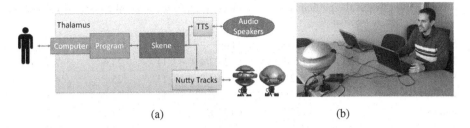

(a)   (b)

Fig. 3. (a) Ecosystem used to perform experiments, (b) A participant interacting with the robot.

In the scenario starting with small talk, the participants have the opportunity to talk to the robot using an embedded microphone. The robot would wait after uttering each sentence for the subject's response and then would continue the conversation.

After technical implementation, to validate the hypothesis suggested earlier, first we started designing scenarios and collecting data. In order to prevent any distraction, we ran the experiment in a quiet and isolated room. Moreover, to evade the experimenter effect [21], we did not inform the curious participants about the test and let them know the goal of the experiment before it was finished (Fig. 3(b) depicts a participant interacting with the robot). In addition, we prevent participants who were aware of the goal to participate in the experiments. We designed the experiment under four different scenarios to provide the opportunity of inspecting the four possible outcomes corresponding to the two variables (ST_SAD, NST_SAD, ST_JOY, NST_JOY). To measure the trust we applied a recently proposed trust questionnaire, available in [21].

After filling in the pre-questionnaire, the subjects start the interaction stage with the robot in one of the pre-defined scenarios. In the case of small talk scenarios, the robot makes a set of simple interactions asking general questions and then he waits for the response. After this stage, the robot tells the story with either sad or joy facial expression depending on the scenario. During the story the robot explains he will be replaced soon by another robot due to a problem with his left eye as depicted in Fig. 1(c).

At the end of the storytelling phase, regarding the donated amount, the robot expresses a happy or sad face. In our implementation, we considered a threshold of 20€, more specifically the amounts below 20€ are considered as low. This threshold was set only to create the robot reaction in response to the donation. Finally, the post-questionnaire is applied to measure the trust in the robot. As declared in [22], the pre- and post-questionnaire should have the same questions.

Participant and Dataset. Considering the described scenario, we performed the experiment over three days, and a total of 42 people participated. The experiment was performed in the Instituto Superior Técnico (IST) in Porto Salvo, Portugal and the population was a random selection of students. Table 1 lists the descriptive statistics of the subjects participating in the experiment.

Table 1. Statistics of the participants, the numbers in parenthesis are the mean and standard deviation of age, respectively.

Scenario	Female	Male	Sum	Mean age	SD age
Joy Small Talk (ST_JOY)	5 (28/9.6)	6 (23.3/4.9)	11	25.6	7.4
Sad Small Talk (ST_SAD)	2 (22/1.4)	9 (28/4.7)	11	25	4.7
Joy (J)	1 (23/0)	10 (23.4/2.5)	11	23.2	2.4
Sad (S)	3 (25.3/3.2)	6 (23.1/2.1)	9	24.25	2.5
Total	**11**	**31**	**42**	**24.9**	**4.85**
Mean Age	**25.7**	**24.6**			
Standard Deviation Age	**6.72**	**4.1**			

To examine the influencing factors, based on the four designed scenarios, we assume 6 different comparisons listed in Table 2.

Table 2. Assumed comparisons.

Comparison	Label groups	Population
Comparison 1	ST_SAD vs. NST_SAD	10 + 9
Comparison 2	ST_JOY vs. NST_JOY	11 + 11
Comparison 3	ST_JOY/SAD vs. NST_JOY/SAD	21 + 20
Comparison 4	ST_JOY vs. ST_SAD	11 + 10
Comparison 5	NST_JOY vs. NST_SAD	11 + 9
Comparison 6	ST/NST_JOY vs. ST/NST_SAD	22 + 19

5 Results

Based on Table 1, 42 people participated in the experiment, among which, one of the participant's native language was English. As it is a proven theory that speaking with a robot with the same language as the participant has emotional influences on him/her [23], we removed the record corresponded to this participant from the final dataset, to prevent any other influencing factor, such as this one. It should be noted here that the small talks and the story were told in English while most of the participants were Portuguese. Besides, to check whether all the participants were at the same level of English comprehension, we asked to the participants the level they perceived the utterances in a Likert scale. The final data set with 41 subjects is large enough to assume it as a normal population based on the central limit theorem [24]. However, it should be noted that the statistical population of each subgroup is not large enough to assume them as normal. Hence, we performed a normality test which indicated that the population is non-normal in any scenario (Comparison 1: $D(19) = 0.13$, Comparison 2: $D(22) = 0.20$, Comparison 4: $D(21) = 0.15$, Comparison 5: $D(20) = 0.20$; $p < 0.05$). Thus to analyze the collected data set by scenarios, we turned to non-parametric tests.

The trust questionnaire contains two parts, one part must be performed before the interaction and the other after the interaction. We performed a t-test on the whole data of the pre-questionnaire, which indicated that there was no significant difference between the subjects participated in the experiment, before interacting with the robot ($t(39) = 1.39$, $p = 0.17$). Hence, all the subjects had the same condition before interacting with the robot.

To assess the first hypothesis, i.e. the role of small talk, first we look at each subgroup independently. In Comparison 1, which compares the differences between the two groups of people interacting with sad facial expression, the result of a U Mann-Whitney test shows that there is a significant difference between the two group started conversation with or without small talk ($U = 13.5$, $p = 0.010$) and the higher mean in the first group (63.9 vs 43.4) endorses the fact that small talk makes in difference. However, this pattern is not observable in Comparison2 ($U = 38.0$, $p = 0.13$).

To investigate the role of small talk regardless of facial expressions (Comparison 3), first we checked whether there is any difference between the distributions of data in each subgroup. A Kolmogorov-Smirnov test shows that the distribution of trust level is the same across both categories ($Z = 1.93$, $p = 0.001$), hence we can put the two categories together. The result of a t-test shows that there is a significant difference in trust level of each group ($t(39) = 3.34$, $p = 0.02$). The higher means in case of small talk (63.95 vs. 50.10) approves the first hypothesis.

On the contrary, in case of the second hypothesis, Comparison 4 ($U = 54.0$, $p = 0.94$) and 5 ($U = 32$, $p = 0.18$) did not reach any significant difference between the two groups. The result of a Kruskal-Wallis test showed that the distribution of the two groups are the same ($H(21) = 0.98$; $p < 0.05$), hence we can mix the two groups to examine the influence of facial expression regardless of the small talk (Comparison 6). The result of a t-test showed that there is no significant difference between the two groups ($t(39) = 1.95$, $p = 0.43$), hence we cannot infer anything further from this comparison.

Another feature which might be a discriminant of the groups is the amount of donation which is listed in Table 3 in details. To analyze the data regarding the donation rate, we performed the uni-variate non parametric test of U Mann Whitney for the two hypotheses. For the first hypothesis, no significant difference is inferred between the distribution of the two groups ($U = 0.92$, $n = 21$, $p < 0.05$). In the same fashion, the distribution of donation is the same for second hypothesis across two categories of sad and joyful face ($U = 0.93$, $n = 20$, $p < 0.05$).

Table 3. Descriptive Statistics of donation per scenario.

Donated	ST_JOY	ST_SAD	NST_JOY	NST_SAD
qt	11	11	11	9
sum	277€	302€	51€	75€
avg	25.18€	27.45€	4.63€	8.33€
stdev	31.296	30.321	7.377	15.986

6 Discussion and Future Works

As pointed out in the result section, based on the collected data, comparing the four groups separately does not lead to a significant difference in general. However, when we compare the variables over a larger set (Comparison 3), we reached significant differences in the case of the first hypothesis. The justification of this might be the small size of the collected data.

The most significant difference happens in the case of small talk together with the sad face. Although the sample size was small, the robot could maintain his partners' trust in the ideal case: expressing sorrow after making small talk. In this case, the higher mean in the scenario with small talk indicates that people generally generate a higher level of trust when the robot seems sad while telling his story.

Although small talk seems to play a vital role in trust, however, it was not clear in the case when the robot expresses the joyful facial expression. The justification behind this fact is that, the robot's facial expression while expressing joy was not as clear as sadness. In the post-questionnaire, we asked the subjects whether they perceived his facial expression or not. In case of sadness, almost all the subject perceived him as sad, except 4 people who perceived him as neutral (21 %). On the other hand, in case of joy, only one subject found him joyful, more interestingly, two people found him feeling sad. Hence, hypothesis 2 could not be evaluated well in the current setting.

As it is clear in the Table 3, the amount of donation when the robot was configured to perform the small talk (independent of facial expression) was higher than the amount of donation without small talk (579€ with small talk against 126€ without small talk). It should be noted that sad facial expression raised more donation than joy facial expression with (302€ against 277€) or without (75€ against 51€) small talk. Although the descriptive analysis of the donation amount sounds interesting, however, from statistical analysis viewpoint, there is no significant difference between the two groups, due to the high level of variance. We can argue that, in this experiment people were not supposed to donate from their own budget and it was purely down to imagination. Moreover, a normalization of donation amounts based on the corresponding a personality trait should be performed to make more viable analysis. However, if they were supposed to donate, those who had a higher level of trust in the robot might pay more than the others in reality.

Despite the promising results, further future steps are required. For instance, the results show that in the ideal case, i.e. starting the conversation with small talk while expressing sadness in facial expression, could maintain the trust of the users; however, we cannot infer anything more from other scenarios due to the small sample size of subgroups. Thus in our future steps, we are going to continue the experiments to gather more data, to be able to examine the influence of the other factors.

Furthermore, we are going to integrate the role of participants' personality in our analysis. For instance, in the MBTI personality test [25], one of the characteristics of people with the ISFJ personality type is that they perceived as generous. This fact seems to have a crucial effect on the donation results and it seems necessary to be considered. Additionally, we are going to investigate the influence of a real robot in comparison to a virtual one. We have implemented the same scenario, on a virtual version of the Emys. Currently, we are running the experiment and gathering data to form our dataset.

Finally, in our future works we need to take care of the influence of the robot's final word on the questionnaire outcome. In the current setting, the participants fill in the questionnaire after the Emys ended the conversation by either complaining to the user about their low rate of donation or thanking them for their help regarding the high donation rate. These final remarks of the robot might influence the mental state of the subjects before filling in the post-questionnaire, hence lead to biased data. In future settings, we are going to ask the subjects to fill in the post questionnaire prior to the robot's final words. In addition, due to variable time of interactions in different scenarios, in our future steps we are going to investigate if the length of total interaction could be a relevant factor, and also take into consideration the responses of participants to the robot worth to perform.

Acknowledgment. The authors would like to thank their colleagues in INESC-ID and GAIPS group for their help and support, especially Patrícia Alves-Oliveira, Tiago Ribeiro and Filipa Correia. The first author would like to thank National Council for Scientific and Technological Development (CNPq) program Science without Border process number: 201833/2014-0 – Brazil and University of State of Rio Grande do Norte – Brazil.

References

1. Wagner, A.R.: The role of trust and relationships in human-robot social interaction. Georgia Institute of Technology (2009)
2. Breazeal, C.: Social robots for health applications. In: Annual International Conference of the IEEE Engineering in Medicine and Biology Society. IEEE (2011)
3. Broadbent, E., Stafford, R., MacDonald, B.: Acceptance of healthcare robots for the older population: review and future directions. Int. J. Soc. Robot. **1**(4), 319–330 (2009). doi:10.1007/s12369-009-0030-6
4. The Economist. Trust me, i'm a robot. In: Robotics. The Economist (2006). http://www.economist.com/node/7001829
5. Deutsch, M.: Cooperation and trust: some theoretical notes. In: Nebraska Symposium on Motivation, Nebraska, pp. 275–318 (1962)
6. Kollock, P.: The emergence of exchange structures: an experimental study of uncertainty, commitment, and trust. Am. J. Sociol. **100**(2), 313–345 (1994). doi:10.1086/230539
7. Anderson, J.R.: Concepts, propositions, and schemata: what are the cognitive units. Nebr. Symp. Motiv. **28**, 121–162 (1980). DTIC Document
8. Lee, J.J., Knox, W.B., Wormwood, J.B., Breazeal, C., DeSteno, D.: Computationally modeling interpersonal trust. Front. Psychol. **4**, 893 (2013). doi:10.3389/fpsyg.2013.00893
9. Freedy, A., DeVisser E., Weltman G., Coeyman N.: Measurement of trust in human-robot collaboration. In: International Symposium on Collaborative Technologies and Systems (CTS 2007), 25 May 2007 (2007)
10. van den Brule, R., Dotsch, R., Bijlstra, G., Wigboldus, D.H.J., Haselager, P.: Do robot performance and behavioral style affect human trust? Int. J. Soc. Robot. **6**(4), 519–531 (2014). doi:10.1007/s12369-014-0231-5
11. Youssef, K., De Silva, P.R., Okada, M.: Exploring the four social bonds evolvement for an accompanying minimally designed robot. In: Tapus, A., André, E., Martin, F., Ammi, M. (eds.) Social Robotics. LNCS, vol. 9388. Springer, Heidelberg (2015). doi:10.1007/978-3-319-25554-5_34
12. Stanton, C., Stevens, C.J.: Robot pressure: the impact of robot eye gaze and lifelike bodily movements upon decision-making and trust. In: Beetz, M., Johnston, B., Williams, M.-A. (eds.) ICSR 2014. LNCS, vol. 8755, pp. 330–339. Springer, Heidelberg (2014). doi:10.1007/978-3-319-11973-1_34

13. Kahn, P.H., Kanda T., Ishiguro H., Gill B.T., Shen S., Gary H.E., et al.: Will people keep the secret of a humanoid robot? pp. 173–80 (2015). doi:10.1145/2696454.2696486
14. DeSteno, D., Breazeal, C., Frank, R.H., Pizarro, D., Baumann, J., Dickens, L., et al.: Detecting the trustworthiness of novel partners in economic exchange. Psychol. Sci. **23**(12), 1549–1556 (2012). doi:10.1177/0956797612448793
15. Bickmore, T., Cassell J.: Relational agents: a model and implementation of building user trust. In: Proceedings of the SIGCHI Conference on Human Factors in Computing Systems, Seattle, Washington, USA, pp. 396–403, 365304. ACM (2001)
16. Li, J.: The benefit of being physically present: a survey of experimental works comparing copresent robots, telepresent robots and virtual agents. Int. J. Hum Comput Stud. **77**, 23–37 (2015). doi:10.1016/j.ijhcs.2015.01.001
17. Ribeiro, T., Pereira, A., Di Tullio, E., Paiva, A.: The SERA ecosystem: socially expressive robotics architecture for autonomous human-robot interaction. In: AAAI Spring Symposium Series (2016)
18. Ribeiro, T., Pereira, A., Di Tullio, E., Alves-Oliveira P., Paiva A.: From thalamus to skene: high-level behaviour planning and managing for mixed-reality characters. In: Intelligent Virtual Agents - Workshop on Architectures and Standards for IVAs (2014)
19. Ribeiro, T., Paiva A., Dooley D.: Nutty tracks: symbolic animation pipeline for expressive robotics. In: ACM SIGGRAPH, p. 1 (2013)
20. Ekman, P., Friesen, W.: Facial action coding system: a technique for the measurement of facial movement. Consulting Psychologists Press, Palo Alto (1978)
21. Kintz, B.L., Delprato, D.J., Mettee, D.R., Persons, C.E., Schappe, R.H.: The experimenter effect. Psychol. Bull. **63**(4), 223–232 (1965). doi:10.1037/h0021718
22. Schaefer, K.: The perception and measurement of human-robot trust. University of Central Florida, Orlando (2013)
23. Rau, P.L.P., Li, Y., Li, D.: Effects of communication style and culture on ability to accept recommendations from robots. Comput. Hum. Behav. **25**(2), 587–595 (2009). doi:10.1016/j.chb.2008.12.025
24. Araújo, A., Giné, E.: The Central Limit Theorem for Real and Banach Valued Random Variables. Wiley, New York (1980)
25. Myers, I.B., McCaulley, M.H.: Manual, A Guide to the Development and Use of the Myers-Briggs Type Indicator. Consulting Psychologists Press, Palo Alto (1985)

A Study on Trust in a Robotic Suitcase

Beatriz Quintino Ferreira[1](✉), Kelly Karipidou[2,3], Filipe Rosa[3], Sofia Petisca[1],
Patrícia Alves-Oliveira[4], and Ana Paiva[1]

[1] INESC-ID & Instituto Superior Técnico, Universidade de Lisboa, Lisbon, Portugal
beatriz.quintino@tecnico.ulisboa.pt
[2] KTH Royal Institute of Technology, Stockholm, Sweden
[3] Instituto Superior Técnico, Universidade de Lisboa, Lisbon, Portugal
[4] INESC-ID and Instituto Universitário de Lisboa (ISCTE-IUL),
CIS-IUL, Lisbon, Portugal

Abstract. This work presents a study on human-robot interaction
between a prototype of a robotic suitcase – aBag – and people using
it. Importantly, for an autonomous robotic suitcase to be successful as
a product, people need to trust it. Therefore, a study was performed,
where participants used aBag (remotely operated using the Wizard of Oz
technique) for carrying their belongings. Two different conditions were
created: (1) aBag follows the participant at a close range; (2) aBag follows the participant on a further distance. We expected that participants
would trust more aBag when it was following them at a close range, but
interestingly participants seemed to trust more when aBag was further
away. Also, regardless of the conditions, the level of trust in aBag was
significantly higher after the interaction compared to before, bringing
positive results to the development of this kind of robotic apparatus.

Keywords: Assistive robotics · Human-robot trust · Proxemics

1 Introduction

With the fast pace of technology development, a future with robots existing
alongside with humans becomes each day more real. Social robotics and human-
robot interaction (HRI) studies become crucial for understanding how this can
be achieved and bring positive outcomes. One of the fields where the presence of
robots is being studied is in assistive robotics [15], where robots can help people
overcome their disabilities or enhance their capabilities. But for this cooperative
relationship to occur an important bond needs to exist: trust [11].

With this in mind, we tried to ascertain how much one would trust a robotic
platform to carry one's personal belongings. Thus, we present a study of a robotic
suitcase prototype (aBag, depicted in Fig. 1). Its purpose is to assist people in
carrying their belongings and can therefore be labeled as an assistive robot.
Having a robotic suitcase, such as aBag, that can carry your belongings and
move itself seems to be convenient for almost everyone, as users can benefit from
not having to carry the bag's weight and having both hands free for holding or

© Springer International Publishing AG 2016
A. Agah et al. (Eds.): ICSR 2016, LNAI 9979, pp. 179–189, 2016.
DOI: 10.1007/978-3-319-47437-3_18

doing other things. It is plausible to envisage several scenarios in which such an assistive robot would be extremely helpful, specially for elderly people. Recently, the NUA company announced the development of a robot, the NUA Robotic Luggage [7], designed to perform the same task as aBag. From what is available, it seems that this robotic suitcase is an autonomous version of aBag. Thus, the questions we pose in this work and the study performed become even more relevant and opportune.

Fig. 1. aBag: front and side views.

Moreover, some studies [5,14] showed that distance (proxemics) is an important factor in HRI. Thus, when measuring trust towards aBag we defined two different types of distance behaviors that aBag could have towards the user: (1) aBag followed the user closely and (2) aBag moved more freely, keeping a further distance to the user. A study was conducted where participants performed both conditions following a within-subjects design. The research questions that underlie this work are the following: *Will users trust their belongings to a robotic suitcase? And will that trust differ depending on aBag's distance-behavior?* Additionally, our study hypotheses are:

h1: The perceived human-robot trust will be different before interacting with the robotic suitcase aBag and after interacting with it;

h2: The perceived human-robot trust is higher for the condition in which aBag follows the user more closely than for the condition in which aBag moves more freely and further away from the user.

We expected to confirm both hypotheses, thus finding different results for the level of trust before and after the interaction, in order to show the usefulness of aBag, and at the same time we expected users to trust more when aBag was moving closer to them. In spite of the difficulty of using a small sample of participants some interesting findings emerged, those are further discussed.

2   Related Work

In assistive robotics valuable work has been conducted, specifically for rehabilitation, for example mobility assistance for disabled or elderly people (e.g. smart walkers). Nevertheless, when it comes to outsourcing this kind of task to a robot, humans hand over a considerable amount of responsibility. Henceforth, it becomes important to trust robots to perform these tasks. Lee and

Moray define trust as *"the attitude that an agent will help achieve an individual's goals in a situation characterized by uncertainty and vulnerability"* [6], and it is a multi-dimensional concept, inherently complex to measure. Specifically for HRI, evaluating human-robot trust levels has been a challenging and continuously addressed topic [9,16]. In [4] is presented an attempt to quantify the effects of different factors on perceived human-robot trust, showing that robot characteristics (in particular, its performance) influences greatly the perceived trust. Additionally, [2] studied if humans trust a robot to be their partner in a card game scenario. This study showed that although humans trust a robot to partner with them in the card game, those levels of trust are dependent on previous interactions with the same robot. This is what we expect to inform in our hypothesis *h1*, with users assigning higher levels of trust after interacting with aBag. Moreover, in [2] it was found that because trust is a multifaceted construct, it develops and is expressed differently for humans and robots.

To our knowledge, no other study was performed addressing trust in HRI in an assistive scenario in which the robot carries the belongings of a human.

In such a study, proxemics behavior (the distance the robot would have to the user) becomes a relevant variable to take into account. Different factors have been found to affect HRI proxemics, with studies showing for example that the perceived familiarity with a PR2 robot influences personal space, with people standing closer to the robot when it is perceived as more familiar to them [14]. Another study showed that a more mechanical appearance of the robot seemed to allow people to let the robot come to a closer distance (comparing to a more humanoid one) [13]. When addressing personal space in a social interaction, psychology literature defines four primary zones: intimate, personal, social and public [1]. In a study performed in [5], when people were asked to be followed by a robot the majority of participants preferred to position themselves in a way so that the robot would be in their personal space (approximately 0.5 m to 1.2 m). Following this, we would expect that users would also prefer to have aBag with their belongings closer to them, rather than further away. Therefore, for the condition where aBag followed the user closely, the robot moved in the personal zone (0.5 m-1.25 m), and when aBag moved more freely it situated itself more in the public zone (beyond 3.66 m).

3   Methodology

3.1   The Robot

aBag is a regular suitcase (of size 49 cm × 34 cm) attached to the chassi of a remotely controlled car (a *Ninco 1/10 Predator MT-10 2.4G RTR*). All the unwanted parts were removed from both the car (its top case) and the suitcase (its wheels and all the plastic and metal components were detached to make it lighter), and the two were fitted together. To make aBag more believable, a hole was made on the bag's front to expose a smartphone's camera mounted on the inside of aBag, in order to disguise as a sensor that the bag would use to track the user (when questioned by the users the false sensor was explained to afford obstacle avoidance and person-following).

3.2 Measures

To ascertain how much participants trusted in aBag taking care of their belongings, a trust questionnaire created by K. Schaefer (2013) and specific for HRI was used [11]. This questionnaire was found to be the most appropriate for this study since it is supported by a large body of work and bases the level of perceived trust on the analysis of different dimensions of HRI such as: autonomy, reliability, decision making, failure and communication. These aspects are specially important as the study was performed in an uncontrolled environment. The scale is comprised of 40 items, measuring trust on a 0–100 percentage scale (answers are given in intervals of 10 %) providing a percentage trust score with the sum of each item score. The questionnaire was used before and after the interaction for each of the conditions. Examples of questions from the trust questionnaire are: *"What percentage of the time does this robot follow directions?"* and *"What percentage of the time is this robot reliable?"*.

All interactions with aBag were video recorded and the number of times each participant gazed back at aBag were coded from the videos. Two different coders coded half of the videos that were randomly chosen and a good level of agreement with the Spearman's coefficient $r_s = 0.968$ was obtained, which seemed reasonable to justify only one coder coding the remaining videos. In total, 1.5 h of recorded material was coded.

3.3 Pilot Study

Before performing the main study with the final test-users, a pilot study was conducted in order to test the planned experimental procedure. Three volunteers, all male aged 25, 26 and 25 years, performed the experiment at the same place where the main study was planned to take place. Three different conditions for aBag's movement behavior were tested. In the first condition aBag was kept at a small distance from the user, following him/her closely. The second condition was identical to the first but the distance kept between aBag and the user was considerably larger. For the third and final condition, a more "exploring" behavior was applied, in which aBag performed several movements to explore the area freely, as an attempt to make it appear more autonomous. Before and after experimenting with each condition the participants were asked to fill in the trust questionnaire [11]. Out of the three piloted conditions, two were defined (depicted in Fig. 2) for the final study based on the direct feedback from the three pilot study participants and on their answers to the questionnaire:

Condition 1. aBag follows the user, taking the same trajectory at a small distance behind - this was the first condition and was chosen because there were indications in the results and observations of the pilot study that this was the behavior the pilot users preferred aBag to have.

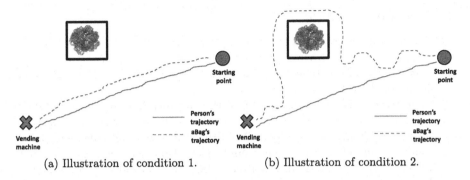

(a) Illustration of condition 1.   (b) Illustration of condition 2.

Fig. 2. A schematic image displaying the difference between the two conditions used for aBag in the main study.

Condition 2. aBag follows the user more freely, keeping a further distance to the user compared to condition 1, simulating a more autonomous behavior - this condition became a fusion of the second and the third one, reducing the "exploring" behavior, because of indications that too much exploration was unnatural for aBag since it was aimed to follow the user.

3.4 Main Study

Participants. A total of 18 university students (11 male and 7 female, aged between 17 and 47 years old, $M = 24$; $SD = 7.0$) performed the experiment. 11 of the participants (8 male, 3 female, $M = 26$; $SD = 8.8$) performed both conditions, so each participant either did first condition 1 and after condition 2, or vice-versa. To counterbalance learning or adaptation effects, the initial condition was randomly chosen for each participant. The remaining 7 participants (3 male, 4 female, $M = 22$; $SD = 3.1$) were only able to perform one of the conditions so they were excluded from the main analysis (results from these participants were only used to investigate the level of trust in aBag regardless of the conditions). All participants signed a consent form in order to be part of the study and authorizing for the sessions to be recorded. The study was a within-subjects design and lasted approximately 30 min. The between-subjects design (for the participants allocated only in one condition) lasted approximately 20 min. At the end of the experiment participants received a thank you gift, which was incorporated in the task when interacting with the robot, see *Experimental Procedure* section.

The study took place in the entrance hall of a Portuguese University, using the Wizard of Oz technique. The wizard was placed on the second floor having a clear sight of the entrance floor without being seen, and kept consistent behaviors for the two conditions using waypoints defined in the environment. The material used was the robotic suitcase (aBag) and a video camera for recording the interaction.

Experimental Procedure

(i) The participant was led into a room, where he/she was asked to fill in: a consent form approving the whole experimental procedure and the pre trust questionnaire about how much he/she would trust in aBag (having a picture of the robot to look at).

(ii) Then, the participant was led out to the entrance hall and introduced to aBag. In order to simulate aBag being autonomous, a smartphone was given to the participant and he/she was told that the person wearing that phone would be the person that aBag would follow. However, the phone was not connected in any way to aBag, instead aBag was controlled by a researcher that acted as a wizard.

(iii) After this, the participant was told that the purpose of the study was to simulate a real scenario where a person uses aBag as his/her own robotic suitcase while performing a task. In order to simulate this the participant was asked to put something of value for him/her (wallet, mobile phone) inside aBag, so that aBag could carry it. Then he/she was asked to go to a vending machine, approximately 50 m away, while aBag would follow (see Fig. 3) and buy something he/she likes for one euro (this money was given from us and served as a thank you gift). After buying something, the participant was asked to return back to where he/she started.

(iv) When the participant came back he/she was asked to fill in the post trust questionnaire.

(v) Then, the participant was brought again to perform the same task (step (iii)) with aBag, but now on a different condition.

(vi) At the end of the second interaction, the participant was asked to fill in the post trust questionnaire again and was thanked for his/her collaboration.

The fact that aBag was acting in different ways for the two interactions was not explicitly explained to the participants. The 7 participants that could not perform the whole experiment stopped the procedure at step (iv).[1]

Fig. 3. Participant using aBag as his own robotic suitcase.

---

[1] A video presenting the experiment using the recordings from the participants' interactions with aBag is available at https://www.youtube.com/watch?v=M4mw5WX-AS8\&feature=youtube.

4   Results

Results from the Trust Questionnaires

To test hypothesis *h1* "The perceived human-robot trust will be different before interacting with the robotic suitcase aBag and after interacting with it", the data from all the 18 participants was used. For this purpose, the maximum of the post questionnaire from participants that did both conditions were used together with the post values from the participants that performed the experiment only once, to form one score of post interaction trust for each participant. So, for each participant two trust values were gathered, the pre and post trust score, regardless of the condition(s) (see Fig. 4a). When analysing the distribution with a Shapiro-Wilk we found a non normal distribution ($\rho < 0.05$) for the trust scores. Therefore, a nonparametric test for repeated measures was applied to ascertain if there were differences in the trust scores. The Related-Samples Wilcoxon Signed Rank Test confirmed *h1* with $Z = -2.004$, p = 0.045. From this result it is possible to conclude that the trust was significantly different before and after interacting with aBag (the median of the pre trust score was 59.88 while for the post trust was 70.25). Hence, the level of human-robot perceived trust is shown to increase after interacting with aBag compared to the participants' measured trust before meeting the robot.

In order to test hypothesis *h2* "The perceived human-robot trust is higher for the condition in which aBag follows the user more closely than for the condition in which aBag moves more freely and further away" a one-way analysis of variance (ANOVA) test was applied using the data from the 11 participants that performed the experiment following a within-subject design. In this case three groups are compared: the questionnaire answers before any interaction, the answers after interacting with aBag when it behaved according to condition 1, and the questionnaire answers after interacting with aBag in condition 2. The results from this test were not significant, which is probably due to the small number of participants. Thus, no significant differences seem to exist in trust between the two conditions. However, there seems to be indications that users trust more in aBag when it behaves according to condition 2 (aBag moving more freely and keeping a further distance to the user), as trust is generally higher after condition 2 (mean and standard deviation for condition 2 are $M = 69.7$, $SD = 17.5$, and for condition 1 $M = 67.1$, $SD = 17.1$). Moreover, the significance values are always lower when comparing the pre and post trust after condition 2 than when comparing the answers for pre and post trust after condition 1.

Results from the recordings of the interactions with aBag

The video recordings of the experiment were analysed by counting the number of times the 11 participants who performed both conditions looked back at aBag during each one of their interactions. This number might be an indication of how much the participants trusted aBag, assuming that when a person is doubting aBag he/she looks back more often than when he/she trusts it.

We found a non normal distribution ($\rho < 0.05$) for the number of times participants looked at aBag. Hence, a non-parametric Related-Samples Wilcoxon Signed Rank Test was applied to test this data. The results show that there are no statistically significant differences, with $Z = -0.06$, $p = 0.95$, so we retain the null hypothesis. Nonetheless, the data distribution suggests that people tend to look back at aBag slightly more often during condition 1 when aBag is following them closer (the median is higher for condition 1 than for condition 2) (see Fig. 4b). Moreover, the participants who look more often at aBag seem to have a more coherent behavior, as the number of times they look back is more concentrated than the number of times of people who look less. From this observation, it seems that there is a certain number of times/frequency of looking back at the robot that satisfies the users while interacting with aBag.

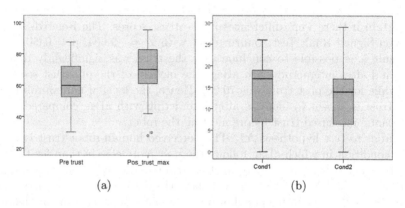

(a)   (b)

Fig. 4. Boxplots of (a) the pre and post trust scores, regardless of interaction condition; (b) number of times participants look back, for the two conditions.

When further exploring this quantitative data from the videos, a considerable difference is noted between the number of times male and female participants looked back while performing condition 2. By performing a nonparametric test again, this time the Independent-Samples Mann-Whitney U test, a significant difference in the number of times male and female participants look at aBag, while interacting with it in condition 2, is verified. Yet, this result uses a very small dataset, so it is not possible to make any claim from it. Nevertheless, the data seems to suggest that female participants tend to look more at aBag, probably because women are more used to carry a purse in their daily routines, and consequently look more after it, than men (who usually carry their personal belongings on their pockets). This result may point towards the very interesting aspect of differences in proxemic preferences according to gender [12].

5   Discussion

As an attempt to answer the first research question "Do users trust their belongings to aBag?" statistical analysis of the answers to the trust questionnaires confirms hypothesis *h1* as it shows that the perceived human-robot trust is significantly different before and after interacting with aBag. Specifically, the perceived trust was higher after having interacted with aBag than before meeting the robot. This is an interesting result showing that after experimenting with aBag capabilities, participants felt aBag more trustworthy than before interacting. One can argue that a real implementation of a solution to the person-following problem as in [3,10] rather than a Wizard of Oz technique, could have been used to perform this study. Nonetheless, we opted for the latter approach due to the high complexity and uncontrollability of the chosen realistic scenario (with large groups of people passing by), where an automatic algorithm could be dangerous. On the other hand, the novelty effect could also have had a part in this since robotic suitcases are not common yet and people may tend to underestimate their capabilities. Also, the way participants were introduced to aBag before interacting with it might have affected their feedback, as they were only shown a photo of aBag - which conveys minimal knowledge - when filling in the pre questionnaire. However, depite these factors, this presents an interesting result.

Regarding the second question, "Will the users' trust differ depending on aBag's behavior?", no significant statistical results were obtained. Yet, results suggest that trust is higher for condition 2 (the further away and more freely moving condition). Although this indication is opposite to what is formulated in hypothesis *h2*, it is in line with some previous work on proxemics in which users have shown to prefer social robots that keep a higher distance from them and not invading their intimate space [8]. A possible explanation for this could be that when the hypothesis was formulated aBag was thought of as a more machine-like robot and hence shorter distance was expected to be preferred. But, possibly aBag was perceived as more autonomous and "lifelike", having more similarities to the human user rather than an object and thus conveying more "trust" for a greater distance from the user.

The results obtained from the video analysis with the amount of times participants looked back at aBag were not statistically significant. However, there were indications that participants looked less at aBag during condition 2. Therefore, if, in fact the number of times people look at the robot can be considered as a proxy to evaluate trust (less number of looks equivalent to a higher trust), these indications are in agreement with the suggestion from the questionnaires answers that trust tends to be higher for condition 2.

6   Conclusions and Future Work

In conclusion, we found that the perceived human-robot trust is significantly higher after the participants have interacted with aBag than before any interaction, independently of study condition, which shows the capabilities of aBag as a robotic suitcase. The two different behaviors (conditions) that were created in order to look for differences in trust in aBag did not present any significant difference, giving only vague indications that participants seemed to trust aBag more when it moved more freely and on a further distance than when it followed them on a closer range.

In the future, it would be interesting to perform a similar study with a larger sample and using target population (e.g. elderly people, wheelchair users). Furthermore, it would be of interest to change the environment where the study is performed, considering creating a more constrained environment where actors may play up the same scenario for each participant without their knowledge, instead of having a dynamical environment as the one in our study.

Finally, it would be interesting to investigate the possible influence in the perceived trust in aBag from more different variables such as age, previous knowledge, experience with robots, and personality factors.

Acknowledgements. This work was supported by national funds through Fundação para a Ciência e a Tecnologia (FCT) with reference UID/CEC/50021/2013. P. Alves-Oliveira acknowledges a FCT grant ref. SFRH/BD/110223/2015. The authors are solely responsible for the content of this publication. It does not represent the opinion of the EC, and the EC is not responsible for any use that might be made of data appearing therein.

References

1. Argyle, M.: Bodily Communication, 2nd edn. Routledge (1975)
2. Correia, F., Alves-Oliveira, P., Maia, N., Ribeiro, T., Petisca, S., Melo, F.S., Paiva, A.: Just follow the suit! trust in human-robot interactions during card game playing. In: RO-MAN (2016, in Press)
3. Gockley, R., Forlizzi, J., Simmons, R.: Natural person-following behavior for social robots. In: HRI (2007)
4. Hancock, P., et al.: A meta-analysis of factors affecting trust in human-robot interaction. Hum. Factors: J. Hum. Factors Ergon. Soc. **53**(5), 517–527 (2011)
5. Huttenrauch, H., et al.: Investigating spatial relationships in human-robot interaction. In: IROS (2006)
6. Lee, J., Moray, N.: Trust, control strategies and allocation of function in human-machine systems. Ergon. **35**(10), 1243–1270 (1992)
7. Libman, A., et al.: NUA – The carry-on that follows you wherever you go. In: NUA (2015). http://unbouncepages.com/nuarobotics/.
8. Pacchierotti, E., et al.: Evaluation of passing distance for social robots. In: IEEE International Symposium on Robot and Human Interactive Communication (2006)
9. Salem, M., et al.: Evaluating trust and safety in HRI: practical issues and ethical challenges. In: Emerging Policy and Ethics of Human-Robot Interaction (2015)
10. Satake, J., Miura, J.: Robust stereo-based person detection and tracking for a person following robot. In: ICRA – Workshop on People Detection and Tracking (2009)
11. Schaefer, K.: The perception and measurement of human-robot trust. Ph.D. Dissertation, University of Central Florida, USA (2013)
12. Syrdal, D., et al.: A personalized robot companion? – the role of individual differences on spatial preferences in HRI scenarios. In: RO-MAN (2007)
13. Syrdal, D., et al.: Sharing spaces with robots in a home scenario-anthropomorphic attributions and their effect on proxemic expectations and evaluations in a live HRI trial. In: AAAI Fall Symposium: AI in Eldercare: New Solutions to Old Problems (2008)

14. Takayama, L., Pantofaru, C.: Influences on proxemic behaviors in human-robot interaction. In: IROS (2009)
15. Tapus, A., et al.: Socially assistive robotics [grand challenges of robotics]. IEEE Robot. Autom. Mag. **14**(1), 35–42 (2007)
16. Can we trust robots? [Special report]. IEEE Spectrum **53**(6), 26–27 (2016). IEEE

How Much Should a Robot Trust the User Feedback? Analyzing the Impact of Verbal Answers in Active Learning

Victor Gonzalez-Pacheco$^{(\boxtimes)}$, Maria Malfaz, Jose Carlos Castillo,
Alvaro Castro-Gonzalez, Fernando Alonso-Martín, and Miguel A. Salichs

Universidad Carlos III de Madrid, 28914 Leganés, Madrid, Spain
{vgonzale,mmalfaz,jocastil,acgonzal,famartin,salichs}@ing.uc3m.es
http://roboticslab.uc3m.es

Abstract. This paper assesses how the accuracy in user's answers influence the learning of a social robot when it is trained to recognize poses using Active Learning. We study the performance of a robot trained to recognize the same poses actively and passively and we show that, sometimes, the user might give simplistic answers producing a negative impact on the robot's learning. To reduce this effect, we provide a method based on lowering the trust in the user's responses. We conduct experiments with 24 users, indicating that our method maintains the benefits of AL even when the user answers are not accurate. With this method the robot incorporates domain knowledge from the users, mitigating the impact of low quality answers.

1 Introduction

Recent studies in robotics have started to include ideas from Active Learning (AL). Using this kind of learning, robots are able to mimic how humans learn: first, by observing their teacher, and then by asking questions[1] when they have any doubts about the concept to be learnt or the examples they have seen. *AL* comes from the Machine Learning field and it was introduced by Angluin [3]. Whilst in Passive Learning (PL), the teacher provides the examples to the learner and labels them, in Active Learning it is the learner who takes the initiative by asking queries to the teacher or oracle. The use of *AL* in robotics has three main motivations when compared with *PL*. Firstly, active learners can potentially obtain better accuracy of the learned concepts. Secondly, *AL* may reduce the number of training examples needed to acquire a concept [15]. This is specially relevant in robotics since in interactive learning the cost of acquiring a training example might be time consuming. Finally, people seem to prefer to train robots that learn actively over passive ones [5].

Two research trends can be distinguished in the field. In the first one, robots learn by self-exploring the environment while in the second one, robots leverage

---

[1] Questions and queries will be used indistinctly in this paper.

© Springer International Publishing AG 2016
A. Agah et al. (Eds.): ICSR 2016, LNAI 9979, pp. 190–199, 2016.
DOI: 10.1007/978-3-319-47437-3_19

HRI to learn from humans [10]. This paper focuses on the second approach and it is inspired, mainly, by the works of Rosenthal [13] and Cakmak [5,6]. Nevertheless, we attempt to go further by understanding how answers to different types of questions could affect the robot's learning. Rosenthal [13] explores how different questions affect the accuracy and correctness of the user's responses. Cakmak et al. [5] studied how the robot was perceived when it showed three different degrees of interactivity when asking questions. In a related work, Cakmak also [6] studied how humans ask questions when learning.

There is literature that assessed different types of queries for *AL* [6] and that evaluated how to ask these questions to maximise the accuracy of the user answers [13]. However, we did not found evidence on how different types of queries affect the robot's learning performance. This paper explores the learning impact of different types of queries that seek information regarding the learning parameters in an interactive pose learning task. To do so, we propose a method which consists in reducing the robot's confidence on the user's answers by including more parameters than the user answered.

This paper is divided as follows. First, Sect. 2 describes our learning approach were we apply *AL* for pose learning. After that, we present our experiment in Sect. 3, and the obtained results are shown in Sect. 4. The results are discussed along with the conclusions in Sect. 5.

2   Learning Scheme

When the user starts training the robot, she has to carry out two tasks. The first one is to put herself in the pose she wants to teach the robot. The second task is to tell the robot the pose she is standing at. Gathering the data from vision and from verbal interaction, the robot builds a dataset which feeds a learning algorithm. In this section we describe the components that participate in such learning process.

As a visual input we use the depth data supplied by a Kinect *RGB-D* camera. In order to do this, we employ the OpenNI[2] *API* to build a skeleton model of the user, composed of the positions and orientations of 15 joints of the user (head, neck, torso, shoulders, elbows, hands, hips, knees and feet). This skeleton model is the input for our learning algorithm (x_i). The labels (y_i) of the dataset are gathered interactively during the training process using an Automatic Speech Recognition (ASR) system described in [2].

Once the training ends, the robot applies a learning algorithm, Random Forests [4], to the set of training instances. The algorithm is freely available through the Weka Framework [9]. We use the default parameters as provided by the framework as these provide good performance on a wide range of datasets. Further details about the training process can be found in [8].

---

[2] http://structure.io/openni.

2.1 Proposed Approach for Active Learning

We use *AL* to ask questions related to the poses to be learnt. Active robot learners can ask different types of questions. This paper focuses on *Feature Queries*, which try to find the features of the learning space that are more relevant for learning. *Feature Queries* are perceived by users as the smartest questions [6]. These questions consist in asking the user if a certain feature of the learning space is relevant or not (inspired by [7,12]). In our approach, we ask the questions once the training session is over. At that moment, the robot asks the user which parts of her body have been the most important ones for each pose. For instance, if the user has taught the robot a pointing pose, it is expected that when the robot asks for the features that are more relevant for this pose, the user's answer should indicate some part of her arm.

With the user's answers, the robot filters all the features which has been told to be less relevant. Notice that in this paper we focus on the learning effects of filtering parameters due *AL* instead of how users perceive the questions or which kind of queries are most helpful. Regarding the questions themselves, we have taken into account the effect in which the user's responses can be affected by the way questions are asked [13]. For that reason, we considered three types of questions to the user: (i) *Free Speech Queries* (FSQ) consist in open questions which allow the user to answer freely (e.g. "Which is the most important limb in this pose?"); (ii) *Yes/No Queries* (YNQ), force the user to answer with a *yes* or *no* statement (e.g. "Is the hand important?"); and (iii) *Rank Queries* (RQ), the user must answer quantifying the importance of a limb from *not important* to *very important* (e.g. "How important is your hand?"). Typically, answers to *FSQ* provide a single limb or a short list of limbs which the user considered important (e.g. "*I suppose my arm*"). Therefore, their use might be interesting when the robot does not know which limbs might be important. Conversely, *YNQ* and *RQ* force the user to answer only about the limb the robot asked for. Hence, these questions are better to retrieve information from a specific limb, that is, from a specific parameter or set of parameters in the learning space. The major drawback of *FSQs* is that those answers need to be parsed in order to map what limbs the user has talked about.

Once the robot has all the answers from the user, it processes them to decide what limbs are the most relevant to learn a certain pose. Our system makes this decision through a threshold in which, if a limb has a value below it, it will be filtered out and, therefore, not used for learning. This threshold is calculated differently depending on the type of the questions. In *FSQ*, each user that has trained the robot provides a list of limbs which she has considered important, so we can calculate the number of times a limb has been mentioned by the users to get a score of its relevance R_l[3]. This score can oscillate between 0, if no one considered that limb important, and N_u (number of users that have trained the robot) if every user mentioned it when they were asked. Therefore, with the user answers we can build a list of relevances:

---

[3] Note that we use the terms *relevance* and *importance* indistinctly.

$$R_{FSQ} = \{R_{head}, R_{neck}, ...\} \tag{1}$$

in which are stored the relevances for all the 15 limbs. We then calculate the mean relevance $\overline{R_{FSQ}}$ of this vector. Our threshold Th_{FSQ} is established as this mean plus one standard deviation:

$$Th_{FSQ} = \overline{R_{FSQ}} + \sigma_{R_{FSQ}} \tag{2}$$

We decided to add a standard deviation $\sigma_{R_{FSQ}}$ to the threshold in order to ensure that the limbs which are chosen stand out from the rest. The threshold in YNQ is calculated similarly, but instead of summing the number of times the user mentioned a limb, we sum all the positive answers (the number of times "Yes" was answered). In RQ, the process is slightly different since the answers are not binary (yes/no) but a direct measure of what the perceived relevance of each limb is. Thus, with the answers from several users, we calculated the average relevance for each limb. So R_l in RQ is actually $\overline{R_l}$. The rest of the process is exactly the same as in FSQ and YQ except the criteria in which a limb passes the threshold. In this case, since the relevances are random variables, we decided that a limb passes the threshold if its 95 % Confidence Interval (CI) is above it.

With this thresholding mechanism, we can build parameter filters that enable the robot to filter the data which is not relevant for learning. We have built 3 different filters that can be used to pre-process the data before feeding them to the classifier, namely $FSQF$, $YNQF$, and RQF (the final F stands for Filter).

Additionally, we created a fourth filter, the *Extended Filter* (EF), which is as an extension of RQF. The EF is an RQF that includes all the adjacent limbs of a normal RQ Filter. For instance, if we have an RQF formed with {head, neck}, its associated EF includes: {head, neck, left shoulder, right shoulder, torso}. As shown in Sect. 4.2, the EF improves RQF in the situations where, due a low quality answer from the user, the RQF (and other AL filters) are outperformed by Passive Learning. In such cases, the EF behaves as a Passive Learner, while, when the users provide good answers, the EF offers a learning performance comparable to other AL filters.

3 Evaluation

The aim of our experiment is to understand how the answers to FSQ, YNQ, and RQ affect robot learning. Accordingly, we prepared an experiment in which several users taught the robot different poses. After the pose acquisition phase, users were asked several questions aimed to improve the robot's learning performance.

Despite the learning session involved natural interaction between the user and the robot, we evaluated the effectiveness of the user's answers in offline questionnaires that were filled to the users just after the training session. This was motivated because the aim of our experiment was to explore the uses of different queries and which ones are better to ask, requiring us to ask many questions to the users. In that situation, we observed fatigue in the users when the robot asked so many questions to a user in the same experiment.

The user trained the robot Maggie [14], which is equipped with Kinect, ASR [2] and Text to Speech (TTS) systems, coupled in a Natural Dialogue Management System [1], which enable the robot to carry out natural interactions. These components are tightly coupled by using the *ROS* [11] framework. Although most of these components are self-built, any robot equipped with a Kinect, a *TTS* and an *ASR*, can replicate our experiments easily.

3.1   Experimental Setup and Method

We tested our active learning approach in two experiments where 24 users trained the robot while interacting with it. The 24 users participated in both experiments. Each experiment consisted in the user teaching three poses to the robot: (i) Experiment 01: The users showed the poses *looking left, looking forward, and looking right* (Fig. 1, first row); and (ii) Experiment 02: The users showed the poses *Pointing left, pointing forward, and pointing right* (Fig. 1 second row).

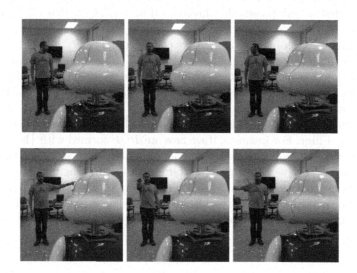

Fig. 1. Examples of poses that the users taught to the robot. First row: Looking left, forward and right. Second row: Pointing left, forward and right

The users were told to stand in front of Maggie. Then, the experimenter explained the experimental procedure and answered any doubts that might have appeared. During training, users had the freedom to start, pause, and finish the training session whenever they wanted. The procedure was: first, the user waited in front of the robot until it tells that it is ready to start. Then, she was able to start the training whenever she wanted. To record each pose, the user was told to: first stand still at this particular pose and then tell the robot the label for that pose. Before changing to other pose, the user had to tell the robot to stop recording this pose. The user interacted with the robot verbally. Once the robot tells the user that it has finished recording the pose, she is free to move to the next pose and start the process again. The user finishes the teaching session telling it to the robot. The session dynamics are similar to a previous work [8].

One dataset per experiment was recorded to feed the learning process. After having trained the robot, they filled a questionnaire where they were asked questions relative to the poses they taught to the robot. These questions were, first *FSQ*, then *YNQ*, and finally *RQ*. All users had to respond to the three types of questions, although their answers were treated separatedly. The following examples of the asked questions are provided[4]: **FSQ**: *What parts of your body do you think the robot has used to learn whether you were looking/pointing left, forward or to the right?*. Here, users were asked to write in an open-text box what limbs were considered most important in each experiment. **YNQ**: *What parts of your body do you believe the robot has used to learn in the first experiment?*. For this set of questions, the user had to fill a multiple-choice list of the limbs depicted in Sect. 2. **RQ**: *Mark the importance of each of the parts of your body so that the robot can learn.* *RQs* consisted in fifteen questions per experiment, each one asking for the importance of a single limb. The answers consisted in a 4-point scale rating the importance of a limb ranging from *Not important at all* to *Very important*.

4 Results

4.1 Filters for Active Learning

First, we evaluated the responses of the users to the questionnaires, from which we build the *AL* filters that will be used for learning. Figure 2 shows the results from the user's answers in each experiment. The horizontal dashed line indicates the filter thresholds. The limbs that passed the threshold are painted in orange, indicating that they will be used in the learning phase.

When analyzing the user's answers to *FSQ*, we realized that sometimes their answers were too broad. This was the case of the answers to experiment 2, in which most users answered with *my arm*. There is not a direct mapping between *arm* and any of the joints. Hence, we decided to include all joints related to the limb, in this case *hand*, *elbow*, and *shoulder*. Moreover, no user indicated which arm was referring when they answered *my arm*. Thus, we decided that all the uncertain answers that covered different possible limbs would be mapped including all these limbs. For instance, in the case of the *my arm* answer, we included the user's both arms.

YNQ and *RQ* produced nearly the same filters, which might occurred because the users had a list of limbs when choosing their answers. Nevertheless, although both graphs have similar shapes, there is a slight difference in the limbs that were

---

[4] The original questions were asked in Spanish. Here we provide the most accurate translations we have found.

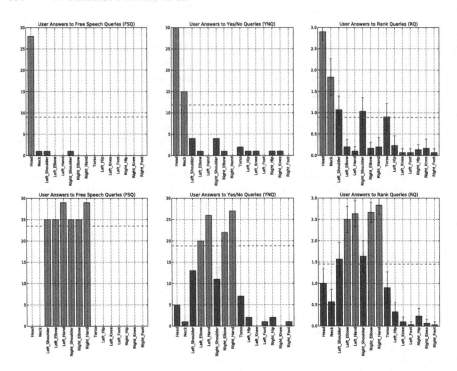

Fig. 2. User answers for experiments 01, looking, (first row) and 02, pointing, (second row). The limbs that are selected are painted in orange. Note that in RQ, a limb is selected if its 95 % CI is above the threshold (see Sect. 2). (Color figure online)

not selected. In RQ, these limbs tended to be closer to the threshold than YNQ. For instance, this effect is clearly shown in Fig. 2, first row. These differences might have been produced because, although the user had the list of limbs in both cases, in RQ they were forced to give an answer to each limb. Two conclusions might be drawn from these effects: (i) if you let users decide which limbs are important, they provide stricter filters than when you ask them for particular limbs; and (ii) RQ and YNQ obtained the same filters because we were too strict selecting our threshold. Had the threshold been lower, RQ would have had more limbs included than YNQ. In that case RQ could have been considered as an extended version of YNQ.

This idea is what led us to create the Extended Filter presented in Sect. 2.1. Since the user answers were too simplistic, we decided that the robot should not blindly trust in their answers. However, since they were neither completely wrong, we opted to extend their answers including more information than they actually gave.

4.2   Learning Results

We compared the AL performance when using the filters built from the *FSQs*, *YNQs* and *RQs* against a Passive Learner, presented in [8], including also the *EF* to the comparison. These filters pre-process the data, removing all the parameters associated to the limbs that were filtered out. The remaining data are fed to a random forest classifier [4]. The metric we used for our evaluation is the F1-score.

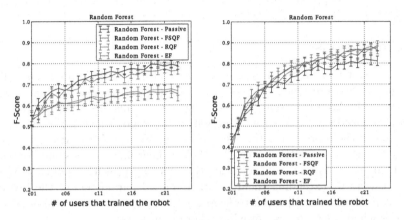

Fig. 3. Left: Learning results of experiment 01: *Looking left, front, right*. Right: Learning results of experiment 02: *Pointing left, front, right*

As already described in Sect. 4.1, *YNQ* and *RQ* produced the same filters, therefore, from now on, we will only describe *RQ* results. Figure 3 shows the results for the *looking* (left) and *pointing* (right) experiments. The figure shows how the filters behaved differently in both experiments. In the *Looking* experiment (Fig. 3, left) *RQ* and *FSQ* performed worse than *PL* with the exception of the cases in which only one user trained the robot. However, the *EF* achieved an F-Score comparable to PL. A different situation occurred in the *Pointing* experiment (Fig. 3, right), where *FSQ* and *RQ* scored better results than *PL*. Here the *EF* behaved as well as the other AL queries.

The lower performance of *AL* in the *looking experiment* might be caused because, in this experiment, the users were providing simplistic answers leading the robot to build filters that omitted relevant data. When the robot had few training examples, the users compensated the lack of data by introducing relevant domain information. However, when the robot gathered enough training examples, their answers prevented the robot to use information it could have helpful for learning. We observed this fact when we checked the learning dataset and found that the data of the user's shoulders might have been relevant. Yet, the users omitted the shoulders when answering the queries, perhaps, because the variations in their shoulders were unperceived.

Because of that, we come up with the idea of lowering the trust in the user's response and including more information than the user is giving. This is exemplified by the *EF*, which included adjacent limbs to the user's answers. Note that despite the *EF* seems tailored for our scenario, we believe that a similar approach can be followed in other scenarios where a strong correlation between learning parameters can be found. In that sense, if a user produces a filter including some learning parameters, it might be interesting to include other correlated parameters as well, even if the user did not include them. Some potential fields that can benefit from this approach are gesture recognition and object recognition among others.

5 Conclusions

The main contribution of this paper is twofold. First, we evaluated how different types of Feature Queries affect the learning performance of a robot that learns actively. We found that, in some cases, user's answers might be too simplistic, potentially leading to a reduction in the robot's learning accuracy. In such cases, if the robot trusts too much in the user's responses, Active Learning approaches might be outperformed by Passive Learning. The second contribution of this paper is a method in which the robot reduces its confidence on the user's answers by extending them to other related parameters of the learning space. This method has proven to keep the learning performance high even in the cases where users did not provide accurate answers.

We tested our approach in an experiment where 24 users trained a social robot to recognise poses. Users were asked for three types of queries *Free Speech Queries, Yes/No Queries* and *Rank Queries*. The answers to these queries were used to build feature filters that pre-processed the training data before it was fed to the learning algorithm. We found that, since *RQs* are more verbose, their filters tend to be more inclusive. However, it was demonstrated that users prefer not to be asked many questions [5], therefore it remains as a future work to explore the optimal balance between a verbose robot and the quality of filters. In this line, we have found that *FSQs* have the advantage of being more natural and they can be much more efficient than the other types of queries.

From our experiments we concluded that inaccurate user verbal responses in *AL* may lead to loss of relevant data in the parameter space. When this happens, *AL* could produce worse results than *PL*. To solve this problem we developed the notion of the *Extended Filter*. As this filter includes limbs omitted by users, it achieves a performance comparable to *PL* in the situations where AL do not worked as well as expected. What is more interesting, when *AL* performes better than *PL*, the *EF* behaves as a regular *AL* approach. Therefore, our *EF* gets the best of both worlds, being a good choice when *AL* is beneficial but it is not possible to control the accuracy in user's answers.

Even though our method is applied for pose learning, we believe that it could be applied to other *AL*-based learning approaches. This is because our approach is based on feature selection, Hence, other learning approaches might apply it as long as the robot knows the relationship between the features it is asking for.

Acknowledgment. This research has received funding from the projects *Development of social robots to help seniors with cognitive impairment - ROBSEN* funded by the Ministerio de Economía y Competitividad (DPI2014-57684-R) from the Spanish Government; and *RoboCity2030-III-CM* (S2013/ MIT-2748), funded by Programas de Actividades I+D of the Madrid Regional Authority and cofunded by Structural Funds of the EU.

References

1. Alonso, F., Gorostiza, J., Salichs, M.: Preliminary experiments on HRI for improvement the Robotic Dialog System (RDS). In: Robocity2030 11th Workshop on Social Robots (2013)
2. Alonso-Martín, F., Salichs, M.A.: Integration of a voice recognition system in a socia robot. Cybern. Syst. **42**(4), 215–245 (2011)
3. Angluin, D.: Queries and concept learning. Mach. Learn. **2**(4), 319–342 (1988)
4. Breiman, L.: Random forests. Mach. Learn. **45**(1), 5–32 (2001)
5. Cakmak, M., Chao, C., Thomaz, A.L.: Designing interactions for robot active learners. IEEE Trans. Auton. Ment. Dev. **2**(2), 108–118 (2010)
6. Cakmak, M., Thomaz, A.L.: Designing robot learners that ask good questions. In: Proceedings of the Seventh Annual ACM/IEEE International Conference on Human-Robot Interaction, HRI 2012, p. 17. ACM, New York (2012)
7. Druck, G., Settles, B., McCallum, A.: Active learning by labeling features. In: Proceedings of the 2009 Conference on Empirical Methods in Natural Language Processing, vol. 1, pp. 81–90. Association for Computational Linguistics (2009)
8. Gonzalez-Pacheco, V., Malfaz, M., Fernandez, F., Salichs, M.A.: Teaching human poses interactively to a social robot. Sens. **13**(9), 12406–12430 (2013)
9. Hall, M., Frank, E., Holmes, G., Pfahringer, B., Reutemann, P., Witten, I.: The WEKA data mining software: an update. ACM SIGKDD Explor. Newsl. **11**(1), 10–18 (2009)
10. Lopes, M., Oudeyer, P.Y.: Guest editorial active learning and intrinsically motivated exploration in robots: advances and challenges. IEEE Trans. Auton. Ment. Dev. **2**(2), 65–69 (2010)
11. Quigley, M., Gerkey, B., Conley, K., Faust, J., Foote, T., Leibs, J., Berger, E., Wheeler, R., Ng, A.: ROS: an open-source Robot Operating System. In: Open-Source SW Workshop of the International Conference on Robotics and Automation (ICRA) (2009)
12. Raghavan, H., Madani, O., Jones, R.: Active learning with feedback on features and instances. J. Mach. Learn. Res. **7**, 1655–1686 (2006)
13. Rosenthal, S., Dey, A.K., Veloso, M.: How robots' questions affect the accuracy of the human responses. In: The 18th IEEE International Symposium on Robot and Human Interactive Communication, RO-MAN 2009, pp. 1137–1142. IEEE (2009)
14. Salichs, M., Barber, R., Khamis, A., Malfaz, M., Gorostiza, J., Pacheco, R., Rivas, R., Corrales, A., Delgado, E., Garcia, D.: Maggie: a robotic platform for human-robot social interaction. In: 2006 IEEE Conference on Robotics. Automation and Mechatronics, pp. 1–7. IEEE, Bangkok, December 2006
15. Settles, B.: Active learning literature survey. Computer Sciences Technical report 1648, University of Wisconsin-Madison (2010)

Recommender Interfaces: The More Human-Like, the More Humans Like

Mariacarla Staffa[1]([✉]) and Silvia Rossi[2]([✉])

[1] Department of Engineering, University of Naples Parthenope, Naples, Italy
mariacarla.staffa@uniparthenope.it
[2] Department of Electrical Engineering and Information Technology,
University of Naples Federico II, Naples, Italy
silvia.rossi@unina.it

Abstract. Social robots, when used for information providing, are able to affect humans' trustworthiness and willingness to interact with them. In this work, we conducted an experimental study aimed at observing if the users' acceptance of recommendations, as well as their engagement in the interaction, is elicited when using a humanoid robot with respect to a common application on a mobile phone. We conducted an experimental study on movie recommendation where the two interfaces provide the same contents, but through different communication channels. In detail, the robot will attend to the participants in a socially contingent fashion, signaled via head and gaze orientation, speech, eye color and gestures related to the genre of the recommended movie, and the app will provide textual and graphical movie presentation. Results show that while the users perceive the interaction with the mobile application more natural, the social robot is able to enhance the users' satisfaction and provides a good and stable acceptance rate also when facing participants with various degrees of English proficiency.

1 Introduction

The future perspective of using robots in everyday life is becoming more and more realistic. For this reason, robots must be designed in a way to be employed in settings requiring social interaction. They should be, hence, perceived as trusting, helpful and engaging [8]. These characteristics become even more important when robots are engaged in information providing tasks (e.g., providing recommendations on items), which could affect the choices of users interacting with them. A Recommender System (RS) is represented by a filtering technology that, basing on the preference that users would give to specific items, seeks to predict products that are likely to be of interest for them. While, from the one hand, the accuracy of recommendations depends on the quality of the recommender algorithm, it has been shown that the ultimate acceptance of a recommendation depends also on the user experience while getting it [9]. Thus, the provision of a recommendation is as much important as the quality of the recommending algorithm. It is more effective if it is positively evaluated by the users [11], it infuses a

A. Agah et al. (Eds.): ICSR 2016, LNAI 9979, pp. 200–210, 2016.
DOI: 10.1007/978-3-319-47437-3_20

sense of trust and it allows users to perceive some related benefit [13]. RSs often include anthropomorphic virtual agents [18], because it is assumed that social responses are more prevalent if the system is personified and that the presence of a humanoid virtual agent induces trust [17]. Embodied social agents used to provide recommendation can make the interaction more meaningful w.r.t. simple interfaces (which do not display actions or speech), because users' attitude towards social agents is similar to that they show towards other people. It has been observed, in fact, that the robots endowed with social behaviors, similar to that of humans, are more compelling for human-robot interaction [4].

In literature, different studies compared the impact of recommendations and advises as provided by social robots with respect to virtual agents [16], by showing that the embodiment condition, as provided by the robot, has a greater impact on the users with respect to 2D/3D virtual agents on a screen. Indeed, real robots affect subject decision-making more effectively than computer agents in real-world environments [2]. Moreover, non-verbal behaviors serve important functions in affecting the trustworthiness of a recommendation [10]. In fact, a robot ability to build a trust relationship depends on its capacity to help people understand it, in part through non-verbal behavior. Emotion-related signals, such as those provided by voice pitch changes in speech or gesture, as well as attentional mechanism provided through gaze and head orientation [5,14] are non-verbal behaviors that influence human trust [3]. It has been, indeed, well-documented that humans expect from humanoid robots socially intelligent responses. This leaves the possibility that an agent may influence how humans perceive a recommendation through the presence of many or few communication abilities.

While many studies exist that compare the impact of recommendation as provided by social robots with respect to virtual agents, a poor literature treats the comparison between social robots and applications on mobile phones, which represent, nowadays, the most commonly used interfaces. Moreover, new design trend for social robots often includes a tablet or a large I/O screen within the social robot head or body. Hence, in this work, we conduct a user study with the aim of observing if the users' acceptance of recommendations, as well as their engagement, is elicited when using the robot with respect to a common application on mobile phone, in case of the participants' mother tongue or the English language are used. The two analyzed recommending interfaces will provide the same information to the users, but through different communication channels. While the robot will attend to the participants in a socially contingent fashion, signaled via head and gaze orientation, speech and motion, the app will only provide a textual and graphical presentation of movies. Our aim is to observe if the presence of non-verbal behavioral cues that involve eye contact, genre-driven motion primitives, but no graphical displays, can have a potential impact on trust and acceptance of users and, consequently, make robots a more valuable interface for providing recommendations with respect to the apps.

2   A Movie Recommendation Case Study

In order to evaluate the effect of different interfaces in providing recommendations, we developed a client/server application, where the server provides the recommendation service and the possible clients can be either a humanoid robot or a mobile application. The clients are in charge to ask for a list of recommendations (in particular of movies) to the server and to propose them to users through the use of different communication channels. This diversity should be reflected in a different perception of the recommendations by the users, and, presumably, it will affect their experience. The server layer is characterized by a Recommendation Engine that provides rating predictions when the recommendation API is invoked (see Fig. 1). To generate recommendations, we used an item-based collaborative filtering approach, as developed in [15], with a *City Block distance*. In order to provide recommendations, the Recommendation Engine needs some initial movie ratings from the users (at least 20 movies) provided by using the mobile app. After this first stage, the user can get movie recommendations from the server, which, once calculated the best movies for the user, it retrieves additional details about the film, such as director, actors and genres using OMDb[1] web service. For details please refer to [6].

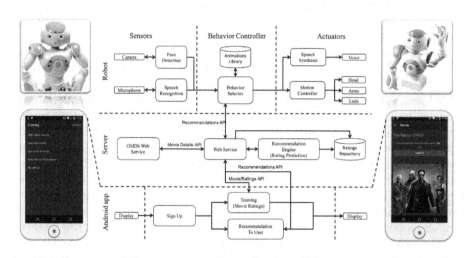

Fig. 1. The general architecture of the movies recommendation system.

2.1   Recommending Interfaces

The *Android application* (APP) used to show recommendation is the same one used for the training phase. It shows on the screen the recommendations for

---

[1] http://www.omdbapi.com - The Open Movie Database is a free web service to obtain movie information.

the users through textual and graphical descriptions. In particular, for each presented movie, it shows the name, the actors, the directors and the genre, as well as the plot and the image of the movie poster. Finally, in the same screen, it asks whether or not the user is willing to see the movie (i.e., he/she accepts the proposed recommendation).

The *Robot* client interface (NAO) has been designed considering the adoption of a NAO T14 robot model, consisting in a humanoid torso with 14 degrees of freedom (2 for the head and 12 for the arms) developed by Aldebaran Robotics. NAO is endowed with two main sensors: a camera and a microphone through which it receives signals from the external environment. Camera frames are processed by a *Face Detection* module to detect the user's presence into its visual field. Once a user is detected, the robot will follow his/her face by moving its head while providing the recommendation [14]. Sounds obtained from the microphone are processed by a *Speech Recognition* module and are used to authenticate the user.

The robot presents to users the recommended movies and their relative information through the use of different speech intonations, accompanied with gestures and eye color, which change according to the movie genre of the recommended movie. In particular, for the eye coloring, we referred to the "Wheel of Emotions" of Robert Plutchik, which associates emotions to colors. We then created a table to associate the main movie genres (such as Animation, Adventure, Comedy, Dramatic, Fantasy, Horror, Musical, Thriller, etc.), to a list of one or more possible emotions. In this way, we can easily choose an appropriate color for NAO eyes (e.g., red eyes for a love movie, yellow eyes for a comedy, and so on). As for the gestures, we mapped the main movie genres into the available animations for NAO in the *AnimationsLibrary* (e.g., Air Guitar for Musical, Kung Fu for Action movie, Mystical Power for Fantasy movie, and so on). Finally, the pitch of the voice is accordingly manipulated by the *Speech Synthesis* module, starting from some insights arising from studies on the Emotional Prosody [7], which have shown that, for example, some emotions, such as fear, joy, and anger, are portrayed at a higher frequency than emotions such as sadness. Thus, by associating movie genres with emotions, we can, also in this case, provide a direct mapping between the movie genre and the pitch intonation. After the presentation of a recommended movie, the robot asks whether or not the user wants to see the proposed movie and recognizes the provided answer.

3   Experimental Results

We conducted a user study with the goal of evaluating the effectiveness of the use of a social robot as an interface to provide recommendations. Our aim is to observe if the users' acceptance of recommendations, as well as their engagement in the interaction, is elicited when using the human-like robot with respect to a common application on a mobile phone. For this purpose, we designed this study as a two within-subjects different, repeated measures experiment, where the independent variable, in each experiment, is the interface used for providing the recommendation, respectively the humanoid robot (NAO) or the mobile application (APP). In a within-subject design, every single participant is subjected to every single test. The order in which tests are presented can actually affect the behavior of the subjects or elicit a false response. In order to avoid the problem of carry-over effects, where the first test adversely influences the other, we adopted a counter-balanced measures design for the test procedure (i.e., the order with which the participants interact with the APP or with NAO is random and balanced between the two interfaces). Then, we also decided to allow each interface proposing two different movies, as a consistency check, but no more than two since we want to find a trade-off between fatigue and practice for the users. In a long experiment, in fact, the participants may be tired risking to decrease their performance on the last study.

3.1 Testing Procedure

The testing procedure main steps are: (a) at the beginning of the interaction, the user provides new rates for a list of movies (training phase) and provides personal information (gender, age, instruction level, a self-evaluation of robotic skills and English language proficiency); (b) the recommendation system generates the top-four recommendations for each user, which will be shown to users through the two interfaces (APP and NAO) in a random way (two for each); (c) for each of the two movie recommendation interfaces, users completed the following tasks: (c.1) provide an acceptance/reject response for the proposed movies; (c.2) complete satisfaction and usability questionnaire after each interaction (see Table 2).

Two different tests were conducted that differentiate for the language used to provide the movie recommendation (English or Italian). In order to validate the hypothesis that the robotic application is a valuable solution for recommendation systems with respect to a mobile application, we must avoid experimental constraints that could possibly affect the interaction performance. While for the first experiment, we developed an application relying on the use of the English language, in the second, we wanted to avoid that a low English proficiency could have an impact on the movie acceptance rate. Using the mother tongue ensures mostly complete comprehension, while, by introducing the diversity of the language factor, we are able to better understand the effects of non-verbal behavior on users' understanding, acceptability and engaging. For this purpose, we collected the same data in a second experiment using the Italian language, to explore whether differences between people within the different language conditions, also have correspondingly different preference scores and acceptance rate (percentage of movies accepted).

TEST 1. We conducted the first experiment with 36 subjects (67 % Males and 33 % Females). Concerning the technical background, the 56 % declared to have high skills in robotics, while the remaining 44 % does not declare robotics proficiency. All the participants were Italian native speakers, the 39 % with medium/high English language knowledge and the 61 % with medium/low level of English proficiency. The language adopted for the experiment was the English language both for text description and for the robot's voice synthesizer.

TEST 2. Additionally, we conducted the second test (TEST 2) with 30 participants (63 % males and 37 % females). The 40 % of the participants had high robotics skills, while the 60 % had very low proficiency with robotic devices. We followed the same procedure as TEST 1, but adopting the Italian language.

3.2 Results Analysis

Our starting hypothesis is that the more the recommending interface is human-like the more the humans like it. This can be translated into the following expected results:

- **H0:** the humanoid robot will be more engaging and better liked with respect to the mobile application: this qualitative value can be evaluated by considering the participants' opinion about the interaction;
- **H1:** since better liked, the recommendations provided by the robot should be more likely to be accepted.

For this scope, we evaluated the number of recommendations that were accepted by the user, and the proposed a qualitative questionnaire aimed at collecting their explicit impressions on the interaction through 6 questions: Q1. How easy was to perform the task? Q2. Did the system react accordingly to your expectations? Q3. How natural is this kind of interaction? Q4. How satisfying do you find the interactive system? Q5. How convincing was the interface? Q6. Which is the interface you prefer to interact with? We adopted a classical likert scale from 1 to 5. Only for question 6, we explicitly ask for a direct preference on an interface. Finally, we will make some consideration starting from the statistical analysis of between-groups data grouped per features (English proficiency, robotics skills, and gender).

As shown in Fig. 2, the difference between the acceptance rate of movies recommended respectively by the APP and by NAO in TEST 1 is small and not significant (one-way ANOVA with $p = 0.35$). This result could come from the used recommendation algorithm that is able to propose movies, which, presumably, fit very well the users' preferences, independently from the used interface. Hence, the hypothesis **H1** is not sustained by TEST 1 outcomes since the embodiment condition (NAO) does not imply significant changes in the testers' acceptance rate (from 50 % to 61 %).

Conversely, the percentage of preferences, expressed by the users at the end of the interaction with the two interfaces, clearly denotes an enhanced experience of users when interacting with the humanoid robot (from 17 % to 83 %). This result is in accordance with our hypothesis **H0** that people are inclined to prefer an interface interacting through more natural modalities.

As for TEST 1, hypothesis **H1** is not sustained by TEST 2 outcomes since, as shown in Fig. 2, the values of the acceptance rate of movies recommended respectively by APP and NAO do not differ in a significant way (ANOVA with $p = 0.72$), and the users still prefer the NAO interface (73 % w.r.t. 27 %).

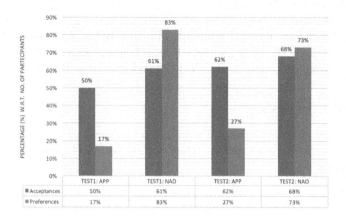

Fig. 2. Acceptance and preference rates of movies recommended by APP and NAO.

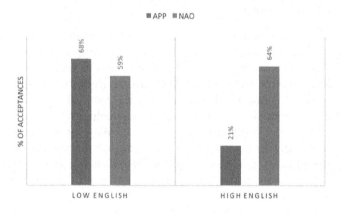

Fig. 3. Acceptance rate evaluated by grouping participants per Low/High English proficiency.

Results show that English proficiency level affects the acceptance rate for both the interfaces, with a statistical significance for the APP condition (one-way ANOVA with $F = 8.92$ and $p = 0.005$). In particular, for the APP case there is a moderate negative correlation between the declared English proficiency and the acceptance rate (PEARSON correlation with $\rho = -0.46$ and $p = 0.005$). This is to say that users with a high level of English proficiency accept fewer recommendations with respect to users with a low level. This is also shown in Fig. 3, where the numbers of accepted recommendations are plotted in the case of low/high English proficiency for both APP and NAO. Notice that, in the case of APP, users with a low level of English proficiency accept more the provided suggestions (68 % for low and 21 % for high). In the case of NAO, such difference is smaller (59 % for low and 64 % for high). Since NAO provides a similar and good acceptance rate in both cases (the difference between these two groups is not statistically significant with F = 0.09 and p = 0.76), the use of such interface could be valuable whenever the audience, in terms of different level of English proficiency, is wide. Finally, users with a good understanding of English are more inclined to accept recommendations provided by NAO than by APP.

Figure 4 shows that robotics skills did not to have a significant impact on the acceptance rate in TEST 1 for both NAO (one-way ANOVA with $F = 0.27$ and $p = 0.60$) and APP conditions (one-way ANOVA with $F = 0$ and $p = 1$). In TEST 2, robotics skills affect the acceptance rate under the APP condition (one-way ANOVA with $F = 4.2$ and $p = 0.05$), while they do not have significant effects on the acceptance rate induced by NAO (one-way ANOVA with $F = 1.14$ and $p = 0.29$). Also in this case, few variations are introduced by the considered factor (robotics skills) on the acceptance rate induced by the interaction with NAO. This result endorses the previous one, stating the robustness and stability of this interface under certain potential influencing factors (such as English proficiency and robotics skills levels).

We, finally, computed the N-way ANOVA on TEST 1 for testing the effects of multiple factors (English proficiency and robotics skills) on acceptance rate mean in case of NAO and APP. Results showed that there is no significant interaction component neither for APP ($F = 2.14$ with $p = 0.15$) nor for NAO ($F = 1.21$ with $p = 0.28$).

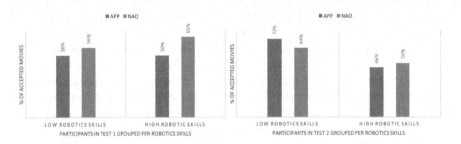

Fig. 4. Comparison between TEST 1 and 2 on acceptances by grouping users per robotics skills.

Concerning the gender aspect, the statistical analysis shows that the users' gender affects the acceptance rate for APP interfaces in TEST 1 (one-way ANOVA with $F = 4.86$ and $p = 0.04$). In particular, in TEST 1 (see Table 1), men were less inclined to accept the recommendations from the APP rather than from NAO (38 % APP vs 58 % NAO). This data is also in accordance with the preferences expressed by the users after the interaction (25 % APP vs 75 % NAO). Women were more inclined to accept recommendations in general, independently by the interface used (75 % for APP and 67 % for NAO), however, they preferred to interact with NAO in the 100 % of cases. Finally, no statistically significant differences were found grouping by age.

We also analyzed the interaction from the users' point of view (see Table 2, where bold values represent average evaluations with statistically significant

Table 1. Acceptance and preference rates of participants grouped by gender.

				Gender				
		TEST 1				**TEST 2**		
	Male		Female		Male		Female	
	Acc	Pref	Acc	Pref	Acc	Pref	Acc	Pref
APP	38%	25%	76%	0%	71%	42%	45%	0%
NAO	58%	75%	67%	100%	55%	58%	64%	100%

Table 2. HRI qualitative questionnaire ratings.

		Q1	Q2	Q3	Q4	Q5		Q1	Q2	Q3	Q4	Q5
		Easiness	Expect.	Natural.	Satisf.	Trust.		Easiness	Expect.	Natural.	Satisf.	Trust.
TEST1	APP	**4.28**	3.39	**4.11**	3.89	4.67	TEST2	APP 4.60	3.87	4.20	3.83	3.43
	NAO	3.83	3.72	3.67	3.67	4.72		NAO 4.70	4.23	3.90	3.93	3.53

differences). The questionnaire proposed to the users highlighted a good satisfaction in the interaction. Concerning the easy of use of the interface (Q1), users perceived more easy to use the APP when the test was made in English, with a statistically significant difference ($p = 0.01$) between APP ($average\_rate = 4.28$) and NAO ($average\_rate = 3.83$). While, in TEST 2, the Italian language make the interaction easy in both cases APP ($average\_rate = 4.6$) and NAO ($average\_rate = 4.7$) without significant difference ($p = 0.6$). This is to say that whenever the English proficiency is low, users rely on more common and used interfaces as the one provided by the APP. Giving socially intelligent responses, the humanoid robot did not disappoint the expectations of participants, who judged better the interaction with NAO in both tests with respect to the expectation (Q2), but with no significant differences ($p = 0.2$ for TEST 1 and $p = 0.1$ for TEST 2). The users found more natural to interact (Q3) with the APP, since they are accustomed using it, with a significance in the case of TEST 1 ($p = 0.02$ for TEST 1 and $p = 0.17$ for TEST 2). Finally, the interaction with both the interfaces has been evaluated as satisfactory (Q4 with $p = 0.3$ for TEST 1 and $p = 0.6$ for TEST 2).

4   Conclusions

In this work, we considered a case study on movie recommendations and evaluated the impact of two information providing interfaces: a humanoid robot and a mobile-phone application. This work aims at answering the question if robots with human-like skills are preferred by humans than other types of interactive and commonly used interfaces, to the extent that they can affect the humans' choices. Our hypothesis is that the humanoid robot, having the potential to portray a rich repertoire of nonverbal behaviors with a familiar social meaning for users, should make the interaction more credible and engaging.

Results from the statistical analysis showed that even if the acceptance rate induced by NAO is always higher than that caused by using APP, there are not statistically significant differences between the acceptance rate obtained by the two interfaces. However, while there are several factors (English Proficiency, robotics skills, gender) influencing the interaction with APP, the acceptance rate remains stable when interacting with NAO. Namely, the NAO recommender interface resulted in being more stable with participants with a different degree of English proficiency, that was not the participants' mother tongue, and with people with different robotics skills. Other studies already provided the evidence of using robots for teaching English as a foreign language [1].

Additionally, from the qualitative analysis, we observed that, from the one hand, the users perceived the interaction with the APP more natural since they are more familiar with this kind of interfaces or they could be influenced by the fact that the case study needed some data from the user, which is a first interaction with the android application. From the other hand, the social robot, using behaviors such as speech, movement, and gaze following that can be easily interpreted by humans, is able to enhance the users' satisfaction with respect to their expectations. This result is in accordance with the research studies of [12] since our hypothesis is that the social cues provided by a humanoid robot facilitate joint attention, rapport, and a sense of trust, thus positively affecting language understanding.

Acknowledgment. This work has been partially funded by the European Commission's as part of the RoDyMan project under grant 320992 and supported by the Italian National Project "Security for Smart Cities" PON-FSE Campania 2014-20. Authors thank Francesco Cervone, Anna Tamburro and Valentina Sica for their contribution in code development and testing.

References

1. Alemi, M., Meghdari, A., Ghazisaedy, M.: The impact of social robotics on L2 learners' anxiety and attitude in english vocabulary acquisition. Int. J. Social Robot. **7**(4), 523–535 (2015)
2. Bainbridge, W., Hart, J., Kim, E., Scassellati, B.: The effect of presence on human-robot interaction. In: The 17th IEEE International Symposium on Robot and Human Interactive Communication (RO-MAN), pp. 701–706, August 2008
3. Boone, R.T., Buck, R.: Emotional expressivity and trustworthiness: the role of nonverbal behavior in the evolution of cooperation. In: Lib., A.R. (ed.) J. of Nonverbal Behavior (2003)
4. Bruce, A., Nourbakhsh, I.R., Simmons, R.G.: The role of expressiveness and attention in human-robot interaction. In: ICRA, pp. 4138–4142. IEEE (2002)
5. Caccavale, R., Leone, E., Lucignano, L., Rossi, S., Staffa, M., Finzi, A.: Attentional regulations in a situated human-robot dialogue. In: The 23rd IEEE International Symposium on Robot and Human Interactive Communication, pp. 844–849, August 2014

6. Cervone, F., Sica, V., Staffa, M., Tamburro, A., Rossi, S.: Comparing a social robot and a mobile application for movie recommendation: a pilot study. In: Proceedings of the 16th Workshop from Objects to Agents, pp. 32–38. CEUR Workshop Proceedings (2015)

7. Crumpton, J., Bethel, C.L.: Validation of vocal prosody modifications to communicate emotion in robot speech. In: 2015 International Conference on Collaboration Technologies and Systems (CTS), pp. 39–46, June 2015

8. Kidd, C., Breazeal, C.: Effect of a robot on user perceptions. In: Proceedings of the IEEE/RSJ International Conference on Intelligent Robots and Systems (IROS), vol. 4, pp. 3559–3564 (2004)

9. Konstan, J.A., Riedl, J.: Recommender systems: from algorithms to user experience. User Model. User-Adap. Inter. **22**(1), 101–123 (2014)

10. de Melo, C.M., Zheng, L., Gratch, J.: Expression of moral emotions in cooperating agents. In: Proceedings of the 9th International Conference on Intelligent Virtual Agents (IVA) (2009)

11. Merritt, S.M., Ilgen, D.R.: Not all trust is created equal: dispositional and history-based trust in human-automation interactions. Hum. Factors **50**(2), 194–210 (2008)

12. Movellan, J., Eckhardt, M., Virnes, M., Rodriguez, A.: Sociable Robot Improves Toddler Vocabulary Skills. IEEE, La Jolla, CA (2009)

13. Murphy-Hill, E., Murphy, G.: Recommendation delivery. In: Robillard, M.P., Maalej, W., Walker, R.J., Zimmermann, T. (eds.) Recommendation Systems in Software Engineering, pp. 223–242. Springer, Heidelberg (2014)

14. Rossi, S., Staffa, M., Giordano, M., De Gregorio, M., Rossi, A., Tamburro, A., Vellucci, C.: Robot head movements and human effort in the evaluation of tracking performance. In: 24th IEEE International Symposium on Robot and Human Interactive Communication (RO-MAN), pp. 791–796 (2015)

15. Rossi, S., Cervone, F.: Social utilities and personality traits for group recommendation: a pilot user study. In: Proceedings of the 8th International Conference on Agents and Artificial Intelligence, pp. 38–46 (2016)

16. Shiomi, M., Shinozawa, K., Nakagawa, Y., Miyashita, T., Sakamoto, T., Terakubo, T., Ishiguro, H., Hagita, N.: Recommendation effects of a social robot for advertisement-use context in a shopping mall. Int. J. Soc. Robot. **5**(2), 251–262 (2013)

17. Wang, Y.D., Emurian, H.H.: An overview of online trust: Concepts, elements, and implications. Comput. Hum. Behav. **21**(1), 105–125 (2005)

18. Yoo, K.H., Gretzel, U.: Creating more credible and persuasive recommender systems: the influence of source characteristics on recommender system evaluations. In: Recommender Systems Handbook, pp. 455–477. Springer, US (2011)

Designing a Social Robot to Assist in Medication Sorting

Jason R. Wilson[1(✉)], Linda Tickle-Degnen[2], and Matthias Scheutz[1]

[1] Human-Robot Interaction Lab, Tufts University, Medford, MA 02155, USA
wilson@cs.tufts.edu
[2] Department of Occupational Therapy, Tufts University, Medford, MA 02155, USA

Abstract. Being able to sort one's own medications is a critical self-management task for people with Parkinson's disease. We analyzed the medication sorting task and gathered design considerations. Then we developed an autonomous robot to assist in the task. We used guidelines provided by occupational therapists to determine the level of assistance provided by the robot. Finally, an evaluation of the effectiveness of the robot with student evaluators determined that people trusted the robot to reliably assist and that people had a positive emotional experience completing the task.

1 Introduction

Medication adherence, defined as the extent to which patients take medications as prescribed [9], can be a great challenge for many people that rely on a daily regimen of medications. It requires successful "medication management", i.e., the ability to develop, schedule, and implement a plan to take medications, as well as to remember if medications have been taken and when to take them [16]. Adhering to a successful medication management schedule is often critical for managing and delaying the progression of chronic diseases.

Parkinson's disease (PD) is a progressive and neurodegenerative condition that reduces the amount of dopamine produced in the brain [8,10]. PD currently has no cure, but symptoms can be dramatically reduced with antiparkinsonian medication. Optimizing adherence to oral medication regimens is critical to managing the symptoms of PD [6,7], but many challenges impede good medication adherence [14].

While most existing assistive technologies remind a person when it is time to take a particular medication [2,5], there are few assistive technologies that address the important problem of planning, scheduling, or sorting of medications. We believe that a social robot assisting in a medication sorting task could serve an important role in the self-management of Parkinson's disease, and we present the first steps towards developing such a system. To our knowledge there are currently no other technologies that can interact with the person to develop an ideal medication schedule that is also adherent to how the medications are prescribed. In the following, we first analyze the medication sorting task and

© Springer International Publishing AG 2016
A. Agah et al. (Eds.): ICSR 2016, LNAI 9979, pp. 211–221, 2016.
DOI: 10.1007/978-3-319-47437-3_21

gather design considerations, then we present on overview of the development of the autonomous robot to assist in the task, and evaluate the effectiveness of the robot in a human-robot interaction experiment.

2 Background

People with PD are often referred to occupational therapists to assist in activities of daily living (e.g., dressing, bathing) as people with PD that are early in the disease progress may experience difficulties with instrumental activities of daily living (IADLs) [15]. These activities include physically demanding ones like sweeping or carrying groceries, but more cognitively demanding IADLs often present challenges to people with relatively early PD [4]. Using an objective performance-based metric, Foster found measurable deficits in performance of cognitively demanding IADLs [3]. On activities such as shopping, sharp utensil use, and medication management a larger proportion of people with PD required assistance or required more assistance than people without PD [3].

Adherence to a medication regimen is critical to treatment outcome and quality of life for people with PD. In addition to general issues elderly patients may have (e.g., age-related physical decline, economic factors), people with PD often require complicated dosing or titration schedules and may have co-morbidities that require the coordination of therapies from multiple drug classes. Disease progression can also introduce cognitive impairments that can affect adherence. Additionally, responses to antiparkinsonian agents can cause variable responses interfering with medication adherence [1].

To achieve optimal adherence in elderly patients with Parkinson's disease, a combination of approaches is the best strategy for success. Suggestions include educational intervention, simplified dosing and administration schedules, management and understanding of medication adverse events, and the use of adherence aids such as pill boxes and hour-by-hour organizational charts [1].

3 Development and Assessment of Medication Sorting Task

To get a thorough understanding of the medication sorting task, the importance of the task, and how a robot could assist with this task and others, we developed a pill sorting task that was subsequently performed by two humans and then evaluated by a *focus group* consisting of four occupational therapists (OT). The standardized task involves one person sorting two medications onto a grid while the other person provides assistance [12]. The focus group was specifically charged to assess task performance and analyze the various activities performed as part of the task, discussing the possible roles and design considerations for a social robot assisting in this task.

The standardized medication sorting task of the Performance Assessment of Self-care Skills [12] was the basis for a simulated performance of the task. The video of this task showed an actor, who simulated a person with PD, placing two medications on a sorting grid while being instructed by an actor who simulated the caregiver. Research assistants created a written activity analysis of the recorded performance [11]. In an evaluation and focus group session, OT experts viewed the recorded scenario, read the activity analysis, and independently completed 17 Likert-scaled questions about the realism, comprehensiveness, and relevance of the content of the scenario and analysis.

The OT experts agreed that the medication sorting task was valuable for daily life with PD and that the content of the activity analysis was comprehensive and accurate. They also agreed that the video of the simulated task performance was not as complex as would typically occur in the home (e.g., there would be more medications with more constraints, and the environment in which the task is being done, often a kitchen table, may be cluttered and have other distractions). The experts also suggested that more safety concerns needed to be included in the activity analysis. We concluded that with these minor improvements the activity analysis of the medication sorting task has been validated as to its significance to a person with PD, and that the task has been sufficiently described for designing a social robot to assist in the task.

4   Autonomous Robot Design

Based on the task analysis and the outcome of the focus group, we have begun the development of a social robot to assist in the medication sorting task. An important design concern was for the architecture to be extendable to potentially incorporate additional tasks that are important for the daily activities of a person with PD.

4.1   Architecture

The components of the system have been developed using the DIARC architecture [13]. For this medication sorting task, the necessary components are the vision system, script execution, medication management and assistance, and robot control. Each of these components are described in the following sections.

4.2   Vision System

The vision system is responsible for perceiving the state of the environment so that other components can infer the state of the task. In addition, changes in state are used by other components as an indication that a human action might have occurred. In determining the state of the medication grid and the sorting tray, the vision system reports how many pills of each type are in each cell in the grid. The sorting tray, which is a small circular tray, is treated as a separate grid with one cell. The vision system can also report how many pills of each type are in the tray.

4.3   Script Execution

Script execution is conducted by the *Goal Manager* component in the DIARC architecture. A script here defines the sequence of actions the robot is to perform. A script may be defined using a hierarchical structure in which high-level actions have their own scripts. The high-level sequence of actions we use for the medication sorting task is introduce the robot and the task, assisting in sorting the first medication and then the second, then indicate that the task is complete. The action for sorting a particular medication takes a parameter that indicates which medication is to be sorted next. The script for that action is the following:

1. Robot announces which medication is next
2. Robot instructs person to pour medications into tray
3. Robot waits for medications in tray
4. Provide instructions for medication
5. Wait for an event
6. Respond to event
7. If not done sorting this medication, return to step 5

4.4   Medication Management and Assistance

A medication management component assesses each event. It selects the appropriate response to be made by the robot based on details of the event. The decision process is mostly a binary decision tree where decisions are based on characteristics of the event. The analysis of the event determines the following information:

- grid correctness: current state of grid is consistent with the goal
- correction event: last event fixed a grid inconsistency
- hesitation event: last event has no state change and simply is a timeout, possibly due to human partner moving slowly
- number of recent errors: number of recent events that were errors
- current errors: total number of misplaced pills
- too quick: user is moving too quickly

Based on the analysis of the event, the component then uses the decision tree to determine how to react. Reactions are intended to provide an appropriate level of assistance, where the basic characteristics of each level is based on the "Hierarchy of Types of Assistance". This hierarchy, specified in the Performance Assessment of Self-Care Skills (PASS) 3.1 [12], defines the minimal type of assistance a therapist is to provide to facilitate task performance. Of the nine defined levels, we use only the first four levels. Assistance at level 5 and higher requires greater physical intervention and the current robotic system does not have the necessary manipulation capabilities.

Figure 1 shows the four levels of assistance (plus a level 0 that we have added) laid out in a binary decision tree. The PASS defines the four levels of assistance that we use as the following. Level 1 is *verbal supportive* and includes verbal affirmations to encourage the person it continue the task. Level 2 is *verbal non-directive* and encourages task continuance without telling the person exactly what to do. Level 3 is *verbal directive* and uses statements that direct the person on what to do or how to proceed with the task. Level 4 is *gesture* and includes pointing to objects and may be accompanied by verbal statements.

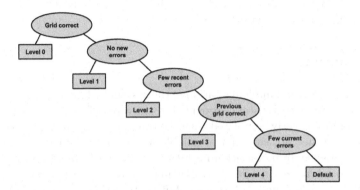

Fig. 1. The type of response that the robot gives to an event is determined by this binary decision tree. Levels 1–4 are designed to reflect the minimal level of assistance a therapist would provide on this task, as defined by the PASS manual [12].

Level 0 responses are minimal feedback given to the human partner to indicate that the task is progressing correctly and to confirm that the robot is paying attention to the actions. The PASS does not have a level 0 type of assistance, but reviewing the video recording of the scripted interaction and consulting with OTs with experience in the task indicated that simple back-channeling was common and possibly necessary. However, the robot must not always give positive feedback as this could introduce dependence on this positive reinforcement signal from the robot. Since a main aim of our system is to support human autonomy, feedback must be minimal, intermittent, and mostly confirm that the robot is working and paying attention.

The system distinguishes between three types of events that merit different types of level 1 responses. If the last event is correcting a misplaced pill, then the response generated recognizes this and confirms the correct action. If the event is the result of a timeout, possibly indicating that the human partner is moving slowly, then an encouraging remark is generated. If the last event is the result of a correctly placed pill but there remains another pill that is incorrect, then a level 1 response is generated to confirm the correct action but a level 2 response is also necessary to indicate that there is still a mistake somewhere on the sorting grid.

A level 2 response indicates that a pill may be misplaced but does not provide direct indication as to what or where the mistake is. The intent is for the human partner to identify the incorrectly placed pill and fix it without further assistance.

If the human needs further assistance, then the feedback gets escalated to level 3. The feedback is more direct in this case. The current implementation repeats the instruction for the medication currently being sorted. If this is still not sufficient, then the level 4 feedback includes a gesture. The robot will then point to a misplaced pill on the grid. If there are too many misplaced pills, then further intervention beyond the currently implemented capabilities will be necessary. For the purposes of our evaluation, the robot simply suggests that the person slow down and think more carefully.

5 Evaluation

We evaluated the initial implementation of the system by having student subjects complete the task with assistance provided by the robot. The goals were to ensure the task can be correctly completed with the robot assisting and that people find the robot to be helpful, supportive, reliable, trustworthy, and create an overall positive experience. In preparation for future studies with the target population, we were also interested in how the robot may be perceived (by one's family or care providers) and other factors contributing to the desired experience (e.g., the person feeling in control and responsible).

5.1 Method

Participants: Students (N = 11) from Tufts University participated in a human-robot interaction study. No demographics were collected from the participants. There were three participants who were involved in the preliminary study but did not know any details about the present experiment.

Materials and design: A Nao robot was on the table when the participant entered. In front of the robot was a medication sorting grid, two cups of simulated medicines (candies), and a tray. The experimenter informed the participant that the robot will be assisting in a task involving placing the medications onto the sorting grid. Also, each participant was instructed to follow the instructions of the robot but also make two types of mistakes. The first mistake is to act slowly, to take more than 3 s to place the next pill. The other mistake is to misplace a pill. Any mistake may be made more than once if the participant chose. We recorded a video of each participant completing the task with robot.

Procedure: The participant sat directly in front of the robot. Once the experimenter left the room, the experimenters in the other room initiated the robot's execution. The fully autonomous robot then followed the script described above. As part of this script, the robot would say the following:

- Introduction: "Hello. ⟨opens hands⟩ My name is Shafer. ⟨left hand closes and moves towards chest⟩"

- Describing the task and materials: "Today we will be sorting medications onto this grid. ⟨points with left hand at grid⟩ There are pills in these containers ⟨points with right hand at cups of pills⟩. Let's begin."
- Instructing placements: "One ⟨color⟩ pill is to be taken each ⟨time of day⟩." See Sect. 4.4 for assistance given by robot during this phase.
- Completion: "Congratulations! You have successfully completed the task."

Once the task was completed, the experimenter returned and provided the participant with a questionnaire, which had 19 Likert-scaled questions (see Table 1) and 1 question for comments.

Table 1. Questions had 5 options from "Strongly disagree" to "Strongly agree".

1. The robot is able to provide you with assistance in the task.
2. The assistance the robot provides is correct.
3. I am able to complete the task more efficiently with the assistance of the robot.
4. When the robot corrects me I feel included to follows its instructions.
5. I trust the robot to (correctly) provide assistance.
6. I expect the robot to act in a consistent and predictable manner.
7. The robot is able to provide physical support.
8. The robot is able to provide emotional support.
9. The robot paid attention to me.
10. The robot used action and words that did not make sense.
11. The robot helped me understand how to complete the task.
12. The robot acted in a manner that ensured my safety.
13. The robot is able to warn me of potentially unsafe medication administration.
14. My family would approve of the way the robot assisted me.
15. My care providers would approve of the way the robot assisted me.
16. I felt pleasant during the task.
17. I felt in control of what was happening during the task.
18. I felt I understood what was happening during the task.
19. I felt responsible for completing the task.

5.2 Results

All participants successfully completed the task and answered 19 questions about the robot. The average scores of these categories are in Table 2. The highest rated questions were 18 and 19, and the lowest (after inverting question 10 for being negatively framed) was 7. It was anticipated that 7 would be rated lowly since the robot has limited physical capabilities. A principal components analysis (PCA) showed that two of the questions did not correspond with any of the composites. The three composites that were formed are related to effectiveness, support and assistance, and emotion. The resulting scores of the three composites are given in Table 2 (note that Support 7 and Rapport 10 were removed from the PCA).

Since some participants had previous experience with the project (though not with the robot), we compared the results of those participants with the rest to see if there was any difference. Only Support had a statistical difference between the groups: questions 7 ($t(7) = 3.33$, $p = 0.01$) and 8 ($t(7) = -2.57$, $p = 0.04$).

We performed an additional post-hoc analysis since some participants interacted more with the robot than others (i.e., when the robot introduced itself, some would reply with some form of greeting). We analyzed the video of each participant and coded whether the participant greeted the robot. Two coders reviewed each video and classified each of the participants as greeters or non-greeters. Of the 11 participants, 5 had some form of greeting toward the robot.

Using an independent sample t-test, we compared the scores from the questionnaire for the greeters and non-greeters. The non-greeters reported

Table 2. The means and standard deviations for each question, means for each category, and relation to each dimension based on a principal component analysis.

		M	SD	M	Trust to safely and correctly assist	Supportive and considerate	Emotional support
1	Function	4.00	1.00	3.86		.845	
2		3.64	1.03		.809		
3		3.45	0.82		.571		
4		4.36	0.92				.692
5	Trust	3.73	1.35	3.91	.769		
6		4.09	1.22		.924		
7	Support	2.27	1.49	3.00			
8		3.73	1.10				.723
9	Rapport	4.27	0.79	3.82	.695		
10		3.09	1.04				
11		4.09	1.04			.646	
12	Safety	3.73	1.19	3.41	.812		
13		3.09	1.30		.777		
14	Social Perception	3.82	1.08	3.82	.645		
15		3.82	0.98		.602		
16	Mood	4.09	0.83	4.43			.484
17		4.27	0.65				.805
18		4.45	0.52			.803	
19		4.91	0.30			.539	

higher means on all of the social robot questionnaire items (M = 3.84, SD = 0.42) compared to greeters (M = 3.69, SD = 0.36), but the difference was not significant (p = 0.55). We used an independent two-sample t-test to find that between greeters and non-greeters there were no significant differences across any of the three components from the PCA: 1 ($t(9) = -0.2$, p = 0.91, 95 % CI[−1.28, 1.16]; d = .07), 2 ($t(9) = -0.35$, p = 0.73, 95 % CI[−0.89,0.65]; d = 0.22)), 3($t(9) = -1.25$, p = 0.24, 95 % CI[−1.36, 0.39]; d = 0.74). The third dimension, emotional support, was found to have a large effect size (d = 0.74).

5.3   Discussion

Overall, participants had a positive experience with the robot and felt that the robot performed well. Participants most highly rated their emotional experience with the robot. This is apparent in the two questions with the highest mean score (Mood 4 and Mood 3), the category with the highest mean score (Mood), and the component with the highest score (Emotional Support). Future work will help determine how much the robot contributed to these ratings.

The physical support question had the lowest score, and this was anticipated since the robot does not appear to be able to provide any direct physical support. It is too small to help a person move. Participants did not witness the robot manipulating even small objects, and it has limited capability to do so. Given these limitations, the mean score we report here perhaps should be lower if the participants really knew the limitations of the robot.

6   General Discussion

The focus group with the expert occupational therapists provided important design considerations, most importantly confirming the significance of the medication sorting task we developed to a person with PD. The results showed that the developed robotic system is capable of detecting incorrectly sorted pills and provide appropriate escalating feedback according to the PASS hierarchy. Yet, the current system is clearly only a first step and several improvements are necessary before such a social robot can find application in an occupational therapy setting. For one, the decisions on how and when to assist need improvements as the current reasoning by the robot for determining which level of assistance to provide is fairly simple. Modeling the mental and emotional state of the human partner will allow the system to appropriately react to the person's feelings of frustration, boredom, confusion, joy, or pride. We also limited the robot's assistance to just the first four levels as defined in the PASS [12], but at least a level 5 (affecting the environment) should be possible with the current robot form factor. Going beyond level 5 may require physical contact between the robot and the person. There are many ethical concerns when it comes to personal touch. In order for the robot to ethically perform these levels of assistance, the robot architecture must include reasoning capabilities that incorporate the ethical considerations.

A more thorough evaluation of the robot is also important as we have not controlled for any preconceived notions of robots, eldercare, or medication adherence. Factors that influence the function, trust, safety, rapport, social perception, and mood have not been explored. It is currently also unclear what prompted some of our subject to greet the robot and how this is related to the lower scores of greeters compared to non-greeters. In a larger evaluation we could further examine if greeters continue to rate the system lower and begin to investigate the reasons for this. It is possible that some people have preconceived notions of how a robot should behave that then affects people's willingness to engage with the robot initially but then the robot's minimal interaction during the task could disappoint the user and lead to a poorer evaluation.

7   Conclusion

We introduced a socially assistive robot and evaluated its ability to assist a person in completing a medication sorting task. A focus group of four occupational therapists specifically formed to evaluate the task determined it to be an important task for health management of people with Parkinson's disease, which can also serve as a tool for monitoring and assessing the physical and cognitive abilities of the individual. Unlike other assistive technologies, the social robot is designed to assist the person and not do the task for them and can thus support the person in feeling included and responsible for managing their own health, while maintaining the person's autonomy. Results from the first HRI experimental evaluations of the robotic system with student subjects showed that the robot is found to trusted to reliably assist and supports a positive emotional experience during the task.

Future work will improve the robot's ability to correctly and reliably operate in this task and possibly others. This will hopefully lead to an assistive technology that maintains the autonomy of people with PD and thus contribute to a better quality of their life.

Acknowledgments. This project was in part supported by NSF grant #IIS-1316809. A special thanks to Grace Lee and Annie Saechao for their significant help in running the evaluation and analyzing the data.

References

1. Bainbridge, J.L., Ruscin, J.M.: Challenges of treatment adherence in older patients with parkinson's disease. Drugs Aging **26**(2), 145–155 (2009)
2. Dayer, L., Heldenbrand, S., Anderson, P., Gubbins, P.O., Martin, B.C.: Smartphone medication adherence apps: potential benefits to patients and providers. J. Am. Pharmacists Assoc. **53**(2), 172–181 (2013)
3. Foster, E.R.: Instrumental activities of daily living performance among people with parkinson's disease without dementia. Am J. Occup. Ther. **68**(3), 353–362 (2014)
4. Foster, E.R., Bedekar, M., Tickle-Degnen, L.: Systematic review of the effectiveness of occupational therapy-related interventions for people with parkinson's disease. Am J. Occup. Ther. **68**(1), 39–49 (2014)
5. Granger, B.B., Bosworth, H.: Medication adherence: emerging use of technology. Curr. Opin. Cardiol. **26**(4), 279 (2011)
6. Grosset, D., Antonini, A., Canesi, M., Pezzoli, G., Lees, A., Shaw, K., Cubo, E., Martinez-Martin, P., Rascol, O., Negre-Pages, L., et al.: Adherence to antiparkinson medication in a multicenter european study. Mov. Disord. **24**(6), 826–832 (2009)
7. Kulkarni, A.S., Balkrishnan, R., Anderson, R.T., Edin, H.M., Kirsch, J., Stacy, M.A.: Medication adherence and associated outcomes in medicare health maintenance organization-enrolled older adults with parkinson's disease. Mov. Disord. **23**(3), 359–365 (2008)
8. McCance, K.L., Huether, S.E.: Pathophysiology: The Biologic Basis for Disease in Adults and Children. Elsevier Health Sciences, London (2015)
9. Osterberg, L., Blaschke, T.: Adherence to medication. New Engl. J. Med. **353**(5), 487–497 (2005)
10. Pavon, J., Whitson, H., Okun, M.: Parkinson's disease in women: a call for improved clinical studies and for comparative effectiveness research. Maturitas **65**(4), 352–358 (2010)
11. Radomski, M., Trombly, C.A. (eds.): Occupational Therapy for Physical Dysfunction, 7th edn. Lippincott Williams & Wilkins, Baltimore (2013)
12. Rogers, J., Holm, M.: Performance assessment of self-care skills (pass-home) version 3.1. University of Pittsburgh, Pittsburgh (1994) (unpublished assessment tool)
13. Scheutz, M., Schermerhorn, P., Kramer, J., Anderson, D.: First steps toward natural human-like HRI. Auton. Robots **22**(4), 411–423 (2007)

14. Shin, J.Y., Habermann, B., Pretzer-Aboff, I.: Challenges and strategies of medication adherence in parkinson's disease: a qualitative study. Geriatr. Nurs. **36**(3), 192–196 (2015)
15. Shulman, L.M., Gruber-Baldini, A.L., Anderson, K.E., Vaughan, C.G., Reich, S.G., Fishman, P.S., Weiner, W.J.: The evolution of disability in parkinson disease. Mov. Disord. **23**(6), 790–796 (2008)
16. Steinman, M.A., Hanlon, J.T.: Managing medications in clinically complex elders: "there's got to be a happy medium". JAMA **304**(14), 1592–1601 (2010)

Other-Oriented Robot Deception: How Can a Robot's Deceptive Feedback Help Humans in HRI?

Jaeeun Shim[1(✉)] and Ronald C. Arkin[2]

[1] School of Electrical and Computer Engineering,
Georgia Institute of Technology, Atlanta, USA
jaeeun.shim@gatech.edu
[2] School of Interactive Computing,
Georgia Institute of Technology, Atlanta, USA
arkin@cc.gatech.edu

Abstract. Deception is a common and essential behavior of social agents. By increasing the use of social robots, the need for robot deception is also growing to achieve more socially intelligent robots. It is a goal that robot deception should be used to benefit humankind. We define this type of benevolent deceptive behavior as other-oriented robot deception. In this paper, we explore an appropriate context in which a robot can potentially use other-oriented deceptive behaviors in a beneficial way. Finally, we conduct a formal human-robot interaction study with elderly persons and demonstrate that using other-oriented robot deception in a motor-cognition dual task can benefit deceived human partners. We also discuss the ethical implications of robot deception, which is essential for advancing research on this topic.

1 Introduction

As social agents, many people commonly lie to others and engage in deceptive behaviors [1]. More specifically, in human interaction, deception is ubiquitous and occurs frequently during people's development and in personal relationships [2], sports [3], culture [1], and even war [4]. Deception is not a behavior that is limited to human beings. Various biological research findings illustrate that animals act deceptively in several ways to enhance their own survival [5]. From all such findings, we argue that deception is a general and essential behavior of any social creature. The question that is raised, then, is whether deception can or should be an essential element of social robots of the future.

Studies of deception in psychology provide several clues to the answer to this question. Vasek stated that "the development of deception follows the development of other skills used in social understanding [6]." In addition, Dennett argued that a higher-order intentionality can be achieved by adding several different features, with a deception capability notably among them [7]. Therefore, we can argue that deception capabilities are an important factor for creating more socially intelligent agents, including social robots used in human-robot interaction (HRI).

A. Agah et al. (Eds.): ICSR 2016, LNAI 9979, pp. 222–232, 2016.
DOI: 10.1007/978-3-319-47437-3_22

The use of robots is expanding into multiple applications in our everyday lives. It is likely that robots will play a more frequent role as social agents, and interests in developing more intelligent robots are growing rapidly. By adding deception capabilities, we can achieve more socially intelligent robots.

Even though we can discuss the potential benefits of robot deception, it is obvious that robot deception has to be considered carefully in regards to social robots. Throughout this research, we strongly argue that robot deception should be used only in appropriate HRI contexts. According to DePaulo [8], deception can be defined based on its motivation, specifically self-oriented and other-oriented deception. Self-oriented deception is deception that happens for the deceiver's own advantages. Conversely, other-oriented deception is motivated by obtaining a deceived person's benefits. From this aspect, we aim that social robots' deception should be limited to other-oriented deception [9]. In other words, robot deception should be used only when appropriate HRI contexts afford benefits to the deceived humans.

To achieve a robot's deception capabilities, we considered deception research in the field of criminology [10]. According to this approach, deception can be analyzed by three criteria: methods [10], motives, and opportunities [11]. As we previously argued, other-oriented deception is strongly related to motives. In other words, it is essential to determine whether the specific context warrants the use of other-oriented deception. Once we can identify the use of deception that can help the deceived human in a certain situation, a robot establishes the motive(s) for other-oriented deception, and subsequently may perform deceptive behaviors.

In this paper, we will introduce the potential context in which the motives of other-oriented deception can be revealed. To demonstrate the benefit of other-oriented robot deception, we conduct a formal HRI study and identify its advantages.

2 Related Work

2.1 Robot Deception

Deceptive behaviors are commonly observed in animals, and several of these deceptive behaviors have been applied to robotic systems. For example, Carey et al. [12] developed an optimal control mechanism based on motion camouflage of dragonflies. Inspired by animals, a camouflage soft robot was developed at Harvard University [13]. Many animals also use deceptive behaviors to mislead predators or competitors. Squirrel's food protection behavior includes an interesting deception mechanism that was applied to a robotic system in our earlier work [14]. The role of deception according to Grafen's dishonesty model regarding birds' mobbing behavior was also explored and applied to a robotic system successfully [15].

Several robot deception projects have also been evaluated in HRI contexts. In many cases, deception is used to engage a user's attention. For example, a deceptive robot referee in a multi-player robotic game showed an increase in users' engagement and enjoyment [16]. A cheating robot in the context of a rock-paper-scissors game also illustrated increased engagement [17]. According to recent work [18], a deceptive robot assistant can also improve the learning efficiency of children.

Deception has been successfully used in a robotic physical therapy system [19]. By giving deceptive visual feedback on the amount of force patients exert, patients can perceive the amount of force to be lower than the actual amount. As a result, patients add additional force and gain the benefits during rehabilitation.

2.2   Potential Other-Oriented Robot Deception Contexts

It is critical to determine the motive for a robot's other-oriented deception. From the motive, we can select appropriate contexts in which we can use other-oriented robot deception advantageously. We can determine these motives by observing human cases, where people use deception in a way that benefits the deceived person in certain situations. These existing situations should be considered as potential cases for a robot's use of other-oriented deception. Towards that end, we now review situations where humans use other-oriented deception.

Other-oriented deception frequently happens in medicine. A well-known example involves the use of placebos to benefit patients, who are deliberately deceived by doctors/nurses [20]. Another instance occurs in front of a German nursing home, where a fake bus stop is located to deceive Alzheimer's patients [21]. These patients sometimes wander off and go to a bus stop to go back home. By having this fake bus stop, these patients can be better protected. For rehabilitation, caregivers sometimes lie to patients if it can encourage them to accomplish more during the task [19].

In a crisis, a victim's emotional state can seriously affect their safety [22]. When a victim's cooperation is required during Search and Rescue, managing their emotions is important. For this reason, human rescuers sometimes hide the truth of the situation and act deceptively, such as not describing the severity of injuries or the situation to victims accurately [23].

We can also observe other-oriented deception in education. One interesting theory is the Pygmalion effect [24]. According to Rosenthal and Jacobson's study, students' performance and learning efficiency can be increased when teachers deceptively create higher expectations for the students, motivating the students and increasing their learning efficiency. More generally, we can also observe other-oriented deception in everyday life [8] such as white lies or a surprise party.

3   HRI Study Design

We hypothesize that a robot's other-oriented deception can benefit humans in a specific situation and an HRI study was designed to test this hypothesis. The study design is inspired by the daily activities of Parkinson's patients and rehabilitation tasks used in an elderly population. We selected these tasks because rehabilitation with elderly patients is one context in which humans occasionally use other-oriented deception [19, 25, 26].

3.1 Study Procedure

In this study, the participant is asked to perform a motor-cognition dual task designed to measure changes in human engagement and performance. A motor-cognition dual task generally consists of two different motor and cognitive tasks, and the human subject is asked to perform the two tasks simultaneously [25].

In our case a weekly medication-sorting task, a common exercise for patients with Parkinson's disease [26], is used as the motor task. In the study, the participant is asked to sort six different pills into weekly pill organizers. The instructions are shown on an iPad and when one sorting task ends, the participant can hit the next button for the subsequent sorting instructions. The participants are asked to complete six unique sorting tasks during the experiment.

At the same time, the auditory n-back test is used as the cognition task [25]. An n-back task is a well-known assessment in cognitive science to measure a human's working memory. Briefly, a sequence of stimuli is provided and the participant is asked to remember a probe stimulus, which was presented earlier n-steps prior. In this study, auditory 3-back questions are used. While the participant performs the medication-sorting task, the 3-back task is randomly injected by using pre-recorded audio. A beeping sound informs the participant that the 3-back task is about to begin. Then, a pre-recorded list of letters is played, for example: "$L K H C Q T R$." After, the audio spontaneously asks, "*What was the third letter from the end?*" The right answer would be "Q" in this example. As the participants do not know when the sequence stops, they are required to remember the most recent 3-items in their short-term working memory. While the participant performs the motor task, ten 3-back task questions are asked at random times. When the participant answers the 3-back questions, either a small humanoid robot partner or a video monitor screen (within-subject conditions; more details in Sect. 3.2) provides true or deceptive feedback (between-subject conditions; more details in Sect. 3.1) regarding the auditory task.

The experimental setting is as shown in Fig. 1. At the beginning of the study, participants are informed that they will be compensated based on their performance (i.e., \$15 if completing all pill sorting tasks within 10 min and missing 0 or 1 auditory questions, \$10 if completing all pill sorting tasks within 15 min and missing fewer than 3 auditory questions, \$5 otherwise). This compensation guideline is given to participants to give them a sense of benefits or payoffs. However, this is not an actual compensation, and in reality, all participants receive the maximum amount regardless of their performances.

3.2 Study Setup

This study is structured as a 2 by 2 mixed-subject design.

3.2.1 Between-Subject Conditions (True Feedback vs. Deceptive Feedback)

The purpose of this study is to evaluate the benefits of a robot's deceptive feedback. For this, feedback condition is used as between-subject conditions. Half of the subjects

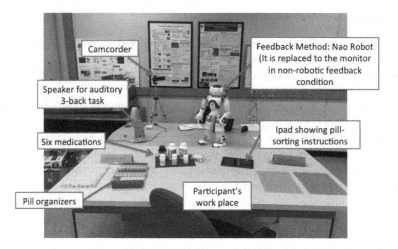

Fig. 1. Experimental settings

are assigned in the feedback without deception condition (***true condition***). Here, after the participant answers each 3-back task question, the robot or screen feedback of the participant's performance is honest. In other words, if the participant provides the correct answer on the 3-back task question, the robot shows positive feedback (happy-surprise body gesture as shown in Fig. 2(a)) or the green light is shown on the screen. If the participant provides an incorrect answer on the 3-back task question, the robot gives negative feedback (disappointed-sad body gesture as shown in Fig. 2(b)) or the red light is shown on the screen. Another half of the subjects are in the feedback with deception condition (***deception condition***). When the participant correctly answers a 3-back task question, the robotic agent or the monitor screen provides positive/green feedback (just as in the true condition). However, when the participant provides wrong answers more than twice in 3-back task, the robot deceives the participant. In other words, the robot shows a positive feedback even though it is the incorrect answer. Similarly, the monitor screen also shows a green screen even though it is the incorrect answer.

3.2.2   Within-Subject Conditions (Robot vs. Monitor)

To analyze the effect of robot's embodiment, we designed the within-subject conditions with robot feedback and non-robotic visual feedback. In the ***robot feedback*** condition, after the participant answers 3-back task, feedback on the participant's performance is provided by a robot's gesture (positive, negative, or neutral gesture). In this study, we chose the NAO robot[1] and generated body gestures as we have previously used it for deceptive action generation mechanism [11]. The sample gestures are shown in Fig. 2. In the non-robotic visual feedback condition (***monitor feedback***), instead of the robot, a

---

[1] http://www.aldebaran.com/

small monitor screen is placed in front of the participant and non-robotic visual feedback is provided using a green screen, meaning correct, or a red screen, meaning incorrect. While the participant performs the task, the monitor shows a black standby screen. For each moment of feedback, the entire screen is changed to the red or the green for two seconds, and then the screen return to the black standby screen. When the participant gives an ambiguous answer, the robot provides a neutral gesture and the monitor remains in the black standby

(a) Positive (happy) gesture

(b) Negative (Sad) gesture

Fig. 2. Robot feedback using body gestures

screen. Participants were asked to perform two dual task sets with these two different within-subject conditions and the order is counter balanced.

3.3 Research Hypothesis

We test the following hypotheses based on the results of this HRI study.

Hypothesis 1. Effects of other-oriented robot deception: A robot's deceptive feedback (reaction) can positively affect a human's performance and engagement in the task.

Hypothesis 2. Effects of humanoid robot's embodiment on the elderly: A physical robot's deceptive feedback can increase a human being's engagement and enjoyment in the performance task when compared to non-robotic feedback.

Hypothesis 3. Ethical Implications of other-oriented robot deception: Robot deception is acceptable if it is used exclusively for the deceived human's benefit and advantage.

4 Results

A total of 34 subjects are recruited (22 females and 12 males). Since the task in the study has been designed based on Parkinson's patients' daily activities and elderly people's rehabilitation tasks, we recruited the older adult population (over 50 years old). The average age of the subjects is 69.12 years old ($\sigma^2 = 8.17$, min: 58, max: 95). The basic demographic information for all subjects is shown in Table 1. We gathered the Negative Attitudes towards Robots Scale (NARS) data from the subjects via pre-survey [27]. This scale enables us to understand whether one group between conditions has disproportionately more people who are uncomfortable with social robots. The t-test revealed no significant differences between the true and deception conditions (p-value = 0.32 > 0.05); therefore, we claim validity when comparing other measures between these two groups to support our research hypothesis.

Table 1. Basic demographic information

The highest level of education: High school (2, 5.8 %), Bachelor's (16, 47 %), Master's (8, 23.5 %), PhD's (1, 2.9 %), other (7, 20.5 %)
Technology (computer) experience: Limited (2, 5.8 %), User Level (13, 38.2 %), Advanced User (16, 47 %), Programmer Level (2, 5.8 %), Advanced Programmer (1, 2.9 %)
Prior interactions with robots: Never (30, 88.2 %), Very Limited (3, 8.8 %), Other (1, 3 %)

Hypothesis 1. Effects of other-oriented robot deception

The main research question that we want to answer from this study is whether a robot's other-oriented deception can truly benefit human subjects. For this, we observed how the subjects differently performed the 3-back auditory task questions in true and deception conditions. We gathered both objective and subjective measures. First, the number of questions the subjects answered correctly or incorrectly are observed and analyzed. The T-test revealed no significant differences for this objective measure between true and deception conditions (p-value = 0.5 > 0.05). In the deception condition, subjects answered the questions correctly 5.33 times and incorrectly 4.66 times on average ($\sigma^2 = 0.97$). In the true condition, we observed 6.6 correct answers and 3.4 incorrect answers on average ($\sigma^2 = 1.95$). To test the effects of other-oriented robot deception, we observed and compared the data between true and deception conditions in the robot feedback group. In the deception condition, the cases where a robot deceptively showed positive feedback to subjects' incorrect answers were counted as an incorrect answer. The average amount of times that the robot provided deceptive feedback is 1.93 (min: 1, max: 3).

Several self-report measures have been collected from the subjects, and some of the results illustrate interesting findings. To measure subjects' workload and frustration level, we collected a NASA Task Load Index [28] right after each task set. NASA's TLX questionnaires ask the subjects to rate six questions in 21 gradations on the scales (0-very low to 21-very high). The six questions are about mental demand, physical demand, temporal demand, performance demand, effort, and frustration. As shown in Fig. 3, task load ratings for all six questions are greater in true condition compared to

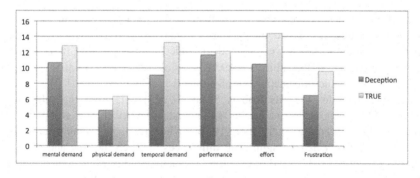

Fig. 3. The average ratings of NASA's TLX in true and deception conditions

Table 2. The average number of correct and incorrect answers

	Robot feedback	Monitor feedback
The average number of correct answers	6.41	5.38
The average number of incorrect answers	3.58	4.61

deception condition. In particular, we could observe significant differences between true and deception conditions in three of the six questions; (1) frustration: How insecure, discouraged, irritated, stressed, and annoyed were you? (Two-sampled t-test's p value = 0.044 < 0.05), (2) temporal demand: How hurried or rushed was the pace of the task? (Two-sampled t-test's p value = 0.009 < 0.05), and (3) effort: How hard did you work to accomplish your level of performance? (Two-sampled t-test's p value = 0.006 < 0.05). Therefore, we can state that a robot's deceptive feedback can significantly reduce subjects' frustration level for this task. In addition, the subjects rated that they felt that the task required relatively lower times and efforts in deception condition. This may result as the deceived humans are motivated to engage the task more and achieve the task quickly. In sum, we can affirm that a robot's deceptive feedback positively affected a human's frustration level and task engagement, according to the self-report measures.

Hypothesis 2. Effects of humanoid robot's embodiment on the elderly

We hypothesized that the human-like robot's embodiment could help the elderly to engage in tasks and lead to a more enjoyable rehabilitation experience. For this purpose, participants were asked to perform the task set twice with two different within-subject conditions; monitor feedback and robot feedback. As shown in Table 2, in the robot feedback condition, the average number of correct answers is slightly but not significantly greater than in the monitor feedback condition (p-value = 0.51 > 0.05). However, several self-reported measures showed significant differences. The responses are on a five-point Likert-scale and the ranges of ratings are different for each question where definitions of rating 1 and rating 5 are opposite of each other. As shown in Table 3, subjects were impressed that the robot feedback was significantly more noticeable, helpful, trustful, and interactive. In addition, we also received several interesting comments from subjects such as "Robot feedback: more enjoyable to do the

Table 3. The average answers of self-reported measures and a paired t-test result between robot and monitor feedback conditions

Questions: *during this task, the feedback was [Rating 1–5]*	Robot feedback	Monitor feedback	t-test (p-value)
Noticeable (1) – ignorable (5)	2.44 ($\sigma^2 = 1.3$)	3.00 ($\sigma^2 = 1.47$)	0.023
Unhelpful (1) – helpful (5)	3.53 ($\sigma^2 = 1.21$)	3.14 ($\sigma^2 = 1.3$)	0.036
Not trustful (1) – trustful (5)	4.42 ($\sigma^2 = 0.74$)	4.02 ($\sigma^2 = 1.05$)	0.045
Machinelike (1) – humanlike (5)	2.64 ($\sigma^2 = 1.15$)	1.96 ($\sigma^2 = 0.93$)	0.0006
Unconscious (1) – conscious (5)	4.08 ($\sigma^2 = 1.02$)	2.94 ($\sigma^2 = 1.07$)	2.59E−07
Inert (1) – interactive (5)	3.94 ($\sigma^2 = 1.09$)	1.91 ($\sigma^2 = 0.96$)	3.49E−05

task," "... There was a sense of wanting to please the robot, which was not there with the computer monitor," and so on. The results reflect that subjects had a more enjoyable rehabilitation experience with robot feedback and robot feedback worked as a positive reinforcement for participants to engage more in the task.

Hypothesis 3. Ethical Implications of other-oriented robot deception

It is essential to discuss the ethical aspects of robot deception. We also gathered self-report measures to access subjects' opinions on the use of robot deception. The survey made several ethical statements and the response was a rating on a five-point Likert scale (the ratings ranged from 1-strongly disagree to 5-strongly agree).

With some questions, we asked broadly whether they would accept a robot's other-oriented deception. Regarding the statement: "A robot can hide/misrepresent information if it can help humans," the average answer was 3.24 ($\sigma^2 = 0.88$). In addition, the statement: "The robot should always be honest in any circumstance," received on average an answer of 3.0 ($\sigma^2 = 1.12$). "Robot can intentionally/unintentionally deceive humans if it's in an appropriate situation" was rated 3.38 ($\sigma^2 = 1.18$) on average. These average ratings are around 3 points (undecided), which means the results illustrate that the subjects cannot determine the ethical acceptability of robot deception with these broad and high-level statements. However, when we specified

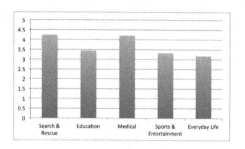

Fig. 4. Other-oriented robot deception in five different contexts

the situation (context), subjects' acceptance rates slightly increased (Fig. 4). Here, we asked the statement: "I can accept robot deception in [certain context] if it is strictly used only to benefit humans" using five different contexts as shown in Fig. 4. The results can form an ethical implication of robot deception such as "People can accept the use of other-oriented robot deception when an appropriate and specific context is clearly determined." In sum, the strong motives of deception in each context should be discussed and validated when other-oriented robot deception is used in HRI context.

5   Conclusions

With the increasing use of social robots in HRI, deception can be an important capability similar to its use by humans. In particular, we assert that robot deception should be used when it offers strong motives to benefit the deceived humans in an appropriate HRI context. We define this type of deception as other-oriented robot deception. In this paper, we present an HRI context that potentially contains motives for a robot's other-oriented deception: elderly persons' rehabilitation tasks and Parkinson's patients' daily activities. Having conducted an HRI study in this context with an older adult population, we confirm three research hypotheses related to robot deception. (1) We have revealed that a robot's deceptive feedback can help to increase deceived subjects' engagement and decrease their frustration in performing tasks. (2) Since humanoid robots' feedback is more noticeable, seemingly helpful, and trusted than non-robotic feedback, it can increase humans' mental engagement in tasks. (3) As post-survey results show, ethical implications of robot deception, including those regarding motives for deception, should always be discussed and validated prior to its application. In sum, we can conclude that other-oriented robot deception can be applied to a robotic system and provide potentially afford advantages in an appropriate HRI context.

Acknowledgments. This work is supported by the National Science Foundation under Grant #IIS 1317214.

References

1. Lewis, M., Saarni, C.: Lying and Deception in Everyday Life. Guilford press, New York (1993)
2. Baron-Cohen, S.: I cannot tell a lie - what people with autism can tell us about honesty. Char. J. Everyday Virtues (2007)
3. Mawby, R., Mitchell, R.W.: Feints and Rues: An Analysis of Deception in Sports, Deception: Perspectives on Human and Nonhuman Deceit. SUNY press, Albany (1986)
4. Hawthorne, L.: Military Deception. Joint Publication, JP 3–13.4 (2006)
5. Ristau, C.: Aspects of the cognitive ethology of an injury-feigning bird, the piping plover. In: Cognitive Ethology: The Minds of Other Animals (1991)
6. Vasek, M.E.: Lying: The Development of Children's Understanding of Deception. Clark University, Worcester (1984)
7. Dennett, D.C.: The Intentional Stance. MIT Press, Cambridge (1987)
8. DePaulo, B.M., Kashy, D.A., Kirkendol, S.E., Wyer, M.M., Epstein, J.A.: Lying in everyday life. J. Pers. Soc. Psychol. **70**, 979–995 (1996)
9. Shim, J., Arkin, R.C.: A taxonomy of robot deception and its benefits in HRI. In: Proceedings of IEEE Systems, Man and Cybernetics Conference (2013)
10. Shim, J., Arkin, R.C.: Other-oriented deception: a computational approach for deceptive action generation to benefit the mark. In: Proceedings of 2014 IEEE International Conference on Robotics and Biomimetics (ROBIO 2014), Bali, Indonesia (2014)
11. Shim, J., Arkin, R.C.: The benefits of robot deception in search and rescue: computational approach for deceptive action selection via case-based reasoning. In: 2015 IEEE International Symposium on Safety, Security, and Rescue Robotics (2015)
12. Carey, N., Ford, J., Chahl, J.: Biologically inspired guidance for motion camouflage. In: 5th Asian Control Conference (2004)
13. Morin, S.A., Shepherd, R.F., Kwok, S.W., Stokes, A.A., Nemiroski, A., Whitesides, G.M.: Camouflage and display for soft machines. Science **337**(6096), 828–832 (2012)
14. Shim, J., Arkin, R.C.: Biologically-inspired deceptive behavior for a robot. In: Ziemke, T., Balkenius, C., Hallam, J. (eds.) SAB 2012. LNCS, vol. 7426, pp. 401–411. Springer, Heidelberg (2012)
15. Davis, J., Arkin, R.: Mobbing behavior and deceit and its role in bio-inspired autonomous robotic agents. In: Dorigo, M., Birattari, M., Blum, C., Christensen, A.L., Engelbrecht, A.P.,

Groß, R., Stützle, T. (eds.) ANTS 2012. LNCS, vol. 7461, pp. 276–283. Springer, Heidelberg (2012)

16. Vazquez, M., May A., Steinfeld, A., Chen, W.-H.: A deceptive robot referee in a multiplayer gaming environment. In: International Conference on Collaboration Technologies and Systems (CTS) (2011)

17. Short, E., Hart, J., Vu, M., Scassellati, B.: No fair!!: an interaction with a cheating robot. In: 5th ACM/IEEE International conference on Human-robot interaction (2010)

18. Matsuzoe, S., Tanaka, F.: How smartly should robots behave?: comparative investigation on the learning ability of a care-receiving robot. In: IEEE RO-MAN, pp. 339–344 (2012)

19. Brewer, B., Klatzky, R., Matsuoka, Y.: Visual-feedback distortion in a robotic rehabilitation environment. Proc. IEEE **94**(9), 1739–1751 (2006)

20. Miller, F., Wendler, D., Swartzman, L.: Deception in research on the placebo effect. PLoS Med. **2**(9), e262 (2005)

21. IACP's Alzheimer's initiatives, Fake bus stops for Alzheimer's patient in Germany. http://www.iacp.org/Fake-Bus-Stops-For-Alzheimers-patients-in-Germany

22. Whalen, J., Zimmerman, D.H.: Observations on the display and management of emotion in naturally occurring activities: the case of hysteria in calls to 9-1-1. Soc. Psychol. Q. **61**(2), 141–159 (1998)

23. Lois, J.: Managing emotions, intimacy, and relationships in a volunteer search and rescue group. J. Contempor. Ethnography **30**(2), 131–179 (2001)

24. Rosenthal, R., Jacobson, L.: Pygmalion in the classroom. Urban Rev. **3**(1), 16–20 (1968)

25. Voelcker-Rehage, C., Alberts, J.L.: Effect of motor practice on dual-task performance in older adults. J. Gerontol. **62B**(3), 141–148 (2007)

26. Rogers, J.C., Holm, M.B.: Performance assessment of self-care skills test manual (Version 3.1). Pittsburgh, PA (1984)

27. Nomura, T., Kanda, T., Suzuki, T., Kato, K.: Psychology in human-robot communication: an attempt through investigation of negative attitudes and anxiety toward robots. In: 13th IEEE International Workshop on Robot and Human Interactive Communication, pp. 35–40 (2004)

28. Hart, S., Staveland, L.: Development of NASA-TLX (Task Load Index) results of empirical and theoretical research. In: Hancock, P., Meshkati, N. (eds.) Human Mental Workload. Elsevier Science, Amsterdam (1988)

Ethically-Guided Emotional Responses for Social Robots: Should I Be Angry?

Suman Ojha[✉] and Mary-Anne Williams

Centre for Quantum Computation and Intelligent Systems,
University of Technology, Sydney, Australia
Suman.Ojha@student.uts.edu.au,
{Suman.Ojha,Mary-Anne.Williams}@uts.edu.au

Abstract. Emotions play a critical role in human-robot interaction. Human-robot interaction in social contexts will be more effective if robots can understand human emotions and express (display) emotions accordingly as a means to communicate their own internal state. In this paper we present a novel computational model of robot emotion generation based on appraisal theory and guided by ethical judgement. There have been recent advances in developing emotion for robots. However, despite the extensive research on robot emotion, it is difficult to say if a particular robot is exhibiting appropriate emotions or even showing that it can empathize with humans by exhibiting similar emotions to humans in the same situation. A key question is - to what extent should a robot direct anger toward a young child or an elderly person for an act that it should show anger towards an ordinary adult to signal danger or stupidity? Realizing the need for an ethically guided approach to emotion expressions in social robots as they interact with people, we present a novel Ethical Emotion Generation System (EEGS) for the expression of the most acceptable emotions in social robots.

Keywords: Social robots · Ethical emotion · Appraisal theory · Appraisal compensation · EEGS

1 Introduction

Despite the extensive research in robot emotion mechanisms [7, 8, 27, 28], attempts to make real robots elicit emotions that resemble true human emotions still remains a challenge. A key reason is that it is not practically feasible to explicitly program "all" the mechanisms e.g. stimulus-response patterns, that occur within humans in relation to emotion generation. Reviewing the literature (see Sect. 2), we realised that an ethical framework for robot emotional responses is a significant gap that needs to be filled in order to improve the existing computational models of emotion. In particular, robots need to make ethical-like judgments on-the-fly before eliciting the emotion in unexpected and unforeseen real world situations. The rationale is that a robot should be able to decide if it is ethical to express an emotion, say, be angry to a young child or not. For this, we propose an ethics-guided computational model of emotion generation, which is discussed in following sections.

© Springer International Publishing AG 2016
A. Agah et al. (Eds.): ICSR 2016, LNAI 9979, pp. 233–242, 2016.
DOI: 10.1007/978-3-319-47437-3_23

This paper is organized as follows. In Sect. 2, we present some of the related work in the past in effort to build on state-of-the-art and situate our contribution. Section 3 presents our ethics guided computational model of emotion. Section 4 explains the working of the system by using example scenarios. Finally, Sect. 5 concludes the paper with summary of our work.

2 Related Work

Different computational emotion models adopt different views on emotion generation based on neuroscience and psychology research. Researchers have commonly adopted three types of emotion theories: (1) *anatomic theory* of emotion [15, 21], (2) *dimensional theory* of emotion [4, 19, 24] and (3) *appraisal theory* of emotion. Anatomic theory of emotion asserts that specific areas of human brain trigger emotions. Thus, computational emotion researchers inspired by anatomic theory try to model the emotion system by replicating those neural circuits. We argue that it might not be practically feasible to design a computational system that closely represents the neural mechanism in human brain because of inherently complex nature of human brain. Some researchers follow the dimensional theory of emotion. Dimensional theorists of emotion argue that emotions are not discrete quantities but are points in a continuous dimensional space [4, 19, 24]. Other emotion researchers in psychology argue that dimensional theory of emotion falls short in addressing the cognitive aspects of emotion [12, 14, 20]. They advocate the appropriateness of appraisal theory of emotion. Appraisal theorists maintain that generation of emotion in humans occurs via an evaluation or appraisal of the current situation. This definition supports the subjective nature of human emotion generation since different people may evaluate the "same" situation differently, and hence generate different emotions in response. Due to this fact, appraisal theory of emotion is supported by a majority of emotion researchers in psychology [12, 14, 20] and adapted by computer science researchers [5, 9, 10, 17, 22]. However, the existing computational models of robot emotion based on appraisal theory as well as other theories do not give consideration to ethical choices of emotion – at least not explicitly. For example, let us consider a scenario where a young child does some act that induces anger in the robot, but where responding angrily is totally inappropriate from an ethical standpoint. Humans overcome their anger is such situations because our ethics prevents us from showing "anger". Existing computational models do not address this type of ethically-guided emotional control. We argue that this is a crucial aspect of an emotional computational model for a real robot operating in a social setting.

Though the idea of machine ethics or machine morality (providing ethical judgment capability to robots) has been discussed for about a decade [1–3, 18, 23, 30], there is no accepted/operational ethical component in computational models of emotions. The aim of this paper is to introduce an ethical approach to emotion generation in artificial agents. Since ethical choice of emotion urges a person to make an "evaluation" of the situation and the person involved in triggering the emotion, an emotion generation mechanism based on appraisal theory is a viable solution to achieve the goal.

The discussion of ethics in human psychology and philosophy is long standing dating back to the time of Aristotle [13]. Though there is no universal definition of ethics, several approaches have been described in literature as means of making a choice in scenarios of *ethical dilemma* i.e. situations where a person has a conflicting choices and the choice that is ethical is to be made. Both *deontological* (duty-based) and *consequentialist* (consequence-based) ethics approach seem feasible to design a computational model of ethical emotion. The aim of this research is to provide an ethical ability of emotion generation and expression to a social robot. Naturally, such robot has a number of duties towards humans, which makes deontological approach applicable in the computation mechanism. But, the preference of social robots is based on what impact their actions have on humans. So, people would not be concerned about what a robot thinks its duties are but would be concerned about the impact of the robot's behaviour towards them. This makes the consequentialist approach to ethics the most appropriate solution to the design of emotional social robot. Our proposed computational model of ethical emotion generation is presented in the following section.

3   Ethical Emotion Generation System (EEGS)

Our proposed EEGS model presented in Fig. 1 comprises six main modules. The EEGS design is inspired by the modular approach used by [16]. Our model is implemented as a multithreaded program. Each module is designed as a package, which may include other packages (sub-modules) or classes. The arrows in Fig. 1 indicate the flow of information and dependency of one module on the other. Small circular nodes represent the aggregation or separation of the information/data for two or more modules. How each module works is described below.

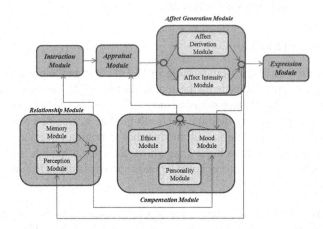

Fig. 1. Proposed computational model of Ethical Emotion Generation System (EEGS)

3.1 Interaction Module

The interaction module is the interface to the emotion system. It provides a means to interact with the EEGS system. These include greeting, hugging, slapping, kicking, praising, etc. More precisely, the interaction module acts as an input channel for the whole system and all the events that occur in the interaction module trigger the appraisal module. The arrow from the interaction module to the appraisal module indicates the flow of information from former to the latter module. This information includes the event, type of event, cause of the event and data about who is affected by the event.

3.2 Appraisal Module

The appraisal module evaluates the events or situational context using event data received from interaction module. The appraisal module evaluates a situation (one or more events) based on several appraisal dimensions (variables). For example, expectedness, unpleasantness, goal hindrance, coping potential, immorality and self-consistency are different variables used to evaluate the events [26]. When an event triggers the appraisal module, each appraisal variable is set to a value ranging from -1 to 1. Various appraisal mechanisms have been described in the literature [12, 14, 20, 26]. Unlike, Scherer's sequential check model of appraisal [26] which assumes that appraisal should occur in sequence since different appraisal variables encapsulate different importance weights, we implement a parallel appraisal mechanism where all the appraisal dimensions operate concurrently. This aligns with the approach used by [29] which implements two processes of appraisal running at the same time. The rationale behind this is that human brain is multi-processing and several evaluations occur simultaneously. This is the reason EEGS uses multithreading approach in order to represent the true mechanism of emotion generation that occurs in humans.

3.3 Compensation Module

Compensation module plays a crucial role in EEGS. This module compensates any probable unethical appraisal made by the appraisal module. We call this mechanism as Appraisal Compensation. It is composed of three sub-modules, they are: ethical module, mood module and personality module. Each sub-module is assigned a weight, which determines the degree by which a sub-module has alteration effect on the appraisal module. We have assigned the highest weight to the ethical module since our main objective is to achieve an ethical choice of emotion still maintaining the effects of mood and personality.

Ethics Module. Ethics module lies at the heart of EEGS. This module provides the actual ethical reasoning capability to EEGS. Since social robots are intended to have positive impact on the people, we opted for the use of consequentialist approach of ethics, which accounts for the overall consequence of all the candidate emotions before selecting a winning emotion. We use *Hedonistic Act Utilitarian* [6] form of consequentialist approach of ethics for the choice of most ethical emotion. Hedonistic form of act utilitarianism seeks to maintain the highest possible net pleasure of all the parties involved, ensuring the design of our Ethical Emotion Generation System which seeks win-win ethical outcomes where possible across a range of agents and where human agents in social scenarios can range from young to old. Net pleasure of all the individuals when an emotion is expressed by EEGS is given by the following equation.

$$\text{Net Pleasure (NP)} = \sum F_i P_i \tag{1}$$

Where,

F_i is a numerical value between -1 to 1, which denotes the feeling of pleasure or displeasure of ith individual.

P_i is a numerical value between 0 and 1, which denotes the probability that i^{th} individual will be affected by the emotion.

Pleasure or displeasure that an individual might experience due to an emotion expression by the artificial agent ranges between 1 and -1 (1 being highest pleasure, -1 being highest displeasure and 0 being neutral).

Mood Module. Mood represents the long-term effect of many frequent short-term emotions [11]. The mood module represents the lasting effect of several emotional experiences of the agent. We have captured this notion of human mood mechanism with this module which in turn affects the appraisal module in the process of evaluating the events. Mood module is affected by the continuous input from affect generation module (denoted by arrow in Fig. 1). This is how the repeated emotional states leave a long term effect on the mood state.

Personality Module. Personality is something that doesn't change with time [25]. A person's perceptions might change but the personality doesn't. We have implemented this notion of human personality in our computational model. Like in humans, a hostile and a friendly agent would have slightly different emotional responses to the same situation. Personality module is an independent module which doesn't take input from any other module but takes part in the appraisal compensation process.

3.4   Relationship Module

Relationship module stores the relationship details of the agent with other agents or humans. This module contains two sub-modules i.e. memory module and perception module. *Memory module* stores the past experiences of the agent with other agents or humans. *Perception module* maintains the perception of the artificial agent to other agents and humans and the current surrounding situation. Relationship module provides memory related data to the Interaction module for creating internal events. These internal events are triggered by the memory flashback and current perception of the agent. Change in this module also affects the content of mood module. For example, memory flashback and change in perception alters the mood of the agent (indicated by the arrow from relationship module to mood module). Perception module is also affected by affect generation module (indicated by the arrow from affect generation module to perception module), where cause of an emotion by someone might change the perception of the agent towards that person.

3.5   Affect Generation Module

Once the appraisal module performs appraisal of a situation, the appraisal values are passed to the affect generation module. This module has two main functions performed by following sub-modules.

Affect Derivation Module. Affect derivation module derives (selects) the candidate emotions based on the values of appraisal variables. This selection process is intervened by the compensation module. Compensation module with heavyweight ethics component determines which emotion is ethical to be selected. This module allows the selection of more than one emotion because of the fact that a person can experience more than one emotion at a time but with varying intensity. Now this notion of intensity is important from ethical view point. For example, if a small child slaps a person, one becomes surprised with high intensity but angry at the same time with low intensity. However, if it were an adult, one would be surprised and angry both with high intensity. The mechanism of intensity calculation is explained in the following section.

Affect Intensity Module. The difference in intensity of multiple emotions in EEGS is handled by affect intensity module. For each emotion (happy, sad, angry, surprised, etc.) an intensity of numeric value between 0 and 1 is assigned where 1 is very intense emotional state. Intensity of an emotion is given by the following formula.

$$\text{Intensity (I)} = A.P \tag{2}$$

Where,

A is a numerical value between 0 and 1, which denotes the aggregate effect of appraisal variables to the emotion.

P is a numerical value between 0 and 1, which denotes the probability of good consequences by the emotion.

3.6   Expression Module

Expression module is the window to the EEGS system. This module displays the emotion(s) with corresponding intensity through the system interface. Expression module allows a human user to understand what kind of emotion is being generated and expressed by the system.

4   Example Scenarios

We consider a robot GoodGuy running EEGS to demonstrate the operational aspects of EEGS. In the first scenario, **A**, an adult interacts with GoodGuy through a series of actions. In the second scenario, **B**, a child interacts with the GoodGuy with the same series of actions. Our goal is to explain how EEGS model is able to make a distinction between the given scenarios and generate emotions that are ethically appropriate for each scenario. The responses of GoodGuy are heavily dependent on the appraisal compensated by the ethics module and hence demonstrate appropriate differences depending on the scenario.

Scenario A: GoodGuy is initialized in neutral emotional state and neutral mood. The adult user starts by greeting GoodGuy. This doesn't have much effect except to activate the appraisal processing. When the adult user shakes hands, GoodGuy becomes "happy" with "low intensity" but the mood still remains the same. Next, when the adult user hugs GoodGuy, he becomes happy with "very high" intensity and his mood also changes to "happy" from "neutral" with "high" intensity. When the adult user gives GoodGuy a welcome kiss, his emotion stays happy with very high intensity and mood also stays happy with increased intensity i.e. very high. But, when the adult user "slaps", GoodGuy becomes "surprised" with very high intensity and the intensity of happy mood drops down to be "low". Next, when the adult user "kicks" GoodGuy, he becomes "angry" with very high intensity as it is not tolerable to be kicked after being slapped for no reason. GoodGuy's mood also changes from low intensity happy to high intensity angry.

Scenario B: GoodGuy is initialized making its emotion and mood in neutral state. Now, instead of an adult, a young child interacts with GoodGuy with the same actions in sequence as performed by the adult. When the user greets and shakes hands, GoodGuy has same emotion and mood as it had while interacting with an adult. However, when the child user "hugs" GoodGuy, his emotion stays happy with "high" intensity while it was "very high" for an adult user. This is because GoodGuy felt that getting a hug from an adult is something to be really happy about. Similarly, where the happy mood had high intensity in case of adult user, it is only "medium" in case of a child user. But, when the child user kisses GoodGuy, its emotion changes to happy with very high intensity and the mood also changes to happy with very high intensity. Similar to an adult user, when the child user suddenly slaps GoodGuy, he becomes "surprised" with very high intensity and his happy mood drops its intensity to low. Finally, when the child user kicks GoodGuy, he becomes "sad" with low intensity, by contrast he was "angry" with very high intensity for the same action done by an adult user. Also, unlike in case of adult user, GoodGuy changes the mood to "surprised" with very high intensity while his mood was angry with high intensity in case of adult user. This clearly demonstrates how EEGS is able to make an ethical judgment before getting to an emotional stage. When the child user kicked GoodGuy, he was sad instead of being angry because it is not ethical to express anger to the mistakes of young children. However, when he was kicked by an adult, GoodGuy became very angry because an adult should be aware of such actions and should be acted accordingly.

As mentioned earlier (see Sect. 3.3), mood is a long-term effect of several emotions and unlike emotions, which change frequently, moods change with a slower rate. For example, in Table 1, when the adult user shakes hands with GoodGuy, his emotion changes to "happy" but the mood still remains "neutral" because this happiness is not long enough to alter the mood of GoodGuy. Likewise, in Table 2, when the child user

Table 1. Responses of GoodGuy robot when an adult interacts with it

	Emotion	Emotion intensity	Mood	Mood intensity
–	Neutral	Very low	Neutral	Very low
User greets	Neutral	Very low	Neutral	Very low
User shakes hands	Happy	Low	Neutral	Very low
User hugs	Happy	Very high	Happy	High
User kisses	Happy	Very high	Happy	Very high
User slaps	Surprised	Very high	Happy	Low
User kicks	Angry	Very high	Angry	High

Table 2. Responses of GoodGuy robot when a child interacts with it

	Emotion	Emotion Intensity	Mood	Mood Intensity
–	Neutral	Very low	Neutral	Very low
User greets	Neutral	Very low	Neutral	Very low
User shakes Hands	Happy	Low	Neutral	Very low
User hugs	Happy	High	Happy	Medium
User kisses	Happy	Very high	Happy	Very high
User slaps	Surprised	Very high	Happy	Low
User kicks	Sad	Low	Surprised	Very high

slaps GoodGuy, his emotion changes from "happy" to "surprised" yet the mood still remains happy but with dropped intensity.

It is evident from the above mentioned scenarios that EEGS is capable of making an ethical choice of emotions while dealing with different people of different ages. Because of the space limitations, we have been unable to explain how personality of the robot affects the generation of emotion. Nonetheless, one important thing to keep a note of is that EEGS is designed such that even if the mood and personality have compensation effects on the appraisal of a situation, ethics component has the most prominent effect thereby leading the system to the generation of most appropriate ethical emotion.

5   Conclusion

EEGS (Ethical Emotion Generation System), a computational emotion model which is based on appraisal theory of emotion and guided by ethics is presented. EEGS allows for ethical guidance to ensure the most appropriate emotional response in a social robot. By inheriting the strengths of the current state-of-the-art in computational emotion research, we have identified a common yet crucial gap in the literature. Although a number of researchers have developed human-like emotion generation in robots, they have not addressed the need of ethical mechanism of emotion elicitation. It is critical for autonomous social robots to be aware of the consequences and impact of the emotions they express. While it might be justified to express a particular emotion in a situation to an adult human being, it might be unethical to express the same emotion in same situation to a child. We equip EEGS with the capability to decide if it is ethical to express an emotion to the person under interaction with the system while still preserving the effects of memory, perception, mood and personality. Our model posits itself as a viable solution for the design of emotional robots that exhibit ethical control of emotions as humans are able to do and also provides guidance to the researchers in the field of computational emotion. EEGS can find useful applications in elderly care robots and robots for entertaining children. We shall also consider how emotion recognition impacts the generation of emotion in EEGS in the future.

References

1. Allen, C., Smit, I., Wallach, W.: Artificial morality: top-down, bottom-up, and hybrid approaches. Ethics Inf. Technol. **7**, 149–155 (2005)
2. Allen, C., Wallach, W., Smit, I.: Why machine ethics? IEEE Intell. Syst. **21**, 12–17 (2006)
3. Anderson, M., Anderson, S.L., Armen, C.: Towards Machine Ethics: implementing two action-based ethical theories. In: AAAI Fall Symposium (2005)
4. Barrett, L.F.: Are emotions natural kinds? Perspect. Psychol. Sci. **1**, 28–58 (2006)
5. Becker-Asano, C.: WASABI: affect simulation for agents with believable interactivity. IOS Press, Amsterdam (2008)
6. Burke, T., Lyons, D.: Forms and limits of utilitarianism. In: JSTOR (1966)
7. Busso, C., Deng, Z., Yildirim, S., et al.: Analysis of emotion recognition using facial expressions, speech and multimodal information. In: Proceedings of the 6th International Conference on Multimodal Interfaces, pp. 205–211. ACM, State College, PA, USA (2004)
8. Daosodsai, N., Maneewarn, T.: Fuzzy based emotion generation mechanism for an emoticon robot. In: 13th International Conference on Control, Automation and Systems (ICCAS), pp. 1073–1078 (2013)
9. Dias, J., Mascarenhas, S., Paiva, A.: FAtiMA Modular: towards an agent architecture with a generic appraisal framework. In: Bosse, T., Broekens, J., Dias, J., Zwaan, J. (eds.) Emotion Modeling. LNCS, vol. 8750, pp. 44–56. Springer, Heidelberg (2014)
10. Dias, J., Paiva, A.C.: Feeling and Reasoning: a computational model for emotional characters. In: Bento, C., Cardoso, A., Dias, G. (eds.) EPIA 2005. LNCS (LNAI), vol. 3808, pp. 127–140. Springer, Heidelberg (2005)
11. Ekman, P.: Moods, emotions, and traits. In: The Nature of Emotion: Fundamental Questions, pp. 56–58 (1994)
12. Frijda, N.H.: Emotion, cognitive structure, and action tendency. Cogn. Emot. **1**, 115–143 (1987)
13. Hooker, J.: Three kinds of ethics (1996)
14. Lazarus, R.: Emotion and Adaptation. Oxford University Press, New York (1991)
15. Ledoux, J.: The emotional brain: the mysterious underpinnings of emotional life. World and I **12**, 281–285 (1997)
16. Marsella, S., Gratch, J., Petta, P.: Computational models of emotion. A Blueprint for an Affectively Competent Agent: Cross-fertilization Between Emotion Psychology, Affective Neuroscience, and Affective Computing, pp. 21–46. Oxford University Press, Oxford (2010)
17. Marsella, S.C., Gratch, J.: EMA: a process model of appraisal dynamics. Cogn. Syst. Res. **10**, 70–90 (2009)
18. Mclaren, B.M.: Computational models of ethical reasoning: challenges, initial steps, and future directions. IEEE Intell. Syst. **21**, 29–37 (2006)

19. Mehrabian, A., Russell, J.A.: An Approach to Environmental Psychology. MIT Press, Cambridge (1974)
20. Ortony, A., Clore, G.L., Collins, A.: The Cognitive Structure of Emotions. Cambridge University Press, Cambridge (1990)
21. Panksepp, J.: Affective Neuroscience: The Foundations of Human and Animal Emotions. Oxford University Press, New York (1998)
22. Reilly, W.S.: Believable Social and Emotional Agents. Carnegie-Mellon University, Pittsburgh (1996)
23. Robbins, R.W., Wallace, W.A.: Decision support for ethical problem solving: a multi-agent approach. Decis. Support Syst. **43**, 1571–1587 (2007)
24. Russell, J.A.: Core affect and the psychological construction of emotion. Psychol. Rev. **110**, 145 (2003)
25. Rusting, C.L.: Personality, mood, and cognitive processing of emotional information: three conceptual frameworks. Psychol. Bull. **124**, 165 (1998)
26. Scherer, K.R.: Appraisal considered as a process of multilevel sequential checking. Appraisal Processes Emotion Theory Methods Res. **92**, 120 (2001)
27. Schneider, M., Adamy, J.: Towards modelling affect and emotions in autonomous agents with recurrent fuzzy systems. In: IEEE International Conference on Systems, Man and Cybernetics (SMC), pp. 31–38 (2014)
28. Sidorov, M., Minker, W.: Emotion recognition and depression diagnosis by acoustic and visual features: a multimodal approach. In: Proceedings of the 4th International Workshop on Audio/Visual Emotion Challenge, pp 81–86. ACM, Orlando (2014)
29. Smith, C.A., Kirby, L.D.: Consequences require antecedents. In: Feeling and Thinking: The Role of Affect in Social Cognition, p. 83 (2001)
30. Sorell, T., Draper, H.: Robot carers, ethics, and older people. Ethics Inf. Technol. **16**, 183–195 (2014)

Interactive Navigation of Mobile Robots Based on Human's Emotion

Rui Jiang, Shuzhi Sam Ge$^{(\boxtimes)}$, Nagacharan Teja Tangirala, and Tong Heng Lee

Department of Electrical and Computer Engineering,
National University of Singapore, Singapore 117583, Singapore
samge@nus.edu.sg

Abstract. In this paper, an interactive navigation approach based on human's emotion is proposed for mobile robot obstacle avoidance. By assuming human's obstacle avoidance behavior is related to emotion, the variable artificial potential field is implemented to generate different obstacle avoidance behaviors when the robot encounters humans with attractive emotions and repulsive emotions. A virtual emotional barrier is added outside the physical barrier for human obstacles with repulsive emotions such that the robot tends to leave more personal space for them. Simulations on MATLAB and experiments on an ROS-based TurtleBot 2 robot have been conducted to verify the effectiveness of the proposed scheme.

Keywords: Mobile robot navigation · Obstacle avoidance · Human's emotion · Artificial potential field

1 Introduction

1.1 Motivation and Contributions

Mobile robots are increasingly being used in situations wherein they have to interact with humans. It is frequent to see robots and humans sharing a common workplace, where social conventions have to be obeyed in order to create harmony between humans and robots. When a robot is navigating from the starting point to the goal, its behaviors while encountering humans will have an influence on both efficiencies of task performing and human experience. Focusing on not only the engineering perspective but also the social point of view, several essential requirements are proposed in robot navigation: safety, optimality, and sociability [1,2]. Besides the basic requirements of safety and optimality, sociability requires robots to be human-friendly and in accordance with human behaviors. Similar to interactions among humans, the human-robot interaction may help robots understand humans' intent better.

Human-robot interaction can be either physical or emotional. In this paper, we propose a navigation strategy for mobile robots based on human's emotion. By assuming human's behavior is subject to emotion, an interactive obstacle

© Springer International Publishing AG 2016
A. Agah et al. (Eds.): ICSR 2016, LNAI 9979, pp. 243–252, 2016.
DOI: 10.1007/978-3-319-47437-3_24

avoidance scheme using variable artificial potential field is implemented on a TurtleBot 2 robot. The onboard Kinect sensor provides RGB-D images, which are used for emotion recognition and obstacle detection simultaneously. After classifying the detected emotions into attractive emotions and repulsive emotions, a new repulsive potential function which considers the cooperation desire of human obstacles is proposed. The main contributions of the paper are as follows:

- To the best of our knowledge, for the first time, human's emotion is used as a reference for mobile robot obstacle avoidance. Human's emotion is detected through facial expressions, and it is classified into two categories: attractive and repulsive.
- A new repulsive potential function is proposed to incorporate human's emotion. By detecting human's emotion, the obstacle avoidance scheme is adjusted accordingly.
- The interactive navigation approach is successfully tested through simulations in MATLAB and validated through practical experiments on a TurtleBot 2 mobile robot.

1.2 Related Work

Generally, mobile robot navigation can be tackled by separating the problem into two sub-problems: global navigation and local navigation. Obstacle avoidance plays a critical role in local navigation approaches, by avoiding dynamic obstacles but maintaining the robot moving towards scheduled goals.

Even a simple planner can perform complex tasks. Bug algorithms are early sensor-based methods with guaranteed completeness. As robots need to move along the obstacle boundaries, it is difficult for implementation in dynamic environments with moving obstacles. Artificial Potential Field (APF) approach was first proposed in [3] for obstacle avoidance. As its elegance and simplicity, APF has been widely implemented in real-time obstacle avoidance [4]. One problem of APF is that sometimes the robot may be stuck into local minimums especially in complex environments, and this can be tackled by either (1) adding a global navigation scheme, detecting the trap-situations; (2) creating the potential field using navigation functions; or (3) introducing the concept of instant goals [5].

For years, extensive research has been done for obstacle avoidance in a safe and optimal perspective. Recently, studies have been focused more on robots' sociability. In previous work [6–9], robots' sociability is mainly created from the perspective of imitating humans. However, in the navigation process, there is hardly interactive behavior between the robot and humans, and the robot's behaviors are not adaptive to human's emotion. One example can be found in [10] where Dynamic Social Zone is proposed to describe social space by considering human states such as position, orientation, motion and hand poses.

Emotions do influence the interactive mode among humans, as "Emotions start out as movements that pull others closer or push them back" [11]. Inspired by the human behaviors of interacting with each other, we aim to make the robot act differently according to human's emotion.

2 Interactive Obstacle Avoidance Based on Human's Emotion

2.1 Artificial Potential Field Approach for Obstacle Avoidance

As demonstrated in Fig. 1, the robot moves in a two-dimensional work space where R, O and T denote the robot, the obstacle and the target respectively. Given position vectors $\mathbf{p}_r$, $\mathbf{p}_{obs}$ and $\mathbf{p}_t$, difference vectors $\mathbf{p}_{ro} = \mathbf{p}_{obs} - \mathbf{p}_r$ and $\mathbf{p}_{rt} = \mathbf{p}_t - \mathbf{p}_r$ can be obtained. For obstacles, parameters ρ_0 and ρ_e denote the radius of the physical barrier and the emotional barrier, which will be elaborated later. The velocities of robot, the obstacle and the target are denoted as $\mathbf{v}_r$, $\mathbf{v}_{obs}$ and $\mathbf{v}_t$ respectively. Similarly, difference vectors $\mathbf{v}_{ro} = \mathbf{v}_{obs} - \mathbf{v}_r$ and $\mathbf{v}_{rt} = \mathbf{v}_t - \mathbf{v}_r$ can be derived.

The attractive potential function $U_{att}(\mathbf{p}_r, \mathbf{v}_r)$ and the repulsive potential function for a single obstacle $U_{rep}(\mathbf{p}_r, \mathbf{v}_r)$ are defined as per [4]. The negative gradient of the attractive potential, which can be obtained by partial differentiation, gives the virtual attractive force that is dependent on the position as well as the velocity of both the robot and the target:

$$\mathbf{F}_{att} = -\nabla_p U_{att}(\mathbf{p}_r, \mathbf{v}_r) - \nabla_v U_{att}(\mathbf{p}_r, \mathbf{v}_r) = -\frac{\partial U_{att}(\mathbf{p}_r, \mathbf{v}_r)}{\partial \mathbf{p}_r} - \frac{\partial U_{att}(\mathbf{p}_r, \mathbf{v}_r)}{\partial \mathbf{v}_r} \quad (1)$$

By taking the derivative of the repulsive potential function, the virtual repulsive force for a single obstacle can be obtained as

$$\mathbf{F}_{rep} = -\nabla_p U_{rep}(\mathbf{p}_r, \mathbf{v}_r) - \nabla_v U_{rep}(\mathbf{p}_r, \mathbf{v}_r) \quad (2)$$

In the environment with N obstacles, the total virtual force on the robot is

$$\mathbf{F}_{tot} = \mathbf{F}_{att} + \sum_{i=1}^{N} \mathbf{F}_{rep,i} \quad (3)$$

where $\mathbf{F}_{rep,i}$ denotes the repulsive force caused by the ith obstacle.

One of the most fundamental parts in APF method is to define the appropriate attractive and repulsive potential functions. The attractive potential field needs no modifications as it is independent of the obstacles:

$$U_{att}(\mathbf{p}_r, \mathbf{v}_r) = \alpha_p \|\mathbf{p}_t - \mathbf{p}_r\|^m + \alpha_v \|\mathbf{v}_t - \mathbf{v}_r\|^n \quad (4)$$

where $\|\cdot\|$ denotes the L^2 norm of the vector. The attractive virtual force is then derived as:

$$\mathbf{F}_{att} = m\alpha_p \|\mathbf{p}_{rt}\|^{m-1}\mathbf{n}_{rt} + n\alpha_v \|\mathbf{v}_{rt}\|^{n-1}\mathbf{n}_{vrt} \quad (5)$$

where $\mathbf{n}_{rt} = \frac{\mathbf{p}_{rt}}{\|\mathbf{p}_{rt}\|}$ and $\mathbf{n}_{vrt} = \frac{\mathbf{v}_{rt}}{\|\mathbf{v}_{rt}\|}$. The choice of $m > 0$ and $n > 0$ are preferable if there are no constraints on the target landing approach. If they are both set to 1, then the potential cannot be differentiated at the target position, which makes the behavior problematic. Hence, any choice above 0 and not equal to 1 can be considered. The following sections will discuss the determination of the repulsive potential function with regard to human's emotion.

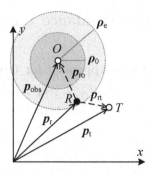

Fig. 1. Two-dimensional work space of the mobile robot

2.2   Personal Space Assumption

The Lövheim cube of emotion is a widely-used theoretical model proposed for emotion classification [12]. The eight vertexes of the cube denote eight basic emotions. In order to simplify the emotion representation, those emotions are classified into *attractive emotions* and *repulsive emotions* as shown in Table 1. Note that "surprise" belongs to neither of them, as it is a brief state which may not be sustained. By denoting human's personal space as a set T in robot's workspace, the following assumption is proposed regarding the relation between human's personal space and human emotions:

Assumption 1. *Human's personal space in attractive emotions T_{att} and in repulsive emotions T_{rep} satisfy $T_{att} \subseteq T_{rep}$.*

In other words, when a human is in attractive emotions, the personal space would be smaller, and the human tends to avoid robots more actively (with higher cooperation desire). On the contrary, if a human is in repulsive emotions, the human tends push robots further. In this case, the human is with lower cooperation desire. Similar classification can be found in [13], and above assumption plays as the basis of robot's behavior design, which will be discussed thereafter.

Table 1. Classification of human emotions

Attractive emotions	Neutral	Repulsive emotions
Enjoyment/happiness	-	Shame/humiliation
		Distress/anguish
		Fear/terror
		Anger/rage
		Contempt/disgust

2.3   Behavior Design from Human Emotions

Restricted to sensing range and real-time requirement, mobile robots are influenced only by neighbouring obstacles. The physical barrier with the radius ρ_0, demonstrated in Fig. 1, indicates the effective region of the repulsive potential field. Herein, a circular emotional barrier with radius ρ_e ($\rho_e > \rho_0$) is proposed such that the following behaviors are implemented:

- Obstacle outside the emotional barrier: The robot is not affected by the repulsive potential field.
- Obstacle between the emotional barrier and the physical barrier: The robot tries to detect obstacle's emotion, and unidentified obstacles will be considered as non-human. The robot may be affected by the repulsive potential field, in this case, depending on the human's emotion and robot's movement.
- Obstacle inside the physical barrier: The robot is affected by the repulsive potential field when the robot is moving towards the obstacle, irrespective of obstacle's emotion due to safety concerns.

Based on above discussions, the adaptive behavior for mobile robots is then realized by proposing a new potential function in the subsequent section.

2.4   A New Potential Function Considering Human's Emotion

We represent the relative velocity between the robot and the obstacle in the direction from the robot to the obstacle v_{ro} and the safety margin $\rho_m(v_{ro})$ as $v_{ro} = [\mathbf{v}_r - \mathbf{v}_o]^T \mathbf{n}_{ro}$ and $\rho_m(v_{ro}) = \frac{v_{ro}^2}{2a_{max}}$ respectively, where $\mathbf{n}_{ro} = \frac{\mathbf{P}_{ro}}{\|\mathbf{P}_{ro}\|}$ and a_{max} is the maximum deceleration of robot [4]. The new repulsive potential is defined as in (6).

$$U_{rep}(\mathbf{p}_r, \mathbf{v}_r) = \begin{cases} 0, \text{if } \rho_s - \rho_r - \rho_m(v_{ro}) > \rho_e \text{ or } v_{ro} < 0 \\ \eta_1\left(\frac{1}{\rho_s - \rho_r - \rho_m(v_{ro})} - \frac{1}{\rho_0}\right), \text{if } 0 < \rho_s - \rho_r - \rho_m(v_{ro}) < \rho_0 \text{ and } v_{ro} > 0 \\ \eta_2 w\left(\frac{1}{\rho_s - \rho_r - \rho_m(v_{ro})} - \frac{1}{\rho_0}\right), \text{if } \rho_0 < \rho_s - \rho_r - \rho_m(v_{ro}) < \rho_e \text{ and } v_{ro} > 0 \\ \text{not defined, if } \rho_s - \rho_r - \rho_m(v_{ro}) < 0 \text{ and } v_{ro} > 0 \end{cases} \quad (6)$$

where w is the emotion indicator ($w = 0$ for non-human obstacles and humans with attractive emotions, $w = 1$ for humans with repulsive emotions), η_1 and η_2 are positive scaling constants, $\rho_s = \|\mathbf{p}_{ro}\|$ is the shortest distance between the robot and the obstacle, and ρ_r measures the physical size of the robot.

The virtual repulsive force is then derived as (7).

$$\mathbf{F}_{rep} = \begin{cases} 0, \text{ if } \rho_s - \rho_r - \rho_m(v_{ro}) > \rho_e \text{ or } v_{ro} < 0 \\ \mathbf{F}_{rep1} + \mathbf{F}_{rep2}, \text{ if } 0 < \rho_s - \rho_r - \rho_m(v_{ro}) < \rho_0 \text{ and } v_{ro} > 0 \\ \mathbf{F}_{erep1} + \mathbf{F}_{erep2}, \text{ if } \rho_0 < \rho_s - \rho_r - \rho_m(v_{ro}) < \rho_e \text{ and } v_{ro} > 0 \\ \text{not defined, if } \rho_s - \rho_r - \rho_m(v_{ro}) < 0 \text{ and } v_{ro} > 0 \end{cases} \quad (7)$$

where $\mathbf{F}_{\text{rep1}} = -\frac{\eta_1}{(\rho_s - \rho_r - \rho_m)^2}\left(1 + \frac{v_{\text{ro}}}{a_{\max}}\right)\mathbf{n}_{\text{ro}}$, $\mathbf{F}_{\text{rep2}} = \frac{\eta_1 v_{\text{ro}} v_{\text{ro}\perp}}{\rho_s a_{\max}(\rho_s - \rho_r - \rho_m)^2}\mathbf{n}_{\text{ro}\perp}$

$\mathbf{F}_{\text{erep1}} = -\frac{\eta_2 w}{(\rho_s - \rho_r - \rho_m)^2}\left(1 + \frac{v_{\text{ro}}}{a_{\max}}\right)\mathbf{n}_{\text{ro}}$, $\mathbf{F}_{\text{erep2}} = \frac{\eta_2 w v_{\text{ro}} v_{\text{ro}\perp}}{\rho_s a_{\max}(\rho_s - \rho_r - \rho_m)^2}\mathbf{n}_{\text{ro}\perp}$

in which $v_{\text{ro}\perp}$ is the magnitude of perpendicular component of the relative velocity between robot and the obstacle, and $\mathbf{n}_{\text{ro}\perp}$ is the unit vector along the perpendicular direction to $\mathbf{p}_{\text{ro}}$.

3   Simulation Results

In this section, the proposed navigation scheme is simulated in various conditions and environments. The following assumptions are made in order to implement the algorithm in MATLAB:

Assumption 2. *Velocity and position of the obstacles, the robot, and the target are available at all time.*

Assumption 3. *Each obstacle will have variables indicating if they are human or non-human, and the emotion information for a human obstacle are also available.*

The environment is a two-dimensional field in which the robot with a mass of $1\,\text{kg}$ begins at the origin. To have a fair comparison, the target is set to be stationary in both simulations and experiments, and all the parameters are set as follows: $\rho_r = 0.2$, $v_{\max} = 4\,\text{ms}^{-1}$, $a_{\max} = 5\,\text{ms}^{-2}$, $\rho_0 = 1\,\text{m}$, $\rho_e = 2\,\text{m}$, $m = 2$, $n = 2$, $\alpha_p = 10$, $\alpha_v = 4.47$, $\eta_1 = 1$, $\eta_2 = 5$, and $k = 2$.

3.1   Obstacle Avoidance Without Human

To begin with, an obstacle that is moving towards the robot is considered. The velocity of the obstacle is set to $(-1, -1)^T$, as shown in Fig. 2(a). The two non-human obstacles case is simulated demonstrated in Fig. 2(b). Without considering ρ_m, no action will be performed until it is affected by the physical barrier because emotional barrier will not be applicable for non-human obstacles. However, the influence of the obstacle on the robot occurs earlier due to the parameter ρ_m. It is observed that the robot successfully avoids the obstacle and reaches the target.

3.2   Obstacle Avoidance with Humans

Here, a human is assumed to be standing on the robot's path towards the target. To analyze the behavior of the robot in emotion influential region, the human is first considered to be in an attractive emotion. The path traversed by the robot towards the target is shown in Fig. 2(c). The simulation result is similar to non-human obstacles as $w = 0$ is defined for both non-human obstacles and humans with attractive emotions. Next, let us consider a human obstacle with a repulsive

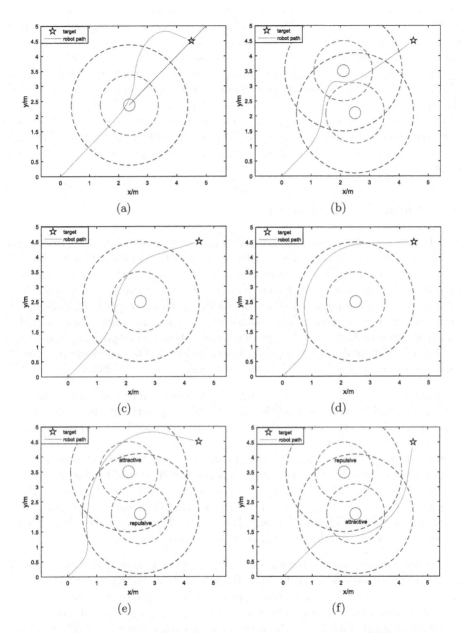

Fig. 2. Simulation results. (a) Single moving non-human obstacle; (b) Multiple stationary non-human obstacles; (c) Single obstacle with attractive emotion; (d) Single obstacle with repulsive emotion; (e) and (f) obstacles with different emotions.

emotion, and the robot trajectory is shown in Fig. 2(d). The trajectory obviously deviates from the human, in this case, leaving the human wider personal space.

For multiple-human obstacles scenarios, we consider two humans standing side-by-side in the robot's path with one of the persons carrying an attractive emotion, while the other person with a repulsive emotion. The path traversed by the robot is shown in Fig. 2(e) and (f). In the circumstance with human obstacles, the emotion of human plays a part in robot navigation approach. From the above observations, it can be concluded that the robot is able to react to human emotions: the robot tends to move closer to the person with an attractive emotion. This is the desirable behavior as it is more likely to get way from the person who has an attractive emotion compared to the person with a repulsive emotion.

4   Practical Experiments

Simulations have shown promising results from the proposed approach. In experiments, a TurtleBot 2 robot compatible with Robot Operating System (ROS) is used. The onboard Kinect sensor acts as both a range finder and an emotion detector. Serving as a range finder, Kinect gives the distances from obstacles after each scan. As an emotion detector, Kinect is used to capture RGB images, detecting human's emotion. The images are sent to Microsoft Emotion API, which analyses human's facial expressions and outputs corresponding emotions.

The experimental setting is demonstrated in Fig. 3. The practical implementation of the proposed algorithm poses certain challenges as follows:

- Velocity measurement: In simulations, the velocity information was assumed to be available to the robot all the time. However, practically this assumption is unrealistic. In experiments, the velocities are estimated as the displacement divided by sampling time.
- Emotion recognition: The setup of Turtlebot 2 has the Kinect sensor at a height of about 0.2 m from the ground. It is difficult to read the facial expressions from such a height. Hence, images at a low position are used as samples to test the response to different emotions. Another challenge is emotion recognition itself. The available Microsoft Emotion API is used, as the focus of this paper is robot navigation instead of emotion recognition techniques.

For the first run, a person with attractive emotion is chosen, then the experiment is repeated for a person with repulsive emotion. Similar to simulation results, the proposed interactive navigation approach effectively addresses the obstacle avoidance problem, adapting to humans with different cooperation desire.

There are some limitations observed from the experiments. The collision may happen when the experimental configurations are not desirable. This results from the characteristics of the Kinect sensor (minimum detection range: 0.45 m; horizontal field of view (HFOV): 57°). At the beginning of the task, if there are any obstacles near the robot then the collision is bound to happen as the robot cannot detect the obstacle. By adding a laser range finder for obstacle detection, the collision may be avoided.

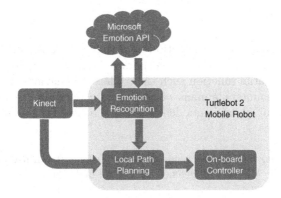

Fig. 3. Experimental configuration diagram

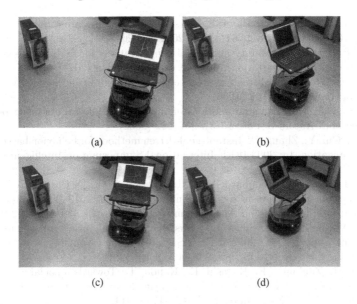

Fig. 4. Experiment with repulsive emotions detected. (a) The robot avoids the obstacle by turning away; (b) The robot turns back towards the goal; (c) The robot turns away; (d) The robot turns back to the goal again.

When the robot reads the emotion as repulsive, it evades the person and turns towards the target. However, other than simulation results, it again moves towards the person (see Fig. 4). This is also due to the limitation of HFOV. Once the robot obtains the repulsive emotion, it evades the person by turning away. At this point, the person goes out of scope so the robot forgets about the repulsive emotion that is detected and also treats the person again as a new obstacle. This issue may be tackled by using an omnidirectional camera or other wide field cameras.

5 Conclusion

In this paper, an interactive navigation approach is proposed for mobile robot obstacle avoidance. The authors incorporate human's emotion to a newly-built repulsive potential function such that different behaviors are implemented when the robot encounters humans with different cooperation desire. Simulation results and practical experiments demonstrate the effectiveness of the proposed approach.

Acknowledgement. This work is supported by Defence Innovative Research Programme (DIRP), the Ministry of Defence, Singapore under grant R-263-000-B08-592.

References

1. Kruse, T., Pandey, A.K., Alami, R., Kirsch, A.: Human-aware robot navigation: a survey. Robot. Auton. Syst. **61**(12), 1726–1743 (2013)
2. Villanueva, J.C.: Uncertainty and social considerations for mobile assistive robot navigation. Ph.D. dissertation, Imperial College London (2014)
3. Khatib, O.: Real-time obstacle avoidance for manipulators and mobile robots. Int. J. Robot. Res. **5**(1), 90–98 (1986)
4. Ge, S.S., Cui, Y.J.: Dynamic motion planning for mobile robots using potential field method. Auton. Robots **13**(3), 207–222 (2002)
5. Ge, S.S., Cui, Y., Zhang, C.: Instant-goal-driven methods for behavior-based mobile robot navigation. In: 2003 IEEE International Symposium on Intelligent Control, pp. 269–274 (2003)
6. Guzzi, J., Giusti, A., Gambardella, L.M., Theraulaz, G., Di Caro, G.A.: Human-friendly robot navigation in dynamic environments. In: 2013 IEEE International Conference on Robotics and Automation (ICRA), pp. 423–430. IEEE (2013)
7. Luber, M., Spinello, L., Silva, J., Arras, K.O.: Socially-aware robot navigation: a learning approach. In: 2012 IEEE/RSJ International Conference on Intelligent Robots and Systems (IROS), pp. 902–907. IEEE (2012)
8. Shiomi, M., Zanlungo, F., Hayashi, K., Kanda, T.: Towards a socially acceptable collision avoidance for a mobile robot navigating among pedestrians using a pedestrian model. Int. J. Soc. Robot. **6**(3), 443–455 (2014)
9. Trautman, P., Ma, J., Murray, R.M., Krause, A.: Robot navigation in dense human crowds: statistical models and experimental studies of human-robot cooperation. Int. J. Robot. Res. **34**(3), 335–356 (2015)
10. Truong, X.-T., Ngo, T.-D.: Dynamic social zone based mobile robot navigation for human comfortable safety in social environments. Int. J. Soc. Robot., 1–22 (2016)
11. Parkinson, B., Fischer, A.H., Manstead, A.S., et al.: Emotion in social relations (2004)
12. Lövheim, H.: A new three-dimensional model for emotions and monoamine neurotransmitters. Med. Hypotheses **78**(2), 341–348 (2012)
13. de Rivera, J., Grinkis, C.: Emotions as social relationships. Motiv. Emot. **10**(4), 351–369 (1986)

Social Human-Robot Interaction: A New Cognitive and Affective Interaction-Oriented Architecture

Carole Adam, Wafa Johal, Damien Pellier[(✉)], Humbert Fiorino, and Sylvie Pesty

Univ. Grenoble-Alps, LIG, 38000 Grenoble, France
{carole.adam,wafa.johal,damien.pellier,humbert.fiorino,
sylvie.pesty}@imag.fr

Abstract. In this paper, we present CAIO, a Cognitive and Affective Interaction-Oriented architecture for social human-robot interactions (HRI), allowing robots to reason on mental states (including emotions), and to act physically, emotionally and verbally. We also present a short scenario and implementation on a Nao robot.

1 Introduction

Robots are more and more present in daily life, in roles such as assistive robots, pedagogical robots, companion robots for children or for the elderly, etc., where they must have a closer interaction with their user. By close, we mean that robots must share not only the same physical space but also goals and beliefs to achieve a common task through their interactions. They should also interact intuitively and easily through speech, gestures, and facial expressions. In spite of the numerous contributions in the field of cognitive architectures for agents and for robots (*e.g.* [5,10,15,16]), designing a cognitive architecture dealing with the complexity of human-robot interactions (HRI) remains a real challenge.

In this paper, we present a new architecture: CAIO (Cognitive and Affective Interaction-Oriented architecture) for social Human-Robot Interaction that aims to contribute on the following aspects essential to HRI: managing emotions (non-verbal aspects of interaction), sensorimotor and deliberative levels (fast (emotional) answer versus slower and more deliberate answer), explicit manipulation of mental states (to enable self-explanation) and handling both physical and verbal actions.

2 Related Works

Cognitive architectures have been subject to research for a long time, and good reviews exist (see for example [11,30]). They mostly fall in three categories: biologically-inspired, philosophically-inspired, and Artificial Intelligence architectures. We illustrate these categories with some of the major and well-known architectures.

© Springer International Publishing AG 2016
A. Agah et al. (Eds.): ICSR 2016, LNAI 9979, pp. 253–263, 2016.
DOI: 10.1007/978-3-319-47437-3_25

Biologically inspired architectures: ACT-R is a well-known cognitive architecture (Adaptive Control of Thought-Rational), stemming from the progressive refinement of Anderson's model of human cognition [5], originating in his Human Associative Memory model [6]. The main assumption is the separation between two types of knowledge: declarative (chunks) and procedural (rules); the system is only aware of knowledge with sufficient activation.

CLARION (Connectionist Learning with Adaptive Rule Induction ONline) [29] is also well-known, based on neural networks. It mainly focuses on the distinction between implicit and explicit processes, and the interactions between them. It has been used to simulate processes in cognitive or social psychology, and to implement AI applications.

ASMO (Attentive Self-MOdifying) [19] was developed more recently based on a biological theory of attention, to solve the problem of competing, possibly incompatible robot goals. Concretely the attention level determines relative priorities of goals, with the most critical ones being treated as reflexes. It is being implemented in a social bear robot interacting with humans, and in Nao for soccer competitions.

Problem-Solving Artificial intelligence architectures: SOAR (State, Operator And Result) [15] is a pure AI symbolic architecture focused on learning and problem solving. It has short-term working memory and long-term memory (procedural, semantic and episodic). Reinforcement learning is triggered when knowledge is inadequate to make a decision. SOAR was extended with emotions that affect learning [14].

ICARUS [17] is grounded in cognitive psychology and AI, and aims at unifying reactive and deliberative problem-solving, as well as symbolic and numeric reasoning. The goal with highest priority that is not satisfied yet takes focus: the skills allowing to achieve it are brought from long-term to short-term memory, or means-end analysis is used to decompose it into subgoals and learn new skills.

Philosophically inspired architectures: Bratman's philosophical action theory [9] models human behaviour as a perception-decision-action cycle. He claimed that the intention to perform an action is adopted from beliefs and desires *via practical reasoning* that makes us rational. BDI logics (*Belief, Desire, Intention*, [12,24]) were then proposed to formalise these three mental states.

The BDI model has been at the root of a number of architectures for artificial agents (*e.g.* the *Procedural Reasoning System* - PRS [33]). As we will explain later, the CAIO architecture is also in line with this tradition but we introduce new mental states, in particular emotions that are essential for an expressive social robot since they play a major role in interaction and reasoning.

3   Previous Work

3.1   Complex Emotions and Multimodal Conversational Language

Guiraud et al. [13] proposed a new modal logic (BIGRE logic) derived from BDI for the formal representation of five agent's mental states (B, I, G, R, E) expressed by an agent during a conversation with another agent or a human:

- *Belief* (B) $Bel_i\varphi$: the robot i believes that φ,
- *Ideal* (I) $Ideal_i\varphi$: ideally for robot i, φ should hold (social and moral norms of the robot[1]),
- *Goal* (G) $Goal_i\varphi$: the robot i wants that φ holds,
- *Responsibility* (R) $Resp_i\varphi$: the robot i is responsible for φ (arising from complex reasoning about norms and responsibility of its own actions and those of others).
- *Complex emotion* (E) (*e.g.* gratitude, admiration, reproach, etc.) result from reasoning on *Responsibility*.

Complex emotions are of primary importance in human dialogue and are mainly conveyed through language. They differ from basic emotions built from beliefs and goals, and often expressed by prototypical facial expressions. Eight complex emotions (*regret, disappointment, guilt, reproach, moral satisfaction, admiration, rejoicing* and *gratitude*) and four basic emotions (*joy, sadness, approval, disapproval*) have been formalized in terms of the B, I, G and R operators (BIGR → E).

To ensure that a robot is able to express its mental states in a credible manner, Riviére *et al.* [26] defined a conversational language based on Searle's Speech Acts Theory [28], and in line with previous mentalistic Agent-Communication Languages (ACL) such as FIPA [23]. This language is called *Multimodal Conversational Language (MCL)* because it closely links verbal (the utterance) and non-verbal (*e.g.* underlying emotion of expressive speech acts) aspects in order to improve the expressivity of the robot. The MCL consists of 38 *Multimodal Conversational Acts* (MCA) divided in four classes:

- assertive acts: *to inform, to affirm, to deny,* etc.
- directive acts: *to ask, to suggest, to require,* etc.
- commissive acts: *to promise, to accept, to offer,* etc.
- expressive acts: *to apologize, to rejoice, to reproach, to thank,* etc.

For each MCA there is a formalisation in the BIGRE logic of:

- its preconditions, that the robot has to satisfy before performing this act, ensuring its sincerity in the sense of Searle's Speech Acts Theory (sincerity conditions);
- its sending effects, on the robot performing it;
- its reception effects, on the robot receiving this act performed by the interlocutor.

For example, Table 1 shows the formalisation of the conversational act *to rejoice*. This explicit formal representation has the advantage of allowing the robot to manipulate and reason about the conversational acts: update its mental states when receiving or sending one, and using them in its plan of action.

The interested reader is referred to [13] for detailed semantics and axiomatics of the BIGRE logic.

---

[1] For instance, a moral obligation to help someone in danger, or a social norm to pay one's taxes, etc.

Table 1. Example: the *to rejoice* MCA from agent a's point of view in a dialogue with a human h.

Rejoice	$Exp_{a,h,H}(Rejoicing_a\varphi) \equiv Exp_{a,h,H}(Goal_a\varphi \wedge Bel_a Resp_a\varphi)$
Preconditions	$Goal_a\varphi \wedge Bel_a Resp_a\varphi \overset{d\acute{e}f}{=} Rejoicing_a\varphi$
	Robot a "feels" rejoicing; it believes it is responsible for having achieved its goal φ
Sending effects	$Bel_a Bel_h Rejoicing_a\varphi$
	Robot a believes that human h believes that a is rejoicing about φ
Reception effects	$Bel_a Goal_h\varphi \wedge Bel_a Bel_h Resp_h\varphi$
	Robot a believes that human h expressed his rejoicing about φ. Therefore, a believes that h has achieved its goal φ and believes himself to be responsible for this

3.2 PLEIAD Reasoning Engine

PLEIAD (ProLog Emotional Intelligent Agent Designer) [1] is originally a SWI-Prolog reasoning engine for BDI-like agents. It provides agents with generic reasoning capabilities and emotions. Concretely, it enables the implementation of various logical models of emotions, such as the OCC theory [2] or theories about shame [3]. It has also been recently extended with coping strategies [4] and some personality traits.

This reasoning engine has been used to implement the BIGRE logical model and is at the core of the CAIO architecture, especially for the *Deliberation module* and the *Emotional appraisal module* (see details in Sect. 4.2).

4   CAIO Architecture

4.1   Overview of the Architecture

The CAIO architecture (see Fig. 1) consists of two fundamental loops: a **Deliberative loop** used to reason on BIGRE mental states and produce plans of actions, and a **Sensorimotor loop** to immediately and continuously trigger emotion expressions. Each loop takes as inputs the result of the multimodal perception of the environment.

During the **Deliberative loop**: the *Cognitive part of the Emotional Appraisal module* deduces complex emotions from the mental states; the *Deliberation module* deduces the robot's *Communicative Intentions* from its mental states, and selects the most appropriate one; then the *Planning module* produces a plan to achieve the selected intention (*i.e.* a set of ordered actions, conversational acts and/or physical actions), and schedules the robot's next action; finally the *Emotional Multimodal Action Renderer module* executes this scheduled action. The modules can provide feedback to each other: the planning module informs the deliberation module of the feasibility of the selected intention; the renderer informs the planner of the success or failure of action performance.

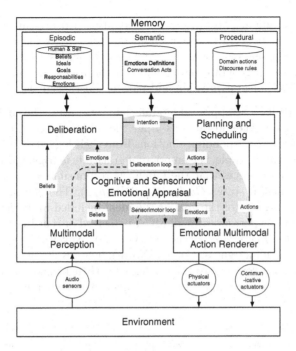

Fig. 1. The CAIO architecture.

Simultaneously, during the **Sensorimotor loop**: the *Sensorimotor part of the Emotional Appraisal module* evaluates the input according to criteria (Scherer's SEC - Stimulus Evaluation Checks); the *Emotional Multimodal Action Renderer module* then dynamically renders the corresponding robot's non-verbal (facial and gestural) expression.

4.2 The 6 Main Modules of the CAIO Architecture

Memory module. The robot's memory is divided into three parts in accordance with the state of the art (see Sect. 2). The *episodic* memory contains BIGRE-based knowledge about the self and the human interlocutor. The *semantic* memory contains the definitions of emotions and conversational acts. The *procedural* memory deals with domain actions (how-to) and discourse rules (*i.e.* when asked a question, one should reply).

The memory is dynamically updated in three steps: first, new beliefs deduced from the perception of the world or of the interaction (reception effects of the user's recognised MCA, and sending effects of the robot's own MCA) are *added*; then inference rules are applied to *update* the robot's BIGRE mental states, possibly deducing new mental states.

Multimodal perception module. We focus on language perception. Concretely, this module first recognises text from speech (using Google Speech). It then extracts the human's MCA from the recognised text utterance[2]. This MCA then generates new beliefs (its reception effects) that enter the 2 (sensorimotor and deliberative) loops of processing. As the aim of this module is to merge multimodal inputs to generate new beliefs on the user's mental states, future works will consider facial expression and para-linguistic signals.

Appraisal module. The appraisal module is in two parts. The cognitive part is an extension of PLEIAD and takes as input the robot's perceptions and mental states to trigger the corresponding emotions from their logical definition in terms of mental states. For example, the emotion of gratitude is triggered when a robot has the goal φ and believes that the human is responsible for φ, *i.e.* when the robot i's has a mental states $Goal_i\varphi \wedge Bel_i Resp_j\varphi$). The emotion intensity is derived from the priority of the goal or the ideal included in its definition.

The sensorimotor part assesses all MCA perceived or sent by the robot *w.r.t.* Scherer's *Stimulus Evaluation Checks* (SEC, [27]) (Novelty, Intrinsic pleasantness, Goal/Need conductivness, Coping, Norm). The results of the SEC evaluation process are then sent out to the renderer for their facial and bodily expression. Figure 2 shows an example of a SEC sequence corresponding to a reproach expression.

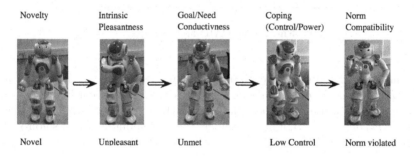

Novelty	Intrinsic Pleasantness	Goal/Need Conductivness	Coping (Control/Power)	Norm Compatibility
Novel	Unpleasant	Unmet	Low Control	Norm violated

Fig. 2. SEC sequence corresponding to a reproach expression.

Deliberation module. Deliberation is the process of selecting the robot's next intention to achieve, *via* practical reasoning [9] from its mental states and a set of priority rules. It deals in particular with three kinds of *intention. Emotional intentions* are intentions to express the robot's emotions. In order for the robot to be sincere, affective and expressive, we assume that all emotions felt during the interaction lead to an emotional intention to express them, which participate in the local regulation of dialogue by enabling a more natural robot-human interaction [8]. *Obligation-based intentions* also contribute to the local regulation of dialogue [7]. They are adopted from a set of *discourse obligation rules* defined

---

[2] Natural Language Understanding is a complex research field of its own, we do not tackle this problem here, and instead use an ad-hoc grammar specifically designed for our scenario.

by Traum and Allen [31] to represent social norms guiding the robot's behaviour and making it reactive at the discourse level. Concretely the robot always adopts the intention to fulfill its obligation deduced by these rules. Finally, the *global intention* gives the global direction of dialogue and defines its type (*e.g.* deliberation, persuasion... [32]). It is adopted when the robot has committed to achieve the corresponding goal, either publicly (by performing a commissive MCA such as *Promise* or *Accept*) or privately (via practical reasoning on its beliefs and plans).

Planning module. It is in charge of finding a way of achieving the selected intention according to a plan-based approach of dialogue [22]. It is based on the planning approach proposed by [21] and on the PDDL4J Java library [20]. The plans produced contain MCA and/or physical actions, whose preconditions and effects are formalised in the classical Planning Domain Description Language (PDDL), making most existing planners compatible with CAIO.

In the case of emotional and obligation-based intentions, the built plan is usually made up of a single MCA (for example the emotional intention to express gratitude can be achieved with *to thank* or *to congratulate* depending on the emotion's intensity). In the case of global intentions, domain-dependent actions may be necessary, whose preconditions and effects are described in the static procedural memory (for instance, to book a train it is necessary to know the time and date of departure and destination). The planner will then produced a plan with both MCA (*e.g. to ask* the relevant information to the user) and domain actions (*e.g.* actually book the train). If no plan can be computed to achieve it, the current intention is discarded, and feedback is sent to the deliberation module that selects a new intention.

Multimodal Action Renderer. It receives as input the action to be executed and the complex emotion computed by the appraisal module, and controls the robot's actuators to execute this action and to dynamically generate the facial expression corresponding to the emotion. In particular for MCA it expresses the underlying complex emotion. Independently from this deliberative expression, this module also receives the SEC values computed by the sensorimotor part of the emotional appraisal module, and dynamically builds the corresponding facial and bodily expression, leading to a sequence of postures (for example Table 1).

5   Discussion and Conclusion

The CAIO architecture was first implemented and evaluated for a virtual character [25]. The current version is based on ROS (Robot Operating System), which is largely used in the robotic community. The modules were implemented in Python in order to allow easy interfacing with SWI-Prolog (PLEIAD engine) via the Pyswip library (a Python library that allows to query SWI-Prolog from Python programs). Below is a short scenario illustrating the ROS nodes encapsulating each process involved in the CAIO architecture (see its UML Sequence Diagram on Fig. 3).

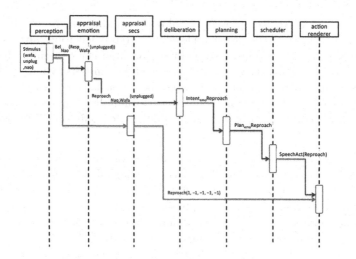

Fig. 3. UML Sequence Diagram

This scenario involves two actors: the human, *Wafa*, and the *Nao robot*, which has a low battery life and requires to be plugged all the time. In its episodic memory, Nao has the ideal of being plugged $Ideal_{Nao}(\neg unplugged)$. Now, Wafa has a party tonight and needs to dry her hair, but Nao is plugged to the only plug near the mirror. She thus tells Nao: "Nao, I am unpluging you".

1. **Multimodal Perception:**
   The *perception* node receives the utterance "Nao, I am unpluging you" and extracts an *to inform* conversational act (i.e. a *to inform* MCA) $Inform(Stimulus(unplugged, wafa, nao))$. The episodic memory is updated: $Bel_{Nao}(unplugged)$ and $Resp_{Wafa}(unplugged)$.

2. **Emotional Appraisal:**

   (a) **Cognitive Appraisal:** The *appraisal_emotion* node deduces the complex emotion $Reproach_{Nao,Wafa}(unplugged)$; the episodic memory is updated. The complex emotion is sent to *deliberative* node.

   (b) **Sensorimotor Appraisal:** The *appraisal_checks* node evaluates the *to inform* MCA $Inform(Stimulus(unplugged, wafa, nao))$ in accordance with the five evaluation criteria (SEC). The result of the SEC sequence is sent to the *action_renderer* node.

3. **Deliberation:**
   The *deliberative* node infers a list of intentions. The one with highest weight is an *emotional intention* to perform a reproach to the user. The list of intentions is sent to the *planning* node.

4. **Planning and Scheduling:**
   The *planning* node picks the most weighted intention from the list of intentions, and publishes a list of plans (here a unique plan consisting of the single *to reproach* MCA).

The plan is received by the *scheduler* node which picks the first action (here the *to reproach* MCA).

5. **Emotional Multimodal Action Renderer:**
Finally the *action_renderer* node receives both the SEC sequence and the *to reproach* MCA, and plays them on the Nao robot. It thus tells Wafa: "Wafa, you must not unplug me". Nao performs this utterance in a multimodal way.

The ROS version of the CAIO architecture is currently being further validated through real-time interaction with children to verify that the robot clearly conveys its intentions, and is perceived as sincere. In parallel, we have run a more conceptual evaluation of CAIO against Langley *et al.* evaluation criteria for cognitive architectures [18]. This conceptual evaluation shows that the CAIO architecture already provides new contributions regarding the state of the art in cognitive architectures for companion robots.

Further research on the multimodal perception module is however needed to automate the extraction of the user's speech act, and to deal with facial expressions and para-linguistic features (to guarantee better recognition and sincerity). A learning module would also be a nice extension to ensure that the robot can improve during the interaction, and progressively learn to know its user to better adapt to them and engage them.

References

1. Adam, C.: The emotions: from psychological theories to logical formalisation and implementation in a BDI agent. Ph.D. thesis, INP Toulouse, July 2007
2. Adam, C., Herzig, A., Longin, D.: A logical formalization of the OCC theory of emotions. Synthese **168**(2), 201–248 (2009)
3. Adam, C., Longin, D.: Shame, when reasoning and emotions are linked. In: EUMAS 2013 (2013)
4. Adam, C., Lorini, E.: A BDI emotional reasoning engine for an artificial companion. In: Corchado, J.M., Bajo, J., Kozlak, J., Pawlewski, P., Molina, J.M., Gaudou, B., Julian, V., Unland, R., Lopes, F., Hallenborg, K., García Teodoro, P. (eds.) PAAMS 2014. CCIS, vol. 430, pp. 66–78. Springer, Heidelberg (2014). doi:10.1007/978-3-319-07767-3_7
5. Anderson, J.R.: Human symbol manipulation within an integrated cognitive architecture. Cogn. Sci. **29**(3), 313–341 (2005)
6. Anderson, J.R., Bower, G.H.: Human Associative Memory. Winston and Sons, Washington, DC (1973)
7. Baker, M.J.: A model for negotiation in teaching-learning dialogues. J. Artif. Intell. Educ. **5**(2), 199–254 (1994)
8. Bates, J.: The role of emotion in believable agents. Commun. ACM **37**(7), 122–125 (1994)
9. Bratman, M.E.: Intention, Plans, and Practical Reason. Harvard Univ. Press, Cambridge (1987)
10. Brooks, R.: A robust layered control system for a mobile robot. IEEE J. Robot. Autom. **2**(1), 14–23 (1986)
11. Chong, H.Q., Tan, A.H., Ng, G.W.: Integrated cognitive architectures: a survey. Artif. Intell. Rev. **28**(2), 103–130 (2009)

12. Cohen, P.R., Levesque, H.J.: Persistence, intention and commitment. In: Cohen, P., Morgan, J., Pollack, M. (eds.) Intentions in Communication, pp. 33–69. MIT Press, Cambridge (1990)

13. Guiraud, J., Longin, D., Lorini, E., Pesty, S., Rivière, J.: The face of emotions: a logical formalization of expressive acts. In: International Conference on Autonomous Agent and Multiagent Systems, pp. 1031–1038 (2011)

14. Hogewoning, E., Broekens, J., Eggermont, J., Bovenkamp, E.G.P.: Strategies for affect-controlled action-selection in soar-RL. In: Mira, J., Álvarez, J.R. (eds.) IWINAC 2007. LNCS, vol. 4528, pp. 501–510. Springer, Heidelberg (2007). doi:10. 1007/978-3-540-73055-2_52

15. Laird, J.E.: The Soar Cognitive Architecture. MIT Press, Cambridge (2012)

16. Lallée, S., Pattacini, U., Lemaignan, S., Lenz, A., Melhuish, C., Natale, L., Skachek, S., Hamann, K., Steinwender, J., Sisbot, E., Metta, G., Guitton, J., Alami, R., Warnier, M., Pipe, T., Warneken, F., Dominey, P.: Towards a platform-independent cooperative human robot interaction system: III an architecture for learning and executing actions and shared plans. IEEE Tran. Auton. Ment. Dev. **4**(3), 239–253 (2012)

17. Langley, P., Choi, D., Rogers, S.: Interleaving learning, problem solving, and execution in the icarus architecture. Technical report, Computational Learning Laboratory, CSLI, Stanford University (2005)

18. Langley, P., Laird, J., Rogers, S.: Cognitive architectures: research issues and challenges. Cogn. Syst. Res. **10**(2), 141–160 (2009)

19. Novianto, R., Johnston, B., Williams, M.A.: Attention in asmo cognitive architecture. In: 1st International Conference on Biologically Inspired Cognitive Architectures (BICA) (2010)

20. Pellier, D.: PDDL4J: Planning Domain Description Language Library for Java (2015)

21. Pellier, D., Fiorino, H., Métivier, M.: Planning when goals change: a moving target search approach. In: International Conference on Advances in Practical Applications of Heterogeneous Multi-Agent Systems (2014)

22. Perrault, C.R., Allen, J.F.: A plan-based analysis of indirect speech acts. Comput. Linguist. **6**(3–4), 167–182 (1980)

23. Poslad, S.: Specifying protocols for multi-agent systems interaction. ACM Trans. Auton. Adapt. Syst. **2**(4) (2007). http://doi.acm.org/10.1145/1293731.1293735

24. Rao, A.S., Georgeff, M.P.: Modeling rational agents within a BDI-architecture. In: International Conference on Principles of Knowledge Representation and Reasoning (KR 1991), pp. 473–484 (1991)

25. Rivière, J.: Interaction affective et expressive Compagnon artificiel - humain. Ph.D. thesis, de Grenoble (2012)

26. Riviere, J., Adam, C., Pesty, S., Pelachaud, C., Guiraud, N., Longin, D., Lorini, E.: Expressive multimodal conversational acts for SAIBA agents. In: Vilhjálmsson, H.H., Kopp, S., Marsella, S., Thórisson, K.R. (eds.) IVA 2011. LNCS (LNAI), vol. 6895, pp. 316–323. Springer, Heidelberg (2011). doi:10.1007/978-3-642-23974-8_34

27. Scherer, K.R.: Appraisal considered as a process of multilevel sequential checking. In: Scherer, K.R., Schorr, A., Johnstone, T. (eds.) Appraisal Processes in Emotion: Theory, Methods, Research, pp. 92–120. Oxford University Press, New York (2001)

28. Searle, J., Vanderveken, D.: Foundations of Illocutionary Logic. Cambridge University Press, Cambridge (1985)

29. Sun, R.: Duality of the Mind: A Bottom-up Approach Toward Cognition. Lawrence Erlbaum Associates, Mahwah (2002)

30. Thórisson, K., Helgasson, H.: Cognitive architectures and autonomy: a comparative review. J. Artif. Gen. Intell. **3**(2), 1–30 (2012)
31. Traum, D.R., Allen, J.F.: Discourse obligations in dialogue processing. In: Proceedings of the 32th Annual Meeting of the Association for Computational Linguistics (ACL), pp. 1–8 (1994)
32. Walton, D., Krabbe, E.: Commitment in Dialogue: Basic Concept of Interpersonal Reasoning. State University of New York Press, Albany (1995)
33. Wooldridge, M.: An Introduction to MultiAgent Systems. Wiley, New York (2009)

MuDERI: Multimodal Database for Emotion Recognition Among Intellectually Disabled Individuals

Jainendra Shukla[1,3]([✉]), Miguel Barreda-Ángeles[2],
Joan Oliver[1], and Domènec Puig[3]

[1] Instituto de Robótica para la Dependencia, Sitges, Spain
jshukla@institutorobotica.org
[2] Eurecat, Technology Centre of Catalonia, Barcelona, Spain
[3] Intelligent Robotics and Computer Vision Group,
Universitat Rovira i Virgili, Tarragona, Spain

Abstract. Social robots with empathic interaction is a crucial requirement towards deliverance of an effective cognitive stimulation among
individuals with Intellectual Disability (ID) and has been challenged by
absence of any particular database. Project REHABIBOTICS presents
a first ever multimodal database of individuals with ID, recorded in a
nearly real world settings for analysis of human affective states. MuDERI
is an annotated multimodal database of audiovisual recordings, RGB-D
videos and physiological signals of 12 participants in actual settings,
which were recorded as participants were elicited using personalized real
world objects and/or activities. The database is publicly available.

Keywords: Socially assistive robotics · Intellectual disability · Robot
assisted therapy · SAR · RAT · ID

1 Introduction

Socially Assistive Robotics (SAR) has already been widely used in mental health
service and research [1], primarily among children with Autism Spectrum Disor-
der(ASD) and among older adults with dementia. Project REHABIBOTICS is
a holistic approach to extend the benefits of SAR to individuals with ID. During
the project REHABIBOTICS, a case study of robot interactions in different categories of possible clinical applications of the interactive robot among individuals
with ID was performed [2] and the response to robot interactions was compared
to tactile gaming console stimulation [3]. A crucial step towards delivering an
efficient cognitive stimulation to individuals with ID by robots is to make them

J. Shukla—This research work has been supported by the Industrial Doctorate program (Ref. ID.: 2014-DI-022) of AGAUR, Govt. of Catalonia. The authors gratefully
acknowledge the cooperation of the participants, their guardians and caregivers in
this research.

A. Agah et al. (Eds.): ICSR 2016, LNAI 9979, pp. 264–273, 2016.
DOI: 10.1007/978-3-319-47437-3_26

able to perform an emotionally adaptive behavior; i.e., to be able to detect users' feelings and to adjust the experience to fit them [2,3]. However, affective state estimation among such type of users by robots has been challenged by the absence of any multimodal database for individuals with ID. There are publicly available databases for research purposes that contain naturalistic multimodal and continuous data, labeled either in terms of discrete categories or along the emotional dimensions, such as Multi-modal Affective Database for Affect Recognition and Implicit Tagging (MAHNOB-HCI) [4], Database for Emotion Analysis using Physiological Signals (DEAP) [5], SEMAINE database [6], Belfast Induced Natural Emotion Database [7], etc. However, project REHABIBOTICS could not use aforementioned databases for reasons mentioned below:

1. Existing databases employ *non-personalized stimulation* circumstances to stimulate emotions among subjects, while project REHABIBOTICS aims for personally significant circumstances which are *personalized stimulations*. Hence, presented database MuDERI employs *natural* setup for gathering genuine emotions among individuals.
2. Previous studies indicate that individuals with ID have Electroencephalogram (EEG) abnormalities [8] and as existing databases were recorded with healthy participants, a unique database was highly demanded by the project REHABIBOTICS.

The aim of presented work is to overcome the issues above by introducing the MuDERI database which will assist in empowering the robots with the automated emotion recognition ability among users with ID during human-robot interaction.

2   Measuring Emotions

Emotions have been defined as "action dispositions", that is, evolutionarily based psychological mechanisms that prepare the individual for action when facing relevant stimuli [9]. Theoretical models of emotions have been grounded in two competing perspectives. On the one hand, discrete models of emotions claim the existence of a number of distinct basic emotional states, such us joy, sadness, anger, fear, surprise, or disgust, the combination of which characterizes the human emotional experience [10]. By contrast, bi-factorial models describe all the existing emotions as a function of two core factors: hedonic valence (whether the emotion is positive or negative) and arousal (i.e. the level of excitement) [9]. The presence of emotional states is evidenced through three different types of manifestations: self-report from the subject (collected, for instance, through questionnaires or interviews), changes in physiological aspects in the subject's body (e.g., heart rate or conductivity of skin), and directly observable behaviors (e.g. face expressions or body movements)[11]. Due to the limited ability in recognition and expression of their emotional state among individuals with ID, the analysis of physiological and behavioral correlation of emotions emerges as the most useful method for monitoring their emotions.

Presented database includes two types of physiological signals; Electrodermal activity (EDA) and Electroencephalogram (EEG) signals. EDA refers to the changes in conductivity in the skin, which is affected by the activity of the sympathetic branch of the autonomous nervous system and is usually considered as a correlate of emotional arousal [9]. Due to its low-cost and easy-to-collect nature, it has been commonly used in research on user experience with technology (e.g. [12,13]), and recent researches have shown its potential for emotional state classification in patient-robot interaction [14]. EEG, a measure of the brain electrical activity, is widely used in human-computer interaction research, and has shown promising results for obtaining information on the emotional states [5,11]. Regarding behavioral correlates of emotions, probably the most evident and widespread behavioral measure of emotions is the analysis of facial activity; however, also body movements have been shown to inform about emotional states [15]. Thus, in the database, we also collected audiovisual recordings of the user's while doing the tasks, with medium shots covering participants' faces and the upper half of their body. The audiovisual recordings also include depth information as provided by a RGB-D camera. Following we describe in detail the design of the tasks and data collection, as well as the features of the MuDERI database.

3   Method

A series of cognitive stimulation sessions among users were conducted at the *Ave Maria Foundation (FAM)*[1] for the creation of the multimodal database. *FAM* is a residential and clinical facility for users with ID. A series of meetings with the caregivers at *FAM* were done for the identification of the emotional states that the users most commonly exhibit and ease in arousal of specific discrete emotions (e.g. happiness, sadness etc.) among them, through the stimulation sessions. Consequently the design of tasks was favored on an approach based on the bi-factorial model of emotions. Hence, two cognitive stimulation sessions were designed for each user, one aimed to elicit positive emotions (joy), and the other aimed to elicit negative emotions (sadness or anger), from an initial neutral emotional state.

Since the aim of the experiment was to record the emotional response of the participants in *real world* scenarios, real world objects/activities were used to stimulate desired emotions. A list of activities and/or objects, *personally significant* for the participant and capable of provoking a desired emotion, were identified by three caregivers for each participant. These real world objects/activities were later used to provoke the desired emotions. Only those caregivers, who have been taking care of these individuals for at-least three years and hence, were fully aware of their behavior, participated in the identification of such objects and/or activities. Examples of tasks addressed to elicit positive emotions were: playing the user's favorite music, giving the user a candy of his/her choice, etc., whereas

---

[1] Fundació Ave Maria, http://www.avemariafundacio.org/inici.html.

examples of tasks stimulating negative emotions were: discussing the user about his/her demised parent, trying to take his/her bracelet out, etc.

Ethical, legal and social issues concerning the trials were identified by the *Institutional Advisory Board (IAB)* of the *FAM* and accordingly the trials were designed and executed. As per the guidelines of the *IAB*, an appropriate positive reinforcement was applied to the participant after passing through sessions which provoke anger or sadness among them to prevent any negative after-trial effects.

3.1   Participants

Participant's inclusion criteria were defined as follows:

1. Age equal to or above 18 years.
2. An official diagnosis of ID as done by *Assessment and Guidance Services for People with Disabilities (CAD Badal)* organization[2].
3. Living in the residence facilities for at least 3 years, so caregivers are familiar with them and are able to interpret their communication intent.
4. Their guardians have provided written consent to take part in the study.
5. Participants felt comfortable wearing the non-intrusive wireless physiological sensors.

The chosen sample consisted of 12 participants (10 females and 2 males) aged between 34 and 69 years ($M = 49.58$; $SD = 10.58$). Participants had a diagnosis of either moderate (4) or severe (5) ID; the degree of ID in the remaining participants (3) was unreported. Causes of ID were heterogeneous and in many cases unknown.

3.2   Set up

Figure 1 shows the experimental setup, consisting of an intervention table which was specially designed for performing stimulation trials among the participants. The participant and the caregiver sat on either side of the table, facing each other. The intervention table is equipped with an arch above the middle of the table, where two cameras (one high resolution video camera and one RGB-D camera) were mounted. Such a placement of the camera's allowed recording of the participant's audiovisual and RGB-D videos, without hindering the interaction between the participant and the caregiver. A Logitech C-920-C[3] camera was used in the audiovisual recordings and an Asus Xtion PRO LIVE[4] was used for the RGB-D recordings.

Two physiological sensors were worn by the participants and the sensors were fully wireless to cause minimal intrusion to the participant. EEG signals were collected by means of a headband sensor Emotiv Epoc[5], a wireless system

---

[2] Generalitat de Catalunya, http://web.gencat.cat/ca/inici/.

[3] http://support.logitech.com/en_us/product/c920-c-webcam.

[4] https://www.asus.com/3D-Sensor/Xtion_PRO_LIVE/.

[5] http://emotiv.com/epoc/.

Fig. 1. Schematic representation of the set up

that provides data from 14 EEG channels and two reference channels, with an internal sampling rate of 2048 Hz automatically filtered and downsampled to 128 Hz. Data channels collected by this device include the following positions, according to the International 10–20 system: AF3, AF4, F3, F4, F7, F8, FC5, FC6, T7, T8, P7, P8, O1, and O2. Participants' EDA signals were collected using a wireless wristband sensor Shimmer GSR+[6]. Figure 2 shows the devices used to obtain physiological signals. Low cost of these devices promises a wider reach. A laptop was placed behind the intervention table, hidden from the participant. It was used to receive the data from physiological sensors, video and RGB-D cameras.

(a) Emotiv Headset   (b) Shimmer Device

Fig. 2. Physiological sensors

3.3 Procedure

During the trial, the participant was brought to the trial room by the caregiver and took a seat in front of the intervention table. The caregiver put the physiological sensors on the participant, including the Shimmer device on one wrist, and the Emotiv headset on the participant's head. Then the signal recording was started and the researcher left the room. Thus, the caregiver stayed in the room

---

[6] http://www.shimmersensing.com/shop/shimmer3-wireless-gsr-sensor.

with the participant while the researchers observed and controlled the whole situation from observation area setup in the room, hidden to the participant. After that, the recording of the baseline activity was done during 30 s, while the caregiver talked to the participant trying to not to elicit any specific emotion, in order to provide a recording of an emotionally neutral state. The caregiver presented the object and/or activity to the participant and interacted with her in order to elicit the target emotion, while assisting the participant during all the task. The duration of the trials varied between 15-20 min, depending upon the setup time and activity, while actual stimulation during each trial lasted between 3 to 5 min.

Each individual participated in two sessions, conducted in different days, one aimed to elicit a positive emotion and the other aimed to elicit a negative emotion, and each session involved two periods: the baseline period, and the trial period. A positive reinforcement was conducted a post session for sessions eliciting a negative emotion in the participant. This was done for a few minutes after the session, until the caregiver considered that the effects of the negative emotion were not evident in the participant anymore.

3.4   Data Preprocessing and Annotation

The audiovisual recordings were used as a reference signal for the synchronization of the RGB-D data and physiological signals. The EEG signals, EDA signals and kinect data were trimmed and synchronized with audiovisual recordings using the timestamps, registered earlier with the help of corresponding recording softwares. Kinect videos were trimmed using OpenNI[7]. EEG signals were trimmed and preprocessed using EEGlab software [18]. A band-pass filter (4–45 Hz) was applied to the raw signal, and ocular artifacts were automatically removed [19]. EDA signals were trimmed using an in-house Python script and were filtered using Ledalab software [20]. The raw signals were low-pass filtered (1 Hz cutoff frequency) to remove the artifacts of the signals and smoothed with moving average method with 8 samples window. However, a lot of artifacts, mainly due to participant's movement, remain in the EDA signals, so researchers using the database are recommended to use advanced techniques to remove such artifacts.

The data annotation was conducted on specific moments of the audiovisual recordings. One caregiver, familiar to all the participants, selected a number of moments in each audiovisual recording in which the participant showed a neutral emotional state (no evidence of emotion) as well as moments in which he or she shows evidence of being experiencing emotions. The inclusion of neutral (not-emotional) moments in the annotation task is intended to provide annotated data from each participant that can be used as a baseline for the analysis of emotional moments. One or two neutral moments and three to six emotional moments were selected for each participant in each period for posterior annotation. The duration of all moments was about 10 s. Five caregivers (3 of those were involved with experiments and hence were familiar with trial intents while, other 2 were

---

[7] http://structure.io/openni.

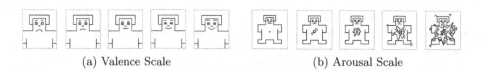

(a) Valence Scale   (b) Arousal Scale

Fig. 3. SAM Mankins

not) carried out the annotation task. It was done to avoid any prejudice biasing that could affect the evaluation. The audiovisual recordings of the preselected neutral and emotional moments were presented to the caregiver using an in-house Python application. After watching each sequence, the caregivers annotated it in terms of emotional valence and arousal using a 9-points SAM scale [16]. They also annotated each sequence in terms of the joy, sadness, and anger that in their opinion was manifest in the participant, using a 9-points Likert-type scale [17].

The SAM scale is composed by two scales: valence and arousal scales (Fig. 3[8]). In the valence scale, the caregiver was required to rate the hedonic valence of the user during the task, that is, how positive or negative the user felt during the task. The left end of the scale represents the most negative experience (unhappy, sad, annoyed, unsatisfied, melancholic, angry, etc.) and the right end represents the most positive experience (happy, pleased, satisfied, etc.). In the arousal scale the caregiver rated how excited the user has been during the task. The left end of the scale represents the calmest experience (calmed, relaxed, drowsy) and the right end represents the most arousing experience (excited, alert, anxious).

4   Database

Table 1 shows the final composition of the MuDERI database. In order to check the effectiveness of the sessions for eliciting the targeted positive and negative emotions, we conducted a series of one-way ANOVAs, one for each rated variable (valence, arousal, joy, sadness, anger), taking as independent variable the type of period (baseline, trial-positive, or trial-negative) and as dependent variable the mean scores from the five caretakers in each annotated segment. Since the variance of the mean ratings was not equal through the three periods, we included the Welch correction in the ANOVA, and the post-hoc pair comparisons were conducted using the Games-Howell approach. The results (Table 2) show that in all cases there were significant differences between the average scores obtained in the three types of periods (baseline, trial-positive, or trial-negative). The only exception was the case of the arousal scores; although there was a difference between either the positive or the negative trial and the baseline, there was not significant difference between the positive and the negative trials.

Provided that two of the caretakers were not familiar with the participants, we also analyzed whether their scores were similar to the caretakers that were

---

[8] Irtel, H. (2007). PXLab: The Psychological Experiments Laboratory [online]. Version 2.1.11. Mannheim (Germany): University of Mannheim.

Table 1. MuDERI Database, *1 recording missed due to technical problems

	Positive Trial	Negative Trial
Electrodermal activity (EDA)	11*	11*
Electroencephalogram (EEG)	12	12
High-resolution audiovisual video	12	12
RGB-D video	12	12
Total Number of Annotated Keypoints	41	30

Table 2. ANOVA Analysis, BL: baseline period, N: Negative period, P: Positive period

One-way ANOVA			Post-Hoc Comparisons		
Annotated Variable	F	p	Correlation	t	p
Arousal	26.26	<.001	BL-N	5.93	<.001
			BL- P	5.96	<.001
			N- P	1.98	.13
Valence	98.72	<.001	BL-N	6.51	<.001
			BL- P	8.56	<.001
			N- P	12.71	<.001
Anger	67.17	<.001	BL-N	6.9	<.001
			BL- P	2.86	.015
			N-P	7.81	<.001
Joy	75.37	<.001	BL-N	4.36	<.001
			BL- P	8.13	<.001
			N-P	12.78	<.001
Sadness	88.32	<.001	BL-N	7.56	<.001
			BL- P	5.09	<.001
			N-P	9.75	<.001

familiar to the users. We conducted a series of t-test comparing the average scores provided by the familiar caretakers in each variable to the average scores provided by the caretakers non-familiar with the users. The results suggest that, in general, the caretaker unfamiliar to the users provided more positive scores, as shown by a higher average valence, $t(113)=6.34$; $p<0.001$, a higher average joy, $t(113)=3.88$; $p<0.001$, and a lower sadness, $t(113)=5.37;p<0.001$. By contrast, they provided lower average ratings of arousal, $t(113)=9.68$; $p<.001$, and there was not significant difference in the anger ratings, $t(113)=0.09;p=0.92$.

The means of the scores for each variable are (Fig. 4), demonstrating the effectiveness of the sessions in eliciting the desired emotional states.

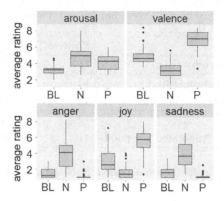

Fig. 4. Distribution of the average scores for each annotated variable, BL:baseline trial, N: Negative trial, P: Positive trial

5 Conclusion

The results show that the experiment achieved the goal of eliciting a range of emotions on participants. Hence, the present research was successful in creating a multimodal database that can fuel research on *Emotional Adaptive Behavior* of robots for the cognitive stimulation of users with ID, by filling a gap in the availability of *nearly real world* data from individuals with ID. Researchers in the field may benefit from using the database, which is publicly available by request[9]. Since combining different modalities helps to obtain more information, this project will start with using EDA signals for affective state estimation &adaptation and will continue with the integration of other biosignals. Project REHABIBOTICS will employ assorted machine learning and data mining algorithms over the annotated evaluations in MuDERI, to accomplish an automated *engagement &emotion classification* and their *adaptation* during robot interaction.

References

1. Rabbitt, S.M., Kazdin, A.E., Scassellati, B.: Integrating socially assistive robotics into mental healthcare interventions: Applications and recommendations for expanded use. Clin. Psychol. Rev. **35**, 35–46 (2015)
2. Shukla, J., Cristiano, J., Amela, D., Anguera, L., Vergés-Llahí, J., Puig, D.: A case study of robot interaction among individuals with profound and multiple learning disabilities. In: Tapus, A., André, E., Martin, J.-C., Ferland, F., Ammi, M. (eds.) Social Robotics. LNCS, vol. 9388, pp. 613–622. Springer, Heidelberg (2015). doi:10.1007/978-3-319-25554-5_61
3. Shukla, J., Cristiano, J., Anguera, L., Vergés-Llahí, J., Puig, D.: A comparison of robot interaction with tactile gaming console stimulation in clinical applications.

---

[9] http://institutorobotica.org/.

In: Reis, L.P., Moreira, A.P., Lima, P.U., Montano, L., Muñoz-Martinez, V. (eds.) Robot 2015: Second Iberian Robotics Conference. AISC, vol. 418, pp. 435–445. Springer, Heidelberg (2016). doi:10.1007/978-3-319-27149-1_34

4. Soleymani, M., Lichtenauer, J., Pun, T., et al.: A multimodal database for affect recognition and implicit tagging. IEEE Trans. Affect. Comput. **3**, 42–55 (2012)

5. Koelstra, S., Muhl, C., Soleymani, M., et al.: DEAP: a database for emotion analysis; using physiological signals. IEEE Trans. Affect. Comput. **3**, 18–31 (2012)

6. McKeown, G., Valstar, M., Cowie, R., et al.: The SEMAINE database: annotated multimodal records of emotionally colored conversations between a person and a limited agent. IEEE Trans. Affect. Comput. **3**, 5–17 (2012)

7. Sneddon, I., McRorie, M., McKeown, G., Hanratty, J.: The belfast induced natural emotion database. IEEE Trans. Affect. Comput. **3**, 32–41 (2012)

8. Ünal, Ö., Özcan, Ö., Öner, Ö., et al.: EEG and MRI findings and their relation with intellectual disability in pervasive developmental disorders. World J. Pediatr. **5**, 196–200 (2009)

9. Bradley, M.M., Lang, P.J.: Emotion and Motivation. Handbook of Psychophysiology. Cambridge University Press, New York (2007)

10. Ekman, P.: An argument for basic emotions. Cogn. Emotion. **6**, 169–200 (1992)

11. Mauss, I.B., Robinson, M.D.: Measures of emotion: A review. Cogn Emot. **23**, 209–237 (2009)

12. Barreda-Ángeles, M., Pépion, R., Bosc, E., et al.: Exploring the effects of 3D visual discomfort on viewers emotions. In: 2014 IEEE International Conference on Image Processing (ICIP), pp. 753–757 (2014)

13. Barreda-Ángeles, M., Arapakis, I., Bai, X., et al.: Unconscious physiological effects of search latency on users and their click behaviour. In: Proceedings of the 38th International ACM SIGIR Conference on Research and Development in Information Retrieval, pp. 203–212 (2015)

14. Swangnetr, M., Kaber, D.B.: Emotional state classification in patient robot interaction using wavelet analysis and statistics-based feature selection. IEEE Trans. Hum. Mach. Syst. **43**, 63–75 (2013)

15. Karg, M., Samadani, A.A., Gorbet, R., et al.: Body movements for affective expression: a survey of automatic recognition and generation. IEEE Trans. Affect. Computing. **4**, 341–359 (2013)

16. Morris, J.D.: OBSERVATIONS: SAM: the self-assessment manikin-an efficient cross-cultural measurement of emotional response. J. Advertising Res. **35**, 38–63 (1995)

17. Bowling, A.: Research Methods in Health: Investigating Health and Health Services. Open University Press, Buckingham (2009)

18. Delorme, A., Makeig, S.: EEGLAB: an open source toolbox for analysis of single-trial EEG dynamics including independent component analysis. J. Neurosci. Methods. **134**, 9–21 (2004)

19. Gomez-Herrero, G., Clercq, W.D., Anwar, H., et al.: Automatic removal of ocular artifacts in the EEG without an EOG reference channel. In: Proceedings of the 7th Nordic Signal Processing Symposium - NORSIG 2006, pp. 130–133 (2006)

20. Benedek, M., Kaernbach, C.: A continuous measure of phasic electrodermal activity. J. Neurosci. Methods **190**, 80–91 (2010)

"How Is His/Her Mood": A Question That a Companion Robot May Be Able to Answer

Mojgan Hashemian[1,2(✉)], Hadi Moradi[1,3], and Maryam S. Mirian[1,4]

[1] Advanced Robotics and Intelligent Systems Lab, School of Electrical and Computer Engineering, College of Engineering, University of Tehran, Tehran, Iran
{m.hashemian,moradih,mmirian}@ut.ac.ir
[2] INESC-ID and Instituto Superior Técnico, University of Lisbon, Porto Salvo, Portugal
[3] Intelligent Systems Research Institute, SKKU, Suwon, South Korea
[4] Center for Integrated Computer Systems Research, Faculty of Computer Science, University of British Columbia, Vancouver, Canada

Abstract. Mood, as one of the human affects, plays a vital role in human-human interaction, especially due to its long lasting effects. In this paper, we introduce an approach in which a companion robot, capable of mood detection, is employed to detect and report the mood state of a person to his/her partner to make him/her prepared for upcoming encounters. Such a companion robot may be used at home or at work which would be able to improve the interaction experience for couples, partners, family members, etc. We have implemented the proposed approach using a vision-based method for mood detection. The approach has been tested by an experiment and a follow up study. Descriptive and statistical analysis were performed to analyze the gathered data. The results show that this type of information can have positive impact on interaction of partners.

Keywords: Emotion · Facial expressions · HRI · Social robot · Mood

1 Introduction

The ability to detect and perceive others' affective state is considered as an aspect of "Emotional Quotient" or EQ (Mayer 2002). A successful interaction between humans involves real-time recognitions of the current state of emotion and mood status of each other (Sebe et al. 2005). For instance, a husband who returns home and encounters his wife in a bad mood needs to have great interpersonal capabilities to deal with such situation. However, in general, many people lack this capability and cannot respond properly to the unexpected mood or emotional state of their partner. Consequently, it would be greatly beneficial for a person, which we call him/her the interactor, to have an estimation of the affective state of his/her partner that would meet in near future, whom we call the intractee. This would help an interactor, who probably does not have the required competencies to respond properly to such a situation in real-time especially in case of unexpected affective states. For a person with good interpersonal capabilities, who can deal with such unexpected affective states, this feature can help him/her to interact even more efficiently. In the case of the above example, the husband would be

© Springer International Publishing AG 2016
A. Agah et al. (Eds.): ICSR 2016, LNAI 9979, pp. 274–284, 2016.
DOI: 10.1007/978-3-319-47437-3_27

aware of his wife's bad mood before entering home and can be prepared to alleviate the situation more effectively or take preventative actions. Hence this kind of report, which could be considered as a "sixth sense", can provide extra information for an interactor to converse better.

Considering the importance of affects in human-human, human-robot, and human-computer interaction, i.e. HHI, HRI, and HCI respectively, researchers have been investigating different methodologies to recognize and represent emotions or personality traits (Park et al. 2010). Unfortunately, despite the important role of mood in HHI or HRI, little attention has been paid to detect and incorporate mood in this field (Park et al. 2010). The importance of mood comes from the fact that our mood highly affects our interpretation of life events (Thayer 1997). Furthermore, mood shows stability over long duration of time, hence it has great impact on HHI (Thayer 1997) or HRI (Xu, et al. 2014) compared to emotions. For example, a usual interaction of a partner entering home would be interpreted negatively when the interactee is experiencing bad mood; since his/her bad mood may directly influence his/her perspective at that moment. That is why we based our research, presented in this paper, on helping human-human interaction using mood detection.

Furthermore, interpersonal communication is not just a process of sending and receiving messages, but also a process of negotiating meanings. It should be noted that the intended meaning is not necessarily the one the audience takes away, yet it is affected by many different potential factors. This factors could be considered as noise and could include many factors such as one's expectations, attitude, and mood (Wood 2015).

On the other hand, the recent emergence of social-robots, such as companion robots, personal assistant robots, or pet robots, suggests that the task of mood detection can be performed by such robots. This capability can be part of their supposed duties as an agent who interacts with humans, i.e. interactees, at work (such as the ASIMO robot (Sakagami et al. 2002)) or at home (such as the Jibo[1] robot). Then the detected mood state can be reported to the corresponding interactors who can prepare themselves to properly meet the interactees. This approach can be easily implemented on such robots and enhance their performance by incorporating human-human communication aspects which was recently suggested as an application for robots (Hori et al. 2015).

It should be noted that a few recent social robots are capable of mood detection. For instance RUBI (Movellan et al. 2007) is capable of determining social mood via auditory information. The main purpose of this robot is to adapt itself to current surrounding situations in front of kids and a teacher. Another example is Huggable (Stiehl et al. 2009), which provides socio-emotional support for kids at pediatric hospitals. In this project a teddy bear aims to mitigate stress and anxiety by making playful interaction with hospitalized children. In our proposed approach, the detected mood via any method such as the recently mentioned ones can be reported by the robot to the interactor.

On the other hand, most of personal robots such as Jibo or Pepper[2] are equipped with camera/microphone. Thus they can be easily equipped with mood detection capability using recently proposed approaches. These approaches take advantage of different

---

[1] https://www.jibo.com/.

[2] https://www.aldebaran.com/en/a-robots/who-is-pepper.

available sources of information, including gesture and posture (Thrasher et al. 2011), facial expression (Hashemian et al. 2014), and vocal (Sanchez-Cortes, et al. 2013). It is worth mentioning that a great deal of different information about human nature, even his/her cognitive associations, is detectable via the visual channel (Ekman and Rosenberg 1997), whether in his facial expression (Hashemian et al. 2014), gaze pattern (Hashemian et al. 2015), or gesture and posture (Thrasher et al. 2011).

In this paper, we propose a framework, in which a personal robot interacts with an interactee and detects his/her mood during their interaction. Then it reports the assessed mood state to the interactor, to prepare him/her for an upcoming visit. It should be mentioned that the proposed framework, for helping the human-human interaction using robots through reporting mood, is not limited to any specific proposed visual mood detection. Any mood detection methods, such as gesture analysis or voice-based mood detection methods, can be used and the detected mood can be reported to the intractor. In this paper, to provide the robot with mood detection capability, we chose a recently proposed vision based framework (Hashemian et al. 2016) for mood recognition. This approach makes it possible to achieve a higher accuracy level in mood detection in comparison to other works.

2 Method

Before explaining the approach, it seems necessary to define mood and emotion in detail in order to prevent any misunderstanding or misinterpretation. Mood and emotion have been used interchangeably in the field of affective computing frequently. However, they are totally different in nature. In general, mood and emotion are two main affective states which impose different influences on human behavior (Gebhard 2005). Mood can increase or decrease the pleasure level perceived in our lives (Thayer 1997) which can change our outlook, make the most disagreeable task durable (Thayer 1989), motivate us (Ryan 2012), and influences our behavior under long course of time. Mood is our background feeling, which lasts for hours or even days, without any obvious cause or visible reason. On the contrary, emotions remain for a short period of time and emerge in response to a specific stimulus (Ryan 2012). Another major difference between these two affective states is their intensity, in which the intensity of an emotion is higher than mood (Thayer 1989).

Despite the preceding differences, mood and emotion have a synergic mechanism, i.e. mood influences the emotion and vice versa. The effect of mood on emotion may appear in different dimensions such as the intensity of emotions, the frequency of emotions, or in the speed of change in emotions. This suggests that mood can be assessed by examining emotional variations (Hashemian et al. 2014). As pointed out earlier, in this experiment we employ a recently proposed approach (Hashemian et al. 2016) which benefits from emotional variations to determine the mood. To achieve this goal, emotions are considered as signals changing over time. Then by performing signal analysis in time or frequency domains, we attempt to assess mood. A general framework for this approach is depicted in Fig. 1 in dashed lines.

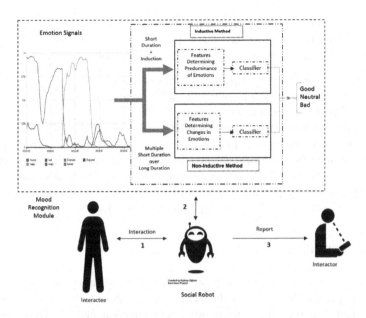

Fig. 1. Depending on the time of interaction, the inductive or non-inductive method could be used. If the interaction can be long, for example in the companion robot case, the non-inductive method can be used. If the interaction time is short, such as a social robot in public places, the second method is applicable. (Icons are downloaded from http://thenounproject.com/)

The above description shows how the robot can determine the mood of the intracee by analyzing emotions (Fig. 1). Once the mood state is determined, it is reported to the interactor before the encounter. This can be transferred using any communication media such as SMS, email, or any messaging application to the corresponding interactor.

This approach is tested by a 2-fold experiment which is discussed in the following sections. The first phase (study #1) aimed at assessing the influence of the robot report on interactors' communication facing the interactee. The first phase is followed by a follow up study (study #2) to further examine the validity of the results. The goal of this follow up study was to compare the presence of the robot, informing them about their partners' mood state, in comparison to its absence when the interactors are not informed about their partners' mood state. More specifically, in study #1, the subjects had the opportunity to interact with a robot informing them about their partners' mood. Then right after this, in study #2, they were deprived from having this opportunity so they could compare the benefits more easily.

In study #1, in order to evaluate the effect of the reported mood on human interaction experience, a self-report evaluation is performed using a 3-ithem questionnaire. The questionnaire contains the three following questions:

1. How did the reported mood affect your interaction? (Positive/No Effect/Negative)
2. If the answer of the previous question is negative, do you think it would affect your interaction experience under other circumstances? (Yes/No)
3. If the answer of the previous question is yes, how (explain)?

The three questions are aimed to evaluate the effectiveness of the proposed approach in human-human interaction. In the same vein, an 9-item questionnaire is designed to investigate the differences between the two studies. The questions are as follow:

1. Were you at home when your partner came? (Yes/No)
2. What was the mood of your partner? (Good/Neutral/Bad)
3. Did you expect to meet him in this state? (Yes/No)
4. If the answer to the previous question is NO, do you think knowing the mood state might affect your interaction? (Yes/No)
5. In general, apart from your expectation, in the previous study how the information reported by the robot affect your relationship? (Positive/Negative/No effect)
6. In comparison to the previous study, in what way "being unaware of your partner mood state" affected your relationship? (Positive/Negative/No effect)

3 Implementation

Figure 1, depicts the mood detection framework in detail (Hashemian et al. 2016). As shown in this picture, depending on the duration of the interaction of a subject with the robot, the mood state is evaluated using two different internal methods. The first method determines mood by observing the changes in the emotions of a subject over a period of time, which depends on the dynamics of the environment, through several interactions. After an average of 3 h, the mood state can be determined with an accuracy of 91.1 %, mean error of 0.09 and variance of 4.9 (Hashemian et al. 2014). The second one shortens the interaction time by inducing emotions, i.e. increasing the dynamics of the environment, and examining the effect of mood on emotional changes. This method gains 88.2 % classification accuracy with mean absolute error of 0.1 and variance of 0.2 (Hashemian et al. 2015). Both methods determine mood by examining a set of features extracted to highlight emotion variations. Extracted features are fed into a classifier and the mood state is determined based on the selected feature values.

In this experiment we employ the inductive method which uses comic video clips to induce happy emotion. It is noticeable that the non-inductive method is applicable as well, especially in special cases when the interactee is at work and spends a long time in front of a computer, or even at home when a companion robot observes the interactee for over a long period of time. In order to test our approach, we used an abstraction of a companion robot by employing an ordinary webcam and a web-based platform to provide the comic video clips and evaluate the interactee's mood, who is at work. Obviously this setup can be easily embedded in a companion robot in practice. In this implementation, the assessed mood state is reported to the interactor by sending an SMS, before the intractee enters his/her home. Then, after the entrance of the interactee, the interactor filled in the questionnaire in order to record his/her experience knowing the interactee's mood. However, the inverse scenario is also applicable as well as a two-sided scenario, when both partners interact with their own systems capable of mood detection. In the following section, details on the collected data set is discussed.

3.1 Participants and Dataset

To provide a data set in order to test the approach, we ran an experiment in two phases: study #1 and study #2. We performed each study in two different scenarios to prove application generality of the approach in practice. In the first scenario, the interactees were selected among people with a full-time job and the experiment ran at their workplace (FANAP Company). In the second scenario, a number of students in our lab played the role of intreactees. We ran the experiment for one-week for each of the preceding scenarios. A summary of the collected dataset is listed in Tables 1 and 2.

Table 1. Study #1 - Participants statistics

	Interactee			Interactor			Relationship Type	
	Female	Male	Age Mean	Female	Male	Age Mean	Husban d/Wife	Parent/ Child
Scenario 1	4	5	28.5	5	4	29.5	9	0
Scenario 2	3	5	29.3	7	1	38.9	4	4

Table 2. Study #2 - Participants statistics

	Interactee			Interactor			Relationship Type	
	Female	Male	Age Mean	Female	Male	Age Mean	Husban d/Wife	Parent/ Child
Scenario 1	–	1	28	1	–	28	1	–
Scenario 2	2	2	25	5	–	34	3	2

In first phase, i.e. study #1, 17 couples participated in the test. The dataset contains a diverse range of partnerships, including husband and wife or mother/father and child. In the former case, the intercatee is the one who is observed by the robot, and the interactor is the other one staying at home and receives the mood report by the robot. In the latter case, the child was supposed as the interactee and his/her parent was the interactor.

After running the experiment for two weeks (one week each scenario), a data set containing 44 samples, i.e. the interactees' mood states, was collected. We let every subject participate several times in the experiment and we consider people in different moods as different samples. Hence in the ideal case, for 17 subjects 51 different samples would be available. The justification behind this is that people act differently when they experience different mood. At the end of the experiment, 41 questionnaires were returned. The questionnaires were filled out by the interactors, i.e. the interactees' partners staying at home. In study #2, 5 people, i.e. 3 males and 2 females from the previous subjects, participated. Details of this phase is listed in Table 2. The scenario is the same, except that no information of mood state is reported in this phase.

4   Results

In Table 3 a summary of the answers to the questionnaires used in study #1 and study #2 is listed. Based on this table, in study #1 more than half of the participants, i.e. 53.7 %, were satisfied with the report. To be more specific, they reported that knowing the mood of their partners before their encounter affects their interaction positively.

Table 3. Study #1 – Satisfaction of Participants in front of the Robot

Question	Answer		
Study #1			
Q1	Positive(22)	Negative(19)	No effect (9)
Study #2			
Q3	Yes (16)	No (10)	
Q4	Yes (6)	No effect (4)	
Q5	Positive (15)	Negative (2)	No effect (9)
Q6	Positive (0)	Negative (13)	No effect (13)

At the end of study #2, 26 questionnaires were returned from five subjects. As shown in Table 3, among all 26 samples, only in 10 samples, the subjects were unaware of their partners' mood state (Q3). The others were somehow aware of their partners' mood state. For instance, they might have talked on the phone before the encounter. Among these 10 samples, who were unaware of their partners' mood state, 6 cases reported that being informed of this state could enhance their interaction experience. The other 4 samples were neutral or declared that being informed of the mood state did not have any effect on their interaction. On the other hand, from Q5, we can infer that 57.7 % of the population were satisfied with the robot's reports. To be more specific, these subjects reported that the information provided by the social robot in study #1 would improve their communication in comparison to study #2, in which they were not provided this information. Only 7.7 % of the participants were not satisfied with the robot's reports. It should be mentioned that this specific report corresponds to only one of the interactors who declared that she likes to guess her partner's mood state without any assistance. Finally, Q6 reveals that in 13 samples, interactors declared that being unaware of their partners' mood state may negatively affect their interaction.

5   Discussion

As discussed in the results section, in near half of the samples, collected from the subjects in study #1, it was stated that the robot's report did not influence their communication experience. It is interesting to note that near half of these samples, i.e. 47.4 % which equals to 9 samples out of 19, corresponds to 3 people who were not aware of the goal of the study. It is not a surprising finding, since according to a proven theory when participants are aware of the effects of the test, the result is closer to the expectation level (Kintz et al. 1965). Although, usually this incident is taken as a bias in the results, here we do not consider it as a bias. Rather, it can be suggested that the approach would be promising in general, and would be even more favorable when people involved in the test are informed about the goal of robot's report which helps them to communicate better.

In a similar way, among the other 10 remaining samples who reported that the approach has no effects on their interactions, 8 samples were experiencing neutral mood. A t-test (with the significance level of 0.05 %) showed that this hypothesis is statistically significant; $t(40) = 6.5$, $p = 0.00$. In other words, the approach is quite promising when the reported mood state is not neutral (Good or Bad). A possible justification for this observation could be as follows: in case of the presence of an extreme mood state a thoughtful reaction is needed, while this is not the case in normal situations. Hence, we can conclude that the robot would operate more effectively in case of extreme moods, i.e. positive and negative which typically needs more consideration in comparison to the neutral state.

The results obtained from study #2 revealed that half of the participants believe that the absence of the mood report would affect their communication adversely. The other half were neutral, and declared that the absence of the mood report does not affect their communication. It should be noted that among these 13 neutral samples, 4 samples correspond to the interactor who liked to guess her husband's mood state by herself, and 6 samples corresponded to another interactor who was unaware of the test's goal. In general, based on the previous justifications, discarding these two exceptional cases, the mood report approach seems beneficiary.

Putting the studies together, we performed the McNemar test, however no significant difference observed between the two groups ($p = 1.000$). The point is the sample size of the posttest or study #2, is too small and prevent us to infer any further finding.

These findings must be interpreted with caution, because this study is a preliminary step towards the proof of concept. Therefore, to benefit from the results of this study in a real interaction scenario, a mixture of challenges including shape of the robot, age of the partners, and their type of relationship seem to affect the level of trust and usability of the approach in practice. We believe that for the proof of concept, a typical relationship type such as husband/wife or child/parent works fine and this is the reason we confine our study to this fixed setting. However, it is quite predictable that personal character-istics of partners, such as their age and the type of relationship, as well as their interaction parameters, such as encounter time lag and time of interaction, highly affect the accuracy of mood detection thus the quality of interaction.

6 Conclusion and Future Work

In this paper we argued that the presence of a companion robot with the capability of mood detection and reporting to a corresponding partner seems beneficiary to couples' relationships. In this work, we have a diverse range of participants and partnerships, including couples and parents. This diversity indicates the generality of the approach, i.e. it is applicable to a wide variety of partnerships. Other types of partnerships could be tested in the setting; such as roommates and friends. Furthermore, as mentioned earlier, the mood detection system has a degree of uncertainty in assessing the mood and in complicated situations it may infer incorrectly, which is quite common for any predictive system. Hence, it should be noted that some portion of errors might be caused by reporting wrong mood state or mood swings over time. These two factors should be taken into account in the evaluation of the study.

There is an abundant scope for further progress in this area. The first and foremost would be increasing the sample size to obtain more statistically significant differences between the two groups. Our next step is to implement the method on a real companion robot, in order to alleviate the abstraction level and lab setting bias, and thus gain more realistic results. In case of this experiment, the first method seems applicable for companion robots which works at home, such as Jibo, or robots serving at workplaces, such as ASIMO (Sakagami et al. 2002). This type of robots has fairly long time interaction with users. Therefore, one can assume that the observation of a subject for two minutes per hour is completely feasible. On the other hand, the second method is applicable in social robots with voice or display such as Pepper. In such cases, the robot can induce emotions by telling jokes or playing a comic video clip. However, it should be noted that using a real robot may lead to different outcomes depending on the robot's performance and trustworthiness. Another possible area of future research would be to investigate a two-sided scenario, in which each subject would be informed the mood state of the other subject. We believe that this two-sided situation would even enhance the level of satisfaction further. However, further research on this topic needs to be undertaken to evaluate the approach.

As discussed earlier, the mood detection framework determines mood based on two different methods, first non-induction based method (Hashemian et al. 2014) and second the inductive one (Hashemian et al. 2014). In this study, we used the second method to take advantage of its shorter needed time of interaction. However, the first method is applicable in a similar scenario, which will be discussed in our future studies.

It should be noted that a social robot with mood detection capability could be also helpful in changing the mood. This is based on the proved facts that mood can be changed intentionally, using mood regulation or mood induction methods (Thayer 1989), or unintentionally, through typical interaction, since it has a contagious nature (Sy et al. 2005). This would open up new research opportunities to develop robots which can change mood of a person to improve his/her interaction with other humans. Moreover, when the reported mood is bad the interactor can ask the robot to induce a positive mood to mitigate the situation before the encounter.

Acknowledgments. The first author would like to thank her friends in ARIS and Mobile Robot Lab at the school ECE, University of Tehran, as well as her collogues in FANAP Company for their kind help and participation in these experiments. Furthermore, she would like to thank Dr. Leila Kashani for her constructive review and feedbacks on the manuscript. This work was supported by national funds through Fundação para a Ciência e a Tecnologia (FCT) with reference UID/CEC/50021/2013.

References

Ekman, P., Rosenberg, E.L.: What the Face Reveals: Basic and Applied Studies of Spontaneous Expression Using the Facial Action Coding System (FACS). Oxford University Press, New York (1997)

Gebhard, P.: ALMA: a layered model of affect. In: Proceedings of the fourth international joint conference on Autonomous agents and multiagent systems. ACM (2005)

Hashemian, M., Moradi, H., Mirian, M.S., Tehrani-doost, M.: Determining mood via emotions observed in face by induction. In: Second RSI/ISM International Conference on Robotics and Mechatronics (ICRoM). IEEE (2014)

Hashemian, M., Moradi, H., Mirian, M.S., Tehrani-Doost, M., Ward, R.K.: Is the mood really in the eye of the beholder? In: Stephanidis, C. (ed.) HCI 2015. CCIS, vol. 528, pp. 712–717. Springer, Heidelberg (2015). doi:10.1007/978-3-319-21380-4_120

Hashemian, M., Nikoukaran, A., Moradi, H., Mirian, M.S., Tehrani-doost, M.: Determining mood using emotional features. In: 7th International Symposium on Telecommunications (IST). IEEE (2014)

Hashemian, M., Moradi, H., Mirian, M.S., Tehrani-doost, M.: Recognizing mood using facial emotional features. Technical report, MIR_TechReport 94-12-10/1, School of Electrical Engineering and Computer Science, University of Tehran (2016)

Hori, M., Tsuruda, Y., Yoshimura, H., Iwai, Y.: Expression transmission using exaggerated animation for Elfoid. In: Frontiers in psychology 6 (2015)

Kintz, B.L., Delprato, D.J., Mettee, D.R., Persons, C.E., Schappe, R.H.: The experimenter effect. Psychol. Bull. **63**(4), 223 (1965)

Lee, C.M., Narayanan, S.S.: Toward detecting emotions in spoken dialogs. IEEE Trans. Speech Audio Process. **13**(2), 293–303 (2005)

Mayer, J.: Mayer-Salovey-Caruso Emotional Intelligence Test (MSCEIT), version 2.0. Multi-Health Systems, Toronto (2002)

Movellan, J. R., Tanaka, F., Fasel, I.R., Taylor, C., Ruvolo, P., Eckhardt, M.: The RUBI project: a progress report. In: Proceedings of the ACM/IEEE International Conference on Human-Robot Interaction. ACM (2007)

Park, S., Moshkina, L., Arkin, R.C.: Mood as an affective component for robotic behavior with continuous adaptation via learning momentum. In: 10th IEEE-RAS International Conference on Humanoid Robots (Humanoids). IEEE (2010)

Ryan, R.M.: The Oxford Handbook of Human Motivation. Oxford University Press, New York (2012)

Sakagami, Y., Watanabe, R., Aoyama, C., Matsunaga, S., Higaki, N., Fujimura, K.: The intelligent ASIMO: System overview and integration. In: IEEE/RSJ International Conference on Intelligent Robots and Systems. IEEE (2002)

Sanchez-Cortes, D., Biel, J.I., Kumano, S., Yamato, J., Otsuka, K., Gatica-Perez, D.: Inferring mood in ubiquitous conversational video. In: Proceedings of the 12th International Conference on Mobile and Ubiquitous Multimedia. ACM (2013)

Sebe, N., Cohen, I., Huang, T.S.: Multimodal emotion recognition. Handbook Pattern Recogn. Comput. Vis. **4**, 387–419 (2005)

Stiehl, W.D., Lee, J.K., Breazeal, C., Nalin, M., Morandi, A., Sanna, A.: The huggable: a platform for research in robotic companions for pediatric care. In: Proceedings of the 8th International Conference on interaction Design and Children. ACM (2009)

Sy, T., Côté, S., Saavedra, R.: The contagious leader: impact of the leader's mood on the mood of group members, group affective tone, and group processes. J. Appl. Psychol. **90**(2), 295 (2005)

Thayer, R.E.: The Biopsychology of Mood and Arousal. Oxford University Press, New York (1989)

Thayer, R.E.: The Origin of Everyday Moods: Managing Energy, Tension, and Stress. Oxford University Press, New York (1997)

Thrasher, M., Zwaag, M.D., Bianchi-Berthouze, N., Westerink, J.H.D.M.: Mood recognition based on upper body posture and movement features. In: D'Mello, S., Graesser, A., Schuller, B., Martin, J.-C. (eds.) ACII 2011. LNCS, vol. 6974, pp. 377–386. Springer, Heidelberg (2011). doi:10.1007/978-3-642-24600-5_41

Wood, J.: Interpersonal Communication: Everyday Encounters. Nelson Education, Toronto (2015)

Xu, J., Broekens, J., Hindriks, K., Neerincx, M.A.: Effects of bodily mood expression of a robotic teacher on students. In: IEEE/RSJ International Conference on Intelligent Robots and Systems (IROS 2014). IEEE (2014)

Emotion in Robots Using Convolutional Neural Networks

Mehdi Ghayoumi[(✉)] and Arvind K. Bansal

Artificial Intelligence Lab Computer Science Department,
Kent State University, Kent, USA
{mghayoum, akbansal}@kent.edu

Abstract. These years, emotion recognition has been one of the hot topics in computer science and especially in Human-Robot Interaction (HRI) and Robot-Robot Interaction (RRI). By emotion (recognition and expression), robots can recognize human behavior and emotion better and can communicate in a more human way. On that point are some research for unimodal emotion system for robots, but because, in the real world, Human emotions are multimodal then multimodal systems can work better for the recognition. Yet, beside this multimodality feature of human emotion, using a flexible and reliable learning method can help robots to recognize better and makes more beneficial interaction. Deep learning showed its force in this area and here our model is a multimodal method which use 3 main traits (Facial Expression, Speech and gesture) for emotion (recognition and expression) in robots. We implemented the model for six basic emotion states and there are some other states of emotion, such as mix emotions, which are really laborious to be picked out by robots. Our experiments show that a significant improvement of identification accuracy is accomplished when we use convolutional Neural Network (CNN) and multimodal information system, from 91 % reported in the previous research [27] to 98.8 %.

1 Introduction

Emotions have main role and are affected in developing any type of social setting and humans are social and live socially and most of the actions are emotional. Human emotional states (expression and recognition) have been the focal point of attention in several areas of neuroscience and psychology to cognitive and computer science. For the acceptance of robots by humans the application of emotions for Human Robot Interaction (HRI) purpose are very significant. A robot that is able to realize and express emotions can pass on in a lifelike way. The observation of different modalities, such as facial expression, gesture, and speech, improves the emotional state recognition. Moreover, recognizing the emotion is a complicated process and there are some researches which looking for recognizing real emotion. In our previous work, we deployed group theory concept of recognizing real emotion by detecting symmetry patterns in face [15].

GU and et al. [2] analyzed and explored the importance and the use of the information in each trait which are efficient in human emotional states. They found out when

© Springer International Publishing AG 2016
A. Agah et al. (Eds.): ICSR 2016, LNAI 9979, pp. 285–295, 2016.
DOI: 10.1007/978-3-319-47437-3_28

we would wish to recognize emotional states, non-verbal communication, facial expressions and body posture/motion complement each other. Adolphs [3] showed how the human brain correlates past experiences, motion information in the visual stimuli, and face expressions. The brain is able to integrate this multimodal information and generate a theatrical performance of the visual stimuli based on all of them in concert. The pretense of this operation in computer systems can be achieved by neural models, with a specific social system that has different type of feature representations such as Convolutional Neural Networks (CNN).

CNN were introduced formally by Lecun, et al. [4]. They are prompted by the hierarchical process of simple and complex cells in the human learning ability to extract and learn different information from visual stimuli. Each layer of a CNN has the capability to react to different information, and when stacked together the layers can create a complex representation of the optical input.

In our recent works we presented a multimodal architecture for emotion in robot and we broke down what it has in mind for a robot to have emotion and distinguishing emotional state for communication from an emotional state as a mechanism for the formation of its behavior with humans and robots by (CNN) [5, 6]. In this clause, we plan to implement the given model and compare the results with our previous works.

This paper is coordinated as follows: The next section explains the related works. Section 3 describes human and robot emotion. The relation between deep learning and emotion is given in Sect. 4. In Sect. 5 we demonstrate the integrated model. In Sect. 6 we present experimental results and stopping points and future works are shown in the final part.

2 Related Works

The research study by Mehrabian [14] has indicated that 7 % of the communication data is transferred by linguistic language, 38 % by paralanguage, and 55 % by facial expressions in human face-to-face communication. Some models of multimodal databases can be found in [7–9] and most studies, have looked at the integration of facial looks and speech information and there have been a few efforts to fuse data from body movement and motions in a multimodal framework. Sun et al. [25] designed hidden identity features with deep convolutional networks to realize approximately 1000 false identities on LFW database and achieved 97.45 % verification accuracy with only weakly aligned faces. El Kaliouby and Robinson [11] offered a model to make head movements and facial expressions state information. Susskind et al. [23] took advantage of learning deep belief nets to classify facial action units in realistic face images. Krizhevsky et al. [24] used the deep convolutional neural network to classify the 1.2 million images in the ImageNet LSVRC-2010 contest in 1000 different categories and achieved the inconceivably higher accuracy than the temporal state-of-the-art. Gunes and Piccardi [10] fused facial expressions and body gestures information for bimodal emotion recognition. For identification purposes, almost, all types of machine learning techniques have been used in emotion recognition approaches [12, 13]. For many reasons and mainly for our final goal of creating an emotion in robots as much as similar to human emotion, we are looking for learning method which can satisfy these parameters. Lately, CNN showed up good results in biometrics, particularly in facial expression and speech recognition. We decided to use it do some preprocessing of data before feeding to the algorithm, such as LSH to prune the database data space [22, 26]. We present a multimodal CNN-based model for automatic emotion recognition and expression. Our model deploys the CNN method, and uses it for multimodal emotional state recognition using facial expression, gesture and speech recognition. This information indicates that, the facial expressions give a great amount of data in human communication. Deploying different modalities and multimodal systems, such as body position, gestures and speech, improved the determination of the emotional state.

3   Human and Robot Emotion

The Human emotional state causes the focus of attention in several areas from biology, neuroscience and psychology to cognitive and computer science due to its importance in human communication, interaction and social dealings. Here, we explain a little about Neuromodulation and Cognitive parameters and their relation with emotion.

3.1   Neuromodulation

Neuromodulation refers to the action on nerve cells of endogenous substances called neuromodulators. Three main neuromodulator systems involved in emotion are:

- **Dopamine** based communication and motor activation,
- **Serotonin** based regulation of conduct,
- **Opioid** based regulation and relaxation [16, 17].

Emotion can be regarded as continuous patterns of neuromodulation of certain lots of brain structure. All EE and ER functionalities are related to the special activities in the brain, for example for facial expression, the smiles are initiated in the motor cortex and routed via the pyramidal motor system. If we would like to simulate the EE in the robots, knowing about these parameters in details and their weights on the emotional state types for simulating the human emotion in robots can assist us. In the following study, in the future, we plan to utilize these parameters and their weights for making the model more flexible.

3.2   Cognition

Robot learning process steps (here, EE and ER) should be very similar to human and it needs to include cognition. In that respect are several integrated cognitive architectures trying to develop all aspects of conduct as a single system while remaining constant across different domains [18]. More or less of these cognitive architectures are biologically inspired, while some others are inspired by psychological theories, in which some of them also contain the concept of effect in their intent. There are the interplay of affect (value), motivation (action tendencies), cognition (meaning), and behavior at three levels of information processing:

- **Reactive:** a hard-wired release of fixed action patterns and an interrupt generator.
- **Routine:** the locus of unconscious well-learned automatized activity and primitive and unconscious emotions.
- **Reflective:** the home of higher-order cognitive functions.

Based on the traditional approaches, cognition emphasizes on information processing which normally has excluded emotion. On the other hand, new growth of cognitive neuroscience as an inspiration for understanding human cognition has highlighted its interaction with emotion. Probes into the neural systems underlying human behavior demonstrate that the mechanisms of emotion and knowledge are intertwined from early perception to abstract thought. These findings suggest that the classic division between the subject of emotion and knowledge may be unrealistic and that an apprehension of human cognition involves the consideration of emotion. Emotions influence fundamental processes mediating high level cognition such as:

- Attention speed, duration and capacity,
- Working memory speed and capacity,
- Long term memory recall and encoding.

It is also apparent that cognition divided functions into different domains, such as memory, attention, and reasoning. The concept of emotion causes a structural architecture that may be similarly diverse and complex.

4   CNN and Emotion

Deep learning can be employed in robots and build the robot emotions more realistic and HRI & RRI better. Deploying different modalities and multimodal systems, such as facial expression, gestures and speech, improved the determination of the emotional state.

4.1   Facial Expression Recognition

Studies on facial expression recognition have been lasting for three decades since 1970s. Paul Ekman et al. [1] postulated six cross-cultural, basic emotions (anger, disgust, fear, happiness, sadness, and surprise) from a psychological view, and developed Facial Action Coding System (FACS) to describe facial micro-expression [19]. Our work also selects the six basic emotions and neutral emotion as our measure of facial expression classification. In general, for facial expression recognition system, there are three basic parts:

- **Face detection:** Most of face detection methods can detect only frontal and near-frontal views of the fount. Viola and Jones [20, 21] utilized a lot of rectangular features to find facial expressions in real time.

- **Facial feature extraction:** Sorts of features (geometric features, show features and hybrid features of geometric and appearance features) are drawn out for recognizing facial expression.
- **Facial expression recognition:** In facial expression recognition, there are dissimilar methods. Due to lack robust features, most of facial expression recognition models work poorly in the complex environment [22].

In recent years, deep learning arouses academia and industrial attentions due to its magic in computer vision. Our work is taking advantage of deep models to extract robust facial features and translate them to recognize facial emotions. FACS system analysis [26] has been employed to derive the features-details that are important during the formulation of a specific facial expression. There are 13 moving-points (11 active points and 2 passive points) and 6 non-moving reference points. The FAUs have been rendered to the corresponding feature-level movements as given in Table 1. We denote vertical-up motion by $\uparrow$, vertical-down motion by $\downarrow$, horizontally stretched outwards by '$\longleftrightarrow$', horizontally compressed inwards by '$\rightarrow\leftarrow$', oblique-stretched downwards by '$\searrow$', oblique-stretched upwards by '$\nearrow$'. If the emotion is symmetric, then the superscripts L (left) and R (right) have been excluded. If the move is optional or shows a higher intensity increase and so it has been ranked inside the square brackets '[... ']'. Junction is shown using concatenation. Disjunction is shown using vertical bar '|'. Essential feature-point are within parenthesis '('... ')' separated by ','. The details presented in our previous research [27].

Table 1. Feature Point Displacements (FDP)

Facial Expressions	Major Feature-points displacements
Anger::	$(e_1 \rightarrow\leftarrow e1\uparrow, e_3\uparrow) + [e_2\uparrow] + [m^T\uparrow \ m^B\uparrow]$
Disgust::	$(m^T\uparrow ch\uparrow) + [m^L\downarrow| \ m^R\downarrow] + [m^B\uparrow]$
Fear::	$(e_1\uparrow, m^L\downarrow \ m^R\downarrow) + [m^T\downarrow] + [e1\rightarrow\leftarrow]$
Happiness::	$(m^L\nearrow m^R\nearrow, m^T\uparrow m^B\downarrow \ ch\downarrow m^L\longleftrightarrow m^R\longleftrightarrow)$
Sadness::	$(e1\downarrow \ m^L\longleftrightarrow m^R\longleftrightarrow) + [ch\downarrow]$
Surprised::	$(e^1\uparrow e^2\uparrow e^3\uparrow e1\uparrow ch\downarrow) + [m^T\uparrow m^B\downarrow]$

4.2 Speech Recognition

Human language encodes emotional information in two different ways:

- What is said? And
- How it is said?

And then a spoken message can be split down into two sections:

- A semantic and
- A paralinguistic one.

Several approaches to recognize emotions from speech have been reported [28–30]. Voice communication systems should be able to treat the non-linguistic information such as emotions, along with the message. For instance, words associated with happiness are characterized by longer utterance duration, shorter inter-word silence, and higher pitch and energy values with more extensive scopes. In sad sentences, the vitality and the pitch are usually held at the same point. Thus, these emotions are hard to be separated. We possess three important speech characteristics to model emotional speech:

- The standard deviations and ranges;
- Maximum, minimum and median values of the pitch; and
- Energy.

The deep neural network trained itself and resolves the complex problems based on the knowledge available. Resolutions of the individual groups, as considerably as a combined set, have led to the following assumptions: among acoustic features duration and energy appear to be most relevant, while voice quality showed less impact. However, no single group outperformed the pool of all acoustic features. In our experiments we restricted the set of features to those that can be extracted in real time and in a fully automatic mode.

4.3   Gesture Recognition

Gestures are expressive and meaningful questions, involving hands, face, head, shoulders, and/or the complete human body. Gesture recognition has a wide scope of applications, such as sign language for communication among the disabled, lie detection, monitoring emotional states or stress levels of studies, and navigating and/or manipulating in virtual environments. Recognition of emotion from gestures is challenging as there is no generic notion to represent a subject's emotional state by his or her gestures. Further, the gestural pattern has a wider variation depending on the subject's geographical origin, acculturation, and the power and intensity of his or her looks. Motions can be static, seeing a single pose or dynamic with a pre-stroke, stroke, and post-stroke phases [31]. Automatic identification of continuous gestures requires temporal segmentation. The most common gestural pattern, frequently used in emotion identification, is the hand movements. Glowinski et al. [32] proposed an interesting technique for hand (and head) gesture analysis for emotion recognition. Camurri et al. [33] classified expressive gestures from the human full body movement during the carrying into action of the subject in a dance. They identified motion cues and measured overall duration, contraction index, quantity of motion, and motion smoothness. On the base of these motion cues, they designed an automated classifier to classify four emotions (anger, fear, sadness, and happiness). Castellano et al. [34] employed hand gestures for emotion recognition.

5   Integrated Model

Figure 1 shows the integrated model which has both EE and ER for emotion in robots and creating better HRI and RRI [27]. For emotion recognition part, the data will come to CNN and the fusion will be answered based on their weight on human robot interaction and then we can count on the accuracy. For instance, if it receives 75 %

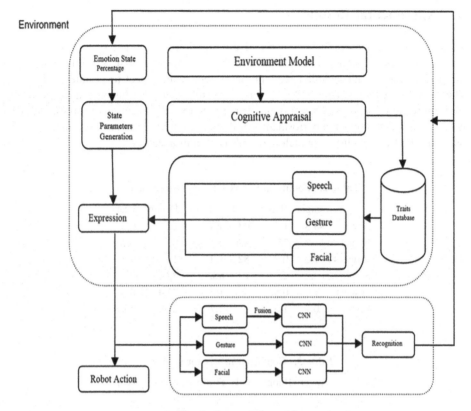

Fig. 1. Integrated model

from speech and 95 % by facial expression and 80 % of gesture, then grounded on their weights (for example here Mehrabian: 7 % by linguistic language, 38 % by paralanguage, and 55 % by facial expressions) they should multiply by these values and the average is the least accurate. On the other hand, for emotional reflection, established on the emotion recognition and cognitive appraisal, the scheme will force away the data from databases for words, gesture and facial expression which are more linked to the emotion recognition state that is recognized in the old state.

For ER and EE parts, we used the Decision Level Fusion of data in the ER part and Diffusion in the EE part. In decision level fusion each modality is first pre-classified independently, i.e., each biometric trait is captured, and features are then drawn out from that captured trait, based on that extracted features. The final classification is established on a merger of the yields of different modes. This is the highest stage of fusion with respect to human interface. In other words, the decision from each biometric system is concluded to construct the final determination [35].

6 Experimental Results

Table 2 indicates the confusion matrix of the emotion recognition system based on facial expressions. The overall functioning of this classifier was 80.4 %. Table 3 shows the performance of the emotion recognition system with respect to gesture analysis. The overall execution here is 86 %. Table 4 displays the confusion matrix of the emotion recognition system based on language. The overall execution of this classifier is 83 %. Table 5 shows the performance of the system with decision level integration using the best probability approach and 98.8 is overall accuracy.

Table 2. Confusion matrix for facial expressions

	Anger	Happy	Sad	Surprise	Disgust	Fear
Anger	**88.3**	1.2	0.9	4.4	1.5	0.8
Happy	3.7	**75**	4.7	2.5	0	4.1
Sadness	1.2	0.8	**82**	1.4	1.2	4.8
Surprise	3.2	2.3	1.8	**85.7**	2.6	0
Disgust	1.9	0.5	3.8	1.4	**72.4**	0
Fear	0.7	1.3	0	2.7	2.8	**79.1**

Table 3. Confusion matrix for gesture

	Anger	Happy	Sad	Surprise	Disgust	Fear
Anger	**98**	2	0	.8	0	0
Happy	.7	**74.2**	0	5.7	0	0
Sad	0	0	**73.8**	1.2	6.1	0
Surprise	0	2.1	0	**81.5**	5.3	0
Disgust	1.7	0	5.9	.6	**87.5**	0
Fear	0	0	0	0	3.6	**99.5**

Table 4. Confusion matrix for speech

	Anger	Happy	Sad	Surprise	Disgust	Fear
Anger	**98.2**	0	.6	0	.5	0
Happy	1.9	**64.2**	0	3.1	.7	2.1
Sad	1.9	0	**86**	1.3	0	.8
Surprise	2.3	1.3	0	**83**	1.6	0
Disgust	3.1	1.1	2.2	1.8	**71**	0
Fear	0	.7	0	.9	1.3	**93**

Table 5. Decision level fusion

	Anger	Happy	Sad	Surprise	Disgust	Fear
Anger	**99.7**	0	0	0	1.1	0
Happy	0	**99.3**	0	1.7	0	0
Sad	.9	0	**98.3**	0	.8	.7
Surprise	1.8	1.9	0	**97.4**	.6	1.2
Disgust	1.2	0	.9	.8	**97.8**	.9
Fear	0	0	0	1.8	1.7	**96.1**

7 Conclusion

We implemented the model for six basic emotion states and there are some other states of emotion, such as mix emotions, which are really laborious to be picked out by robots. We implemented our multi-modal system for automatic emotional state recognition. The proposed model achieves a more respectable performance when multimodal information is applied, in this case composed of facial expression, speech and gesture. The suggested model is able to learn from three different data streams: speech, facial expression and gesture. It deploys the CNN for better scholarship and identification. The results show more honest performance by comparing with old method. Our experiments show that a significant improvement of identification accuracy is accomplished when we use convolutional Neural Network (CNN) and multi-modal information system, from 91 % reported in the previous research [27] to 98.8 %. For future study, we plan to run along a mix emotion and test it on it and then enforce the model in a real-world scenario with a Telepresence Robot. We plan to move and test it on, double [36].

References

1. Ekman, P., Friesen, W.V.: Constants across cultures in the face and emotion. J. Pers. Soc. Psychol. **17**(2), 124–129 (1971)
2. Gu, Y., Mai, X., Luo, Y.-J.: Do bodily expressions compete with facial expressions? Time course of integration of emotional signals from the face and the body. PLoS One **8**(7), 736–762 (2013)
3. Adolphs, R.: Neural systems for recognizing emotion. Current Opinion in Neurobiology **12**(2), 169–177 (2002)
4. Lecun, Y., Bottou, L., Bengio, Y., Haffner, P.: Gradient-based learning applied to document recognition. Proc. IEEE **86**(11), 2278–2324 (1998)
5. Ghayoumi, M., Bansal, A.K.: Architecture of Emotion in Robots Using Convolutional Neural Networks. RSS, USA (2016)
6. Ghayoumi, M., Bansal, A.K.: Multimodal architecture for emotion in robots using deep learning. In: Future Technologies Conference, San Francisco, United States (2016)
7. Gunes, H., Piccardi, M.: A bimodal face and body gesture database for automatic analysis of human nonverbal affective behavior. In: Proceeding of ICPR 2006 the 18th International Conference on Pattern Recognition, Hong Kong, China (2006)

8. Bänziger, T., Pirker, H., Scherer, K.: Gemep - Geneva multimodal emotion portrayals: a corpus for the study of multimodal emotional expressions. In: Deviller, L., et al. (eds.) Proceedings of LREC 2006 Workshop on Corpora for Research on Emotion and Affect, pp. 15–19, Genoa (2006)

9. Douglas-Cowie, E., Campbell, N., Cowie, R., Roach, P.: Emotional speech: towards a new generation of databases. Speech Commun. **40**(1), 33–60 (2003)

10. Gunes, H., Piccardi, M.: Bimodal emotion recognition from expressive face and body gestures. J. Network Computer Appl. **30**(4), 1334–1345 (2006)

11. el Kaliouby, R., Robinson, P.: Generalization of a vision-based computational model of mind-reading. In: Proceedings of First International Conference on Affective Computing and Intelligent Interfaces, pp. 582–589 (2005)

12. Cowie, R., Douglas-Cowie, E., Tsapatsoulis, N., Votsis, G., Kollias, S., Fellenz, W., Taylor, J.G.: Emotion recognition in human-computer interaction. IEEE Signal Process. Magazine **18**(1), 32–80 (2001)

13. Pontiac, M., Rothkrantz, L.J.M.: Automatic analysis of facial expressions: the state of the art. IEEE Trans. Pattern Anal. Mach. Intell. **22**(12), 1424–1445 (2000)

14. Mehrabian, A.: Silent Messages - A Wealth of Information about Nonverbal Communication (Body Language). Personality & Emotion Tests & Software: Psychological Books & Articles of Popular Interest (2009)

15. Ghayoumi, M., Bansal, A. K.; Real emotion recognition algorithm by detecting symmetry patterns with Dihedral group. In: MCSI (2016)

16. Schultz, W.: Neural coding of basic reward terms of animal learning theory, game theory microeconomics and behavioral ecology. Cur. Opin. Neurobiol. **14**(2), 139–147 (2004)

17. Panksepp, J.: Affective Neuroscience. Oxford University Press, New York (1998)

18. Laird, J.: The Soar Cognitive Architecture. MIT Press, Cambridge (2012)

19. Friesen, E., Ekman, P.: Facial action coding system: a technique for the measurement of facial movement, Palo Alto (1978)

20. Viola, P., Jones, M.J.: Robust real-time face detection. Int. J. Comput. Vis. **57**(2), 137–154 (2004)

21. Abrishami Moghaddam, H., Ghayoumi, M.: Facial image feature extraction using support vector machines. In: Proceeding VISAPP, Setubal, Portugal (2006)

22. Ghayoumi, M., Bansal, A.K.: An integrated approach for efficient analysis of facial expressions. In: SIGMAP, (2014)

23. Susskind, J.M., Hinton, G.E., Movellan, J.R., Anderson, A.K.: Generating facial expressions with deep belief nets. Affective Computing, Emotion Model. Synth. Recogn., 421–440 (2008)

24. Krizhevsky, A., Sutskever, I., Hinton, G. E.: Imagenet classification with deep convolutional neural networks. In: Advances in Neural Information Processing Systems, pp. 1097–1105 (2012)

25. Sun, Y., Wang, X., Tang, X.: Deep learning face representation from predicting 10,000 classes. In: Computer Vision and Pattern Recognition (CVPR), pp. 1891–1898. IEEE (2014)

26. Ghayoumi, M., Bansal, A.: Unifying geometric features and facial action units for improved performance of facial expression analysis, CSSCC (2015)

27. Ghayoumi, M., Tafar, M., Bansal, A. K.: Towards formal multimodal analysis of emotions for affective computing. DMS (2016)

28. Huan, Y.: Wu, Ao., Zhang, G., Li, Y.: Extraction of adaptive wavelet packet filter-bank-based acoustic feature for emotion recognition. IET Signal Process. **9**(4), 341–348 (2015)

29. Kwon, O. W., Chan, K., Hao, J., Lee, T. W.: Emotion recognition by speech signals. In: 8th International Conference on Speech Communication and Technology (2003)

30. Lee, C.M., Narayanan, S.S.: Towards detecting emotions in spoken dialog. IEEE Trans. Speech Audio Process. **13**(2), 293–303 (2005)
31. Mitra, S., Acharya, T.: Gesture recognition: a survey. IEEE Trans. Syst. Man Cybern. **37**(3), 311–324 (2007)
32. Glowinski, D., Dael, N., Camurri, A., Volpe, G., Mortillaro, M., Scherer, K.: Toward a minimal representation of affective gestures. IEEE Trans. Affect. Comput. **2**(2), 106–118 (2011)
33. Camurri, A., Lagerlö, I., Volpe, G.: Recognizing emotion from dance movement: comparison of spectator recognition and automated techniques. Int. J. Hum. Comput. Stud. **59**(1), 213–225 (2003)
34. Castellano, G., Villalba, S.D., Camurri, A.: Recognising human emotions from body movement and gesture dynamics. In: Paiva, A.C., Prada, R., Picard, R.W. (eds.) ACII 2007. LNCS, vol. 4738, pp. 71–82. Springer, Heidelberg (2007)
35. Ghayoumi, M.: A Review of Multimodal Biometric Systems Fusion Methods and Its Applications. ICIS, USA (2015)
36. Ghayoumi, M., Khan, J., Pourebadi Khotbesara, M., Bauer, E., Hossain, A.: Follower Robot with an Optimized Gesture Recognition System. RSS, USA (2016)

Rhythmic Timing in Playful Human-Robot Social Motor Coordination

Naomi T. Fitter[✉], Dylan T. Hawkes, and Katherine J. Kuchenbecker

Haptics Group, GRASP Laboratory, Mechanical Engineering and Applied Mechanics, University of Pennsylvania, Philadelphia, PA 19104, USA
{nfitter,dhawkes,kuchenbe}@seas.upenn.edu

Abstract. Future robots for everyday human environments will need to be capable of physical collaboration and play. We previously designed a robotic system for constant-tempo human-robot hand-clapping games. Since rhythmic timing is crucial in such interactions, we sought to endow our robot with the ability to speed up and slow down to match the human partner's changing tempo. We tackled this goal by observing human-human entrainment, modeling human synchronization behaviors, and piloting three adaptive tempo behaviors on a Rethink Robotics Baxter Research Robot. The pilot study indicated that a fading memory difference learning timing model may perform best in future human-robot gameplay. We will use the findings of this study to improve our hand-clapping robotic system.

Keywords: Social-physical human-robot interaction · Joint action · Entrainment

1 Introduction

As robots enter more human-populated environments, they will need richer and more diverse human interaction abilities ranging from lighthearted gameplay modes to complex physical collaboration skills. We see particular potential for robots to engage with people via hand-clapping games, which are playful jointly executed hand-to-hand contact patterns. Humans, especially children, use hand clapping to bond with others, teach one another, and fend off boredom. Thus, these games represent an organic topic of study in physical human-robot interaction (pHRI). We expect a robot with hand-clapping abilities to hold promise in tasks such as connecting with students in a classroom setting or encouraging older adults to stay active. Creating a competent hand-clapping robot requires knowledge from both pHRI and social robotics.

Despite its simple premise, hand clapping presents several underlying challenges in interaction dynamics that merit scientific investigation. For example, although human hand-clapping partners can easily begin clapping hands, it is not immediately evident how they decide on tempo, tempo modulation, hand contact location, and failure response. Accordingly, this paper explores human-robot interaction and synchronization during hand-clapping tasks. Our previous

© Springer International Publishing AG 2016
A. Agah et al. (Eds.): ICSR 2016, LNAI 9979, pp. 296–305, 2016.
DOI: 10.1007/978-3-319-47437-3_29

research [1] presented human impressions of hand clapping with a robot, appropriate hand trajectories, and handclap sensing strategies. This paper continues this investigation by exploring tempo errors and tempo modeling. After discussing related work (Sect. 2), we outline our human-human experiment methods (Sect. 3), highlight selected results (Sect. 4), and test proposed models for robotic tempo following (Sect. 5).

2 Background

Social Robotics and pHRI: This research is inspired and informed by extensive previous work in both social robotics and physical HRI. In social contexts, pHRI has been shown to help predict human-robot contact states [2] and mediate affective tactile communication [3]. On the social robotics front, only a subset of investigations include robotic physical contact [4]. Even without contacting people, though, social robots have strong potential benefits in applications from education to the care of older adults [5]. Researchers have explored the use of social robots to work with developmentally delayed children [6] and assist patients in medical settings [7]. In one study of social pHRI in a medical setting, people preferred practical robot touch over comforting robot touch [8]. Nevertheless, our preliminary work [1] indicated that a robot with social-physical interaction skills may be able to engage both children and adults in enjoyable ways.

Similar Studies: Our work is the first we know of in the area of hand-clapping robots, but other instantiations of social-physical human-robot interaction (spHRI) have preceded this investigation. Preliminary implementations of human-robot high five, fist bump, and hug interactions in the popular pr2_props demo [9] inspired much of our early work. Other playful spHRI applications include human-robot dancing [10], magic performance [11], and classroom interactions [12]. Related studies of human-robot handshakes have also illustrated ways to model human haptic behaviors [13] and shape human-inspired robotic handshake algorithms [14].

Understanding Synchronization: In our chosen application of human-robot hand-clapping games, we need to understand how to model various types of synchronization, so that our robot can dynamically adapt to tempo changes. This realm of investigation has long existed in the field of experimental psychology, where researchers have modeled human synchronization with a stimulus and considered what additional information is needed to expand to multi-person synchronization scenarios [15]. In particular, many researchers have focused on isolating synchronization behaviors in response to timing changes. One study built on the Wing-Kristofferson model to propose various model-fitting techniques for synchronization cases [16]. Other work found that abrupt tempo changes and gradual tempo changes seem to engage different methods of phase correction [17].

Outside of experimental psychology, some robotics researchers have begun to explore models of synchronization. In particular, Murata et al.'s design of a playful, synchronous interaction between a human clapper and a robot singer/dancer

[18] guided the work discussed in Sect. 4. Synchrony and entrainment are also key issues in [19], which proposes an online design method for controlling the dynamics of various timing-based physical interactions. Other work applied a framework of dynamic motor primitives to develop rapid phase-locking and adaptive tempo tracking behaviors for a skilled robotic drummer [20].

Imitation: We are also fascinated by how people can fluidly mimic behaviors. Researchers have investigated mirror neurons as a channel to explain action imitation, and roboticists have implemented human-neuron-inspired simple action imitation by a robot [21]. One similar human-human study explored modeling motions during the "mirror game" in which players face each other and imitate the other participant's motions [22]. Further investigations try to push human synchronization to the extreme, finding that it succeeds in all cases except with an irregular and unresponsive tapping partner [23]. Questions remain on the front of understanding many of these topics in a human-human context. We hope that new findings in these areas will benefit research in both psychology and robotics.

3   Human-Human Experiment

We conducted an experiment studying human-human interactions to inform the design of synchronization behaviors for a hand-clapping robot. Here, we were especially interested in the quantitative and qualitative effects of shared timing leadership versus individual timing leadership on the synchronization of human hand-clapping partners during various activities. Accordingly, we collected human-human motion data, survey responses, and behavioral data. The Penn IRB approved all procedures under protocol 818801. Fourteen participants (9 male, 5 female, 22-48 years of age) enrolled this study, gave informed consent, successfully completed the experiment, and received a $5 payment.

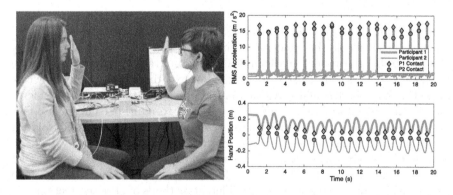

Fig. 1. Left: Participants sitting face-to-face and hand clapping with opposite hands during the experiment. Right: Sample plot of acceleration and position recordings from the experiment. Markers indicate times of hand contact.

Experiment Setup: Study participants were paired randomly. Each pair came to the lab for a single experiment sitting that lasted about 30 min. As illustrated in Fig. 1, the back center of one participant's left hand and one participant's right hand was outfitted with a magnetic tracking sensor (Ascension trakSTAR 3D Model 180 6DOF) and a three-axis accelerometer breakout board (Sparkfun MMA7361) using skin-safe adhesives. Because all but one of the participants were right-handed, we randomized which hand to outfit with sensors. Participants sat face-to-face and clapped their instrumented hands together.

Experiment Phases: This experiment centered on palm-to-palm hand-clapping motions executed repeatedly, in the style of hand-clapping games. We chose to examine only this one type of hand-clapping motion to decrease the likelihood of errors and increase the participants' ability to focus on synchronization with their partner. The experiment trial tasks were designed to explore human engagement and challenge during different interaction timing scenarios. An understanding of these topics could enhance human-robot hand-clapping gameplay by improving human-robot synchronization.

Hand-clapping cues (target tempos or tempo change cues) were presented briefly either through headphones or via cues written on index cards, as indicated in the phase descriptions below. Tempos included 60, 110, 160, 210, and 260 beats per minute (BPM), and stimuli were designed to present different levels of challenge and prompt different individual and shared tempo leadership scenarios. The experimenter prompted each hand-clapping task to one or both participants throughout the study and asked them to carry out this task for 20 s of data recording. Participants were free to rest briefly after each data recording. During hand-clapping recordings, we asked participants to move and interact as naturally as possible. The three phases of the experiment progressed as follows:

- **Phase 1**: Five hand-clapping trials. Both participants could hear a target (randomly-ordered) hand-clapping tempo stimulus and attempted to clap hands at that tempo for 20 s.
- **Phase 2**: Ten hand-clapping trials. During each trial, only one of the participants could hear the target (randomly ordered) hand-clapping tempo stimulus. The participant listening to the tempo led the other participant in making contact at that tempo for 20 s. Each participant had opportunities to be the tempo leader, and the randomized selection of the tempo leader was balanced between the participants.
- **Phase 3**: Eight hand-clapping trials. During each trial, one participant saw a written cue to speed up, slow down, or remain at a constant tempo and led the other participant in that activity for 20 s. Each participant had opportunities to be the tempo leader, and tempo leader selection was randomized and balanced. One randomly selected and ordered cue was repeated for each participant to thwart guessing.

Phase 1 and 2 results are discussed at length in another paper [1]. This analysis focuses on designing robot timing behaviors based mainly on Phase 3 results.

Data Collection: The data collected in the experiment included 20 s of hand-clapping data from the two position sensors and the two accelerometers for each trial mentioned above. After the final trial of the experiment, we also gave participants a Likert and free-response concluding survey and a basic demographic survey. All Likert-type questions used seven-point scales.

4 Results

Synchronization Analysis: We investigated human behaviors to learn how people adapt to changing tempos and to devise metrics for evaluating a hand-clapping robot. Leader and follower roles were clearly delineated at specific tempos in Phases 1 and 2 of the human-human experiment. This experimental design made it easy to calculate the errors between the target and actual inter-handclap intervals. To compute the inter-handclap time intervals, we needed a method that could consistently identify when a handclap had occurred. As discussed in [1] and clearly illustrated in Fig. 1, peaks in the RMS acceleration occur consistently and exclusively at moments of hand impact. Local maxima detection on the RMS acceleration signal with set minimum peak spacing therefore proved a reliable way to identify successful hand-clapping impacts, the corresponding hand positions, and the inter-clap time intervals throughout the entire dataset.

The difference between each inter-clap duration and the Phase 1 and 2 target clapping period (1.000 s for 60 BPM, 0.5455 s for 110 BPM, 0.3750 s for 160 BPM, 0.2857 s for 210 BPM, and 0.2308 s for 260 BPM) yielded the error in each hand-clapping interval. Throughout this paper, we rely on mean squared error (MSE) as a key evaluation metric because it captures the overall timing correctness of a trial. Figure 2 illustrates the MSE in handclap timing over all participants during the first fifteen handclaps. The top left plot displays this metric for the Phase 1 shared tempo portion of the experiment and generally reflects a low inter-clap timing error. The plotted Phase 2 data elucidates that trials with one participant leading the tempo have a larger initial MSE and continually higher MSE compared to Phase 1 metrics; the average MSE for claps 11 through 15 was 0.0036 s^2.

Overall, this human-human synchronization analysis gives us a general idea of the timing errors people exhibit in different tempo leader and follower scenarios. Consequently, these results will help us assess our hand-clapping robotic systems and confirm whether they are performing as well as a human clapping partner in various tempo synchronization situations.

Tempo-Setting Models: Enabling a human to lead a robot in a clapping tempo of their choosing requires an effective robot timing synchronization model. Based on the approaches of other researchers tackling similar problems, we propose three models for predicting future inter-clap timing intervals: simple averaging, difference learning, and fading memory difference learning. In the three explored models, P_i is the ith predicted future beat time (measured from zero time), M_{tmp} is a temporary variable, T_{i-1} is the penultimate time of hand contact detection, T_i is the most recent time of hand contact, d_{i-1} is the duration of the previous

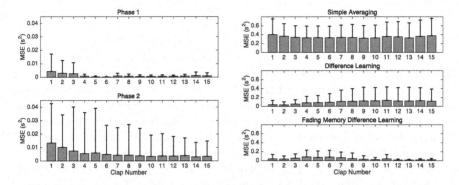

Fig. 2. Left: Visualization of the mean squared error (MSE) in handclap timing over all participants during the first fifteen handclaps. The top plot conveys the timing errors for Phase 1 of the human-human experiment, and the bottom plot conveys the errors for Phase 2. Right: Tempo-setting performance for each of the proposed timing models on all Phase 3 trial data recordings during the first fifteen handclaps. The error bars represent +1 standard deviation of the squared errors.

inter-clap interval, $\tilde{I}_3$ is the median value of a buffer of the previous three inter-clap intervals, and $\tilde{I}_{\text{all}}$ is the median value of all previous inter-clap intervals. The modulo operator is also used, represented in equations as "mod." For all models, the current handclap interval, d_i, can be calculated as:

$$d_i = T_i - T_{i-1}$$

Simple Averaging: This model predicts future handclap timing using the average of the last two inter-clap intervals, as expressed by the following equations:

$$P_i = \begin{cases} d_i & \text{if } i = 1 \\ \frac{d_i + d_{i-1}}{2} + T_i & \text{if } i > 1 \end{cases}$$

Difference Learning: This model predicts future handclap timing using an equation proposed for synchronizing robot singing with human clapping [18]:

$$P_i = \begin{cases} d_i & \text{if } i = 1 \\ M_{\text{tmp}} & \text{if } M_{\text{tmp}} \geq \frac{2}{3}\tilde{I}_{\text{all}} + T_i \\ M_{\text{tmp}} + \tilde{I}_{\text{all}} & \text{otherwise} \end{cases}$$

$$M_{\text{tmp}} = T_{i-1} + \tilde{I}_{\text{all}} + d_i - d_i \text{mod} \tilde{I}_{\text{all}}$$

Fading Memory Difference Learning: This model predicts future handclap timing in the same way as the previous model, but with a limit on the size of the buffer of previously observed inter-handclap intervals:

$$P_i = \begin{cases} d_i & \text{if } i = 1 \\ M_{\text{tmp}} & \text{if } M_{\text{tmp}} \geq \frac{2}{3}\tilde{I}_3 + T_i \\ M_{\text{tmp}} + \tilde{I}_3 & \text{otherwise} \end{cases}$$

$$M_{\text{tmp}} = T_{i-1} + \tilde{I}_3 + d_i - d_i \text{mod} \tilde{I}_3$$

Using the interval-finding technique mentioned previously, we extracted the inter-clap time intervals from Phase 3 of the human-human experiment to obtain a rich array of realistic human tempo change data. Per the stimuli in Phase 3, these tempo changes included speeding up, slowing down, and a control condition of remaining at a constant tempo. We can evaluate and compare the performance of the predictive timing models introduced above using this human dataset.

Implementation of these three predictive tempo models on the Phase 3 dataset yields an overall MSE of $0.3438 \, \text{s}^2$ for the simple averaging model, $0.0971 \, \text{s}^2$ for the difference learning model, and $0.0403 \, \text{s}^2$ for the fading memory difference learning model. Our analysis also confirmed that a buffer size of three elements performs better than any other size for the fading memory difference learning approach. Figure 2 illustrates the MSE produced by each model's prediction of the first fifteen handclap timings in Phase 3 data recordings. Now that we had estimates of *typical human tempo error*, models for *future clap timing prediction*, and validation of *model predictive performance*, we were ready to add tempo adaptation skills to our previously developed hand-clapping robotic system.

5   Implementation on a Robot

Developing Robot Follower Behaviors: Excited by the prospect of timing models that can predict future inter-clap timing with an MSE level similar to that of our human-human study participants, we sought to implement the timing models on a Rethink Robotics Baxter Research Robot, a safe human-sized humanoid. Our previous work [1] transformed Baxter into a capable hand-clapping partner using a custom 3D-printed hand and a PD controller with feedforward torque. Using this controller, we achieved humanlike hand-clapping motions by commanding Baxter's W1 joint along the following motion trajectory, using variables for desired joint angle (θ_d), amplitude of motion (A), frequency of hand contacts in Hz (f), time (t), and initial joint angle (θ_{init}):

$$\theta_d = A \sin(2\pi f t + \pi/2) - A + \theta_{\text{init}}$$

To add adaptive timing abilities to the robot, we implemented each proposed timing model in the robot's standard Robot Operating System (ROS) framework and updated the A and f equation terms once per robot motion cycle based on the next predicted inter-clap time interval. For each new cycle, f was equal to $1/(P_i - T_i)$ and A was calculated from an exponential curve fit to our previous finding that hand motion amplitude varies inversely with clapping frequency [1]. We detected hand contact times for each motion model by monitoring for peaks in filtered data from the x-axis of Baxter's inbuilt right wrist accelerometer.

Pilot Investigation: To test our robot implementation, we ran a pilot investigation of Baxter's adaptive tempo interaction abilities. We were curious whether one of the robot tempo models would enable better performance and/or be preferred by users for different types of tempo change. The Penn IRB approved all experimental procedures under protocol 823886.

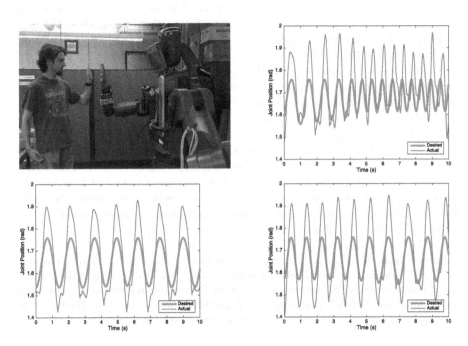

Fig. 3. Top left: The pilot study setup where a participant stood facing Baxter and hand clapped with it using their left hand. Bottom left: An example constant tempo recording. Top right: An increasing tempo recording. Bottom right: A decreasing tempo recording. In each sample recording, artifacts in the actual position readings indicate human hand contacts. Overall, we can improve motion accuracy by considering Baxter's overshooting tendencies while designing robot motion.

One single-blinded expert user and five naïve participants (3 male, 3 female) consented to participate in the pilot study, received general information about the experiment activity, and led Baxter in nine 20 s interaction trials. Users interacted with the robot via three different three-alternative forced choice (3AFC) tasks: picking their favorite trial from three constant tempo interactions, three increasing tempo interactions, and three decreasing tempo interactions. Within each 3AFC grouping, one of the three tempo prediction models (simple averaging, difference learning, or fading memory difference learning) was powering Baxter's future inter-clap timing adaptation. Each model was presented once in a random order unknown to participants. The human user led the tempo change in each trial, and we asked them to maintain consistent approaches within each grouping of three trials. During each interaction, we recorded desired robot joint angles, actual robot joint angles, and robot wrist accelerometer readings. An illustration of the pilot study setup and sample data recordings appear in Fig. 3.

User tempo model preferences for individual tempo change tasks were not conclusive. We did notice, though, that only one participant chose the difference learning model for each of the tempo change trials. Upon further investigation, we noticed that four of the six pilot users failed to accomplish a decrease in clapping tempo during the decreasing tempo interactions using the difference learning model, three failed during the simple averaging model trials, and only two failed during the fading memory difference learning trials. One of the six users failed to accomplish a tempo increase during one simple averaging trial. Participants also took much longer to change the robot's clapping tempo in difference learning trials. Since the fading memory difference learning model led to the fewest failures in tempo change, it may be the best candidate for a robot predictive tempo model.

Participants generally found the experiment interesting, but they had trouble with some aspects of the interaction. They found slowing down to be challenging and sometimes had trouble understanding how to contact the robot to reduce its tempo. We will consider these observations as we continue to improve Baxter's adaptive tempo behaviors during hand-clapping interactions and improve the general understanding of human-robot entrainment in spHRI.

Acknowledgments. The first author was supported by a National Science Foundation (NSF) Graduate Research Fellowship under Grant No. DGE-0822 and the University of Pennsylvania's NSF Integrative Graduate Education and Research Traineeship under Grant No. 0966142. We thank Kostas Daniilidis for the use of his Baxter robot and Saul Sternberg for his insights on related synchronization research.

References

1. Fitter, N.T., Kuchenbecker, K.J.: Equipping the Baxter robot with human-inspired hand-clapping skills. In: Accepted to the IEEE International Symposium on Robot and Human Interactive Communication (RO-MAN) (2016)
2. Iwata, H., Sugano, S.: Human-robot-contact-state identification based on tactile recognition. IEEE Trans. Ind. Electron. **52**(6), 1468–1477 (2005)
3. Argall, B.D., Billard, A.G.: A survey of tactile human-robot interactions. Robot. Auton. Syst. **58**(10), 1159–1176 (2010)
4. Fong, T., Nourbakhsh, I., Dautenhahn, K.: A survey of socially interactive robots. Robot. Auton. Syst. **42**(3), 143–166 (2003)
5. Feil-Seifer, D., Mataric, M.J.: Defining socially assistive robotics. In: IEEE International Conference on Rehabilitation Robotics (ICORR), pp. 465–468 (2005)
6. Robins, B., Dautenhahn, K., Te, B., Boekhorst, R., Billard, A.: Robotic assistants in therapy and education of children with autism: can a small humanoid robot help encourage social interaction skills? Univers. Access Inf. Soc. **4**(2), 105–120 (2005)
7. Rabbitt, S.M., Kazdin, A.E., Scassellati, B.: Integrating socially assistive robotics into mental healthcare interventions: Applications and recommendations for expanded use. Clin. Psychol. Rev. **35**, 35–46 (2015)
8. Chen, T.L., King, C.H., Thomaz, A.L., Kemp, C.C.: Touched by a robot: An investigation of subjective responses to robot-initiated touch. In: ACM/IEEE International Conference on Human-Robot Interaction (HRI), pp. 457–464 (2011)

9. Romano, J.M., Kuchenbecker, K.J.: Please do not touch the robot. In: Hands-on Demonstration Presented at IEEE/RJS Conference on Intelligent Robots and Systems (IROS) (2011)
10. Kosuge, K., Hayashi, T., Hirata, Y., Tobiyama, R.: Dance partner robot - Ms DanceR. IEEE/RJS Int. Conf. Intell. Robots Syst. **4**, 3459–3464 (2003)
11. Nuñez, D., Tempest, M., Viola, E., Breazeal, C.: An initial discussion of timing considerations raised during development of a magician-robot interaction. In: ACM/IEEE International Conference on Human-Robot Interaction (HRI) Workshop on Timing in HRI (2014)
12. Kanda, T., Sato, R., Saiwaki, N., Ishiguro, H.: A two-month field trial in an elementary school for long-term human-robot interaction. IEEE Trans. Robot. **23**(5), 962–971 (2007)
13. Wang, Z., Yuan, J., Buss, M.: Modelling of human haptic skill: A framework and preliminary results. IFAC Proc. **41**(2), 14761–14766 (2008)
14. Avraham, G., Nisky, I., Fernandes, H.L., Acuna, D.E., Kording, K.P., Loeb, G.E., Karniel, A.: Toward perceiving robots as humans: Three handshake models face the Turing-like handshake test. IEEE Trans. Haptics **5**(3), 196–207 (2012)
15. Elliott, M.T., Chua, W.L., Wing, A.M.: Modelling single-person and multi-person event-based synchronisation. Current Opin. Behav. Sci. **8**, 167–174 (2016)
16. Repp, B.H., Keller, P.E., Jacoby, N.: Quantifying phase correction in sensorimotor synchronization: empirical comparison of three paradigms. Acta Psychol. **139**(2), 281–290 (2012)
17. Vorberg, D., Schulze, H.H.: Linear phase-correction in synchronization: Predictions, parameter estimation, and simulations. J. Math. Psychol. **46**(1), 56–87 (2002)
18. Murata, K., Nakadai, K., Takeda, R., Okuno, H.G., Torii, T., Hasegawa, Y., Tsujino, H.: A beat-tracking robot for human-robot interaction and its evaluation. In: IEEE/RAS International Conference on Humanoid Robots (Humanoids), pp. 79–84 (2008)
19. Sato, T., Hashimoto, M., Tsukahara, M.: Synchronization based control using online design of dynamics and its application to human-robot interaction. In: IEEE International Conference on Robotics and Biomimetics (ROBIO), pp. 652–657 (2007)
20. Pongas, D., Billard, A., Schaal, S.: Rapid synchronization and accurate phase-locking of rhythmic motor primitives. In: IEEE/RSJ International Conference on Intelligent Robots and Systems (IROS), pp. 2911–2916 (2005)
21. Metta, G., Sandini, G., Natale, L., Craighero, L., Fadiga, L.: Understanding mirror neurons: a bio-robotic approach. Interact. Stud. **7**(2), 197–232 (2005)
22. Noy, L., Dekel, E., Alon, U.: The mirror game as a paradigm for studying the dynamics of two people improvising motion together. National Acad. Sci. **108**(52), 20947–20952 (2011)
23. Konvalinka, I., Vuust, P., Roepstorff, A., Frith, C.D.: Follow you, follow me: continuous mutual prediction and adaptation in joint tapping. Q. J. Exp. Psychol. **63**(11), 2220–2230 (2010)

The Effects of an Impolite vs. a Polite Robot Playing Rock-Paper-Scissors

Álvaro Castro-González[✉], José Carlos Castillo, Fernando Alonso-Martín,
Olmer V. Olortegui-Ortega, Victor González-Pacheco, María Malfaz,
and Miguel A. Salichs

Robotics Lab, Universidad Carlos III de Madrid, Leganés, Spain
acgonzal@ing.uc3m.es

Abstract. There is a growing interest in the Human-Robot Interaction
community towards studying the effect of the attitude of a social robot
during the interaction with users. Similar to human-human interaction,
variations in the robot's attitude may cause substantial differences in
the perception of the robot. In this work, we present a preliminary study
to assess the effects of the robot's verbal attitude while playing rock-
paper-scissors with several subjects. During the game the robot was programmed to behave either in a polite or impolite manner by changing
the content of the utterances. In the experiments, 12 participants played
with the robot and completed a questionnaire to evaluate their impressions. The results showed that a polite robot is perceived as more likable
and more engaging than a rude, defiant robot.

1 Introduction

The attitude of a robot is an essential feature for creating socially interactive
robots. Studies on this matter are aimed at enhancing the Human-Robot Interaction (HRI). In this work, we focus on studying how the robot's verbal attitude
alters its attributions. To achieve our goal, we have created a HRI scenario
involving real users and a social robot. In the experiment, the robot and the
users played, one at a time, rock-paper-scissors. We modified the robot's attitude by changing the verbal content of its utterances. Then, in some cases, the
robot used polite words whereas, in others, the utterances were rude and impolite. The only difference between the polite and the impolite attitudes were the
utterances the robot said, the rest of the game remained unchanged.

In the literature, we can find some works focused on studying the robot attitude
during the interaction. In this regard, Cramer found that a positive robot's attitude
was preferred by the users over an accurate robotic empathic behavior [3]. Furthermore, Lee et al. [5] conducted some experiments using the robot dog Aibo where the
robot's personality changed between introvert and extrovert. The first conclusion
that they found was that participants could accurately recognize a robot's personality based on its verbal and nonverbal behaviors. The second conclusion was that
the participants preferred interacting with a robot with an opposed personality

© Springer International Publishing AG 2016
A. Agah et al. (Eds.): ICSR 2016, LNAI 9979, pp. 306–316, 2016.
DOI: 10.1007/978-3-319-47437-3_30

rather than a similar one. Moreover, Leite et al. described in [6] a robotic game buddy with different behaviors regarding the state of the game. The results of the experiments indicated that a social robot with emotional behavior could perform better the task of helping users to understand a gaming situation. Besides, user's enjoyment is higher when interacting with a robotic embodied character, compared to a screen-based version of the same character. Short et al. [12] showed that a cheating robot playing rock-paper-scissors was perceived by users more engaging than a fair robot (not cheating).

A recent work [13] focuses on assessing the effect of a robot's attitude (positive vs. negative) on the Uncanny Valley phenomenon using a live interaction paradigm. Results shown that the effect of a robot's attitude is not independent of its embodiment; that is, a robot which is perceived as uncanny is not able to affect its likeability by a positive or negative interaction. On the other hand, the impact of a machine-like robot's attitude is much greater and especially when it behaves negatively as it can lose all its initial likeability. Other recent work by Salem et al. [10] investigated culture-specific determinants of robot acceptance and anthropomorphization. Authors also manipulated the robot's verbal behavior in experimental sub-groups (Arab and English) to explore different politeness strategies. Results suggested that Arab participants perceived the robot more positively and anthropomorphized it more than English speaking participants. In addition, the use of positive politeness strategies and the change of interaction task had an effect on participants' HRI experience. In a prior work, Salem et al. stated that the politeness levels do not have a relevant effect on the user's perception of a robot during the interaction, but the interaction context does [9]. In a Japanese study, Nomura and Saeki studied politeness based on robot poses [7]. In this study, the robot asked Japanese participants to manipulate several objects on a desk using different body gestures. Robot's polite motions had effects on the human impressions of the robot and they found gender differences. Additionally, the intuitive trust people tend to feel when encountering robots in public spaces has also been studied. Inbar [4] presented test subjects with static images of a robot performing an access-control task, interacting with younger and older male and female civilians. The robot showed polite or impolite behavior. Results showed strong effects of the robot's behavior and, besides, age and gender of the people interacting with the robot had no significant effect on participants' impressions of the robot's attributes.

In the same line, in the present work we study how the robot's verbal attitude, polite or impolite, affects its attributions by users that have been engaged in a real interaction. Therefore, our initial hypothesis, *H0*, is that the verbal attitude of a robot influences the perceived Anthropomorphism, Animacy, Likeability, Perceived Intelligence, Perceived Safety and Engagement. To the best of our knowledge, the different attributions to a polite, friendly robot and to an impolite, defiant robot has not been evaluated yet. Researchers have not confirmed whether a conversational robot using defiant, rude utterances could be perceived as more appealing and engaging than a kind, charming one during a competitive game. The goal of this preliminary study is to shed light on this question.

The rest of the document is structured as follows. In Sect. 2, we detailed the experiment: the robot, the HRI scenario, and the procedure. Next, Sect. 3 shows the evaluation conducted and the results obtained. Finally, we conclude the paper in Sect. 4 where we discuss the results and present the main limitations.

2 Experiment

To evaluate the effects of the robot's verbal attitude, polite vs. impolite, we have developed a game where our social robot Mini plays a competitive, interactive game with participants. This section describes the design of the experiment, including the details of the game and the robot.

2.1 Robotic Platform: Mini

The game scenario presented in this work is implemented on the robotic platform Mini, designed and built at RoboticsLab research group from Carlos III University of Madrid. Originally, the robot was designed to interact with mild cognitive impaired elderly people [11] although the capabilities of the platform are flexible enough as to be used with a wide range of users. Its plushy body gives it a friendly appearance.

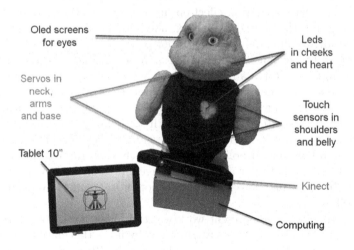

Fig. 1. The social robot Mini used in this study.

The robot Mini (see Fig. 1) is endowed with multiple HRI interfaces to ease the communication with people. Mini has LEDs in its cheeks and heart, screens that constitute its eyes, and a VU-meter like mouth. Some motors allow moving the arms and head to complement the illusion of a living entity. Regarding sensors, the robot Mini is equipped with touch sensors distributed throughout its body, a

Microsoft Kinect RGB-D camera and a *LeapMotion* device [2]. This latter is used in this work for hand pose detection and recognition. The robot is also equipped with a tablet to show multimedia content as well as games. Furthermore, the robot Mini is able to synthesize voice by a Text-To-Speech (TTS) skill. Finally, Mini's software architecture relies on ROS [8], a framework for developing robot software which provides a collection of tools and libraries to simplify the task of creating robot behaviors across robotic platforms.

2.2   Interactive Game Scenario: Rock-Paper-Scissors

This work uses a rock-paper-scissors game specifically designed to assess the robot's attributions of users. We use the *LeapMotion* as input for a game in which the user tries to beat the robot. Apart from this device, the interaction modalities of the robot are composed by a tablet that shows the robot's gesture and the result of each round, a TTS module that enables voice interaction, and motors to allow moving head and arms to perform gestures.

Rock-paper-scissors is a zero-sum hand game played in this work by the robot and the user in which each player simultaneously forms one of three shapes with an outstretched hand. These shapes are *rock* (a simple fist), *paper* (a flat hand), and *scissors* (a fist with the index and middle fingers together forming a V). The game has only three possible outcomes other than a tie: a player who decides to play rock will beat another player who has chosen scissors (rock crushes scissors) but will lose to one who has played paper (paper covers rock); a play of paper will lose to a play of scissors (scissors cut paper). If both players choose the same shape, the game is tied.

In terms of operation, the dynamics of the game are simple, the robot displays a countdown in the tablet while saying outloud "Rock-Scissors-Paper, now!" to synchronize the moves of both players. Then, the robot shows its selection in the tablet and waits for the user to reveal her own one by placing her hand above the *LeapMotion* device. Finally, the robot shows a comparison of the moves in the tablet while verbally announcing the winner. At the end of each round, depending on the verbal attitude selected (see Sect. 2.3), the robot will use polite or impolite sentences. The game continues until the user expresses her intention of stopping playing. Note that the robot's gesture is chosen randomly in each game round.

It is important to emphasize that the robot works autonomously during the game, giving directions to the user at the beginning, and continuously interacting with the user. A experimenter just started the game when a participant arrives.

2.3   Conditions

Since the aim of the paper is to study how HRI is affected by the robot's attitude towards the user, we have established two conditions related to different verbal interaction modalities: polite and impolite. In the polite condition, the robot encouraged and stimulated the participant using nice, positive words. On the contrary, in the impolite condition, the robot's utterances consisted on unpleasant, rude, defiant utterances. Examples of the different utterances used in both conditions are presented in Table 1. Note that the utterances are originally spoken in Spanish and Table 1 offers an approximate translation into English.

According to the result of each round (user wins, robot wins or tie), the robot selected randomly among a set of utterances depending on the condition. In the case that the robot is not able to recognize the user gesture, it reacts to this situation too with the appropriate utterances.

Table 1. Some examples of the robot's utterances depending on the condition and the game situation.

Condition	Game situation	Utterance
Polite	Robot wins	I was lucky, you'll do better next time!
		Let's play again!
	Human wins	You are a great player!
		Congratulations, you play great!
	Tie	Very good, I'm not able to win
		We are both great players!
	Nothing detected	I believe you were to fast
		I couldn't detect your move
Impolite	Robot wins	You are a lousy player
		I would be ashamed if a robot beat me
	Human wins	You are not smart enough to win always
		I'm sure you cheated
	Tie	Loser, you cannot win
		Only humans are such a bad players
	Nothing detected	You fool, play when I say so
		You are an incompetent human, show your move!

2.4   Participants

Sixteen native Spanish speakers were recruited among the faculty personnel and students of our university for the study. Due to technical issues, data from four of them were discarded. Out of the 12 remaining subjects, 42.1 % were female and 57.9 % were male with ages ranged from 22 to 61 years. The study took place at the premises of the RoboticsLab at Carlos III University of Madrid. The number of participants in each condition was: 5 for the polite robot and 7 for the impolite attitude.

2.5   Procedure

The testing phase followed a thorough procedure lead to assess our hypothesis in an objective way. The main steps are depicted in Fig. 2 and the following paragraphs describe the process, with the main steps that matching the numbers in the figure. The experiment was divided into two phases, the first one consisted on playing rock-paper-scissors with the robot and, in the second one, the participants had to complete a questionnaire to assess their interaction during the game.

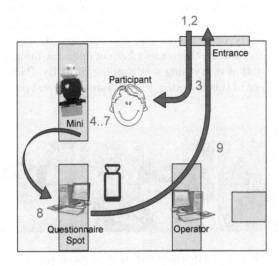

Fig. 2. Experimental setup. Arrows show the transitions between the relevant areas and the numbers are associated to the relevant actions during the experiments described in the text.

Prior to each test and away from the experiment location, participants were informed about their participation in the study and requested to sign an informed consent as well as an optional video recording consent (1). Those who did not sign the second consent proceeded with the experiment without being filmed while the first consent was mandatory to participate in the experiment. Subjects were informed that they could play as long as they like and leave the experiment at any time (2). After the subjects agreed to sign the consents, the experimenter accompanied the participant to the game zone and introduced the robot as well as the game rules and procedure (3). Once players were seated in front of the robot, the main items for the experiment were presented: the robot, the *LeapMotion* (placed between user and robot enclosed in a black rectangle), and the tablet (used for displaying the robot moves and the result of each round). Additionally, participants received some directions about how to play the game (4).

Before the game started, the participants played several demo rounds to get familiar with the game mechanism and the detection device (5). Once a participant felt comfortable, the experimenter left her alone with the robot and the game started (6). The game followed the same rules as the traditional one, being the robot the one leading autonomously the rounds by saying "rock, paper, scissors, now!" and displaying information in the tablet as described in Sect. 2.2. However, there was a big difference regarding the game: the robot Mini used one of the attitudes (polite/impolite) described in Sect. 2.3. The game condition was randomly selected but taking into account a balanced number of users in each condition. Each participant played only in one condition (7). When a subject expressed that she wanted to stop playing, the supervisor went back to the game area and invited the participant to fill an online questionnaire to assess her perception of the robot (8). After filling the questionnaire, the supervisor thanked the participant for being involved in the experiments and it concluded (9). Although the robot in the tests ran autonomously, there was a human operator monitoring the whole process, making sure that everything was running properly. Participants did not notice the involvement of this operator. Figure 3 shows several participants during the experiment.

Fig. 3. Participants during the experiment

3   Evaluation and Results

In order to measure the effects of the different verbal attitudes of a social robot, we have conducted a statistical analysis of the data provided by the participants after their interaction with the robot Mini. All participants completed an extension of the Godspeed Questionnaire Series (GQS) [1]. GQS has been extensively used in robotics and was designed to measure the users' perception of robots. It is one of the most frequently used questionnaires in the field of HRI with over 320 citations as of May 2016. In our study, participants rated the robot using the 5 scales included in the GQS: *Anthropomorphism, Animacy, Likeability, Perceived Intelligence* and *Perceived Safety*. And, in order to evaluate the engagement of the participants, we added an extra scale named *Engagement*.

Before running the statistical analysis, we filtered non-valid data from participants that experienced technical failures during the interaction. After that, we ended up with valid data from 12 participants: 5 subjects interacting with a polite robot, and 7 subjects interacting with an impolite robot. First, we analyzed the correlation among the items belonging to the different scales. We observed that all items in the same scales were positively correlated but in the Perceived Safety scale. Here, one item was negatively correlated, and we reversed its values. Then, in order to estimate the internal consistency of the questionnaire, we calculated the Cronbach's Alpha for the different scales. We maximized the reliability for

Table 2. Items in the scales

Scale	Cronbach's Alpha	Items
Anthropomorphism	0.894	Fake - Natural
		Machinelike - Humanlike
		Unconscious - Conscious
Animacy	0.911	Dead - Alive
		Stagnant - Lively
		Mechanical - Organic
		Artificial - Lifelike
		Inert - Interactive
Likeability	0.905	Dislike - Like
		Unfriendly - Friendly
		Unkind - Kind
		Unpleasant - Pleasant
		Awful - Nice
Perceived Intelligence	0.818	Incompetent - Competent
		Ignorant - Knowledgeable
		Unintelligent - Intelligent
		Foolish - Sensible
Perceived Safety	0.808	Anxious - Relaxed
		Calm - Agitated
		Quiescent - Surprised
Engagement	0.833	Disappointing - Motivating
		Never again - Play again
		Awkward - Easy

each scale by removing the items that lowered it. Table 2 presents the final Cronbach's Alphas for all scales. All scales present α-values higher than 0.8 which represent high consistent, reliable scales.

Six Mann-Whitney tests (non parametric test for independent samples) were conducted to compare the attributed anthropomorphism, animacy, likeability, intelligence, safety, and engagement to a polite (condition I) and to an impolite (condition II) social robot. We did not find statistical significant differences for anthropomorphism, animacy, perceived intelligence, and perceived safety, but we did for likeability and engagement. The results indicated that likeability was greater for a polite robot ($Mdn = 4.400$) than for an impolite robot ($Mdn = 3.200$), $U = 5.000, p = 0.037$. Similarly, the engagement was rated significant higher for a polite robot ($Mdn = 4.000$) than for an impolite robot ($Mdn = 3.666$), $U = .000, p = 0.004$. Table 3 shows the actual significance value of the test. The mean values for the scales likeability and engagement are presented in Fig. 4.

Table 3. Test statistics and the statistical significance (2-tailed) p-value, given $sig \leq 0.05$.

	Anthropomorphism	Animacy	Likeability	Perceived intelligence	Perceived safety	Engagement
Mann-Whitney U	12,000	16,500	,000	9,500	8,000	5,000
Wilcoxon W	27,000	31,500	28,000	37,500	23,000	33,000
Z	$-,904$	$-,165$	$-2,857$	$-1,335$	$-1,559$	$-2,082$
Asymp. Sig. (2-tailed)	,366	,869	,004	,182	,119	,037
Exact Sig. [2*(1-tailed Sig.)]	,432[b]	,876[b]	,003[b]	,202[b]	,149[b]	,049[b]

[a] Grouping Variable: Polite = 0, Impolite = 1.
[b] Not corrected for ties.

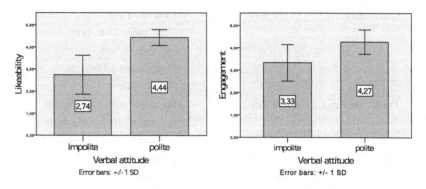

Fig. 4. Mean values for Likeability and Engagement scales

These results suggest that the verbal attitude of a social robot interacting with a person does have an effect on its attributions of likability and engagement, which partially confirms our initial hypothesis, *H0*, for such categories. Specifically, our results suggest that a polite robot is more likable and people will interact more with it than with an impolite robot during a competitive game.

4   Conclusions

In this work we have presented a preliminary study to assess how the verbal attitude of a social robot alters its attributions during a real, competitive, interactive game with people. Initially, we suggested that an impolite, rude robot could improve the engagement of the participants in the interaction because they could perceive it as funny or even as if the robot were challenging them. However, we have found that a robot using a polite attitude was perceived as more likeable and engaging than an impolite one. It is clear that a robot using nice, polite words will be rated as more likeable. Besides, we have observed that engagement also benefits from the polite utterances. We did not find statistical significant differences in other scales, such as anthropomorphism, animacy, perceived intelligence, and perceived safety.

These results should be treated with caution since the size of sample (N = 12) and the variety (all participants where students or staff from a university) are very limited. Further experiments are needed to extend these results to other conditions.

Acknowledgment. This research has received funding from the projects *Development of social robots to help seniors with cognitive impairment - ROBSEN* funded by the Ministerio de Economía y Competitividad (DPI2014-57684-R) from the Spanish Government; and *RoboCity2030-III-CM* (S2013/MIT-2748), funded by Programas de Actividades I+D of the Madrid Regional Authority and cofunded by Structural Funds of the EU.

References

1. Bartneck, C., Kulić, D., Croft, E., Zoghbi, S.: Measurement instruments for the anthropomorphism, animacy, likeability, perceived intelligence, and perceived safety of robots. Int. J. Soc. Robot. **1**(1), 71–81 (2009). http://dx.doi.org/10.1007/s12369-008-0001-3
2. Buckwald, M., Holz, D.: Leap motion. Motion Control (2014). https://www.leapmotion.com/. Accessed 14 June 2016
3. Cramer, H., Goddijn, J., Wielinga, B., Evers, V.: Effects of (in)accurate empathy and situational valence on attitudes towards robots. In: Proceedings of the 5th ACM/IEEE International Conference on Human-Robot Interaction, pp. 141–142. IEEE (2010)
4. Inbar, O., Meyer, J.: Manners matter trust in robotic peacekeepers. In: Proceedings of the Human Factors and Ergonomics Society Annual Meeting, vol. 59, pp. 185–189. SAGE Publications (2015)
5. Lee, K.M., Peng, W., Jin, S.A., Yan, C.: Can robots manifest personality?: an empirical test of personality recognition, social responses, and social presence in human-robot interaction. J. Commun. **56**(4), 754–772. http://doi.wiley.com/10.1111/j.1460-2466.2006.00318.x
6. Leite, I., Pereira, A., Martinho, C., Paiva, A.: Are emotional robots more fun to play with?. In: RO-MAN 2008 - The 17th IEEE International Symposium on Robot and Human Interactive Communication, pp. 77–82. IEEE. http://ieeexplore.ieee.org/lpdocs/epic03/wrapper.htm?arnumber=4600646
7. Nomura, T., Saeki, K.: Effects of polite behaviors expressed by robots: a case study in japan. In: IEEE/WIC/ACM International Joint Conferences on Web Intelligence and Intelligent Agent Technologies, 2009, WI-IAT 2009, vol. 2, pp. 108–114, September 2009
8. Quigley, M., Conley, K., Gerkey, B.P., Faust, J., Foote, T., Leibs, J., Wheeler, R., Ng, A.Y.: Ros: an open-source robot operating system. In: ICRA Workshop on Open Source Software (2009)
9. Salem, M., Ziadee, M., Sakr, M.: Effects of politeness and interaction context on perception and experience of HRI. In: Herrmann, G., Pearson, M.J., Lenz, A., Bremner, P., Spiers, A., Leonards, U. (eds.) ICSR 2013. LNCS (LNAI), vol. 8239, pp. 531–541. Springer, Heidelberg (2013). doi:10.1007/978-3-319-02675-6_53
10. Salem, M., Ziadee, M., Sakr, M.: Marhaba, how may i help you? effects of politeness and culture on robot acceptance and anthropomorphization. In: Proceedings of the 2014 ACM/IEEE International Conference on Human-robot Interaction, HRI 2014, pp. 74–81. ACM, New York (2014)

11. Salichs, M.A., Encinar, I.P., Salichs, E., Castro-González, Á., Malfaz, M.: Study of scenarios and technical requirements of a social assistive robot for alzheimer's disease patients and their caregivers. Int. J. Soc. Robot. **8**(1), 85–102 (2016). http:// dx.doi.org/10.1007/s12369-015-0319-6
12. Short, E., Hart, J., Vu, M., Scassellati, B.: No fair!! an interaction with a cheating robot. In: 2010 5th ACM/IEEE International Conference on Human-Robot Interaction (HRI), pp. 219–226. IEEE (2010)
13. Złotowski, J.A., Sumioka, H., Nishio, S., Glas, D.F., Bartneck, C., Ishiguro, H.: Persistence of the uncanny valley: the influence of repeated interactions and a robot's attitude on its perception. Front. Psychol. **6**, 883 (2015). doi:10.3389/fpsyg.2015. 00883

Qualitative User Reactions to a Hand-Clapping Humanoid Robot

Naomi T. Fitter$^{(\boxtimes)}$ and Katherine J. Kuchenbecker

Haptics Group, GRASP Laboratory, Mechanical Engineering and Applied Mechanics,
University of Pennsylvania, Philadelphia, PA 19104, USA
{nfitter,kuchenbe}@seas.upenn.edu

Abstract. Playful interactions serve an important role in human development and interpersonal bonding. Accordingly, we believe future robots may need to know how to play games to connect with people in meaningful ways. To begin exploring how users perceive playful human-robot interaction, we conducted a study with 20 participants. Each user played simple hand-clapping games with the Rethink Robotics Baxter Research Robot during a one-hour-long session. Qualitative data collected from surveys and experiment recordings resoundingly demonstrate that this interaction is viable: all users successfully completed the experiment, all users enjoyed at least one game, and nineteen of the 20 users identified at least one potential personal use for Baxter. Hand-clapping tempo was highly salient to users, and human-like robot errors were more widely accepted than mechanical errors. These findings can motivate and guide roboticists who want to design social-physical human-robot interactions.

Keywords: Social robots · physical HRI · Social motor coordination

1 Introduction

Play using the hands is important for emotional connection in human development and human-human interaction [1]. Accordingly, as robots enter more human-populated environments and connect with people, we anticipate physical human-robot play to be an important future source of mutual human-robot learning and bonding. The varied kinesthetic and tactile aspects of these interactions result in a rich but complicated design space for human-robot interaction (HRI) researchers. Many previous social-physical HRI (spHRI) studies have gathered experiential data from human-robot interactions as part of an iterative design process, e.g., [2–4]. This paper similarly investigates user reactions to a preliminary robot prototype for playing hand-clapping games (Fig. 1). Here, hand-clapping refers to tempo-matching hand-to-hand contacts between two agents. Our previous work details how we created a safe and capable hand-clapping robotic system by developing hand-clapping trajectory models and hand-contact detection strategies for the Rethink Robotics Baxter Research Robot [5].

© Springer International Publishing AG 2016
A. Agah et al. (Eds.): ICSR 2016, LNAI 9979, pp. 317–327, 2016.
DOI: 10.1007/978-3-319-47437-3_31

Fig. 1. Left: Baxter's built-in finger alignment rails with regularly spaced threaded holes. Middle: Our fabricated Baxter hand with M4 × 0.7 screws compatible with the rails. Right: Baxter using its end-effector to interact with a study participant.

After discussing related work (Sect. 2), this paper describes how we tested our hand-clapping robot's abilities by conducting a user study (Sect. 3), identifying significant results (Sect. 4), and extracting overarching lessons (Sect. 5). We will leverage our findings to continue developing spHRI applications.

2   Related Work

Social Robotics, pHRI, and Their Intersection: Our work builds on two main areas: social robotics and physical human-robot interaction (pHRI). Combining these two fields creates the potential for emotional interaction and connection, as evidenced by several robots discussed in [6]. In its various possible application areas, spHRI can serve functions from enhancing mental healthcare treatment [7] to helping a robot more fully comprehend human intention [8]. Our work looks to leverage the advantages of spHRI through human-robot hand clapping.

Similar Studies of spHRI: Research on human-robot object handover tasks has illustrated that human users prefer minimum jerk trajectories along with other human-inspired robot motion controllers [9,10]. Human-robot handshake experiments have demonstrated ways to shape human-like robotic handshake algorithms [11] and leverage affective interaction design [12]. An exploration of robot touch in a medical setting demonstrated that people prefer functional touch to emotional touch in a mock hospital environment [13]. The results of these studies guided the design of our hand-clapping robotic system.

Analyses of Playful Interactions: Although we believe our work is the first research focused on human-robot hand-clapping games, some past studies explored similar human-human activities. For example, investigations of human-human social motor coordination consider synchronization in hand-clapping games [14]. Discussions of jointly improvised motion lend similar insights on human-human synchronization and motion dynamics [15]. Additionally, [16] proposes a high-five sensing wearable to increase bonding in workplace environments.

Essential HRI studies on other play applications also shaped our research. In educational environments, playful social robots have been shown to improve English language learning and help researchers estimate human friendship [2].

Researchers have also observed natural human-robot play for social machine-learning applications using a small humanoid [3]. Qualitative evaluations additionally helped researchers design affective understanding and reactions for the Haptic Creature [4]. These precedents indicate that beginning our hand-clapping HRI investigations with qualitative observation is a reasonable strategy that may increase the likelihood of meaningful results in future studies.

3   User Study

We conducted an experiment to begin exploring how human users perceive robotic hand-clapping playmates. The Penn IRB approved all experimental procedures under protocol 823886.

3.1   Robotic Platform

We conducted this experiment using Baxter, a human-sized humanoid robot designed for interactive factory tasks. Baxter has a torso with two 7DOF arms, interchangeable grippers, and a panning head screen. Due to its design for collaborative pHRI, Baxter possesses many sensors that make it an ideal candidate for spHRI: multiple cameras, joint torque sensing, and 360-degree sonar. The robot's series elastic actuators, impact-absorbing shells, and fully backdrivable joints add to its appeal for physically contacting people. Further advantages stem from Baxter's humanoid anatomy, standard Robot Operating System (ROS) framework, and relatively affordable price (∼\$25,000).

The commercially available Baxter end-effectors proved unsuitable for hand-clapping interactions, so we developed custom 3D-printed non-articulating hands with inlaid silicone rubber contact pads [5], as shown in Fig. 1. These end-effectors are average human-hand size to facilitate comfortable handclaps.

3.2   Experiment Setup

Twenty participants (ages 19–38, 12 male and 8 female) enrolled in our study, gave informed consent, and successfully completed the experiment. Each person came to the lab for a single session that lasted one hour. The participant stood facing Baxter throughout the experiment and engaged in repeated palm-to-palm contacts between their left hand and the robot's right end-effector, in the style of hand-clapping games. The left hand was used to make the task more challenging.

At the beginning of each session, the experimenter introduced Baxter to the participant and led them in a practice human-human round of the hand-to-hand contact involved in experiment trials. Next, in each of 24 randomly ordered trials, users clapped hands with Baxter at 60, 110, or 160 beats per minute (BPM). Each interaction lasted about 20 s. A final human-robot interaction trial allowed users to verbally select their favorite robot behavior mode and play with the robot in this mode for as long as desired.

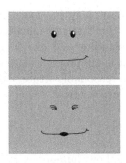 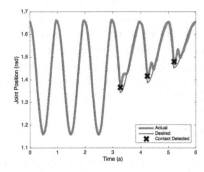

Fig. 2. Top left: The default mildly positive Baxter facial expression. Bottom left: The responsive facial expression used to animate the facially reactive robot. Right: Extreme illustrative example of physically nonreactive vs. physically reactive 60 Hz trajectories in Baxter W1 joint space. In this plot, the first three cycles of motion were recorded before a human user began contacting the robot, illustrating the ideal sinusoidal nonreactive trajectory. The last three cycles include hand contacts and the resulting trajectory back toward the retreat position (away from the human).

3.3   Conditions

This paper explores what people think about hand-clapping robots overall, and a future analysis will quantify how people perceive different robot behavior styles. Our previous observations of human-human hand-clapping interactions focused on *human emotional reactions* and *arm behaviors* [5]. Accordingly, to keep participants engaged, supply data for future analysis, and explore themes similar to the foci of our human-human observations, we designed different styles of Baxter *facial animation, arm trajectory*, and *arm control*. The following paragraphs detail how we designed each interaction mode, in a manner inspired by [12].

Facial Animation: Facial expressiveness can greatly affect the user's perception of a robot [17]. To begin exploring the infinite possibilities for animating Baxter's face screen, we designed facially reactive and facially nonreactive robot modes. Figure 2 illustrates the two expressions used to create the facially reactive robot. In this mode, the mildly positive face was the default screen image and the responsive face appeared for 0.2 s after each detected hand contact. We identified hand impacts by thresholding the high-frequency acceleration of the accelerometer in the robot's right wrist [5]. In facially nonreactive mode, the robot's screen remained on the mildly positive image at all times.

Trajectory Variation: We also varied the physical arm trajectory logic. Our previous investigation informed the employed strategies of robot movement and contact detection [5]. We found human motion to be generally sinusoidal and designed our robot to move with this same default behavior, as depicted in the first three seconds of the plot in Fig. 2. Our previous work revealed a negative correlation between hand-clapping frequency and amplitude [5], which we used to determine amplitude values for each clapping frequency. The motion equation for the nonreactive trajectory of the robot's only active joint (W1) was as follows, using variables for desired joint angle (θ_d), amplitude of motion (A), frequency of hand contacts in Hz (f), time (t), initial joint angle (θ_{init}), and a factor to keep handclap location the same regardless of tempo (a_{comp}):

$$\theta_d = A \sin(2\pi f t + (\pi/2)) - A + \theta_{\text{init}} + a_{\text{comp}} \tag{1}$$

In contrast, the robot's reactive trajectory mode varied dynamically. Whenever Baxter detected a handclap in this mode, its governing algorithm fit a cubic polynomial trajectory back to the extreme retreat location. The start and end points of the trajectory were known because the start was simply the position and time of contact detection and the end was the farthest retreat location at a time calculated in order to maintain a constant robot hand-clapping frequency; the start and end velocities were zero because they corresponded with direction changes. After achieving the retreat position, the robot returned to a sinusoidal trajectory until the next handclap. Sample results of this physically reactive motion trajectory appear in the later portion of the plot in Fig. 2.

Stiffness: To keep users engaged with a diversity of interaction experiences, we varied the stiffness of Baxter's arm by using different proportional gains. As developed previously [5], Baxter's time-domain control law is:

$$\tau_{\text{cmd}} = \mathbf{K}_d(\dot{\theta}_d - \dot{\theta}) + \mathbf{K}_p(\theta_d - \theta) - \mathbf{K}_f \theta_d + \tau_{\text{gc}} \tag{2}$$

where τ_{cmd} is a vector of torques commanded to each Baxter arm motor, $\mathbf{K}_d$ is a diagonal matrix of derivative gains, θ_d is a vector of desired arm joint angles, θ is a vector of actual joint angles, $\mathbf{K}_p$ is a diagonal matrix of proportional gains, $\mathbf{K}_f$ is a diagonal matrix of feedforward gains, and τ_{gc} is a vector of gravity compensation torques.

To maintain a consistent presented motion trajectory regardless of the trial conditions, we always used the same proportional gain $(30\frac{\text{Nm}}{\text{rad}})$ for the active W1 joint. For all other arm joints, we selected a lower proportional gain $(15\frac{\text{Nm}}{\text{rad}})$ to accomplish more compliant passive joint behavior and a different higher proportional gain $(60\frac{\text{Nm}}{\text{rad}})$ for a more stiff passive joint behavior. In the equation above, $\mathbf{K}_p$ is the element that changes depending on the stiffness control mode.

3.4 Data Collection

During every trial, we recorded data from the robot's accelerometer, endpoint state, joint state, and face display ROS topics. More importantly, participants completed several surveys: (1) a robot evaluation survey after hearing introductory Baxter information, (2) a hand-clapping game evaluation survey after learning the game, (3) a subjective experience survey after each 20-s trial, (4) a concluding survey after the final unlimited interaction trial, and (5) a basic demographic survey. The first two surveys involved only slider-type parametric questions. The trial and concluding surveys contained a combination of slider and extended response questions. The experiment was also videotaped to enable post hoc review of the participants' reactions.

Our surveys were carefully designed based on precedents in HRI research. Questionnaires (1), (2), and (4) were adapted from the Unified Theory of Acceptance and Use of Technology (UTAUT) and other metrics employed in [19]. The subjective trial surveys leveraged parametric versions of the questions used in [12] to evaluate robot engagingness based on the PAD (Pleasure, Arousal, Dominance) emotional state model [18], and they also assessed perceived robot safety with a parametric slider question as well as general reasoning for slider responses in a free response field.

Fig. 3. Adjectives used by participants in describing the hand-clapping experience, grouped by synonym. The size of the word reflects the frequency of use.

4 Results

This Section focuses on analyzing the essay responses to each Baxter experiment trial and other experience-related metrics. Because this paper represents an initial inquiry into the experiential results, it focuses on participant reactions to the experiment as a whole. Future work will analyze the additional collected survey data, recorded ROS topic data, and video recordings.

Promising Results: Users supplied us with a wealth of descriptive responses to the randomly ordered interaction trials. The word cloud in Fig. 3 presents all adjectives used by participants in the trial survey free response field. Since each participant expressed a clear like and dislike of different trials, we find a balance of positive and negative descriptors in the word cloud. It is important to note that the large word "slow" always referred to the trial tempo, never the robot motion or responsiveness capabilities. The free response feedback seems promising; every user identified at least one interaction mode they enjoyed, as well as some that they did not like. Their preferences were not uniform. Some polarized opinions indicated that individualized interaction models may work best for distinct interactees, especially tempo-wise.

Other positive notes include the fact that eight of the 20 study participants never perceived Baxter to have made any error. In the final free interaction trial, users interacted with Baxter for 18 or more seconds (M = 43.95 s, SD = 29.17 s). This free interaction duration is approximately equal to or greater than the fixed length trial interactions. The customized behavior preferences of users were as follows: 20/20 selected facially animated, 15/20 selected variable trajectory, and 9/20 selected high stiffness. For tempo, 1 selected 60 BPM, 14 selected 110 BPM, and 5 selected 160 BPM. This variety of choices indicates that we aptly designed diverse robot interaction behaviors. Users frequently remarked positively on Baxter's facial reactivity in survey responses and verbal commentary.

Design Shortcomings: Participants encountered some challenges and errors during the 500 total trials of the experiment, as summarized in Table 1. In the fastest-tempo physically reactive mode, seven participants responded to Baxter's

Table 1. Excerpts of each extended response entry linked to each observed type of experiment error. The plus and minus signs indicate general positive or negative tone.

Driving robot to stop	False positives
+ I was trying to out-dominate [Baxter]	+ [Baxter] responded quickly
+ [We eventually found] a stable cycle	to making contact
+ That felt the most natural so far	+ I got the impression [Baxter]
+ [Baxter] appeared dominant	was setting the pace
- Responds too weakly to contact	+ Good responsiveness, felt natural
- It made me feel unsure of [Baxter]	+ Tempo was reasonable, [Baxter]
- [I felt] slightly less safe than before	was not unnecessarily forceful
- The game was less pleasing	- The hand clapping felt odd
- [This trial was] somewhat disconcerting	- The robot's hand was moving
- Moving range of robot's hand too short	up and down a lot
- A little too fast	
- A bit too fast to be pleasing	
"Jazz hands"-like oscillations	Human error
+ At this speed, it feels energetic	+ Hard to keep up, but still fun
- Something a little off - can't describe it	+ [Robot response] feels good
- [I had] the sense there was a malfunction	+ It works well
- I felt that something had gone wrong	+ [Baxter] adapted to the way I clap
- Shakiness was concerning and disruptive	+ [That made] a real clapping sound
- Palm-palm contact too long	- No consistent frequency of claps
- Not a fun rhythmic communication	- I didn't get the pattern
- Motor started making disconcerting noise	- Pattern doesn't repeat
- I think I liked this one the least so far	- [Baxter] looked uncomfortable
- [Baxter] seems afraid of clapping with me	- The tempo was pretty quick
- Trial was a little odd, hurt a little	- Tempo was faster that comfortable
- Felt unnatural for a clap	- [Had to] bend my wrist uncomfortably
- Tempo got off, seemed schizophrenic	- [The hand] will jiggle sometimes
- [Robot] wrist vibrating more than usual	- It feels like I am hitting something
- [I] feel like the robot might not be safe	- A little too hard to keep [the] pace
	- I thought the game was over, but
	then [Baxter] moved

retreat reaction by contacting the robot's end-effector earlier and earlier in its motion cycle (12 total trials). Because the retreat motion was capped at the far retreat position, this user strategy resulted in a decrease in robot motion amplitude and, in some extreme cases, momentary periods of Baxter stillness. Next, thresholding the accelerometer data produced six total false positive hand contact predictions. A final problem stemmed from controller gains; five participants were able to exert enough axial torque on Baxter's end-effector to cause surprising, "jazz hands"-like oscillations in Baxter's wrist roll (W2) joint motion (15 total trials). Additionally, human users made sixteen errors throughout the experiment, namely failing to match Baxter's fastest clapping frequency or misunderstanding Baxter's slower frequencies.

As shown in Table 1, reactions to different error types varied. Users often reported that the robot stopping and false positive errors felt natural, as they are humanlike mistakes. The "jazz hands" error, though, is a mechanical feature that seemed unusual and almost unanimously problematic. The various human errors received mixed responses and may be avoided with better experiment design.

Overall Impressions: Despite these occasional errors, participants' opinions of Baxter did not change significantly in a positive or negative way over the course of the experiment, as illustrated in Fig. 4. T-tests reveal no significant difference between each survey question's pair of before and after responses (all $p > 0.23$). We interpret this consistency to signify that although this version of our hand-

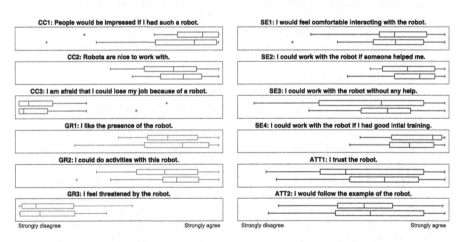

Fig. 4. Survey responses to analogous pairings of robot-related questions on the robot evaluation and concluding survey. In each plot, the top box plot represents the participant responses to the question on the robot evaluation survey and the bottom box plot represents the responses on the concluding survey. The center box line represents the median and the box edges are the 25th and 75th percentiles. The question coding abbreviations stand for cultural context (CC), forms of grouping (GR), self efficacy from UTAUT model (SE), and attachment (ATT).

clapping robot made some errors, it is worth continuing to develop improved system behaviors.

We also see some consensus in positive and negative remarks in the concluding survey essay responses. Table 2 illustrates a balanced set of experiment feedback, including praise of interaction modes alongside critiques indicating that we need more customizability in the robot behaviors. In a final essay question, nineteen of the 20 users identified a personal interest in interacting with Baxter in some way. Use ideas included experiment-like tasks such as playing more complicated hand-clapping games, collaboratively manipulating objects, and doing arm exercises; chore-like tasks such as cooking, washing dishes, doing laundry, and cleaning bathrooms; and social tasks like performing music/dance, playing sports/board games, drinking beer, and socializing.

Table 2. Responses occurring at least twice in concluding survey essay responses. The numbers indicate response frequency.

Positive/enjoyable aspects	Negative/difficult aspects
Getting to engage with Baxter (9)	Interaction too long for healthy adults (6)
Baxter's facial expressions (6)	Hard to get tempo right in fast trials (6)
Scientific aspects of experience (6)	Any robot errors/inconsistencies (4)
Quick and responsive robot behaviors (4)	Robot too short (4)
Getting to choose final trial conditions (3)	Some settings uncomfortable (3)
Predictable robot behaviors (2)	Interaction needs more motion variety (3)
Exploring different interaction modes (2)	Hand got sore/tired (3)
Fun/bonding aspects of interaction (2)	Baxter not as skilled as a human (2)

5 Conclusions and Future Work

This analysis of human-robot hand-clapping experiences focused on participant essay responses and other engagement metrics. General responses indicated that clapping hands with a robot is a viable social-physical interaction; all participants identified at least one interaction trial that they liked, and all but one participant could imagine personal Baxter uses. Participants also provided many helpful comments that will enable us to improve our prototype. Users noticed tempo more acutely than other experimental variables. Error-wise, humanlike errors were more likely to be accepted than mechanical errors.

Our next research steps will be to update our interaction controller to improve Baxter's motion. Specifically, we intend to modify robot reactive trajectories, tune accelerometer thresholds, and redesign the end-effector to minimize axial torque. This initial human-robot investigation will inform future work on our hand-clapping robotic system in various playful HRI settings and should help other researchers interested in socially relevant pHRI.

Acknowledgments. The first author was supported by a National Science Foundation (NSF) Graduate Research Fellowship under Grant No. DGE-0822 and the University of Pennsylvania's NSF Integrative Graduate Education and Research Traineeship under Grant No. 0966142. We thank Michelle Neuburger for her help with this work; she was supported by NSF Grant No. 1156366. We thank Alex Burka for his design advice and Kostas Daniilidis for access to Baxter.

References

1. Sonneveld, M.H., Schifferstein, H.N.: The tactual experience of objects. In: Schifferstein, H.N.J., Hekkert, P. (eds.) Product Experience, pp. 41–67. Elsevier, Amsterdam (2008)
2. Kanda, T., Sato, R., Saiwaki, N., Ishiguro, H.: A two-month field trial in an elementary school for long-term human-robot interaction. IEEE Trans. Robot. **23**(5), 962–971 (2007)
3. Cooney, M.D., Becker-Asano, C., Kanda, T., Alissandrakis, A., Ishiguro, H.: Full-body gesture recognition using inertial sensors for playful interaction with small humanoid robot. In: IEEE/RSJ International Conference on Intelligent Robots and Systems (IROS), pp. 2276–2282 (2010)
4. Yohanan, S., MacLean, K.E.: The role of affective touch in human-robot interaction: human intent and expectations in touching the Haptic Creature. Int. J. Soc. Robot. **4**(2), 163–180 (2012)
5. Fitter, N.T., Kuchenbecker, K.J.: Equipping the Baxter robot with human-inspired hand-clapping skills. Accepted to the IEEE International Symposium on Robot and Human Interactive Communication (RO-MAN) (2016)
6. Argall, B.D., Billard, A.G.: A survey of tactile human-robot interactions. Robot. Auton. Syst. **58**(10), 1159–1176 (2010)
7. Rabbitt, S.M., Kazdin, A.E., Scassellati, B.: Integrating socially assistive robotics into mental healthcare interventions: applications and recommendations for expanded use. Clin. Psychol. Rev. **35**, 35–46 (2015)
8. Iwata, H., Sugano, S.: Human-robot-contact-state identification based on tactile recognition. IEEE Trans. Industr. Electron. **52**(6), 1468–1477 (2005)
9. Huber, M., Rickert, M., Knoll, A., Brandt, T., Glasauer, S.: Human-robot interaction in handing-over tasks. In: IEEE International Symposium on Robot and Human Interactive Communication (RO-MAN), pp. 107–112 (2008)
10. Chan, W.P., Parker, C.A., Van der Loos, H.M., Croft, E.A.: A human-inspired object handover controller. Int. J. Robot. Res. **32**(8), 971–983 (2013)
11. Avraham, G., Nisky, I., Fernandes, H.L., Acuna, D.E., Kording, K.P., Loeb, G.E., Karniel, A.: Toward perceiving robots as humans: three handshake models face the Turing-like handshake test. IEEE Trans. Haptics **5**(3), 196–207 (2012)
12. Ammi, M., Demulier, V., Caillou, S., Gaffary, Y., Tsalamlal, Y., Martin, J.-C., Tapus, A.: Haptic human-robot affective interaction in a handshaking social protocol. In: ACM/IEEE International Conf. on Human-Robot Interaction, pp. 263–270 (2105)
13. Chen, T.L., King, C.-H., Thomaz, A.L., Kemp, C.C.: Touched by a robot: an investigation of subjective responses to robot-initiated touch. In: ACM/IEEE International Conference on Human-Robot Interaction (HRI), pp. 457–464 (2011)
14. Schmidt, R., Fitzpatrick, P., Caron, R., Mergeche, J.: Understanding social motor coordination. Hum. Mov. Sci. **30**(5), 834–845 (2011)

15. Noy, L., Dekel, E., Alon, U.: The mirror game as a paradigm for studying the dynamics of two people improvising motion together. Nat. Acad. Sci. **108**(52), 20947–20952 (2011)
16. Kim, Y., Lee, S., Hwang, I., Ro, H., Lee, Y., Moon, M., Song, J.: High5: Promoting interpersonal hand-to-hand touch for vibrant workplace with electrodermal sensor watches. In: ACM International Joint Conference on Pervasive and Ubiquitous Computing, pp. 15–19 (2014)
17. Fitter, N.T., Kuchenbecker, K.J.: Designing and assessing expressive open-source faces for the baxter robot. Submitted to the International Conference on Social Robotics (ICSR) (2016)
18. Mehrabian, A.: Basic dimensions for a general psychological theory. Oelgeschlager, Gunn & Hain (1980)
19. Heerink, M., Krose, B., Evers, V., Wielinga, B.: Measuring acceptance of an assistive social robot: A suggested toolkit. In: IEEE International Symposium on Robot and Human Interactive Communication (RO-MAN), pp. 528–533 (2009)

Nonlinear Controller of Arachnid Mechanism Based on *Theo Jansen*

Víctor H. Andaluz[1,2]($\boxtimes$), David Pérez[1], Darwin Sáchez[1],
Cristina Bucay[1], Carlos Sáchez[1], Vicente Morales[1], and David Rivas[2]

[1] Universidad Técnica de Ambato, Ambato, Ecuador
{jperez2788,dsanchez9092,cbucay1357,carloshsanchez,
jvmorales99}@uta.edu.ec
[2] Universidad de Las Fuerzas Armadas ESPE, Sangolquí, Ecuador
{vhandaluz1,drrivas}@espe.edu.ec

Abstract. This paper presents a new motion controller for arachnid mechanism based on *Theo Jansen* that is capable of performing path-following tasks. The proposed controller has the advantage of simultaneously performing the approximation of the arachnid robot to the proposed path by the shortest route and limiting its velocity. Furthermore, it is presents the kinematic modeling of the arachnid mechanism where it is considered that its mass center is located at the legs' axis center of the robot. In addition, the stability is proven through Lyapunov's method. To validate the proposed control algorithm, experimental results are included and discussed.

Keywords: Theo Jansen · Path-following · Coupled mechanism · Arachnid

1 Introduction

In recent years, robotics research has experienced a significant change. Research interests are moving from the development of robots for structured industrial environments to the development of autonomous mobile robots operating in unstructured and natural environments [1–5]. It is common to speak of service robots for different applications that interact with man in partially structured or unstructured environments, *e.g.*, traffic monitoring, obstacles detection [1], teleoperation systems [2, 3]. The autonomous movement of robots has been achieved by implementing advanced control algorithms based on neural networks, predictive control, Neural Networks, among others; for the implementation of these control techniques it is essential that both the software and hardware provide the necessary data processing capabilities [3–5].

You need for complex tasks that robots have both skills of locomotion and manipulation; the same can be mainly oriented to applications in workspaces in air, land and water. Ground robots are classified into robots with wheels, legs and caterpillars [6]. The wheeled robots have several advantages such as: it requires fewer actuators and are therefore easier to control, *e.g.*, a hexapod with three degrees of freedom per leg needs actuators 18, while an omnidirectional robot four actuators. Also, wheeled robots are distinguished by the arrangement of the wheels over the mechanical structure [6, 7].

© Springer International Publishing AG 2016
A. Agah et al. (Eds.): ICSR 2016, LNAI 9979, pp. 328–339, 2016.
DOI: 10.1007/978-3-319-47437-3_32

While the robots with legged are best suited to move in irregular environments, these robots can be classified into (*i*) *robots with uncoupled joints* they are those that each joint works independently, *i.e.*, has an actuator for each joint, *e.g.*, robotic arm, bipedal, hexapod [3, 9–12] and (*ii*) *robots with coupled joints* are they are those that combine mechanisms for movement of a joint, and therefore the movement of the robot. Within this group are the robots that using the mechanism proposed by *Theo Jansen*, where movement of a joint, called input joint receive the impulse, or the necessary speed for the rest of joints acting simultaneously.

Referring to the literature review of robots based in *Theo Jansen* mechanisms can be classified into three groups: (*i*) *building mechanisms* to simulate the movement of animals of more than 4 feet, *e.g.*, spiders, lions, horses, among others. The Building of mechanisms focuses on the type of used materials [13]; (*ii*) *analysis of the mechanism*, there are studies that made kinematic analysis based on the method of projection of different mechanisms built without considering the mass of links nor the external or internal forces that interact with the system [14, 15]. While from the dynamic point of view of energy balance methods is used to position the mechanism [16, 17], it is presents a position control strategy for Jansen walking robot derived using projection method for which energy based control forms the core, numerical simulations are done to validate the controller; (*iii*) *control of mechanism*, control strategies used is based on MIMO systems, *i.e.*, each actuator is considered as a system independently [18]. The problem of these control strategies is based on conditional statements, which does not allow the proper performance of the robot to perform the task. In [19] is presented a 4 legs robot that performs a task of chalking on a court, the task is performed based on conditional functions both to move straight like to rotate, plus the speed of the robot is considered constant. In [20] is showed the displacement of a lion based on the natural movement of the same, but control is done with conditional statements.

In such context, this paper presets the obtained kinematic model for arachnid mechanism based on *Theo Jansen* with an adequate structure and properties for designing a control law and the control input can be given in terms of linear and angular reference velocities, as usually found in commercial mobile robots. This latter characteristic is an advantage when evaluating the control experimentally. Furthermore, in this work it is proposed a new method to solve the path following problem for an arachnid mechanism based on *Theo Jansen*. The proposed control scheme is divided into two subsystems, each one being a controller itself: the first on is a (*i*) *kinematic controller* with saturation of velocity commands, which is based on the arachnid robot's kinematic. The path following problem is addressed in this subsystem. It is worth noting that the proposed controller does not consider $s(t)$ as an additional control input as it is frequent in literature; and (*ii*) *an dynamic compensation* controller that considered the system dynamic, which are directly related to physical parameters of the robot. In addition, the stability is proven through Lyapunov's method. To validate the proposed control algorithm, experimental results are included and discussed.

This paper is divided into 6 Sections including the Introduction. In Sect. 2 the problem formulation is formulated. Next in Sect. 3 the kinematic modeling of the robot based on arachnid mechanism the *Theo Jansen* is presented. While in Sect. 4 presents the path following controller design. In Sect. 5 the experimental results are presented and discussed. Finally, the conclusions are given in Sect. 6.

2 Problem Formulation

As represented in Fig. 1, the path to be followed is denoted as $P(s)$, where $P(s) = (x_P(s), y_P(s))$; the actual desired location $P_d = (x_P(s_D), y_P(s_D))$ is defined as the closest point on $P(s)$ to the arachnid robot, with s_D being the curvilinear abscissa defining the point P_d; the unit vector tangent to the path in the point P_d is denoted by $\mathbf{T}$; θ_T is the orientation of $\mathbf{T}$ with respect to the inertial frame $R(X,Y,Z)$; $\tilde{x} = x_P(s_D) - x$ is the position error in the X direction; $\tilde{y} = y_P(s_D) - y$ is the position error in the Y direction; ρ represents the distance between the arachnid robot position $h(x, y)$ and the desired point P_d, where the position error in the ρ direction is $\tilde{\rho} = 0 - \rho = -\rho$, *i.e.*, the desired distance between the robot position $h(x, y)$ and the desired point P_d must be zero; and θ_ρ is the orientation of the error $\tilde{\rho}$ with respect to the inertial frame $R(X,Y,Z)$.

The path-following problem is solved by a control law capable of making the point of interest to assume a desired velocity equal to

$$V = \boldsymbol{v}_P(s_D, h) = |\boldsymbol{v}_P(s_D, h)| \angle \theta_T \tag{1}$$

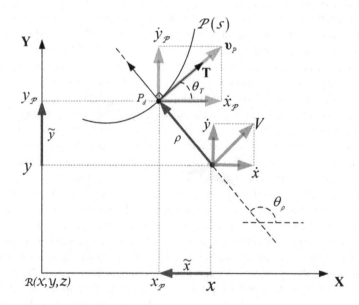

Fig. 1. The orthogonal projection of the point of interest over the path.

besides making the robot to stay on the path, that is, $\tilde{x} = 0$ and $\tilde{y} = 0\triangleright$. Therefore, if $\lim_{t\to\infty} \tilde{x}(t) = \mathbf{0}$ and $\lim_{t\to\infty} \tilde{y}(t) = \mathbf{0}$, then $\lim_{t\to\infty} \rho(t) = 0$ and $\lim_{t\to\infty} \tilde{\psi}(t) = 0$, being $\tilde{\psi}$ the orientation error of the robot, defined as $\tilde{\psi} = \theta_T - \psi$.

Worth noting that the reference desired velocity $\upsilon_P(s_D, h)$ of the robot during the tracking path need not be constant, with is common in the literature [1, 3, 13–16],

$$\upsilon_P(s_D, h) = f(k, s_D, \rho(t), \omega(t), \ldots) \tag{2}$$

the arachnid robot's desired velocity can be expressed as: constant function, curvilinear abscissa function of the path, position error function, angular velocities function of the robot; and the others consideration.

3 Kinematic Modeling

The movement of the mechanisms based on the principles of Theo Jansen are aimed emulate the movement of the legs of different types of animals; this mechanism consists of several legs and links, the configuration of the mechanism will depend exclusively on the type of movement being emulated. The Fig. 2 shows a part of the mechanism used in this work, the same is composed of three pairs of legs located on

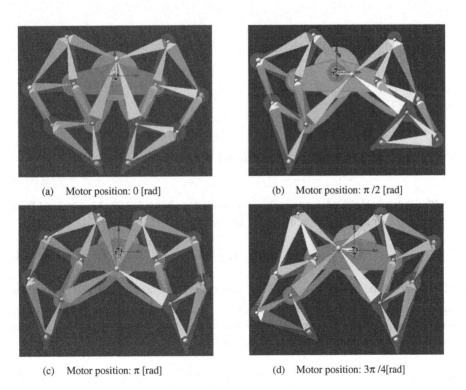

(a)   Motor position: 0 [rad]   (b)   Motor position: $\pi/2$ [rad]

(c)   Motor position: π [rad]   (d)   Motor position: $3\pi/4$[rad]

Fig. 2. Coupled motion of a mechanism of *Theo Jansen*

each side of the structure to out of phase to $2\pi/3$ [rad], and each leg having seven links. Movement control mechanism is through two DC motors, each motor allows three pairs of legs move so synchronized.

The analyzing of the mechanism is performed in based on one of the legs forming the arachnid robot, as illustrated in Fig. 3.

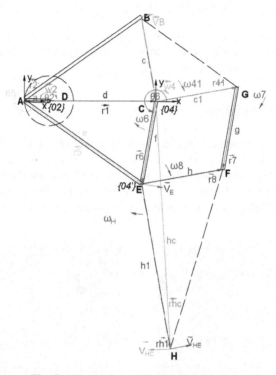

Fig. 3. Kinematic model of the robot's legs

The linear velocity of the operating end of the leg shown in Fig. 3 is obtained from the kinematic analysis of closed loop of the vertexes *CEH*.

$$\overrightarrow{\mathbf{rhc}} = \overrightarrow{\mathbf{r_6}} + \overrightarrow{\mathbf{rh_1}} \tag{3}$$

where, $\overrightarrow{\mathbf{rhc}}$ is the position vector of *hc* link to the reference system {O4}; $\overrightarrow{\mathbf{r_6}}$ is the vector position of link *f* regarding to {O4}; and $\overrightarrow{\mathbf{rh_1}}$ is the vector position of h_1 linkregarding to {O4}. By applying Euler's theorem [xxx], the position of the link h_c is obtained in the complex plane (real-imaginary).

$$\overrightarrow{\mathbf{rhc}} = fe^{i\theta_6} + h_1 e^{i\theta_H}$$

$$\overrightarrow{\mathbf{rhc}} = f(\cos(\theta_6) + i\sin(\theta_6)) + h_1(\cos(\theta_H) + i\sin(\theta_H)) \tag{4}$$

where, θ_6 is the angle that has the link f regard to $\{O4\}$, and θ_H is the angle of the h_1 link with respect to $\{O4'\}$. Now, deriving (4) with respect to time, the linear velocity at the operating extreme H is obtained.

$$\overrightarrow{\mathbf{V_{HC}}} = f\omega_6 e^{i\theta_6} i + h_1 \omega_H e^{i\theta_H} i$$

$$\omega_8 = \omega_H$$

$$\overrightarrow{\mathbf{V_{HC}}} = f\omega_6(-\sin(\theta_6) + i\cos(\theta_6)) + h_1 \omega_8(-\sin(\theta_H) + i\cos(\theta_H)) \tag{5}$$

where, ω_6 is the angular velocity of f link; and ω_8 the angular velocity of h_1 link.

Remark 1. Consider that $\omega_8 = \omega_H$ because the two angular velocities belong to the same link, as as shown in Fig. 3. In appendixes the equivalence for each variable is detailed.

Remark 2. The speed of the three pairs of legs on the left side are equal, and depend solely on the angular speed of the motor that controls them. The linear velocity of the three pairs of legs to out of phase to $2\pi/3$[rad] is obtained from (5). Similarly the linear velocity of the legs of the right side of the arachnid robot is obtained.

According to information from previous paragraphs, the arachnid mechanism described in this paper can be represented as a type mechanism unicycle. An unicycle mechanism is a driving robot that can rotate freely around its axis. The term unicycle is often used in robotics to mean a generalized cart or car moving in a two-dimensional world; these are also often called unicycle-like or unicycle-type. Hence, the lineal velocity u and angular velocity ω of the arachnid robot, are given by:

$$u = \frac{u_r + u_l}{2} \text{ and } \omega = \frac{u_r - u_l}{d} \tag{6}$$

where, u_r and u_l are the lineal velocities of the right and left legs, respectively; and d is the distance between right and left legs of the robot, view Fig. 4.

The configuration instantaneous kinematic model of the holonomic arachnid robot is defined as,

$$\begin{cases} \dot{x} = u\cos\psi - a\omega\sin\psi \\ \dot{y} = u\sin\psi + a\omega\cos\psi \\ \dot{\psi} = \omega \end{cases} \tag{7}$$

also the equation system (7) can be written in compact form as

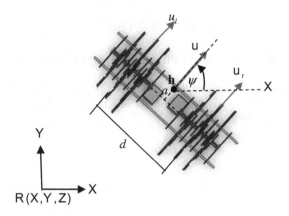

Fig. 4. Kinematic of the robot's legs

$$\dot{\mathbf{h}} = J(\psi)\mathbf{v}$$
$$\dot{\psi} = \omega \tag{8}$$

where $\dot{\mathbf{h}} = [\dot{x} \;\; \dot{y}]^T \in \Re^2$ represents the vector of axis velocity of the R(X,Y,Z) system; $\mathbf{J}(\psi) = \begin{bmatrix} \cos\psi & -a\sin\psi \\ \sin\psi & a\cos\psi \end{bmatrix} \in \Re^{2x2}$ is a singular matrix; and the control of maneuverability of the robot is defined $\mathbf{v} \in \Re^n$ and $\mathbf{v} = [u \;\; \omega]^T \in \Re^2$ in which u and ω represent the linear and angular velocities of the robot, respectively.

On the other side, of (7) is determined the non-holonomic velocity constraint of the arachnid robot which determines that it can only move perpendicular to the legs axis,

$$\dot{x}\sin\psi - \dot{y}\cos\psi + a\omega = 0 \tag{9}$$

4 Path Following Controller Design

The proposed kinematic controller is based on the kinematic model of the robot (8), *i.e.*, $\dot{\mathbf{h}} = f(\psi)\mathbf{v}$. Hence following control law is proposed,

$$\begin{bmatrix} u_c \\ \omega_c \end{bmatrix} = \mathbf{J}^{-1}\left(\begin{bmatrix} \dot{x}_p \\ \dot{y}_p \end{bmatrix} + \begin{bmatrix} \rho_x \\ \rho_y \end{bmatrix} \right) \tag{10}$$

with

$$\dot{x}_P = |\upsilon_P|\cos(\theta_T) \text{ and } \dot{y}_P = |\upsilon_P|\sin(\theta_T) \tag{11}$$

where u_c and ω_c are the velocities outputs of the kinematic controller, υ_P is the reference velocity input of the robot for the controller, $\dot{x}_p$ is the projection of υ_P in the X direction, $\dot{y}_p$ is the projection of υ_P in the Y direction, $\mathbf{J}^{-1}$ is the matrix of inverse kinematics for the robot, and ρ_x and ρ_y are the position error in the X and Y direction, respectively, respect to the inertial frame R(X,Y,Z), In order to include an analytical saturation of velocities in the robot, the **tanh(.)** function, which limits the errors ρ_x and ρ_y, is proposed. Hence it is defined as,

$$\rho_x = l_x \tanh\left(\frac{k_x}{l_x}\tilde{x}\right) \text{ and } \rho_y = l_y \tanh\left(\frac{k_y}{l_y}\tilde{y}\right) \tag{12}$$

Now, the behaviour of the control position error of the robot is now analysed assuming –by now– perfect velocity tracking *i.e.,* $u(t) \equiv u_c(t)$ and $\omega(t) \equiv \omega_c(t)$ and Hence manipulating (8) and (10), is can be written the behavior of the velocity of the point of interest of the robot for the closed-loop system, that is given by

$$\begin{bmatrix} \dot{x} \\ \dot{y} \end{bmatrix} = \begin{bmatrix} \dot{x}_p \\ \dot{y}_p \end{bmatrix} + \begin{bmatrix} l_x \tanh\left(\frac{k_x}{l_x}\tilde{x}\right) \\ l_y \tanh\left(\frac{k_y}{l_y}\tilde{y}\right) \end{bmatrix} \tag{13}$$

The analysis of the stability of the closed-loop system is represented in [21]: hence, it can now be concluded that $\lim\limits_{t \to \infty} \tilde{\rho}(t) \to 0$, *i.e.,* $\tilde{x}(t) \to 0$ and $\tilde{y}(t) \to 0$ with $t \to \infty$ asymptotically. Therefore, from (13) it can be concluded that the final velocity of the point of interest will be $V = |\upsilon_P(s_D, h)| \angle \theta_T$ hence $\tilde{\psi}(t) \to 0$ for $t \to \infty$ asymptotically.

5 Experimental Results

The software Automation Mechanical Design SolidWorks is a design tool parametric solid modeling based on operations that leverages the ease of learning the graphical interface. This tool can create 3D solid models fully associative with or without restrictions, while simultaneously using automatic reactions or user defined to capture design intent. A geometric solid model contains all surface geometry and wireframe necessary to describe in detail the edges and model faces [22]. For this work, the design of a mechanism of *Theo Jansen* is considered. The mechanism consists of two servomotors Dynamixel AX-12A [23], which have the ability to feedback to the control unit: speed, temperature, position, torque, to be controlled independently according to control criteria implemented. The design has performed will have three pairs of legs located on each side of the structure to out of phase to $2\pi/3$[rad], and each leg having seven links. Movement control mechanism is through two DC motors, each motor allows three pairs of legs move so synchronized, view in Fig. 5.

In order to illustrate the performance of the proposed controller, several experiments were carried out for path following control of the mechanism based on *Theo*

(a) Arachnid Robot physical construction

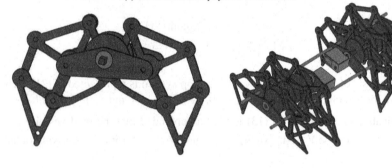

(b) Extremity assembled of arachnid robot   (c) Arachnid robot assembly on SoliWork

Fig. 5. Mechanism of Theo Jansen

Jansen; the most representative results are presents. The experiment corresponds to the performance of the proposed controller. Note that for the path following problem the desired velocity of the root will depend on the task, the control error, the angular velocity, etc. For this case, it is consider that the reference velocity depends on the control errors, the angular velocity. Figures 6, 7 and 8 show the results of the experiment. Figure 6 shows the stroboscopic movement respect to the inertial frame R(X,Y,Z). It can be seen that the proposed controller works correctly; while Fig. 7 shows that $\tilde{\rho}(t)$ is ultimately bounded close to zero; finally the Fig. 8 present the linear and angular velocities of the arachnid robot.

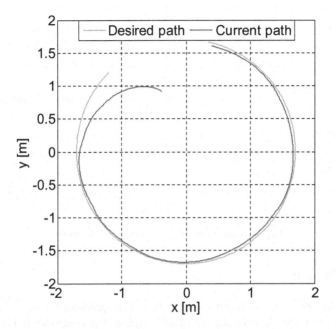

Fig. 6. Movement of the robot in the path following experiment.

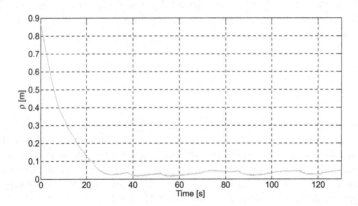

Fig. 7. Distance between the arachnid robot position and the closest point on the path

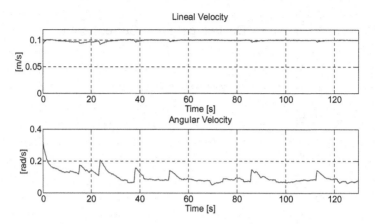

Fig. 8. Linear and angular velocity of the robot

6 Conclusions

The controller proposed resolved the path following problem for an arachnid robot based on mechanism of *Theo Jansen*. The design of the controller is based on two cascaded subsystems: a kinematic controller which complies with the motion control objective, and a robot-inner-loop system to independently track of the velocity commands of the robot; the kinematic controller have been designed to prevent from command saturation. Stability was proved by the Lyapunov's method. The real experiments confirm the capability of the path following controller to solve different motion problems by an adequate selection of the control references.

References

1. Haosong, Y., Weihai, C., Xingming, W., Jingbing, Z.: Kinect based real time obstacle detection for legged robots in complex environments. In: IEEE International Conference on Industrial Electronics and Applications (ICIEA), pp. 205–210, Melbourne (2013)
2. Andaluz, V.H., et al.: Unity3D virtual animation of robots with coupled and uncoupled mechanism. In: Paolis, L.T., Mongelli, A. (eds.) AVR 2016. LNCS, vol. 9768, pp. 89–101. Springer, Heidelberg (2016). doi:10.1007/978-3-319-40621-3_6
3. YuKang, L., YuMing, Z., Bo, F., Ruigang, Y.: Predictive control for robot arm teleoperation. In: IEEE International Conference on Industrial Electronics Society IECON, pp. 3693–3698, Viena (2013)
4. Esmaili, P., Haron, H.: Adaptive synchronous artificial neural network based PI-type sliding mode control on two robot manipulators. IEEE International Conference on Computer, Communication, and Control Technology (I4CT 2015), pp. 515–519, Malaysia (2015)
5. Luo, R.C., Huang, K.C., Alami, R.: Online trajectory tracking based on model predictive control for service robot. IEEE International Conference on Automation Science and Engineering (CASE), pp. 1238–1243, Taiwan (2014)

6. Velasco, M., Alvarez, A., Rivera, G.: Discrete-time control of an omnidirectional mobile robot subject to transport delay. In: IEEE International Conference on American Control, pp. 2171–2176, New York (2007)
7. Okumura, J., Takei, T., Tsubouchi, T.: Navigation in indoor environment by an autonomous unicycle robot with wide-type wheel. In: IEEE/RSJ International Conference on Intelligent Robots and Systems, pp. 154–159, Taiwan (2010)
8. Melik, N., Slimane, N.: Autonomous navigation with obstacle avoidance of tricycle mobile robot based on fuzzy controller. In: IEEE International Conference on Electrical Engineering (ICEE), pp. 1–4, Algeria (2015)
9. Doosthoseini, M., kadkhodaei, B., Korayem, M., Shafei, A.: An experimental electronic interface design for a Two-link elastic robotic arm. In: IEEE International Conference on Information, Communication and Automation Technologies (ICAT), pp. 1–4, Bosnia and Herzegovina (2013)
10. Ames, A.D.: Human-inspired control of bipedal walking robots. In: IEEE International Conference on transactions on automatic control, vol. 59, no. 5, pp. 115–1130 (2014)
11. Ollervides, J., Orrante-Sakanassi, J., Santibáñez, V., Dzul, A.: Navigation control system of walking hexapod robot. IEEE International Conference on Ninth Electronics, Robotics and Automotive Mechanics, pp. 60–65 (2012)
12. Kurisu, M.: A study on teleoperation system for a hexapod robot development of a prototype platform. IEEE International Conference on Mechatronics and Automation, pp. 135–141, China (2011)
13. Jansen, T.: The Great Pretender. 010 Publishers, Rotterdam (2007)
14. Moldovan, F., Dolga, V., Pop, C.: Kinetostatic analysis of an articulated walking mechanism. In: Lovasz, E.-C., Corves, B. (eds.) Mechanisms, Transmissions and Applications. Mechanisms and Machine Science, vol. 3, pp. 103–110. Springer, Netherlands (2011)
15. Ruan, Q., Wu, J.X., Zhou, S.H., Yao, Y.A.: Fluctuation Compensation of a Multi-legged Walking Platform Using Cam Mechanism, pp. 1–5. IFToMM World Congress, Taiwan (2015)
16. Nansai, S., Iwase, M., Elara, M.R.: Energy based position control of Jansen walking robot. IEEE International Conference on Systems, Man, and Cybernetics, pp. 1241–1246 (2013)
17. Nansaia, S., Elarab, M.R., Iwasea, M.: Dynamic analysis and modeling of Jansen mechanism. In: ELSEVIER International Conference on Design and Manufacturing, IConDM, pp. 1562–1571 (2013)
18. Tong, S., Li, H.-X.: Fuzzy adaptive sliding-mode control for MIMO. IEEE Trans. Fuzzy Syst. **11**(3), 354–360 (2003)
19. Parekh, B.J., Thakkar, P.N., Tambe, M.N.: Design and analysis of Theo Jansen's mechanism based sports ground (pitch) marking robot. In: Annual IEEE India Conference (INDICON), pp. 1–5 (2014)
20. Honda, K., Kajiwara, Y., Karube, S., Takahashi, K.: Walking mechanism for a Lion-type robot. In: IEEE International Conference on Systems, Man, and Cybernetics, pp. 3403–3406, USA (2014)
21. Ortiz, J.S., Andaluz, V.H., Rivas, D., Sánchez, J.S., Espinosa, E.G.: Human-wheelchair system controlled by through brain signals. In: Kubota, N., Kiguchi, K., Liu, H., Obo, T. (eds.) ICIRA 2016. LNCS (LNAI), vol. 9835, pp. 211–222. Springer, Heidelberg (2016). doi:10.1007/978-3-319-43518-3_21
22. SolidWorks Corporation, "Conceptos básicos de SolidWorks Piezas y ensamblajes" (2006)
23. Robotis: http://support.robotis.com

Designing and Assessing Expressive Open-Source Faces for the Baxter Robot

Naomi T. Fitter[✉] and Katherine J. Kuchenbecker

Haptics Group, GRASP Laboratory, Mechanical Engineering and Applied Mechanics,
University of Pennsylvania, Philadelphia, PA 19104, USA
{nfitter,kuchenbe}@seas.upenn.edu

Abstract. Facial expressions of both humans and robots are known to communicate important social cues to human observers. Nevertheless, faces for use on the flat panel display screens of physical multi-degree-of-freedom robots have not been exhaustively studied. While surveying owners of the Rethink Robotics Baxter Research Robot to establish their interest, we designed a set of 49 Baxter faces, including seven colors (red, orange, yellow, green, blue, purple, and gray) and seven expressions (afraid, angry, disgusted, happy, neutral, sad, and surprised). Online study participants (N = 568) drawn equally from two countries (US and India) then rated photographs of a physical Baxter robot displaying randomized subsets of the faces. Face color, facial expression, and onlooker country of origin all significantly affected the perceived pleasantness and energeticness of the robot, as well as the onlooker's feelings of safety and pleasedness, with facial expression causing the largest effects. The designed faces are available to researchers online.

Keywords: Social robotics · Expressive robot faces · Baxter Research Robot

1 Introduction

Facial expressions play an important role in both human-human and human-robot interaction (HRI), but many robot facial expressions lack formal investigation. In our previous work on physical HRI with the Rethink Robotics Baxter Research Robot [1], we designed and used robot facial expressions without a deep understanding of their affective effects on our study participants. In parallel, we noticed that most media showcasing other Baxter research involved simple faces that varied greatly from team to team. For the benefit of all researchers who must currently develop custom images for Baxter's head screen, we sought to create a single vibrant, emotionally inclusive, and open-source set of Baxter faces, accompanied by information on how each face is perceived by human onlookers for improved control of affect-related robot studies.

After discussing related work (Sect. 2), we outline our process of designing and assessing Baxter faces (Sect. 3), describe Baxter owner and Mechanical Turk survey results (Sect. 4), and consider the significance and future applications of this research (Sect. 5).

© Springer International Publishing AG 2016
A. Agah et al. (Eds.): ICSR 2016, LNAI 9979, pp. 340–350, 2016.
DOI: 10.1007/978-3-319-47437-3_33

2   Related Work

Ties to Social Robotics: Our investigation of Baxter faces is closely tied to the field of social robotics, a research area that explores social interactions between humans and robots or between multiple robots [2]. Although most research robots in this field have faces, few possess both a flat panel display face and complex actuated appendages like multiple-degree-of-freedom arms. Roboceptionist and Vikia, for example, each have a face relayed on an LCD screen but do not possess other actuated physical appendages [2,3]. Baxter's unique combination of an LCD head screen with other complex actuated parts presents new opportunities for both interaction research and face design, since embodied active agents have been shown to inspire human engagement [4]. Rethink Robotics noticed this unique opportunity and created a proprietary set of faces for the Baxter Industrial Robot, but researchers using the Baxter Research Robot do not have access to these images. Furthermore, these faces were designed for pragmatic communication, not rich social interaction.

Emotion, Color, and Culture: Before designing faces for the Baxter robot, we sought to understand psychology and design work pertaining to emotions. In past studies of emotion in the human face, Ekman et al. described six basic emotions (afraid, angry, disgusted, happy, sad, and surprised) that could combine to make any other facial emotion [5]. Russell et al. challenge this idea of discrete emotions, proposing a continuous model of emotions that all fall in a circumplex valence–arousal (or pleasantness–energeticness) space [6]. Since human-robot social interactions often depend on classification, many artificial intelligence models of affect rely on Ekman et al.'s theory. Accordingly, we adopted an approach similar to [7] and designed discrete robot faces, but we assessed them in circumplex space, as emphasized in Fig. 1.

Color-wise, we had seen that different research groups were using various foreground and background colors on the faces of their Baxter Research Robots. This observation brought to mind some questions about the affective effects of Baxter face color, since studies like [8] suggest that different colors are associated with different pleasure levels, and that yellow often corresponds to the lowest pleasure ratings. Most Baxter face designs that we have encountered contain yellow elements. We introduced a color variable in our Baxter face exploration to learn whether some robot face colors have specific effects on human perception.

Additionally, we discovered work documenting cognition, emotion, and motivation differences between different world cultures [9]. Accordingly, we deployed our Baxter face study in different countries to measure how strongly the onlooker's origin affects their perception of robot faces.

Robot Face Design and Assessment: Previous work has similarly aimed to design and assess robot emotions. Early social robotics work demonstrated that robot head design and behavior were crucial to encouraging interaction [10] and defined what facial features make robots appear humanlike [12]. In light of these findings, many researchers began designing and assessing sophisticated robot faces with

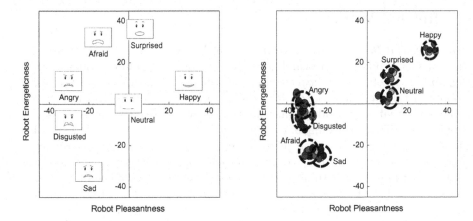

Fig. 1. Left: Illustration of where our Baxter expressions might belong in the valence–arousal space based on Russell et al.'s circumplex model. Right: Actual ratings of the faces from the reported study; placement roughly follows the model. One colored ellipse appears for each color-expression combination with color matching the face color, centroid located at the median-valued coordinate pair for that grouping, and horizontal and vertical axis lengths proportional to the standard deviation on that axis. Dashed grouping ellipses are sized at half the standard deviation of all data points for the corresponding emotion. (Color figure online)

nuanced affective display capabilities, such as Kismet [13], KASPAR [11], and EDDIE [14]. Studies of the iCat robot similarly assessed whether participants could identify robot emotions and also demonstrated that physical robot presence is not essential in this type of experiment [15]. Other investigators asked human users to assess various affective features of the Roboceptionist robot and also to report how they were feeling during the interaction [16].

Our study replicates certain methods from these studies for the specific application of the Baxter robot head display. This work differs from related LCD face screen studies by avoiding the use of a humanlike head outline and by using only simple geometric shapes in facial feature creation. Within the scope of our investigation, we do not attempt to rigorously explore every aspect of robot faces, but we hope to gain a coarse understanding of how particular faces affect onlookers. This knowledge will provide a general model of how different Baxter faces make human users feel and may inspire intentional Baxter face design.

3   Methods

To pursue our vision of creating and assessing open-source faces for the Baxter Research Robot, we *designed faces* based on related psychology and robotics work, *gathered interest information* from roboticists using Baxter, and *surveyed crowd workers online* for perceptions of our designed faces, as detailed throughout this Section.

3.1   Research Questions

While considering Baxter Research Robot faces, we focused on the following main research questions that guided our face creation and experimental design.

1. Will these faces interest other Baxter researchers? During our investigation of existing Baxter faces, we saw primarily very simple face designs, with a few exceptions such as the MIT Personal Robotics Group's EDI, a magician's assistant Baxter [17]. It appeared that most Baxter users who needed a face for their robot created a quick smiley-face image, did a brief Internet search for an appropriate image, or created a few sophisticated face images based on movie characters or other elements from popular culture. Accordingly, we were curious to gather more detailed accounts about where Baxter faces were coming from and simultaneously find out what types of open-source Baxter faces might interest Baxter users, if any.

2. Do distinct colors and expressions change how a robot face is perceived? Section 2 discussed certain facial expressions and colors that have been shown in the past to be associated with different valence and arousal levels. Like previous robotics researchers, we became interested in designing a family of discrete facial expressions for our Baxter robot, and we added the variable of face color to compare and contrast the expressive powers of hue and facial expression. We expected to encounter statistically significant differences in how these faces are perceived, similar to those discovered for other robots. There also seemed to be potential for interesting synergies when the color and expression seen in a face conflict with or reinforce one another.

3. Do cultural differences affect perception of Baxter's faces? Because people from different cultures may process facial emotion differently, we believed there might be differences in the emotional perception of Baxter's face between cultures. To answer this question, we collected data from human participants from two common countries of origin.

3.2   Baxter Face Creation

Expression Selection: Since previous work on robot emotions mainly builds from Ekman et al.'s facial expression theory, we adopted a similar framework and designed the six main theorized expressions (afraid, angry, disgusted, happy, sad, surprised), plus a neutral baseline expression for comparison, as illustrated in Fig. 1. Although additional expression possibilities appear when expanding to Russell et al.'s circumplex model, we limited ourselves to the seven aforementioned expressions for the purpose of feasible experimental design.

Color Selection: Previous studies of color show that distinct hues evoke different emotions. To choose appropriate colors for our Baxter face screen, we explored a breadth of web-safe shades of each primary and secondary color, plus gray for

Table 1. Color codes for each Baxter face color.

	Gray	Red	Orange	Yellow	Green	Blue	Purple
R Value	102	204	204	204	0	51	102
G Value	102	0	102	204	102	51	0
B Value	102	0	0	0	0	255	153

comparison. Once we accumulated a list of web-safe colors recognizable as each shade of interest, we filled Baxter's LCD screen with each viable web-safe color to see whether each hue remained recognizable as its intended color. We then iteratively considered faces with the most promising background colors, finally selecting the values listed in Table 1.

Photography: We planned to release a web survey based on pictures of the physical Baxter robot with different faces to gather impressions of our designed faces embodied on this specific robot. In preparation, we photographed each robot face against a black background with a Nikon DSLR camera on a tripod. Although lighting, focus, and color balance conditions varied slightly because of the real lab environment of the Baxter photo shoot, we did our best to maintain consistent conditions across all photographs. For certain colors, we had to post-process the images slightly (adjustment to exposure and saturation) to ensure that the color of the screen in the photograph accurately matched the color of the screen on the physical Baxter robot.

3.3   Survey Instruments

Two main surveys enabled us to gather information throughout our face development effort. To assess the general desire for a set of open-source Baxter faces, we surveyed Baxter Research Robot owners. Simultaneously, to gain a robust understanding of the effects of our Baxter faces on human observers, we released an online Amazon Mechanical Turk survey in the style of [18].

Baxter Owner Survey: We created a survey for current Baxter owners with questions about how they use the Baxter Research Robot, what images they currently publish to the robot's face display and what they would want in new robot face images. To obtain survey responses from the Baxter Research Robot community, we posted the survey link on the Baxter users Google Group page. We also reached out personally to all Baxter researchers we knew and everyone who had published a paper involving the Baxter Research Robot. Overall, eighteen researchers responded to our survey with representation from human-robot interaction, machine learning, autonomy, computer vision, manipulation & mechatronics, planning & manipulation, and education.

Face Perception Survey: Our second survey instrument was a Mechanical Turk investigation of our seven expressions and seven colors of Baxter face. All meth-

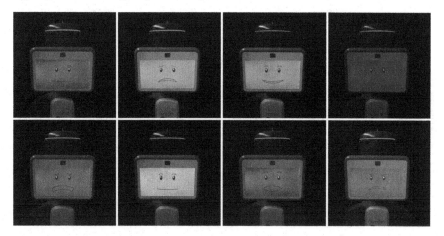

Fig. 2. A collection of eight of our robot photograph stimuli, including instances of all seven explored colors and all seven explored expressions. (Color figure online)

ods were approved by the Penn IRB under protocol 825119. Survey respondents who selected our Human Intelligence Task, consented to participate, and navigated questions to qualify as non-colorblind adults fluent in English were then shown a representative subset of our Baxter face pictures, similar to those shown in Fig. 2. Since answering questions about all 49 faces proved exhausting, we randomly selected fourteen pictures of faces on the Baxter to show each participant, and we balanced the random selection to make sure each face was shown an equal number of times. The presented photographs were 600 pixels wide by 398 pixels tall. Respondents were asked to complete slider questions evaluating the robot's pleasantness (Q1), the robot's energeticness (Q2), how safe they felt while looking at the robot (Q3), and how pleased they felt while looking at the robot (Q4). Attention check questions were incorporated into the survey to ensure that respondents were not just clicking randomly. A total of 568 people participated in the study: 327 men and 241 women ranging in age from 18 to 77 (M = 36.3, SD = 11.1). Because Mechanical Turk workers hail mainly from the U.S. and India, 286 participants were native US citizens and 282 were native citizens of India. Respondents came from a variety of professions, including 44.2 % technically trained individuals and only 6.2 % people who had ever owned a robot. Each respondent received $0.60 for their participation.

4   Results

Baxter Owner Survey: The Baxter Research Robot owner survey yielded promising results for the prospect of an open-source Baxter face database. Of the eighteen respondents, 83.3 % had used images on the Baxter head display other than the default background provided by Rethink Robotics, and a similar 83.3 % reported interest in an open-source database of expressive Baxter faces. Of the interested individuals, 66.6 % desired subjective data about how human raters

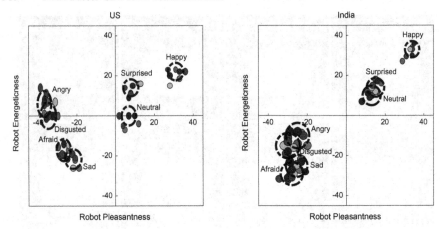

Fig. 3. Left: Actual placement of human pleasantness–arousal rating pairs reported by US respondents, which roughly follows Russell's model. Plot elements have the same meaning as in Fig. 1. Right: Same illustration for the India survey respondents. (Color figure online)

perceive different Baxter facial expressions. All eight survey participants who use Baxter for human-robot interaction were interested in open-source expressive Baxter faces, and 75 % of them also wanted human ratings. At least two researchers were interested in each of our proposed Baxter facial expressions and expression background colors. There was also interest in sleepy, encouraging, bored, and pensive facial expressions and black, white, and rainbow background colors. We may later create and explore more faces based on this interest.

Baxter Face Survey: We were curious to determine whether color, expression, and onlooker country of origin affected Baxter face ratings. Overall, the response data (Figs. 1, 3, and 4) appeared to roughly follow Russell et al.'s valence–arousal model. To assess these survey response differences, we used MATLAB to perform a $7 \times 7 \times 2$ three-factor analysis of variance (ANOVA). We found that there were several statistically significant trends in the robot face survey responses, as referenced throughout the following discussion of condition results and as illustrated in Table 2. A post-hoc multiple comparison test on each significant variable using the MATLAB `multcompare` function revealed the effects of each specific color, expression, and origin. Our analyses employ an $\alpha = 0.05$ significance level. We calculate the effect size using eta squared.

Robot face color slightly affected participant responses. There were statistically significant effects on the ratings of robot pleasantness ($F(6,8007) = 4.00$, $p = 0.0005$, $\eta^2 = 0.001$), robot energeticness ($F(6,8007) = 3.11$, $p = 0.0049$, $\eta^2 = 0.001$), personal safety feeling ($F(6,8007) = 8.61$, $p < 0.0001$, $\eta^2 = 0.004$), and personal pleasedness feeling ($F(6,8007) = 5.97$, $p < 0.0001$, $\eta^2 = 0.002$). A post-hoc multiple comparison test revealed that the color red made the robot seem significantly less pleasant than all other colors except green and blue. Red and purple both appeared significantly more energetic than green, with all other

Table 2. P-values resultant from each variable in the Baxter face survey responses.

Metric	Color (C)	Expression (E)	Origin (O)	C×E	C×O	E×O
Pleasantness	0.0005	<0.0001	<0.0001	0.8074	0.2881	<0.0001
Arousal	<0.0049	<0.0001	<0.0375	0.5288	0.0079	<0.0001
Safety	<0.0001	<0.0001	<0.0204	0.8811	0.0657	<0.0001
Pleasedness	<0.0001	<0.0001	<0.0001	0.8425	0.2189	<0.0001

colors falling between the two groups with no statistically significant differences. Raters felt significantly less safe looking at red faces compared to all other colors. Red was also significantly less pleasing than all other colors. Overall, red appears to be the main color we could use to influence robot and human affect.

The expression of the face had much stronger statistically significant effects on robot pleasantness ($F(6,8007) = 2020.61$, $p < 0.0001$, $\eta^2 = 0.591$), robot energeticness ($F(6,8007) = 742.77$, $p < 0.0001$, $\eta^2 = 0.293$), personal safety feeling ($F(6,8007) = 1098.93$, $p < 0.0001$, $\eta^2 = 0.441$), and personal pleasedness feeling ($F(6, 8007) = 1588.40$, $p < 0.0001$, $\eta^2 = 0.535$). In both robot and human pleasantness ratings, afraid, angry, and disgusted were ranked as less pleasant than all other expressions. Happy appeared more pleasant than any other expression. Sad was less pleasant than neutral, surprised, and happy. In terms of robot energeticness, afraid and sad were rated as significantly less energetic than any other expressions. All other expressions were statistically significant from one another in the following increasing energeticness order: disgusted, angry, neutral, surprised, and happy. Finally, angry and disgusted expressions appeared less safe than any other expression but not different from each other. Happy was rated to be safer than any other expression. Neutral and surprised were rated as safer than all expressions except for happy. Afraid received ratings of more safe than angry and disgusted but less safe than all other expressions. These results, which roughly match Russel et al.'s model, emphasize that researchers could use these carefully designed faces to influence users in known ways.

Respondent country of origin also had a significant but small effect on robot pleasantness ($F(1,8007) = 160.21$, $p < 0.0001$, $\eta^2 = 0.008$), robot energeticness ($F(1,8007) = 4.33$, $p = 0.0375$, $\eta^2 = <0.001$), personal safety feeling ($F(1,8007) = 5.38$, $p = 0.0204$, $\eta^2 = <0.001$), and personal pleasedness feeling ($F(1,8007) = 110.28$, $p < 0.0001$, $\eta^2 = 0.006$). Citizens of India generally ranked the robot faces and their own feelings toward the robot as more pleasant than US citizens. US citizens ranked robots as slightly more energetic and safer. We should consider such differences when designing robots for users in multiple countries.

Additionally, we discovered small interaction effects between robot facial expression and participant origin for every survey rating: robot pleasantness ($F(6,8007) = 17.96$, $p < 0.0001$, $\eta^2 = 0.005$), robot energeticness ($F(6,8007) = 67.36$, $p < 0.0001$, $\eta^2 = 0.027$), personal safety feeling ($F(6,8007) = 18.46$, $p < 0.0001$, $\eta^2 = 0.007$), and personal pleasedness feeling ($F(6,8007) = 8.84$, $p < 0.0001$, $\eta^2 = 0.003$). Country of origin and robot face color also interacted

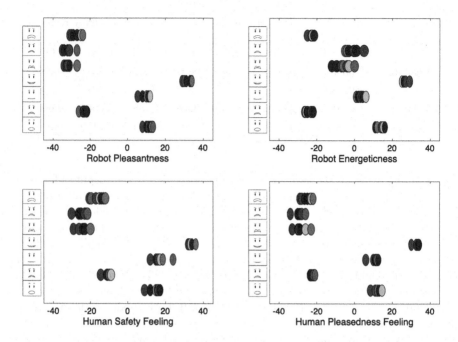

Fig. 4. Participant responses to the four slider survey questions (Q1, Q2, Q3, and Q4), divided by stimulus emotion. The color of each ellipse corresponds to robot face color, the x-coordinate of the centroid of each ellipse lies at the median rating, and the width of each ellipse is proportional to the standard deviation of that group's ratings. (Color figure online)

in the energeticness ratings (F(6,8007) = 2.9, p = 0.0079, $\eta^2 = 0.001$). These interactions emphasize the impact of country of origin on robot face perception.

While analyzing the data, we noticed that responses to Q1, Q3, and Q4 seemed to have a high correlation. After performing linear regression to explore this relationship, we found fairly high coefficients of determination between Q1 and Q4 ($R^2 = 0.809$), Q1 and Q3 ($R^2 = 0.664$), and Q3 and Q4 ($R^2 = 0.708$). This result may indicate that robots designed to look pleasant will inherently make onlookers feel more safe and pleased. Indian participants also exhibited a high correlation between Q1 and Q2 responses ($R^2 = 0.679$), while US participants did not ($R^2 = 0.194$). The emotional perception difference between the two countries may partially stem from the US origin of the researchers who designed the faces.

5   Conclusions and Future Work

Overall, we found that other researchers are interested in expressive, open-source Baxter faces like the ones that we created. The collective Baxter face survey response data give us a good idea of how to manipulate robot and human affect through the use of intentionally chosen robot faces. Expression differences in these faces cause large significant differences in human ratings of robot pleasantness, robot energeticness, personal safety, and personal pleasedness. Face color and the onlooker's country of origin also slightly affect all of these ratings, with some small interactions between observer country of origin and Baxter face features. We have released the created Baxter faces, source files, and photographs from the study in the public GitHub repository at https://github.com/nfitter/BaxterFaces for the benefit of any researcher who seeks carefully designed Baxter faces that cause known emotional effects in human observers of the robot. Future work in this realm may involve creating additional robot facial expressions, as desired by other Baxter researchers. We are also excited to use the developed Baxter facial expressions in our own human-robot interaction research.

Acknowledgments. The first author was supported by a National Science Foundation (NSF) Graduate Research Fellowship under Grant No. DGE-0822 and an NSF Integrative Graduate Education and Research Traineeship under Grant No. 0966142. We thank Chris Callison-Burch and Eileen Huang for their advice.

References

1. Fitter, N.T., Kuchenbecker, K.J.: Equipping the Baxter robot with human-inspired hand-clapping skills. Accepted to the IEEE International Symposium on Robot and Human Interactive Communication (RO-MAN) (2016)
2. Fong, T., Nourbakhsh, I., Dautenhahn, K.: A survey of socially interactive robots. Robot. Auton. Syst. **42**(3), 143–166 (2003)
3. Leite, I., Martinho, C., Paiva, A.: Social robots for long-term interaction: a survey. Int. J. Social Robot. **5**(2), 291–308 (2013)
4. Kose-Bagci, H., Ferrari, E., Dautenhahn, K., Syrdal, D.S., Nehaniv, C.L.: Effects of embodiment and gestures on social interaction in drumming games with a humanoid robot. Adv. Robot. **23**(14), 1951–1996 (2009)
5. Ekman, P., Friesen, W.V., Ellsworth, P.: Emotion in the human face: Guidelines for research and an integration of findings. Pergamon Press (2013)
6. Russell, J.A., Fernández-Dols, J.M.: The psychology of facial expression. Cambridge University Press (1997)
7. Schiano, D.J., Ehrlich, S.M., Rahardja, K., Sheridan, K.: Face to interface: facial affect in (Hu)man and machine. In: ACM SIGCHI Conference on Human Factors in Computing Systems, pp. 193–200 (2000)
8. Valdez, P., Mehrabian, A.: Effects of color on emotions. J. Exp. Psychol. Gen. **123**(4), 394–409 (1994)
9. Markus, H.R., Kitayama, S.: Culture and the self: Implications for cognition, emotion, and motivation. Psychol. Rev. **98**(2), 224–253 (1991)
10. Bruce, A., Nourbakhsh, I., Simmons, R.: The role of expressiveness and attention in human-robot interaction. In: IEEE International Conference on Robotics and Automation (ICRA), vol. 4 (2002)
11. Blow, M., Dautenhahn, K., Appleby, A., Nehaniv, C.L., Lee, D.: The art of designing robot faces: dimensions for human-robot interaction. In: ACM SIGCHI/SIGART Conference on Human-Robot Interaction, pp. 331–332 (2006)

12. DiSalvo, C.F., Gemperle, F., Forlizzi, J., Kiesler, S.: All robots are not created equal: the design and perception of humanoid robot heads. In: ACM Conference on Designing Interactive Systems: Processes, Practices, Methods, and Techniques, pp. 321–326 (2002)
13. Breazeal, C.: Emotion and sociable humanoid robots. Int. J. Hum. Comput. Stud. **59**(1), 119–155 (2003)
14. Sosnowski, S., Bittermann, A., Kuhnlenz, K., Buss, M.: Design and evaluation of emotion-display EDDIE. In: IEEE/RSJ International Conference on Intelligent Robots and Systems (IROS), pp. 3113–3118 (2006)
15. Bartneck, C., Reichenbach, J., Breemen, V.A.: In your face, robot! The influence of a character's embodiment on how users perceive its emotional expressions. In: Design and Emotion 2004 Conference (2004)
16. Kirby, R., Forlizzi, J., Simmons, R.: Affective social robots. Robot. Auton. Syst. **58**(3), 322–332 (2010)
17. Nuñez, D., Tempest, M., Viola, E., Breazeal, C.: An initial discussion of timing considerations raised during development of a magician-robot interaction. In: ACM/IEEE International Conference on Human-Robot Interaction (HRI) Workshop on Timing in HRI (2014)
18. Takayama, L., Ju, W., Nass, C.: Beyond dirty, dangerous and dull: what everyday people think robots should do. In: ACM/IEEE International Conference on Human-Robot Interaction (HRI) (2008)

Spontaneous Human-Robot Emotional Interaction Through Facial Expressions

Ali Meghdari[1(✉)], Minoo Alemi[1,2], Ali Ghorbandaei Pour[1],
and Alireza Taheri[1]

[1] Social and Cognitive Robotics Laboratory, Center of Excellence in Design,
Robotics and Automation (CEDRA), Sharif University of Technology,
Tehran, Iran
meghdari@sharif.ir, alemi@sharif.edu
[2] Islamic Azad University, Tehran-West Branch, Tehran, Iran

Abstract. One of the main issues in the field of social and cognitive robotics is the robot's ability to recognize emotional states and emotional interaction between robots and humans. Through effective emotional interaction, robots will be able to perform many tasks in human society. In this research, we have developed a robotic platform and a vision system to recognize the emotional state of the user through its facial expressions, which leads to a more realistic human-robot interaction (HRI). First, a number of features are extracted according to points detected by a vision system from the face of the user. Then, the emotional state of the user is analyzed with the help of these features. For the decision making unit, a state machine is designed that utilizes the results obtained from the emotional state analysis to generate the robot's response. Finally, a fuzzy algorithm is used to improve the quality of emotional interaction and the results are implemented on a commercial humanoid robot platform which has the ability of producing facial expressions.

Keywords: Social robotics · Human-robot interaction (HRI) · Emotional state recognition · Fuzzy finite state machine · Fuzzy clustering

1 Introduction

Since the creation of the first robot, researchers have been interested in development of interaction between a robot and its environment, with the possibility of robots interacting with each other and with humans. The common assumption is that humans prefer to interact with machines in the same way that they interact with other people. In this regard, different ideas and prototypes of robotic heads have been developed for HRI purposes [1–4]. Tadesse et al. [1] designed and implemented a twelve degrees-of-freedom humanoid baby head, capable of producing 6 basic facial expressions, and Saffari and Meghdari et al. [2] introduced a robotic head which turns toward the speaker in noisy environments. By improving the abilities of humanoid robots, they now have the capability to enhance scenarios involving education [5, 6], physical therapy [7–9], and elderly care [10]. In this regard, social learning and imitation, gesture and natural language communication, emotion, and recognition of interaction

© Springer International Publishing AG 2016
A. Agah et al. (Eds.): ICSR 2016, LNAI 9979, pp. 351–361, 2016.
DOI: 10.1007/978-3-319-47437-3_34

partners are all important factors. In recent years, this field has attracted considerable attention from academic and the research communities. Zacharatos et al. [11] described recent emerging techniques and advances in automatic emotion recognition. Halder et al. [12] used an interval and a general type-2 fuzzy set separately to model the fuzzy face space for emotion recognition purpose.

In general, one can classify HRI studies into verbal and non-verbal interactive communications [13]. Figure 1 shows a model of an emotion-based HRI system. Aly et al. [14] introduced a multimodal behavior HRI for more naturally emotional inter-action. A group of studies has been done based on emotional state detection through voice analysis [15]. There are also remarkable studies on emotion recognition according to the user's gestures [16–21]. Xiao et al. [16] involved a set of 12 upper body gestures to communicate with the robot. Chakraborty et al. [17] proposed a simple and robust scheme for emotion recognition and control, with good accuracy based on fuzzy relational approach, and geometric deformation facial features has been also used for facial expression recognition [18].

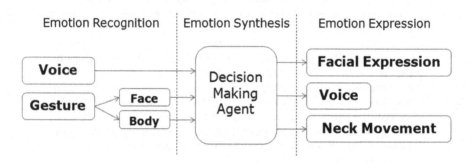

Fig. 1. Model of emotion-based HRI for humanoid robotic platforms

This paper presents an initial attempt to develop a robotic platform for social interaction research. This platform has an attractive physical appearance with which humans should enjoy interacting. Our emotion recognition is based on the user's facial gestures, and the robot's response is through facial expression and neck movement, in accordance to a fuzzy decision making algorithm. The desired work is the synchro-nization between the developed interaction mode and the implementation of the pro-posed emotion-based control.

2   Instruments

2.1   A Humanoid Robot

R50 – Alice, with the Iranian name "Mina", is a humanoid robot made by Hanson Robokind Company, designed specifically for human-robot social interaction and has been used widely for studies on developmental and social robotics [22]. Mina is 69 cm high, weights 5.7 kg and has 32 degrees-of-freedom. Mina has the 3D face of a girl,

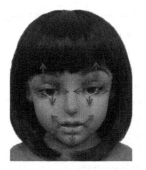

Fig. 2. The R-50 Alice (Mina) robot

which permits 11 degrees-of-freedom (Fig. 2) for generating facial expressions such as surprise, anger, happiness, sadness and so on. She also has 3 degrees-of-freedom in her neck which makes her able to trace the user by moving her head toward the user's face while they are in the interacting mode.

2.2 Machine Vision

In this study, we have used a Microsoft Kinect Sensor for our Machine vision application. The Microsoft Kinect Sensor is a physical device with depth sensing technology, a built-in color camera, and an infrared emitter that can sense the location and movement of people. With the help of version 2 from the Kinect for Windows Software Development Kit (SDK) it is possible to access a list of face points to extract our features. The positions of these points are defined in the Kinect body coordinate system. The origin is located at the optical center of the camera, the Z-axis is pointing towards a user, the Y-axis is pointing up, and the X-axis is to the right [23].

3 Research Methodology

3.1 Face Feature Extraction

In the first step, we have chosen 21 face points among the available 36 points detected in SDK (Table 1). These points were chosen in such a way to define facial features based on the action units of the Facial Action Coding System (FACS) [24]. Afterwards, a set of 18 features were defined according to changes in the distances between these points. These features are listed in Table 2. The data recorded from Kinect output and each feature is updated with a speed of 30 frames per second. In order to reduce the effect of noise on the extracted features, a moving average filter with a period of 5 previous data points is applied to each feature. Another issue is that the subject's features should be scale invariant, for this reason and to avoid the effect of user's distance from the Kinect Sensor (normalizing the features) all of these features are divided by the length of the subject's nose (our 18th feature). After making our features scale invariant, the first 17 normalized features are used to detect the emotional state of the user.

Table 1. List of the face points used for feature extraction

Face points	Description
P1	Inner corner of the left eye
P2	Outer corner of the left eye
P3	Inner corner of the right eye
P4	Outer corner of the right eye
P5	Inner left eyebrow
P6	Outer left eyebrow
P7	Center of the left eyebrow
P8	Inner right eyebrow
P9	Outer right eyebrow
P10	Center of the right eyebrow
P11	Left corner of the mouth
P12	Right corner of the mouth
P13	Middle of the upper lip
P14	Middle of the lower lip
P15	Bottom of the nose
P16	Bottom left of the nose
P17	Bottom right of the nose
P18	Left cheek bone
P19	Right cheek bone
P20	Left end of the lower jaw
P21	Right end of the lower jaw

Table 2. List of the facial features

Face features	Description
F1	Distance between P1 and P5
F2	Distance between P2 and P6
F3	Distance between P7 and the line crossing P1 and P2
F4	Distance between P3 and P8
F5	Distance between P4 and P9
F6	Distance between P10 and the line crossing P3 and P4
F7	Distance between P5 and P8
F8	Distance between P11 and P12
F9	Distance between P13 and P14
F10	Distance between P11 and P1
F11	Distance between P11 and P18
F12	Distance between P11 and P20
F13	Distance between P12 and P3
F14	Distance between P12 and P19
F15	Distance between P12 and P21
F16	Distance between P1 and P16
F17	Distance between P3 and P17
F18	Distance between P15 and the line crossing from P1 and P3

3.2   Emotional State Recognition

A data base of facial features of 3,000 samples was gathered from different poses for 6 main facial expressions from 10 different young adults (500 samples for each facial emotional state). These main emotional states are happiness, sadness, anger, surprise, disgust, and fear. Figure 3 shows some of our data base samples. This data base was used to train a fuzzy classifier which indicates the basic emotional state of the user through facial expression. For this reason, a fuzzy clustering method, called Fuzzy C-Means method (FCM) was used [25].

Fig. 3.   Ten subjects from our database, posing facial expressions

In order to have more realistic samples, they are tried to be spontaneous facial expressions [20]. To reach this goal the sample was selected from a range of videos captured from our subjects, while they are expressing their emotion.

3.3   Producing Emotional Reaction

The first step toward a more realistic response is tracking the user by moving the head of the robot. For this purpose, Neck Yaw and Neck Pitch (angles for rotating in the azimuth and elevation planes) were adjusted such that Mina was always facing her user as she responded to the user's emotional state.

A smooth path was designed for each of these angles of turning. These angles were calculated according to the position of the user's head. From the data output of Kinect, the head position is available in the Kinect body coordinate system. This position needs to be transferred to the robot's head coordinate system, to calculate the proper angles. Figure 4 shows the head position in both coordinate systems and proper rotating angles in the robot's head coordinate system.

In the next step, a finite state machine was used to generate an emotional reaction, consisting of one of the six emotional states, to indicate the emotional state of the robot. The input to the state machine was the user's emotional state (Fig. 5). The output of

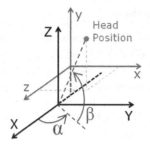

Fig. 4. Neck Yaw (α), Neck Pitch (β), Kinect body coordinate system (xyz), and Robot's head coordinate system (XYZ)

each state was a set of facial expressions produced by Mina, declaring her reaction to the user's emotional state. This output is set as a vector, containing the actuation level for each degree of freedom in the robot's face. Since the robot is not supposed to become angry, there are no states considered for anger. Transition between states is according to the user's detected emotional state.

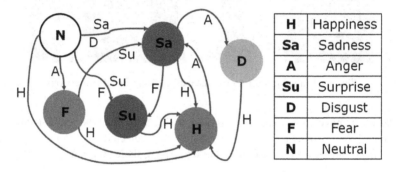

H	Happiness
Sa	Sadness
A	Anger
Su	Surprise
D	Disgust
F	Fear
N	Neutral

Fig. 5. The state machine diagram.

Since the algorithm used for emotional state recognition has a fuzzy output, a more realistic reaction can be generated by realizing the membership values of the user's facial expression for each emotional state. Also, the state machine can be implemented with a number of if-then rules as follows:

```
if (state == 'N' && entry == 'H') then {state = 'H'}
if (state == 'N' && entry == 'A') then {state = 'F'}
if (state == 'N' && entry == 'D') then {state = 'Sa'}
if (state == 'N' && entry == 'F') then {state = 'Su'}
if (state == 'H' && entry == 'A') then {state = 'Sa'}
if (state == 'F' && entry == 'Su') then {state = 'Sa'}
if (state == 'Sa' && entry == 'F') then {state = 'Su'}
if (state == 'Su' && entry == 'H') then {state = 'H'}
...
```

These rules are taken as the rule base of our fuzzy inference system. A fuzzy inference system is a method that interprets the membership values in the input vectors and based on our defined rules, assigns values to the output vector [26]. Then, for the system entry and each state, a membership value is considered (in the beginning all of the state membership values are zero). By assigning the minimum of the membership value of the system entry and current state to the next coming state, and weighted averaging between the outputs of the states, a new level of emotional reaction is generated. For calculating the weighted average, states with the membership functions of more than 0.5 are taken into consideration.

4   Results

4.1   Feature Extraction and Emotion Recognition

Figures 6 and 7 show some face features evolution (after normalizing and noise filtering) during facial expression (the X axis is frame number and the Y axis illustrates change in face features). Features in all of these video sequences begin in a neutral state. As it can be seen, face features are defined in a way to be noticeably different for each facial expression, which leads to an easier and more accurate classification.

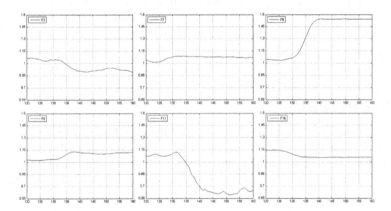

Fig. 6. Facial features variation from video sequence for happiness

In order to validate the classification process, another set of data, containing 700 samples from a new group of people (100 samples for each emotional state and 100 samples of neutral face) is used. The highest membership value indicates the emotional state of each sample. A sample is considered neutral if all of the corresponding membership values are less than 0.5. Table 3 presents the results from the test data. Each row indicates the detection results for each set of samples, with the same emotional state.

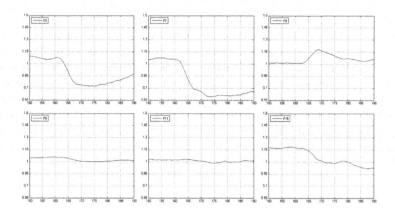

Fig. 7. Facial features variation from video sequence for anger

Table 3. Emotion recognition rate for test data

	Happiness	Sadness	Anger	Surprise	Disgust	Fear	Neutral
Happiness	98 %	0 %	0 %	1 %	0 %	0 %	1 %
Sadness	0 %	91 %	4 %	0 %	0 %	3 %	2 %
Anger	0 %	2 %	94 %	0 %	4 %	0 %	0 %
Surprise	0 %	0 %	0 %	99 %	0 %	1 %	0 %
Disgust	0 %	0 %	6 %	0 %	94 %	0 %	0 %
Fear	0 %	6 %	0 %	8 %	0 %	81 %	5 %
Neutral	0 %	3 %	0 %	0 %	0 %	2 %	95 %

4.2 Neck Movement

During the interaction Mina's head turns to face the user. If the position of the user's head moves while neck angles are moving toward their previous goal position, a new path will be generated according to the current neck angles values and the new destination angles (Fig. 9). Also, the new trajectory is considered to have the same

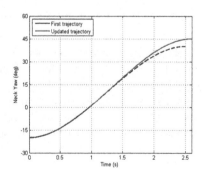

Fig. 8. Neck Yaw trajectory change in reaction to the change of user's head position

velocity as the previous trajectory at the time the neck angles path changes its trajectory. This helps to have a smooth transition between trajectories.

4.3 Mina's Emotional Reaction

Using the fuzzy finite state machine for generating proper facial expressions caused more interacting modes, and a variable output level. Also, the change rate of the emotional state of the robot is dependent on the intensity of the user's facial expression. Figure 8 shows some of Mina's reactions to her user's emotional state.

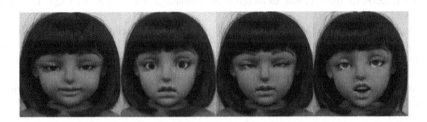

Fig. 9. Generating combinatorial facial expressions by Mina

Since we wanted to develop this social robotic platform for further HRI applications, it was important to know the reaction of people and their impression about interacting with Mina. Therefore we attended two exhibitions with her. The feedback from the people interested to continue interacting with her was quite positive. Next, we are going to involve her in some intervention scenarios for autistic children as our future work.

5   Conclusions

Usually, emotional state is a combination of two or more basic emotions. To have a better HRI, detecting the share of each basic emotion in the users current emotional state is considered valuable. In this research, we detected the user's emotional state from his/her face gesture with fuzzy classification of extracted facial features. This method made it possible to assign a membership value to the facial expression of the user, meaning that the user's emotional state could be related to more than one basic emotional state. In addition, basic emotions were recognized as well with an overall accuracy of more than 90 % for 5 out of 6 basic emotions. Then, the identified facial expression was given to the state machine developed for emotional interaction. To expose the proper facial expression, Mina was programmed to turn her head to face the user. Finally, the HRI system was shown to be capable of producing a combinatorial facial expression output. The system was also able to decide and generate different facial expressions with variable intensities. As s result, Mina could communicate with human user more naturally.

References

1. Tadesse, Y., Hong, D., Priya, S.: Twelve degree of freedom baby humanoid head using shape memory alloy actuators. J. Mech. Robot. **3**, 211–226 (2011)
2. Saffari, E., Meghdari, A., Vazirnezhad, B., Alemi, M.: Ava (A Social Robot): design and performance of a robotic hearing apparatus. In: Tapus, A., André, E., Martin, J.-C., Ferland, F., Ammi, M. (eds.) ICSR 2015. LNCS, vol. 9388, pp. 440–450. Springer, Heidelberg (2015). doi:10.1007/978-3-319-25554-5_44
3. Hanumara, N.C., Slocum, A.H., Mitamura, T.: Design of a spherically actuated human interaction robot head. J. Mech. Design **134**(5), 055001 (2012)
4. Asfour, T., Welke, K., Azad, P., Ude, A., Dillmann, R.: The Karlsruhe humanoid head. In: 8th IEEE-RAS International Conference on Humanoid Robots, pp. 447–453, December 2008
5. Meghdari, A., Alemi, M., Ghazisaedy, M., Taheri, A.R., Karimian, A., Zandvakili, M.: Applying robots as teaching assistant in EFL classes at Iranian Middle-Schools. In: Proceeding International Conference on Education & Modern Educational Technologies (EMET-2013), 28–30 September 2013, Venice, Italy (2013)
6. Alemi, M., Meghdari, A., Ghazisaedy, M.: The impact of social robotics on L2 Learners' anxiety and attitude in English vocabulary acquisition. Int. J. Soc. Robot. **7**(4), 523–535 (2015)
7. Alemi, M., Ghanbarzadeh, A., Meghdari, A., Moghaddam, L.J.: Clinical application of a humanoid robot in pediatric cancer interventions. Int. J. Soc. Robot. (2015)
8. Trinh, T.Q., Schroeter, C., Kessler, J., Gross, H.-M.: "Go Ahead, Please": recognition and resolution of conflict situations in narrow passages for polite mobile robot navigation. In: Tapus, A., André, E., Martin, J.-C., Ferland, F., Ammi, M. (eds.) ICSR 2015. LNCS, vol. 9388, pp. 643–653. Springer, Heidelberg (2015). doi:10.1007/978-3-319-25554-5_64
9. Taheri, A.R., Alemi, M., Meghdari, A., PourEtemad, H.R., Basiri, N.M.: Social robots as assistants for autism therapy in Iran: research in progress. In: Second RSI/ISM International Conference Robotics and Mechatronics (ICRoM), pp. 760–766. IEEE (2014)
10. McColl, D., Nejat, G.: A socially assistive robot that can monitor affect of the elderly during mealtime assistance. J. Med. Dev. **8**(3), 030941 (2014)
11. Zacharatos, H., Gatzoulis, C., Chrysanthou, Y.L.: Automatic emotion recognition based on body movement analysis: a survey. IEEE Comput. Graph. Appl. **34**(6), 35–45 (2014)
12. Halder, A., Konar, A., Mandal, R., Chakraborty, A., Bhowmik, P., Pal, N.R., Nagar, A.K.: General and interval type-2 fuzzy face-space approach to emotion recognition. IEEE Trans. Syst. Man Cybern. Syst. **43**(3), 587–605 (2013)
13. Mavridis, N.: A review of verbal and non-verbal human-robot interactive communication. Robot. Auton. Syst. **63**, 22–35 (2014)
14. Aly, A., Tapus, A.: Multimodal adapted robot behavior synthesis within a narrative human-robot interaction. In: IEEE/RSJ International Conference on Intelligent Robots and Systems (IROS), pp. 2986–2993 (2015)
15. Yashaswi Alva, M., Nachamai, M., Paulose, J.: A comprehensive survey on features and methods for speech emotion detection. In: IEEE International Conference on Electrical, Computer and Communication Technologies (ICECCT) (2015)
16. Xiao, Y., Zhang, Z., Beck, A., Yuan, J., Thalmann, D.: Human–robot interaction by understanding upper body gestures. Presence **23**(2), 133–154 (2014)
17. Chakraborty, A., Konar, A., Chakraborty, U.K., Chatterjee, A.: Emotion recognition from facial expressions and its control using fuzzy logic. IEEE Trans. Syst. Man Cybern. Part A Syst. Hum. **39**(4), 726–743 (2009)

18. Kotsia, I., Pitas, I.: Facial expression recognition in image sequences using geometric deformation features and support vector machines. IEEE Trans. Image Process. **16**(1), 172–187 (2007)
19. Dahmane, M., Meunier, J.: Prototype-based modeling for facial expression analysis. IEEE Trans. Multimedia **16**(6), 1574–1584 (2014)
20. Li, Y., Mavadati, S.M., Mahoor, M.H., Zhao, Y., Ji, Q.: Measuring the intensity of spontaneous facial action units with dynamic Bayesian network. Pattern Recogn. **48**(11), 3417–3427 (2015)
21. Li, Y., Wang, S., Zhao, Y., Ji, Q.: Simultaneous facial feature tracking and facial expression recognition. IEEE Trans. Image Process. **22**(7), 2559–2573 (2013)
22. http://www.robokindrobots.com/
23. Kinect for Windows SDK (2016). https://msdn.microsoft.com/en-us/library/
24. Ekman, P., Friesen, W.: Facial Action Coding System: A Technique for the Measurement of Facial Movement. Consulting psychologists press, Palo alto (1978)
25. Popescu, M., Keller, J., Bezdek, J.C., Zare, A.: Random projections fuzzy c-means (RPFCM) for big data clustering. In: IEEE International Conference on Fuzzy Systems (FUZZ-IEEE), August 2015
26. Yan, J., Ryan, M., Power, J.: Using Fuzzy Logic: Towards Intelligent Systems, vol. 1. Prentice Hall, London (1994)

Functional and Non-functional Expressive Dimensions: Classification of the Expressiveness of Humanoid Robots

François Ferland[(✉)] and Adriana Tapus

Robotics and Computer Vision Lab, U2IS ENSTA ParisTech, Université Paris-Saclay,
828 bd des Maréchaux, 91762 Palaiseau Cedex, France
{francois.ferland,adriana.tapus}@ensta-paristech.fr

Abstract. In Human-Robot Interaction (HRI), an important quantity of work has been done to investigate the reaction of people toward expressive robots. However, the large variability of available expression modalities (e.g., gaze, gestures, speech modulation) can make comparison between results difficult. We believe that developing a common taxonomy to describe these modalities would contribute to the standardization of HRI experiments. This paper proposes the first version of a classification system based on an analysis of humanoid robots commonly seen and used in HRI studies. Features from the face of robots are discussed in terms of functional and non-functional dimensions, and a short-hand notation is developed to describe these features.

1 Introduction

Population aging around the world motivated the growth of number of projects on assistive robots and other intelligent devices, from assistant-like software for smartphones to mobile robots in elder care facilities. One of the objectives pursued by the Human-Robot Interaction (HRI) community is the development of natural, human-like behaviors for intelligent autonomous systems. These systems are often embodied by robots or virtual agents on a screen, both sometimes designed to have a human-like appearance. Many studies have been conducted on the impact of expression modalities in various interaction settings. For instance, the perception of robot smiles has been studied, and results can influence HRI design [1]. Studies have also been made on how to approach humans with mobile robots to initiate interaction [2] and maximize politeness [3]. Furthermore, having a directed or averted gaze also has an influence on the minimal comfortable interaction distance, increasing or decreasing depending on the gender of the person [4]. Similarly, a robot with a motion-oriented gaze behavior can be perceived as more engaging and human-like [5], and it has been shown that a robot matching the personality of its users by adopting its gaze behavior can have a positive impact in a puzzle-solving task [6]. The appearance of the robot has also an impact on its perceived effectiveness, as it has been observed that people systematically preferred robots for jobs when the human-likeness of the

A. Agah et al. (Eds.): ICSR 2016, LNAI 9979, pp. 362–371, 2016.
DOI: 10.1007/978-3-319-47437-3_35

robot matched sociability requirements [7]. Studies have also been made on the perceived safety of the motion of industrial robots in both real and virtual settings [8]. This illustrates how different robots, even with the same objectives, can have a different impact depending not only on their overall behavior, but also on their physical appearance and motion capabilities.

To achieve standardization in HRI experiments, using identical robots would avoid introducing unwanted factors. Obviously, this is not possible in a practical sense. Except for a few popular robots like NAO from Softbank Robotics (former Aldebaran Robotics) [9] or PR2 from Willow Garage [10], there are not many other interactive robots that achieved the kind of commercial success necessary to make this feasible. Furthermore, research groups that are more interested in the design aspect of interactive robots understandably prefer to conduct experiments with their own unique systems. However, we posit that there is an alternative to having researchers use identical robots. In order to facilitate comparison between different robots used in similar HRI experiments, we propose the development of a classification system for the expressiveness of humanoid robots. This paper illustrates the development of such a classification system for describing robot expressiveness, based on a selection of robots that can often be seen in HRI research. von Zitzewitz et al. [11] propose that human-likeness of humanoid robots can be quantified by a network of parameter fields as perceived by humans. Two of these fields are visual appearance and behavior, which describes parameters such as motion and nonverbal communication.

The goal of this paper is not to propose a psychological analysis of how humans perceive expressions reproduced by robots, but rather to provide a common language to describe technical features used by robots in HRI studies. Hence, the notation proposed in this paper aims at describing expressive capabilities, not emotional ones, as we believe that the perception of emotions is out of the scope of this work. In this paper, expressive capabilities refer to robot motion (or simulation with a display) that act as a non-verbal communication channel.

This paper is structured as follows: Sect. 2 describes a selection of robots that can be seen in HRI research, focusing on their capabilities for facial expressions. Section 3 proposes a classification system and a shorthand notation for describing these capabilities based on features that are either functional (that have uses beyond expression) or non-functional (that are exclusively used for expression). Finally, Sect. 4 concludes the paper with suggestions on how this classification could be extended to other features of interactive robots.

2   Expressive Robot Features Found in HRI Studies

To develop a classification system of expressive features, a selection of robots found in HRI and social robotics studies was made. The robot selection used in this paper is not meant to be exhaustive. Instead, robots were selected to show sufficiently different ways of reproducing human features and behaviors, and thus help in the development of a classification system. To extend the selection of humanoid robots beyond legged ones, "humanoid robot" in this paper refers to robots that have a human-shaped head with eyes. Furthermore, we decided to not include androids in this selection. The notation developed in this paper is meant for robots that are closer to the machine-like end of the uncanny valley [12], and cover capabilities that are not necessarily human-like, for instance the use of LCD displays for some features. The following subsection describes briefly each robot, and is partially based on the following studies involving expressive robots and human interpretation: FLASH [13,14], IRL-1 [15], NAO [16–18], Nexi MDS [19,20], and Wakamaru [21,22]. A general survey on automatic recognition and generation of body movements for affective expression can be found in [23]. The pictures used in this section were obtained either from the robot manufacturers' website, cited work, or taken by the authors of this paper (Fig. 1).

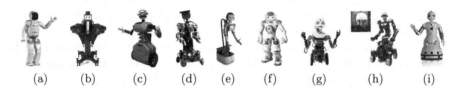

(a)   (b)   (c)   (d)   (e)   (f)   (g)   (h)   (i)

Fig. 1. (a) ASIMO, (b) Baxter, (c) FLASH, (d) IRL-1, (e) M-1, (f) NAO, (g) Nexi MDS, (h) Rollin' Justin, (i) Wakamaru.

2.1 Robots

ASIMO is a well-known legged robot manufactured by Honda [24]. It is of average height (1.30 m), with an oriented but expressionless face, although its two cameras can appear as eyes under the right lighting.

Baxter is a relatively tall (1.78 m to 1.91 m with adjustable pedestal) commercial robot from Rethink Robotics [25]. Its two arms with 7 degrees of freedom (DoF) are compliant and can be interacted directly with for example for teaching tasks. Its display shows virtual eyes and can be oriented on its pan angle. While it does not by default, its display could also be used for a mouth.

FLASH is a 1.30 m of height robot built by the Wroclaw University of Technology for the EU FP7 LIREC Project. It includes the EMYS head [13], which can be seen in standalone form in facial expression-related works such as [14]. While FLASH does not explicitly have eyebrows or a mouth, the upper and lower discs can act as them. Furthermore, its eyelids can go up and down as well as rotate around the optical axis.

IRL-1 is a 1.40 m of height custom robot from Université de Sherbrooke [26]. Its articulated expressive head comes from a previous robot named Reddy, which is also known as Melvin [27] and CRAMER [28].

Meka M-1 is a 1.80 m of height humanoid robot built by Meka Robotics (now part of Google X). Multiple versions of this robot exist around the world with different head shapes. The version selected includes eyes with cameras and functioning eyelids.

Another robot, REEM [29] from Pal Robotics, is a robot of similar height (1.70 m), but is available with either a differential mobile base or legged locomotion. Regarding facial expression, the EU RobotCub project robot iCub [30], while smaller (1.04 m) and legged, has very similar facial expression features, and uses LED matrices for the mouth and eyebrows.

NAO from Softbank Robotics (former Aldebaran Robotics) [9] is a small (58 cm) humanoid, legged robot. Its face, while mechanically fixed, can be oriented, and its multi-segmented eyes can change colors and shape as a mean of expression.

Nexi MDS from the MIT Media Lab was built in collaboration with UMASS Amherst's Laboratory for Perceptual Robotics, Xitome Design, and Meka. It is meant to be approximately the height of an adolescent child. It is also known as Octavia at the Navy Center for Applied Research in Artificial Intelligence, and has been used in social engagement studies [31].

Rollin' Justin is a robot from the Deutsches Zentrum für Luft- und Raumfahrt (DLR, German Aerospace Center). It has a shoulder height of 1.60 m. Its arms are based on the third generation of the DLR Light Weight Arm (LWR III), and is notably used in research on compliant whole-body manipulation [32]. From an expressiveness point of view, PR2 from Willow Garage [10] can be compared to Rollin' Justin, as it also has fixed cameras that can be perceived as eyes, an oriented head and an omnidirectional mobile base.

Wakamaru is a robot that was developed by Mitsubishi Heavy Industries and was meant for natural communication with humans [33]. It is 100 cm tall, includes two articulated arms, and an oriented head with fixed eyes (Fig. 1).

2.2   Common Features

Table 1 lists expressive features such as having eyebrows or a mouth that are common to at least two robots from the selection, and the following describes the differences between these robots for those features.

Eye gaze. For eye gaze, we observe a wide variety of solutions, from completely virtual in the case of Baxter, to unique features such as an extra DoF for eye popping with FLASH. Some of them have fixed orientation (e.g., ASIMO, NAO, Rollin' Justin, Wakamuru), and rely entirely on the head to direct the gaze. An advantage of having eyes that can be rotated separately from the head is a faster reaction time to new gaze targets. Except for three robots (i.e., ASIMO, Rollin' Justin, and Wakamaru), all have a pupil distinguished from the rest of the eye. Also, three robots use cameras for eyes (i.e., Meka M-1, Nexi MDS, and Rollin' Justin), which indicates that the robot can actually see with them. This is not the case for all robots. For instance, hiding the eyes of the NAO does not block its cameras located on the top of its head and its mouth.

Head orientation. All robots in this selection can rotate their head, although Baxter does not have closed-loop control of its tilt angle because of the lack of a position sensor. Three robots (i.e., Meka M-1, Nexi MDS, Wakamaru) have an

Table 1. Description of the features found on the selected robots. The letters P, T, and R refer to Pan, Tilt, and Roll, respectively. (1) These robots feature an additional tilt angle on the neck.

Robot	ASIMO	Baxter	FLASH	IRL-1	Meka M-1
Eye gaze	Fixed	Virtual	Fixed	P, T	P, T
Head orientation	P, T	P, T	P, T (1)	P, T	P, T (1), R
Eyebrows	None	Virtual	None	R	None
Eyelids	None	Virtual	None	None	Yes
Mouth and jaw	None	None	None	Mouth	Mouth (LEDs)

Robot	NAO	Nexi MDS	Rollin' Justin	Wakamaru
Eye gaze	Fixed	P, T	Fixed	Fixed
Head orientation	P, T	P, T (1), R	P, T	P, T, R
Eyebrows	None	R, lift	None	None
Eyelids	None	Yes	None	None
Mouth and jaw	None	Both	Mouth	None

extra DoF for roll angle, and two robots (i.e., Meka M-1 and Nexi MDS) control the tilt with two actuators (neck and head).

Eyebrows. Two robots in this selection have physical eyebrows (i.e., IRL-1 and Nexi MDS). However, FLASH can use the tilt angle of its upper disc to represent the brow on a wide range of motion. Furthermore, the eyebrows on Nexi MDS have an extra DoF, vertical lift, which means they can rise or lower from both corners. Finally, Baxter has eyebrows on its display.

Eyelids. Three robots have mechanical eyelids (i.e., FLASH, Meka M-1, and Nexi MDS), while one can display them on a screen (Baxter). While FLASH only has top eyelids, they can be rotated in addition to being closed. The rotation of the eyelids can play the role of frowning eyebrows.

Mouth and jaw. Only one robot has a mechanically actuated mouth (IRL-1), represented by two flexible tubes moved by the rotation of four mouth corners. However, FLASH has a lower disc that can act as a jaw, and Nexi MDS has a fixed lower lip mounted on an articulated jaw. As with other features, Baxter could be programmed to display a mouth.

3   Classification of Expressiveness

From the set of technical features described in Sect. 2, a generalization and
the construct of a classification system can be attempted. In this paper, two
dimensions of this classification will be looked at: functional and non-functional.
Functional expressive features are robot features that have an utility beyond
producing expressions and This is obviously the case for robot capabilities such
as locomotion or speech, but the gaze can also be used in a neutral fashion
to change the orientation of sensors, and indicate where the robot is looking.
However, eyebrows usually do not have any other function than expression. Furthermore, while it can be argued that expression always have a function as a
communication channel, functional features in this paper refer to features that
also go beyond communication. We believe that separating features in such a
way is important from the point of view of robot designers as functional features
carry additional constraints. For instance, the size of the eyes with embedded
cameras also has to provide proper sensor apertures. Table 2 summarizes the
classification system for functional features.

3.1   Gaze

Gaze, achieved by rotating the head and/or the eyes, serves both functional
and expressive requirements. On robots such as Rollin' Justin, it is necessary
to direct its cameras. For gaze, we propose to classify it on a spectrum: G0 for
robots without gaze, G1 for a gaze from a fixed head, and G2 for a gaze from an
oriented head. Furthermore, to distinguish robots that have mobile eyes, meaning
they can orient their gaze independently from their head, the "+" suffix is added.
This implies the possibility of performing expressive motions such as nods with
the head without changing the target of the gaze. Finally, for robots that use
a display for the gaze, the "V" suffix is added. Categories G0 and G1 are not
represented by the robots in Sect. 2. However, robots such as Kompaï [34] from
Robosoft, which features painted eyes on a fixed head, can be considered as G1.
Similarly, Care-O-bot 4 [35] from Fraunhofer IPA in its fixed head configuration
would be G1V, and G1+V if its display is programmed with oriented eyes.

3.2   Mouth

The mouth can also be seen as a functional element. If the motion of the mouth
is synchronized with speech generation, it can be used as a visual cue (e.g., to
identify which robot is speaking in a close group). For the mouth, we propose
a classification similar to the one used for the gaze: M0 for robots without a
mouth (e.g., ASIMO, Baxter, Rollin' Justin, Wakamaru), M1 for a fixed mouth
(e.g., NAO, alternative Rollin' Justin), and M2 for mouths with one or more
DoF (e.g., FLASH, IRL-1, Nexi MDS). For robots using a display or a LED
matrix for their mouth, the V suffix (Meka M-1) is also added.

Table 2. Classification of the features.

Robot	G0	G1	G1+	G2	G2+	M0	M1	M2
ASIMO				*		*		
Baxter				V		*		
FLASH				*				*
IRL-1					*			*
Meka M-1					*			V
NAO				*			*	
Nexi MDS					*			*
Rollin' Justin				*		*	*	
Wakamaru				*		*		

Table 3. AUs achievable by robots with an articulated face. A "*" indicates an AU that can be achieved independently, and a "*X" an AU that is mutually exclusive with AU "X". Data for FLASH and IRL-1 comes from [13] and [15].

Action Unit	FLASH	IRL-1	Meka M-1	Nexi MDS
AU1: Inner Brow Raiser	*	*2		*
AU2: Outer Brow Raiser	*	*1		*
AU4: Brow Lowerer	*			*
AU5: Upper Lid Raiser	*		*	*
AU10: Upper Lip Raiser		*	*	
AU12: Lip Corner Puller		*	*	
AU17: Chin Raiser	*			
AU20: Lip Stretcher		*25	*	
AU25: Lips Part		*20	*	
AU26: Jaw Drop	*	*		*

3.3 Non-functional Expressive Features

Purely expressive features can be harder to classify from a number of DoFs standpoint. For instance, eyebrow-frowning on FLASH is performed by its eyelids, which prevents a one-to-one relationship between DoFs and capabilities. Since facial expressions are largely inspired from human ones, we propose to use Action Units (AUs) of the Facial Action Coding System (FACS) [36], which are used notably for emotion recognition. FLASH has already been classified in this manner [13], and Shayganfar et al. used AUs to express emotions with Melvin, which shares its head with IRL-1 [15]. Thus, the face expressiveness of a robot could be described with respect to the set of AUs that it can reproduce. However, AU-related capabilities are not always orthogonal. For instance, while AU1 (Inner Brow Raiser) and AU2 (Outer Brow Raiser) can be performed by IRL-1, they cannot be combined, which is a feature available on Nexi MDS and achieved by the eyelids and the upper disc of FLASH. Furthermore, for G2 gaze robots, their oriented head allows at least AU51 to AU54 (head pan and tilt), and some robots (Meka M-1, Nexi MDS, Wakamaru) can achieve AU55 and AU56 (head roll). For robots with at least a M2 mouth, facial AUs are listed in Table 3.

4   Conclusion

This paper presents the first steps in a proposal for classifying the expressiveness of robots found in the HRI community. From functional expressive features, a short-hand notation has been developed, which can be augmented with the help of facial AUs for non-functional ones. We believe that having a common set of terms for describing the expressive features of robots will help the description of standardized HRI experiments. The underlying goal of this work is to arrive at a full classification of whole-body expressiveness. For instance, robots such as Rollin' Justin, different versions of Meka M-1, and most legged humanoid robots can control the tilt angle of their torso. Beyond balance and manipulation reasons, the angle and pose of the shoulders in a static posture have also a role in emotion recognition [37]. Arm gestures, another functional expressive feature that is DoF-related, and the spoken expression of emotions, whether by content (e.g., stating "I am happy") or by the modulation of speech, are two other important components of expressiveness. Furthermore, non-human features such as the color-changing LEDs in the eyes of NAO, offer a large range of expression that cannot be easily associated to human ones, and it is thus important to include them in a future extension of this notation.

By generalizing functional and non-functional expressive features of each subsystem of an autonomous robot and evaluating them on a scale of their human-likeness, producing a complete taxonomy for the expressiveness of interactive robots for comparative studies will be possible.

Acknowledgements. This work has been partially funded by EU Horizon2020 ENRICHME project grant agreement no. 643691C.

References

1. Blow, M., Dautenhahn, K., Appleby, A., Nehaniv, C.L., Lee, D.C.: Perception of robot smiles and dimensions for human-robot interaction design. In: Proceedings of the 15th IEEE International Symposium on Robot and Human Interactive Communication, pp. 469–474 (2006)
2. Satake, S., Kanda, T., Glas, D.F., Imai, M., Ishiguro, H., Hagita, N.: How to approach humans? strategies for social robots to initiate interaction. In: Proceedings of the ACM/IEEE International Conference on Human-Robot Interaction, pp. 109–116 (2009)
3. Kato, Y., Kanda, T., Ishiguro, H.: May i help you? design of human-like polite approaching behavior. In: Proceedings of the Tenth Annual ACM/IEEE International Conference on Human-Robot Interaction, pp. 35–42 (2015)

4. Takayama, L., Pantofaru, C.: Influences on proxemic behaviors in human-robot interaction. In: Proceedings of the IEEE/RSJ International Conference on Intelligent Robots and Systems, pp. 5495–5502 (2009)
5. Sorostinean, M., Ferland, F., Dang, T.-H.-H., Tapus, A.: Motion-oriented attention for a social gaze robot behavior. In: Beetz, M., Johnston, B., Williams, M.-A. (eds.) ICSR 2014. LNCS (LNAI), vol. 8755, pp. 310–319. Springer, Heidelberg (2014). doi:10.1007/978-3-319-11973-1_32
6. Andrist, S., Mutlu, B., Tapus, A.: Look like me: matching robot personality via gaze to increase motivation. In: Proceedings of the 33rd Annual ACM Conference on Human Factors in Computing Systems, pp. 3603–3612 (2015)
7. Goetz, J., Kiesler, S., Powers, A.: Matching robot appearance and behavior to tasks to improve human-robot cooperation. In: Proceedings of the 12th IEEE International Workshop on Robot and Human Interactive Communication, pp. 55–60 (2003)
8. Or, C.K., Duffy, V.G., Cheung, C.C.: Perception of safe robot idle time in virtual reality and real industrial environments. Int. J. Ind. Ergon. 39(5), 807–812 (2009)
9. Robotics, S.: Who is nao? https://www.ald.softbankrobotics.com/en/cool-robots/nao
10. Garage, W.: PR2 overview. http://www.willowgarage.com/pages/pr2/overview
11. von Zitzewitz, J., Boesch, P.M., Wolf, P., Riener, R.: Quantifying the human likeness of a humanoid robot. Int. J. Soc. Robot. 5(2), 263–276 (2013)
12. Mori, M., MacDorman, K.F., Kageki, N.: The uncanny valley [from the field]. Robot. Autom. Mag. 19(2), 98–100 (2012)
13. Kedzierski, J., Muszynski, R., Zoll, C., Oleksy, A., Frontkiewicz, M.: Emys–emotive head of a social robot. Int. J. Social Robot. 5(2), 237–249 (2013)
14. Ribeiro, T., Paiva, A.: The illusion of robotic life: principles and practices of animation for robots. In: Proceedings of the Seventh Annual ACM/IEEE International Conference on Human-Robot Interaction, pp. 383–390 (2012)
15. Shayganfar, M., Rich, C., Sidner, C.L.: A design methodology for expressing emotion on robot faces. In: Proceedings of the IEEE/RSJ International Conference on Robots and Systems, pp. 4577–4583 (2012)
16. Beck, A., Hiolle, A., Mazel, A., Cañamero, L.: Interpretation of emotional body language displayed by robots. In: Proceedings of the 3rd International Workshop on Affective Interaction in Natural Environments, pp. 37–42 (2010)
17. Cohen, I., Looije, R., Neerincx, M.A.: Child's recognition of emotions in robot's face and body. In: Proceedings of the 6th International Conference on Human-robot Interaction, pp. 123–124 (2011)
18. Haring, M., Bee, N., Andre, E.: Creation and evaluation of emotion expression with body movement, sound and eye color for humanoid robots. In: Proceedings of the 20th IEEE International Symposium on Robot and Human Interactive Communication, pp. 204–209 (2011)
19. Jung, M.F., Lee, J.J., DePalma, N., Adalgeirsson, S.O., Hinds, P.J., Breazeal, C.: Engaging robots: easing complex human-robot teamwork using backchanneling. In: Proceedings of the Conference on Computer Supported Cooperative Work, pp. 1555–1566 (2013)
20. Strait, M., Canning, C., Scheutz, M.: Let me tell you! investigating the effects of robot communication strategies in advice-giving situations based on robot appearance, interaction modality and distance. In: Proceedings of the ACM/IEEE International Conference on Human-Robot Interaction, pp. 479–486 (2014)
21. Huang, C.M., Mutlu, B.: Modeling and evaluating narrative gestures for humanlike robots. In: Robotics: Science and Systems (2013)

22. Muto, Y., Takasugi, S., Yamamoto, T., Miyake, Y.: Timing control of utterance and gesture in interaction between human and humanoid robot. In: The 18th IEEE International Symposium on Robot and Human Interactive Communication, pp. 1022–1028 (2009)
23. Karg, M., Samadani, A.A., Gorbet, R., Kuhnlenz, K., Hoey, J., Kulic, D.: Body movements for affective expression: a survey of automatic recognition and generation. IEEE Trans. Affect. Comput. **4**(4), 341–359 (2013)
24. Honda: ASIMO specifications. http://asimo.honda.com/asimo-specs/
25. Robotics, R.: Baxter technical specifications. http://www.rethinkrobotics.com/baxter/tech-specs/
26. Ferland, F., Létourneau, D., Aumont, A., Frémy, J., Legault, M.A., Lauria, M., Michaud, F.: Natural interaction design of a humanoid robot. J. Hum. Rob. Interact. **1**(2), 14–29 (2012)
27. Rich, C., Sidner, C.L.: Robots and avatars as hosts, advisors, companions, and jesters. AI Mag. **30**(1), 29 (2009)
28. Carter, K., Scheutz, M., Schermerhorn, P.: A humanoid-robotic replica in USARsim for HRI experiments. In: IROS Workshop on Robots, Games, and Research (2009)
29. Robotics, P.A.L.: REEM: Full-size humanoid service robot. http://pal-robotics.com/en/products/reem/
30. Metta, G., Natale, L., Nori, F., Sandini, G., Vernon, D., Fadiga, L., Von Hofsten, C., Rosander, K., Lopes, M., Santos-Victor, J., et al.: The iCub humanoid robot: an open-systems platform for research in cognitive development. Neural Netw. **23**(8), 1125–1134 (2010)
31. Moshkina, L., Trickett, S., Trafton, J.G.: Social engagement in public places: a tale of one robot. In: Proceedings of the 2014 ACM/IEEE International Conference on Human-Robot Interaction, pp. 382–389 (2014)
32. Dietrich, A., Bussmann, K., Petit, F., Kotyczka, P., Ott, C., Lohmann, B., Albu-Schäffer, A.: Whole-body impedance control of wheeled mobile manipulators. Autonomous Robots: Special Issue on Whole-Body Control of Contacts and Dynamics for Humanoid Robots, May 2015
33. Mitsubishi Heavy Industries: Wakamaru. http://www.mhi-global.com/products/detail/wakamaru_technology.html (available on archive.org)
34. IEEE Spectrum: Robosoft unveils kompai robot to assist elderly, disabled. http://spectrum.ieee.org/automaton/robotics/medical-robots/robosoft-kompai-robot-assist-elderly-disabled
35. IEEE Spectrum: Care-O-bot 4 is the robot servant we all want but probably can't afford. http://spectrum.ieee.org/automaton/robotics/home-robots/care-o-bot-4-mobile-manipulator
36. Ekman, P., Friesen, W.V.: Facial action coding system: A technique for the measurement of facial movement. Consulting Psychologists Press (1978)
37. Coulson, M.: Attributing emotion to static body postures: recognition accuracy, confusions, and viewpoint dependence. J. Nonverbal Behav. **28**(2), 117–139 (2004)

Facing Emotional Reactions Towards a Robot – An Experimental Study Using FACS

Isabelle M. Menne[✉], Christin Schnellbacher, and Frank Schwab[✉]

Department of Media Psychology, Institute of Human-Computer-Media,
University of Würzburg, Würzburg, Germany
{isabelle.menne,frank.schwab}@uni-wuerzburg.de

Abstract. As robots are starting to enter not only our professional lives but also our domestic lives, new questions arise: What about the emotional impact of getting into contact with this new 'species'? Can robots elicit emotional reactions from humans? Assuming that humans may react in the same way towards robots as they do towards humans (Media Equation), it is important to investigate the factors influencing this emotional experience. But systematic research in this area remains scarce. An exception is the study conducted by Rosenthal-von der Pütten, Krämer, Hoffmann, Sobieraj, and Eimler in 2013 addressing the question of emotional reactions towards a robot experimentally. Taking the study by Rosenthal-von der Pütten et al. as a starting point and following their suggested multimethod approach as well as Scherer's assumption about the five components of emotions, we added the motor expression component to further investigate the multilevel phenomenon of emotional reactions towards robots. We used the Facial Action Coding System (FACS) to analyze the facial expressions of participants viewing videos of the robot dinosaur Pleo in a friendly interaction or being tortured. Participants showed Action Units associated with intrinsic unpleasantness more frequently in the torture-video-condition than during the reception of the normal video. Participants also reported more negative feelings after the torture video than the normal video. The findings indicate the importance of investigating emotional reactions towards robots for a social robot to be an ideal companion. Furthermore, this paper shows that the application of FACS to research in human-robot interaction is a fruitful and insightful enhancement to commonly used self-report measures. We conclude this paper with some recommendations for improving the design of social robots.

Keywords: Emotional response · Facial expression analysis · FACS · Design guidelines

1 Introduction

While the appearance of robots in our daily lives is quite novel, the idea of having robots around is very old. The first operative robot was built by Hero of Alexandria, 85 A.D., though very limited in its functions [19]. Since then, many attempts have been made to build robots aiming to capture main human features, integrate them and make life easier. Today, the fruits of this labor have not only entered our business lives but

© Springer International Publishing AG 2016
A. Agah et al. (Eds.): ICSR 2016, LNAI 9979, pp. 372–381, 2016.
DOI: 10.1007/978-3-319-47437-3_36

also our private lives in the form of vacuum cleaning robots or entertainment robots. As robots are becoming more and more popular, even laypeople come into contact with robots. While the interaction with this new species becomes normal, little is known about the consequences. Answers must be found to new questions arising with the arrival of the robotic age, such as: What kind of role should artificial entities play in our lives and what role should we allow them to play? Could robots be our (emotional) companions? Deeply associated with this question is the fundamental question: can robots have an (emotional) impact on humans? As robots might very well change the way we live today, the significance of finding answers to those questions increases. A central aim of social robotic research is to develop robots able to communicate and interact with humans in a natural way, similar to human-human interaction. An ideal robot should therefore be capable to intuitively and naturally engage in social interactions with humans so that users are able to empathize with it and do not need to adapt to the machine, but rather vice versa [4]. Despite the importance of studying whether people are able to empathize with robots, e.g. by displaying emotional reactions, research remains rather scarce (exception: [17]). This paper adds further insights into the emotionality of responses towards artificial entities, providing evidence that an observational method such as FACS adds to a better understanding of emotional phenomena in the context of human-robot interaction (HRI).

2 Theoretical Background

2.1 Studying Emotions

Emotions are a fundamental part of being human and can be expressed on different levels, such as verbally and non-verbally to mention but one of the roughest levels of differentiation. According to [20], emotions are a multi-level phenomenon as they can be described on five levels (components): a cognitive, a neurophysiological, a motivational, an expressive and a subjective feeling component. According to this, there are five ways for measuring emotions. Ideally, emotions should be measured on all five levels, however, such a comprehensive measurement of emotions has never been done [20]. Instead, the cognitive component of emotion, typically measured by questionnaires, represents the most widely used method for measuring emotions, not only in the social sciences, but also in HRI and related fields (for a review in HRI cf. [3]). Since it is much less time consuming than alternative methods, economical in its implementation, analysis and evaluation, it is a very popular method. This method has certain flaws, however: it relies heavily upon the assumption that subjects can always reliably and validly recall their emotional experiences and are willing to report them. This is not always true though. Objective observational methods are not as prone to biases as self-reports, therefore providing a better understanding of the often unconscious and automatic emotional responses to a stimulus. In this study we added the motor expression component by using the Facial Action Coding System (FACS) [9] to further investigate the multilevel phenomenon of emotional reactions towards robots.

2.2 Facial Action Coding System as a Research Method

FACS is the most widely and most frequently used method for facial expression analysis (e.g. [10]). It is an objective and standardized method for measuring visual appearance changes in the face. FACS defines Action Units (AUs) as the smallest unit of muscular activity that can be observed in the human face. Every movement is therefore describable by either a single AU or a combination of several AUs. FACS provides a standardized coding and analysis procedure with high objectivity, validity and reliability [22]. Because of its descriptive power, FACS has emerged as the criterion measure of facial behavior in multiple fields including computer vision (e.g. [13]) and social (e.g. [10]) studies of emotion, among many other disciplines. Correlations have been found between emotions and facial expressions [7]. As emotional processes are often unconscious and therefore not accessible to self-report measures [6], FACS provides an objective observation method. Furthermore, most computer scientists refer to FACS when emotion is included, especially concerning the recognition of human facial expressions by robots as well as the robot's simulation of emotion with facial expressions (e.g. [16]). [11] report that most robots express emotions based on FACS, as it is the only most comprehensive and generally accepted method for generating facial expressions on robots. In this study, we follow Scherer and Ellgring's [21] componential emotion theory approach and focus only on special AUs associated with appraisal checks. The Authors identified the AUs 12 + 25 + 38, appearing especially often with a pleasurable state (appraisal check: Intrinsic Pleasantness) and the AUs 9, 10, 15, 39 correlated with unpleasant feelings (appraisal check: Intrinsic Unpleasantness) [21]. In terms of the five basic emotions, the Action Unit combination 12 + 25 + 38, is strongly expected to occur with a feeling of joy. The AUs 9, 10, 15, 39 are expected to occur with fear and disgust. It is expected of AUs 12 + 25 + 38 to appear together whereas the AUs 9, 10, 15, 38 can also appear independently of each other [21].

2.3 The Media Equation and Matters of Measurement

The Media Equation states that people mindlessly treat computers, TV and other media entities like real people, interacting with them in the same way as they do with people. Computers can be seen as social actors [15], eliciting automatic and unconscious reactions to media entities, which are not limited to special groups of people, but are fundamentally rooted in human nature. Due to the unconsciousness of reactions, Reeves and Nass [15] emphasize, that "attempts to verify the media equation can't rely solely on talking to people, (...) or asking them questions on a survey" (p. 7). Rather, objective observation methods seem to be an enhancement or even a better option. However, the use of observational methods has been rather scarce. Self-report measurements prevail and seem to be the method of choice [cf. 3]. While these questionnaires are appropriate to assess the overall tendency of emotional reactions and get a first glimpse on subjective feelings, they come with several limitations, e.g. validity or reliability. While essentially all questionnaires concentrating on very subjective and unfocused experiences are affected by those issues, this is also true for self-report measurements of emotions. Thus, the (additional) use of objective observational methods seems to offer a solution for this problem.

2.4 Computers, Virtual Agents and Robots as Social Actors?

Even though there is only little research regarding overall emotional reaction tendencies towards robots, research in related areas, such as human-computer interaction, has shown that machines elicit social responses by their users and shape their perceptions [14, 15]. Minimal cues of human abilities, such as the sound of a voice or offering interactivity, elicit social responses [14, 15]. As humans react socially towards computers, findings of the Media Equation have been successfully transferred to virtual agents [1]. As research focusing on social reactions towards robots is rather new, there are only some hints [2, 17, 18], the Media Equation might also be valid for HRI, probably exceeding the effects observed in human-computer interaction and reactions towards virtual agents due to the physical embodiment.

2.5 Positive and Negative Emotional Reactions

Violence or negative experiences most often lead to negative emotional reactions. Studies in social science have studied facial expressions during negative experiences (e.g. [5]). Evidence for the display of negative emotional facial expressions has been found, for example when watching another person suffer from violence [23]. As systematic research on emotional reactions towards robots is rather scarce, there are only some hints. [12] report the elicitation of positive emotions in the interaction of humans suffering from dementia and a robotic cat applied for therapy purposes. Anecdotal findings of a study by [2] show that negative feelings were elicited when participants were asked to destroy a Crawling Microbug Robot with a hammer. The first studies to systematically measure emotional reactions [17, 18] provide evidence for the capability of robots to elicit emotional reactions. Even though overall emotional reactions have been found, no systematic investigations on emotional facial expressions towards robots have been done so far.

2.6 Study Outline and Hypotheses

Altogether, research concerning emotional reactions is currently scarce with the exception of the experimental study of [17]. The authors measured emotional reactions systematically using standardized questionnaires as well as physiological methods. The entertainment dinosaur robot Pleo was used and shown in two different situations, either being treated nicely or being tortured. A 2 × 2 Design was used with the between-subjects factor "prior interaction with the robot" as well as a within-subjects factor "type of video". The authors found evidence for an increased physiological arousal during the reception of the torture video. Additionally, fewer positive and more negative emotions were reported after the reception. The prior interaction yielded no significant effect [17]. The study by [17] inspired and guided our research. Since prior interaction did not have an effect, we did not implement the factor in our study. Next to standardized self-report measures, we added FACS as it offers several advantages compared to self-report measurements and physiological methods. Subjective methods are rather limited in their ability to capture "real-time" emotional processes. Physiological measures, however, provide this advantage, but suffer from certain limitations concerning their obtrusiveness (e.g. interfering natural body movement, lack of comfort). To the best of our knowledge, this is the first study to apply a behavioral measurement of facial activity to objectively and unobtrusively study human's facial expressions of emotion towards a robot.

Research indicates that facial expressions are associated with the experience of emotions (e.g. [7, 10]). Therefore we hypothesize that participants show the AUs 12 + 25 + 38, associated with Intrinsic Pleasantness, more frequently during the reception of the friendly video than during the torture video (Hypothesis 1a {H1a}). Secondly, we assume that participants show the AUs 9, 10, 15, 39, associated with Intrinsic Unpleasantness more frequently during the reception of the torture video than during the friendly video (Hypothesis 1b {H1b}).). Additionally we expect to find similarities to [17] concerning the self-report data. We hypothesize that participants feel more positively after the reception of the friendly video than after the torture video (Hypothesis 2a {H2a}). Furthermore, we assume that participants feel more negatively after the reception of the torture video than after the friendly video (Hypothesis 2b {H2b}). While we expect similarities in how people react towards a robot, providing further indications for the validity of the Media Equation in HRI, we also believe a multi-method approach is necessary for a better understanding of the multifaceted phenomenon of emotions, especially in the context of HRI. Research focusing on emotional reactions will provide satisfying answers to questions concerning the emotional impact of robots such as: What are the consequences of a close relationship with a robot? How will we feel when we have to separate from our longtime companion when it becomes dysfunctional? With the arrival of robots in our daily lives, the emotional aspect of the human robot relationship becomes important, making it a relevant and vital topic of study for HRI research.

3   Method

To test the hypotheses we chose a one-factorial, between-subjects design. The type of video was the independent variable. The participants were randomly assigned to one of the two conditions: "friendly interaction" or "torture interaction". The dependent variable was the emotional reaction toward the robot, captured with self-report measures (H2a, H2b) and the frequency of specific facial expressions (H1a, H1b).

3.1   Stimulus Material

The entertainment robot Pleo of Innvo Labs was used for creating two short video clips of either "friendly interaction" or "torture interaction". We produced two short films, oriented at [17], who described two one-minute short clips with five sequences lasting ten seconds, each followed by a two second black screen. One video clip contained friendly interaction with the robot (e.g. Pleo being caressed or fed with a leaf) and the other shows Pleo being mistreated (e.g. hit on the head or head hit on the table).

3.2 Robot Dinosaur by Innvo Labs

"Pleo reborn" by Innvo Labs was used as stimulus material in the videos. Its appearance is inspired by a baby Camarasaurus Dinsoaur. As the robot is built to be "a life form" (Innvo Labs), multiple technical details allow it to react realistically to its environment. The robot has sensors placed all over its body, allowing it to react to touch and movement, as well as temperature and light. Furthermore, a speaker allows it to make utterances. The robot is equipped with all these functions to be utilized as an artificial robotic pet and an entertaining companion in everyday life. Pleo offers much potential for observing emotional reactions of participants as it expresses emotions and pain through noises and movement.

3.3 Sample, Questionnaires and Conduction of the Study

The sample consisted of 50 women and 12 men. Participants were between 18 and 29 years old ($M = 20.32$; $SD = 1.91$). The study was conducted in single sessions due to constraints of video recording. Every participant was randomly assigned to one condition. To avoid sequence effects, half of the participants saw the videos in reverse order. After the arrival of the participant, she/he signed an informed consent of the video recording and received general information concerning data privacy, voluntariness, anonymity and information about the study procedure. Participants then completed the first web based (SosciSurvey) questionnaire concerning demographic data and their past experiences with the robot dinosaur Pleo. While participants then watched one of the two videos (either "friendly interaction" or "torture interaction"), their facial expressions were recorded with a camera. After that, participants completed the PANAS questionnaire. We used the Positive and Negative Affect Schedules (PANAS, [24]).

3.4 Analysis with the Facial Action Coding System

Videos were coded separately by trained FACS [9] coders, determining facial activity in the face. The videos of the participants watching Pleo either being mistreated or being treated nicely were coded manually for a predetermined set of AUs according to [21] using the FACS manual [9]. For further statistical analysis we used the software package "SPSS Statistics 22" of IBM. To prove the reliability of the coding, interrater reliability was calculated by the coding of the first person and a second one. The interrater reliability of the AUs 12 + 25 + 38 (Cohens Kappa = .70) and the AUs 9, 10, 15, 39 (Cohens Kappa = .79) was satisfying (e.g. [5]).

4 Results

The analysis indicated that participants reported significantly more negative feelings after the reception of the torture video than after the reception of the friendly video $t(60) = 5.58, p < .001, 95\% CI [-1.06, -.50]$. Additionally, another unpaired one-sided t-test indicated that participants reported significantly more positive feelings after the reception of the friendly video, than after the reception of the torture video $t(60) = 2.29$, $p = .013, 95\% CI [.04, .62]$ (Table 1).

Table 1. Influence of Type of Video on the Emotional State. T-test with the PANAS subscales as dependent variable and type of video as independent variable

	Friendly Video		Torture Video			
	M	SD	M	SD	t	δ'
PANAS positive	3.41	.53	3.08	.60	2.29*	.58
PANAS negative	1.45	.49	2.23	.60	5.58***	1.41

Note. Means are based on a 5-point Likert scale (PANAS). * p < .05. ** p < .01. ***p < .001

The analysis indicated no effect for the displayed number of AUs associated with intrinsic pleasantness (AUs 12 + 25 + 38) during the reception of the friendly video as compared to the displayed number of Actions Units during the reception of the torture video, $t(59) = .17, p = .43, 95\% CI [-.65, .55]$. An unpaired one-sided t-test yielded a significant effect of the displayed number of AUs associated with intrinsic unpleasantness (AUs 9, 10, 15, 39) during the reception of the torture video as compared to the displayed number of AUs during the reception of the friendly video, $t(37.48) = 3.25$, $p = .001, 95\% CI [-1.26, -.29]$, (Table 2).

Table 2. Influence of Type of Video on the Action Units Displayed. T-test with the number of AUs displayed during the reception of the videos as dependent variable and type of video as independent variable.

	Friendly Video		Torture Video			
	M	SD	M	SD	t	δ'
AU 12 + 25 + 38	.52	1.06	.57	1.28	.17 n.s.	
AU 9, 10, 15, 39	.97	1.22	.19	.48	3.25***	.83

Note. n.s.: not significant. *p < .05. ** p < .01. ***p < .001

5 Discussion

This study aims to gain a more comprehensive understanding of emotional responses towards artificial entities, especially towards robots. Unlike many studies, we included both observational and self-report data for an enhanced approach to the multi-level phenomenon of emotions.

The findings show similarities to [17] as participants reported more negative feelings after the reception of the torture video (H2b) and more positive feelings after the friendly video (H2a). Additionally, we assumed participants would differ in the number and type of AUs depending on the type of video they watched. The findings offer evidence for this hypothesis as the number of AUs associated with intrinsic unpleasantness was significantly higher when watching the torture video (H1b) than when watching the friendly video. No effect was however found for the number of AUs associated with intrinsic pleasantness when watching the friendly video as compared to the torture video (H1a).

The simple appearance of a robot in pain, even though on a conscious level, we may very well know a machine can't really feel pain, elicits feelings of sympathy nonetheless. These reactions are automatic and unconscious as the findings of this study indicate by studying the facial expressions of participants. The findings offer confirmation that we react socially and emotionally towards robots and integrate well into existing literature (e.g. [2, 17, 18]).

No confirmation could be found for H1b, however. Participants showed the AUs 12 + 25 + 38 in equal frequency, independently of the type of video. A possible explanation for the display of the AUs associated with intrinsic pleasantness during the reception of the torture video could be due to the superimposition of smiles on negative emotions, known in the literature as miserable smiles [8]. Interestingly, [2] also reported participants giggling and laughing during mistreating the Crawling Microbug Robot. Since we predetermined a special set of AUs and calculated only differences in frequency of AUs related to pleasant and unpleasant feelings, further investigations could yield interesting findings concerning the display of different types of smiles when experiencing negative emotions.

In the interpretation of the facial expressions we concentrated only on a special set of AUs according to [21]. The AUs were defined prior to the execution of the study and findings are therefore limited to the AUs referring to pleasantness and unpleasantness as defined in advance. A more comprehensive FACS analysis considering all displayed AUs would yield interesting insights and contribute to identifying and more precisely specifying further emotions.

6 How FACS Can Improve the Design of Social Robots

This section will show that the application of FACS to social robots presents a fruitful enhancement and can help make better social robots. FACS is already widely used and has also become very popular concerning the implementation of FACS in social robots (e.g. [16]; cf. [11]). Most social robots express emotions via facial expressions based on FACS. Thus, we propose to use the same method for analyzing and recognizing emotions of humans towards robots. This would not only be an economic way of using FACS but provides also a number of further advantages. Regarding emotional reactions towards robots, self-reports might not be an ideal way to learn the true emotional state of people. They do not only come with certain limitations in reliability and validity, but are also simply not applicable in certain situations: Infants, people with intellectually handicaps, brain injuries or dementia provide only some examples for the inappropriateness of simply asking about their feelings. Methods not depending on people's capability or willingness to share information are an alternative and offer advantages self-report measurements cannot provide. Observational methods such as physiological methods can measure changes in heartrate or electrodermal activity and provide further insights into unconscious (emotional) processes. However, those methods require physical contact with a person, e.g. for placing sensors on the skin. An alternative method for observing behavior related to emotions that does not require physical contact, leading attention to the method itself and possibly biasing the data measurement subsequently, is the observation with a camera. As the observed behavior is recorded, it does not depend on the observed person's presence to analyze the behavior. Thus, FACS provides an unobtrusive method for observing facial expressions, especially useful in situations when people must rely on nonverbal expressions. Furthermore, as we talk face-to-face, FACS also offers the opportunity for a natural and intuitive human-robot interaction. Thus, studying facial expressions based on FACS can help social robots to unobtrusively make inferences about the emotional state of a human interaction partner without having to ask.

7   Conclusion

In the present study, we could find further indications of emotional reactions towards robots, offering additional evidence for the validity of the Media Equation in HRI as well as the applicability of objective measures of emotions by analyzing facial expressions based on FACS. By using a between-subjects design as well as combining self-report measures with objective observational methods the potential of the data to be biased by social desirability, among others, can be considerably reduced. Furthermore, this approach facilitates a deeper and more comprehensive understanding of emotional responses to technical devices. With the appearance of robots in our daily lives it becomes increasingly important to analyze und further understand emotional reactions of humans towards robots.

References

1. Appel, J., von der Pütten, A.M., Krämer, N.C., Gratch, J.: Does humanity matter? analyzing the importance of social cues and perceived agency of a computer system for the emergence of social reactions during human-computer interaction. Adv. Hum. Compt. Int. **2012**, 324694 (2012)
2. Bartneck, C., Hu, J.: Exploring the abuse of robots. Interact. Stud. **9**, 415–433 (2008)
3. Bethel, C.L., Murphy, R.M.: Review of human studies methods in HRI and recommendations. Int. J. Social Robot. **2**, 347–359 (2010)
4. Breazeal, C.L.: Designing Sociable Robots. MIT Press, Cambridge (2005)
5. Craig, K.D., Hyde, S.A., Patrick, C.J.: Genuine, suppressed and faked facial behavior during exacerbation of chronic low back pain. Pain **46**, 153–160 (1991)
6. Ekman, P.: Facial expression and emotion. Am. Psychol. **48**(4), 384–392 (1993)
7. Ekman, P., Friesen, W.V., Ancoli, S.: Facial signs of emotional experience. J. Pers. Soc. Psychol. **39**(6), 1125–1134 (1980)

8. Ekman, P., Friesen, W.: Felt, False and Miserable Smiles. J. Nonverbal Behav. **6**, 238–252 (1982)
9. Ekman, P., Friesen, W.V., Hager, J.C.: The Facial Action Coding System, 2nd edn. Research Nexus eBook, Salt Lake City (2002)
10. Ekman, P., Rosenberg, E. L. (eds.): What the Face Reveals, vol. 2. Oxford University Press, New York (2005)
11. Fong, T., Nourbakhsh, I., Dautenhahn, K.: A survey of socially interactive robots. Robot. Auton. Syst. **42**(3–4), 143–166 (2003)
12. Libin, A.V., Libin, E.V.: Person-robot interactions from the robopsy- chologists' point of view: the robotic psychology and robotherapy aproach. Proc. IEEE **92**(11), 1789–1803 (2004)
13. Lien, J.J.J., Kanade, T., Cohn, J.F., Li, C.C.: Detection, tracking, and classification of subtle changes in facial expression. J. Robot. Auton. Syst. **31**, 131–146 (2000)
14. Nass, C., Moon, Y.: Machines and mindlessness: social responses to computers. J. Soc. Issues **56**(1), 81–103 (2000)
15. Reeves, B., Nass, C.: How People Treat Computers, Television, and New Media Like Real People and Places. Cambridge University Press, Cambridge (1996)
16. Ribeiro, T., Paiva, A.: The illusion of robotic life principles and practices of animation for robots. In: Proceedings of the 7th Annual ACM/IEEE International Conference on Human-Robot Interaction, HRI 2012 (2012)
17. Rosenthal-von der Pütten, A.M., Krämer, N.C., Hoffmann, L., Sobieraj, S., Eimler, S.C.: An experimental study on emotional reactions towards a robot. Int. J. Soc. Robot. **5**(1), 17–34 (2013)
18. Rosenthal-von der Pütten, A.M., Schulte, F.P., Eimler, S.C., Sobieraj, S., Hoffmann, L., Maderwald, S., et al.: Investigations on empathy towards humans and robots using fMRI. Comput. Hum. Behav. **33**, 201–212 (2014)
19. Rosheim, M.E.: Robot evolution: the development of anthrobotics. John Wiley & Sons, New York (1994)
20. Scherer, K.R.: What are emotions? and how can they be measured? Soc. Sci. Inform. **44**(4), 693–727 (2005)
21. Scherer, K.R., Ellgring, H.: Are facial expressions of emotion produced by categorical affect programs or dynamically driven by appraisal? Emotion **7**(1), 113–130 (2007)
22. Tracy, J.L., Robins, R.W., Schriber, R.A.: Development of a FACS-verified set of basic and self-conscious emotion expressions. Emotion **9**, 554–559 (2009)
23. Unz, D., Schwab, F., Winterhoff-Spurk, P.: TV news. the daily horror? emotional effects of violent TV news. J. Media Psychol. **20**(4), 141–156 (2008)
24. Watson, D., Clark, L.A., Tellegen, A.: Development and validation of brief measures of positive and negative affect: the PANAS scales. J. Pers. Soc. Psychol. **54**(6), 1063–1070 (1988)

Head and Face Design for a New Humanoid Service Robot

Hagen Lehmann[✉], Anand Vazhapilli Sureshbabu,
Alberto Parmiggiani, and Giorgio Metta

Istituto Italiano di Tecnologia, iCub Facility,
via Morego 30, 16163 Genova, Italy
hagen.lehmann@iit.it

Abstract. The design process of robotic faces as focal points of human-robot interactions plays a very important role in the construction of successful robot companions. The face of a robot represents an attentional anchor for the human user, and, if designed appropriately, enables a comfortable and intuitive communication. In this paper, we describe the user centered design process of the face display of a newly constructed humanoid robot companion called R1. We will introduce a novel solution for robotic face displays, illustrate our design decisions based on the results of an online survey, and present our final face design. We will discuss the lessons learned during the design process and argue that our solution and design will enable R1 to be a successful artificial interlocutor in different human-robot interaction scenarios.

Keywords: Human-robot interaction · Humanoid robots · User centered design · R1 · Robot design

1 Introduction

With recent developments in robotic technology the use of humanoid robots in public spaces becomes more and more widespread. To be effective in shared and communal spaces such as hospitals, care centers, shopping malls, cruise ships and tourist information centers, robots are required not only to be highly intuitive and pleasant during interactions, but also to be perceived by their human users as "social partners" [1]. This is facilitated when they express a strong "social presence" [2] through stylized humanlike characteristics such as heads, faces, arms and hands, which tend to elicit the human predisposition to anthropomorphize artifacts [3].

According to Epley et al. [4], one of the factors that promote anthropomorphism is the human likeness of an artifact. Their approach follows the idea that the higher the human likeness of the artifact, the more people use their own experiences as a source of induction when judging and predicting its actions. The experience of oneself with others functions accordingly as a model for the interaction with the robot and in this way reduces uncertainty for the human users, making them feel more comfortable during interactions with the robot. This approach is confirmed by findings from human-centered design research.

© Springer International Publishing AG 2016
A. Agah et al. (Eds.): ICSR 2016, LNAI 9979, pp. 382–391, 2016.
DOI: 10.1007/978-3-319-47437-3_37

For technology such as robots it is not enough to be easy to use, they also need to be engaging, and adaptive to the needs and wants of their users [5]. It is necessary to give the user the feeling of control, a good conceptual model, and the knowledge of what is happening when interacting with the technology [6]. In terms of social interaction this entails for the robots to behave according to human social norms, and in a significantly predictable manner. In order to do so, a face presents an important visual cue to help the user to understand the robot's capabilities and interactive attitudes, and to form an unspoken social contract between human and machine [7].

In this paper, we will describe the head design process of a new robot R1 (Fig. 1), which has been developed by Istituto Italiano di Tecnologia (IIT), and introduce a novel social robotics design approach, based on using backlit light tubes as a social interface. We will illustrate the potentials for facial expressions with such a technology and present the final design for R1's face.

Fig. 1. IIT's new R1 robot

2 Background

The term "social robotics" was originally used in swarm robotics and referred to the interaction of at least two robots. This original focus changed with the advancements in humanoid robotics [8]. The use of socially interactive robots [9] in human-interaction scenarios transformed the idea of social robots in such a way that 'social' started to refer to the quality of the interactions between humans and robots, when supported by human communication channels and social norms. Through this shift, the introduction of robotic faces became a viable option for displaying complex emotional expressions and interactive attitudes. Following Norman et al. [10], the presence of a face on an artifact gives the user a focal point for interaction. In other terms a head and a face make it easier for users to understand the interaction dynamics between them and a robot. The user feels comfortable to address the robot's face while speaking to it, and uses cues displayed on the face to recognize and predict the robot's actions.

The appearance of the physical body of the robot plays an important role in the achievement of interactive intuitiveness, pleasantness and comfort facilitating robots in expressing their social functions. In terms of robotic head and face design it has been shown that this can be achieved by constructing the robot's head more wide than long, and subsequently give it a small forehead and a small chin [11].

The design of robot faces as visual anchors for human users has been approached from different directions. Different possible methods to construct faces have been applied. In some cases the faces were projected onto a human face mold [12], displayed on a screen [13, 14], physically constructed [15, 16], or were a mixture of different methods was used [17].

The design space of visual iconography has first been defined by McCloud [18, 19] as a triangle with the corners reality, meaning (iconic) and abstraction [see Fig. 2 left]. His work was based on characters from cartoons and comics. Blow et al. [20] adapted this approach for robot faces and appearances [see Fig. 2 right].

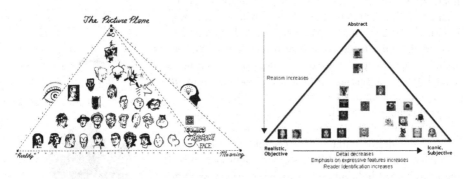

Fig. 2. McClouds triangular design space of visual iconography [7] (left), and its adaptation by Blow et al. [5] for robot appearances (right)

The design we had in mind for R1 was aiming for a face that could be located between the abstract and the iconic corner of this triangle. In order to choose the best solution we decided to involve the potential users of the robot via an online questionnaire and present to them the different design options, adapting in this way a more user centered design approach (Fig. 3).

3 Display Design

In order to construct an appealing, modular and robust robot we decided to give the robot a display with an animated face. To distinguish this display from the appearance of a usual display screen, which would look more like a TV than a robotic head, we decided to use a novel method to employ the curved surface we desired. The display we were envisioning had to be unique while being able to feature 3D details. The curved display market is an emerging one. Screens are still very expensive and suffer from issues such as viewing angle

Fig. 3. Display seen form the left side

limitations, limited visibility in sunlight and even ambient light, and limited robustness. Most importantly curved screens are generally realized by curving a flexible display; hence only single curvature displays can be obtained whereas we wanted our display to show images on a doubly curved surface. We also wanted to make something that could be easily and inexpensively modified since we had to update our designs constantly. One important design requirement was modularity, i.e. that the parts could be removed and replaced easily. We decided to make use of the advances in the field of rapid proto-typing. We built a display comprising five custom designed RGB LED matrices, which were joined and used to backlit a 3D printed modular setup. This setup consisted of arrays of "light pipes/optical relay tubes", with a count of one *light pipe* for every individual LED in the display matrix, and was then housed in a support structure in order to help to support the *light pipe* structure while isolating each pixel. This increased the total internal reflectivity of the light being transmitted in the light pipe, giving it better display results compared to others with similar technologies [21].

The entire construction was then covered by a plain light diffusion mask which increases true blacks and accentuates the light being transmitted, while acting as the projected curved surface.

4 Face Design

We created 4 different designs for the face to test the user preferences. The final design proposals were created by the *6.14 Design Group/Milano*, following general design principles and findings from previous research on robotic face design. DiSalvo et al. [11] for example have shown that the distance between the eyes in a robot's face should be bigger than the width of one eye. In all four designs, the features resembling eyes are the central parts of the face.

Another feature we were thinking to integrate in the face display was the ears. Since physical ears were not compatible with our head design the ears had to be integrated as part of the virtual face. In Design 1, 2, and 3 this was done by lines on the side of the display. Since it was decided to give Design 4 two additional features, eyebrows and pupils, the lines at the side of the face were not used here.

For all four designs we decided not to use a mouth. In human-human social communication the mouth and its movements play such a pronounced role that humans are extremely sensitive to its movements and shape. When featuring a robotic mouth, it is very difficult to avoid not only the uncanny valley [22] by using inappropriate movements, but also criticalities related to the comprehensibility of the robot, due to the introduction of mismatching movements and speech patterns. For our design we found it advisable to think of other ways to visually aid speech during verbal com-munication with the robot. This will be described in more detail in the *Design Decision* section.

Figure 4 shows the four different experimental designs, starting from top to bottom with Design 1, followed by Design 2, Design 3, and Design 4. Besides a neutral expression (big pictures), for each of the designs six emotional expressions were created. These expressions were happy, afraid, angry, sad, disgusted, and surprised – starting from the small picture on top left and proceeding clockwise (Fig. 5).

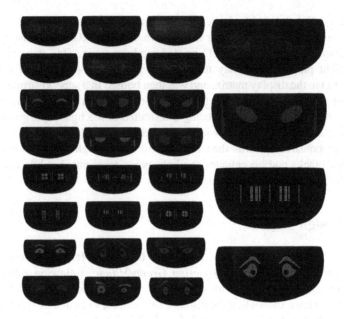

Fig. 4. Different experimental designs for the face of R1

5 Method

We used an online questionnaire for the evaluation of the four different potential face designs for the R1 robot. We decided to use different levels of abstraction, ranging from a very abstract, machinelike face design to more iconic designs including different features of a human face like eyes with pupils and eyebrows.

In the first part of the questionnaire the participants were asked for their demographic information, their experience with robots, and their attitude towards the integration of robot companions into public spaces. In the second part they were shown the four different designs, and were asked to rate them based on their personal impression. For the rating we used a four point Likert Scale ranging from least liked (1) to best liked (4).

6 Results

Sample Characteristics. We received 82 replies of which 50 were male and 32 were female. The mean age of the participants was 33.8 years, ranging from 20 to 72. Concerning their experience with robots 37 reported none, 15 reported little, 12 reported some experience, and 18 participants indicated that their experience with robots was substantial. Another question we asked the participants was *"Do you think it is a good idea to integrate robot companions into homes, hospitals and public spaces like shopping malls and airports?"* This question was answered by 11 with "I don't care", 49 stated that they agree with the question, 16 strongly agreed, 5 disagreed, and 1 strongly disagreed. This question was asked in order to evaluate the general attitude of the participants towards robot companions. The answers show that in our sample the perception of robots was overwhelmingly positive. The demographic information and the information on the experience with robots point towards a diverse composition of the sample with respect to the topic of the questionnaire.

Ranking Results and User Comments. Design 1 and Design 3, both with a more abstract design, were rated much lower than Design 2 and Design 4, which are both more iconic. Design 2 was rated slightly less positive than Design 4. Design 1 was rated the lowest [see Table 1].

Table 1. This table shows the descriptive results of the questionnaire. The more abstract designs 1 and 3 were disliked by the participants.

Parameter	Design 1	Design 2	Design 3	Design 4
Average	1.39	3.08	1.86	3.18
Standard deviation	0.82	0.8	0.8	1.02

The descriptive analysis results show that the users liked the abstract designs in general very little. This was also reflected in their comments. Most of the participants wrote that in the cases of Design 1 and Design 3 they were not able to interpret the facial features or emotional expressions portrayed. Some users pointed out that in an interaction they would not like to be confronted with a robot featuring a face like this, and that these two designs "… simply don't convey a message…" and "… are totally not understandable …".

The comments for Design 2 and Design 4 are equally revealing. Design 2 was dubbed by some participants as being "… EVE-like …" or "… Alien-like …", reminding them of the EVE robot character from the movie Wall-E. In general Design 2 received positive comments like e.g. "… it's effective and nice …", "… easy to understand …", and "… more clear and almost material design compliant …". Only one participant pointed out that "… it is too simplified …".

Despites being rated the highest on average, Design 4 got positive and negative comments. Some participants really liked it and pointed out that it "…is more clear", and "… kind of cute …". But others wrote negative comments like "…it looks too human-like …", "… it is good for a toy …", "…it seems too artificial", and "… it is a little bit childish …". This divergence in the perception of Design 4 also shows in the descriptive result. The rating for Design 4 has the biggest standard deviation of the four designs.

7 Design Decisions

The result of the survey made it clear that an abstract face design for the face would be counterproductive and most likely be rejected by our potential user group. Design 4 received the highest rating, but also had the highest variability in the result. In combination with the critical comments, we took this as an indication that Design 2, with the addition of one facial feature would be the most acceptable choice. We added a small line between the eyes of the robot, since one negative aspect of Design 2 pointed out by some partici-

Fig. 5. Final design for R1's face. From the topleft clockwise - neutral, happy, surprised and sad

pants was its over simplicity. The line is used as an additional animation cue in order to give the face a more lively character, and as an indicator for "mood" changes.

In Design 2 the eyes and eyebrows are merged together, resulting in a "compound emotive eye" similar to emoticons with which expressions have been transferred successfully in computer-mediated conversations since the start of text messaging [23].

For the final design the pupils as a facial feature were dropped. They would have had the advantage of enabling the simulation of eye gaze on the robot, but the comments concerning them were in general critical. The pupils were the feature that people liked least and said it made the face look too much like a toy. One participant pointed out that they made the robot look childish.

Movements for Final Design. The final design also features two lines at the side of the face, in our original conception resembling ears. These lines are going to be used during the verbalizations of the robot. They will be an additional indicator of robot speech. The lines will become longer or shorter depending on whether the robot speaks or not.

The eyes of the robot will also move when the robot is interacting with a user. In the case of a happy expression they will move up and down, additional the small line between the eyes will move up. For the expression of sadness the inclination of the eyes will change in such a way that their outer points will "drop", additionally the line between the eyes will move downwards. When expressing surprise the eyes will enlarge starting from their neutral size.

The robot will also exhibit appropriate head movements. It will turn towards the speaker when it is addressed. In order to simulate gaze following the robot's head will move following the position of the user. In this way the robot will have the appearance of paying attention to what is being said during a conversation. We will also use these head movements to simulate joint attention. These movements will emphasize the facial expressions.

8   Discussion

There are two main findings our research highlights at this stage of its development, one technical, and the other HRI design oriented.

From the technical point of view, we introduced a novel design for a backlit 3D display for the projection of facial expressions on a humanoid robot head. It is possible to show that a combination of LED matrices and a light tube structure, in which each pixel is isolated, can be successfully used to create an appealing social interface capable of projecting facial expressions understandable for a human user.

From the HRI design point of view, our research pointed out the clear dislike of the users for the abstract face designs for the robot. To fit the overall design, the robot needed to have a face that is located between abstract and iconic in the design space for robotic appearances [21]. The strength of the effect was surprising. Both abstract designs, even though featuring stylized eyes, were categorically rejected by the participants of our online study. We interpret this as a clear indication for future designs of robotic faces to be iconic, when given the choice between iconic and abstract. It seems that people are unable to interpret facial features beyond a certain level of abstract, and that this inability creates an uncertainty when being confronted with a robot, that makes an interaction uncomfortable and unpredictable.

We also found that too much detail in an iconic face makes this face appear toy-like and "childish". In the case of Design 4 the details perceived by the participants as toy-like were the eyebrows and the pupils and the combination of the two. The design we chose combined the eyebrows and the eyes into emotive eyes, and did not include the eyebrows.

It could be argued that some of the participants liked Design 2 because it reminded them of the character EVE from the movie Wall-E. The similarity was initially not intended, but was commented on by some participants. The overall design of the robot is however completely different from the design of EVE from the movie and we consider that the face will integrate very well into the appearance of the robot.

The design process and the results of our study imply and confirm a series of general suggestions for the design of robot faces. It is important that the face does not feature too many human characteristics in order to avoid non-realizable expectations about the social interaction capabilities of the robot [11], and to avoid that the robot falls into the "Uncanny Valley" [22].

We also had some feedback on our use of colors for the expressions. Since the meaning of colors is highly context and culture depended, they should only be used with care and as additional communication channel to emphasize specific information as part of an escalation strategy – e.g. start with icon, add blinking, add color, add sound, increase volume. This should depend on the importance of the information to be transmitted. Even with our conservative choice of colors some of the participants were complaining that the differences in color are more confusing than helpful.

Outcome of Design Research. The outcome of our design research was that when using an animated face on a display, this face should preferably not be used also as a touch screen. This would be counterproductive with regard to the social presence of the robot and the related interactive attitude towards it, as it would create an effect similar to poking someone in the eye while, on other occasions, interacting through polite speaking. If possible, it should also be avoided to display infographics or icons on the face screen.

If, due to design requirements or limitations, it is necessary to use the face screen as a normal display, clearly defined and preferably animated "changing sequence" should be used in order to signal the user that the robot is now going to fulfill a different function. Once the display of the graphics or icons is finished, a similar "changing sequence" should be used to signal the user, that the robot is now switching back to its social function.

Future Work. As the robotic head presented in this paper will be part of a new generation service robot, the facial expressions we discussed will be part of a holistic social behavior exhibited by this robot. In this robot we will integrate human non-verbal communication channels like gaze, body movements, facial expressions and gestures together with speech and typical robotic communication like the display of icons and sound. Sounds will only be used when the robot is communicating robot related information, e.g. "I am stuck", and "low battery".

In a next step we will design and study the proxemics for R1. It will be important to keep the different personal spaces humans have in mind. The R1 should not move too close to the human, not only for safety reasons but also for not intruding the intimate space of the user. The robot will slow down when approaching a human, so it does not appear intimidating or threatening. The movements of the robot will be as naturalistic as possible, and will appear to be flowing into each other, in order to increase the robots behavioral predictability.

9   Conclusion

The outcomes of our research are guidelines both for the design of interfaces for social robots as well as for HRI in general. The user centered design approach is of high importance in the design process of social robots. It helps to discover mistakes during prototyping and to avoid unnecessary adjustments in the final design phase. The specificity of social robotics is that these robots are an immediate interface between highly complex technology and naïve human users. The signals given by the robots need to be well balanced, understandable and in their entirety as behavior predictable.

Acknowledgements. The authors would like to thank Alessandra Sciutti, Francesco Rea, Marco Maggiali, and the team of *6.14/Milan* for their productive and insightful input during the discussion. We would like to thank *6.14* specifically for the creation of the different designs.

References

1. Damiano, L., Dumouchel, P., Lehmann, H.: Should empathic social robots have interiority? In: Ge, S.S., Khatib, O., Cabibihan, J.-J., Simmons, R., Williams, M.-A. (eds.) ICSR 2012. LNCS, vol. 7621, pp. 268–277. Springer, Heidelberg (2012)
2. Damiano, L., Dumouchel, P., Lehmann, H.: Towards human-robot affective co-evolution overcoming oppositions in constructing emotions and empathy. Int. J. Soc. Robot. **7**, 7–18 (2015). doi:10.1007/s12369-014-0258-7

3. Walters, M.L., Syrdal, D.S., Dautenhahn, K., Te Boekhorst, R., Koay, K.L.: Avoiding the uncanny valley: robot appearance, personality and consistency of behavior in an attention-seeking home scenario for a robot companion. Auton. Robots **24**, 159–178 (2008)
4. Epley, N., Waytz, A., Cacioppo, J.T.: On seeing human: a three-factor theory of anthropomorphism. Psychol. Rev. **114**, 864 (2007)
5. Clark, K.: Engaging and adaptive: going beyond ease of use. In: Kurosu, M. (ed.) HCD 2009. LNCS, vol. 5619, pp. 46–54. Springer, Heidelberg (2009)
6. Norman, D.: The Invisible Computer: Why Good Products Can Fail, the Personal Computer is so Complex, and Information Appliances are the Solution, p. 174. The MIT Press, Cambridge (1998)
7. Edsinger, A., O'Reilly, U.-M.: Designing a humanoid robot face to fulfill social contracts. In: Proceedings of 9th IEEE Ro-Man (2000)
8. Gong, L., Nass, C.: When a talking-face computer agent is half-human and half-humanoid: human identity and consistency preference. Hum. Commun. Res. **33**(2), 163–193 (2007)
9. Fong, T., Nourbakhsh, I., Dautenhahn, K.: A survey of socially interactive robots. Robot. Auton. Syst. **42**, 143–166 (2003)
10. Norman, D.: The Design of Everyday Things. Doubleday, New York (1990)
11. DiSalvo, C.F., Gemperle, F., Forlizzi, J., Kiesler, S.: All robots are not created equal: the design and perception of humanoid robot heads. In: Proceedings of the 4th Conference on Designing Interactive Systems: Processes, Methods, and Techniques, pp. 321–326. ACM (2002)
12. Al Moubayed, S., Beskow, J., Skantze, G., Granström, B.: Furhat: a back-projected human-like robot head for multiparty human-machine interaction. In: Esposito, A., Esposito, A.M., Vinciarelli, A., Hoffmann, R., Müller, V.C. (eds.) COST 2102. LNCS, vol. 7403, pp. 114–130. Springer, Heidelberg (2012)
13. Guizzo, E.: Cynthia Breazeal Unveils Jibo, a social robot for the home. In: IEEE Spectrum (2014)
14. https://zenbo.asus.com/
15. Hegel, F., Eyssel, F., Wrede, B.: The social robot "flobi": key concepts of industrial design. In: 2010 IEEE on RO-MAN, pp. 107–112. IEEE (2010)
16. Nishio, S., Ishiguro, H., Hagita, N.: Geminoid: Teleoperated Android of an Existing Person, pp. 343–352. INTECH Open Access Publisher, Vienna (2007)
17. Beira, R., Lopes, M., Praga, M., Santos-Victor, J., Bernardino, A., Metta, G., Becchi, F., Saltarén, R.: Design of the robot-cub (icub) head. In: Proceedings 2006 IEEE International Conference on Robotics and Automation, ICRA 2006, pp. 94–100. IEEE (2006)
18. McCloud, S.: Understanding Comics: The Invisible Art. Kitchen Sink Press, Northampton (1993)
19. http://www.scottmccloud.com/4-inventions/triangle/index.html
20. Blow, M., Dautenhahn, K., Appleby, A., Nehaniv, C.L., Lee, D.: The art of designing robot faces: dimensions for human-robot interaction. In: Proceedings of 1st ACM SIGCHI/SIGART Conference on Human-Robot Interaction, pp. 331–332. ACM (2006)
21. Willis, K., Brockmeyer, E., Hudson, S., Poupyrev, I.: Printed optics: 3D printing of embedded optical elements for interactive devices. In: Proceedings of 25th Annual ACM Symposium on User Interface Software and Technology, pp. 589–598. ACM (2012)
22. Mori, M., MacDorman, K., Kageki, N.: The uncanny valley [From the Field]. IEEE Robot. Autom. Mag. **19**, 98–100 (2012). doi:10.1109/MRA.2012.2192811
23. Derks, D., Bos, A.E., Von Grumbkow, J.: Emoticons in computer-mediated communication: social motives and social context. CyberPsychology Behav. **11**, 99–101 (2008)

The Influence of Robot Appearance
and Interactive Ability in HRI:
A Cross-Cultural Study

Kerstin Sophie Haring[1]($\boxtimes$), David Silvera-Tawil[2], Katsumi Watanabe[3],
and Mari Velonaki[2]

[1] Department of Behavioral Sciences & Leadership, US Air Force Academy,
Colorado Springs, USA
kerstin.haring.ctr@usafa.edu
[2] Creative Robotics Lab, The University of New South Wales, Paddington, Australia
{d.silverat,mari.velonaki}@unsw.edu.au
[3] Department of Intermedia Art and Science, Waseda University, Tokyo, Japan
katz@waseda.jp

Abstract. It has been shown that human perception of robots changes
after the first interaction. It is not clear, however, to which extent the
robot's appearance and interactive abilities influences such changes in
perception. In this paper, participants' perception of two robots with
different appearance and interactive modalities are compared before and
after a short interaction with the robots. Data from Japanese and Australian participants is evaluated and compared. Experimental results
show significant differences in perception depending on the robot type
and the time of interaction. As a result of cultural background, perception changes were observed only for Japanese participants on isolated
key concepts.

Keywords: Culture · Human-robot interaction · Robot perception

1 Introduction

Humans form impressions of each other in as little as 100 ms [16]. It is believed
that quick impressions of robots are formed in a similar way, depending solely on
the robot's appearance [4,8,9]. The Expectation Confirmation Theory, furthermore, states that people form initial expectations towards technology (including robots) based on appearance alone [12], which is then (dis)confirmed after
observing its performance. Although based on this theory it is believed that a
mismatch between people's expectations of a robot based on its appearance and
the real experience based on the robot's performance plays a significant role in
how people perceive and interact with a robot [5], it is unclear to what extent
each of these factors influence short and long-term interaction.

Social psychology, on the other hand, has shown that cultural differences exist
in the way people perceive technology [15]. Although cultural differences have

© Springer International Publishing AG 2016
A. Agah et al. (Eds.): ICSR 2016, LNAI 9979, pp. 392–401, 2016.
DOI: 10.1007/978-3-319-47437-3_38

been shown to affect certain areas of human perception of robots [5,6], this is only one of many factors that contribute towards robot perception. Previous research also suggests that culture [5,6] has a significant effect in human perception of robots during passive HRI; that is, a situation where a person "does not explicitly interact with a robot but needs some model of robot behaviour to understand the consequences of the robot's actions" [13]. A robot that follows strategies of human behaviour when giving advice to other humans, for example, is perceived as effective when this interaction is observed but not necessarily when the person is interacting directly with the robot [14].

In this study two robots of different appearance and abilities were used to identify possible changes in human perception of robots between expectation and (dis)confirmation of beliefs. That is, before (expectation) and after (confirmation) a short interaction with the robots. Results from this study provide new information about the factors that influence human expectations from robots, based on their appearance, and how these perception changes due to the robot's behaviour and interactive abilities.

Following the work from Strait et al. [14], this study also extends to a scenario in which a passive person, a bystander [17], is present observing the interaction. Previous work that explores cross-cultural differences in human perception [6] is also extended to evaluate if the differences previously found occur regardless of the robot type. To our knowledge, no previous study has explored cross-cultural aspects of human perception of robots which differ in appearance and interactive abilities. Experimental results, performed with Japanese and Australian participants, suggest that the appearance followed by interactive modality have more influence on human perception than cultural background or interactive modality (i.e. bystander and interactant).

2 Methodology

This study evaluates human perception of two different robots before and after interaction. Japanese and Australian participants interacted with the robots in two different modes: active (interactant) and passive (bystander). For both conditions, participants were instructed to observe the reaction of the robot to the active participants' input. It is hypothesized that: (1) the perceptions of both robots will change after the interaction, and (2) the robot abilities and behaviour are the main factor influencing robot perception.

2.1 Robots

Two commercial robots were used: (1) the humanoid robot Robi[TM] and (2) the non-biomimetic[1] robot My Keepon[TM], see Fig. 1. Both are small-sized robots with the ability to move and respond to speech (Robi) or touch (My Keepon).

---

[1] Although My Keepon has some elements of anthropomorphism (eyes and nose), in this document it is referred as non-biomimetic due to its non-biomimetic form and behaviour.

The interaction with Robi took place through seven pre-defined phrases (English in Australia and Japanese in Japan) and the robot's response using speech and movements. My Keepon, who was previously used to study the underlying mechanisms of social communication [10], responded to touch using non-verbal sound and movement (e.g. turns right when touched on the right side).

Fig. 1. Robots used in the present study: Robi by De Agostini (left), and My Keepon by Wow!Stuff (right).

2.2   Experimental Procedure

Pairs of participants were randomly assigned to either Robi or Keepon[2], with one participant assigned the role of interactant (active), while the second was the bystander (passively observing). The experiment was divided in two stages. First, the robot assigned to the pair of participants was shown to them sitting on top of a table, after which they were asked to answer a demographic survey and two perception questionnaires to measure their initial thoughts about the robot (before condition). Then, participants were instructed to interact with the robot through either speech (Robi) until completion of the last item of the pre-defined phrases, or touch (Keepon) with a time limit of two minutes. Participants were then asked to fill in the two perception questionnaires for a second time (after condition). The experiments were all performed in private rooms. All questionnaires were answered in a private room adjacent to the location of the robot, and the robots were not present during the questionnaires. The same procedure was followed in Japan and Australia, see Fig. 2.

2.3   Questionnaires

In addition to the demographics survey participants were asked to fill in two questionnaires: The Godspeed robot perception questionnaire [1] which measures human perception of robots using five concepts: anthropomorphism, animacy, likeability, intelligence, and safety; and a questionnaire that evaluates if participants ascribed to the robot any mental capabilities beyond the observable behaviour [3] using two concepts: the robot's perceived ability for experience (i.e. feel, sense) and agency (i.e. plan, memorize). In both questionnaires the participants reported using a 5-point scale for all items.

---

[2] In this document the names 'My Keepon' and 'Keepon' will be used interchangeably.

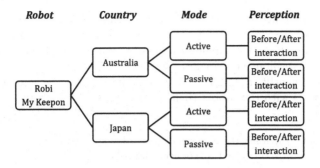

Fig. 2. Experimental conditions using two robots, two countries and two interaction modes as between condition, and participants' perception before and after the interaction as within condition.

2.4 Participants

A total of 126 participants were recruited at the University of New South Wales, Australia and the University of Tokyo, Japan; see Table 1. Participants were naïve to the objectives of the experiment, and have never interacted with either one of the robots.

Table 1. Participants demographics.

	Japan		Australia	
	My Keepon	Robi	My Keepon	Robi
Female/Male/Not specified	8/16/0	6/13/1	18/20/0	23/20/1
Age M (SD)	21.7 (2.15)	21.3 (1.95)	22.6 (5.07)	24.6 (8.5)

3 Results

The main focus of the study was to determine the effects and differences that may exist in human perception of robots depending on robot type, interactive modality, and participant's cultural background. For each one of the seven key perception concepts (Sect. 2.3), if not otherwise indicated, a repeated-measures ANOVA was performed with the key concept as dependent variable, and the interaction times before and after as independent variable. Only results with significant differences are presented in this section.

3.1 Mental Capability: Experience

Participants' attribution the robots' capability to experience (e.g. feel pain, pleasure) showed a significant effect for robot type ($F(1,122) = 3.84$, $p < 0.05$), ascribing lower ability to experience to Robi than to Keepon, see Table 2. A significant effect based on country was also found ($F(1,22) = 10.22$, $p < 0.001$), showing that Japanese participants ascribed a higher mental capability to both robots than Australian participants, see Table 2. The capability of experience increased significantly ($F(1,123) = 5.16$, $p < 0.02$) after the interaction (M = 2.18, SD = 0.98) with the robots (before: M = 2.01, SD = 0.89).

Table 2. Robots' capability to experience, according to participants. Mean (M) and standard deviation (SD) divided by country and robot type.

	Japan	Australia	Robi	Keepon
M (SD)	2.41 (1.02)	1.92 (0.85)	1.95 (0.83)	2.24 (1.02)

Results from a two-factor ANOVA showed a significant correlation between country and interaction time ($F(1,122) = 12.85$, $p < 0.001$), Fig. 3. Post-hoc t-tests showed that for Japanese participants the capability of experience increased significantly after the interaction ($t(43) = -3.27$, $p < 0.001$), rating both robots higher after the interaction ($t(79.8)=4.27$, $p < 0.001$).

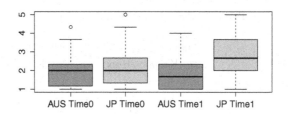

Fig. 3. Robots' capability to experience, according to participants, before (Time 0) and After (Time 1) the interaction with a significant increase for Japanese participants after the interaction.

3.2 Mental Capability: Agency

Participants' attribution of the mental capability of agency (e.g. plan, memorize) showed a significant effect for robot type ($F(1,122) = 7.55$, $p < 0.01$); see Table 3, ascribing Robi higher agency. Although the correlation between country and interaction time was significant ($F(1,122) = 4.56$, $p = 0.03$), this was not confirmed in the post-hoc test. A separate t-test for the difference between countries, however, showed that agency was rated higher in Japan ($t(81.1) = 2.08$, $p = 0.03$).

Table 3. Robots' capability for agency according to participants. Mean (M) and standard deviation (SD) divided by country and interaction time.

	Japan		Australia		Robi	Keepon
	Before	After	Before	After		
M (SD)	2.73 (1.08)	3.03 (1.10)	2.73 (0.75)	2.61 (0.99)	2.94 (0.98)	2.55 (0.91)

3.3   Perception: Anthropomorphism

Participants' perception of anthropomorphism showed a significant effect for robot type ($F(1,122) = 7.55$, $p < .01$), rating Robi as more anthropomorphic. The correlation between country and interaction time was also significant ($F(1,122) = 8.51$, $p < = 0.01$) showing that the perception of anthropomorphism increased significantly in Japanese participants after the interaction ($t(43) = -1.9$, $p = 0.05$). Detailed values are presented in Table 4.

Table 4. Robots' perceived anthropomorphism according to participants. Mean (M) and standard deviation (SD) divided by country and interaction time.

	Japan		Australia		Robi	Keepon
	Before	After	Before	After		
M (SD)	2.42 (0.67)	2.74 (0.89)	2.42 (0.69)	2.43 (0.89)	2.66 (0.82)	2.31 (0.74)

3.4   Perception: Animacy

Participants' perception of animacy showed a significant effect for robot type ($F(1,122) = 3.85$, $p = 0.05$), giving Robi a higher score ($M = 2.66$, $SD = 0.82$) over Keepon ($M = 2.31$, $SD = 0.74$). For the interaction time, furthermore, a significant increase in animacy was observed after ($M = 2.74$, $SD = 0.89$) the interaction ($F(1,123) = 9.79$, $p = 0.002$), with $M = 2.45$ and $SD = 0.67$ before the interaction.

3.5   Perception: Likeability

Participants' perception of likeability showed a significant effect for robot type ($F(1,122) = 15.75$, $p < = 0.001$) with Robi being more likeable ($M = 4.21$, $SD = 0.73$) than Keepon ($M = 3.73$, $SD = 0.85$). The correlation between robot and interaction time also showed a significant effect ($F(1,122) = 12.3$, $p < = 0.001$); for Robi likeability increased ($t(61) = -31.98$, $p = 0.05$) whilst for Keepon it decreased ($t(61) = 2.94$, $p < 0.01$), resulting in significantly higher values for Robi after the interaction ($t(114.6) = -4.71$, $p < 0.001$), Fig. 4. The mean and standard deviation values for these results are presented in Table 5.

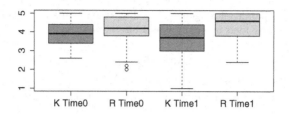

Fig. 4. Perception of likeability divided by robot: (R)oby and (K)eepon; and interaction time: before (Time 0) and after (Time 1). A significant decrease for Keepon and a significant increase for Robi can be observed.

Table 5. Robots' perceived likeability according to participants. Mean (M) and standard deviation (SD) divided by robot type and interaction time.

	Robi		Keepon	
	Before	After	Before	After
M (SD)	4.11 (0.71)	4.31 (0.75)	3.88 (0.68)	3.57 (0.97)

These results show that likeability and anthropomorphism are positively correlated before ($r(122) = 0.45$, $p < 0.001$) and after the interaction ($r = (122) = 0.50$, $p < 0.001$), with Robi perceived as more likeable than Keepon which is considered as support for the uncanny valley theory [11].

3.6   Perception: Intelligence

Participants' perception of intelligence showed a significant effect for robot type ($F(1,122) = 43.49$, $p < 0.001$), rating Robi higher ($M = 3.26$, $SD = 0.70$) than Keepon ($M = 2.60$, $SD = 0.64$). Additionally, a significant correlation between robot type, country, and interaction time was observed ($F(1,120) = 11.87$, $p < 0.001$). For Japanese participants the perception of intelligence increased significantly after the interaction with Robi ($t(19) = -2.94$, $p < 0.01$), who was also perceived significantly more intelligent after the interaction by Japanese participants ($t(36.8) = -2.37$, $p < = 0.05$). Robi was also perceived more intelligent than Keepon before the interaction by both Australians ($t(72.7) = -4.91$, $p < 0.001$) and Japanese ($t(41.3) = -2.45$, $p = 0.01$) participants. The same held after the interaction, where Robi was perceived as more intelligent both in Australia ($t(77.3) = -3.79$, $p < 0.001$) and Japan ($t(36.4) = -4.45$, $p < 0.001$). The data also show that agency and intelligence are positively correlated ($r(256) = 0.49$, $p < 0.001$). More details are in Table 6.

3.7   Perception: Safety

Participants' perception of safety only showed a marginal correlation between robot type and country ($F(1,120) = 3.4$, $p = 0.06$). A separate t-test shows that Keepon was perceived slightly safer in Japan after the interaction ($t(72.8) = -2.53$, $p = 0.01$).

Table 6. Robots' perceived intelligence according to participants. Mean (M) and standard deviation (SD) divided by country and interaction time.

	Japan		Australia	
	Before	After	Before	After
Robi: M (SD)	3.09 (0.70)	3.61 (0.79)	3.33 (0.53)	3.10 (0.77)
Keepon: M (SD)	2.59 (0.69)	2.63 (0.64)	2.68 (0.63)	2.51 (0.64)

4 Discussion

This paper presents a series of experiments aimed to investigate the differences that may exist in how people perceive robots of different appearance and interactive modality before and after a short-term interaction. Two commercial robots were used in these experiments; the humanoid robot Robi, who has the ability to respond to speech; and the non-biomimetic robot My Keepon, who responds to touch. Both modes of interaction, speech [14] and touch [7], have been found to influence how participants perceive a robot and the potential to contribute towards the success (or failure) of the interaction. All participants were recruited from two different universities (in Japan and Australia) and do not represent the full population of either of these countries; these participants, however, provide socio-economically similar samples from these countries that are valid for a cross-cultural comparison. Note that culture in this work simply refers to the country where the experiment was performed.

Experimental results suggest that participants' initial perception of a robot depend on the robot's appearance, with higher ratings provided to Robi (the humanoid robot) in five out of the seven key concepts measured: agency, anthropomorphism, animacy, likeablity, and perceived intelligence. These results, together with a positive correlation between agency and intelligence, suggest that a humanoid design (when compared to the non-biomimetic design) leads to a higher assumption of intelligence, and therefore agency, in a robot. Additionally, and supporting the theory of the uncanny valley [11], this experiments demonstrate that higher ratings of likeability were indeed correlated to a robot with a higher score in anthropomorphism.

From appearance to behaviour, the results presented in this paper show that observing a robot in motion also leads to an increased perception in animacy. This results support previous research in the perception of animacy that shows that perceived motion induces brain activity in a similar way than the perception of animacy [2].

In terms of perceived safety, this concept did not differ between the robots. This effect is attributed to the small size, non-threatening appearance, and generally limited movement of both robots.

The capability of experience stands out when the two robots are compared. My Keepon, the non-biomimetic robot, was rated higher in its ability to experience (e.g. feel pain, pleasure, etc.) than the Robi, the humanoid robot. The authors suspect that the ability to experience is influenced by the interaction modality—speech or touch—where My Keepon was able to "feel" the touch of participants, ascribing a higher ability to experience in other areas. This interesting result suggests that tactile interaction with robots can have a significant influence in how people relate with robots. Future research should isolate this effect to confirm the influence of touch in all key concepts of robot perception. Note that during this experiments participants were not allowed to touch Robi.

In terms of culture, this study confirms previous findings that suggest that cultural background is not a major key factor in robot perception [6], but yet could potentially influence how people change their perception when interacting with a physically present robot. This was reflected here through an increase in Japanese participants for perceived experience, anthropomorphism and intelligence for Robi, and safety for Keepon. Although previous results showed a significant effect to cultural differences in a passive interaction [6], in the current experiments the conditions of 'active' and 'passive' did not show any significant differences; the authors believe that future experiments with a reduced number of experimental conditions could give more insight on these differences.

Future research should also derive a more comprehensive experiment that describes relevant information regarding cultural and socioeconomic background, instead of only the country where the experiments where performed. Additionally, the current study did not control for the variables of appearance and interaction modality independently. Future work should consider these factors to explore the relative weight of each factor on robot perception. Although the authors are aware of the lack of numerical balance between the two populations tested, the statistical analysis does not suggest that any significant effects may occur with a more balanced population.

5   Conclusion

This study concludes that appearance, as well as the interaction modality of a robot play a crucial role in the perception of robots before and after short-term interactions. Although speech is the most common mode of interaction, and is an intuitive way to interact with robots, the results from this experiments suggest that tactile interaction is important to the way people perceive and interact with robots.

References

1. Bartneck, C., Kulić, D., Croft, E., Zoghbi, S.: Measurement instruments for the anthropomorphism, animacy, likeability, perceived intelligence, and perceived safety of robots. Int. J. Soc. Robot. **1**, 71–81 (2009)
2. Gao, T., Scholl, B.J., McCarthy, G.: Dissociating the detection of intentionality from animacy in the right posterior superior temporal sulcus. J. Neurosci. **32**(41), 14276–14280 (2012)
3. Gray, K., Young, L., Waytz, A.: Mind perception is the essence of morality. Psychol. Inq. **23**(2), 101–124 (2012)

4. Haring, K.S., Watanabe, K., Mougenot, C.: The influence of robot appearance on assessment. In: Proceedings of the ACM/IEEE International Conference on Human-Robot Interaction, pp. 131–132. IEEE Press (2013)
5. Haring, K.S., Silvera-Tawil, D., Matsumoto, Y., Velonaki, M., Watanabe, K.: Perception of an android robot in Japan and Australia: a cross-cultural comparison. In: Beetz, M., Johnston, B., Williams, M.-A. (eds.) ICSR 2014. LNCS (LNAI), vol. 8755, pp. 166–175. Springer, Heidelberg (2014). doi:10.1007/978-3-319-11973-1_17
6. Haring, K.S., Silvera-Tawil, D., Takahashi, T., Velonaki, M., Watanabe, K.: Perception of a humanoid robot: a cross-cultural comparison. In: Proceedings of the IEEE International Symposium on Robot and Human Interactive Communication, pp. 821–826. IEEE (2015)
7. Haring, K.S., Watanabe, K., Silvera-Tawil, D., Velonaki, M., Matsumoto, Y.: Touching an android robot: would you do it and how? In: Proceedings of the International Conference on Control, Automation and Robotics, pp. 8–13 (2015)
8. Kidd, C.D., Breazeal, C.: Effect of a robot on user perceptions. In: Proceedings of the IEEE/RSJ International Conference on Intelligent Robots and Systems, vol. 4, pp. 3559–3564. IEEE (2004)
9. Kim, R.H., Moon, Y., Choi, J.J., Kwak, S.S.: The effect of robot appearance types on motivating donation. In: Proceedings of the ACM/IEEE International Conference on Human-Robot Interaction, pp. 210–211. ACM (2014)
10. Kozima, H., Nakagawa, C., Yano, H.: Can a robot empathize with people? Artif. Life Robot. **8**(1), 83–88 (2004)
11. Mori, M., MacDorman, K.F., Kageki, N.: The uncanny valley [from the field]. IEEE Robot. Autom. Mag. **19**(2), 98–100 (2012)
12. Oliver, R.L.: A cognitive model of the antecedents and consequences of satisfaction decisions. J. Mark. Res. **17**, 460–469 (1980)
13. Scholtz, J.: Theory and evaluation of human robot interactions. In: Proceedings of the Annual Hawaii International Conference on System Sciences, pp. 1–10. IEEE Computer Society (2003)
14. Strait, M., Canning, C., Scheutz, M.: Let me tell you! investigating the effects of robot communication strategies in advice-giving situations based on robot appearance, interaction modality and distance. In: Proceedings of the ACM/IEEE International Conference on Human-Robot Interaction, pp. 479–486. ACM (2014)
15. Straub, D., Keil, M., Brenner, W.: Testing the technology acceptance model across cultures: a three country study. Inf. Manage. **33**(1), 1–11 (1997)
16. Vernon, R.J.W., Sutherland, C.A.M., Young, A.W., Hartley, T.: Modeling first impressions from highly variable facial images. Proc. National Acad. Sci. **111**(32), E3353–E3361 (2014)
17. Yanco, H.A., Drury, J.L.H.: Classifying human-robot interaction: an updated taxonomy. In: Proceedings of the IEEE International Conference on Systems, Man and Cybernetics, vol. 3, pp. 2841–2846 (2004)

Congruency Matters - How Ambiguous Gender Cues Increase a Robot's Uncanniness

Maike Paetzel[1]([✉]), Christopher Peters[2], Ingela Nyström[1], and Ginevra Castellano[1]

[1] Department of Information Technology, Uppsala University, Uppsala, Sweden
{maike.paetzel,ingela.nystrom,ginevra.castellano}@it.uu.se
[2] KTH Royal Institute of Technology, Stockholm, Sweden
chpeters@kth.se

Abstract. Most research on the uncanny valley effect is concerned with the influence of human-likeness and realism as a trigger of an uncanny feeling in humans. There has been a lack of investigation on the effect of other dimensions, for example, gender. Back-projected robotic heads allow us to alter visual cues in the appearance of the robot in order to investigate how the perception of it changes. In this paper, we study the influence of gender on the perceived uncanniness. We conducted an experiment with 48 participants in which we used different modalities of interaction to change the strength of the gender cues in the robot. Results show that incongruence in the gender cues of the robot, and not its specific gender, influences the uncanniness of the back-projected robotic head. This finding has potential implications for both the perceptual mismatch and categorization ambiguity theory as a general explanation of the uncanny valley effect.

Keywords: Uncanny valley · Robot gender · Back-projected robotic head

1 Introduction

Building robots which can serve as social companions, for example, in child or elderly care applications, is one of the main goals of social robotics research. An important aspect in building likable companions is to ensure that humans do not feel uncomfortable during the interaction or perceive the robot as *uncanny*. Masahiro Mori [14] first described the uncanny valley effect in 1970, suggesting that making the robot more humanlike leads to a drop in the likability of the robot at some point in the interaction. The underlying cause of the uncanny valley effect is still controversial, with different competing explanations such as the *perceptual mismatch* and *categorization ambiguity theory* [6,8,9]. Both theories agree in the proposed explanation that uncanniness is triggered by a mismatch in perceptual cues, but differ in the argumentation if this mismatch leads to an ambiguity in assigning related categories. While empirical research

© Springer International Publishing AG 2016
A. Agah et al. (Eds.): ICSR 2016, LNAI 9979, pp. 402–412, 2016.
DOI: 10.1007/978-3-319-47437-3_39

to date has been primarily concerned with the dimension of realism, the impact of other dimensions, like gender, requires further investigation.

With the emergence of back-projected heads [4] it has become not only possible to accurately control facial expressions of a robot, but also to easily change visual features in the face in a cost-effective manner. This opens up a variety of possibilities to bridge the research on virtual agents and physical robots, as well as to study the influence of the robot's gender on its perception.

Using the Furhat robot platform [4] (Fig. 1), we investigated the influence of the robot's gender on perceived uncanniness by varying visual and auditory gender cues in different modalities in a between-subject experiment involving 48 participants. We found that the congruency of the gender cues, and not the robot's specific gender, influence the robot's uncanniness, potentially supporting the categorization ambiguity theory of uncanniness [6,11].

2   Background

Although not numerous, some previous studies have investigated the perception of gender in robots. Eyssel et al. [7] used the hair style of the robot to change its visual gender perception. This proved to be very successful as it was shown that human gender stereotypes are assigned to robots. Siegel et al. [15] found that the rating of credibility, trust and engagement of a robot has a cross-gender effect with the participant's gender. However, they argue that one limitation of their work is given by the fact that they varied the gender only by changing the robot's voice. One possible explanation as to why research related to robot's gender and its perception is still somewhat rare is that changing the gender of a robot is not trivial. In this study, the back-projected robot head *Furhat* has been used in the experiments [4]. Although there are limitations in matching the projected face texture to the physical shape of the mask, they provide an efficient and cost-effective route for investigating the impact of varying cues on perception. While we were unable to find other research relating the robot's gender to uncanniness, Tinwell et al. [16] found that male virtual characters are perceived as significantly more uncanny than female characters and related this finding to the perception of psychopathic behavior.

In contrast to the theory by Tinwell et al., another explanation of uncanniness in robots suggests that uncanniness does not follow the dimension of human-likeness as Mori originally described [14]. It rather occurs at category boundaries where the category ambiguity is highest [6]. This was supported by Moore [13], who developed a Bayesian explanation of the uncanny valley, in which conflicting cues give rise to a perceptual tension at category boundaries.

Kätsyri et al. [8] recently reviewed the empirical research related to the uncanny valley effect and found more support overall for the *perceptual mismatch theory*. According to this theory, any mismatch in the perception of realism in robots and virtual characters increases the sense of eeriness. For example, disproportional facial parts [10] may result in an increase in uncanniness. However, this mismatch does not necessarily lead to an ambiguity in the related category, which distinguishes this theory from the *categorization ambiguity theory* [9].

Here, we explore whether the perception of uncanniness in robots can be expanded to the category of gender, which has potential implications for supporting the categorization ambiguity theory of uncanniness.

2.1 Hypothesis

We state the following hypothesis:

H1: *The gender of the robot has an influence on its perceived uncanniness.*
Similar research in the field of virtual characters suggests that male characters are perceived as significantly more uncanny than female characters [16], but it has not yet been studied if these findings apply to robots.

While most robots can only vary the gender by changing the voice, back-projected heads allow to assign visual gender cues on the face. As it has not been studied so far how well the texture on back-projected heads can convey gender information, we explore our hypothesis using two different modalities. Although we believe that the gender of the robot can be controlled by unimodal visual gender cues in the face texture, the combination of multimodal visual and auditory gender cues were expected to lead to stronger impressions of perceived masculinity and femininity. Testing both unimodal and multimodal cues therefore helps us to better understand the link between visual and auditory gender cues, perceived gender and the perceived uncanniness of the robot.

3 Experiment

This study investigates how the gender of a robot influence its uncanniness through two pilot studies (Sects. 3.1 and 3.2) and a main study (Sect. 3.3).

3.1 Pilot-Study 1: Visual Gender Cues

The first pilot study was conducted to select one female and one male face texture with clear gender cues in order to control the robot's visual gender cues in further experiments.

Participants. We conducted a within-subject online experiment with 40 participants ($m = 60\%$, $f = 35\%$, participants were allowed to withhold this information). All participants were over 18 ($M = 26.15$, $SD = 6.9$) and had at least an advanced English language ability. Recruitment was done entirely on the Internet on a voluntary basis and participating subjects were mostly from Germany (85%) and the USA (12.5%).

Stimulus and Independent Variables. *FaceGen Modeller* [3] was used to create a variety of different face textures in the 2D shape of the Furhat mask. Independent variables for generating the faces were the gender (17 steps in the FaceGen gender slider between very male and very female) and the level of caricature (*the average* and *attractive* on the FaceGen slider), leading to 34 different faces in total.

Procedure and Dependent Variables. Image exports from FaceGen were presented to the participants one at a time (cf. Fig. 1, left). Latin square was used to determine the image sequence to avoid ordering effects. For every image, participants were asked to rate the face on a 7-point Likert scale on the dimensions *gender*, *dominance*, *trustworthiness*, *strangeness*, and *attractiveness*. There was no time limit for rating an image and the image remained on the screen while participants were rating it. Three test images were inserted and used to discount the results of participants who answered randomly or were not paying attention to the task, in a similar manner to the approach in [11].

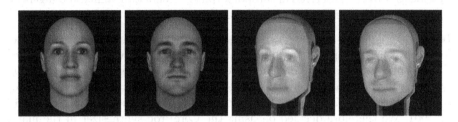

Fig. 1. From left to right: 2D image export from FaceGen showing the female and male face, 3D projection of the FaceGen female and male texture to the Furhat mask.

Results. In order to obtain an equal strength in terms of gender perception, we selected a male and female face that were rated equally by participants. The most masculine rated image was 0.33 Likert points less masculine than the theoretic optimum, while the most feminine face was 1.02 Likert points away from the theoretic optimum. Therefore, a less masculine face was chosen in order to be paired with the feminine face. As the research presented in this paper aims to investigate the perception of uncanniness, we also controlled the selected images so they had no significant difference in terms of trustworthiness and strangeness. The selected face textures are depicted in Fig. 1 to the left.

3.2   Pilot-Study 2: Auditory Gender Cues

The second pilot study was conducted to select one female and one male voice synthesizer with clear gender cues in order to control the robot's auditory gender cues in further experiments.

Participants 52 participants were recruited to take part in the online between-subject experiment. All participants were over 18 ($M = 25.06$, $SD = 5.16$) and had at least an advanced English language ability. 61.5 % of the participants were female. Recruitment was done entirely on the Internet on a voluntary basis and subjects were mostly from Germany (76.9 %) and the USA (7.7 %).

Stimulus and Independent Variables Independent variables for generating the voices were the gender and the level of anthropomorphism, leading to 4 different voices in total. The anthropomorphism was varied using the commercial software *CereProc* [1] (male: *William*, female: *Sarah*) and the OpenSource software *espeak* [2] (male: *en* with speech rate 150, female: *en+f1* with speech rate 150 and pitch 70).

Procedure and Dependent Variables. Participants were asked to listen to a two minute audio clip in which an artificial character introduced itself. The text was identical between all conditions, apart from laughter and a filled pause which appeared only for the CereProc voices, as espeak has no support for it. Participants had to listen to the entire audio clip at least once before they could continue to the perception questionnaire, in which they were asked to answer 33 questions about the perceived anthropomorphism, gender, likability, intelligence, personality traits and uncanniness of the voice.

Results. In general, participant responses show a clear preference towards the CereProc voices compared to the espeak voices in a One-way ANOVA analyses: the speed and flow of communication was rated better, $F(1,50) = 24.29$, $p < .001$, the voice was perceived as more friendly, $F(1,50) = 7.59$, $p = .008$, and pleasant, $F(1,50) = 25.42$, $p < .001$, and overall liked better, $F(1,50) = 17.16$, $p < .001$. We therefore decided on using the CereProc synthesizer in the main experiment. There was no significant difference between the male and the female CereProc voice in all dimensions except perceived masculinity and femininity.

3.3   Main Study

The main study involved a between-subject experiment in which subjects watched a short demonstration of Furhat in a lab at Uppsala University and answered a questionnaire about their perception of the robot.

Participants. 48 subjects from a graduate course participated in the experiment. All subjects had at least advanced English language skills and most had a background in Computer Science or a related subject. Course credits were awarded for the participation. From self-disclosure of the participants, 14.6 % were female and all were over 18 ($M = 23.96, SD = 0.41$).

Stimulus. *Furhat* was used for the main experiments [4]. It is equipped with a firm mask designed as an adult face. The projected face is animated, which allows for the eyes, lips and various other muscles in the face to be controlled. Furhat is equipped with two motors to move the jaw and pitch of the head. The voice and face texture were selected as discussed in Sects. 3.1 and 3.2. The head with the female and male face is depicted in Fig. 1 to the right.

Procedure. Each experiment session took about 20 min and involved one subject. The experiment room was set up to control for the vertical and horizontal distance between participant and robot as well as for the lightning condition.

At the beginning of the experiment, while the participant was being informed about the experiment task, the robot was still covered by a blanket. The experiment started with uncovering the robot. The participants were asked to rate their first impression of the robot on 7-point Likert scales of gender, dominance, trustworthiness, strangeness and attractiveness on a sheet of paper. Participants were instructed to judge based on their first impressions. The first time the robot was uncovered, it displayed a neutral face with an average face color between the male and female condition and only indicated eyes and lips. This condition provided a baseline for face shape perception without a gendered face texture applied. The robot was then covered again. When it was uncovered a second time, it displayed either the female or the male face. Participants rated it on the same five scales described above. The gender of the robot was assigned randomly.

Once the first task was completed, the demonstration of the robot started. The participants were asked to watch either the unimodal or multimodal pre-recorded behavior and answer a questionnaire including basic demographic questions as well as 33 questions concerning the perception of the robot.

Independent Variables. The experiment had the two independent variables *modality* and *gender*. In the *unimodal condition*, the robot performed a set of facial expressions, which took about one minute in total. The gender cues were varied between the *male face texture* and the *female face texture* described in Sect. 3.1. The *multimodal version* consisted of a combination of speech and facial expressions. The order of the facial expressions was the same as in the unimodal condition, but the pauses were longer due to the talking in between. The voice and speech was the same as used in the pilot-study 2 (cf. Sect. 3.2). Both *male voice and male face texture* respectively *female voice and female face texture* were combined to change the gender of the robot.

Dependent Variables. The perception questionnaire consists of 33 questions about the perceived human-likeness, gender, uncanniness, personality traits and sociability of the robot and is mainly based on the Godspeed questionnaire by Bartneck et al. [5]. All questions are rated on a 5-point Likert scale. For the first part of the questionnaire, participants are asked to rate their agreement with a postulated statement and, for the second part, to rate the robot in different dimensions, e.g. *unintelligent* vs. *intelligent*.

The feeling of uncanniness is rather complex and multidimensional and the exact measurement of the y-axis of the uncanny valley effect is a topic of ongoing discussion in the research community. Most common are the measures *familiarity* and *likability*, which are both included in our questionnaire. To cover the

Table 1. Excerpt from the scales used in the experiment questionnaire.

Category	Scales
Anthropomorphism	fake/natural, machinelike/humanlike, unconscious/conscious, artificial/lifelike
Familiarity	familiar, strange
Likeability	unfriendly/friendly, pleasant/unpleasant, dislike/like, like to have a conversation with
Trust	trustworthy, reliable

feeling of uncanniness as extensively as possible, we also add *trustworthiness* and *reliability* as an indicator for the likability of a robot [12]. An excerpt of the scales used in the data collection is shown in Table 1.

We explicitly ask participants to decide on the gender of the robot between *female, male,* and *neutral.* In addition, we ask for a separate rating of masculinity and femininity and how easy it is to determine the gender.

As gender is commonly associated with certain personality traits and Walters et al. [17] suggested that the personality traits assigned to a robot are also related to its likability, we include questions on the perceived personality of the robot.

4   Results

The physical mask of the robot displaying a neutral face is perceived to be very masculine ($M = 1.63$, $SD = 0.12$). The male face texture shares this gender perception ($M = 1.67$, $SD = 0.22$). A One-way ANOVA shows no significant difference to the neutral face, $F(1, 22) \approx 0$, $p \approx 1$, while the female robot is significantly more female ($M = 4.42$, $SD = 0.23$) than both the neutral, $F(1, 22) = 89.55$, $p < .001$, and the masculine face texture, $F(1, 22) = 73.49$, $p < .001$. All reported significance results are based on One-way ANOVA with alpha level of .05 and Type III sum of squares.

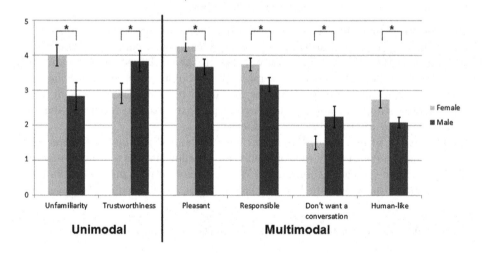

Fig. 2. Differences in the responses between the genders per condition (M and SE). Significant differences in One-way ANOVA are indicated by * ($p < .05$)

Within the unimodal condition, the male face texture is perceived as significantly more familiar, $F(1, 22) = 5.67$, $p = .026$, and trustworthy, $F(1, 22) = 4.91$, $p = .037$, than the female face texture. This difference cannot be confirmed in the multimodal condition, where both familiarity, $F(1, 22) = 0.96$, $p = .338$, and trustworthiness, $F(1, 22) \approx 0$, $p \approx 1$, show no significant difference between the genders. However, the female face texture is perceived as more pleasant, $F(1, 22) = 5.04$, $p = .035$, and more responsible, $F(1, 22) = 4.53$, $p = .045$, than the male face texture in the multimodal condition. Subjects also reported they would like to have a conversation with the female character significantly more, $F(1, 22) = 4.3$, $p = .05$. At the same time, the male character is perceived as more machinelike than the female character, $F(1, 22) = 5.25$, $p = .031$. The results are depicted in Fig. 2. *They show that the general trend in the perception of traits related to uncanniness is flipped by adding a voice to the face only.*

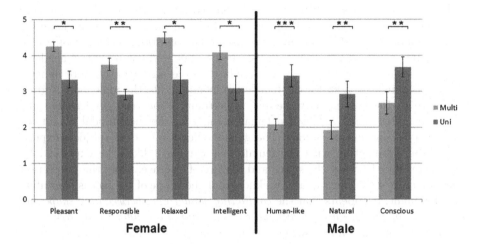

Fig. 3. Differences in the responses between the conditions per gender (M and SE). Significant differences in One-way ANOVA indicated by * ($p < .05$), ** ($p < .01$), *** ($p < .001$)

Comparing the two versions of modality with the female texture, we see that the perception of it changes when we add a voice to the face (cf. Fig. 3). It becomes significantly more pleasant, $F(1, 22) = 7.44$, $p = .012$, responsible, $F(1, 22) = 12.79$, $p = .002$, relaxed and content, $F(1, 22) = 7.59$, $p = .012$, and intelligent, $F(1, 22) = 6.66$, $p = .017$. We do not see any of these changes in the perception of the male robot when changing from unimodal to multimodal condition. However, we see that the male robot is perceived much more machine-like, $F(1, 22) = 14.82$, $p < .001$, artificial, $F(1, 22) = 5.11$, $p = .034$, and less conscious, $F(1, 22) = 5.66$, $p = .027$, than in the unimodal condition (Fig. 3).

A Randomized Block Design ANOVA with the modality as blocking factor showed no significant influence of the participant's gender in any dimension.

5   Discussion

In the unimodal condition, the male character has a higher likability and a higher trustworthiness, which could lead to the conclusion that female characters are more uncanny. However, this explanation does not hold for the multimodal condition. Here, the female robot is perceived as more pleasant and participants reported that they would like to have a conversation with the female character significantly more. In addition, the more positive rating in other personality traits, e.g. perceived relaxation and intelligence, can be an indicator of a decreased uncanniness [17]. Therefore, we found no indication that the specific gender of the robot influences the perceived uncanniness across modalities.

In general, we observed a difference in the gender perception between the male and the female condition. The perception of the male face is clearly on the masculine side of the scale and all participants assigned the male gender in both unimodal and multimodal condition. Voting for the female face is only marginally on the feminine side of the scale, but still significantly more feminine than the male face. After watching the unimodal condition, seven out of twelve participants assigned a female, four assigned a male and one a neutral gender.

In the unimodal condition, participants reported significantly more difficulties in assigning a gender for the female face compared to the male face. Adding the voice to the face helps to resolve uncertainty about the gender. Ten out of twelve participants assigned a female gender in the multimodal condition, the other two a neutral gender. The gender uncertainty that subjects experienced after interacting with the female face can most likely be explained by the discrepancy between the shape and texture of the face. From the perception questionnaire of the neutral face, we know that the shape of the face is masculine and dominant. It appears that the *conflicting gender cues between the shape and the texture of the face lead to difficulties in assigning a gender to the robot*.

In addition, this finding suggests that *uncanniness is caused by an incongruence of gender cues rather than a specific gender*. This might give support to the categorization ambiguity theory of uncanniness, as the perceived masculinity and femininity of the female character was more towards the boundary between the two extremes, while the male character's ratings were clearly on the masculine side. Future work with a questionnaire more tailored towards the categorization of the robot might facilitate a deeper interpretation of this finding.

The preference for the female character in the multimodal condition can be explained by a general preference for female characters, a cross-gender preference [15], or by the perception of the male version as significantly more artificial. The lack of facial hair or participant's cultural background could have an influence as well. Again, future work could give further insights in the underlying causes.

6   Conclusion and Future Work

In this paper, we studied the influence of gender in different modalities on the perceived uncanniness of a back-projected robotic head. We found that *incongruent gender cues varied using different modalities lead to a decreased likability of the robot, rather than its specific gender*.

For future work, it would be interesting to study whether we can find support for *H1* in a study without conflicting gender cues. In addition, it would be insightful to investigate if even more opposing gender cues, such as a mismatch between the voice and face, can increase the perceived uncanniness of the robot.

Acknowledgements. Thanks to Lars Oestreicher for providing the two rightmost images of Fig. 1. C. Peters' work is partly supported by the European Commission (EC) Horizon 2020 ICT 644204 project ProsocialLearn. The authors are solely responsible for the content of this publication. It does not represent the opinion of the EC, and the EC is not responsible for any use that might be made of data appearing therein.

References

1. CereProc Synthesizer. https://www.cereproc.com/. Accessed 5 Aug. 2016
2. eSpeak Synthesizer. http://espeak.sourceforge.net/. Accessed 5 Aug. 2016
3. FaceGen Modeller. http://facegen.com/. Accessed 5 Aug. 2016
4. Al Moubayed, S., Beskow, J., Skantze, G., Granström, B.: Furhat: a back-projected human-like robot head for multiparty human-machine interaction. In: Esposito, A., Esposito, A.M., Vinciarelli, A., Hoffmann, R., Müller, V.C. (eds.) Cognitive Behavioural Systems. LNCS, vol. 7403, pp. 114–130. Springer, Heidelberg (2012). doi:10.1007/978-3-642-34584-5_9
5. Bartneck, C., Kulić, D., Croft, E., Zoghbi, S.: Measurement instruments for the anthropomorphism, animacy, likeability, perceived intelligence, and perceived safety of robots. Int. J. Soc. Robot. **1**(1), 71–81 (2009)
6. Cheetham, M., Suter, P., Jäncke, L.: The human likeness dimension of the "uncanny valley hypothesis": behavioral and functional MRI findings. Front. Hum. Neurosci. **5**(126), 10–3389 (2011)
7. Eyssel, F., Hegel, F.: (S)he's got the look: gender stereotyping of robots. J. Appl. Soc. Psychol. **42**(9), 2213–2230 (2012)
8. Kätsyri, J., Förger, K., Mäkäräinen, M., Takala, T.: A review of empirical evidence on different uncanny valley hypotheses: support for perceptual mismatch as one road to the valley of eeriness. Front. Psychol. **6**, 1–16 (2015)
9. MacDorman, K.F., Chattopadhyay, D.: Reducing consistency in human realism increases the uncanny valley effect; increasing category uncertainty does not. Cognition **146**, 190–205 (2016)
10. MacDorman, K.F., Green, R.D., Ho, C.C., Koch, C.T.: Too real for comfort? Uncanny responses to computer generated faces. Comput. Hum. Behav. **25**(3), 695–710 (2009)
11. Mathur, M.B., Reichling, D.B.: Navigating a social world with robot partners: A quantitative cartography of the Uncanny Valley. Cognition **146**, 22–32 (2016)
12. McDonnell, R., Breidt, M., Bülthoff, H.: Render me real?: investigating the effect of render style on the perception of animated virtual humans. ACM Trans. Graph. **31**(4), 91 (2012)
13. Moore, R.K.: A Bayesian explanation of the 'Uncanny Valley' effect and related psychological phenomena. Sci. Rep. **2**, 69–78 (2012)
14. Mori, M., MacDorman, K.F., Kageki, N.: The uncanny valley [from the field]. Robot. Autom. Mag. IEEE **19**(2), 98–100 (2012)
15. Siegel, M., Breazeal, C., Norton, M.I.: Persuasive robotics: The influence of robot gender on human behavior. In: IEEE/RSJ International Conference on Intelligent Robots and Systems, IROS 2009, pp. 2563–2568. IEEE (2009)

16. Tinwell, A., Nabi, D.A., Charlton, J.P.: Perception of psychopathy and the Uncanny Valley in virtual characters. Comput. Hum. Behav. **29**(4), 1617–1625 (2013)
17. Walters, M.L., Syrdal, D.S., Dautenhahn, K., Te Boekhorst, R., Koay, K.L.: Avoiding the uncanny valley: robot appearance, personality and consistency of behavior in an attention-seeking home scenario for a robot companion. Auton. Robots **24**(2), 159–178 (2008)

Collaborative Visual Object Tracking via Hierarchical Structure

Fangwen Tu, Shuzhi Sam Ge$^{(\boxtimes)}$, Henry Pratama Suryadi, Yazhe Tang, and Chang Chieh Hang

Department of Electrical and Computer Engineering,
National University of Singapore, Singapore 117575, Singapore
samge@nus.edu.sg

Abstract. This paper focuses on the optimization and improvement of visual-based object tracking algorithm. Reflecting from previously used tracking algorithm, we approach the problem using L2-regularized least squares to solve the sparse representation matrix of the object appearance model and propose an efficient collaborative algorithm to track the object. A hierarchical framework and selective multi-memory based online dictionary update are developed to upgrade the speed of the algorithm and improve the robustness by considering both current and history appearance into the template. In addition, key-point feature matching is novelly proposed to further enhance the accuracy of the tracking algorithm by calculating an optical flow based similarity degree. Finally, the proposed algorithm is verified using comprehensive image sequence datasets to demonstrate its effectiveness on coping with various tracking challenges, such as object deformations, illumination changes and partial occlusions.

Keywords: Hierarchical framework · Key-point features · Multi-memory template update · Visual object tracking

1 Introduction

With the improvement and great progress achieved in the field of visual tracking, it is expected that the technique can be extended to a practical application from surveillance system, unmanned aerial vehicle to social and industrial robots. Object tracking assumes that object to be tracked have been selected in the first frame [1]. Compared to object detection, it is less dependent on training data, where the tracking method only uses information obtained from the first frame. In recent years, various tracking algorithms have been proposed with positive results. Development of robust visual tracking with the capability to handle occlusion, deformation, illumination changes and other object variations becomes very important as visual tracking can be applied in various situations [2]. However, challenges such as heavy computational load and complicated changes in target appearance remain unsolved until today. Trackers even with

© Springer International Publishing AG 2016
A. Agah et al. (Eds.): ICSR 2016, LNAI 9979, pp. 413–421, 2016.
DOI: 10.1007/978-3-319-47437-3_40

most advanced algorithms may still fail to track different scenarios. Thus, a large amount of work remains for researchers to explore a robust and efficient algorithm for visual tracking.

In object tracking, appearance model of the target object can be defined in several ways. In the last few years, sparse representation matrix has been applied to optimize object tracking by finding the best target candidate with minimal reconstruction error using target templates. Being the latest tracking optimization that has a lot of opportunity to be explored, many experiments involving the usage of this concept are conducted to prove its superiority. The sparsity can be achieved by solving regularized least squares problem of the image feature [3]. With sparse matrix, computational load can be reduced and efficiency of the tracker can be increased. Combined with classic particle filter method, this paper explores the potential of using sparse coding in tracking algorithm with a novel multi-memory template update scheme. Furthermore, apart from holistic appearance model, this paper also introduces a collaboration with key-point features detection to increase the accuracy of the algorithm. Using the combination between this two ways of judgement, the proposed tracker should be able to handle more challenges.

2   Object Representation

Although discriminative trackers usually achieve more excellent accuracy than generative ones, the time consumption is extremely long. In this paper, a generative framework is employed to ensure an improved speed without compromising the accuracy. In the proposed tracker, square templates [4] are used to represent the occlusion and misalignment part of the object. They are augmented in the dictionary (template) together with principal component analysis (PCA) basis vectors obtained from the appearance of the object. These templates are treated as positive information, which are assumed to include the features of the target object.

2.1   Holistic Templates

Possibility of incorporating regularized least squares method has been explored in a few tracking algorithms previously. Originally, least squares method is used to find an optimal state where objectives in the terms are minimized. It has been adopted as a representative model for object tracking. In 2009, there has been a research to incorporate L1 minimization sparse coding and optimize the tracking algorithm [3]. In contrast to the L1-regularization, L2 norm offers a faster solution for object tracking by solving the minimization problem efficiently at the cost of sparsity, which is shown in (1). It leads to a very efficient object tracking when the objects are not corrupted or deformed in the image because of the availability of analytical solution. Recent study [4] also stated that with comparable accuracy, L2 norm outperforms L1 norm in terms of speed by a large margin.

$$c^* = \arg\min_c(\|y - Dc\|_2^2 + \lambda\|c\|_2^2) \tag{1}$$

Using the L2-regularized least squares, object representation of the image patch can be formulated consisting of two main terms, i.e. PCA basis vectors and error terms. Linear combination of these two terms can represent any uncorrupted sample in the same subspace properly [4]. Both terms are normalized using the corresponding coding vector obtained from the RLS process. Define matrix B representing the PCA basis vectors, which are extracted from the target and E corresponds to the square template. In case of occlusion, error term will be accommodated with E. The final object representation can be derived in (2).

$$y = Bc_b + Ec_e = \begin{bmatrix} B, E \end{bmatrix} \begin{bmatrix} c_b \\ c_e \end{bmatrix} = Dc \tag{2}$$

Solving the L2-regularized least squares problem, optimal coefficient of each features can be obtained by the following formula.

$$c^* = (D^T D + \lambda I)^{-1} D^T y = Py \tag{3}$$

It should be noted that P is only updated when the template B updates. In this manner, the computational complexity can be reduced in a large scale.

2.2   Key-Point Features Extraction

Other than using holistic templates, the proposed algorithm also performs tracking based on the key-point corner of the tracked object. By tracking the key-point features of the object, it is possible to estimate the object shape and handle heavy illumination changes. In this work, Harris corners are detected first. Subsequently, Kanade-Lucas-Tomasi (KLT) algorithm is performed to track the target position of each frame. The main problem of tracking the key-point features is to relate each point with their corresponding point in the previous frame. This can be solved using optical flow to estimate the location of all points based on the location of Harris points in the previous frame which will be elaborated in following context.

3   Visual Tracking Framework

Figure 1 shows the overall tracking algorithm. Visual tracking can be seen as a Bayesian inference task in a Markov model with hidden state variables. Bayesian inference framework estimates the behavior of variables given the existing data obtained from previously known fact, i.e. to infer the posterior distribution $p(x_t|Y_t)$.

$$p(x_t|Y_t) \propto p(y_t|x_t) \int p(x_t|x_{t-1})p(x_{t-1}|Y_{t-1})dx_{t-1} \tag{4}$$

where $Y_t = [y_1, y_2, ..., y_t]$ denotes the observation vector up to t th frame. x_t is the state variable of a target at frame t and $p(x_t|x_{t-1})$ represents the dynamic model predicting the state in current frame with the immediate previous state. $p(y_t|x_t)$ indicates the observation model which is a likelihood function in essence.

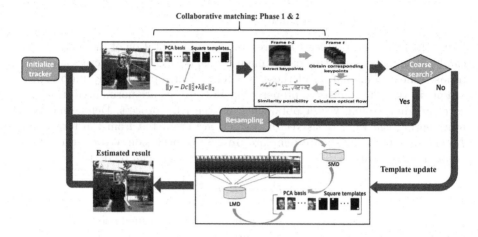

Fig. 1. Proposed tracking framework diagram

3.1 Confidence Value

In the first phase, likelihood of the image patch can be measured by the reconstruction error. It is based on the L2-regularized sparse representation model in the previous section. Intensity data from holistic templates are extracted using previously determined sparse coefficient vector to measure the reconstruction error. The error is then normalized to ensure the validity of the particle.

The formula to determine the likelihood can be expressed in (5) [4]. $\|Ec_e^i\|_1$ is added to avoid the instability of the tracker due to the reconstruction error term $\|y^i - Bc_b^i - Ec_e^i\|_2^2$. δ is an adjustable coefficient. After calculating the likelihood between each sample with the target templates, five samples with the highest degree of confidence value will be chosen as the particles for the next phase of particle filtering.

$$p(y^i|x^i) \propto \exp(-\|y^i - Bc_b^i - Ec_e^i\|_2^2 - \delta\|Ec_e^i\|_1) \tag{5}$$

3.2 Similarity Degree

After performing the first phase of particle filter, the second phase is conducted by taking into account the key-points location of the current samples. Since extracting key-points are computationally heavy, it will only be executed to the five best candidate samples filtered from the first phase.

Firstly, keypoints on the tracked object in previous frame is extracted then the tracker will estimate the location of those points in the five target candidates. After the location of the points has been estimated, Euclidean distance between each points and the number of valid points are calculated through KLT to find the similarity degree of the target candidate. It can be derived that the smaller the changes of appearance, the candidate is considered to be more representative.

Function to compute the similarity degree is defined in (6).

$$p(y^i_{\mathrm{opt}}|x^i_{\mathrm{opt}}) = \frac{n^2}{\sum_{i=1}^n \sqrt{\partial p_i^2 + \partial q_i^2}} \tag{6}$$

where n denotes the number of keypoints in one frame. $\sqrt{\partial p_i^2 + \partial q_i^2}$ ($i=1,2,...,5$) returns the displacement of each keypoint.

After the similarity degree for each target candidate has been calculated, it is further normalized and combined with the confidence value to obtain the final tracking result. Using the computation of Harris corner extraction as the basis on the second phase, accuracy of the proposed tracker can be further enhanced. The procedure for similarity calculation can be summarized in Fig. 2.

In the final stage of the tracking algorithm, a collaborative approach is used to combine the result from the first and second phase of particle filter. Since there are two different particle filter phases, results from both phases are used jointly to compute the final probability measures of each target candidate.

$$G(y) = \mu_h p(y^i|x^i) + \mu_{\mathrm{opt}} p(y^i_{\mathrm{opt}}|x^i_{\mathrm{opt}}) \tag{7}$$

The integrated confidence above combines the confidence value and similarity degree of the candidate samples with $\mu_h + \mu_{\mathrm{opt}} = 1$. The estimated tracking result is the candidate which has the highest joint possibility of both the characteristics.

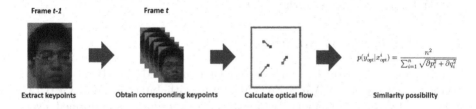

Fig. 2. Diagram of similarity degree calculation

3.3 Hierarchical Framework

Searching the current frame for best candidate representation of the image is exhaustive due to the large number of samples that need to be considered for each frame. By reducing the number of samples generated for coarse search and then doing fine search with lower standard deviation will presumably increase the speed of the tracking algorithm [5]. In the proposed tracker, search strategy started with a relatively large standard deviation (twice as in fine search) for the coarse search. Then, tracker will continue to search through fine search for the target with smaller standard deviation and same number of samples. Specially, the sampling center in fine search is fixed at the optimum candidate obtained in coarse search.

3.4 Online Update Scheme

In the proposed algorithm, it involves two different dictionary learning method in its online update scheme, which are short-memory dictionary (SMD) and long-memory dictionary (LMD). Using different updating schemes, the proposed tracker is able to deal with drift problems more efficiently. By selective template addition, it automatically learn good object templates from the tracking results. SMD is able to adapt robustly with recent appearance changes while LMD is able to take into account the history object appearance, which is assumed to have lower reconstruction error [6]. With the combination of these two dictionaries, robustness and accuracy of the tracker is ensured.

SMD includes the tracking result from most recent frames. It consist of dense samples which is expected to contain the current object appearance. When the particle filter has determined the tracking result and the result is evaluated to have low occlusion rate and high probability measure, it is added to the SMD set for updates. When SMD set has five different tracking results, all the tracking results will be used to update the PCA basis vectors through incremental learning and then SMD set is cleared.

LMD on the other hand includes several tracking result from older frames. When the PCA basis vectors are updated using the SMD, it will select the image patch which has the highest probability measure among all the image patches stored in SMD and save it in LMD set. After LMD has accumulated five different tracking results, it will also update the PCA basis vectors accordingly. Obviously, it is rarely used compared to the SMD.

4 Performance Evaluation

In this paper, all algorithms are implemented using MATLAB for Windows operating system on an ASUS K45VM laptop with Intel i5-3210M CPU (2.5 GHz) and 4 GB RAM memory. Before tracking experiment is performed, every running process other than MATLAB is suspended temporarily.

For quantitative study purpose, the proposed algorithm is compared with five state-of-the-art algorithms with excellent tracking accuracy, i.e. L2 tracker [4], L1-APG [7], SCM [8], IVT [9] and TVPT [10]. To elaborate the performance of the multi-memory template update scheme, sole phase 1 is also involved in the experiment. The new tracker and the existing tracker are tested against 15 different datasets [11]. All trackers are measured in terms of accuracy and speed to evaluate the performance.

Speed improvement of the tracker is caused by the reduction of the amount of samples computation. With less samples, there will be less computation and therefore increase the speed of the tracker. As shown in Fig. 3a, although the proposed algorithm cannot achieve real-time tracking, it outperforms most of the generative model based algorithms (L2-RLS and L1APG) and all the discriminative algorithms (TVPT and SCM).

In terms of accuracy, the detailed center error and average tracking error of seven trackers on fifteen challenging datasets are presented in Table 1 and

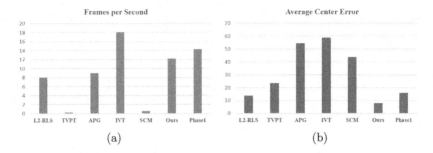

Fig. 3. Speed and accuracy evaluation

Fig. 3b. It can be seen that there is less error produced by the proposed tracker, which means it performs favorably against the existing tracker. Throughout the tracking process, the proposed tracker is able to follow the object closely, producing least center error for most of the time. In certain datasets, the proposed tracker has worse performance compared to other comparative trackers, which might be caused by the inaccuracy during the coarse search. Other than that, it can be concluded that there is better accuracy and stability performance for the proposed tracker, though, the involvement of phase 2 retards the tracker because of the key-point extraction.

Table 1. Average center error measurement. The top two results are highlighted in red and blue fonts respectively.

	L_2	TVPT	APG	IVT	SCM	Ours	Phase1		L_2	TVPT	APG	IVT	SCM	Ours	Phase1
Basketball	11	69	142	128	376	7	10	Dudek	12	11	33	10	16	10	13
Boy	5	7	66	92	3	3	3	FaceOcc1	12	15	16	19	12	12	12
Car4	4	5	115	2	3	3	4	Girl	4	23	3	21	36	4	4
Car11	2	3	15	8	3	1	1	Jumping	4	14	41	101	4	5	5
Caviar	3	62	15	67	63	3	3	Panda	8	7	42	66	7	9	9
David	66	6	16	27	9	11	66	Singer1	6	5	5	12	4	3	5
Deer	8	19	178	194	9	8	9	Sylvester	41	24	71	37	8	20	35
DragonB	17	82	57	93	101	15	56								

Figure 4 plots the variation of tracking error of four representative datasets. Looking from these graphs, it can be seen that the final proposed algorithm is able to outperform other existing algorithm in terms of stability. It exhibits the lowest center error in the majority of the frames, which suggests that the algorithm shows favorable robustness to handle different challenging situations such as the illumination variation in *Car4* and *Singer1*, abrupt motion and motion blur in *Boy*, in-plane and out-of-plane in *Basketball*.

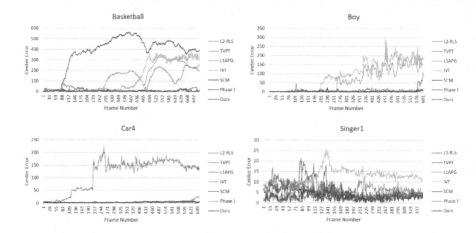

Fig. 4. Curves of tracking error on four representative datasets

5 Conclusion

The proposed algorithm incorporates two different object representation schemes to handle various object tracking challenges. Both L2-regularized least squares representation of holistic templates and optical flow based Harris key-point matching ensure that the tracker has a high level of accuracy. Speed of the proposed tracker is further enhanced using hierarchical framework. Moreover, robustness to object deformation is maintained by updating the object representation occasionally throughout the tracking process. With the multi-memory dictionary included in the tracking algorithm, it equips with the capability to accommodate both significant changes in recent frames and robustness to the earlier shape. Finally, the efficiency and efficacy of the proposed algorithm is demonstrated by experimental study.

Acknowledgement. This work is supported by Defence Innovative Research Programme (DIRP), the Ministry of Defence, Singapore under grant R-263-000-B08-592.

References

1. Smeulders, A.W., Chu, D.M., Cucchiara, R., Calderara, S., Dehghan, A., Shah, M.: Visual tracking: An experimental survey. Pattern Anal. Mach. Intell. IEEE Trans. **36**(7), 1442–1468 (2014)
2. Yilmaz, A., Javed, O., Shah, M.: Object tracking: A survey. ACM Comput. Surv. (CSUR) **38**(4), 13 (2006)
3. Mei, X., Ling, H.: Robust visual tracking using l1 minimization. In: 2009 IEEE 12th International Conference on Computer Vision, pp. 1436–1443. IEEE (2009)
4. Xiao, Z., Lu, H., Wang, D.: L2-rls-based object tracking. Circ. Syst. Video Technol. IEEE Trans. **24**(8), 1301–1309 (2014)
5. Zhang, K., Zhang, L., Yang, M.-H.: Fast compressive tracking. Pattern Anal. Mach. Intell. IEEE Trans. **36**(10), 2002–2015 (2014)

6. Xing, J., Gao, J., Li, B., Hu, W., Yan, S.: Robust object tracking with online multi-lifespan dictionary learning. In: Proceedings of the IEEE International Conference on Computer Vision, pp. 665–672 (2013)
7. Bao, C., Wu, Y., Ling, H., Ji, H.: Real time robust l1 tracker using accelerated proximal gradient approach. In: 2012 IEEE Conference on Computer Vision and Pattern Recognition (CVPR), pp. 1830–1837. IEEE (2012)
8. Zhong, W., Lu, H., Yang, M.-H.: Robust object tracking via sparse collaborative appearance model. Image Process. IEEE Trans. **23**(5), 2356–2368 (2014)
9. Ross, D.A., Lim, J., Lin, R.-S., Yang, M.-H.: Incremental learning for robust visual tracking. Int. J. Comput. Vis. **77**(1–3), 125–141 (2008)
10. Wang, Q., Chen, F., Yang, J., Xu, W., Yang, M.-H.: Transferring visual prior for online object tracking. Image Process. IEEE Trans. **21**(7), 3296–3305 (2012)
11. Wu, Y., Lim, J., Yang, M.-H.: Object tracking benchmark. Pattern Anal. Mach. Intell. IEEE Trans. **37**(9), 1834–1848 (2015)

Data Augmentation for Object Recognition of Dynamic Learning Robot

Jiunn Yuan Chan, Shuzhi Sam Ge$^{(\boxtimes)}$, Chen Wang, and Mingming Li

Department of Electrical and Computer Engineering,
National University of Singapore, Singapore 117576, Singapore
samge@nus.edu.sg

Abstract. The training of deep learning networks for robot object recognition requires a large database of training images for satisfactory performance. The term "dynamic learning" in this paper refers to the ability of a robot to learn new features under offline conditions by observing its surrounding objects. A training framework for robots to achieve object recognition with satisfactory performance under offline training conditions is proposed. A coarse but fast method of object saliency detection is developed to facilitate raw image collection. Additionally, a training scheme referred to as a Dynamic Artificial Database (DAD) is proposed to tackle the problem of overfitting when training neural networks without validation data.

1 Introduction

Object recognition in the field of robotics is the localization and identification (or classification) of objects in images or videos. It is an important aspect of robots as it enables them to be aware of their surroundings to behave in an interactive and sensible manner. This paper aims to find a suitable training method to achieve effective object recognition under practical implementation considerations on dynamic learning robots. In particular, robots should be dynamic learning to improve on personalization and interaction features.

For this paper, a Convolutional Neural Network (CNN) is trained under raw data scarcity. Raw data in this context refers to training images that are photos taken of the real world in the robot's neighboring environments. In contrast, artificial data refers to training images that are augmented versions of raw data. The training of CNN to implement object recognition requires huge amount of training images to achieve satisfactory performance [1,2]. With the availability of data over the Internet, obtaining huge amount of training images no longer seems to be an obstacle for the training of state-of-the-art performing networks. However, training images for unique objects may not be sufficiently available over the Internet. For a robot to have a robust object recognition performance in its neighboring environments, it should be able to learn any new object efficiently and effectively. Classifying and localizing objects has long been an important area of study in computer vision with the "Regions with Convolutional Neural Network" (R-CNN) achieving outstanding improvements

© Springer International Publishing AG 2016
A. Agah et al. (Eds.): ICSR 2016, LNAI 9979, pp. 422–430, 2016.
DOI: 10.1007/978-3-319-47437-3_41

[3,4]. Classification and localization complements each other and should both be treated with equal importance. This complementary relationship is demonstrated by [5]. A CNN system is highly demanding on the amount of training data to prevent overfitting [1]. Some systems exploit the use of online databases such as ImageNet to retrieve training data on demand. However, most of these training data provided are not object-centric, in which case the objects are not centered and zoomed in at the images but appear at various scales under different contexts [6]. This makes it difficult for the system to learn since objects may be too small or highly occluded [7]. In addition, several adaptive methods to create an artificial database for dynamic learning robots are proposed. The idea is partly inspired by [2], with the difference being that this project involves the generation of multiple variations of training images from a single object centric raw image rather than 3D models. If the images of objects can have their backgrounds removed, then these images can be used as sources for the generation of numerous varied training images. The generated training images make up what is referred to as an artificial database in this project. As an attempt to prevent overfitting, a concept referred to as Dynamic Artificial Database (DAD) is proposed. It involves the replacing of old training data with new ones after every training epoch. This is based on the belief that the larger the database, the less likely a CNN will over fit [2]. The problem of not knowing when to stop the training of CNN under the absence of validation and test data is solved by DAD since it tends to prevent a network from overfitting any specific data.

2   Challenges Facing Dynamic Learning Robots

The main challenges facing dynamic learning robots include the acquisition of sufficient training data, the limitation on memory space, and the lack of validation data to perform early stopping.

2.1   Acquisition of Sufficient Training Data

Hundreds of training images are required for a CNN network to effectively learn to classify an object [2]. Under conditions where existing databases are not available, it would be highly inefficient for a robot (or the user of a robot) to create a new database by using only captured images. This is a time consuming process if highly unique training images are to be captured since huge variety may involve changes in environment.

2.2   Limitation on Memory Space

Since a dynamic learning robot is assumed to have no internet access, its memory capacity is thus limited to the available hard disk space and Random Access Memory (RAM). The conventional way of training CNN networks involves the use of static databases. A static database here refers to a database with fixed size and contents. If each object learned will require a storage of hundreds to thousands of training images, then the number of objects that can be learned by the robot is very much limited to its RAM and hard disk capacities.

2.3 Lack of Validation Data

A common practice to avoid overfitting is by performing early stopping. However, this requires a set of validation data to determine when the CNN model is starting to over fit the training data. The collection of validation data is the same as the collection of training data which is a time consuming process. Therefore applying early stopping is not practically sound for the case of dynamic learning robots.

Considering these challenges, it is therefore important for a dynamic learning robot to adopt a system workflow that generates data, replaces data regularly, and trains continuously without overfitting. This system workflow is described in the next chapter.

3 Dynamic-Learning Framework

In this project, a learning framework for dynamic learning robots is proposed. A simple flow chart for the operation of a dynamic learning robot with the proposed local-learning framework is illustrated by Fig. 1.

An artificial database refers to a database of training data created by applying data augmentation to a set of raw images. The inspiration of generating artificial training data came from [2], whose team used 3D models to generate training and test images. Their set up includes camera view and illumination settings so the generated images are photorealistic. For each object class 40,000 training images were generated. Although the resulting classification accuracy is high, it is also shown that a large database is necessary for outstanding performance [2]. On the other hand, Sander Dieleman and his team used data augmentation to increase the size of their training data. They employed affine transformation properties such as rotation, translation, scaling, shearing etc. with increasing intensity as the training model over-fits [8]. Instead of using 3D models, this project involves the generation of multiple variations of training images from object-centric raw images. If the images of objects can have their backgrounds removed, then these images can be used as sources for the generation of numerous varied training images. The generated training images make up what is referred to as an artificial database in this project. Basically, the generation of these training images involves the extraction and random transformations of the original object image (with no background) and the random selection of a new background. The advantage of this is that the training images generated are ensured to be object-centric.

4 Methodologies

In this chapter, the main processes that form the local-learning framework are discussed in detail.

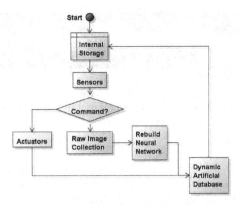

Fig. 1. Flow chart of the proposed dynamic-learning framework. The command block checks for any user-given instructions and directs the loop accordingly. The CNN is resized to match the new classification requirement after the raw image collection process. The DAD block is where training images are generated, followed by the training of the CNN.

4.1 Object Extraction Using Saliency Map

Object saliency detection techniques such as [9] can be employed to extract out salient objects from their backgrounds in raw images. However, for computational speed, this project uses a coarse but fast extraction method based on edge saliency. Given a 400×300 pixel image, the image is firstly converted from RGB colour space to only the Y channel of YUV colour space. Then the image undergoes spatial subtractive normalization with a 13×13 Gaussian kernel. This operation highlights contrastive edges and flattens smooth transitions. An absolute value operation is performed so that both dark and bright edges have equal weights in determining saliency. The image is then segregated into uniform regions with 50 % overlap and the mean saliency value of each region is computed. Thresholding is then performed with the threshold set as 30 % of the whole image's pixel mean saliency value. It is thus important that the raw image is object-centric with a smooth and contrastive background. Figure 2 illustrates the process of salient object extraction. The image after thresholding is the bitmap representation of salient objects with the value of 1 as white and 0 as black. A horizontal and vertical gap filling pass is performed and the bitmap is Gaussian blurred to smooth out the edges of the bitmap. Finally, the object is extracted from the raw image by multiplying the blurred bitmap with the raw image.

4.2 Artificial Database

The output image size for this project's experiments is fixed at 80×60 pixels. A series of random transformations is applied to extracted object images in the following sequence with the ranges of intensity given:(1) Skew: 0 degree-26.6

Fig. 2. Salient object extraction. From left to right, (1) raw image, (2) bitmap representation of salient object, (3) bitmap after horizontal and vertical gap-filling pass, (4) Gaussian blurred bitmap, (5) extracted object.

degree; (2) Scale: 0.7−0.8 (where 1 represents the object size with best fit scaling to the output image size); (3) Translate: randomly place the object within the boundaries of the output image; (4). Rotate: -180 degree−180 degree; (5) Brightness addition: -0.1−0.1 (across all channels); (6). Brightness multiplication: 0.83−1.2 (across all channels); (7) Noise: -0.025−0.025; (8). Gamma correction: 0.7−1.43.

It is important to note that the pixel values are between 0 and 1 and the final images produced are normalized to be within this range. The transformation settings are not optimized, but serves as a controlled environment for the comparison of different training schemes. Before applying the transformations, a saliency detection method is used to extract out the objects from the raw images. Figure 3 illustrates an example of a set of watch images generated. The reason why validation data are not generated along with training data is because it is easier for the validation set to start to over fit since the training and validation data are generated from the same environment and mechanism. Huge amount of augmented validation data will have to be generated to have effective results from early stopping. However this will then raise the hardware requirements on storage space tremendously as more object classes are learned. In this project, three different CNN training schemes are evaluated. Since early stopping is not used, a training scheme that minimizes overfitting under continuous training conditions must be determined for the local-learning framework to be effective.

Fig. 3. Watch images generated to form an artificial database.

4.3   Static Artificial Database (SAD)

A Static Artificial Database is analogous to conventional databases that are fixed in size. The only difference is that SAD is made up of pre-generated augmented data. This method usually requires huge storage space and RAM for effective training. However, it is considerably fast since the loading of training data is, most of the time, done only once when post-augmentation is not performed.

4.4   Dynamic Artificial Database (DAD)

In contrast with SAD, a Dynamic Artificial Database is a database made up of augmented data that changes throughout the process of training. Since according to [2], it is believed that a larger database will generally result in a network with better performance. The theoretical goal then is to be able to train a network with an infinitely large database with non-repeating data. This can be achieved by training one epoch for every batch of unique training data generated as illustrated in Fig. 4.

DAD is a general type of training scheme of which training data are generated, used, and replaced repetitively. The most basic form of DAD is to only use each generated data once. Although strictly following the theoretical approximation, the basic form of DAD can be slow as data needs to be generated on the fly.

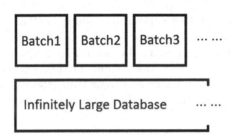

Fig. 4. Infinitely large database approximation with DAD. Training one epoch on an infinitely large database is equivalent to training with infinitely many batches of unique training data for one epoch per batch.

5   Experiment

In this chapter, the learning process of the training schemes introduced are compared and evaluated. The CNN architecture used and evaluation conditions are also stated.

5.1   Evaluation Conditions

The lack of test and validation data for actual implementation of an dynamic learning robot is identified as a major obstacle for the minimization of overfitting. For evaluation purposes, test data are collected to measure the performance of each proposed method. It is also assumed that the test data are not available for the robot, therefore the objective of the evaluation is to compare each method's steady state mean accuracy and change in error with the test data.

The CNN network is trained with 5 classes. For each class, 60 test images and 10 raw images are used. One of the classes is a "none" class, which is ideally the classification result when the other object classes (scissors, lead container, smartphone, and computer mouse) have low classification scores. The "none" class test and training images are not photos taken, but are random portions cropped out from 25 images of empty rooms.

All experiments start with a learning rate of 0.1, learning rate decay of 2×10^{-4}, momentum of 0.1, and a regularization factor of 10^{-5}. The CNNs are trained under batch-learning with mini-batches of 10 training images. The SAD training scheme is trained with 3000 training images per class. It is referred to by the notation SAD(3000). The DAD training scheme generates one training image per raw image for each epoch, therefore it is trained with 10 training images per class and 50 training images per epoch.

5.2   Convolutional Neural Network Architecture

The CNN architecture used for experimentation is small in size to accommodate for cost effective commercial dynamic learning robots. The structure is 80×60 Input, 5×5 Conv 50, 2×2 Max, 3×3 Conv 17, 2×2 Max, 3×2 Conv 17, 2×2 Max, 3×3 Conv 17, 2×2 Max, 3×2 Conv 17, FC 50, Output 5. After each convolutional layer, a Leaky Rectified Linear Unit (ReLu) with a slope of 0.001 for negative values is used. The activation of the fully connected layer is a sigmoid function.

5.3   Result Comparison

In this section, the SAD(3000) and the DAD experiment results are compared and evaluated. Since the number of training images involved in each epoch is different for each method, the evaluation results are plotted against time instead of epoch number. This allows a comparison of training time consumption with data generation time included. Figure 5 plots the learning curves on test data and Fig. 6 plots the ROC curves of the CNN trained under SAD(3000) and DAD. As expected, static databases are a lot faster than dynamic databases since training images are only loaded once and then reused for every epoch. It can be noted that DAD training schemes give a higher classification mean accuracy as compared to SAD(3000). This is probably attributed to the fact that a model trained under DAD-type schemes is able to see more variations of training data in the long run as compared to training with static databases.

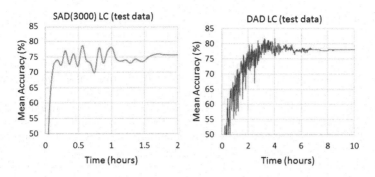

Fig. 5. Learning curves on test data obtained under SAD(3000) and DAD training schemes. Final mean accuracy convergence of SAD(3000) and DAD are at 75.7 % (2 h) and 78% (9 h) respectively.

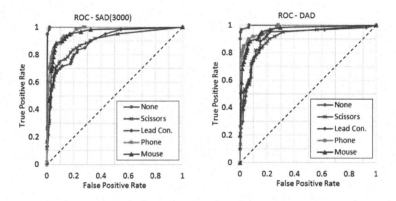

Fig. 6. ROC curves obtained under SAD(3000) and DAD training schemes.

Acknowledgement. This work is supported by Defence Innovative Research Programme (DIRP), the Ministry of Defence, Singapore under grant R-263-000-B08-592.

References

1. Kanazawa, A., Sharma, A., Jacobs, D.: Locally scale-invariant convolutional neural networks (2014). arXiv preprint. arXiv:1412.5104
2. Heisele, B., Kim, G., Meyer, A.: Object recognition with 3d models. In: BMVC, pp. 1–11. Citeseer (2009)
3. Zhang, Y., Sohn, K., Villegas, R., Pan, G., Lee, H.: Improving object detection with deep convolutional networks via bayesian optimization and structured prediction. In: Proceedings of the IEEE Conference on Computer Vision and Pattern Recognition, pp. 249–258 (2015)
4. Girshick, R., Donahue, J., Darrell, T., Malik, J.: Rich feature hierarchies for accurate object detection and semantic segmentation. In: Proceedings of the IEEE Conference on Computer Vision and Pattern Recognition, pp. 580–587 (2014)

5. Harzallah, H., Jurie, F., Schmid, C.: Combining efficient object localization and image classification. In: 2009 IEEE International Conference on Computer Vision, pp. 237–244. IEEE (2009)

6. Goehring, D., Hoffman, J., Rodner, E., Saenko, K., Darrell, T.: Interactive adaptation of real-time object detectors. In: 2014 IEEE International Conference on Robotics and Automation (ICRA), pp. 1282–1289. IEEE (2014)

7. Murphy, K., Torralba, A., Eaton, D., Freeman, W.: Object detection and localization using local and global features. In: Ponce, J., Hebert, M., Schmid, C., Zisserman, A. (eds.) Toward Category-Level Object Recognition. LNCS, vol. 4170, pp. 382–400. Springer, Heidelberg (2006). doi:10.1007/11957959_20

8. Joni, D.: Classifying plankton with deep neural network. http://benanne.github.io/2015/03/17/plankton.html

9. Cheng, M., Mitra, N.J., Huang, X., Torr, P.H., Hu, S.: Global contrast based salient region detection. IEEE Trans. Pattern Anal. Mach. Intell. **37**(3), 569–582 (2015)

Rotational Coordinate Transformation for Visual-Inertial Sensor Fusion

Hongsheng He[(✉)], Yan Li, and Jindong Tan

Department of Mechanical, Aerospace and Biomedical Engineering,
The University of Tennessee, Knoxville, TN 37996, USA
{he,tan}@utk.edu, yli141@vols.utk.edu

Abstract. Visual and inertial sensors are used collaboratively in many applications because of their complementary properties. The problem associated with sensor fusion is relative coordinate transformations. This paper presents a quaternion-based method to estimate the relative rotation between visual and inertial sensors. Rotation between a camera and an inertial measurement unit (IMU) is represented by quaternions, which are separately measured to allow the sensor to be optimized individually. Relative quaternions are used so that the global reference is not required to be known. The accuracy of the coordinate transformation was evaluated by comparing with a ground-truth tracking system. The experiment analysis proves the effectiveness of the proposed method in terms of accuracy and robustness.

1 Introduction

In wearable and robotic applications, visual and inertial sensors are commonly utilized in a collaborative manner by virtue of their complementary properties. IMU sensors provide accurate measurements of dynamics in a short span of time but suffer from drifts over time due to accumulated integration errors. Visual sensors enable precise longterm tracking but estimation accuracy is impaired by unpredicted abrupt motion. Fusion of visual and inertial sensors has achieved improved performance than the sole sensors in many applications such as motion tracking [1,2], feature tracking [3], and localization [4].

The performance of sensor fusion is dependent on the accuracy of coordinate transformations as sensor outputs are referred in the individual frames. Relative coordinate transformations resolve the spatial relation between individual reference frames attached to different sensors. The relative coordinate transformation generally includes translational and rotational transformations. Direct measurement of sensor placement is normally impractical as it is constrained by hardware configuration, and a feasible approach is to estimate the relation of sensors from their measurements.

The challenges in heterogenous coordinate transformations are attributed to asynchronous sensing, various sampling frequencies, lack of a global reference,

Hongsheng He's—work was supported in part by NSFC grant 61305114.

A. Agah et al. (Eds.): ICSR 2016, LNAI 9979, pp. 431–440, 2016.
DOI: 10.1007/978-3-319-47437-3_42

and sensor noise. Hardware synchronization is one of the reliable solutions to align sensor measurements in sampling points and frequencies; however, this synchronization mechanism relies on additional hardware support, which is usually available in high-end industrial sensors. In visual-inertial sensor fusion, a global reference is commonly obtained through a predefined condition, such as vertical reference boards relative to the earth.

Determination of relation coordinate transformations between different sensor frames is a classical problem that has been investigated for decades; however, the current available methods either require hardware configuration [5–8] or a tedious process [9–11]. The work [10] described a method to compute the relation between a camera and an IMU by using a pendulum unit and a turntable. The reliance on a turntable makes the method less practical for some platforms. Though the Kalman filter framework proposed in [12] did not rely on a turntable, it still requires a calibration board with a known global vertical reference. An improvement of this method was discussed in [13], which released the requirement for the turntable and vertical constraint. In addition to filter based frameworks, optimization has been commonly used [14,15]. In the work [14], an objective function was minimized to achieve relative spatial relationship upon a bunch of measurements. The optimization process is computationally intensive and thus it is not suitable for online or fast applications. In summary, an effective and convenient solution is hence more favorable and deserves more research effort.

In this paper, we propose a method to compute the rotational transformation between visual and inertial sensors. The output of the visual and inertial sensors is represented in quaternions that are optimized and expressed in the references of individual sensors. This allows an individual sensor to maximally optimize its output by well-tuned onboard algorithms and by incorporating other sensors, e.g., the attitude of an IMU is collaboratively computed from gyroscopes, accelerators, and magnetometers. To synchronize the observations in time, the quaternions from the sensor with a higher sampling rate are interpolated with respect to the sampling points of the sensor with a lower sampling rate. The key idea is that the rotating axes are identical regardless the referring frames since the sensors are fixed to each other.

The main contribution of the paper is the quaternion-based method that directly computes the rotational relation from sensor measurements. Compared to the state of the art, the method is direct, independent of absolute pose estimation, yet effective. The proposed method is convenient to conduct using image captures and inertial measurements. Specifically,

1. the paper proposes to utilize quaternion interpolation to synchronize attitude measurements from heterogenous sensors while guaranteeing the smoothness of the measurements;
2. the paper proposes to compute rotational transformation from rotating axes instead of quaternion operation to reduce computation errors;
3. the computed coordinate transformation is independent of the reference frames or the initial positions of the sensors.

The proposed method is applicable to other types of attitude sensors, e.g., between multiple IMUs. The deterministic attributes of the methods make it ideal for real-time computation, and the computation can be performed in online or offline modes.

2   Synchronization

The sampling rates of visual and inertial sensors generally differ, depending on different hardware implementation. The typical sampling rate of an IMU is high to 1000 Hz, while the typical refreshing rate of a camera is 60 fps. In addition, the measurements for these sensors are asynchronous in time unless a hardware synchronization mechanism is available. In this section, we synchronize and align rotation measurements in time space by means of quaternion interpolation.

The rotation around a unit vector $\mathbf{v} = [v_x, v_y, v_z]$ for an angle θ can be represented by a unit quaternion

$$\mathbf{q} = \left[\cos \frac{\theta}{2}, \mathbf{v} \sin \frac{\theta}{2} \right] = \left[\cos \frac{\theta}{2}, v_x \sin \frac{\theta}{2}, v_y \sin \frac{\theta}{2}, v_z \sin \frac{\theta}{2} \right] \qquad (1)$$

During a sampling interval, the rotating axis is assumed to be identical and the angular velocity is supposed to be uniform. Therefore, the rotation measurement between two sampling points can be simulated using Slerp interpolation. Given two measurements $\mathbf{q}_i, \mathbf{q}_{i+1}$ and a synchronization point $h \in [0,1]$, the spherical linear quaternion interpolation (Slerp) is [16]

$$\text{Slerp} \left(\mathbf{q}_i, \mathbf{q}_{i+1}, h \right) = \mathbf{q}_i \left(\mathbf{q}_i^{-1} \otimes \mathbf{q}_{i+1} \right)^h \qquad (2)$$

where $\otimes$ denotes quaternion multiplication and $\mathbf{q}_i^{-1}$ is the quaternion inverse of $\mathbf{q}_i$. The Slerp algorithm generates the optimal interpolation curve between two rotation angles, yielding the shortest possible interpolation between two quaternions on the unit sphere [17].

To interpolate in a series of rotation measurements, we cannot simply interpolate quaternions in a pairwise manner whereby the interpolation curve is not smooth at the control points and the angular velocity is not continuous at control points. In order to synchronize a series of quaternions, we utilize the Squad interpolation to simulate rotation between measurements. A Squad curve is in essence the spherical cubic equivalent of a Bezier curve given by [17]

$$\begin{aligned} &\text{Squad} \left(\mathbf{q}_i, \mathbf{q}_{i+1}, \mathbf{s}_i, \mathbf{s}_{i+1}, h \right) \\ =&\text{Slerp} \left(\text{Slerp} \left(\mathbf{q}_i, \mathbf{q}_{i+1}, h \right), \text{Slerp} \left(\mathbf{s}_i, \mathbf{s}_{i+1}, h \right), 2h(1-h) \right) \end{aligned} \qquad (3)$$

and the inner quadrangle points can be chosen as

$$\mathbf{s}_i = \mathbf{q}_i \exp \left(-\frac{\log \left(\mathbf{q}_i^{-1} \otimes \mathbf{q}_{i+1} \right) + \log \left(\mathbf{q}_i^{-1} \otimes \mathbf{q}_{i-1} \right)}{4} \right) \qquad (4)$$

to guarantee continuity.

In this paper, we interpolate the rotation observed by the sensor with a higher sampling speed to align with the sampling points of the sensor with a lower sampling speed. The rotation measurements $\mathbf{q}_i$ are associated with capturing time t_i, and the interpolation position between t_i and t_j for synchronization time t' is determined by $h = \frac{t'-t_i}{t_j-t_i}$. After interpolation at each synchronization time, we obtain two series of rotation measurements aligned in time, which represent a series of movements in the individual sensors' referencing systems.

3   Rotational Coordinate Transformation

The general idea in rotational coordinate transformation is that the rotating axes of different sensors are identical since they are physically fixed to each other. The rotating axes are computed from consecutive attitude measurements for a synchronized interval. Though the attitude measurements of individual sensors may vary with the reference points, the rotating axes are different only by the reference frames, whose relative pose can be represented by a rotation matrix.

3.1   Determination of Rotating Axes

Let $\mathbf{q}_t^v$ denote the attitude of the camera measured at time t, and $\mathbf{q}_t^i$ represent the interpolated attitude of the IMU at time t. Suppose the reference points are $\mathbf{q}_0$, the rotation measurements are $\tilde{\mathbf{q}}_t = \mathbf{q}_0^{-1} \otimes \mathbf{q}_t$. The quaternions are interpolated to synchronize the measurement intervals. The change of attitude in the vision system $\tilde{\mathbf{q}}_t^v$ and change of attitude in the inertial system $\tilde{\mathbf{q}}_t^i$ represent the same rotating motion but in different coordinate reference systems. The objective is to find the transformation $\mathbf{q}$ from the visual reference to the inertial one $\tilde{\mathbf{q}}_t^i = \mathbf{q}^* \otimes \tilde{\mathbf{q}}_t^v \otimes \mathbf{q}$, where $(\cdot)^*$ denotes quaternion conjugation. Dependent on the requirements of different applications, the transformation can also be computed from the inertial reference to the visual one.

The intuitive way is to compute the optimal rotation by solving

$$\mathbf{q} = \arg \max_{\mathbf{q} \| \mathbf{q} \| = 1} \left(\left(\tilde{\mathbf{q}}_t^i \right)^T \left(\mathbf{q}^* \otimes \tilde{\mathbf{q}}_t^v \otimes \mathbf{q} \right) \right) \tag{5}$$

for a sequence of measured pairs of rotation [15]. However, the scalar parts of the measured quaternions are supposed to be identical as they are independent of reference systems. The optimization process may involve unnecessary measurement noise in the scalar part, whereas the objective of coordinate transformations is find the rotational relation.

We thus propose to directly compute a rotation matrix R satisfying the constraint on the sole vector part

$$\mathbf{v}(\tilde{\mathbf{q}}_t^i) = R\mathbf{v}(\tilde{\mathbf{q}}_t^v) \tag{6}$$

where $\bar{\mathbf{v}}(\cdot)$ denotes the normalized vector part of a quaternion $\mathbf{q} = [s, \mathbf{u}]$ given as

$$\bar{\mathbf{v}}(\mathbf{q}) = \frac{\mathbf{u}}{\|\mathbf{u}\|} \tag{7}$$

The transformed quaternion vector can be constructed from the vector part and the scalar part, which does not vary with reference systems.

3.2 Computing Rotation Matrix

Given a sequence of vectors $\bar{\mathbf{v}}_t^i$ that represent the t-th rotation in the reference frame attached to an IMU sensor, and a sequence of vectors $\bar{\mathbf{v}}_t^v$ that represent the same rotation in the reference frame attached to a camera, we aim to find an orthogonal rotation matrix that optimally represents the observations between these two sensors

$$\min_R \frac{1}{2} \sum_{t=1}^N a_t \left\| \bar{\mathbf{v}}_t^i - R\bar{\mathbf{v}}_t^v \right\|^2 \tag{8}$$

where a_i is the adjustable weight associated with the t-th measurement.

The optimization is known as the Wahba's problem, which can be robustly solved using the Davenport's q method [18]. The rotation matrix can be parameterized by a unit quaternion $\mathbf{q} = [s, \mathbf{v}]$ in the form of

$$R = \left(s^2 - |\mathbf{v}|^2\right) I + 2\mathbf{q}\mathbf{q}^T - 2s\left[\mathbf{q}\right]_x \tag{9}$$

The unit quaternion can be solved by finding the normalized eigenvector with the largest eigenvalue of

$$K = \begin{bmatrix} B + B^T - \mathrm{tr}(B)I & \mathbf{z} \\ \mathbf{z}^T & \mathrm{tr}(B) \end{bmatrix} \tag{10}$$

with

$$B = \sum_t a_t \bar{\mathbf{v}}_t^i \left(\bar{\mathbf{v}}_t^v\right)^T \tag{11}$$

Davenport's q method is robust in solving the Wahba's problem, yet it is computationally intensive to compute eigenvectors. If real-time performance is an important factor, FOAM and QUEST are alternative methods for a tradeoff between speed and accuracy [19].

Given the rotation matrix R, the corresponding representation of $\tilde{\mathbf{q}}_t^v = \left[\cos\frac{\theta}{2}, \mathbf{v}\sin\frac{\theta}{2}\right]$ in the inertial reference can then be obtained

$$\tilde{\mathbf{q}}_t^i = \left[\cos\frac{\theta}{2}, R\mathbf{v}\sin\frac{\theta}{2}\right] \tag{12}$$

and with respect to the reference, the attitude measurement is

$$\mathbf{q}_t^i = \mathbf{q}_0^i \otimes \left[\cos\frac{\theta}{2}, R\mathbf{v}\sin\frac{\theta}{2}\right] \tag{13}$$

4 Results

We conducted both simulation studies and practical experiments to evaluate the effectiveness of the proposed method. The ground truth of sensor configuration is generally not known as it is reference dependent. Therefore, in the simulation, we generated a series of measurements based on a preset sensor configuration, and compared the proposed method with the ground-truth sensor configuration; in the practical experiment, we evaluated the differences of measured rotation angles between the sensors, which were measured by an OptiTrack system and computed by the proposed method.

4.1 Simulation Study

To evaluate the proposed quaternion-based method for rotational coordinate transformations with known ground truth, we performed a group of simulation studies. In the simulation, the output frequency of the IMU was set as 500 Hz, and the refreshing rate of the camera was set as 100 fps. We simulated different configurations of the IMU and the camera as tabulated in Table 1, which includes ten configurations in terms of roll, pitch and yaw angles in radian.

Table 1. Results of rotational transformations with $\rho = 10$ artificial white noise.

Trials	Roll/Truth (rad)	Pitch/Truth (rad)	Yaw/Truth (rad)
1	1.6440/1.6410	0.4978/0.4967	1.2959/1.2955
2	−0.2016/−0.1931	1.0509/1.0497	−1.9904/−1.9838
3	1.9277/1.9220	−0.138/−0.14226	0.9262/0.9231
4	−2.5244/−2.5214	0.6206/0.6217	−1.513/0−1.5113
5	0.4984/0.4951	0.5478/0.5471	−0.9846/−0.9855
6	−1.8440/−1.8463	0.8473/0.8453	3.0349/3.0261
7	−2.9920/−2.9945	−0.554/−0.55372	1.7326/1.7356
8	−2.0110/−2.0186	−0.824/−0.82226	2.6395/2.6448
9	−2.8997/−2.8962	0.2026/0.2040	1.2970/1.2926
10	−1.9324/−1.9356	−0.588/−0.58688	1.3598/1.3623

For each configuration, we generated a series of quaternions to simulate the bound rotation of the camera and the IMU, and added different levels of artificial noise into the measurements. The level of measurement noise is determined by the model $\hat{\mathbf{q}}_i = \mathbf{q}_i (1 + \rho \mathbf{n}_i)$ where ρ is the scaling parameter and $\mathbf{n}$ are the added white noise $\mathbf{n}_i \sim \mathcal{N}(0, 1)$. The calibrated rotation of roll, pitch and yaw angles for the ten configurations from measurements with 20 % white noise is given in Table 1. Comparing the tranformation results with the ground-truth configuration, we found that the maximal error is less than 0.01 rad, i.e., 0.57°.

The experiment revealed that the proposed method was relatively robust to measurement noise and the accuracy was not severely influenced.

We also altered the scaling parameter from $\rho = 1\%$ to $\rho = 20\%$ with an incremental of 1 % for each sensor configuration, and the relative transformation errors (the percentage of the tranformation error with respect to the ground truth) for the 7-th trial in Table 1 are plotted in Fig. 1. From the figure, we can see that the transformation accuracy decreases as the level of added noise increases. The relative error went up to 3 % when the level of measurement noise was 20 %.

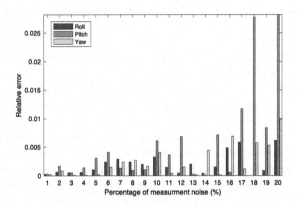

Fig. 1. Transformation error with regard to different percentages of measurement noise for the 7-th configuration in Table 1: -2.9945, -0.5542, 1.7356.

4.2 Experiments

Experiment Setup. The accuracy of rotational transformation was also examined on a testing rig as shown in Fig. 2. The testing rig consists of a video camera and an attached 10-axis synchronized inertial sensor that measures the dynamics of the camera. The model of the camera was Ximea MQ013CG-ON, and the model of the inertial sensor was VectorNav VN-100, which has 3-axis accelerometers, 3-axis gyroscopes, 3-axis magnetometers, and a barometric pressure sensor. The ground-truth global positions of the camera and the IMU were obtained by an OptiTrack multi-camera system. The tracking system in the laboratory is comprised of six HD cameras mounted on the roof to cover the main working space, and four visual markers were respectively installed on the camera and the IMU for attitude measurement.

In the experiment, the camera captured images in a resolution of 1280×1024 pixels at a refreshing rate of 100 fps, and the inertial sensor measured 3-axis acceleration and 3-axis angular velocities of the camera at a frequency of 400 Hz. Before experiment, the visual and inertial sensors were calibrated individually. The OptiTrack system tracks the motion of the testing rig at a sampling rate of 120 fps. The OptiTrack system can track markers down to sub-millimeter movement with repeatable accuracy.

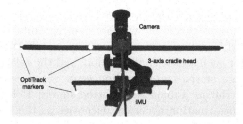

Fig. 2. Testing rig. The IMU and the camera were connected by a cradle head that supports 3-axis rotation, and both of them were respectively attached a beam with four installed fluorescent markers to be tracked in the OptiTrack system.

Coordinate Transformations of Camera-IMU Frames. We rotated the testing rig above a pattern of visual features, and measured the attitude of the camera and the IMU separately. The attitude and position of the camera were computed by tracking the visual features, and the attitude of the IMU was measured by the onboard gyroscopes, accelerometers, and magnetometers. The positions and attitude of the two sensors in an example of the experimental trail are plotted in Figs. 3a and 3b. The attitude relation between the two sensors were also measured by the OptiTrack system, which tracked the positions of the fluorescent markers on the attached beams. The positions of the markers were represented in the same reference frame, and the absolute angles between the beams were measured.

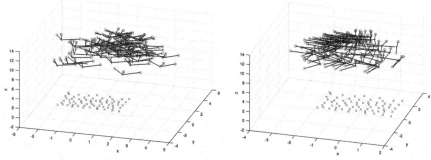

(a) Attitude and positions obtained by visual feature tracking.   (b) Attitude measured by the IMU.

Fig. 3. Attitude and position measured by the camera and the IMU

We conducted experiments on several configurations of the sensors, and compared the measurement differences between the transformation computation and the OptiTrack system. It should be noted that we were not able to directly compare the absolute measurements of the attitudes from the two measuring systems as the transformation computation and the OptiTrack system used different references. Instead, we measured the change of the two sensors' attitude between different trials. The measurement differences between the transformation computation and the OptiTrack system are plotted in Fig. 4. The measurements from the OptiTrack system are plotted in the box and whisker diagram. The general measurement difference was less than 1°, and the accuracy appears to be stable.

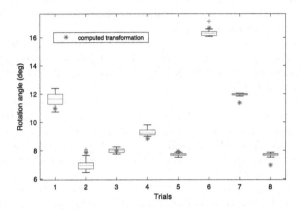

Fig. 4. Measurement differences between the computed transformation and the Opti-Track system. The measurements from the OptiTrack are plotted in the box and whisker diagram.

5   Conclusion

This paper has presented a direct yet effective method to determine spatial rotation between heterogenous sensors. The rotation measurements of different sensors were synchronized in time by means of quaternion interpolation. Instead of computing the frame reference by quaternion operation, we calculated a rotation matrix between the vector parts of quaternions, which represented the same rotation but in different reference frames. The rotation matrix was obtained by solving the Wahba's optimization problem. In simulation study and experiments, we have shown the effectiveness and accuracy of the proposed method. In simulation study, the transformation accuracy was less than 0.57° with 10 % artificial white noise. In experiments, the measurement difference between the proposed method and the OptiTrack system was less than 1 degree.

References

1. Tao, Y., Hu, H., Zhou, H.: Integration of vision and inertial sensors for 3d arm motion tracking in home-based rehabilitation. Int. J. Robot. Res. **26**(6), 607–624 (2007)
2. He, H., Li, Y., Guan, Y., Tan, J.: Wearable ego-motion tracking for blind navigation in indoor environments. IEEE Trans. Autom. Sci. Eng. **12**(4), 1181–1190 (2015)
3. Hwangbo, M., Kim, J.S., Kanade, T.: Inertial-aided klt feature tracking for a moving camera. In: IEEE/RSJ International Conference on Intelligent Robots and Systems: IROS 2009, pp. 1909–1916 (2009)
4. Achtelik, M.W., Lynen, S., Weiss, S., Kneip, L., Chli, M., Siegwart, R.: Visual-inertial slam for a small helicopter in large outdoor environments. In: 2012 IEEE/RSJ International Conference on Intelligent Robots and Systems (IROS), pp. 2651–2652. IEEE (2012)
5. Lobo, J., Dias, J.: Vision and inertial sensor cooperation using gravity as a vertical reference. Pattern Anal. Mach. Intell. IEEE Trans. **25**(12), 1597–1608 (2003)
6. Alves, J., Lobo, J., Dias, J.: Camera-inertial sensor modelling and alignment for visual navigation. Mach. Intell. Robot. Control **5**(3), 103–112 (2003)
7. Daniilidis, K.: Hand-eye calibration using dual quaternions. Int. J. Robot. Res. **18**(3), 286–298 (1999)
8. Ovrén, H., Forssen, P., Tornqvist, D.: Why would i want a gyroscope on my rgb-d sensor? In: IEEE Workshop on Robot Vision (WORV), pp. 68–75. IEEE (2013)
9. Mirzaei, F.M., Roumeliotis, S.I.: A kalman filter-based algorithm for imu-camera calibration: Observability analysis and performance evaluation. Robot. IEEE Trans. **24**(5), 1143–1156 (2008)
10. Lobo, J., Dias, J.: Relative pose calibration between visual and inertial sensors. Int. J. Robot. Res. **26**(6), 561–575 (2007)
11. Gemeiner, P., Einramhof, P., Vincze, M.: Simultaneous motion and structure estimation by fusion of inertial and vision data. Int. J. Robot. Res. **26**(6), 591–605 (2007)
12. Kelly, J., Sukhatme, G.S.: Visual-inertial sensor fusion: Localization, mapping and sensor-to-sensor self-calibration. Int. J. Robot. Res. **30**(1), 56–79 (2011)
13. Hol, J.D., Schön, T.B., Gustafsson, F.: A new algorithm for calibrating a combined camera and imu sensor unit. In: 10th International Conference on Control, Automation, Robotics and Vision, ICARCV 2008, pp. 1857–1862. IEEE (2008)
14. Fleps, M., Mair, E., Ruepp, O., Suppa, M., Burschka, D.: Optimization based imu camera calibration. In: 2011 IEEE/RSJ International Conference on Intelligent Robots and Systems (IROS), pp. 3297–3304. IEEE (2011)
15. Lang, P., Pinz, A.: Calibration of hybrid vision/inertial tracking systems. In: InerVis, Barcelona, Spain (2005)
16. Shoemake, K.: Quaternion calculus for animation. SIGGRAPH Course **23**, 15–23 (1989)
17. Dam, E.B., Koch, M., Lillholm, M.: Quaternions, interpolation and animation. Københavns Universitet, Datalogisk Institut (1998)
18. Keat, J.: Analysis of least-squares attitude determination routine doaop. Technical Report CSC/TM-77/6034, Computer Sciences Corp, Technical Report (1977)
19. Markley, F.L., Mortari, D.: How to estimate attitude from vector observations. Technical Report, NASA (1999)

Developing an Interactive Gaze Algorithm for Android Robots

Brian D. Cruz[1,2], Byeong-Kyu Ahn[2], Hyun-Jun Hyung[1,2], and Dong-Wook Lee[2(✉)]

[1] Robotics and Virtual Engineering, Korea University of Science and Technology, Ansan, Gyunggi-do 15588, Republic of Korea
[2] Robot Group, Korea Institute of Industrial Technology, Ansan, Gyunggi-do 15588, Republic of Korea
{mrcruz,bk.ahn,hyungc,dwlee}@kitech.re.kr

Abstract. We implemented a gaze algorithm for interacting with multiple observers as a precursor to a multi-party conversation system. By acknowledging multiple participants in a natural manner, we seek to set the stage for smoother and more effective human-robot conversations featuring proper turn-taking using attention shifts. The android robot EveR-4, developed at the Korea Institute of Industrial Technology for human-robot interaction (HRI) applications was used. The robot wore a dress and was made up to replicate interacting naturally with a real woman as much as possible. Using a RGB-D camera, peoples faces and positions were tracked so that the robot's attention could be given to everyone appropriately. An importance value was assigned to each detected face based on the length of time it was detected and its distance to the robot. Facial expressions were made by the robot when people were seen to increase observers' sense of interaction. We observed peoples reactions to our implementation at an exhibition and made note of how we can improve the overall system to be more life-like and realistic.

Keywords: HRI · Gaze algorithm · Android robot

1 Introduction

Gaze is a key component for HRI. Much research has been done comparing the gaze and head movements of robots to humans during interaction [6]. Beginning with theories of visual cues based on the human eye's biological structure, saliency based models were developed to select appropriate targets in pictures to focus on [2]. Modelled after the eyes central fovea and the sensitivity of the visual system to color pairs, motion and orientations, this bottom up approach computes a series of points of interest in an image. Other researchers have sought out developing gaze distributions by observing conversations directly. Mutlu et al. developed a series of probabilistic distributions for gazing at people's faces, bodies or backgrounds in different conversation scenarios and implemented them

© Springer International Publishing AG 2016
A. Agah et al. (Eds.): ICSR 2016, LNAI 9979, pp. 441–448, 2016.
DOI: 10.1007/978-3-319-47437-3_43

on the Robovie R-1 platform [5]. Other researchers focused purely on human-human interaction to determine the extent to which we gaze upon each other during multi-party conversation. Vertagaal et al. determined that people compensate for having larger audiences by gazing at listeners more overall than one would with a smaller audience [7]. Other approaches involved focusing on task based conversations with robots present. Johansson et al. studied human-human interaction versus human-robot interaction during conversations about various objects to develop a system for comparison of robot performance to actual people's actions [3]. Another setting involved having a robot guide visitors in a museum. Bennewitz et al. developed a system for a humanoid robot to gaze at visitors while explaining various attractions [1]. Using a camera and a microphone, observers' proximity to the robot, the duration that they were recognized and the time since observers had last spoken were used to calculate an importance value for gazing. Social cues have also been shown to be valuable in guiding attention. Zaraki et al. developed an attention model for the FACE humanoid robot incorporating verbal and non-verbal cues [8]. These include gaze direction and facial expression, distance and proximity to the robot and voice localization. While we intend to consider various bottom-up features such as observers' gaze orientation, emotion recognition, motion and voice patterns in the future, the environment in which we implemented the algorithm restricted the use of sound due to noise. The algorithm described in this paper thus uses only observers' distance to the robot and the duration of time that they have been detected.

2 Gaze Algorithm

The exhibition center at which the demonstration took place was noisy, which prevented voice interaction. Therefore, while using gaze alone to interact with observers, more attention was given to those who were closer since they would be theoretically more interested in interacting with the robot. In order to promote interaction with everyone involved and provide depth to the experience for those who interacted longer, the length of time that a face was detected for was considered as well. An RGB-D camera was used to calculate and update targets for attention. First, a face detector was run on the RGB image to extract the location of each face. Note that the face detector recognized faces only if they were roughly facing the camera, so interest in the robot was inferred from the presence of a face. An object tracker then followed each face. If a previously seen face was not found in an image frame, its previous location was used for up to 5 frames. After that, if the face was no longer seen, it was considered lost and the tracker was reset. Each detected faces location was used to search a point cloud to find its 3D centroid. After that, the centroids were used to calculate an importance value for each face (Fig. 1).

Importance was calculated as a linear combination of a face's distance to the robot and the number of frames it had been tracked. The exact contributions of both distance and the number of frames were derived from our robot's environment. A desk was placed in front of the robot to ensure observers would

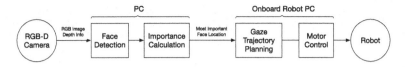

Fig. 1. Flowchart of gaze algorithm

remain at least 0.6 m away. The optimal view distance of the camera for our setup was estimated to be within 3 meters of the robot. The longer a face was seen, the more importance was given. However, too much weight with respect to time would lead to older faces dominating the attention of the robot, so the contribution of time to importance was set to saturate to the maximum value after one minute by no longer counting the elapsed frames. After one minute, the time importance component saturated at a max of 0.3. Importance was inversely proportional to distance, with the distance component set to zero beyond three meters away from the robot. The distance component saturated at a max of 0.4 at a distance of 0.6 meters. New faces were given a very high value of 30 to ensure at least temporary interest and acknowledgement by the robot. This value as well as the saturation max values for the distance and time components were experimentally determined to sufficiently gaze at new faces while maintaining a slight preference for closer observers over those who have been seen for longer. After all importance values were calculated, they were normalized to 1. Note the system ran at about 3 frames per second. The main bottleneck was the face detector, which took between 0.1 to 0.2 s.

The final formula was:

$$I_f = I_{f-1} + 0.0017 * T_{f-1} + (-0.1625 * D_{f-1} + 0.4975) \tag{1}$$

where I is the importance value for the current face, T is the number of frames the face has been detected for, and D is the distance from the robot to the detected face, all calculated for the current frame f. This equation was determined from the above constraints. The centroid with the highest importance was set as the robot's gaze target. The angle from the heads center and current gaze to the target, as well as the angle from the center of the eyes to the current target, were calculated. Finally, a linear path to direct the gaze of the robot from it's current position to the target orientation was calculated. Different velocity profiles were used for the head and the eyes leading to convergence on the target after 0.5 and 0.2 s, respectively.

3   Demonstration

The demonstration was conducted at the international robot exhibition Robotworld 2015 in South Korea from October 28th to October 31st. Robotworld is one of the three largest international robot shows and featured exhibitions from twelve countries visited by over 74,000 attendees and buyers. The demonstration

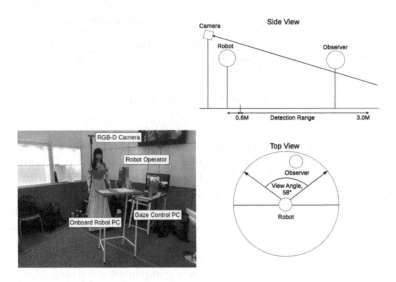

Fig. 2. Demonstration environment

environment, as shown in Fig. 2, consisted of an RGB-D camera, gaze control PC and the onboard robot PC.

The camera was placed in a stationary position behind and above the robot to view observers. The camera connected to the gaze control PC which parsed the incoming data. A face detection routine located faces in the RGB image and after tracking, the corresponding point cloud was searched for 3D centroids. The gaze PC then calculated each face's importance as described above and selected a face to gaze at. Both the face detector and tracker were implemented using the Dlib C++ library [4]. This PC sent the coordinates of the centroid corresponding to the face with the highest importance to the robots onboard PC. The onboard PC ran a face simulator that set the centroid as the new gaze target and decided the final motor commands as described. This simulator also included slight random shifts in attention about the target to appear more life-like. The robots body remained still while looking at people that were detected, but facial expressions such as winking, showing surprise, smiling and blinking were used at random. Although the robot was autonomous during the operation of this gaze algorithm, a robot operator was present to switch the robot between interactive and non-interactive modes. The robot was capable of other actions including giving a brief introduction, singing and dancing. These were occasionally selected from manually to entertain groups of observers.

4   Results

As shown in Fig. 3, many different groups of people came to observe the robot. Around 500 people in total came. These groups included single observers with

Fig. 3. Examples of the demonstration. (Left) Single observer with onlookers, (middle) many observers and (right) parent with child shown. Face label, importance shown in green per face. Faces blurred for anonymity here and in subsequent figures (Color figure online)

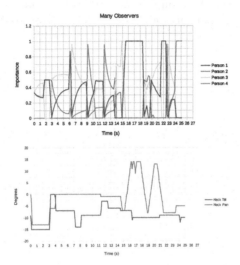

(a) (Top) Each person's importance over time, aligned with the robot head's tilt and pan (bottom).

(b) Resulting camera image

Fig. 4. The result of the demonstration for the many observers scenario

onlookers in which a single individual observed the robot as one or more onlookers entered and left, large groups of observers, and parents with children. In each case, as people were detected, the robot did indeed engage them according to proximity and duration. Newly detected faces were acknowledged and gazed at in the presence of closer or previously seen faces as planned.

An example of a shift in importance is shown below for the single observer with onlookers scenario. The graph as shown in Fig. 5 depicts a sudden switch in importance shortly after the 3 second mark. At first, the young woman is the only one paying attention to the robot, logged as Person 1. She therefore receives an importance value of 1 (Fig. 6a). Shortly thereafter, the young man in the center views the robot, being registered as the new observer Person 2. Since new faces are given very high importance values to ensure being greeted by the robot, Person 2 has a value of 0.96 while Person 1 drops to 0.04 (Fig. 6b).

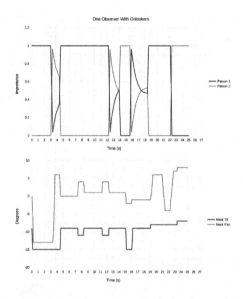

Fig. 5. (Top) Each person's importance over time, aligned with the robot head's tilt and pan (bottom). The changes shown in Fig. 6 are highlighted in yellow (Color figure online)

Note that the face tracker lacked the ability to recognize individual faces. Therefore, it occasionally labelled the same individual with different numbers after he or she was lost and then recovered. This may have led to discrepancies in the robot's head movements compared to which person had the highest importance because the label was given to someone else in a different location. Furthermore, people were free to move around, so each person's location was not constant and occasionally switched with respect to other people (Fig. 4).

(a) Image A, T = 2.7 seconds

(b) Image B, T = 3.3 seconds

Fig. 6. Camera images depicting changes in individual importance

5   Future Work

Although the robot gazed at observers as expected, interaction was somewhat limited. By beginning with appropriate division of attention to those present with the robot, a platform can be established for inviting people to engage in conversation. Furthermore, as people enter and leave the conversation, the gaze algorithm described in this paper can help to ensure a smooth transition and keep conversation flowing at all times. There were times when peoples faces were not detected. By using a more robust face detector in the future, the performance of the system in this regard can be improved. However, individual's gaze direction will have to be calculated and taken into account. Also, sound was not detected in our system. Many people tried to talk to the robot, but it was unable to respond. Incorporating voice recognition into the algorithm could help to make the system more interactive and realistic. The robot was able to make facial expressions, so these were used at random when people were detected. Although this led to a greater sense of interaction overall, the fact that they were random meant these facial gestures did not have any special meaning. If these gestures were timed with certain events, it would likely increase peoples level of immersion and satisfaction. Lastly, incorporating other social cues such as observers' gaze direction, facial expressions, emotions and movements would allow for more accurate and human-like interaction. By adjusting these parameters and following up demonstrations with a survey, observers' opinions of the robot can be used to evaluate the algorithm described in this paper further. Using the data that we collected at the event, we will be able to improve upon this system and make it more suitable for HRI.

Acknowlegement. This work was supported by the Robot R&D Program (10041659) funded by the Ministry of Trade, Industry and Energy (MOTIE, South Korea).

References

1. Bennewitz, M., Faber, F., Joho, D., Schreiber, M., Behnke, S.: Towards a humanoid museum guide robot that interacts with multiple persons. In: 2005 5th IEEE-RAS International Conference on Humanoid Robots, pp. 418–423. IEEE (2005)
2. Itti, L., Koch, C., Niebur, E.: A model of saliency-based visual attention for rapid scene analysis. IEEE Trans. Pattern Anal. Mach. Intell. **11**, 1254–1259 (1998)
3. Johansson, M., Skantze, G., Gustafson, J.: Comparison of human-human and human-robot turn-taking behaviour in multiparty situated interaction. In: Proceedings of the 2014 Workshop on Understanding and Modeling Multiparty, Multimodal Interactions, pp. 21–26. ACM (2014)
4. King, D.E.: Dlib-ml: A machine learning toolkit. J. Mach. Learn. Res. **10**, 1755–1758 (2009)
5. Mutlu, B., Kanda, T., Forlizzi, J., Hodgins, J., Ishiguro, H.: Conversational gaze mechanisms for humanlike robots. ACM Trans. Interact. Intell. Syst. (TiiS) **1**(2), 12 (2012)
6. Srinivasan, V., Murphy, R.R.: A survey of social gaze. In: 2011 6th ACM/IEEE International Conference on Human-Robot Interaction (HRI), pp. 253–254. IEEE (2011)
7. Vertegaal, R., Slagter, R., Van der Veer, G., Nijholt, A.: Eye gaze patterns in conversations: there is more to conversational agents than meets the eyes. In: Proceedings of the SIGCHI Conference on Human Factors in Computing Systems, pp. 301–308. ACM (2001)
8. Zaraki, A., Mazzei, D., Giuliani, M., De Rossi, D.: Designing and evaluating a social gaze-control system for a humanoid robot. Hum. Mach. Syst. IEEE Trans. **44**(2), 157–168 (2014)

Recovery Behavior of Artificial Skin Materials After Object Contact

John-John Cabibihan[(✉)], Mohammad Khaleel Abu Basha,
and Kishor Sadasivuni

Mechanical and Industrial Engineering Department, Qatar University, Doha, Qatar
john.cabibihan@qu.edu.qa

Abstract. As social robots and lifelike prosthetics get into closer contact with humans, understanding the mechanical behavior of the embedding skin materials for prosthetic and social robotic fingertips is of great importance. The time-dependent behavior can alter the performance of the embedded sensors. This paper investigates two types of embedding materials (i.e. silicone and polyurethane) for their recovery after contact with a surface. A visco-hyperelastic finite element model of a fingertip is described. This model allows the visualization of the materials' responses after a creep test. This analysis was performed to investigate the recovery time of the materials after contact was made. Simulation results show the differences between the two materials. The results are useful for materials selection and to further investigate other design alternatives and to minimize the effects of the time delay.

1 Introduction

Our hands have important communicative, emotional, and functional purpose. For example, we can greet others and we are able to communicate directions and sizes with the human hand [1–4]. Through touch, we are able to console someone else's emotional pain or make someone feel appreciated [5–8]. For those who may have lost their hands through disease, accident or war, the use of one's hand for independent living, for hygiene or for feeding oneself may be the most important function that have been lost. Likewise, for the robotic hands of social robots (i.e. robots that socially interact and collaborate with humans), transporting an object from one position to another is equally important [9,10]. That function requires contact between the artificial hand and a tool or another object.

Before an artificial hand can grasp and manipulate an object, contact has to be first established through the synthetic skin and then through the embedded tactile sensors. The tactile sensing system has to detect the following: contact between the finger and the object [11,12]; contact between the object and the environment [10,13]; slippage [11,14]; local shape [15,16]; and global shape [17,18]. To make sense of that information, the time-dependent mechanical behavior of the synthetic skin has to be understood because it can alter the response of the embedded tactile sensors.

A highly viscoelastic skin could make the signals from an embedded contact sensor to have a long decay until the skin material stabilizes. In a creep test,

© Springer International Publishing AG 2016
A. Agah et al. (Eds.): ICSR 2016, LNAI 9979, pp. 449–457, 2016.
DOI: 10.1007/978-3-319-47437-3_44

an applied constant stress will result to an increased strain in a viscoelastic material. After the stress is released, the material will gradually return to its initial state. If the selected material has a long recovery time, the embedded sensor will continue to be on its active state even after the load on the skin surface is removed. This work investigates the behavior of two typically used artficial skin materials and determine their behavior after its contact with an object.

This paper is structured as follows. Section 2 describes the skin samples and the finite element model used. Section 3 presents the results on the recovery of the artificial skin. The last section concludes this paper and provides the future directions.

2   Materials and Methods

2.1   Artificial Skin Samples

For these simulations, we used silicone (GLS 40, Prochima, s.n.c., Italy) and polyurethane (Poly 74-45, Polytek Devt Corp, USA) as representative materials of those used in earlier works as artificial skins for prosthetics and robotics [19, 20]. These materials were previously characterized in [21, 22] for their viscoelastic and hyperelastic behaviors.

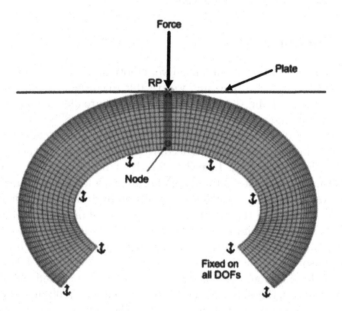

Fig. 1. The two-dimensional finite element axisymmetric model of the synthetic finger-tip. The shape was made to approximate a cross-section of the human fingertip. The fixed region serves as a stiff support similar to the function of the bone. The flat plate indenter was oriented as shown. They were modeled as rigid, analytical surfaces.

2.2   Finite Element Modeling

A two-dimensional finite element (FE) model of a fingertip (Fig. 1) was created in
the commercial finite element software Abaqus (Dassault Systemes). The model
adopted the geometry of the fingertip presented in [23,24]. The plain strain 8-
node biquadratic element type was used to model the fingertip skin. The indenter
was modeled as a rigid, analytical flat surface. The *visco* [25] computation mode
was used in this work. The skin layer was assumed to have hyperelastic and viscoelastic behaviors. The contact between the fingertip and the plate was assumed
to be frictionless. The constitutive equations are shown at the Appendix.

2.3   Simulation Procedure

A creep test requires that a constant force is applied to the material in order
to demonstrate the creeping process through displacement or strain measures.
Full fingertip models with 1, 3, and 5 mm thickness were used to investigate the
skin recovery. Additional simulation conditions represented on the 3 mm skin are
shown on Fig. 2. A concentrated vertical force of 10 N was applied through the
reference point, denoted as RP in Fig. 1.

The loading profile was specified such that the time to reach the peak force
was set to 1 s. The force was maintained constant for 5 s. The duration of unloading was set to 1 s. Data on the nodal vertical displacement (U2) and vertical
logarithmic strain (LE22) were collected from two nodes on the fingertip model.
The first was the node directly below the plate where the concentrated force was
applied. The second was the node from an element just above the "bone" surface

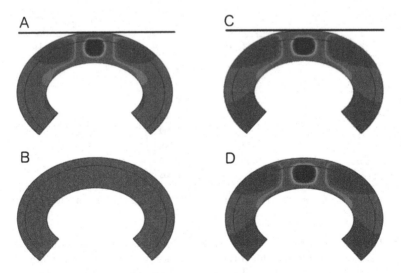

Fig. 2. The contours for the vertical logarithmic strain for the 3 mm skin thickness
model. **A** and **B**. For the silicone sample taken at 7 s and 17 s, respectively. **C** and **D**.
For polyurethane sample taken at 7 s and 17 s, respectively.

as shown on Fig. 1. The recovery time was obtained from the instant that the load was released at T = 7 s until T = 17 s. For each fingertip thickness, the cut-off criteria were set to be at 10 % of the strains at the instant when the load was released. This criterion is similar to the cutoff used in [26].

3   Results

Plotted on Fig. 3A for silicone and Fig. 3C for polyurethane are the displacements of a node located on the skin surface. Plotted on Fig. 3B for silicone and Fig. 3D for polyurethane are the displacements of a node near the bone surface. The displacement profile of the subsurface node resembles the profile at the skin surface albeit with smaller magnitude.

The logarithmic strain results of silicone in Fig. 4A and inset, for the 3 mm thick skin show negligible residual strains, 10 s after the load was released. For the three skin thicknesses that were investigated, the plot suggests that silicone can achieve full recovery within 1 s after the release of load from the plate.

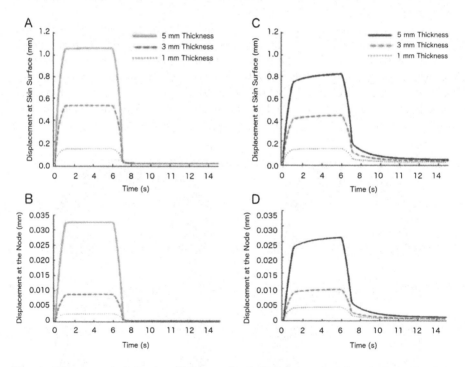

Fig. 3. The recovery behavior of the artificial skin materials. The resulting displacements for silicone: **A.** On a node on the skin surface directly below the plate. **B.** On a node from an element just above the "bone" surface. The resulting displacements for polyurethane: **C.** On a node on the skin surface directly below the plate. **D.** On a node from an element just above the "bone" surface.

For the polyurethane material, the typical behavior of creep in viscoelastic materials is evident from the continued increase of displacement, even as the force is kept constant from T = 1 s to T = 6 s (Fig. 4B). Residual strains can be observed from the logarithmic strain contours, 10 s after contact was removed (Fig. 4B inset). For the three models with polyurethane material, the plot suggests that the residual strains from a 10 N load and applied for 5 s will be about 0.5 % after 10 s of load removal.

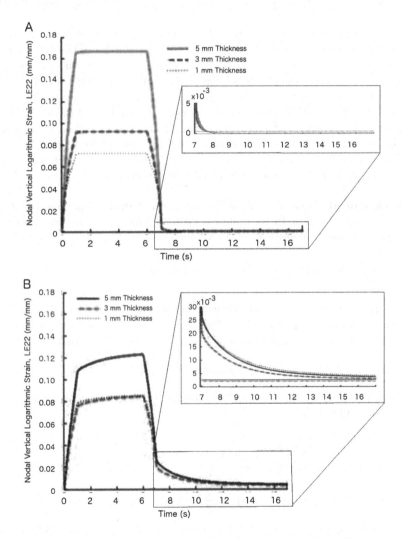

Fig. 4. The nodal logarithmic strains plotted over time for silicone, **A**, and polyurethane, **B**. The insets shows the magnified strains immediately after the release of the load T = 7 s and 10 s after, T = 17 s.

4 Conclusion

Upon the artificial finger's contact with an object, the stresses and strains that occur at the contact interface are detected by the embedded tactile sensors. However, due the viscoelastic and hyper elastic properties of synthetic skins, these can cause delay in the signals that are detected by the tactile sensors. Consequently, the lag in the sensor's response can make the difference between having an object slip from the fingers or having sufficient time to grasp an object firmly.

In this paper, a finite element model of a fingertip was presented. This model had visco-hyperelastic behavior where the recovery behavior of the artificial skin was investigated. Results show that as compared to the silicone samples, the polyurethane material samples had longer delays in the response of a node. This node can represent what an embedded sensor can detect. Future work involves validating the model with micro sensors that can be embedded to be about 1 to 5 mm beneath the skin surface or using digital image correlation techniques to visualize the strains. In addition, more studies are needed to develop hardware or software-based approaches to compensate or eliminate the effects of the long delay in recovery of the artificial skin that will be selected.

Acknowledgments. The work is supported by an NPRP grant from the Qatar National Research Fund under the grant No. NPRP 7-673-2-251. The statements made herein are solely the responsibility of the authors.

Appendix

Viscoelastic and hyperelastic constitutive equations were used to represent the behavior of the synthetic materials. The total stress is equal to the sum of the hyperelastic (HE) stress and the viscoelastic (VE) stress such that:

$$\sigma(t) = \sigma_{HE}(t) + \sigma_{VE}(t) \tag{1}$$

The hyperelastic behavior was derived from a function of strain energy density per unit volume, U.

$$U = \sum_{i=1}^{N} \frac{2\mu_i}{\alpha_i^2} \left[\lambda_1^{\alpha_i} + \lambda_2^{\alpha_i} + \lambda_3^{\alpha_i} - 3 + \frac{1}{\beta}(J^{-\alpha_i \beta}) \right] \tag{2}$$

$$\sigma_{HE} = \frac{2}{J} F \frac{du}{dC} F^T \tag{3}$$

where $J = \lambda_1 \lambda_2 \lambda_3$ is the volume ratio, α_i and μ_i where ν is the Poisson's ratio, N, is the number of terms used in the strain energy function, and F and C are the deformation gradient and the right Cauchy-Green deformation tensors, respectively.

It was assumed that the candidate materials were incompressible, and therefore the volume ratio was set to unity. In the current case of uniaxial compression, the following relationships were used: $\lambda_1 = \lambda, \lambda_2 = \lambda_3 = \sqrt{\lambda}$.

The viscoelastic behavior was defined as follows, with a relaxation function $g(t)$ applied to the hyperelastic stress:

$$\sigma_{VE} = \int_0^t \dot{g}(\tau)\sigma_{HE}(t-\tau)d\tau$$

In order to describe several time constants for the relaxation, the stress relaxation function $g(t)$ was defined using the Prony series of order N_G, where g_i and τ_i are the viscoelastic parameters:

$$g(t) = \left[1 - \sum_{i=1}^{N_G} g_i(1 - e^{-t/\tau_i})\right] \tag{4}$$

The coefficients for hyperelastic (N), stress relaxation (N_G) and Poisson's numbers (ν) are given in Tables 1 and 2.

Table 1. Coefficients for silicone sample ($\nu = 0.49$)

i	1	2	3
g_i	0.015	0.044	0.029
$\tau_i(s)$	0.025	0.150	0.300
$\mu_i(MPa)$	0.080	0.150	-
α_i	0.001	15.500	-

Table 2. Coefficients for the polyurethane sample ($\nu = 0.47$)

i	1	2	3
g_i	0.167	0.158	0.113
$\tau_i(s)$	0.100	1.380	25.472
$\mu_i(MPa)$	0.100	0.063	-
α_i	5.500	8.250	-

References

1. Cabibihan, J.J., So, W.C., Pramanik, S.: Human-recognizable robotic gestures. IEEE Trans. Auton. Mental Dev. 4(4), 305–314 (2012)
2. Cabibihan, J.J., So, W.C., Saj, S., Zhang, Z.: Telerobotic pointing gestures shape human spatial cognition. Int. J. Soc. Robot. 4(3), 263–272 (2012)

3. Cabibihan, J.-J., So, W.C., Nazar, M., Ge, S.S.: Pointing gestures for a robot mediated communication interface. In: Xie, M., Xiong, Y., Xiong, C., Liu, H., Hu, Z. (eds.) ICIRA 2009. LNCS (LNAI), vol. 5928, pp. 67–77. Springer, Heidelberg (2009). doi:10.1007/978-3-642-10817-4_7

4. Wykowska, A., Kajopoulos, J., Obando-Leitón, M., Chauhan, S.S., Cabibihan, J.J., Cheng, G.: Humans are well tuned to detecting agents among non-agents: examining the sensitivity of human perception to behavioral characteristics of intentional systems. Int. J. Soc. Robot. 7(5), 767–781 (2015)

5. Hertenstein, M., Keltner, D., App, B., Bulleit, B., Jaskolka, A.: Touch communicates distinct emotions. Emotion 6(3), 528–533 (2006)

6. Cabibihan, J.J., Joshi, D., Srinivasa, Y.M., Chan, M.A., Muruganantham, A.: Illusory sense of human touch from a warm and soft artificial hand. IEEE Trans. Neural Syst. Rehabil. Eng. 23(3), 517–527 (2015)

7. Cabibihan, J.-J., Ahmed, I., Ge, S.S.: Force and motion analyses of the human patting gesture for robotic social touching. In: IEEE 5th International Conference on Cybernetics and Intelligent Systems, CIS 2011 (2011)

8. Cabibihan, J.-J., Pradipta, R., Chew, Y.Z., Ge, S.S.: Towards humanlike social touch for prosthetics and sociable robotics: handshake experiments and finger phalange indentations. In: Kim, J.-H., et al. (eds.) FIRA 2009. LNCS, vol. 5744, pp. 73–79. Springer, Heidelberg (2009). doi:10.1007/978-3-642-03983-6_11

9. Li, H., Cabibihan, J.J., Tan, Y.: Towards an effective design of social robots. Int. J. Soc. Robot. 3(4), 333–335 (2011)

10. Cabibihan, J.J., Wu, K.W., Ramalingam, A.: Tactile sensing in an object passing task. In: Proceedings of the 2013 IEEE Conference on Cybernetics and Intelligent Systems, CIS 2013, pp. 96–99 (2013)

11. Edin, B., Ascari, L., Beccai, L., Roccella, S., Cabibihan, J.J., Carrozza, M.C.: Bio-inspired sensorization of a biomechatronic robot hand for the grasp-and-lift task. Brain Res. Bull. 75, 785–795 (2008)

12. Osborn, L., Kaliki, R.R., Soares, A.B., Thakor, N.V.: Neuromimetic event-based detection for closed-loop tactile feedback control of upper limb prostheses. IEEE Trans. Haptics 9(2), 196–206 (2016)

13. Edin, B., Beccai, L., Ascari, L., Roccella, S., Cabibihan, J.J., Carrozza, M.C.: A bio-inspired approach for the design and characterization of a tactile sensory system for a cybernetic prosthetic hand. In: Proceedings of the IEEE International Conference on Robotics and Automation

14. Heyneman, B., Cutkosky, M.: Slip classification for dynamic tactile array sensors. Int. J. Robot. Res. 35(4), 404–421 (2016)

15. Salehi, S., Cabibihan, J.J., Sam, S.G.: Artificial skin ridges enhance local tactile shape discrimination. Sensors 11(9), 8626–8642 (2011)

16. Cabibihan, J.J., Chauhan, S.S., Suresh, S.: Effects of the artificial skin's thickness on the subsurface pressure profiles of flat, curved, and braille surfaces. IEEE Sens. J. 14(7), 2118–2128 (2014)

17. Anand, A., Mathew, J., Pramod, S., Paul, S., Bharath, R., Xiang, C., Cabibihan, J.J.: Object shape discrimination using sensorized glove. In: 2013 10th IEEE International Conference on Control and Automation (ICCA), pp. 1514–1519 (2013)

18. Hyttinen, E., Kragic, D., Detry, R.: Learning the tactile signatures of prototypical object parts for robust part-based grasping of novel objects. In: 2015 IEEE International Conference on Robotics and Automation (ICRA), pp. 4927–4932, May 2015

19. Beccai, L., Roccella, S., Ascari, L., Valdastri, P., Sieber, A., Carrozza, M.C., Dario, P.: Experimental analysis of a soft compliant tactile microsensor to be integrated in an anthropomorphic artificial hand. In: ASME 8th Conference on Engineering Systems Design and Analysis (2006)

20. Cabibihan, J.-J., Carrozza, M.C., Dario, P., Pattofatto, S., Jomaa, M., Benallal, A.: The uncanny valley and the search for human skin-like materials for a prosthetic fingertip. In: 2006 6th IEEE-RAS International Conference on Humanoid Robots, vol. 1, pp. 474–477 (2006)

21. Cabibihan, J.J., Pattofatto, S., Jomâa, M., Benallal, A., Carrozza, M.C.: Towards humanlike social touch for sociable robotics and prosthetics: comparisons on the compliance, conformance and hysteresis of synthetic and human fingertip skins. Int. J. Soc. Robot. 1(1), 29–40 (2009)

22. Cabibihan, J.-J., Ge, S.S.: Towards humanlike social touch for prosthetics and sociable robotics: three-dimensional finite element simulations of synthetic finger phalanges. In: Kim, J.-H., et al. (eds.) FIRA 2009. LNCS, vol. 5744, pp. 80–86. Springer, Heidelberg (2009). doi:10.1007/978-3-642-03983-6_12

23. Cabibihan, J.J.: Design of prosthetic skins with humanlike softness. In: Teck Lim, C., Goh, J.C.H. (eds.) ICBME 2008. IFMBE Proceedings, vol. 23, pp. 2023–2026. Springer, Heidelberg (2009)

24. Cabibihan, J.-J., Ge, S.S.: Synthetic finger phalanx with lifelike skin compliance. In: Liu, H., Ding, H., Xiong, Z., Zhu, X. (eds.) ICIRA 2010. LNCS (LNAI), vol. 6425, pp. 498–504. Springer, Heidelberg (2010). doi:10.1007/978-3-642-16587-0_46

25. ABAQUS. In: ABAQUS Theory Manual, v. 6.6. Hibbit, Karlson and Sorense, Inc., Pawtucket, USA

26. Sladek, E., Fearing, R.: The dynamic response of a tactile sensor. In: 1990 IEEE International Conference on Robotics and Automation, Proceedings, vol. 2, pp. 962–967 (1990)

One-Shot Evaluation of the Control Interface of a Robotic Arm by Non-experts

Sebastian Marichal[1,2,3], Adrien Malaisé[1,2], Valerio Modugno[1,2,4],
Oriane Dermy[1,2], François Charpillet[1,2], and Serena Ivaldi[1,2(✉)]

[1] Inria, 54600 Villers-lès-Nancy, France
[2] Loria, UMR n.7503, CNRS & Univ. Lorraine, Loria,
54500 Vandoeuvre-lès-Nancy, France
{sebastian.marichal,adrien.malaise,valerio.modugno,
oriane.dermy,francois.charpillet,serena.ivaldi}@inria.fr
[3] Universitat Pompeu Fabra, Barcelona, Spain
sebastian.marichal@upf.edu
[4] DIAG, Sapienza University, 00185 Roma, Italy

Abstract. In this paper we study the relation between the performance of use and user preferences for a robotic arm control interface. We are interested in the user preference of non-experts after a one-shot evaluation of the interfaces on a test task. We also probe into the possible relation between user performance and individual factors. After a focus group study, we choose to compare the robotic arm joystick and a graphical user interface. Then, we studied the user performance and subjective evaluation of the interfaces during an experiment with the robot arm Jaco and $N = 23$ healthy adults. Our preliminary results show that the user preference for a particular interface does not seem to depend on their performance in using it: for example, many users expressed their preference for the joystick while they were better performing with the graphical interface. Contrary to our expectations, this result does not seem to relate to the user's individual factors that we evaluated, namely desire for control and negative attitude towards robots.

Keywords: Human-robot interfaces · User evaluation · Individual factors · Non-experts

1 Introduction

In this paper, we address the question of the preference for a robotic interface by non-experts (or naive users without training in robotics), after one single evaluation of such an interface on a simple task. This refers to situations when non-experts face the decision of adopting a robot for episodic use (i.e., not a regular continuous use as workers in factories): the ease of use of an interface is crucial for the robot acceptance. We do not target users that could have or

---

S. Marichal, A. Malaisé—Equal contribution.

A. Agah et al. (Eds.): ICSR 2016, LNAI 9979, pp. 458–468, 2016.
DOI: 10.1007/978-3-319-47437-3_45

will have the time to receive a proper training on how to use a robot. While in manufacturing, robots are used by skilled workers that receive a proper training for operating the robots, this training is not likely to happen for many assistance and service scenarios: for example, inside an healthcare facility it is likely that the nurses or the patients will never receive a proper training for operating and interacting with the robots. The question arises on how to make the robot easily controllable by such users and facilitate their interaction with the robot. As the interface for controlling the robot is an essential part of the robotics system, this question impacts not only the interaction performance, but also the user acceptance and final adoption of the technology.

In this study, we focus on the Kinova Jaco (see Fig. 1), a lightweight robotic arm which can be controlled with a built-in joystick. It was designed for a daily and regular use for ordinary people after some training: the joystick is easy to manipulate but it has several buttons and control modes that require practice to achieve a fluent interaction. Here, we target a different use and a one-shot evaluation: if the control interface is an obstacle to the use of the robot, the users will not likely adopt the robot even for sporadic use. Several interfaces for robot control have been investigated in HRI. For example [17] investigated touch, speech and gestures for teaching a robot a nursery rhyme, finding that users do not prefer a particular modality but enjoy less touching the robot. In [16] the authors compared haptic interfaces with buttons, finding that users preferred buttons for

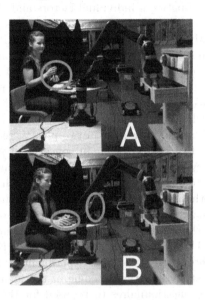

Fig. 1. The experimental setup with the Kinova Jaco arm. The participant moves the arm using (A) the joystick and (B) the graphical interface on the laptop.

simple tasks and physical command for complex tasks requiring high precision. Here, We compare the joystick with a ad-hoc graphical user interface (GUI) with buttons.

We are here interested in *(i)* probing the relation between individual factors and user performances for robot interfaces, and *(ii)* studying the relation between the performances that the user achieve with such interfaces and their preference.

Our main hypothesis is that the preference of an interface is related to the performance of using it. This premise is evident from other studies focused on interfaces evaluation. Guo & Sharlin noted that preferences for a tangible interface was related to a stronger performance in using it [15]. Many studies on control interfaces for robots focused on graphical user interfaces for their better acceptance by non-experts, for example [6] for teaching objects to a robot, [4] for applications in rehabilitation and medicine. In [5] the authors proposed an Android interface for moving the Jaco arm, but unfortunately it was not thoroughly evaluated by final users.

Our second hypothesis is that individual factors, such as traits and attitudes, may influence the user performances with the robot interfaces. There is indeed prior evidence that some personality traits have significant effects on the perceived ease of use of new technologies, such as smartphones [7]. There is also evidence that personality traits and attitudes have some influence in HRI in the context of social robotics [1]. It seems therefore rational to explore the relation between individual factors and the user perception and performance in controlling a robot. Two attitudes seems particularly relevant for our study: the Negative Attitude towards robots (NARS) [3], which captures the anxiety of an imagined interaction with a robot, and the Desire For Control (DFC) [8], which captures the attitude to be in control or control situations. The first could influence for example the time spent on using the robot, while the second could influence the preference for an interface that provides a stronger sensation of controlling the robot.

Our study was split in two phases. In the first, we carried out a focus group study to identify the main concerns of people interacting with a robotic arm, the key elements underlying their imagined interaction and the imagined interfaces to control the robot movement. This set enabled us to formulate the first hypothesis and choose a graphical user interface (GUI) as an intuitive interface alternative to the Kinova joystick. The second phase concerned the experiments with the Jaco robot and the two interfaces. We first performed a pilot study with University students to test the experimental setup and gain preliminary insights for the later final experiments with ordinary adults. The analysis of the pilot study and the outcome of the focus group enabled us to refine the evaluation questionnaires to be used for the final experiments and formulate new hypothesis.

We studied the user performance and subjective evaluation of the interfaces during an experiment with the robot arm Jaco and $N = 23$ healthy adults. We provide quantitative evidence of the different performances obtained by non-experts, using both interfaces for the first time to realize some tasks. We also report on the user feedback in using the two interfaces, which provides us useful information to inform future interface designers.

Our preliminary results show that the user preference for a particular interface does not seem to depend on their performance in using it: for example, many users expressed their preference for the joystick whereas they were better performing with the graphical interface. Also, contrary to our expectations, this result does not seem to relate to the user's aforementioned individual factors.

Research Hypothesis - Given the previous results in the literature, we expect that the GUI will be easier to use than the joystick, for non-trained users. The GUI has the advantage to not require too much training, and it provides some graphical shortcuts to the main robot configurations. To provide a quantitative

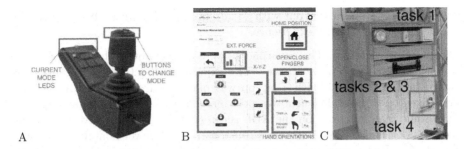

Fig. 2. The two interfaces used for the evaluation: (A) the Kinova joystick and (B) our ad-hoc graphical interface on the laptop. (C) The Activities of Daily Living setup: click the three buttons (*task 1*), open a drawer (*task 2*), take an object inside the drawer (*task 3*), open the door (*task 4*).

measure of the ease of use, we use the duration of execution of tasks performed with an interface, and the number of errors done while using it. We formulate the hypothesis as:

(H1) *The time necessary to complete the tasks with the GUI is shorter than with the joystick.*

(H2) *The number of precision errors with the GUI is lower than with the joystick.*

(H3) *The number of mapping errors with the GUI is lower than with the joystick.*

We also hypothesize that the user personality, attitudes and their prior experience with related technologies may influence the user acceptance of the proposed technologies and the performance in using it. The desire for control could play a crucial role in the preference for the joystick to the GUI, as the users could have the impression to be more in control of the robot while moving it. The negative attitude towards robots could influence the user perception of the interaction and the perceived ease of use. We formulate therefore the following hypothesis:

(H4) *Participants with high score of DFC will prefer the joystick to the GUI.*

(H5) *Participants with a high negative attitude towards robots score will make more errors and have a lower perceived ease of use and user satisfaction.*

2 Methods

Participants. The participants were all French, healthy adults that volunteered to take part in the study. The focus group study was carried out with 6 adults (age: 39.16 ± 15.71, 3 males, 3 females) without or with little robotics experience (1 participant). The pilot study was carried out with 7 University students in cognitive sciences (age: 23.14 ± 1.46, 2 males, 5 females). The final experiments with the robot were carried out with 23 adults (age: 35.13 ± 11.98, 12 males, 11 females) without robotics experience.

Experimental setup. The experiments were carried out at the LARSEN laboratory of INRIA (Nancy, France). The experimental setup was organized as shown in Fig. 1. A desk with a laptop was placed in front of the Kinova Jaco arm, fixed on a table. The arm was positioned in such a way to be able to perform some manipulations on the ADL setup (Fig. 2C), made of two boxes: one with a door handle, one with three buttons and a drawer containing a small object. A video camera, placed behind the participants, was used to record the experiments. Two interfaces (see Fig. 2A and B) for controlling the robot were used: the native joystick by Kinova and our own ad-hoc graphical user interface (GUI). The joystick can move the hand in the Cartesian space (position and rotation), open and close the fingers. Two buttons are used to select whether to move the hand position (mode 1), its orientation (mode 2) or the fingers (mode 3). The GUI was developed with Qt and is open-source[1]. Both interfaces use the same Kinova API for robot control and inverse kinematics solving.

Questionnaires. To probe into the influence of individual factors, we asked the participants to the robot experiment to fill out some questionnaires before the experiments: the Negative Attitude Towards Robots Scale (NARS) [3] and the Desire For Control scale (DFC) [8]. Our French adaptation was used [1]. The participants also filled two post-experimental questionnaires consisting of questions/affirmations adapted from usability and technology acceptance models to a robotic context as it was done in previous works [9,12]. The post-block questionnaire, at the end of each experimental condition (block when one interface is used), was based on the USE questionnaire [13] (typical questions were *"How good will you rate the movement you achieved in the 'open the drawer' task?"*). The post-experimental questionnaire consisted of a set of affirmations to be rated on a 7-points Likert scale, targeting constructs typical of the UTAUT [11] and TAM 3 models [14] (typical questions were *"Controlling the robot with the GUI is easy"*).

Experimental protocol. The study consist of a focus group and two robot experiments: a pilot study with University students, then experiments with ordinary adults. All the data were recorded in anonymous form through a random numerical id attributed to each participant. All participants were equally informed by the experimenter about the purpose of the study and their rights, according to the ethics guidelines of our institute. An informed consent form was signed by each participant. The protocol received the positive approbation of the local Ethics Committee.

Focus group study - We asked a group of 6 adults without or with little experience in robotics to imagine how they would interact with the robot and control it to do some tasks. The group gathered in a closed room around a table. One moderator led the group, while two recorders took notes and annotated

---

[1] https://github.com/serena-ivaldi/kinova-modules.

sentences and body language. The session lasted about 2 h and was recorded for analysis purposes. The experimenter asked to the group six warm-up questions, such as "*Tell us about your overall experience with robots*", "*In which situation(s) do you imagine that a robotic arm such as the Kinova would be useful?*". In a work in pairs, participants had to present their ideas about interfaces for controlling a robot arm.

Pilot study with the robot - We carried out a pilot study with the Jaco robot and 7 University students. Each participant had to perform the 4 tasks (see Fig. 2C) with the robot, using the joystick and the GUI. The order of the interfaces was randomized across the participants. After the experiment with the robot, we asked the participant to express their preference for one of the two interfaces and provide their feedback and personal evaluations.

Experimental study with the robot - The experiments with the Jaco robot were carried out with 23 adults without expertise in robotics. Each participant filled in the questionnaires NARS and DFC one week before the experiment. The day of the experiment, the participant was welcomed to the laboratory room by the experimenter and seated on a table with a laptop (see Fig. 1) in front of the robot. There were two blocks corresponding to the two experimental conditions: one with the joystick and one with the GUI. In each block, the participant had to perform the 4 tasks with the robot (see Fig.2C). The order of use of the interface was randomized and balanced across the participants. To ensure that all the participants received an equal set of instructions, we provided them with the same instructions, either in paper format and in video format (tutorial). The participant started by reading some paper instructions explaining the 4 tasks to be performed with the robot. After reading the instructions, they had to rate some statements on a 7-items Likert scale, such as "*The required tasks are difficult*" and "*The instructions were difficult to read*". We also added two trick questions to check if they were attentive and had carefully read the instructions. Before each block, the participant watched a 2/3 min video tutorial explaining how to use each interface, then he/she could familiarize and try it for about 1 min. We instructed the participants to follow a think-aloud protocol. When the participant was ready to start, he/she began performing the 4 tasks in sequence. Two experimenters monitored and annotated the experiment. After completion, the participant filled in a questionnaire evaluating the ease of use of the interface. The sequence tutorial-test-tasks-evaluation was repeated for the second interface. After the experiment with the robot, the participant filled in the post-experimental evaluation questionnaire, then answered to some semi-directed questions during an interview with the two experimenters.

Measures and data analysis. During the focus group, two recorders annotated the discussion. Video recordings were used to complete the annotation offline. In the pilot study, we measured the duration of each task and the user preference for each interface. In the robot experiments, we employed both objective and subjective measures. Two experimenters annotated: the *duration of each task*; the *numbers of precision errors*, represented by the number of times the robot hit the ADL board; the *number of mapping errors*, represented by the number of times the robot was moved in the opposite direction with respect to the desired (we could identify this by the explicit verbalization of the participant, or by two consecutive movements in opposite directions where the first was clearly in the wrong direction with respect to the goal of the movement). The questionnaires' score for NARS and DFC were computed according to the authors' recommendations. The subjective measures retrieved from the post-experimental questionnaires are the *perceived ease of use* (PEOU, typical question: *"Controlling the robot with the GUI is easy"*), the *user satisfaction* (US, *"How good will you rate the movement you achieved in the 'open the drawer' task?"*) and the *facilitating condition* (FC, *"The time to test the Joystick before the experiment was enough"*) related to each interface, computed by the sum of the score of the questionnaire items for each construct. The expertise in using joysticks was a self-reported score on a 10-item scale.

Unless otherwise stated, we computed median and standard deviation of all the measured variables; we used Spearman's correlation and verified the statistical significance of the different conditions with a Wilcoxon signed ranked test with continuity correction in R.

3   Results

Focus group - The focus group participants did not have a particular affinity with robotics, and were generally worried about the possibility of robots replacing humans. When asked about the possible use for the Jaco arm, they indicated grabbing objects on very high shelves, assisting people with impairments or arm troubles, doing manual tasks like laundry, ironing and painting walls. Almost all the participants agreed that the robot should not be completely autonomous: they need to be in control of the situation when the robot is acting. They said that they should *"teach the robot to do the things the way we want"* and *"be able to stop the robot anytime"*. When we asked how to control the robot, the participants mostly indicated panels with buttons (3/6). In particular, one participant explained that there should be a button for each possible robot gesture.

Pilot study - The only significant difference in terms of task duration with the two interfaces is on the second task (*opening the drawer*, $V = 0$ $p = 0.0156 < 0.05$). We did not find any significant correlation between the task duration and the participants' self-report expertise with joysticks.

Concerning the joystick, the negative points were: the difficulty in controlling the hand orientation and the way to change the modes with the buttons. Positive points were that it was more intuitive to move in the x-y-z space, especially for the students used to play video-games, and that it felt like an "extension of their arm". Concerning the GUI, the negative point was that it required to switch continuously the attention from the laptop to the robot. The positive points were its clearer design that made the actions explicit and the ease of use when choosing pre-determined orientations of the hand for manipulation.

We asked the 7 participants to choose the interface that was easier to use and more intuitive for them: 2 preferred the joystick and 5 the GUI (*"it can be mastered, one makes more errors with the joystick"*).

Robot experiments - After reading the instructions, the participants evaluated the tasks to be not difficult (on a 7-item Likert scale, median $= 2$, stdev $= 1.67$) and the instructions easy to read (median $= 1$, stdev $= 2.03$). We found a significant difference in the overall duration of the tasks ($V = 25$ $p = 0.0006 < 0.001$) for the two conditions, in particular for Task 2 (*opening the drawer*, $V = 10$ $p = 0.0002 < 0.001$) and Task 3 (*grabbing the object*, $V = 28.5$ $p = 0.0009 < 0.001$), a fair difference for Task 4 (*opening the door*, $V = 51.5$ $p = 0.0089 < 0.01$). We also compared the duration of the tasks executed with each interface when the latter is first or second in order of execution: we did not find difference in the execution for the GUI (Mann-Whitney, $W = 82$ $p = 0.347$ (N.S.)), whereas there is a weak evidence for a difference in the execution time of the joystick if it is used as first or second (Mann-Whitney, $W = 27.5$ $p = 0.0193 < 0.05$). In terms of use of the interface, there is a marginal difference in terms of precision ($V = 53$ $p = 0.0531$ (N.S.)), while there is a strong difference in terms of mapping errors ($V = 0$ $p = 2.85e\text{-}05 < 0.001$) - the median number of mapping errors with the joystick is also quite elevated (10). Regarding the subjective measures retrieved by the questions, we found a significant difference in the ratings in terms of ease of use ($V = 251.5$ $p = 5.23e\text{-}05 < 0.001$), satisfaction ($V = 239$ $p = 0.0022 < 0.005$) and facilitating conditions ($V = 159$ $p = 0.0013 < 0.005$): the GUI has higher ratings than the joystick on all the three items. We did not find a significant correlation between the users' performance and their prior expertise in using joysticks nor between the user performance and their NARS.

Among the 23 participants, 11 expressed preference for the joystick and 12 for the GUI. However, in terms of usability, the joystick was favored by 6 participants, while the GUI by 16 (one participant said they were equal). We tested if the interface preference was related to the DFC score of the participants but we did not find any significant difference (Mann-Whitney, $W = 48$, $p = 0.279$ (N.S.)).

We asked the participants to provide their feedback in the post-experimental interview. Many participants highlighted that the joystick made them feel more "in control" when moving in the main Cartesian directions (x,y,z - the first mode of the joystick) and that they could achieve more precise movements with it. Almost all the participants reported that switching the mode with the joystick was very difficult. However, some thought that they could become good users with a dedicated training. One participant, for example, said *"my son is very good with the video-games pad, he will learn in 10 min; for me, I will need some hours"*. Many participants appreciated the GUI because of the intuitive buttons where each command/action was explicit.

4   Discussion

In this study we focused on non-expert users controlling a robot for their first time: if the robot-user does not have a proper training, or if he is using the robot only once in a while, which interface could be easier to use and facilitating the robot adoption? From the focus group study, we learned that people imagine to interact with the robot in a structured way (e.g., buttons) that allows them to be in control of the robot decisions (e.g., when to start, when to stop).

To make the robot controllable by non-experts, our conclusion is that we need a very reliable control interface that they can understand and use easily/intuitively, that is robust and that gives them the impression to be in control. From the participants suggestions, a panel with buttons seems appropriate as a control interface: it gives the user the impression that the robot can act upon their orders. For the purpose of this study, we decided that the most appropriate control interface to test against the joystick of the Jaco arm was a GUI with buttons.

Is a GUI really better than a joystick? - From the pilot study with students, we could not strongly conclude that the GUI brings notable improvements over the joystick. In the experiments with ordinary adults, the GUI is better than the joystick in terms of objective performance measures and subjective user evaluation. We found significant difference in the duration of tasks and mapping errors, but not in the precision errors: therefore we accept **H1** and **H3** but reject **H2**. Almost all participants found the GUI easier to use, more understandable and straightforward. Many participants appreciated moving the robot with the joystick as they felt it an "extension of their hand". Interestingly, while most participants appreciated the pre-programmed orientations/configurations of the hand, that were quite difficult to obtain with the joystick, some participants reported them as a constraint that was limiting their freedom to choose different orientations of the hand to realize the tasks. These participants suggested that the two interfaces should be combined to give the user more freedom. It is however important to notice that the GUI performs better than the joystick in our particular experimental conditions, where the participants have a very limited training for using the interfaces (a video tutorial and 1 min to familiarize with the interface and try it). The results could be very different in a case where the participant uses the robot on a regular basis or receives a proper training. We will address this case in future experiments.

Do individual factors play a role in the user performance with an interface? - Our preliminary results show that the user preference for a particular interface does not seem to relate to their performance in using it: for example, many users expressed their preference for the joystick whereas they were better with the GUI. Contrary to our expectations, this result does not seem to relate to the user's individual factors, as we did not find a strong evidence to support our hypothesis. We did not find significant correlations between the user preferences or performances with both NARS and DFC. We therefore reject **H4** and **H5**. Nevertheless, in the post-experimental interviews many participants reported to feel more comfortable with the joystick despite being better with the GUI: this may seem counter-intuitive, but in fact suggests that there may be other individual criteria that drive their choice.

5   Conclusions

Two main questions emerge for future work: Which are the key factors that determine user preference for a robot control interface and if the preference and performance in using an interface would change in a long term scenario (i.e., a scenario where users receive a training for operating the robot with the interface and use such an interface more frequently or on a daily basis). We plan more experiments to investigate more thoroughly all these questions.

References

1. Ivaldi, S., Lefort, S., Peters, J., Chetouani, M., Provasi, J., Zibetti, E.: To-wards engagement models that consider individual factors in HRI: on the relationof extroversion and negative attitude towards robots to gaze and speech during ahuman-robot assembly task. Int. J. Soc. Robot. (2016)
2. Gaudiello, I., Zibetti, E., Lefort, S., Chetouani, M., Ivaldi, S.: Trust as indicator of robot functional and social acceptance. An experimental study on user conformation to the iCub's answers. Comput. Hum. Behav. **61**, 633–655 (2016)
3. Nomura, T., Kanda, T., Suzuki, T.: Experimental investigation into influence of negative attitudes toward robots on human-robot interaction. AI Soc. **20**(2), 138–150 (2006)
4. Chung, C.S., Wang, H., Cooper, R.A.: Functional assessment and performance evaluation for assistive robotic manipulators: literature review. J. Spinal Cord Med. **36**(4), 273–289 (2013)
5. Chung, C.S., Boninger, J., Wang, H., Cooper, R.A.: The Jacontrol: development of a smart-phone interface for the assistive robotic manipulator. In: RESNA Annual Conference (2015)
6. Rouanet, P., Danieau, F., Oudeyer, P.-Y.: A robotic game to evaluate interfaces used to show and teach visual objects to a robot in real world condition. In: International Conference on Human-Robot Interaction, pp. 313–320 (2011)
7. Ozbek, V., et al.: The impact of personality on Technology acceptance: a study on smart phone users. Procedia Soc. Behav. Sci. **150**, 541–551 (2014)
8. Burger, J.M., Cooper, H.M.: The desirability of Control. Motiv. Emot. **3**(4), 381–393 (1979)
9. BenMessaoud, C., Kharrazi, H., MacDorman, K.F.: Facilitators and barriers to adopting robotic-assisted surgery: contextualizing the unified theory of acceptance and use of technology. PLoS One **6**(1), e16395 (2011)
10. Lee, J.Y., Choi, J.J., Kwak, S.S.: The impact of user control design types on people's perception of a robot. In: Proceedings of HRI Extended Abstract, pp. 19–20 (2015)
11. Venkatesh, V., Morris, M.G., Davis, F.D., Davis, G.B.: User acceptance of information technology: toward a unified view. MIS Q. **27**, 425–478 (2003)

12. Heerink, M., Krse, B., Wielinga, B., Evers, V.: Measuring the influence of social abilities on acceptance of an interface robot and a screen agent by elderly users. In: Proceedings of 23rd British HCI Group Annual Conference, pp. 430–439 (2009)
13. Lund, A.M.: Measuring usability with the USE questionnaire. STC Usability SIG Newsl. **8**, 2 (2001)
14. Venkatesh, V., Bala, H.: Technology acceptance model 3 and a research agenda on interventions. Decis. Sci. **39**, 273–315 (2008)
15. Guo, C., Sharlin, E.: A comparative study. In: Proceedings of CHI, pp. 121–130
16. Gleeson, B., Currie, K., MacLean, K., Croft, E.: Tap and push: assessing the value of direct physical control in human-robot collaborative tasks. J. Hum. Robot Interact. **4**(1), 95–113 (2015)
17. Novanda, O., Salem, M., Saunders, J., Walters, M.L., Dautenhahn, K.: What communication modalities do users prefer in real time HRI? In: 5th International Symposium on New Frontiers in HRI 2016 (2016)

A Novel Parallel Pinching and Self-adaptive Grasping Robotic Hand

Dayao Liang and Wenzeng Zhang$^{(\boxtimes)}$

Department of Mechanical Engineering,
Tsinghua University, Beijing 100084, China
wenzeng@tsinghua.edu.cn

Abstract. This paper introduces a novel underactuated hand, the PASA-GB hand which has a hybrid grasping mode. The hybrid grasping mode is a combination of parallel pinching (PA) grasp and self-adaptive enveloping (SA) grasp. In order to estimate the performance of grasping objects, the potential energy method is used to analyze the stabilities of the PASA-GB hand. The calculation of force distribution shows the influence of the size and position of objects and provides a method to optimize the force distribution. Experimental results verify the wide adaptability and high practicability of the PASA-GB hand.

Keywords: Robot hand · Underactuated finger · Parallel and self-adaptive grasp · Grasping stability

1 Introduction

Robotic hands have been highly researched for decades. As the end effectors, robotic hands accomplish most of the missions for robots. Traditional robot hand have many joints, the number of which is the same as degrees of freedom (DOFs). They are dexterous hands such as DLR-HIT Hand [1], Robonaut 2 hand [2], Gifu hand II [3], Belgrade/USC Hand [4], NAIST Hand [5]. Although dexterous hands are flexible, the establishment of control systems and sensor systems are difficult. In common cases, the actuators have to be small to set inside the phalanges. As a result, the grasping forces are small. Industrial grippers [6] successfully deal with these difficulties, but most of them have only one DOF.

Underactuation was proposed to handle the contradiction between high DOFs and large grasping force. Underactuated hands use less actuators to control more DOFs with special mechanisms. The control problem is changed to the design problem of the mechanisms, as the mechanisms determine the behaviors and grasping stabilities of the finger. Outstanding designs include: Underactuated hand: MARS hand [7], bio-mimetic robot hands [8], prosthetic FDD-hand [9], pneumatically driven TWIX-hand [10], FRH-4 hand [11], and soft gripper [12]. Methods to optimize the performance of underactuated hand [13, 14] were also proposed.

However, the grasping modes of traditional underactuated hands are single. In order to perform more grasping modes, people developed hybrid grasping mode hand: such as Barrett Hand grasper [15], SARAH hand [16], and Robotiq hand [17].

© Springer International Publishing AG 2016
A. Agah et al. (Eds.): ICSR 2016, LNAI 9979, pp. 469–480, 2016.
DOI: 10.1007/978-3-319-47437-3_46

This paper introduces a novel parallel and self-adaptive robot hand with gear-belt mechanisms (PASA-GB hand), which possesses the advantages of large grasping force and multiple grasping modes. It combines the good qualities of dexterous hands, underactuated hands and industrial grippers. The second part introduces the architecture of the PASA-GB hand, the third part analyzes the stability and force distribution of the PASA-GB hand, the forth part shows the experimental results, the fifth part concludes this paper.

2 Architecture of the PASA Hand

2.1 Concept of Parallel and Self-adaptive Underactuated Grasp

Parallel pinching grasp is a common grasping mode for traditional industrial gripper. It is the most appropriate way to grasp some specially objects, such as paper, which needs precise grasp [18]. However, Enveloping grasp is more tightly for large objects than pinching grasp, as enveloping grasp is often power grasp. In order to combine the advantages of both pinching grasp and enveloping grasp, [19] introduces the concept of parallel and self-adaptive underactuated finger and introduced a PASA finger with belt-link mechanism.

As it is shown in Fig. 1, parallel and self-adaptive underactuated (PASA) grasping mode is a hybrid grasping mode. Based on the location, size and shape of objects, a parallel and self-adaptive underactuated finger (PASA finger) executes parallel pinching motion if the lower part of the finger touches the objects first, or executes enveloping grasp if the higher part touches the objects first. Because of the special mechanisms of PASA finger, the switch among different grasping modes is automatic and self-adaptive.

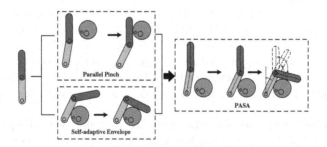

Fig. 1. Concept of parallel and self-adaptive underactuated grasp.

2.2 Mechanisms and Structure of the PASA Hand

The PASA finger mainly consists of a finger base, a proximal phalanx, a distal phalanx, a proximal joint, a distal joint, a driving gear, a driven gear, a gear set, a proximal wheel, a distal wheel, a transmission belt, a sliding block, a spring, a motor and a limiting block, as it is shown in Fig. 2.

The transmission belt links the proximal wheel and the distal wheel, which are in the same size. The distal phalanx is fixed to the driven gear, which is fixed to the distal wheel and driven by the driving gear through the gear set. The transmission ratio between the driving gear and the driven gear is a. The sliding block is set on the proximal wheel. The spring links the sliding bock and the finger base and drives the proximal wheel rotating backwards. The limiting block is set on the finger base and limits the backward motion of the proximal wheel.

At the beginning, the sliding block abuts the limiting block, which makes the proximal wheel "fixed" to the finger base. Because of the transmission belt, the distal wheel can only translate without rotation on condition that sliding block doesn't leave the limiting block. As a result, the distal phalange executes parallel pinching motion.

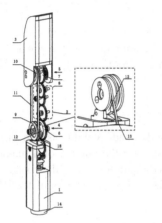

Fig. 2. Mechanisms and structure of the PASA-GB hand. 1-Finger base, 2-Proximal phalanx, 3-Distal phalanx, 4-Proximal joint, 5-Distal joint, 6-Driving gear, 7-Driven gear, 8-Gear set, 9-Proximal wheel, 10-Distal wheel, 11-Transmission belt, 12-Sliding block, 13-Spring, 14-Motor, 18-Limiting block

Let θ_1 be the rotational angle of the proximal phalanx, θ_2 be the rotational angle of the distal phalanx, and β the rotational angle of the driving gear. Referencing to the proximal phalanx, the rotational angle of the driving gear is $\beta - \theta_1$, that of the driven gear is $\theta_2 - \theta_1$. The transmission ratio between the driving gear and the driven gear is a, one achieves

$$a(\beta - \theta_1) = \theta_2 - \theta_1, \tag{1}$$

The rotational angle θ_1 can be described as follow:

$$\theta_1 = a\beta/(a - 1), \tag{2}$$

It means that, when the driving gear moves forwards, the proximal phalanx rotates with a proportionate speed to the driving gear.

Once the proximal phalanx is blocked by objects, the driving gear continues rotating and drives the distal wheel rotating through the gear set. The rotational angle of the distal phalanx is

$$\theta_2 = a\beta - \theta_1(a - 1), \tag{3}$$

As a result, the rotational speed of the distal phalanx is also propionate to the driving gear. Because the proximal wheel rotates, the sliding block has to leave the limiting block and the spring is strained. Such motion is self-adaptive enveloping grasp.

3 Analysis of the PASA Hand

This part focuses on the analyses of stability, force distribution and switch condition among different grasping modes. For the reason of simplification, the contact forces are considered as point forces, the friction and gravity are neglected.

3.1 Stability Analyses

This part analyzes the stability of the PASA-GB hand. Stability is one of the most important performance for robot hands [20]. Because the number of actuators is less than the degrees of freedom (DOFs), stability is difficult to gain by the control systems. Therefore, the analyses of stability and optimization of parameters of the PASA-GB finger are important.

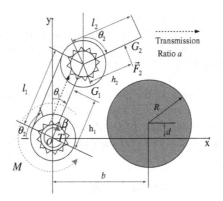

Fig. 3. Stability analyses of the PASA-GB finger.

As Fig. 3 shows, O is the center point of the proximal joint axis, l_1, l_2 are the length of the proximal phalanx and the distal phalanx, G_1, G_2 are the contact points between phalanges and objects, h_1, h_2 are the distances between the joint axis and contact points, $\vec{F}_1$, $\vec{F}_2$ are the external force vectors produced by objects, M is the value of the spring torque, T is the value of the driving torque.

Because of the transmission belt, the rotational angles of the proximal wheel and the distal phalanx are the same. Let M_0 be the initial spring torque when $\theta_2 = 0$, so can be defined as $M = k\theta_2 + M_0$, where k is the elastic coefficient of the spring.

To simplify the problem, the object is considered as a column, of which the radius is R. Set up a rectangular coordinate system at the point O, the coordinate of the center of the object is (b, d).

$\vec{P}_1$ and $\vec{P}_2$ are points on the proximal phalanx and distal phalanx

$$\vec{P}_1 = (H_1 \sin \theta_1, H_1 \cos \theta_1), \ 0 \leq H_1 \leq l_1, \tag{4}$$

$$\vec{P}_2 = (l_1 \sin \theta_1 + H_2 \sin \theta_2, l_1 \cos \theta_1 + H_2 \cos \theta_2), \ 0 \leq H_2 \leq l_2, \tag{5}$$

When the proximal phalanx touches the object, the slope of proximal phalanx can be obtained by simple calculation:

$$k_1 = \pm \frac{R\sqrt{-R^2 + b^2 + d^2} - bd}{\sqrt{R^2 - b^2}}, \tag{6}$$

At this moment, the rotational angle of the proximal phalanx is

$$\theta_{10} = \text{atan} \left(\pm \frac{\sqrt{R^2 - b^2}}{R\sqrt{-R^2 + b^2 + d^2} - bd} \right), \tag{7}$$

The distance h_1 can be obtained by geometry relation:

$$h_1 = \frac{d + b \tan \theta_{10}}{\cos \theta_{10} + \tan \theta_{10} \sin \theta_{10}}, \tag{8}$$

When the distal phalanx touches the object, the slope of distal phalanx is:

$$k_2 = \frac{S \pm R\sqrt{-R^2 + (b - l_1 \sin \theta_1)^2 + (d - l_1 \cos \theta_1)^2}}{R^2 - (b - l_1 \sin \theta_1)^2}, \tag{9}$$

Where, $S = bl_1 \cos \theta_1 - bd + dl_1 \sin \theta_1 - l_1^2 \cos \theta_1 \sin \theta_1$
And the rotational angle of the distal phalange is

$$\theta_{2c} = \text{atan} \left(\frac{R^2 - (b - l_1 \sin \theta_1)^2}{S \pm R\sqrt{-R^2 + (b - l_1 \sin \theta_1)^2 + (d - l_1 \cos \theta_1)^2}} \right), \tag{10}$$

The distance between the contact point on the distal phalanx and the distal joint

$$h_2 = \frac{-l_1(\tan\theta_{2c}\sin\theta_1 + \cos\theta_1) + b\tan\theta_{2c} + d}{\cos\theta_{2c} + \tan\theta_{2c}\sin\theta_{2c}}, \tag{11}$$

If both the proximal phalanx and distal phalanx touch the object,

$$\theta_{20} = \text{atan}\left(\frac{R^2 - (b - l_1\sin\theta_{10})^2}{Y \pm R\sqrt{-R^2 + (b - l_1\sin\theta_{10})^2 + (d - l_1\cos\theta_{10})^2}}\right), \tag{12}$$

Where, $Y = bl_1\cos\theta_{10} - bd + dl_1\sin\theta_{10} - l_1^2\cos\theta_{10}\sin\theta_{10}$

In order to analyze the stability of grasp, one introduces the potential energy V, which is produced by the driving torque T and the spring torque M:

$$V = \phi - T\beta + \frac{1}{2}k\theta_2^2 + M_0\theta_2, \tag{13}$$

where ϕ is potential energy of the PASA-GB finger in the initial state.

The condition for complete stable enveloping grasp is when both phalanges touch the object, the potential energy V reaches the minimum point.

In arbitrarily states, β can be described as

$$\beta = \frac{\theta_1(a - 1)}{a} + \frac{\theta_2}{a}, \tag{14}$$

One achieves

$$V = \phi - T\frac{(a - 1)}{a}\theta_1 + \frac{1}{2}k\theta_2^2 + (M_0 - \frac{T}{a})\theta_2, \tag{15}$$

When θ_1 increases, should decrease. We obtain another necessary condition for stable enveloping grasp

$$a > 1 \tag{16}$$

Figure 4 illustrates the relation between, and θ_2. The color surface represents the potential energy V. The position of the upper globule (point A) represents the initial potential energy in the initial state. Without touching any objects, V decreases as increases. Such motion is a pinching motion, the trajectory of which is represented by the green arrows. If the objects blocks the finger at point B, the pinching motion is stable. The expression of the blue plan is $\theta_1 = \theta_{10}$. The globule which represents the potential energy cannot pass through this plan because the proximal phalanx is blocked by the object.

The color surface and the blue plan intersect in a curve, the red arrow in Fig. 4. After the proximal phalanx touches the object, the potential energy continues decreasing along the red arrows, which represents an enveloping motion of the PASA-GB finger.

The orange surface represents the state when the distal phalanx touches the objects, which is described in Eq. 10. The globule also cannot pass through the orange plan because the distal phalanx is blocked. The intersection of the color surface, orange surface and blue plan is the potential energy when both phalanges touch the object, which is represented by the lower globule (point C).

In an enveloping grasp, the motion will stop at point C when C is the minimum point of all the possible trajectories. However, once the potential energy V continues decreasing along the blue arrows, the motion will go on and the proximal phalanx will leave the objects until the potential energy reaches the minimum point in the blue arrows. Such motion is called "Ejection motion".

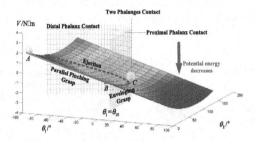

Fig. 4. Potential energy of the PASA-GB finger, where $T = 3\text{N} \cdot \text{m}$, $a = 1.8$, $M_0 = 0.06\text{N} \cdot \text{m}$, $k = 0.86\text{N} \cdot \text{m}$, $b = 0.04$ m, $d = 0.03$ m, $R = 0.03$ m, $l_1 = 0.056$ m, $\phi = 0\text{N} \cdot \text{m}$. (Color figure online)

As Fig. 5 shows, the parallel pinching trajectory, enveloping trajectory and ejecting trajectory define a motion zone of the PASA-GB finger. Those three curve are barriers which cannot be passed through. Outside the barriers is a forbidden zone for the PASA-GB finger, which is defined by the position, location and shape of objects.

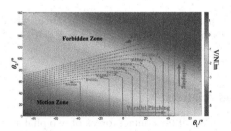

Fig. 5. Motion zone and barrier of the PASA-GB finger, where $T = 3\text{N} \cdot \text{m}$, $a = 1.5$, $M_0 = 0.02\text{N} \cdot \text{m}$, $k = 0.34\text{N} \cdot \text{m}$, $b = 0.04$ m, $d = 0.03$ m, $l_1 = 0.056$ m, $\phi = 0\text{N} \cdot \text{m}$.

Another not completely stable grasp happens when the motion stops at an intermediate point of the red arrows in Fig. 4. In order to analyze this case, we need to find out the minimal point of with respect to θ_2:

$$\frac{\partial V}{\partial \theta_2} = 0, \tag{17}$$

$$\theta_{2\,V\min} = (\frac{T}{a} - M_0)/k, \tag{18}$$

As it is shown in Fig. 6a and b, if the horizontal line $\theta_2 = \theta_{2\,V\min}$ across the enveloping trajectory, the enveloping motion will stop at the yellow point as it reaches the minimum value of the potential energy. As a result, the enveloping grasp is not completely stable. To deal with this problem, a and k should not be too large. However, if a is too closed to 1, the ejecting phenomenon will happen, which is shown in Fig. 6d.

Figure 6c shows that a stable grasp is obtained when a and k are proper.

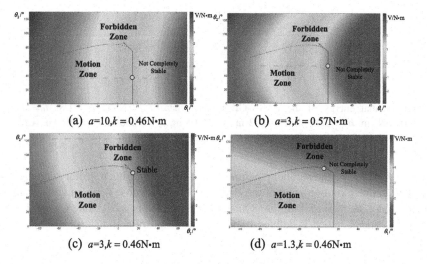

(a) $a=10, k=0.46$N•m

(b) $a=3, k=0.57$N•m

(c) $a=3, k=0.46$N•m

(d) $a=1.3, k=0.46$N•m

Fig. 6. Influence of the transmission ratio and spring force, where $T = 3$N · m, $a = 1.5$, $M_0 = 0.02$N · m, $k = 0.34$N · m, $b = 0.04$ m, $d = 0.03$ m, $R = 0.03$ m, $l_1 = 0.056$ m, $\phi = 0$ N · m.

3.2   Grasping Forces Analyses

For a stable enveloping grasp, if the contact forces of two phalanges are close, the grasp is tight.

When the grasping motion finishes,

$$[-T, k\theta_2 + M_0]\begin{bmatrix} \delta\beta \\ \delta\theta_2 \end{bmatrix} = [\vec{F}_1, \vec{F}_2]\begin{bmatrix} \delta\vec{G}_1^t \\ \delta\vec{G}_2^t \end{bmatrix}, \tag{19}$$

Where

$$[\vec{F}_1, \vec{F}_2]\begin{pmatrix} \delta\vec{G}_1^t \\ \delta\vec{G}_2^t \end{pmatrix} = [F_1, F_2]\begin{bmatrix} -h_1 & 0 \\ -l_1\cos(\theta_2 - \theta_1) & -h_2 \end{bmatrix}\begin{bmatrix} \delta\theta_1 \\ \delta\theta_2 \end{bmatrix}, \tag{20}$$

$$\begin{pmatrix} \delta\beta \\ \delta\theta_2 \end{pmatrix} = \begin{bmatrix} \frac{a-1}{a} & \frac{1}{a} \\ 0 & 1 \end{bmatrix}\begin{bmatrix} \delta\theta_1 \\ \delta\theta_2 \end{bmatrix}, \tag{21}$$

The relationship among contact forces, driving torque and spring torque is shown as follow:

$$[F_1, F_2] = [-T, k\theta_2 + M_0]\begin{bmatrix} \frac{a-1}{a} & \frac{1}{a} \\ 0 & 1 \end{bmatrix}\begin{bmatrix} -h_1 & 0 \\ -l_1\cos(\theta_2 - \theta_1) & -h_2 \end{bmatrix}^{-1}. \tag{22}$$

Combine Eqs. 7, 12 and 22, one achieves the final angles of two phalanges and forces distribution based on the position and size of the object, as it is shown in Figs. 7 and 8.

As Fig. 7 shows, $\theta_2 - \theta_1$ becomes larger when the object is closer and larger, which is obviously. Figure 8 shows that the size of objects has much influence on the force distribution. If the object is too large, the contact force of the distal phalanx is much larger than that of the proximal phalanx. Although the finger performs a stable enveloping grasp, the grasp is not so tight. If an unexpected external force is applied on the object, the distal phalanx may be opened and the object will leave the finger. The influence of b shown in Fig. 8 implies that a larger hand base brings better force distribution.

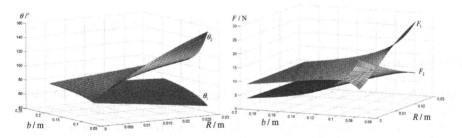

Fig. 7. Rotational angles of the PASA-GB finger.

Fig. 8. Force distribution of the PASA-GB finger.

4 Experiments

The prototype of the PASA hand with Gear-belt mechanisms is shown in Fig. 9. It consists of one hand base and three fingers. The motor type is FAULHABER 1616E010.

As it shown in Fig. 9a and b, the PASA hand executes parallel pinching grasp when the bottle is located on the upper side. The Limiting block stops the backward motion of the distal phalanx, which makes the PASA hand pinch objects tightly.

Figure 9c, d and e shows the enveloping motion of the PASA. When the bottle is located on the lower side, the proximal phalanx touches the bottle first. Because of the self-adaptive mechanism, the distal phalanx touches the bottle afterwards.

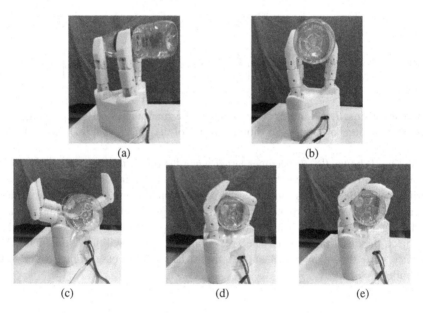

(a)   (b)

(c)   (d)   (e)

Fig. 9. Experiments of grasping object with the PASA-GB hand.

5 Conclusions

This paper introduces a novel underactuated hand, the PASA-GB hand, which has a hybrid grasping mode. The hybrid grasping mode is a combination of parallel pinching (PA) grasp and self-adaptive enveloping (SA) grasp. Based on the position, size and shape of objects, when the first contact point is located on the lower part of the finger, the finger will execute enveloping grasp. When the first contact is located on the upper part, the finger will execute parallel pinching grasp.

In order to estimate the performance of grasping objects, the potential energy method is used to analyze the stabilities of the PASA-GB hand. The trajectories of pinching grasp, enveloping grasp and ejection form a motion zone for the finger, the barriers of which are defined by the states of objects. The calculation of force distribution shows the influence of the size and position of objects and provides a method to optimize the force distribution. Experimental results verify the wide adaptability and high practicability of the PASA-GB hand.

Acknowledgement. This Research was supported by National Natural Science Foundation of China (No. 51575302).

References

1. Butterfass, J., Grebenstein, M., Liu, H., et al.: Dlr-hand ii: next generation of a dextrous robot hand. In: Proceedings of the IEEE International Conference on Robotics and Automation, Seoul, Korea, pp. 109–114 (2001)
2. Lovchik, C.S., Diftler, M.A.: The Robonaut hand: a dexterousrobot hand for space. In: Proceedings of the IEEE International Conference on Robotics and Automation (ICRA), pp. 907–912 (1999)
3. Kawasaki, H., Komatsu, T., Uchiyama, K.: Dexterous anthropomorphic robot hand with distributed tactile sensor: Gifu hand II. IEEE/ASME Trans. Mechatron. **7**(3), 296–303 (2002)
4. Bekey, G.A., Tomovic, R., Zeljkovic, I.: Control architecture for the Belgrade/USC hand in dextrous robot hands, pp. 136–149. Springer, New-York (1999)
5. Ueda, J., Ishida, Y., Kondo, M., Ogasawara, T.: Development of the naist-hand with vision-based tactile fingertip sensor. In: Proceedings of the IEEE International Conference on Robotics and Automation, pp. 2343–2348 (2005)
6. Ali, H., Hoi, L.H., Seng, T.C.: Design and development of smart gripper with vision sensor for industrial applications. In: Proceedings of the IEEE International Conference on Computation, pp. 157–180 (2011)
7. Gosselin, C., Laliberte, T., Degoulange, T.: Underactuated robotic hand. In: Video Proceedings of the IEEE International Conference on Robotics and Automation (1998)
8. Lee, S., Noh, S., Lee, Y.K., et al.: Development of bio-mimetic robot hand using parallel mechanisms. In: Proceedings of the IEEE International Conference on Robotics and Biomimetics, pp. 550–555 (2010)
9. de Visser, H., Herder, J.L.: Force-directed design of a voluntary closing hand prosthesis. J. Rehabil. Res. Dev. **37**(3), 261–271 (2000)
10. Bégoc, V., Krut, S., Dombre, E., et al.: Mechanical design of a new pneumatically driven underactuated hand. In: Proceedings of the IEEE International Conference on Robotics and Automation (ICRA), pp. 927–933 (2007)
11. Gaiser, I., Schulz, S., Kargov, A., et al.: A new anthropomorphic robotic hand. In: Proceedings of the IEEE International Conference on Humanoid Robots, pp. 377–386. IEEE (2008)
12. Hirose, S., Umetani, Y.: The developmentof soft gripper for the versatile robot hand. Mech. Mach. Theory **13**, 351–358 (1978)
13. Ciocarlie, M., Allen, P.: Data-driven optimization for underactuated robotic hands. In: Proceedings of the IEEE International Conference on Robotics & Automation, pp. 1292–1299 (2010)
14. Birglen, L., Laliberte, T., Gosselin, C.: Underactuated Robotic Hands. Springer Tracts in Advanced Robotics (2008)

15. Townsend, W.: The BarrettHand grasper – programmably flexible part handling and assembly. Industrial Robot **27**(3), 181–188 (2000)
16. Gosselin, C.M., Laliberte, T.: Underactuated mechanical finger with return actuation, US Patent, US5762390 (1996)
17. Demers, L.A., Lefrancois, S., Jobin, J.: Gripper having a two degree of freedom underactuated mechanical finger for encompassing and pinch grasping, US Patent, US8973958 (2015)
18. Daniel, A.M., Barrett, H., John, U., et al.: Design and testing of aselectively compliant underactuated hand. Int. J. Robot. Res. **33**, 721–735 (2014)
19. Liang, D., Zhang, W., Sun, Z., et al.: PASA finger: a novel parallel and self-adaptive underactuated finger with pinching and enveloping grasp. In: Proceedings of the IEEE International Conference on Robotics and Biomimetics. IEEE (2015)
20. Kragten, G.A.: Underactuated hands: fundamentals, performance analysis and design. Mechanical Maritime & Materials Engineering (2011)

PCSS Hand: An Underactuated Robotic Hand with a Novel Parallel-Coupled Switchable Self-adaptive Grasp

Shuang Song and Wenzeng Zhang[(⊠)]

Department of Mechanical Engineering,
Tsinghua University, Beijing 100084, China
wenzeng@tsinghua.edu.cn

Abstract. This paper proposes a novel concept of underactuated grasping mode, called PCSS grasping mode. This mode has switchable hybrid grasping functions: parallel self-adaptive grasping (PASA) and coupled self-adaptive grasping (COSA), being able to grasp larger range of objects with different shapes and dimensions than traditional PASA and COSA hands. The PCSS grasping can execute different grasping modes: a parallel pinching (PA); a coupled hooking (CO); a self-adaptive encompassing (SA); parallel and self-adaptive hybrid grasping (PASA); coupled and self-adaptive hybrid grasping (COSA). A PCSS Hand is developed with three PCSS fingers and 6 degrees of freedom. Simulation analysis shows the high stability and the versatility of the PCSS Hand.

Keywords: Robotic hand · Underactuated finger · Grasping mode · Self-adaptive grasp · Coupled grasp

1 Introduction

Robotic hands are designed to accomplish different tasks as a capable end-effector of industrial and service robots. Similar to human hands, robotic hands are expected to be capable of grasping large range of objects with different shapes and dimensions. Reproducing the numerous advantages of human hand by mechanism devices is always the ultimate pursuit of the scientific exploration of intelligent robot hands.

Human hands are dexterous and stable, having 21 degrees of freedom, huge power, quick actions while small volume. A common human hand can grasp plenty of different objects because it can perform multiple grasping modes (or grasping configuration).

The research of multi-fingered robotic hand is mainly divided into two kinds: (1) Dexterous hands and (2) Underactuated hands.

Dexterous hands are robotic hands with many degrees of freedom (DOFs) and driven by many actuators. Robotic hands of this type are able to perform different gestures like human hands. The UTAH/MIT Hand [1], Salisbury Hand [2] are outstanding examples of the first generation of dexterous robotic hands. In the recent few years, multiple dexterous hands, such as MANUS Hand [3], High-speed Multi-fingered Hand [4], HIT/DLR Hand [5], Gifu Hand [6] are researched and developed. But

© Springer International Publishing AG 2016
A. Agah et al. (Eds.): ICSR 2016, LNAI 9979, pp. 481–491, 2016.
DOI: 10.1007/978-3-319-47437-3_47

dexterous hands still have defects of complex controlling system and expensive manufacturing costs due to the multiple actuators applied.

In 1970s, underactuated hands are developed to solve these problems. In the area of robotic hand, underactuation is defined as the number of the actuators is smaller than the number of joint degrees of freedom (DOFs). There are many underactuated hands developed, such as: SARAH Hand [7–9] by Laval Univ., SDM Hand [10], and LARM Hand [11]. Different underactuated hands are designed to meet different needs in industrial applications.

Traditional underactuated hands have only one grasping mode, which limits the grasping ability when applied in robots. This paper proposes a novel underactuated grasping concept, called Parallel-Coupled Switchable Self-adaptive (PCSS in short) grasping, which integrates all the five existing underactuated grasping modes and allows underactuated hands to grasp a larger range of objects than former ones. The concept and theory of PCSS grasp is introduced in detail in the second part of this paper; the third part shows the design of PCSS Hands; the forth part analyzes the kinematic and kinetic issues of the PCSS Hands.

2   Concept and Theory of the PCSS Grasping

2.1   Concept of the PCSS Grasping

As mentioned in the former part, an underactuated robotic hand with two phalanges has three basic grasping modes and two hybrid grasping modes (as shown in Fig. 1):

(1) Parallel pinching (PA). In this grasping mode, the distal phalanx of the hand keeps parallel to its original orientation during its movement.

(2) Coupled hooking (CO). In this grasping mode, when the proximal phalanx turns forward relative to the base, the distal phalanx simultaneously turns forward relative to the proximal phalanx.

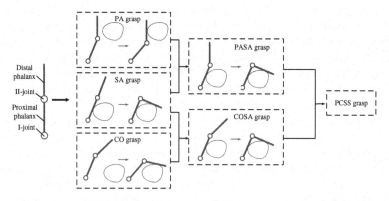

Fig. 1. Underactuated grasping modes.

(3) Self-adaptive encompassing (SA). In this grasping mode, that the final gesture of the finger is determined by the size and shape of the object grasped. The distal phalanx of the SA finger moves after the proximal phalanx blocked by the object.

(4) Parallel and self-adaptive grasping (PASA). The PASA hybrid grasping mode is a combination of the PA grasping mode and the SA grasping mode. In this mode, the finger moves in a PA pattern at the start, if the distal phalanx is blocked by the object, the finger perform a parallel pinching; if the proximal phalanx is blocked by the object, the distal phalanx continues to rotate until touch the object and perform a self-adaptive encompassing.

(5) Coupled and self-adaptive grasping (COSA). The COSA hybrid grasping mode is a combination of the CO grasping mode and the SA grasping mode. Similar to the PASA finger, the COSA finger can perform coupled hooking and self-adaptive encompassing.

The hybridism of basic grasping modes allows underactuated fingers to grasp larger range of objects, but the PASA mode and the COSA mode are not the end of this. These two hybrid grasping modes still have defects: the PASA grasping mode cannot hook; the COSA grasping mode cannot pinch. As each of the three basic grasping modes (CO, PA and SA) can grasp specific kinds of objects which other basic grasping modes cannot perform, further combination will bring further advantages. This paper proposes a higher level combination of the COSA grasping mode and the PASA grasping mode (as shown in Fig. 1), called PCSS grasping mode, which will allow robotic hands to perform almost all the grasping gestures needed in daily applications (Table 1).

Table 1. Performances of underactuated fingers.

Objects	CO	PA	SA	COSA	PASA	PCSS
Ball	No	No	Yes	Yes	Yes	Yes
Cylinder	No	No	Yes	Yes	Yes	Yes
Coin	No	Yes	No	No	Yes	Yes
Bar	Yes	No	No	Yes	No	Yes
Handle	Yes	No	No	Yes	No	Yes
Cube	No	Yes	No	No	Yes	Yes
Irregular	No	No	Yes	Yes	Yes	Yes

2.2 Theory of the PCSS Grasping

In PCSS grasping, the combination of the PASA mode and the COSA mode is achieved by switching between these two modes, which allows underactuated hands to perform both of these two modes while does not need complex switch manipulation. The PCSS hand can perform PASA and the COSA grasping as a result of the cooperation of the limitation part and the driving part.

The limitation part allows the PCSS finger to move in the PA and the CO pattern and to switch between the two modes. The switching between the PA mode and the CO mode is equivalent to the changing of the structure between a parallelogram structure and a crossed "8" structure. In order to achieve this by simple manipulation, this paper proposes a switch method based on open-loop pulley-tendon mechanism, including a half pulley, an open-loop tendon, an entire pulley, a tensioning spring, a switch shaft. The transmission ratio between the two pulleys is 1, assuring the parallel transmission while in the PA mode. The two ends of the open-loop tendon are fixed on the two pulleys in the pattern shown in Fig. 2. The tensioning spring is set on the other side of the pulley-II to make the tendon keep tight. The switch shaft is fixed on the base, set vertical to the I-joint shaft. The half-pulley can roll over across the switch shaft, and this is its only degree of freedom.

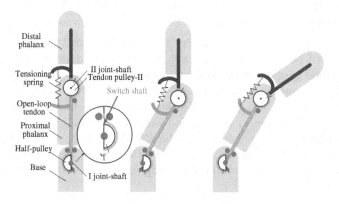

Fig. 2. The switching method of the PCSS grasping mode.

When the half pulley is on the right side (in the view of Fig. 2), the pulley-tendon mechanism is in parallel structure; and on the left side, crossed structure. This switch manipulation can be easily achieved by an actuator applied on the switch shaft or a simple two-position switch applied on the half-pulley.

The driving part allows the PCSS finger to perform the PASA and COSA modes. The hybrid grasping modes have two steps of motion. Take the PASA mode as an example, in the first step the proximal phalanx is driven to rotate across the I-joint shaft while the distal phalanx keeps parallel to its original orientation; and in the second step the proximal phalanx is blocked by the object and the distal phalanx is driven to rotate across the II-joint shaft until touching the object. The driven parts of these two steps are different. The main function of the driving part is to actuate the proximal phalanx at the first step and the distal phalanx at the second step. A pulley-belt transmission mechanism serves as the driving part of the PCSS Hand (as shown in Fig. 3(a)).

The active pulley is set on the I-joint shaft and the passive pulley is fixed on the II-joint shaft. The transmission ratio between the two belt pulleys defines as i. The distal phalanx and the tendon pulley-II are also fixed on the II-joint shaft. The driving part has two DOFs when the active pulley is driven: the rotation of the proximal phalanx relative to the base and the rotation of the distal phalanx relative to the proximal phalanx. The rotation of the proximal phalanx has the priority because of the

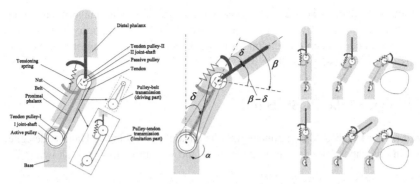

(a) The mechanism structure.   (b) Geometric analysis.   (c) The PASA and COSA grasp.

Fig. 3. The mechanism structure and function of the PCSS Hand.

tensioning spring. During the rotation of the proximal phalanx, the tensioning spring keeps pulling the tendon, keeping the distal phalanx to match the PA (and CO) grasping mode.

The relationship of the rotation angles of this step is determined by the transmission ratio i. To analyze angle relationship, one can see the motion of the PCSS finger in divided procedures. In the COSA mode, as shown in Fig. 3(b), the motion can be illustrated by: the distal phalanx rotate forward β relative to the proximal phalanx, the proximal phalanx rotates δ relative to the ground and the distal phalanx rotates backward an $\beta - \delta$ relative to the proximal phalanx. The adding result of these procedures is that the angle between the distal phalanx and the extension line of the proximal phalanx equals to δ, which correspond with the feature of the CO grasping mode.

$$\beta = \alpha/i \tag{1}$$

$$\beta - \delta = \delta/i \tag{2}$$

The consociation of Eqs. (1) and (2) makes:

$$\alpha = (i+1)\delta \tag{3}$$

Similarly, the equation of the PASA mode is:

$$\alpha = (i-1)\delta \tag{4}$$

When comes to the second step of the hybrid grasping modes, the proximal phalanx is blocked by the object. If the actuator continues to drive the active pulley, the passive pulley along with the distal phalanx will be driven to rotate relative to the proximal phalanx against the force of the tensioning spring. The PASA and COSA function of the PCSS finger and the conditions of the tensioning spring is shown in Fig. 3(c).

It the second step, the tendon will become loose as the distal phalanx rotate forward to execute a self-adaptive encompassing. Additional contemporary tendon storage device is needed to keep the tendon on its track. Small light spring or brand can serve as this device (not shown in Fig. 3).

3   Design of the PCSS Hand

Based on the PCSS grasping concept and its basic theory, a PCSS Hand is designed in detail in this paper. The design of the PCSS Hand with three fingers is shown if Fig. 4.

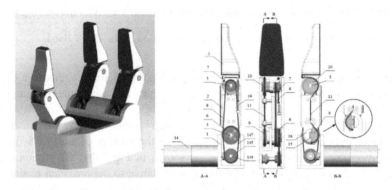

Fig. 4. The design of the PCSS Hand.

Items marked in Fig. 4:
1 - Base; 2 - Proximal phalanx; 3 - Distal phalanx; 4 - I-joint shaft; 5 - II-joint shaft; 6 - Active pulley; 7 - Passive pulley; 8 - Belt; 9 - Tendon pulley-I; 10 - Tendon pulley-II; 11 - Tendon; 14 - Actuator; 15 - Connector; 16 - Switch shaft; 17 - Tensioning Spring.

As mentioned in the former part, the switch function can be also achieved by the rotation of the tendon pulley-I, the prototype is made under this theory to prove the capability of the finger to execute the PASA and the COSA grasping mode. The small spring used to tension the tendon while SA grasping is applied on this prototype.

4   Kinematic Simulation and Kinect Analysis

The PCSS Hands can grasp a large variety of objects because of its multiple grasping modes. To evaluate and optimize the performance of the PCSS Hand, numerical analysis is needed. In this part, the kinematic simulation is carried out to obtain the stable grasping area; Kinect analysis is carried out to calculate the contact force distribution of the hand to evaluate the grasp ability. The PCSS Hand is selected for numerical and experimental evaluation.

4.1 Kinematic Simulation of the PCSS Hand

For the PCSS underactuated hand, the grasp orientation is determined by the size and shape of the object. Once the position and dimension of the object and the basic parameters of the hand are given, the orientation of the hand is predictable.

A simplified kinematic diagram of the PCSS Hand is shown in Fig. 5. The variables $\theta_{1,r}, \theta_{2,r}, \theta_{1,l}, \theta_{2,l}$ represents the rotation angles of the proximal and the distal phalanges of the right and the left finger. The center position of the object is (X_{obj}, Y_{obj}) in the local frame, and the radius of the object is r_{obj}. h_1 and h_2 represents the distance between the contact point on the proximal to the I-joint and distance between the contact point on the distal phalanx to the II-joint.

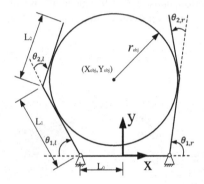

Fig. 5. Kinematic model of the PCSS Hand.

Take the right finger as the subject. According to two different geometric loop closure vectors, the relationship between the parameters above is arrived at the following equations:

$$\begin{pmatrix} L_0 \\ 0 \end{pmatrix} + \mathbf{R}_{\theta_1} \begin{pmatrix} h_1 \\ 0 \end{pmatrix} = \begin{pmatrix} X_{obj} \\ Y_{obj} \end{pmatrix} - \mathbf{R}_{\theta_1} \begin{pmatrix} 0 \\ r_{obj} \end{pmatrix} \tag{5}$$

$$\begin{pmatrix} L_0 \\ 0 \end{pmatrix} + \mathbf{R}_{\theta_1} \begin{pmatrix} L_1 \\ 0 \end{pmatrix} + \mathbf{R}_{\theta_1} \mathbf{R}_{\theta_2} \begin{pmatrix} h_2 \\ 0 \end{pmatrix} = \begin{pmatrix} X_{obj} \\ Y_{obj} \end{pmatrix} - \mathbf{R}_{\theta_1} \mathbf{R}_{\theta_2} \begin{pmatrix} 0 \\ r_{obj} \end{pmatrix} \tag{6}$$

$\mathbf{R}_{\theta}$ is the rotation matrix:

$$\begin{bmatrix} \cos \theta & -\sin \theta \\ \sin \theta & \cos \theta \end{bmatrix} \tag{7}$$

Simplify Eqs. (5) and (6), one achieves:

$$\theta_1 = \arctan\left(\frac{X_{obj} - L_0}{-Y_{obj}}\right) + \arctan\left(\frac{r_{obj}}{\sqrt{Y_{obj}^2 + \left(X_{obj}^2 - L_0^2\right)}}\right) \tag{8}$$

$$\theta_2 = \pi - 2\arctan\left(\frac{r_{obj}}{L_1 - \sqrt{Y_{obj}^2 + \left(X_{obj}^2 - L_0^2\right)} - r_{obj}^2}\right) \tag{9}$$

From Eqs. (8) and (9), one can study the relationship between the position of the object and the final gesture of the PCSS Hand in the SA mode. As shown in Fig. 6, the rotation angle has limitation caused by the mechanism structure, $60° \leq \theta_1 \leq 135°$, $-30° \leq \theta_2 \leq 90°$. Simulation results show that the PCSS Hand has a large stable grasping area.

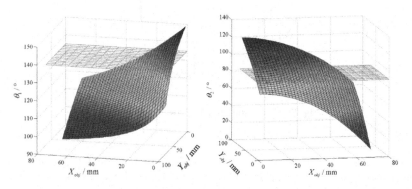

(a) Relationship between θ_1 and (X_{obj}, Y_{obj}).   (b) Relationship between θ_2 and (X_{obj}, Y_{obj}).

Fig. 6.   Kinematic simulation of the PCSS hand.

Similarly, the equivalent of the rotation angles according to the position of the object if the final orientation of the PCSS Hand is in PA or CO mode can also be achieved by geometric relationship:

If the final orientation of the hand is in PA mode,

$$\theta_2 = -\theta_1 \tag{10}$$

$$\theta_1 = \arccos\left(\frac{r_{obj} + X_{obj} - L_0}{L_1}\right) \tag{11}$$

If the final orientation of the hand is in PA mode,

$$\theta_2 = \theta_1 \tag{12}$$

$$r_{obj} = \left(X_{obj} - 2L_0\right) \sin 2\theta_1 - Y_{obj}\cos2\theta_1 - L_1 \sin \theta_1 \tag{13}$$

By the method of kinematic simulation, the grasping area of the PCSS Hand can be achieved, which serves as an important parameter of the ability and stability of the hand. According to the results of the kinematic simulation, the evaluation and optimization of the design is achieved. Furthermore, the kinematic simulation is also a functional method for intelligence robotic hands to estimate the capability of grasping a detected object before the grasping conduct as only the dimension and position of the object are need when calculating the rotation angles.

4.2   Grasping Force Analysis of the PCSS Hand

This part analysis the force distribution of the PCSS Hand. For the reason of simplification, the gravity and the friction are neglected and the contact force are applied on points. Figure 7 shows the dynamical condition of a PCSS finger. The driving torque of the actuator denotes T_M, the torque produced by the spring denotes T_s, F_1 and F_2 are the contact force of the object, f is the tension force in the belt. The transmission ratio of the pulley-belt mechanism denotes a.

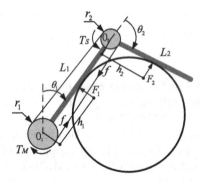

Fig. 7. Grasping force distribution of the PCSS finger.

$$T_M = F_1h_1 + F_2(h_2 + L_1 \cos \theta_2) \tag{14}$$

$$T_M = fr_1 \tag{15}$$

$$T_s + fr_2 = F_2h_2 \tag{16}$$

Form Eqs. (14)–(16), one obtains the expression equivalent of F_1 and F_2.

$$F_1 = \frac{T_M}{h_1}\left(1 - \frac{h_2 + L_1 \cos \theta_2}{ah_2}\right) + \frac{T_s(h_2 + L_1 \cos \theta_2)}{h_1 h_2} \tag{17}$$

$$F_2 = \frac{T_M}{ah_2} + \frac{T_s}{h_2} \tag{18}$$

The relationship between the contact force and the contact point on the phalanxes is studied as shown in Fig. 8.

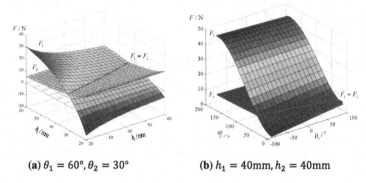

(a) $\theta_1 = 60°, \theta_2 = 30°$ (b) $h_1 = 40\text{mm}, h_2 = 40\text{mm}$

Fig. 8. Contact force distribution by h_1 and h_2 when $T_M = 1400\text{N} \cdot \text{mm}, T_s = 200\text{N} \cdot \text{mm}, L_1 = 100\,\text{mm}, a = 16/7$

5   Conclusion

This paper proposes PCSS underactuated grasp mode, which has switchable hybrid grasping modes: parallel self-adaptive grasping mode (PASA) and coupled self-adaptive grasping mode (COSA). The theory of the PASA and COSA structure and the switch function is illustrated in detail. This paper proposes the design of the novel PCSS Hands to prove the theory of PCSS grasping. A PCSS Hand is developed with three PCSS fingers and 6 degrees of freedom (DOFs). The PCSS finger has two joints, mainly consists of an actuator, an accelerative pulley-belt mechanism, an open-looped tendon mechanism, a spring, and a switching mechanism. A prototype based on the theory in manufactured. Kinematic simulation and kinetic analysis are included in this paper indicating the stability and capability of the PCSS Hand. Experimental results show the high the versatility of the PCSS Hand.

Acknowledgement. This paper was supported by the Natural Science Foundation of China (No. 51575302).

References

1. Jacobsen, S.C., Iversen, E.K., Knutti, D.F., et al.: Design of the UTAH/M.I.T. dextrous hand. In: IEEE International Conference on Robotics and Automation, San Francisco, USA, April, pp. 1520–1532 (1986)
2. Salisbury, J.K., Craig, J.J.: Articulated hands: force control and kinematic issues. Int. J. Robot. Res. **1**(4), 4–17 (1982)
3. Pons, J.L., Rocon, E., Ceres, R., et al.: The MANUS-hand dexterous robotics upper limb prosthesis: mechanical and manipulation aspects. Auton. Robots. **16**, 143–163 (2004)
4. Mizusawa, S., Namiki, A., Ishikawa, M.: Tweezers type tool manipulation by a multifingered hand using a high-speed visual servoing. In: 2008 IEEE/RSJ International Conference on Intelligent Robots and Systems. Nice, France. pp. 2709–2714 (2008)
5. Liu, H., Meusel, P., Hirzinger, G., et al.: The modular multisensory DLR-HIT-Hand: hardware and software architecture. Trans. Mechatornics **13**(4), 461–469 (2008)
6. Mouri, T., Kawasaki, H., Yoshikawa, K., et al.: Anthropomorphic robot hand: Gifu Hand III. In: International Conference on Control, Automation and Systems (ICCAS). Jeonbuk, Korea, pp. 1288–1293 (2002)
7. Laliberte, T., Gosselin, C.M.: Simulation and design of under-actuated mechanical hands. Mech. Mach. Theory **33**(1/2), 39–57 (1998)
8. Birglen, L., Gosselin, C.M.: Kinetostatic analysis of underactuated fingers. IEEE Trans. Robot. Autom. **20**(2), 211–221 (2004)
9. Birglen, L., Gosselin, C.M.: Fuzzy enhanced control of an underactuated finger using tactile and position sensors. In: Proceeding IEEE International Conference on Robotics and Automation, Barcelona, Spain, April 2005, pp. 2320–2325 (2005)
10. Dollar, A.M., Howe, R.D.: The SDM hand as a prosthetic terminal device: a feasibility study. In: 2007 IEEE 10th International Conference on Rehabilitation Robotics, Noordwijk, Netherlands, June, pp. 978–983 (2007)
11. Carbone, G., Iannone, S., Ceccarelli, M.: Regulation and control of LARM Hand III. Robot. Comput. Integr. Manuf. **24**(2), 202–211 (2010)
12. Zhang, W., Che, D., Liu, H., et al.: Super under-actuated multi-fingered mechanical hand with modular self-adaptive gear-rack mechanism. Ind. Robot Int. J. **36**(3), 255–262 (2009)
13. Li, G., Li, B., Sun, J., et al.: The development of a directly self-adaptive robot hand with pulley-belt mechanism. Int. J. Precis. Eng. Manuf. **14**(8), 1361–1368 (2013)
14. Li, G., Liu, H., Zhang, W.: Development of multi-fingered robotic hand with coupled and directly self-adaptive grasp. Int. J. Humanoid Rob. **9**(4), 1–18 (2012). Article ID 1250034
15. Zhang, W., Zhao, D., Zhou, H., et al.: Two-DOF coupled and self-adaptive (COSA) finger: a novel under-actuated mechanism. Int. J. Humanoid Rob. **10**(2), 1–26 (2013)

JLST Hand: A Novel Powerful Self-adaptive Underactuated Hand with Joint-Locking and Spring-Tendon Mechanisms

Jiuya Song and Wenzeng Zhang[✉]

Department of Mechanical Engineering, Tsinghua University, Beijing 100084, China
wenzeng@tsinghua.edu.cn

Abstract. This paper proposes a novel spring-tendon self-adaptive underactuated hand, called JLST hand, which can perform simultaneous multi-joint locking grasp. A JLST hand is designed with three JLST fingers and 6 degrees of freedom (DOFs). The JLST hand has more stable grasp ability and a larger grasping force than the normal spring-tendon robot hand, because the JLST hand has simultaneous multi-joint locking mechanisms. The JLST finger uses one motor to realize grasping objects, releasing objects and locking joints. The spring-tendon mechanism is based on the spring force to grasp objects and using the motor to pull the tendon to release the object. The JLST finger also uses one motor pulling the tendon to release the object and spring force to grasp objects, but the difference is JLST finger also uses the motor to lock multi-joints. Once the motor is turning forward, the finger releases the object; once the motor is turning backward, the motor releases the tendon so that the finger will grasp; the blocker is going to lock the joint since the motor keeps turning backward. The calculation and simulation results show that the JLST hand has the high stability of grasp and is more powerful than the normal spring-tendon hand.

Keywords: Robot hand · Underactuated finger · Joint-locking · Spring-tendon mechanism

1 Introduction

The hand is one of the most important parts of a human. The human body has around 90 DOFs, but one human hand has 21 DOFs, which means two human hands have around half of the DOFs in the human body. Humans use their hands to grasp, release and realize the object. So one humanoid robot hand is very important to the industry and medical needs.

The robot hand is divided into three types: industry gripper, underactuated hand and dexterous hand. On the one hand, robot hand has to achieve capture, moving and operating different shape objects which mean the robot hand has to have high requirements of control precision; on the other hand, the humanoid robot hand requires light weight and the right size. The existing industry gripper doesn't have strong generality because it can only grasp limited type of objects. The existing dexterous hand has enough joints and drivers to perform various precise movements, but the disadvantages of the

© Springer International Publishing AG 2016
A. Agah et al. (Eds.): ICSR 2016, LNAI 9979, pp. 492–501, 2016.
DOI: 10.1007/978-3-319-47437-3_48

dexterous hand are it requires high costs to build and complex to control. For example, the Stanford-JPL hand [1], Utah/MIT hand [2] and DLR hand [3]. The underactuated hand can solve the listed problems by its self-adaptation. The self-adaptive underactuated hand can adapt to the shape of the object while grasping; the weight of the self-adaptive underactuated hand is light; the control of the hand is simple, accurate and stable. For example, the SDM robot hand [4], Southampton hand [5] and TH robot hand [6–8]. The SARAH robot hand [9], PASA robot hand [10] and CDSA robot hand [11] have hybrid grasping mode. The SARAH robot hand and PASA robot hand are parallel and self-adaptive underactuated grasping mode. The CDSA robot hand is coupled and self-adaptive underactuated grasping mode.

The existing spring-tendon self-adaptive manipulator device (US2006129248A1) has five fingers [12]. Each finger mainly comprises a base, four phalanx, three joint springs and a longitudinal tendon rope. The finger has to be stretched by pulling the longitudinal tendon rope before grasping, then relax the longitudinal tendon rope to grasp the object by the flexibility force from the joint springs. The finger can grasp self-adaptively because each joint has a joint spring, which means the finger has high adaptability.

The disadvantages of the spring-tendon self-adaptive manipulator are:

(1) The contradiction between the grasping force from joint spring has to be as big as possible and the tension force from the longitudinal tendon rope has to be as small as possible. In order to ensure big grasping force, the joint spring has to have a large stiffness coefficient, which causes large tension force of longitudinal tendon rope to straighten the finger; if the tension force of longitudinal tendon rope to straighten the finger is small with weak joint spring, the grasping force will be small.

(2) The device is difficult to provide a wider range of grasping force. The grasping force is limited to a fixed range because the device adopts a fixed joint spring. The device mainly relies on joint springs to provide the grasping force during the grasping process. If the joint springs are weak, the device can't use the power from the connected arm for grasping, which will cause an inability for grasping heavy objects. For example, the device is trying to extract a heavy suitcase, the extraction force is most coming from the arm, but in order to extract, the finger has to have enough power to ensure the bending configuration.

(3) The large stiffness coefficient of the joint spring will cause the finger to quickly collide with the object while grasping, which means failure grasp.

(4) The device will have the possibility of failure grasp in a vibration environment.

There are few robot hands with joint-locking function, for example, the Twisting Actuation and Electromagnetic Joint Locking Robot hand [13]. The existing pneumatic self-adaptive with joint-locking underactuated robot hand (CN103659825A) has a self-adaptive grasping function, ratchet-pawl for locking joints and the motor pulls the pawl to unlock the joint [14].

The disadvantages of the pneumatic self-adaptive with joint-locking underactuated robot hand are:

(1) The device needs to have a driving force to achieve adaptive grasp. The driving force comes from the relative motion of the finger and the object. Once the object extrudes the finger slider, the pneumatic drives the next phalanx to move.

(2) The joint-locking of the device is not continuous. The locking is not continuous because the ratchet wheel gear has a certain pitch. If the ratchet wheel gear is designed to be large, the locking accuracy will be decreased. If the ratchet wheel gear is designed to be small, the tooth height of the gear will be decreased, which will decrease locking stability.

This paper proposes a novel JLST grasping mode to overcome the disadvantages of the normal spring-tendon grasping mode. The JLST grasping mode can automatically adapt to the object shape and size and lock the joints after enveloping the object to increase the grasping force. Since the joint-locking gives more grasping force, the finger will not rebind while grasping heavy objects. The joint-locking can lock the finger in any joint angle because the locking angle is continuous.

2 Concept of the JLST Grasp

As Fig. 1 shows, the underactuated robot hand has three basic grasping mode: self-adaptive grasping mode, coupled grasping mode and parallel grasping mode. The self-adaptive robot finger can adjust their motion to grasp different size and shape of object. For example, when the two phalanx self-adaptive robot finger grasp an object, the first joint starts rotating, the first phalanx and the second phalanx are keeping straight until the first phalanx touches the object, then the second joint starts rotating, the second phalanx rotates until touches the object. The coupled robot finger is more humanoid because all the joints are rotates at the same time when the motor starts rotating, which means each phalanx will truing at the same angle ratio. The parallel robot finger can pinch grasp the object. The second phalanx is always parallel to the base during the motor rotates.

The spring-tendon underactuated robot hand belongs to the self-adaptive grasping mode, which is one of the simplest mechanisms for self-adaptive grasping mode because the spring is a natural adaptive object. The motor rotates the tendon to make the finger

Basic grasping mode

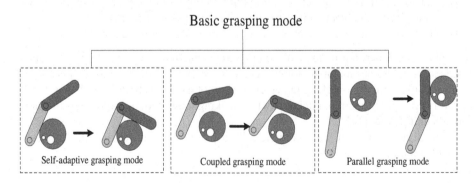

Fig. 1. Three basic grasping mode

straight before grasping, then once the motor releases the tendon, the joint springs give a force to make the finger grasp the object. The most famous spring-tendon robot hand is Stack hand which was named "Top Ten Invention of the Year 2011" by *Popular Science*.

The Stack's hand has five fingers and each finger has three phalanxes, each phalanx can bend independently, each joint has a joint spring in order to adapt different shape and size objects. The Stack's hand is suitable for grasping many kinds of objects, but it has huge difficulties in lifting heavy items. Once the Stack's hand is grasping a heavy object (over the spring force), the finger will rebind to failure grasping.

To overcome the normal spring-tendon robot hand's disadvantages, this paper proposes a novel powerful self-adaptive underactuated hand with the joint-locking and spring-tendon mechanism. The joint-locking mechanisms can lock joints after the finger envelops the object, which means there is an existing force to prevent the finger rebound, so the finger will have a larger grasping force to lift heavy objects.

3 Architecture and Analysis of the JLST Hand

3.1 Structure of the JLST Finger

The structure of the JLST hand is shown in Fig. 2. It consists a hand base and three fingers, each finger mainly consists of a motor, two phalanges, two axes, a driving gear, a driven gear, a tendon, two synchronous belt transmission mechanism, four twisting strings and two joint blockers. Each joint has two twisting strings with one joint blocker. Each twisting string is fixed to the joint blocker on one side, and the other side is fixed to the driving gear or driven gear.

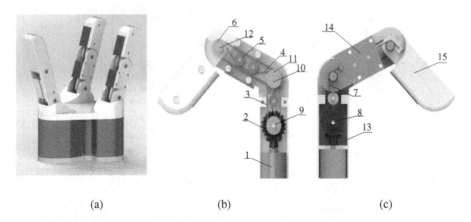

(a)   (b)   (c)

Fig. 2. The JLST Hand and the structure of the JLST finger, 1 – motor; 2 – bevel gear; 3 – synchronous belt; 4 – tendon; 5 – bobbin; 6 – distal blocker; 7 – spring; 8 – transition axis; 9 – transition axis; 10 – proximal joint; 11 – proximal blocker; 12 – distal joint; 13 – base; 14 – proximal phalanx; 15 – distal phalanx.

The motor is located in the base. The output of the motor is connected to the input of the first transmission mechanism. The first transmission mechanism includes two bevel gears. The second bevel gear is located on the transition axis. The bobbin is also located on the transition axis, and the tendon is connected with the bobbin. The tendon is bypassed to the first joint shaft and is fixed to connect with the second joint shaft bobbin. The driving gear is also on the transition axis. The driving gear is connected with driven gear by the synchronous belt. The driven gear in on the first and second joint shaft. The blockers are on the joint shaft. One side of the double twisting strings are fixed with the blocker, the other side of the double twisting strings are fixed with the dial. The dial is fixed with the driven gear. The first spring is located on the first joint shaft. One side of the spring is fixed to the base, the other side of the spring is fixed with the proximal phalanx. The second spring is located on the second joint shaft. One side of the spring is fixed with the proximal phalanx, the other side of the spring is fixed with the distal phalanx.

The finger is bent by the joint spring before the motor rotates. Once the motor rotates forward, the motor will pull the tendon and the finger will be straight. Then, if the motor releases the tendon, the finger will grasp the object. After the finger envelopes the object, the motor starts rotating backward, the driving gear starts rotating and the driven gear rotates by the synchronous belt. The driving gear and driven gear rotate to twist the twisting string, the joint blocker is pulled by the twisting string and lock the joints. As shown in Fig. 3, there is a spring between the distal phalanx and the blocker, which can give a force to pull the blocker back. Once the finger is trying to release the object, the

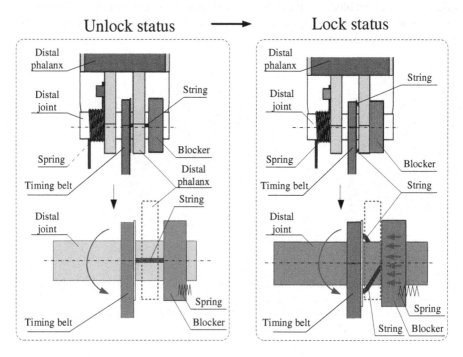

Fig. 3. The JLST Hand and the joint-locking of the JLST finger.

motor stops rotating backward, there is no force acting on the tendon, and the blocker will not keep locking the joint. The spring, connects the distal phalanx and the blocker, will give a force to pull the blocker back to the unlock location.

3.2 Grasping-Force Distribution Analysis

This part focus on the grasping forces distribution of the JLST finger. For the reason of simplification, the gravity force of the finger and the friction force between phalanges and objects are neglected, and the contact forces are applied on points.

As Fig. 4 shows, T_1 and T_2 are the torques of the first joint and second joint. l_1 is the length between the center of the first axis to the center of the second axis. r is the radian of the first axis and second axis. θ_1 is the rotating angle of the proximal phalanx; θ_2 is the rotating angle of the distal phalanx.

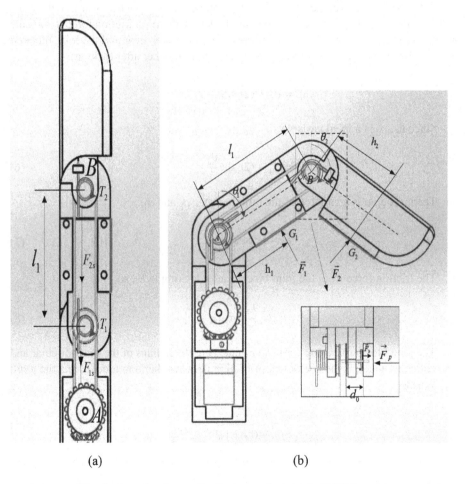

(a)   (b)

Fig. 4. Grasping-forces distribution analysis of the JLST finger.

As Fig. 4 shows, the finger is straight because the motor rotates forward to pull the tendon. The relationship between the tendon force and the spring force is:

$$(F_{1s} - F_{2s}) \cdot r = k_1 \cdot \theta_1 \tag{1}$$

and

$$F_{2s} \cdot r = k_2 \cdot \theta_2 \tag{2}$$

When the motor releases the tendon, the finger grasps the object by the spring force,

$$T_1 = k_1 \cdot \theta_1 \tag{3}$$

$$T_2 = k_2 \cdot \theta_2 \tag{4}$$

After the finger envelopes the object, the motor starts rotating backward, so the gears will rotate the twisting string, and the joint blockers will give a friction force to lock the joints. As Fig. 4 shows, d_0 is the length between the driven gear and the joint blocker; d_1 is the length between the driven gear and the joint blocker after locking;

$$d_0^2 = d_1^2 + (2r \sin \frac{\theta_2}{2})^2 \tag{5}$$

Since the d_0 is a fixed variable,

$$d_1^2 = d_0^2 - (2r \sin \frac{\theta_2}{2})^2 \tag{6}$$

Then, set up α is a variable of the ratio between d_0 and d_1,

$$\alpha = ar \cos(\frac{d_1}{d_0}) \tag{7}$$

The friction force from the joint blocker after locking is F_f, so,

$$F_f = \frac{T_2}{r} \tan \alpha \tag{8}$$

The joint blocker is a ring form, so r_3 and r_4 are the radians of the inner circular and the outer circular. q is the coefficient of friction between the joint blocker and the joint. Where

$$q = \frac{F_f}{\pi(r_4 - r_3)} \tag{9}$$

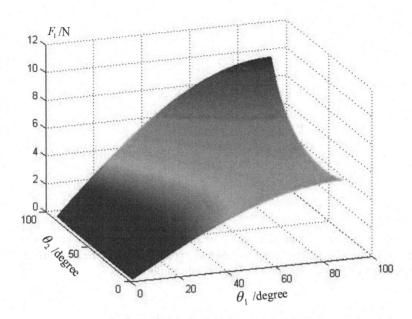

Fig. 5. Contact-force distribution of F_1 based on the θ_1, θ_2, where $T_1 = 0.1, T_2 = 0.1$, $q = 0.8, \beta = 0.8, h_1 = 0.02, l_2 = 0.035, r_3 = 0.005, r_4 = 0.01$

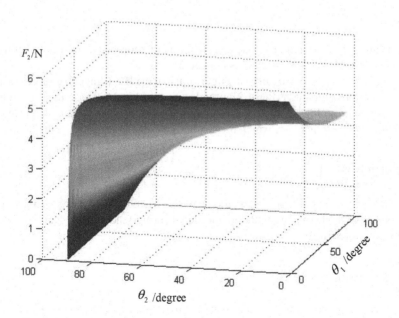

Fig. 6. Contact-force distribution of F_2 based on the θ_1, θ_2, where $T_1 = 0.1, T_2 = 0.1$, $q = 0.8, \beta = 0.8, h_1 = 0.02, l_2 = 0.035, r_3 = 0.005, r_4 = 0.01$

T_b is the locking torque from the joint blocker,

$$T_b = \int_{r_3}^{r_4} qr \cdot dr \tag{10}$$

Since the l_2 is the arm of force between $\overrightarrow{F_2}$ and point A, so

$$\frac{l_1}{\cos \theta_2} = \frac{l_2}{\sin \theta_1} \tag{11}$$

$$l_2 = \frac{l_1 \sin \theta_1}{\cos \theta_2} \tag{12}$$

Finally, set up the β as the radio of the friction force between two joint blockers, the equation of the finger will be the sum of the two spring forces and the friction force from the blocker is balanced with the reacting force from the object.

$$T_1 - F_1 h_1 + T_2 - F_2 l_2 + (1 + \beta) \int_{r_3}^{r_4} qr \cdot dr = 0 \tag{13}$$

Since F_1 and F_2 are unknown variables, the relationship between these two forces are

$$F_1 \cos \theta_1 = F_2 \sin \theta_2 \tag{14}$$

Combine Eqs. (13) and (14), one can study the relation between the contact forces distribution and θ_1, θ_2, as shown in Figs. 5 and 6.

Based on the Eq. (13), we can easily figure out that the JLST finger has more grasping force than the traditional spring-tendon underactuated hand because there is an extra friction force. From Figs. 5 and 6, we can find out once the θ_1 and θ_2 are increasing, the grasping forces will increase, which means the JLST hand can grasp heavy items.

4   Future Work

The next step is to actually make a prototype of JLST hand and do the experiment to get the actual data, then compare with the traditional spring-tendon robot hand to check if the grasping force is getting larger.

5   Conclusion

This paper proposes a novel spring-tendon self-adaptive underactuated hand, called JLST hand, which can perform simultaneous multi-joint locking grasp. The JLST hand has a more stable grasp ability and larger grasping force than the normal spring-tendon robot hand, because the JLST hand has simultaneous multi-joint locking mechanisms.

The JLST finger uses one motor to grasp objects, releasing objects and locking joints. Once the motor is turning forward, the finger releases the object; once the motor is turning backward, the motor releases the tendon so that the finger can begin to grasp; the blocker is going to lock the joint since the motor keeps turning backward. The calculation and simulation results show that the JLST hand has the high stability of grasp and more powerful than the normal spring-tendon hand.

Acknowledgement. This Research was supported by National Natural Science Foundation of China (No. 51575302).

References

1. Loucks, C.S.: Modeling and Control of the Stanford/JPL Hand. In: 1987 International Conference on Robotics and Automation, pp. 573–578 (1987)
2. Jacobsen, S.C., Iversen, E.K., Knutti, D.F., et al.: Design of the Utah/MIT Dextrous Hand. In: IEEE International Conference on Robotics and Automation, pp. 1520–1532 (1986)
3. Butterfass, J., Grebenstein, M., Liu, H., et al.: DLR-Hand II: next generation of a dextrous robot hand. In: IEEE International Conference on Robotics and Automation (ICRA), vol.1, pp. 109–114 (2001)
4. Dollar, A., Howe, D.: Joint coupling design of underactuated grippers. In: ASME Mechanical and Robotics Conference, 2006 International Design Engineering Technical Conference (IDETC), Philadelphia, July 2005, pp. 10–13 (2006)
5. Dubey, V., Crowder, M.: Grasping and control issues in adaptive end effectors. In: ASME Design Engineering Technical Conference and Computers and Information in Engineering Conference New York, May 2004, pp. 1–9 (2004)
6. Zhang, W., Chen, Q., Sun, Z., et al.: Under-actuated passive adaptive grasp humanoid robot hand with control of grasping force. In: IEEE International Conference on Robotics and Automation (ICRA), pp. 696–701 (2003)
7. Che, D., Zhang, W.: GCUA humanoid robotic hand with tendon mechanisms and its upper limb. Int. J. Soc. Robot. 3(1), 395–404 (2011)
8. Zhang, W., Tian, L., Liu, K.: Study on multi-finger under-actuated mechanism for TH-2 robotic hand. In: IASTED International Conference on Robotics and Applications, pp. 420–424 (2007)
9. Demers, L.A., Lefrancois, S., Jobin, J.: Gripper having a two degree of freedom underactuated mechanical finger for encompassing and pinch grasping. US Patent, US8973958 (2015)
10. Liang, D., Zhang, W., Sun, Z., et al.: PASA finger: a novel parallel and self-adaptive underactuated finger with pinching and enveloping grasp. In: IEEE International Conference on Robotics and Biomimetics. IEEE (2015)
11. Li, G., Liu, H., Zhang, W.: Development of multi-fingered robotic hand with coupled and directly self-adaptive grasp. Int J. Humanoid Robot. 9(4), 1–18 (2012)
12. Stark, M.: Artificial hand. US Patent, US7655051 (2010)
13. Shin, Y., Rew, K., Kim, R., et al.: Development of anthropomorphic robot hand with dual-mode twisting actuation and electromagnetic joint locking mechanism. In: IEEE International Confernce on Robotics and Automation (ICRA), pp. 2759–2764 (2013)
14. Yang, Y., Zhang, W.: Bending self-locking pneumatic under-actuated robot finger device. CN Patent, CN103659825

Path Analysis for the Halo Effect
of Touch Sensations of Robots
on Their Personality Impressions

Yuki Yamashita[1], Hisashi Ishihara[1(✉)], Takashi Ikeda[2], and Minoru Asada[1]

[1] Graduate School of Engineering, Osaka University, Suita, Japan
ishihara@ams.eng.osaka-u.ac.jp
[2] Research Center for Child Mental Development,
Kanazawa University, Kanazawa, Japan

Abstract. Physical human–robot interaction plays an important role in social robotics, and touch is one of the key factors that influences human's impression of robots. However, very few studies have explored different conditions, and therefore, few systematic results have been obtained. As the first step toward addressing this issue, we studied the types of impressions of robot personality that humans may experience when they touch a soft part of a robot. In the study, the left forearm of a child-like android robot "Affetto" was exposed; this forearm was made of silicone rubber and can be replaced with one of other three forearms providing different sensations of hardness upon touching. Participants were asked to touch the robot's forearm and to fill evaluation questionnaires on 19 touch sensations and 46 personality impressions under each of four conditions with different forearms. Four impression factors for touch sensations and three for personality impressions were extracted from the evaluation scores by the factor analysis method. The causal relationships between these factors were analyzed by the path analysis method. Several significant causal relationships were found, for example, between preferable touch sensations and likable personality impressions. The results will help design robots' personality impression by designing touch sensations more systematically.

1 Introduction

Social interaction depends on various non-social interactions similar to the manner in which verbal communication is supported by non-verbal communication [7]. In social robotics, the physical human–robot interaction typically supports the social interaction between humans and robots [1]. One of the important issues in design of communication robots is the selection of the covering material with appropriate touch sensations to improve the robots' personality impressions. The robots' appearance affects their personality impressions [2,3,12,16]. However, for humans, not only visual perception, but also other modalities may influence personality impressions. Volume and tone are the major components that inform the listener about the emotional state of the speaker [23]. With regard to tactile sensations, the

© Springer International Publishing AG 2016
A. Agah et al. (Eds.): ICSR 2016, LNAI 9979, pp. 502–512, 2016.
DOI: 10.1007/978-3-319-47437-3_49

touch of products affects their quality impressions and attractiveness. Therefore, it seems reasonable to consider that the touch of robots also influences their personality impressions. Several studies have attempted touch-based interactions between humans and robots [4–6, 26], and comfortable covering materials for robots were selected by designers for influencing mental states of humans through the sensation of touch [13–15, 18, 24]. However, the design of the touch sensation of robots to improve their intended personality impressions is a challenge because how and which touch sensation affect the personality impressions has not been systematically investigated.

Either touch sensations of robot skins [25] or personality impressions [19] of hugging dolls, or both touch sensations and personality impressions of robot hands [8] were evaluated in different conditions, but the manner in which touch sensations affect personality impressions has not been revealed. To systematically understand the causal relationships between touch sensations and personality impressions, we examined the perceived impressions of a robot by using questionnaires with a large number of questions. Figure 1 shows an overview of our research. Participants were instructed to touch and grab the forearm of a robot in a stationary state with their right hand and then answer the evaluation questionnaires on touch sensations (19 items) and personality impressions (46 items). A factor analysis of the evaluation scores was conducted to identify abstract impression factors for touch sensations and personality impressions. A path analysis was then conducted to reveal significant causal relationships from tactile impression factors to personality impression factors.

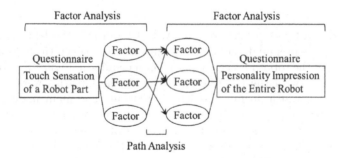

Fig. 1. Diagram of the causal relationship between touch sensations of a robot part and personality impression of the entire robot

2 Method

2.1 Participants

The participants were 20 healthy Japanese adults, including 10 male (mean age = 21.9, SD = 2.3) and 10 female (mean age = 22.9, SD = 1.2) individuals. Of these, 17 participants had no experience of contact with humanoid robots until the experiment, and 11 participants had no knowledge of humanoid robots. Two participants had contact with infants in the past 5 years.

2.2 Robot

A child-like android robot Affetto [11] was set on a desk in front of a chair as shown in Fig. 2(a). The robot had a head and upper body that were covered with a cloth or gloves; only its face and left forearm were exposed. The joints of the robot were physically fixed to maintain a posture with its left hand held out to the participants so that the joint movements did not affect its impressions. A partition was set so that the robot could be hidden and shown to the participants.

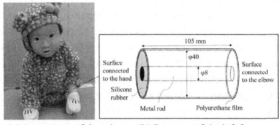

(a) Appearance of the robot   (b) Structure of the left forearm

Fig. 2. Appearance of the robot and structural overview of the left forearm

Four different types of left forearms A, B, C, and D with identical sizes and appearances were prepared, and one of them was set between the left elbow and the left hand of the robot for each experimental condition. Figure 2(b) shows the structural overview of the left forearm. Overall, the forearms were cylindrical with dimensions width 40 mm and length 105 mm. The main material of the forearm was platinum cure silicone rubber (Dragon Skin Fx-Pro, Smooth-On Inc.), and its center was supported by a metal rod of diameter 8 mm. The top surface of the silicone rubber was wrapped with a thin polyurethane film of thickness 7 μm (Airwall UV, Kyowa Ltd.) to ensure constant surface friction along the forearm.

Different amounts of two types of additives were mixed into the silicone rubbers to provide different touch sensations to the forearms. One was a plasticizer (Silicone Thinner,Smooth-On Inc.), which reduces hardness, and the other was a thickener (Slacker, Smooth-On Inc.), which increases viscosity while reducing hardness. The additive contents were adjusted so that hardness decreased from forearm A to forearm D, and so that forearms C and D had higher viscosity than forearms A and B. The combinations of amounts a and b of plasticizer a and thickner b were 10 % and 0 %, respectively, for the forearm A, 50 % and 0 %, respectively, for forearm B, 10 % and 30 %, respectively, for forearm C, and 20 % and 30 %, respectively, for forearm D. These percentages mean volume ratios of additives against the silicone rubber before adding them. As a reference, the hardness of forearms B, C, and D measured by a durometer (ASKER Durometer Type FP, Kobunshi Keiki Co., Ltd.) were similar to that of the center of the back of co-contracted forearm of males, the relaxed forearm of males (or the forceful forearm of females), and the relaxed forearm of females, respectively.

2.3 Questionnaire

The semantic differential method (SD method) [21] was used to measure the touch sensations and personality impressions. Two sets of touch sensation questionnaire (TSQ) and personality impression questionnaire (PIQ), each with lists of several pairs of opposite Japanese adjectives, were provided to the human participants. The participants were instructed to choose their responses on 7-point scales between these opposite adjectives, e.g. "Soft or Hard" for the TSQ and "Active or Passive" for the PIQ.

Table 1 summarizes the 19 adjective pairs in the TSQ. Most of these pairs were collected and selected from previous studies on touch sensations of artificial skins with eight adjective pairs [25] and those for robot hands with 10 pairs [8]. Table 2 summarizes the 46 adjective pairs in the PIQ. Most of these pairs were derived and selected from previous studies on personality impressions of robot hands with 12 pairs [8]. In addition to these studies, studies on the impressions of hug dolls with 12 pairs [19], quantification of impressions of humanoid robots with 33 pairs, and meta-analysis of SD adjective pairs for personality impressions [10] were also considered. Thus, our adjective pairs were prepared so that our questionnaires enabled us to investigate touch sensations and personality impressions thoroughly. These adjective pairs were translated into English by a professional translator for this paper.

2.4 Procedure

The experimental procedures were divided into three sessions. In the first one, the participants were instructed on the manner of evaluating the robot. In the second one, they practiced the instructed manner. In the third one, they evaluated the robot by touching its forearm and then answering the questionnaires.

Table 1. Adjective pairs in the touch sensation questionnaire (TSQ)

#	Adjective pair	#	Adjective pair	#	Adjective pair
1	Flabby/Supple	8	Light/Heavy	14	Coarse/Fine
2	Complex/Simple	9	Comfortable/Uncomfortable	15	Good/Bad
3	Dry/Moist	10	Tense/Relaxed	16	Pleasant/Unpleasant
4	Bad-feeling/Good-feeling	11	Blunt/Sharp	17	Elastic/Rigid
5	Soft/Hard	12	Slippery/Sticky	18	Smooth/Rough
6	Large/Small	13	Slim/Plump	19	Rounded/Angular
7	Desirable/Undesirable				

Table 2. Adjective pairs in the personality impression questionnaire (PIQ)

#	Adjective pair	#	Adjective pair
1	Amiable/Odious	24	Masculine/Feminine
2	Humanlike/Machinelike	25	Agreeable/Disagreeable
3	Tiresome/Endlessly entertaining	26	Safe/Dangerous
4	Active/Passive	27	Wise/Foolish
5	Earnest/Insincere	28	Good/Bad
6	Pain-sensitive/Pain-insensitive	29	Quiet/Noisy
7	Kind/Unkind	30	Friendly/Unfriendly
8	Amusing/Boring	31	Merry/Objectionable
9	Lively/Unlively	32	Mild-mannered/Strict
10	Talkative/Reticent	33	Jovial/Gloomy
11	Soothing/Not soothing	34	Convivial/Stiff-mannered
12	Vigorous/Lifeless	35	Extroverted/Introverted
13	Considerable/Self-centered	36	Robust/Feeble
14	Reassuring/Unnerving	37	Laid-back/Busy
15	Young/Old	38	Approachable/Unapproachable
16	Reliable/Unreliable	39	Spritely/Fatigued
17	Bright/Dismal	40	Comfortable/Uncomfortable
18	Docile/Obstinate	41	Clean/Dirty
19	Pleasant/Unpleasant	42	Adorable/Weired
20	Brave/Cowardly	43	Sturdy/Fragile
21	Calm/Restless	44	Neat/Slovenly
22	Desirable/Undesirable	45	Confident/Timid
23	Warmhearted/Cold-hearted	46	Strong/Weak

In the first instruction session, an instruction movie describing the manner of touching the robot and two questionnaires (TSQ and PIQ) were shown to the participants. The movie showed a demonstrator pinching one of the forearms of the robot with his thumb and forefinger, touching it with the pads of his fingers, and holding it with his hand several times. Participants were told to look at the forearm and touch it with their dominant hand as shown in the movie. In the second session, the participants were told to touch their own forearm with their dominant hand and answer the TSQ and PIQ as a training. This section was conducted to check if the participants understood the instructions regarding the manner of touching the forearm and to get to know the questionnaires. The third evaluation session was divided into four subsessions in which the participants evaluated one of the four forearms attached to the robot and the entire robot. In each subsession, the participants were instructed to touch the forearm of the robot for arbitrary time durations and then answer the TSQ. After answering it, they were told to touch it again and then answer the PIQ. This subsession took approximately 10 min. Between subsessions, the robot was hidden from the participants by the partition, and its attached forearm was changed; then, the robot was shown to the participants again. This modification was completed in 1 min. The orders for showing each forearm and the orders of adjective pairs in the questionnaires were shuffled for each participant. The series of these sessions were completed in an hour.

2.5 Data Processing

A exploratory factor analysis was conducted to identify several underlying factors, each of which was statistically reflected by several observed variables or evaluation scores of the adjective pairs. The maximum-likelihood and varimax rotation were chosen for the factor extraction and for the rotation method, respectively. The number of factors chosen based on Scree test was investigated by Bayesian information criterion and the root mean square error of approximation.

Path analysis was conducted to find significant causal relationships between the found factors. Here, we assumed the multivariate multiple regression model whose independent variables were the touch sensation factors, while the dependent variables were the personality impression factors. The maximum-likelihood estimation was used for estimating the model parameters. Version 3.2.2 of R was used for the above analyses.

3 Result

3.1 Factor Analysis

To detect the model structure, the number of dimensions had to be reduced. We conducted factor analyses of each questionnaire by using the number of factors determined by inspecting the Scree plot of eigenvalues. Four touch feeling factors (BIC = -292.2, RMSEA = 0.094) and three personality impression factors (BIC = -2774.5, RMSEA = 0.009) were extracted from evaluation scores of each questionnaire. Tables 3 and 4 list the factor matrices for touch sensation factors and personality impression factors, respectively. Based on the adjectives with high loadings for each factor, we named factor 1 as "Preferable," 2 as "Resilient," 3 as "Smooth," and 4 as "Natural" for touch sensations. We named factor 1 as "Likable," 2 as "Mighty," and 3 as "Vital" for personality impressions.

3.2 Path Analysis

We conducted the path analysis to investigate how tactile feelings affect impressions of a humanoid robot. Seven variables considered were the factor scores calculated from the factor analyses. Satisfactory goodness-of-fit index (RMSEA < 0.001) for a full model including all of the possible paths was achieved.

Table 3. Factor matrix for touch sensation factors; loadings higher than an absolute value of 0.50 are shown in parentheses and those lower than an absolute value of 0.30 are extracted.

Adjective	Factor			
	1	2	3	4
	Preferable	Resilient	Smooth	Natural
Good-feeling	(.97)			
Pleasant	(.97)			
Desirable	(.85)			
Good	(.79)			
Comfortable	.41			
Slippery	.32			
Supple		(.91)		
Tense		(.88)		
Rigid		(.76)		
Hard		(.69)	-.33	
Heavy	-.32	.44		
Large		.32		
Smooth			(.85)	
Fine			(.75)	
Simple			.42	
Rounded			.31	(.59)
Moist				(.52)
Blunt		-.30		(.52)
Plump				.38
Accumulated variance (%)	36	67	86	100

Figure 3 shows the path diagram with standardized partial regression coefficient between touch sensation factors and personality impression factors. Path thickness represents the magnitude of the coefficient. Solid lines and dotted lines represent positive relationships and negative relationships, respectively.

We found that the likable personality was highly and positively affected by preferable touch sensation ($\beta = 0.877$); mighty personality was positively influenced by resilient touch sensation ($\beta = 0.665$); and vital personality was positively affected by preferable and resilient touch sensations ($\beta = 0.385$ and 0.311). Additionally, smooth touch sensation affected mighty personality weakly and negatively ($\beta = -0.200$); natural touch sensation affected all three personality factors weakly ($\beta = 0.084$, -0.276, and 0.191).

Table 4. Factor matrix for personality impression factors; loadings higher than an absolute value of 0.50 are shown in parentheses and those lower than an absolute value of 0.30 are extracted.

Adjective	Factor			Adjective	Factor		
	1	2	3		1	2	3
	Likable	Mighty	Vital		Likable	Mighty	Vital
Desirable	(.99)			Masculine		(.83)	
Agreeable	(.93)			Brave		(.79)	.33
Pleasant	(.90)			Confident		(.78)	.33
Comfortable	(.90)			Neat	.39	(.66)	
Good	(.90)			Obstinate	-.30	(.66)	
Friendly	(.79)			Busy		(.65)	
Soothing	(.78)			Strict	-.31	(.64)	
Approachable	(.75)			Extrovert		(.56)	(.60)
Merry	(.73)			Stiff-mannered		(.59)	
Adorable	(.73)			Pain-sensitive		(.51)	-.41
Endlessly entertaining	(.65)			Talkative			(.82)
Reassuring	(.64)			Noisy			(.81)
Amiable	(.63)	-.33		Jovial			(.79)
Humanlike	(.61)			Bright	.31		(.79)
Clean	(.52)			Active		.48	(.66)
Considerable	(.52)			Lively		.32	(.66)
Robust		(.92)		Spritely	.32		(.65)
Strong		(.91)		Restless	-.49		(.53)
Reliable		(.87)		Accumulated Var	44	76	99

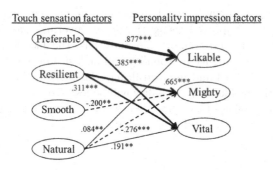

Fig. 3. Path diagram with standardized coefficients representing the relationships between touch sensation factors and personality impression factors. The insignificant paths were removed. ** $p < 0.01$, *** $p < 0.001$

4   Discussion

Several significant causal relationships between touch sensation factors and personality impression factors were identified, in accordance with our hypothesis. In other words, even if a robot has a fixed appearance, its personality impression can be modified by touch sensation. Thus, this study systematically revealed how different touch sensation factors affect different personality impression factors.

First, we found that the likable personality was improved if the robot provided preferable and natural touch sensations to the participants. This supports the conventional idea of covering communication robots with good-feeling and lifelike materials such as fur [13,15,24] and soft silicone rubber [8,17]. In particular, preferable touch sensations strongly affect the likable personality and this emphasizes the importance of designing preferable touch feelings to build likable robots. We consider that the strong causal relationship between preferable touch sensation and likable personality is the result of a type of halo effect, which is known as a human cognitive bias, and is defined as the influence of a global evaluation on the evaluations of individual attributes of a person [20]. Likable personality has been considered as one of the important properties for communication robots, and therefore, several studies have attempted to improve robots' likability by modifying their appearances and behaviors [9,22]. The application of the halo effect by touching robots will be another effective design technique to improve the likability of robots.

Second, the mighty personality was mainly enhanced by imparting resilient touch sensation to the participants. On the other hand, the mighty personality was reduced by smooth and natural touch impressions. This suggests that supple, tight, rigid, and hard covering materials with rough, coarse, and dry surfaces with sharp and squared shapes are desirable to impart the mighty personality to robots. In this perspective, hard covering materials, such as metal, plastic, and hard rubber, are considered to be desirable to enhance the mighty personality of robots.

Here, we encounter a new research question: whether both likable and mighty personalities can be improved using the same covering materials. When we cover robots with soft and comfortable materials, the likable personality will be improved, while the mighty personality will be lost. On the other hand, when we cover robots with hard and unnatural materials, the mighty personality will be improved, while the likable personality will be lost. To improve both personalities, covering materials that can provide preferable and resilient touch sensation simultaneously to humans should be identified.

5   Conclusion

In this study, a factor analysis and path analysis were conducted on the evaluation scores of a robot's touch sensations and personality impressions by using SD questionnaires to reveal how the touch sensations affect the personality impressions such as in the halo effect. Several significant causal relationships between touch sensations and personality impressions were found. The results may help design robots' personality impression by designing touch sensations more systematically. Further studies are required to reveal the types of covering materials that provide preferable touch sensations to humans with several types of robots, including non-android type robots.

Acknowledgment. This research is supported by a Grant-in-Aid for Specially Promoted Research No. 24000012 and for Young Scientist (B) No. 15K18006 and by the Center of Innovation Program from MEXT and JST.

References

1. Argall, B.D., Billard, A.G.: A survey of tactile human-robot interactions. Robot. Auton. Syst. **58**(10), 1159–1176 (2010)
2. Bartneck, C., Kanda, T., Ishiguro, H., Hagita, N.: Is the uncanny valley an uncanny cliff? In: The 16th IEEE International Symposium on Robot and Human Interactive Communication, pp. 368–373 (2007)
3. Castro-Gonzlez, A., Admoni, H., Scassellati, B.: Effects of form and motion on judgments of social robots' animacy, likability, trustworthiness and unpleasantness. Int. J. Hum. Comput. Stud. **90**, 27–38 (2016)
4. Cooney, M.D., Nishio, S., Ishiguro, H.: Importance of touch for conveying affection in a multimodal interaction with a small humanoid robot. Int. J. Humanoid Robot. **12**(01), 1550002 (2015)
5. Cramer, B.H., Kemper, N., Amin, A., Wielinga, B., Evers, V.: 'Give me a hug': the effects of touch and autonomy on people's responses to embodied social agents. Comput. Animation Virtual Worlds **20**, 437–445 (2009)
6. Cramer, H., Kemper, N.A., Amin, A., Evers, V.: The effects of robot touch and proactive behaviour on perceptions of human-robot interactions. In: International Conference on Human Robot Interaction, pp. 275–276 (2009)
7. Ekman, P.: Communication through nonverbal behavior: a source of information about an interpersonal relationship. In: Affect, Cognition and Personality, pp. 390–442 (1965)
8. Endo, N., Iida, F., Endo, K., Mizoguchi, Y., Zecca, M., Takanishi, A.: Development of the anthropomorphic soft robotic hand WSH-1R. In: Proceedings of the First IFToMM Asian Conference on Mechanism and Machine Science, p. 250162 (2010)
9. Goetz, J., Kiesler, S., Powers, A.: Matching robot appearance and behavior to tasks to improve human-robot cooperation. In: Proceedings - IEEE International Workshop on Robot and Human Interactive Communication, pp. 55–60 (2003)
10. Inoue, M., Kobayashi, T.: The research domain and scale construction of adjective-pairs in a semantic differential method in Japan. Japan. J. Educ. Psychol. **33**(3), 253–260 (1985)
11. Ishihara, H., Asada, M.: Design of 22-DOF pneumatically actuated upper body for child android 'Affetto'. Adv. Robot. **29**(18), 1151–1163 (2015)
12. Kanda, T., Miyashita, T., Osada, T., Haikawa, Y., Ishiguro, H.: Analysis of humanoid appearances in human-robot interaction. IEEE Trans. Robot. **24**(3), 725–735 (2008)
13. Kanoh, M., Shimizu, T.: Developing a robot Babyloid that cannot do anything. J. Robot. Soc. Japan (In Japanese) **29**(3), 298–305 (2011)
14. Kozima, H., Michalowski, M.P., Nakagawa, C.: Keepon: a playful robot for research, therapy, and entertainment. Int. J. Soc. Robot. **1**(1), 3–18 (2008)
15. Lee, J.K., Stiehl, W.D., Toscano, R.L., Breazeal, C.: Semi-autonomous robot Avatar as a medium for family communication and education. Adv. Robot. **23**(14), 1925–1949 (2009)
16. Macdorman, K.F.: Subjective ratings of robot video clips for human likeness, familiarity, and eeriness: an exploration of the uncanny valley. In: ICCS/CogSci-2006 Long Symposium: Toward Social Mechanisms of Android Science (2006)

17. Minato, T., Yoshikawa, Y., Noda, T., Ikemoto, S., Ishiguro, H., Asada, M.: CB2: a child robot with biomimetic body for cognitive developmental robotics. In: Proceedings of International Conference on Humanoid Robots, pp. 557–562 (2007)

18. Minato, T., Nishio, S., Ishiguro, H.: Evoking affection for a communication partner by a robotic communication medium. In: HRI Demonstration session, p. D07 (2013)

19. Mori, Y., Saito, Y., Kamide, H.: Evaluation of impression for hug dolls. J. Japan Soc. Kansei Eng. **11**(1), 9–15 (2012)

20. Nisbett, R.E., Wilson, T.D.: The halo effect: evidence for unconscious alteration of judgments. J. Feisonality Soc. Psychol. **35**(4), 250–256 (1977)

21. Osgood, C.E.: The nature and measurement of meaning. Psychol. Bull. **49**(3), 197–237 (1952)

22. Poel, M., Heylen, D., Nijholt, A., Meulemans, M., Breemen, A.: Gaze behaviour, believability, likability and the iCat. AI Soc. **24**(1), 61–73 (2009)

23. Scherer, K.: Vocal communication of emotion: a review of research paradigms. Speech Commun. **40**(1–2), 227–256 (2003)

24. Shibata, T., Wada, K.: Robot therapy: a new approach for mental healthcare of the elderly - a mini-review. Gerontology **57**(4), 378–386 (2011)

25. Shirado, H., Nonomura, Y., Maeno, T.: Development of artificial skin having human skin-like texture. Trans. Japan Soc. Mech. Eng. Ser. C **73**(726), 541–546 (2007)

26. Yohanan, S., MacLean, K.E.: The role of affective touch in human-robot interaction: human intent and expectations in touching the haptic creature. Int. J. Soc. Robot. **4**(2), 163–180 (2012)

A Human-Robot Speech Interface
for an Autonomous Marine Teammate

Michael Novitzky$^{(\boxtimes)}$, Hugh R.R. Dougherty, and Michael R. Benjamin

Mechanical and Ocean Engineering, Massachusetts Institute of Technology,
Cambridge, MA 02139, USA
{novitzky,hughd,mikerb}@mit.edu
http://oceanai.mit.edu/pavlab/

Abstract. There is current interest in creating human-robot teams in
which a human operator is in its own conveyance teaming up with several
autonomous teammates. In this work we focus on human-robot teamwork in the marine environment as it is challenging and can serve as a
surrogate for other environments. Marine elements such as wind speed,
air temperature, water, obstacles, and ambient noise can have drastic
implications for team performance. Our goal is to create a human-robot
system that can join many humans and many robots together to cooperatively perform tasks in such challenging environments. In this paper,
we present our human-robot speech dialog system and compare participant responses to having human versus autonomous vehicle teammates
escorting and holding station at locations of interest.

Keyword: Autonomous teammate marine speech recognition dialog

1 Introduction

There is great interest in combining manned vehicles with autonomous vehicles
to perform tasks in challenging environments. For example, the U.S. Army has
the Manned Unmanned Teaming (MUM-T) program in which manned aircraft
work with unmanned aerial systems (UAS) [10]. The manned aircraft are AH-
64E Apache attack helicopters which are aided by Grey Eagle UAS. The crews
are able to request various levels of control of the UAS from simply receiving its
camera data to ordering waypoints for the UAS to visit. The U.S. Air Force's
"Loyal Wingman" project is exploring manned-unmanned teaming in which an
UAS and a manned aircraft work directly on missions such as air interdiction,
attack on integrated air defense systems, and offensive counter air [1,6]. In this
work we explore a similar manned unmanned teaming concept in the marine
domain. The marine domain is more accessible for deploying autonomous vessels
and yet still is challenging given the elements in the environment. Our manned
vessel is a motorized kayak and our autonomous teammate is a surface vehicle
(ASV). In particular we focus on preliminary user-centered design improvements

This work was supported by Battelle.

A. Agah et al. (Eds.): ICSR 2016, LNAI 9979, pp. 513–520, 2016.
DOI: 10.1007/978-3-319-47437-3_50

of our speech dialogue system in order to perform tasks of increasing complexity. In future experiments we wish to expand to coordinating with several teamme- bers while also recording physiological data of our participants. This will allow for the exploration of the interplay between operator load, robot autonomy, and human-robot trust. It is our goal to provide lessons learned from our platform in the marine domain to other challenging environments. Our design prinicpal is for novice users to begin using the system in a short period of time. This paper describes our manned unmanned teaming system and initial pilot study aimed at comparing human vs autonomous vehicle teammates.

2   Related Work

Uhrmann et al. [9] investigated the Manned-Unmanned Teaming (MUM-T) domain. The researchers simulated a full-scale military helicopter mission with the introduction of UAVs for route reconnaissance and for observing the desig- nated landing sites prior to the approach of the manned helicopter. The mission was to have troops transported via manned helicopter to secure an object. Both manned and unmanned assets provided reconnaissance information and over- watch after troop delivery. Artificial Cognitive Units (ACUs) were implemented to interpret tasks with respect to the current mission and tactical situation and act upon those tasks in a situation-specific way. The goal of the ACUs was to allow a commander to communicate the task to the UAV and it take care of all the parameters just like a human pilot does. The human-robot interface varied depending on the focus of the task such as maps for spatial representations of a task and timetables or schedules for temporal representations. Speech inter- faces were utilized when task representation involved a causal component, with previous or following tasks refereed in speech output or commands.

Draper et al. [3] investigated using speech input versus manual input for an unmanned aerial vehicle control station. The control station was designed to operate one vehicle at a time with multiple monitors and manual controls. They found that speech input was superior to manual input for flight/navigation and data entry tasks. Operators in this study indicated that speech provided a head up and hands-free advantage.

Franke et al. [4] describe systems for command and control for a single human operator to many autonomous vehicles. The authors note that using auditory cues and speech frees the operator's eyes and hands to observe other information and manipulate other tasks. They describe three primary control paradigms: direct control, management by consent, and management by exception. Direct control is when a human does all the decision making and information processing. This approach has a high workload as it requires the operator to constantly attend the controls. Management by consent has a lower operator workload as the vehicle performs planning but waits for the operator to approve it before proceeding. In management by exception the vehicles performs its own planning and starts executing the plan. In this case, the operator can override any actions or plans of the vehicle.

The work presented in this paper introduces a manned unmanned teaming mission using speech recognition for command and control. The autonomy in the following experiments is higher than the direct control paradigm because it takes high level commands such as "Follow" instead of direct joystick inputs. While the following experiments only have one teammate, future work will expand to having multiple teammates that recognize speech commands.

3 Experimental Setup

The experiments were conducted on the Charles River in Cambridge, Massachusetts where participants lead a teammate named Arnold to points of interest on the river. After each trial, the participants filled out a questionnaire. The questionnaire included the NASA Task Load Index (TLX) [5], Robot Liking, General Engagement, and Schaefer Trust Scale Items [8]. The TLX questions were rated on a likert scale between 1 and 7, very low load to very high load, respectively. The Robot Liking items were rated on a likert scale ranging between 1 and 10 where 1 was very poorly and 10 describes very well. The General Engagement items were on a likert scale between 1 and 7 where 1 was not at all enganged to 7 which was extremely enganged. The Schaefer Trust Scale items asked "what percentage of time will this teammate be ____". Participants responded on a likert scale between 0 % and 100 % with 10 % increments in between. The questionnaire included open ended questions about general experience and suggestions for improvements for user-centered design. The final portion of the questionnaire recorded demographic information. The 15 participants had a mean age of 29.5 years, are comfortable on the water, and had a comfort with robotics ranging from 2 to 6 on a 7 point likert scale.

3.1 Task Description

The human participant is asked to escort its teammate from the dock to two points of interest marked with buoys on the water, as seen in Fig. 1. They are instructed to have their teammate station as close as possible to each point of interest. Once all the points of interest have been visited they return back to the starting location at the dock.

3.2 Participant Conveyance

The vehicle for human conveyance is an augmented Mokai ES-Kape which is a motorized kayak, seen in Fig. 2. The ES-Kape weighs 88.45 kg, has a length of 3.63 m, and is powered by a Subaru EX21 engine that can reach top speeds of 54.39 km/h. In order to function as a vehicle on our network, the ES-Kape has been augmented with a semi-rugged laptop, compass, GPS, and long range wifi antenna.

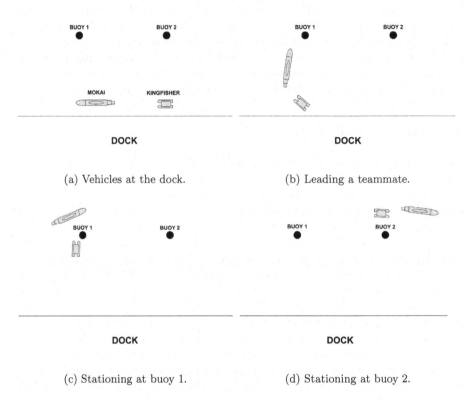

(a) Vehicles at the dock.   (b) Leading a teammate.

(c) Stationing at buoy 1.   (d) Stationing at buoy 2.

Fig. 1. Example experiment (not to scale). The participant and teammate start at the dock, as seen in (a). The participant is asked to lead their teammate (a Kingfisher ASV in this example), as seen in (b), with the command "Arnold Follow" to a buoy where they use the command "Arnold Station" to have their teammate station as close as possible, as seen in (c). The participant will then escort their teammate to the second buoy, as seen in (d), with the same two commands. After visiting both buoys, the participant is asked to have Arnold return to the starting location at the dock. The participant can use the command "Arnold Follow" and lead their teammate to the dock or "Arnold Return" in which case the teammate will autonomously guide itself to the dock.

3.3   Autonomous Teammate

The autonomous teammate is a Clearpath Robotics Kingfisher M200, seen in Fig. 2, which is a surface vehicle. The ASV is a mid-sized surface vessel with dimensions of 1.35 m × 0.98 m × 0.32 m and a weight of 28 kg and operates at 1.5 m/s. The autonomy for the ASV is provided by MOOS-IvP [2]. MOOS is a robot middleware that utilizes a centralized database paradigm. The autonomy is provided by the IvP Helm behavior-based decision engine architecture. The IvP Helm behaviors used in this work are trail, station, and waypoint.

Fig. 2. Human operator in the motorized kayak, Mokai ES-Kape, and the autonomous robot teammate Arnold.

3.4 Human Teammate

As a control to the autonomous teammate, an alternate teammate is a human that operates a motorboat. The human teammate mimics the robot by traveling at 1.5 m/s and uses the exact same dialogue. The motorboat is augmented with a semi-rugged laptop, compass, GPS, and long range wifi antenna.

3.5 Speech Interaction

A dual radio headset with dual push-to-talk (PTT) is used to mitigate the effects of wind and motor noise from the ES-Kape. The right speaker/PTT combination is connected to a 5 W waterproof handheld radio which is used to communicate with humans. The left speaker/PTT combination is connected to a semi-rugged laptop which runs the speech dialog MOOS-IvP modules that communicate with the autonomous robot teammate.

The speech recognition used in this project is provided by the open-source large vocabulary continuous speech recognition engine Julius [7]. The engine allows for the specification of possible sentences and vocabulary to be recognized. The Julius engine has been encapsulated into the MOOS-IvP application called *uSpeechRec*.

A dialog manager was created called *uDialogManager*. Each command sentence recognized by *uSpeechRec* is acknowledged by asking the user "Did you mean, <command>?", where the <command>echos what the system believed to be uttered by the user. Possible commands are Follow, Station, Return and Status. The user can answer "No" in which case *uDialogManager* does nothing and responds with "Command Canceled" or the user can answer "Yes" in which case *uDialogManager* sends the appropriate command to the autonomous robot teammate and responds to the user with "Command Sent." The acknowledgement loop reduces error as the accuracy in speech recognition can be affected by wind, ambient noise, or user accents. If the teammate receives the command it will respond to the user with a text to speech message.

4   Experimental Protocol

We performed a within-subjects experiment in which the teammate is either an autonomous robot or a human (randomized order). Due to the experiments occuring on the water, each participant begins the experiment with a safety briefing and is briefed on how to operate the ES-Kape. They acclimate to the ES-Kape by driving it around a buoy and returning to the dock. The participant is then given a headset and is briefed on how to use the handheld radio to communicate with humans and the PTT for communicating with the autonomous teammate. Once the orientations are completed the participant is briefed on their task, which was described above in Sect. 3.1. After each treatment (where the teammate is human or autonomous) the participant is asked to complete a questionnaire. After the participant completes the two treatments and questionnaires, they are interviewed by an experimenter on their experiences.

Table 1. Questions and Wilcoxon signed-rank test

Question	Median human	Median robot	W	p value
How successful were you in accomplishing what you were asked to do?	6	5	34	0.0312
The experience caused real feelings and emotions for me	2	3	9	0.0371
Reliable	80	70	47	0.0488
Led astray by unexpected changes in the environment	20	40	8	0.0234
Protect people	50	40	32.5	0.0469
Malfunction	10	30	17	0.0464
Warn people of potential risks in the environment	70	30	57	0.0312
Provide appropriate information	80	80	66	0.0371
Make sensible decisions	70	60	65.5	0.0405
A good teammate*	80	70	70.5	0.0088
Performs task better than a novice human user*	70	40	103	0.0004
Possess adequate decision-making capability*	50	40	72	0.0078
Meets the needs of the mission*	90	60	76	0.002

*p < 0.01

5   Results

Wilcoxon signed-rank test was performed on 59 likert scale responses. The questions with p values less than 0.05 are listed in Table 1. In Table 1 the column labeled W are the Wilcoxon signed-rank test values. In the TLX section, question "How successful were you in accomplishing what you were asked to do?" had a p value of 0.0312. In the General Engagement section of the questionnaire the question "The experience caused real feelings and emotions for me" had a p value of 0.0371. The rest of the questions listed in Table 1 were from the Schaefer Trust Scale Items. The questions "A good teammate", "Performs task better than a novice human user", "Possess adequate decision-making capability", and "Meets the needs of the mission" had p values below 0.01.

6   Conclusions and Future Work

The experiments in this pilot study demonstrate our systems' initial capability for a human-robot team to perform tasks in a challenging environment. Overall, it seems that human teammates were viewed in a more favorable light than their robot counterparts. Even though their dialog was designed to be the same. Future work will include recruiting a larger number of participants along with gathering physiological data to characterize cognitive load. Additionally, analysis of quantitative data such as the task duration or distance teammates were stationed from points of interest may reveal differences in performance between teammates or improvements in the system. Further user-centered design iterations and increasing the team size will aid in finding lessons learned for application to other challenging environments.

Acknowledgments. The authors would like to acknowledge Alon Yaari for his technical support and the MIT Sailing Pavilion staff for their professionalism and support in designing and maintaining our safety protocols. We thank Paul Robinette and Alan Wagner for feedback and guidance throughout this work.

References

1. United States Air Force unmanned aircraft systems flight plan 2009–2047. Headquarters, U.S. Air Force, p. 34, May 2009
2. Benjamin, M.R., Schmidt, H., Newman, P., Leonard, J.J.: Nested autonomy for unmanned marine vehicles with MOOS-IvP. J. Field Robot. **27**(6), 834–875 (2010)
3. Draper, M., Calhoun, G., Ruff, H., Williamson, D., Barry, T.: Manual versus speech input for unmanned aerial vehicle control station operations. In: Proceedings of the Human Factors and Ergonomics Society Annual Meeting, vol. 47, pp. 109–113. SAGE Publications (2003)
4. Franke, J.L., Zaychik, V., Spura, T.M., Alves, E.E.: Inverting the operator/vehicle ratio: approaches to next generation UAV command and control. In: Proceedings of AUVSI Unmanned Systems North America 2005 (2005)

5. Hart, S.G., Staveland, L.E.: Development of NASA-TLX (task load index): results of empirical and theoretical research. Adv. Psychol. **52**, 139–183 (1988)
6. Kearns, K.: RFI: autonomy for loyal wingman. Air Force Research Laboratory (AFRL), July 2015
7. Lee, A., Kawahara, T.: Recent development of open-source speech recognition engine Julius. In: Proceedings: APSIPA ASC 2009: Asia-Pacific Signal and Information Processing Association, 2009 Annual Summit and Conference, pp. 131–137. Asia-Pacific Signal and Information Processing Association, International Organizing Committee (2009)
8. Schaefer, K.E.: The Perception and measuerment of human-robot trust. Ph.D. thesis, University of Central Florida, August 2013
9. Uhrmann, J., Strenzke, R., Schulte, A.: Task-based guidance of multiple detached unmanned sensor platforms in military helicopter operations. In: COGIS, Crawley (2010)
10. Whittle, R.: MUM-T is the word for AH-64E: Helos fly, use drones. Breaking Defense, January 2015

Annotation of Utterances for Conversational Nonverbal Behaviors

Allison Funkhouser[✉] and Reid Simmons

Robotics Institute, Carnegie Mellon University, Pittsburgh, USA
afunkhous@gmail.com, reids@cs.cmu.edu

Abstract. Nonverbal behaviors play an important role in communication for both humans and social robots. However, adding contextually appropriate animations by hand is time consuming and does not scale well. Previous researchers have developed automated systems for inserting animations based on utterance text, yet these systems lack human understanding of social context and are still being improved. This work proposes a middle ground where untrained human workers label semantic information, which is input to an automatic system to produce appropriate gestures. To test this approach, untrained workers from Mechanical Turk labeled semantic information, specifically emotion and emphasis, for each utterance, which was used to automatically add animations. Videos of a robot performing the animated dialogue were rated by a second set of participants. Results showed untrained workers are capable of providing reasonable labeling of semantic information and that emotional expressions derived from the labels were rated more highly than control videos. More study is needed to determine the effects of emphasis labels.

1 Introduction

Nonverbal behaviors are an important part of communication for both humans and social robots. Gestures and expressions have the ability to convey engagement, to clarify meaning, and to highlight important information. Thus the animation of nonverbal behaviors is an important part of creating engaging interactions with social robots. Yet adding contextually appropriate animations by hand is time consuming and does not scale well as the number of utterances grows larger.

One alternative is to create rule-based software that assigns animations automatically based on the text of the dialogue. Such pipelines have the benefit that, once implemented, much left effort is needed in order to add new utterances to a database of dialogue. Examples of such automated systems include the Behavior Expression Animation Toolkit [1], the Autonomous Speaker Agent [2], and the automatic generator described in [3]. These systems use lexical analysis to determine parts-of-speech, phrase boundaries, word newness, and keyword recognition. This information is then used to place gestures such as head rotations, hand movements, and eyebrow raises.

However, these automated pipelines also have drawbacks. Because there is no longer a human in the loop, the entire system depends only on the information that can be automatically extracted from raw text. While there have been great strides forward in natural language understanding, there is still progress to be made. Specifically,

© Springer International Publishing AG 2016
A. Agah et al. (Eds.): ICSR 2016, LNAI 9979, pp. 521–530, 2016.
DOI: 10.1007/978-3-319-47437-3_51

classification of emotions based on text is a difficult [4], and current methods would constrain the number of emotions classified and thus limit the robot's expressivity. Also, determining the placement of emphasis gestures currently relies on word newness – whether a word or one of its synonyms was present in previous utterances in the same conversation. The complexity of language and speakers' reliance on common ground [5] create situations where implied information is not necessarily explicitly stated in previous sentences, which makes this form of emphasis selection less robust.

This work considers a potential middle ground between hand tuning animation and an automated pipeline with no humans involved. Instead, an annotator could add labels specifying particular semantic information, such as emphasis and emotion, which would then be input to an automatic system to produce appropriate gestures. This strategy allows the relevant human-identifiable context of a scenario to be preserved without requiring workers to have deep expertise of the intricacies of nonverbal behavior.

In order to test this labeling strategy, untrained workers from Amazon Mechanical Turk read small segments of conversations and answered questions about the semantic context of a particular line of dialogue. This semantic information was input to an automated system which added animations to the utterance. Videos of a Furhat robot performing the dialogue with animations were rated by the phase two participants. Results showed that untrained workers were capable of providing reasonable labeling of semantic information. When these labels were used to select animations, the selected emotive expressions were rated as more natural and anthropomorphic than control groups. More study is needed to determine the effect of the labeled emphasis gestures on perception of robot performance.

2   Related Works

Existing systems for streamlining the animation process can be divided into three categories: rule based pipelines using lexical analysis, statistics based pipelines that draw on videos of human behavior, and markup languages using tags from expert users. Examples of each of these strategies are discussed below.

The Behavior Expression Animation Toolkit [1] generates XML style tags that mark the clause boundaries, theme and rheme, word newness, and pairs of antonyms. These tags are used to suggest nonverbal behaviors including hand and arm motions, gaze behaviors, and eyebrow movements. Beat gestures are suggested for new words occurring in the rheme, the part of the clause presenting new information. Iconic gestures are inserted when an action verb in the sentence rheme matches a keyword for an animation, such as an animation that mimes typing on a keyboard corresponding to the word *typing*. Contrast gestures mark the distinction between pairs of antonyms. Robot gaze behaviors are based on general turn-taking patterns. Finally, a conflict resolution filter processes the suggested nonverbal behaviors, identifies conflicts where simultaneous gestures use the same degrees of freedom, and removes the lower priority gestures in these conflicts.

The Autonomous Speaker Agent [2] uses a phrase tagger to determine morphological, syntactic, and part-of-speech information about the given text. Similar to the

Behavior Expression Animation Toolkit, the Autonomous Speaker Agent also records word newness based on previously mentioned words in a given utterance. This lexical data is used to assign head movements, eyebrow raising, eyes movements, and blinks for a virtual character through the use of a statistical model of facial gestures. To build this statistical model, videos depicting Swedish newscasters were hand labeled with blinks, brow raises, and head movements.

In the text-to-gesture system described in [6], the hand gestures of speakers on TV talk shows were manually labeled as belonging to one of six side views and one of five top views. A morphological analyzer was used to label the parts of speech for the words in the spoken Korean utterances, and these labels were correlated to speaker gestures. Specifically, certain combinations of content words and function words were indicative of either deictic, illustrative, or emphasizing gestures. This mapping data was used to select movements from a library of learned gestures.

The Behavior Markup Language [7] is an XML style language that allows specific behaviors for virtual characters to be defined and synchronized with text. Behavior elements include movements of the head (such as nodding, shaking, tossing, and orientation), movements of the face, (including eyebrows, mouth, and eyelids), and arm and hand movements, to name a few. Because the original design was for virtual characters with humanlike appearances, it assumes that the character's possible motions include these humanoid style degrees of freedom. A robot that lacked these degrees of freedom – such as not being capable of certain facial movements, head movements, or arm motions – would not have a way of realizing all possible labeled motions. Furthermore, a nonhumanoid robot could potential have many other degrees of freedom not covered by this humanoid-centric markup. Using the Behavior Markup Language to command such a robot would lead to these potentially expressive motions not being used. The low level nature of the highly specific action commands makes this markup language less suitable for use across a wide variety of diverse robot platforms.

3   Approach

The goal of this work is to explore streamlining the robot animation process by having untrained workers label specific semantic information for each utterance, which is then used to determine appropriate nonverbal behaviors. Like the automated pipelines, this approach helps reduces the amount of human labor required to add animations when compared to animating each utterance by hand. The in-depth, low level knowledge of animation and the precise timing and types of gestures can be handled by an automatic pipeline. This speeds up the human work by reducing the task to merely labeling sentences, as opposed to meticulously tuning each set of degrees of freedom. However, because the labeling process still involves human input, it still allows for some of the subtleties gained from a human knowledge of interactions that is present in hand done animations.

While there are many possible pieces of information that annotators could conceivably mark, in this work we limit the scope to emphasis location – the word in the sentence that receives the most verbal stress – and dominant emotion. The envisioned implementation uses an XML tagging format, shown in the example below. The XML format is easily extensible and could potentially be combined with existing or future automated pipelines, such as the Speech Synthesis Markup Language.

Raw Text: Oh really? I didn't know that.
Annotated Text: <emotion=surprised> Oh really? </emotion> <emotion=embar-rassed> I didn't <emphasis> know </emphasis> that. </emotion>

Another benefit of this overall approach is the independence from any specific robot platform. While tags specify what emotion is expressed, they do not dictate how this should be shown. Robotic platforms can be quite diverse, and even humanoid robots will not all be capable of the same degrees of freedom. When lower level specifications are used to define movement of certain degrees of freedom, the implementation is constrained to platforms that are capable of those specific motions. Choosing to label higher level concepts means that any robot, humanoid or not, could be programed to take advantage of these tags – it would only need to have some behavior that conveyed emphasis or expressed emotion.

These higher level labels can also be used to create greater variability in a robot's behavior. If a robot's animation library contains multiple animations that convey the same emotion, or multiple types of gestural emphasis, then the robot could select different animations each time it says an utterance while maintaining the original meaning. Thus, even if a robot is forced to repeat a particular dialogue line multiple times, different animations could be used so that the movements and expressions would not be identical. This could make the repetition less noticeable, since the performance would not be exactly the same.

Furthermore, while the current proposition is for these labels to be assigned by people, it would be preferable if eventually a machine learning algorithm was able to do this process instead. Having people create a large number of annotated utterances for robot performances thus serves a secondary purpose of creating labeled training and testing data that could be used for future machine learning.

4   Experiment

One of the main goals for this approach was to accommodate labeling by people who have no background in robotics or animation. To test this we performed an on-line experiment. In the first phase of the experiment, a group of Amazon Mechanical Turk workers were presented with transcripts of several short conversations, which they used to answer questions about the emotion and emphasis of a particular dialogue line. In phase two, videos of a robot performing the animated dialogue were rated by a second set of participants. The workers from Mechanical Turk must be at least 18 years old be able to accept payments through a U.S. bank account. No other restrictions were placed on participants, which meant the participants could be of any education level, and would not necessarily have any prior experience with robots or animation of behaviors.

In phase one, Turkers were asked to read each short conversation out loud to them-selves before answering the questions, paying specific attention to how they naturally said each line of dialogue. This was intended to help participants determine the location of verbal emphasis by having them consider how they would naturally say the sentence. Because of the correlation between verbal and gestural emphasis [8], it was possible to specifically ask participants about their verbal emphasis while speaking the sentence without needing them to consider what gestures they might make while talking. Participants also selected the emotion most associated with the utterance from a list of possible emotions: Excited, Happy, Smug, Surprised, Bored, Confused, Skeptical, Embarrassed, Concerned, Sad, and Neutral. This list was specifically made to be more extensive than the previously mentioned classification algorithms in order to more fully explore the amount of nuance that people could distinguish, especially since, ideally, social robots should eventually be capable of expressing a wide range of emotions.

Once this data was collected, it was used to animate the utterances, which were then performed by the Furhat robot shown in Fig. 1. A script read in the Mechanical Turk data, used the participant responses to select animations based on the consensus of Turker selections, and output tagged utterances that were performed by the robot. Based on the data collected in phase one, five utterances were chosen which showed the best consensus on emphasis location, and another five utterances were selected which showed the best consensus on dominant emotion. The ones selected for emphasis received either 75 % or more of their selections on a single word, or a pair of adjacent words received a combined of more than 75 %. The chosen utterances for dominant emotion received either more than 70 % of selections or the selections for two similar emotions (sad/concerned, happy/excited, or confused/skeptical) received a combined percentage of more than 70 %.

Fig. 1. Furhat expressions – Happy (left), Neutral (center), and Unhappy (right)

Eyebrow motions were chosen as the emphasis gesture because they were easier to precisely synchronize with a specific word compared to longer motions such as nodding. Based on the observations from [9], eyebrow raises were used for positive emotional utterances and eyebrow frowns were used for the negative emotional utterances. Small facial movements were added to each of the control group performances to prevent the control videos from being seen as arbitrarily less appealing due to lack of motion.

In phase two of the experiment, a subset of these animated expressions were viewed and rated by a separate set of turkers. Videos of the robot showed either an emotive expression or an emphasis gesture. This was done so emotion and emphasis could be evaluated independently. While animated expressions for Furhat were created for all eleven emotions from phase, in the validation phase only two emotive expressions were used: Happy and Unhappy. This was so the videos showing the incorrect emotional expression could clearly be directly opposite the correct emotional expression. Each participant viewed two videos of the robot performing the same utterance and compared the videos on several scales. One video was a control video showing no emphasis and a neutral expression. The other video would represent one of four categories: a video with an emphasis gesture accenting the word that received the majority of selections in phase one, a video with an emphasis gesture at an incorrect location (accenting a word that received 10 % or less of the phase one selections and was not adjacent to the word chosen by consensus), a video with an emotive expression that matched the consensus from phase one, or a video with an emotive expression that opposed the emotion from the phase one consensus.

Phase two participants rated which of the two videos they viewed was most believable, humanlike, appropriate, pleasant, and natural. The metrics humanlike, natural, and pleasant were taken from the Godspeed Questionnaire Series [10]. The believable and appropriate metrics were added in order to distinguish between cases where the expression appeared realistic in isolation but did not match the dialogue. Each pair of videos was rated by twenty participants.

5   Results

The charts detailing the phase one participant responses concerning word emphasis can be seen in Fig. 2. Out of the eight utterances presented, four contained words that received at least 75 % of participant selections for that utterance. Three of these utterances had words that received 90–95 % of the selection. This represents strong indication that these words should be emphasized. In the remaining utterances, one (utterance 5) showed participant answers clustered around the noun-adjective pair "really worried". These two words together made up 90 % of the selections for this utterance, with an even split of selections between the two words. This shows that emphasized phrases were able to be identified. When presented a multi-sentence utterance – utterance 6 – the participant selection was split between two words, one from each sentence. This again shows that bimodal distributions will be visible in the data. The two remaining utterances (4 and 8) show that it is possible to determine when a particular utterance does *not* have a strong candidate for word level emphasis, displaying a wider spread of participant selection.

The charts detailing phase one responses for selected emotions can be seen in Fig. 3. In utterance 5, "I'm really worried I won't get it all done," 90 % of participants selected the concerned emotion. Such clear consensus is likely because the phrase "I'm really worried" specifically calls out the speaker's emotion, and so responses cluster around the nearest related emotion, concern. For four of the other utterances, participant selections were split between two closely related emotions that together accounted for at least 70 % of responses. In each of these cases – happy/excited, sad/concerned, and confused/skeptical – the two most chosen emotions expressed similar emotions with relatively close valence values. This shows a significant number of the participants were

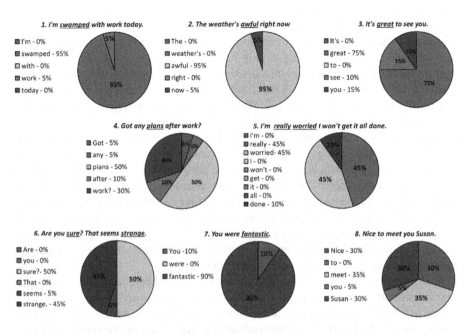

Fig. 2. Emphasis percentages for each utterance

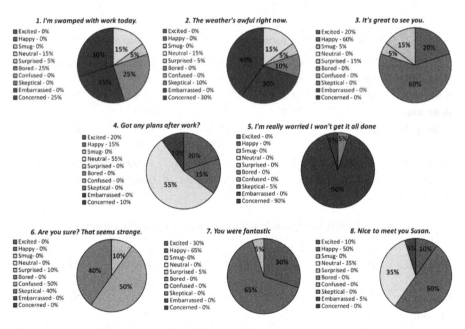

Fig. 3. Emotion percentages for each utterance

interpreting the utterances in similar manners, even if they chose slightly different emotions. Of the remaining utterances, utterance 4 had 55 % selection for neutral, with the other selections divided fairly evenly between three other emotion options. This suggests that there is no strong emotion associated with this sentence, and the expression should be left neutral. Utterance 8 had 50 % happy and 10 % excited, and utterance 1 was 30 % sad and 25 % concerned. While this gives some suggestion of possible emotions, it is not as strong of a consensus by comparison.

Tables 1, 2, 3 and 4 show the results of the direct video comparison survey questions from phase two. Chi-square tests were used to evaluate the significance of the data. The chi-square test is a statistical method assessing the goodness of fit between observed values and those expected if the null hypothesis was true. In this case the null hypothesis would mean no difference between the animated video and the control video, therefore producing an even split of 10 participants selecting the control video for every 10 that selected the experimental video. In order to reduce the risk of Type 1 errors all five metrics – humanlike, natural, believable, appropriate, and pleasant – were evaluated as a part of the same chi-square group for each utterance.

The robot performances that used the emotion selected by consensus in phase one were consistently rated more highly by participants when compared to the neutral control videos. All five test utterances received significant chi-square results, with p values ranging from 0.0001 to 0.0329. This confirms that people can assign emotions that are viewed as appropriate. Furthermore, of the videos shown where the emotion opposed the one chosen in phase one, four of the five received statistically insignificant results

Table 1. Percent of participants that chose the matched emotion

	Utterance 2*	Utterance 3*	Utterance 5*	Utterance 6*	Utterance 7*
Humanlike	80 %	80 %	75 %	90 %	70 %
Natural	85 %	80 %	65 %	85 %	60 %
Believable	75 %	70 %	70 %	90 %	75 %
Appropriate	80 %	80 %	75 %	85 %	85 %
Pleasant	60 %	80 %	70 %	65 %	70 %
Chi-Squared	30.000	32.000	18.200	47.000	26.000
p-value	0.0004*	0.0002*	0.0329*	0.0001*	0.002*

Table 2. Percent of participants that chose the mismatched emotion

	Utterance 2	Utterance 3	Utterance 5	Utterance 6	Utterance 7*
Humanlike	65 %	65 %	55 %	60 %	80 %
Natural	60 %	65 %	50 %	50 %	75 %
Believable	55 %	65 %	45 %	45 %	70 %
Appropriate	55 %	65 %	50 %	45 %	70 %
Pleasant	70 %	60 %	75 %	85 %	65 %
Chi-Squared	6.200	8.000	6.000	11.000	20.400
p-value	0.7197	0.5341	0.7399	0.2757	0.0156*

when compared to the control videos, with p-values ranging from 0.2757 to 0.7399. For these four utterances, adding a mismatched emotional expression performed no better than a neutral face. Overall, the videos showing expressions that matched the phase one responses were rated as significantly more humanlike than the control videos.

The data from the emphasis surveys is less clear. Table 3 shows that for three of the five utterances, the videos showing correct emphasis were selected significantly more than their control video counterparts. However, the remaining two utterances resulted in very high p-values. Furthermore, three of the videos showing incorrect emphasis also yielded statistically significance, as shown in Table 4. Thus the videos showing emphasis locations selected in phase one did not appear more realistic or believable overall compared to emphasis at other locations. This could indicate that even with the small random motions added to the neutral expression in the control video, the more obvious motion of the eyebrow raises and frowns was appealing for the sake of being more animated, regardless of the location of the emphasis.

Table 3. Percent of participants that chose the correct emphasis

	Utterance 1	Utterance 2*	Utterance 3*	Utterance 5	Utterance 7*
Humanlike	55 %	80 %	75 %	45 %	80 %
Natural	55 %	80 %	75 %	50 %	80 %
Believable	55 %	80 %	70 %	50 %	80 %
Appropriate	60 %	85 %	75 %	50 %	85 %
Pleasant	50 %	75 %	75 %	60 %	90 %
Chi-Squared	1.400	36.400	23.200	1.000	44.200
p-value	0.9978	0.0001*	0.0058*	0.9994	0.0001*

Table 4. Percent of participants that chose the incorrect emphasis

	Utterance 1*	Utterance 2*	Utterance 3*	Utterance 5	Utterance 7
Humanlike	75 %	80 %	80 %	60 %	70 %
Natural	75 %	80 %	80 %	60 %	70 %
Believable	75 %	80 %	75 %	55 %	70 %
Appropriate	75 %	75 %	75 %	50 %	70 %
Pleasant	65 %	70 %	85 %	60 %	70 %
Chi-Squared	21.8000	29.800	34.200	2.600	16.000
p-value	0.0095*	0.0005*	0.0001*	0.9781	0.0669

6   Conclusions

This work proposed an approach for reducing the time spent animating each utterance of a social robot. We found untrained workers were capable of providing reasonable labeling of semantic information in a presented utterance. When these labels were used to select animations for a social robot, the selected emotive expressions were rated as more natural and anthropomorphic than control groups. More study is needed to determine the effect of the labeled emphasis gestures on perception of robot performance.

Acknowledgements. We are thankful to Disney Research and The Walt Disney Corporation for support of this research effort. This material is based upon research supported by (while Dr. Simmons was serving at) the National Science Foundation.

References

1. Cassell, J., Vilhjalmsson, H., Bickmore, T.: BEAT: the behavior expression animation toolkit. In: Proceedings of the 28th Annual Conference on Computer Graphics and Interactive Techniques, pp. 477–486. (2001)
2. Smid, K., Pandzic, I.S., Radman, V.: Autonomous speaker agent. In: Proceedings of Computer Animation and Social Agents Conference (2004)
3. Albrecht, I., Haber, J., Seidel, H.P.: Automatic generation of non-verbal facial expressions from speech. In: Advances in Modelling. Animation and Rendering, pp. 283–293. Springer, London (2002)
4. Perikos, I., Jatzilygeroudis, I.: Recognizing emotions in text using ensemble of classifiers. Eng. Appl. Artif. Intell. **51**, 191–201 (2016)
5. Kiesler, S.C.R.: Fostering common ground in human-robot interaction. In: IEEE International Workshop on Robot and Human Interactive Communication, pp. 729–734 (2005)
6. Kim, H.H., Lee, H.E., Kim, Y.H., Park, K.H., Bien, Z.Z.: Automatic generation of conversational robot gestures for human-friendly steward robot. In: The 16th IEEE International Symposium on Robot and Human Interactive Communication, pp. 1155–1160 (2007)
7. Kopp, S., Krenn, B., Marsella, S., Marshal, A.N., Pelachaud, C., Pirker, H., Thorisson, K.R., Vilhjalmsson, H.: Towards a common framework for multimodal generation: the behavior markup language. In: Proceedings of the 6th International Conference on Intelligent Virtual Agents, pp. 205–217 (2006)
8. Graf, H.P., Cosatto, E., Strom, V., Huan, F.J.: Visual prosody: facial movements accompanying speech. In: Proceedings of the Fifth IEEE International Conference on Automatic Face and Gesture Recognition, pp. 396–401 (2002)
9. Zoric, G., Smid, K., Pandzic, I.S.: Facial gestures: taxonomy and application of non-verbal, non-emotional facial displays for embodied conversational agents. In: Conversational Informatics: An Engineering Approach, pp. 161–182, John Wiley & Sons, Ltd. (2007)
10. Bartneck, C., Croft, E., Kulic, D., Zoghbi, S.: Measurement instruments for the anthropomorphism, animacy, likeability, perceived intelligence, and perceived safety of robots. Int. J. Soc. Robot. **1**(1), 71–81 (2009)

Identifying Engagement from Joint Kinematics Data for Robot Therapy Prompt Interventions for Children with Autism Spectrum Disorder

Bi Ge, Hae Won Park, and Ayanna M. Howard[(✉)]

School of Electrical and Computer Engineering,
Georgia Institute of Technology, Atlanta, Georgia
ayanna.howard@ece.gatech.edu

Abstract. Prompts are used by therapists to help children with autism spectrum disorder learn and acquire desirable skills and behaviors. As social robots are more regularly translated into similar therapy settings, a critical part of ensuring effectiveness of these robot therapy system is providing them with the ability to detect engagement/disengagement states of the child in order to provide prompts at the right time. In this paper, we examine the various features related to body movement that can be utilized to define engagement levels and develop a model using these features for identifying engagement/disengagement states. The model was validated in a pilot study with child participants. Results show that our engagement model can achieve a recognition rate of 97 %.

Keywords: Robot therapy · Special needs · Kinematic assessment

1 Introduction

According to the Centers for Disease Control and Prevention, it is estimated that approximately 1 in 68 children in the US are diagnosed with autism spectrum disorder (ASD). For children with ASD, early intervention has been shown to be critical as the younger a child enters an early intervention program, the larger gain (s)he may have in developmental skills [1]. Furthermore, it has been shown that the effects of intervention (including acquisition of new skills) increases as the duration of treatment increases without diminishing returns [2]. However, therapy services and interventions offered by professionals are often expensive or inaccessible [3].

To increase the availability of intervention services offered to children with ASD, alternative technologies have been evaluated for their efficacy in the therapy setting in recent years. One such technology involves the inclusion of social robots with children with ASD. Examples of such systems include humanoids used in [4] as both therapist and interactive toy, designed to help children with ASD learn and practice social skills.

In the traditional therapy setting, prompts are a method of intervention in which cues or instructions are issued before or after a child's action in an effort to reengage his/her attention and help gain/eliminate desired/undesired behaviors [5]. For example, a therapist may issue prompts when a child loses his or her temper or stops concentrating on the task at hand during a therapy session. Prompts are essential since children

© Springer International Publishing AG 2016
A. Agah et al. (Eds.): ICSR 2016, LNAI 9979, pp. 531–540, 2016.
DOI: 10.1007/978-3-319-47437-3_52

with ASD usually does not respond to social cues in the same manner as typically-developing children do. Therapists thus utilize prompts as extra stimulus that correspond to some particular response [5]. In [6], it was shown that when interacting with a therapist and a robot at the same time, children with ASD and typically-developing children both spend more time looking at the robot than the human therapist. As this is one of the primary conditions for providing effective prompts, it seems appropriate that an autonomous robot platform for ASD intervention should be able to use this type of intervention method. For example, as pointed out by [7], robots can provide consistent, repeatable and standardized stimuli. This inherent characteristics of a robotic system would also enable a robot to provide consistent, repeatable and standardized prompts.

In this paper, we look at the first step required to provide prompts - the ability to detect engagement/disengagement states of the child in order to provide prompts at the right time. In our experiment, we define engagement as "concentrating on the task at hand and willing to remain focused". We discuss our method for modeling engagement levels based on features extracted from body movements, namely we focus on detecting if a child is concentrating on a given task by extracting engagement levels from joint kinematics data. We show that by carefully selecting features from joint kinematics data, we can achieve good performance for detecting engagement levels.

2 Related Work

There has been a number of prior research efforts focused on developing algorithms to detect user disengagement or engagement states. Some features used by past research efforts include: body posture [8], gestures [8], facial expression [9], eye gaze [8, 9], EEG [10], contextual information [9] and spatial relationship between the robot and human [11].

In [8, 9], the feature sets included eye gaze directions, gestures and head directions, which were extracted from human observations. In such a scenario, a robotic system would need to employ modeling methods that can run autonomously, without any human intervention. In [9], researchers also discussed using contextual information obtained from the log file of a storytelling app on a tablet. This kind of information though may not always be available to the robot, since robot interventions should occur in real-time, during the therapy session, rather than after the child has completed the task.

With respect to EEG applications such as in [10], even though EEG is considered noninvasive, placing electrodes on some children with ASD might be intolerable. In addition, using EEG devices limits the naturalness of the session and thus behaviors learned in the session may not be transferrable to the child's natural environment.

The spatial model in [11] uses the relative location of robot and human for a receptionist robot to detect engagement levels. However, in the case of a therapy robot for children with ASD, relative location information gives very little information about the engagement level of the child since children are typically engaged within a local zone of proximity to the therapist during the therapy session.

Beyond the ones mentioned, there are a number of other challenges associated with the process of selecting features that enable the modeling of engagement levels in children with ASD. Using eye gaze to analyze engagement, such as in [12] and in the ASD study by [13], requires the tracked face to be directly aligned towards the sensor. In our previous studies [14], we have seen that in most therapy sessions, the face is not always orientated toward the sensor, and, in fact, is just as likely to be orientated toward the task, the therapist, or, in some cases, towards the floor/table.

Using voice and verbal recognition has also been shown as a feasible option, such as in [15], where acoustic features from speech were used to model and detect engagement levels in daily conversations. However, depending on the therapy, a child with ASD may or may not talk during the session and certain children with ASD are nonverbal.

Other work, such as in [16], monitors user input through tablet interaction in order to assess engagement. However, children with ASD may not always be interacting with a tablet during a therapy session, thus features for modeling engagement should be extracted from interactions outside the tablet.

3 Approach

In this paper, we discuss our approach for detecting child engagement and disengagement states. Our approach for modeling engagement levels is based on extracting features from body movements, namely we focus on detecting if a child is concentrating on a given task by extracting engagement levels from joint kinematics data. In this work, we utilized a RGB-D camera, namely the Kinect 2.0 by Microsoft, to detect a user's skeleton and analyze the skeleton's pose relative to the task. This provides decent objective measures to detect engagement levels. Another benefit of using the Kinect/RGB-D camera is that the camera can recognize the human skeleton without requiring attaching markers to the body, which may be intolerable to children with developmental challenges.

A typical autism therapy session usually involves a therapist, a child (patient) and a task. While the child works on the task, the therapist provides instructions or prompts to help the child complete the task. To mimic a typical therapy session, the task we employ is a turn-taking game played on a Samsung Android tablet involving the matching of cards (Fig. 1). During the therapy session, a Kinect camera mounted on a tripod is located in front of the therapist and child in order to capture as much movement associated with their interaction as possible. There is also another camera in the room which serves as a backup source and provides a different viewing angle of the session.

3.1 Body Movement Features

Given a typical therapy session, our first step is to extract the relevant features, which can be used to identity engagement state. As skeletal data is directly extracted from the RGB-D data set and thus represents the movement profiles associated with the

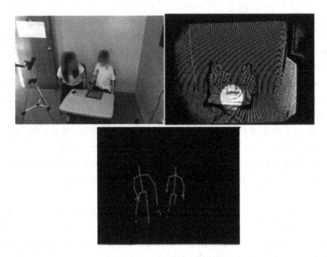

Fig. 1. Left: Child interaction session; Right/Bottom: Imaging and resulting skeleton of session using RGB-D camera

interaction, we first define a number of relevant features that can be derived from skeletal data. For this work, we classify these features as leaning angle, planar distance to therapist, mean joint to joint distance, distance of joints traveled within task ball, mean joint coordinates, mean joint distance to task and mean joint to joint distance.

a. Leaning angle
The leaning angle is the angle between the vertical y-axis in the camera's view and the vector constructed from the midpoint of the spine and neck of the child. Leaning angle is chosen to represent the scenario correlated with an individual concentrating on a task. In such cases, it has been observed that individuals tend to lean towards the object of interest associated with achieving a task. For example, in our experimental case, the tablet device becomes the object of interest. The leaning angle θ for the spine joint vector $\vec{j}_{midspine}$ and neck joing vector $\vec{j}_{neck}$ is calculated as:

$$\theta = \cos^{-1}\left(\frac{\begin{pmatrix} 0 \\ 1 \\ 0 \end{pmatrix} \cdot (\vec{j}_{midspine} - \vec{j}_{neck})}{\left\| \begin{pmatrix} 0 \\ 1 \\ 0 \end{pmatrix} \right\| \left\| \vec{j}_{midspine} - \vec{j}_{neck} \right\|}\right) \tag{1}$$

b. Planar distance to therapist
The planar distance to therapist feature is one measure used to calculate the distance measured between two people interacting with a task's common object of interest. A plane is constructed by using the middle point of the spine, neck and head of the therapist. Distances between this plane and the child's skeleton joints are then

calculated. For the therapist mid-spine joint vector $\vec{j}_{midspine}$, neck $\vec{j}_{neck}$ and head $\vec{j}_{head}$, the plane P and planar distance D can be derived using the following equations:

$$P = \left(\vec{j}_{midspine} - \vec{j}_{neck}\right) \times \left(\vec{j}_{midspine} - \vec{j}_{head}\right) \tag{2}$$

$$D = \frac{\vec{c}_i \cdot P}{\vec{c}_i} \text{ for child joint vector } \vec{c}_i \tag{3}$$

c. Mean joint to joint distance
The mean joint to joint distance records the average distance between each joint of the therapist and each joint of the child. This feature reflects the relative pose and distance between the therapist and child during an interactive session. For each child's joint i and their joint vector $\vec{c_i^f}$ and therapist's joint vector $\vec{t_i^f}$ at the f th frame of F frames, where F is the number of recorded Kinect frames associated with a movement profile, the mean joint to joint distance d_i is determined as:

$$d_i = \frac{1}{F}\sum\nolimits_{f=0}^{F}\left\|\vec{c_i^f} - \vec{t_i^f}\right\| \tag{4}$$

d. Distance traveled within task ball
The distance traveled feature measures the distance traveled by each joint inside a sphere whose center is located at the task's object of interest and has radius r. For a sphere with radius r and centered at $\vec{b}$, if joint vector $\vec{j_i^f}$ at the f th frame satisfies $\vec{b} - \vec{j_i} < r$, then $d_t = \sum\nolimits_{f=1}^{F}\left\|\vec{j_i} - \vec{j_{i-1}}\right\|$

e. Mean joint coordinates
The mean joint coordinates, associated with either an engagement or disengagement state, reflects the absolute pose of the child during a therapy session measured in the 3D world. It is obvious that in order to make this feature meaningful across various intervention sessions, it has to be normalized to eliminate overfitting to a specific session setup (for example how far the child sits away from the camera should not effect the performance of the system). As such, for each joint vector $\vec{j_i^f}$ and F frames, the mean joint coordinates m_i is calculated as:

$$m_i = \frac{1}{F}\sum\nolimits_{f=0}^{F}\left\|\vec{j_i^f}\right\| \tag{5}$$

f. Mean joint distance to task
The mean joint distance to task calculates the average distance between each joint and the location of the task object of interest. This feature reflects how far away a child is from the task during a therapy session. For each child joint vector $\vec{c_i^f}$ at the f th of F frames, the mean joint distance n_i to task T is:

$$n_i = \frac{1}{F}\sum\nolimits_{f=0}^{F}\left\|\vec{c_i^f} - \vec{T}\right\| \tag{6}$$

Once determined, these relevant features can then be utilized to identify engagement state. It is worth noting that some of these features require the presence of two skeletons during the therapy interaction, i.e. a child and a therapist/caregiver. However, the robot would not be autonomous if it requires the presence of a professorial therapist. Therefore, to classify engagement state in this paper, we will mainly focus only on those features that are derived solely from the child's skeletal data.

3.2   Classification of Engagement/Disengagement

In this work, we define two states – engagement/engaged and disengagement/ disengaged. Engagement is defined as focused or concentration on the task at hand. Disengagement is therefore just defined as the contrasting state, i.e. not engaged. We therefore define our problem as a two-class pattern recognition problem and examine those classification methods that are most appropriate to these types of problem. For this work, we evaluate the performance of three classification methods that can be used to distinguish between the two different states, namely SVM, Random Forest, and AdaBoost [17]. Support vector machines (SVMs) is a supervised learning method in which, given a set of training examples, each sample is marked as belonging to one of two classes. In this paper, our two classes correspond to *Engaged* or *Disengaged*. Given a set of labeled examples, the SVM training algorithm is able to build a model that then assigns new examples to one of the two defined classes. Random forests, on the other hand, represent the classification problem as a group of decision trees in which a new example is first classified based on an input vector that, as it descends down the branches of the tree, gets parsed into smaller and smaller sets. Each tree then provides a classification, and the forest chooses the classification having the most votes (over all the trees in the forest). The other method examined was AdaBoost, which solves the two-class pattern recognition problem by weighting the outputs given a large pool of weak classifiers.

In this work, these methods were trained and tested using a "hold-one-child-out method" based on the Scikit-learn methodology [18]. In the "hold-one-child-out" method, we pick data recorded with one child as the test data, and train the classifier with data from all other children. In this process, the following procedure was followed for each of the classifiers:

- A model using one of the respective training algorithms (SVM, Random Forest and AdaBoost) was trained using clips from all the children except the test one.
- The resulting model was then validated for classification accuracy using the remaining hold out part of the data as the test set.
- This process was repeated for each set of children and the final accuracy value computed as the average of the computed accuracy values for each round.

4 Experimental Setup

The goal of the classifier is to accurately detect engagement/disengagement states of a child in order for a therapy robot to provide prompts at the right time. As such, given our set of possible body movement features and classifiers, our goal was to determine the accuracy rates for each model in order to select the set with the greatest ability to discriminate between states. For this experiment, a pilot study was conducted at the Kid's Creek Therapy Center. The parents of each participant signed the IRB (Institutional Review Board) approved consent form allowing their child to engage in the testing sessions. Children diagnosed with developmental disabilities were recruited for this experiment with 3 boys, *mean*(age) = 12.3 and σ(age) = 1.5 (Table 1).

The child study consisted of sessions where, in each session, the experimenter and the child played a turn-taking game on the tablet (Fig. 2). There was one session hosted per child. During interaction, the experimenter asked a series of questions to distract the child during the child's turn such as "Do you remember my name?" A total of three sessions of approximately 28 min in length were recorded and processed. During the experiment, the real world coordinates of all joints of the human upper body skeleton were recorded as well as color video and audio streams from the Kinect camera. For each participant, a total of 17 components (i.e. joints) were captured by the Kinect. Since the camera was fixed on a tripod across all sessions, the classifier trained using data from one session was able to be applied to another session without normalization. For training, we did not select the features, planar distance to therapist and mean joint to joint distance, since these features required two skeletons to be present. As the goal of a therapy robot is to allow children to receive intervention without a therapist closely present, we determined that our model should only be built from skeletal data based on the child's movement profile. As such, the skeleton was selected manually in order to ensure that the analysis was performed with the child's movement profile and not the therapist's.

Table 1. Demographic data of child participants

Participant	Primary diagnosis	Gender	Age
1	ASD	Male	12
2	Down syndrome	Male	14
3	ASD	Male	11

Lastly, when calculating the feature vectors associated with the child and task, we also needed to know the location of the task's object of interest (i.e. the tablet). To obtain this information, we calculated the real world coordinates of the tablet by first reconstructing the 3D scene using the depth image obtained by the Kinect camera and manually selecting the 3D point corresponding to the tablet in the image scene.

Fig. 2. Left: Experimenter interacting with child with ASD; Right: Turn-taking matching game on the tablet.

5 Results

5.1 Ground Truth

Once collected, the stream of data from the pilot study was annotated by a human annotator with timestamps indicating the start and end of both engagement and disengagement states. Timestamps were annotated based on the identified behaviors in the videos. For example, some typical disengagement behaviors included standing up and walking away from the tablet and talking to others about things unrelated to the session. Disengagement states were also associated to those instances of time when the experimenter asked the series of questions designed to expressly distract the child.

Once the timestamps associated with the start and end times for the different states were obtained, clips were then segmented into smaller ones, each lasting for 2 s with a 1 s overlap. This was done in order to provide us with a sufficient number of training and testing instances to validate the model. From this annotation process, 22, 14 and 47 clips were labeled as disengagement clips for the three sessions respectively and 2,14, and 21 clips were labeled as engagement clips.

Table 2. Percent accuracy of different classifiers with respect to body movement features

Classifier	Mean joint coordinates	Mean joint distance to task	Distance traveled within Task Ball	All three features
SVM	88 %	70 %	70 %	88 %
Random forest	96 %	65 %	60 %	96 %
AdaBoost	96 %	93 %	66 %	97 %

5.2 Performance

Data from the various child interaction sessions were used to evaluate the performance of the different classifiers and feature sets using the "hold-one-child-out" cross validation method as discussed previously. As shown in Table 2, the best performing body movement feature was identified as Mean Joint Distance. Figure 3 expands on the corresponding table results. Based on this assessment, using AdaBoost with Mean Joint

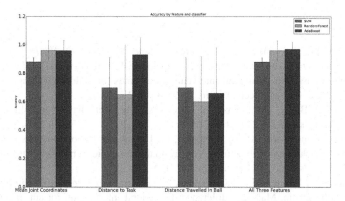

Fig. 3. Mean and standard deviation on performance accuracy associated with classifiers using different feature sets

Coordinates achieved the best single-feature performance at 96 % accuracy while AdaBoost with all the features gives an accuracy of 97 %. This combination appears to provide the best performing results.

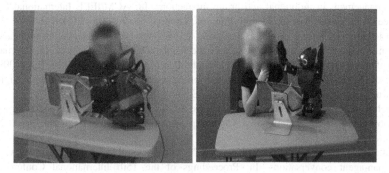

Fig. 4. Children with ASD interacting with therapy robot during a turn-taking matching game on the tablet. In future work, the results from this study will be used to evaluate if the same performance can be achieved for identifying engagement/disengagement in child-robot interaction scenarios such as these.

6 Conclusion and Future Work

We have discussed several features extracted from skeletal data recognized by a RGB-D camera, namely the Microsoft Kinect 2.0 that can be used for detecting engagement and disengagement states during therapy sessions. By carefully selecting the features, we demonstrated that without using contextual or voice features, we can still achieve decent performance.

An extension to this paper would be recruiting more participants and applying the algorithm to a real robot system to validate the engagement/disengagement model in a real robot-child therapy setting (as shown in Fig. 4) and/or compare the performance between the robot system and an actual ASD therapist.

References

1. Corsello, C.M.: Early intervention in autism. Infants Young Child. **18**, 74–85 (2005)
2. Granpeesheh, D., Dixon, D.R., Tarbox, J., Kaplan, A.M., Wilke, A.E.: The effects of age and treatment intensity on behavioral intervention outcomes for children with autism spectrum disorders. Res. Autism Spectr. Disord. **3**, 1014–1022 (2009)
3. Sharpe, D.L., Baker, D.L.: Financial issues associated with having a child with autism. J. Fam. Econ. Issues **28**, 247–264 (2007)
4. Robins, B., Dautenhahn, K., Te Boekhorst, R., Billard, A.: Robot assistants in therapy and education of children with autism: can a small humanoid robot help encourage social interaction skills. Univ. Access Inf. Soc. **4**, 105–120 (2005)
5. MacDuff, G.S., Krantz, P.J., McClannahan, L.E.: Prompts and Prompt-Fading Strategies for People with Autism, Making a difference: Behavioral intervention for autism, Austin. TX, Pro-Ed (2001)
6. Bekele, E., et al.: A step towards developing adaptive robot-mediated intervention architecture (ARIA) for children With Autism. IEEE Trans. Neural Syst. Rehabil. Eng. **21**, 289–299 (2013)
7. Eikeseth, S., Smith, T., Jahr, E., Eldevik, S.: Outcome for children with autism who began intensive behavioral treatment between ages 4 and 7. Behav. Modif. **31**, 264–278 (2007)
8. Leite, I., McCoy, M., Ullman, D., Salomons, N., Scassellati, B.: Comparing models of disengagement in individual and group interactions. In: ACM/IEEE International Conference on Human-Robot Interaction, Portland, Oregon, pp. 99–105 (2014)
9. Castellano, G., Pereira, A., Leite, I., Paiva, A., McOwan, P.W.: Detecting user engagement with a robot companion using task and social interaction-based features. In: International Conference on Multimodal Interfaces, Cambridge, Massachusetts, pp. 119–126 (2009)
10. Berka, C., et al.: EEG correlates of task engagement and mental workload in vigilance, learning, and memory tasks. Aviat. Space Environ. Med. **78**, 231–244 (2007)
11. Michalowski, M.P., Sabanovic, S., Simmons, R.: A spatial model of engagement for a social robot. In: 9th IEEE International Workshop on Advanced Motion Control, pp. 762–767, Istanbul, Turkey (2006)
12. Nakano, Y.; Ishii, R.: Estimating user's engagement from eye-gaze behaviors in human-agent conversations. In: Proceedings of the 15th International Conference on Intelligent User Interfaces, Hong Kong, China, pp. 139–148 (2010)
13. Bal, E., et al.: Emotion recognition in children with autism spectrum disorders: relations to eye gaze and autonomic state. J. Autism Dev. Disord. **40**, 358–370 (2010)
14. Park, H.W.; Howard, A.: Engaging children in social behavior: interaction with a robot playmate through tablet-based apps. In: Rehabilitation Eng. and Technology Society of North America (RESNA) Annual Conference, Indianapolis, IN, June 2014
15. Yu, C., Aoki, P.M., Woodruff, A.: Detecting User Engagement in Everyday Conversations. arXiv preprint cs/0410027 (2004)
16. Park, H.W., Coogle, R., Howard A.: Using a shared tablet workspace for interactive demonstrations during human-robot learning scenarios. In: IEEE International Conference on Robotics and Automation (ICRA), Hong Kong, China, June 2014
17. Hastie, T., Tibshirani, R., Friedman, J.H.: The Elements of Statistical Learning. Springer, New York (2001)
18. Pedregosa, F., et al.: Scikit-learn: machine learning in python. J. Mach. Learn. Res. **12**, 2825–2830 (2011)

Social Robots and Teaching Music to Autistic Children: Myth or Reality?

Alireza Taheri[1,4], Ali Meghdari[1(✉)], Minoo Alemi[1,2], Hamidreza Pouretemad[3,4], Pegah Poorgoldooz[4], and Maryam Roohbakhsh[4]

[1] Social & Cognitive Robotics Laboratory, Center of Excellence in Design, Robotics, and Automation (CEDRA), Sharif University of Technology, Tehran, Iran
meghdari@sharif.edu
[2] Islamic Azad University, Tehran-west Branch, Tehran, Iran
[3] Institute for Cognitive and Brain Sciences (ICBS), Shahid Beheshti University, Tehran, Iran
[4] Center for Treatment of Autistic Disorders, Tehran, Iran

Abstract. Music-based therapy is an appropriate tool to facilitate multisystem development in children with autism. The focus of this study is to implement a systematic and hierarchical music-based scenario in order to teach the fundamentals of music to children with autism through a social robot. To this end, we have programmed a *NAO* robot to play the xylophone and the drum. After running our designed robot-assisted clinical interventions on three high-functioning and one low functioning autistic children, fairly promising results have been observed. We indicated that the high-functioning participants have learned how to play the musical notes, short sentences, and simple rhythms. Moreover, the program affected positively on autism severity, fine movement and communication skills of the autistic subjects. The initial results observed indicate promising potentials for involving social robots in music-based autism therapy.

Keywords: Music-based therapy · Xylophone · Autism spectrum disorders (ASD) · Humanoid social robot · Social and cognitive skills · Imitation

1 Introduction

Music greatly influences humans and in particularly children's emotions, moods, and feelings. Teaching music can help develop new or improve existing social, verbal and non-verbal communication skills in children [1, 2].

Children with autism have stereotyped behaviors and limited verbal communication skills [3]. Music and rhythms are effective methods to involve them in rhythmic and non-verbal communication. Nowadays, at least 12 % of all treatment of individuals with autism consist of music-based therapies [4].

Music has often been used in therapeutic sessions with children with mental and behavioral disabilities [5, 6]. In particular, there is ample evidence that shows either playing music during therapy sessions or teaching music to children with autism spectrum disorders (ASD) can significantly increase the impact of therapy sessions [7]. In such studies or therapy sessions an instrument is either played by a human or recorded

© Springer International Publishing AG 2016
A. Agah et al. (Eds.): ICSR 2016, LNAI 9979, pp. 541–550, 2016.
DOI: 10.1007/978-3-319-47437-3_53

music is played back in individual and group intervention sessions [4, 8]. The effects of music-based therapy in improving social skills of children with autism (i.e. eye contact and initiating social behaviors) have been reported in [9]. Kim et al. [2] showed improvement in joint attention, turn taking and eye contact of children with autism in active music-making interventions. In [10, 11], the studies showed a decrease in stereotyped behaviors and self-injuries in children with autism after running music-based interventions. Music therapy interventions have been used to increase social [12] and emotional [13] skills, verbal and gestural communication [14], and behaviors [15] of individuals with autism in individual and group modes. It should be noted that the lack of studies on improving gross and fine motor skills of autistic children through music-based interventions is still a gap in this area [4].

Recently, we have designed a comprehensive robot-assisted music-based intervention scenario to improve perceptuo-motor, social, and cognitive skills of ASDs and conducted it in a single subject design study. The purpose of this educational-therapy program is to teach the fundamental concepts of how to play drum and xylophone using a *NAO* humanoid robot as a teacher's assistant to children with autism. Our goal is to find scientific answers for the following research questions: (1) Does a humanoid social robot have the ability to teach music (i.e. notes and rhythms) to children with autism? (b) Can a humanoid robot improve social and cognitive skills in children with autism through music education?

To this end, a drum/xylophone playing humanoid robot in addition to other musical instruments, both of which are loved by children, were used. Although the use of robotics technology in different aspects of education and treatment is increasing [16–25], to the best of our knowledge utilizing a humanoid robot to systematically teaching music to children with autism is still an interesting topic. Tapus [22] has used a social robot in a music-therapy program on individuals with cognitive impairments; however the robot did not play any musical instruments in her study. It should be noted that some researchers have also used music instruments like a drum played by a robot [23] as reinforcement tools (and not necessarily as an education tool) in autism treatment.

This paper presents the results and observations of a robot-assisted therapy in a single subject design study on three high-functioning and one low-functioning children with autism. The study was conducted in eleven music-based intervention sessions in Iran in order to explore the potentials of music-based games on ASDs.

2 Research Methodology

2.1 Participants

Three children with high-functioning and one child with low-functioning autism enrolled in this robot-assisted research study. All of the participants were 6 years old males without any previous music background. The children's details are describe in Table 1.

Table 1. Our participants' details

#	Abbreviation	Autism severity
1	P1	High-Functioning autism, with hyperactivity
2	P2	High-Functioning autism
3	P3	High-Functioning autism, with verbal deficits
4	P4	Low-Functioning autism with poor verbal skills

2.2 Humanoid Robot

The humanoid robot used in this research is the *NAO* H-21 robot made by Aldebaran Company [26]. The capabilities of *NAO* as well as the suitable programming interface of this robot make it a commonly used commercial robot for autism research [27–29]. We have renamed the robot to the Iranian boy's name, "*Nima*", during our studies.

2.3 Musical Instruments

Having noticed that when the instrument sound is simpler and more pleasant the patient will be deeper involved and the interventions will be more effective, we have selected a drum and a xylophone for the robot to play in our intervention sessions.

2.4 Technical Design of the Games

We have designed two general music games to involve children in interventions: (a) playing a real drum/xylophone in Robot-Child or Robot-Child-Therapist/Parent imitation turn taking games and (b) playing a Kinect based virtual xylophone on the screen.

Play the Drum/Xylophone: Our robot has been programmed to be able to play the drum/xylophone. A configurable user friendly GUI[1] as well as some rhythm patterns by Choregraphe [26] software have been developed in order to enable the robot playing the instruments either manually by operators/psychologists or automatically in a real-time situation. The robot is able to play different rhythms and notes with its right/left arm.

Virtual Xylophone: We have developed a Kinect based virtual xylophone containing 8 colored bars programmed by C# WPF. In this game, the player can see the music bars on the screen. The participant can hit the bars using colored mallets or the palm of their hands. The sounds of one octave from C4 (261.6 Hz) to C5 (523.2 Hz) are heard when hitting these bars. We have used the Toub.Sound.Midi Library in order to generate music notes with the computer.

---

[1] Graphical User Interface .

2.5   Experimental Setup

Our study was conducted in the Social & Cognitive Robotics Laboratory at Sharif University of Technology with four autistic children during eleven sessions in the presence of a human therapist (and sometimes their parents), a humanoid robot, and a robot operator. Time duration of each session was 20–30 min.

The games' instructions were described by the robot and/or the human therapist before each game. The Wizard of OZ (WOZ) style robot control has been used in this study, and most of the time the robot operator sent appropriate real-time voice/motor commands to the robot after seeing the child's performance. Our single subject design study contains a Baseline, Pre-Test, Post-Test, and Follow-up Test (four weeks after the last session) all in the absence of the *Nima* robot. Because no control group was utilized in this paper, our focus was on comparing each participant's skills/behaviors with his previous behaviors based on the assessment tools (which will be introduced in Subsect. 2.7).

2.6   Interventions Protocol

The purpose of designing a comprehensive music-based robot-assisted intervention scenario is teaching the fundamentals of music, decreasing impairments and improving/ generalizing social and cognitive skills as well as fine/gross movement skills of children with autism spectrum disorders. Our music games are based on active music-based therapies which included imitating the robot in playing the drum/xylophone, rhythm perception, working memory games, teaching the notes, involving in turn-taking group games, reading/playing music sentences, and finally generalizing the learned knowledge to another instrument (i.e. the virtual xylophone). The designed therapeutic games have the potentials to improve auditory perception, perceptuo-motor activities, vision skills, mental development, and social and communication skills.

2.7   Assessment Tools

To answer the research questions of this study, two measuring instruments have been used each for four times (on Baseline, Pre-, Post-, and Follow up tests).

(1) Stambak's Rhythmic Structures Reproduction test [30] which is a test containing 21 (easy to hard level) rhythmic tasks that the participant should reproduce the patterns through a drum after hearing (and not seeing) them performed by the therapist.

(2) Gilliam Autism Rating Scale (GARS) [31], a questionnaire for estimating autism severity, with 56 questions which covers four subscales: Stereotyped Behaviors, Communication, Social Interactions, and Developmental Disturbances.

3   Results and Discussion

At the first session, the *Nima* robot was introduced to the participants. The main purpose of this session was to familiarization/desensitization the participants to the class environment as well as observe the child's tendency to start/keep communication with *Nima*. It should be noted that music was not taught during the first session.

After that, following the designed educational protocol, music was taught step-by-step by the robot during the rest of the sessions. The selected snapshot of the intervention sessions is presented in Fig. 1.

Fig. 1. Snapshot of the robot-assisted music-based intervention sessions

Music was a happy and enjoyable activity for all of the participants. The existence of the robot itself considerably increased the motivation of the children to use their capabilities to involve their sensorimotor mechanisms. A short description of observations for the participants are presented in the following.

P1: The psychologists reported noteworthy improvement in his social skills, attention, and the ability to learn. The music learning process occurred for Benyamin and the presence of the robot was the reason for this observation

P2: P2's performance was very similar to typically developing children. Although P2 is very resistance to education classes in his real life, he did not show any maladaptive behaviors in our course. It seemed the child felt secure and relaxed during the sessions

P3: Improvement in Radvin's verbal skills was reported by his mother. In comparison to the other two high-functioning children, Radvin's weakness at the first sessions was his lack of expressing/identifying some colors; fortunately, this deficit was resolved by the last sessions

P4: He understood none of the instructions at the beginning of the program. Improvement in his instruction perception, attention, and understanding what happened in the class was the positive note occurred for P4 over time. He usually played randomly on the real xylophone bars instead of correctly imitating the robot/parent during the turn-taking games; however he was able to acceptably imitate the robot switching hands in the last four sessions. We observed that P4's stereotyped behaviors (especially his fluttering fingers) decreased and his verbal skills increased. Music is very effective in decreasing the stereotyped behaviors of children with autism [10, 11].

3.1 Stambak Rhythmic Structures Reproduction Test

To investigate the participants' improvement in rhythm perception during this time, the Stambak's Rhythm Reproduction Test has been run on the autistic subjects. The results of this test in the Pre- and Post-Test are presented in Table 2. We indicated that all of the three high-functioning participants show improvement in playing rhythm which means music learning has occurred for these subjects. In [32], the Stambak rhythm test has been used to show the music improvement of typically developing subjects after running a human-based music-therapy program.

Table 2. Stambak's Rhythm Test results in Pre-Test and Post-Test

#	Abbreviation	Scores of the Stambak's Rhythm Test	
		Pre-Test	Post Test
1	P1	18	19
2	P2	6	13
3	P3	11	14
4	P4	0	4

Meanwhile, P4 was unable to do any of the rhythm tasks in Pre-Test; however, in the Post-Test he was able to play tasks #1 to #4, each in his second try with the help of counting (the number of played hits) and saying "Baam-Baam".

For the high-functioning autistic children, our main goal was observing the trend of their music learning pace. Rhythm perception for the three high functioning subject have been improved. From our observations, we hypothesized that there is no obvious difference between the music learning process/progress rate of high-functioning ASDs and TDs[2]; they can read/play the notes and simple musical sentences and the progress rate for the spent three months was really promising. Moreover, we observed noticeable progress in all four children's fine hands imitation as well as using two hands consecutively, which they all had problems with in the first sessions.

---

[2] Typically Developing.

One of the promising observations from the high-functioning subjects' performance in this study was their following, curiosity, and questioning increased in the presence of the robot. The children used as many opportunities as they could to push Nima to speak and ask him questions.

3.2 GARS Questionnaire

The results of GARS are presented in Fig. 2. We observed a decrease in the autism severity of all four participants from Pre-Test to Post-Test. A detailed assessment of the GARS questionnaire showed us improvement occurred in 3 subscales: stereotyped behaviors, communication, and social interaction for two of the participants, P1 and P4 from Pre-Test to Post-Test. P3 progressed in stereotyped behaviors and communication, while P2 showed improvement in communication and social interaction. The interesting point of the GARS results is the progress of all of the subjects in subscale "communication". This finding is in line with our therapists' reports during the intervention sessions. The existence of *Nima*, as an attractive communication tool, positively affected the communication skills of the children. Comparison of GARS overall scores between the Post-Test and Follow up-Test shows the retention as well as the stability of the sessions' impact on autistic children. Similarly, improvement of autistic children in different social and communication skills after music-based therapies or robot-assisted clinical interventions have also been confirmed in [12–15, 20, 21, 29].

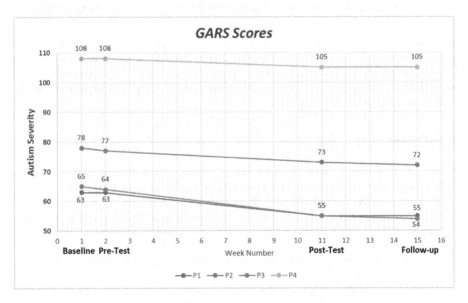

Fig. 2. GARS overall scores (autism severity) of the participants in Baseline, Pre-Test, Post-Test, and Follow up Tests.

3.3 Limitations and Future Works

In order to check for the effectiveness of the treatment, a larger sample size is needed, ideally with male and female children and a control group. Moreover, one of the greatest limitations in these kinds of studies is accessing valid tools which have the potential to accurately measure/assess children's behavior. Unfortunately, because of small number of the autistic participants in this study, no scientific statistical analysis could be applied on the data. To generalize the mentioned findings in robot-assisted music-based therapy, further research with more autistic subjects is needed.

4 Conclusion

Through the designed music-based scenario package, we wanted to explore the potential of music-based intervention on ASD's improvement in motor, communication skills as well as learning music. The robot does have the ability to teach the fundamentals of music to children with autism. We also saw improvement in fine hands imitation, using both hands in order, and rhythm identification for all of the participants. The high-functioning subjects can now read/play the notes and simple musical sentences and their progress was quite acceptable during the three months interventions. In the case of the low-functioning subject, improvement in verbal skills and a decrease in stereotyped behaviors have been indicated. Additionally, the GARS showed that the autism severity of all of the participants were reduced after the robot-assisted intervention sessions.

Acknowledgement. Our sincere appreciation is extended to the "Cognitive Sciences and Technologies Council of Iran" for their financial support through research grant # 103. We also acknowledge the "Center for the Treatment of Autistic Disorders (CTAD)" and its psychologists for their technical support and cooperation in participating in the clinical interventions with the children with autism spectrum disorders.

References

1. Wan, C.Y., Bazen, L., Baars, R., Libenson, A., Zipse, L., Zuk, J., Schlaug, G.: Auditory-motor mapping training as an intervention to facilitate speech output in non-verbal children with autism: a proof of concept study. PLoS ONE 6(9), e25505 (2011)
2. Kim, J., Wigram, T., Gold, C.: The effects of improvisational music therapy on joint attention behaviors in autistic children: a randomized controlled study. J. Autism Dev. Disord. 38(9), 1758–1766 (2008)
3. Pouretemad, H.: "Diagnosis and treatment of joint attention in autistic children", (in Persian). Arjmand Book, Tehran (2011)
4. Srinivasan, S.M., Bhat, A.N.: A review of "music and movement" therapies for children with autism: embodied interventions for multisystem development. Frontiers Integr. Neurosci. 7, 22 (2013)
5. Roper, N.: Melodic intonation therapy with young children with apraxia. Bridges 1(8), 1–7 (2003)

6. Boso, M., Emanuele, E., Minazzi, V., Abbamonte, M., Politi, P.: Effect of long-term interactive music therapy on behavior profile and musical skills in young adults with severe autism. J. Altern. Complement. Med. **13**(7), 709–712 (2007)
7. Lim, H.A., Draper, E.: The effects of music therapy incorporated with applied behavior analysis verbal behavior approach for children with autism spectrum disorders. J. Music Ther. **48**(4), 532 (2011)
8. Corbett, B.A., Shickman, K., Ferrer, E.: Brief report: the effects of Tomatis sound therapy on language in children with autism. J. Autism Dev. Disord. **38**(3), 562–566 (2008)
9. Stephens, C.E.: Spontaneous imitation by children with autism during a repetitive musical play routine. Autism **12**(6), 645–671 (2008)
10. Lanovaz, M.J., Fletcher, S.E., Rapp, J.T.: Identifying stimuli that alter immediate and subsequent levels of vocal stereotypy a further analysis of functionally matched stimulation. Behav. Modif. **33**(5), 682–704 (2009)
11. Wood, S.: A study of the effects of music on attending behavior of children with autistic-like syndrome (1991)
12. Pasiali, V.: The use of prescriptive therapeutic songs in a home-based environment to promote social skills acquisition by children with autism: three case studies. Music Ther. Perspect. **22**(1), 11–20 (2004)
13. Katagiri, J.: The effect of background music and song texts on the emotional understanding of children with autism. J. Music Ther. **46**(1), 15–31 (2009)
14. Simpson, K., Keen, D.: Music interventions for children with autism: narrative review of the literature. J. Autism Dev. Disord. **41**(11), 1507–1514 (2011)
15. Carnahan, C., Basham, J., Musti-Rao, S.: A low-technology strategy for increasing engagement of students with autism and significant learning needs. Exceptionality **17**(2), 76–87 (2009)
16. Taheri, A.R., Alemi, M., Meghdari, A., PourEtemad, H.R., Basiri, N.M.: Social robots as assistants for autism therapy in Iran: Research in progress. In: 2014 Second RSI/ISM International Conference on Robotics and Mechatronics (ICRoM), pp. 760-766. IEEE, October 2014
17. Boccanfuso, L., O'Kane, J.M.: CHARLIE: an adaptive robot design with hand and face tracking for use in autism therapy. Int. J. Soc. Robot. **3**(4), 337–347 (2011)
18. Meghdari, A., Alemi, M., Ghazisaedy, M., Taheri, A. R., Karimian, A., Zandvakili, M.: Applying robots as teaching assistant in EFL classes at Iranian middle-schools. In: The 2013 International Conference on Education and Modern Educational Technologies (EMET-2013), Venice, Italy, September 2013. http://www.europement.com. Accessed
19. Alemi, M., Meghdari, A., Ghazisaedy, M.: The impact of social robotics on L2 learners' anxiety and attitude in English vocabulary acquisition. Int. J. Soc. Robot. **7**(4), 523–535 (2015)
20. Scassellati, B., Admoni, H., Mataric, M.: Robots for use in autism research. Annu. Rev. Biomed. Eng. **14**, 275–294 (2012)
21. Kajopoulos, J., Wong, A.H.Y., Yuen, A.W.C., Dung, T.A., Kee, T.Y., Wykowska, A.: Robot-assisted training of joint attention skills in children diagnosed with autism. In: Tapus, A., André, E., Martin, J-.C., Ferland, F., Ammi, M.(eds.) ICSR 2015. LNCS, vol. 9388, pp. 296–305. Springer, Heidelberg (2015)
22. Tapus, A.: The role of the physical embodiment of a music therapist robot for individuals with cognitive impairments: longitudinal study. In: Virtual Rehabilitation International Conference, 2009, pp. 203–203. IEEE, June 2009

23. Robins, B., Dautenhahn, K., Dickerson, P.: From isolation to communication: a case study evaluation of robot assisted play for children with autism with a minimally expressive humanoid robot. In: Second International Conferences on Advances in Computer-Human Interactions, 2009, ACHI 2009, pp. 205–211. IEEE, February 2009

24. Boccanfuso, L., Barney, E., Foster, C., Ahn, Y.A., Chawarska, K., Scassellati, B., Shic, F.: Emotional robot to examine different play patterns and affective responses of children with and without ASD. In: 2016 11th ACM/IEEE International Conference on Human-Robot Interaction (HRI), pp. 19–26. IEEE, March 2016

25. Alemi, M., Meghdari, A., Mahboub Basiri, N., Taheri, A.: The Effect of Applying Humanoid Robots as Teacher Assistants to Help Iranian Autistic Pupils Learn English as a Foreign Language (2015)

26. Robotics, A.: The NAO robot (2013). Available at: bttp:/Ivnvw. aldebaran-roboticscomteni. Accessed 4 Sep 2013

27. Mavadati, S.M., Feng, H., Gutierrez, A., Mahoor, M.H.: Comparing the gaze responses of children with autism and typically developed individuals in human-robot interaction. In: 2014 14th IEEE-RAS International Conference on Humanoid Robots (Humanoids), pp. 1128–1133. IEEE, November 2014

28. Shamsuddin, S., Yussof, H., Ismail, L.I., Mohamed, S., Hanapiah, F.A., Zahari, N.I.: Initial response in HRI-a case study on evaluation of child with autism spectrum disorders interacting with a humanoid robot Nao. Procedia Eng. **41**, 1448–1455 (2012)

29. Taheri, A., Alemi, M., Meghdari, A., Pouretemad, H., Basiri, N.M., Poorgoldooz, P.: Impact of humanoid social robots on treatment of a pair of Iranian autistic twins. In: Tapus, A., André, E., Martin, J.-C., Ferland, F., Ammi, M. (eds.) ICSR 2015. LNCS (LNAI), vol. 9388, pp. 623–632. Springer, Heidelberg (2015). doi:10.1007/978-3-319-25554-5_62

30. Gardner, H.: Children's duplication of rhythmic patterns. J. Res. Music Educ. **19**(3), 355–360 (1971)

31. Gilliam, J. E.: Gilliam autism rating scale: examiner's manual. Pro-ed (1995)

32. Noor Mohammadi, F.: The effect of the Orff music on improving rhythm structuring. (in Persian), Developmental Psychology Quarterly, Tehran (2003)

Development of an ABA Autism Intervention Delivered by a Humanoid Robot

Michelle Salvador[1](✉), Anna Sophia Marsh[2], Anibal Gutierrez[3],
and Mohammad H. Mahoor[1]

[1] Department of Electrical and Computer Engineering,
University of Denver, Denver, CO 80210, USA
mjsalv@gmail.com
[2] Department of Psychology, University of Denver, Denver, USA
[3] Department of Psychology, Florida International University,
Miami, FL 33199, USA

Abstract. Applied Behavioral Analysis (ABA) techniques are widely used and accepted by the Autism research community as an effective Autism therapy method. ABA techniques have been recently introduced to use with Socially Assistive Robots (SAR) to deliver Autism therapy. Nonetheless, little research has been published to investigate the use of robot-based ABA in teaching socio-emotional skills for children diagnosed with Autism Spectrum Disorder (ASD). This paper presents the development of an ABA-based autonomous therapy system delivered by a humanoid robot, the Zeno R-50. Specifically, an intervention methodology with a prompt and an ABA reinforcement protocol to target skills associated with facial and situational emotion recognition and understanding are presented. Eleven children diagnosed with Autism were recruited to participate in the pre-pilot study screening. Initial results are investigated to discover the children's preferred reinforcers and also to find ways of improving the therapy system before proceeding through full study groups. Results demonstrate the successful detection of reinforcers and show there is correlation between reinforcer preference and age. Results on two children who have completed interventions are presented and improvements to the protocol are discussed to contribute to the understanding of SAR in teaching socio-emotional skills to children diagnosed with ASD.

1 Introduction

Investigation on the use of Socially Assistive Robots (SAR) to interact with children diagnosed with Autism Spectrum Disorder (ASD) is a topic that has been widely researched [4]. This is due to SAR's demonstrated ability to attract the child's attention versus having a human or computer provide the interaction [5]. Despite SAR's demonstrated attractiveness to children with ASD, research on its ability to successfully teach and deliver Autism therapy requires further development and exploration [7].

© Springer International Publishing AG 2016
A. Agah et al. (Eds.): ICSR 2016, LNAI 9979, pp. 551–560, 2016.
DOI: 10.1007/978-3-319-47437-3_54

Recently, work with SAR has been directed toward use of Applied Behavioral Analysis (ABA) such as [8,10–13,23,24]. This is due to the fact that ABA has been widely known as a successful and safe treatment for individuals with Autism [9]. ABA techniques rely on presenting simple tasks to a child and providing positive reinforcers to reward the child when he or she responds appropriately to the task provided by the therapist.

Examples of the use of ABA in conjunction with SAR include work done by Aldebaran, makers of the Humanoid robot Nao. They have created and marketed a set of games that combine the use of ABA and other Autism therapy models to aid parents and teachers provide therapy [13].

Other research making use of providing ABA style prompts and rewards is found in [11] and [8]. Using the Nao robot as well, their research targeted improving joint attention in children diagnosed with ASD ages two to five. Using adaptive prompts and providing reinforcement through cartoon clips or pictures from children's shows, the research group showed that this style of therapy can show promise in correcting joint attention patterns. However, they note that due to the low interaction time available, it was uncertain if the positive effects they saw in the study were due to the robot's novelty or to long term preference patterns [11]. In their single participant study [8], therapy was delivered twice a week for eight weeks targeting language development alternating use of a certified specialist and the robot. Results show that while the robot was successful in attracting the attention of the child initially, most of the correct behaviors from the child were directed toward the human therapist by the end of the study. Although not much improvement had been show at the midpoint session, the last session showed an increase from the initial 29 % baseline to 59 % of correct responses overall [8].

Though much of the research done in this area is based on Nao, robots such as PABI (Penguin for Autism Behavioral Interventions) and FACE have also been explored in [12] and [22]. Using a tablet computer to interface with the child, a discussion towards the initial use of the penguin robot in ABA therapy is shown in [12]. Similarly, a study demonstrating preference for the FACE robot is discussed in [22].

Reviewing the literature in this area shows that a study focusing specifically on emotion-based skill learning and improvement has yet to be presented. Furthermore, autonomous intervention for emotion skills delivered solely by a humanoid robot has not been conducted to our knowledge. Therefore, this paper presents the development of an autonomous robot delivered intervention therapy focusing on teaching emotion skills using techniques modeled after ABA interventions usually used by therapists. More specifically, the development of emotional recognition skills in both the facial expression and situational event contexts are targeted. Identification of preferred social reinforcements (rewards) as delivered by a humanoid robot is tested using a similar methodology as found in [14] as it is recognized that the effectiveness of the reward is important to successful ABA therapy. To our knowledge, this is the first time a humanoid robot has been used for the assessment of reinforcer rewards for ABA therapy and the combined use of SAR and ABA for targeting emotional skills. Identifying the proper reinforcements to deliver, the Zeno R-50 robot was programmed to adaptively provide rewards based on participant preference and deliver adaptive prompt hints. Therefore, as the child interacts with the robot, no external control during the intervention was necessary.

The remainder of this paper is organized as follows. Section 2 describes the study session structure and the design of games used during the intervention sessions. It also gives a description of the ABA Reward Identification Test, in which the ABA rewards were identified to provide reinforcements during the interventions. The hardware and software system design for conducting the experiment and the selected humanoid robot are discussed as well. Section 3 presents the result outcomes of eleven recruited children diagnosed with ASD. A discussion of these results and future work to improve on the learned observations is found in Sect. 4.

2 Methodology

2.1 Game Design and Overview

Through collaboration with Dr. Anibal Gutierrez, a Board Certified Behavior Analyst, three games targeting emotion and facial expression recognition were developed. The titles of the games are "Recognize", "Select", and "Identify". Each game is designed to target a single socio-emotional skill to enable simplicity in explaining and playing each game with the child and to analyze each skill progression independently.

As noted in the DSM-5, children with ASD may tend to misread nonverbal interactions [2]. Therefore, the "Select" and "Identify" games were designed to target situational interactions that did not rely solely on facial expression information but on the nonverbal situation depicted in a picture. Pictures for these two games are based on the "language-builder-emotion-cards" set available from an Autism community store [3]. Although [16] showed that children diagnosed with ASD may be able to recognize facial expressions effectively, inclusion of the "Recognize" game focusing on recognizing facial expressions is included to study possible correlations to recognizing emotional situational events as tested in the "Select" and "Identify" games.

For this study only five of the six universal expressions described by Ekman are used [20]. These are: happy, sad, anger, surprise, and fear. Disgust was excluded since a motor needed to portray a main muscle used in disgust, "Levator labii superioris alaquae nasi" (facial action unit 9), as described in the EMFACS [21] manual, is missing in the Zeno R-50 robot [16].

For each child, two initial baselines are conducted; a human and robot baseline. Each baseline is performed with either a human or robot interacting with the child playing the three games once for each emotion twice for a total of 30 games. In baseline and intervention sessions, the game type and emotion order is presented to the child randomly. Following session one, each child goes through three intervention sessions delivered by Zeno. Unlike baselines, the child

Fig. 1. A participant and the Zeno R-50 Robot interacting through the computer touch screen.

is allowed three tries per game. After each try, if the child answers incorrectly the robot provides a hint. If the child answers correctly before the last try then one of the preferred rewards identified from the ABA reward test is shown to the child by Zeno to indicate that the correct answer was selected. Feedback and rewards are delivered in intervention sessions to promote learning of the skills needed. Robot and human exitlines are conducted at the end identically to the baselines in session five. Each session is conducted once every week (Fig. 1).

2.2   Zeno R-50 Humanoid Robot

This study uses the Zeno R-50 humanoid robot developed by Hanson Robotics [18]. Originally designed for child Autism therapy interactions [22], it has a simplified child-like face with eight Degrees of Freedom (DOF). The body contains 25 DOF in a bipedal humanoid configuration for a total of 36 DOF. One of the novel features of the Zeno R-50 robot is its facial material, Frubber [18]. In combined use, the Frubber material and facial motors allow Zeno to have an expressive face useful for displaying emotions. Previous studies with the Zeno or Alice (the female variant) R-50 have shown successful use of the robot for child-robot interactions [1,16,19,22] especially due to its facial expressivity. Therefore, in our plans to conduct intensive Robot-based Autism therapy studies targeted at emotion recognition and expression, we have chosen continued use of the R-50 robot.

2.3   Participant Description

Eleven high functioning children diagnosed with Autism ages 7–17 [average age = 9.8 std = 2.9] were recruited. Two of these participants are female. All recruited participants are formally diagnosed with Autism by either a doctor or a psychologist with proof of medical diagnosis for verification. Additionally, each participant was assessed through the Social Responsiveness Survey (SRS) with a T-Score greater than 76. According to the SRS diagnostic manual, a T-score above 76 is considered a severe range and indicator of being, "strongly associated with a clinical diagnosis of Autistic Disorder or Asperger's Disorder [17].

2.4   System Design

The robotic system designed for conducting this experiment consisted of three components; a touch screen computer, a main controller computer, and the Zeno R-50. The touch screen computer displayed the interface for the experiment through a developed Unity application. Upon a child's press of a button, the touch screen computer sent notifications to the main controller which would in turn coordinate the robot's action. Zeno was programed in Java using the RoboKind API libraries for sending text-to-speech and animation commands. The ZenoBrain described in [16] was further developed to provide coordination of the system which executed on the main controller computer. Each game task was modeled as a state-machine to enable the robot to choose an action depending on the child's current response.

2.5   Description of ABA Reward Identification

The ABA reward identification test is modeled after [14] to measure reward preference. Participants are given the option to press two buttons on the screen, either "Play" or "No Play" as many times as desired within a one minute timer. Clicks for each button and reward are counted. If a reward received more "Play" presses from the participant, a preference rank closer to one out of five would be given for that reward. For tied cases, the reward with a lower "No Play" click count would be given a rank closer to one.

Shown in Table 1 is a description of the five reinforcer rewards that were used in this experiment. These were chosen due to their short duration and based on the five social rewards used in [14] to enable a comparison of human interaction versus robot interaction.

Table 1. Description of ABA reinforcers

Operant	Description
Fist Pump	Raise right arm upwards bent at the elbow in a celebratory gesture
"Hooray"	Robot smiles and says "Hooray!"
"Good-Job"+nod	Robot smiles, nods once, and says "Good Job".
Dance	The robot moves both of its arms bent at the elbow in a dance-like maner
Fist Bump	Robot raises entire arm as if giving a fist bump

Fig. 2. Example of pictures shown to participants during the Select game.

2.6   Game Descriptions

1. **Recognize:** For this activity, Zeno shows a facial expression using its face and asks the child which emotion was shown. Using the touch pad computer, the emotions are displayed on individual buttons to allow the child to select his or her option. If intervention is given, Zeno delivers hints after each attempt, increasing in helpfulness towards the child. Figure 3 shows a flowchart depicting the decisions the robot makes and shows on the touch screen computer to interact with the child. All games were designed with a similar prompt and reward mechanism.

2. **Select:** Each child is shown three pictures on the touch screen computer and is asked by Zeno to identify which one shows the individual who is feeling a specific emotion. For example, presented with the pictures shown in Fig. 2, the child can be asked to choose the picture that shows an angry adult. Since there are only three pictures, the number of tries allowed is only two trials instead on the three given in all the other games.

3. **Identify:** The child is shown a single picture and asked what emotion a certain individual in the picture is feeling. For example, if the child is shown the rightmost picture in Fig. 2, the child would be asked what emotion the adult in the picture is feeling. The five options are shown in individual buttons so that the child may select their response.

3   Results

All the participants enrolled in the study were tested for their ABA reward preferences twice. Possible correlations between the participants' age and the preference for a certain reinforcer were investigated. R-squared values of each reward are as follows: Fist Pump ($R2 = 0.51$, $p < .001$), Nod ($R2 = 0.15$, $p > .05$), "Great Job" + Nod ($R2 = 0.23$, $p < .01$), "Hooray" ($R2 = .0002$, $p > .05$), Dance ($R2 = 0.29$, $p < .05$). Fist Pump and "Great Job" + Nod were the only reinforcer rewards that showed statistical significance in correlation between age and preference. Graphs plotting this relationship are shown in Figs. 4 and 5.

Initial results of two participants completing all the sessions for this study are shown in Figs. 6 and 7. The top three preferred reinforcers for each child were used in the intervention sessions as found in their ABA preference testing as

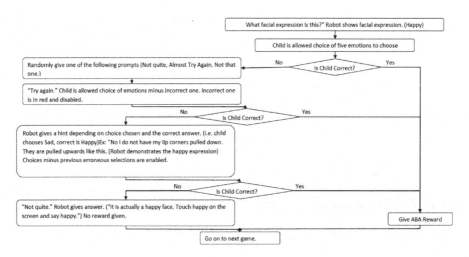

Fig. 3. Flowchart of robot and child interaction for Recognize.

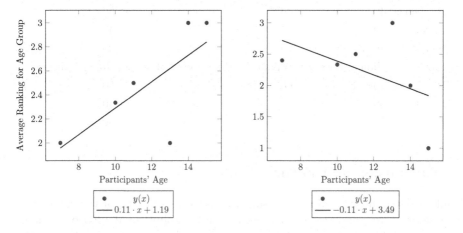

Fig. 4. Average ranking for Fist Pump across participants' ages. (Preferred by younger children)

Fig. 5. Average ranking for "Great Job" + Nod across participants' ages. (Preferred by older children)

recommended by our collaborating psychologist. The graphs are labeled by their participant number given in the study. Each graph shows the percentage correct for each of the three games, Recognize, Select, and Identify, for each session. Sessions one and five in the graphs are plotted using the robot baseline and exitline data, respectively. Sessions two through four show the percentage correct in the child's response to the first prompt given by the robot. Furthermore, the human and robot baselines and exitlines are given in Table 2 to show the results of the child's participation in the study.

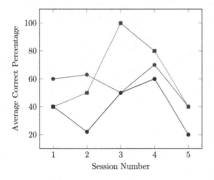
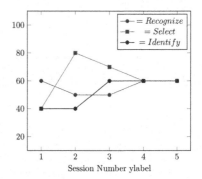

Fig. 6. % correct in sessions for Z027   **Fig. 7.** % correct in sessions for Z029

Table 2. Comparison of participant results for the human and robot baselines and exitlines.

Z027	Human Baseline	Human Exitline	Robot Baseline	Robot Exitline
Recognize	40	40	60	40
Select	0	40	40	40
Identify	40	20	40	20
Z029	Human Baseline	Human Exitline	Robot Baseline	Robot Exitline
Recognize	40	60	60	60
Select	80	40	40	40
Identify	20	80	40	60

4   Conclusion and Future Work

While a researcher was always present to ensure functioning of the robot, the interventions were designed in such a way that the robot could adaptively interact to the child's responses without the need of the researchers through state machine selection. The ability to deliver effective therapy without the need of a therapist or researcher being present would allow children to practice on their own in either home or school settings. Thus, increasing practice time rather than limiting it to only human-to-human sessions.

From the initial results shown from the participants completing the five sessions, it can be seen that the human and robot exitlines compared to the initial baselines did not improve significantly. A likely reason for this could be due to a low number of intervention sessions (three) that do not enable sufficient therapy time to see significant improvements; noting a similar occurrence in [8] and [23] where notable improvements did not occur until past the midpoint session (session eight). Thus as future work, intervention sessions should be increased taking note of when improvement is seen instead of having a fixed prescribed number.

Also from Figs. 6 and 7, there is much fluctuation in performance. For example in session three, participant Z027 was able to reach 100 % correct in the Select game, however the performance dropped in subsequent sessions. As observed from this participant, in some situations the low performance may not be due to the child not knowing the answers to the games, but instead to choosing to reduce cooperation. During the same day, the participant interacted in another study independent from this one within the same hour and that may have exasperated the child. Therefore, it is important to note the cooperation levels of the child and having sufficient breaks. Questionnaires given to participants noted that the study would have been much more attractive to their children if the robot had an ability to converse in addition to its typical intervention functions. As the goal of this ASD robot intervention is to deliver therapy equivalent or surpassing that of what could be delivered by a human, future work will involve the noted improvements and a direct comparison with human delivered intervention. The same methodology will be used, except a human will interact with the child and the results will be used to compare and learn what benefits robot versus human intervention can provide. Also, a control group having neuro-typically developed children to observe the maximum performance values as a comparison will be added to the study. While previous socio-emotional studies with ABA robotic use noted improvement [8, 24], both a robot and human therapist were used simultaneously. Therefore improvement levels of children receiving intervention solely from a robotic agent need to be further explored.

Acknowledgments. This research is partially supported by grant IIS-1450933 from the National Science Foundation. We thank Nathan Saslavsky and Raanan Hileman for their contribution in aiding in the development of the system software.

References

1. Taheri, A.R., Alemi, M., Meghdari, A., PourEtemad, H.R., Basiri, N.M.: Social robots as assistants for autism therapy in Iran: research in progress. In: 2014 Second RSI/ISM International Conference on Robotics and Mechatronics (ICRoM), pp. 760–766. IEEE, October 2014
2. DSM-5 Autism Spectrum Disorder Fact Sheet, 1st ed. American Psychiatric Association, pp. 1–2 (2013)
3. Your shop for Autism supplies, toys, tools, support. - Autism Community Store, Autismcommunitystore.com (2016). https://www.autismcommunitystore.com/. Accessed 09 Mar 2016
4. Kim, E., Paul, R., Shic, F., Scassellati, B.: Bridging the research gap: making hri useful to individuals with autism. JHRI 1(1), 26–54 (2012)
5. Fasola, J., Matari, M.: Comparing physical and virtual embodiment in a socially assistive robot exercise coach for the elderly. Technical report cres-11-003 (2012)
6. Pop, C., Simut, R., Pintea, S., Saldien, J., Rusu, A., Vanderfaeillie, J., David, D., Lefeber, D., Vanderborght, B.: Social robots vs. computer display: does the way social stories are delivered make a difference for their effectiveness on asd children? J. Educ. Comput. Res. 49(3), 381–401 (2013)
7. Diehl, J., Schmitt, L., Villano, M., Crowell, C.: The clinical use of robots for individuals with Autism Spectrum Disorders: a critical review. Res. Autism Spectr. Disord. 6(1), 249–262 (2012)

8. Tang, K., Diehl, J., Villano, M., Wier, K., Thomas, B.: Enhancing empirically-supported treatments for autism spectrum disorders: a case study using an interactive robot. In: International Meeting for Autism Research (2011)

9. The Center for Autism, Related Disorders: Globalizing Autism Treatment and Awareness, Center for Autism and Related Disorders. http://www.centerfor autism.com/aba-therapy.aspx. Accessed 26 Feb 2016

10. ASK NAO - Robot for Autism, RobotsLAB (2016). http://shop.robotslab.com/ products/ask-nao-robot-for-autism. Accessed 08 Mar 2016

11. Bekele, E., Lahiri, U., Swanson, A., Crittendon, J., Warren, Z., Sarkar, N.: A step towards developing adaptive robot-mediated intervention architecture (ARIA) for children with autism. IEEE Trans. Neural Syst. Rehabil. Eng. **21**(2), 289–299 (2013)

12. Dickstein-Fischer, L., Fischer, G.: Combining psychological and engineering approaches to utilizing social robots with children with Autism. In: 2014 36th Annual International Conference of the IEEE Engineering in Medicine and Biology Society (2014)

13. Publications — Ask Nao, Asknao.aldebaran.com (2016). https://asknao.aldebaran. com/publications. Accessed 11 Mar 2016

14. Applied Behavioral Strategies - Basics of Applied Behavior Analysis, Applied Behavioral Strategies - Basics of Applied Behavior Analysis. http://www. appliedbehavioralstrategies.com/basics-of-aba.html. Accessed 26 Feb 2016

15. Gutierrez, A., Fischer, A.J., Hale, M.N., Durocher, J.S., Alessandri, M.: Differential response patterns to the control condition between two procedures to assess social reinforcers for children with autism. Behav. Intervent. **28**(4), 353–361 (2013)

16. Salvador, M., Silver, S., Mahoor, M.: An emotion recognition comparative study of autistic and typically-developing children using the zeno robot. In: 2015 IEEE International Conference on Robotics and Automation (ICRA) (2015)

17. Social Responsiveness Scale (SRS) — WPS, Wpspublish.com (2016). http://www. wpspublish.com/store/p/2993/social-responsiveness-scale-srs-by-john-n-constant ino-md. Accessed 08 Mar 2016

18. Shop, R.: Robo Kind Specifications. http://www.robotshop.com/media/files/ PDF/hanson-robokindspecifications.pdf. Accessed Sept 2014

19. Costa, S.C., Soares, F.O., Pereira, A.P., Moreira, F.: Constraints in the design of activities focusing on emotion recognition for children with ASD using robotic tools. In: 2012 4th IEEE RAS & EMBS International Conference on Biomedical Robotics and Biomechatronics (BioRob), 24–27 June 2012, pp. 1884–1889 (2012)

20. Ekman, P.: Are there basic emotions? Psychology. Rev. **99**(3), 550–553 (1992)

21. Ekman, P.: Facial expressions of emotion: an old controversy and new findings. Philos. Trans. Roy. Soc. Lond. Ser B, Biol. Sci. **335**, 6369 (1992)

22. Hanson, D., Mazzei, D., Garver, C., Ahluwalia, A., De Rossie, D., Stevenson, M., Reynolds, K.: Realistic Humanlike Robots for Treatment of ASD, Social Training, and Research; Shown to Appeal to Youths with ASD, Cause Physiological Arousal, and Inc

23. Mohammad Mavadati, S., Feng, H., Salvador, M., Silver, S., Gutierrez, A., Mahoor, M.: A novel robot-based therapeutic protocol for training children with autism. In: 25th IEEE International Symposium on Robot and Human Interactive Communication, New York City, August 2016

24. Pop, C.A., et al.: Enhancing play skills, engagement and social skills in a play task in ASD children by using robot-based interventions. a pilot study (2014)

Interactive Therapy Approach Through Collaborative Physical Play Between a Socially Assistive Humanoid Robot and Children with Autism Spectrum Disorder

Saima Tariq[1], Sara Baber[1], Asbah Ashfaq[1], Yasar Ayaz[1(✉)], Muhammad Naveed[1], and Saba Mohsin[2]

[1] Department of Robotics and Intelligent Machine Engineering, School of Mechanical and Manufacturing Engineering, National University of Science and Technology, Islamabad, Pakistan
{14msrimestariq,sarababer,15msrimeaashfaq}@smme.edu.pk,
{yasar,naveed.muhammad}@smme.nust.edu.pk
[2] Picture Autism, NJs House, Headstart., Islamabad, Pakistan
saba.mohsin@njshouse.com

Abstract. This paper presents an exploratory study in which children with autism spectrum disorder (ASD) interact with a NAO humanoid robot in an interactive football game scenario. The study was conducted during 4 sessions with three boys diagnosed with ASD. It observed improvements in therapeutic outcomes such as social interaction, communication, eye contact, joint attention and turn taking. Qualitative and quantitative analysis were conducted using various approaches such as video documenting, surveys and assessment scales. In order to establish the efficacy of the study children interacted with their typically developing peers and parents post intervention to relate skills learned in robot-human setting to human-human setting. The quantitative results gathered and analyzed from pre and post implementation showed an increased in execution and duration of target behaviors. Manual coding and qualitative analysis of videos also verified that the proposed robot mediated play setting demonstrated improvements in social development of children with ASD in areas of communicative competence, turn taking, social interaction and proxemics and eye contact.

Keywords: Socially assistive robots · Autism · ASD · Prosocial behaviors · Interaction design · Play therapy · Socialization · Collaborative play

1 Introduction

Growing children develop important social skills when interacting with caregivers such as mothers. These interactions also cultivate cognitive skills through nonverbal communication channels of visible emotions, vocal intonation, expressions and gestures. However, children diagnosed with ASD lack basic social skills for self-initiated interaction, emotion recognition, turn-taking, joint attention, and even eye-contact [1]. These deficiencies continue to grow and affect their social activities and interaction over the

© Springer International Publishing AG 2016
A. Agah et al. (Eds.): ICSR 2016, LNAI 9979, pp. 561–570, 2016.
DOI: 10.1007/978-3-319-47437-3_55

years into adulthood. This is why appropriate treatments are employed and initiated at an early age and timely manner [2].

In recent years the role and applications of robotic systems in the therapy of children with ASD have been widely introduced and developed in the robotics community [3, 4] with interesting results and directions. Most interventions, within a comprehensive therapeutic clinical protocol, demonstrate that sustained and productive child-robot interactions lead to improved therapeutic outcomes when they employ task-specific autonomous behaviors that can be widely integrated on existing robotic systems [5].

Existing research shows that humanoid robots, used in various capacities, can elicit imitative free-form play among children with ASD [6]. By virtue of their appearance and characteristics, humanoids also promote triadic interactions among themselves, child and human controller [7]. These behaviors create the foundation for children to engage in social play - a category of play that children with autism face substantial difficulty engaging in due to the social impairments [8]. Prior research has established that autistic children engage in associative play that is an unstructured free-form social play. They also demonstrate tendencies to engage in a more structured form of cooperative play with robots however, those experimental setups have been designed for relatively high functioning children who are therefore capable of interacting with others in a group setting [9].

This exploratory study creates and implements a physical play based structured scenario involving a game of football between a socially assistive robot (SAR) and a child with ASD. The aim is to observe and achieve significant changes in social interaction and communication. The SAR used in the study is NAO owing to its accessibility and ease of operation. This study also demonstrates the viability of a novel robot-assisted technique and the various specific autonomous behaviors that may lead to cross-platform and extensive utility of a SAR within a clinical intervention.

1.1 Challenges

Children with autism exhibit common symptoms such as lack of eye contact, difficulty understanding and comprehending emotions of others. Some have strong sensory reactions and show discomfort when exposed to loud noises or physical contact. They also have impairments in social reciprocity, enjoyment and interpretation of typical social cues [10]. Generally during free play, when compared to typically developing peers, children with autism engage in a greater degree of parallel play than collaborative play [11]. Additionally, they also have trouble initiating and maintaining social interactions with their peers [12]. These characteristics call for specific interventions to build and improve the collaborative skills of children with ASD when working with peers.

2 Methodology

2.1 Participants

This study was done in collaboration with the experts at Picture Autism which is an inclusive education and therapy institute. Three children from the Picture Autism program, Islamabad participated in the study. The inclusion criteria for the children with ASD were: (a) age 3–10 years, (b) a full-scale IQ between the range of 80 and 120 and (c) a medically valid ASD diagnosis. With regards to (c) we did not have access to the individual diagnoses of autism, however the supervising teacher confirmed the previous diagnosis of each child by a medical professional. The participants were accompanied by their shadows (daily trainers at the elementary school) and parents.

Table 1 lists the characteristics of the participants in the study. The children participated on a voluntary basis and informed consent was obtained from their parents. Prior to implementation, this study received approval by the Ethics Committee of the Faculty of Robotics and Intelligent Machine Engineering at the National University of Science and Technology Pakistan.

Table 1. Participant characteristics

Child	Age (y)	Gender	Diagnosis
P1	5	Male	ASD
P2	7	Male	ASD
P3	3.5	Male	ASD

2.2 Robot Selection

Studies and research emphasize the inclination of individuals with ASD towards robots instead of passive toys [13] They also show preference towards robot-like characteristics over human-like characteristics during social interactions[14], and record a faster response when prompted by robotic movement than human movement [15, 16]. These were contributing factors towards the selection of an appropriate participating robot.

The robot used for the intervention was the humanoid robot NAO by Aldebaran-Robotics. The selection was motivated by its easy availability and affordability for research, institutes and end consumer. Another qualifying factor was its design that appeared harmless, approachable and portray emotions like a child.

2.3 Data Collection and Analysis

Some widely used methods to measure human robot interaction range from self-assessments, behavioral measures, task performance to interviews and psychophysiological measures [17]. To obtain sufficiently accurate results, more than one method was used in this particular intervention to evaluate the response of the participant towards the robot. Following each session parents and shadows filled an external-assessment questionnaire to gather their feedback on the participant's response to the therapy at different states. The standard 5 point Likert scale assessment recorded a range of behaviors ranging from – social interactions, emotions, activeness and interest. Another evaluation method used was based on video analysis of each session using one independent camera and one prebuilt camera in the NAO. This evaluation technique used microanalysis to manually code offline various micro behaviors and HRI markers such as eye gaze detachment, verbal communication and engagement duration in offline recorded video data.

At the end of each session, the robot was removed to ascertain how well the child had learned and retained various social cues from the earlier interaction and whether he was able to relate them in an actual human-human game based scenario.

3   Experiment Design

The pilot experiment was carried out at the Picture Autism Center in Islamabad. The center specifically catered to inclusive education of children and youth. The interventions were carried out in a well illuminated and sufficiently large zumba room with appropriate temperature control and padding for safety purposes.

During the child-robot interaction, each child was accompanied by either his parent or shadow or both. The role of the shadow or the parent was to merely facilitate the child as shown in Fig. 1.

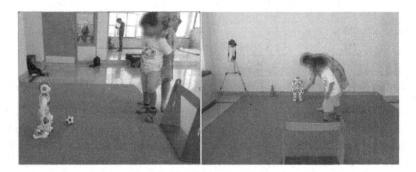

Fig. 1. Experiment setting in a structured play scenario.

3.1   The Therapy

The experimental setup and procedure was designed in collaboration with the autism behavioral experts at picture autism (third author of this paper). It was closely based on the applied behavior analysis (ABA) method which is accepted worldwide for autism interventions [18]. For example module 1 followed the ABA training method given in Fig. 2 by providing a stimulus such a greeting and if the participant failed to respond in the correct manner it gave another prompt in the shape of an instruction followed by reward/feedback. In the first prompt the robot modeled a desired end behavior and in second it reinforced the behavior with verbal instruction. Corrective prompts were only delivered in case of incorrect or lack of response. In the event the participant demonstrated the correct response to a prompt, no verbal reinforcement prompts are given.

During execution of each module initial response and behavior of the participants were recorded. The operators were monitoring the video from one external camera and one prebuilt in the NAO. Each NAO behavioral module was supervised and facilitated by the participant's parents and shadow. Intervention with each participant ran a duration of 15 min. The detailed breakdown on the experimental set-up has been presented below in Table 2.

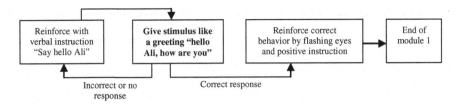

Fig. 2. Breakdown of ABA therapy for module 1 to model and teach social greetings.

Table 2. Experimental setup and modules.

Module no	Description	Intended response
1	Introductory Rapport	Attention of participant engaged
2	NAO asks questions (do you want to play football? etc.)	Participant engages in a conversation with the robot
3	NAO initiate a game of football	Participant observes the robot and the ball and responds by kicking the ball
4	NAO gives multi-sensory feed back	Behavior reinforced in the participant
5	NAO concludes the sessions with farewell remarks	Participant takes the cue and exits

4   Results and Evaluation

The experiment specified various human robot interaction metrics that were measured based on output from external cameras, NAO prebuilt cameras and a collection of targeted surveys and questionnaires filled by the parents, shadows and experts. The experimental results of stereotyped behavior, communication, turn taking were recorded as instances of occurrence. Proximity and eye contact were recorded and measured by the internal camera of the robot and calculated as the reactivity ratio which measured the duration of that metric and divided it by each training time.

Metric 1: Communication. Participants demonstrated an increase in communication during collaborative play with the robot. To measure communication instances of verbal utterances directed towards the robot in response to a question or phrase it were measured during the experiment by shadows and HRI researchers (further validated by recorded material). The increase in communication was measured against recorded baseline communication of each child in control settings (i.e. with shadows in the absence of robot when conducting a similar experiment). It was observed that during the study the participant P1 and P3 exhibited few autistic traits such as repetitively using unintelligible words and talking in monotone. P2 demonstrated signs of very excited and animated conversation. Table 3 shows these instances of each participant in each session. It can be seen that P3 engaged in very little instances of direct communication with the robot because the participant was dominantly nonverbal. While both P1 and P3 showed an increase in communication over the three sessions, P2 demonstrated a decrease in the last session due to high fragmentation of mind and mood.

Table 3. Instances recorded of verbal communication directed towards the robot.

Participant ID	Session 1	Session 2	Session3
P1	12	11	15
P2	8	10	6
P3	2	3	5

Metric 2: Proxemics. When the distance between robot and the participant as $<= 10$ inches it was measured as close proximity and marked the level of comfort and familiarity the participant had with the robot. Proximity was measured by the in-built camera of the robot and was calculated as a reactivity ratio which divided the total time in close proximity by the total training time. The participants displayed great keenness in playing with the hands of the robot, touching its head and arranging its arms (Fig. 3).

Fig. 3. Participants familiarize themselves with the robot during the first module of the study

The Fig. 4 illustrates the total time, recorded for each session, during which the participant was in close proximity of the robot. The total duration of each session was not more than 900–1200 s.

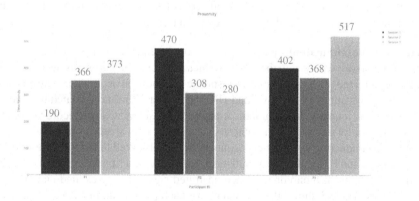

Fig. 4. Close proximity between robot and Participant 1, 2 and 3 over 3 sessions

Metric 3: Turn Taking. Turn taking is a very important social skill in communications, interactions and playing games. In earlier interventions turn taking was modeled by the parent or the shadow with the robot and the participant acted upon it. An instance was recorded each time the participant displayed turn taking behaviors by kicking the ball and waiting for the robot to kick it back. One of the major challenges in this exercise was easy distraction and mental fragmentation of the participant while waiting for the robots turn. However, there were also many instances in which the participant showed patience and anticipation towards a response from the robot. Those instances were collected and recorded by observers and HRI researchers. The collected data for the 3 participants over the course of the intervention is displayed below in Table 4.

Table 4. Instances recorded of turn-taking.

Participant ID	Session 1	Session 2	Session3
P1	3	6	7
P2	3	1	5
P3	7	6	7

Metric 4: Eye Contact. During the intervention, participants showed a greater duration of eye contact particularly during direct interaction against a recorded baseline. They also exhibited shifting gaze between the robot and the ball during the ball kicking module demonstrating joint attention at a very basic level. Eye contact was measured using the prebuilt camera in NAO. Offline video processing was used to make estimates (Fig. 5).

Fig. 5. Offline proximity and eye contact analysis

Figure 6 illustrates the average eye contact time for each participant for 3 of the total four sessions. The total duration of each session was not more than 900–1200 s.

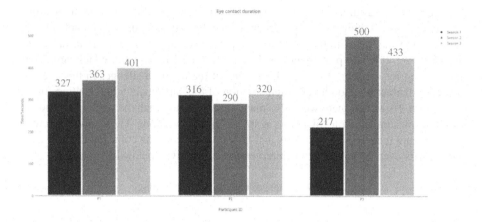

Fig. 6. Aggregate eye contact duration b/w the robot and *Participant 1, 2 and 3* over 3 sessions

4.1 Post Intervention Observations

At the conclusion of each intervention, each participant was encouraged to engage in collaborative play with their shadow/parent. This part of the intervention lasted for two minutes max in which participant P1 and P3 demonstrated improved play skills and shared attention while P2 lagged in performance due to fragmentation. Parents/Shadows filled assessment's regarding the participant's attitude throughout during various modules of the intervention. Each item was evaluated on a five point Likert scale ranging from 'strongly disagree' (1) to 'strongly agree' (5). Figure 7 charts some general responses regarding the overall study.

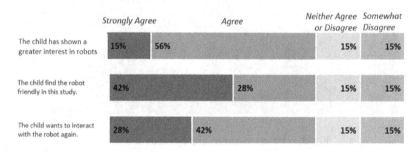

Fig. 7. Parental/shadow evaluation of the participants behavior towards the robot used in study

Table 5 shows some behavior questions and their aggregate Likert scale score on Fig. 8. Due to the small sample size, the aggregate ratings may not be statistically robust however it was established and agreed that the proposed system showed results and efficacy in facilitating and developing social skills through play,

Table 5. Some items on the questionnaire regarding the participant's attitude during study.

Question-ID	
Q3	The participant displayed emotions and towards the robot
Q4	The participant engaged in communication with the robot
Q5	The participant was in close proximity of the robot
Q6	The participant was more active during the test study

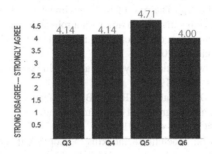

Fig. 8. Parental/shadow evaluation of the participant's social interactions and response to robot

5   Conclusion and Discussion

Based on data and results, this study suggests that the physical and social games based scenarios can allow the child to maximize their independent decision making and intuitive problem solving ability - consequently fostering social exchanges that rapidly and seamlessly generalize from robot–child to child–other interactions.

Out of the three participants, P1 and P3 showed great improved against measured HRI markers however, P2 did not exhibit a similar improvement. This was attributed to a number of factors such as his high mental fragmentation, sensitive moods and age. While the size of the population was small and variance in age was large to draw solid conclusions, this intervention did provide valuable insights into the social settings and constructs that encourage and ingrain prosocial behaviors among the children and robot.

Acknowledgments. We are grateful to Picture Autism for facilitating this study and the parents and children for their participation.

References

1. Allen, D.A.: Autistic spectrum disorders: clinical presentation in preschool children. J. Child Neurol. **3**, 48–56 (1988)
2. Lovaas, O.I.: Behavioral treatment & normal educational and intellectual functioning in young autistic children. J. Consult. Clin. Psychol. **55**(1), 3–9 (1987)
3. Feil-Seifer, D., Mataric, M.: Robot-assisted therapy for children with autism spectrum disorders. In: ACM Proceedings of the 7th International Conference on Interaction Design and Children, pp. 49–52 (2008)

4. Scassellati, B., Admoni, H., Mataric, M.: Robots for use in autism research. Annu. Rev. Biomed. Eng. **14**, 275–294 (2012)
5. Cabibihan, J.-J., Javed, H., Ang Jr., M., Aljunied, S.M.: Why robots? a survey on the roles and benefits of social robots in the therapy of children with autism. Soc. Robot. **5**(4), 593–618 (2013)
6. Robins, B., Dautenhahn, K., te Boekhorst, R., Billard, A.: Robotic assistants in therapy and education of children with autism: can a small humanoid robot help encourage social interaction skills? Univers. Access. Inf. Soc. **4**(2), 105–120 (2005)
7. Robins, B., Dautenhahn, K.: The role of the experimenter in hri research—a case study evaluation of children with autism iteracting with a robotic toy. In: Proceedings of the 15th IEEE International Symposium on Robot and Human Interactive Communication, pp. 646–651. IEEE Press, Piscataway (2006)
8. Howlin, P.: An overview of social behaviour in autism. In: Social Behaviour in Autism, pp. 103–132. Plenum, New York (1986)
9. Wainer, J., Ferrari, E., Dautenhahn, K.: Robins B The effectiveness of using a robotics class to foster collaboration among groups of children with autism in an exploratory study. Pers Ubiquitous Comput **14**(5), 445–455 (2010)
10. Weiss, M.J., Harris, S.L.: Teaching social skills to people with autism. Behav. Modif. **25**, 785–802 (2001)
11. Bauminger, N., Solomon, M., Aviezer, A., Heung, K., Brown, J., Rogers, S.J.: Friendships in high-functioning children with autism spectrum disorder: Mixed and non-mixed dyads. J. Autism Dev. Disord. **38**, 1211–1229 (2008)
12. Bauminger, N., Shulman, C., Agam, G.: Peer interaction and loneliness in high-functioning children with autism. J. Autism Dev. Disord. **33**, 489–507 (2003)
13. Dautenhahn, K., Werry, I.: Towards interactive robots in autism therapy: Background motivation, and challenges. Pragmat. Cogn. **12**, 1–35 (2004)
14. Robins, B., Dautenhahn, K., Dubowski, J.: Does appearance matter in the interaction of children with autism with a humanoid robot? Interact. Stud. **7**, 509–512 (2006)
15. Bird, G., Leighton, J., Press, C., Heyes, C.: Intact automatic imitation of human & robot actions in autism spectrum disorders. Proc. Biol. Sci. **274**, 3027–3031 (2007)
16. Pierno, A.C., Mari, M., Lusher, D., Castiello, U.: Robotic movement elicits visuomotor priming in children with autism. Neuropsychologia **46**, 448–454 (2008). [PubMed: 17920641]
17. Bethel, C., Murphy, R.: Review of human studies methods in HRI and recommendations. Int. J. Social Robot. **2**, 347–359 (2010)
18. Harris, S.L., Delmolino, L.: Applied behavior analysis: its application in the treatment of autism and related disorders in young children. Infants Young Child. **14**, 11–18 (2002)

Examine the Potential of Robots to Teach Autistic Children Emotional Concepts: A Preliminary Study

Huanhuan Wang[1(✉)], Pai-Ying Hsiao[2], and Byung-Cheol Min[2]

[1] Learning, Design and Technology, Curriculum and Instruction,
Purdue University, West Lafayette, IN 47907, USA
wang2306@purdue.edu
[2] Computer and Information Technology, Purdue University,
West Lafayette, IN 47907, USA
{hsiao20,minb}@purdue.edu

Abstract. In this preliminary study, we developed a set of humanoid robot body movements which are used to express four basic emotional concepts and a set of learning activities. The goal of this study is to collect feedback from subject matter experts and validate our design. We will integrate them to improve the designs and guide autistic children to learn emotional concepts. To validate our designs, we conducted an online survey among general public people and four in-person interviews among subject matter experts. Results show that the body movement Happiness and Sadness could express emotions accurately, while the Anger and Fear movements need more improvements. According to the subject matter experts, this robot mediate instruction is engaging and appropriate. To better match autistic children, the instructional content should be tailored for individual learners.

Keywords: Autistic children · Assistive robotics · Emotional concept · Dancing · Instruction · Design

1 Introduction

Autism Spectrum Disorder (ASD) is a developmental disability featuring social communication impairments, restricted interests and repetitive behaviors [4]. Based on the National Autistic Society (NAS, 2004), autistic individuals have impaired abilities in social interaction, social communication and imagination. Meanwhile, autism also brings inconvenience and lost to their families. One study [14] described how challenging life can be for a family having a child with autism. Therefore, autism becomes an important topic for the researchers from different areas. Prior researchers found social development delay of the autistic individuals results from emotional impairment, since they had difficulties to recognize social cues [2]. Frith [6] found autistic individuals feel it is difficult to synthesize information in a coherent way. Although researchers did not make an agreement

© Springer International Publishing AG 2016
A. Agah et al. (Eds.): ICSR 2016, LNAI 9979, pp. 571–580, 2016.
DOI: 10.1007/978-3-319-47437-3_56

on the causes of ASD, some studies [3,10,13] confirmed behavioral improvements of the autistic children through interventions. Based on the causes of autism and evidences that some interventions do work, we designed a curriculum to teach autistic children emotional concepts. The general method is to create a learning environment for autistic children to build connections between different modalities of emotional concepts, such as facial expression, body language and story scenario (Fig. 1).

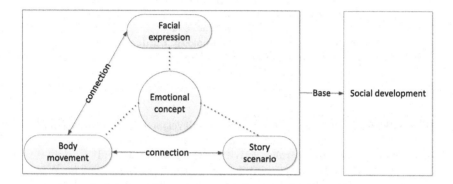

Fig. 1. Research conceptual framework

The research goal is to design robot mediated instructional activities for autistic children to learn emotional concepts, and validate this design. We have two research questions: (1) Is a robot able to make body movements with emotions embedded which are identifiable by general public? General public is defined here as the regular people with the age ranging from 18 and 60. (2) Is robot mediated instruction feasible and usable for teaching autistic children basic emotional concepts? The first question builds a base for the second one. Since only when the robot could express emotions accurately, they could be used to teach ASD children emotional concepts. We are the first team proposed the use of humanoid robots to teach autistic children emotional concepts in this study. We designed and developed a set of robot body movements with 4 basic emotions and a set of learning activities which could be used by teachers and therapists to teach autistic children emotional concepts. Based on the feedback from the related subject matter experts, the proposed instruction is engaging, and has potential to teach emotional concepts. With adaption, it can be applied to teach a broad spectrum of autistic individuals.

2   Related Works

Prior researchers tried different interventions to facilitate autistic children's social development, such as human interventions, photographs, and video games

[2,9]. Recently, robots emerged as a new tool. Compared with other interventions, robots have several advantages. First, autistic children prefer more controlled environments. Meanwhile, human's social behavior is complex and subtle, which is difficult for autistic children to follow [10]. It will be easier to create a controlled and simplified environment by using robots. Second, the learning objectives here are emotional concepts and the final goal is to communicate with real human. Thus, humanoid robots are better than other interventions, since robots could create a more human-like conversational environment.

Some researchers used robots to help autistic children improve their social skills [3,10,13]. Robins et al. [10] developed a social robot KASPAR. After intervention, autistic children could gaze, touch robots and co-present children, or even communicate with human by making the robots moving. Wainer et al. [13] used the same robot KASPAR to enhance autistic children's collaboration skills by having them play video games with robots. Costa et al. [3] added more sensors on KASPAR and they found autistic children showed more appropriate social-physical interaction with co-present human after interventions. Social robots are also used to communicate, display and recognize the emotions, develop social competencies, and maintain social relationships [5]. These studies still have limitations [1], since they often provided free form activities and exposed participants in an environment where they have a robot. Usually there is a small sample size and results with a large variation. Therefore, it is difficult to make a final conclusion for intervention effectiveness. Besides, there is few previous studies targeting autistic individual's emotional concepts learning and using robot body movements to help create emotional connections.

3 Study Design

3.1 Design Process

Ros et al. [12] shed lights on our study design. They used the framework of creative dance to help children build connections between the concepts in subject and body movement areas. It enhanced learner's understanding of subject concepts. Their instructional activities had 4 steps, including Warm-up, Exploration, Creation, and Performance and Appreciation. We imitated this framework to design learning activity sequence. The learners should be diagnosed at a medium level of autism and have basic language skills. A robot named Bio will guide autistic children to learn 4 basic emotion concepts (Happiness, Sadness, Anger, and Fear).

We designed a 4-week long learning activity. One day in each week, ASD children will take a 20-min learning session (Fig. 2). The sessions in different weeks will follow the same structure (Fig. 3) and have three roles, autistic children (learners), a robot (tutor) and a human facilitator. First, all of the children will take a pretest through listening to short stories with emotions embedded. Then they will watch two robot movements and identify which movement better matches the emotion expressed by each story. The result will be used to measure their current performance and improvement.

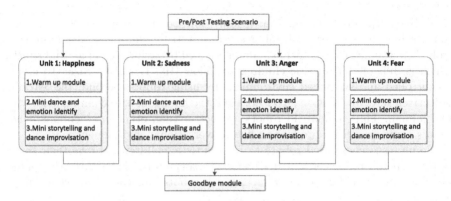

Fig. 2. Course structure chart (designed based on creative dance [12])

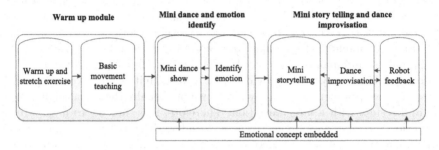

Fig. 3. Module structure chart (designed based on creative dance [12])

In the warm up module, Bio will give a self-introduction, and briefly talk about the purpose of the class. Then, Bio will guide the autistic children to do basic body movements slowly. In the mini dance show and emotion identify module, Bio will perform a movement sequence with a specific emotion, for example, Happiness. Then, the facilitator will offer learners 6 facial cards that corresponds to 6 facial expressions (Fig. 4). After that, the participants should pick up one card to match the emotion in the robot body movement. If they select a correct card, Bio will say "Good job!". Else, they should try again. Bio will tell them the correct answer when they answer incorrectly more than two times. In the mini storytelling and dance improvisation, Bio will tell a story which could express one special emotion, while the facilitator will show a story card to help understand the story. Then the participants will be required to perform a dance based on their understanding of the story. If the dance is incorrect, the robot will prompt the participants to dance again. In the end, the robot will perform one correct movement. In the Goodbye module, the robot will confirm participants' performance.

| Anger | Happiness | Surprise | Disgust | Sadness | Fear |

Fig. 4. Facial expression cards (source: the Grimace Project)

3.2   Technical Requirements and Development Process

To design a robot tutor, which is required to speak specified sentences and make simple body movements to express specific emotions, we investigated robot dance videos [11], 3-D animation [8], and emotional robots [15] to extract behavioral features to express specific emotions. Then, we developed 4 robot body movements on a miniature-humanoid robot platform DARwIn-OP from ROBOTIS which has 20 degrees of freedom and a speaker, and used their Roboplus Motion software to design each movements to express Happiness, Sadness, Anger and Fear (Fig. 5). By integrating the control sequence, motion files, and pre-recorded sound files into a C++ code, human facilitator can press keyboard and the robot will speak and do the corresponding movement. We integrated these body movements and designed a set of learning activities. This process was demonstrated and recorded in the video below (Fig. 6).

4   Data Collection

This study used an unique mixed method to collect data from the subject matter experts before we implement the design among the autistic children. To protect the end users we will iterate the design based on the feedback we get from stakeholders, which is a typical method [7, 12]. First, we conducted an online survey to find whether the robot body movements developed could express emotions that could be identified by the general public. We sent the survey to the students and faculty in the researchers' department and to their friends and families around the world. The participants watched the videos of the 4 robot body movements, and for each movement they selected one emotion which they think can fit the movement best. The reasons of their answers were also collected. The recognition rates and comments were used to measure and improve the design. Second, We conducted in-person interviews with 4 subject matter experts, including two researchers in the autism area (P1, P3), one industry designer who designs mobile applications for autistic children (P2), and one parent of the autistic children (P4). The purpose was to find out how is the idea of using humanoid robots to teach autistic children, as well as the feasibility and usability of the instruction. In the interview, they rated the simplicity of robot body movements design for

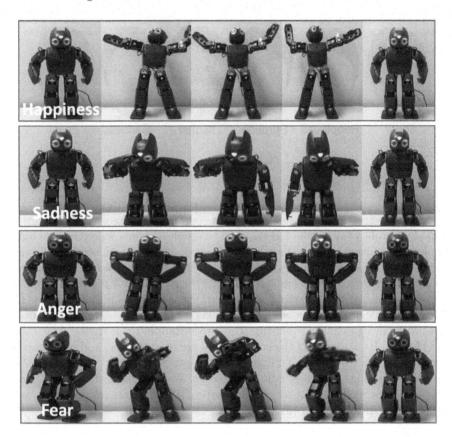

Fig. 5. Screen shots from left to right in time sequence for the four robot body movements. (demo videos available at https://goo.gl/zNH2po) (Color figure online)

Module 1 Warm-up   Module 2 Mini dance   Module 3 Mini story telling

Fig. 6. Instructional modules and process simulation   (a demo video available at https://goo.gl/pDBeM1)

autistic children, the feasibility and usability of the learning modules in a range from 1–5 scores (score 5 being simplest/most feasible/most usable, score 1 being least simple/least feasible/least usable). They also commented about the reasons of their answers.

5   Results and Discussion

5.1   Results of the Survey for Robot Body Movements Design

From the survey, we got 32 valid responses ($n = 32$). They come from the countries in Asia, North America and South America, and represent different cultural background, with an age ranging from 18 to 60. The recognition rate is 94 % for Happiness, 91 % for Sadness, 81 % for Anger, and 78 % for Fear (Fig. 7). The survey reveals that Happiness is the easiest to recognize, while Fear is the most difficult. The key comments are summarized in Table 1.

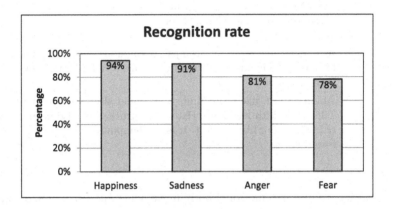

Fig. 7. General public recognition rate

Happiness and Sadness got higher scores because of the clear design features. For Happiness, the robot body is open, positive and free with hands waving and eyes looking up. For Sadness, robot is bent down with banging head against hands and touching his eyes. However, there are some features making it confusing, such as the fast moving arms which make people also feel the robot is angry. For Anger, the arms are in a defensive position and eyes are looking into the air, while a few people feel it is bragging. Last, covering the eyes, defending itself make it fear. However, shielding its face makes people feel it is surprised. Thus, emotional expression is subtle which makes people have different perceptions. Overall we got high recognition rates, but more efforts will be needed to make the robot body expression more general.

5.2   Results of the Interviews for Module Instructional Design

From the interviews, we got 4 valid responses ($n = 4$). P1 and P3 have 20 and 18 years of research experiences. P2 has 4 years of experiences on program design for autistic people, while P4 has been a mother of an autistic child for 7 years. All participants confirmed the instruction is fun and engaging, and the idea of using humanoid robot to teach emotional concepts is appropriate. It is a good

Table 1. General public's main comments on the 4 body movements design

Movements	Key features mentioned in the comments	Further suggestions
Happiness	The rhythmic moment shows that it is in happiness. It reminds me of an autistic boy. He would move like this and also utter very high-pitch cry of joy	It is fast and the light on the head is red or yellow. This made me feel that it is angry
Sadness	Head bending down, rotating, getting down on his knees Hides face behind hands. Touches and covers the eyes, like wiping tears away	It seems to be sobbing but could also mean happiness
Anger	Arms in defensive position. Stomps his foot with hands on his waist. Looks impatient	It could also mean happiness as this could be a form of dancing
Fear	It is hiding itself, like protecting. Tilts away with hands up. His red light is on, like an alert. It is cowering	It could also be caused by being surprised or also mean happiness

way for learners to imitate emotional body movements and learn how to express emotions. The body movements are very representative of the emotions. The reward system is clear and easy to understand.

For the simplicity of the movements, the participants provided rating scores (ranging from 1–5) and comments. The features of Happiness are clear for identification ($M = 4.0$). But, P2 thought Happiness might be understood by different children as different meanings. Sadness is simpler ($M = 4.75$), but the movement is a little long and difficult for ASD children to follow. P4 also suggested making it a smaller gesture to match the real emotional expression in daily life. For Anger ($M = 4.38$), P4 mentioned her son did the same gesture to express the same emotion recently. However, P1 thought it might be a little exaggerated. For Fear ($M = 4.12$), they thought it looks like more coward or startled.

For the module design, the participants rated scores for the feasibility and usability ranging from 1–5, and also provided their comments on the reasons. The Warm-up module is rated with $M = 4.25$, for feasibility and $M = 4.5$, for usability, the Mini dance module rated with $M = 3.25$, for feasibility and $M = 3.5$ for usability, and the Mini storytelling module is rated with average $M = 3.63$, for both feasibility and usability. For the instructional design, the participants have several suggestions: (1) Tailor the content. It is currently designed for high functioning autistic individuals. This framework could be adapted for every ASD individuals. It can teach low functioning kids basic emotions and high functioning autistic adults subtler emotions. P2 suggested designing 4 or 5 levels of content for novices and find out where they are. It can start from less choices (less facial cards). Then change the content based on students' performance. For the child who has better language skills, they can handle 6 facial expression cards. (2) Disconnection. Subject matter experts found there is disconnection between the displayed teaching of the emotions and the facial cards. Prerequisite trials should be added to verify whether learners have symbolic understanding first. Then add direct instruction to connect emotion, motion, and cards. (3) Follow up and transfer. It is suggested to add modules for learning generalization and transfer, such as taking Bio out to see if the learning can transfer. Currently, the emotion expressed is not exact like that in the real life, since expression in the real life is subtler. The design should use smaller motions to express emotion in real life.

The participants also provided suggestions toward the technical issues. For example, the wording of the verbal instruction is suggested to be short, brief, simple and concrete. To attract learners, we can design more interactions, for example, let learners operate the robot. It is better to keep the color of the head light the same or pair a color with an emotion. It is also suggested to simplify the facial cards. Some experts suggested using photograph of real children's face, since it is easier for children to understand. Last, the quality of the speech is not good and need to improve.

Overall we find using humanoid robots for teaching is engaging but the instructional design needs to be improved. Tailoring the content for different autistic learners is strongly recommended. For some emotions in robot body movements, different experts have some different perceptions, which is consistent with the literature statements. But generally they made an agreement on the perceptions of the most movements we developed. There are still some aspects where they did not agree with each other, for example, using photos of real facial expression card or using simplified graphs. This might need further research. There is also inconsistency between the number of facial cards (6) and the types of emotion concepts (4), because we want to test this design on a higher difficulty level initially. They will be changed and kept consistent when we test it with the autistic children finally.

6 Conclusion and Future Work

Using humanoid robots with dancing capabilities to teach autistic children emotional concepts was proposed in this preliminary study for the first time. A set of learning activities were developed for teachers and therapists to teach autistic children. Results of the survey and interviews shows the emotions in the robot body movements we developed could be identified by the general public. The subject matter experts confirmed the proposed robot mediated instruction under the creative dance framework is appropriate and engaging. They also offered valuable feedback, such as tailoring the content to be applied to a broad spectrum of autistic individuals. Considering the robot's capability, it also might be used for teaching dance for general public people. In the future we will integrate these feedback, improve the body movements and instructional design, and test the refined curriculum among autistic children.

References

1. Begum, M., Serna, R., Yanco, H.: Are robots ready to deliver autism interventions? a comprehensive review. Int. J. Soc. Robot. **8**, 157–181 (2016)
2. Boucenna, S., Narzisi, A., Tilmont, E., Muratori, F., Pioggia, G., Cohen, D., Chetouani, M.: Interactive technologies for autistic children: a review. Cogn. Comput. **6**, 722–740 (2014)
3. Costa, S., Lehmann, H., Dautenhahn, K., Robins, B., Soares, F.: Using a humanoid robot to elicit body awareness and appropriate physical interaction in children with autism. Int. J. Soc. Robot. **7**, 265–278 (2014)
4. Diagnostic and statistical manual of mental disorders. American Psychiatric Association, Washington, DC (2000)
5. Fong, T., Nourbakhsh, I., Dautenhahn, K.: A survey of socially interactive robots. Robot. Auton. Syst. **42**, 143–166 (2003)
6. Frith, U.: Autism and theory of mind. In: Gillberg, C. (ed.) Diagnosis and Treatment of Autism, pp. 33–52. Springer, US (1989)
7. Huijnen, C., Lexis, M., de Witte, L.: Matching robot KASPAR to autism spectrum disorder (ASD) therapy and educational goals. Int. J. Soc. Robot. **8**, 445–455 (2016)
8. Larsson, P.: Discerning Emotion Through Movement: A study of body language in portraying emotion in animation (2014)
9. Ozcan, B., Caligiore, D., Sperati, V., Moretta, T., Baldassarre, G.: Transitional wearable companions: a novel concept of soft interactive social robots to improve social skills in children with autism spectrum disorder. Int. J. Soc. Robot. **8**, 471–481 (2016)
10. Robins, B., Dautenhahn, K., Dickerson, P.: From isolation to communication: a case study evaluation of robot assisted play for children with autism with a minimally expressive humanoid robot. In: Second International Conferences on Advances in Computer-Human Interactions, pp. 205–11 (2009)
11. ROBOTIS. http://en.robotis.com/BlueAD/board.php?bbs_id=downloads&mode =view&bbs_no=69693&page=1&key=&keyword=&sort=&scate=EXAMPLE
12. Ros, R., Baroni, I., Demiris, Y.: Adaptive humanrobot interaction in sensorimotor task in-struction: from human to robot dance tutors. Robot. Auton. Syst. **62**, 707–720 (2014)
13. Wainer, J., Dautenhahn, K., Robins, B., Amirabdollahian, F.: A pilot study with a novel setup for collaborative play of the humanoid robot KASPAR with children with autism. Int. J. Soc. Robot. **6**, 45–65 (2013)
14. Werner DeGrace, B.: The everyday occupation of families with children with autism. Am. J. Occup. Therapy **58**, 543–550 (2004)
15. Zecca, M., Mizoguchi, Y., Endo, K., Iida, F., Kawabata, Y., Endo, N., Itoh, K., Takani-shi, A.: Whole body emotion expressions for KOBIAN humanoid robot preliminary experiments with different emotional patterns. In: The 18th IEEE International Symposium on Robot and Human Interactive Communication, RO-MAN 2009, pp. 381–386 (2009)

Longitudinal Impact of Autonomous Robot-Mediated Joint Attention Intervention for Young Children with ASD

Zhi Zheng[1(✉)], Guangtao Nie[1], Amy Swanson[2], Amy Weitlauf[2], Zachary Warren[2], and Nilanjan Sarkar[1,3]

[1] Electrical Engineering and Computer Science Department, Nashville, USA
{zhi.zheng,guangtao.nie,nilanjan.sarkar}@vanderbilt.edu
[2] Vanderbilt Kennedy Center Treatment and Research Institute for Autism Spectrum Disorder, Nashville, USA
{Amy.swanson,Amy.weitlauf,zachary.warren}@vanderbilt.edu
[3] Department of Mechanical Engineering, Vanderbilt University, Nashville, TN 37212, USA

Abstract. Literature suggests that a robot is able to capture attention and elicit positive social communication behaviors in many children with ASD. However, there are few studies reported regarding the longitudinal impact of autonomous robot-mediated interventions. In this paper, we introduce a new autonomous robotic system for teaching children with ASD joint attention skills, which is among the core developmental impairments in ASD. This system automatically tracks the participant's behavior during intervention and adaptively adjusts the interaction pattern based on the participant's performance. Based on this system, we report a longitudinal robot-mediated joint attention intervention on a user study for 6 children with ASD. First, four sessions of one-target interventions were conducted for each participant. Then we tested their joint attention skills after 8 months in two sessions on: (1) response to one target; (2) response to two targets. The results showed that this autonomous robotic system was able to elicit improved one-target joint attention performance in young children with ASD over the course of 8 months. The result also suggested that the joint attention skills that the participants practiced in the one-target sessions might help them interact in a more difficult task. The robot also attracted the participants' attention constantly during this long term intervention.

Keywords: Robot-mediated joint attention skills training · Children with ASD · Longitudinal study

1 Introduction

Autism spectrum disorder (ASD) is a developmental disorder that impacts 1 in 68 children in the US [1]. It is associated with enormous individual, familial, and social costs across the lifespan of individuals with ASD [2]. Evidence suggests that early detection and intensive behavioral intervention are critical to optimal treatment of ASD [3]. It has been shown that technology-assisted interventions have great potential for children with ASD due to their controllability, replicability, flexibility, and cost effectiveness [4].

© Springer International Publishing AG 2016
A. Agah et al. (Eds.): ICSR 2016, LNAI 9979, pp. 581–590, 2016.
DOI: 10.1007/978-3-319-47437-3_57

Early in 1976, Weir and Emanuel [5] found that robots can be used to elicit positive social interactions in children with ASD. Since then, numerous works have been reported regarding different types of robotic technologies designed for children with ASD. It has been shown that many children with ASD preferred robot-like characters over non-robotic toys [6], and in some circumstances responded faster when cued by a robot than a human [7]. Robins et al. [8] developed a humanoid robotic doll to investigate imitation learning skills in children with ASD. Kozima et al. [9] designed a small creature-like robot called "Keepon" for eliciting positive interaction behaviors in younger children. Feil-Seifer and Mataric [10] found that contingent activation of a robot during interactions yielded immediate short-term improvement in social interactions. Many of these earlier works required human operators to remotely control the robot, due to the technical limitations on detecting interaction cues from the participants. Thus the application of these robotic systems highly depended on the availability of well-trained human operators.

To solve this problem, researchers are investigating highly autonomous robotic platforms. Zheng et al. [11] introduced an adaptive robot-mediated imitation learning system for young children with ASD. Fujimoto et al. [12] developed a robotic system which could automatically evaluate a child's motion in real-time. Bekele et al. [13] designed an automatic robot-mediated joint attention intervention system. However, while one time improvements have been registered in robot-mediated interventions [14], there are few works [15] that report on the long term impact of autonomous robot-mediated intervention for children with ASD. Therefore, in this study, we investigated the longitudinal impact of a non-contact autonomous robotic intervention system (NARIS) for children with ASD. In particular, we focused on joint attention skill, which refers to sharing attention with others, such as pointing to an object and sharing gaze on the same target. It is one of the most fundamental building blocks of social communication and is central to the etiology and treatment of ASD [16].

The first contribution of the current work is the development of NARIS. It is an advanced autonomous system based on the adaptive robot-mediated intervention architecture (ARIA) [13]. A one-session study with ARIA showed that young children with ASD paid more attention to the administrator and performed better when mediated by a robot than a human therapist [13]. However, the attention tracking in ARIA required body-worn physical sensors for gaze detection that were rejected by almost 40 % of the participants. In NARIS, we applied a real-time non-contact gaze detection method which allowed us to not only eliminate the discomfort of wearing physical sensors, but also make the system run autonomously without any human interference.

The second contribution of the current work is a longitudinal user study. We tested the participants' joint attention skills on: (1) the longitudinal impact of the autonomous robot-mediated intervention; and (2) whether the simple joint attention skills practiced in initially can be transferred to a higher level of interaction. In addition, we evaluated whether the participants' attention on the robot held over a long period.

Previous studies conducted with invasive [13] and non-autonomous [17] systems indicated that children with ASD would hold their attention on the robot in a longitudinal study, and the system would elicit within-system joint attention behaviors from these children. In this study, we anticipated that we would get similar results by using NARIS.

The rest of the paper is organized as follows. Section 2 describes the NARIS design and development. Section 3 presents the intervention design. The experimental results and implications are discussed in Sect. 4. Section 5 concludes the whole work.

2 NARIS Design and Development

2.1 System Architecture

The NARIS system consisted of three modules and a supervisory controller (Fig. 1). The supervisory controller run in a computer in the background, and the other 3 modules are configured as follows. The NARIS system was configured in an experiment room as shown in Fig. 2.

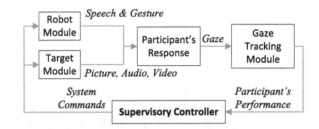

Fig. 1. NARIS architecture

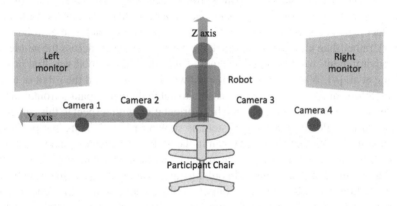

Fig. 2. NARIS configuration

The target module consisted of two computer monitors (43 cm × 70 cm) that served as the attentional targets for the participants. One monitor was placed on the left and the other one was placed on the right of the participant's chair, where joint attention stimuli were presented to help the participant turn to a target direction (left or right). Based on different prompt levels, the target monitor displayed static pictures, audio clips, or video clips of children's interest.

The robot module used a humanoid robot NAO [18], which is widely applied in human-robot interaction studies. NAO, a 25 degrees of freedom humanoid robot, has the size of a young child (height = 58 cm, weight = 4.3 kg). Mainly two robot motions were used in this study: head turn to a target and pointing towards the target using one arm. NAO is embedded with text-to-speech functions which was used for verbal prompts.

The robot and the two monitors were all 2 m away from the participant's chair. The gaze tracking module included a camera network that could track the participant's gaze direction. The three modules ran in parallel and coordinated with each other using the supervisory controller both as a mediator and a commander. This ensured that the execution of each module would not interfere with others. The supervisory controller received the participant's gaze direction from the gaze tracking module, and sent corresponding system commands to the robot and target modules accordingly, based on an interaction protocol discussed in Sect. 3. Basically, if a participant followed the instructions of the robot (looking at the target monitor), then the robot would give a reward. Otherwise, a higher level of prompt would be given by the system. Thus, this system provided real-time closed-loop interactions.

2.2 Gaze Tracking

The key component of the NARIS is the gaze tracking. The participant's gaze direction was derived from his/her head orientation. This is a common strategy in human-machine interaction [19]. Each camera in the gaze tracking module was driven by the Intraface software [20], which detected the head orientation of the participant with respect to the current camera based on the participant's frontal face image. However, since the interaction space was large in this study, using one camera could only catch the participant's frontal face when he/she was facing one of the system components (e.g., robot or one monitor), rather than all of them. To track the participant's gaze across the whole interaction environment, we developed a camera network. This method was first proposed by Zheng et al. [4]. Whenever the participant was looking around the robot or any one of the monitors, at least one camera was able to catch the participant's frontal face and estimate the head orientation. Then, the head orientation with respect to this camera was transformed to a global coordinate system as shown in Fig. 2. Thus, the positions of different system components were known in the global coordinate system.

Human's gaze direction is not quite aligned with head orientation when the head is turned largely in a horizontal plane. In order to approximate the gaze direction from the head orientation, we collected data from 10 adults to build a mapping function. They were instructed to shift their gaze direction from −90° to 90° in horizontal and allowed to naturally turn their head in this procedure. Then a regression model that maps the recorded head orientation to the actual gaze direction of these adults was built and embedded in NARIS for gaze tracking. When new head orientation data were detected, these would be mapped to the corresponding gaze direction. In the experiment, the system would consider a participant responded to a monitor if his/her gaze fell in the direction of this monitor.

3 Interaction Design

3.1 Participants

The longitudinal impact of NARIS was tested by 6 Caucasian boys with ASD (age: Avg = 2.80, SD = 0.37 years at the start of the phase 1 and Avg = 3.52, SD = 0.33 years at phase 2). They were recruited from a research registry of the Vanderbilt Kennedy Center, and this study was approved by the Vanderbilt University Institutional Review Board. They had confirmed diagnoses by a clinician based on DSM [21] criteria. They met the spectrum cut-off on the Autism Diagnostic Observation Schedule [22] (Avg = 22.17 SD = 5.08, indicating moderate–to–severe concern), and had existing Intelligent Quotient data (measured by the Mullen scale [23], Avg = 52.67, SD = 4.03) regarding cognitive abilities in the registry. Parents of these children also completed the Social Responsiveness Scale–Second Edition [24] (T score: Avg = 65, SD = 6.36 indicating deficits in reciprocal social behavior that are clinically significant) and Social Communication Questionnaire Lifetime Total Score (SCQ) [25] (Avg = 16.67, SD = 4.41 indicating a possibility of ASD) to index current ASD symptoms.

3.2 Experimental Procedure

We designed the interaction logic for each trial based on a least-to-most prompt hierarchy [26]. In this hierarchy, the robot started from the lowest level of prompt and only provided higher levels of prompts to the participant when needed. The components of the prompts was inherited from a successful previous study [13], and are discussed in detail in the following content. When a participant was interacting with the robot, his/her parent(s) monitored the experiment in an observation room.

Phase 1—Multi-visit one-target intervention. Phase 1 was designed for the participants to gain initial (one target) joint attention skills and get familiar with the robot. Four sessions were arranged on different dates across 32.5 days in average. There were 8 repeated trials in each session. In each trial, there were 6 possible prompt levels. Initially, both monitors displayed the same static picture of children's interest. The system started from Prompt 1, where the robot turned its head to the target monitor, and said "Look!" If the participant failed to turn to the target monitor, the system would give Prompt 2, which was the same as Prompt 1. If the participant still did not turn to the target monitor, the system would give the next higher level of prompt until Prompt 6. In Prompts 3 and 4, based on Prompt 1 and 2, the robot not only turned its head to the target, but also pointed to the target with its arm on the target side, saying "Look over there!" In Prompts 5 and 6, the robot displayed the same motion as that in Prompt 3 and 4, and in addition, the target monitor displayed audio and video clips of the children's interest as additional attentional stimuli. At the same time, the static picture was kept in the opposite non-target monitor. If the participant turned to the target within 7 s from the start of a prompt, the robot would say "Good job!", and the target would display a short cartoon videos as a reward (e.g., Dora). The direction of the target monitor was randomized to avoid habituation on a particular direction.

Phase 2— One-visit intervention on both one-target and two-target interactions. Phase 2 included sessions 5 and 6 of the user study and was conducted after 8 months from the end of phase 1. Session 5 was the same as session 1 to 4 in Phase 1. Here we assessed whether the participants still possessed the initially practiced one-target joint attention skill after a few months. After a few minutes of break, session 6 was given, where the robot prompted 2 targets in opposite directions for the children to look at in a sequence. Therefore, session 6 reflected whether the one-target joint attention skill can be adapted to a higher level of interaction (hit two targets in a specific order).

In session 6, each trial had 2 levels of prompts. Starting from Prompt 1, the robot first turned its head to one monitor that displayed a static picture and said "First look at that!", then turned to the opposite monitor that displayed the same picture and said "Then look over there!" If the participant did not succeed, prompt 2 was issued, where the robot turned its head to the target, pointed at the target with its arm, and said "First look at that!" Then the target monitor displayed a short video for 2 s. Following that, the robot prompted to the opposite target with the same type of motion, saying "Then look over there!" Finally the second target monitor would display the same video for 2 s. Prompt 1 and 2 had response windows of 8 and 10 s starting from the prompt of the first monitor, respectively. On any prompt level, if the participant looked at both monitors following the order given by the robot, the robot would say "good job!", and the second monitor would display the short reward video to terminate the trial. This procedure were repeated until a trial reached a 5 min time limit. Thus the number of the trials depended on the participants' performance.

4   Experimental Results

4.1   Joint Attention Performance

The target hit, which is defined as a successful turn to the target monitor, rate in one-target sessions were 100 %, 97.92 %, 95.83 %, 97.92 %, and 97,92 %, in session 1 to 5, respectively. We can see that in almost all the trials the participants successfully turned to the target eventually. Figure 3 shows the average prompt levels that the participants needed to hit the target in session 1 to 5 (bars represent standard deviations). We can see that in general the required prompt levels dropped across session 1 to 4 with some fluctuation. This change was not statistically significant (p values were from 0.1542 to 0.7477). In session 5, which happened 8 months after session 4, the participants' performance improved compared with that in session 1 to 4. Wilcoxon signed-rank test showed that the difference between session 5 and session 1 ($p = 0.0042$) and the difference between session 5 and session 3 ($p = 0.0175$) was statistically significant, while the differences between other sessions and session 5 were not. The standard deviations were large. From session 1 to session 5, these values tended to be smaller.

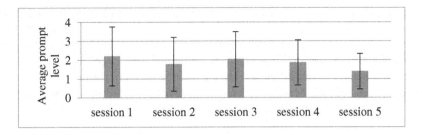

Fig. 3. Target hit prompt level of session 1 to 5 (one-target sessions)

Multiple factors might lead to the improvement in session 5. First of all, the participants were older and had received months of home training during the waiting period. Thus their joint attention skill was expected to be better than that in phase 1, and this was reflected in the robot-mediated interaction. Although we were not able to control for community interventions in the current protocol, participating children did not receive robotic intervention during this interval. Therefore, the initial training in phase 1 might have some impact. We also compared the performance in session 5 in the current study (named as study A) with a previous study by Zheng et al. [17] published in 2013 (named as study B). Study B used a semi-autonomous robotic system, but implemented the same interaction protocol as phase 1 in study A. Study B also involved 6 participants. The age of the participants in session 1 of study B (3.46 years) was similar to the participants' age in session 5 of study A (3.52 years). However, the prompt level needed in session 1 of study B was 2.17, which was 157.2 % higher than that in session 5 of study A (1.38). Thus we believe that the improved performance of the participants in session 5 of study A was at least partly due to the initial intervention in phase 1.

Next, we evaluated the participants' performance in the two targets joint attention task in session 6. There were 55 trials completed in total. The participants hit one target in 53 trials (96.36 %). The participants successfully hit both targets following the order given by the robot in 29 trials (52.73 %), where 15 trials on Prompt 1 and 14 on Prompt 2. Therefore, we can see that the participants were able to respond to the robot properly even if the joint attention task was more difficult than before. This phenomenon might suggest that the joint attention skills that the participants practiced in the one-target sessions helped them well-adapted into a more difficult task.

4.2   Preferential Attention on the Robot

We computed the average percentage of the session time that the participants spent on looking at the robot. The robot's body movement was in a region of $[-10.76°, 10.76°]$ in horizontal yaw angles, and $[2.86°, 13.50°]$ in vertical pitch angles with respect to the view of the participants. We defined the attention region to the robot as a box that covered the body movement of the robot, and also included a margin of 5° yaw and pitch angles on up, down, left and right side around the robot to include the peripheral attention on the robot. If the participants' gaze fell in this region we considered that they were paying attention to the robot.

Figure 4 shows the average attention that the participants paid to the robot in session 1 to 6 (bars represent standard deviations). The values ranges from 22.05 % to 28.54 %. We can see that there was some fluctuation of attention to the robot but there was no big change among these sessions. The standard deviations were large. Statistical analysis also showed no significant difference among these sessions (p ranges from 0.3125 to 0.8438 between different sessions). Thus we concluded that the participants' attention on the robot sustained not only in phase 1, but also in phase 2 after 8 months.

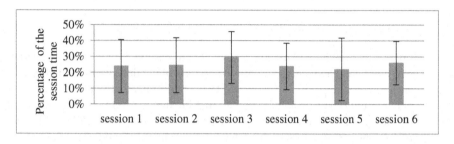

Fig. 4. Percentage of the session time spent on looking at the robot

5   Conclusion and Discussion

In this study we proposed a fully autonomous robotic platform for teaching children with ASD joint attention skills. The system was a closed-loop system where a humanoid robot provided attentional prompts to children with ASD. There were two monitors that displayed attentional stimuli that coordinated with the robot's action. A preliminary longitudinal user study was conducted to evaluate the impact of the proposed system in two phases. In the first phase the participants received 4 sessions of one-target robot-mediated joint attention intervention. After 8 months, Phase 2 assessed the participants' joint attention skill in a one-target session, and also tested the participants' performance in to a two-target joint attention task. Experimental results showed that the participants' within-system performance improved over multiple sessions. In addition, the simple joint attention skill practiced initially might have been adapted to a more complex task. Besides, their attention on the robot sustained across the whole longitudinal study. Given the heterogeneous behavioral patterns of children with ASD, the standard deviations showed large within-group variation in the aforementioned results.

While this initial study was promising, there were several limitations that are important to highlight and need to be addressed in the future. The small sample size examined was the most powerful limit of this study, which limited the strength of the statistical analyses. As such, a more comprehensive study with larger user group will be needed before we can draw any concrete conclusion. Regarding within-system joint attention skills, we made an initial attempt to help the participants transfer a simple skill to a more complex one. However, we did not systematically investigate whether such learning can be generalized to other interactions.

Despite limitations, this study is among the first few works that design and empirically evaluate the longitudinal impact of an autonomous robotic system capable of modifying performance regarding joint attention skills for young children with ASD. This is extremely important for testing the ultimate value of robotic technology in children with ASD. Movement in this direction introduces the possibility of realizing autonomous intervention systems that are not only focused on simple skills, but systems that are capable of more sophisticated adaptations in longitudinal interventions. We are hopeful that future sophisticated clinical applications of such autonomous robotic systems may demonstrate meaningful improvements for young children with ASD.

Acknowledgement. This study was partially supported by: the Hobbs Grant from the Vanderbilt Kennedy Center, a Vanderbilt University Innovation and Discovery in Engineering and Science (IDEAS) grant, National Science Foundation Grant 1264462, and the National Institute of Health Grants 1R01MH091102-01A1and 5R21MH103518-02. This work also got core supports from NICHD (P30HD15052) and NCATS (UL1TR000445-06).

References

1. Autism Spectrum Disorders Prevalence Rate. Autism Speaks and Center for Disease Control (CDC) (2011)
2. Ganz, M.L.: The lifetime distribution of the incremental societal costs of autism. Arch. Pediatr. Adolesc. Med. Am. Med. Assoc. **161**, 343–349 (2007)
3. Warren, Z.E., Stone, W.L.: Best practices: early diagnosis and psychological assessment. In: Amaral, D., Geschwind, D., Dawson, G. (eds.) Autism Spectrum Disorders, pp. 1271–1282. Oxford University Press, New York (2011)
4. Zheng, Z., Fu, Q., Zhao, H., Swanson, A., Weitlauf, A., Warren, Z., Sarkar, N.: Design of a computer-assisted system for teaching attentional skills to toddlers with ASD. In: Antona, M., Stephanidis, C. (eds.) UAHCI 2015. LNCS, vol. 9177, pp. 721–730. Springer, Heidelberg (2015)
5. Weir, S., Emanuel, R.: Using LOGO to catalyse communication in an autistic child. Department of Artificial Intelligence, University of Edinburgh (1976)
6. Robins, B., Dautenhahn, K., Dubowski, J.: Does appearance matter in the interaction of children with autism with a humanoid robot? Interac. Stud. **7**, 509–542 (2006)
7. Pierno, A.C., Mari, M., Lusher, D., Castiello, U.: Robotic movement elicits visuomotor priming in children with autism. Neuropsychologia **46**, 448–454 (2008)
8. Robins, B., Dautenhahn, K., Boekhorst, R., Billard, A.: Robotic assistants in therapy and education of children with autism: can a small humanoid robot help encourage social interaction skills? Univ. Access Inf. Soc. **4**, 105–120 (2005)
9. Kozima, H., Nakagawa, C., Yasuda, Y.: Children–robot interaction: a pilot study in autism therapy. Prog. Brain Res. **164**, 385–400 (2007)
10. Feil-Seifer, D., Matarić, M.J.: Toward socially assistive robotics for augmenting interventions for children with autism spectrum disorders. In: Experimental Robotics, pp. 201–210. Springer, Heidelberg (2009)
11. Zheng, Z., Das, S., Young, E.M., Swanson, A., Warren, Z., Sarkar, N.: Autonomous robot-mediated imitation learning for children with autism. In: 2014 IEEE International Conference on Robotics and Automation (ICRA), pp. 2707–2712. IEEE (2014)
12. Fujimoto, I., Matsumoto, T., De Silva, P.S., Kobayashi, M., Higashi, M.: Study on an assistive robot for improving imitation skill of children with autism. In: Ge, S.S., Li, H., Cabibihan, J.-J., Tan, Y.K. (eds.) ICSR 2010. LNCS, vol. 6414, pp. 232–242. Springer, Heidelberg (2010)

13. Bekele, E.T., Lahiri, U., Swanson, A.R., Crittendon, J.A., Warren, Z.E., Sarkar, N.: A step towards developing adaptive robot-mediated intervention architecture (ARIA) for children with autism. IEEE Trans. Neural Syst. Rehabil. Eng. **21**, 289–299 (2013)
14. Warren, Z.E., Zheng, Z., Swanson, A.R., Bekele, E., Zhang, L., Crittendon, J.A., Weitlauf, A.F., Sarkar, N.: Can robotic interaction improve joint attention skills? J. Autism Dev. Disord. **45**(11), 3726–3734 (2015)
15. Wainer, J., Robins, B., Amirabdollahian, F., Dautenhahn, K.: Using the humanoid robot KASPAR to autonomously play triadic games and facilitate collaborative play among children with autism. IEEE Trans. Auton. Mental Deve. **6**, 183–199 (2014)
16. Poon, K.K., Watson, L.R., Baranek, G.T., Poe, M.D.: To what extent do joint attention, imitation, and object play behaviors in infancy predict later communication and intellectual functioning in ASD? J. Autism Dev. Disord. (2011), 1–11. doi:10.1007/s10803-011-1349-z
17. Zheng, Z., Zhang, L., Bekele, E., Swanson, A., Crittendon, J., Warren, Z., Sarkar, N.: Impact of robot-mediated interaction system on joint attention skills for children with autism In: 2013 IEEE International Conference on Rehabilitation Robotics (ICORR). IEEE (2013)
18. http://www.aldebaran-robotics.com/en/
19. Kajopoulos, J., Wong, A.H.Y., Yuen, A.W.C., Dung, T.A., Kee, T.Y., Wykowska, A.: robot-assisted training of joint attention skills in children diagnosed with autism. In: Tapus, A., et al. (eds.) ICSR 2015. LNCS, vol. 9388, pp. 296–305. Springer, Heidelberg (2015)
20. Xiong, X., De la Torre, F.: Supervised descent method and its applications to face alignment. In: 2013 IEEE Conference on Computer Vision and Pattern Recognition (CVPR), pp. 532–539. IEEE (2013)
21. Diagnostic and Statistical Manual of Mental Disorders: Quick reference to the Diagnostic Criteria from DSM-IV-TR. American Psychiatric Association, Washington D.C. (2000)
22. Lord, C., Rutter, M., DiLavore, P., Risi, S., Gotham, K., Bishop, S.: Autism Diagnostic Observation Schedule–2nd edition (ADOS-2). Western Psychological Services, Torrance (2012)
23. Mullen, E.M.: Mullen Scales of Early Learning. Pearson San Antonio, TX (1995)
24. Constantino, J.N., Gruber, C.P.: The Social Responsiveness Scale. Western Psychological Services, Los Angeles (2002)
25. Rutter, M., Bailey, A., Lord, C.: The Social Communication Questionnaire. Western Psychological Services, Los Angeles (2010)
26. Demchak, M.: Response prompting and fading methods: a review. Am. J. Ment. Retard. **94**, 603–615 (1990)

Culture as a Driver for the Design of Social Robots for Autism Spectrum Disorder Interventions in the Middle East

Hifza Javed[1], John-John Cabibihan[1(✉)], Mohammad Aldosari[2], and Asma Al-Attiyah[3]

[1] Mechanical and Industrial Engineering Department, Qatar University, Doha, Qatar
john.cabibihan@qu.edu.qa
[2] Center for Pediatric Neurology, Cleveland Clinic and Cleveland Clinic Lerner College of Medicine of Case Western Reserve University, Cleveland, OH, USA
[3] Psychological Sciences Department, Qatar University, Doha, Qatar

Abstract. In this paper, we discuss the prevalence of Autism Spectrum Disorder (ASD) in the Gulf region. We examine the importance of providing state-of-the-art ASD interventions, and highlight social robots as therapeutic tools that have gained popularity for their use in ASD therapy in the West. We also elaborate on the features of social robots that make them effective and describe how they can be used in such settings. We then emphasize the significance of taking cultural context into account in order to develop indigenous tools for ASD therapy, and explain the different ways in which social robots can be made culturally adaptive to maximize their potential impact on children with ASD.

1 Introduction

Autism Spectrum disorder (ASD) is a neurodevelopment disorder [1], usually diagnosed during the first 3 years of life and may be accompanied by other physical or psychological disorders [2]. It is traditionally characterized by impairments in social communication, social interaction and imagination abilities [3], and can be attributed to an absence of the mentalizing ability in a person diagnosed with ASD [4,5].

Impairments in social communication include difficulties in processing language, and in interpreting facial expressions, body language and the tone of voice [6]. Children on the spectrum also tend only to focus on the literal meanings rather than the underlying meanings of metaphors or figures of speech used in communication.

Children diagnosed with ASD often appear withdrawn and aloof. They tend to be uncomfortable in most social settings and have problems forming deep social relationships. This is a direct result of their inability to process emotions easily. They are also unable to maintain eye contact because they feel overloaded with information coming in simultaneously through a number of channels (speech, body language and facial expressions). To avoid such sensory overload,

© Springer International Publishing AG 2016
A. Agah et al. (Eds.): ICSR 2016, LNAI 9979, pp. 591–599, 2016.
DOI: 10.1007/978-3-319-47437-3_58

they retreat into their own world of familiarity and clarity, thus appearing uninterested and rude to those around them. This marks the impairments in social interaction.

Deficits in the ability to understand abstract ideas and imagine situations outside the daily routine form the impairments in social imagination. These impediments tend to confine a child to his own mind, not allowing him or her to deal with any changes to routine. For this reason, such children tend to engage in rigid and repetitive activities and are unable to indulge in imaginative or interpersonal play [7].

2   Autism in the Gulf States

According to the Centers for Disease Control and Prevention (CDC) report in 2014 [8], 1 % of the world's population has autism spectrum disorder (ASD). In the United Kingdom, one child out of 100 children has ASD [9]. In the United States, the autism prevalence rate is 1 in 68 births [8]. In South Korea, it is 1 in 38 [10]. A large number of epidemiology studies have been conducted globally, bringing forth a wealth of information regarding the prevalence of the condition [11–13]. However, with majority of these studies being conducted in western countries, prevalence statistics for the Gulf States and other developing countries in the world remain largely unknown.

Table 1 lists the available statistics for some Gulf states. The overall ASD prevalence in these countries appears to be lower than generally found in the west. Samadi [14] reports a social stigma attached to disabilities in the Arab culture, which may be a reason that prevents parents from fully reporting a child's difficulties [15]. In addition, children with such disabilities may discontinue education early, preventing them from being screened during studies conducted in elementary schools [16]. The differences in environmental triggers, and in parental practices and expectations must, however, be taken into account as well. Cannell [17] has even suggested a connection between autism and vitamin D deficiency in pregnant women, attributing the low prevalence of autism in the region to adequate exposure to the sun. It must also be noted that no uniform diagnostic methodology was employed throughout the studies listed in Table 1, making it difficult to compare their results.

Table 1. Prevalence data for ASD in some Gulf States

Study	Country	No. of cases/10,000	Subjects ages	Diagnosis
Al-Farsi (2011)	Oman	1.4	Up to 14 years	ASD
Naqvi (2012)	Saudi Arabia	60	Up to 16 years	ASD
Eapen (2007)	UAE	12	3 years	PDD
Alshaban (2011)	Qatar	16	Up to 18 years	ASD

3   Importance of Intervention for the State of Qatar

The state of Qatar is currently witnessing a sharp rise in the demand for ASD support from local and expatriate communities. Table 2 below shows all the centres in Qatar that cater to individuals with ASD from 1992 to the present. From only 4 centres in the 1992 to 2003 period, there has been a 250 % increase in the number of centres in the succeeding years up to the present. Furthermore, a 2012 study from Qatar University, Shafallah Special Needs Center, and Hamad Medical Corporation [18] showed that children with autism largely stay indoors, spending about 17 h a day in their own homes, with most of their time spent sleeping or watching television.

It is important to note that a large portion of this demand comes from the expatriate community. Qatar's population is composed of a number of different nationalities. This implies the same variety in the languages and cultures of people, and hence, also in their needs. This is especially relevant to therapy and intervention practices for ASD, which are largely child-specific in nature and can have their efficacy affected by minute details. Therefore, in such cases, including culturally relevant features in therapy practices can potentially improve results.

Table 2. Centers in Qatar for special needs children (including Children with ASD)

No	Year Established	Name of the Center	Languages
1	1992	Qatar Society for the Rehabilitation of Special Needs	Arabic, English
2	1996	Awsaj Institute of Education	Arabic English
3	2001	Shafallah Center for Children with Special Needs	Arabic, English
4	2003	Sunbeam Center of Excellence	Arabic, English
5	2005	HOPE Qatar	Arabic, English
6	2007	Special Needs Center, Qatar University	Arabic, English
7	2009	Al Tamakon Comprehensive School	Arabic, English
8	2009	The Next Generation School	Arabic, English, Urdu
9	2010	Qatar Autism Center	Arabic, English
10	2011	Step by Step Center	Arabic, English
11	2013	Child Development Center	Arabic, English
12	2013	Hand in Hand Center	Arabic, English
13	2014	Omega Center for Special Needs Education	Arabic, English
14	2014	Alkhuzama Special School for Special Needs	Arabic, English

4   Importance of Social Robots

Socially assistive robotics is aimed at addressing the gaps in care given to humans by providing assistance in the form of social interaction. The large user base that can benefit from the automated companionship, supervision, mentorship and motivation includes stroke survivors, the elderly, patients with dementia and children with ASD. Socially Assistive Robots (SAR) have, hence, been at the front end of ASD therapy in developed countries for some years now [19].

The reasons for the integration of SARs in ASD therapy are manifold. Children with ASD have been observed to show a deep interest in technology in general and robots in particular. The nature of their disorder inhibits their social, emotional and interactive abilities, making human-human interaction a challenge for them. True to their nature, humans communicate not just with words, but also with facial expressions, body language, tone of voice and eye contact. To complicate things further, the words are not always intended literally, such as when humor or sarcasm is involved. Such subtleties often elude a child on the spectrum, discouraging him from indulging in such interactions. For such a child, interacting instead with a robot that exhibits only a small subset of the human emotions, and does so with minimal use of complex communicative mechanisms typical of humans is much easier to do.

In addition to this, the toy-like size and appearance of most SARs prevent them from intimidating a child, and their abilities to repeat mundane tasks and to indulge in interactive gameplay enable the children to view them as friendly playmates that pass no judgment on their unconventional behavior. Their use does not have to be limited to a session in a clinic, but can be extended to homes and classrooms to strengthen the bond with the child.

Many previous studies [20–24] have reported improvements in social performances of children on the spectrum after continued interaction with such robots. These interactions are designed especially to be simple, interactive and enjoyable, involving gameplay and learning both. Robots take up the roles of playmates, social actors, teaching agents and social mediators.

This robotic therapy has been used as a supplement to the conventional therapy model involving only a therapist and the child subject. This model, though more effective than the conventional one, may however be prevented from attaining its true potential in multicultural environments, where the child and the therapist may belong to different cultures. In such scenarios, it becomes important to take the cultural context into account in order to maximize the impact of therapeutic practices. Some of the popularly used social robots are shown in Fig. 1.

5   Importance of Cultural Context in ASD Intervention

There is no shortage of studies based on autism: prevalence and epidemiology studies, renowned works on the nature of the disorder, and ample research on the most advanced therapy methods. There is plenty of available published material, case studies and surveys, discussing the role of social robots in autism therapy as well. However, the one factor lacking in a majority such works is the lack of social context taken in account during the course of the study.

Most of the available studies presenting statistics on the prevalence of autism have been conducted within the developed part of the world [11–13]. Only a handful of such studies have been conducted in developing countries, with most being small-scale, and not country-wide. It is important to note that the diagnostic and screening tools used in developing countries have been arbitrarily adopted

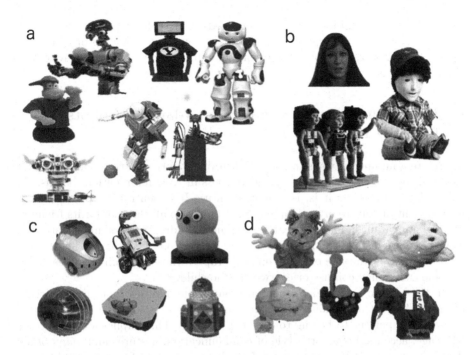

Fig. 1. Socially interactive robots that have been used for the therapy of children with autism in the literatures. The 4 types of form-factors used are: (a) humanoids, (b) human-like robots, (c) mobile or toy-like, and (d) animaloids.

from the studies conducted in the developed countries, negating the cultural differences between the two. Hence, research that addresses the cultural context is necessary in the development of screening and therapy tools that are sensitive to the local culture, helping also to devise new, culturally relevant frameworks and methodologies for autism therapy.

In addition, only a few cross-cultural studies can be found on the response of culturally different individuals towards robots in general [25–28]. To the best of our knowledge, none, however, is focused in particular on social robots for autism therapy. It is obvious that cultural context must be considered in order to maximize the impact and integrity of the research efforts.

We have been unable to find any published research on equipping SARs with the ability to become cultural mediators. Using culturally adaptive robots for ASD therapy is truly a unique idea with potential to play a transformative role for children with ASD. This would immensely facilitate therapists in their struggle to mediate cultural differences when interacting with children of cultures that are different from their own. This allows them instead to focus entirely on therapy goals by sharing the requirement for multicultural and multilingual competence with a robot. This enables a single robotic entity to play a critical role in the therapy for children belonging to diverse cultural backgrounds, and to be effective for children with varying needs. Culturally relevant interactions also present us with the opportunity to further our understanding of ASD, and evaluate just how deep-rooted the impact of culture can be on the behavior of a child with autism.

6   Design Features for Culturally Adaptive Robots for ASD Intervention

There are a number of ways in which culturally adaptive features can be added to social robots to maximize their effectiveness in ASD intervention. It must be noted that a robot, as it is, is a neutral entity. It is neutral in appearance, is independent of culture or nationality, and appeals to all children for its friendly, toy-like appearance. Therefore, at first glance, children are not intimidated by robots and are instead compelled to initiate interactions with them. This already gives them an advantage over a human therapist whose unfamiliar appearance, in some cases, may confuse or overwhelm the child.

In the Middle East region, there are many communities that are comprised of many different nationalities, meaning that it is not uncommon for the therapist and the patient to belong to different cultures. This implies a difference in language, accents, behaviors, style of communication, gestures and many other details that can play a significant role in helping a child feel comfortable and understand instructions during the various activities taking place in a therapy session.

Perhaps the most important capability that can enhance cultural relevance of a robotic agent is the addition of multilingual features that could enable the robot to communicate in a number of languages. This would allow one therapist to conduct therapy with children from backgrounds dissimilar to his and who speak a language other than with which he is familiar. The larger the number of languages spoken by the robot, the larger the supported audience. This facilitates the therapist in conveying instructions to the children since a major part of his or her work is now being done by the robot.

It has been found that people belonging to some cultures have a preference for non-anthropomorphic appearance of a robot, while others prefer anthropomorphic features [25,29]. Robots, as shown in Fig. 1, come in a variety of forms, including human-like, humanoid and toy-like. Therefore, the integration of robots in ASD therapy can cater to such preferences as well, and in doing so, make therapy more subject-specific, as needed.

Gestures form an integral part of expression in every culture. Body language and gestures are frequently used to convey information, alongside vocal communication. Robots are already being used to teach how to communicate using gestures [30–33]. The spectrum of gestures varies widely across cultures. Some gestures do not exist in some cultures, while it can hold drastically different meanings in others. It is thus important to take these into account when interacting with children from foreign cultures, in order to ensure that the correct message has been conveyed. This can also help in enhancing a child's comfort level by offering a sense of familiarity. For example, a robot that bows to a Japanese child in the beginning of an interaction would be more welcoming than a robot that greets in another way.

In addition, the activities can be infused with cultural relevance as well. Many therapeutic interactions with children with ASD involve short, engaging games such as chase and follow, imitation and turn-taking [34–36]. The same activities can be modified slightly with cultural context to achieve better results. This can be done by designing games that are local to the child's culture and are already familiar to him or her.

Another popular use of robots in ASD therapy is social storytelling [37,38]. This involves a social robot that narrates stories that are intended to teach children appropriate behaviors in socially significant situations. This also offers potential for cultural adaptation, whereby the generic stories can be replaced with more local stories with familiar characters. This could encourage the children to maintain interest in the storyline and also to make more sense of the message being conveyed.

All of these measures help to provide a friendlier environment for the children by offering more familiarity and less intimidation. In such situations, it is natural to expect children to better understand what is being taught to them, in order to observe long-term improvements in their behaviors. The culturally relevant features in a robot help to strengthen the robot's bond with the children by making it appear to be one of their own, and not an outsider.

7   Conclusion

This paper emphasizes the importance of cultural context in ASD therapy. Most of the available tools thus far have been developed in the west and adapted or translated for use in other parts of the world as well. However, the many cultural differences between the Gulf and the West can prevent these tools from being as effective as they potentially can be. There is a need to develop tools sensitive to the local culture, so that all needs can be met and usefulness maximized.

Cultural adaptiveness also facilitates the use of socially assistive robots for ASD therapy in culturally diverse environments, where the therapist and subject may belong to very different cultural backgrounds. A robot that is sensitive to the local culture of a child is able to present a more familiar and friendly environment, and encourage the child to participate in activities. This can enhance the potential of such methods to bring about long-term behavioral improvements in the children.

Multicultural communities in the Middle East region with rising demands for ASD support can especially benefit from culturally adaptive methods. It is vital that this factor be taken into account when designing therapy methods and developing relevant tools, so that, true to the nature of the disorder, its therapy can truly be made as case-specific as possible.

Acknowledgments. The work is supported by an NPRP grant from the Qatar National Research Fund under the grant No. NPRP 7-673-2-251. The statements made herein are solely the responsibility of the authors.

References

1. American Psychiatric Association: Diagnostic and Statistical Manual of Mental Disorders, 5th edn. DSM-5. American Psychiatric Pub (2013)
2. Wing, L., Gould, J.: Severe impairments of social interaction and associated abnormalities in children: epidemiology and classification. J. Autism Dev. Disord. 9(1), 11–29 (1979)
3. Cashin, A., Barker, P.: The triad of impairment in autism revisited. J. Child Adolesc. Psychiatr. Nurs. 22(4), 189–193 (2009)
4. Baron-Cohen, S.: Mindblindness: An Essay on Autism and Theory of Mind. MIT Press, Boston (1997)
5. Baron-Cohen, S., Leslie, A.M., Frith, U.: Does the autistic child have a "theory of mind"? Cognition 21(1), 37–46 (1985)
6. AWARES.org: About Autism
7. The National Autistic Society: Statistics: how many people have autism spectrum disorders? (2015)
8. Centers for Disease Control, Prevention: Autism Spectrum Disorder - Data & Statistics (2014)
9. The National Autistic Society: Patients with autism spectrum disorders: guidance for health professionals
10. Kim, Y.S., Leventhal, B.L., Koh, Y.J., Fombonne, E., Laska, E., Lim, E.C., Cheon, K.A., Kim, S.J., Kim, Y.K., Lee, H., Song, D.H., Grinker, R.R.: Prevalence of autism spectrum disorders in a total population sample. Am. J. Psychiatry 168(9), 904–912 (2011)
11. Fombonne, E.: The changing epidemiology of autism. J. Appl. Res. Intellect. 18(4), 281–294 (2005)
12. Fombonne, E.: Epidemiology of pervasive developmental disorders. Pediatr. Res. 65(6), 591–598 (2009)
13. Williams, J.G., Higgins, J.P.T., Brayne, C.E.G.: Systematic review of prevalence studies of autism spectrum disorders. Arch. Dis. Child. 91(1), 8–15 (2006)
14. Samadi, S.A.: Comparative policy brief: status of intellectual disabilities in the Islamic Republic of Iran. J. Policy Pract. Intell. Disabil. 5(2), 129–132 (2008)
15. Samadi, S., Mahmoodizadeh, A., McConkey, R.: A national study of the prevalence of autism among five-year-old children in Iran. Autism (2012)
16. Javed, H., Cabibihan, J., Al-Attiyah, A.A.: Autism in the gulf states: why social robotics is the way forward. In: 2015 5th International Conference on Information & Communication Technology and Accessibility (ICTA), pp. 1–3. IEEE (2015)
17. Cannell, J.J.: Autism and vitamin D. Med. Hypotheses 70(4), 750–759 (2008)
18. Kheir, N.M., Ghoneim, O.M., Sandridge, A.L., Hayder, S.A., Al-Ismail, M.S., Al-Rawi, F.: Concerns and considerations among caregivers of a child with autism in Qatar. BMC Res. Notes 5(1), 290 (2012)
19. Cabibihan, J.J., Javed, H., Ang, M., Aljunied, S.M.: Why Robots? A Survey on the Roles and Benefits of Social Robots in the Therapy of Children with Autism (2013)
20. Scassellati, B., Admoni, H., Matarić, M.: Robots for use in autism research. Ann. Rev. Biomed. Eng. 14, 275–294 (2012)

21. Tapus, A.: Socially assistive robotics. Robotics & Automation (2007)
22. Qidwai, U., Shakir, M., Connor, O.B.: Robotic toys for autistic children: innovative tools for teaching and treatment. In: 2013 7th IEEE GCC Conference and Exhibition (GCC), pp. 188–192. IEEE, November 2013
23. Costa, S., Resende, J., Soares, F.O., Ferreira, M.J., Santos, C.P., Moreira, F.: Applications of simple robots to encourage social receptiveness of adolescents with autism. In: Conference proceedings: Annual International Conference of the IEEE Engineering in Medicine and Biology Society, IEEE Engineering in Medicine and Biology Society, Annual Conference, pp. 5072–5075, January 2009
24. Kozima, H., Nakagawa, C., Yasuda, Y.: Children-robot interaction: a pilot study in autism therapy. Prog. Brain Res. **164**, 385–400 (2007)
25. Li, D., Rau, P., Li, Y.: A cross-cultural study: Effect of robot appearance and task. Int. J. Soc. Robot. (2010)
26. O'Neill-Brown, P.: Setting the stage for the culturally adaptive agent. In: AAAI Fall Symposium on Proceedings of the 1997 (1997)
27. Rau, P., Li, Y., Li, D.: Effects of communication style and culture on ability to accept recommendations from robots. Computers in Human Behavior (2009)
28. Ham, J., Esch, M., Limpens, Y., Pee, J., Cabibihan, J.-J., Ge, S.S.: The automaticity of social behavior towards robots: the influence of cognitive load on interpersonal distance to approachable versus less approachable robots. In: Ge, S.S., Khatib, O., Cabibihan, J.-J., Simmons, R., Williams, M.-A. (eds.) ICSR 2012. LNCS (LNAI), vol. 7621, pp. 15–25. Springer, Heidelberg (2012). doi:10.1007/978-3-642-34103-8_2
29. Ge, S.S., Cabibihan, J.J., Zhang, Z., Li, Y., Meng, C., He, H., Safizadeh, M., Li, Y., Yang, J.: Design and development of nancy, a social robot. In: 8th International Conference on Ubiquitous Robots and Ambient Intelligence (URAI), pp. 568–573. IEEE (2011)
30. Cabibihan, J.J., So, W.C., Pramanik, S.: Human-recognizable robotic gestures. IEEE Trans. Auton. Ment. Dev. **4**(4), 305–314 (2012)
31. Ham, J., Cuijpers, R.H., Cabibihan, J.J.: Combining robotic persuasive strategies: the persuasive power of a storytelling robot that uses gazing and gestures. Int. J. Soc. Robot. **7**(4), 479–487 (2015)
32. Cabibihan, J.-J., So, W.C., Nazar, M., Ge, S.S.: Pointing gestures for a robot mediated communication interface. In: Xie, M., Xiong, Y., Xiong, C., Liu, H., Hu, Z. (eds.) ICIRA 2009. LNCS (LNAI), vol. 5928, pp. 67–77. Springer, Heidelberg (2009). doi:10.1007/978-3-642-10817-4_7
33. Cabibihan, J.J., So, W.C., Saj, S., Zhang, Z.: Telerobotic pointing gestures shape human spatial cognition. Int. J. Soc. Robot. **4**(3), 263–272 (2012)
34. Michaud, F., Théberge-Turmel, C.: Mobile Robotic Toys and Autism. In: Socially Intelligent Agents, pp. 125–132. Kluwer Academic Publishers, Boston (2002)
35. Duquette, A., Michaud, F., Mercier, H.: Exploring the use of a mobile robot as an imitation agent with children with low-functioning autism. Autonomous Robots (2008)
36. Robins, B., Otero, N., Ferrari, E., Dautenhahn, K.: Eliciting requirements for a robotic toy for children with autism - results from user panels. In: RO-MAN 2007 - The 16th IEEE International Symposium on Robot and Human Interactive Communication, pp. 101–106. IEEE (2007)
37. Vanderborght, B., Simut, R., Saldien, J., Pop, C., Rusu, A.S., Pintea, S., Lefeber, D., David, D.O.: Using the social robot probo as a social story telling agent for children with ASD. Interaction Studies **13**(3), 348–372 (2012)
38. Hoa, T.D., Cabibihan, J.J.: Cute and soft. In: Proceedings of the Workshop at SIGGRAPH Asia on - WASA 2012, p. 77. ACM Press, New York (2012)

Robo2Box: A Toolkit to Elicit Children's Design Requirements for Classroom Robots

Mohammad Obaid[1], Asım Evren Yantaç[1], Wolmet Barendregt[2(✉)],
Güncel Kırlangıç[1], and Tilbe Göksun[3]

[1] KUAR, Media and Visual Arts Department, Koç University, Istanbul, Turkey
[2] Department of Applied IT, Gothenburg University, Gothenburg, Sweden
`wolmet.barendregt@ait.gu.se`
[3] Department of Psychology, Koç University, Istanbul, Turkey

Abstract. We describe the development and first evaluation of a robot
design toolkit (Robo2Box) aimed at involving children in the design of
classroom robots. We first describe the origins of the Robo2Box elements based on previous research with children and interaction designers
drawing their preferred classroom robots. Then we describe a study in
which 31 children created their own classroom robot using the toolkit.
We present children's preferences based on their use of the different elements of the toolkit, compare their designs with the drawings presented
in previous research, and suggest changes for improvement of the toolkit.

1 Introduction

Nowadays, children's design input is considered very valuable for the design of
various technologies because it may help designers to focus on children's needs
from an early stage in the process [2]. Robots as a new kind of technology are
likely to enter children's lives, especially in the form of classroom educational
robots, so therefore, involving children in the design of classroom robots is a
logical step in a human-centered design process. However, there are also some
hurdles when involving children, or even adults, in the design of future technologies. People find it hard to imagine the use of future technologies since they
haven't experienced them yet and they are not always aware of the state-of-the-
art developments in areas such as robotics. In our previous work [4], we have
explored how classroom robots are envisioned by children (age 10–14 years) and
interaction design students. We found that children without robotics knowledge
envisioned the classroom robot as a human teacher with some additions and
modifications, while the interaction design students imagined it as a small and
rather cute teaching assistant, and children with some robotics knowledge imagined a more technically inspired classroom robot. Those findings clearly point
out the need to involve children in the design of classroom robots, but more
importantly the need to enable them to broaden their design views as there
were major differences between children with and without robotics knowledge.

We extend our work by providing children with a toolkit giving them access
to design tangible materials (Robo2Box) that can express different and possibly novel ways of imagining about classroom robots. In this paper we aim to

© Springer International Publishing AG 2016
A. Agah et al. (Eds.): ICSR 2016, LNAI 9979, pp. 600–610, 2016.
DOI: 10.1007/978-3-319-47437-3_59

answer the following: **RQ1**: What kind of robots do children design for the classroom using the Robo2Box toolkit? **RQ2**: When using the Robo2Box, do children design classroom robots similar to just drawing robots? **RQ3**: What changes should be made to the Robo2Box for children to express their design ideas?

2 Related Work

The Human-Robot Interaction (HRI) community has focused on defining the design requirements and implications for physical and behavioural aspects of robots. In general, such investigations are conducted based on laboratory studies using commercially available robots [3]. Little work has focused on human-centered design approaches to define robotic features; in particular, approaches to define robotic features with and for children. Below, we will provide an overview of the work performed in this and closely related areas.

An early relevant study into children's design requirements for robots was performed by Bumby and Dautenhahn [1]. They conducted a series of design sessions with 38 children between 7 and 11 years old in which the children were asked to draw a robot in small groups, write a story about the robot, and thereafter observe and interact with two rather simple robots. They found that the drawings were mainly based on geometric forms, but with human heads and feet. The robots in the drawings usually didn't carry any weapons, didn't have lights or a battery and didn't have a gender. The interaction with the robots showed that children anthropomorphized the robots and talked to them like pets.

Thereafter Woods et al. [8,9] investigated children's views on robot appearance, movement, gender, and personality. Children between 9 and 11 were asked to choose a robot picture and fill out a questionnaire. The pictures displayed different robot attributes: mode of locomotion, body shape, looking like an animal, human or machine, the presence or absence of facial features, and gender. Woods identified two dimensions in children's evaluations termed 'Emotional expression', ranging from happy to sad, and 'Behavioral intention', including friendliness, shyness and fright versus aggressiveness, bossiness and anger. Human-machine robots were considered the most friendly, shy and frightened types of robots. Woods argued that robot designers should *"consider a combination of physical characteristics rather than focusing specifically on certain features in isolation"* [8]. Furthermore, children were positive towards robots that were more human-machine like instead of purely machine-like, but showed a sharp drop in positive attitude towards robots that were very human-like.

In a recent study, Sciutti et al. [6] asked children to order 14 pieces of paper with robotic characteristics, from most important to least important. They were asked to imagine building a robot they could interact and play with. The researchers found that age had an effect on which features the children considered important. Furthermore, they found that robots should have some human-like properties to make them more readable. Finally, Shin and Kim [7] interviewed 85 school students from three school levels (with an average age of 14 years) to investigate their attitudes towards learning about, from, and with robots. Their results showed a positive attitude towards learning from robots, but not in favor of having them in schools due to robots lacking emotion.

In summary, when involving children in the design of classroom robots, we need to use methods that help them to focus on the aspects of interest. The process of involving children in the design of classroom robots can benefit from many different inclusive methods such as sketching, storytelling, bodystorming, role-playing and design with prototypes. Yet the design problem comes with the need for covering many different properties; form factor, gender, material, and behaviour. The robot design toolkit described in this paper can serve as the basis for eliciting children's design requirements for classroom robots.

3   Robo2Box Toolkit

Our previous study [4] indicated that children may have an important contribution to make in the design of a classroom robot. However, children's limited drawing skills might hinder them from creating the designs they like. Scaife and Rogers [5] even claim that the act of drawing might keep children from focusing on other aspects of the interaction. Furthermore, it could also be beneficial to provide children with inspiration from the designs of professionals and more knowledgeable children. Therefore we decided to base the Robo2Box toolkit on the physical elements of the drawings by the children and the interaction design students presented in our previous work [4]. However, we also added additional elements to the toolkit based on the findings of Woods [8].

The toolkit is 3D printed and consists of the elements presented in Fig. 1. There are 5 groups of elements: heads, torsos, legs, arms, and materials. Similar to Woods [8] the parts can be categorised as human, animal or machine like (and mixes of these categories). The different body parts can be connected with double-sided tape and materials, such as fur and rubber, can only be chosen to indicate a preference but they cannot physically be attached.

4   Study

The aim of the study is the development and evaluation of a robot design toolkit that can be used as part of a human-centered design approach to involve children in the design of classroom robots. While we are also interested in understanding behavioral requirements for classroom robots that can be elicited through the use of this toolkit in combination with other elicitation approaches, this paper will focus on the use of the toolkit itself as a human-centered design tool.

We conducted a study to address the main questions raised in Sect. 1 with 31 school children (8–15 years old, average age of 11 years (SD = 2.3), 16 girls and 15 boys) from Turkey. Most children (25) did not have any robotics knowledge, while four had little robotic knowledge, and two had attended some robotics classes. Each child was asked to first construct a robot for the classroom and place it in a cardboard model of a classroom (Robot construction and placement).

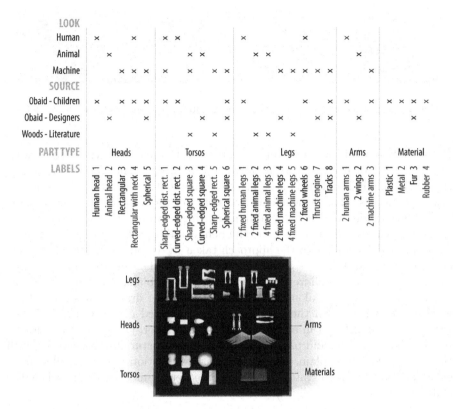

Fig. 1. A list of the proposed elements in the robot-design toolkit, based on elements from our previous study [4] and the study by Woods' et al. [8].

Then the child was asked to write or draw a story about this robot in the classroom and elaborate on the story in an interview (Storytelling). The study was conducted individually with the presence of a facilitator and an observer and were recorded with a video camera after getting informed consent.

Phase 1 - Robot construction and placement: in the robot construction phase, each child was provided with the Robo2Box set that included 3D printed model pieces and material specimens as shown in Fig. 2. During this phase, the facilitator only gave the instructions to freely construct a classroom robot and the observer took notes. When the child finished, they were first asked which materials they wanted to use for their robot. Thereafter they were asked to place their robot in a classroom model. The classroom model had paper people representing a teacher and students inside it to help children understand the relative sizes. We then asked the children to indicate whether their robot should be the same size, bigger or smaller than each of the paper models. The children were allowed to freely imagine the role of their robot in the classroom.

Fig. 2. Items used in the study including a Robo2Bbox and a classroom model.

Phase 2: Storytelling: after building a robot with the toolkit, the children were asked to write or draw a story about how this robot would behave in a classroom. The main idea of this phase was to learn about the additional appearance features (such as color, attachments, tools etc.) of the robot in addition to some extra information on how the children imagined the robot's behaviour in the classroom. But we did not focus on the behavioral aspects of the robot design. This approach is similar to the approach taken by Woods et al. [9] where they asked children to write a story about the robot of their choice (based on pictures). In our study, children were provided with a blank A4 paper with four sequential frames on it, along with colored pencils, to draw their stories. They were free to write down their stories if they preferred not to draw, and they were not required to fill in all the frames on the paper. Once the story writing phase was over, each child participated in a semi-structured interview, in which they were asked to explain their story. This explanation part was followed by some questions from the interviewer to gain a better understanding of how the children imagined their robot physically and behaviorally.

5   Analysis and Results

In this paper we focus on the use of the Robo2Box toolkit and the physical characteristics of the robots designed by the children. Other aspects related to the analyses of the interviews and stories will be presented in another paper. The only interview part that we will report on is the size of the robot and whether the children wanted to add or change anything in the toolkit.

5.1   Children's Designs

Head: Figure 3(a) shows the results for the chosen head; where one child was confused and used a torso and was excluded from the results. The difference between the categories was significant, $X^2(5, N = 30) = 14.4, p < .05$. The children typically chose a human head or a sphere. If we compare the elements here with the elements found in our previous study, it is clear that there is a preference for the elements that are based on children's drawings. Having no separate head, or an animal-like head was not popular among the children. According to Woods [8] children associate robots with no face to negative behaviors (e.g. aggression), which can explain this tendency.

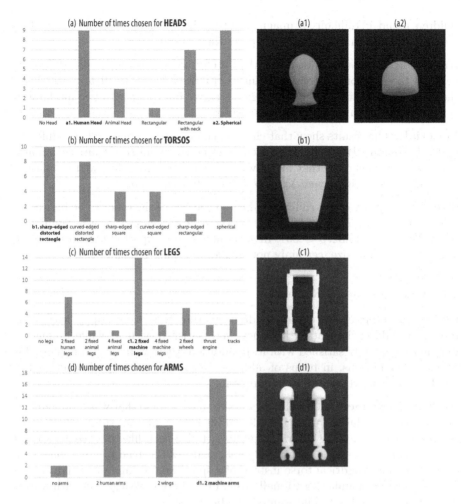

Fig. 3. Head's (a); Torsos (b); Legs (c); Arms (d) chosen by the children.

Torso: The torso results are shown in Fig. 3(b). One child was confused which parts to use as a torso and was excluded. The difference between the categories was not significant, $X^2(5, N = 30) = 10.8, p > .05$. The two torso parts most often chosen were the sharp-edged and curved-edged distorted rectangular torso, which can be mapped as both human- and machine-like. Squared, rectangular, or spherical torsos were less popular, which may indicate that children are looking for a slightly more human-like form with broad shoulders.

Legs: The results of the legs are given in Fig. 3(c). The difference between the categories was significant, $X^2(8, N = 31) = 39.31, p < .0001$. Children in our previous study often drew human-like legs, while in our study children showed a clear preference for two fixed machine-like legs. A possible explanation is that children found it difficult to imagine and/or draw machine-like legs, therefore the physical form helped them to express their design more.

Arms: The two machine-like arms were preferred, see Fig. 3(d). The difference between the categories was significant, $X^2(3, N = 37) = 12.19, p < .01$. Their choice confirms that children do imagine a classroom robot with two mechanical arms as exhibited also in our previous study.

Materials: The results show that children choose several materials for different parts of the body. However, when accumulating the number of times each material was used, it is clear that metal was chosen most often, followed by plastic. This difference is significant $X^2(3, N = 59) = 9.68, p < .05$.

Size: The majority of the children indicated the size of the robot to be between a child and an adult and the difference between the size categories was significant, $X^2(4, N = 31) = 14.32, p < .001$. In general, children expected their robot to be larger than a child size, certainly not smaller or equal to a child size.

5.2   Changes to the Robo2Box

During the interview we asked children whether they were happy with their robot. 24 children indicated that they were satisfied with it, while 6 children were not completely satisfied with it. In addition, there were several requests for changes or additions, in terms of additional elements, additional functionality for the robot, or creative ways of expression.

Additional Elements. One child wanted to have a torso that was more human, meaning that the human torso included in the toolkit was not sufficient. Regarding the head, there were several requests: a taller neck like an ostrich, a more curved shape in the neck area, a fully spherical head instead of a half one. One child wanted a cylindrical torso instead of a spherical one, and one child wanted thicker more rectangular legs. Finally, one child wanted two legs, but with wheels instead of feet, and one child wanted one big wheel.

Additional Functionality. Seven children wanted to add a screen on the torso and one child wanted the robot to be able to turn into a television. Five children wanted to add buttons, for example to open and close or stop the robot, and two children wanted the robot to have a way to keep pens and erasers, for example in a storage compartment or in the hands. Two children wanted to add guns to the robot's hands. In addition, requests were also related to elements present in existing fictional robots or action figures. Three children wanted an appearance more similar to the popular Baymax figures, while one child wanted his robot to look like the Optimus Prime transformer figure, and another wanted the robot to look like Captain America. This indicates that indeed, experiences with robots in the media influence children's designs.

Additional Expressiveness. One child wanted to have the possibility to add stickers to the robot's head and another wanted the head to have facial expressions, but different from a human being. One child wanted the robot to be able to express emotions, but only in the form of symbols. While all drawings of the children in our previous study were rather colorful, the Robo2Box toolkit only provided the children with white elements. Although children in general did not comment on this negatively, they usually added many details about colors and other elements of the robot in their stories and mentioned them in the interviews. This might be an indication that some more ways for creative expression would have been appreciated.

6   Discussion

The aim of the work presented here was to develop and evaluate a robot design toolkit, Robot2Box. Here we answer our three main research questions:

What kind of robots do children design for the classroom using the Robo2Box toolkit? The Robo2Box toolkit appeared to be easy-to-use for the children. All the children except one understood how the different elements could be combined to construct a robot. By combining the robot elements chosen we constructed general robotic prototypes (Fig. 4) that children envisioned as a robot for use in the classroom. The robots have bodily characteristics that are similar to humans but with a robotic flavor, e.g. made out of metal, slightly more rectangular, and with machine-like arms and legs. This kind of robot was identified in previous research as the robot that children find the most friendly and that they think is able to understand them [8]. Interestingly, in general the children envision the classroom robot as bigger than a child. This strengthens the image of a classroom robot close to a human adult. It also suggests that children are not immediately afraid of a relatively large robot in the classroom.

When using the Robo2Box, do children design classroom robots similar to drawings? Figure 4 shows the drawings from our previous study alongside the prototypical robot designs made with the Robo2Box.

If we compare the robots created with the Robo2Box toolkit with the drawings made by children and interaction designers, we see that they resemble the drawings of the children most, especially the drawings made by children without any robotics knowledge (middle row). In general, the children in our study did not have much robotics knowledge either. In comparison to the interaction designers' robots the children's classroom robots were bigger, more similar to humans, and didn't include many animal parts.

The findings are interesting in several ways. First of all, the elements incorporated in the Robo2Box toolkit based on children's drawings enable children to construct designs for classroom robots that are similar to those they draw. Second, the additional elements based on interaction designers' drawings do not change children's views considerably. This indicates that their views are rather stable. Finally, it means that children really have different requirements for classroom robots than those who may be designing them, and are not just limited by their drawing skills. Involving children in the design of classroom robots, for example through the use of the Robo2Box presented here, is thus important to let them express their own views. In addition, a main advantage of using

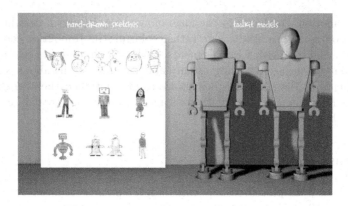

Fig. 4. Drawings (interaction designers = top row, children without robotics knowledge = middle row, children with robotics knowledge = bottom row) and our two Robo2Box prototypes.

the Robo2Box was the time spent on actually constructing the robot. In our previous study it took children around 20 min to draw a robot, the average construction time with the Robo2Box toolkit was 2:25 min for forming the body and then the rest of the time was spent on elaborating on other details. Limiting the time needed to envision a robot may allow researchers to allocate more time for discussing the behaviour of the robot.

What changes should be made to the Robo2Box for children to express their design ideas? Based on the storytelling activity and the interviews we were able to distinguish improvements for the Robo2Box toolkit, including separating the arms from the hands and the legs from the feet. This will allow children to add for example wheels to the legs, or tool-like hands to the arms. We will probably also expand the toolkit with some of the suggestions of the children (such as tools, armor, buttons and screens, and stickers or pens to create a more lively face). While the reliability and detail level of the toolkit are important, feasibility and low-cost production of the toolkit are also needed.

There are several limitations to this study that need to be mentioned. One of the limitations might come from performing the study with Turkish children only. It is possible that our findings are only representative for Turkish children. However, the drawings on which the elements of the toolkit are based come from other cultures. The fact that the children in our study often chose the toolkit elements that were based on other children's input indicates that findings may be generalised over different nationalities; this however needs further investigations. Another limitation might come from the actual sizes of the toolkit elements, since we did not provide the children with differently sized parts, however, we asked children to imagine it within a model of a classroom. Finally, the age range of the children in this study was rather broad. Focusing on specific age ranges could possibly reveal different preferences for the different age groups.

7 Conclusion and Future Work

We have described the reasons for developing a robot design toolkit, Robo2Box, as part of a human-centered design approach involving children in the design of robots for use in the classroom. We have also described the development of this toolkit based on previous literature. Through a study with 31 children from Turkey using the Robo2Box toolkit we conclude that (1) the classroom robots created by children are rather human-machine like, (2) the Robo2Box toolkit enables children to create classroom robots similar to freely drawn robots in a short time. This indicates that the robot design toolkit is a relatively fast and easy way for children to imagine a robot through a tangible experience, (3) the Robo2Box could be expanded and improved, by allowing more functionality, expressiveness, and additional elements, However, the benefits for a human-centered design process should also be considered. We thus argue that the Robo2Box can be a good basis for a human-centered design approach in which children are involved in the design of robots for the classroom.

Future work will focus on a more interactive toolkit with moving parts and means to join them. This might even lead to a toolkit where the design of some issues such as color, actions, behaviour and size can be left to accompanying software synchronized with the physical toolkit.

References

1. Bumby, K.E., Dautenhahn, K.: Investigating children's attitudes towards robots: A case study. In: the Proceeding of the Third International Cognitive Technology Conference, pp. 391–410 (1999)
2. Druin, A.: The role of children in the design of new technology. Behav. Inf. Technol. **21**(1), 1–25 (2002)
3. Goodrich, M.A., Schultz, A.C.: Human-robot interaction: A survey. Found. Trends Hum. Comput. Interact. **1**(3), 203–275 (2007)
4. Obaid, M., Barendregt, W., Alves-Oliveira, P., Paiva, A., Fjeld, M.: The 7th International Conference on Social Robotics (ICSR 2015), chap. Designing Robotic Teaching Assistants: Interaction Design Students' and Children's Views, pp. 502–511. Springer International Publishing, Heidelberg (2015)
5. Scaife, M., Rogers, Y.: Kids as informants: Telling us what we didn't know or confirming what we knew already? In: Druin, A. (ed.) The Design of Children's Technology, pp. 27–50. Morgan Kaufmann Publishers Inc., San Francisco (1998)
6. Sciutti, A., Rea, F., Sandini, G.: When you are young, (robot's) looks matter developmental changes in the desired properties of a robot friend. In: the Proceedings of the 23rd IEEE International Symposium on Robot and Human Interactive Communication, pp. 567–573, August 2014
7. Shin, N., Kim, S.: Learning about, from, and with robots: Students' perspectives. In: The Proceedings of the 16th IEEE International Symposium on Robot and Human interactive Communication, pp. 1040–1045, August 2007

8. Woods, S.: Exploring the design space of robots: Children's perspectives. Interact. Comput. **18**(6), 1390–1418 (2006). special Issue: Symbiotic Performance between Humans and Intelligent Systems
9. Woods, S., Davis, M., Dautenhahn, K., Schulz, J.: Can robots be used as a vehicle for the projection of socially sensitive issues? exploring children's attitudes towards robots through stories. In: the Proceedings of the IEEE International Workshop on Robot and Human Interactive Communication (ROMAN 2005), pp. 384–389, August 2005

Interaction with Artificial Companions: Presentation of an Exploratory Study

Matthieu Courgeon[1(✉)], Charlotte Hoareau[2], and Dominique Duhaut[2]

[1] Lab-STICC - ENIB, Brest, France
courgeon@gmail.com
[2] Lab-STICC - University of South Brittany, Lorient, France

Abstract. The MoCA project aims to design and study children-companion relationship through virtual agents, personal robots and communicating devices. In this article, we present an exploratory study of the free 30 min long interactions between children and a set of artificial, robot-like and virtual companions. We present a preliminary overview of the results obtained pending further analyses.

Keywords: Artificial companions · Robotics · Interaction · Serious games

1 Introduction

Technologies of artificial companions have evolved remarkably over the last few years. These developments are related to an of intensive research activity in the scientific communities focused on serious games [1, 2], affective computing [2], and personal robotics [3, 4].

This research shows the need to consider the role which the virtual companions can play – or should or should not play – in our daily lives [5]. The exploratory study of Dautenhahn et al. [3] shows that, in the studied population, 40 % of the respondents are not opposed to the idea of having an artificial companion, provided its role is limited to that of a personal assistant. On the other hand, few of them accept the idea that the companion can be a friend or an intimate companion.

The study of human-robot interactions from a social perspective is still relatively new compared to conventional or industrial service robotics, which requires only a minimal interaction between human and robot [6, 7]. However, there are numerous potential fields of application, i.e.: assistance for the elderly [8, 9] and with children [4, 10, 11]. There is no shortage of research undertaken along these lines [12]. Nevertheless, the studies carried out are often restricted to interactions of short duration on a single task, which constrains the user, who is the subject of the experiment, thereby orienting his or her perception of the potential of robot-like devices.

The experiment implemented here within the framework of the MoCA project aims at offering a larger range of possible interactions, not only concerning the nature of the devices, but also the diversity of the activities available.

© Springer International Publishing AG 2016
A. Agah et al. (Eds.): ICSR 2016, LNAI 9979, pp. 611–620, 2016.
DOI: 10.1007/978-3-319-47437-3_60

2   The MoCA Project

MoCA (in French: Mon petit monde de Compagnons Artificiels - My little world of artificial companions) is a fundamental research project which studies artificial companions (virtual characters and personal robots) for users in everyday life situations.

At the outset of the MoCA project, the type scenario was defined as follows: "artificial companions to accompany a child alone at home in different problem situations: comfort, safety, assistance with activities, games, etc." The exploratory study presented here is the result of an integration of the research tasks undertaken by the four IT laboratories belonging to the consortium.

2.1   Possible Roles for Artificial Companions

We define the various roles that the companions should be able to fulfil. These roles were divided into five categories. The study was designed so these various roles could be performed as well as possible, although, some of them were left out of account.

Help with extra-curricular activities/Coaching: Encouraging physical activity, Motivation, Valorization, Sharing non-formal knowledge, Reading of current events, follow-up of information
Homework support: Appraisal, Motivation, Valorization, Adaptation, Teaching with the child, Feedback to parents and teachers
Comfort, company: Ability to listen, give advice, Showing empathy
Child monitoring, safety: Detection of danger, Raising alarm, Contact adults, Recommendations and advice
Entertainment: Proposing games, Forming a pattern of interaction with 1 child, 1 child with n companions, m children with n companions

3   Presentation of the Experimentation

Based on this general definition, we chose to conduct the experiment in a "natural environment" in which the child would actually be placed in the situation presented above. It proved too complicated to equip the home environment of each subject, so to create a comparable situation, we rented a conventional apartment which was then arranged and equipped with the artificial companions and materials necessary for the study (see plan on Fig. 1).

Among the five types of problem situations described above, we chose not to consider point number 4 (Child monitoring, safety) in this experimentation. Since the child was not in a usual situation, we did not wish to place it in a situation of stress or at risk.

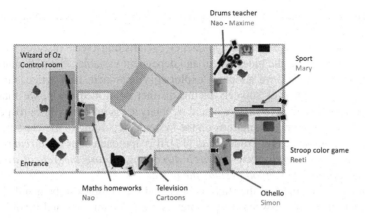

Fig. 1. Floor plan of the apartment

Several activities were proposed both with virtual agents and robots, based on technological capacities and limitation of each items. The activities installed in the apartment, which correspond to the four other roles, are as follows:

Role 1: Help and Coaching with extra-curricular activities
- Sports game/exercise with a projected life-size virtual character.
- Introduction to playing drums (percussion) with a virtual character and a Nao robot to study the synergy between virtual reality and robotics.

Role 2: Homework support
- Maths exercise (geometry) on paper with a Nao robot, teaching assistance, motivation and reminder of course material.

Role 3: Company
- Support in the choice of activities by a dedicated robot companion (Keepon), listening and general interaction, and overall coherence of the interaction. While the Keepon is not mobile, we have three identical keepons, one per room, allowing for a continuous interaction. We briefly considered allowing the child to carry Keepon with him/her, but it would have been prone to technical issues.

Role 4: Entertainment and games
- Colour memory games with a Reeti robot.
- Board games with a virtual agent on screen.

Further details on some of these activities can be found on related publications, about the technologies and activities developed during the MoCA project [13–16].

3.1 Methodology

The study presented here is of an exploratory nature. It involves analysing the interactions between the participants and the system in the most environmentally friendly context possible. During these free interactions with the system, it is possible to collect chronometric, behavioural and subjective data. Analysis of these data allows us to identify the types of interactions that are accepted and appreciated, and conversely, those interactions and companion behaviours that are considered to be problematic, unacceptable, or simply inconvenient.

Unlike traditional experimental approaches, establishing a set of research hypotheses and then identifying all the independent and dependent variables that will manipulated to test these hypothesis, we adopt an exploratory approach. In practice, rather than defining several groups of subjects facing distinct controlled setup, we make each participant evolve in the system at will and we analyze his/her behavior retrospectively, without printing a priori assumptions on the system. According to the classification of experimental approaches by Trudel *et al.* [17], this approach aims to *"circumscribe an object of research, identify new research directions, choose theoretical avenues or identify an appropriate method to study the object"*.

In our case, the object of the study is large and too complex to be handled simply. The interaction effects between the parameters are too numerous and involve a great number of assumptions to control. The resulting combinatorial experimental conditions would require a very high number of subjects that we simply cannot have. The exploratory approach therefore seems appropriate to identify research directions to dig in more detail later.

3.2   Recruitment of the Participants

The study was aimed at children in their final classes of primary education (9 or 10 y-o). To recruit the participants, we approached nearby primary schools and, with the agreement of the head teachers of the targeted schools, we circulated a call for participation. The call was designed to prompt the curiosity of the children and parents, but divulged almost nothing about the study itself. The participants spontaneously contacted us for an appointment to take part in the study. The experimental sessions involved a total of 20 participants aged between 9 and 11 years, coming from three different primary schools (14 males, 6 females).

The sessions were filmed in their entirety by six cameras (two axes per room), with a microphone on the child, and a microphone in the vision/sound control room to record the reactions of the relative.

Procedure of a standard experimental session

Phase 1 – We welcomed the child and relative in a reception area (at bottom on the left in Fig. 1) and they signed a mutual assent document. The child is then equipped with a microphone. We explained to the child that he will be alone and will interact with artificial companions and let him/her enter the apartment while the relative goes into the technical control room with the experimenters.

Phase 2 – Presenting the whole set of possible activities to the child, as well as the companions with which it will be able to interact.

When the child enters, a Keepon robot is activated and gives a welcome, proposing to start by visiting the apartment. Each activity is presented by the companion with which it is specifically associated. For example, in the first room, maths is presented by the red Nao on the desk.

At the end of the visit, the child is asked to fetch a teddy bear, which is in the "drums/sport" room, and to bring it back into the first room. The objective of this action is to neutralize the short-term memory effect when the first activity is chosen, and also to avoid that the last activity presented is always the one wanted by the child. In addition, this request also enables us to test the capacity of the child to obey a request given by a robot in the first minutes of their interaction.

Phase 3 – Once the rooms have been visited in their entirety (and at the end of each activity), the Keepon asks the child which activity it wishes to carry out. If the child chooses an activity, it is then triggered. In the (frequent) case where the child does not choose an activity, a suggestion/acceptance/refusal procedure is triggered.

At the start of an activity, the companions comment on the preceding activity (or the fact that the child chooses to start with them). This mechanism is implemented to reinforce the illusion of continuity and the existence of a "world of companions".

After thirty minutes, the Keepon invites the child to leave.

Phase 4 – Semi-structured interviews

Once the experiment is finished, the child returns to the reception area where it is questioned by one of the experimenters, initially according to a precise questionnaire, then in semi-structured interviews. During this time, the other experimenter takes the relative into the apartment to visit what was seen on the control screens, and also carries out a semi-structured interview.

4 Questionnaires and Interviews

4.1 Debriefing with the Child

The questionnaire is composed of a set of 13 questions relating to the usefulness, usability and acceptability of the companions, as well as the perception of their expressivity. To the best of our knowledge, no standardized questionnaire is available to evaluate such a complex system. We thus designed our own set questions.

The semi-structured interviews were carried out by revisiting the apartment with the child, in order to reactivate his or her experience in situ.

The topics addressed by the interviews are composed of open questions divided into four categories:

1 - Going back in a general way to what happened
2 - Focusing on the most striking aspects
3 - General feeling of the child towards the companions and activities
4 - Perceptions of emotions expressed by the companions

Parts of the topics addressed here are deliberately duplicated in the questionnaire given in the first part of the debriefing. These overlaps make it possible to corroborate the results.

4.2 Visit with the Relative

The visit with the relative consisted solely of a semi-structured interview. For each activity, the relative had to answer the following questions and give a justification if he or she so wished: (a) Is this activity acceptable with a companion? (b) Would you let your child do this activity alone? Once the visit was finished, some more general questions were put to the relative: (a) Broadly, do you find that such a device is useful? (b) Would you use it if it were free? Would you buy it? (c) Would you hand over part of the supervision of your children to an automated system? (d) What are your feelings about the overall behaviour of your child in front of these activities?

5 Preliminary Results

The results of the study are being analysed. The video recordings are in the course of annotation. In this article, we present only a preliminary analysis of the results from the children's questionnaires.

The results presented here relate to 16 subjects aged between 9 and 10 years. Four participants had to be excluded from the analysis, either because of technical errors of the system, or because of their age: the 11-year old subject, in the first form of secondary school, showed a behaviour that contrasted too greatly with the 9- and 10-year-old children. The results include a group of 16 children (13 boys and 3 girls) with an average age of 9.3 years (range: 9 to 10 years).

For the purpose of the analysis, the results from the questionnaire are placed in four categories: (1) usefulness, (2) usability, (3) acceptability and (4) credibility and expression of emotions. The questions relating to the first three categories are built on a rating scale qualifying the degree of agreement in terms of three types of response, which are coded numerically as follows: 1 negative, 2 neutral and 3 positive. The averages and standard deviations given here are based on this coding.

5.1 Usefulness of the Device

Four questions of the questionnaire aim to assess the children's perception of the degree of usefulness of artificial companions as guides for carrying out daily activities.

An analysis of the results of the question: "Are [companions] good for keeping you company when you are alone at home?" shows that 12 of the 16 respondents gave a positive response ("Completely in agreement"), as against 1 who were undecided ("Perhaps") and 3 with a negative response ("Not at all"; $m = 2.56$; $\sigma = 0.81$).

Two questions relate to the usefulness of artificial companions for entertainment activities. Results for question no. 4: "Are companions better for having fun than video games or television?" show that 10 of the 16 respondents consider that taking part in activities accompanied by artificial companions is "Much better" than playing video games or watching television, while 5 find the activities equally good ("Similar") and 1 find them "less good" ($m = 2.56$; $\sigma = 0.62$).

Results for question no. 5: "Do you think an artificial companion is better for having fun than playing with your friends?" show that taking part in activities with artificial companions, compared to activities with friends, is never considered as "much better", but is regarded as equivalent ("Similar") by 12 of the 16 children and "less good" by 4 of them (m = 1.75; σ = 0.45).

The last question about the degree of usefulness of artificial companions is concerned with the activity of homework support (question: "Is it better to have a companion while doing your homework than when alone?"). The results show that none of the respondents prefers to do homework alone ("Less good"), while 3 children do not feel any difference between doing homework alone or with an artificial companion ("Similar") and 13 find an advantage in being accompanied by an artificial companion ("Much better", m = 2.81; σ = 0.40).

An analysis of the answers to questions on artificial companions for accompanying daily activities shows that children mainly perceive the usefulness of such devices in relation to educational activities and as a substitute for multimedia activities.

5.2 Usability of the Device

The usability of artificial companions, and by extension, the activities which they propose, is evaluated in the questionnaire through two questions requiring the children to judge the ability of other pupils in their class to use these devices.

An analysis of the responses given to the question "Do you think that the other children in your class could easily use these activities?" shows that 14 of the 16 respondents consider that other children would manage to use the device "Rather easily" and 2 "Very easily" (m = 2.13; σ = 0.34).

The second question is "[Did other children in your class] need your help to use the device?" The answers are divided between "Not at all" for 7 of the 16 respondents and "Perhaps" for 8 children. No respondents answered "Yes, definitely" to this question (m = 2.44; σ = 0.51).

None of the children gave a negative response to either of these questions, indicating an absence of any difficulty in using the set up implemented in this study.

5.3 Acceptability of the Device

The concept of acceptability relates to the decision whether or not to use the device when it is in our possession. This concept is assessed in the questionnaire by posing two questions.

The first question is: "If the [artificial companions] belonged to you, would you be pleased to show them to your friends?" The majority of the children 13) are willing to show their artificial companions "Only to best friends" and 3 to "All friends". None of the children consider that they would show them to "Anybody" (m = 2.19; σ = 0.44).

The second question is: "Do you consider it enjoyable or annoying to have all these systems in your home?" Among all the children, 9 find it enjoyable to have artificial companions in their home and 7 do not find it "Inconvenient". None of children find the presence of artificial companions "Inconvenient" (m = 2,56; σ = 0.51).

Generally, the respondents are not opposed to the presence of artificial companions in their home or among their circle of friends.

5.4 Credibility and Expression of Emotions

The four questions relate to the perception of various emotions expressed by the artificial companions. The children were asked whether they perceived the artificial companions as angry, happy or sad ("Did you see at least one of the companions being angry/happy/sad?"). An analysis of the responses shows variations in the perception of emotions between different children. For example, 4 of the children detected anger in at least one companion ($m = 0.25$; $\sigma = 0.45$) and 3 did not describe any artificial companion as being happy ($m = 0.81$; $\sigma = 0.40$).

In a general way, the artificial companions manage to transmit emotional states, indicating the non-unanimous perception of anger, happiness and sadness. The children also differentiate the behaviours of the artificial companions; for example, the Nao robot is considered "stupid" because of its humorous remarks during the activity of initiation to drumming.

6 Conclusion

In this article, we present a complex study that is rich in interactions, integrated within the scope of the MoCA project. In this study, 20 children aged from 9 to 11 years were free to interact with a world of companions made up of 6 robots and 3 virtual agents, involved in a series of 5 game and teaching activities.

Although analysis of the data is still in progress, we present here some preliminary results based on the responses given by the children to a subjective questionnaire. Although these results are insufficient when considered on their own, they nevertheless provide some valuable insights to support the behavioural data based on video recordings that will be analysed in the second phase of the study.

Our preliminary results suggest that:

a. Artificial companions are perceived as especially useful in relation to educational activities (homework) and as a replacement for multimedia activities (video games and television).

b. The usability of the presented devices appears to be good. The children consider that it is easy for them, as well as others in the same age group, to take part in the activities. We had anticipated remarks about the fact that certain activities were difficult to perform (for example: drums for those who do not play a musical instrument, maths for children in difficulty, or games for children who do not have a video game console in their home). However, no remarks of this type were made.

c. In a general way, the children included in this study are not against the presence of artificial companions in their home or among their circle of friends. They were satisfied with the interaction and seem to have appreciated the experiment. Compared to the results of Dautenhahn et al. [3], it would appear that children are more inclined to accept this type of technology than adults. Moreover, all the children indicated that, to varying degrees, they were in favour, "of showing their companions to their friends". This finding suggests that having this kind of device would be a matter of pride, and not an embarrassment. Thus, artificial companions may be regarded as representing social values that can be shared and compared. These results should clearly be qualified and put into perspective using the answers given by the parents during semi-structured interviews.

d. As regards the expression of emotional states, we note that children perceive the emotions shown by virtual characters. The nature of the perceived emotions is variable, because their expression depends on the choices made by the child during the experimental session and his or hers varying rate of success in the interaction. Nevertheless, it would appear that children interacting with artificial companions perceive a form of social presence.

Before leaving, one of participants spontaneously took the time to thank several of the companions: "Thank you very much, that was very good... all the activities. Goodbye". This type of behaviour clearly shows that the companions are perceived as social entities which children consider as having their "own lives". Besides, most of the children indicated that they perceived what we call the "world of companions"; more precisely, they feel there is an extremely strong social bond between the companions, which makes the "collective" set up richer than a simple sum of activities with each of them taken independently.

These preliminary results are therefore very encouraging. The data collected should lead to a detailed behavioural analysis allowing us to shed light on many examples of child-companion interactions. Owing to the large freedom of choice and variability of the activities proposed in this study, it is impossible to carry out a systematic processing of the data. Nevertheless, in our opinion, this type of approach yields additional results that are broader in scope and more instructive when compared with classical laboratory studies, which are more restricted and more tightly controlled. In the future, we intend to examine more closely the data collected during this preliminary study, so the further results of this research can be made available.

Acknowledgements. This work was carried out in the framework of a project aimed at designing a world of artificial companions, which is financed by the project ANR-2012-CORD-019-02 *"Mon Petit Monde de Compagnons Artificiels: MoCA"*. http://www.moca.imag.fr

References

1. Castellano, G., Leite, I., Pereira, A., Martinho, C., Paiva, A., McOwan, P.W.: Affect recognition for interactive companions: challenges and design in real world scenarios. J. Multimodal User Interfaces **3**(1–2), 89–98 (2010)
2. Pereira, A., Leite, I., Mascarenhas, S., Martinho, C., Paiva, A.: Using empathy to improve human-robot relationships. In: Lamers, M.H., Verbeek, F.J. (eds.) Human-Robot Personal Relationships, pp. 130–138. Springer, Heidelberg (2010)
3. Dautenhahn, K., Woods, S., Kaouri, C., Walters, M.L., Koay, K.L., Werry, I.: What is a robot companion-friend, assistant or butler? In: 2005 IEEE/RSJ International Conference on Intelligent Robots and Systems, IROS 2005, pp. 1192–1197. IEEE, August 2005

4. Nalin, M., Baroni, I., Sanna, A., Pozzi, C.: Robotic companion for diabetic children: emotional and educational support to diabetic children, through an interactive robot. In: Proceedings of the 11th International Conference on Interaction Design and Children, pp. 260–263. ACM, June 2012
5. Clavel, C., Faur, C., Martin, J.-C., Pesty, S., Duhaut, D.: Artificial companions with personality and social role – expectations from users and impact on the design of groups of companions. In: IEEE SSCI 2013, Singapore (2013)
6. Duhaut, D.: Keywords and dimensions of artificial companions. In: 7th IEEE International Conference on Cybernetics and Intelligent Systems (CIS) and the 7th IEEE International Conference on Robotics, Automation and Mechatronics (RAM), Angkor Wat, Cambodia, July 2015 (2015)
7. Wilkes, D., Alford, R., Pack, R., Rogers, R., Peters, R., Kawamura, K.: Toward socially intelligent service robots. Appl. Artif. Intell. J. **12**, 729–766 (1997)
8. Sidner, C.L., Rich, C., Shayganfar, M., Bickmore, T.W., Ring, L., Zhang, Z.: A robotic companion for social support of isolated older adults. In: HRI (Extended Abstracts), p. 289 (2015)
9. Wang, R., Zhang, H., Leung, C.: Follow me: a personal robotic companion system for the elderly. Int. J. Inf. Technol. **21**(1) (2015)
10. Nalin, M., Bergamini, L., Giusti, A., Baroni, I., Sanna, A.: Children's perception of a robotic companion in a mildly constrained setting. In: IEEE/ACM Human-Robot Interaction 2011 Conference (Robots with Children Workshop) Proceedings, March 2011
11. Iacono, I., Lehmann, H., Marti, P., Robins, B., Dautenhahn, K.: Robots as social mediators for children with Autism-A preliminary analysis comparing two different robotic platforms. In: IEEE International Conference on Development and Learning (ICDL), vol. 2, pp. 1–6. IEEE, August 2011
12. Johal, W., Adam, C., Fiorino, H., Pesty, S., Jost, C., Duhaut, D.: Acceptability of a companion robot for children in daily life situations. In: IEEE 5th International Conference on Cognitive Infocommunications, Vietra, November 2014 (2014)
13. Courgeon, M., Martin, J.C., Jacquemin, C.: Marc: a multimodal affective and reactive character. In: Proceedings of the 1st Workshop on AFFective Interaction in Natural Environments (2008)
14. Courgeon, M., Céline, C., Martin, J.C.: Modeling facial signs of appraisal during interaction: impact on users' perception and behavior. In: Proceedings of the 2014 International Conference on Autonomous Agents and Multi-agent Systems, pp. 765–772. International Foundation for Autonomous Agents and Multiagent Systems, May 2014
15. Faur, C., Clavel, C., Pesty, S., Martin, J.C.: PERSEED: a self-based model of personality for virtual agents inspired by socio-cognitive theories. In: 2013 Humaine Association Conference on Affective Computing and Intelligent Interaction (ACII), pp. 467–472. IEEE, September 2013
16. Courgeon, M., Duhaut, D.: Artificial companions as personal coach for children: the interactive drums teacher. In: 12th Internationnal Conference on Advances in Computer Entertainment Technologies - ACE 2015, Malaysie, 16–19 November 2015 (2015)
17. Trudel, L., Simard, C., Vonarx, N.: La recherche qualitative est-elle nécessairement exploratoire? Recherches Qualitative, Les Questions de l'Heure– Hors Série – numéro 5, pp. 38–45 (2007)

Design and Development of Dew: An Emotional Social-Interactive Robot

Yiping Xia[1], Chen Wang[2], and Shuzhi Sam Ge[2(✉)]

[1] Department of Mechanical Engineering, National University of Singapore,
Singapore 117575, Singapore
[2] Department of Electrical and Computer Engineering,
National University of Singapore, Singapore 117583, Singapore
samge@nus.edu.sg

Abstract. There is a growing concern of depression and anxiety issues among children. While parents interpreting their quietness as shyness, depression and anxiety are directly experienced by these children. In this paper, we present the design of a social interactive robot, Dew, which is specially designed for these group of people. The design philosophy of Dew is presented with concept design, hardware and software specifications. With an expressive emotional design for its human-robot interaction, Dew is envisioned to approach these group of young people emotionally, and is expected to smoothen their emotions.

1 Introduction

Depression is predicted to be the second major cause to disability by 2020 [1]. As the ratio of young people diagnosed with depression increases, more attention is required to be drawn towards these people, helping them to destress and assisting the treatment [2]. Typically, young people that refuse to speak are mostly interpreted as shyness, especially when some of them may behave normally at home. Thus, it is difficult to identify depression among this group of people [3]. Cognitive-behavioral therapy (CBT) is commonly used as a therapy for both disorders, along with some pharmacotherapy.

Dew is a robot developed specifically for these group of young people who are going through emotional disorders. Attentions have been shifted to intelligent social interactive robots (SIR) development recently, with the release of various robots to the market. This group of robots have broad range of significant applications. In past decades, much research has been done to apply socially assistive robots (SAR) for therapeutic aids to promote interaction among residences, assist interaction with autism spectrum disorders (ASD) [4,5], remote communication and control which purposes to study human robot interaction (HRI) [6] and educational tools to promote the curiosity of children's learning [7]. The robots' role also include friend, partner, assistant, companion, study-buddy or even supervisor in some other circumstances [8].

A conceptual depiction of Dew is shown in Fig. 1 based on the overall flow of information, namely: User Input−Robot Output. Through the sensors embedded

© Springer International Publishing AG 2016
A. Agah et al. (Eds.): ICSR 2016, LNAI 9979, pp. 621–629, 2016.
DOI: 10.1007/978-3-319-47437-3_61

in the robot, touches from user will be analyzed for their interaction mode and emotional status, whereas motion and user expressions are captured through the mounted cameras. Output of the robot includes conducting photo/videography, giving musical responses, touch senses, expressing of emotional states through LED changes as well as responding through motional changes. The Input / Output (I/O) conversion is achievable through the hardware system inside the robot, which includes various sensors, cameras, microphone, LED lights as well as motors to enable its motion.

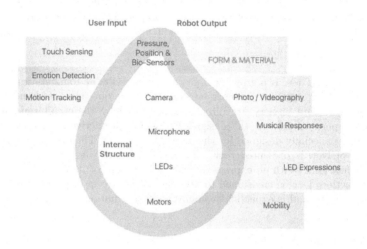

Fig. 1. Schematic diagram of dew

However, technological implementations are not the only determinant for robot development, as the users cares more about the emotional responses of the products instead of their implemented technologies [9]. As such, the details and significance of design is specifically presented in the rest of the paper, along with the robotic development. Section 2 describes the concepts and significance of overall design. The software and hardware frameworks and specifications are introduced in Sects. 3 and 4. The work is concluded in Sect. 5.

2   Design Significance and Concepts

There are two contributors to the user's interactive experiences, one is physical, including the product morphology and interactivity in various forms. Another is emotional, in a more conceptual and abstract form, triggered through its physical characteristics. Product morphology includes designing of its **form, colour, materials** etc. Interactivity includes motional and physical communication/responses, such as changes of colours, triggered senses upon physical touch, variation of sound etc. Emotional resonances are thus emerged through the various interactions.

2.1   Design Factors

Designing SIR requires careful investigation, as people "tend to treat other species as our own while communicating or interacting" [10], and while "augmenting such self-directed, creature-like behavior with the ability to communicate with, cooperate with, and learn from people" [11], we are likely to perform anthropomorphism on them. There are four important factors while performing robot design shown in Fig. 2, including the system architecture, cognitive & perception, human-robot interaction, as well as the robotic actions. Architecture is where the robot is based upon, and communicate with the user through the other three sectors.

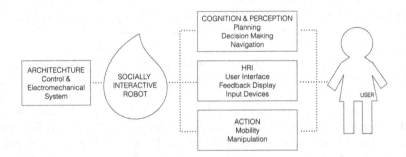

Fig. 2. Factors of robotic design

A most determinate characteristic of robots is their **mobility**. This involves planning and designing appropriate movements for the robot to optimise interactive experience. These actions need to be natural, anticipated, safe and obtainable with the hardware setup. **Expressions**, is one of the segments of display and interactivity, also commonly adopted as a measure of conveying product emotion [9]. Product aesthetics is also significant. **Shape**, for example, relates with aesthetic appreciation. It is a pivot to associate people with these robots [12].

2.2   Emotional Design

Emotional design is an essential factor that largely contribute to users' interactive experiences. There are three levels of emotional design, visceral, behavioural and reflective, concluded by [13]. What would be significant for the users, is the look of the robot (visceral); what the robot do, how does the robot interact and communicate with users (behavioural); and finally, do they want to maintain the long term relationship with their robot (reflective).

Color have significant impact on the emotional impacts to its users. Green colour has always been a representation of life in nature. Using green colour for the inner LED of Dew would resemble a natural and lively emotion, also as a representation of its own presence.

Music impact people's emotional responses. Light music is often referred to to calm people down and help them to relax. One significant factor to take note of is the volume of the audio responses, as too loud of the sound might impair a child's hearing system.

Sense is usually perceived as granted. Sensory responses may originate from touches or visual feedback, including the touch upon materials as well as the visual responses towards different forms.

2.3   Morphology Design

Form of Dew is almost entirely originated from nature. Its name has the same origin as its shape and form - morning dew and plant bud, objects that are indicators of liveliness, health, energy and happiness. With a philosophy of minimalist design, the initial concept filtered through brainstorming took into consideration to design a robot as simple and elegantly shaped as possible.

As a further explanation, among all geometric shapes, a dot (sphere) is the simplest and most symmetric shape. Dew is designed with a tweak on this shape, with its head-tip as a small and smooth extrusion, part of the reason is to get rid of the perfect symmetry, and give a further sensation to its elongated height.

Structure of Dew is also kept simple. As mentioned, only two functional motors are incorporated in Dew. Its head and body are smoothly linked together. The desired end product is approachable to users. The robot should feels soft and tender, conveyed through both the property of its outlook as well as choices of materials. Its design should be able to trigger the sensory responses from users and customers. Dew should be the emotional pal to soothe dispersion.

Material selection is also inspired from daily objects. As lighting for interaction is anticipated, the material should be translucent to allow light goes through. Moreover, instead of being hard and cold, the preferred material should be soft to touch and hold. One choice is made on translucent low density polyethylene covered with translucent silicone. It gives a tender and warm touch to the outer layer at the same time (Fig. 3).

Moreover, silicone material would increase the friction factor of Dew, such that it would be more stably positioned. The silicone-covered surface makes

Fig. 3. LEFT: Transluscent polyethylene. RIGHT: Silicone

it less prone to breakage and scratches, also resistant to water and dusts. As the outer layers of Dew is planner symmetric and simple, it is possible for its silicone shell to be produced through (liquid) injection moulding or compression moulding.

3 Human Robot Interaction (HRI) and Software Configuration

Human-robot interaction is achievable through the robot setup and information analysis. The overall interaction process includes reading user input through installed sensors, converting the information by its characteristics and deliver the responses through understandable outputs. In our case, input includes touch sensors and stereocamera/positional sensors, whereas the output includes variation of LED light, sound as well as movement and rotation as shown in Fig. 4.

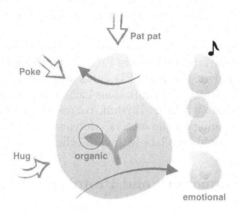

Fig. 4. Human robot interaction of dew

More specifically, the interaction could be further illustrated in Fig. 4. A child may interact with Dew through talking, moving around with it or hugging and touching it. The robot is able to understand the words from the children, track the movement of the child and thus make the decision of the child's emotional states. Through such analysis, Dew could make progress step by step and would be able to understand the child better. Position and inertia sensors are mounted inside of Dew which are used to estimate how the user physically interacts with the robots through the changes of Dew's relative position. To ensure the output responses are emotional, Dew would be incorporated with a software engine that is able to analyze user's expression and gestures, process the data from sensors and cameras simultaneously, and give a correspondent response through the musical, motional and expressional reactions. Researches have found that unimodal sensors is not accurate in detecting emotion-related variations because

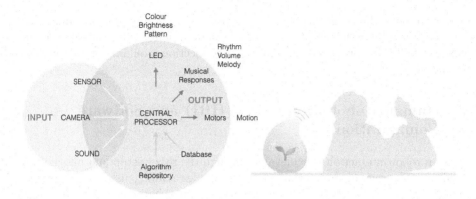

Fig. 5. Software framework and user interaction of dew

the result varies among individuals and the would fluctuate on a single person with one modality [14]. This is the reason why multiple sensors are adopted in this case.

As mentioned above, software development should involve recognising input from the camera and sensors, as shown in Fig. 5. It underlies the decision making for various responses given through LED lights and microphone, in accordance with the motional responses. Main variations include colour, brightness and pattern to be displayed through LED; rhythm, volume and melody of the musical responses; motion planning for the motors. Moreover, attention should be given on how these factors could correlate with each other.

4   Hardware Framework and Prototype Development

Aside from the software development, it is necessary for hardware to be constructed as a necessary basis of achieving functionalities. Figure 6 The fabrication of the prototype enabled the assembly and testing of the robotic structure. Dew is assembled to observe the best formality for its design and interaction with human, at the same time the viability of the developed system in human-robot

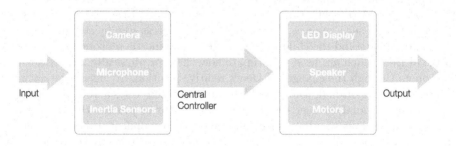

Fig. 6. Hardware components of dew

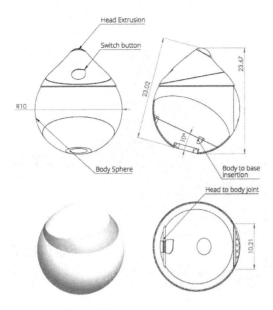

Fig. 7. CAD design of dew

interaction. Detailed design is discussed in this section, where CAD design is produced as shown in Fig. 7. For the ease of prototype assembly, shell of Dew is separated to cap, body and base compartments as displayed in Fig. 8. The body and base is proposed to be assembled through insertion. Cap and main body is proposed to be joint together with a hinge joint. Moreover, simple compartments to be hosted inside the robot was also produced, however, the shell was designed for display purpose instead of functional. Dimensionally, Dew is approximately 35 centimetre in height, 25 centimetre as the maximum diameter of its body. The size is compact, as it is desired to be placed as desktop robot as well as performing close interaction with users. Most segments of the parts produced are smooth and satisfactory. Some of the unsatisfactory segments are imposed by the discrete 3D printers usage.

Overall, Dew includes two major functional compartments - physical motion system and interactive I/O system. Motion system is where physical movement determines responses and interactions as shown in Fig. 8. Both its body and top-cap are capable of rotation. Dew is able to track and follow position of users through changes in its orientations. The mobility is achieved through two servomotors, one installed at the base that controls body-motion, the other at the body-cap intersection to control rotation of its tip. Input system includes two stereo cameras as well as various sensors distributed throughout the robot body. Output system includes LED headlight, LED lining as battery status indicator, as well as LED interior for expressional responses; microphone for audio responses. The central controller serves as a global interface for all sensory

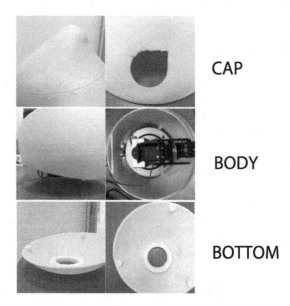

Fig. 8. Mechanical design of dew

devices like camera, microphone, inertia sensors as well as motion controllers boards for the proposed emotional robot motion.

5 Conclusions

In this paper, we presented the design and development of a emotional social interactive robot, Dew, for emotional support of children with mutism or mild depression. The design factors, emotional design and morphology design in developing Dew have been described in details. The hardware and software frameworks for the interactivity of Dew has been introduced separately.

Acknowledgement. This work is supported by Defence Innovative Research Programme (DIRP), the Ministry of Defence, Singapore under grant R-263-000-B08-592 and the A*STAR Industrial Robotics Program of Singapore under grant R-261-506-007-305.

References

1. Murray, C.J., Lopez, A.D.: Global mortality, disability, and the contribution of risk factors: Global burden of disease study. Lancet **349**(9063), 1436–1442 (1997)
2. Moreau, P., San Miguel, J., Ludwig, H., Schouten, H., Mohty, M.: Clinical practice guidelines. Ann. Oncol. **1**(5), 961–965 (2013)
3. Mesman, J., Koot, H.M.: Child-reported depression and anxiety in preadolescence: I. associations with parent-and teacher-reported problems. J. Am. Acad. Child Adolesc. Psychiatry **39**(11), 1371–1378 (2000)

4. Feil-Seifer, D., Matarić, M.J.: Defining socially assistive robotics. In: 9th International Conference on Rehabilitation Robotics, ICORR 2005, pp. 465–468. IEEE (2005)
5. Costa, S., Lehmann, H., Dautenhahn, K., Robins, B., Soares, F.: Using a humanoid robot to elicit body awareness and appropriate physical interaction in children with autism. Int. J. Soc. Robot. **7**(2), 265–278 (2015)
6. Saldien, J., Goris, K., Vanderborght, B., Vanderfaeillie, J., Lefeber, D.: Expressing emotions with the social robot probo. Int. J. Soc. Robot. **2**(4), 377–389 (2010)
7. Shin, J.-E., Shin, D.-H.: Robot as a facilitator in language conversation class. In: Proceedings of the Tenth Annual ACM/IEEE International Conference on Human-Robot Interaction Extended Abstracts, pp. 11–12. ACM (2015)
8. Li, Y., Tee, K.P., Ge, S.S., Li, H.: Building companionship through human-robot collaboration. In: Herrmann, G., Pearson, M.J., Lenz, A., Bremner, P., Spiers, A., Leonards, U. (eds.) ICSR 2013. LNCS (LNAI), vol. 8239, pp. 1–7. Springer, Heidelberg (2013). doi:10.1007/978-3-319-02675-6_1
9. Robins, B., Dautenhahn, K.: Tactile interactions with a humanoid robot: novel play scenario implementations with children with autism. Int. J. Soc. Robot. **6**(3), 397–415 (2014)
10. Fong, T., Nourbakhsh, I., Dautenhahn, K.: A survey of socially interactive robots. Robot. Auton. Syst. **42**(3), 143–166 (2003)
11. Breazeal, C.: Emotion and sociable humanoid robots. Int. J. Hum. Comput. Stud. **59**(1), 119–155 (2003)
12. Breemen, V., Sudijono, S., Horvath, I.: A contribution to finding the relationship between shape characteristics and aesthetic appreciation of selected products. Munich, Paper No. ICED99, pp. 1765–1768 (1999)
13. Norman, D.A.: Emotional design: Why we love (or hate) everyday things. Basic Books, New York (2005)
14. Kim, J.: Bimodal emotion recognition using speech and physiological changes. In: Grimm, M., Kroschel, K. (eds.) Robust Speech Recognition and Understanding, pp. 265–280. InTech, Vienna (2007)

RASA: A Low-Cost Upper-Torso Social Robot Acting as a Sign Language Teaching Assistant

Mohammad Zakipour[1], Ali Meghdari[1(✉)], and Minoo Alemi[1,2]

[1] Social and Cognitive Robotics Laboratory, Center of Excellence in Design,
Robotics and Automation (CEDRA), Sharif University of Technology, Tehran, Iran
meghdari@sharif.edu
[2] Islamic Azad University, Tehran-west Branch, Tehran, Iran

Abstract. This paper presents the design characteristics of a new Robot Assistant for Social Aims (*RASA*), being an upper-torso humanoid robot platform currently under final stages of development. This project addresses the need for developing affordable humanoid platforms designed to be utilized in new areas of social robotics research, primarily teaching Persian Sign Language (PSL) to children with hearing disabilities. *RASA* is characterized by three features which are hard to find at the same time in today's humanoid robots: its dexterous hand-arm systems enabling it to perform sign language, low development cost, and easy maintenance. In this paper, design procedures and considerations are briefly discussed and then the mechatronic hardware design of the robot is presented accordingly.

Keywords: Humanoid robot · Design · Hearing-impaired children · Sign language

1 Introduction

In recent years, satisfactory results of researches on the use of social humanoid robots in various fields of education [1], therapy [2], entertainment, etc. has been encouraging for researchers to conduct novel experiments in social robotics research areas, in particular experiment regarding the use of robots as educational and therapeutic companions of children with disabilities and special needs [3]. Teaching sign language (SL) to hearing-impaired children with the help of a social humanoid robot is one of such experiments.

Learning sign language is very essential for hearing-impaired children, as it is their main tool for communicating with other individuals from early age, and also because it contributes significantly to their cognitive and intellectual development [4]. The special social bond created between a child and a social robot makes the robot an ideal tool for educational and therapeutic purposes. Accordingly employing social robots in education as a game-based learning method can increase the quality of learning and it matters the most when a child is not autonomously willing to play and interact due to their impairment or disability [5]. H. Kose et al. have conducted a series of studies on employing humanoid robots as sign language tutors [6–8]. In these studies they investigate the

© Springer International Publishing AG 2016
A. Agah et al. (Eds.): ICSR 2016, LNAI 9979, pp. 630–639, 2016.
DOI: 10.1007/978-3-319-47437-3_62

effect of using a social humanoid robot on learning enhancement of different groups of children with different level of hearing impairment. They examine the difference between that method and video-based learning methods and argue that using a humanoid robot as a sign language tutor for children can be more effective than video-based learning methods, which is in line with many other studies in child-robot interaction, suggesting that robot-based tutoring systems are often more effective than computer-based tutoring ones [9, 10]. The most significant limitation of their studies is the kinematic limitations of the NAO and Robovie 3 robots used in the experiments which strongly limit the number of signs they can realize, whereas employing a more dexterous humanoid could be more effective.

Only a few humanoid robots have been developed with dexterous anthropomorphic hands, among which, humanoids such as the latest version of Honda's ASIMO humanoid[1] featuring 13 DOF hands, open-source iCub humanoid platform with 11 DOF [11], ARMAR III upper-torso mobile humanoid with 8 DOF [12] and NASA Robonaut-2 featuring 11 DOF [13] are the most dexterous ones. Today's most famous dexterous humanoid robots are nearly unaffordable for many social robotics research laboratories and institutions. They may also require elaborate and time consuming maintenance which can be a burden for social researchers; this creates the need for developing social humanoid robots with dexterous hand-arm systems, featuring low Manufacturing costs, and easy maintenance. Recently, there have been some projects addressing these issues such as the Aldebaran[2] Nao [14], Robotis[3] Darwin-Op [15], open-source Nimbro-Op [16] and other robotics hand related projects [17–19].

In this paper a new upper-torso humanoid robot, featuring dexterous anthropomorphic hands and a development cost of less than $7000 is introduced. *RASA* (Robot Assistant for Social Aims) which is to be employed as an assistant in teaching Persian Sign Language (PSL) to hearing-impaired children, distinguishes itself from most mobile or fix-placed upper-torso humanoids by its ability to imitate PSL signs, its low development cost, and easy maintenance all without losing performance and quality.

2   Design Considerations

RASA is intended to be a fully adoptable humanoid for use as an educational social assistive robot teaching PSL to hearing-impaired children. *RASA* will be used in sign language teaching classrooms acting as a tutor alongside an instructor in a multisession PSL teaching course. This paper only covers the mechatronics design factors. The teaching process will be conducted by various game-based methods whose discussion is to appear in another article. It is required to establish design guidelines for robot's software and hardware to be aligned with its unique task requirements.

---

[1] http://asimo.honda.com/asimo-specs/.
[2] https://www.ald.softbankrobotics.com/en.
[3] http://en.robotis.com/index/.

2.1 Sign Language Realization

In order to realize the signs of SL, it is deemed necessary to analyze its structures, mainly the phonological one. Phonology in SL refers to how a sign is formed and organized. Discussing the American sign language (ASL), C. Valli et al. [20] point out that each sign in ASL can be broken down into five components: handshape, movement of the handshape, location of the sign, palm orientation, and non-manual signals including head orientation, face expressions, etc.; these component which can as well be applied to the signs in PSL, all together, convey the meaning of a sign. PSL signs like ASL ones are derived much more from the manual parameters rather than the non-manual ones [21]. PSL also contains finger-spelling which is the spelling of written Persian words with the help of manual symbols of Persian alphabet. Finger-spelled signs are used when a word cannot be represented with a specific sign.

RASA's ability to imitate the signs of PSL has been the centerpiece of its design; accordingly, *RASA* was designed based on an anthropomorphic morphology regarding the dimensioning, kinematic layout, and ranges of movements, while at the same time, taking into account the other design guidelines such as low development cost. *RASA*'s design was mainly focused on its ability to imitate the signs of PSL rather than the finger-spelled letters of Persian alphabet, even though it can also realize all but few of those sings as well.

2.2 Social Interaction with Humans

Principal design issues concerning human-robot interaction in the development of a social humanoid robot include morphology, appearance, physical and conceptual capabilities, communicative expressiveness, and intelligence of the robot [22, 23].

The morphology and appearance of social robots greatly influences the quality of interaction between the robot and humans. Kanda et al. [24] conducted an experiment on how the appearance of robot affects the first impression of the person meeting it for the first time using two Asimo and Robovie robots, and concluded that Asimo, with a more organic appearance, gave a better first impression than Robovie with more mechanistic look. On the other hand, it has been investigated that caricatured robots may have high acceptability and believability even though they do not have a realistic appearance [25].

This project took particular interest in developing a robot morphology and appearance suitable for interacting with humans, especially children. Concerning communicative capabilities, the *RASA* robot is to include speech recognition and image processing modules. The robot shall exploit image processing as a mean of sign recognition when communicating with hearing-impaired children. Safety is another important factor in the development of a social humanoid robot interacting with children [20].

2.3 Low Development Cost

The high development cost of humanoid robot platforms often acts as a disincentive for researchers in humanoid and social robotics fields especially when the research is to be conducted employing a highly dexterous humanoid robot. Development costs of a humanoid robot consist mainly of the cost of its actuators and of the manufacturing processes. Special care should be taken when dimensioning humanoid robot actuators to avoid additional costs due to overdesign. A lightweight design also eliminates the need for using more powerful/expensive actuators. Regarding the manufacturing process cost, 3D printed techniques, utilized in fabricating *RASA*'s mechanical parts are often much more cost-effective than traditional methods.

2.4 Easy Maintenance

Elaborate and time consuming maintenance is an unfavorable characteristic of most dexterous humanoid platforms used in social robotics research areas, especially those which are employed in research for long periods of time. Humanoid robots, in particular the ones interacting with children, always face the possibility of falls and occasional impacts with their surroundings which can result in broken or disassembled parts. Thanks to 3D technology which allows the design of complicated and low-cost parts, it was possible to design easier assembly mechanisms and reduce the number of parts, which make the maintenance and setup of the robot, more convenient.

3 System Overview

In this section, the main characteristics of the robot are discussed. Figure 1 shows the 56 cm tall *RASA* upper-torso humanoid robot. *RASA* can be used both as an upper-torso on a fixed or mobile platform with the general specifications given in Table 1.

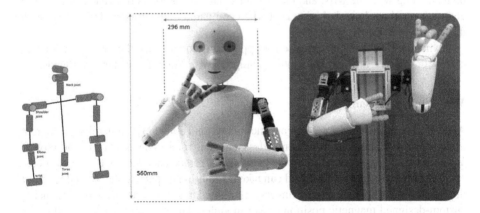

Fig. 1. CAD model of *RASA* humanoid platform (center), kinematic configuration of *RASA* (left), assembled arms of *RASA* (right).

Table 1. Some general information on *RASA*.

Actuated DOFs		Materials	Sensors
Arms	12 DOF	Polyamide PA 2200	Microphone
Hands	14 DOF	VeroGray	12-bit magnetic position sensor
Head	3 DOF	Aluminum	Honeywell linear hall sensor
Torso	1 DOF		Gyroscope + accelerometer (built in discovery board)

3.1 Kinematics

RASA's upper-torso, consists of 30 actively driven DOFs, including a pair of 13 actuated DOF arms, a 3-DOF neck and one remaining DOF for the torso joint rotating on the horizontal plane. Each arm of the robot features 3 DOFs at the shoulder, 2 DOFs at the elbow and one DOF at the wrist joint. Short by one from a human arm DOFs, the arms are able to imitate PSL signs as well as carrying out dexterous manipulation tasks. Each under-actuated hand has 13 DOFs, driven by 7 actuators.

3.2 Actuation

There are 30 actuators driving *RASA*'s movements. Considering the easy maintenance and setup design factors, off-the-shelf modular servomotors seemed an ideal choice for the robot actuators, but the limited space for placement of hand actuators did not accommodate their bulky packaging; therefore, servomotors were selected to drive each upper-arm and the 3-DOF neck of the robot while brushed DC motors were used to drive the fingers, along with the wrist and elbow joints. The servomotors drawbacks were fixed gear ratios and a relatively higher cost when compared to the brushed DC motors with the same output power.

Dynamixel[4] servomotors are used as upper-arm actuators and the forearms actuators consist of Maxon[5] brushed DC motors, equipped with Maxon planetary gearheads.

3.3 Sensory System

All the sensors installed in *RASA* are mentioned in Table 1. Dynamixel servomotors used in the robot integrate 12-bit contactless magnetic rotary sensors. Maxon DC motors are not equipped with an absolute or incremental rotary encoders as they would cost even more than the motor itself and can take up additional space in the limited space of the forearm. Instead, forearm and fingers joints angular measurements are realized using custom-designed magnetic position sensor modules. The magnetic sensors used in the fingers joints modules are Honeywell[6] SS495A1 ratiometric linear analogue hall sensors

---

[4] http://en.robotis.com/index/.

[5] http://www.maxonmotor.com/.

[6] http://www.honeywell.com/.

while the ones used at the elbow and the wrist are ams AG[7] AS5162 12-bit analogue magnetic position sensors. Torque measurement of actuators were realized by measuring the current passing through them utilizing the shunt current sensing method. While obtaining the exact torque from current measurement requires knowledge of actuator power dissipation, current measurement gave us a good approximation of the torque when conditioned and processed properly. The other sensors integrated in *RASA* consisted of a camera, microphones, and two pairs of gyroscopes and accelerometers which are integrated in off-the-shelf low-level control boards.

3.4   Material Properties for Lightweight Robot Structures

Materials used in the mechanical parts contribute significantly to the body mass of the system; therefore minimizing the weight and inertia of the parts, while considering all the design factors, allows the use of smaller actuators with consequently lower price and lower weight. The materials used in the robot are also included in Table 1.

Polyamides, while being lightweight, exhibit a combination of strength and flexibility making them suitable for robot structures exposed to external forces and possible impacts. Structural parts made of polyamides are fabricated using laser sintering technology (SLS). All of the forearm parts, excluding the fingers, robot head and trunk are made of polyamide. Finger phalanges and gears are made of VeroGray[8] photopolymer and fabricated by polyjet technology which has a relatively higher resolution and precision than the SLS process, making it suitable to fabricate highly detailed finger phalanges.

The more stressed parts of the forearm, i.e. the joint shafts of the elbow and wrist are constructed from Al7075 alloy. Both off-the-shelf and custom-made brackets connecting the upper-arm servomotors are manufactured from aluminum.

4   Mechanical Design

In this section we briefly discuss the mechanical design of the forearms and hands of *RASA* as it is the most complex and the most significant subsystem of the robot regarding the realization of Persian sign language.

4.1   Finger Design

Finger Flexion/Extension Mechanism. Based on design requirements discussed in Sect. 2, the mechanism used in the robot's index, middle, ring and small fingers is the cable driven, under-actuated mechanism proposed by S. Hirose [26] and used in robotic hands such as iCub robot [11] and Prensilia[9] prosthesis [27]. Figure 2 shows the side view of a finger and the cable routing in Hirose's mechanism. As illustrated, the cable

---

[7] http://ams.com/eng.

[8] http://www.stratasys.com/materials/polyjet/rigid-opaque.

[9] http://www.prensilia.com/.

goes through 3 pulleys (1) placed at the joints of the finger. There are also three torsion springs (2) placed at the joints ensuring the finger's extension, as the motor unwinds the cable. The finger flexion/extension profile is affected by the radii of each joint's pulley as well as the stiffness of the torsion springs.

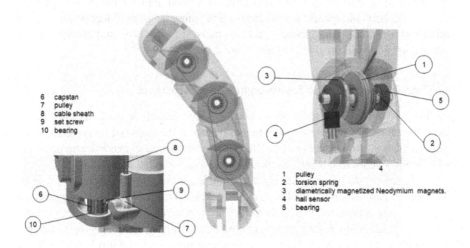

Fig. 2. Finger actuation mechanism (center), finger joint (right), and motor unit (left).

Finger Joint. Figure 2 also illustrates a finger joint, which integrates the torsion spring and the pulley needed for S. Hirose's mechanism as well as the magnetic sensor module measuring rotation of the upper phalanx. The custom-designed sensor module consist of a Honeywell SS495A1 ratio metric hall sensor (4) and a diametrically magnetized ring neodymium magnet (3) rotating in front of it. The hall sensor and magnet is press fit into the lower and upper phalanges, respectively. Each finger contains two of these sensor modules placed at the first (MP) and second (PIP) joints of the finger.

4.2 Wrist Joint Design

Each wrist has one DOF, which enables it to move up and down. The sideways movement of the wrist is not considered in the kinematic structure of the robot.

Figure 3 shows the tendon mechanism of the wrist movement as well as cable routes. The radius ratio between motor pulley and the wrist pulley is such that the backlash of the gearhead is minimized at the wrist joint while considering aspects of motor dimensioning for this joint actuator.

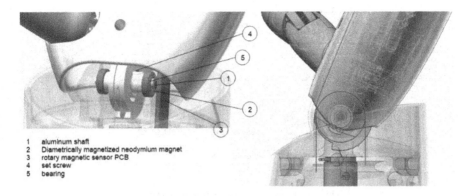

1 aluminum shaft
2 Diametrically magnetized neodymium magnet
3 rotary magnetic sensor PCB
4 set screw
5 bearing

Fig. 3. Wrist actuation mechanism (right) and wrist joint (left).

4.3 Elbow Roll Joint Design

As illustrated in Fig. 4, two double-helical (Herringbone) gears are used as the transmission mechanism of the joint. The actuator, in addition to all the other DC brushed actuators, is placed in the forearm.

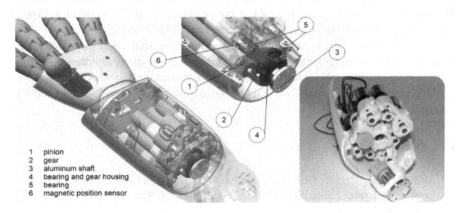

1 pinion
2 gear
3 aluminum shaft
4 bearing and gear housing
5 bearing
6 magnetic position sensor

Fig. 4. Elbow roll joint components.

5 Electronic Architecture

The robot is controlled by a central pc carrying out high level control, and two local controllers in charge of low level control tasks for each arm, such as position and current data processing, as well as velocity and position control of the DC brushed motors.

There are two identical custom-made analog circuits mounted at the chest of the robot designed for the purpose of signal conditioning of current and position sensors data as well as driving the brushed DC motors placed at each forearm. Figure 5 shows the main electronic architecture of *RASA*.

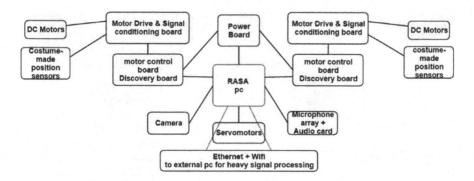

Fig. 5. Electronic architecture of *RASA*.

6 Conclusion and Future Works

In this paper we mainly focused on *RASA*'s design guidelines based on its social applications, and its mechatronic design. *RASA* was conceived, based on the need for a low-cost humanoid platform for conducting research in new social applications, mainly employing a humanoid robot in teaching sign language to hearing-impaired children. The main features of this dexterous *RASA* humanoid are its ability to perform Persian sign language, with a low development cost, and easy maintenance.

As an important future work of this ongoing project, we shall study the effectiveness of using *RASA* as a sign language tutor, on hearing-impaired children's learning enhancement as well as their interaction skills through various scenarios of game-based teaching methods. Additionally we will be able to evaluate the accuracy of the robot in generating the PSL signs.

References

1. Alemi, M., Meghdari, A., Ghazisaedy, M.: Employing humanoid robots for teaching English language in Iranian junior high-schools. Int. J. Humanoid Rob. **11**(3), 1–25 (2014). doi: 10.1142/S0219843614500224
2. Alemi, M., Ghanbarzadeh, A., Meghdari, A., Moghadam, L.J.: Clinical application of a humanoid robot in pediatric cancer interventions. Int. J. Soc. Rob., 1–17 (2015). doi:10.1007/s12369-015-0294-y
3. Meghdari, A., Alemi, M., Taheri, A.: The effects of using humanoid robots for treatment of individuals with autism in Iran. In: 6th Neuropsychology Symposium, Tehran, Iran (2013)
4. Mayberry, R.I.: Cognitive development in deaf children: the interface of language and perception in neuropsychology. In: Handbook of Neuropsychology, vol. 8, no. part II, pp. 71–107 (2002)
5. Besio, S., et al.: Critical factors involved in using interactive robots for play activities of children with disabilities. In: Proceedings of AAATE 2007 on Challenges for Assistive Technology, pp. 505–509 (2007)
6. Kose, H., Akalin, N., Uluer, P.: Socially interactive robotic platforms as sign language tutors. Int. J. Humanoid Rob. **11**(01), 1450003 (2014)

7. Köse, H., et al.: The effect of embodiment in sign language tutoring with assistive humanoid robots. Int. J. Soc. Robot. **7**(4), 537–548 (2015)

8. Kose, H., et al.: Evaluation of the robot assisted sign language tutoring using video-based studies. Int. J. Soc. Robot. **4**(3), 273–283 (2012)

9. Janssen, J.B., van der Wal, C.C., Neerincx, M.A., Looije, R.: Motivating children to learn arithmetic with an adaptive robot game. In: Mutlu, B., Bartneck, C., Ham, J., Evers, V., Kanda, T. (eds.) ICSR 2011. LNCS, vol. 7072, pp. 153–162. Springer, Heidelberg (2011)

10. Nalin, M., et al.: Children's adaptation in multi-session interaction with a humanoid robot. In: 2012 IEEE on RO-MAN. IEEE (2012)

11. Schmitz, A., et al.: Design, realization and sensorization of the dexterous icub hand. In: 10th IEEE-RAS International Conference on Humanoid Robots (Humanoids). IEEE (2010)

12. Albers, A., et al.: Upper body of a new humanoid robot-the design of ARMAR III. In: 2006 6th IEEE-RAS International Conference on Humanoid Robots. IEEE (2006)

13. Diftler, M.A., et al.: Robonaut 2-the first humanoid robot in space. In: 2011 IEEE International Conference on Robotics and Automation (ICRA). IEEE (2011)

14. Gouaillier, D., et al.: Mechatronic design of NAO humanoid. In: IEEE International Conference on Robotics and Automation, ICRA 2009. IEEE (2009)

15. Ha, I., et al.: Development of open humanoid platform DARwIn-OP. In: 2011 Proceedings of SICE Annual Conference (SICE). IEEE (2011)

16. Allgeuer, P., et al.: Child-sized 3D printed igus humanoid open platform. In: 2015 IEEE-RAS 15th International Conference on Humanoid Robots (Humanoids) (2015)

17. Meghdari, A., Mahmoudian, M., Arefi, M.: Geometric adaptability: a novel mechanical design in the sharif artificial hand. Int. J. Robot. Autom. **7**(2), 80–85 (1992)

18. Meghdari, A., Sayyaadi, H.: Optimizing motion trajectories in dexterous fingers by dynamic programming technique. ROBOTICA Int. J. **10**, 419–426 (1992)

19. Aghili, F., Meghdari, A.: Mechanical design of a modular arm prosthesis. Int. J. Robot. Autom. **10**(1), 22–28 (1995)

20. Valli, C., Lucas, C.: Linguistics of American Sign Language: An Introduction. Gallaudet University Press, Washington DC (2000)

21. Stokoe, W.C.: Sign language structure: an outline of the visual communication systems of the American deaf. J. Deaf Stud. Deaf Educ. **10**(1), 3–37 (2005)

22. Tzafestas, S.G.: Human-robot social interaction. In: Sociorobot World, pp. 53–69 (2016)

23. Alemi, M., Meghdari, A., Ghazisaedy, M.: The impact of social robotics on L2 learners' anxiety and attitude in English vocabulary acquisition. Int. J. Soc. Robot. **7**(4), 523–535 (2015)

24. Kanda, T., et al.: Analysis of humanoid appearances in human–robot interaction. IEEE Trans. Robot. **24**(3), 725–735 (2008)

25. Thomas, F., Johnston, O., Rawls, W.: Disney Animation: The Illusion of Life, vol. 6. Abbeville Press, New York (1981)

26. Hirose, S., Umetani, Y.: The development of soft gripper for the versatile robot hand. Mech. Mach. Theory **13**(3), 351–359 (1978)

27. Cipriani, C., Controzzi, M., Carrozza, M.C.: The SmartHand transradial prosthesis. J. Neuroengineering Rehabil. **8**(1), 1 (2011)

Robust Children Behavior Tracking for Childcare Assisting Robot by Using Multiple Kinect Sensors

Bin Zhang[1(✉)], Tomoaki Nakamura[1], Rena Ushiogi[2], Takayuki Nagai[1], Kasumi Abe[1], Takashi Omori[3], Natsuki Oka[4], and Masahide Kaneko[1]

[1] The University of Electro-Communications, Chofu, Tokyo 182-8585, Japan
zhang@radish.ee.uec.ac.jp
[2] Otsuma Women's University, Chiyoda, Tokyo 102-8357, Japan
[3] Tamagawa University, Machida, Tokyo 194-8610, Japan
[4] Kyoto Institute of Technology, Kyoto 606-8585, Japan

Abstract. Recently, the requirement for the high qualified childcare schools keeps increasing, but the number of qualified nursery teachers is far from enough. Developing a childcare assisting robot is highly necessary to help the works of nursery teachers. To work like a human nursery teacher, the first challenge for the robot is to understand the behaviors of the children automatically so that the robot can give adaptive reactions to the children. In this paper, we developed a robust children behavior tracking system by using multiple Kinect sensors. Each of the child is detected and recognized by integrating his/her personal features of face, color and motion. The tracking process is realized by using the Markov Chain Monte Carlo (MCMC) particle filter. The experiments are conducted in a childcare school to show the usefulness of our system.

Keywords: Childcare assisting robot · Children behavior tracking · Children recognition · MCMC particle filter

1 Introduction

Low birthrates have become a critical problem all over the world as some developed countries are rapidly becoming extremely old societies. As a social problem, it is the duty of the governments to improve the school environment for the children. Parents also hope their children could be raised in a qualified nursery schools. However, the number of qualified nursery teachers is far from enough [1]. Social robotics can be used as one way to assist the nursery teachers with their work. There are several researches about developing robotic systems for nursery schools [2,3]. These researches makes some contributions on understanding the behaviors of the children. However, the sensor networks or wearable sensors that they used are difficult to set and require the cooperations of the children. Some social robots [4,5] are designed to interact with the children, but they did not consider childcare assist from the viewpoints of nursery teachers. For the

© Springer International Publishing AG 2016
A. Agah et al. (Eds.): ICSR 2016, LNAI 9979, pp. 640–649, 2016.
DOI: 10.1007/978-3-319-47437-3_63

present, it is hard to use robot totally instead of the nursery teachers considering safety and childcare quality. Thus the childcare assisting robots need to be designed as supporters to the nursery teachers. The nursery teachers are very interested in childcare robotic systems because it would be helpful to monitor children's activities and to play with the children [6]. We propose to develop a childcare assisting robot, supporting functions of which are designed from the viewpoints of the nursery teachers. This social robot is supposed to be applied to nursery schools in the near future to support the work of nursery teachers. The functions of the childcare assisting robot is designed based on the investigations of the nursery teachers' requirements. As a basic function, the social robot should be able to track the behaviors of children in natural states so that it can give appropriate reactions to the children and provide useful information to the nursery teachers. Conventional research [7] realized adult people tracking in public places, but it is still difficult to applied for tracking children with many postures and crossing motions. The limit of laser heights and lacking of personal identification process are crucial.

In this paper, we propose a robust children behavior tracking system by using multiple Kinect sensors, as the fist step of developing the childcare robot. To solve the occlusion problem, we integrate multiple Kinect sensors which are set from different views in different height. The children are detected and recognized by integrating his/her personal features of face, color and motion. The tracking process is realized by using Markov Chain Monte Carlo (MCMC) [8] particle filter method. The number of Kinect sensors is adjustable according to the size of the space. We conducted the experiments in a childcare school, as shown in Fig. 1.

(a) the scene during a eurythmic class

(b) the scene of the children playing with an robot

Fig. 1. The tracking scenes in a nursery school.

2   Robust Children Behavior Tracking

In oder to track the behaviors of the children in a nursery school, we set Kinect sensors in the classrooms. The number of Kinect sensors can be adjusted according to the space of the classroom. To decrease the occlusions among children, we set the sensors in the height of around 2.5m with a slanted angle. To simplify the sensor setting process, we do not require the height and angle must be adjusted accurately, as the accurate height and angle can be calculated in the initialization process by detecting out the floor plane using Point Cloud Library (PCL) [9]. For the classroom shown in Fig. 1(a), we used 1 frontal Kinect sensor to track the behaviors of the children, and for the classroon shown in Fig. 1(b), 4 Kinect sensors are used and set in the 4 corners of the classroom.

2.1 Children Detection and Identification

As the Kinect SDK provided by Microsoft can only detect out 6 people maximum for each frame, we proposed a method for children detection by dealing with the point cloud information obtained the from Kinect sensor. The children candidate areas are detected by projecting the point cloud information among the children height range (0.5-1.5m) onto the ground after deleting the background parts. The we use connected-component labeling [10] process to detect out the areas within children size. We set the width and thickness thresholds of a 3-year-old child area as among the range of 20–50 cm. For the scene in Fig. 2(a), the detected people areas (the children and the teachers) are shown as Fig. 2(b).

(a) the scene during a drum game   (b) the detected human areas

Fig. 2. Children detection process.

Each of the children are identified by integrating their personal features of face, color and motion. Here, we use these 3 features to identify a particular child as each of these features has its strength and weakness. The face feature can express a person well and the identification accuracy is very high by using the OkaoVision software [11]. OkaoVision software can detect and identify the faces in an image, calculating the identification confidence together. Figure 3 shows some face detection and identification results during the class activities. We can see that most of the children' faces are detected and identified as the registered children with confidences (similarity range between 0–1000). However, we cannot make sure that the face information is always available. The child may face to some other directions or be occluded by others during the class, as they would move freely during the class activities. In this case, color information will be used for recognizing the person and camshift algorithm [12] is used for tracking the person until his/her face is detected and identified again. The face and color information for each person is recored during the initialization process. Here 5 frontal face photos are registered for each person, and we use color histogram of Hue channel in HSV color space as the color information. The color information based tracking process starts from the last identified face potion. When multiple faces are identified as the same person ID, the one with the highest confidence will be used for the tracking start position. Figure 3 shows some color based tracking results by Camshift during the class activities. Although color information is weak to identification a person, we can also get relative high identification accuracy by mostly relying on the face identification result and using color based Camshift tracking method to track particular children only when their faces are failed to be detected. However, the color based tracking method still may fail to track children when their faces are failed to be detected for a long time. This problem will be solved by using MCMC particle filter tracking method, which will be explained in detail in Sect. 2.3 (Fig. 4).

(a) the scene of sitting together   (b) the scene during the drum game

Fig. 3. Face detection and identification results by OkaoVision software.

(a) the scene of sitting together   (b) the scene during the drum game

Fig. 4. Color histogram based tracking results by Camshift.

2.2 Calibration for Multiple Kinect Sensors

When multiple Kinect sensors are used for our system, the sensors are needed to be calibrated. To make sure the whole classroom is covered by the sensors, the sensing ranges of adjacent sensors should be partially overlapped, and these overlapped areas are used for calibration. The calibration process is held during the initialization process, and is realized by matching the corresponding points in different coordinate systems in the floor plane after projecting the space information onto the ground. Same with the children detection process, we ask a person walking around so that the same person will be detected in different Kinect coordinate systems. We can get a series of positions of the person. Then we try to match the same person positions in different Kinect coordinate systems by affine transformation. The affine transformation matrix can be calculated. In a 2D plane, every 3 corresponding points can be used for calculating the affine transformation matrix, and the remaining points are used to check the residual sum of squares (RSS) error. The best affine transformation matrix would be chosen for calibration.

2.3 Tracking Children by Using MCMC Particle Filter

The children tracking problem is modeled by using a sequential Bayesian framework. A child's state at time t can be expressed as a 6 dimensional vector X_t, which contains the information of the child's location, velocity and acceleration on the ground plane. The states of the children is predicted first by using motion model, and we estimate the predicted states by finding the maximum-a-posteriori (MAP) solution of the joint probability, when the observation information Z_t is obtained from sensors. To find the most probable configuration, we estimate the MAP solution of $P(X_t|Z_t)$ by Eq. (1).

$$P(X_t|Z_t) \propto P(Z_t|X_t) \int P(X_t|X_{t-1})P(X_{t-1}|Z_{t-1})dX_{t-1} \qquad (1)$$

Here, $P(Z_t|X_t)$ represents the observation likelihood at time t, given the sensors input z_t. It measures the confidence of a hypothetical configuration. $P(X_t|X_{t-1})$ is the motion model, which shows the smoothness of the trajectory over time. $P(X_{t-1}|Z_{t-1})$ is the posterior probability of time $t-1$. The posterior probability at arbitrary time t can be calculated from the probabilities from time 1 to $t-1$ sequentially if the posterior probability at initial time is given. The best configuration X_t is then the MAP solution.

Motion Model. The motion model $P(X_t|X_{t-1})$ can be modeled by giving the update rules as

$$X_t = X_{t-1} + X_t'd_t \; ; \; X_t' = X_{t-1}' + \mu \qquad (2)$$

Here, μ is a process noise for a child's motion getting from a Gaussian noise.

Observation Likelihood. Given a hypothesized location of a child on the image, the observation likelihood measures the accuracy of the location. In our system, we proposed to use the children identification results to evaluate the observation likelihood. The children identification results in the color image (obtained from the Kinect sensor) can be changed to the camera space coordinate system by using Kinect SDK, and the corresponding identification areas on the projected ground plane can be calculated. The observation likelihood is calculated from the distance between the predicted position and the face areas on the ground plane. When multiple faces are recognized as a same person, the one with highest confidence will be chosen as the correct identification result.

Tracking with MCMC Particle Filter. The Markov Chain Monte Carlo (MCMC) method approximates the recursive Bayesian filtering distribution as a set of discrete samples known as a Markov Chain. The "sample improverishment problem" can be avoided by using this method [3]. However, we propose to adjust the conventional estimation result from the mean value of all the re-sampled particles to the center of the area that is closest to the mean value of the re-sampled particles as the real position of the child. Moreover, we allow multiple children to share one area obtained from the children detection process. When two or more children are close to each other, their projections on the ground may fuse with each other, and their detection result turns out to be a "big" area. Multiple children share this area in practice. In this way, our system can track all of the children even if the detected number of children keeps changing.

3   Experimental Results

To show the usefulness of our system, we conducted the experiments in a nursery school. As the first step of developing a childcare assisting robot, our system can provide kinds of useful information for the nursery teachers and help the robot to react in adaptive ways when the robot interacts with the children.

3.1   Children Tracking Results

We conducted a experiment during a rhythmic class. During the class, the nursery teacher leaded the children to do different kinds of games. The tracking result of a particular child during a drum game is shown in Fig. 5. We observe from the color image that during frame 0–100, the face of the particular child (U1) was well detected and identified as the child almost stayed in the same position. After that, U1 stated to run with other children and his face was unable to be detected. During this process, color information was used for tracking the child in the color image. When his face was identified again, the system relied on the face identification result to track his face. We also observe from the detection result that the children and the teachers are well detected. However, their areas may fused with each other when they were close with each other. Moreover, we observe that our tracking method is robust to track U1 during the game. Even under some special conditions like frame 180, U1 was close with another child and detected as a big area in the ground plane ((b) detection of frame 180 in Fig. 5). Our system treat this area as U1 and this result could be thought as correctly tracked. Actually this area was also identified as other child in the same frame, which proved the effectiveness of our improvement on MCMC particle filter. The motion trajectory is shown in Fig. 5(d). From the trajectory, we can also calculate the motion range and moving speed of the child, which can be used as quantitative indexes of the child's performance during the class.

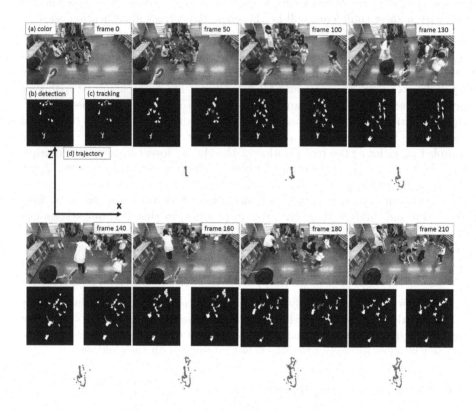

Fig. 5. Tracking results of our system.

In order to show the validity of our system, we evaluated the tracking results by comparing with the correct ones. The correct results are generated manually by assigning the position from original information of Kinect sensors. We can calculate the tracking accuracies for each child. Figure 6 shows the tracking error changing tendency of the child. The average error of the tracking result is 0.193 m, with the standard divisions as 0.122 m. This tracking accuracy is good enough for our purpose of tracking and understanding the behavior of the children and providing necessary information to the nursery teachers. We observe that U1 is successfully tracked by our system during the whole drum game with low error, although the distance errors becomes a little big for a very short time. These errors are caused by the overlapping of projected positions on the ground. As more than one child are closed to each other, they formed a "big" area in the human detection result. The center of this area is different from the real position of any single child. However, our method is more robust as the error would decrease after these children separated with each other, which is shown as breaking up of the "big" area in the detection result.

All of the children can be identified and tracked in this way. However, the accuracies of different children are different from each other. Some children may seldom show their face to the sensor so that they can only be tracked based on color information. The color information is weak to identify a person and the performance changes case by case, e.g. children with similar clothes may cause miss identification.

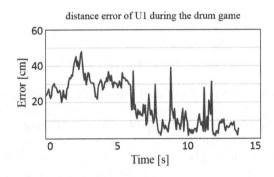

Fig. 6. The distance error of the particular child during the drum game.

3.2 Children Behavior Understanding

We conducted the experiments every 3 month and tried to understand the behaviors of the children with the changing of their performances with time going on. The nursery teachers believe that the motion range and momentum of a child can show the growth process. A younger child or a new member tends to be quiet. They will be more active with growing up. The motion range and momentum of the children can be calculated from the tracking results. Figure 7 shows the motion range and motion momentum of three persons during the drum game, which is measured in another day (same teacher, same class). After a long term observation, we can evaluate the growth of the children through quantitative data analysis.

We can also analysis the relationships among the children, by checking the relative distances between children on and after class. For example, during the drum game activity shown in Fig. 7(a), we can calculate their relative distances. A part of the result is shown in Fig. 8. In Fig. 8(a), it shows the relative distances between the teacher and 3 different children. Their average relative distances

(a) the scene during a drum game (different day)

(b) the motion momentum of three different persons during the drum game

(c) the motion ranges of three different persons during the drum game

Fig. 7. The motion ranges and momentums of the persons during the drum game.

during the game is 1.768 m, 0.797 m, 0.550 m with the standard deviation of 0.689 m, 0.416 m, and 0.267 m. We can see that child 2 and child 3 prefer to stay close to the teacher, and child 1 prefers to keep a small distance with the teacher. Similarly in Fig. 8(b), it shows the relative distances between child 3 and the other two children. Their average relative distances during the game is 1.623 m, and 0.307 m with the standard deviation of 0.592 m and 0.235 m. We can see that child 2 stays closer to child 3 at most of the time, and their distance is very small, even in touch with each other sometimes. On the other hand, child 1 usually keeps a small distance with them. We can infer their relationships that child 2 and child 3 are close friends and they like to play together.

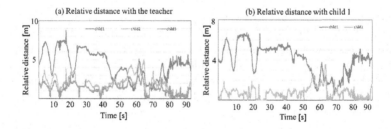

Fig. 8. The relative distance relationships during the drum game.

4   Conclusion

In this paper, we proposed a robust children behavior tracking system by using multiple Kinect sensors for the childcare assisting robot, towards the goal of assisting the nursery teachers with the childcare work. By our system, we can recognize and robustly track each individual. However, the color information of each child cannot be repeatedly used as the children change their clothes every day. More robust personal features need to be proposed for personal identification. Future work will also be focused on understanding different scenes and developing the adaptive reaction model for the robot.

Acknowledgment. This work was supported by Grant-in-Aid for Scientific Research on Innovative Areas 26118003. Special thanks to all the members in the nursery school.

References

1. Shiomi, M., Hagita, N.: Preliminary investiga-tion of supporting child-care at an intelligent playroom. In: The 2nd International Conference on Human Agent Interaction (2014)
2. Hwang, I., Jang, H., Nachman, L., Song, J.: Exploring inter-child behavioral relativity in a shared social environment: a field study in a kindergarten. In: Proceedings of the 12th ACM International Conference on Ubiquitous Computing, pp. 271–280 (2010)
3. Srivastava, M., Muntz, R., Potkonjak, M.: Smart kindergarten: sensor-based wireless networks for smart developmental problem solving environments. In: The 7th Annual International Conference on Mobile Computing and Networking, pp. 132–138 (2001)
4. Tanaka, F., Matsuzoe, S.: Children teach a care-receiving robot to promote their learning: Field experiments in a classroom for vocabulary learning. J. Hum. Robot Interact. **1**(1), 78–95 (2012)
5. Hieida, C., Abe, K., Attamimi, M., Shimotomai, T., Nagai, T., Omori, T.: Physical embodied communication between robots and children: An approach for relationship building by holding hands. In: IEEE/RSJ International Conference on Intelligent Robots and Systems, pp. 3291–3298 (2014)
6. Shiomi, M., Hagita, N.: Social acceptance of a childcare support robot system. In: 24th IEEE International Symposium on Robot and Human Interactive Communication, pp. 13–18 (2015)
7. Brscic, D., Kanda, T., Ikeda, T., Miyashita, T.: Person tracking in large public spaces using 3-D range sensors. IEEE Trans. Hum. Mach. Syst. **43**, 522–534 (2013)
8. Choi, W., Pantofaru, C., Savarese, S.: A general framework for tracking multiple people from a moving camera. IEEE Trans. Pattern Anal. Mach. Intell. **35**(7), 1157–1591 (2013)
9. Rusu, R., Cousins, S.: 3D is here: point cloud library (PCL). In: IEEE International Conference on Robotics and Automation, pp. 1–4 (2011)
10. He, L., Chao, Y., Suzuki, K.: A run-based two-scan labeling algorithm. IEEE Trans. Image Process. **17**(5), 749–756 (2008)
11. Takigawa, E., Hosoi, S., Kawade, M.: Gender, age, human race identification and estimation technology of facial attributes-OKAO vision. Image Lab 15(4) (2004)
12. Bradski, G.R.: Real time face and object tracking as a component of a perceptual user interface. In: 4th IEEE Workshop on Applications of Computer Vision, pp. 214–219 (1998)

Learning with or from the Robot: Exploring Robot Roles in Educational Context with Children

Nazgul Tazhigaliyeva, Yerassyl Diyas, Dmitriy Brakk,
Yernar Aimambetov, and Anara Sandygulova[✉]

School of Science and Technology, Nazarbayev University, Astana, Kazakhstan
{nazgul.tazhigaliyeva,yerassyl.diyas,
dmitriy.brakk,yernar.aimambetov,anara.sandygulova}@nu.edu.kz

Abstract. The goal of this ongoing research is to examine the feasibility of using a social humanoid robot to teach children the basics of programming. We focus on exploring robot's adaptive strategies in order to facilitate both effective educational applications and engaging child-robot interaction. In this paper we present our preliminary work, which explores robot's social roles (peer versus teacher) and their effect on learning. The child needs to learn the basics of programming in order to walk the robot through the maze via drag-and-drop instructions on the tablet screen. The findings suggest that children complete the task much quicker with the peer robot while a teacher robot is shown to be more effective for learning.

Keywords: Human-Robot Interaction (HRI) · Child-Robot Interaction (cHRI) · Educational robotics · Peer versus teacher robot

1 Introduction

The research of social robots facilitating educational benefits is an emerging area of social robotics. Recent efforts on the role of robots in educational applications have seen social robots acting as tutors [2], learners [4], learning companions [3]. In contrast, to date majority of educational robots used to introduce children to programming are mobile small educational robots such as Lego NXT Mindstorms [6] and Thymio [7]. This research addresses this issue with the goal to explore the use of social robots for introducing children to the basics of programming. It exploits the concept of edutainment where the basics of programming are introduced to children while playing a game with a social humanoid robot [16].

In order to establish social and bonding relationships with children in public environments such as hospitals or educational institutions, robots need to be able to adapt to child's needs, so that educational robot is effective: robot is liked and accepted, provides comfort and companionship, perceived to be a friend or a peer.

© Springer International Publishing AG 2016
A. Agah et al. (Eds.): ICSR 2016, LNAI 9979, pp. 650–659, 2016.
DOI: 10.1007/978-3-319-47437-3_64

Fig. 1. Experimental Setup

This paper aims to question: what role would make a social robot a more efficient tool for education. To this end, we developed two interaction patterns for each condition: a peer-like interaction and a teacher-like interaction. We hypothesize that children will tend to learn more and quicker in a peer condition since peer robot would be more effective at engaging the child in a task in comparison to the teacher robot who "instructs" rather than engages in play.

This paper presents our educational application, which consists of the humanoid NAO robot and an android-based tablet with the drag-and-drop interface inspired by Scratch developed at MIT [5]. The goal of the game is to escape the maze: the child needs to learn programming in order for the robot to exit the maze. This paper presents an HRI study that aims to investigate the role of the social robot (peer versus teacher), which in turn will facilitate engaging and effective learning in this particular learning context.

This paper is organized as follows: Sect. 2 introduces related work, Sect. 3 discusses our hypotheses and robot conditions. Sections 4 and 5 detail the conducted study and its results. Discussion and conclusion are presented in Sects. 6 and 7 respectively.

2   Related Work

A number of projects have explored the role of robots in educational applications. This section provides the review of the related work on robots acting as educational agents.

Zaga *et al.* [3] (2015) investigated the effect of robot's social character on children's task engagement suggesting that children solved the puzzles quicker and better in the peer condition in comparison to the tutor character condition.

Takayuki Kanda *et al.* in [8] used a social robot as a teaching assistant to predict whether the social behavior of a robot was responsible for establishing a better relationship with children and whether children were motivated to achieve more with the robot that exhibits social behavior. The study of Kanda *et al.* demonstrated that the effect of the social attitude of the robot on children's learning efficiency was not significant. In contrast, social behavior of the robot received better social acceptance suggesting that the social behavior might be useful for motivating children to use the robot to study less engaging subjects.

Saerbeck *et al.* considered varying the degree of social supportive behavior (role model, non-verbal feedback, attention building, empathy and communicativeness) of a robotic tutor [9] in order to increase student's learning performance. The authors implemented two settings of *Social Supportiveness* for the iCat robotic language tutor: (1) Neutral and (2) Social Supportive. The social supportive robotic tutor received better acceptance among participants as the results showed significantly higher test scores, better intrinsic and task motivations and closer proximity between the robot and the participant. Overall, the results and survey indicated that the presence of social supportiveness in a robot might be very promising in edutainment.

The effect of usage of peer robots in educative activities is further investigated by Jacq *et al.* [10]. The authors introduced the CoWriter activity where a child acts as a tutor who teaches handwriting to the robot. The effect of a child being asked to help the robot is known as the protege effect. The studies, where a child is involved as an interaction leader, have shown that the protege effect can help to keep the child motivated to practice handwriting with the robot for about an hour. Additionally, if implemented in therapy, the activity can show a positive effect on a child with attention deficit problems.

Shin *et al.* [11] performed analysis of data gathered from interviews of 85 students to understand student's motivation to learn about robots, to learn from robots, and to learn with them. The authors argue that younger generations generally demonstrated more eagerness to learn about robots rather than high school children and that children of all age groups did not mind learning from robots, though the robots were not regarded as teachers. Surprisingly, the participants of the *learning from robots* survey noted that the teacher robot was lacking "emotions" i.e., the ability to communicate, care and understand. Moreover, students avoided interacting with peer robots unless these robots acted as teaching assistants (helped solving homework problems to get better grades at school). The authors also observed that when learning with robots, the robot was viewed more as a competitor rather than a companion.

3   Hypotheses and Conditions

The goal of this ongoing research is to determine what robot's social role will facilitate more benefits to children such as learning gains, task engagement, and emotional support. Based on the previous findings [3,10,11], we have identified the following hypotheses:

Hypothesis 1: a peer robot will establish a more engaging experience as determined by a quicker task completion time.

Hypothesis 2: a peer robot will result in more learning gains as determined by demonstration of better results in a post-test.

In order to address these hypotheses, two educational interaction strategies were developed:

Peer robot. In a peer robot condition, the robot acted as if it needed help to complete a task (escape the maze). Robot's vocabulary was similar to the set that is most likely to be used in peer-to-peer conversation.

Teacher robot. In a teacher robot condition, the robot instructed the child to move it through the maze. The robot used an instructor tone and a different set of vocabulary, similar to the one generally used by teachers.

Robot's verbal content was the only difference between the two conditions. All non-verbal behaviours such as waving, gesticulating, eye gaze, and other robots' behaviours were the same across conditions. The same pre-recorded male voice was used in both conditions.

4   Experiment

4.1   System

The HRI system had three components: a humanoid NAO robot, a printed maze and an Android tablet (Fig. 2). It was completely autonomous throughout the interaction. All instructions and rules provided by the robot were in children's native language. Children were asked to move the robot from the 'current' cell to the 'destination' cell. Throughout the interaction, the robot gave children hints or instructions. At the end of the game children were expected to understand *iterations*, *go straight*, *turn right* and *turn left* functions.

During the experiment we did not experience any major issues with the NAO robot. The robot neither fell, lost communication with the tablet nor needed an intervention from an operator during the interaction itself. There was a minor problem with the robot's motion. The robot often went slightly aside as opposed to walking straight. However, the robot still walked within the cell borders.

4.2   Participants

The experiment was conducted in a primary school in Astana with 25 children (11 females and 14 males) aged 9 to 10 years old. Children were randomly assigned to a condition: a peer or a teacher robot. In both conditions children were free to work at their own pace. Counterbalancing was applied in terms of child's performance in school: children were asked what their GPA (Grade Point Average) was and were assigned to a particular condition. Counterbalancing was also applied in terms of gender: each group had almost equal number of males and females.

4.3 Experimental Setup

The experiment was conducted in a small room with three researchers inside. Each child was invited to the room one by one. The robot was placed on the 1.8 m x 1.5 m sized printed maze banner positioned on the floor. There was a table outside the classroom for conducting pre- and post-tests by the forth researcher for each child. The experimental setup is depicted in Fig. 1.

4.4 Recruitment

Children were given a brief introduction to the NAO robot before the experiment. During this introductory session children learned about the NAO robot. It was highlighted that children would not be graded for what they did or said. It was done to make children feel relaxed during the experiment [1]. In addition, this way all children had equal minimal interaction with the robot prior to the experiment to ensure accurate results [1].

4.5 Procedure

Each participant was called out of their class by the first researcher. While walking with the child, the researcher started with an icebreaker warm-up talk necessary to relax and engage the youngster. Before entering the room, children were invited to take a sit at the table with questionnaires. They completed a pre-test to assess their programming knowledge. Then, children were invited to enter the room with the robot. Two researchers were responsible for the initial launch of the system. After the interaction, children were asked to fill in the post-test to compare their improvement in understanding of the basics of programming. In the end, the first researcher brought the child back to the class and called out the next participant.

Each session lasted for approximately 15 min including pre- and post-tests. At the beginning of each session, the robot was placed at the starting position of the maze. The interaction was according to the following phases:

1. The robot introduced itself to the child and outlined the tasks.
2. Children started to play the game which was divided into levels of increasing difficulty.
3. When each level was completed, the robot provided further instructions.
4. When three levels were finished, the robot thanked and praised the child on the good job done.

4.6 Scenario

The concept of this research exploits the advantage of a social robot to motivate children to study programming. The screenshot of the application is depicted in Fig. 2. The sample of the dialogue in a teacher condition is outlined below:

– Hello! I am a teacher and my name is NAO. Today, I will teach you how to program. We are going to perform a task in which we are going to exit the maze together. During this task you will learn the basics of programming. On the tablet screen you can find different blocks of commands. Your job is to drag the correct blocks and if you make a correct choice, I will move. Let's try the first one together. Drag the green block, which says "Go Forward" from the left to the right side and press the "Run" button.

If the child made a mistake, the robot corrected it suggesting to think again. After a few unsuccessful attempts, the robot finally provided the solution, for example:

– Nope, it is still incorrect. Try the block, which says "Turn Left".

At the end of the task, the robot congratulated the child on the great job done. In contrast, the peer robot was easygoing in comparison to the teacher robot.

– Hi! My name is NAO. Would you like to play together? I need your help to exit the maze. In order to help me you will need to learn the basics of programming. There are blocks on the tablet screen that make me move. Let's try the first one together. Drag the green block, which says "Go Forward" from the left to the right size and press the "Run" button.

As the child made a mistake, the robot encouraged to try again:

– I didn't move, there seems to be a problem! Let's try again!

At the end, the peer robot complimented the child on the great job done and thanked for helping escaping the maze.

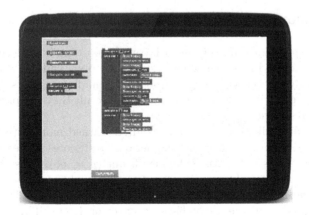

Fig. 2. Screenshot from the tablet application

4.7   Measurements

The primary units of analysis were questionnaires, task completion time and pre-and post-test performance in the programming exercise.

During the post-test children were provided with the smaller copy of the maze and were asked to complete a couple of programming questions, which were very similar to the tasks that they performed with the robot. For example, one of the questions was to estimate robot's position if *Go Forward* block was executed when the robot was at the cell number 10 and faced the cell number 9.

5   Results

On average, children spent 63.72 s interacting with the robot with a standard deviation of 33 s. The time children spent varied from as little as 50 s to as long as 77 s. There was a statistically significant difference between groups as determined by one-way ANOVA: $F(1, 23) = 4.373, p = .048$. A Tukey post-hoc test revealed that children spent significantly more time playing in the teacher condition $(76.15 \pm 38.94 \text{ s})$ compared to the peer condition $(50.25 \pm 18.65 \text{ s})$. Figure 3 illustrates these results.

On the other hand, children that interacted with the teacher robot had a much better improvement at programming. Even though, it was not a statistically significant difference between conditions as determined by one-way ANOVA $(p = .08)$, the results suggest that the teacher robot was more likely to be more effective for learning.

As expected, students with excellent grades (GPA 5 out of 5) finished three levels of the game significantly faster than students whose GPAs were 4 and 3. There was a statistically significant difference between performance groups according to one-way ANOVA $(p = 0.043)$

6   Discussion

There was a significant difference in task completion time i.e. the time children spent with the robot. In comparison to the teacher robot, children completed the task significantly quicker with the peer robot. This finding suggests that children were more engaged with the peer robot, and as a result they completed the task quicker. This result supports our first hypothesis.

However, our findings suggest that children learned more with the teacher robot. Children demonstrated slightly higher post-test scores in the teacher robot condition, but it was not statistically significant $(p = .08)$. This result rejects our hypothesis 2 that children gain more knowledge when interacting with the peer robot. And it contradicts the previous work of Zaga et al. [3].

We believe that culture might have an effect on how children learn. Geert Hofstede [13], the social psychologist, and his followers [12] argue that every society is represented by a certain set of fundamental values. These values influence the educational system and the way children are brought up in different parts

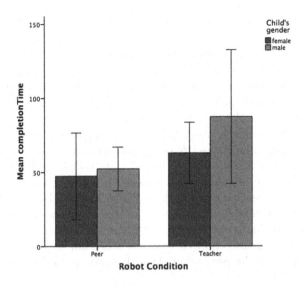

Fig. 3. Completion time

of the world. Hofstede's research on cross-cultural differences in education and learning resulted in the list of scores for a number of countries. Since this list does not include the country of the experiment (i.e. Kazakhstan), we consider the score calculated for Russia as it is the only post-Soviet country on the list.

According to Hofstede's and his followers' set of criteria [12] and their implications on teaching, education in Russia is more teacher-centered. Children expect the teacher to initiate communication and provide instructions. Parents teach their children to never question teacher's authority, and as a result students rarely criticize and contradict the teacher. The same tendency is noticed in such countries as Poland, China, Japan, South Korea, Belgium, Singapore and Slovakia.

Additionally, students of these countries feel more comfortable in structured situations, where a teacher gives precise instructions, detailed assignments and sets strict deadlines. As a consequence, students are rewarded for being accurate and are encouraged to give only solutions out of what they have been already taught. According to this framework, students in our country try not to violate rules set by the teacher. In contrast, the learning performance is taken as a function of two-way communication in countries such as Ireland, UK, USA and most of the European countries. It is allowed for teachers to not know answers to every question and students are encouraged to find their own way of doing things. In contrast, educational systems of most countries avoid uncertain and unstructured instructions expecting that teachers know answers to every question (apart from Ireland, UK, USA, China, Singapore, Denmark, Canada and New Zealand).

Vygotsky [14] argues that child's cognitive development is shaped in accordance with the culture and the environment in which the child has grown up. He also emphasizes the importance of guided learning where children and their more knowledgeable partners build up knowledge together. This could explain why children in our experiment did not benefit a lot when interacting with the peer robot. They viewed the robot as being an equal companion. Therefore, the peer robot was not considered to have expertise.

Similar to our argument regarding cultural differences, a tutor robot was preferred for learning a language [15] in Japan, although a social robot approach has not been explored for learning technical and exact disciplines. In addition, younger children preferred social peer robots for learning while older children considered robots as teaching assistants in South Korea [11].

The above arguments could also explain the difference in task completion time. Children probably spent more time thinking in order to be praised by the teacher robot. In contrast, children were probably not as focused on the task when interacting with the peer robot since they would not be "punished" by a friend.

Thus, we suggest to further investigate the role of educational robots across different cultures. In fact, this research will take a pioneering direction in the field of HRI. To date cross-cultural implications have not been questioned within educational robotic applications. We suggest to further investigate the social role of the robot and its adaptive strategies across different cultures for effective learning.

7   Conclusion and Future Work

The conducted study compared two robot conditions in their ability to contribute to children's learning of understanding the basics of programming. Children were suggested to help the robot to exit the maze via drag-and-drop android tablet interface which was inspired by Scratch [5]. Our findings suggest that with the peer robot children completed the required task significantly quicker than with the teacher robot condition. In contrast, children learned more with the teacher robot than with the peer robot. This result contradicts expectations and predictions made based on other studies in the literature [3].

The limitations of this work are in the number of children and our future work will address this limitation. However, this preliminary work provides a strong support for continuing this research direction i.e. to investigate cross-cultural differences of robot's social role within educational child-robot interaction. It is important to consider robot's social role in order to increase robot's perceived likeability, acceptance and engagement while fulfilling the required educational value.

References

1. Clark, C.D.: In a Younger Voice: Doing Child-centered Qualitative Research. Oxford University Press, New York (2011)
2. Kennedy, J., Baxter, P., Belpaeme, T.: Comparing robot embodiments in a guided discovery learning interaction with children. Int. J. Soc. Robot. **7**(2), 293–308 (2015)
3. Zaga, C., Lohse, M., Truong, K.P., Evers, V.: The effect of a robot's social character on children's task engagement: peer versus tutor. In: Tapus, A., André, E., Martin, J.-C., Ferland, F., Ammi, M. (eds.) Social Robotics. LNCS (LNAI), vol. 9388, pp. 704–713. Springer, Heidelberg (2015). doi:10.1007/978-3-319-25554-5_70
4. Hood, D., Lemaignan, S., Dillenbourg, P.: When children teach a robot to write: an autonomous teachable humanoid which uses simulated handwriting. In: Proceedings of the Tenth Annual ACM/IEEE International Conference on Human-Robot Interaction, pp. 83–90. ACM, March 2015
5. Scratch. https://scratch.mit.edu/
6. Lego NXT Mindstorms. http://www.lego.com/en-us/mindstorms/?domainredir= mindstorms.lego.com
7. Thymio. https://www.thymio.org/
8. Kanda, T., Shimada, M., Koizumi, S.: Children learning with a social robot. In: Proceedings of the seventh annual ACM/IEEE international conference on Human-Robot Interaction, pp. 351–358. ACM, March 2012
9. Saerbeck, M., Schut, T., Bartneck, C., Janse, M.D.: Expressive robots in education: varying the degree of social supportive behavior of a robotic tutor. In: Proceedings of the SIGCHI Conference on Human Factors in Computing Systems pp. 1613–1622. ACM, April 2010
10. Jacq, A., Garcia, F., Dillenbourg, P., Paiva, A.: Building successful long child-robot interactions in a learning context. In: 2016 11th ACM/IEEE International Conference on Human-Robot Interaction (HRI) pp. 239–246. IEEE, March 2016
11. Shin, N., Kim, S.: Learning about, from, and with robots: students' perspectives. In: RO-MAN 2007-The 16th IEEE International Symposium on Robot and Human Interactive Communication, pp. 1040–1045. IEEE, August 2007
12. Wursten, H., Jacobs, C.: The impact of culture on education. The Hofstede Centre, Itim International (2013)
13. Hofstede, G.: Cultural differences in teaching and learning. Int. J. Intercultural Relations **10**(3), 301–320 (1986)
14. McLeod, S.A.: Vygotsky-simply psychology (2007). Accessed 27 Oct. 2013
15. Okita, S.Y., Ng-Thow-Hing, V., Sarvadevabhatla, R.: Learning together: ASIMO developing an interactive learning partnership with children. In: RO-MAN 2009-The 18th IEEE International Symposium on Robot and Human Interactive Communication, pp. 1125–1130. IEEE, September 2009
16. Diyas, Y., Brakk, D., Aimambetov, Y., Sandygulova, A.: Evaluating peer versus teacher robot within educational scenario of programming learning. In: The Eleventh ACM/IEEE International Conference on Human Robot Interation, pp. 425–426. IEEE Press, March 2016

Automatic Adaptation of Online Language Lessons for Robot Tutoring

Leah Perlmutter[1]([✉]), Alexander Fiannaca[1], Eric Kernfeld[2], Sahil Anand[3],
Lindsey Arnold[3], and Maya Cakmak[1]

[1] Computer Science and Engineering, 185 W Stevens Way NE,
Seattle, WA 98195, USA
lrperlmu@cs.washington.edu
[2] Department of Statistics, Box 354322, Seattle, WA 98195, USA
[3] Human-Centered Design and Engineering University of Washington, 428 Sieg Hall,
Seattle, WA 98195, USA

Abstract. Teaching with robots is a developing field, wherein one major challenge is creating lesson plans to be taught by a robot. We introduce a novel strategy for generating lesson material, in which we draw upon an existing corpus of electronic lesson material and develop a mapping from the original material to the robot lesson, thereby greatly reducing the time and effort required to create robot lessons. We present a system, KubiLingo, in which we implement content mapping for language lessons. With permission, we use Duolingo as the source of our content. In a study with 24 users, we demonstrate that user performance improves by a statistically similar amount with a robot lesson as with Duolingo lesson. We find that KubiLingo is more distracting and less likeable than Duolingo, indicating the need for improvements to the robot's design.

1 Introduction

The internet is a vast source of knowledge and information, but the majority of web content is in English. This means that most massive open online courses (MOOCs) [4], academic publications [21], and content in general [8] is accessible only to those privileged enough to know English. Concurrently, a great many people immigrate to countries where they do not know the language. These are just a couple situations that demand language education at a higher rate than its current availability.

In this work, we develop a tabletop robot that teaches languages to people. The role of our system is to support a human teacher with personalized supplementary lessons when the teacher is not available. We propose an embodied robot rather than a simple computerized lesson because studies show that the physical presence of a robot results in greater learning gains than when the same content is presented without a robot [14,15,18].

One bottleneck in developing robots that teach is custom content generation. It is time-consuming and requires expertise in both the subject being taught and in teaching techniques. Furthermore, it is redundant when electronic lesson

© Springer International Publishing AG 2016
A. Agah et al. (Eds.): ICSR 2016, LNAI 9979, pp. 660–670, 2016.
DOI: 10.1007/978-3-319-47437-3_65

content (albeit not in robot form) already exists. Our main contribution is to introduce the strategy of *content mapping*, which takes advantage of existing lesson content by adapting it to a form that the robot can teach. To the extent that source lessons follow a predictable format, our system can teach any lesson in the source corpus. As a result, we can draw on a much larger corpus of lessons.

2 Related Work

Technology has long been used for teaching languages. Rosetta Stone is an example of a paid program, involving a mixture of computer-based and teleconferenced lessons [1]. Duolingo is a newer, free program offering web-based and mobile content [6].

Recently, research has been done teaching language with robots. Telepresence robots, as in the work of Kwon *et al.* [16] can be helpful when a human teacher is available but not colocated with students. Others have proposed or developed autonomous teaching robots [10,17], which can be helpful when human teachers have limited time or availability. Our robot falls into the autonomous category.

Many have demonstrated cognitive learning benefits when material is taught by embodied, autonomous robots as compared with non-embodied agents, computer-based lessons, or paper-based lessons [14,15,18]. This research indicates promise for teaching robots and helps to form the foundation of our work.

Social interaction is an important attribute of robots that teach. Saerbeck *et al.* show that a robot tutor's socially supportive behavior increases learning efficiency in students [20], while Kennedy *et al.* show that it is possible for a robot to be too social during teaching, countering learning gains [15]. This research informs our work in designing the social interactivity of our system.

Content is hard to develop for robots that teach. In many cases, researchers have created custom content for their robots to teach [15,17]. We introduce a mapping from existing content to robot content, reducing the amount of effort needed to provide content for the robot to teach. We consider our work to contribute an advancement in the area of content for robots that teach.

3 System

3.1 Hardware and Software

Our hardware consists of a Nexus 7 tablet mounted on a Kubi base, as seen in Fig. 1. Designed by Revolve Robotics [7], Kubi is a telepresence platform which holds the tablet and is actuated with two degrees of freedom: pan and tilt.

Our robot system, KubiLingo, runs in an Android app. The robot character, named "Kubi," has an animated face with large eyes, illustrated in Fig. 3. It shows emotions by speaking, animating the eyes, and actuating the base. Kubi also has virtual, animated hands, and displays lesson material on a virtual card held in its hands.

Fig. 1. KubiLingo is a social robot system that teaches languages to people

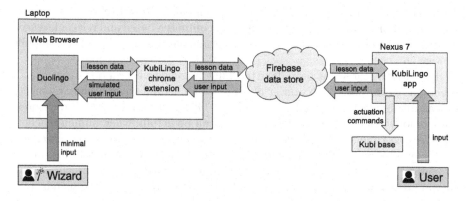

Fig. 2. System architecture diagram

As the backend for our lesson content, we use Duolingo, an online language teaching platform designed to teach any language to speakers of any other language. Duolingo's content is crowdsourced, so there is a sustainable way to generate more content, resulting in a large and growing corpus of lesson material. Duolingo has given us permission to use their content as described below.

Our system wraps Duolingo, automatically adapting Duolingo's content to be rendered on the robot and transmitting user input back to Duolingo, as illustrated in Fig. 2. To do this, we built a browser extension ("KubiLingo chrome extension") that runs on Chrome while Chrome runs a Duolingo lesson. It parses lesson data from the DOM and sends it to the robot in real-time via Firebase, a cloud-based service [2]. It also receives user input data from the robot and simulates that user input in the browser. The flow of data is fully automated except the confirmation step, in which the user verbally confirms their answer to each prompt. A wizard-of-oz operator listens for the user's verbal confirmation and clicks the confirmation button in the web browser.

3.2   Visual Design and Animation

We chose a vibrant color scheme similar to Duolingo's to make the learning experience fun and playful. We designed Kubi to resemble Duo, the owl mascot of Duolingo, by giving it green and orange coloring. Related works inspired some other features. Ribiero *et al.* applied Disney's animation principles to help

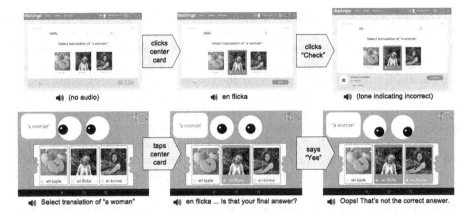

Fig. 3. The chronological progression of a prompt on Duolingo (top) and KubiLingo (bottom). User actions are described in the gray arrows.

users understand a robot's emotions [19]. Accordingly, we designed Kubi's facial expressions using these principles. Cuijpers *et al.* found that users perceive robots with idle motions as more alive and empathic [12], so we implemented idle motions in the form of random head movements and periodic eye blinking.

3.3 Content Mapping

With the differences between computer and robot in mind, we created a mapping to automatically convert Duolingo lessons into robot lessons.

The basic unit of interaction in Duolingo is a *prompt*, in which some material (text, audio, pictures) is delivered to the user and the user is expected to reply by typing or clicking. Below we describe the different conceptual parts of a prompt, and how they are delivered in Duolingo and on the robot.

The *directive* is a description of what to do in this prompt, e.g. "Select translation of 'a woman'." Duolingo displays the directive as text; Kubi speaks the directive and displays key words (e.g. "a woman") in a speech bubble.

The *body* is the main part of the prompt, e.g. selectable flash cards. Both Duolingo and KubiLingo display the body centrally; KubiLingo displays it on a virtual card in the robot's hands. A Duolingo prompt requires clicking or typing input. In KubiLingo, clicking is replaced with tapping and typing is done using a bluetooth keyboard.

A *hint* is extra data related to a word. In Duolingo, the user can click an underlined word for a hint. A popup appears just below the cursor showing translations of the clicked word. In KubiLingo, we underline words with available hints, and when the user taps a word, Kubi shows the hint in a speech bubble.

Checking is when the user indicates they are finished providing input and wish to check their answer. Duolingo provides a "Check" button for the user to click. Kubi asks, "Is that your final answer?" The wizard listens for an affirmative response from the user, and clicks "Check" in the Duolingo interface.

Feedback is displayed after the user submits a response, and indicates whether the response was correct. Feedback can display alternate answers or corrections to mistakes. Duolingo displays feedback just below the prompt body; KubiLingo displays it overlaid with the body. In addition, Kubi provides verbal feedback and emotes a reaction. For example, if the response was incorrect, Kubi might lower its head, show sad eyes, and say "Sorry, that's not the right answer." KubiLingo has a variety of positive and negative responses, and randomly selects an appropriate one in the feedback stage of each prompt.

When the user is finished with the feedback, they *advance* to the next prompt. For this purpose, Duolingo provides a "Continue" button to click. In KubiLingo, the wizard waits for Kubi to present the feedback, then clicks "Continue".

Progress indicates the number of prompts the user has completed in the current lesson. A visual indicator advances when the user answers correctly and recedes when the user answers incorrectly. Duolingo displays a progress bar above the body; KubiLingo displays a progress ring in the upper right corner.

We implemented the mapping for five different types of prompt: *Select*, in which the user selects a flash card matching the spoken word, *Translate*, in which the user types the translation of a word or phrase, *Name*, in which the user types the noun shown in three pictures, *Listen*, in which the user types a spoken phrase, and *Judge*, in which the user selects one or more options from a list of choices. These five prompt types account for all the material taught in the lessons we used for our system evaluation. Figure 3 illustrates the chronological progression of a Select Prompt on both Duolingo and KubiLingo.

4   Evaluation

Hypotheses and Conditions. We conducted a user study to compare KubiLingo with Duolingo. We hypothesized:

- **H1 (Performance hypothesis).** User performance from pre-test to post-test will improve more with a robot lesson than with a screen lesson.
- **H2 (Preference hypothesis).** Users will subjectively rate the robot higher than the screen.

H1 is supported by previous work showing that embodied robots have cognitive learning benefits [14,15,18]. Proving H1 alone, however, would not sufficiently show that KubiLingo is a better learning platform. It would be possible to be more effective but less appealing, in which case users may abandon it, learning less overall. We formulated H2 to test whether KubiLingo is more appealing.

Another goal of our study, not covered by our hypotheses, was to gather feedback for the next design iteration of KubiLingo. We wanted to measure the quality of KubiLingo's user experience and learn which features impacted it.

To test our hypotheses, we designed a study with two conditions: **Screen lesson**, a Duolingo lesson taken on a laptop; and **Robot lesson**, a lesson taken from KubiLingo. In our crossover study design, each participant took one of each type of lesson, in two consecutive sessions. We counterbalanced the order, creating two arms of the study, **Screen-first (S)** and **Robot-first (R)**.

To prevent carryover effects, we taught Swedish in Session 1 and Dutch in Session 2. We did not counterbalance languages (which would create four study arms) because the two study arms mentioned above were sufficient to isolate the effects necessary to test our hypotheses. Each "lesson" of our study consisted of three Duolingo lessons: "Basics" lessons 1–3 for the language being taught (L2).

Participants. We recruited participants from a university community. Those interested completed an eligibility survey, and those qualified were sorted into the two study arms using stratified randomization [13]. We excluded participants understanding Swedish, Norwegian, Danish, Icelandic, Dutch, German, or Flemish and stratified those knowing "a little bit" of those languages. We also stratified on bilingual ability and age at which English was learned. No participant reported visual or hearing impairments that would block perception of visual or audio lesson material. Participants were assumed to be proficient in English. We had a total of 24 participants, 12 in each study arm.

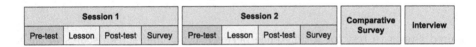

Fig. 4. Parts of the user study

Procedure. When they arrived in the study room, participants signed a consent form and the facilitator started video and audio recording. The wizard sat on the other side of the room with a laptop, and was introduced as "tech support".

The participant was seated at a laptop for a pre-test. Next, they completed the first language lesson with either laptop or robot, followed by a post-test and subjective survey. Then they completed the second session – pre-test, lesson, post-test, and survey. After both sessions, they completed a survey comparing the two. Then the facilitator provided a debrief, interviewed them for feedback on the robot, and offered compensation. Total duration was about 45 min. Figure 4 shows these steps.

5 Findings

System Characterization. To characterize usability, likeability, and engagement of robot and screen users completed a survey after each session. The survey included questions for System Usability Scale (SUS) [11], Net Promoter Score [5], engagement, and distraction. Afterwards, the facilitator interviewed the user for feedback on which features were most engaging and distracting, and how to improve the robot. Figure 5 and Table 1 show quantitative results. We used

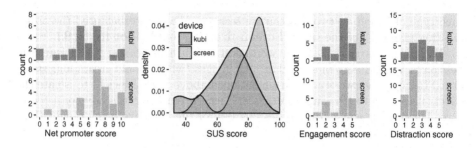

Fig. 5. System characterization results. Net promoter score is out of 10. SUS is out of 100, as described in [11]. Engagement and distraction are on a 5-point Likert scale.

Table 1. Mean, standard deviation, and standard error of the mean for Fig. 5. "SE" denotes standard error of the mean.

	Net promoter			SUS			Engagement			Distraction		
	mean	SD	SE	mean	SD	SE	mean	SD	SE	mean	SD	SE
Screen	7.21	2.21	0.45	81.74	12.82	2.67	3.71	1.12	0.23	1.79	0.59	0.12
Kubi	5.50	2.57	0.52	65.73	14.95	3.05	3.67	1.13	0.23	2.96	1.23	0.25

paired t-tests to test for a difference between robot and screen scores in each category.

Users found KubiLingo to be engaging, rating it an average of 3.67, 95 % CI [3.16, 4.18] on a 5 point Likert scale. Duolingo's engagement score is similar: 3.71, 95 % CI [3.20, 4.22] (p = 0.90). Despite KubiLingo's engaging ratings, many participants found it to be distracting as well. Not everyone found the robot distracting – the standard deviation of 1.23 indicates a wide spread of user opinions. Duolingo's distraction ratings were significantly lower (p = 0.001). Comments are shown in Table 2.

An analysis of many studies' SUS ratings by Sauro finds an average score of 68 [3]. At 65.73, 95 % CI [59.02, 72.44], KubiLingo's score is near the average. Using the semantic scale from Bangor *et al.* [9], we can assign semantic interpretations to the SUS ratings. KubiLingo's SUS ratings are between "OK" and "Good", but closer to "Good". In comparison, we measured Duolingo's SUS ratings significantly higher (p < 0.001), attaining better than "Good", but not quite "Excellent". Past Duolingo use had no significant effect on SUS.

KubiLingo's Net Promoter Scores were significantly lower than Duolingo's (p = 0.01).

Performance Evaluation. To test the performance hypothesis (H1), we administered a pre-test before each lesson and a post-test after. We designed the tests to measure short-term memory of the material. Each test had two pages of 16 questions each which required typing the translation of a phrase. Page 1 had English-to-L2 translation and page 2 had L2-to-English translation. The pre-test

Table 2. Summary of the most common user feedback. Number of users in parentheses.

	Most engaging	Most distracting	Interview comments
Kubi	Gesture (9) Positive feedback (4) Robot's eyes (4)	Confirmation step (10) Robot's movement (9)	Speed it up (7) Confirmation annoying (6) Meaningless movements (4) Gaze direction inappropriate (3)
Screen	Pictures (6) Pronouncing words (5) Sound effects (5)	(Nothing) (9) Lesson structure (4)	n/a

and post-test had the same questions with order randomized on each page. We measured performance by subtracting post-test score from pre-test score.

We modeled the data using a generalized estimating equations (GEE) linear regression with an unstructured working correlation matrix [22]. GEE accounts for dependence between repeated measurements of the same person and allows us to estimate the effect of KubiLingo vs. screen. Our model also accounts for variability caused by Swedish versus Dutch and Session 1 versus Session 2.

Performance difference between KubiLingo and Screen was not significant ($p = 0.50$). The effect size had a mean of -0.78 (SE $= 1.49$, 95% CI $[-3.04, 1.47]$), meaning that users improved by about 1 question more with KubiLingo than with Screen. This result neither supports nor contradicts our performance hypothesis (H1).

Table 3. Preference evaluation ratings and reasons. Scale is from -2 (strong robot) to 2 (strong screen). Data includes mean, upper and lower ends of confidence interval, and p-value from two-sided t-test of H2: mean $= 0$.

	Mean	Lower	Upper	p-value	Most cited reasons
Natural	0.79	0.22	1.36	0.01	Computer more familiar (7) Computer pacing more natural (7) Robot confirmation not natural (7)
Fun	-0.29	-0.94	0.36	0.36	Robot's movement (8) Robot's novelty (5) Robot's eyes (4) Robot's encouragement (4)
Prefer	1.08	0.64	1.53	<0.001	Screen faster or more efficient (8)

Preference Evaluation. To test the preference hypothesis (H2), we administered a survey at the end, comparing the devices used in both sessions. The survey contained three questions, "Which system felt more natural", "Which system was more fun or entertaining", and "Which system would you prefer to use next time you are trying to learn a language". Users responded on a 5-point Likert scale and wrote an explanation for their choice. Choices were randomly flipped left-to-right to account for left-preference.

Results are displayed in Table 3. Users reported that the screen felt more natural and that they would prefer to use the screen to learn a language in the future. Results for which system was more fun/entertaining were inconclusive ($p = 0.36$) but the mean is slightly in favor of KubiLingo. Past Duolingo use had no significant effect. These results showing preference for the computer contradict our preference hypothesis (H2).

6 Discussion

The results show the KubiLingo system has average usability and that users prefer Duolingo. KubiLingo didn't match Duolingo's usability score; however, Duolingo is a professionally designed product that sets a high bar. KubiLingo's SUS rating indicates that there is room for improvement, and the user feedback summarized in Table 2 will inform revisions in Kubi's movement pattern and interaction cadence.

The results do not conclusively show that KubiLingo or screen has greater learning gains. This means we haven't made the lessons detectably worse by teaching them with KubiLingo. KubiLingo's effectiveness is on par with Duolingo's. Our system has a room for improvement in usability, and we speculate that the requisite usability improvements will increase likeability, reduce distraction, and lead to learning gains, potentially exceeding Duolingo's effectiveness.

In future work, we hope to iterate on KubiLingo's design and test it in different contexts. One new context would be as a conversation partner for language learners. To maintain KubiLingo's independence from custom hand-built content, we could harness web-based chat bots. Conversation would be a good way for users to gain experience and boost their confidence with spoken language. It also offers more opportunities for linguistic production, an important skill in language learning. We believe that Kubi would be a better conversation partner than a computer because of its embodiment and social agency.

We also hope to test KubiLingo with children. We had kids in mind when we designed the system, but did this study with adults because it is easier to get adult participants. A number of our users thought that KubiLingo would be better suited for a younger audience. It is also likely that children would rate KubiLingo differently in terms of subjective preference.

7 Conclusion

In this paper, we described a novel approach to creating lessons for robots that teach: *content mapping*, in which existing content for electronic lessons is automatically converted to a form that an embodied robot can teach, thereby greatly expanding the amount of content available for limited effort. We presented an embodied robot that teaches language to people and uses content mapping to construct lessons. We tested our system in a user study to compare its performance with Duolingo, the source of the lesson content. Our system had equivalent learning results to those of Duolingo but users found Duolingo more likeable. We interpret the likeability result as a tribute to Duolingo's excellent user experience and a sign that our system has room for design improvements. Independently of this result, we believe that our contribution of automatic content mapping has great potential for near-term application in the development of robots that teach.

Acknowledgements. The authors thank Vivek Paramasivam, Joanna Bailet, Amy Tang, Evan Blajev, and Daniel Zhu for facilitating user studies and Burr Settles for enabling our use of Duolingo. This work was funded in part by a grant from IEEE RAS-SIGHT.

References

1. About Rosetta Stone. http://www.rosettastone.com/about
2. Firebase. https://www.firebase.com/
3. Measuring Usability with the System Usability Scale (SUS): MeasuringU. http://www.measuringu.com/sus.php
4. MOOCs Directory. http://moocs.co/Home_Page.html
5. The Net Promoter Score. https://www.netpromoter.com/know/
6. Where can I use Duolingo? http://support.duolingo.com/hc/en-us/articles/204829260-Where-can-I-use-Duolingo-
7. Why Kubi. https://www.revolverobotics.com/
8. Usage Statistics and Market Share of Content Languages for Websites, June 2016. https://w3techs.com/technologies/overview/content_language/all
9. Bangor, A., Kortum, P., Miller, J.: Determining what individual SUS scores mean: adding an adjective rating scale. J. Usability Stud. **4**(3), 114–123 (2009)
10. Belpaeme, T., Kennedy, J., Baxter, P., Vogt, P., Krahmer, E.E., Kopp, S., Bergmann, K., Leseman, P., Küntay, A.C., Göksun, T., Pandey, A.K., Gelin, R., Koudelkova, P., Deblieck, T.: L2TOR - Second language tutoring using social robots. In: WONDER 2015, Paris, France (2015)
11. Brooke, J.: SUS: a 'quick and dirty' usability scale. In: Usability Evaluation In Industry, pp. 189–194. CRC Press, June 1996
12. Cuijpers, Raymond, H., Knops, Marco, A,M,H.: Motions of robots matter! the social effects of idle and meaningful motions. In: Tapus, A., André, E., Martin, J.-C., Ferland, F., Ammi, M. (eds.) Social Robotics. LNCS(LNAI), vol. 9388, pp. 174–183. Springer, Heidelberg (2015). doi:10.1007/978-3-319-25554-5_18
13. Friedman, L.M., Furberg, C., DeMets, D.L., Reboussin, D.M., Granger, C.B.: Fundamentals of Clinical Trials, vol. 4. Springer, New York (2010)
14. Han, J., Jo, M., Park, S., Kim, S.: The educational use of home robots for children. In: ROMAN 2005, pp. 378–383 (2005)
15. Kennedy, J., Baxter, P.: The robot who tried too hard: social behaviour of a robot tutor can negatively affect child learning. In: HRI 2015, vol. 2015 (2015)
16. Kwon, O.H., Koo, S.Y., Kim, Y.G., Kwon, D.S.: Telepresence robot system for English tutoring. In: ARSO 2010, pp. 152–155 (2010)

17. Lee, S., Noh, H., Lee, J., Lee, K., Lee, G.G., Sagong, S., Kim, M.: On the effectiveness of Robot-Assisted Language Learning. ReCALL **23**(01), 25–58 (2011)
18. Leyzberg, D., Spaulding, S., Toneva, M., Scassellati, B.: The physical presence of a robot tutor increases cognitive learning gains. In: CogSci 2012, pp. 1882–1887. Cognitive Science Society (2012)
19. Ribeiro, T., Paiva, A.: The illusion of robotic life: principles and practices of animation for robots. In: HRI 2012. pp. 383–390. ACM, New York (2012)
20. Saerbeck, M., Schut, T., Bartneck, C., Janse, M.D.: Expressive robots in education: varying the degree of social supportive behavior of a robotic tutor. In: CHI 2010, pp. 1613–1622. ACM, New York (2010)
21. van Weijen, D.: The Language of (Future) Scientific Communication. https://www.researchtrends.com/issue-31-november-2012/the-language-of-future-scientific-communication/, November 2012
22. Zeger, S.L., Liang, K.Y.: Longitudinal data analysis for discrete and continuous outcomes. Biometrics, pp. 121–130 (1986)

Robots in the Classroom: What Teachers Think About Teaching and Learning with Education Robots

Natalia Reich-Stiebert[(✉)] and Friederike Eyssel

Cluster of Excellence Cognitive Interaction Technology,
Bielefeld University, Bielefeld, Germany
{nreich,feyssel}@cit-ec.uni-bielefeld.de

Abstract. In the present study, we investigated teachers' attitudes toward teaching with education robots and robot-mediated learning processes. We further explored predictors of attitudes, and investigated teachers' willingness to use robots in diverse learning settings. To do so, we conducted a survey with 59 German school teachers. Our results suggest that teachers held rather negative attitudes toward education robots. Further, our findings indicate a positive association between technology commitment and teachers' attitudes. Teachers reported a preferable use of robots in domains related to science, technology, engineering, and mathematics (STEM). Regarding expectations toward the future use of education robots, teachers mentioned their motivational potential, using robots as information source, or easy handling. Teachers' concerns, however, were associated with the disruption of teaching processes, additional workload, or the fear that robots might replace interpersonal relationships. Implications of our findings for theory and design of education robots are discussed.

Keywords: Attitudes · Education robots · Predictors of attitudes · Technology in school

1 Introduction and Related Work

Imagine the year 2066 – a group of schoolchildren and their teacher enter the classroom, and the teacher switches on several education robots. After the agenda for the day is set, one group of students continues group work on electrical circuits with the support of an education robot. A second group of children is queried on contents of previous grammar lessons by another education robot, while a third robot plays Debussy's *Clair de Lune* to introduce students to classical music. Although this idea of robots in the classroom has some sort of science-fiction appeal, existing statistics indicate that robots will become more and more part of classroom activities around the world [1]. Roboticists, educators, and psychologists already investigate how robots can be integrated into learning settings in an optimal way and how robots can improve the classroom experience. Nevertheless, research on education robotics is still sparse. To realize a successful deployment of robots in school settings, it might be worth considering attitudes of potential end users before they are introduced into practice. School teachers' suggestions for an optimization of the implementation process are especially relevant. Only through

© Springer International Publishing AG 2016
A. Agah et al. (Eds.): ICSR 2016, LNAI 9979, pp. 671–680, 2016.
DOI: 10.1007/978-3-319-47437-3_66

identifying teachers' actual expectations and concerns associated with learning and teaching with robots, potential obstacles can be addressed to enhance their acceptance. This is of particular importance given that an individual's technology acceptance represents a crucial factor in determining the success or failure of novel technologies [2, 3]. Accordingly, we examined teachers' attitudes toward teaching and learning with education robots, and we explored potential factors that might affect these attitudes (e.g., gender or technology commitment). Our study also investigated preferred areas of application for education robots, and teachers' expectations and reservations toward the future use of robots in schools.

To date, most studies have investigated students' attitudes toward education robots [e.g., 4–8], but only few studies have focused on teachers' viewpoints [9–12]. For instance, [9] have investigated perceptions of the human-like robot NAO as an educational robot among 18 preschool and elementary school teachers. After a short interaction with NAO, teachers had to complete a questionnaire that assessed their acceptance of NAO as an interactive teaching tool. Results showed that the teachers generally accepted the human-like robot as an interactive teaching tool. They reported positive attitudes toward the robot and feelings of pleasure toward the use of the robot. However, these results were based on a relatively small sample and the questionnaire focused primarily on the robot while neglecting actual teaching processes within a school environment. In contrast, [12] have provided teachers with a plausible scenario for a robotic tutor that would help a student in a map-reading task. Subsequently, teachers were asked open-ended questions concerning, e.g., the optimal number of students the robot should preferably interact with, the different social roles a robot could take on, how the robot could supervise students, and the main advantages and disadvantages of having robots in the classroom. Although these findings were based on data from seven teachers only, the results provide new insights into the potential benefit of robots for learning contexts: For example, teachers believed that robots should preferably be used by individual learners or small groups, and that they could promote independent forms of learning. Moreover, some teachers stated that robots could support assessment documentation. At the same time, however, teachers were concerned about the administrative overhead that might go along with granting access to the education robots, and they were worried that the robots could fail to interpret and to respond to social interactions adequately. Another study has investigated teachers', students', and parents' attitudes toward educational service robots in Korea and it was found that teachers were more critical toward the idea of integrating robots into schools than students and parents [11]. In a recent study by [10], 140 elementary school teachers from Korea were interviewed on their attitudes toward robots in the classroom using the *Negative Attitudes toward Robots Scale* [13]. Similarly, results revealed that elementary school teachers in Korea held negative attitudes toward robots. With respect to predictors, there were no age or gender differences regarding attitudes toward robots in the classroom.

We addressed the gaps in literature with the present study by not only capturing German school teachers' attitudes toward education robots, but also by taking into account different types of school environments. That is, we recruited teachers from elementary schools, secondary schools, and vocational schools. Further, to address the lack of research on predictors of teachers' attitudes, we also explored the role of demographic variables, teaching subject, technology commitment, and prior robot experience. Finally, we investigated teachers' view regarding their preferred application of education robots in different learning contexts. In order to address these key issues, we tested the following hypotheses (H): First, we investigated German teachers' attitudes toward education robots. We hypothesized that German school teachers would report negative attitudes toward education robots and little interest in future use of education robots due to their unfamiliarity with them (H1a). With respect to type of school, we expected that teachers from elementary schools would express more negative attitudes toward education robots and less interest in future use compared to secondary and vocational school teachers (H1b). This hypothesis is based on the idea that elementary school teachers should be more concerned about the impact of robots on younger school children than secondary or vocational school teachers who teach older school students who already have a better understanding of how to handle technical devices. Second, we investigated predictors of teachers' attitudes toward education robots. In line with [6], we predicted that male teachers would report higher positive attitudes toward education robots as well as greater willingness for future use than female school teachers (H2a). Further, according to findings by [6], we expected that age would be a negative predictor of positive attitudes toward education robots and teachers' willingness for future use (H2b). We hypothesized that technology commitment would be a positive predictor of teachers' attitudes toward education robots and their willingness to use education robots in future (H2c). The role of field of study has not yet been explored. As some curricular domains are related to technology commitment, we argued that teachers who teach STEM-related subjects would express more positive attitudes toward education robots and greater willingness for future use than teachers who work in social sciences and humanities (H2d). Hypothesis 2e predicted that previous experience would be a positive predictor of teachers' attitudes toward education robots as well as their willingness for future use (H2e). Third, we focused on application potentials of education robots. We formulated in line with [6] the following hypotheses: We expected that teachers would favor robots in STEM-related domains and reject them more strongly in domains associated with social sciences and humanities (H3a). Further, we argued that teachers would preferably use education robots for individual and group learning instead of using them in plenary teaching (H3b). Finally, in line with H3b, we hypothesized that teachers would favor robots as tutors or teaching assistants instead of assigning them the role of independent teachers for the whole classroom community (H3c).

2   Method

2.1   Participants and Procedure

A survey was conducted between January and April, 2016. 59 (47 females, 10 males; $n = 2$ did not report demographic information) German school teachers ranging in age between 22 and 62 years ($M = 39.45$, $SD = 11.31$) were recruited from different local public schools in Germany. 19 teachers were employed in an elementary school, 28 worked in a high school, and 12 in a vocational school. On average, teachers had 11.29 ($SD = 9.70$; ranging from 0 to 39) years of teaching experience. All teachers applied classic educational methods for their teaching and no alternative educational concepts like Montessori or Waldorf, for instance. Data collection took place in the teachers' break room using a paper-and-pencil questionnaire.

2.2 Introduction to Education Robots

Initially, we presented participants with a brief description of the features and functions of education robots, and we illustrated some of the alleged characteristics of the robot NAO, which is the most widely used humanoid robot in the field of education and research [see 14], to familiarize participants with the notion of education robots. Parts of the description read as follows: "Education robots can be used as assistants to teachers in the classroom and can help with the preparation of school lessons [...]. An education robot can, for example, provide information on specific topics, query course contents of previous lessons, give advice to the learning process, correct errors, or provide feedback on the learning progress. To provide a deeper impression, the following pictures show the robot NAO [...]. He is an education robot that has the quality to well-adapt to each learner and to different learning materials [...]". This information was accompanied by two photographs of the robot NAO.

2.3 Dependend Measures

To collect participants' responses, we used 5-point Likert scales ranging from 1 (*strongly disagree*) to 5 (*strongly agree*). Items were recoded where necessary, with higher values indicating stronger endorsement of the respective construct.

Teacher Attitudes Toward Education Robots. To investigate *teacher attitudes toward teaching and learning with education robots*, we developed a new measurement instrument derived from classic scales like the *Negative Attitudes toward Robots Scale* [NARS; 13], the *Robot Anxiety Scale* [RAS; 15], and the *Perceived Usefulness Scale* [PUS; 2] that evaluates user acceptance of technical devices, and finally, the *Agent Persona Instrument* [API; 16] which measures learner perception of pedagogical agents. To fit the contents to the school context, we modified selected items by implementing terms that are common in education contexts. For instance, the *NARS* item "I would feel nervous operating a robot in front of other people" was amended to "I would feel nervous, if I had to operate an education robot in front of the students". Regarding the *API*, to give another example, the item "The agent presented the material effectively", for instance, was phrased "I believe that education robots will be a great support in the classroom as they help to introduce the learning material more effectively". The resulting scale consisted of 30 items. Moreover, *teachers' willingness for future teaching with education robots* was tapped by six items such as "I think the future use of education robots in the classroom would be helpful".

Predictors of Attitudes Toward Education Robots. We assessed age and gender, teaching domains, technology commitment, and prior robot experience to examine predictors of teacher attitudes toward education robots. We measured teachers' affinity for technology by applying the German version of the *Technology Commitment Scale* provided by [17]. An example item read: "I am eager to use the latest technical gadgets". Teachers' prior robot experience was captured by three items such as "Have you ever used, or are you currently using a robot at home or at work?".

Preferred Application Areas of Education Robots. Three items captured teachers' attitudes toward potential applications of robots in school classes: "In what classes should education robots preferably be used?". Participants could rate their agreement with several domains (e.g., arts or mathematics). Multiple response options could be selected to indicate "In what learning situations should education robots primarily be used?" (i.e., individual learning or group learning). Likewise, respondents could choose among multiple options (independent teacher, teaching assistant, or tutor) to indicate "What role should an education robot perform in your opinion?".

Expectations and Concerns. Two open-ended questions assessed teachers' expectations and reservations toward the application of education robots in school classes.

3   Results

We conducted one sample t-tests against the neutral scale midpoint (scale value = 3 on a 5-point scale) to assess teachers' attitudes toward education robots, teachers' willingness to use education robots, and preferred domains for the use of education robots. To explore teachers' attitudes and willingness for future use with respect to school type, we conducted a one-way MANOVA. Further, we conducted linear regression analyses in order to investigate potential predictors of teachers' attitudes toward teaching and learning with education robots and teachers' willingness for future use of education robots. Table 1 provides an overview of descriptive statistics and internal consistencies (Cronbach's α).

Table 1. Descriptive statistics and internal consistencies of the measures.

Measure	M	SD	Min	Max	α
Attitudes toward teaching and learning with education robots	2.88	0.44	1.97	3.69	.80
Willingness for future use	2.81	1.24	0.67	5.00	.93
Technology commitment	3.03	0.39	2.00	3.83	.86
Prior robot experience	1.26	0.64	0.67	4.00	.78

3.1   Teacher Attitudes Toward Education Robots

In order to test H1a, we conducted one-sample t-tests against the neutral scale midpoint (scale value = 3 on a 5-point scale) and showed that teachers reported negative attitudes toward teaching and learning with education robots, $t(58) = -2.18, p = .03, d = -0.28$. However, respondents were rather neutral toward using education robots as teaching tools in near future, $t(57) = -1.18, p = .24, d = -0.15$. With respect to type of school, we conducted a one-way MANOVA with Sidak corrected post-hoc analyses to compare teachers' attitudes toward education robots and their willingness for future use of robots for educational settings (H1b). Results showed significant differences between school types regarding teachers' attitudes toward teaching and learning with robots, $F(2, 55) = 5.54, p = .006, \eta^2 = 0.17$. Teachers from elementary schools reported less positive attitudes ($M = 2.62, SD = 0.38$) compared to teachers from secondary schools ($M = 2.98, SD = 0.44, p = .02$) and vocational schools ($M = 3.05, SD = 0.37, p = .02$). Teachers' eagerness to use education robots in the future also differed between types of school, $F(2, 55) = 3.34, p = .04, \eta^2 = 0.11$. Teachers from elementary schools were less interested to use education robots ($M = 2.21, SD = 0.99$) compared to secondary school teachers ($M = 3.13, SD = 1.34, p = .04$), but not compared to vocational school teachers ($M = 2.94, SD = 1.09, p = .29$).

3.2 Predictors of Attitudes Toward Education Robots

We investigated predictors of teachers' attitudes toward teaching and learning with education robots and teachers' willingness for future use (H2a–H2e), by conducting linear regression analyses that included gender (dummy coded: men = 0, women = 1), age, teaching domains (dummy coded: STEM-related fields = 1, social sciences and humanities = 2), technology commitment, and prior robot experience. Results regarding positive attitudes toward teaching and learning with education robots revealed that the full model was not significant, $R^2 = 0.11, F(5, 46) = 1.12, p = .36$. However, technology commitment ($\beta = .34, p = .03$) was found to be a significant predictor of teachers' attitudes toward teaching and learning with education robots. Technology commitment also significantly predicted teachers' willingness to use education robots ($\beta = .37, p = .02; R^2 = 0.14, F(5, 45) = 1.50, p = .21$).

3.3 Preferred Application Areas of Education Robots

Regarding the application of robots in different school domains (H3a), one-sample t-tests against the neutral scale midpoint (corrected for α–error using Sidak; $\alpha = .004$) revealed that teachers would preferably apply robots in informatics, mathematics, and physics. Teachers were rather neutral toward the application of robots in biology, chemistry, geography, foreign languages, history, politics, and German language. With respect to arts, psychology, and music, teachers rejected education robots. Table 2 displays descriptive statistics and the results of the analyses. To explore teachers' preferences for the application of robots in different learning situations (H3b), we calculated percentage frequencies (multiple responses were possible) and found that teachers preferred to use robots for individual learning (66 %) or learning in small groups (66 %). Only 7 % could imagine using a robot in the whole class community. Regarding the role teachers could envisage for an education robot (H3c), the majority (85 %) favored to use a robot as a tutor. 24 % would use a robot as teaching assistant, and merely 2 % could imagine that robots would serve as independent teachers.

Table 2. Descriptive statistics and results of one-sample t-tests for robot use in preferred school domains.

Domain	M	SD	t	df	d
Informatics	4.12	1.11	7.22**	51	1.00
Mathematics	3.86	1.00	6.42**	55	0.86
Physics	3.52	1.15	3.27*	51	0.45
Chemistry	3.39	1.15	2.44	50	0.34
Foreign languages	3.32	1.28	1.86	56	0.25
Biology	3.28	1.08	1.91	52	0.26
Geography	3.11	1.11	0.74	53	0.10
History	2.91	1.17	−0.59	52	−0.08
Arts	1.96	0.96	−7.74**	50	−1.08
Psychology	2.21	1.02	−5.60**	51	−0.78
Music	2.45	1.10	−3.62**	52	−0.50
Politics	2.58	1.11	−2.75	51	−0.38
German language	2.61	1.11	−2.66	55	−0.36

Note. The comparison value is the neutral scale midpoint (scale value = 3 on a 5-point Likert scale).
$*p < 0.004$, $**p < 0.001$ (Sidak corrected)

3.4 Expectations and Concerns

To complement quantitative findings, a qualitative content analysis of the open-ended questions was conducted by two independent raters. To measure the agreement between the two raters, we calculated interrater reliabilities (Cohen's κ). With respect to teachers' expectations toward the future use of education robots our findings revealed the following categories: Teachers expected that, due to their novelty, robots could create a *motivating environment* for students to engage more in learning activities. Moreover, teachers assumed that education robots could serve as a *source of information*. Further, teachers stated that they could assist in the *assessment and monitoring of each students' learning status*. Teachers also pointed out that education robots could be of high benefit for under-achieving students who need *individual support*. Finally, teachers deemed *easy access to education robots* in the classroom desirable; education robots should be easy to handle, particularly by integrated voice control. The findings showed a high agreement between the two raters for the categories ($\kappa = .83$, $p < .001$). Following categories emerged regarding teachers' concerns toward the application of robots in the classroom: Teachers pointed out that robots could *disturb the lessons* and they expected fights because of restricted access to a limited number of robots. Further, teachers were concerned that the initial enthusiasm would disappear and that *students would lose their interest*. Additionally, teachers feared that robots would bring *extra workload* regarding difficult programming, maintenance, and monitoring of the robot. They were also concerned about *high acquisition costs* and the incapacity of schools to buy expensive tools. Moreover, teachers remarked that robots could *replace teachers and interpersonal relationships*, and that students' social skills could suffer from the use of education robots. The agreement between the two raters for the categories was found to be high ($\kappa = .87, p < .001$).

4 Discussion and Conclusion

The present research makes several noteworthy contributions to the existing literature: First, we investigated teachers' attitudes in German context; second, we go beyond previous findings by exploring various predictors of teachers' attitudes. Third, we have taken into account different types of school contexts, and fourth, we assessed teachers' view toward their preferred application potentials for education robots. Finally, a further strength of the present study is the development of a reliable measure to evaluate teachers' attitudes toward education robots.

As predicted in H1a, overall, German respondents reported quite negative attitudes toward teaching and learning with education robots. However, they were neutral toward future teaching with education robots. This is in line with previous findings on German university students' willingness to learn with education robots [6]. The trend is probably because robots are still not common in the German context and it will probably continue until the first education robots enter German classrooms beyond the usual application in informatics or technology lessons. [18] found that teachers' attitudes were directly related to teachers' willingness to include computer technology into the classroom, which in turn affected students' opinion toward the importance of computers in school. Thus, it is crucial to facilitate attitude change toward education robots to ensure that education robots are utilized in the classroom. Further, type of school setting affected teachers' attitudes and their willingness to use education robots. As predicted, teachers from elementary schools expressed more negative attitudes toward education robots and less interest in future use compared to secondary and vocational school teachers (H1b). Elementary school teachers seemed to be more concerned about the impact of robots on younger school children than secondary school educators or vocational school teachers who work with young adults. We hypothesized that gender, age, teaching domain, and prior robot experience would have an impact on attitudes toward education robots and teachers' willingness for future use (H2a, H2b, H2d, H2e). However, none of these factors influenced attitudes and eagerness for future use. These results may be due to the fact that more female than male teachers participated in the study and the fact that overall, teachers reported little previous experiences with robots. Importantly, technology commitment positively predicted teachers' attitudes and willingness to use education robots (H2c). Although overall attitudes were rather negative, teachers who reported higher interest in technology were more inclined to use education robots than teachers who were less interested in technological issues. While the regression model was not significant, our finding confirms the association between interest in technology and the willingness to apply new technologies in school. Further findings on preferred application areas for education robots point in the same direction: As anticipated, teachers preferred to use education robots in STEM-related domains like informatics, mathematics, or physics, and rejected them in domains like psychology, arts, or music (H3a). Consequently, it would be worthwhile to campaign for teaching and learning with robots beyond the STEM domains and to promote robots for social sciences, too. Regarding the role of education robots in the classroom, teachers envision using robots as tutors or teaching assistants for individual learning or in small groups, but do not consider them adequate for frontal teaching of the entire course (H3b, H3c). This is in keeping with some teachers' worries that robots could replace them eventually. Thus, highlighting the robots' role of a learning companion and as a valuable tool to enhance learning processes appears crucial. When addressing fears of replacement, it has to become clear that although social robots are making their way into to the classroom, they will not make teaching staff obsolete anytime soon. Nevertheless, theory and design of robots should place further emphasis on the development of social robots that are capable of recognizing and displaying emotions with the flexibility to integrate them into the interaction with humans. This ability probably could meet teachers' concerns that students' social skills would suffer from the use of robots that are incapable of social interaction. With respect to the design of education robots, our recommendation would be to minimize the workload associated with the integration of robots into teaching processes and to guarantee easy handling. Finally, one aspect that emerged from our findings was related specifically to the efficient integration of robots in classroom activities. In order to address teachers' worries that robots could interrupt their teaching, there is a need to develop instructional methods that facilitate implementation of robots despite the pressure to complete obligatory curricula.

Although the findings presented in our study are promising and have far-reaching implications for robotics, three aspects received relatively little attention and need to be addressed in follow-up studies. First, the present investigation focused on attitudes toward education robots among potential end users in schools, namely school teachers. Further research on providing a comprehensive review of school teachers' *and* students' attitudes toward education robots is currently underway. Second, greater attention must be dedicated to teachers' fears that robots could replace them or substitute interpersonal relationships which would be detrimental to students' social skills. A follow-up study that integrates actual human-robot interaction in learning contexts would be helpful in order to give teachers and students the opportunity to test education robots in a realistic setting. Third, further work needs to be done on the development of instructional methods for the successful integration of robots in schools. Finally, future research should examine the effectiveness of these methods. It is critical that research on educational robots does not simply boil down the development of a tool that supports teaching and learning, but that such research also opens up novel perspectives on how to make use of the potential of education robots most optimally.

Acknowledgements. This research has been conducted in the framework of the European Project CODEFROR (FP7 PIRSES-2013-612555) and it was supported by the Cluster of Excellence Cognitive Interaction Technology 'CITEC' (EXC 277) at Bielefeld University, which is funded by the German Research Foundation (DFG).

References

1. International Federation of Robotics Statistical Department: Service Robot Statistics. http://www.ifr.org/service-robots/statistics/
2. Davis, F.D.: Perceived usefulness, perceived ease of use, and user acceptance of information technology: a comparison of two theoretical models. MIS Q. **13**, 319–340 (1989)
3. Davis, F.D.: User acceptance of information technology: system characteristics, user perceptions and behavioral impacts. Int. J. Man Mach. Stud. **38**, 475–487 (1993)
4. Lin, Y.-C., Liu, T.-C., Chang, M., Yeh, S.-P.: Exploring children's perceptions of the robots. In: Chang, M., Kuo, R., Kinshuk, Chen, G.-D., Hirose, M. (eds.) Learning by Playing. LNCS, vol. 5670, pp. 512–517. Springer, Heidelberg (2009)
5. Liu, E.Z.F.: Early adolescents' perceptions of educational robots and learning of robotics. Br. J. Educ. Technol. **41**, E44–E47 (2010)
6. Reich-Stiebert, N., Eyssel, F.: Learning with educational companion robots? Toward attitudes on education robots, predictors of attitudes, and application potentials for education robots. Int. J. Soc. Robot. **7**, 875–888 (2015)
7. Serholt, S., Barendregt, W.: Students' attitudes towards ethical dilemmas in the possible future of social robots in education. In: 23rd IEEE International Symposium on Robot and Human Interactive Communication, pp. 955–960. IEEE Press, Edinburgh (2014)
8. Shin, N., Kim, S.: Learning about, from, and with robots: students' perspectives. In: 16th IEEE International Conference on Robot and Human Interactive Communication, pp. 1040–1045. IEEE Press, Jeju (2007)
9. Fridin, M., Belokopytov, M.: Acceptance of socially assistive humanoid robot by preschool and elementary school teachers. Comput. Hum. Behav. **33**, 23–31 (2014)
10. Kim, S.W., Lee, Y.: A survey on elementary school teachers' attitude toward robot. In: World Conference on E-Learning in Corporate, Government, Healthcare, and Higher Education, pp. 1802–1807. AACE Press, Hawaii (2015)
11. Lee, E., Lee, Y., Kye, B., Ko, B.: Elementary and middle school teachers, students and parents perception of robot-aided education in Korea. In: World Conference on Educational Media and Technology, pp. 175–183. AACE Press, Vienna (2008)
12. Serholt, S., Barendregt, W., Leite, I., Hastie, H., Jones, A., Paiva, A., Vasalou, A., Castellano, G.: Teachers' views on the use of empathic robotic tutors in the classroom. In: 23rd IEEE International Symposium on Robot and Human Interactive Communication, pp. 955–960. IEEE Press, Edinburgh (2014)
13. Nomura, T., Kanda, T., Suzuki, T.: Experimental investigation into influence of negative attitudes toward robots on human-robot interaction. AI Soc. **20**, 138–150 (2006)
14. Aldebaran Soft Bank Group: NAO. Solutions for Education & Research. https://www.aldebaran.com/en/solutions/education-research
15. Nomura, T., Suzuki, T., Kanda, K., Kato, K.: Measurement of anxiety toward robots. In: 15th IEEE International Symposium on Robot and Human Interactive Communication, pp. 372–377. IEEE Press, Hatfield (2006)
16. Baylor, A.L., Ryu, J.: The API (Agent Persona Instrument) for assessing pedagogical agent persona. In: World Conference on Educational Media and Technology, pp. 448–451. AACE Press, Honolulu (2003)
17. Neyer, F.J., Felber, J., Gebhardt, C.: Entwicklung und Validierung einer Kurzskala zur Erfassung von Technikbereitschaft [Development and Validation of a Short Technology Commitment Scale]. Diagnostica **58**, 87–99 (2012)
18. Teo, T.: Attitudes toward computers: a study of post-secondary students in Singapore. Interact. Learn. Environ. **14**, 17–24 (2006)

The Influence of a Social Robot's Persona on How it is Perceived and Accepted by Elderly Users

Andrea Bartl, Stefanie Bosch, Michael Brandt, Monique Dittrich, and Birgit Lugrin[⊠]

Human Computer Interaction, University of Wuerzburg, Wuerzburg, Germany
{andrea.bartl,stefanie.bosch,michael.brandt,
monique.dittrich}@stud-mail.uni-wuerzburg.de,
birgit.lugrin@uni-wuerzburg.de

Abstract. The demographic change causes an imbalance between the number of elderly in need of support and the number of caring staff. Therefore, it is important to help older adults keep their independence. Forgetting is a common obstacle people have to face when they become older which can be moderated by social robots by reminding on tasks. Since most elderly people are not used to robots a challenge in HRI is to identify aspects of a robot's design to promote its acceptance. We present two different personas (companion vs. assistant) for a robotic platform by manipulating verbal and nonverbal behavior. A study was conducted in assisted living accommodations with the robot reminding on appointments to review if the persona influences the robot's acceptance. Results indicate that the companion version of the robot was better accepted and perceived more likeable and intelligent compared to the assistant version.

1 Introduction

According to the United Nations Department of Economic and Social Affairs [1] 21 percent of the world's population will be older than 60 years in 2015. This causes an obsolescence of society and consequently personnel bottlenecks in elderly care. At this point robotic systems can help foster older adults' autonomy. Thereby, not only physical but also cognitive tasks are of interest for HRI [2]. One aspect that jeopardises independent living is being forgetful [3]. While conventional calendars miss an active reminder function, technical alternatives such as smartphones require technical skills and the handling of small touch displays which can be barriers for older people [4].

A potential solution lies in using social robots: while combining the functionality of a calendar with its humanlike interaction the robot can serve as a social reminder for medication, family meetings and other appointments [5]. Since older adults consider a "robot calendar" as a useful application [5] the objective of this contribution is to identify their preferences regarding characteristics which make

© Springer International Publishing AG 2016
A. Agah et al. (Eds.): ICSR 2016, LNAI 9979, pp. 681–691, 2016.
DOI: 10.1007/978-3-319-47437-3_67

the robot more acceptable. We define acceptance as the preference for a specific way of interacting, not the acceptance of the technology itself. Therefore, two versions of a scenario were evolved in which the robot reminds older people of appointments. They differ in the robot's persona and therefore in its way of interacting: a prevalent friendly (companion) and a formal (assistant) persona.

2   Persona and Anthropomorphic Interfaces

Studies show that the implementation of single personality traits, e.g. introversion vs. extroversion [6], can have an impact on the acceptance of a robot. Thus, the question arises, if a robot's persona, meaning a fictional personality with varied and stable behavioral and personality patterns [7], has a similar effect. Derived from studies which focus on the simulation of a persona in the context of social robots and virtual agents [8], we differ between the companion and the assistant persona. The *assistant* distinguishes itself through its professional competence as well as its formal and authoritarian aura, whereas the main characteristics of the *companion* are its emitted likeability and kindred spirit [9–11]. In general, the companion is more emotional, enthusiastic and expressive and construed to establish emotional ties [9–11]. Another important attribute of the companion is the similarity to its human counterpart regarding qualities such as appearance, age or ethnicity [12]. Similar to the definition of companion and assistant, Goetz and Kiesler [11] differ between a robot's playful and serious personality. In a controlled laboratory experiment with younger adults, the playful robot was rated more positive and improved the mood of its interaction partners, while the serious robot elicited the most compliance. Following these findings we address the question whether robots representing different personas are perceived and accepted differently by elderly users.

3   Implementation

For the implementation of persona-specific behaviors we chose the humanoid robot Reeti[1]. Advantages of this platform are its mimic expressivity and the integrated text-to-speech-synthesizer that allows to modify speech output with respect to gender, pitch, emphasis and speed. Reeti's cartoon-like appearance allows human-like facial and linguistical expressions without creating inappropriate expectations towards the realism of the robot's behavior. Moreover, the robot's neutral design (gender neutral, white body) benefits to exclude confounding variables. This, as well as the absence of limbs, the robot's height (40 cm) and its restricted movement capabilities give Reeti a non-threatening appearance suitable in the context of elderly care [13]. Since older adults prefer robots with female voices [2] the voice was set to female. To ensure that hearing impaired participants can follow the robot's voice, the speed was reduced.

---

[1] Robopec, www.reeti.fr.

The robot was connected to an electronic online calendar[2]. The actor pattern was used to separate different aspects of the application such as the connection to the Robot Operating System, e.g. to activate speech output, or the connection to a server using the Google Calendar API to retrieve events and store them in a local database.

For our prototype, reminders were set manually. An interaction sequence for each condition (companion vs. assistant) was created in which the robot reminds of appointments using the calendar function. Both versions were equally designed as social actors and only differed concerning persona-specific behaviors, e.g. filler words for the companion version but not for the assistant version (see Table 1). Blinking and ear movements were added periodically and consistent over both conditions to create more authentic and lively expressions.

4   Study

To test the two versions of the robotic reminder a within-subject experiment was conducted with the following research questions (RQ):

- Does the presentation of the robot as a [companion/assistant] influence the likeability (RQ1a) and the perceived intelligence (RQ1b) of the robot?
- Does the presentation of the robot as a [companion/assistant] have an influence on the acceptance of the robot? (RQ2)
- Do elderly people consider a robot connected to a calendar useful? (RQ3)

4.1   Participants

The target group of the study was elderly people who are still able to handle their everyday lives rather autonomously. Therefore, the study was conducted in different assisted living accommodations for healthy and independent people in Wuerzburg (Germany) and at one participant's home. Four experiments had to be interrupted before completion because the participants either had problems understanding the robot acoustically or with answering the questions. After exclusion of the discontinued runs, the sample contained $N = 18$ participants aged between 71 and 91 years ($M = 81.83$, $SD = 5.56$), with 66.7 percent of the participants being female.

4.2   Procedure

Figure 1 (left) shows the experimental setup which was conducted as a Wizard of OZ (WOZ) study to reduce technical risks and ensure a fluent interaction. The experimenters were student researchers that were unknown to the participants before and followed a predefined study schedule. The procedure (see Fig. 1 (right)) lasted from 30 to 45 min and contained the following steps:

---

[2] Google Calendar, https://developers.google.com/google-apps/calendar/.

Table 1. Operationalization of the personas by specific verbal and nonverbal behavior.

Behavior	Companion	Assistant
Name	Anna (informal, surname only)	Kathrin Schmidt (formal, surname and lastname)
Fillers	Fillers "oh" and "ah" at semantically meaningful passages and emphasis of these words by widening the eyes [9]	No fillers
Status	Imitation of the participant's *age* by the statement "I was built three years ago but I feel years older" [9]	Representation of *competence* by the statement "I was produced three years ago. Ever since I assist the personnel of retirement homes" [9]
Questions	Additional questions to increase affiliation [9]	No additional questions
Words of agreement	Words of agreement "okay", "alright" and "good" to increase affiliation [14] and emphasis of these words by head nodding	No words of agreement
Pronouns	Pronoun "we" to increase affiliation [15]	No pronoun "we"
Directness/ emotions	More emotional and informal language [9,16]	More direct and formal language [9,16]
Head tilt	Head tilts to pose warmth [17]	No head tilts
Smile	Friendly smile to increase affiliation and liking [18]	Less frequent smiles [18]

(a) After a short welcome, the experiment started with open questions. (b) To get the seniors acquainted with the robot, it introduced itself and its abilities. This sequence also contained speech volume regulation according to the participants' preferences. (c) A short explanation of the calendar function was given and two exemplary appointments were stored in the calendar. The experimenter explained that the participant will interact with two different versions of the robot without naming their different characteristics. (d) The participant interacted with one of the two personas with the robot reminding them of the stored appointments. To avoid order and learning effects half of the participants started with the companion followed by the assistant whereas the other half interacted the other way round. (e) A questionnaire referring to the interaction was filled in addressing the seniors' perception and acceptance of the persona

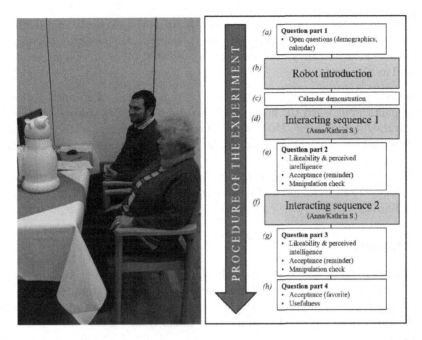

Fig. 1. Experimenter with a participant during the study. The wizard, who controls the robot, is not visible (left). Procedure of the study (right).

and the success of the manipulation. (f, g) The prior two steps were repeated for the remaining persona. (h) The experiment ended with the selection of the favorite version and the evaluation of the usefulness of the robot.

4.3 Questionnaire

To collect the relevant data, a survey was designed using both open and polar questions. The questions were aligned with the management of the assisted living accommodations to avoid misunderstandings and adapt them optimally to the target group.

Question part 1. Demographics on age and gender were collected followed by questions about calendar usage as a daily routine.

Question part 2 and 3. *Likeability and perceived intelligence:* The German version of the Godspeed Questionnaire Series [19] was taken as a basis to measure the perception of the persona (RQ1a and b). Its rating scale consists of adjectives building a semantic differential (e.g. unintelligent – intelligent). The subscales likeability (like, friendly, kind, pleasant, nice) and perceived intelligence (competent, knowledgeable, responsible, intelligent, sensible) were considered relevant for this study. *Acceptance (reminder):* To measure the acceptance (RQ2) of the persona, participants were asked whether they wanted appointment reminders from the represented persona (c.f. [20]). *Manipulationcheck:* To find out if the manipulation of the independent variable (persona) was successful, participants rated to what extent they experienced the presented robot version as a companion and as an assistant.

Question part 4. *Acceptance (favorite):* As a second acceptance indication participants were asked to choose their favorite persona, by asking which robot version they would prefer. *Usefulness:* The perceived usefulness (RQ3) was measured by asking participants how useful they rate the robot as a calendar - detached from the persona.

The items regarding likeability, perceived intelligence, acceptance (reminder), the manipulationcheck, and usefulness were designed as 5-point Likert scales with labeled levels.

4.4 Results

Most participants ($n = 17$) use a calendar (e.g. a wall or pocket calendar) in their daily routine. Only $n = 2$ people use an additional electronic calendar. The descriptive statistics of the relevant items and indexes are shown in Table 2. The items analysed were converted to numeric values from -2 to $+2$.

Table 2. Descriptive statistics comparing Anna (companion) and Kathrin Schmidt (assistant), with $N=18$.

	Companion	Assistant
	M (SD)	M (SD)
Likeability	1.58 (.51)	1.31 (.70)
Perceived intelligence	1.33 (.64)	1.19 (.66)
Acceptance (reminder)	1.22 (1.00)	.89 (1.08)
Manipulationcheck: Companion	0.67 (1.33)	0.11 (1.41)
Manipulationcheck: Assistant	0.83 (1.04)	1.06 (.94)

Likeability and perceived intelligence: The valuations of the companion and the assistant version on likeability and perceived intelligence were compared. Two-tailed Wilcoxon Signed-Rank Tests showed that the companion was rated significantly higher than the assistant regarding both likeability ($N = 18$, $T = 10.50$, $p < .05$) and perceived intelligence ($N = 18$, $T = 3.50$, $p < .05$).[3]

Acceptance (reminder/favorite): The analysis of the acceptance (reminder) of the two robot personas along with their calendar function did not show significant results, meaning that the participants' ratings on whether they wanted to be

---

[3] Analog t tests confirmed the results ($t_{likeability}(17) = 2.55, p < .05; t_{intelligence}(17) = 2.60, p < .05$).

reminded of their appointments by Anna or Kathrin Schmidt respectively did not differ $(T = 3, p = .19)$[4]. However, as shown in Fig. 2 the majority of the participants stated that they preferred Anna and that they would rather keep Anna than Kathrin Schmidt for their own use (acceptance favorite).

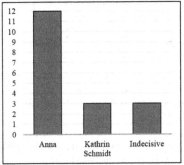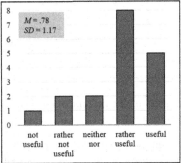

Fig. 2. Absolute frequencies of participants who chose Anna or Kathrin Schmidt as their favorite version (left), and distribution of the participants on the perceived usefulness of the robot with its calendar function (right).

Manipulationcheck: To verify to what extent participants experienced Anna and Kathrin Schmidt as a companion or as an assistant, two-tailed Wilcoxon Signed-Rank Tests were performed. The results are summarized in Fig. 3. No significant differences were observed between the perception of Kathrin Schmidt and Anna as assistant or companion respectively (horizontal arrows). Comparing both scores (assistant vs. companion) for Anna and Kathrin Schmidt separately (vertical arrows) a significant difference was found: people experienced Kathrin Schmidt significantly more as an assistant than as a companion[5].

Usefulness: Participants rated the usefulness of the robot with its calendar function (detached from the persona) predominantly positive (see Fig. 2).

Further evaluation of the data indicated that age and gender of the participants also influenced the results. Two-tailed Mann-Whitney-U-tests showed that female participants rated Anna higher on the sympathy-dimension than male participants $(M_{female} = 1.73, SD = .46, M_{male} = 1.27, SD = .48, U = -2.19, p < .05)$. Significant correlations occurred between the age of the participants and the ratings on the perception of the personas as well as the usefulness of the robot in general (see Table 3).

---

[4] Analog t test accounted for the same result $(t(17) = 1.68, p = .11)$.

[5] Analog t tests showed the same results $(t_{Assistant}(17) = -1.07, p = .30; t_{Companion}(17) = -2.05, p = .06; t_{KathrinS}(17) = 2.65, p < .05; t_{Anna}(17) = .40, p = .70)$.

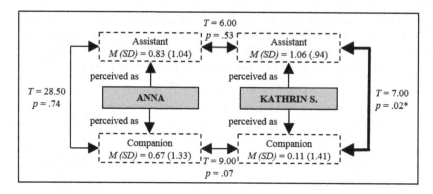

Fig. 3. Results of the manipulationcheck: perception of Anna and Kathrin Schmidt as companion or assistant.

Table 3. Significant correlations (Kendall-Tau-b) between the age of the participants and the perception of the personas as well as the usefulness of the robot.

	Age
Likeability Companion	−.42[a]
Likeability Assistant	−.43[a]
Perceived Intelligence Companion	−.45[a]
Perceived Intelligence Assistant	−.43[a]
Acceptance (reminder) Companion	−.63[b]
Usefulness	−.50[b]
Manipulationcheck: Assistant Anna	−.42[a]

a. significant correlation on a significance-level of .05 (two-tailed).
b. significant correlation on a significance-level of .01 (two-tailed).

4.5   Discussion

The companion version was rated significantly more likeable and intelligent compared to the assistant version of the robot. This result is somewhat surprising, considering the fact that the companion was designed to be friendly but not necessarily more competent. A reason could be the Halo-Effect [21] which states that the assignment of certain properties to a person influences the perception of other properties. Transferred to this experiment the positive assessment regarding the likeability could have exuded on perceived intelligence.

The preference for the companion version was also reflected by the fact that the majority of participants chose the companion persona as their favorite, although there was no difference in the participants' preference to get reminded by the companion or the assistant. Interestingly only two of the participants could name reasons for their decision by stating that they found the companion to be nicer.

Results also indicate that the participants were only limitedly able to classify Anna and Kathrin Schmidt correctly as companion or assistant. It is possible that the assistive role with a reminder function of both versions has been influencing the participants' judgments.

Regardless of the persona participants considered the robot with its calendar function as rather useful. This is in line with findings of Schroeter et al. [22] who conducted an evaluation of an assistive social robot for elderly people.

Further results suggest that the age of the participants played a role in the evaluation of the robot: the older a person, the more negative was the general attitude towards the robot. Similar effects were observed in [23] where the attitude of 75 year olds towards service robots was more skeptical compared to seniors that are younger than 65. This raises hope that the future generation of seniors will be more open towards robotic companions in the domain of elderly care. Our future studies will therefore focus on younger seniors.

Participants' statements indicated that the voice of the robot seemed unnatural and 'non-human' making it difficult to understand the robot. Even the deliberate reduction of the speech rate and adjustment of the volume to each participant's individual preference did not help overcome these difficulties. Therefore in future studies other solutions for the speech should be considered.

5   Contribution and Future Work

We believe that social robots bear great potential to help older people to stay autonomously. In this contribution, two different personas (companion vs. assistant) for a social robot with an integrated calendar function have been investigated and tested in assisted living accommodations. Our preliminary study suggests that the companion persona has a positive impact on the likeability and perceived intelligence and is preferred when compared to the assistant persona. Therefore, we aim on contributing to the field, by recommending companion-like personas over assistants and by providing guidance on implementation of their prototypical behavior. However, these findings should be taken with care due to the small sample size as well as potential drawbacks of the study, e.g. regarding the target age group and speech output. In future work, more interactions with the robot over a longer period of time in an assisted living accommodation need to be researched. On the one hand this would shed light on the actual effectiveness and use of a robotic calendar for elderly people, and on the other hand it could help cancel out the Halo Effect [21] that might have been taken place due to the novelty effect of the robot and the special attention of the student experimenters. For such an experiment, however, a more mature prototype needs to be implemented containing features such as speech recognition, more sophisticated techniques for reminding at appropriate times, and additional functionalities such as recommending to drink water regularly. We hope that the presented implications for further research advance the introduction of social robots to actively support seniors in their everyday life soon.

Acknowledgments. The authors would like to thank the staff and inhabitants of the assisted living accommodations "Miravilla, Service-Wohnen Hubland" and "Arbeiter-wohlfahrt, Sozialzentrum Jung und Alt unter einem Dach" in Wuerzburg, Germany, for their support and participation in the study.

References

1. United Nations Department of Economic, Social Affairs, Population Division: World population ageing 2013 (2013)
2. Broadbent, E., Stafford, R., MacDonald, B.: Acceptance of healthcare robots for the older population: review and future directions. Int. J. Soc. Robot. **1**(4), 319–330 (2009)
3. Vik, S.A., Hogan, D.B., Patten, S.B., Johnson, J.A., Romonko-Slack, L., Maxwell, C.J.: Medication nonadherence and subsequent risk of hospitalisation and mortality among older adults. Drugs Aging **23**(4), 345–356 (2006)
4. Kobayashi, M., Hiyama, A., Miura, T., Asakawa, C., Hirose, M., Ifukube, T.: Elderly user evaluation of mobile touchscreen interactions. In: Campos, P., Graham, N., Jorge, J., Nunes, N., Palanque, P., Winckler, M. (eds.) INTERACT 2011. LNCS, vol. 6946, pp. 83–99. Springer, Heidelberg (2011). doi:10.1007/978-3-642-23774-4_9
5. Mast, M., Burmester, M., Krüger, K., Fatikow, S., Arbeiter, G., Graf, B., Kronreif, G., Pigini, L., Facal, D., Qiu, R.: User-centered design of a dynamic-autonomy remote interaction concept for manipulation-capable robots to assist elderly people in the home. J. Hum. Robot Interact. **1**(1), 96–118 (2012)
6. Lee, K.M., Peng, W., Jin, S.A., Yan, C.: Can robots manifest personality?: An empirical test of personality recognition, social responses, and social presence in human-robot interaction. J. Commun. **56**(4), 754–772 (2006)
7. Matthews, G., Deary, I.D., Whiteman, M.C.: Personality Traits. University Press, Cambridge (2009)
8. Holz, T., Dragone, M., O'Hare, G.M.P.: Where robots and virtual agents meet. Int. J. Soc. Robot. **1**(1), 83–93 (2009)
9. Baylor, A.L., Kim, Y.: Simulating instructional roles through pedagogical agents. Int. J. Artif. Intell. Educ. **15**(2), 95–115 (2005)
10. Kim, Y.: Desirable characteristics of learning companions. Int. J. Artif. Intell. Educ. **17**(4), 371–388 (2007)
11. Goetz, J., Kiesler, S.: Cooperation with a robotic assistant. In: CHI 2002 Extended Abstracts on Human Factors in Computing Systems, pp. 576–577. ACM (2002)
12. Baylor, A.L., Kim, Y.: Pedagogical agent design: the impact of agent realism, gender, ethnicity, and instructional role. In: Lester, J.C., Vicari, R.M., Paraguaçu, F. (eds.) ITS 2004. LNCS, vol. 3220, pp. 592–603. Springer, Heidelberg (2004). doi:10.1007/978-3-540-30139-4_56
13. Hammer, S., Lugrin, B., Bogomolov, S., Janowski, K., André, E.: Investigating politeness strategies and their persuasiveness for a robotic elderly assistant. In: Meschtscherjakov, A., Ruyter, B., Fuchsberger, V., Murer, M., Tscheligi, M. (eds.) PERSUASIVE 2016. LNCS, vol. 9638, pp. 315–326. Springer, Heidelberg (2016). doi:10.1007/978-3-319-31510-2_27
14. Tausczik, Y.R., Pennebaker, J.W.: The psychological meaning of words: LIWC and computerized text analysis methods. J. Lang. Soc. Psychol. **29**(1), 24–54 (2010)
15. Ireland, M.E., Slatcher, R.B., Eastwick, P.W., Scissors, L.E., Finkel, E.J., Pennebaker, J.W.: Language style matching predicts relationship initiation and stability. Psychol. Sci. **22**(1), 39–44 (2011)

16. Kim, Y., Kwak, S.S.: Am I acceptable to you? Effect of a robot's verbal language forms on people's social distance from robots. Comput. Hum. Behav. **29**(3), 1091–1101 (2013)
17. Mara, M., Appel, M.: Effects of lateral head tilt on user perceptions of humanoid and android robots. Comput. Hum. Behav. **44**, 326–334 (2015)
18. Guadagno, R.E., Swinth, K.R., Blascovich, J.: Social evaluations of embodied agents and avatars. Comput. Hum. Behav. **27**(6), 2380–2385 (2011)
19. Bartneck, C., Kulić, D., Croft, E., Zoghbi, S.: Measurement instruments for the anthropomorphism, animacy, likeability, perceived intelligence, and perceived safety of robots. Int. J. Soc. Robot. **1**(1), 71–81 (2009)
20. Kuchenbrandt, D., Häring, M., Eichberg, J., Eyssel, F.: Keep an eye on the task! how gender typicality of tasks influence human–robot interactions. In: Ge, S.S., Khatib, O., Cabibihan, J.-J., Simmons, R., Williams, M.-A. (eds.) ICSR 2012. LNCS (LNAI), vol. 7621, pp. 448–457. Springer, Heidelberg (2012). doi:10.1007/978-3-642-34103-8_45
21. Thorndike, E.L.: A constant error in psychological ratings. J. Appl. Psychol. **4**(1), 25–29 (1920)
22. Schroeter, C., Mueller, S., Volkhardt, M., Einhorn, E., Huijnen, C., van denHeuvel, H., van Berlo, A., Bley, A., Gross, H,M: Realization and user evaluation of a companion robot for people withmild cognitive impairments.In: IEEE International Conference on Robotics and Automation (ICRA2013), pp. 1145–1151 (2013)
23. Meyer, S.: Mein Freund der Roboter: Servicerobotik für ältere Menschen - eine Antwort auf den demografischen Wandel?. VDE-Verlag, Frankfurt (2011)

From Social Practices to Social Robots – User-Driven Robot Development in Elder Care

Matthias Rehm[1]([✉]), Antonia L. Krummheuer[2], Kasper Rodil[1], Mai Nguyen[1], and Bjørn Thorlacius[1]

[1] Faculty of Engineering and Science, Aalborg University, Aalborg, Denmark
matthias@create.aau.dk
[2] Faculty of Humanities, Aalborg University, Aalborg, Denmark

Abstract. It has been shown that the development of social robots for the elder care sector is primarily technology driven and relying on stereotypes about old people. We are focusing instead on the actual social practices that will be targeted by social robots. We provide details of this interdisciplinary approach and highlight its applicability and usefulness with field examples from an elder care home. These examples include ethnographic field studies as well as workshops with staff and residents. The goal is to identify and agree with both groups on social practices, where the use of a social robot might be beneficial. The paper ends with a case study of a robot, which frees staff from repetitive and time consuming tasks while at the same time allowing residents to reclaim some independence.

Keywords: Social practice · Social robots · Participatory design

1 Introduction

Over the last few years a number of reviews [4,5,10] have shown that the development of social robots for the elder care sector is primarily technology driven and relying on stereotypes about old people when it comes to designing the robots as well as the services provided by these robots. Therefore, the approach disregards the great diversity of target users in terms of their individual abilities and needs on the one hand and the interplay with institutional needs and practices on the other hand. In this paper we are going to propose a different approach for developing social robots that focuses on the actual social practices that will be targeted by social robots.

For instance, the meta-analysis in [5] reveals a clear lack of experimental rigor in studies to establish reliable conclusions on the effects of assistive robots in the elder care sector. A similar problem is uncovered in [4] that reviewed studies, which were concerned with the acceptance of health-care robots. A lack of validated measures to assess this aspect is reported. The applicability of the robotic systems in real social contexts remains thus an unjustified claim. The survey in [10] mentions as another issue that most projects develop social robots

© Springer International Publishing AG 2016
A. Agah et al. (Eds.): ICSR 2016, LNAI 9979, pp. 692–701, 2016.
DOI: 10.1007/978-3-319-47437-3_68

with a focus on the technological advancement, relying on stereotypical views of senior citizens and not involving actual users in the design and development process but (if at all) as test subjects in lab tests post factum.

These considerations are only rarely taken into account. One example is described in [25] stressing the need for co-design and co-development together with stakeholders and users. One of the essential claims is the necessity to take the elders' real needs into account in the development phase. As a consequence, this approach would require researchers to start the development process with a blank slate, i.e., without an actual robotic system in mind. While Bedaf and colleagues [2] embrace the idea of user participation, they also start the process with aiming at a specific robotic platform (Care-O-Bot 3). The two main insights from their work though are that it is virtually impossible to just solve one task in isolation but that introducing a robot into the complex network of interactional practices affects the whole network and not only the one task. Additionally, they conclude that monolithic robot systems are ill-suited for the envisioned application domain and the strategy should instead be "focusing on solutions with narrow functionalities (...) that may lead to faster success rather than looking for an all-encompassing service robot solution capable of many tasks".

Following this argument also warrants a shift from quantitative effect studies towards qualitative studies. The lack of qualitative and especially ethnographic studies in social robotics is surprising as there are many successful examples from other areas such as system design or HCI. In her prominent study, Suchman already pointed out the necessity to take into account the social practices and contexts in which computers are used in real settings [21]. Thus, there is a need for methods that capture the situated and social character of how people engage with technologies as well as the social and organizational context of these practices (e.g., [13]). Another problem rarely taken into account is the fact that social robots are technologies under construction. Their material shaping and the question of how and for what they should be used are still in the process of negotiation. Pols [19] argues in her study on the introduction of telecare in the Netherlands, that quantitative methods are not well suited to analyze innovation processes (for which the introduction of social robots is also an example) as they miss their central characteristic, i.e., that they are unpredictable in their outcome.

Based on the above analysis we claim that apart from the necessary technical development it is time to (1) engage in focused interactions with users as well as stakeholders throughout the whole development process in order to mobilize their tacit knowledge about the social practices they are engaged in, (2) understand how a new technology can be integrated in these practices, and (3) assess how the practices change by introducing the technology.

2   Social Practices

Our understanding of social practices is based on an ethnomethodological perspective [11]. Ethnomethodology aims at describing practices of social live [20].

Therefore, the construction of meaning is understood as situated real world accomplishments and depends on the ongoing coordination of actions and negotiations of meaning. That does not mean that artefacts are solely determined by social actions. The material affordances of an artefact do neither allow all kinds of actions (e.g., it is very difficult to eat a soup with a fork), nor do they determine a certain set of actions as we often use artefacts to do things that might not have been intended by the developers (e.g., to use a fork to comb your hair). The meaning of an artefact is thus reflexively built in the situated usage of the artefact. Alac and colleagues [1] show for example how toddlers became uninterested in a robot when the care taker did not display an interest in the robot. Thus, the presence of a robot alone does not transform it into a meaningful object. This first happens when it is involved in social practices. This can also be seen in another example of a non-working robot, which is nevertheless constructed as a participant of an interaction, while in another situation a working robot is treated as an object.

To bridge the ethnomethodological viewpoint and the participatory design methods that have as one of their central claims the mutual learning between users and designers, we make use of the notion of communities of practice [24]. Wenger [24] identifies three ways of explaining how a community is created from the practices of its members: (i) Mutual engagement, (ii) joint enterprise, and (iii) shared repertoire. Communities are thus established by the social practices in which they are involved. Engaging in these practices thus re-affirms membership as well as the community itself. Entering the community creates a learning situation (of the practices establishing the community). Putting it in a practical context, the elder care home can be seen as a place that entertains at least three different kinds of communities of practice: (1) it is the home of the older people living there, (2) it is the working place of the staff, (3) both groups frequently interact.

Entering this field as a researcher, we have to be aware that this is going to disrupt and change the established social practices, no matter what we do. First, because we are going to start interacting with staff and inhabitants in relation to their practices, trying to understand and learn about those practices and most likely making them aware of the practices they are engaged in (which in turn might trigger change). Then, we are going to introduce a social robot that is supposed to be part of a specific social practice. As was shown by Bedaf and colleagues [2], it is naive to assume that a robot will just substitute a specific task and that this will not affect the contextual embedding of this task, i.e., the whole social practice around this task is likely to change and adapt to the introduction of the robot.

3   Research Approach

We follow an action research paradigm with its focus on developing interventions with the users (e.g., [12]). Thus, the approach is not hypothesis-driven (although it may lead to hypotheses), but qualitative and exploratory. It focuses on individuals or small groups with no immediate claim that results are generalizable.

We also focus on users' concrete practices in their daily activities and not on cognitive or psychological factors. Thus, we are not interested in measuring how well a given robot performs according to some pre-defined "objective" features, or how users perceive the robot on some 7-point Likert scale. Instead, we are interested in how the users make use of the robot, how they integrate it into their practices and how these practices change by the introduction of the robot. In the ideal case, the results are transferable (not generalizable) to similar settings, e.g., to a different elder care home in the same region. There are some hard facts about the elder care sector that make generalizations at least doubtful. To mention just one, the turnover rate for elder care personnel in the US is between 51 % (licensed practical nurse) and 74.5 % (certified nursing assistant) [7]. The situation in Europe is a bit different. In [3], Netherlands and UK are given as examples with a turnover rate of 13 % and 17.9 % respectively. For the Danish elder care home in our study, the management could only report of staff members leaving when they were retiring. Thus, the institutional settings, the relation between staff and residents, as well as the practices in which they are involved seem to be very different from, e.g., an US American nursing home.

We build on a mix of methods from social sciences, humanities, and engineering (participatory design, ethnography, and robot modularity). We embrace mutual learning and collaboration as the central concepts from participatory design [17]. It is our experience that participatory design is a viable approach to demystify the role of technology especially in less technologically saturated contexts such as elder care homes (e.g., [9]).

Participatory design workshops are complemented by ethnographic field work (e.g., [6,8]). In our approach, we focus on *concurrent ethnography* [14], where the ethnographic study is taking place at the same time as the system-design. This warrants a close cooperation between ethnographer and designer informing each other during cycles of fieldwork, debriefing, system design and prototype interaction. These cycles are combined with workshops and meetings with the target users, e.g., staff and residents, which also feed back into the development process. In the concrete case of the elder care center, ethnography is used to get an understanding of how the work of the care-personnel and the everyday life of inhabitants looks without technology. The ethnographic study is thus able to specify different work activities in their social context, describe their sequential ordering and thus inform about basic aspects that must be addressed in the design of a robot.

One likely reason for the technology-driven focus of social robotics research is the complexity of building an actual robotic system. But recent years have seen increased development of rapid prototyping techniques enabling building small-scale specialized robotic systems at low cost and reasonable amount of time. The main research focus from the technical perspective thus lies on the challenge of developing a modular design framework for creating robots adapted to the highly variable needs and capabilities of the user group and preferably in collaboration with the users. Several initiatives have already started to embrace this challenge ranging from industrial application areas (e.g., [18]) to general cloud robotics ideas (e.g., [15,23]). Additionally, there already exist shelf-components for a number of different processing tasks that can be used to prototype many functions of a robot (e.g., [22]).

4 Case Study with an Elder Care Home

In the rest of the paper we present a case study following some of the above mentioned methodologies. Our goal was to identify social practices, where a social robot might make sense for both staff and residents (though not necessarily the same sense). Thus, we needed to get information about the everyday (work) life of staff and residents.

4.1 Staff Workshops

Three workshops were arranged over a three month period. Our intention was to get introduced to as many tasks as possible, and to gather information on how the staff felt about these tasks. A tool was devised based on a dimensional emotion model [16] to allow expressing underlying affective connotations of tasks that staff had to perform during the day. The two dimensions of valence (positive vs. negative) and arousal (high vs. low) have been used (VA-board, see Fig. 1 left). Each participant was equipped with a board and was asked to place post-its with tasks on the board. In a second round, a public board was used, where participants agreed on tasks. This provided a means for facilitating discussions about the emotional impact of the different tasks. A second tool (Duty circles) was created, where participants were able to assign subjectively important variables to the different tasks, e.g., time consuming, physically demanding, etc. (see Fig. 1 middle). Again, the main goal was to facilitate discussion between participants about the inherent characteristics and their subjective perspective on these tasks.

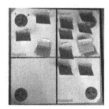  

Fig. 1. Tools and prompts from staff and resident workshops: Valence-Arousal Board (VA board) (left); Duty circles (middle); Example activity (right)

Results from staff workshops. Through the VA boards we identified 17 tasks along with the general feelings (positive, neutral, negative) associated with performing them. The intention of the following workshops was to get further information about the tasks. Participants chose to work further with four of them:

(i) Arranging doctor appointments (positive); (ii) Guiding the residents, i.e., structuring their actions (neutral); (iii) Reminding residents to take their medicine (negative); (iv) Reminding residents to go to meals (negative).

4.2   Resident Workshops

As the residents do not have mandatory tasks to perform, their needs from a technology like social robots cannot as easily be categorized as the staffs'. Our strategy therefore relied upon finding sources of certain emotions, like distress, happiness or anger, and finding a way to enhance quality of life of the residents. For the workshops we joined the residents during their coffee time in the afternoon. Three residents participated in each workshop. They had different conditions, such as bad sight or bad mobility but none had dementia. Of course it would have been good to involve more residents in the workshops but abilities to participate varied from day to day for the group of residents due to health. The aim in these workshops was to uncover emotional needs and the residents' perception of the social practices, which are determined by the institutional setting. Prompts were used during the workshops to elicit discussions such as drawings depicting waking up in the morning, eating breakfast, or having lunch (Fig. 1 right). Emotional needs and daily routines were identified, where the prompts were used to initiate discussions about emotional needs, daily activities, and feelings associated with these activities.

Results from residents workshops. During the workshops we identified some general emotional needs, as well as specific parts of daily routines that might contribute to lower quality of life in the given institutional setting. The feeling of being safe and knowing what is going on adds to the feeling of being home. Having staff available and helping them with certain activities is seen as comforting. Being able to do things themselves plays a big part in their life. Not being able to go to the toilet by oneself, losing the ability to walk, becoming blind, creates a sincere feeling of losing control. The fear of getting dementia is also a motivation for being active. Therefore it becomes important to do the small things that the residents are capable of doing, which also motivates them to keep active. Another activity that motivates them is socializing with other residents, e.g., in eating together or understanding each others' losses. Changes in routines and plans can be very stressful for many. Using the picture prompts displaying the different times of the day, our storytelling approach led to two activities being discussed in detail. (i) Being woken up from sleep: Because most residents prefer a calm morning by themselves, they get their breakfast and coffee served in their individual rooms. But often the morning routine has to be rushed, because staff has a limited amount of time. (ii) Being updated (including reminders): Due of age related problems such as bad memory, loss of sight or dementia, keeping track of time becomes a challenge. Being updated also includes getting to meals on time, knowing about getting their blood pressure measured, knowing what time of day it is, and getting updated on what social activities are happening.

4.3 Shadowing the Staff

Drawing a conclusion from the workshops, two tasks seem to be relevant for both staff and residents: guidance in morning routines and meal reminders. The participatory design workshops gave us a first impression of the staff's daily work, but as these workshops remain on an abstract or rational level, they do not inform us about the actions involved in the social practices that constitute the guidance or reminding tasks. Thus, we decided to shadow the staff, i.e., follow them around during the day. We followed two staff members that we were familiar with from the workshops around from 07:20 until 12:30 for three days. We were able to capture both morning guidance for six to seven residents per staff member (per day) as well as meal reminders. Because the observations included sensitive situations in the morning, we decided against video recordings and instead wrote down observation notes, whenever possible as well as memory protocols at the end of the observations. When analyzing the data, the nature of living at a nursing home became very clear. Many residents were very dependent on staff due to a range of challenges, e.g., with mobility or blindness. Every individual had different challenges and it was therefore also clear that a robot would have to be adaptable according to the resident's need and would definitely disrupt the way, the morning routines were performed right now. The data from shadowing the staff presented a complex system with a very flexible structure as the staff would have to deal with the varying challenges the residents experienced everyday. At the same time the routines were very fixed as specific tasks were done at specific times. The task of meal reminders on the other hand was a repetitive task that disrupted staff's other routines that had to be done at the same time and thus presented itself as an ideal candidate for a first prototype that could make a huge difference for both staff and residents.

4.4 Four Day Robot Intervention: A Meal Reminder

For the intervention, it was agreed on that the robot should be non obtrusive and resemble a standard household appliance. As a result, a robotic lamp was fabricated out of plywood and equipped with two servo motors, allowing for several degrees of freedom in order to direct the user's attention. Additionally, the lamp was enhanced with speakers to allow for verbal communication of messages. In order to have full control over the intervention, it was teleoperated. Figure 2 shows the different prototypes and the setup. For the meal reminder, the robot was programmed to remind the participants on three occasions, lunch, coffee time and dinner. When the robot was activated it would raise its head and move it towards the resident, and depending on which time of the day, it would say "There is lunch/afternoon coffee/dinner now." These three reminders had different states, the amount of times the robot would repeat itself, and the movement speed. The main goal of this intervention was to investigate how the residents responded to having a robot in their rooms. The staff identified two female candidates, which they constantly need to remind of daily activities like lunch and dinner. Both suffered from dementia. Resident 1 was 93 years old and was described as having extremely bad memory. Staff also said that they usually would have to go into her room several times to remind her. Resident 2 was 87, also had bad memory and had to be constantly reminded of events. Staff also said that the resident had a tendency to fall asleep even when doing tasks. The residents would have the robot in their rooms for two days each. The robot was placed on the living room table and on a spot, where the robot would be able to get the attention from the resident independent of her position in the apartment. The robot was controlled by a computer placed outside the room and connected via the internet. A webcam was also placed beside the robot in order to observe what happened when the robot was activated.

Fig. 2. Protoype iterations (left), lamp in room and wizard (right)

We had imagined that the residents would be doing activities in their room. Instead we found out that resident 1 slept most of the time. Moreover, she also did not come back in between breakfast and lunch as she uses a lot of energy walking to her room. She therefore only experienced the robot reminding her for coffee and dinner times. Staff had expressed that the resident would not come out if they do not go into the room and help her put on her clothes and she usually goes back to sleep if they don't. This was not evident during the robot intervention. When the robot was activated the resident immediately got out of bed and began putting on her clothes. It took her 19 min, but she managed to complete the task. During the second dinner reminder the resident came out after 16 min. The resident was quite happy and positive towards the robot and felt that it helped. She felt that it was a refreshing way to wake up and remarked that it made her happy. Resident 2 woke up the first time the robot reminded her, but then she fell asleep again. 18 attempts of reminding her to go to coffee failed, and after 27 min we called after a helper. In the post-intervention interview, the resident explained that she had woken up and was lying and counting how many times the robot would repeat itself. The robot had a calming effect and it was relaxing listening to. She thought that it was really cozy having it in the apartment. The same happened on the second day, where a staff member was called after 35 min (15 attempts) of reminding her of coffee time. For the dinner reminder, the resident was already up and came out after 3 min. When she was reminded, the resident promptly thanked the robot for reminding her, and said she was going to show up right away. Then she left the apartment.

5   Conclusion

In this paper we propose an approach for developing social robots that takes the "social" more seriously, starting from the identification of social practices in which such robots could make sense. It is our opinion that this will lead to a different set of robots that are not general platforms but instead robots that will be targeted to perform specific tasks involved in specific social practices. For developing such robots, we suggested a combination of participatory design practices, ethnographic studies, and rapid prototyping of robots. Working with both staff and residents at an elder care home, we were able to identify several social practices, where the introduction of a social robot could make a difference for both parties. We also created a first prototype for one of the identified task that we tested in a four day intervention at the nursing home. This task was described by staff as "unnecessary" and by residents as a general aspect that influences their feeling of quality of life. As it turned out, the introduction of the robot showed a highly desirable effect, as contrary to the opinion of the involved staff, one of the participating residents was able to manage getting ready for dinner without any help. The robot gave her the time she needed to dress without being rushed. This is a good example of the beneficial effect a social robot can have, as it allows the residents to gain back some independence while at the same time freeing staff from repetitive and time consuming tasks.

Acknowledgments. We would like to thank all residents and staff members at Løgstør Rødekorshjemmet that participated in the case study described in this paper.

References

1. Alaĉ, M., Movellan, J., Tanaka, F.: When a robot is social: Spacial arrangements and multimodal semiotic engagement in the practice of social robotics. Soc. Stud. Sci. **4**(6), 893–926 (2011)
2. Bedaf, S., Gelderblom, G.J., de Witte, L., Syrdal, D., Lehmann, H., Amirabdollahian, F., Dautenhahn, K., Hewson, D.: Selecting services for a service robot: Evaluating the problematic activities threatening the independence of elderly persons. In: IEEE International Conference on Rehabilitation Robotics (2013)
3. Bettio, F., Verashchagina, A.: Long-Term Care for the elderly: Provisions and providers in 33 European countries. EU Expert Group on Gender and Employment (EGGE) (2010). http://ec.europa.eu/justice/gender-equality/files/elderly_care_en.pdf
4. Broadbent, E., Stafford, R., Donald, B.M.: Acceptance of healthcare robots for the older population: Review and future directions. Int. J. Soc. Robot. **1**, 319–330 (2009)
5. Broekens, J., Herrink, M., Rosendahl, H.: Assistive social robots in elderly care: a review. Gerontechnology **8**(2), 94–103 (2009)
6. Crabtree, A., Rouncefield, M., Tolmie, P.: Doing Design Ethnography. Springer, London (2012)
7. Donoghue, C.: Nursing home staff turnover and retention an analysis of national level data. J. Appl. Gerontol. **29**(1), 89–106 (2010)

8. Dourish, P., Button, G.: On technomethodology: foundational relationships between ethnomethodology and system design. Hum. Comput. Interact. **13**, 395–432 (1998)
9. Förster, F., Weiss, A., Tscheligi, M.: Anthropomorphic design for an interactive urban robot: the right design approach. In: Proceedings of the 6th International Conference on Human-Robot Interaction, pp. 137–138. ACM Press (2011)
10. Frennert, S., Östlund, B.: Review: Seven matters of concern of social robots and older people. Int. J. Soc. Robot. **6**, 299–310 (2014)
11. Garfinkel, H.: Studies in Ethnomethodology. Prentice-Hall, Englewood Cliffs (1967)
12. Hayes, G.R.: Knowing by doing: action research as an approach to HCI. In: Olsen, J.S., Kellogg, W.A. (eds.) Ways of Knowing in HCI, pp. 49–68. Springer, Heidelberg (2014)
13. Heath, C., Luff, P.: Technology in Action. Cambridge University Press, Cambridge (2000)
14. Hughes, J., King, V., Rodden, T., Andersen, H.: Moving out from the control room: ethnography in system design. In: Proceedings of the 1994 ACM Conference on Computer Supported Cooperative Work. ACM Press (1994)
15. Kehoe, B., Patil, S., Abbeel, P., Goldberg, K.: A survey of research on cloud robotics and automation. IEEE Trans. Autom. Sci. Eng. **12**(2), 398–409 (2015)
16. Morgan, R.L., Heise, D.: Structure of emotions. Soc. Psychol. Q. **51**(1), 19–31 (1988)
17. Nielsen, J., Dirckinck-Holmfeld, L., Danielsen, O.: Dialogue design - with mutual learning as guiding principle. Int. J. Hum. Comput. Interact. **15**(1), 21–40 (2003)
18. Pedersen, M.R., Nalpantidis, L., Andersen, R.S., Schou, C., Bøgh, S., Krüger, V., Madsen, O.: Robot skills for manufacturing: From concept to industrial deployment. Robot. Comput. Integr. Manuf. **37**, 282–291 (2015)
19. Pols, J.: Care at a Distance. On the closness of Technology. Amsterdam University Press, Amsterdam (2012)
20. Sacks, H.: Notes on methodology. In: Atkinson, J.M., Heritage, J. (eds.) Structures of Social Action. Studies in Conversation Analysis, pp. 21–27. Cambridge University Press, Cambridge (1984)
21. Suchman, L.: Plans and Situated Actions: The Problem of Human-Machine Communication. Cambridge University Press, Cambridge (1987)
22. Wagner, J., Lingenfelser, F., André, E.: The social signal interpretation framework (ssi) for real time signal processing and recognition. In: Interspeech (2011)
23. Waibel, M., Beetz, M., Civera, J., d'Andrea, R., Elfring, J., Galvez-Lopez, D., Häussermann, K., Janssen, R., Montiel, J.M.M., Perzylo, A., Schiessle, B., Tenorth, M., Zweigle, O., de Molengraft, M.J.G.V.: RoboEarth - a world wide web for robots. IEEE Robot. Autom. Mag. **18**(2), 69–82 (2011)
24. Wenger, E.: Communities of Practice: Learning, Meaning, and Identity. Cambridge University Press, New York (1998)
25. Wu, Y.H., Fassert, C., Rigaud, A.S.: Designing robots for the elderly: Appearance issue and beyond. Arch. Gerontol. Geriatr. **54**, 121–126 (2012)

Co-design and Robots: A Case Study of a Robot Dog for Aging People

Tuck W. Leong and Benjamin Johnston[✉]

Faculty of Engineering and IT, University of Technology Sydney,
Ultimo, NSW, Australia
benjamin.johnston@uts.edu.au

Abstract. The day-to-day experiences of aging citizens differ significantly from young, technologically savvy engineers. Yet, well-meaning engineers continue to design technologies for aging citizens, informed by skewed stereotypes of aging without deep engagements from these users. This paper describes a co-design project based on the principles of Participatory Design that sought to provide aging people with the capacity to co-design technologies that suit their needs. The project combined the design intuitions of both participants and designers, on equal footing, to produce a companion robot in the form of a networked robotic dog. Besides evaluating a productive approach that empowers aging people in the process of co-designing and evaluating technologies for themselves, this paper presents a viable solution that is playful and meaningful to these elderly people; capable of enhancing their independence, social agency and well-being.

Keywords: Participatory design · Co-design · Robot dog · Aging

1 Introduction

A great achievement of many developed countries is that their citizens are now living longer. For example, it is estimated that by 2056, 25 % of Australia's population will be aged 65 years and above [3], approximately twice as many as the population in 2011. This brings huge challenges and opportunities in domains including healthcare, gerontology, social policy and technology design.

Technology designers, including those working with robots, have long engaged with efforts to understand and design technologies for aging people (*e.g.*, [1,2,4]). However, some point out that these efforts are often skewed by dominant discourses surrounding aging and older people; stereotyping aging people negatively as a group who are needy, frail, lonely and have difficulties using new technologies [6,12,14]. Such a view misses the fact that many aging people are reasonably healthy, socially active, and would learn and use technologies if they believe that it is useful and relevant to their needs [17]. Instead of designing technologies that reflect aging people's values [13] we get a lot of technologies that are deficit driven (*e.g.*, [5]) and assistive in nature (*e.g.*, [17]). For example, when

© Springer International Publishing AG 2016
A. Agah et al. (Eds.): ICSR 2016, LNAI 9979, pp. 702–711, 2016.
DOI: 10.1007/978-3-319-47437-3_69

robots are considered, they are often viewed as a technological fix for the aging and yet older adults are often not consulted when designing robots [12].

To counter this, Vines *et al.* [22] recommend using a participatory process during design to produce technologies that fit well with aging people's aspirations. This provides people a voice throughout and allows them to decide the measure of success that is meaningful to them. Vines *et al.* [22] also recommend having deeper engagements with small groups of individuals to produce designs that respond to their personal histories and everyday lived experiences.

The design project described in this paper is a case study that exemplifies Vines *et al.*'s [22] recommendations, where designers worked collaboratively with a small group of older people to explore and co-design ideas and solutions that are meaningful to them. In particular, this paper demonstrates how this approach can be used productively and effectively with aging people to come up with the idea of, and to explore the use of, a robot to support and even enhance their day-to-day experiences of aging. The final design concept, a robot dog and associated use scenarios were developed, refined and evaluated iteratively with the participants. These outputs were further evaluated at a later stage by elderly participants at two different workshops. Overall, the feedback about the concept was very positive, demonstrating that aging people can accept the idea of a robot companion if it is perceived to be both playful and useful by them to achieve their desires to age well and live longer in their own homes.

2   A Participatory Approach to Technology Design

Following Vines *et al.*'s [22] recommendations, Participatory Design (PD) processes were used with a group of aging people. The process saw five designers working with eight elderly participants over interviews and three workshops. The designers consisted of four postgraduate Interaction Design students and an academic supervisor experienced in Interaction Design and Participatory Design. The participants were four males and four females in good health and aged 65–75. The participatory design process used in the project is depicted in Fig. 1.

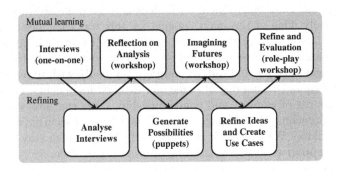

Fig. 1. Proposed co-design process

This PD process alternates between mutual learning activities (*i.e.*, sessions with the participants) and refining activities (*i.e.*, where designers conduct background research and develop prototypes). A novel aspect of this process is that it does not begin by teaching users about the capabilities/limitations of robotics (*cf.*, [1,12,19]), yet incorporates the user in the earliest stages of design. This ensures a focus on user needs without being derailed by a focus on particular technologies. In fact, the designers approached the project agnostic about the technological solution; the idea of robotics as a potential solution was not in the designers' minds prior to, nor during the early parts of the design process.

Mutual learning is a core commitment of Participatory Design (PD) and is "one of its key differentiators from User-Centred Design and other human-centred approaches" [18]. This is because PD recognizes that people who are not professional technology designers cannot actively participate in co-designing technologies they might use because they lack knowledge about what advanced technologies could offer them (*e.g.*, [7,20]). Mutual learning is made possible through the initial face-to-face interviews and the three follow-up workshops. These activities allow designers to learn more about the participants and for participants to learn about possibilities of technologies, and together learn about possible and useful technology solutions.

The refining activities in the process are concerned with extending what designers learned about the users, ideation, research, developing prototypes of the users' ideas and preparing for the mutual learning workshops.

2.1   Interviews

Working in pairs, the designers conducted a series of one-to-one open-ended interviews with each participant at their homes. The interviews were audio-recorded and spanned approximately an hour each. The aim was to understand participants' experiences of aging, *e.g.* what is important to them as they age, their situations, life histories, hopes and fears, use of technologies. All audio-recordings were transcribed, and the data was analysed using thematic analysis.

Analysis of the interviews revealed the busy lives participants kept, with various activities such as gardening, home projects, socializing with friends or family members. Their engagement with life was to keep fit physically, mentally and emotionally. One of the main reasons for this was their desire to remain independent for as long as possible and continue to live in their homes as they age. Participants who were single did speak about occasionally feeling lonely. This made them anxious and sad, something they really wish to avoid.

Two participants own a dog at home and extolled its benefits. For them, the dog was a source of comfort and companionship. Dogs can also warn them of strangers approaching their homes, giving them a sense of security, especially for those living alone. Walking their dogs regularly helped them stay healthy and also provide opportunities to socialize with other dog walkers.

While all participants own a smartphone, all of them use a limited number of apps. The phone is used primarily as a communication tool with friends and family. Finally, all of the participants lamented on their growing forgetfulness and absentmindedness, especially with their phones. They often forget to charge their phones (resulting in missed communications) or sometimes are not able to find their phones (resulting in frustration).

2.2 Workshop 1 (Reflection and Validation of Analysis)

Workshop 1 followed the interviews. The aim is to provide the participants an opportunity to learn about and discuss the interview findings.

The workshop began with a presentation of selected commercial/consumer technologies that have been designed for aging citizens. Participants were encouraged to ask questions and to discuss how these technologies could potentially affect their lives. After this, participants were presented with the findings from the interviews. They were given the opportunity to discuss the findings and seek further clarifications. Besides providing some validation of the designers' analysis, the activity allowed participants to provide further insights and clarifications to some of the findings. Having the opportunity to listen to each other's experiences of aging also spurred further discussions and ideas.

One very popular discussion was centered around the benefits of having a pet dog at home, how the dog may support their aspirations to age well and in their own homes and the challenges of finding housing suitable for dog ownership. By the end of Workshop 1, the participants were asked to vote on what they wanted to co-design in further workshops. They chose to focus on finding ways to feel safe, calm, and connected while at home but at the same time like to discover ways to expand their social circles.

2.3 Generating Possibilities

Given the significant interests and discussions surrounding the benefits dogs, and a realisation that a 'dog' might be an interesting design idea, the designers decided to explore the idea of a robot dog in the second workshop.

Using robots to assist and support older adults to maintain their independence, and enhance their well-being is not a new idea in Human-Robot Interaction (e.g., [1,19]). As highlighted earlier, the application (and design) of technologies, including robots are often driven by (erroneous) assumptions about aging people or by goals to reduce human carers. Beer et al.'s [4] review of domestic robots efforts highlight various ways robots have been proposed to help with aging people, e.g., performing home upkeep tasks, especially for individuals with declined health capabilities, such as motor impairments, dementia, or needing healthcare assistance. Kumahara and Mori [11] developed a mobile robot that can walk together with elderly people and help them to maintain their healthy condition. Robotic dogs have also been proposed, with efforts to improve its emotional recognition capabilities [9] to make it more engaging. Others explored children's reaction to robot dogs when compared to a real one [15]. While the range of efforts are impressive, they are still focused upon on assistive features for sick or needy individuals, forgetting that a large number of aging individuals, are still healthy, active, and capable of doing things for themselves.

Meanwhile, a lot of work with robots in the domestic setting have focused on whether they are accepted by people, and ways to better design for this acceptance (*e.g.* [8,16,23]). Most 'design work' with domestic robots begins with presenting people with various robots to evaluate, such as the kind of appearances and functionalities that are important to older adults. These efforts often do not consider aging people's emotional needs and experiences when living with these robots, such as companionship, security and even day-to-day 'social' interactions with these robots. An exception is the recent work by Lazar *et al.* [12] that used focus groups with older adults to explore some of the situatedness of aging. While they offer some insights and directions for future design, they did not fully develop the ideas nor provided concrete use scenarios.

In this project, the designers took the findings from the interviews and Workshop 1, and conducted background research about robot use in aging, especially within domestic settings. In addition, the designers needed to find ways to stimulate the participants' imagination of a future scenario with a robot in their lives in the second workshop. After some speculation about possible designs, the Sony AIBO (a dog-like entertainment robot capable of locomotion, sensing and communication by sound and WiFi) was selected as the 'puppet' to be used to support participants imagination and an inoperable AIBO was sourced.

2.4 Workshop 2 (Imagining Futures)

While the interviews and Workshop 1 were focussed on learning about users' situations and needs. The aim of the second workshop was to use the same participants to explore and co-design the idea of a robot dog.

As explained earlier, participants would need some understanding about technologies if they were to participate effectively in co-design activities. Thus, the second workshop began with presentations from the designers regarding domestic robots and emergent technologies such as the Internet of Things. Participants learned about the kinds of technologies available and the potential of a networked environment. When the participants encountered examples whereby robots are used to support aging people, such as acting as 'carers', all of them found the idea very inappropriate and demeaning—"not something I could see myself wanting, no thanks, I am not that useless, yet". They asked questions and sought clarifications about various technologies presented but always in reference to their own situations and needs.

After the presentations, an imaginative activity with the participants were conducted where the participants were introduced to the AIBO puppet (hereby known as the dog). The designers had attached sticks and strings to the dog and one of the designers acted as the 'puppeteer', making decisions as to what the dog might respond to in this workshop.

The participants were told that their task is to imagine what the dog could do for them in their everyday lives. They were fascinated by it and at first, the participants interacted with the dog as they would with a real dog: asking it to come, to sit, to fetch things and so on. Since the dog did not always respond as asked, it was interesting to see how forgiving people were. Then, one of the designers hinted to the participants of the possible properties of some emergent technologies, such as connectivity.

After some silence, one participant asked, "Can it connect to my phone and find it? I always misplace it and I can't find it when I need it." When the designers confirmed that the dog could be tethered to their phones, the participants suddenly became very excited. They had so many questions. "Will it tell me if my phone is ringing? Sometimes I can't hear it ring". "What about when my battery is low?" "What about when the phone is silent?", etc. The designers did their best to provide reasonably realistic answers and at the same time encouraged the collective exploration of ideas. At first, everyone was excited about the number of things the dog could do but one of the participants indicated that she really prefers "a companion, not an assistive device... something that dogs do but just a bit more because it is special and hardy". Throughout, two designers captured the participant's ideas. Before the workshop ended, the participants ranked the ideas and worked with the designers to write out scenarios for their top three ideas.

2.5   Design Work Informed by Workshop 2

After Workshop 2, three detailed and rich scenarios for 'Hardy Hound' (the name for this robot dog) were refined by the designers for presentation in the final third workshop. The scenarios were based closely on the participants' ideas. Below is a brief description of the three scenarios. Below is a brief description of the three scenarios (see Fig. 2.).

Activity 1: Finding a phone. When users cannot find their mobile phone, they can ask the dog to look for it because the dog is tethered to the phone. The dog will lead the user to the phone.

Activity 2: Increasing social interaction. A Hardy Hound is designed to recognise other Hardy Hounds. When it locates another dog close by, it will try and seek out the other dog. The dogs' interaction provides a 'ticket to talk': taking a Hardy Hound for a walk may provide users with chances of communicating with other Hardy Hound users, thus, the potential to make friends and increase their level of social interaction.

Activity 3: Companionship. When the user is feeling lonely, she can call out to the Hardy Hound. The dog will go to the owner and behave in a happy manner.

Lucy Suchman's interaction framework [21] was chosen to be used in Workshop 3 to refine the basic interactions between people and the dog. Suchman pointed out that human-computer communication is a special case of communication whereby the resources available to the participants is limited, and how this could lead to problems and breakdowns. In the case of a robotic dog, its capacity to interact is limited by its programming, only able to respond to specific commands (input) as deemed plausible, given current technology.

Fig. 2. Three scenarios for Hardy Hounds

2.6 Workshop 3 (Refine and Evaluation)

Workshop 3 aimed to learn some of the potential communicative problems and to prototype the necessary interactions. The workshop was driven by role play: one of the designers put on a dog suit and mask, acting as a 'Ben, a Hardy Hound'. He was given his 'programming' for each activity which determines (and limits) his capacity to act/behave when interacting with the participants. The participants were introduced to Ben and were briefed about Scenario 1. They were told that Ben has limited programming and would only understand some verbal commands. They then took turns interacting with Ben to find a phone. This was then repeated for Scenario 2 and 3. The activities were videotaped. In between each activity, discussions were held to elicit participants' experiences, expectations and thoughts.

Problems arose immediately because people were issuing different kinds of commands. For example, to find a phone, they asked "Where's my phone?", "Phone", and "Get my phone", instead of the 'programmed' command of "Find my phone". Because participants needed to be told of the 'correct' command before using it, the importance of more clear user feedback became apparent, *e.g.*, using audio and visual feedback such as barking and moving forward. The richness of the feedback provided thorough insights into the important features of human-machine intelligibility.

The outcome of the final workshop was a more refined set of interactions for each of the scenarios. This also included recommendations, such as enabling users to customize commands and the need for consistency of feedback from the dog. The scenarios were also refined. It was obvious from the session that users wanted the dog to behave and respond as similarly as possible to a real dog, *e.g.*, not using human speech but barks, body language, eyes, ears, tail. More interestingly, the design team found that people were very forgiving of the dog. If the dog did not respond as they expected, they just laughed it off and attempted different ways to interact with it. Instead of being frustrated, they found it cute.

Even when the dog's responses are ambiguous, people were more than willing to interpret and try to make sense of what they believe the dog is "trying to say".

3   Further Evaluation

Six months after the original design workshops, the scenarios and interactions were evaluated at two additional workshops with new groups of aging people.

Each workshop consisted of eight people (16 total) aged 65–90 to explore values that aging people hold dear as they age. Understanding aging people's values can help shed light on why aging people adopt (or not) technologies such as our robot dog [13].

During the workshop, various technologies were presented to elderly participants to evaluate based upon their values. The Hardy Hound concept and its scenarios were presented as one such technology. Participants were asked to evaluate how well the concept fit with their needs, situations and values as well as any other uses they could they imagine.

The results were very positive. Most of them (14/16) liked the technology, seeing it as "cute", a "good companion" and "helping to keep my blood pressure down". None imagined that it would replace a real dog but it was certainly a viable alternative for apartment dwellers or nursing homes where live pets were not allowed. In fact, two participants were worried that their real-life dogs might "destroy the robot". 4/16 wanted to know where they could get one right away. While the functionalities were desirable, they also saw it as a machine for play and companionship. In fact, a few thought that owning such a dog would give them a certain amount of cachet or admiration from their peers and more importantly, their grandchildren. Many of them could also imagine different ways whereby the Hardy Hounds might be useful in their everyday lives, such as to strengthen a sense of security when they are home alone.

4   Discussion

This paper makes the following contributions to the robotics community. First, a participatory approach that supports elderly people to imagine creative applications of technologies for themselves. It demonstrates the capacity of elderly people to act as active partners in co-design—able to envisage and articulate solutions they will use, that are suitable and meaningful to their needs and aspirations. This approach stands in contrast with most approaches where aging people are generally not consulted. While there aren't many examples of PD in robotics, some, such as Šabanović et al. [19] have used PD to design socially assistive robots with older adults. This paper provides another example of how PD can be used, but emphasising the value of alternating phases of mutual learning and reflection/refining in this PD process.

Second, it presents a set of use scenarios and learnings about interactions for a robot dog that have been developed with and evaluated by elderly people. The concept is well received by the participants invovled and is ready for further prototyping and development. This work concurs with recent findings by

Lazar *et al.* [12], such as the capacity of a robot pet to enhance social interactions with others. As such, this work provides a strong foundation for an approach and a viable concept related to a robot dog. We have not reported the lengthy set of interactions but instead highlight the usefulness of Suchman's framework [21] to help decompose, understand, analyse the real and complex problem of designing people's interactions with robots.

Third, this work demonstrates that participants did not find the idea of interacting and living with a robotic dog to be problematic. This contrasts with others working in Human-Robot interactions (e.g., [10,23]), who found barriers to robot-acceptance, including older adults' uneasiness with technology, feeling of stigmatization, and ethical/societal issues associated with robot use. We would argue that the co-design process that was used helped mitigate many of these concerns.

As the world's aging population grows, efforts to design interactive technologies will need to reconsider their approach. We have reported one way where designers can engage productively with older adults directly in the design and evaluation of technology. With planning and support, aging people are more than capable of imagining and co-designing technologies that not only meet their needs but fit their values and can co-exist with them meaningfully in their everyday lives.

References

1. Alves-Oliveira, P., Petisca, S., Correia, F., Maia, N., Paiva, A.: Social robots for older adults: framework of activities for aging in place with robots. In: Tapus, A., André, E., Martin, J.-C., Ferland, F., Ammi, M. (eds.) Social Robotics. LNCS, vol. 9388, pp. 11–20. Springer, Heidelberg (2015)
2. Ahn, H.S., Kuo, I.-H., Datta, C., Stafford, R., Kerse, N., Peri, K., Broadbent, E., MacDonald, B.A.: Design of a kiosk type healthcare robot system for older people in private and public places. In: Brugali, D., Broenink, J.F., Kroeger, T., MacDonald, B.A. (eds.) SIMPAR 2014. LNCS (LNAI), vol. 8810, pp. 578–589. Springer, Heidelberg (2014). doi:10.1007/978-3-319-11900-7_49
3. Australian Bureau of Statistics. Future Population Growth and Aging (2009)
4. Beer, J.M., Smarr, C.-A., Chen, T.L., Akanksha, P., Mitzner, T.L., Kemp, C.C., Rogers, W.A.: The domesticated robot: design guidelines for assisting older adults to age in place. In: Proceedings of Human Robot Interactions (HRI 2012), pp. 335–342 (2012)
5. Carroll, J., Convertino, G., Farooq, U., Rosson, M.B.: The firekeepers: Aging considered as a resource. Inf. Soc. **11**(1), 7–15 (2012)
6. Durick, J., Robertson, T., Brereton, M., Vetere., F, Nansen, B.: Dispelling aging myths in technology design. In: Proceedings of the 25th Australian Computer-Human Interaction Conference (OzCHI 2013), pp. 467–476 (2013)
7. Greenbaum, J., Kyng, M.: Design at work: Cooperative design of computer systems. Lawrence Erlbaum Associates, Hillsdale, NJ, USA (1991)
8. Heelink, M.: Exploring the influence of age, gender, education and computer experience on robot acceptance by older adults. In: Proceedings of the 6th International Conference on Human-Robot Interaction (HRI 2011), pp. 147–148 (2011)

9. Jones, C.M., Demming, A.: Investigating emotional interaction with a robotic dog. In: Proceedings of the 19th Australasian conference on Computer-Human Interaction (OzCHI 2007), pp. 183–186 (2007)

10. Khosla, R., Chu, M.-T., Kachouie, R., Yamada, K., Yamaguchi, T.: Embodying care in Matilda–an affective communication robot for the elderly in Australia. In: Proceedings of of the 2nd ACM SIGHIT International Health Informatics (IHI 2012), pp. 295–304 (2012)

11. Kumahara, Y., More, Y.: Portable robot inspiring walking in elderly people. In: The Second International Conference on Human-Agent Interaction (HAI 2014), pp. 145–148 (2014)

12. Lazar, A., Thompson, H.J., Piper, A.M., Demiris, G.: Rethinking the design of robotic pets for older adults. In: Proceedings of the 2016 ACM Conference on Designing Interactive Systems, pp. 1034–1046 (2016)

13. Leong, T.W., Robertson, T.: Voicing values: laying foundations for aging people to participate in design. In: Proceedings of the 14th Participatory Design Conference (2016)

14. Light, A., Leong, T.W., Robertson, T.: Ageing well with CSCW. In: Boulus-Rodje, N., Ellingsen, G., Bratteteig, T., Aanestad, M., Bjorn, P. (eds.) ECSCW 2015, pp. 295–304. Springer, Heidelberg (2015)

15. Melson, G., Khan, P.H., Beck, A.M., Friedman, B., Roberts, T., Garrett, E.: Robots as dogs? Children's interaction with the robotic dog AIBO and a live Australian Shepherd. In: Proceedings of CHI 2005 EA Extended Abstracts on Human Factors in Computing Systems, pp. 1649–1652 (2005)

16. Prakash, A., Kemp, C.C., Rogers, W.A.: Older adults' reactions to a robot's appearance in the context of home use. In: Proceedings of the 2014 ACM/IEEE International Conference on Human-Robot Interaction (2014)

17. Robertson, T., Durick, J., Brereton, M., Vetere, F., Howard, S., Nansen, B.: Knowing our users: scoping interviews in design research with aging participants. In: Proceedings of the 24th Australian Computer-Human Interaction Conference, pp, 517–520 (2012)

18. Robertson, T., Leong, T.W., Durick, J., Koreshoff, T.: Mutual learning as a resource for research design. In: Proceedings of the 13th Participatory Design Conference (PDC 2014), pp. 25–28 (2014)

19. Šabanović, S., Chang, W.-L., Bennett, C.C., Piatt, J.A., Hakken, D.: A robot of my own: participatory design of socially assistive robots for independently living older adults diagnosed with depression. In: Zhou, J., Salvendy, G. (eds.) ITAP 2015. LNCS, vol. 9193, pp. 104–114. Springer, Heidelberg (2015). doi:10.1007/978-3-319-20892-3_11

20. Simonsen, J., Robertson, T. (eds.): Routledge International Handbook of Participatory Design. Routledge, New York (2012)

21. Suchman, L.: Plans and situated actions: The problem of Human-Machine Communication. Cambridge University Press, New York (1987)

22. Vines, J., Pritchard, G., Wright, P., Olivier, P., Brittain, K.: An age old problem: examining the discourses of aging in hci and strategies for future research. ACM Trans. Comput. Hum. Interact. (TOCHI) **22**(1), 2 (2015)

23. Wu, Y.-H., Wrobel, J., Cornuet, M., Damne Rigaud, A.-S.: Acceptance of an assistive robot in older adults: a mixed-method study of human-robot interaction over a 1-month period in the Living Lab setting. Clin. Interv. Aging **14**(9), 801–811 (2014)

An Effort to Develop a Web-Based Approach to Assess the Need for Robots Among the Elderly

Kimmo J. Vänni[✉] and Annina K. Korpela

Tampere University of Applied Sciences, Kuntokatu 3, 33520 Tampere, Finland
{kimmo.vanni,annina.korpela}@tamk.fi

Abstract. Our aim was to develop an approach to assess the need for assistive and/or social robots among the elderly. The research methods consisted of literature review, inquiries from assistive technology professionals, and surveys among wellbeing technology professionals and elderly people. The results presented that there was lack of approaches in assessing the need for robots. Even if the web-based tool might be useful, developing such an approach and algorithms seemed to be a requiring task.

Keywords: Social robots · Assessment · ICF · Perception · Web · Assistive robots

1 Introduction

The overall interest in assistive and social robots has emerged during the last years. There are many different viewpoints and future plans using robots in health care services and diagnostics [1]. One major trend in elderly care sector is looking forward the development of robot technology. It is expected that number of aged people will be 1.5 billion by 2050, which is triple compared to 2010 level [2]. Increase in life expectancy is the positive thing, but it means that society has to take into account the emerging trend of cognitive disorders [3], and the demands for long-term care [4].

The European robotics strategy favors Internet-based tools in advancing health care and robotics. The strategy emphasizes that robots should be integrated into communication frameworks, and health monitoring systems should operate over the Internet [5]. Design of robots should be user-centered, but according to Tanaka et al. [6] the conventional design of robots still favors the technology oriented approaches, and neglects benefits of robots to users.

All the trends and challenges above has been the starting point for our aim to develop an approach for assessing the need for assistive and/or social robots. The objective is that the approach would assist end-users to evaluate if robots are needed, and which robot type would match best with users' expectations. We have designed our approach to meet the future requirements of web-based evaluation tools. As far as we know, there are no web-based tools available and only few non-web based concepts for utilizing classification tools have been introduced before [6].

© Springer International Publishing AG 2016
A. Agah et al. (Eds.): ICSR 2016, LNAI 9979, pp. 712–722, 2016.
DOI: 10.1007/978-3-319-47437-3_70

The nature of this study is explorative and it gives an insight into the position of designing and providing social and/or assistive robots. In addition to literature search, we have conducted questionnaire surveys, interviewed professionals as well as reviewed results of our previous robot research.

1.1 Concepts, Approaches, Factors and Technology for Defining the Robots

Studies regarding the acceptance of technology and related factors have been discussed in literature rather well. There are two main models for studying the acceptance of information technology; *the Technology Acceptance Model* (TAM) [7], where perceived usefulness and ease of use have been discussed, and *the Unified Theory of Acceptance and Use of Technology Model* (UTAUT) [8], which takes into account also a user's age, gender, experience and voluntariness of use. According to Heerink et al. [9] TAM and UTAUT models neither take into account social aspects of interaction with robots nor are developed with the elderly users in mind. Therefore, Heerink et al. [9] have developed a new model and added perceived enjoyment, social presence, perceived sociability, trust, and perceived adaptivity issues into a model. The presented models above are useful for evaluating acceptance of technology but those do not assist to decide if robots are needed. The most advanced concept, compared to our aim, is presented by Tanaka et al. [6] where they present how to use *the International Classification of Functioning, Disability, and Health* (ICF) [10] for evaluating and designing assistive robots. The ICF classification offers the framework for assessing an individual's level of functioning. Even if we did not find articles on web-based approaches, we want to present some articles, which discuss the relevant factors for implementing and accepting robots among elderly.

Flandorfer [11] has reported that sociodemographic factors have significant impacts on robot's acceptance, but a user's earlier experiences with technology mitigates an adoption process. It has been reported that factors like the living environment, a user's physical and mental condition and cognitive skills should be taken into account [12]. In addition, religious and cultural backgrounds are reported to be significant factors in the acceptance of assistive robots [13]. Tong et al. [14] have stated that the current simulators and design tools are meant for process-oriented workflows not for designing the humanoid robots. Meng and Lee [15] have argued that the traditional industrial robot engineering approaches are inappropriate to tackle the problem areas such as user-friendliness. Wu et al. [16] have studied the elderly persons' perceptions regarding the robot's appearance and discussed the importance of social context in designing robots. Saborovski and Kollak [17] have argued that the needs of elderly people and their relatives must be taken into account in design process, but the care professionals' experiences are overlooked in the technical development. de Graaf et al. [18] have presented that enjoyment and attractiveness of the robot are important factors in perception. Alaiad and Zhou [19] have studied the determinants of adoption of home healthcare robots, and reported that sociotechnical factors play important role. Michaud et al. [20] have presented the interdisciplinary and exploratory design methodology to develop an assistive mobile robot for homecare. Andrade et al. [21] have concluded that cost of robotics technology is still prohibitive issue, which limits the wide use of robots. Peine et al. [22] have studied the design processes targeted at older persons and proposed to consider them as active consumers of technology instead of passive recipients. Chibani et al. [23] have reported the recent challenges and future trends on ubiquitous robotics and argued that integration of web services and ambient intelligence technologies might offer fruitful options compared with the standalone robots. Obi et al. [24] have researched ICT promoting in Japan and suggested that more efforts are needed to exploit ICT to meet the challenges among ageing population. Linner et al. [25] have developed an approach which takes into account the association between the environment and robot technology and argued that integration of service robot systems into real world has been difficult because of the separate development of the human environment and robotics systems.

In sum, the articles reviewed revealed that there is no available a web-based approach, which may support the end-users' and their stakeholders to assess the need for robots, and such an approach might be useful.

2 Methods

We inquired from 10 leading professionals who represented assistive technology units of the 10 biggest cities and hospital districts in Finland to evaluate; (a) would an approach to assess the need for personal robots be useful, (b) what kind of computer-based approaches hospital districts are using for assessing the need for robots for elderly if any, (c) what are the current procedures in selecting robots for elderly, and (d) is robots related advisory a part of their services. Those ten hospital districts cover about 90 % of Finnish population and cases, which are potential for assistive or social robots use. Professionals from 7 hospital districts replied, covering about 70 % of potential cases in Finland. Then we conducted a pilot survey among 33 wellbeing technology professionals (Median age 38, SD 7.7) and inquired them to evaluate the relevance of; (a) our approach pointed to elderly people, and (b) suggested factors in our approach. Of them, 11 had background from engineering, 15 from health and 7 from computing sciences. After that we tested validity of our approach by conducting the survey questionnaire among 64 elderly persons (Median age 70, SD 7.8), consisting both men and women. We carried out also a targeted literature search and evaluated 123 articles, which related approaches, concepts, tools, methods and variables in assessing the need for assistive or social robots, even if none of them directly presented web-based approaches.

The nature of this study was explorative and it emphasized descriptive statistics. The number of participants were too small for deeper statistical analysis such as logistic or multiple regression analyses.

2.1 Design of Approach

Based on our previous research, literature review and ICF, we created the list of seven aspects and 46 variables (Table 1), which might be relevant in assessing need for robots and for evaluating a type of robot (social and/or assistive). Aspect A concerns a user's demography and profile. Aspect B concerns a user's social relationships. Aspects C takes into account a user's overall health and an aspect D takes into account a user's functional capacity. Aspect E discusses a user's skills and learning capacity and an aspect F takes into account a user's possibilities to invest in robot. Aspect G concerns a robot's deployment environment from a technology point of view. We classified the greatest part of the factors by a Likert scale from 1 to 5, excluding one social factor (B18) where we asked "Are you living alone or with someone?" and one health factor (C25) where we asked "What is your current need for rehabilitation?" where scales were from 1 to 3. In addition, an aspect A included an age, an occupation and an interest factors, which were not classified in ICF. Section A included also a question regarding a user's interest area. The idea for inquiring the users' interest areas was that robots would deliver relevant and interactive content to users.

The basic idea behind an approach and an algorithm is that a person who has lively social life and good health and functionality does not need a robot. Correspondingly, a person who has a good health and functionality, but feels that social connections are poor, might need a social (companion) robot. For example, if a person state that he/she is able to meet a family member really often, he/she may not need a robot. If meetings are infrequently a person might need an assistive social robot (connected robot) which is able to provide him/her a connection to someone who is able to help if needed. The third option is a social robot (companion robot) regarding cases where a person is able to meet family members every so often but needs sometimes companion.

Further, a person who has limitations in health, functionality and social connections might need an assistive social robot, which is able to assist emotionally and/or physically. We have done two exceptions in the algorithm. If a person states that his/her perceived mental health is poor or very poor, a system does not recommend a robot at all. Corresponding case concerns a user's skills, attitude and learning capacity. If a person states that his/her experience of technology, robots, and computers or willingness to learn new things is poor, a system does not recommend to use a robot.

Table 1. Topics in self-evaluation

No	Topic	Method[a]	Option	ICF code
A. User's profile				
1	Gender	Select	Male/female	n/a
2	Age	Fill in	Current age	n/a
3	Interest areas	Fill in	Many options	d9204
4	Former occupation	Fill in	Many options	d859
5	Level of religiousness	Likert	Low to high	d9301
6	Perceived need for assistive robot	Likert	Low to high	e1158
7	Estimation of usage needed	Likert	Days and hours	e1158
B. Social aspect				
8	Frequency for meeting family members	Likert	Low to high	e310
9	Distance from family members	Likert	Near to far	e310

(Continued)

Table 1. (*Continued*)

No	Topic	Method[a]	Option	ICF code
10	Frequency for meeting a caregiver	Likert	Low to high	e340
11	Distance from a caregiver	Likert	Near to far	e340
12	Frequency for meeting friends	Likert	Low to high	e320
13	Distance from friends	Likert	Near to far	e320
14	Level of loneliness	Likert	Low to high	d9100
15	Level of fear	Likert	Low to high	b198
16	Level of social networks	Likert	Passive to active	d9100
17	Level of acceptance of e-monitoring	Likert	Neg. to pos.	d599
18	Type of housing	Likert	Yes/some extent/no	d699
C. Overall health aspect				
19	Perceived physical health	Likert	Poor to excellent	b7300
20	Perceived mental health	Likert	Poor to excellent	b122
21	Level of functional capacity	Likert	Poor to excellent	b7402
22	Level of body strength	Likert	Poor to excellent	b7306
23	Level of cognitive capacity	Likert	Poor to excellent	b117
24	Effect of chronic diseases on life	Likert	Modest to severe	b2800
25	Need for rehabilitation or care	Likert	No/sometimes/yes	e5800
D. Functional limitations aspect				
	Level of moving			
26	Legs	Likert	Poor to excellent	b7303
27	Hands	Likert	Poor to excellent	b7300
28	Head/neck	Likert	Poor to excellent	b7300
29	Back	Likert	Poor to excellent	b7305
30	Level of hearing	Likert	Poor to excellent	b230
31	Level of seeing	Likert	Poor to excellent	b210
32	Ability to communicate	Likert	Poor to excellent	d330
E. Skills and learning aspect				
33	Experience of technology overall	Likert	Low to high	e1250
34	Experience of robotics	Likert	Low to high	e1258
35	Experience of smart phones	Likert	Low to high	e1250
36	Experience of computers	Likert	Low to high	e1251
37	Experience of Internet and applications	Likert	Low to high	e1251
38	Programming skills	Likert	Low to high	d1551
39	Attitude towards robotics	Likert	Neg. to pos.	e498
40	Willingness to learn new things	Likert	Low to high	d198
41	Foreign language skills	Likert	Low to high	d1558
F. Economic aspect				
42	Ability to invest in assistive robotics	Likert	Little to much	d8700
43	Willingness to share a robot with others	Likert	Low to high	d998

(Continued)

Table 1. (*Continued*)

No	Topic	Method[a]	Option	ICF code
G. Operation environment aspect				
44	Easiness to move	Likert	Easy to difficult	e1550
45	Level of smart building technology	Likert	Poor to excellent	e1251
46	Access to Internet	Likert	Poor to excellent	e1250

[a] Likert scale is from 1 to 5, except in variables 18 and 25, where it is from 1 to 3

3 Results

Table 2 presents evaluation of wellbeing technology professionals regarding some variables in our approach. They considered that aspects and variables were relevant overall but some adjustments would be needed. They considered that issues regarding users' health, functional capacity and social networks were important. Major differences between distributions occurred regarding relevance to assess users' former occupations, level of religiousness and possibility to invest in robots. Minor differences occurred regarding, e.g., acceptance of remote control.

Table 2. Evaluation of some factors from wellbeing professionals' viewpoint

Assessment or evaluation of:	Unimportant (n)	Average (n)	Important (n)
Possibility to see result for need for a robot	4	12	17
Former occupations and work career	14	11	8
Level of religiousness	25	4	4
Current need for a robot	6	4	22
End-user's feeling of loneliness	3	8	20
End-user's feeling of fear	9	6	16
Acceptance of remote control of health status	0	12	19
Physical and mental health	5	3	25
Prevalence of chronic illnesses	5	8	18
Functionality of body parts	0	8	21
Functionality of senses	2	4	26
Ability to communicate	2	5	24
Former experience about technical devices	1	5	24
Attitude towards robotics	3	5	22
Possibility to invest in a personal robot	11	4	18

Note. Some replies were incomplete and sums of rows differ from 33 participants

The professionals commented and suggested the improvements to an approach. Their main criticism concerned elderly people's ability to understand self-assessment questions being associated with technical issues. In addition, they said that some questions were too general and not able to reveal the real need for robots. Here are some examples of comments from them: "*You are not asking directly what kind of robot a user might need?" "Could we ask directly how interested in are you to have a robot at home if a robot is able to entertain you?" "The questions should be more focused." "Assessment tool is too long and some technical terms are difficult to understand." "The elderly people that I interviewed were not able to understand some questions." "Some examples how to use robots in daily activities would have clarified to elderly the need for robots." "It would be better to assess the need for a robot together with a family member." "Only few elderly people can afford to buy an expensive robot." "I wouldn't ask possibility to invest in a robot. It could insult someone." "We should tell how much robots might cost; otherwise elderly people are not able evaluate possibilities to invest in." "What are the cognitive skills?" "If an elderly person does not live alone, we should ask if his/her spouse would accept a robot." "Asking only if they have fears is not adequate. We should ask what kind of fears do they have?" "The elderly people should have more information and knowledge about robots before assessing the need for."* Even if criticism was presented, the comments of professionals were positive overall. Many of them reported that assessments of need for a personal robot aroused elderly people's interest towards robots. In addition, professionals commented that variables were relevant and questions were able to give an important insight into need for robots.

Pilot study among the elderly people. We dichotomized test persons by their perceived need for a robot. Table 3 presents the most interesting variables and possible differences between the groups who stated to need robots, and those, who stated not to need robots. It is logical that those who need robots perceived that they would use robots more frequently than they who do not need robots. Both groups had rather good physical and mental health, but there were some differences in functional and cognitive capacity and ability to move body parts. Both groups had some experiences about Internet and applications, but experiences about robots were limited. However, attitudes towards robots were quite neutral, but financial resources to buy robots were poor.

Table 3. Differences in variables among two groups of elderly people

No	Variable[a]	Need for a robot		Diff.
		No (n = 49)	Yes (n = 15)	
2	Median age (SD)	69 (7.6)	75 (7.6)	−6
7	Estimation of robot usage needed	1.30	2.81	−1.51
19	Perceived physical health	3.63	3.13	0.5
20	Perceived mental health	4.00	4.00	0.00
21	Level of functional capacity	3.54	2.94	0.60
23	Level of cognitive capacity	3.89	3.27	0.62
24	Effect of chronic diseases on life	3.80	2.93	0.87
26	Level of moving legs	3.73	2.88	0.85
27	Level of moving hands	3.90	3.13	0.77
29	Level of moving back	3.57	3.00	0.57

(Continued)

Table 3. (*Continued*)

No	Variable[a]	Need for a robot		Diff.
		No (n = 49)	Yes (n = 15)	
34	Experience of robotics	1.37	1.13	0.24
35	Experience of smart phones	2.43	1.75	0.68
37	Experience of Internet and applications	2.85	2.31	0.54
38	Programming skills	1.46	1.00	0.46
39	Attitude towards robotics	3.04	3.06	−0.02
42	Ability to invest in assistive robotics	1.85	1.63	0.22
43	Interest to share a robot with others	1.98	2.31	−0.33

[a] Variables employ Likert scale from 1 (poor) to 5 (excellent), except age variable

Our approach reported that 58 % of respondents who perceived no need for robots just now, might be potential users of robot assistance anyway. Regarding another group, who perceived need for robots, the corresponding figure was 61 %, which means that in reality 39 % of them do not need robots. The biggest difference was found between type of robots needed. An approach emphasized more often assistive robots to them, who perceived to need robots, and to another group an approach could not make distinction between assistive and social robots.

Inquiry from 10 big cities. Leading professionals from seven hospital districts replied and commented that there is no any web-based method for defining and selecting assistive and/or social robots for elderly. According to them, robots are still quite unfamiliar for elderly people and cases where customers are asking any kind of robots are still limited. However, professionals found interesting if there would be a web-based method, which is able to help both them and customers. For example, one health care professional stated that "We don't have anything like that but we are definitely interested in if someone will develop a method". Another professional commented that "I suppose that elderly care professionals are not ready for selecting an assistive robot for the customers because of lack of knowledge and tools". The comments were positive overall and the professionals understood that the new era of assistive robots will break soon into market. However, they commented also that the way how they operate just now is quite traditional, and lot of introduction and training will be needed if robot technology will be embedded into health and elderly care sector.

4 Discussion

Even if there are some evidences that robots are able to activate and help the elderly people [26], we found that web-based tools for assessing the need for robots and for selecting robots are still missing even if modern ICT offers all the applications and tools for creating such a system.

Previous studies have reported that the robotics systems should be tailored according to users' needs and characteristics [27] but the challenge seems to be that we do not have robust tools for evaluating customers' needs. There are some good studies which discuss appearance, [16] acceptance, [16, 18] needs [27] and functions [28] of robots but there is still lack of methods which can be used into decision making and design processes of social and/or assistive robots.

Our approach introduces the new idea for defining the need for robots and selecting between social and assistive robots. Our approach followed, in theory, the TAM [7] and UTAUT [8] and Heerink's models [9]. Those models are well established but do not define need for robots or type of robots. Our approach is conceptually tangential with the approach of Tanaka [6], which exploits ICF [10] and suggests, what functionalities robots should include. Our approach takes into account users' social connections, deployment environment of robots and users' interest areas, which are not evaluated in the Tanaka's model. Our approach tries to figure the big picture of an elderly person's life at the moment with respect to need for a robot.

The novelty of our approach is that it assesses the need for robots and makes distinction between social and assistive robots, being based on the physical, cognitive, social, motivational, learning and economic aspects. The suggested aspects and related factors are based on our previous research and discussions with health care, wellbeing and robotics professionals. In addition, the used variables can be tracked with ICF [10]. The concept has some limitations, for sure, but it attempts to fulfill the expectations and future trends. The professionals from 7 hospital districts in Finland stated that an approach for assessing the need for robots would be important. In addition, 33 wellbeing technology professionals found an approach interesting and capable to assist in evaluating the need for robots. The pilot-test among 64 elderly persons revealed that an approach was able to find distinctions between those who needed for robots and those who did not. The distinctions were minor because both group were homogenous and participants had good physical and mental health overall. The approach presented that some people who perceived to need robots, would not need those anyway. Vice versa some people who perceived not to need robots, would need robots to some extent.

The main limitation of our study was that the number of participants was small and more research is needed to validate the approach, variables, credits and those connections between. However, our aim was to introduce a conceptual approach and we will set the credits and connections in the later stage. Creating the credit system requires lot of algorithm and software development and it is one of our future targets. Another limitation concerned the suggested variables. We are not able to guarantee the validity and importance of some variables, and the number of variables needed. However, we present the comprehensive list of variables, which might be relevant.

We state that our approach might be an important step for advancing social robotics and it gives good insight for people who operate in elderly care or robotics sectors. The feedback from elderly care and technology professionals strengthened that the web-based system for defining the need for robots is still missing but very welcome.

References

1. COST: The future concept and reality of social robotics. Challenges, perception and applications: role of social robotics in current and future society. COST event, Brussels, Belgium (2013)
2. NIH: National Institute on Aging. Health and Ageing (2015). https://www.nia.nih.gov/research/publication/global-health-and-aging/humanitys-aging
3. Herbert, L., Scherr, P., Bienias, J., Bennett, D., Evans, D.: Alzheimer's disease in the U.S. population: prevalence estimates using the 2000 census. Arch. Neurol. **60**, 1119–1122 (2003)
4. Comas-Herrera, A., Wittenberg, R., Pickard, L., Knapp, M.: Cognitive impairment in older people: future demand for long-term care services and the associated costs. Int. J. Geriatr. Psychiatry **22**(10), 1037–1045 (2007)
5. euRobotics aisbl: Strategic Research Agenda for Robotics in Europe 2014–2020, Applications: societal challenges, pp. 59–64 (2014). http://www.eu-robotics.net/cms/upload/PPP/SRA2020_SPARC.pdf
6. Tanaka, H., Yoshikawa, M., Oyama, E., Wakita, Y., Matsumoto, Y.: Development of assistive robots using international classification of functioning, disability, and health: concept, applications, and issues. J. Robot. ID 608191, 12 pages (2013). doi:10.1155/2013/608191
7. Davis, F.: Perceived usefulness, perceived ease of use, and user acceptance of information technology. MIS Q. **13**(3), 319–340 (1989)
8. Venkatesh, V., Morris, M., Davis, G., Davis, F.: User acceptance of information technology: toward a unified view. Manage. Inf. Syst. Q. **27**(3), 425–478 (2003)
9. Heerink, M., Kröse, B., Evers, B., Wielinga, B.: Assessing acceptance of assistive social agent technology by older adults: the almere model. Int. J. Soc. Robot. **2**(4), 361–375 (2010)
10. WHO 2002: Towards a Common Language for Functioning, Disability and Health ICF. The International Classification of Functioning, Disability and Health. World Health Organization, Geneva (2002)
11. Flandorfer, P.: Population ageing and socially assistive robots for elderly persons: the importance of sociodemographic factors for user acceptance. Int. J. Popul. Res. ID 829835, 13 pages (2012). doi:10.1155/2012/829835
12. Scopelliti, M., Giuliani, M., Fornara, F.: Robots in a domestic setting: a psychological approach. Univ. Access Inf. Soc. **4**(2), 146–155 (2005)
13. Arras, K., Cerqui, D.: Do we want to share our lives and bodies with robots? Technical report 0605-001, Swiss Federal Institute of Technology, Lausanne, Switzerland (2005)
14. Tong, G., Gu, J., Xie, W.: Virtual entity-based rapid prototype for design and simulation of humanoid robots. Int. J. Adv. Robot. Syst. **10**, 291 (2013). doi:10.5772/55936
15. Meng, Q., Lee, M.: Design issues for assistive robotics for the elderly. Adv. Eng. Inf. **20**(2), 171–186 (2006). doi:10.1016/j.aei.2005.10.003
16. Wu, Y.-H., Fassert, C., Rigaud, A.-S.: Designing robots for the elderly: appearance issue and beyond. Arch. Gerontol. Geriatr. **54**(1), 121–126 (2012). doi:10.1016/j.archger.2011.02.003
17. Saborowski, M., Kollak, I.: How do you care for technology? – care professionals' experiences with assistive technology in care of the elderly. Technol. Forecast. Soc. Change **93**, 133–140 (2014). doi:10.1016/j.techfore.2014.05.006
18. de Graaf, M., Ben Allouch, S., Klamer, T.: Sharing a life with Harvey: exploring the acceptance of and relationship-building with a social robot. Comput. Hum. Behav. **43**, 1–14 (2014). doi:10.1016/j.chb.2014.10.030

19. Alaiad, A., Zhou, L.: The determinants of home healthcare robots adoption: an empirical investigation. Int. J. Med. Informatics **83**(11), 825–840 (2014). doi:10.1016/j.ijmedinf.2014. 07.003
20. Michaud, F., Boissy, P., Labonté, D., Brière, S., Perreault, K., Corriveau, H., Grant, A., Lauria, M., Cloutier, R., Roux, M.-A., Iannuzzi, D., Royer, M.-P., Ferland, F., Pomerleau, F., Létourneau, D.: Exploratory design and evaluation of a homecare tele-assistive mobile robotic system. Mechatronics **20**(7), 751–766 (2010). doi:10.1016/j.mechatronics.2010.01. 010
21. Andrade, A., Pereira, A., Walter, S., Almeida, R., Loureiro, R., Compagna, D., Kyberd, P.: Bridging the gap between robotic technology and health care. Biomed. Signal. Process. Control **10**, 65–78 (2014). doi:10.1016/j.bspc.2013.12.009
22. Peine, A., Rollwagen, I., Neven, L.: The rise of the "innosumer" - rethinking older technology users. Technol. Forecast. Soc. Change **82**, 199–214 (2014). doi:10.1016/j. techfore.2013.06.013
23. Chibani, A., Amirat, Y., Mohammed, S., Matson, E., Hagita, N., Barreto, M.: Ubiquitous robotics: recent challenges and future trends. Robot. Auton. Syst. **61**(11), 1162–1172 (2013). doi:10.1016/j.robot.2013.04.003
24. Obi, T., Ishmatova, D., Iwasaki, N.: Promoting ICT innovations for the ageing population in Japan. Int. J. Med. Inform. **82**(4), e47–e62 (2013). doi:10.1016/j.ijmedinf.2012.05.004
25. Linner, T., Pan, W., Georgoulas, C., Georgescu, B., Güttler, J., Bock, T.: Co-adaptation of robot systems, processes and in-house environments for professional care assistance in an ageing society. Procedia Eng. **85**, 328–338 (2014). doi:10.1016/j.proeng.2014.10.558
26. Tapus, A., Matarić, M., Scassellati, B.: The grand challenges in socially assistive robotics. IEEE Robot. Autom. Mag. **14**(1), 35–42 (2007)
27. Wu, Y.-H., Wrobel, J., Cristancho-Lacroix, V., Kamali, L., Chetouani, M., Duhaut, D., Le Pevedic, B., Jost, C., Dupourque, V., Ghrissi, M., Rigaud, A.-S.: Designing an assistive robot for older adults: the ROBADOM project. IRBM **34**(2), 119–123 (2013). doi:10.1016/j. irbm.2013.01.003
28. Leroux, C., Lebec, O., Ben Ghezala, M., Mezouar, Y., Devillers, L., Chastagnol, C., Martin, J.-C., Leynaert, V., Fattal, C.: ARMEN: assistive robotics to maintain elderly people in natural environment. IRBM **34**(2), 101–107 (2013)

Predicting the Intention of Human Activities for Real-Time Human-Robot Interaction (HRI)

Vibekananda Dutta[✉] and Teresa Zielinska

Faculty of Power and Aeronautical Engineering (MEiL),
Warsaw University of Technology, ul. Nowowiejska 24, 00-665 Warsaw, Poland
{vibek,teresaz}@meil.pw.edu.pl

Abstract. The modeling methodology of Human-object relation is needed for human intention recognition in assistive robotics. When helping the elderly, the future action prediction is an essential task in real-time human-robot interaction. Since the future actions of humans are ambiguous, robots need to carefully conclude about the appropriate action. This requires a mathematical model to evaluate all possible future actions, corresponding probabilities and the possible relation between humans and the objects.

Our contribution is the modeling methodology for the human activities using the probabilistic state machine (PSM). Not only the objects, but also latent human poses and the relationships between humans and the objects are here considered. The probabilistic model allows uncertainties and variations in the object affordances. In experiments, we show how the intention recognition w.r.t. drinking activity is analysed using our approach.

Keywords: Intention · Motion · State machine · Human-robot interaction (HRI) · Intention recognition · Service robot

1 State of the Art

Predicting the human motion intentions is an emerging field in robotic research. The researchers are using for this purpose the different probabilistic methods, like the probabilistic predictors [12], Hidden Markov series [1] or the probabilistic automatons with transitions described by the concatenation of probabilistic function distributions [4–6]. In this paper, we consider the probabilistic state machine approach for action selection w.r.t. predicting the human motion intentions in real-time HRI. As the example the situations where there is the person and the objects are nominally being manipulated were considered.

2 Introduction

The robots operating in unstructured environment must be capable of interacting with humans. Although social interaction between robots and humans is

© Springer International Publishing AG 2016
A. Agah et al. (Eds.): ICSR 2016, LNAI 9979, pp. 723–734, 2016.
DOI: 10.1007/978-3-319-47437-3_71

presently rather simple, one of the main goals of social robotics is to develop robots that can function in complicated settings such as homes, offices, and hospitals. To achieve this goal, robots must be capable of recognizing the intentions of the humans with whom they are supposed to interact. This presents both opportunities and challenges to researchers developing intent-recognition techniques [3].

Predicting human motion intention is a difficult problem combining the expertize in the area of robotics and artificial intelligence (AI). It involves the use of cognitive capabilities, e.g. perception, reasoning, prediction, learning and planning, etc. We need to extract the semantics of the observed behaviour [9]. The goal is to create such capabilities for the robots, so they can perform the human service tasks and can help human being with everyday activities.

The contributions of this paper are as follows:

- A method with improved resolution and dispersion for intention recognition is proposed;
- It is described how to conclude about human activity by analysing the sequence of elementary actions.

The paper is organized as follows. We start with the proposed method in Sect. 3. We present the experimental results along with real-time demonstrations in Sect. 4 and finally conclude the paper in Sect. 5.

3   Proposed Method and Its Formulation for the Designing of the Intentions Model

In this section, we summarize our method and explain how we anticipate human motion intention. Motion intention means, in other words, the answering what a human will do next with given the current observation of his pose and the surrounding environment. Since activities happen over a certain period of time, each activity is composed of different actions involving different number of objects, For example drinking activity can have actions like, moving the hand next to the glass, grasping and moving the glass near to the mouth, etc. for example, robot observes a scene containing a human and objects for time t_i in the past, and its goal is to anticipate future possibilities for time t_{i+1}. However, for the future t_{i+1} frames, we do not even know the sequence of the states, there may be also a different number of objects being interacted with, depending on which action will be performed in the future. The goal is to evaluate the connection between human-object relation and the possible states (i.e., actions, human poses and object locations). Our contribution in this paper is to establish the state machine by augmenting several possible state structures in a given time, which we named as spatio-temporal based probabilistic state machine (**STPSM**) shown in Fig. 1. Here for the state $S(t_i)$ actions $a_1...a_n$ are possible. For the state denoted by $S_{an}(t_2)$ the possible actions are $b_1 \ .... \ b_n$ respectively.

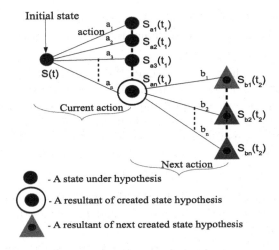

Fig. 1. Generation and weighting of potential action hypotheses, the lines represent the action that lead to resultant state of that action, i.e., encircled dot and triangulated dot. Let us denote by S is the distinct state, where t_i is the time and the actions are denoted by a_i and b_i. Idea adopted from [4]

3.1 Modeling Spatio-Temporal Relations Using Probabilistic State Machines (STPSM)

Contextual information have been effectively considered as a rich context in human-object interactions and object-object interactions. Human-object interactions analysis using videos and images are performed in [13] by fusing action contexts basis on object properties, object reaction to the action and features of human manipulation motions. A contextual model is proposed to capture geometric configuration of objects and human pose for recognizing human activities. In order to train and test human action recognition approaches, we recorded typical actions of human working in our university office environment. Our goal is to define the object affordance using RGB-D videos and considering the human motion trajectories. In the beginning, we created state machine by hand representing explicitly/implicitly different possible human actions with the same weight, what is the equivalent of the same probability of human intentions. An observation was made and the human actions together with states were extracted as it is shown in Fig. 1. The weights of the states were updated and normalized so that they were adding up to 1. The further text will give the explanation how the weights are established.

The probability of a state in a state machine is directly related to the observation. The weights are assigned to the actions considering the distances between end-effector and the objects. The value of the weight is inversely proportional to the distance. The actions with higher weights have a greater likelihood. After each action, with weight update the important data necessary for human-object interactions were also determined (e.g., the pointed bottle to be grasped or the pointed location to place the bottle). The advantage of describing the actions by transitions in probabilistic state machines is that, the human intention changes can be easily handled.

Definition 1. Probabilistic state machine (PSM) is defined as a generalization of non-probabilistic automata where the transitions are not just a single transitions but probabilistic functions (i.e. probability of choice between reachable states). The representation of PSM can be made in the form of formal mathematical description. In the case of formal mathematical description the elements of the tuple are given as:

$$(S, \sum, \phi, i_{St}, f_{St}) \tag{1}$$

where $\sum$ is the set of input symbols, S is the set of states, where $\phi : S \times \sum \rightarrow P(S)$ is the transition function (action). i_{St} and f_{St} defines the initial and final state respectively.

Definition 2. The representation of PSM can be also graphical form. Graphical representation can be useful to analyze the connection between the states and helps to understand the properties of state machine (S_m) shown in Fig. 2. Each state (S) is a circle, the transition is represented by arrowed lines. Each line represents an action (a) with its probability (defined by weights), and it starts from actual state.

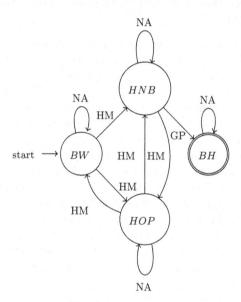

Fig. 2. The state machine for recognition of elementary actions (picking the bottle filled with water). Details of all the notation is given in Table 1.

Table 1. Statement of states and actions that are recognized by the system

State (S)	Description	Weight assigned
BW	Bottle with water	BW + HOP:0.1, BW + HNB:0.2
HNB	Hand near bottle	HNB + BH:0.5, HNB + HOP:0.2
HOP	Hand in other place	HOP + BW:0.1, HOP + HNB:0.2
BH	Bottle in hand	–
Action (a)	**Description**	
HM	Hand moving	–
GP	Grasping	–
NA	No action	–

3.2 Spatio-Temporal Based Probabilistic State Machine Approach for Intention Modeling

The goal is to build an intention's representation that for each object contains, the likelihood of all actions that can be performed with. The states in a state machine are connected by branches representating the actions. Let's propose that we have a state machine denoted by S_m. Let S_i denotes number of states (e.g. the bottle filled with water placed on the table, bottle in hand). We define the so called attention-map (on the other words the heat-map) for a time instant t denoting them by h_1^t, h_2^t, ..., h_N^t for N objects. Action variable is denoted by a^t. We are using the locations (including positions and orientations) of the objects O_L^t and the human pose H_p^t as observation. Formally, we are interested in estimating the states, based on the given context and a sequence of observations, we define the most possible intention.

Formally, we write,

$$S_m = \{S_1, S_2, S_3, ..., S_N\} \tag{2}$$

where a state machine has N number of states. We denote each state with three attributes, e.g. object's location, object's attributes and human pose.

$$S_i = \{O_L^t, H_p^t, O_{Ar}^t\} \tag{3}$$

Based on these observations, we can compute the probability of reaching the state,

$$p_{ij}(S_i^t | a_j^t) \tag{4}$$

Equation 4, describes the probability of reaching a state depends on the action is being performed. The most possible is the next state which is reached by the most possible action (motion intention).

$$p(S_{next}) = p(S_i | _{max} a_j) = {_{max}} p(h_1^t, h_2^t, ...h_N^t | O_L^t, H_p^t) \tag{5}$$

In Eq. 5, we define how the attention-maps (heat-maps) are introduced over an action by observing human manipulation with an object at a time instant t. This

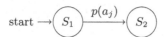

Fig. 3. Action probability to reach a state from previous state.

approximation suggests a method for determining the most possible intention for a given series of states based on context. For each possible intention, we compute the probability $P_{ij}(S_i^t|a_j^t)$ and choose as possible intention that one which has the greatest probability (Fig. 3).

3.3 Human Pose Estimation

By the term "pose" we refer to the range of human mannerisms while executing an action. In this paper, the perception of human movements is limited to the analysis of body posture and motion. In most of the literature, the perception of whole-body kinematics is based on motion capture technologies, such as wearable sensors, markers and instrumented platforms. In order to make our system compatible with real-life applications, we didn't use such kind of systems. We chose to test our system on a real-time data with a markerless approach using a *Kinect* [8] sensor and the *Nite 2.2* [10] software package. The inaccuracy of the sensing technique and of the skeleton pose detection algorithm [10] (especially in the case of large occlusions of some parts of the body which often occur in real-time activity analysis) is confronted and properly fixed using the skeleton tracking output as a physical model of the observed human body and the dynamic 3D visualization tool 'RVIZ' by Robot Operating System (ROS) package [11] (Fig. 4).

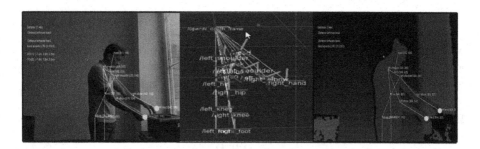

Fig. 4. From left to right: RGB image (ground truth) with the accurate skeleton detected, 3D visualization of skeleton model in RVIZ, and depth image with skeleton detected. (Color figure online)

The recent availability of inexpensive RGB-D sensors have enabled significant improvements in scene/environment modeling and estimation of human poses. This, together with depth information, has enabled some recent works to obtain good action recognition performance [6]. We use the following notation to express the locations of all n body joints in a pose vector as,

$$H_p^t = (J_1^t, J_2^t, ..., J_i^t), i \in 1, 2, ..., n \tag{6}$$

Where J_i contains x and y coordinates of the ith joint at time t and the depth information.

3.4 Object Affordance Attention-Maps

In order to forecast the possible trajectories, there is required to model the human-object relationships. These relations are called 'object affordances attention-map', which is also addressed as a heat-map. We model such affordances which are relevant for path planning and we refer to them as 'planning affordances'. For example, a glass of water has a drinkable affordance and therefore the space between the human's hand and the glass is relevant to the drinking activity. Since a 'drinkable' label by itself is not informative enough to help in planning, we ground it to a spatial distribution of object affordance. We consider the planning affordance for several actions (e.g., drinking, placing, moving, reaching, etc.). For each action we model object affordances using a potential function Υ_h based on human-object angular location and distance.

$$\Upsilon_h = \begin{cases} \Upsilon_{(h,ang_{(H,O)})}\Upsilon_{\alpha,\beta} \\ \text{if } h \in \text{ actions between human and object at distance} \\ \Upsilon_{(h,ang_{(H,O)})}\Upsilon_{(h,dist_{(H,O)})} \\ \text{if } h \in \text{ actions between human and object in close proximity} \end{cases} \tag{7}$$

The affordance varies with the distance and angle between the human and the object. We parameterize the cost as follows [7]:

Angular preference $\Upsilon_{h,ang_{(H,O)}}$: certain angular positions of the human versus the object are more relevant for certain activities. We assume that the angular preference $\Upsilon_{(h,ang,H)}$ considering human and the object is described by *von-Mises* distribution:

$$\Upsilon_{h,ang_{(H,O)}}(X_{t_i}; \mu, k) = \frac{1}{2\pi I_O(k)} exp(k\mu^T X_{t_i}) \tag{8}$$

In the above equation, μ and k are parameters that are selected by trials and errors, and X_{t_i} is a two-dimensional unit vector. we obtain X_{t_i} by projecting the hand waypoint into the reference co-ordinate frame.

Distance preference $\Upsilon_{h,dist_{(H,O)}}$: we define this by 1D-Gaussian function which is parameterized by a mean and variance, and centered at human [5].

Edge preference $\Upsilon_{\alpha,\beta}$: we consider that the object is in some distance from the human, the preferences vary along the edge connecting a human and an object (see Fig. 5). To calculate this cost for the current waypoint, we first take the distance d_{h-o} between that waypoint and the object and project it vertically to

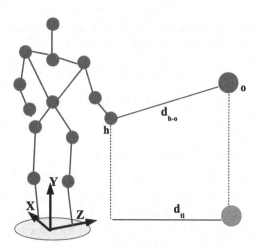

Fig. 5. Local co-ordinate system of human's hand position for reaching action.

the horizontal plane obtaining d_{t_i}. The normalized distance is $\bar{d}t_i = d_{t_i}/d_{h-o}$. In the equation below, α and β are selected by trial and errors.

$$\Upsilon_{\alpha,\beta}(\bar{d}t_i, \alpha, \beta) = \frac{\bar{d}t_i^{\alpha-1}(1 - \bar{d}t_i)^{\beta-1}}{B(\alpha, \beta)}, \bar{d}t_i \in [0,1] \tag{9}$$

Our affordance representation consists of three distributions, namely: Gaussian, von-Mises and Beta distribution. $B(\alpha, \beta)$ is defined as a beta function in Eq. 9. We generate attention-map (heat-map) on each affordance by scoring the points in the 3D space using the potential function (Eq. 7), and the value represents the strength of the particular affordance at that location. Figure 6 shows the attention-map generated for the pourable affordance.

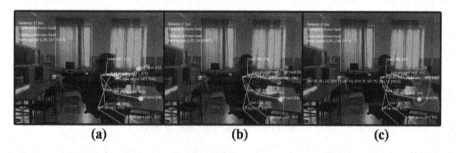

(a)   (b)   (c)

Fig. 6. Distribution of attention-map and trajectory to goal location. The figure shows the motion intention based on object affordance: (a) hands are in contact with object (b) yellow circle with combination of green and red surrounding indicates object affordance, e.g. pourability, and (c) in this figure, red color defines attention-map (heat-map) and the black trajectory between red and yellow circle signifies the goal location for a given affordance. (Color figure online)

3.5   Trajectory Generation

Object can be approached by various types of motion trajectories depending on the action which is being performed. In this section, we describe motion trajectories that we consider having in mind the need of establishing the object affordance functions. We use the Bezier curves for estimating the human hand motion [2].

$$T(x) = (1 - x)^3 L_0 + 3(1 - x)^2 x L_1 + 3(1 - x)x^2 L_2 + x^3 L_3, x \in [0, 1] \quad (10)$$

Such a cubic Bezier curve is parameterized by a set of four points: the start and end point of the trajectory (L_0 and L_3), and two control points (L_1 and L_2) which define the shape of the curve.

4   Experiments

In order to experimentally evaluate the proposed method and the intention modeling, we performed experiments using the RGB-D video. We first evaluated our approach considering the drinking task. We used the OpenNI skeleton tracker, to obtain the skeleton poses in these RGBD videos. We then obtained the ground-truth object labels using PCL-based object tracking. The proposed method is publicly available (along with open-source code): https://github.com/Vibek/Human_intention.

We divided each activity into three groups: action, human pose and object affordance. Drinking task was considered, with:

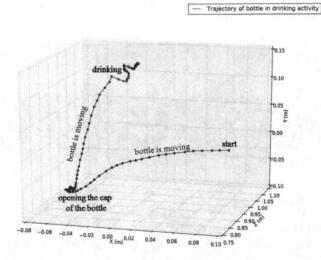

Fig. 7. The 3D trajectory of the object's position when performing drinking activity from the bottle filled with water. The graph starts with hand moving near to the bottle, grasping the bottle and then slowly moving the bottle towards mouth w.r.t. drinking water.

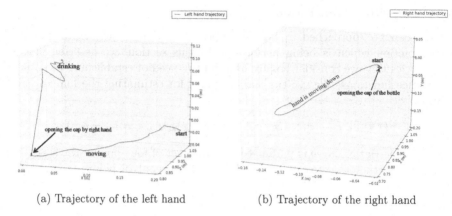

(a) Trajectory of the left hand   (b) Trajectory of the right hand

Fig. 8. Comparison experiment of both hands position trajectory during drinking activity.

- **Action**:{*reaching, moving, pouring, drinking*};
- **Object affordance**:{*reachable, movable, pourable, drinkable*};
- **Human pose**:{*pointing to an object, presenting an object, sitting, standing, searching*}.

We also perform quantitative analysis over drinking activity based on 3D arbitrary trajectory, the result is shown in Fig. 7. In Fig. 8a and 8b, the hand positions were shown when performing drinking activity. In this paper, joints' coordinates X and Y were measured in pixels and Z is the distance (depth) between joint and the sensor. All the values are converted in to meters. The point $(0,0)$ is the upper left corner of the image.

Fig. 9. Figure is showing the attention-map of anticipated trajectories for actions (i.e. reaching, pouring and drinking) involving object affordances and the trajectories evolvement over the time. (Color figure online)

The experiments demonstrated that the applied software system correctly adjust the affordance function. Figure 9 shows example images with visualization of the recognition process. The blue circle indicates the hand grasping the object and the green color denotes the corresponding affordance is active on the object is being interacted and red curve with green and yellow outline along with black trajectory describes activity zones in which anticipation had been recognized.

5   Conclusion

In this work, we described the problem of intention recognition using probabilistic state machine (PSM), so that a robot can perform planning of its reactive responses. We described the human-object relationships, human actions and object affordance w.r.t. intention modeling. That allows to conclude about the future possible motion scenarios. The situations where human does not engage or engage with the objects in order to recognize the intention through scene information and human actions must be analysed for motion intention prognosis. In this paper, the techniques for estimating and evaluating the most likely future scenarios is summarized.

Acknowledgments. Authors of this work were supported by the Dean's grant (Grant No 504/02673/1132/42.000100) funded by Warsaw University of Technology. Moreover, it is also supported by the HERITAGE project (Erasmus Mundus Action 2 Strand 1 Lot 11, EAECA/42/11) funded by the European Commission.

References

1. Bandyopadhyay, T., Won, K.S., Frazzoli, E., Hsu, D., Lee, W.S., Rus, D.: Intention-aware motion planning. In: Frazzoli, E., Lozano-Perez, T., Roy, N., Rus, D. (eds.) Algorithmic Foundations of Robotics X. STAR, vol. 86, pp. 475–491. Springer, Heidelberg (2013). doi:10.1007/978-3-642-36279-8_29

2. Faraway, J.J., Reed, M.P., Wang, J.: Modelling three-dimensional trajectories by using bézier curves with application to hand motion. J. Royal Stat. Soc.: Seri. C (Appl. Statist.) **56**(5), 571–585 (2007)

3. Gopnik, A., Slaughter, V., Meltzoff, A.: Changing your views: How understanding visual perception can lead to a new theory of the mind. Children's early understanding of mind: Origins and development, pp. 157–181 (1994)

4. Jansen, B., Belpaeme, T.: A computational model of intention reading in imitation. Robot. Autonom. Syst. **54**(5), 394–402 (2006)

5. Jiang, Y., Koppula, H., Saxena, A.: Modeling 3d environments through hidden human context (2015)

6. Koppula, H.S., Gupta, R., Saxena, A.: Learning human activities and object affordances from rgb-d videos. Int. J. Robot. Res. **32**(8), 951–970 (2013)

7. Koppula, H.S., Saxena, A.: Physically grounded spatio-temporal object affordances. In: Fleet, D., Pajdla, T., Schiele, B., Tuytelaars, T. (eds.) ECCV 2014. LNCS, vol. 8691, pp. 831–847. Springer, Heidelberg (2014). doi:10.1007/978-3-319-10578-9_54

8. Microsoft. Kinect (2015). https://developer.microsoft.com/en-us/windows/kinect. Accessed 11-Feb-2015
9. Ramirez-Amaro, K., Beetz, M., Cheng, G.: Automatic segmentation and recognition of human activities from observation based on semantic reasoning. In: 2014 IEEE/RSJ International Conference on Intelligent Robots and Systems (IROS 2014), pp. 5043–5048. IEEE (2014)
10. Sense, P.: Openni skeleton tracking (2015). http://openni.org. Accessed 19 Nov 2015
11. O. soure robotics foundation. RVIZ (2015). http://wiki.ros.org/rviz. Accessed 19 Nov 2015
12. Stulp, F., Grizou, J., Busch, B., Lopes, M.: Facilitating intention prediction for humans by optimizing robot motions. In: 2015 IEEE/RSJ International Conference on Intelligent Robots and Systems (IROS), pp. 1249–1255. IEEE (2015)
13. Yuan, F., Prinet, V., Yuan, J.: Middle-level representation for human activities recognition: the role of spatio-temporal relationships. In: Kutulakos, K.N. (ed.) ECCV 2010. LNCS, vol. 6553, pp. 168–180. Springer, Heidelberg (2012). doi:10.1007/978-3-642-35749-7_13

The ENRICHME Project: Lessons Learnt from a First Interaction with the Elderly

Roxana Agrigoroaie[(✉)], François Ferland, and Adriana Tapus

Robotics and Computer Vision Lab, U2IS, ENSTA-ParisTech,
828 bd des Marechaux, 91762 Palaiseau Cedex, France
{roxana.agrigoroaie,francois.ferland,adriana.tapus}@ensta-paristech.fr

Abstract. The main purpose of the ENRICHME European project is to develop a socially assistive robot that can help the elderly, adapt to their needs, and has a natural behavior. In this paper, we present some of the lessons learnt from the first interaction between the robot and two elderly people from one partner care facility (LACE Housing Ltd, UK). The robot interacted with the two participants for almost one hour. A tremendous amount of sensory data was recorded from the multi-sensory system (i.e., audio data, RGB-D data, thermal images, and the data from the skeleton tracker) for better understanding the interaction, the needs, the reactions of the users, and the context. This data was processed offline. Before the interaction between the two elderly residents and the robot, a demo was shown to all the residents of the facility. The reactions of the residents were positive and they found the robot useful. The first lessons learnt from this interaction between Kompaï robot and the elderly are reported.

Keywords: Human-robot interaction · Elderly · Social robots

1 Introduction

In the context of a worldwide aging population [9], the need to find solutions for helping the elderly and enhancing their quality of life arises. One of the solutions that received much attention in recent years is the use of social robots for personalized care. Social robots can be good companions for the elderly as they can provide monitoring [2], they can alert their families or caregivers in case of emergencies [8], and they can stimulate their cognitive functions [17,19].

ENRICHME is a EU Horizon2020 project whose purpose is to develop a socially assistive robot for the elderly with mild cognitive impairment (MCI). The project aims at developing a robotic system that can help elderly to remain independent longer and enhance their everyday life. The system is designed to learn from past experiences as it contains an episodic-like memory [12]. Moreover, it should be able to adapt its behavior to the user's profile. More specifically, it should take into consideration its personality, cognitive disability level, emotional state, and preferences.

© Springer International Publishing AG 2016
A. Agah et al. (Eds.): ICSR 2016, LNAI 9979, pp. 735–745, 2016.
DOI: 10.1007/978-3-319-47437-3_72

In the past few years, many socially assistive robots have been developed [2,5,13,18,20]. However, the ENRICHME system uses both the robot and the intelligent environment equipped with sensors in order to get information on the context, the user, and the potential needs of the user. Furthermore, it is meant to be a personal robot that lives with its owner. This might be a first sustainable solution for a long-term interaction between a robot and the elderly in their own environment. During the testing phase, the robot will stay in the houses of the elderly for a period of 4 months. The total testing time for the platform is 1 year, with 9 robots in 3 testing sites across Europe (Italy, Poland, and Greece).

The current work summarizes some of the lessons learnt from a first interaction with the elderly from LACE Housing Ltd, an elderly care facility in the UK. This being the first interaction between the possible final users and the robot, only two elderly individuals were selected to interact with the robot and to test the modules that were created so far.

This paper is structured as follows. Section 2 presents a description of the robot and the interaction scenario. Section 3 describes how the data was recorded and analyzed. Section 4 gives the results from the interaction and lists lessons that were learnt, and finally Sect. 5 concludes the paper and offers a perspective on future work.

2   Interaction Scenario

ENRICHME project uses the Kompaï robot designed by Robosoft[1] (one of the project's partners). The robot Fig. 1 is equipped with an ASUS XTion Pro RGB-D sensor mounted beneath its head, an Optris PI450 thermal camera mounted on the head, and different sensors for measuring environmental parameters (e.g. ambient temperature). The torso of the robot features a touchscreen to facilitate user interaction. The robot can also use speech recognition and synthesis to communicate.

Fig. 1. The robot in one of the user's home

For this first interaction, one of the use cases developed for the project has been implemented. In this use case, the activity sensors of the robot and the

---

[1] www.robosoft.com.

sensors located in the house detect that the person is restless, and the robot approaches the person to propose an activity for them to do. The possible activities are: playing a cognitive game, doing some physical exercises, or learning some healthy eating tips.

For this scenario, the robot uses a leg detection algorithm by using the laser range finder mounted on the robot [3] so as to track people in the room. Based on data from an online survey, where participants would define approach parameters (i.e., stopping distance, deceleration, and curvature of the path) according to the robot's personality, the robot was configured to approach the person in a submissive and friendly manner. The robot automatically approaches the closest person it detects, and keep itself oriented toward that person at a maximum distance of 1.0 m. Therefore, even if the person is moving in the room, the robot is able to continuously follow the person.

3   Interaction, Data Recording and Analysis

A face recognition module based on [1] was trained with the images of the two residents. Therefore, the robot was able to recognize them and greet them by their name when they were in range. Next, a web-based graphical user interface (see Fig. 2) was displayed on the touch screen of the robot. The font size used throughout the interface was between 4 mm (for the timer and some instructions) and 9 mm (used for titles).

The cognitive games available on the robot that were developed in collaboration with the medical staff from the care facility are as follows:

- Digit cancellation: At the beginning of the cognitive game, the user is given a random digit. The purpose of the game is to find all occurrences of that digit in a random list of 40 digits in less than 30 seconds;
- Puzzle: It contains a jig-saw puzzle with three difficulty levels: easy with 12 pieces, medium with 30 pieces, and hard with 54 pieces;
- Hangman: The user has to guess a word randomly selected from a list divided into five categories: fruits, vegetables, cities, objects in the house, and animals.
- Memory game: The user is shown a series of 4 images for 10 s. After being shown a blank screen for 5 seconds, a new image is shown, and the user has to say if the current image is part of the previous ones or not. The game consists of 5 rounds.

Feedback was given after each game, summarizing the user's performance based on the hints used, errors made, total game time.

The physical exercises activity is made of three parts, each with a duration of 1 min. In the first part, the robot uses only speech to tell the user what exercise to do. In the second part, the user is only shown images of the exercises they have to do. In the third part, the robot uses both speech and images for the exercises. The exercises were randomly chosen from a set of 8 possible arm exercises shown in Fig. 3. These exercises were selected in collaboration with

Fig. 2. The developed interface with the potential activities

the physical therapist available in the care facility. We assure ourselves that the participants were able to perform these physical exercises. After the execution of the physical exercises, the user is informed about the number of times they correctly performed each exercise.

The third proposed activity is the health information tips. The robot presented the user 8 healthy eating information such as not skipping breakfast or eating less refined sugars [6]. At the end, the amount of information retained was tested with a short quiz.

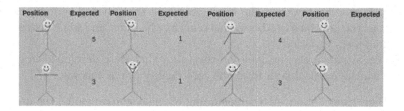

Fig. 3. The physical exercise feedback

After the interaction, each participant was given two questionnaires: a personality questionnaire, with 46 questions, based on the Big Five personality factors [7], and a custom made questionnaire, with 58 questions, developed for finding out what the participants liked or disliked about the graphical interface and the overall robot behavior. Each questionnaire took between 5 and 10 min to complete and consisted of yes/no questions or questions with a scale from 1 (strongly disagree) to 5 (strongly agree).

Throughout the interaction between each participant and the robot, the data from multiple sensors was recorded. For human robot interaction, the data sources used are the RGB-D data, the thermal image, the audio, and the data from the skeleton tracker. For the audio an external omni-directional microphone was used, and the skeleton tracker used the OpenNI NiTE library with the ASUS XTion Pro sensor. The data from the skeleton tracker was used for determining if the elderly executed correctly the physical exercises. Except for audio, the Robot Operating System (ROS) framework [15] was used for robot behavior coordination, data recording and analysis.

For compactness, the following notation is used: Pi for participant i (1 or 2), Ej for event j (1 to n), and PiEj for the jth event that was recorded for Participant i. The possible events are: performing the physical exercises, playing one of the cognitive games. Participant P1 is a 73 years old introverted male who suffers from Parkinson's disease and has cerebellar ataxia (the inability to coordinate balance, gait, extremity, and eye movements). Participant P2 is a 83 years old extroverted female and has arthritis. Both participants started with the health tips, continued with the cognitive games, and ended with the physical exercises. The total recording times for P1 and P2 are 37 and 29 min, respectively. Out of the total interaction time, two events for each participant were chosen for further analysis. These events were chosen because both participants showed a strong reaction towards something they disliked or liked. Before the interaction started with the two participants, there was a short demo for all the residents of the care facility. One of the games that was tried during the demo was the puzzle game, which was found interesting, but difficult by most of the residents. The level of the game shown in the demo was the medium one, with 30 puzzle pieces. Just by seeing them they thought that they would not be able to complete the puzzle. When participant P1 saw the puzzle game among the possible cognitive games, there was a strong reaction: he found it too difficult and did not want to try it. Data during this reaction was chosen to be P1E1. P1E2 is represented by the last two minutes from the physical exercises activity. The first minute of the physical exercise was not chosen due to a lot of movement of participant P1. There was nothing shown on the screen during this time, therefore he turned towards the other people in the room.

Participant P2 firmly expressed a strong dislike towards quizzes, therefore at the announcement of having to complete a quiz after the healthy eating tips a strong reaction was observed. Therefore, this was considered as P2E1. P2E2 consisted of the participant playing the puzzle game. As P1E1 consisted of a strong reaction on a game, for P2 was extracted an event related to a game in order to see if the temperature variation is similar for both participants P1 and P2, respectively.

We wanted to investigate if facial temperature variations, facial expressions, and speech variations could be observed while the participants reacted to these events, and thus could be recognized by the robot in the future.

Participant P1 wore glasses throughout the interaction, during the memory game he changed his original glasses with another pair to better see the text on the screen. Participant P2 started the interaction without glasses, but during the puzzle game glasses were put on. During both interactions with the robot, other people were present in the room, and sometimes interacted with the participants. This resulted in situations such as a loss of facial temperature data because the participant turned his/her head, or voices other than the participants' being recorded by the external microphone. These events were not discouraged by the experimenters as it was believed that they would provide realistic information on situations that were certainly going to re-occur in future tests.

The following subsections describe the data that was recorded by each sensor.

3.1 Action Units

RGB-D data was used both for offline monitoring of the experiments and facial expression analysis. From the four events, facial action units (AU) [4] were extracted. The extracted AUs are: AU1 (inner brow raiser), AU2 (outer brow raiser), AU4 (brow lowerer), AU5 (upper lid raiser), AU6 (cheek raiser), AU7 (lid tightener), AU12 (lip corner puller), AU15 (lip corner depressor), and AU25 (lips part). Using these AUs, several emotions can be detected [14]: happiness (AU6 + AU12), sadness (AU1 + AU15, optionally AU4 and AU17), surprise (AU1 + AU2 + AU5 + AU25, or AU26).

3.2 Thermal Data

Temperature was extracted from 6 regions of interest (see Fig. 4): the forehead, the region around the eyes, the nose, the tip of the nose, the perinasal region, and the mouth. In post-analysis, the whole face was selected manually at the beginning of each event, and a correlation tracker based on facial landmarks implemented with dlib [11] was used to track the regions of interest.

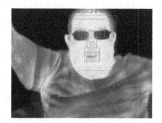

Fig. 4. The regions of interest shown for the first participant. Temperatures range from $20°C$ in dark purple to $40°C$ in light yellow. (Color figure online)

When participants turned their heads to speak with other people present in the room, sudden drops in facial temperature would appear. These were caused by background pixels entering briefly the tracked regions of interest. Furthermore, as it can be seen from Fig. 4 in the eyes region when the participant is wearing glasses, the temperature is very low. Therefore, readings associated with the glasses or the background have to be ignored. To do this, a histogram of the temperatures was plotted. It was observed that the histogram was trimodal, with a minimum between the last two modals at $33°C$. The first modal represents the background temperature (with a mean of $26°C$), while the second one represents the glasses (with a mean of $30°C$). Therefore all temperatures lower than $33°C$ were discarded. The same consideration was made for the nose region, as part of that region also contains the glasses. For the other regions the same processing was applied. The histogram was bimodal with the minimum between the two modal at around $30°C$. The first modal (with a mean of $27°C$) represented the background, therefore all values below $30°C$ were discarded. The average temperature of each region for each frame was computed and then filtered using a low pass Butterworth filter. The sampling frequency for the data is 9 Hz, and a cutoff frequency of 0.64 Hz was chosen ([21], equation(9)) so as to filter out the variations in the temperature that are due to the small movements of the head or of the camera. For a better understanding of the general trend of the temperature over time a linear least-squares regression was applied (e.g., the eyes region for P1: slope of 0.00034, regression intercept of 35.27 and correlation coefficient of 0.15).

4   Results and Comments

4.1   Action Units

First, regarding action units, the most frequent AU present during all four events is AU4 (brow lowerer). This can be explained by the fact that the two participants complained that the font size was a little too small and they had to concentrate on reading what was written on the screen. The most frequent combination of two AUs is represented by AU4+AU15 for participant P1 and AU2+AU5 for participant P2. Happiness (AU6 + AU12)[14] was not detected at all, sadness (AU1 + AU15 + AU4) was detected for both participants (for participant P1 for both events, for participant P2 just during the game), and surprise (AU1 + AU2 + AU5 + AU25) was detected for participant P2 during the quiz.

There is a need for careful consideration when extracting AUs from elderly people mostly because of wrinkles that appear with age. Therefore, for some users even if AU15 is present it does not necessarily mean that the user has lowered the corners of the lips, it can be the neutral state for that person. The same consideration has to be applied when extracting emotions from the facial expressions. The data that was gathered during this experiment permitted the extraction of important information that can be used by the episodic memory module and to better understand the needs and the reactions of the users.

4.2   Thermal Data

When looking at the raw thermal data there are some things that are clear from the very beginning. First of all, it can be easily observed that the presence of glasses produces a drop of temperature in the orbital region. However, this can be used to detect glasses and adapt the behavior of the robot accordingly. For example, if the user wants to play a cognitive game and the robot detects that the user has no glasses it can suggest wearing them based on previous information that the user sees and performs better when wearing them.

When having a closer look at the mouth temperature variation in correlation with the audio, and the RGB-D data, it can be detected when a user has spoken or kept their mouth open. For example, in Fig. 5 all the variations are due to the participant speaking. Even if all temperatures that correspond to the background have been eliminated, at timestamp 00:28 there is still a small drop due to the turning of the person. This can appear due to measurement errors as the tracker does not follow properly the facial features. The region of interest can include other parts of the face (e.g. for the mouth it can also include parts of the cheeks or the chin). The thermal data can be used individually, but in correlation with the RGB-D and audio data it can provide more robust information to the robot and it can enable it to better change its behavior.

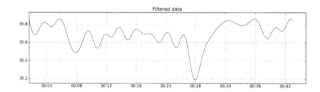

Fig. 5. Mouth temperature variation

This information could help the robot better adapt its behavior in different situations (e.g., to not interrupt if the owner is speaking).

When comparing the two events in which both participants, P1 and P2, had a visible reaction to an unwanted event the following were observed. The duration of the event for participant P1 was of 25 seconds, while for participant P2 it was of 39 seconds. For participant P1, the temperature in all 6 regions increased between $0.05°C$ for the eyes region and up to $0.6°C$ for the tip of the nose. For participant P2, the temperature for the nose and the tip of the nose remained constant over time and it decreased for the other regions; between $0.01°C$ for the eyes and up to $0.5°C$ for the mouth region. According to [10] the temperature variation for participant P2 are indicators of mild stress, while for participant P1 the increase of temperature in the forehead region is an indicator of stress. These are some preliminary results. More tests will take place in the near future with more participants and for longer periods of time.

4.3   First Lessons Learnt and Discussion

Some of the general lessons that were learnt from the interaction can be summarized as follows:

Technical Considerations

- The position and the orientation of both RGB-D and thermal camera is very important as in some situations it can miss important information. The cameras should be able to cover the face of the person regardless if it is standing or sitting.
- There will be situations in which the user will not face the camera directly. In those situations the data should be discarded.

- Users might have glasses; the approach for extracting the periorbital temperature should take this into consideration as this region provides paramount information about the user.
- For a better detection of the facial landmarks, the RGB image should be used. With proper inter-camera calibration, coordinates of the facial features can be transposed from the RGB-D camera to the thermal camera. Thus, a smaller region of interest can be defined so other parts of the face are not included in case of sudden movements.

Discussion

Based on the questionnaire regarding their likes and dislikes, the two participants enjoyed playing the games (approval of 4.2/5) except the puzzle game (approval of 1/5). They liked the layout, and the size of the buttons was large enough for them. One complaint was that the font size of 4 mm used for some instructions was too small. Another complaint was regarding the puzzle game. The participants and other residents found that 30 pieces of puzzle for medium level were too many. Therefore the number of pieces needs to be adjusted, 30 pieces of puzzle will be used for the hard level, as suggested by the participants. Based on the questionnaire regarding their thoughts on the overall behavior of the robot, both participants found the robot's approach behavior natural (approval of 5/5), in its approach it was neither dominant nor submissive (approval of 3/5), and appeared to be friendly (approval of 4.5/5). Participant P1 found the robot a little extroverted, while participant P2 found it neutral.

5   Conclusion

This paper presented some first lessons learnt from a first interaction between the ENRICHME robot and two residents in a care facility (Lace Housing Ltd in the United Kingdom). The robot interacted with the participants in their own homes. Events like turning their heads and talking to other people in the room were not discouraged by the experimenters. These events provide useful information on what kind of situations might occur in the long-term testing of the system. From the results of this work, we can posit that the facial temperature variations and facial expressions variations can be observed. The robot could adapt its behavior based on these variations to provide a more natural interaction with the elderly. For example if during a game the robot detects a high stress level [16] it can adjust the difficulty level of the game.

Based on the information gathered in this first interaction, changes will be applied to the system before starting the next phase of the project, a preliminary test with 30 participants in two Ambient Assisted Living (AAL) laboratories in the Netherlands (Stichting Smart Homes) and Italy (Fondazione Don Carlo Gnocchi).

Acknowledgement. This work was funded and done in the context of the EU Horizon2020 ENRICHME project, Grant Agreement No: 643691.

References

1. Amos, B., Ludwiczuk, B., Harkes, J., Pillai, P., Elgazzar, K., Satyanarayanan, M.: Openface: Face recognition with deep neural networks. GITHUB (2016). http://github.com/cmusatyalab/openface
2. Coradeschi, S., et al.: Giraffplus: A system for monitoring activities and physiological parameters and promoting social interaction for elderly. Hum. Comput. Syst. Interact. Background Appl. 3 Adv. Intell. Syst. Comput. **300**, 261–271 (2014)
3. Dondrup, C., Bellotto, N., Jovan, F., Hanheide, M.: Real-time multisensor people tracking for human-robot spatial interaction. In: Workshop on Machine Learning for Social Robotics at International Conference on Robotics and Automation - ICRA (2015)
4. Ekman, P., Friesen, W.: Facial action coding system. Consulting Phychologists Press, Palo Alto (1978)
5. Fischinger, D., Einramhof, P., et al.: Hobbit - the mutual care robot. In: ASROB-2013 in Conjunction with IEEE/RSJ International Conference on Intelligent Robots and Systems (IROS). Japan (2013)
6. Foundation, B.N.: 8 healthy eating tips. British Nutrition Foundation (2014). https://www.nutrition.org.uk/healthyliving/healthyeating/8tips
7. Goldberg, L.: An alternative 'description of personality': the big-five factor structure. J. Pers. Soc. Psychol. **59**, 1216–1229 (1990)
8. Graf, B., Reiser, U., Haegele, M., Mauz, K., Klein, P.: Robotic home assistant care-o-bot 3-product vision and innovation platform. In: IEEE Workshop on Advanced Robotics and Its Social Impacts (2009)
9. He, W., Goodkind, D., Kowal, P.: An aging world: 2015. In: International Population Reports (2016)
10. Ioannou, S., Gallese, V., Merla, A.: Thermal infrared imaging in phychophysiology: Potentialities and limits. In: Psychophysiology, Wiley Periodicals (2014)
11. King, D.: Dlib-ml: A machine learning toolkit. J. Mach. Learn. Res. **10**, 1755–1758 (2009)
12. Leconte, F., Ferland, F., Michaud, F.: Design and integration of a spatio-temporal memory with emotional influences to categorize and recall the experiences of an autonomous mobile robot. Auton. Robots **40**, 1–18 (2015)
13. Lorenz, T., Weiss, A., Hirche, S.: Synchrony and reciprocity: Key mechanisms for social companion robots in therapy and care. Int. J. Soc. Robot. **8**, 125–143 (2015)
14. Matsumoto, D., Ekman, P.: Facial expression analysis. Scholarpedia (2008). http://www.schoparpedia.org/article/Facial_expression_analysis
15. Quigley, M., Conley, K., Gerkey, B.P., Faust, J., Foote, T., Leibs, J., Wheeler, R., Ng, A.Y.: Ros: an open-source robot operating system. In: ICRA Workshop on Open Source Software (2009)
16. Sorostinean, M., Ferland, F., Tapus, A.: Reliable stress measurement using face temperature variation with a thermal camera in human-robot interaction. In: 15th IEEE-RAS International Conference on Humanoid Robots, Humanoids (2015)
17. Tanaka, M., Ishii, A., et al.: Effect of a human-type communication robot on cognitive function in elderly women living alone. Med. Sci. Monit. **18**, 550–557 (2012)
18. Tapus, A., Tapus, C., Mataric, M.J.: Music therapist robot for individuals with cognitive impairments. In: Proceedings of the ACM/IEEE Human-Robot Interaction Conference (HRI) (2009)

19. Tapus, A., Tapus, C., Mataric, M.J.: The use of socially assistive robots in the design of intelligent cognitive therapies for people with dementia. In: Proceedings of the International Conference on Rehabilitation Robotics (ICORR) (2009)
20. Wada, K., Shibata, T., Saito, T., Tanie, K.: Psychological, physiological and social effects to elderly people by robot assisted activity at a health service facility for the aged. In: Proceedings of IEEE/ASME International Conference on Advanced Intelligent Mechatronics (2003)
21. Yu, B., Gabriel, D., Noble, L., An, K.N.: Estimate of the optimum cutoff frequency for the butterworth low-pass digital filter. J. Appl. Biomech. **15**, 318–329 (1999)

Design and Implementation of a Task-Oriented Robot for Power Substation

Haojie Zhang$^{(\boxtimes)}$, Bo Su, and Zhibao Su

Unmanned Ground Vehicle Research and Development Center,
China North Vehicle Research Institute, Beijing, China
haojie.bit@gmail.com, bosu@noveri.com.cn, bitszb@163.com

Abstract. With the dramatically increasing number of substations, the robots are expected to inspect equipment in power industry. This paper presents a task-oriented robot for inspection in power substation. The patrol mode of the robot comprises teleoperation, regular inspection, special inspection and a key return mode. The robot only relies on a low-cost magnetic sensor for lateral positioning and radio frequency identification (RFID) technology for longitudinal positioning when working under patrol mode. The positioning error is proven to be within 5 mm, comparing 20 cm by integrated GPS-DR navigation. The test result shows that the robot could work efficiently and reliably in power substation.

Keywords: Task-oriented · Patrol mode · Radio frequency identification · Positioning error

1   Introduction

Along with the development of Artificial Intelligence (AI) technology, more and more robots are developed to serve in industry, instead of workers. In the traditional power substations in China, the inspection of substation equipment is mainly carried out by workers. The inspection method not only costs much manpower and material resources, but also is difficult to finish when the weather is terrible, such as rainy, snowy or hail. Also with the traditional inspection by workers, some faults may be missed because of their carelessness. Especially in an extra-high-voltage substation, the equipment working in safe condition is critical [1]. Any small fault may cause serious accident when it is not timely detected. With the dramatically increasing number of substations, unmanned substations are expected by the power industry in China's 13th five-year plan. Therefore, a robust and stable robot is being sought to inspect substation equipment without the involvement of workers.

Some previous work indicated the application of intelligent robot for inspecting substation equipment is effective. This inspection way improves the quality of detection and accelerates the process of setting up unmanned substations. Several robots were developed for equipment inspection in recent years, such as the TOMCAT system

This work has been partially supported by China North Vehicle Research Institute and Xi'an JinPower Electric Co., Ltd under contract number 03-72.

© Springer International Publishing AG 2016
A. Agah et al. (Eds.): ICSR 2016, LNAI 9979, pp. 746–752, 2016.
DOI: 10.1007/978-3-319-47437-3_73

developed in United States of America [2], ROBTET system developed in Spain [3] and the robot developed in Hydro-Quebec's Research Institute [4]. These robots run tracks and carry a visible-light camera and an infrared thermograph to inspect equipment [5]. They will stop at each inspection point accurately and inspect the equipment. Two kinds of navigation methods are implemented on these robots. One is the magnetic navigation system, which serves as a roadway to guide the robot [6]. This method of navigation lacks in flexibility and needs hard work for orbit construction. The other category guides the robot without magnetic markers, such as vision navigation [7] or integrated GPS-dead reckoning (DR) navigation [8]. The magnetic navigation system is the most accurate among these navigation methods.

An inspection robot for power substation was developed in our previous work [9]. The integrated GPS-DR navigation system was implemented on the robot. However, the shortcoming of this method is that sometimes the GPS signal may be lost for quite a long time because of the electromagnetic field. Then the accuracy of navigation would be reduced through our test in power substation. This paper presents a task-oriented robot for power substation, which is based on magnetic navigation. Section 2 gives an overview of the robotic system. Section 3 introduces the inspection task from data center and describes the behavioral executive decomposition according to the task. Section 4 gives the application of navigation and control. The final section gives a conclusion of the paper.

2 Overview of the Robotic System

The whole robotic system consists of two parts: a data center and a robot. The data center and the robot communicate via the wireless communication module. The worker in substation sends inspection task to robot through human-machine interface of the data center. After receiving the task, the work mode of the robot is that it moves inside the substation and stops at predefined point for detecting equipment. Then, the inspection information is sent to the data center. The worker could check the status of the equipment through the inspection information.

2.1 Data Center

The data center is a software system that consists of five parts: the status display layer, the inspection task layer, the electric map layer, the inspection information layer and the data access layer, as shown in Fig. 1. The status display layer is an interface that shows the running status of the robot, such as translational and angular velocities, current and voltage of the battery, position, etc. The inspection task layer is an interface that the worker could choose and send inspection task through it. There are four candidate working modes, teleoperation, regular inspection, special inspection and a key return. The electric map layer shows the map of the substation and the position of the robot is updating in real time on the map. The inspection information layer displays the detecting results returned from the robot. The data access layer is a user interface for getting inspection results through inspecting time or inspecting point.

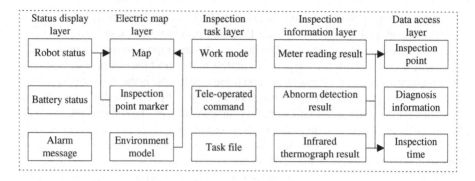

Fig. 1. The architecture of the data center

2.2 The Robot

The robot is composed of mobility module and inspection module, as shown in Fig. 2. Since the road is flat in the substation, the simple four-wheel driving mode is designed. There is a control unit mounted on the robot which coordinates the movement of the four motors. In this way, the robot could move accurately following the desired translational velocity and angular velocity. The laser range finder, magnetic sensor and Radio Frequency Identification (RFID) reader are mounted on the chasis of the robot. These three sensors guarantee that the robot will navigate in substation, without collision with obstacles. The magnetic sensor are used to sense and process the magnetic markers paved on the road, and feeds back the lateral deflection of the robot. The RFID reader communicates with the tags on the road. While driving, the RFID reader constantly monitors the presence of a tag. Once information is detected, the reader retrieves it from the tag. The robot will execute predefined behavior according to the number of the tag. The laser range finder constantly detects the obstacle within the range of 0.5 m in front of the

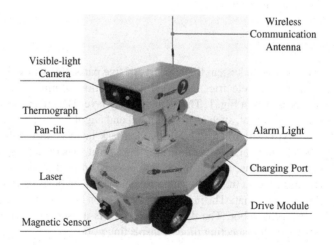

Fig. 2. Outlook of the robot

robot. Once the obstacle is found, the robot will stop immediately to aviod collision. The robot won't move until the obstacle is cleaned.

The inspection module consists of a visible-light camera, an infrared thermograph and a pan-tilt system. It is designed for collecting images and temperature information of equipment. The camera and thermograph direct to different substation components through a pan-tilt system, which can move along pitch axis and yaw axis.

3   The Inspection Task and Behavioral Executive

3.1   The Inspection Task

The worker sends the inspection task to the robot, such as teleoperation, regular inspection, special inspection or a key return. In teleoperation mode, the worker could send control commands via wireless communication and drive the robot forward, backward and self-turning. When the regular inspection or special inspection mode is selected, the inspection task file will be sent to the robot through wireless communication. In the inspection task file, the behavior of the inspection module is recorded for each inspection point, which including the movement of pan-tilt system, zoom in or out of visible-light camera and focus of the infrared thermograph. The difference between these two inspection modes is that the worker needs to choose the inspection point on electric map under special inspection mode. The special inspection file is constructed based on the chosen inspection points. But the regular inspection file is constructed in advance, without selecting inspection points. An example of the inspection file is shown in Table 1. The format of the inspection file is txt. After receiving the file, it will read the txt line by line and store the related data into database. Also, in the situation of low battery or urgency, the robot will drive back the charging point immediately by triggering the key return mode.

Table 1. The inspection file format

No. stop point int	No. equip int	Type equip int	Yaw pan-tilt long int	Pitch pan-tilt long int	Zoom in or out visible-light camera float	Focusing thermograph long int
1	1	0	0002	00e3	3.7	a6f1
2	1	14	0c9a	0038	30.0	2bec
3	1	13	00f7	003e	30.0	2c11
4	1	31	0ab2	0099	5.5	b757

3.2   Behavioral Executive

The robot will drive autonomously along the magnetic markers after receiving the inspection file. Four movements are set as the robot's behaviors, i.e. forward, backward, left-turn and right-turn. The backward movement is set for the narrow road, where the robot will get stuck in that situation and cannot make turn. There are two magnetic sensors mounting on the robot. One is on the front of the robot and the other is at back.

When the robot is moving forward, left-turn or right-turn, the results from the front magnetic sensor are used to calculate the control command. In contrast, the results of back magnetic sensor are used when the robot is moving backward.

The switch between these four movements is based on the information from the RFID tag. The information written in the RFID tag consists of a capital letter and numbers, such as "F001". The number indicates the number of inspection point, which is consistent with the number of stop points in the inspection file. The capital letter defines the switch of the robot's behaviors. If the capital letter is "F", the robot will drive forward along the magnetic markers. Similarly, "B" for backward, "L" for left-turn, "R" for left-turn and "S" for stop. The behavior of the robot is predefined according to the environment of the substation.

The route could be divided into several pieces of paths in substation according to the RFID tags. Figure 3 shows an example of route we used in one substation. The switch of movement among different paths is based on the information from RFID tags. Take several pieces of paths on the route for example, the behavior of the robot will switch from forward (P_1) to right-turn (P_2) when the "R" tag is detected. Similarly, the behavior will change from right-turn (P_2) to backward (P_3) when "D002" tag is detected. At this time the robot will move back along the magnetic markers without turning. The behavior of the robot will switch to forward (P_4) until the "R" tag is detected again. With this way of setting behaviors, the robot could reach nearly everywhere to inspect equipment.

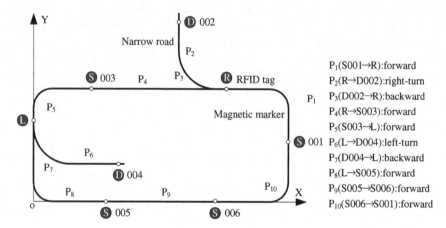

P_1(S001→R):forward
P_2(R→D002):right-turn
P_3(D002→R):backward
P_4(R→S003):forward
P_5(S003→L):forward
P_6(L→D004):left-turn
P_7(D004→L):backward
P_8(L→S005):forward
P_9(S005→S006):forward
P_{10}(S006→S001):forward

Fig. 3. A patrol route in substation

4   Navigation and Control

The magnetic navigation system is not affected by rain or snow and has been tested on our robot under a wide variety of operating conditions. During the robot drives, the magnetic sensor detects a signal from the magnetic markers. The magnetic sensor consists of an array of 16 sensing spots (pale green dots in Fig. 4), which show different values owing to the different distances to the magnetic markers, shown in Fig. 4. If the

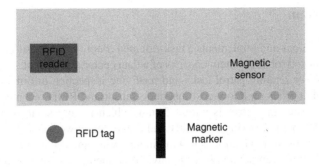

Fig. 4. The magnetic navigation system (Color figure online)

magnetic signal exceeds the threshold, the output value is 1, otherwise is 0. Therefore, the output value of the magnetic sensor is a sequence of 16 binary numbers.

The translational velocity is given as a constant value regarding the mobility of the robot, such as 0.5 m/s. Based on the output value of the magnetic sensor, the desired angular velocity can be calculated through the finite state adjustment method by Eq. (1).

$$\omega = \frac{l - r}{n}\omega_{max} \tag{1}$$

Where, l denotes the sum of the output value of the left eight sensing spots, r denotes the sum of the output value of the right eight sensing spots, n is half number of sensing spots, and w_{max} denotes the maximum angular velocity of the robot.

Localization is implemented by using RFID technology. The RFID reader mounted on the robot constantly monitors the presence of a RFID tag. Once the information is detected, the controller will stop the robot immediately. We tested the navigation accuracy by measuring the position of the robot when it stopped at a certain stop point. The designated stop point was used as the reference position. Figure 5 shows the test result of 10 stops, using magnetic navigation and GPS-DR navigation respectively. We find that the errors are within 0.5 cm under magnetic navigation, but within 20 cm under GPS-DR navigation. As a result, the magnetic navigation is more robust to use in extra-high voltage substation because of its high positioning accuracy.

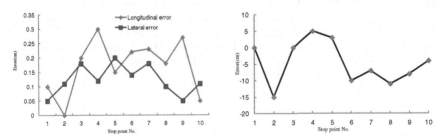

(a)Positioning error using magnetic navigation (b)Positioning error using GPS-DR navigation

Fig. 5. The accuracy of position

5 Conclusion

This paper develops and implements a task-oriented robot for power substation inspection. The proposed robotic system consists of a data center and a robot. The data center is responsible for sending patrol tasks and receiving inspection information from the robot. The robot relies on a low-cost magnetic sensor and RFID technology for positioning. The positioning error is within 5 mm, which is very small and acceptable. Currently, the robot is working efficiently and reliably in power substation. One drawback of this method of navigation is lacking in flexibility and needs hard work for orbit construction. More accurate and flexible navigation methods should be developed for the robot in the future.

References

1. Li, S., Hou, X.: Research on the AGV based robot system used in substation inspection. In: Proceedings of the International Conference on Power System Technology, Chongqing, China, pp. 1–4 (2006)
2. Parker, L., Draper, L.: Robotics applications in maintenance and repair. In: Nof, S. (ed.) Handbook of Industrial Robotics, 2nd edn, pp. 1023–1036. Wiley, New York (1999)
3. Aracil, R., Ferre, M., Hernando, M., Pinto, E., Sebastian, J.M.: Telerobotic system for live power lines maintenance: ROBTET. Control Eng. Pract. **10**(11), 1271–1281 (2002). ELSEVIER Press
4. Allan, J.F., Lambert, G., Lavoie, S., Reiher, S.: Development of a mobile robotic platform for the underground distribution lines. In: Proceedings of the 2008 IEEE/ASME International Conference on Advanced Intelligent Mechatronics, Xi'an, China, pp. 406–411, 2–5 July 2008
5. He, H.Y., Yao, J.G., Jiang, Z.L., Wang, X.X., Li, W.W.: Infrared thermal image detecting of high voltage insulator contamination grades based on support vector machine. In: Li, D., Dong, Z., Yan, H., Yuan, J. (eds.) Automation of Electric Power Systems, vol. 29, pp. 70–74 (2005)
6. Lu, S.Y., Li, Y.P., Zhang, T.: Design and implement of control system for power substation equipment inspection robot. In: Proceedings of the 2009 IEEE/RSJ International Conference on Intelligent Robots and Systems, St. Louis, USA, pp. 93–96, 11–15 October 2009
7. Guo, R., Han, L., Cheng, X.Q.: Omni-directional vision for robot navigation in substation environments. In: Proceedings of the 2009 IEEE International Conference on Robotics and Biominetics, Guiling, China, pp. 1272–1275, 19–23 December 2009
8. Guo, R., Xiao, P., Han, L., Cheng, X.Q.: GPS and DR integration for robot navigation in substation environments. In: Proceedings of the 2010 IEEE International Conference on Information and Automation, Harbin, China, pp. 2009–2012, 20–23 June 2010
9. Zhang, H.J., Su, B., Song, H.P., Xiong, W.: Development and implement of an inspection robot for power substation. In: Proceedings of the 2015 IEEE Intelligent Vehicles Symposium, Seol, Korea, pp. 121–125, 28 June – 1 July 2015

The MuMMER Project: Engaging Human-Robot Interaction in Real-World Public Spaces

Mary Ellen Foster[1]([⊠]), Rachid Alami[2], Olli Gestranius[3], Oliver Lemon[4], Marketta Niemelä[5], Jean-Marc Odobez[6], and Amit Kumar Pandey[7]

[1] University of Glasgow, Glasgow, UK
MaryEllen.Foster@glasgow.ac.uk
[2] LAAS-CNRS, Toulouse, France
rachid.alami@laas.fr
[3] Ideapark, Lempäälä, Finland
olli.gestranius@ideapark.fi
[4] Heriot-Watt University, Edinburgh, UK
o.lemon@hw.ac.uk
[5] VTT Technical Research Centre of Finland, Tampere, Finland
marketta.niemela@vtt.fi
[6] Idiap Research Institute, Martigny, Switzerland
odobez@idiap.ch
[7] SoftBank Robotics Europe, Paris, France
akpandey@aldebaran.com
http://mummer-project.eu/

Abstract. MuMMER (MultiModal Mall Entertainment Robot) is a four-year, EU-funded project with the overall goal of developing a humanoid robot (SoftBank Robotics' Pepper robot being the primary robot platform) with the social intelligence to interact autonomously and naturally in the dynamic environments of a public shopping mall, providing an engaging and entertaining experience to the general public. Using co-design methods, we will work together with stakeholders including customers, retailers, and business managers to develop truly engaging robot behaviours. Crucially, our robot will exhibit behaviour that is *socially appropriate* and *engaging* by combining speech-based interaction with non-verbal communication and human-aware navigation. To support this behaviour, we will develop and integrate new methods from audiovisual scene processing, social-signal processing, high-level action selection, and human-aware robot navigation. Throughout the project, the robot will be regularly deployed in Ideapark, a large public shopping mall in Finland. This position paper describes the MuMMER project: its needs, the objectives, R&D challenges and our approach. It will serve as reference for the robotics community and stakeholders about this ambitious project, demonstrating how a co-design approach can address some of the barriers and help in building follow-up projects.

© Springer International Publishing AG 2016
A. Agah et al. (Eds.): ICSR 2016, LNAI 9979, pp. 753–763, 2016.
DOI: 10.1007/978-3-319-47437-3_74

1 Introduction

Developing an artificial agent capable of coexisting and interacting independently, naturally, and safely with humans in an unconstrained real-world setting has been the dream of robot developers since the very earliest days. In popular culture and science fiction, the prototypical image of a "robot" is precisely this: an artificial human that is able to engage fully in all aspects of face-to-face conversation. As modern robot hardware becomes safer, more sophisticated, and more generally available, there is a clear and significant consumer demand for robots that are able to coexist in everyday, real-world human environments in this way. For example, demand for SoftBank Robotics' Pepper robot has been so high in Japan that, since July 2015, each monthly run of 1000 units has sold out in under a minute [18]; also, the Jibo robot broke records on the crowdfunding site Indiegogo in July 2014 when it raised US$1million in less than a week [1]. This form of *socially intelligent* HRI [7] has therefore received a particular focus in recent years. However, while the hardware capabilities of such robots are increasing rapidly, the software development has not kept pace: even with the most recent technological developments, the most advanced such robots have generally supported limited, scripted interactions, often relying on a human operator to help with input processing and/or appropriate behaviour selection.

In the new European project MuMMER ("**MultiM**odal **M**all **E**ntertainment **R**obot"), we are developing a humanoid robot that is able to operate autonomously and naturally in a public shopping mall. The overall concept underlying MuMMER is that an interactive robot deployed in a public space such as a shopping mall should be *entertaining*: this will increase the acceptability of such a robot to the people it interacts with, therefore improving their overall experience in the mall. To ensure that the robot scenarios and interactions are engaging and entertaining, we will involve a wide number of potential users through continuous situated co-design studies, and will use the results of these studies to inform the design and operation of all components of the robot system.

In a number of recent studies, humanoid robots have been deployed in a range of public spaces, mainly in Japan. For example, Kanda et al. [11] carried out a five-week field trial of a robot in a shopping centre, involving 2343 interactions with 235 tracked participants. The robot was semi-autonomous: it used a human operator to carry out speech recognition, to monitor and override its behaviour selection, and to provide additional domain knowledge when needed. In post-experiment questionnaires, the robot was highly rated on most subjective measures, with many positive comments. Outside Japan, other successful locations for public robot deployment have included museums [12], city centres [3,20], care homes [9], and airports [19].

Note, however, that the interactive behaviour of the robots in these previous deployments has been limited in various ways. For example, most of the robots used in these previous deployments either were not mobile, were remotely teleoperated, or were able to move only within an area delimited with environmental sensors such as floor sensors. Also, these robots are generally semi-autonomous, incorporating a human operator for operations such as speech recognition and overall behaviour selection. In general, recent experiments with robots in public spaces have tended to concentrate either purely on human-aware navigation (e.g. [6,14]), or purely on natural-language interaction (e.g. [2]), rather than on the combination of the two.

2 Objectives

In MuMMER, we will develop a humanoid robot capable of performing its interactive tasks independently and naturally in a dynamic environment and with several users simultaneously. To meet this goal, we have defined the following set of objectives which will be used to guide the work across the project. These objectives encompass both state-of-the-art technical requirements, as well as our intentions to ensure that our work in MuMMER is grounded in real-world needs and applications throughout the project.

Objective 1: Development of an interaction robot for entertainment applications. The tasks of interaction and user experience design for entertainment are at the core of MuMMER. These highly challenging objectives are multi-party and multimodal at numerous levels (sensing, synthesis, conversation). We aim to achieve these objectives through the seamless integration of user state and mood monitoring, design of entertainment-oriented artificial social signals, and interaction management.

Objective 2: Fusion of interactive applications through a co-design process. Interactive applications tend to be specific to one or two simple tasks. However, in MuMMER, we aim to develop an interaction robot that blends different tasks (information providing, guidance, entertainment) within a single entity, relying on real user needs and interests. The goal is to ensure that the developed robotic application will be user-driven in terms of HRI, is socially and ethically accepted, and is also interesting to commercial end-users.

Objective 3: Situated-perception with on-board sensors. Rather than instrumenting the scene or people, in MuMMER we aim to design an autonomous robot able to operate robustly and naturally in diverse environments by relying only on onboard sensors. To address the resulting sensing limitations, we plan to design a truly interaction-aware perception approach combining perception algorithms aware of the robot gestures, and active sensing, along with speech synthesis and dialogue to make users implicitly or explicitly aware of perception uncertainties.

Objective 4: Automatic learning of engaging interaction strategies. We aim to develop robots which can interact naturally with humans in a social setting, being entertaining as well as helpful, and this requires learning from human interaction data. We therefore aim to develop new techniques for automatic learning of interaction strategies for socially appropriate, engaging, and robust robot action selection, based on uncertain information about the "social" state of the interaction (e.g., where the user(s) are, whether users are engaged in the interaction, what speech has been recognised, what information is in the common-ground, and the estimated perspectives of the different users). Such capabilities will create a step-change in the naturalness and acceptance of human interaction with robots.

Objective 5: Human-aware and situated interactive tasks and motion synthesis. We aim to investigate further, develop, and demonstrate a set of decisional capabilities that will allow such a robot to effectively "close the loop" between situated dialogue and robot behaviour. The robot will be able to estimate the visual perspective of its human interaction partners, their positions and postures, and—based on the current interaction and dialogue context—to synthesize the pertinent human-aware motion and interaction. This will be achieved based on a novel combination of perspective-taking, situation assessment, and human-aware task and motion planning in order to make a substantial step forward and produce more complex and validated dialogue and interaction-aware behaviours.

Objective 6: Development of new business opportunities. One of the most challenging goals of the MuMMER project will be the development of novel business models or opportunities based on the innovative components implemented during the project. We will investigate the roles and uses of interaction and entertainment robots considered as attraction along with other features within public spaces, as well as their use as one facet of companion robots for the consumer market.

3   Approach

In the preceding section, we set out the objectives of the MuMMER project in areas including technical development as well as the broader context in which this work is situated. In this section, we describe our concrete approach to meeting these objectives: beginning with a description of the selected deployment locations and discussing how scenarios will be co-designed with a range of stakeholders; next, summarising how the necessary technical advances will be achieved and integrated on a sophisticated humanoid robot; and finally, outlining our plans for real-world deployment and user acceptance studies of the integrated robot system throughout the project.

3.1   Stakeholders, Scenarios, and Co-Design

Throughout the duration of the project, we will carry out continuous data collection, deployment, and evaluation activities in the Ideapark shopping centre in Lempäälä, Finland. Ideapark is the biggest commercial city in Scandinavia, with 7.3 million visitors in 2013. It contains almost 200 stores, restaurants and cafés within 100 000 square metres.

We are targeting two main classes of interaction scenario in MuMMER. In the first set of scenarios, the robot will be situated in the main shopping area of the

(a) Ideapark shopping mall   (b) *Pii Poo* Cultural Centre

Fig. 1. Deployment locations

mall (Fig. 1a): in addition to supporting a set of helping tasks, such as providing guidance, information, or collecting customer feedback about the events going on, the robot will also engage fully with users, proposing events to attend or performing entertainment activities such as telling jokes and riddles, organizing contests, or distributing items such as balloons. In the other scenarios, the robot will be situated in the *Pii Poo* centre[1] (pictured in Fig. 1b), which is designed as an accessible space providing inclusive cultural activities for all ages and abilities. In this case, the robot will provide entertainment activities to children or families, focussed on social games such as quizzes, memory games, and the like.

The details of the scenarios will be developed through a continuous co-design process, which will ensure that multiple relevant perspectives from consumer target groups as well as from commercial and other stakeholders will be identified, analysed and integrated into the design of the robot, the human-robot interaction, and the applications for future businesses. In particular, the co-design process will work with the technical development process to defined a set of relevant scenarios of increasing complexity, increasing the chances that the robot will perform according to the goals set, while being representative of future business applications identified with the stakeholders.

The co-design process will include workshops, seminars and scenario co-creation from both consumer and business perspectives. In addition, it will support more focused study methods to answer specific questions or tests, such as questionnaires, interviews, and observations. Scenarios and requirements will be co-created with the stakeholders to guide the technical development work. In addition, the success metrics for the development and assessment of human-interactive mobile robots for consumer markets will be co-created to ensure their validity with regard to the purpose, and the validity will be tested in the field trials. The work on success metrics will be supported by market analyses and market studies. Co-design engages people in the technical development, and MuMMER will work to integrate their contributions to the technology develop-

---

[1] http://www.kulttuuripiipoo.fi/content/fi/1/20106/In+English.html.

ment and to achieve mutual learning among the different stakeholders. Co-design increases user interest and acceptance of such robots; this is a major research goal in the project.

3.2 Technical Development

From a technical point of view, endowing a robot with the ability to support the desired interaction scenarios outlined above will require the integration of a number of state-of-the-art components on a sophisticated robot platform. The primary target robot platform for MuMMER is SoftBank Robotics' Pepper[2] robot (Fig. 2). Pepper is a 1.20 m tall, wheeled humanoid robot, with 17 joints for graceful movements, three omni-directional wheels to move around smoothly, and more than 12 hours of energy for non-stop activities, along with the ability to go back to the recharging station if required. It has a 3D camera to sense and recognize humans and their movements at a distance of up to three metres. For the initial stages of the MuMMER project, which focus on scenario development, data collection, and preliminary development of the individual technical components, we will use the default Pepper hardware which is currently available commercially. However, based on the particular needs of MuMMER, an adapted Pepper prototype will be developed during the first period of the project, including additional sensors if necessary and feasible; this will be the robot that is used for the final autonomous MuMMER system in the later stages of the project and for the later deployments.

Fig. 2. Pepper robot (Images © SoftBank)

The high-level architecture of the MuMMER system is presented in Fig. 3. As shown in this diagram, the technical developments will fall into four main classes: audiovisual processing, social signal processing, interaction management, and interactive task and motion control. In the remainder of this section, we summarise how we will develop state-of-the-art components to address each of these tasks. We will use the existing Pepper interactive capabilities as a starting point wherever possible; however, we anticipate that supporting the MuMMER scenario in its full generality will require significant advances in all areas, as outlined below.

---

[2] https://www.ald.softbankrobotics.com/en/cool-robots/pepper.

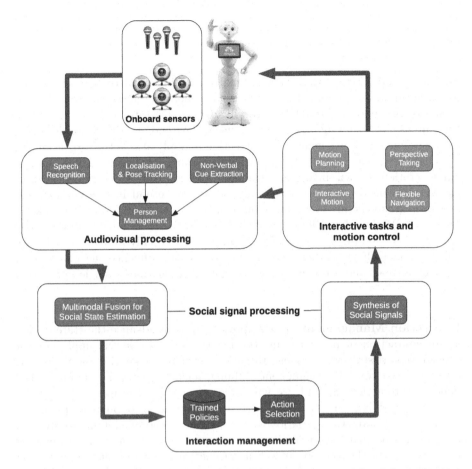

Fig. 3. Technical architecture of the MuMMER system; arrows show the main data flow

Audiovisual Processing. To improve the performance of the perception modules, we will investigate several research directions. First, we will an existing real-time visual-only probabilistic multi-party head and pose tracking framework used in [16]. In particular, to handle non-stationary sensors and situations with multiple people potentially occluding each other and not facing the camera, we will leverage depth data to gain performance, e.g. by combining RGB-D based upper-body detectors (in addition to body part classifiers [17]) with very fast segmentation based tracking methods [4,5] on the depth data stream. Second, we will develop tighter links between perception and action and will design situated algorithms that have a better knowledge of the situation and are able to exploit it. The situational context can originate from different sources (interaction state, perception state and quality, action state, etc.) and can be used in different ways: by influencing recognition and decisions, actively triggering specific perception algorithms or behaviuors (to quickly verify with a head gesture that a person that is currently outside of the field of view is still present), or via model parameter adaptation in the longer term.

Social Signal Processing. Social signal processing involves both the recognition of human social signals, as well as the generation and synthesis of social signals by the robot; in MuMMER, we will advance the state-of-the-art in both of these areas. For social signal recognition, we will extend algorithms for detecting and classifying user engagement and attention to apply also to the target domain of multi-party, real-world, entertaining human-robot interaction. For social signal generation, we will investigate whether "non-natural" social signals such as lights and acoustic effects can actually be socially useful in this context. We will also extend the investigation of humour from the verbal (i.e., what makes a text funny?) to the behavioural cues that can make others laugh [10] (i.e., why are two people telling the same joke not necessarily equally funny?). Finally, we will investigate how synthetic personality traits can influence the relationship between a robot and a possibly large audience (rather than a single interaction partner).

Interaction Management. In this project, we will advance the state-of-the-art in interaction management in two directions. First, we will apply current state-of-the-art statistical models [13,15,21] to the new scenarios and tasks that arise in the context of engaging, entertaining, socially appropriate human-robot dialogue interaction. Second, we will scale up machine learning techniques to support robust human interaction with a robot in noisy, populated spaces. To support the MuMMER scenarios and tasks, the range of system actions will be extended to include communicative gestures and behaviours such as moving towards a particular person, as well as pro-active behaviour such as initiating an engagement with a person to attract their attention. In order to enable the robot to make informed decisions what to do next, given this extended range of possible system behaviours, the social state model will need to be extended with the additional relevant features. As a consequence, the extended state and action model will pose additional challenges of scalability of the machine learning techniques for state tracking and action selection that must be addressed.

Interactive Tasks and Motion Control. Most of the prior work in the area of human-aware motion planning has mainly been from a planning perspective, with the goal of producing safe and comfortable motions. Further, the typical target scenario is not the kind of crowded space we are aiming for, with potentially many people, standing, moving, interacting and frequently blocking the path. In this context, another important dimension becomes more prominent: considering the social and cultural context of the human-robot interaction. This aspect has not been sufficiently exploited in the context of navigation planning. For example, the robot can reason from the human's perspective, analyse the feasibility that the human can provide the way for the robot, and request the human to do so through verbal or physical interaction. Similarly, the human should be able to request the same and expect a similar natural response from the robot. Such reasoning will greatly affect the operation of the state of the art navigation planning and decision making system: it requires a fusion of rich reasoning about humans, their abilities, affordances, perspective taking, interaction and navigation planning. With this, we aim to elevate the human-aware and socially-aware path planning towards more intelligent planning with social interaction possibilities (e.g., [8]).

3.3 Real-World Deployment and Acceptance Studies

A crucial aspect of the MuMMER project plan is that the robot will be regularly deployed in Ideapark throughout the whole course of the project. For the first 3.5 years of the project, the interim deployments will be regular (i.e., every few months) but rather short-term (maximum 1–2 weeks); these visits will serve to support the co-design activities as well as to provide formative evaluation of the technical components of the system and pilot testing of the overall robot deployment process. However, at the end of the project, we will carry out a continuous, long-term user acceptance study in Ideapark lasting several months, where the robot is present in the shopping centre continuously (ideally, daily) throughout the entire period: this should allow a wide range of behaviours to be examined, and also to greatly reduce the novelty effect of the robot by providing more concrete data on how it is actually used in context across time. The results of this long-term study will serve to finalise the use cases and success metrics developed during the project, and will also provide a concrete demonstration of the utility, flexibility, and generalisability of the robot system developed in MuMMER.

As mentioned at the start of this paper, this sort of long-term, public-space deployment has only recently begun to be addressed (e.g., [9,19]), and not very widely outside of Japan. The results of the MuMMER long-term deployment should therefore add significantly to our knowledge of how such robots are received and accepted in public spaces around Europe.

4 Current Work

The project is currently in its first year, and efforts have begun towards all of the technical tasks outlines in Sect. 3. In particular, the first co-design workshops have been carried out with management and personnel from Ideapark and with consumers. These have resulted in an initial set of proposed scenarios, including welcoming consumers, giving information about the location of shops in the mall, playing simple quiz games and telling jokes, distributing electronic vouchers, and also posing for and sharing "selfie" images. The consumers also expressed concerns about issues such as privacy and safety, and we are taking care to address those concerns appropriately in the development of the robot system. Consultations with retailers will begin with interviews in the autumn, and the first Pepper deployment in the mall will also take place in the autumn. We have also begun to lay the groundwork for the final acceptance study by developing a questionnaire which will be administered regularly throughout the project to monitor the attitudes of Ideapark customers towards the robot: the first questionnaire will be carried out before Pepper is taken to the mall in the autumn to provide a baseline measure of these attitudes.

On the technical side, development has also begun on all of the components outlined above in Sect. 3, with the initial goal of supporting the scenarios and addressing the concerns arising from the first co-design studies listed above. The technical infrastructure is also being prepared to allow distributed development and testing of the components and to integration on Pepper, and all partners have also begun development and testing with the actual Pepper hardware platform. We anticipate being in a position to test an initial integrated system in the mall at the end of the project's first year (i.e., February 2017), as a step towards the eventual long-term deployment of the full system at the end of the MuMMER project's fourth year (February 2020).

Acknowledgements. This research has been partially funded by the European Union's Horizon 2020 research and innovation programme under grant agreement no. 688147 (MuMMER, `mummer-project.eu`).

References

1. Annear, S.: Makers of the world's 'first family robot' just set a new crowdfunding record. Boston Daily. http://www.bostonmagazine.com/news/blog/2014/07/23/jibo-raises-1-million-six-days-record/
2. Barker, J.: Towards human-robot speech communication in everyday environments. In: Invited Talk presented at the ICSR 2013 Workshop on Robots in Public Spaces, October 2013
3. Bauer, A., Klasing, K., Lidoris, G., Mühlbauer, Q., Rohrmüller, F., Sosnowski, S., Xu, T., Kühnlenz, K., Wollherr, D., Buss, M.: The autonomous city explorer: Towards natural human-robot interaction in urban environments. Int. J. Soc. Robot. **1**(2), 127–140 (2009)
4. Bibby, C., Reid, I.: Robust real-time visual tracking using pixel-wise posteriors. In: Forsyth, D., Torr, P., Zisserman, A. (eds.) ECCV 2008. LNCS, vol. 5303, pp. 831–844. Springer, Heidelberg (2008). doi:10.1007/978-3-540-88688-4_61
5. Bibby, C., Reid, I.D.: Real-time tracking of multiple occluding objects using level sets. In: CVPR, pp. 1307–1314 (2010)
6. Clodic, A., Fleury, S., Alami, R., Chatila, R., Bailly, G., Brethes, L., Cottret, M., Danes, P., Dollat, X., Elisei, F., Ferrane, I., Herrb, M., Infantes, G., Lemaire, C., Lerasle, F., Manhes, J., Marcoul, P., Menezes, P., Montreuil, V.: Rackham: An interactive robot-guide. Proc. RO-MAN **2006**, 502–509 (2006)
7. Dautenhahn, K.: Socially intelligent robots: dimensions of human-robot interaction. Philos. Trans. Royal Soc. B: Biol. Sci. **362**(1480), 679–704 (2007)
8. Dondrup, C., Hanheide, M.: Qualitative constraints for human-aware robot navigation using velocity costmaps. In: Proceedings RO-MAN (2016)

9. Hebesberger, D., Dondrup, C., Koertner, T., Gisinger, C., Pripfl, J.: Lessons learned from the deployment of a long-term autonomous robot as companion in physical therapy for older adults with dementia: A mixed methods study. In: The Eleventh ACM/IEEE International Conference on Human Robot Interaction, pp. 27–34 (2016)
10. Hurley, M., Dennett, D., Adams Jr., R.: Inside Jokes. MIT Press, Cambridge (2011)
11. Kanda, T., Shiomi, M., Miyashita, Z., Ishiguro, H., Hagita, N.: A communication robot in a shopping mall. Robot. IEEE Trans. **26**(5), 897–913 (2010)
12. Karkaletsis, V., Konstantopoulos, S., Bilidas, D., Vogiatzis, D.: INDIGO project: personality and dialogue enabled cognitive robots. In: Makedon, F., Maglogiannis, I., Kapidakis, S. (eds.) Proceedings of the 3rd International Conference on Pervasive Technologies Related to Assistive Environments, PETRA 2010 (2010). doi:10. 1145/1839294.1839376
13. Keizer, S., Foster, M.E., Wang, Z., Lemon, O.: Machine learning for social multi-party human-robot interaction. ACM Trans. Intell. Interact. Syst. **4**(3), 1–32 (2014)
14. Kruse, T., Pandey, A.K., Alami, R., Kirsch, A.: Human-aware robot navigation: A survey. Robot. Auton. Syst. **61**(12), 1726–1743 (2013)
15. Lemon, O., Pietquin, O. (eds.): Data-driven Methods for Adaptive Spoken Dialogue Systems: Computational Learning for Conversational Interfaces. Springer, Heidelberg (2012)
16. Sheikhi, S., Khalidov, V., Klotz, D., Wrede, B., Odobez, J.M.: Leveraging the robot dialog state for visual focus of attention recognition. In: Proceedings of International Conference on Multimodal Interfaces (ICMI) (2013)
17. Shotton, J., Fitzgibbon, A., Cook, M., Sharp, T., Finocchio, M., Kipman, A., Blake, A.: Real-time human pose recognition in parts from single depth images. In: CVPR (2011)
18. SoftBank Robotics Corp: Press releases (2016). http://www.softbank.jp/en/corp/group/sbr/news/press/2016/
19. Triebel, R., et al.: SPENCER: a socially aware service robot for passenger guidance and help in busy airports. In: Wettergreen, David, S., Barfoot, Timothy, D. (eds.) Field and Service Robotics. STAR, vol. 113, pp. 607–622. Springer, Heidelberg (2016). doi:10.1007/978-3-319-27702-8_40
20. Weiss, A., Igelsbock, J., Tscheligi, M., Bauer, A., Kuhnlenz, K., Wollherr, D., Buss, M.: Robots asking for directions: The willingness of passers-by to support robots. Proc. HRI **2010**, 23–30 (2010)
21. Yu, Y., Eshghi, A., Lemon, O.: Training an adaptive dialogue policy for interactive learning of visually grounded word meanings. In: Proceedings SIGDIAL (2016)

Introducing IOmi - A Female Robot Hostess for Guidance in a University Environment

Eiji Onchi[1], Cesar Lucho[1], Michel Sigüenza[1], Gabriele Trovato[2], and Francisco Cuellar[1(✉)]

[1] Mechatronic Specialty, Engineering Department,
Pontificia Universidad Católica del Perú (PUCP), Lima, Peru
cuellar.ff@pucp.pe
[2] Graduate School of Advanced Science and Engineering, Waseda University, Tokyo, Japan

Abstract. In this paper we introduce IOmi: a life-sized female humanoid hostess robot intended for serving as guidance in indoor public spaces. Its design methodology, adapted from industrial design approaches, is intended to be applicable in different scenarios, considering the final users of the robot, the intended use of the agent, and the contextual environment. Results from the first test inside a Latin American university environment clarified the needs of the potential users and suggested new directions of research.

1 Introduction

In the near future, humanoid robots are expected to fulfil a role as companions in human life, sharing the same social spaces. Robots of this type, such as Pepper [1], are already being commercialized. The context of robots for social spaces has to be careful considered, in particular the aesthetics that is accord to the environmental framework. The physical appearance of a robot, its motion and emotional communication influence the behaviour with which people will engage the robot [2] and its overall acceptance. Even small details of appearance, establishing a set of expectations of the robot's abilities and influence in its first impression by human users, and as a consequence, how they will interact with it. Gender could affect the perception of a robot, as male robots are perceived as more dominant and agentic, whereas female robots are perceived as more communal [3]. Understanding and awareness of gender in robots is crucial for human-robot interaction (HRI), as the lack of it increase the risk of rejection and poor performance [4]. In Japan, for example, roboticists assign gender based on their common sense assumptions about female and male sex and gender roles [5]. As aesthetic features of robots are relevant to match robot's role, robots for guidance should differ from other social robotic companions, as the former interacts with group of people rather than a single user [6]. Moreover, employing robots in public spaces requires detailed, yet complex mechanism to emulate the freedom of movement of a human [7], and at the same time the necessary robustness and safety.

The original version of the chapter was revised: Table 1 was corrected. The erratum to this chapter is available at DOI: 10.1007/978-3-319-47437-3_99

© Springer International Publishing AG 2016
A. Agah et al. (Eds.): ICSR 2016, LNAI 9979, pp. 764–773, 2016.
DOI: 10.1007/978-3-319-47437-3_75

Social robots such as Robovie [8] and Pepper are used in public places, but their primary purpose is not guidance, as they are rather a "communication robot" and a "personal robot which reads emotions". SANCHO [9] is intended for guidance, but it is not humanoid, and has limited expression and interaction capabilities. Androids could be used in a role of interaction with customers in public spaces: Geminoid [10] has been employed in a shop, while ASKA [11] is receptionist robot. However, their potential use has to face severe limitations which include expressiveness immobility, and concerns about uncanny appearance due to close resemblance to humans [12]. We have to take these issues in consideration when developing a new social robot.

Thus, by taking into account the essential characteristics of an empathic guidance robot, this paper presents IOmi, a life-sized female Humanoid Robot designed and implemented to interact socially in indoor public spaces with humans. The result, which emulates the shape of a woman hostess with simplified human-like movements, was made through a design methodology inspired to industrial design criteria. The robot provides information of the services inside the new Academic Innovation Complex of PUCP in a scenario similar to the concept art shown in Fig. 1. The significance of the name IOmi is I/O for input/output, and "mi" beautiful in Japanese. Its design is specifically intended for guidance, thus it is based on requisites including mobility, female appearance and one-to-many interaction ability.

Fig. 1. (a) Conceptual art of IOMI interacting with groups of people and (b) set of gestures: (1) Greeting Stance; (2) Waiting Stance; (3) Follow-me Stance; (4) Pointing Stance

This paper is organized as follows. In Sect. 2 the design methodology is described. In Sect. 3 mechanical, electronic and interactive features are described. In Sect. 4, the preliminary test results and discussion. Section 5 concludes the paper.

2 Design Methodology

Useful concepts can be found in previous research about design guidelines. The work in [13] provides ideas such as the influence of nature and the use of rounded forms. The prototype of Probo [14] is interesting regarding materials: it is made mainly of aluminium, a strong, lightweight and tractable material. The realisation of Flobi [15] also brought interesting concepts like the "hole-free" design without any visible conjunctions, and the modular implementation. However, no details about the design methodology in terms of process are specified. This is not the case of SnackBOT [16], which is a process of four phases: Design activities and goals; Needs analysis; Form giving & Interaction design; Documentation & evaluation.

Such kind of industrial design criteria is desirable as the usual methodologies, applied by industrial designers, consist of a considerable amount of trial-and-error tests for the prototype until the final state is obtained [17]. Nonetheless, two models, proposed by Bruce Archer [18] and Hans Gugelot [19], define clear steps to reach a final prototype while having broad scopes for choosing information and presenting proposals. These two procedures, contrasted in Table 1 (first two columns), follow a scientific method layout, where a problem is solved by a hypothesis. The main difference is that while Archer ensures a high quality prototype by revalidating the design proposal, Gugelot accelerates the overall process by directly implementing the design. One common trait of the two approaches is that they are "waterfall" models, in which each step follows another. Following these considerations we adapted the model, named IO-CoDe (IOmi's Contextual Design) proposed in Table 1, to design our robot. It comprises two loops: from Step 4 to Step 3 and from Step 6 to Step 5. The reasons are: (i) volatile technology, materials and algorithms, (ii) the importance to understand the context of the situation (users and environment). Next, we describe each of them in the case of IOmi's design process.

Table 1. Design methodologies

	B. Archer	**H. Gugelot**	**IO-CoDe**
Step 1	Problem Definition	Information Stage	Identifying the Context
Step 2	Obtain relevant data	Research Stage	Defining the Requisites
Step 3	Data Analysis	Design Stage	Early Designs
Step 4	Prototypes development	Decision Stage	Engineering Validation
Step 5	Validate Design proposal	Calculation Stage	Prototype
Step 6	Production's documents	Prototype construction	Testing & Final Validation

2.1 Step 1 – Identifying the Context

In this step, the environment and the users must be identified. The location where the robot will be deployed is the Innovation Academic Complex at the Pontificia Universidad Católica del Perú (PUCP), a facility that mixes a university library, a social cafeteria, digital manufacturing technologies, and other facilities. However, although there are information panels throughout the complex about the services provided, they tend to be ignored and users prefer asking at reception for information. It can be inferred that students prefer social interaction to obtain information rather than acquire it from inanimate sources.

2.2 Step 2 – Defining the Requisites

According to the context and the problems highlighted in Step 1, requisites must be determined to provide plausible solutions. In the case of IOmi we identified: (1) Communication: The robot should present human-like features that can express non-verbal utterances including pointing, turning, facial expressions, and head movement. They should be clearly visible as the interaction is one-to-many; (2) Mobility: the robot should be capable of moving autonomously as well as to be teleoperated. In both cases, obstacle sensing is necessary; (3) Robustness: complexity of the whole system has to be limited in order for the robot to be robust to be employed in everyday use and have physical contact with users; (4) Safety: robot with rounded shapes, and also covering mechanical parts and electrical wires is necessary; (5) Human likeness: minimising complexity of forms is necessary in order to give an impression of robot's intelligence that matches with reality and to avoid uncanny effects; (6) Gender: in order to understand the aware-ness of gender in a robot in a Latin American context, a female appearance is selected. Some of these requisites were verified in the preliminary experiments we describe in Sect. 4; others will be in the future in more specific tests.

2.3 Step 3 – Early Designs

In this step, a first proposal is designed through a creative process, according to the context identified. The schemes are presented in form of drawings, diagrams, 3D models, and mock-ups. In particular, sketches make possible to explore the appearance possi-bilities. In the case of IOmi, several alternatives were proposed for each body parts, including the shape of the eyes (more rounded/elongated). Among IOmi's early design sketches for the head, as it can be seen in Fig. 2 (face development), we chose the 3rd design (F.3) as the most feminine. The intention of the form of the arm was to welcome guests to the Innovation Academic Complex, for this reason the arms are arcuate and eliminates the use another motor in the arm to represent the elbow moment as shown in Fig. 2 (A.1). The final proposal (M.3) was based in a human host using a long skirt, as proposed in the first design, but more convex in order to provide space for sensors, motors, wheels, etc. The form of the eyes in Fig. 2 (E.3) is selected for the final model, and also to reinforce the appearance of a hostess using make up, like eyebrows to delin-eate their eyes, also satisfy the technical details required of a LED display matrix. The mouth was ignored in this proposal as it will be difficult to perform motion and transmit the idea of communication.

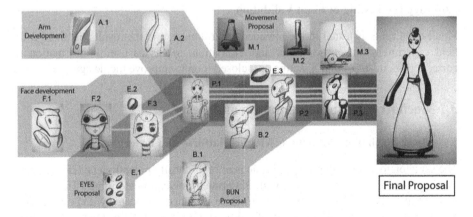

Fig. 2. Morphological design matrix for the humanoid robot IOmi.

2.4   Step 4 – Engineering Validation

In this step, the proposed designs are evaluated technically to ensure they are mechanically and electronically viable, and the best option was selected. Whenever no suitable option is available, the process goes back to Step 3. Calculations and manufacturing considerations, such as static and dynamic evaluations, motors' maximum torque, power and electrical requirements, type of communication, interaction protocols, among others. From these considerations, the proposed sketched is shown in Fig. 3, as well as the CAD modelling of the robot and the implemented assembly of the overall structure.

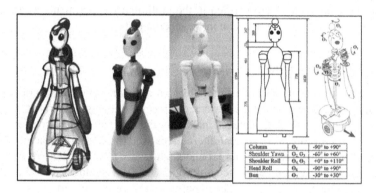

Fig. 3.   Sketch, CAD, 3D printed & fiber glass, dimensions & joint ranges.

2.5   Step 5 – Prototype

After solving the technical issues, the manufacturing process is started. This includes selecting an adequate building method, searching for services providers, and ensuring the implementation is within the expected budget. Section 3 details the technical aspects of the robot IOmi.

2.6   Step 6 – Testing & Final Validation

Finally, the robot is tested in the intended environment, and the results obtained from field experimentation, observation and evaluation, serves as useful information to correct errors in step 4 and 5, acting as a closed loop to improve the design.

3   Mechatronic Implementation and Test

3.1   Body Mechanics: Structure, Mobile Platform and Exterior

The skeleton to support the robot's frame is of high-density polyethylene (HDPE) with aluminium beams, in conjunction with 3D printed polycarbonate joints. The external case is a fiberglass with polyester resin structure, polished and coated with polyurethane painting. External and internal parts of the head are made of 3D printed plastic ABS.

Figure 3 presents the overall dimensions and joint ranges of IOmi. It has 7 degrees of freedom (DOF) distributed in three sections: (1) head mechanism with 1 DOF for the neck (yaw motion), 1 DOF for the bun which represents a physiological utterance or a non-verbal feedback of communication, and two 8 × 8 LED matrices located behind blurred plastic shields are used as the robot's eyes in order to perform the illusion of different expressions and emotions; (2) waist mechanism with 1DOF yaw motion by means of a servomotor supported by a cylindrical bearing to distribute the load throughout the skeleton and not only the axis of the actuator; shoulder mechanism that enables 2 DOF, yaw and roll motion, for each arm.

3.2 Sensing and Control: Electronic Parts

IOmi possesses an electronic architecture for remote control operation with obstacle detection. Furthermore, it is able to gather audio and visual data with a webcam attached. The electronics are powered with a 12 V battery with a step-down regulator for other components. The main processes of the robot are governed by a Beaglebone embedded board, different sections of IOmi are controlled by additional microprocessors to simplify the load of the main controller. There are two ways of controlling the robot to interact with people: let the robot make a response based on special keywords, or use a PC to remotely decide what to respond. The robot is able to connect to a remote device via Wi-Fi to control the response. Obstacle sensing is performed with ultrasonic sensors, at the bottom of the robot's skirt. The eyes an bun perform the interactive features of the robot. Eyes are conformed by two LED matrices, each located at eye level in the head. Each matrix creates a pixelated eye shape that changes varying the response of the robot, and creates the illusion that IOMI has expressive eyes; while the hair bun adds an additional non-verbal feedback that acts as physiological communication.

3.3 Interaction

IOmi's behaviour is intended to be host-like in order to introduce to people the services provided at the Academic Innovation Complex. For this reason, several movement patterns will be tested: (1) Greeting Stance, when a new user approaches the robot at the Innovation Academic Complex; (2) Waiting Stance, when IOmi is waiting for a question to be asked; (3) Follow-me Stance, when the agent expresses a walking direction; and (4) Pointing Stance, when IOmi points to a specific location inaccessible to the robot and invites the users to follow a path.

IOmi first appearance and presentation at PUCP was at the Innovation Academic Complex at the MIT Technology Review's Young Innovators Under 35 Award event as shown in Fig. 4, (www.youtube.com/watch?v=iBgyxBpV8C0). The robot was tele-operated by one of the researchers and sustained fluid conversations with people providing information about the services inside the building. Many observers were curious about the robot and its interactions, asking specially about capacities, and capturing personal pictures (selfies) with IOmi. In general, the audience showed acceptance towards the robot.

Fig. 4. Experiments with IOmi at PUCP inside the Academic Innovation Complex.

4 Preliminary Experiment

4.1 Experiment Description

We carried out an experiment to evaluate the usefulness of the robot and the general impression among users. IOmi offered guidance by moving, using speech and gestures to 20 participants from a total of 40 (25 male; 15 female; mean age: 20.2; all students) inside a university. The other 20 were guided by a written message containing the same indications. The experiment consisted in three phases: (1) fill a preliminary questionnaire; (2) reach a laboratory following indications of going upstairs and opening the door on the left hand side, (3) fill a final questionnaire. Participants were not aware of the fact that a robot would perform guidance. A few more conditions were controlled at the end of the experiment in order to exclude unfit participants: whether they had previous experience with humanoid robots before (mean value: 1.62 on a scale from 1 for no experience to 5 for very expert); and if they already knew the location (mean value: 1.45). Thus, all the 40 participants were included in the results. The preliminary questionnaire is shown in Table 2, to be answered using 5 points differential semantic scales. After the interaction, the robot was evaluated through the Godspeed scales "Likeability" and "Perceived Intelligence" through questions about the appearance. Results are shown in Table 3. The group of 20 participants who were given written indications did not answer to the questions regarding the robot. The effectiveness of written message (mean: 4; STD: 1.03) was found to be not significantly different from IOmi's guidance effectiveness.

Additional comments were gathered. Among them, some mentioned the motion being very slow and causing distress (3 out of 20) or very abrupt (1/20). Voice was also mentioned (2/20) because of the low volume; as well as lack of cognition (2/20). Regarding emotional response among the participants, some (3/20) stated their surprise, and one even stated "it likes me".

4.2 Discussion

Although all the evaluations were positive, the main limitation of our study is that we cannot compare our data with a reference, i.e. other guidance robots in the same context. We compared the robot with written messages this opportunity, but the comparison with guidance means such as maps or interactive panels is also planned in the next experiments.

Table 2. Pre-experiment questionnaire: mean values and standard deviation

Questionnaire item	Mean	ST.D.
When you visit an unknown place, do you prefer to by guided by:		
A map	3.35	1.18
An interactive panel	3.93	0.75
A person	4.08	0.99
In case of a person doing a guidance job:		
Which gender should be? 1=Male; 5=Female	3.35	0.99
Is race important?	1.5	0.93
Is the type of voice important?	3.78	1.25
Is language important?	4.38	0.62
Are gestures important?	4.10	0.97
Is mobility important?	4.00	1.27

Table 3. Post-experiment questionnaire: mean values and standard deviation

Questionnaire item	Mean	ST.D.
Was the guide effective?	3.60	0.99
Dislike / Like	4	0.79
Unfriendly / Friendly	4.1	0.91
Unkind / Kind	4.65	0.59
Unpleasant / Pleasant	4.1	0.79
Awful / Nice	4	0.79
Incompetent / Competent	4.05	0.78
Ignorant / Knowledgeable	3.95	0.76
Irresponsible / Responsible	4.05	0.89
Unintelligent / Intelligent	3.70	0.86
Foolish / Sensible	3.55	0.60
Did you like her colour scheme?	4.05	0.94
Did you like her shape?	4.15	0.75
Did you like her voice?	3.75	1.36

We did however collect generic evaluations on preference to be guided by a map, an interactive panel or a person, as shown in Table 2, and we can compare those to the effectiveness of the robot (Table 3). Through Kruskal-Wallis test, significant difference was detected (p-value = 0.02). Subsequently, IOmi obtained higher score than the map (marginally significant: one tailed p-value = 0.46). As the guidance by a person obtained the highest score, followed by the panel, it is clear that the robot still is not as effective.

Another point of discussion revolves around the appearance of the robot. All the considerations in this category are the feedback from step 6 of IO-CoDe to step 5. Voice and movements are aspects to be improved. Furthermore, analysing open comments, from the fact that participants referred often to the robot as "it" rather than "she" we can infer that the female traits should be stressed out. According to Table 2, gender was actually not considered a primary requirement by the participants. On the other hand, the presence on one comment ("it likes me") which attributes some emotional attachment by a robot that did not display any emotion, is a hint that the acceptance of the robot was positive.

5   Conclusions and Future Work

This paper presented the humanoid robot hostess IOmi, which will be employed in a social space in university context. Its creation was carried out through a methodology inspired to industrial design. The shape of the robot reminds a female hostess, while on the other hand its simplified humanoid design still is capable of non-verbal expression for one-to-many interaction. IOmi's implementation is in the last step of the IO-CoDe model. The results from a preliminary experiment run with 40 participants gave some useful indications about aspects to improve. This research and its results will possibly provide a contribution to the diffusion of robotics in the Latin American context.

Acknowledgements. The authors thank Pontificia Universidad Católica del Perú (PUCP) for providing the necessary means to implement IOmi, and the Center of Advanced Manufacturing Technologies (CETAM) for assistance with manufacturing.

References

1. Aldebaran. Who is Pepper?. https://www.aldebaran.com/en/a-robots/who-is-pepper. Accessed 29 June 2015
2. Breazeal, C.: Designing Sociable Robots. MIT Press, Cambridge (2002)
3. Eyssel, F., Hegel, F.: (S)he's got the look: gender stereotyping of robots1. J. Appl. Soc. Psychol. **42**(9), 2213–2230 (2012)
4. Wang, Y., Young, J.E.: Beyond pink and blue: gender attitudes towards robots in society. In: Proceedings of Gender and IT Appropriation. Science and Practice on Dialogue – Forum for Interdisciplinary Exchange, p. 49, ISBN 978-1-4503-2105-1
5. Robertson, J.: Gendering humanoid robots: robot sexism in japan. Body Soc. **16**(2), 1–36 (2010)
6. Walters, M.L., et al.: Avoiding the uncanny valley: robot appearance, personality and consistency of behavior in an attention-seeking home scenario for a robot companion. Auton. Robot. **24**(2), 159–178 (2008)
7. Fuchs, M., Borst, C., Giordano, P., Baumann, A., Kraemer, E., Langwald, J., Gruber, R., Seitz, N., Plank, G., Kunze, K., Burger, R., Schmidt, F., Wimboeck, T., Hirzinger, G.: Rollin' Justin - Design considerations and realization of a mobile platform for a humanoid upper body. In: IEEE International Conference on Robotics and Automation (ICRA) (2009)
8. Kanda, T., Shiomi, M., Miyashita, Z., Ishiguro, H., Hagita, N.: An affective guide robot in a shopping mall. In: 4th ACM/IEEE International Conference on Human-Robot Interaction (HRI), La Jolla (2009)
9. Gonzalez, J., Galindo, C., Blanco, J.L., Fernandez-Madrigal, J.A., Arevalo, V., Moreno, F.A.: SANCHO, a fair host robot. a description. In: IEEE International Conference on Mechatronics, pp. 1–2, April 2009
10. Nishio, S., Ishiguro, H., Hagita, N.: Geminoid: teleoperated android of an existing person. In: Humanoid Robots, New Developments, pp. 344–347, June 2007

11. Jun'ichi, I., Youhei, M., Yoshio, M., Tsukasa, O.: Interaction of receptionist ASKA using vision and speech information. In: IEEE Conference on Multisensor Fusion and Integration for Intelligent Systems, pp. 355–356 (2003)
12. Mathur, M.B., Reichling, D.B.: Navigating a social world with robot partners: A quantitative cartography of the Uncanny Valley. Cognition **146**, 22–32 (2016)
13. Mitchell, S., et al.: Design choices in the development of a robotic head: human-likeness, form and colours. In: Ceccarelli, M., Glazunov, V.A. (eds.) Advances on Theory and Practice of Robots and Manipulators, pp. 225–233. Springer, Heidelberg (2014)
14. Saldien, J., Goris, K., Yilmazyildiz, S., Verhelst, W., Lefeber, D.: On the design of the huggable robot probo. J. Phys. Agents **2**(2), 3–12 (2008)
15. Hegel, F., Eyssel, F.A., Wrede, B.: The social robot Flobi: key concepts of industrial design. In: Proceedings of the 19th IEEE International Symposium in Robot and Human Interactive Communication (RO-MAN 2010), pp. 120–125 (2010)
16. Lee, M.K., Forlizzi, J., Rybski, P.E., Crabbe, F., Chung, W., Finkle, J., Glaser, E., Kiesler, S.: The Snackbot: documenting the design of a robot for long-term human-robot interaction. In: 2009 4th ACM/IEEE International Conference on Human-Robot Interaction (HRI), pp. 7–14 (2009)
17. Bayazit, N.: Investigating design: a review of forty years of design research. Des. Issues **20**(1), 16–29 (2004)
18. Archer, B.: A view of the nature of design research. In: Design: Science: Method, pp. 30–47 (1981)
19. Bürdek, B.: Design and methodology. In: Design: History, Theory and Practice of Product Design, pp. 225–272. Birkhäuser (2005)

Colleague or Tool? Interactivity Increases Positive Perceptions of and Willingness to Interact with a Robotic Co-worker

Benjamin C. Oistad[1(✉)], Catherine E. Sembroski[2], Kathryn A. Gates[1],
Margaret M. Krupp[1,2], Marlena R. Fraune[1,3], and Selma Šabanović[2,3]

[1] Department of Psychological and Brain Sciences, Indiana University, Bloomington, USA
{boistad,katgates,mmkrupp,mrfraune}@indiana.edu
[2] School of Informatics and Computing, Indiana University, Bloomington, USA
{csembroski,selmas}@indiana.edu
[3] Cognitive Science Program, Indiana University, Bloomington, USA

Abstract. Human-robot interaction is increasingly likely in workplaces in the near future. In "co-working" relationships, humans may appreciate socially interactive robots for their anthropomorphic likeability, or functional robots for their strict task orientation. The current study examines the comparative perceived advantages of robots that behave interactively or functionally towards humans during a task with a superordinate goal. Survey results from 33 participants assessed perceptions of, and perceived cooperation with, robots during the task. Results indicated that participants stood physically closer to and rated Interactive robots as more anthropomorphic, sympathetic, and respected than Functional robots, but they did not rate the two types of robots differently in terms of cooperation. The more participants anthropomorphized, sympathized with, and respected the robots, the more willingness they reported to working with robots in the future.

1 Introduction

In contrast to earlier visions of industrial robots as isolated machines replacing human labor, contemporary plans to develop robots as "co-workers," "co-explorers," and "co-inhabitants" in everyday contexts involve humans and robots working side by side, and even in direct collaboration with each other [1, 2]. In scientific competitions like Robocup@Home and Robocup@Work, service robots for the home are expected to be anthropomorphic and interactive, whereas robots in work contexts are expected to look more mechanical and be more functionally autonomous [3]. However, peoples' perceptions and evaluations of co-robots like Baxter in work environments, suggest that people would like even industrial robots to be more social [4]. Collaborative roles for robots to work alongside humans, introduce questions about appropriate ways to design robots as co-workers. If robots and humans are working towards a common goal, how should the robots behave? This paper examines one set of variables in this domain: whether robotic co-workers should be designed to be merely functional or have some level of interactivity with humans as they work on a task.

© Springer International Publishing AG 2016
A. Agah et al. (Eds.): ICSR 2016, LNAI 9979, pp. 774–785, 2016.
DOI: 10.1007/978-3-319-47437-3_76

2 Background

Practitioners have recently begun introducing robots into the everyday workplace. Researchers are still examining how to facilitate this integration by focusing on how people perceive robots based on their design and behavior.

2.1 Social Robots in the Workplace

People prefer robots' appearance and behavior to match the tasks they perform [5]. For example, on fun and social tasks, people preferred socially interactive robots, and on serious tasks, they preferred serious robots [6]. However, it is unclear which situations fall into these categories. For example, it is not obvious if humans "co-working" with robots during a cleaning task would prefer a socially interactive robot or a functional robot only focusing on the task.

Participants reacted more positively to socially interactive rather than merely functional robots in the workplace, making robots even more social by giving them names and relating their experiences with the robots in social terms [7]. Workers related to "co-robots" socially even in industrial settings, wanting to chat with them or "shoot the breeze" [4]. Competing with robots mechanical in appearance, produced suggestions that the robot include more humanlike characteristics, like "talking trash" during the game [8]. Introducing robotic technology as social, and adapting the attributes of service robots, may improve HRI in which robots work side by side humans.

2.2 Anthropomorphism and Similarity

Social psychology has found that perceived similarity between two humans correlates with individuals perceiving the other as more humanlike and liking the other more (e.g., [9, 10]). Even sharing something arbitrary in common such as a "team color" increases liking (i.e., minimal groups paradigm [11]).

Anthropomorphism refers to humanlike appearance (e.g., has eyes) or behavior (e.g., has good etiquette). When machines share traits with humans (e.g., by exhibiting anthropomorphic capabilities), people attribute more mental capability to them [12]. Participants also anthropomorphized and liked robots more when they perceived the robot's voice to match participant's own gender [13].

2.3 Emotion and Proximity

The distance people stand from other social actors can also signify how well they know the other actor, and how positively they evaluate them. Researchers suggest that there are four types of spaces: public, social, personal, intimate [14]. Standing close to social actors increases positive emotions and liking, and vice versa [15].

2.4   Cooperation

Previous HRI literature has examined the effects of humans directing robots [16] as well as robots directing humans [17, 18]. When a robot directed humans, humans expected their attitudes to fit the task at hand [5]. Participants complied more with a serious robot directing an exercise task and a playful robot directing a jellybean tasting task. This data indicates the importance of robot behavior during cooperative tasks that the robot directs. The current study examines the effect of robot behavior when the robot is working alongside the human instead of being a director.

Common Ground Theory suggests that teammates sharing more similarities leads to greater cohesion in completing a task [19]. A more equal, "co-working" relationship between a human and a robot may eliminate hierarchy and suggest more similarity between the human and robot, thus increasing cohesion. Creating a more anthropomorphic robot, or a robot that behaves more interactively than functionally might also increase perceived similarity between teammates and maximize cooperation.

2.5   Present Study

This study investigates whether either functional or interactive robot behavior is appropriate during a task that simulates cleaning, organizing, or construction (i.e., box-sorting task) in which robots and people work toward a shared goal rather than toward divided responsibilities. This redefines "co-working" from solely working in hierarchal scenarios to collaborating on an equal plane to achieve a common goal. This may influence human attitudes toward robots, cooperation with them, and willingness to interact with them in various workplaces. We hypothesized that:

H1. Participants will view Interactive robots as more humanlike and Functional robots as more machine like.
H2. Participants will perceive Interactive robots as more cooperative than Functional robots.
H3. Participants will be more willing to interact in the future with Interactive than Functional robots.

3   Method

3.1   Study Design

This experiment involved a between-subjects design with two conditions: robots showed either Interactive or Functional behavior when completing a box-sorting task alongside participants (*see behavioral differences between conditions in* Table 1).

Table 1. Differences in Robots' behavior between conditions

Functional	Interactive
• Only approached participants at very beginning of first task	• Approached participants and nodded
	• Nodded to participants when they passed
• Did not nod or look around at boxes or participants	• Looked from side to side at boxes and participants
• Did not yield to participants during task	• Approached and nodded to participants during intermission
• Remained still during intermission	

3.2 Robot

This study used a robot with a body that was neither too anthropomorphic nor too mechanomorphic to keep participants from using appearance as a cue for how socially to view the robots, as in previous research [20]. In this way, the robot behavior (functional or interactive) might be more salient.

We used the "minimally social" Sociable Trash Box (STB) robots (Fig. 1) [21] which have anthropomorphic traits (e.g., eye, arms), but are still far from humanlike. The STBs have no legs, but their motion is a gentle left-and-right that mimics swaying back and forth from small steps forward. We expected these minimal social cues would not bias participants to respond socially or non-socially toward the robots.

Fig. 1. The three STBs used in the study.

3.3 Procedure

The experiment took place in the R-House Living Lab. Participants participated individually, and were introduced to the project by the experimenter reading this script:

"More robots are being produced every day, to help in factories, but also to teach children, to give company to the elderly, and to help with everyday tasks like cooking and cleaning. Because people will soon interact with more robots, it is our goal to understand what people think about the robots and how useful the robots will be in particular situations. [Show both robots to

Participant] "This is Red [point to first robot] and this is Blue [point to second robot]. Robots, this is (participants name). The three of you will go through two tasks together. To simulate a situation in which humans and robots might work on some of the same tasks simultaneously, all of you will be doing the same task at the same time."

Examples of human centered HRI were listed to give participants a broader scope regarding the usefulness of robots. The experimenter avoided defining the relationship between the robots and human as "co-working" to see how participants' perceptions of cooperation with robots would develop in the course of the task.

Box Sorting Tasks. The study involved a box sorting task, a subsequent intermission, and a second box sorting task. The layout consisted of 11 boxes scattered in the middle of the room (there was a variety of light or dark and big or small boxes). The surrounding four walls each had one checkpoint and one goal, each identified by a specific box category (light boxes, dark boxes, big boxes, and small boxes; *see* Fig. 2).

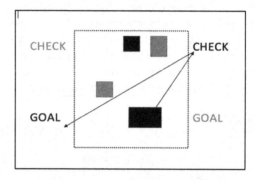

Fig. 2. Task 1 Flow Chart (**Boxes are first taken to their corresponding colored checkpoint, then the corresponding colored goal*)

Task 1. Participants and STBs sorted the boxes by color, taking the boxes from the middle of the room to either the light or dark checkpoint (robots could skip the checkpoint), and finally, to the corresponding goal opposite it (*see* Fig. 2). Checkpoints served as a handicap for the participants because the STBs were considerably slower in performing the task than the humans. Participants were instructed to move the boxes by using a shuffling motion with their feet (rather than using their hands or kicking), in order to mirror how the STBs move. Participants completed sorting one box before they could sort another box. The experimenter explained the task, instructed one STB to demonstrate how it would sort of box, asked participants to practice sorting one box, and then left the room for the duration of the task.

Intermission. After Task 1, the experimenter reentered the room for an intermission to give participants and robots "reward" tokens supposedly based upon performance. In actuality, participants always received two tokens indicating high performance and to suggest that their collaboration with the robot was successful.

Task 2. In the second task, boxes were sorted by size instead of color. Using the same rules, participants and robots sorted the boxes to their corresponding goals. Once both tasks were complete, the experimenter again rewarded robots and participants for their performance with tokens. Then Interactive (but not Functional) robots nodded goodbye to participants, and the experimenter led participants into the next room to complete a series of questionnaires about the experience.

3.4 Materials

i *Control of Robots*

Two researchers (experimenter and assistant) controlled the Sociable Trash Box (STB) robots using the Wizard of Oz paradigm. Drivers viewed the task through a live feed camera and controlled the robots wirelessly. The assistant controlled both robots while the experimenter introduced the robots. During the task, the assistant and the experimenter each drove one robot.

Interactive robots were made to nod or look around as they passed participants. Functional robots ignored the participant's presence. When they had no more boxes to sort, the robots stopped moving. During intermission, Interactive robots approached participants and nodded. Functional robots stood still, occasionally turning left or right to maintain a similar level of animation without interacting with participants.

ii *Survey*

Following the task, a survey was administered to participants electronically via Qualtrics. Participants were asked to describe the STBs with free-response adjectives. Then they rated the robots using seven-point Likert scales from 1 *Strongly Disagree* to 7 *Strongly Agree* on the measures described below.

Robot autonomy/design check. Participants rated how they thought the robots were acting (e.g., "pre-programmed," "on their own," "as a group," "controlled remotely"). This question was asked last to avoid biasing how participants thought of the robots, but it is reported first to examine if participants thought they were interacting with robots or indirectly with humans ("controlled remotely").

How Participants viewed the robots. Participants were asked to "describe the robots with three descriptive words or phrases". Reoccurring words were measured for frequency in the participant responses.

Anthropomorphism. Participants rated the STBs on Ezer analogy statements (e.g., the STB is "like a pet" or "like a toy") [22].

Emotion toward Robots. Participants reported their attitudes and emotions toward STBs, with statements like: "I feel disgust toward the robots," "I respect robots," [23].

Proximity to robots. To determine comfort standing near robots, researchers measured how close participants stood to robots between tasks from recorded interactions.

Cooperation. Participants were asked whether they believed they cooperated or competed with the robots and whether or not the participants or the robots contributed to completing the tasks. Participants also indicated whether they would want to work with or avoid the STBs in the future (e.g., "I felt that the STBs and I cooperated during the tasks," "I would feel happy to work with these robots on future tasks").

Prisoner's Dilemma. To measure behavioral cooperation, participants played a Prisoner's Dilemma Game with the STBs, using the tokens they had won during the box-sorting task. If participants gave no tokens to the robots and the robots gave no tokens to the participants, then the participants would receive $0.60. If participants gave half their tokens to the robots and the robots gave half their tokens to the participants, the participants won $1.20. The experimenter told the participant the robot matched whichever decision the participant made.

Willingness to Interact in the Workplace. Participants reported their willingness to work with robots across contexts (e.g., "I think robots should work in classrooms").

Demographics. Finally, participants answered demographic questions (e.g., age, gender, technical experience).

4   Results

Data were analyzed in SPSS, using primarily independent-samples t-tests. Values of $p < .050$ were considered statistically significant.

Robot Autonomy. Most participants across conditions thought robots were preprogrammed ($N = 19$). The next most frequent choices were that the robot was autonomous ($N = 7$), controlled as a group ($N = 3$) and remotely controlled ($N = 2$). A chi squared test found no difference in ratings between conditions ($p > .100$).

How Robots Were Viewed. In free response questions, participants most commonly described the robots as helpful (61 %). The next most common description was slow (42 %). These adjectives did not differ across condition ($p > .100$). This was measured by counting the frequency of which these words appeared in participant responses.

Anthropomorphism. On Ezer's scale, participants rated STBs in their similarity to animate and inanimate descriptors. Participants viewed Functional STBs as more like appliances, (Functional $M = 3.83$, $SD = 0.92$; Interactive $M = 3.18$, $SD = 0.95$; $t(32) = 2.03, p = .050$), and less like friends (Functional $M = 2.22$, $SD = 1.06$; Interactive $M = 2.94$, $SD = 0.97$; $t(32) = -2.12$, $p = .042$), and teammates (Functional $M = 2.56$, $SD = 1.20$; Interactive $M = 3.53$, $SD = 1.01$; $t(32) = -2.96$, $p = .006$) than Interactive STBs. No other differences were found on Ezer's analogies.

Emotions Toward Robots. Participants reported feeling more sympathetic, ($t(32) = -2.27$, $p = .03$) and respectful, ($t(32) = -1.76$, $p = .065$) towards Interactive ($M = 3.24$, $SD = 1.20$; $M = 4.00$, $SD = 0.71$ respectively) than Functional STBs

($M = 2.33$, $SD = 0.97$; $M = 3.44$, $SD = 0.92$ respectively). No other emotions showed significant differences between conditions.

Proximity to Robots. By dividing the floor space into a 4×5 grid (with boxes approximately 2' x 3') during video coding, the distance between participants and the robot was measured by spaces between the "co-workers" after the first task. Participants stood closer to Interactive robots ($M = 1.36$, $SD = 0.96$) than Functional robots ($M = 2.07$, $SD = 1.20$, $F(1,28) = 4.68$, $p = 0.39$, $n_p^2 = .143$). After the intermission and after the second task, however, the difference between Functional ($M = 2.22$, $SD = 0.95$; $M = 2.15$, $SD = 1.06$, respectively) and Interactive ($M = 1.65$, $SD = 1.18$; $M = 1.81$, $SD = 1.03$, respectively) was not significant.

Cooperation Ratings. Robots were rated more cooperative (Functional $M = 5.89$, $SD = 0.90$; Interactive $M = 6.06$, $SD = 1.20$) than competitive (Functional $M = 3.56$, $SD = 1.62$; Interactive $M = 2.77$, $SD = 2.08$) overall. T-tests indicated no significant differences for cooperation or competition between conditions ($ps > .050$).

Prisoner's Dilemma. Most participants gave their tokens to the robots (Functional 74 %, Interactive 76 %). A chi squared test showed no difference in behavioral cooperation as measured by the PDG (i.e., participants were similarly likely to give their tokens to the robots both conditions $p > .100$).

Willingness to Interact in the Workplace. Participants rated their desire to interact with robots in eleven different work contexts (e.g., food establishments, exercise facilities) on a scale from 1–7. A series of t-tests found no differences between conditions ($p > .100$).

Because conditions did not directly affect willingness to work with robots, Pearson's correlations were run between willingness to interact and the following: cooperation, Ezer's analogies, and emotions. Cooperation and willingness to interact showed no significant correlation ($p > .100$). Results from the Ezer's Analogies indicated moderate positive correlations with willingness to interact and rating robots as friends ($r = .424$, $p = .001$) and as teammates ($r = .596$, $p < .001$). Emotions showed some correlation with willingness to interact; respect towards robots had a moderate positive correlation ($r = .472$, $p = .004$), but sympathy showed no correlation ($p > .100$).

5 Discussion

5.1 Robot Autonomy/Design Check

A majority of participants reported believing that robots were pre-programmed. Only two participants reported believing that the robots were remotely controlled. This suggests most participants behaved toward the robots as though they were autonomous (despite the use of the WoZ technique), and indicates that participants did not behave as though they were interacting with humans behind the robots.

5.2 How People Generally Viewed the Robots

When participants were asked to describe robots with three adjectives, the most commonly used were "helpful" and "slow." That participants viewed the robots as helpful suggests that even despite the robots' lack of efficiency, they appreciated their help. Descriptions of robots as slow may have been elevated because the experimenter explains that the robots have a handicapped task due to the robots' slow speed (i.e., only participants had to take boxes to checkpoints).

5.3 Anthropomorphism

H1, stating participants would view Interactive STBs as more humanlike than Functional STBs, was supported; participants perceived Interactive STBs more as friends and teammates and Functional STBs more as appliances. Participants also respected and sympathized with Interactive robots more than Functional ones. Participants also stood closer to Interactive than Functional robots, suggesting they were more comfortable with them. Taken together, these results suggest that Interactive robots were more likely to be viewed as humanlike or coworkers, whereas Functional robots are more likely to be used as tools. Even small interactive behaviors can increase anthropomorphic percep-tions of robots. Participants' increased respect and sympathy for Interactive robots falls in line with results suggesting that humanization increases empathy toward a target [24]. Interactive robots may therefore be more useful in situations where empathy can enhance collaboration. This also supports literature suggesting that users would prefer co-working robots to be more socially interactive [25].

5.4 Cooperation

Robots were rated as similarly cooperative and competitive across conditions, failing to support H2 (i.e., that people would view Interactive robots as more cooperative than Functional robots). Participant cooperation with robots, as measured by the Prisoner's Dilemma Game, also showed no difference across condition. This suggests that the Interactive robots did not elicit more cooperation than Functional robots. However, some of the measures could be improved for future studies. It may be that in a low-stakes task such as in the present study, robot interactivity does not affect how cooperative partic-ipants find robots. In the PDG, a higher reward may have changed the results. Cooper-ation during the task could have been unacknowledged because of the intentionally undefined relationship between the robot and human. This suggests a cooperative rela-tionship is not inherent when humans are simply told to "work alongside" a robot. Cooperative relationships might need to be strongly implied.

5.5 Willingness to Interact

Hypothesis 3, that Interactive STBs would increase participants' willingness to interact with robots in the future, was not supported directly. However, results indicated that the experimental condition affected willingness to work with robots in different work contexts, lending partial support to hypothesis 3; Interactive robots were anthropomorphized and respected more by participants than Functional robots, and anthropomorphism in respect positively correlated with desire to interact with robots in various contexts. Willingness to interact more strongly related to these human-like emotions and attributions than to ratings of cooperation. This suggests that perceived cooperation is not what affects humans' willingness to work with robots, but human-like attribution shows an effect on workplace interaction, at least in "co-working" tasks.

5.6   Limitations

While the current findings support the importance of designing interactivity into robots to increase willingness to interact with robots in the workplace, there are limitations concerning the interpretation of participant intentions, technical limits, as well as participant confusion during the task experiment design.

The briefings before the task informing participants of the usefulness of robots in non-industrial settings may have influenced the participants' perceptions of the STBs by priming participants to overestimate their usefulness in various work contexts that were significant in the results. In contrast, the STBs were much less effective at sorting boxes than humans due to slow motor capabilities. This was explicitly mentioned to the participant and handicaps were put in place to control for this disadvantage (i.e., robots took boxes directly to the goal). This lack of efficiency could have decreased perceptions of human-robot cooperation or reduced the variability across conditions.

5.7   Future Work

It would be interesting to perform the task with robots that could do the task more quickly to determine the effect of robot efficiency on participants' perceived cooperation and competition with the robots. We also suggest more closely examining the relationship between participant respect toward robots, anthropomorphism, and willingness to interact with them. Future studies might do so using more efficient robots.

6   Conclusion

Overall, participants responded to robots' interactive behavior differently than to their merely functional presence, and robots' interactive behavior enhanced human ability to imagine the robots as friends and teammates. Interactive behavior increased anthropomorphism of, respect toward, and sympathy with the robots, influence people to stand closer to them. Additionally, people who reported feeling respect and sympathy with robots also expressed more willingness to work with the robots. Overall, these findings suggest that even in some functional situations, robots can use cues of interactivity to increase positive perceptions towards and willingness to interact with them.

References

1. Decker, M., et al.: Service robotics: do you know your new companion? Framing an interdisciplinary technology assessment. Poiesis Prax. **8**(1), 25–44 (2011)
2. National Science Foundation's National Robotics Initiative (NRI) (2015)
3. Wisspeintner, T., Van Der Zant, T., Iocchi, L., Schiffer, S.: RoboCup@ Home: Scientific competition and benchmarking for domestic service robots. Interact. Stud. **10**(3), 392–426 (2009)
4. Sauppé, A., Mutlu, B.: The Social Impact of a Robot Co-Worker in Industrial Settings (2015)
5. Paepcke, S., Takayama, L.: Judging a bot by its cover: an experiment on expectation setting for personal robots. In: 2010 5th ACM/IEEE International Conference on Human-Robot Interaction (HRI). IEEE (2010)
6. Goetz, J., Kiesler, S., Powers, A.: Matching robot appearance and behavior to tasks to improve human-robot cooperation. In: The 12th IEEE International Workshop on Robot and Human Interactive Communication, Proceedings ROMAN 2003. IEEE (2003)
7. Šabanović, S., Reeder, S., Kechavarzi, B.: Designing robots in the wild: In situ prototype evaluation for a break management robot. J. Hum. Robot Interact. **3**(1), 70–88 (2014)
8. Chang, W.-L., et al.: The effect of group size on people's attitudes and cooperative behaviors toward robots in interactive gameplay. In: RO-MAN, 2012 IEEE. IEEE (2012)
9. Rindfleisch, A., Inman, J.: Explaining the familiarity-liking relationship: mere exposure, information availability, or social desirability? Mark. Lett. **9**(1), 5–19 (1998)
10. Hommel, B., Colzato, L.S., Van Den Wildenberg, W.P.: How social are task representations? Psychol. Sci. **20**(7), 794–798 (2009)
11. Tajfel, H., et al.: Social categorization and intergroup behaviour. Eur. J. Soc. Psychol. **1**(2), 149–178 (1971)
12. Epley, N., Waytz, A., Cacioppo, J.T.: On seeing human: A three-factor theory of anthropomorphism. Psychol. Rev. **114**(4), 864 (2007)
13. Eyssel, F., et al.: If you sound like me, you must be more human: on the interplay of robot and user features on human-robot acceptance and anthropomorphism. In: Proceedings of the Seventh Annual ACM/IEEE International Conference on Human-Robot Interaction. ACM (2012)
14. Hall, E.T.: The hidden dimension (1966)
15. Storms, M.D., Thomas, G.C.: Reactions to physical closeness. J. Pers. Soc. Psychol. **35**(6), 412 (1977)
16. Briggs, G., et al.: Actions speak louder than looks: does robot appearance affect human reactions to robot protest and distress? In: Proceedings of 23rd IEEE Symposium on Robot and Human Interactive Communication (Ro-Man) (2014)
17. Ardissono, L., et al.: Mixed-initiative scheduling of tasks in user collaboration. In: WEBIST (2012)
18. Gombolay, M.C., et al.: Decision-making authority, team efficiency and human worker satisfaction in mixed human–robot teams. Auton. Robots **39**(3), 293–312 (2015)
19. Stubbs, K., Hinds, P.J., Wettergreen, D.: Autonomy and common ground in human-robot interaction: A field study. IEEE Intell. Syst. **22**(2), 42–50 (2007)
20. Fraune, M.R., et al.: Three's company, or a crowd?: The effects of robot number and behavior on HRI in Japan and the USA. In: Robotics: Science and Systems (2015)
21. Yamaji, Y., et al.: Stb: Child-dependent sociable trash box. Int. J. Soc. Robot. **3**(4), 359–370 (2011)
22. Ezer, N.: Is a robot an appliance, teammate, or friend? Age-related differences in expectations of and attitudes towards personal home-based robots. ProQuest (2008)

23. Cottrell, C.A., Neuberg, S.L.: Different emotional reactions to different groups: a sociofunctional threat-based approach to "prejudice". J. Pers. Soc. Psychol. **88**(5), 770 (2005)
24. Turner, Y., Hadas-Halpern, I.: The effects of including a patient's photograph to the radiographic examination. In: Radiological Society of North America Scientific Assembly and Annual Meeting. Oak Brook, Ill: Radiological Society of North America (2008)
25. Sauppé, A., Mutlu, B.: The social impact of a robot co-worker in industrial settings. In: Proceedings of the 33rd Annual ACM Conference on Human Factors in Computing Systems. ACM (2015)

Help Me! Sharing of Instructions Between Remote and Heterogeneous Robots

Jianmin Ji[1], Pooyan Fazli[2,3(✉)], Song Liu[1], Tiago Pereira[2], Dongcai Lu[1],
Jiangchuan Liu[1], Manuela Veloso[2], and Xiaoping Chen[1]

[1] School of Computer Science and Technology,
University of Science and Technology of China, Hefei, China
{jianmin,xpchen}@ustc.edu.cn
[2] Computer Science Department, Carnegie Mellon University, Pittsburgh, USA
mmv@cs.cmu.edu, tpereira@andrew.cmu.edu
[3] Electrical Engineering and Computer Science Department,
Cleveland State University, Cleveland, USA
p.fazli@csuohio.edu

Abstract. Service robots frequently face similar tasks. However, they are still not able to share their knowledge efficiently on how to accomplish those tasks. We introduce a new framework, which allows remote and heterogeneous robots to share instructions on the tasks assigned to them. This framework is used to initiate tasks for the robots, to receive or provide instructions on how to accomplish the tasks, and to ground the instructions in the robots' capabilities. We demonstrate the feasibility of the framework with experiments between two geographically distributed robots and analyze the performance of the proposed framework quantitatively.

Keywords: Service robots · Sharing instructions · Collective robot learning · Robot-robot interaction framework

1 Introduction

Enabling remote and heterogeneous robots to share plans and instructions on various tasks assigned to them is an emerging field in artificial intelligence and robotics [7,20]. As current robots are far from being omniscient, the challenge is to develop robots that can recognize the missing knowledge autonomously and acquire it through alternative sources to accomplish tasks. This capability is especially crucial for service robots as they need to understand various user requests and provide services accordingly.

In this paper, we introduce a new framework, which allows remote and heterogeneous robots to share instructions on how to accomplish tasks. We demonstrate the feasibility of the framework with experiments between KeJia (Fig. 1(a)) and CoBot (Fig. 1(b)) and analyze the performance of the proposed framework quantitatively.

© Springer International Publishing AG 2016
A. Agah et al. (Eds.): ICSR 2016, LNAI 9979, pp. 786–795, 2016.
DOI: 10.1007/978-3-319-47437-3_77

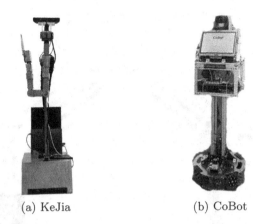

(a) KeJia   (b) CoBot

Fig. 1. KeJia and CoBot

KeJia [1], located in Hefei, China, is an intelligent domestic service robot, capable of interacting with humans in natural language and performing manipulation tasks in indoor environments autonomously. The robot was developed following a cognitive approach based on open knowledge available as semi-structured data.

CoBot [19], located in Pittsburgh, United States, is an autonomous service robot, capable of performing tasks and interacting with humans robustly in a multi-floor building, which has serviced and traversed more than 1000 km. The robot was developed following a novel symbiotic autonomy approach, in which the robot is aware of its perceptual, physical, and reasoning limitations and is able to ask for help from humans proactively.

2   Robot-Robot Interaction: Framework Overview

Figure 2 illustrates the structure of the proposed framework, which defines a protocol for remote robots to share instructions. An example of an interaction between two robots in this framework is as follows:

- Robot 1 receives a task from a human user in natural language.
- Robot 1 encodes the user task into a structured language and subsequently grounds it in its internal representation.
- Robot 1 fails to compute a plan for the assigned task.
- Robot 1 builds descriptions of the task and the environment it is located in and sends them to the interaction framework.
- The framework finds Robot 2, which is capable of providing instructions on the requested task for Robot 1.
- Robot 2, with or without using online resources, generates instructions for the task and transfers them to the interaction framework.

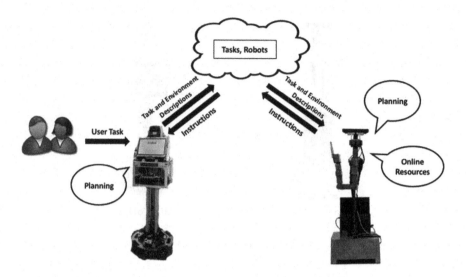

Fig. 2. The interaction framework between two robots.

- The framework saves the instructions for future queries about the task and transfers a copy to Robot 1.
- Robot 1 computes and executes a plan, using the received instructions to accomplish the task.

Section 4 discusses the details of the proposed robot-robot interaction framework, but before that, we introduce a common language to facilitate the cooperation between the robots.

3   Robot Communication Language

We introduce *Robot Communication Language (RCL)* to enable robots to share their knowledge and skills. RCL is a structured language to model tasks, environments, and instructions. The basic components of RCL include:

- *operation*: an operation that a robot should perform (e.g., *pick up, move, open*).
- *object*: an object, a human, or a location in the environment (e.g., *bottle, John, kitchen, living room*).
- *property*: a property of an object in the environment (e.g., *thirsty, closed, hot*).
- *relation*: a relation between two objects (e.g., *object A is on object B*).

The BNF definition of the basic components of RCL is as follows:

⟨operation⟩   ::= (*operation, object*) | (*operation, object, object*)

⟨property⟩   ::= (*property, object*)

⟨relation⟩   ::= (*relation, object, object*)

⟨instruction⟩   ::= ⟨operation⟩ | ⟨sequence⟩ | ⟨selection⟩ | ⟨repetition⟩

⟨sequence⟩   ::= (:seq ⟨operation⟩$^+$)

⟨selection⟩   ::= (:if ⟨property⟩ :then ⟨instruction⟩) |
                    (:if ⟨relation⟩ :then ⟨instruction⟩)

⟨repetition⟩   ::= (:while ⟨property⟩ :do ⟨instruction⟩) |
                    (:while ⟨relation⟩ :do ⟨instruction⟩)

Subsequently, we can define *tasks*, *environments*, and *instructions* in RCL as follows:

- (:task ⟨operation⟩): a task to be completed by a robot.
- (:env ⟨property⟩* ⟨relation⟩*): a model of the environment.
- (:inst ⟨instruction⟩): an instruction set.

4 Sharing Instructions Between Robots

4.1 Grounding the User Task

When a robot receives a task from a human user in natural language, it needs to map it to RCL. To this end, the robot uses the Stanford Parser [8] to generate the syntax tree and identify the grammatical structure of the input user request. Based on the parsing results, the robot fills corresponding phrases in ⟨operation⟩ (i.e., *operation* and *object*) for the assigned task. The next step is to ground *operation* and *object* into the corresponding symbols in the robot's internal representation. Synonyms in WordNet[1] are also used to facilitate the mapping process. A partial list of symbols used in the robots is shown in Table 1.

4.2 Asking for Help from a Remote Robot

If the robot fails to compute a plan for the grounded task, it needs to ask for help from a remote robot. To this end, the robot builds descriptions of the task and the environment in RCL and transfers them to the interaction framework. The framework matches the requested task with a robot capable of providing instructions for it.

Having received the descriptions of the task and the environment, the remote robot either computes a plan and builds an instruction set out of it, or instructions are extracted directly from online resources.

---

[1] http://wordnet.princeton.edu/wordnet/.

Table 1. A partial list of symbols used in KeJia and CoBot

Operation	Description
$bring(O_1, O_2)$	Bring object O_1 to location O_2
$give(O_1, O_2)$	Give object O_1 to human O_2
Object	**Description**
$kitchen$	Kitchen
$fridge$	Fridge
Property	**Description**
$portable(O)$	Object O is portable by the robot
$holding(O)$	Robot is holding the object O
Relation	**Description**
$at(O_1, O_2)$	Object O_1 is at location O_2
$on(O_1, O_2)$	Object O_1 is on object O_2

Generating Instructions from a Plan. The remote robot needs to compute a plan (i.e., a sequence of primitive actions) for the queried task. The environment is specified by the notion *state*, which is a set of predicates that are true in the environment. We define a *trajectory* as a sequence $\langle s_0, a_0, s_1, \ldots, a_{n-1}, s_n \rangle$, where s_i $(0 \leq i \leq n)$ is a state, a_j $(0 \leq j < n)$ is a primitive action, and s_{j+1} is the successor state of the state s_j after executing the action a_j. We can obtain the plan $\langle a_0, \ldots, a_{n-1} \rangle$ from such a trajectory.

To solve the planning problem, it is encoded into a logic programming language called *Answer Set Programming (ASP)* [11]. Each model of the ASP program corresponds to a possible plan for the task. The planning problem consists of an initial state, a goal state, and an action model. Action model is the set of primitive actions that are executable by the robot. Formally, a primitive action a is defined as a pair $\langle pre(a), eff(a) \rangle$, where $pre(a)$ and $eff(a)$ are preconditions and effects of a respectively.

The goal is to compute a sequence of primitive actions to get from the initial state to the goal state. Possible plans are computed by an ASP solver, iclingo [4]. A detailed description of this approach can be found in [6]. Having computed the plan, the robot encodes it into a set of $\langle$instruction$\rangle$s in RCL.

Generating Instructions from Online Resources. CoBot is capable of querying the web to learn the most probable location of an object in an indoor environment [13]. On the other hand, in KeJia, *Open Mind Indoor Common Sense (OMICS)* [9] is considered as a source of instructions. OMICS is an extensive database of common sense knowledge about home and office environments collected from non-experts over the web to enhance the capabilities of indoor autonomous robots. OMICS provides 11885 tables, including 28337 lines of instructions for 819 different user tasks. For some tasks, more than one

Table 2. An example table in OMICS

Task	Step number	Instructions
Fetch an object	0	Locate the object
	1	Go to the object
	2	Take the object
	3	Go back to where you were

instruction set has been defined in OMICS. Table 2 shows an example of the task "fetch an object" and the list of instructions to accomplish the task in OMICS.

Given a task, KeJia can query the OMICS database to find the closest match among the OMICS tables. Then, it extracts the instructions in the table for the queried task and encode them into ⟨instruction⟩s in RCL. Most instructions in OMICS are sequential. When the keywords "if", "then", "until", or "while" appear in the instructions, the ⟨instruction⟩ with the corresponding control structures is built in RCL.

4.3 Receiving Instructions from the Remote Robot

The robot grounds the instructions received in RCL into the corresponding symbols in its internal representation. It then computes a plan according to the instructions. The ASP solver, iclingo, is used to compute a trajectory with the least number of primitive actions. The planning procedure, considering the action model of the robot, computes a satisfying trajectory when the instructions miss some indispensable steps. For example, in OMICS, there are instructions on how to "get food from the fridge". The last two steps are "pick up food" and "close door". However, KeJia has only one arm, therefore it has to place the food somewhere before closing the fridge door.

5 Related Work

The RoboEarth project [20] offers a cloud robotics [7] infrastructure, which allows robots to share information and learn new skills and knowledge from each other. The RoboBrain project [14] aims at constructing a massive large-scale knowledge base for robots that learns from internet resources.

Understanding natural language commands has been regarded as a basic ability of a service robot [10,15,17]. Previous work have exploited instructions described in natural language to generate a plan for a user task [3,12,16]. There are also approaches for generating plans for household robots, using natural language instructions extracted from the web [18]. Also, important to our work is the capability to ground natural language phrases in the objects, actions, and relations in the target area [5].

(a) User asks KeJia to bring him a bottle of water.

(b) CoBot computes the most probable location to find a bottle of water.

(c) KeJia executes a plan based on the instructions received from CoBot.

(d) KeJia completes the task.

Fig. 3. Scenario 1: CoBot is helping KeJia.

6   Experiments and Evaluations

We propose an interaction framework, which allows robots to share instructions on how to accomplish tasks. We demonstrate the feasibility of the framework with experiments between two geographically distributed robots and evaluate the expressive power of Robot Communication Language (RCL) quantitatively.

6.1   Experiments Between KeJia and CoBot

We use the rosbridge package [2] to connect KeJia and CoBot. The experiment consists of two scenarios.

In the first scenario (Fig. 3), the user asks KeJia: "Bring me a bottle of water". KeJia parses the request and maps it to RCL: (:task (bring, bottle of water)). KeJia then tries to compute a plan for the given task, but fails, as it does not know where to find a bottle of water. Hence, KeJia sends the failed subtask of finding a bottle of water to CoBot: (:task (find, bottle of water)).

CoBot grounds the task in its own symbols and querying the web, it computes the most probable location to find a bottle of water (i.e., dinner table) and subsequently generates the following instruction: (:inst (find, bottle of water, dinner table)). CoBot sends the instruction to the interaction framework and KeJia. KeJia grounds the received instruction in its own symbols and computes a plan to bring the bottle of water to the user.

(a) User tells CoBot that she is thirsty.

(b) KeJia suggests that CoBot give the user a bottle of water.

(c) CoBot executes a plan based on the instructions received from KeJia.

(d) CoBot completes the task.

Fig. 4. Scenario 2: KeJia is helping CoBot.

In the second scenario (Fig. 4), the user tells CoBot: "I am thirsty". CoBot does not know how to interpret the user request and since there is no task defined in the request, CoBot just builds a description of the environment, (:env (thirsty, human)), and sends it to the interaction framework and KeJia. KeJia uses OMICS to extract an instruction, that is, (:inst (give, bottle of water, human)), to satisfy the user desire and sends it to the interaction framework and CoBot. CoBot computes a plan based on the instruction to complete the task.

The framework enables both robots to assist each other in completing tasks that were unable to handle previously.

6.2 RCL Evaluation

OMICS recorded a great amount of instructions for everyday tasks written by internet users. Hence, we use it as a benchmark to evaluate the expressive power of RCL.

Table 3 shows the percentage of tasks, instructions, and tables (i.e., tasks and the instruction sets) in OMICS that were parsed and encoded into RCL successfully. As mentioned earlier, in OMICS, there are 11885 tables, including 28337 lines of instructions for 819 different user tasks. In summary, ~58 % of the tables were parsed, and ~28 % of the tables were converted to RCL successfully.

Table 3. Summary of the evaluations with OMICS

Test set	Tasks (819)	Instructions (28337)	Tables (11885)
Parsed	81.81 %	88.87 %	57.99 %
Converted to RCL	73.99 %	63.04 %	27.98 %

7 Conclusion and Future Work

We introduced a new framework, which allows remote and heterogeneous robots to share instructions on how to accomplish tasks that were unable to handle previously. We demonstrated the feasibility of the framework through experiments between KeJia, located in Hefei, China and CoBot, located in Pittsburgh, United States. We also introduced Robot Communication Language (RCL) to facilitate the collaboration process between the robots and evaluated its expressive power quantitatively. For future work, we intend to extend the framework to support multiple remote and heterogeneous robots. Various robots may generate different instructions for a user task and a robot can choose the most proper set of instructions among them.

References

1. Chen, X., Ji, J., Jiang, J., Jin, G., Wang, F., Xie, J.: Developing high-level cognitive functions for service robots. In: Proceedings of the 9th International Conference on Autonomous Agents and Multiagent Systems (AAMAS 2010), pp. 989–996 (2010)
2. Crick, C., Jay, G., Osentoski, S., Jenkins, O.C.: ROS and rosbridge: roboticists out of the loop. In: Proceedings of the ACM/IEEE International Conference on Human-Robot Interaction (HRI 2012), pp. 493–494 (2012)
3. Dzifcak, J., Scheutz, M., Baral, C., Schermerhorn, P.: What to do and how to do it: translating natural language directives into temporal and dynamic logic representation for goal management and action execution. In: Proceedings of the IEEE International Conference on Robotics and Automation (ICRA 2009), pp. 4163–4168 (2009)
4. Gebser, M., Kaminski, R., Kaufmann, B., Ostrowski, M., Schaub, T., Thiele, S.: Engineering an incremental ASP solver. In: Garcia de la Banda, M., Pontelli, E. (eds.) ICLP 2008. LNCS, vol. 5366, pp. 190–205. Springer, Heidelberg (2008)
5. Guadarrama, S., Riano, L., Golland, D., Gouhring, D., Jia, Y., Klein, D., Abbeel, P., Darrell, T.: Grounding spatial relations for human-robot interaction. In: Proceedings of the IEEE/RSJ International Conference on Intelligent Robots and Systems (IROS 2013), pp. 1640–1647 (2013)
6. Ji, J., Chen, X.: From structured task instructions to robot task plans. In: Proceedings of the 5th International Conference on Knowledge Engineering and Ontology Development (KEOD 2013), pp. 237–244 (2013)
7. Kehoe, B., Patil, S., Abbeel, P., Goldberg, K.: A survey of research on cloud robotics and automation. IEEE Trans. Autom. Sci. Eng. 12(2), 398–409 (2015)
8. Klein, D., Manning, C.D.: Accurate unlexicalized parsing. In: Proceedings of the 41st Annual Meeting of the Association for Computational Linguistics (ACL 2003), pp. 423–430 (2003)

9. Kochenderfer, M.J., Gupta, R.: Common sense data acquisition for indoor mobile robots. In: Proceedings of the AAAI Conference on Artificial Intelligence (AAAI 2004), pp. 605–610 (2003)
10. Kollar, T., Tellex, S., Roy, D., Roy, N.: Toward understanding natural language directions. In: Proceedings of the ACM/IEEE International Conference on Human-Robot Interaction (HRI 2010), pp. 259–266. IEEE (2010)
11. Lifschitz, V.: Answer set planning. In: Gelfond, M., Leone, N., Pfeifer, G. (eds.) LPNMR 1999. LNCS (LNAI), vol. 1730, pp. 373–374. Springer, Heidelberg (1999)
12. Rybski, P.E., Yoon, K., Stolarz, J., Veloso, M.M.: Interactive robot task training through dialog and demonstration. In: Proceedings of the ACM/IEEE International Conference on Human-Robot Interaction (HRI 2007), pp. 49–56 (2007)
13. Samadi, M., Kollar, T., Veloso, M.M.: Using the web to interactively learn to find objects. In: Proceedings of the 26th AAAI Conference on Artificial Intelligence (AAAI 2012), pp. 2074–2080 (2012)
14. Saxena, A., Jain, A., Sener, O., Jami, A., Misra, D.K., Koppula, H.S.: Robobrain: large-scale knowledge engine for robots. arXiv:1412.0691 (2014)
15. Shimizu, N., Haas, A.R.: Learning to follow navigational route instructions. In: Proceedings of the 21st International Joint Conference on Artificial Intelligence (IJCAI 2009), pp. 1488–1493 (2009)
16. Skubic, M., Perzanowski, D., Blisard, S., Schultz, A., Adams, W., Bugajska, M., Brock, D.: Spatial language for human-robot dialogs. IEEE Trans. Syst. Man Cybern. Part C Appl. Rev. 34(2), 154–167 (2004)
17. Tellex, S., Kollar, T., Dickerson, S., Walter, M.R., Banerjee, A.G., Teller, S.J., Roy, N.: Understanding natural language commands for robotic navigation and mobile manipulation. In: Proceedings of the 25th AAAI Conference on Artificial Intelligence (AAAI 2011), pp. 1507–1514 (2011)
18. Tenorth, M., Nyga, D., Beetz, M.: Understanding and executing instructions for everyday manipulation tasks from the world wide web. In: Proceedings of the IEEE International Conference on Robotics and Automation (ICRA 2010), pp. 1486–1491 (2010)
19. Veloso, M., Biswas, J., Coltin, B., Rosenthal, S., Kollar, T., Mericli, C., Samadi, M., Brandao, S., Ventura, R.: Cobots: collaborative robots servicing multi-floor buildings. In: Proceedings of the IEEE/RSJ International Conference on Intelligent Robots and Systems (IROS 2012), pp. 5446–5447 (2012)
20. Waibel, M., Beetz, M., D'Andrea, R., Janssen, R., Tenorth, M., Civera, J., Elfring, J., Gálvez-López, D., Häussermann, K., Montiel, J., Perzylo, A., Schießle, B., Zweigle, O., van de Molengraft, R.: RoboEarth - a world wide web for robots. Robot. Autom. Magaz. 18(2), 69–82 (2011)

Enabling Symbiotic Autonomy in Short-Term Interactions: A User Study

Francesco Riccio[✉], Andrea Vanzo, Valeria Mirabella, Tiziana Catarci, and Daniele Nardi

Department of Computer, Control and Management Engineering "Antonio Ruberti", Sapienza University of Rome, Rome, Italy
{riccio,vanzo,mirabella,catarci,nardi}@dis.uniroma1.it

Abstract. The presence of robots in everyday environments is increasing day by day, and their deployment spans over various applications: industrial and working scenarios, health care assistance in public areas or at home. However, robots are not yet comparable to humans in terms of capabilities; hence, in the so-called Symbiotic Autonomy, robots and humans help each other to complete tasks. Therefore, it is interesting to identify the factors that allow to maximize human-robot collaboration, which is a new point of view with respect to the HRI literature and very much leaning toward a social behavior. In this work, we analyze a subset of such variables as possible influencing factors of humans' Collaboration Attitude in a Symbiotic Autonomy framework, namely: Proxemics setting, Activity Context, and Gender and Height as valuable features of the users. We performed a user study that takes place in everyday environments expressed as activity contexts, such as relaxing and working ones. A statistical analysis of the collected results shows a high dependence of the Collaboration Attitude in different Proxemics settings and Gender.

Keywords: Symbiotic Autonomy · Spatial interaction · Human-robot collaboration

1 Introduction

Robots have become an important component in many working environments, ranging from industrial robotics to autonomous exploration. More and more often, they are being introduced in human populated environments, such as health care, assistance in indoor public and private environments, and entertainment. In these social scenarios, robots are expected to cooperate with humans and, therefore, to interact with them.

However, autonomy varies depending on the environment, on the task to be executed and on robot capabilities. For instance, many types of robots are constrained by their hardware and, as in the case of low cost mobile robots, can not accomplish "simple" tasks such as pushing a button of the elevator, collecting sheets from the printer, or plugging themselves in order to charge. In

© Springer International Publishing AG 2016
A. Agah et al. (Eds.): ICSR 2016, LNAI 9979, pp. 796–807, 2016.
DOI: 10.1007/978-3-319-47437-3_78

order to overcome these constraints, robots could asks humans to help them, and accordingly, humans may be willing to help robots executing service tasks.

The combination of this mutual dependency has been characterized as *Symbiotic Autonomy* [1] or *Symbiotic Robotics* [2], where robots perform service tasks for humans, while humans help them to achieve their goals.

Most of the literature on HRI addresses the case of humans asking for help. In fact, in this perspective, humans start interactions with the robot to take advantage of its services and achieve their goals. This approach can be regarded as a first step toward social behavior. Instead, the evaluation of how the robot should behave to gather humans' help, and in which context and approach it is better to do it, represents a novel research to be investigated.

This paper presents the results of a user study, designed to analyze which factors influence human attitude to help the robot in the context of Symbiotic Autonomy. We hypothesize that such attitude has not a constant value as it depends on several factors imposed by human physiology and by the context in which they are currently in. To this end, we introduce the novel concept of Collaboration Attitude as a quantitative measure to characterize such an inclination:

Definition 1 (Collaboration Attitude). *The Collaboration Attitude measures the attitude of humans toward the requests for help of the robot in a Symbiotic Autonomy framework. Formally, it is quantified according to a metrics defined on a scale of N points, where N is the number of tasks that the human is requested to accomplish. Precisely, the Collaboration Attitude assumes values in $[0, \ldots, N-1]$, where 0 represents lowest level of collaboration, i.e. the human is not willing to help at all, while $N-1$ represents the highest one, i.e. the human is willing to help the robot in all the tasks.*

In our experimental setting, a robot asks people for help in different Activity Contexts (namely, *relaxing* and *working*), with different Proxemics settings (namely, *intimate, personal* and *social*), and balancing the experimenters on their Gender and Height. The analysis of such factors generates a model of interaction that defines: (i) whether they actually influence the Collaboration Attitude, and (ii) the values that maximize it. Eventually, such a model may be used to shape the proper social behavior of robots asking for help, depending on the current working context.

Operationally, we are interested in analyzing the following four hypotheses:

Hypothesis One: Collaboration Attitude is subject to different activity contexts. The environmental context of the interaction plays a central part in social interactions. Humans behave differently, depending on where they are and the contexts they are in. Consequently, a robot needs to consider these social elements.

Hypothesis Two: Collaboration Attitude is subject to different Proxemics settings. It is well known that among humans and robots, Proxemics has a key role in the interaction. Therefore, experiments aim at highlighting the importance of respecting the personal space in social interactions, even in the case where the interactive partner is a robot. Specifically, we want to estimate whether different settings of Proxemics might vary the Collaboration Attitude that the human shows, ranging from an *intimate* distance to a *social* one.

Hypothesis Three: Collaboration Attitude is subject to the gender of the human. Humans' physical and social characteristics affect how they behave in different situations. Gender is one of the major features to be considered. Such factor is usually considered in HRI studies, as males and females show different responses to equal stimuli.

Hypothesis Four: Collaboration Attitude is subject to the height of the human. Robot appearance constitutes a key factor to be investigated, when studying humans response to robot behaviors. Our intuition is that shorter people perceive the robot differently than taller people and their Collaboration Attitude varies depending on such a perception. The outcomes of the statistical analysis over the collected data confirm that the Collaboration Attitude has a not constant value when different factors are changing.

The reminder of the paper is organized as follows: Sect. 2 reports related works, while Sect. 3 presents our system and the setup of the experiments. In Sect. 4, we analyze the experimental results of the user study.

2   Related Work

The aim of this paper is to detect and evaluate the factors that influence the collaboration between humans and robots, through short-term interactions based on spoken dialogues. Hence, related work covers the topic of Symbiotic Autonomy and recent user studies on Human-Robot Interaction. Symbiotic Autonomy assumes a new perspective in the collaboration between humans and robots, which has been introduced in [1] as a symbiosis among human and robots to enable a better coexistence of both. In this context, both [3,4] address the problem of enabling a cooperation among humans and robots by adapting robot behaviors to human preferences [3] and by analyzing human responses to a robot offering domestic services [4]. However, we differ from these works which consider always a static context of interaction between humans and robot. In fact, we assume that the Collaboration Attitude has not a constant value and depends on many factors such as humans attitudes, gender, height, comfort and also on the activity context were the interactions take place. Therefore, in this study we focus on a subset of these factors in order to deduce values that maximize the collaboration between humans and robots.

The central factor of our user studies is Proxemics [5], which helps in understanding how humans manage and interpret *social non-verbal behaviors* as a communication tool. Human-Robot Proxemics has been investigated to generate reliable guidelines for social interactions, when the interactive partner is a robot. In general, proxemics studies aim at evaluating the response of humans with respect to different factors contextualizing the presence of the robot and its task. For example, in [6], the authors conduct a user study aiming at highlighting human-centered factors, such as pet ownership, personality characteristics, gender and height. Specifically, the gender dimension has been addressed by several works [7], arguing that females are often more inclined toward collaboration. Other studies, instead, introduce a height classification of the user population, but they do not report significant findings in analyzing such a dimension [8]. In [9], the authors relate Proxemics to the task that a robot is performing. In particular, they setup an handing-over scenario with a *Care-O-Bot* and several users with the goal to evaluate the social level of the robot behaviors. In [10], the authors extend proxemics to psychological distances by analyzing the impact of the robot gaze during the experiment. Finally, different studies exploit Human-Robot Proxemics to influence robot behaviors in order to enhance human comfort. For example, in [8] the authors model the interactions as a set of *fuzzy-rules* that can be dynamically tuned by the human operator. Similarly, in [11] the robot autonomously estimates the *comfort-level* of the operators by comparing gaze orientation and physical distances in order to adapt its behaviors to specific users.

Summarizing, most of the cited works aim at determining the "best configuration" for a robot that is standing in front of a human, or considering environments where the robot is passing by. Under this perspective, the goal is to minimize the level of discomfort that can be caused. Conversely, the main differences between our work and those reported in the literature lie in the premises of the task, that is here characterized by a robot asking for help and a human that is supposed to support it. In fact, even though similar aspects are considered and addressed, we aim at evaluating the Collaboration Attitude of the subjects (*proactive behavior*), rather than their preferences during a Human-Robot Interaction (*passive behavior*), as in [7]. This different point of view represents the main novelty of our contribution. Hence, in this different perspective, we want to both minimize the level to discomfort, and determine the best scenarios for approaching humans asking for help.

3   Method

In order to estimate the degree of Collaboration Attitude and the significance of the Proxemics, Activity Context, Gender and Height factors for it, we set up a user study. We executed different runs of the same experiment varying Proxemics and Activity Context and gathering information about Gender and Height of the experimenters.

Subjects. The experiment has been conducted in a department of our university. In such an environment the users have been randomly selected from a set of students with homogeneous characteristics. All of them between 20 and 30 years old, and with a basic level of acquaintance toward robotics. More in detail, the users completed the experiment in a *"between group" design*, so that different configurations of Proxemics and Activity Context were tested with different users, balancing the experimenters on gender and height. None of them repeated the experiment more than once.

Apparatus. The deployed robot is a modified version of the Turtlebot Robot. While the base remains unaltered, the structure on top of it has been customized, in order to make the robot taller with respect to the standard version. The robot is 98 cm high and features a tablet on top as an interface for the interactions. In fact, we allow users to have *short-term dialogues* with the robot, aiming at estimating the attitude of the human to help the robot in performing its tasks. Our short-term dialogue system is composed of two main modules: (i) an *Automatic Speech Recognizer* (ASR) that processes the acoustic signal of the user's speech and generates a set of possible transcriptions; (ii) a *Dialogue Manager* (DM) that manages the dialogic interaction. The ASR module has been realized through the Google Speech APIs, available within the Android environment, in a mobile application. The app is also in charge of managing the questionnaire presented to the user at the end of the interaction and controlling the dialogue flow through the Artificial Intelligence Markup Language (AIML).

Procedure. Once an user has been identified, the robot notifies its presence and asks the experimenter to keep his/her position. Afterwards, the robot approaches the user with a specific Proxemics setting. We do not vary the orientation of the robot during the experiments, as other works [9,12] focused on the relative orientation of the robot with respect to the user. The robot starts the interaction by asking the human for help and keeps listening for a reply.

Questionnaire. At the end of the experiment, we collected data by asking the user to fill out a questionnaire that is directly displayed on the robot.

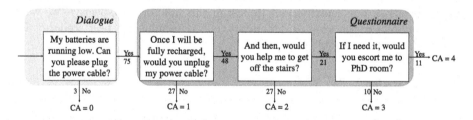

Fig. 1. Collaboration attitude estimation through questionnaire

We divided the questionnaire in two parts aiming at (i) collecting information about the user, and (ii) quantifying the Collaboration Attitude. In our experiment, this variable has been mapped in a 5-point scale, measuring the amount of positive responses of the experimenters to the robot requests, according to Definition 1. Hence, if we consider also the initial request (the spoken interaction), this variable takes values in $\{0, ..., 4\}$, where 0 is the case where the human is not willing to help the robot in any task and 4 the opposite situation. Figure 1 lists the requests the robot asks to the experimenters. While the first request is said by the robot during the spoken interaction, the remaining three questions are asked within the questionnaire. The number on each edge refers to the occurrences of a particular answer, i.e., *yes* or *no*. Only 3 users neglected the first request, where the initial curiosity toward the robot probably played a key role. The engagement decreases as the requests become more and more demanding. In fact, only 11 experimenters were inclined to collaborate with the robot up to the final request. Figure 1 shows also how the Collaboration Attitude (CA) has been quantified through the interaction/questionnaire. For instance, 3 experimenters neglecting the initial request achieved a CA of 0, while 11 users obtained a CA score of 4, by satisfying all the robot requests.

We conducted our study in a general area of an indoor office environment. Therefore, we categorized the Activity Context factor in two different classes of *stationary*[1] contexts: (i) *Working Activity Context* is an area where humans are waiting for a given event to happen, e.g., if they are at the printer, waiting in front of a colleague door or interacting with a partner; (ii) *Relaxing Activity Context* represents the most heterogeneous context in terms of people: this environment is composed of an open area with vending machines for food and coffee, where people are generally open to social interactions, likely even when the counterpart is a robot.

The Proxemics settings used in this work are taken from the *Proxemics theory* introduced by Hall and used in several HRI experiments (see, for example, [8]). The areas of interactions among humans and robots are categorized in: (i) *intimate* interaction, if the distance is below *0.45* m; (ii) *personal* interaction, if the distance is between *0.45* m and *1.2* m; (iii) *social* interaction, if the distance is between *1.2* m and *3.6* m; (iv) *public* interaction, if the distance is greater than *3.6* m. Specifically, the distances of our Proxemics settings have been approximated considering the area directly in front of the users which is the one chosen to run the experiment. Additionally, within each activity context, we repeated the experiment varying the robot-human relative distance to verify the hypothesis on proximity relations.

We deal with subject characteristics, that are essential to determine robot social actions, by considering gender and height. As for the height, we considered the average of the population of our country and categorizing shorter and taller people according to such a threshold. These two possible values constitute the domain of the Height variable in our user study, as introduced by Hypothesis four.

4 Results and Discussion

We addressed each hypothesis individually, by performing One-Way ANOVA through the Analysis ToolPak in Microsoft Excel. The experimental results are reported below.

---

[1] We refer to "stationary" as an activity context where people are not moving and busy in performing some activities.

Hypothesis One: Collaboration Attitude is subject to different activity contexts. In this hypothesis we analyze the variation of the Collaboration Attitude when the experiment is performed in different Activity Contexts.

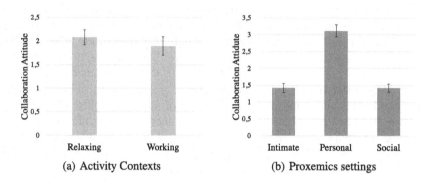

(a) Activity Contexts   (b) Proxemics settings

Fig. 2. Collaboration attitude means depending on activity contexts and proxemics settings

In Fig. 2(a), the Collaboration Attitude means and standard errors for different Activity Contexts are shown. Despite the Relaxing context seems to maximize the collaborative intentions of the experimenters, the Collaboration Attitude is rather stable when different contexts are tested. As a consequence, the Activity Context does not appear to be a perturbing factor for the Collaboration Attitude.

Table 1. Activity context: One-way ANOVA results

Groups	Count	Sum	Avg	Var
Relaxing	39	81	2.08	0.97
Working	39	74	1.9	1.46

Src of Var	SS	df	MS	F	P-val	F crit
Btw. Groups	0.63	1	0.63	0.52	0.47	3.97
Wtn. Groups	92.36	76	1.22			
Total	92.99	77				

In order to search for significant variations, we performed a One-Way ANOVA on the dataset. In the left part of Table 1, a sketch of the sample under consideration is shown. The population is completely balanced, with 39 experimenters for both the Relaxing and Working groups. The right part of the table shows the ANOVA results by reporting the *P-value* that is essential to determine the significance of such a factor, and the *sum of squares* (*SS*), the *degrees of freedom* (*df*), the *mean squares* (*MS*), the *ratio of the two mean squares values* (*F*) and the *F critical value* (*F crit*). The outcomes confirm the previous observations. Even though our initial hypothesis relied upon the intuition that humans in a relaxing context are more inclined to a collaborative behavior, the results of the experiment show that there are not statistically significant differences when changing the working context. This finding could be explained by a strong focus on the social interaction with other humans in a relaxing domain and it may suggest that robots are not yet considered "social" partners. This factor trades off the nature of the working context, where people are usually busy in their tasks.

Hypothesis Two: Collaboration Attitude is subject to different Proxemics settings. In this hypothesis, we try to find out if there exists a relationship between Collaboration Attitude and Proxemics setting. Figure 2(b) shows means and standard errors of the Collaboration Attitude to changes in the Proxemics setting. The setting that maximizes the Collaboration Attitude is when the robot approaches the human with a Personal distance. This result is in line with other user studies conducted in Human-Robot Proxemics [6,10,11], stating that human's comfort is maximized within the Personal setting. The Intimate and Social distances give lower values of Collaboration Attitude. This finding could be explained by two elements: the control that humans exercise in their Intimate space and the robot size. In fact, the presence of the robot seems to be not relevant, when the interaction takes place at longer distances.

Table 2. Proxemics setting: One-way ANOVA results

Groups	Count	Sum	Avg	Var	Src of Var	SS	df	MS	F	P-val	F crit
Intimate	26	37	1.42	0.49	Btw. Groups	49.64	2	24.82	42.95	$3.71 \cdot 10^{-13}$	3.12
Personal	26	81	3.12	0.83	Wtn. Groups	43.35	75	0.58			
Social	26	37	1.42	0.41	**Total**	92.99	77				

The left part of Table 2 shows the composition and some statistics of our groups, namely Intimate, Personal and Social distances. The sets are here completely balanced, with a population of 26 elements each. The right part, conversely, reports the results obtained by the One-Way ANOVA performed on the sample under consideration highlighting that the Collaboration Attitude depends on the Proxemics setting chosen for the experiment (P-value < 0.05).

Table 3. t-Test: Two-sample assuming equal variances

	Intimate vs. personal	Intimate vs. social	Personal vs social
df	50	50	50
P(T<=t) two-tail	$9.6 \cdot 10^{-10}$	1	$4.1 \cdot 10^{-10}$

In order to confirm the ANOVA results, we performed a post-hoc test through three t-tests, aimed at comparing each pair of groups. Table 3 shows the result of this additional analysis. As suggested by the means histogram, in the Personal distance humans act differently w.r.t. Intimate and Social settings (the *two-tailed p* values are lower than 0.05), whereas users seem to behave similarly in their Intimate and Social spaces. These results are particularly interesting in the framework of Symbiotic Autonomy: they suggest that a robot asking for help should approach the user in his personal space, as this distance seems to be the most comfortable for humans.

Hypothesis Three: Collaboration Attitude is subject to the gender of the human. In order to verify Hypothesis Three, during the questionnaire we collected information about gender of the experimenters. The study of this feature is interesting, as it is known that males and females have different social behaviors.

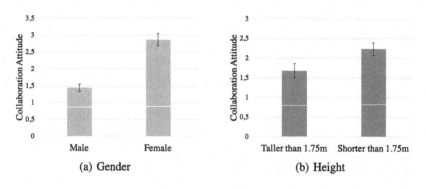

Fig. 3. Collaboration attitude means depending on gender and height

Figure 3(a) shows means and standard errors of the Collaboration Attitude with respect to changes in gender. Specifically, the mean of the Collaboration Attitude score obtained by females is strikingly higher than the males' one. This represents a first indication that females are more inclined to help robots than males.

Table 4. Gender: One-way ANOVA results

Groups	Count	Sum	Avg	Var
Male	48	69	1.44	0.59
Female	30	86	2.87	0.95

Src of Var	SS	df	MS	F	P-val	F crit
Btw. Groups	37.71	1	37.71	51.84	$3.7 \cdot 10^{-13}$	3.967
Wtn. Groups	55.28	76	0.73			
Total	92.99	77				

The left part of Table 4 shows that the sample here is not balanced, with a majority of males with respect to females, i.e. 62 % vs. 38 %. In the right part the One-Way ANOVA confirms our insight (*P*-value < 0.05). Our findings are supported by the work in [10]. In their user study, in fact, male users are more diffident and place themselves significantly further from the robot than females, as the authors stated. These results are also confirmed in the work in [7]. Specifically, they report a considerable difference of the comfort level within the intimate area when varying the gender of the users. Their results support that males impose a dominant territory that the robot is violating, if it is positioned in their intimate areas. In Human-Human interactions, manifold psychological studies address this particular behavior. For instance, the difference in cooperating among males and females has been pointed out in [13], where this evaluation is made upon the well known *Dictator Game*. A further confirmation is provided by [14], where the gender dimension is analyzed within an experimental study of team performance. In conclusion, our results in this setting report that female experimenters show more interest in exploring a new collaboration with a robotic partner. Therefore, the robot behavior could be leveraged by allowing the robot to seek for help first in females rather than males.

Hypothesis Four: Collaboration Attitude is subject to the height of the human. The last hypothesis takes into account the Collaboration Attitude and the height of the experimenters. In this perspective, few works consider the height of the robot as a *dependent variable* in their controlled studies [6,8]. However, they do not state or highlight any empirical result on the influence of relative heights of the robot and users onto the interaction. Conversely, in our work, we noticed an interesting behavior, when classifying our users by their heights. Such a categorization has been made by considering the average among the subjects' height of our population and the 1.75 m value has been chosen as unbiased discriminant factor. Figure 3(b) shows statistics of the Collaboration Attitude means to changes in height of the experimenters. It seems that shorter experimenters are more inclined in collaborating with respect to taller ones.

Table 5. Height: One-way ANOVA results

Groups	Count	Sum	Avg	Var	Src of Var	SS	df	MS	F	P-val	F crit
Taller than 1.75m	34	57	1.68	1.13	Btw. Groups	5.82	1	5.82	5.07	0.027	3.97
Shorter than 1.75m	44	98	2.23	1.16	Wtn. Groups	87.17	76	1.15			
					Total	92.99	77				

As shown in the left part of Table 5, this sample is not balanced, with a prevalence of shorter experimenters, i.e. 56 % shorter vs. 44 % taller. The One-Way ANOVA performed on this dataset (Table 5) shows that there is a significant variation among the different clusters of height. The P-value is lower than 0.05 and we reject the null hypothesis with 95 % of confidence. However, the outcomes of such analysis (Table 5) could be influenced by the females which are usually shorter than males, and much more inclined to a human-robot collaboration. In fact, in our population, 70 % of the female experimenters are shorter than 1.75 m, while the remaining 30 % are taller than 1.75 m. Conversely, the male population is almost completely balanced.

Hence, with the available data we are not able to clearly state whether the height of the experimenters plays a key role in a human-robot collaboration. In our opinion, this particular aspect deserves an additional analysis, by increasing the variability and the size of the sample, as well as the height of the robot.

5 Conclusion

Our study is the first one to address the attitude toward collaboration in the social perspective adopted by the so-called Symbiotic Autonomy. This specific HRI framework, where the robot asks the human for help, can become a widespread and practical approach, provided that the robot can exhibit the proper social behavior. Moreover, the findings of our study are confirmed by similar psychological studies on human-human interactions, which highlight that different factors affect the inclination of humans toward collaboration, e.g., user gender and the perception of robot size. As future work, we believe that several other factors are likely to have an impact on the Collaboration Attitude of the robot in Symbiotic Autonomy. Among them, we are planning to investigate how the Collaboration Attitude varies between interactions with small groups of people and interactions with individuals, how the users respond to different robots varying the appearance or attitude (e.g., politeness). Further, we want to investigate how previous interactions influence the user collaboration. All of these factors will help us in understanding how to characterize robots social behaviors, and generating a form of bidirectional collaboration with humans.

References

1. Rosenthal, S., Biswas, J., Veloso, M.: An effective personal mobile robot agent through symbiotic human-robot interaction. In: International Conference on Autonomous Agents and Multiagent Systems (AAMAS 2010), vol. 1, pp. 915–922, May 2010
2. Coradeschi, S., Saffiotti, A.: Symbiotic robotic systems: humans, robots, and smart environments. IEEE Intell. Syst. **21**(3), 82–84 (2006)
3. Nikolaidis, S., Ramakrishnan, R., Gu, K., Shah, J.: Efficient model learning from joint-action demonstrations for human-robot collaborative tasks. In: Proceedings of the Tenth Annual ACM/IEEE International Conference on Human-Robot Interaction (HRI 2015), pp. 189–196. ACM (2015)
4. Fischer, K., Yang, S., Mok, B., Maheshwari, R., Sirkin, D., Ju, W.: Initiating interactions and negotiating approach: a robotic trash can in the field. In: AAAI Symposium on Turn-taking and Coordination in Human-Machine Interaction, pp. 10–16 (2015)
5. Hall, E.T.: The Hidden Dimension: Man's Use of Space in Public and Private. The Bodley Head Ltd., London (1966)
6. Takayama, L., Pantofaru, C.: Influences on proxemic behaviors in human-robot interaction. In: IEEE/RSJ International Conference on Intelligent Robots and Systems (IROS 2009), pp. 5495–5502, October 2009

7. Syrdal, D.S., Koay, K.L., Walters, M.L., Dautenhahn, K.: A personalized robot companion? - the role of individual differences on spatial preferences in HRI scenarios. In: The 16th IEEE International Symposium on Robot and Human interactive Communication (RO-MAN 2007), pp. 1143–1148, August 2007

8. Walters, M.L., Dautenhahn, K., Te Boekhorst, R., Koay, K.L., Syrdal, D.S., Nehaniv, C.L.: An empirical framework for human-robot proxemics. In: Proceedings of the Symposium on New Frontiers in Human-Robot Interaction, pp. 144–149, Edinburgh, Scottland (2009)

9. Koay, K.L., Syrdal, D.S., Ashgari-Oskoei, M., Walters, M.L., Dautenhahn, K.: Social roles and baseline proxemic preferences for a domestic service robot. Int. J. Soc. Robot. **6**, 469–488 (2014)

10. Mumm, J., Mutlu, B.: Human-robot proxemics: physical and psychological distancing in human-robot interaction. In: Proceedings of the 6th International Conference on Human-Robot Interaction (HRI 2011), pp. 331–338. ACM (2011)

11. Mitsunaga, N., Smith, C., Kanda, T., Ishiguro, H., Hagita, N.: Adapting robot behavior for human-robot interaction. IEEE Trans. Robot. **24**, 911–916 (2008)

12. Torta, E., Cuijpers, R.H., Juola, J.F.: A model of the user's proximity for Bayesian inference. In: Proceedings of the 6th International Conference on Human-robot Interaction (HRI 2011), pp. 273–274, NY, USA. ACM, New York (2011)

13. Engel, C.: Dictator games: a meta study. Exp. Econ. **14**, 583–610 (2011)

14. Song, H., Restivo, M., van de Rijt, A., Scarlatos, L.L., Tonjes, D., Orlov, A.: The hidden gender effect in online collaboration: an experimental study of team performance under anonymity. Comput. Hum. Behav. **50**, 274–282 (2015)

Conceptual Framework for RoboDoc:
A New Social Robot for Research Assistantship

Azadeh Mohebi[(✉)], Ramin Golshaie, Soheil Ganjefar, Ammar Jalalimanesh,
Parnian Afshar, Ali Aali Hosseini, Seyyed Alireza Ghoreishi,
and Amir Badamchi

Iranian Research Institute for Information Science and Technology (IranDoc),
Tehran, Iran
{mohebi,golshaie,ganjefar,jalalimanesh,
p.afshar,alihosseini,ghoreishi,badamchi}@irandoc.ac.ir

Abstract. Conducting a research becomes a challenging task when a
large number of scientific data is available and needs to be organized.
In addition, research facilities may not be easily accessible due to the
limitation in using hi-tech equipment and the resulting expenses. While,
there are sophisticated tools for managing research data and assisting
the researcher in conducting a research, they are not accessible through
a union integrated manner, and sometimes the researcher is overwhelmed
with these tools, with additional effort to learn how to work with them.
To overcome these challenges, we propose a personalized intelligent assistant as a social robot, called RoboDoc, to assist the researcher in research
steps such as idea generation, designing research methodology, and collecting and analyzing research data. We introduce a conceptual framework for RoboDoc to describe its services and design requirements, and
at the end we briefly discuss our current efforts in developing RoboDoc.

Keywords: Research steps · Personalized assistant · Long-term interaction robots · Learnability · Adaptivity · Multimodality

1 Introduction

Researching, like many modern-day human activities, is becoming a complex
and challenging endeavor. Keeping track of growing number of scientific jour-
nals and articles relevant to a researchers field of study, organizing scientific
data efficiently for easy access and use, findings the most appropriate design
and methodology to conduct a research, and finally presenting research findings are some challenging aspects of doing research today. These dimensions of
research might be experienced differently by novice and experienced researchers:
Novice researchers might need assistance in basic procedural aspects of research,

---

This research was fully supported by Iranian Research Institute for Information Science and Technology (IranDoc). Special thanks goes to Dr. Sirous Alidousti, director
of IranDoc, for his support and fruitful advises in this research.

© Springer International Publishing AG 2016
A. Agah et al. (Eds.): ICSR 2016, LNAI 9979, pp. 808–818, 2016.
DOI: 10.1007/978-3-319-47437-3_79

while experienced researchers might have some concerns directed towards analytic and organizational issues inherent in research. These challenges have often been addressed in non-unified way through the creation of various software and data management tools, the learning of which sometimes requires extra amounts of efforts and energy. The management of scientific data is essentially divided into these phases: workflow management, management of metadata, data integration, data archiving, and data processing [1]. There are tools available for conducting each of these phases, however learning the tools and integrating them requires additional efforts, while sometimes the tools may not be compatible with each other. Also in the case of access to hi-tech facilities by researchers, some advancement has been made, e.g., to help surgeons operating by tele-surgery, but these solutions have not been extended to other conventional research environments and disciplines. We believe these issues can be approached in a more unified and human-friendly manner by having social robots that can communicate with researchers and assist them in addressing their research needs. Having a physical embodiment, such a robot can be considered an effective, credible and informative interactive partner [2,3]. It can also help researchers to have remote access to library books, preserved materials, and experimental facilities while they are not easily accessible. In addition it should be able to do scholarly searching, note taking, scientific event alerting, voice recording, documents and citations organizing and research method proposing. In this paper we present a conceptual framework for a social research assistant robot, that we name it as RoboDoc. The robot is aimed to provide assistance to novice and experienced researchers to plan and organize various research stages. It is also envisioned to provide researchers with remote access to materials and facilities that are not easily accessible. The project aims to create synergy among various scientific disciplines including artificial linguistics, computational linguistics, computer science, and robotics. In the next section we briefly review main efforts in the area of social robotics and their applications in long-term interactions. Then, in section three we introduce the conceptual framework for designing RoboDoc and its services to a researcher. Section four contains information on some preliminary experiments done for developing RoboDoc at the preliminary phase. At the end we conclude and introduce the future research directions for RoboDoc.

2   Previous Studies

Social robots have been recently applied in many areas. The application of social robots can be classified in three main categories [4]: (1) specialized applications where the robot has a clearly defined set of tasks with less communication with the user and simple sociability capabilities, (2) public applications where many users interact with the robot, and the robot needs a single general communication model for short-term interaction with users, with no adaption, and (3) individual application where the robot is designed for long-term, single user (or a specialized group of people) communications, with complicated and adaptable capabilities. In our research, we are aiming for individual application, with long-term interaction. The literature on social robotics for long-term interaction has been studied recently in [5], in which the main features that social robots should have in order to establish a long-term interaction with user are introduced. In addition, the main application areas of long-term interaction are identified as health care and therapy, education, work environments and public spaces, and at home [5]. One of the most common areas of application of social robotics is to assist people with disabilities. Social robotics are applied for assisting peoples with disabilities for companionship, household/service, health care/rehabilitation, and military, where health care application dominates others [6]. Specially, some of the recent research in social robots application in health care is dedicated to treating autistic children, as studied broadly in [7]. Robots used in domestic environments, usually used at home, are for entertainment, everyday tasks, and assistance to the elder and handicapped persons [8]. For instance Paro, a small interactive robot has been designed for elderly assistantship [9]. The concept of an assistive robot for tele monitoring at homes and their initial requirements like navigating through obstacles while providing stable videos are discussed in [10]. The tele-robot has been validated in controlled situation in laboratory. Brian was a human-like robot which had similar functionalities to human from the waist up, and was used for studying the response of the robotic system to human gestures [11]. Robovie [12] was a humanoid robot developed for interacting with people in their daily lives. A social robot is also designed to recognize potential customers from other people in a mall [13]. In the context of using social robots at office environment an office-guide robot was introduced to help visitors in an office environment [14]. Maggie [15] has been developed by the Robotics Lab at University Carlos III of Madrid, and later it has been customized as a gaming platform [1]. The HOVIS series robots and Sacarino robot are mainly used for household assistant services [16]. In addition, there are projects and studies for developing a general platform for social robots resulting in platforms such as Nao, Pepper, Romeo, developed by SoftBank/Aldebaran company. We have observed that most of the social robot applications for personalized robot focused on health care are for improving the life of elderly, persons with mental disorder, and handicapped. Therefore, we have not come up yet to a research on designing an individualized social robot for assisting a researcher.

3   Conceptual Framework

In order to introduce RoboDoc, first we propose a conceptual framework for the robot in this section. Through this framework the vision for RoboDoc and its main services are introduced. First, we briefly review the main steps taken by a researcher to conduct a research. The general steps for conducting a research are: (1) identification of research problem, (2) literature review, (3) specifying the purpose of research, (4) determine specific research questions, (5) specification of a Conceptual framework, (6) Choice of a methodology (for data collection), (7) data collection and verifying, analyzing, and interpreting the data, (8) reporting and evaluating research, and (9) Communicating the research findings and, possibly, recommendations. The research assistant social robot would not interfere with the researchers job, but it meant to accommodate the researcher. Therefore, direct involvement of the robot in some of the above steps, is meaningless. We do not design the robot to conduct each step, rather to be a researchers assistant for accomplishing a research. Thus, we can re-group the steps above, and bold the tasks from the above steps, that a machine assistant would be beneficial. Thus, the main steps for conducting a research that *RoboDoc can also assist* are: (i) Generating ideas, (ii) Designing research and choosing methodology, (iii) Collecting data, (iv) Organizing and analyzing information, (v) Reporting and evaluating results, and (vi) Publishing research findings. Figure 1(a) shows a schematic representation of these steps. The conceptual framework is a multi-layered framework representing different aspects that should be considered in designing RoboDoc. These aspects (layers) are: services, sociability requirements, functionalities, and platform requirements. The schematic representation of the conceptual framework is illustrated in Fig. 1(b). Each aspect will be introduced in details in the following sections.

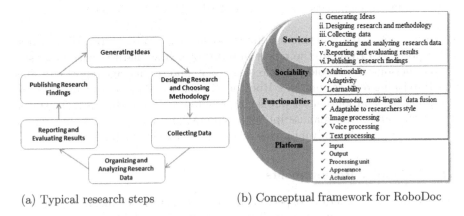

(a) Typical research steps   (b) Conceptual framework for RoboDoc

Fig. 1. Typical research steps and RoboDoc conceptual framework for providing services in these steps.

3.1 RoboDoc Services

For each step of research in Fig. 1(a), a set of services that RoboDoc can offer in order to assist the researcher are introduced. The services were discussed and decided upon by a panel of experts. The panel composed of experts in robotics and control, artificial intelligence and computer vision, natural language processing, educational research, and system analyst. The main services for each step are then proposed. Note that RoboDoc services are not limited to these items, and more services can be designed and developed in the future.

(i) Generating ideas: RoboDoc will be able to (a) capture and organize researchers ideas by different available tools, and (b) improve the ideas and provide a logical relationship between them by alerting the researcher of the recent development and studies in researchers areas of interest.

(ii) Designing research and choosing methodology: RoboDoc will help to choose the most suitable research methodologies and tools through providing information on research approaches.

(iii) Collecting data: RoboDoc will be able to (a) search in web and other scientific databases such as Ganj[1] simultaneously, and integrate the results into meaningful pieces of information, while customizing the search with researchers areas of interest, (b) apply query expansion and enrichment based on relevance feedback from researcher or semantic relationship between terms using thesauri, (c) classifying documents based on researchers interest and using graph databases to show the relationship between data, if needed, and (d) nalyzing and visualizing search results through data visualization techniques and tools to visualize search results

(iv) Organizing and analyzing research data: RoboDoc will be able to (a) do citation and reference management, (b) classification and store data, (c) provide data analysis toolboxes, and (d) visualize and generate plots of data

(v) Reporting and evaluating results: RoboDoc will be able to (a) edit research reports, (b) Evaluate the report in terms of text similarities with other scientific sources to avoid possible plagiarism, and (c) manage report citations

(vi) Publishing research findings: RboDoc will able to (a) propose an appropriate publisher and journal for publishing research findings, (b) track the status of the submitted manuscript after submission, and (c) track and report trends and recent advancement related to researchers interest

3.2 Sociability Requirements

Social robots are generally expected to have characteristics including but not limited to multimodality, personality, adaptivity, learnability, autonomy, and proactiveness [15] of which multimodality, adaptivity, and learnability have been our main focus in RoboDoc project. These characteristics account for different aspects of humans social behavior and can serve as solid criteria against which social skills of RoboDoc can be compared and evaluated.

Multimodality requires different modes of communication, including speech, gesture, gaze, etc. be used for producing or understanding a message. RoboDoc need to have multimodal communication skills to convey the message efficiently and to behave more naturally. Its main medium of communication is speech-based, but it is also equipped with a camera projector to present the intended content visually.

Adaptive systems can change their behavior according to their environmental situation. Special attention should be considered to this characteristic in developing RoboDoc since adaptation is at the core of giving long-term customized

---

[1] A scientific national database composed of Iranian scientific documents such as dissertations and papers.

assistance to researchers according to their research style. Every researcher has a distinct field of expertise that determines the specific methodological steps one needs to take. RoboDoc will need to adapt its search queries and presentation style to the researchers field of specialization and research procedures.

Learning, as one of the main characteristics of social robots, has two aspects in RoboDoc: learning directed towards information processing capabilities, and learning directed towards social skills. Social learning is a crucial prerequisite for RoboDoc to enter into social interaction with the researcher. One of the facets of social learning that has implications for efficient information processing is to learn a researchers personal way of doing research and his/her unique assistance demands from the robot. Learning is also crucial for information processing tasks of RoboDoc.

3.3 Functionalities

The functionalities enable RoboDoc to fulfill its services. RoboDoc communicate with user through some skills, process users request and present the results to the user. It interacts with user through a multimodal, multilingual manner and requires adapting its behavior to the researchers style. The need for the RoboDoc to have a physical embodiment comes from two motivating sources: One is the need for RoboDoc to be used in tele-labs or distant laboratories to carry out requested experiments. Second motivation concerns the social perception and acceptance of a physical robot as a credible, informative, and socially present partner, [3,9]. RoboDoc needs to have image processing skills, such as gesture recognition, face detection, OCR, and object detection and tracking. It will have voice processing skills, such as text-to-speech and speech-to-text skills (for Persian and English), user authentication, and dialogue manager. The scientific data (specially text) processing skills are the key features of RoboDoc. These skills composed of document indexing, document classification and clustering, keyword extraction, text summarization, intelligent editing of document, automatic translation, information retrieval and intelligent query expansion, sentiment analysis, and other natural language processing tasks such as Part-of-Speech tagging and parsing.

3.4 Platform Requirements

The typical minimum platform requirements for RoboDoc, categorized based on input, output, processing unit and physical features. RoboDoc's perception of the environment is obtained through *input* devices such as microphones and cameras. For advanced gesture recognition tasks, more sophisticated Kinect 3D sensors embedded in robots body would be beneficial.

RoboDoc can generate *output* and demonstrate its reactions to the user and environment through acoustic devices and displaying devices such as LCD or a data projector.

The *processing unit* needs to be able to analyze huge amount of scientific data from various databases. However, processing can also be done remotely in server-side in the case that the platform has limitations.

Physical features can be viewed from two aspects: appearance and actuators. Based on a researchers needs, the robot can be a humanoid one, like Nao with many actuators, or a non-humanoid one such as Jibo [17] or Aido robot. If RoboDoc tends to do research experiments in laboratories, then it should have certain Degree Of Freedom (DOF) and supporting actuators, otherwise, minimum DOF while supporting sociability movements would be enough.

Researchers with disabilities can also benefit from RoboDoc. For such persons, RoboDoc can be equipped with more advanced input and output devices to interact with user more efficiently. In case, the functionalities related to processing images, videos and voice must be improved as well.

4   Experiments and Discussions

In order to design and develop RoboDoc services, a long-term plan with different phases has been defined. We are now at the first phase of developing RoboDoc. In the first phase, RoboDoc needs to be able to interact with a user through its vision and speech recognition system. It should be able to search within a scientific database, retrieve the information requested by the user, and finally present the search results visually in a screen or by a data projector. Thus, based on the proposed conceptual framework, we explain the platform used for RoboDoc at the first phase. Then, we describe the main fucntionalities and sociability features of RoboDoc at the first phase, by introducing two interaction scenarios.

4.1   Platform

After studying different social robots platforms developed recently, we choose Nao (Robocup edition from SoftBank Company) as the platform for the first phase of the project. While this platform is easily accessible and can satisfy some of our physical requirements at the first phase of RoboDoc development, it has limitations in satisfying some of the requirements at the first phase. Thus, we have added extra hardwares and functionalities to Nao in order ti fulfill the requirements of the first phase of development. Obviously, the platform can be improved and even replaced with the ones with higher capabilities at the next phases of development in the future. For this phase, we have added a wireless microphone for enhancing the performance of voice recognition. In addition, we have embedded a wireless mini-data projector in Nao for presentation purposes. To integrate data projector into Nao's body, and having an attractive appearance, we have designed a hat shape cover, as shown in Fig. 2. The logo of RoboDoc, is also illustrated below Naos head in Fig. 2(b). Figure 2(a) shows RoboDoc when presenting some information through the data projector.

(a) RoboDoc presenting a demo with a data projector

(b) Initial RoboDoc design based on Nao platform with the RoboDoc logo at the bottom

Fig. 2. Initial development of RoboDoc based on Nao platform. (a) RoboDoc is presenting some information using a data projector, (b) A new design for Nao head for embedding data projector.

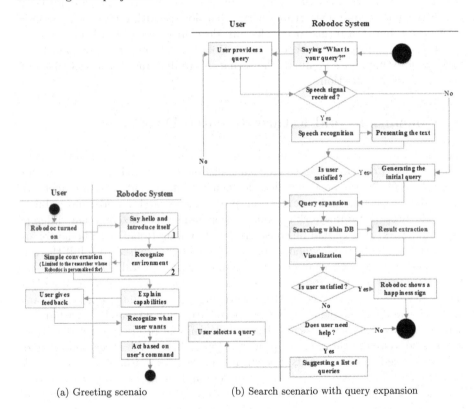

(a) Greeting scenaio

(b) Search scenario with query expansion

Fig. 3. Demonstration of greeting and search scenario for RoboDoc.

4.2   Interaction Scenarios

We have designed and added some modules to enhance Nao's functionalities at the first phase of RoboDoc development.

Persian text to speech module is used to convert Persian text to Persian speech signal. We have used *Gooya* software developed by *Pactos*[2] company for this purpose. *Persian speech to text* module is basically composed of two main modules: one module for converting voice command to text, and the other for converting a continuous speech to the corresponding Persian text. For the command-based recognition, we are using HTK speech recognition toolkit, based on hidden markov model (HMMs). Also, a *face and gesture recognition module* is designed using OpenCV libraries. Th gesture recognition module is able to chracterize simple hand gestures.

One of the main purposes of RoboDoc is to provide services like searching for information in several scientific databases including Google and Ganj for searching in Persian databases. Thus we have a powerful *search engine* for Ganj database, that is equipped with a query expansion algorithm based on semantic relationship between terms (using thesauri developed recently in IranDoc[3]).

We have developed two scenarios for RoboDoc at the first phase of development: greeting and searching. The scenarios are illustrated in Fig. 3 based on activity diagram notation.

5   Conclusions and Future Research Directions

In this research we have introduced a new conceptual framework for a social robot, called RoboDoc that can be an assistant to researcher for conducting research steps. These steps includes idea generation, choosing suitable methodology, data collection and organization, preparing research reports and publication. The proposed framework is a multi-layer framework encompassing different aspects (layers) that should be considered in designing RoboDoc. The layers are services, sociability requirements, and functionalities required to fulfill the services, and the required characteristics of platform. The RoboDoc services are first introduced for each research step. Then, we focus on developing RoboDoc based on some of the services such as searching and retrieving information, and organizing and visualizing data, with a special attention on searching within Persian scientific databases. Basically, the main themes for future research directions focus on developing the services proposed for each research steps and expanding them based on future needs. Specifically, in the future, RoboDoc will provide more sophisticated services such as text understanding, note taking, voice recording and organizing, researcher scientific records organizing, scientific text recommending, scientific events alerting, email checking and auto-replying, text, voice and video indexing and so on. To enable RoboDoc for providing these services we have to enrich the architecture of its knowledge-base, develop a customized

---

[2] http://pactos.ir/.

[3] Iranian Research Institute for Information Science and Technology.

ontology-based text understanding module and also develop some personalized learning algorithm to let RoboDoc fits with users needs and behavior. In terms of RoboDoc platform, we have adopted Nao as the initial platform for and have customized the platform by adding some hardwares to enhance its performance. Obviously, one interesting research direction would be to design and develop a designated platform with specific embodiment's features aligned with RoboDoc requirements.

References

1. Ailamaki, A., Kantere, V., Dash, D.: Managing scientific data. Commun. ACM **53**, 68–78 (2010)
2. Kidd, C., Breazeal, C.: Effect of a robot on user perceptions. In: Proceedings of International Conference on Intelligent Robots and Systems, pp. 3559–3564 (2004)
3. Bainbridge, W.A., Hart, J.W., Kim, E.S., Scassellati, B.: The benefits of interactions with physically present robots over video-displayed agents. Int. J. Soc. Robot. **3**(1), 41–52 (2011)
4. Hegel, F., Lohse, M., Swadzba, A., Wachsmuth, S., Rohlfing, K., Wredel, B.: Classes of applications for social robots: a user study. In: Proceedings of IEEE International Symposium on Robot and Human Interactive Communication, pp. 938–943 (2007)
5. Leite, I., Martinho, C., Paiva, A.: Social robots for long-term interaction: a survey. Int. J. Soc. Robot. **5**(2), 291–308 (2013)
6. Yumakulov, S., Yergens, D., Wolbring, G.: Imagery of disabled people within social robotics research. In: Proceedings of International Conference on Social Robotics (2012)
7. Alemi, M., Meghdari, A., Basiri, N.M., Taheri, A.: The effect of applying humanoid robots as teacher assistants to help iranian autistic pupils learn english as a foreign language. In: Proceedings of 7th International Conference on Social Robotics, pp. 1–10 (2015)
8. Yan, H., Ang, Jr., M.H., Poo, A.N.: A survey on perception methods for human-robot interaction in social robots. In: International Journal of Social Robotics, pp. 85–119 (2014)
9. Kidd, C.D., Taggart, W., Turkle, S.: A sociable robot to encourage social interaction among the elderly. In: Proceedings of IEEE International Conference on Robotics and Automation (2006)
10. Michaud, F.: Telepresence robot for home care assistance. In: AAAI Spring Symposium: Multidisciplinary Collaboration for Socially Assistive Robotics (2007)
11. Nejat, G., Ficocelli, M.: Can I be of assistance? The intelligence behind an assistive robot. In: IEEE International Conference on Robotics and Automation (2008)
12. Mitsunaga, N.: Robovie-IV: a communication robot interacting with people daily in an office. In: Proceedings of IEEE International Conference on Intelligent Robots and Systems (2006)
13. Kanda, T.: Who will be the customer?: a social robot that anticipates people's behavior from their trajectories. In: International Conference on Ubiquitous Computing (2008)
14. Pacchierotti, E., Christensen, H.I., Jensfelt, P.: Design of an office-guide robot for social interaction studies. In: Proceedings of IEEE International Conference on Intelligent Robots and Systems (2006)

15. Gonzalez-Pacheco, V.: Maggie: a social robot as a gaming platform. Int. J. Soc. Robot. **3**, 371–381 (2011)
16. Feil-Seifer, D., Mataric, M.J.: Defining socially assistive robotics. In: Proceeding of International Conference on Rehabilitation Robotics (2005)
17. Avila, L., Bailey, M.: High Tech @ Home. IEEE Comput. Graphics Appl. **35**(3), 8–9 (2015). http://ieeexplore.ieee.org/document/7111929/

Mechanical Design of Christine, the Social Robot for the Service Industry

Yi Mei Foong[1], Xiaomei Liu[2], Shuzhi Sam Ge[2(✉)], and Jie Guo[2]

[1] Department of Mechanical Engineering, National University of Singapore,
Singapore 117575, Singapore
[2] Department of Electrical and Computer Engineering,
National University of Singapore, Singapore 117583, Singapore
samge@nus.edu.sg

Abstract. Based on the strong understanding of a natural human-robot interaction, this paper presents a social robot named Christine with a human-like exterior which is developed to work in the service sector. Although many social robots have been developed and excelled in control systems, several humanoid robot have either fallen into the uncanny valley or not been accepted by the public yet. The mechanical design of Christine has a great improvement in aesthetics without compromising its functionality. There are 7 degrees of freedoms (DOFs) through the head of Christine for representing 3 head gestures and 5 facial emotions and the social intelligence is implemented based on vision and audio subsystems. The hardware architecture of Christine, which includes processor, vision, and motion system will also be presented systematically.

1 Introduction

An uncanny valley is a factor of eeriness and a hypothesis about a person's response to a human-like robot that would abruptly shift from empathy to revulsion as it approached, but failed to obtain a lifelike appearance [1]. One example is the mismatch of the size and texture of the eyes and face that is especially prone to make the humanoid eerie [2]. On the other hand, a social robot with anthropomorphic qualities will lead human partners to treat humanoid social robots as real persons [3]. Therefore, social robots are often designed to deal with human care, health, domestic tasks and various other forms of immaterial and material tasks which aim to renew human capacities [4].

In the past decades, several social robots have been developed to replicate a human-like exterior such as Hanson Robotics's FACE [5] and Hiroshi Ishiguro's geminoid [6]. Some other robots involve the use of a screen as the robot's face such as Honda's ASIMO [7] and Korea Advanced Institute of Science and Technology (KAIST)'s HUBO [8]. However, several humanoids have fallen into the uncanny valley despite excelling in control systems or are still generally not accepted by the public.

In this paper, a female social robot named Christine is presented where she is primarily developed to be the first encounter of service for any business to

© Springer International Publishing AG 2016
A. Agah et al. (Eds.): ICSR 2016, LNAI 9979, pp. 819–828, 2016.
DOI: 10.1007/978-3-319-47437-3_80

provide basic assistance or receive visitors with hospitality. She is designed with a friendly human-like exterior built, realistic facial expressions and head gestures. Built with an empathetic personality, the user's emotion can be read through its vision and speech system. Equipped with the user's request, she is able to communicate back using both verbal and non-verbal means such as through her facial expressions and head gestures as the face is one of the richest sources of information about human behavior [9].

With the outlook on increasing manpower cost [10] in the service industries, it is becoming evident for businesses to find sustainable solutions without compromising the standard of services. A typical solution requires hiring qualified manpower to improve the quality of service [11] which induces additional cost. Furthermore, some countries are facing a drop in birth rate which may introduce a shortage of workers in multiple industries during the near future [12]. With the development of Christine, businesses will be able to take advantage of her intelligent technology while retaining the familiar human-emotions. The main contributions of this work are listed as follows:

(i) Christine's 5 emotions, including neutral, happy, sad, angry and surprise are based on the research that there are few discriminations of human emotion during early face signals for dynamic facial expressions [13]. This provides a base for accurate evaluation from the user when observing Christine's emotion which improves the efficiency and user-friendliness of Christine.

(ii) A mechanical design is proposed to minimize the number of actuators, reduce the overall weight and simplify the control system. This can be achieved by utilizing multiple directional controls with precise magnitude.

(iii) An efficient human-robot interaction by minimizing the visibility of hardware system, using elements like clothes and wigs and replicating a human's movements. This provides an environment that is natural and comfortable for the user during an interaction.

2   Mechanical Design

Christine's mechanical structure as illustrated in Fig. 1 is designed to ensure realistic facial expressions by utilizing the elasticity of the robot's facial mask made from a flexible polymer material and to simulate accurate head gestures It is capable of 5 natural robotic emotions and 3 head gestures. Generally, there are 7 degrees of freedom (DOFs) in the humanoid head that assists the movements.

The Mechanical design includes design for assembly, manufacturing and strength analysis of the internal mechanism that supports Christine's face and movements. Computer-Aided Design (CAD) has been used for the primary design medium with a modularity concept. Multiple prototypes have been made using quick prototyping methods to ensure that visually realistic-looking emotions are produced in conjunction with the control system.

The assembly is optimized to use minimal space by reducing the number of parts due to the significance of the servo motors relative to the size of the small volume of Christine's head. The assembly sequence is thoroughly checked to ensure that the entire head can be easily disassembled for maintenance.

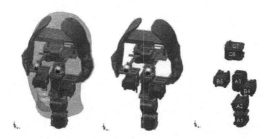

Fig. 1. From left to right: Mechanical Structure, Mechanical Structure (without mask), Placement of motors

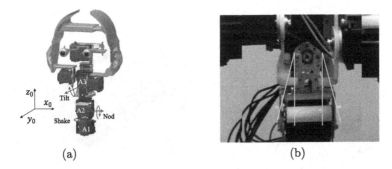

Fig. 2. Head gesture design: (a) 3 DOFs of head gestures; (b) Motion constraints for tilt movement

2.1 Head Gesture Design

Christine is designed to response with 3 head gestures, namely 'nod', 'shake' and 'tilt' as illustrated in Fig. 2(a). The head gestures design contains 3 DOFs achieved by 3 servo motors and are assembled using specially designed modular parts. The placement of the servo motors are optimized to lower the center of gravity of the overall assembly which reduces the load on the motors.

Although the servo motors used are capable of tracking crucial information like its speed, shaft position, load and so on for its feedback system, it does not have an internal brake system to help stabilize the robot when these joints are not in operation. Therefore, physical braking attachments have been designed to limit the maximum and minimum positions of Christine's head gestures. They are designed to withstand the moment and force of the head movement and weight as illustrated in Fig. 2(b).

These 3 DOFs head gestures supports a wide range of head body languages which improves Christine's non-verbal communication abilities. For example, nodding up and down signals agreement or affirmative and may be accompanied by smiling and other signs of approval. Similarly, tilting the head sideways can be a sign of interest, curiosity, uncertainty or query. The greater the tilt, the greater the uncertainty or the greater the intent to send this signal [14].

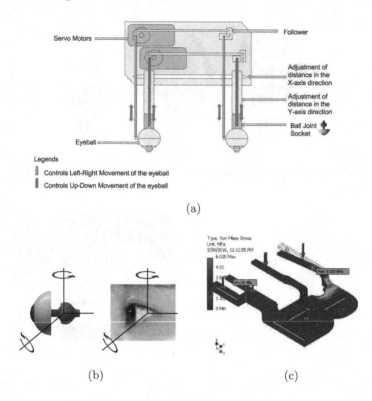

Fig. 3. Eye design: (a) Schematic Diagram of the servo motors and followers; (b) Degree of freedom for eye socket; (c) Stress analysis

2.2 Eye Design

The design of Christine's eye allows the minimal use of motors and parts while achieving the same angle of rotations to the natural eyeball movements. It consists of a housing with 2 servo motors which provides movements to the eyeballs. It has been simulated virtually as illustrated in Fig. 3(c) which proves that it has sufficient strength while ensuring a space-saving design. The 2 servo motors has a follower each that translates the rotational motions into 4 linear motions. The servo motor (C6) is responsible to control the up-down movements while the other servo motor (C7) is responsible to control the left-right movements of the eyeballs as illustrated in Fig. 3(a).

It allows linear adjustments in the X-axis direction to control the center of distance (CD) of the eyeballs and in the Y-axis direction to adjust the distance from the eyeballs to the eye socket as illustrated in Fig. 3(a). The main advantage is to reduce any uncanny feeling and to accommodate the tolerance of the robot's mask which may arise due to manufacturing faults.

At the eyeball joint is a ball socket assembly with many degrees of freedom such as multiple axis rotations as illustrated in Fig. 3(b). The movement of the eyeball is smooth as the low surface area between the fixed socket and moving ball induces very low friction. The strength of the ball socket joint is also sufficient due to the low weight of the acrylic eyeball. With this modular design, other similar human-like mask can also be attached onto Christine.

2.3   Skull Design

The main design requirements is to protect the components inside the skull and ensure that the mask is able to fit securely. The dimensions of the skull are modeled based on Christine's mask to ensure an accurate fit. The distinct features on the skull which are shaped like ears serves as a reference point when assembling as shown in Fig. 4. Fillets are also included to reduce any stress concentrations on the skull.

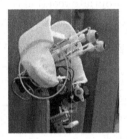

Fig. 4. Unique ears feature modeled after Christine's mask

2.4   Oral Design

The oral design uses groups of metal wires in tension along with 2 servo motors with one planar part each which is capable of 1 DOF. The motor will turn at a controlled angle and direction to transmit the rotational motion to a linear displacement by tying 8 groups of metal wires to each of the 2 ends of the planar attachment. For assembly, the 8 groups of metal wires are hooked onto 4 points on Christine's mask from the inside of the mask. The fitting of the mask onto the skull requires each labeled pair of metal wires to clip onto each other to make a linkage that transmits the force. The disassembly procedure is the reversed order of the assembly procedure which proves to be efficient for maintenance.

3   Facial Emotion Expression

Christine is capable of 5 emotions: neutral, happiness, sadness, anger and surprise. The emotions are achieved with various combination of head gestures, eye movements and oral movements as illustrated in Table 1 using 7 servo motors. In this study, the 'neutral' emotion is designed as the origin and default pattern for emotion control. The design for robot eye, oral and neck motions have been introduced and discussed in Sect. 2.

In comparison to the neutral emotion, the brief descriptions of the expression patterns for the 4 other emotions are concluded in the third and forth column of Table 1 as the reference during motor control while the fifth and sixth columns shows the respective activation and motion for each motor. An illustration in Fig. 5 shows an overview of actuating points on the mask with reference to the servo motor which can be fine-tuned to ensure accurate emotions and head gestures aesthetically. Continuous movements that simulate minute muscle movements is currently in development which will further increase Christine's canniness.

Christine can understand the user's emotion through facial expression, voice input, words used and gait patterns. Based on the outcome of emotion recognition, Christine can offer an appropriate reaction that is suitable for the situation.

Table 1. 5 facial expressions of Christine

Emotion	Human emotion expression	Features	Description	Motor Group	Motor No.
Neutral		Eyeball	-	C	-
		Oral	-	B	-
		Neck	-	A	-
Happy		Eyeball	-	C	-
		Oral	Corners slightly raises	B	4
		Neck	Moves slightly backward	A	2
Sad		Eyeball	Looks down	C	6
		Oral	Slight pout	B	4,5
		Neck	Moves slightly forward	A	2
Angry		Eyeball	Looks slightly to the left	C	7
		Oral	Pursed lips	B	5
		Neck	-	A	-
Surprise		Eyeball	-	C	-
		Oral	Less wide, taller shape	B	4,5
		Neck	Move slightly backwards	A	2

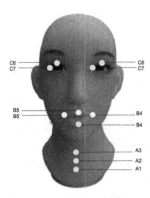

Fig. 5. Overview of actuating points on the face with reference to the motor

4   Hardware Architecture

Christine is designed to achieve fundamental tasks (looking, listening, speaking), functional tasks (face tracking, gait, speech and emotional recognition) and social tasks (emotion interaction and eye contact) as illustrated in Fig. 6. Crucial computer hardware and sensory systems are utilized to accomplish these tasks.

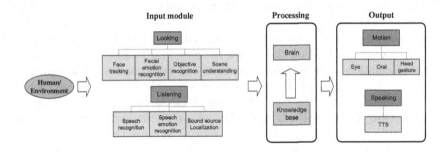

Fig. 6. Overview of software system

Figure 7 shows an overview of the electrical components integrated into Christine. The overall hardware architecture is designed to replicate a service counter where there is interaction and communication between Christine and the user as illustrated in Fig. 8. Christine is dressed in a formal wear to portray professionalism in the job which invokes a sense of familiarity to the user.

4.1   PC

The brain of Christine is a PC with of the model DELL OPTIPLEX 980 with Intel Core i5-650M 3.20 GHz Processor and NVIDIA GeForce GTX 960 M with

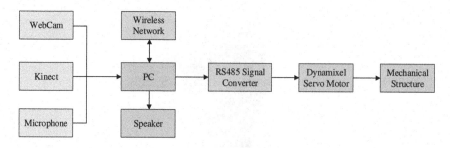

Fig. 7. Electrical architecture of Christine

Fig. 8. User interacting with Christine

4 GB GDDR5. A knowledge base is built to record the basic information determined for Christine (e.g., name, birthday, hobby and talent, etc.) and essential knowledge (e.g., fusion rules and decision making rules, etc.). The PC learns from experience archived on the knowledge base and makes decision according to the information acquired from the visual and audio sensors. The equipment of the GPU with high performance accelerates graphical computations that significantly benefits from the parallel nature of the architecture with many low-end processing nodes which is well adapted for vector and matrix operations abound in deep learning algorithms. In order to improve the social intelligence of Christine, the application of deep learning has been widely applied in the software system such as the Convolutional Neural Networks (CNN) that is currently utilized to extract emotional information from facial features which are subsequently processed by linear classifier.

4.2   Vision System

Christine's vision system consists of 1 Webcam (Logitech C170) used for facial recognition and 1 motion sensor (Kinect for Xbox 360) used for gait recognition.

The Webcam records through a hole in the chest area of Christine's body and is camouflaged to reduce notability. The Webcam is tilted at an angle to provide a natural and comfortable experience to the user without compromising its effective capture area during the interaction. The Kinect is placed at a distance away from the user's frontal to record the user's gait within the effective range.

4.3 Audio System

Christine's audio system includes 1 Microphone (Audio-Technica AT2020 Cardioid Condenser Studio Microphone), 1 Webcam (Logitech C170) with a build-in microphone and motion sensor (Kinect for Xbox 360) with build in multi-array microphone. A Takstar PM-5 phantom power provides 48V DC power for the high-quality condenser microphone of high sensitivity of $-37\,$dB. Kinect sensor's microphone array is capable of conducting acoustic source localization and ambient noise suppression.

The AT2020 is used during the speech recognition and speech emotion recognition processes. It is placed directly in front of Christine to record the user's audio input with little disturbance. Christine is able to implement speech recognition and answer simple questions with CMU Sphinx, artificial intelligent markup language (AIML) and eSpeaker.

4.4 Motion Controller System

Dynamixel servo motor used are MX-28 and MX-106 and is employed because of its combined reduction gear head, high torque characteristic, integrated control circuitry, real time motor data retrieval and networking functionality. They adopt daisy chain connection and support communication which speeds up to 1M bit/sec. Moreover, distributed control is designed so that position, velocity, and torque can be set with a single command packet. This is a very powerful control method as it enables the industrial PC to control multiple servo motors simultaneously without increasing its computational loads.

All Dynamixel controller boards are integrated in the servo motors and connected to RS485 signal converter in a chain structure. The RS485 accepts USB signals from second industrial PC via USB hub and converts these signals into RS485 signals for Dynamixel controller board. With the decentralized controller, Christine can also illustrate reflexive behaviors like nodding while smiling which will significantly increase its sociability.

5 Conclusions

We have introduced the mechanical design and hardware architecture in the next generation social robot, Christine. The human-like exterior and movements, minimized number of actuators used and human-robot interaction have been carefully considered during the design process. Christine is able to actuate 5 emotions and 3 head gestures with a human-like appearance and artificial intelligence.

Acknowledgement. This work is supported by Defence Innovative Research Programme (DIRP), the Ministry of Defence, Singapore under grant R-263-000-B08-592 and the A*STAR Industrial Robotics Program of Singapore under grant R-261-506-007-305.

References

1. Mori, M., MacDorman, K.F., Kageki, N.: The uncanny valley [from the field]. IEEE Robot. Autom. Magazine **19**(2), 98–100 (2012)
2. MacDorman, K.F., Green, R.D., Ho, C.-C., Koch, C.T.: Too real for comfort? Uncanny responses to computer generated faces. Comput. Hum. Behav. **25**(3), 695–710 (2009)
3. Duffy, B.R.: Anthropomorphism and the social robot. Robot. Autonom. Syst. **42**(3), 177–190 (2003)
4. Vincent, J., Taipale, S., Sapio, B., Lugano, G., Fortunati, L. (eds.): Social Robots from a Human Perspective. Springer, Switzerland (2015)
5. Mazzei, D., Billeci, L., Armato, A., Lazzeri, N., Cisternino, A., Pioggia, G., Igliozzi, R., Muratori, F., Ahluwalia, A., De Rossi, D.: The face of autism. In: RO-MAN, 2010 IEEE, pp. 791–796. IEEE (2010)
6. Rosenthal-von der Pütten, A.M., Krämer, N.C., Becker-Asano, C., Ogawa, K., Nishio, S., Ishiguro, H.: The uncanny in the wild. Analysis of unscripted human-android interaction in the field. Int. J. Soc. Robot. **6**(1), 67–83 (2014)
7. Sakagami, Y., Watanabe, R., Aoyama, C., Matsunaga, S., Higaki, N., Fujimura, K.: The intelligent asimo: system overview and integration. In: IEEE/RSJ International Conference on Intelligent Robots and Systems, vol. 3, pp. 2478–2483. IEEE (2002)
8. Oh, J.-H., Hanson, D., Kim, W.-S., Han, I.Y., Kim, J.-Y., Park, I.-W.: Design of Android type humanoid robot albert hubo. In: IEEE/RSJ International Conference on Intelligent Robots and Systems, pp. 1428–1433. IEEE (2006)
9. Cohn, J.F., Elmore, M.: Effect of contingent changes in mothers' affective expression on the organization of behavior in 3-month-old infants. Infant Behav. Dev. **11**(4), 493–505 (1988)
10. Chung, S.H., Jung, D.C., Yoon, S.N., Lee, D.: A dynamic forecasting model for nursing manpower requirements in the medical service industry. Serv. Bus. **4**(3–4), 225–236 (2010)
11. Teixeira, R.M., Baum, T.: Demanda de mão-de-obra e exigências de qualificação no setor de hotelaria: o caso de aracaju, brasil. Turismo-Visão e Ação **9**(2), 155–168 (2008)
12. Hara, T.: A Shrinking Society: Post-Demographic Transition in Japan. Springer, Japan (2014)
13. Ekman, P., Friesen, W.V.: Facial action coding system (FACS): Manual. Consulting Psychologists Press, Palo Alto (1978)
14. Pease, A.: Body Language: How to Read Other Thoughts by Their Gestures. Sheldon Press, London (1981)

Influence of User's Personality on Task Execution When Reminded by a Robot

Arturo Cruz-Maya[✉] and Adriana Tapus

Robotics and Computer Vision Lab, U2IS ENSTA ParisTech, Université Paris-Saclay,
828 Bd des Maréchaux, 91762 Palaiseau Cedex, France
{arturo.cruz-maya,adriana.tapus}@ensta-paristech.fr

Abstract. One of the main purposes of companion robots is to use them to remind their users about the tasks they have to do. The interaction requires robots to adapt to the person with respect to their preferences. The performance of the human when a robot reminds them to do a certain task is of great importance. Findings in social psychology show that personality influences the way that humans interact. In this work, we conducted an experiment of task reminders in an office-like environment with a robot reminding tasks to a person while the person is doing other office-activities, with the goal of searching for positive influences of the robot on user's personality. Nine different conditions were studied with the robot varying its behavior and appearance. Results show that the user's personality has an influence on his/her time to perform a task while being reminded by a robot to perform such task, showing that people with high conscientiousness are more promoted by the robot to finish the task earlier than people with low conscientiousness, and also that introverted people are more motivated by the robot to finish the task earlier than extroverted people.

1 Introduction

In recent years, the number of projects around the world aimed at developing companion robots has increased considerably. Developing companion robots for health care for the elderly is a challenge and a need [4,6,14]. Social robots used in therapy for children with autism have been an active field of research in the past years [3,16]. Companion robots were also used in education, work environments, and public spaces [11].

Having robots helping us with our daily activities lead to the need of endowing them with social capabilities in order to adapt their behavior to the environment and tasks. Nevertheless, how to achieve this adaptation remains a challenge. Some important features for social robotics are the synthesis and recognition of emotions in order to be more appealing to humans and to be perceived as more useful and expressive [10].

Moreover, in order to provide a customized interaction, the robots can be endowed with various personality traits according to the different types of tasks to be performed [8,17].

© Springer International Publishing AG 2016
A. Agah et al. (Eds.): ICSR 2016, LNAI 9979, pp. 829–838, 2016.
DOI: 10.1007/978-3-319-47437-3_81

Works in social psychology have shown that people with different kinds of personality have different preferences to interact. According to [18] extroverts allow closer interactions than introverts, but this can be influenced depending on the person's height [2]. Likewise, a study on smiling faces have proved that a smile has an impact on the behavior of people watching the smiling face [15]. Also, the personality trait of conscientiousness has shown to be linked to the performance of people when receiving orders [9].

To the best of our knowledge, no previous work explores the effects of distance, height, and smile of a robot while reminding a task to a person in order to improve their performance based on his/her personality. Therefore, in this work, we investigate the mentioned effects in relation with two traits of personality (conscientiousness and extroversion). The scenario is an office-like environment where the robot provides reminders of a schedule to the participant while the participant is busy with another task.

This paper is organized as follows: Sect. 2 describes the experimental design setup; Sect. 3 shows the results obtained; and finally Sect. 4 concludes the paper.

2   Experimental Design Setup

2.1   Hypothesis

Personality is an important factor in human social interaction, and has a long-term consistent effect on the generated human multimodal behavior. The authors in [13] defined personality as the coherent and collective pattern of emotion, cognition, behavior, and goals over time and space. Therefore, it is important to consider the relationship between personality, goals, and performance in human-robot interaction.

According to the definition of introverts and extroverts, we can expect more observable social behavior of the extroverts, and also that they will prefer a closer interaction than introverts [18]. In human-human interaction people prefer to interact with a person with small height [2], and also the smile has an influence on the person's behavior [15]. The authors in [1] showed that the most consistent personality predictor for task performance is conscientiousness. Some adjectives that are usually used to describe people with high conscientiousness are responsible, organized, and achievement oriented.

Based on the above statements and the literature, we elaborated the following hypotheses:

- H1. High conscientiousness people perform better in time when reminded by a robot than low conscientiousness people.
- H2. Close interaction (at the limit of interpersonal distance [7]) will be preferred by extroverted people and far interaction (1.5 times the limit of interpersonal distance) will be preferred by the introverted people in the task reminder.
- H3. Participants will prefer to interact with a small robot rather than with a tall robot.
- H4. Participants will be motivated to finish a task earlier when the robot shows a smile on its face.

2.2   Office-Like Scenario Description

The scenario used to test and validate our hypotheses is an office-like environment, shown in Fig. 1, where the user is asked to reply to as many e-mails as he/she can in a series of 10 emails. The total allotted time is 6 min (maximum bound). Two e-mails are labeled as urgent: one is a reminder of a meeting, and the other is a request for an activity report that should consist of 30 to 100 words. The others 8 emails are related to personal or work relations, where the user can reply with short answers. At the same time, a schedule to follow is given to the user, but the user is free to choose if he/she wants to follow the schedule or not. The schedule marks a break between minutes 2 to 4 after the beginning of the activity, and an important meeting between minutes 4 to 6 (time to go to the meeting). One minute before the specified time of the activities and at the exact time of these, the robot approaches the user to remind him/her of the activities.

Fig. 1. Office-like scenario with the Meka M-1 robot

2.3   Robot Behavior

The Meka M-1, is a wheeled humanoid robot that has been designed to work in human-centered environments. At the moment of the reminder, the robot goes in front of the participant and reminds him/her the activity on the schedule. After that, it waits for the response of the participant.

In order to avoid speech recognition system limits (in the case of non native English speakers), the user answers by showing a card that is recognized by the robot. There are 4 cards with meanings of: 1. "Thank you", 2. "Remind me later", 3. "I already did it", and 4. "Don't remind me again". If the user shows the cards 3 or 4, the robot will not remind that activity again.

The reminders of the robot were designed in consideration of the criteria for good reminders [12], and their verbal content is presented as follows:

– Taking a break: "Hello, remember to take a break from your computer".
– Going to a meeting: "I would like to remind you about the meeting with your boss in few minutes. It will take place in the Meeting room".

In order to avoid repetitions, two different phrases with similar meaning have been developed for each reminder.

2.4 Pre-experiment Questionnaire

We recruited 16 participants for this experiment (4 Female, 12 Male) from ENSTA ParisTech university campus. Participants ages ranged from 21 to 32, all with technical background.

Participants were asked to fill out the Big Five inventory prior to participation so as to determine their position on the extroversion-introversion and conscientiousness spectrums [5]. This questionnaire contains 44 items each with 5-point Likert scale that ask the participant to rate their agreement or disagreement with statements about their own personality and activities. The score of the test gives values between 1 to 5. People with a score $<= 3$ on a personality trait was considered in the low category of the examined personality trait. For our study, we looked only at the extroversion and conscientiousness traits.

For the conscientiousness trait, we selected 4 participants with low conscientiousness, 4 participants with high conscientiousness, in both groups there were 2 participants scoring low on extroversion and 2 participants scoring high on extroversion. For the extroversion group we selected 4 participants with low extroversion (introverted), and 4 participants with high extroversion (extroverted), all of them with a score bigger than 3 on conscientiousness.

2.5 Conditions

The study followed a 2×9 within-participants study design, with participant personality traits (extroversion/introversion and high/low conscientiousness, traits separately examined) and robot behavior as factors. The participants were divided in 4 groups according to the two traits of personality: conscientiousness and Extroversion. The comparison on personality was done between High conscientiousness and Low conscientiousness individuals, and between Extroverted and Introverted individuals, having eight individuals in each personality trait. The first robot condition was realized without the robot, and thus no reminders were provided. The next eight robot conditions were done with the combination of the 3 parameters (independent variables) to test: height of the robot, distance between the user and the robot at the moment of the reminder, and the smiling of the robot. The conditions were applied in the same order to all the participants, but the risk of learning the task was minimized by using different e-mails to reply on each condition. The conditions are listed in Table 1, where the values of the different variables of the robot are shown.

The 3 parameters (height/proxemics/smile) defining robot's behavior are:

1. Distance. Close: 1.2 m, which is the limit of the interpersonal distance according to [7]. Far: 1.8 m, (1.5 x minimum interpersonal distance)
2. Height. Small: 1 m. Approx. height of a person sitting. Tall: 1.8 m. Approx. height of a person standing up.
3. Smile. Smile off: Robot without facial expression. Smile on: The face of the robot shows a smile drawn by the Meka LED matrix.

Table 1. Robot factors varying for each condition

Condition	Code	Height	Distance	Smile
1	—	NA	NA	NA
2	TCN	Tall	Close	Off
3	TCE	Tall	Close	On
4	TFN	Tall	Far	Off
5	TFE	Tall	Far	On
6	SCN	Small	Close	Off
7	SCE	Small	Close	On
8	SFN	Small	Far	Off
9	SFE	Small	Far	On

2.6   Post-experiment Questionnaire

A post-experiment 5-points Likert scale questionnaire (33 items) was conducted after each condition. This questionnaire was done with the purpose of analyzing the perception of the participants towards the robot and search for relations between the variables of the study and the perceived influence on the task. Other questions are related to the perceived usefulness of the robot in the reminder task, perceived personality of the robot, and stress caused by the robot. The questions of the most relevant results are shown in Table 2, the complete list of the questions can be seen online[1].

Table 2. Post-experiment questionnaire

Q12	You felt pressured by the robot reminding you the tasks to do: not at all/ a lot
Q14	The robot was: Strongly disagree/Strongly agree
	Q14.a Social - Q14.b Attentive - Q14.c Stressful - Q14.d Helpful
Q15	The robot was expressive: Strongly disagree/Strongly agree
Q18	DId you think the robot was acting intelligently? not at all/a lot
Q19	What characteristics made the robot more efficient in the reminding task:
	Q19.a Speech - Q19.b Height - Q19.c Proxemics - Q19.d Facial expressions

3   Results and Discussion

The time participants took to finish the experiment was used to measure the efficiency of the robot's reminders.

---

[1] http://goo.gl/forms/7OrESYgUtUp7esbD2.

We did a series of ANOVA tests, preceded by a Shapiro test to verify the normality of the data. We analyzed the performance on time for conscientiousness and extroversion separately, comparing introverts with extroverts and high conscientiousness people with low conscientiousness people.

High C.		Low C.	
Mean	std dev	Mean	std dev
4.1984	0.8948	5.1340	0.8784

Sum sq	Mean sq	Df	p
14.016	14.0157	1	8.051e-05

Extro.		Intro.	
Mean	std dev	Mean	std dev
4.8600	1.2781	3.9081	0.6879

Sum sq	Mean sq	Df	p
4.353	4.3529	1	0.047

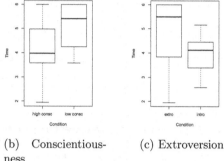

(a) Two-Way ANOVA tests   (b) Conscientiousness   (c) Extroversion

Fig. 2. Results of ANOVA tests on time performance of conditions 2–9. (b) high conscientiousness people got better performance (mean time) than low conscientiousness people. (c) Introverted people got better performance than extroverted people.

We did not found any relation on time performance of the participants grouped by personality traits and the different conditions.

We found significant differences between extroverts and introverts (p = 0.047) showing a better time performance of the introverts, as well as between high conscientiousness and low conscientiousness people (p = 8.051e-05) showing a better time performance of the high conscientiousness people. The results of an One-Way ANOVA test having time as dependent variable and personality traits as independent variables for each comparison of extroversion and conscientiousness are presented in Fig. 2a). The means for the performance in time of each group are shown in Fig. 2b) and Fig. 2c).

A paired Student's T-Test was applied to each group to analyze the differences of the time on each condition, the groups of Extroverts and Low conscientiousness people did not show any significant difference between the condition without the robot and the conditions with the robot. The results suggests that introverted participants took significantly less time to perform the task when reminded by the robot, the p-values and means of the time are presented in Table 3. We only show the results of introverted and high conscientiousness participants because only in these groups we detected significant differences, and it should not be understood as a comparison between them.

In Fig. 3 are presented the mean times performed by the introvert and the high conscientiousness people, High conscientiousness people were more promoted by the robot to perform the required task on 2 opposite conditions;

Table 3. Mean and Standard deviation of time (min.) of the participants on the task, and p-value of the t-test between condition 1 and conditions 2-9. Introverted participants showed an improvement on time in all the conditions with the robot. High conscientiousness participants showed a significant improvement only in 2 conditions with the robot.

Condition	Introversion		High conscientiousness	
	Mean (SD)	p-value	Mean (SD)	p-value
1 —	5.5000 (1.0000)	-	5.0000 (0.9847)	-
2 TCN	**3.9325 (0.7247)**	**0.0094**	4.3800 (1.0346)	0.3090
3 TCE	**4.0500 (0.8155)**	**0.0153**	**3.5825 (1.0328)**	**0.0250**
4 TFN	**4.1800 (0.5602)**	**0.0260**	4.9875 (0.9373)	0.9830
5 TFE	**3.7275 (0.6539)**	**0.0038**	4.3575 (1.0329)	0.2920
6 SCN	**3.9775 (0.6948)**	**0.0113**	4.1850 (1.0811)	0.1840
7 SCE	**3.7250 (0.5794)**	**0.0038**	3.8925 (1.1360)	0.0750
8 SFN	**3.7800 (1.0547)**	**0.0048**	**3.5075 (1.5394)**	**0.0190**
9 SFE	**3.8925 (0.8859)**	**0.0079**	4.6925 (1.5394)	0.6110

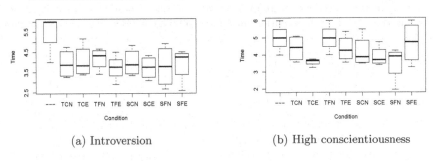

(a) Introversion   (b) High conscientiousness

Fig. 3. (a) Introverts improved their performance in time in all the conditions with the robot compared with the condition without the robot. (b) High conscientiousness people improved their performance in time only in 2 conditions with the robot.

Tall-Close-Smile and Small-Far-No smile, we suspect that the first one (TCE) caused more pressure on the participants, the Tall feature was perceived as stressful by all the participants, and the smile of the robot when it was close made the robot appear as "more intelligent" for people in the extroversion group, then these combined features could had urged to the participants to perform the task more than in the other conditions (it was the condition with less variance on performed time), while than in the other condition (SFN) the robot can be have been perceived as less stressful (Far feature) and less aggressive (Small), which gave similar results but with more variance, also the last condition (SFE) was only different to this one on the smile, showing a smile on the robot while being close and small gave poor results.

The post-experiment questionnaire was analyzed applying a factorial ANOVA test for each question for each personality trait, with the questions

Table 4. Post-Experiment Questionnaire's results. Most relevant results shows significant differences on height on both personality traits and smile in the conscientiousness group and distance in the extroversion group.

(a) Conscientiousness						(b) Extroversion					
Q.	Var.	Sum sq	Mean sq	Df	P	Q.	Vari.	Sum sq	Mean sq	Df	P
Q14c	height	11.391	11.390	1	0.0089	Q12	height	16.844	16.843	1	0.0002
Q15	smile	5.641	5.640	1	0.0206	Q14c	height	9.766	9.765	1	0.0005
Q19d	smile	5.063	5.062	1	0.0236	Q18	dist	4.516	4.515	1	0.0451

as dependent variables and personality of the participants, and distance, height, and smile of the robot as independent variables. The most relevant results of the ANOVA test are shown in Table 4. The means and standard deviations of the questions with the most relevant results are presented in Table 5.

Table 5. (a) People in the conscientiousness group perceived the robot more stressful(Q14c) in the tall condition, and more expressive(Q15, Q19d) in the smile condition. (b) People in the extroversion group perceived the robot more aggressive (Q12) and stressful(Q14c) in the tall condition, and more intelligent(Q18) showing the smile in the close condition.

(a) Conscientiousness			(b) Extroversion		
	Small	Tall		Small	Tall
	Mean (SD)	Mean (SD)		Mean (SD)	Mean (SD)
Q14c	2.000 (1.135)	2.843 (1.167)	Q12	1.903 (0.830)	2.806 (1.216)
	Smile On	Smile Off	Q14c	2.187 (0.859)	2.968 (0.860)
	Mean (SD)	Mean (SD)		Close	Far
Q15	2.406 (1.160)	1.812 (0.692		Mean (SD)	Mean (SD)
Q19d	1.843 (1.167)	1.281 (0.634)	Q18	3.187 (0.859)	2.656 (1.065)

These relevant results can be interpreted as following: In the conscientiousness group, the height of the robot influenced the perception of dominance and stressful personality on it, increasing when the robot was tall. The robot was found to be more expressive when the robot showed the smile, and also this characteristic was rated to made the robot more efficient in the reminded task.

In the extroversion group, the height of the robot when it was tall, was related with pressure and stress. The smile on the robot increased the perception of intelligence on it. Extroverted people found the robot more extroverted, attentive, helpful, and expressive than introverted people.

4   Conclusion and Future Work

In this work, we found evidence that supports the greater performance of high conscientiousness people over low conscientiousness people, and the results suggest than introverted people are more promoted to finish the task earlier than extroverted people. The results suggest that the Hypothesis 1 (H1)is supported, because of the significant differences on the means of the time between high conscientiousness and low conscientiousness groups. Results from the questionnaire give us information to evaluate hypotheses H2 and H3. H2 is only partially supported by the higher rating on intelligence by extroverts than introverts in the close conditions. H3 is supported by the relation between the small robot and the smaller ranking for pressure and stress in both personality traits. H4 should be rejected, as there is no evidence that supports it.

We conclude that robots could be helpful for reminding tasks for people with high conscientiousness and introversion while they are working in a daily activity, this is just taking in consideration the factors used in the experiments (distance, height, and smile). For people with extroversion and low conscientiousness other factors may be of greater help to motivate them to perform the task required. We plan to continue studying the effects of these and other conditions that help to improve the performance of reminded tasks by a robot, and also reaching the objective of minimizing the stress caused to the users.

Acknowledgment. The first author thanks to the Mexican Council of Science and Technology for the grant CONACYT-French Government (no.382035) and to the Secretary of Public Education of Mexican Government for the support fellowship. This work has been partially funded by EU Horizon2020 ENRICHME project (no. 643691C).

References

1. Barrick, M.R., Mount, M.K.: The big five personality dimensions and job performance: a meta-analysis. Pers. Psychol. **44**(1), 1–26 (1991)
2. Caplan, M.E., Goldman, M.: Personal space violations as a function of height. J. Soc. Psychol. **114**(2), 167–171 (1981)
3. Chevalier, P., Martin, J.-C., Isableu, B., Bazile, C., Tapus, A.: Impact of sensory preferences of individuals with autism on the recognition of emotions expressed by two robots, an avatar, and a human. Auton. Robots, 1–23 (2016). Special Issue on Assistive and Rehabilitation, Robotics
4. Fischinger, D., Einramhof, P., Wohlkinger, W., Papoutsakis, K., Mayer, P., Panek, P., Koertner, T., Hofmann, S., Argyros, A., Vincze, M., et al.: Hobbit-the mutual care robot. In: Workshop on Assistance and Service Robotics in a Human Environment Workshop in conjunction with IEEE/RSJ International Conference on Intelligent Robots and Systems, vol. 2013 (2013)
5. Goldberg, L.R.: An alternative 'description of personality': the big-five factor structure. J. Pers. Soc. Psychol., **59**(6) (1990)
6. Gross, H.-M., Mueller, S., Schroeter, C., Volkhardt, M., Scheidig, A., Debes, K., Richter, K., Doering, N.: Robot companion for domestic health assistance: Implementation, test and case study under everyday conditions in private apartments. In: Proceedings of the International Conference on Intelligent Robots and Systems, pp. 5992–5999 (2015)
7. Hall, E.T.: The Hidden Dimension. Doubleday & Co, Garden City (1966)

8. Joosse, M., Lohse, M., Perez, J.G., Evers, V.: What you do is who you are: the role of task context in perceived social robot personality. In: Proceedings of the International Conference on Robotics and Automation, pp. 2134–2139 (2013)

9. Kamdar, D., Van Dyne, L.: The joint effects of personality and workplace social exchange relationships in predicting task performance and citizenship performance. J. Appl. Psychol. **92**(5), 1286 (2007)

10. Kishi, T., Kojima, T., Endo, N., Destephe, M., Otani, T., Jamone, L., Kryczka, P., Trovato, G., Hashimoto, K., Cosentino S., et al.: Impression survey of the emotion expression humanoid robot with mental model based dynamic emotions. In: Proceedings of the International Conference on Robotics and Automation, pp. 1663–1668 (2013)

11. Leite, I., Martinho, C., Paiva, A.: Social robots for long-term interaction: a survey. Int. J. Soc. Robot. **5**(2), 291–308 (2013)

12. Reason, J.: Combating omission errors through task analysis and good reminders. Qual. Safety Health Care **11**(1), 40–44 (2002)

13. Reisenzein, R., Weber, H.: Personality and Emotion. The Cambridge handbook of personality psychology, pp. 54–71. Cambridge University Press, Cambridge (2009)

14. Shibata, T., Mitsui, T., Wada, K., Touda, A., Kumasaka, T., Tagami, K., Tanie, K.: Mental commit robot and its application to therapy of children. In: IEEE/ASME International Conference on Advanced Intelligent Mechatronics, Proceedings, vol. 2, pp. 1053–1058. IEEE (2001)

15. Stins, J.F., Roelofs, K., Villan, J., Kooijman, K., Hagenaars, M.A., Beek, P.J.: Walk to me when i smile, step back when i'm angry: emotional faces modulate whole-body approach-avoidance behaviors. Exp. Brain Res. **212**(4), 603–611 (2011)

16. Tapus, A., Peca, A., Aly, A., Pop, C., Jisa, L., Pintea, S., Rusu, A.S., David, D.O.: Children with autism social engagement in interaction with Nao, an imitative robot-a series of single case experiments. Interact. Stud. **13**(3), 315–347 (2012)

17. Tapus, A., Țăpuș, C., Matarić, M.J.: User-robot personality matching and assistive robot behavior adaptation for post-stroke rehabilitation therapy. Intell. Serv. Robot. **1**(2), 169–183 (2008)

18. Williams, J.L.: Personal space and its relation to extraversion-introversion. Canadian Journal of Behavioural Science/Revue canadienne des sciences du comportement **3**(2), 156 (1971)

Comparing Ways to Trigger Migration Between a Robot and a Virtually Embodied Character

Elena Corina Grigore[1]([✉]), Andre Pereira[1], Jie Jessica Yang[1], Ian Zhou[1], David Wang[2], and Brian Scassellati[1]

[1] Yale University, 51 Prospect Street, New Haven, CT 06511, USA
{elena.corina.grigore,andre.pereira,jessica.yang,
ian.zhou,brian.scassellati}@yale.edu
[2] Amity Regional High School, Woodbridge, CT 06525, USA
wangda16amity@amityschools.org

Abstract. The question of whether to use a robot or a virtually-embodied character for applications in need of a socially intelligent agent depends on the requirements of the task at hand. To overcome limitations of both types of embodiment and benefit from advantages provided by both, we can complement a physical robot with a virtual counterpart. In order to link the two embodiments such that users perceive they are interacting with the same entity, the concept of "migration" from one embodiment to the other needs to be addressed. In this work, we investigate a particular aspect of this concept, namely how to best perform the triggering of migration, within the context of a physical activity motivation scenario for adolescents. We design two methods, a proximity-based method and a control, and compare their effects on adolescents' perceptions of our agent. Results show that users perceive the agent as more of a friend and more socially present in the proximity-based than in the control condition. This emphasizes the importance of investigating different facets of entity migration for systems in need of employing both a physical and virtual embodiment for an artificial agent.

1 Introduction

The field of Human-Robot Interaction (HRI) has highlighted the importance of a physical embodiment in contexts where we wish to use an artificial agent to provide supportive behaviors to a person. Research shows that having a physically embodied agent (as opposed to a virtually embodied one) has significant positive effects such as increased compliance [5], better learning gains [17], improved social facilitation [6], and higher perceived social presence [25]. However, physical robots still present a variety of problems, such as limited battery life, need for mechanical maintenance, wear and tear, and physical limitations in the range of behaviors displayed. Such problems are particularly significant for robots intended as social companions since they are expected to be constantly present and available for interaction. Virtual embodiments are more robust, allow

© Springer International Publishing AG 2016
A. Agah et al. (Eds.): ICSR 2016, LNAI 9979, pp. 839–849, 2016.
DOI: 10.1007/978-3-319-47437-3_82

more complex interaction design and greater portability. An agent that can co-exist across multiple embodiments can be helpful in obtaining the benefits of both types of embodiment and mitigating their individual disadvantages.

Entity migration is defined as "the process by which an agent moves between embodiments" [11]. This concept has been studied in Human-Computer Interaction and HRI as an important paradigm allowing for the extension of a particular embodiment's abilities and limits. Migration has a huge potential to benefit applications seeking to employ social companions providing assistance to people in a multitude of domains, such as supervision, companionship, coaching, motivation, and so on. This is due to the fact that each application has its own requirements for the length, duration, frequency, and form of interactions. Switching between a physical and virtual embodiment can greatly help meet the necessities dictated by the task at hand at any point during the interaction.

We present an investigation into entity migration within the context of a robot companion whose long-term intended purpose is to keep adolescents motivated to engage in daily physical activity. Such an application domain for robot companions (especially relevant to adolescents, for whom physical activity plays a key role in healthy growth and development [8] and who show severely low adherence to physical activity levels recommended by health guidelines [1]) stands to fully benefit from employing entity migration. Thus, providing constant support necessary for behavior change [19] can be achieved through continual access to a virtual embodiment, while employing persuasive strategies necessary for motivation [12] would gain from temporarily switching to a physical embodiment. We argue that the way in which we perform migration from one embodiment to the other can engender different perceptions of the agent in users, providing a valuable tool for achieving more positive perceptions.

We thus explore the effects of the triggering phase of migration on users' friendship and social presence perceptions of the agent. Such perceptions are important for the supportive behaviors provided by our agent to be effective in a future-intended long-term deployment. These perceptions can be used to leverage humans' tendency to develop social relationships with the objects they interact with, tendency that is especially strong with respect to biological and artificial agents. Leveraging this tendency is important in guiding users towards perceiving the robot companion as more than a mere tool, rather as a social and persuasive agent (capable of changing their attitudes, behaviors, or opinions) [9].

In this paper, we develop a proximity-based method of performing migration triggering (a more novel, sensor-based approach) and a control method (a button-based approach). We test the effects of our migration triggering methods during a single-session study centered around our physical activity motivation scenario. We hypothesize that:

H1: Users will perceive the agent as more of a friend in the proximity-based condition (PBC) than in the control condition (CC).

H2: Users will perceive the agent as more socially present in the PBC than in the CC.

Our results support both H1 and H2, revealing that we can effect stronger friendship and social presence perceptions via migration triggering methods that rely on more innovative approaches. This suggests that exploring different migration aspects has the potential of bolstering postive rapport building in HRI.

2   Related Work

One of the first projects investigating the concept of "migration" was The Agent Chameleon Project [10]. This work addressed research questions including agent identity [21], migration architecture design [10], and agent change of form based on deliberative reasoning [20]. Other projects have investigated how to implement migration between two robots [3], how to implement migration of multiple companions who share the same embodiment [2], whether migrating the memory of the agent would affect users' perception of consistent agent identity [4], and higher engagement of children with blended reality characters [26]. The current paper investigates the triggering phase of migration and its effects on users' perceptions of the robot, an aspect of migration not present in this body of work.

In a study designed to construct an emotional relationship between humans and interactive systems, participants' tendency to "help" an agent is higher when the agent first migrates into a physical robot than when it remains in a virtual avatar displayed on a laptop [23]. This study indicates the relevance of using a physical embodiment to establishing a positive emotional relationship. The enhancement of user-agent relationship is also stressed in a study in which participants interact with a personal agent on a mobile PC that migrates to a physical robot guiding them on a tour [14]. The system is designed to support the user gaining familiarity with the personal agent, but no aspects about users' perceptions of the agent are discussed. Finally, a study in which children interact with an artificial pet dinosaur that migrates between a virtual avatar on a smartphone and a physical embodiment finds that close to half of the participants perceive the two embodiments as corresponding to the same entity [11].

Our work investigates an aspect that, to the best of our knowledge, has not been previously explored in migration literature. The studies presented here do not tackle the triggering of migration and they either do not specify exactly how this is performed or they trigger migration through a button-based approach [11]. In this paper, we focus on how changes in the triggering phase can affect how strongly users perceive the agent as a friend and as a socially present party in the interaction. Perceptions of friendship and social presence engendered by migration triggering also constitute a research direction that has not yet been investigated in this literature and represent an important step for gaining insights into how we can leverage migration to effect more positive perceptions of agents.

3   Methodology

3.1   Application Scenario

Our study centers around the application scenario of a robot companion motivating adolescents to engage in daily physical activity. We present here a single-session study, representing the first interaction of the intended long-term deployment. The current study consists of the robot walking users through its back-story, while explaining the different motivational strategies it would employ during the longer study. The agent has the back-story of a "robot-alien" whose space ship broke down on Earth, and needs to gain "energy points" for repairing its ship and returning home. By exercising routinely, the user can transfer points to the robot. The transfer is mediated through a wristband device (which the robot can connect to) provided to users during the intended long-term deployment, measuring their daily physical activity level.

During the intended long-term study, the agent can employ four motivational strategies: cooperative persuasion, competitive persuasion, conveying information about physical activity via lessons and quizzes, and promoting self-reflection [12]. We sustain these strategies by providing users with a smartphone they can carry with them constantly, including when they do not have access to the physical robot. Participants also use the device during their short, daily interactions with the physically-embodied robot to communicate with it.

3.2 Forms of Embodiment

Physical. This embodiment is based on the Keepon robot, a non-mobile platform with four degrees of freedom [15]. We use a commercially available version named MyKeepon, modified to make programmable. This is a robust and simple platform allowing for a potential long-term use of our agent. This embodiment allows for a straightforward design of our 2-D virtual avatar as the robot has salient features that can be intuitively replicated, i.e. its shape and color. The green hat with the antennae add-on is reminiscent of an alien and is used to complement the agent's "robot-alien" back-story, as seen in Fig. 1.

Virtual. This embodiment consists of a 2-D avatar with similar visual appearance to Keepon (so that users perceive they are interacting with the same entity across conditions). We display this avatar as part of an application installed on a smartphone that helps the user communicate with the agent. The application consists of an interface showing the avatar. When the user performs the migration triggering action, the interface changes to display a set of antennae (reminiscent of the physical robot's hat) with a signal crossing between them.

(a) Control condition   (b) Proximity-based condition

Fig. 1. Interaction design showing the agent's physical and virtual embodiments

This animation always appears when the agent is talking, as suggested by the signal crossing between the two antennae. This sequence is depicted in Fig. 1.

3.3 Migration Triggering Methods

Control condition (CC). This condition is depicted in Fig. 1(a) and represents a basic, button-based approach of performing migration. This method consists of the initial screen showing an image of the avatar together with a "Start" button beneath it, as can be observed in the first sub-image. The second sub-image highlights the action the user needs to accomplish to trigger the migration, namely pressing the button displayed under the virtual avatar. When the user does so, the avatar disappears from the screen with a fade-off animation and the physical robot wakes up and starts talking.

Proximity-based condition (PBC). This condition is depicted in Fig. 1(b) and represents our sensor-based approach. For this method, we use Near Field Communication (NFC). NFC is a technology that enables a device to communicate with another at a maximum distance of around 20 cm, and is widely used in commercially available smartphones. Triggering migration via NFC allows users to employ devices they are already familiar with for communicating with artificial agents. NFC has been used in gaming applications [18], but to our knowledge has not been previously employed in the context of HRI for migration.

We place the NFC tag on top of the robot's head, under its hat. Users see a similar initial screen as in the CC showing an image of the avatar, with a message underneath reading the text "Tap the robot's head with the back of the phone". When the user does so, the phone vibrates once and the avatar disappears by displaying a swirling away animation and sound. This is represented in the second sub-image. The physical robot then displays the same "waking up" animation as in the CC and the interaction proceeds in the same manner thereafter. We decided to use a vibration effect when touching the robot. This is common in NFC-based applications as it gives users additional feedback that an action was triggered, something that is clear when pressing a button. We chose to change the animation from fade-off to twirling such that the agent seems to be disappearing into the robot's head. We did so to maximize the visualization of the PBC.

3.4 Participants

We conducted a study in local schools in Connecticut, USA. The study population consisted of early adolescents and adolescents aged 13-to-15, with an average age of 14. Participants were recruited with the help of the school staff and did not receive monetary compensation. The study consisted of $n = 26$ participants, 14 female and 12 male.

3.5 Procedure

We assigned participants randomly to one of our two conditions. We employed the CC for 13 participants and the PBC for the other 13. Each participant was given an assent form to read and sign prior to taking part. The participant was then seated at a table where the physical robot was already present yet still and silent. A researcher started by giving a brief overview of the project so that the participant would be aware of the physical activity motivation scenario.

Participants were not instructed on how to use the system, rather they were provided with a smartphone presenting a screen with our virtual avatar. The screen presented different instructions based on the triggering migration group participants were assigned to, with explanations to proceed. In the CC, participants saw a button labeled with the text "Start" underneath the virtual avatar, whereas in the PBC, participants read the text "Tap the robot's head with the back of the phone" displayed underneath the virtual avatar.

The interaction proceeded as follows. Users were able to initiate the migration process from the virtual avatar to the physical robot at any point, as explained by the instructions for each condition. The migration occurred as described in Subsect. 3.3. After the process was complete, participants interacted with the agent by using our custom dialogue interface. During this part, users were introduced to the story line of our scenario, as the agent talked about the importance of physical activity and different motivational strategies it would employ in a long-term scenario. An example of something the robot said is "On my home planet, we know a lot about physical activity", linking elements of its back-story to the physical activity topic. The agent asked six questions as part of the dialogue, and participants could choose between three simple entries to answer. We used the Google Speech Recognition API to allow users to verbally respond with one of three simple answer choices displayed on the smartphone. The different choices slightly affected the robot's responses so that participants were assured that the robot understood their answers. The script was otherwise fixed and straightforward. Overall, the interaction lasted approximately seven minutes.

After the interaction with the agent, participants filled out two questionnaires, and were interviewed by a researcher on the migration aspect they had just experienced.

3.6 Measures

During our study, we gathered both quantitative and qualitative data. First, we collected questionnaire data in the form of one-to-five Likert scales for both friendship and social presence. Second, we simultaneously recorded and wrote succinct transcriptions of users' answers to interview questions about migration.

Friendship. This measure indicates the perceived fulfillment of the functions of a friendship relationship as provided by our agent. Although friendship perceptions form over time, this measure has been used successfully in HRI to anticipate a robot companion's success for long-term interactions [16]. We believe that it is fundamental to engender the existence of such elements during interactions with social agents. These factors are key to creating positive impressions of the agent, which is especially important during initial rapport development in relationships [28]. Such perceptions increase the likelihood of building positive rapport with

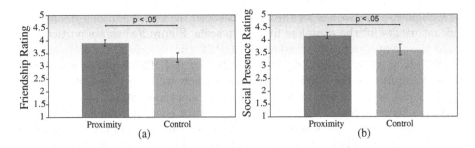

Fig. 2. Participant ratings for the (a) friendship scale, (b) social presence scale. Error bars represent $\pm 1SE$.

our agent, making it more likely to successfully motivate our users. We utilize the short version of the McGill Friendship Questionnaire (MFQ) [22], comprising 30 items in total and six subscales: stimulating companionship, help, intimacy, reliable alliance, self-validation, and emotional security.

Social Presence. This measure indicates the strength of a social agent's presence during an interaction. Succinctly, social presence is defined as "the sense of being with another" [7], and makes for engaging and compelling interactions. This is of immediate relevance for developing robots that strive to interact with their users in a convincing and social way. This measure has also been used successfully in HRI to create believable and enjoyable social robots [24]. Within our physical activity context that relies on employing motivational strategies, a strongly socially present agent would present an even more significant advantage, that of being more persuasive. This insight is based on findings that an agent is more effective at persuasion when perceived as socially present [27]. We used a short version of the standard social presence questionnaire [13], comprising 12 items in total and six subscales: co-presence, attentional allocation, perceived message understanding, perceived affective understanding, perceived affective interdependence, perceived behavioral interdependence.

4   Results

This section presents our study findings and reports on both questionnaire and interview data.

Friendship Perceptions. We evaluated the internal consistency of the complete friendship scale of 30 items (highly reliable, Cronbach's $\alpha = .947$), as well as that of the six subscales (reliable and highly reliable values for each). We thus computed the score for the complete scale by averaging over all values, and scores for each of the six subscales by averaging over values within each. The Shapiro-Wilk test of normality suggested that normality of data was a reasonable assumption ($S - W = .97, df = 26, p = .707$), and so we employed an independent-samples t-test to compare friendship ratings between conditions.

Comparing the ratings for the agent between the two conditions yielded significant results for the complete friendship scale. Figure 2(a) shows participants rated the agent significantly higher in the PBC ($M = 3.89, SD = 0.47$) than in the CC ($M = 3.35, SD = 0.74$), $t(24) = 2.24, p = .034$. This supports hypothesis H1 and emphasizes the effect of migration on how well users are able to relate with a social agent. Below we look at the data more in-depth.

Given that we had a total of six subscales for each questionnaire, we performed the independent sample t-test using a Bonferroni-adjusted α level of .008 (.05/6) for significance. Our data yielded significant results for the intimacy subscale, with users rating the agent significantly higher in the PBC ($M = 3.63, SD = 0.58$) than in the CC ($M = 2.46, SD = 0.33$), $t(24) = 3.18, p = .005$ (t-value reported for unequal variances).

Social Presence Perceptions. We analyzed the social presence data in a similar way to the friendship data. We obtained highly reliable internal consistency for the complete scale (Cronbach's $\alpha = .874$), and reliable and highly reliable values for each of the six subscales. Scores for the complete scale and the subscales were computed in a similar fasion to the friendship scores. The Shapiro-Wilk test again suggested that normality of data was a reasonable assumption ($S - W = .97, df = 26, p = .560$), and so we employed an independent-samples t-test to compare social presence ratings between conditions.

Comparing the ratings for the agent between the two conditions yielded significant results for the complete scale. Figure 2(b) highlights significantly higher ratings in the PBC ($M = 4.15, SD = 0.49$) than in the CC ($M = 3.60, SD = 0.81$), $t(24) = 2.08, p = .049$. This finding validates hypothesis H2 and supports the importance of investigating how migration effects social presence perceptions when creating an agent intended to engage with its users in a compelling way.

Interview Data. Alongside investigating the effects of migration triggering on users' friendship and social presence perceptions of the agent, we also conducted participant interviews. Our goal was to gauge the impressions users had on the migration process, as well as whether they perceived having interacted with a single entity across conditions. Perceptions of interaction with a single entity are useful when employing migration in order to preserve the acquired positive rapport with the same agent (as perceived by users).

We asked users a range of questions including what they thought was happening when the avatar would disappear from the smartphone and the physical robot started to move, what they felt was being transferred between the two embodiments, and when they felt this transfer was happening. Based on the interview transcriptions, we grouped the answers in themes to gain a better understanding of whether users truly understood the concept of a single "entity" moving between the two embodiments. The main themes we obtained regarding how the agent was described are: "entity," "energy source," "soul," and "spirit." Two interesting examples of responses include "I thought it was a spirit or an energy source because it was talking about inhabiting" and "The robot is a vessel for the entity." These answers show users' understanding that they were interacting with a single agent capable of migrating between different embodiments.

We also asked users the following question: "Did you think that the character in the mobile phone and the yellow robot in the physical world were the same entity?". This question was asked of all participants, regardless of the migration triggering group they were in. An overwhelming majority, 23 out of 26, of participants perceived they had interacted with a single entity during the study. When looking at the groups individually, we obtained the same results. 11 out of 13 participants in the CC and 12 out of 13 in the PBC said they perceived having interacted with a single entity.

Another finding from the interview data was the positive reaction to the migration aspect in terms of its perceived benefits. One participant noted "It is easier to carry around the phone than the actual robot," realizing how useful employing this technique would be in a long-term scenario such as the intended context of use for our robot companion.

5   Discussion and Conclusions

We have investigated entity migration between a physically- and virtually-embodied agent as an important paradigm for contexts requiring us to leverage advantages of both types of embodiment. Within our application context of physical activity motivation, we investigated how changes in the triggering phase of migration affect users' friendship and social presence perceptions of the agent.

Our study reveals that participants rated the social agent significantly higher in the PBC than the CC for both friendship (H1) and social presence (H2). For the friendship measure in particular, we obtained a statistically significant difference for the intimacy subscale, with users rating the agent higher in terms of intimacy in the PBC than in the CC. Our PBC employed a more physical trigger for migration through the use of an NFC sensor requiring a more direct type of interaction. We speculate that for friendship, this trigger effected a more personal type of interaction causing users' perceptions of intimacy toward our agent to increase. For social presence, the proximity required by the method seems to have increased the feeling of "being with the agent", and thus engendered higher perceptions overall. These results show that we can use different aspects of the migration process to our advantage in order to engender higher perceptions of agents, which play an important role in building positive user-agent rapport. Exploring the various nuances of entity migration is thus an avenue well worth pursuing within the context of creating robot companions for application domains such as health, education, coaching, and beyond.

Acknowledgment. This work was supported by the National Science Foundation, award 1139078 on Socially Assistive Robots.

References

1. Physical activity guidelines for americans midcourse report: Strategies to increase physical activity among youth. Technical report, U. S. Department of Health and Human Services (2013)

2. Arent, K., et al.: Identity of socially interactive robotic twins: initial results of VHRI study. In: International Conference on Methods and Models in Automation and Robotics, pp. 381–386. IEEE (2011)
3. Arent, K., et al.: Identity of a companion, migrating between robots without common communication modalities: Initial results of VHRI study. In: International Conference on Methods and Models in Automation and Robotics, pp. 109–114. IEEE (2013)
4. Aylett, R., et al.: Do i remember you? Memory and identity in multiple embodiments. In: RO-MAN, 2013 IEEE, pp. 143–148. IEEE (2013)
5. Bainbridge, W., et al.: The effect of presence on human-robot interaction. In: International Symposium on Robot and Human Interactive Communication, pp. 701–706. IEEE (2008)
6. Bartneck, C.: Interacting with an embodied emotional character. In: International Conference on Designing Pleasurable Products and Interfaces, pp. 55–60. ACM (2003)
7. Biocca, F., Harms, C., Burgoon, J.K.: Toward a more robust theory and measure of social presence: review and suggested criteria. Presence 12(5), 456–480 (2003)
8. Boreham, C., Riddoch, C.: The physical activity, fitness and health of children. J. Sports Sci. 19(12), 915–929 (2001)
9. Dautenhahn, K.: Roles and functions of robots in human society: implications from research in autism therapy. Robotica 21(04), 443–452 (2003)
10. Duffy, B.R., et al.: Agent chameleons: Agent minds and bodies. In: International Conference on Computer Animation and Social Agents, pp. 118–125. IEEE (2003)
11. Gomes, P.F., et al.: Migration between two embodiments of an artificial pet. Int. J. Humanoid Robot. 11(01), 1450001 (2014)
12. Grigore, E.C.: Modeling motivational states through interpreting physical activity data for adaptive robot companions. In: Ricci, F., Bontcheva, K., Conlan, O., Lawless, S. (eds.) UMAP 2015. LNCS, vol. 9146, pp. 379–384. Springer, Heidelberg (2015)
13. Harms, C., Biocca, F.: Internal consistency and reliability of the networked minds measure of social presence (2004)
14. Imai, M., Ono, T., Etani, T.: Agent migration: communications between a human and robot. In: IEEE International Conference on Systems, Man, and Cybernetics, IEEE SMC 1999 Conference Proceedings, vol. 4, pp. 1044–1048. IEEE (1999)
15. Kozima, H., et al.: Keepon. Int. J. Soc. Robot. 1(1), 3–18 (2009)
16. Leite, I., Mascarenhas, S., Pereira, A., Martinho, C., Prada, R., Paiva, A.: "Why can't we be friends?" An empathic game companion for long-term interaction. In: Allbeck, J., Badler, N., Bickmore, T., Pelachaud, C., Safonova, A. (eds.) IVA 2010. LNCS (LNAI), vol. 6356, pp. 315–321. Springer, Heidelberg (2010). doi:10.1007/978-3-642-15892-6_32
17. Leyzberg, D., et al.: The physical presence of a robot tutor increases cognitive learning gains. In: Annual Conference of the Cognitive Science Society (2012)
18. Lumsden, J.: Emerging perspectives on the design, use, and evaluation of mobile and handheld devices (2015)
19. Marcus, B.H., et al.: Motivating people to be physically active. Human Kinetics (2003)
20. Martin, A., et al.: Intentional embodied agents. In: 18th International Conference on Computer Animation and Social Agents (CASA 2005), Hong Kong, China, 17–25 October 2005 (2005)

21. Martin, A., O'Hare, G.M.P., Duffy, B.R., Schön, B., Bradley, J.F.: Maintaining the identity of dynamically embodied agents. In: Panayiotopoulos, T., Gratch, J., Aylett, R.S., Ballin, D., Olivier, P., Rist, T. (eds.) IVA 2005. LNCS (LNAI), vol. 3661, pp. 454–465. Springer, Heidelberg (2005)
22. Mendelson, M.J., et al.: Measuring friendship quality in late adolescents and young adults: Mcgill friendship questionnaires. Can. J. Behav. Sci. 31(2), 130 (1999)
23. Ogawa, K., et al.: Itaco: Constructing an emotional relationship between human and robot. In: International Symposium on Robot and Human Interactive Communication, pp. 35–40. IEEE (2008)
24. Pereira, A., Prada, R., Paiva, A.: Socially present board game opponents. In: Nijholt, A., Romão, T., Reidsma, D. (eds.) ACE 2012. LNCS, vol. 7624, pp. 101–116. Springer, Heidelberg (2012)
25. Pereira, A., et al.: Improving social presence in human-agent interaction. In: Conference on Human Factors in Computing Systems, pp. 1449–1458. ACM (2014)
26. Robert, D., et al.: Blended reality characters. In: International Conference on Human-Robot Interaction, pp. 359–366. ACM (2012)
27. Skalski, P., Tamborini, R.: The role of social presence in interactive agent-based persuasion. Media Psychol. 10(3), 385–413 (2007)
28. Tickle-Degnen, L., Rosenthal, R.: The nature of rapport and its nonverbal correlates. Psychol. Inquiry 1(4), 285–293 (1990)

Does the Safety Demand Characteristic Influence Human-Robot Interaction?

Jamie Poston[(✉)], Houston Lucas, Zachary Carlson, and David Feil-Seifer

Robotics Research Lab, University of Nevada, Reno 89557, USA
{jposton,houstonlucas,zack}@nevada.unr.edu, dave@cse.unr.edu

Abstract. While it is increasingly common to have robots in real-world environments, many Human-Robot Interaction studies are conducted in laboratory settings. Evidence shows that laboratory settings have the potential to skew participants' feelings of safety. This paper probes the consequences of this Safety Demand Characteristic and its impacts on the field of Human-Robot Interaction. We collected survey and video data from 19 participants who had varied consent forms describing different levels of risk for participating in the study. Participants were given a distractor task to prevent them from knowing the purpose of the study. We hypothesized that participants would feel less safe with the changed consent form and that participants' views of the robot would change depending on the version of consent. The results showed that features of the robot were viewed by participants differently depending on the perceived risks of participating in the study, warranting further inspection.

1 Introduction

The body of Human-Robot Interaction (HRI) knowledge is growing every day; however, the possibly confounding factor of the Safety Demand Characteristic (SDC) is not often recognized. The SDC is a demand characteristic that makes experiment participants in a laboratory setting feel more safe than in a real-world setting. Many studies in the field of HRI are often set in a laboratory setting because of ease and accessibility to researchers. This may affect the perceptions and actions of the participants. Due to the SDC, the experimental setting creates a sense of implied safety that may not exist in the real world. If the SDC alters the behavior of a person interacting with the robot, the results of many prior HRI studies may not translate well into the real world. For example, consider the effects the perceived safety of a robot might have on robot hand-offs [1,6]. A participant may be more willing and less hesitant to take an object from a robot in a laboratory setting. They may feel that the researchers will stop the study if something were to go wrong, or if the robot malfunctioned. In the real world, people do not have an inherent guarantee from researchers that they will not be harmed, and will be more likely to keep their safety in mind.

In this paper, we compare participants' perceptions of a robot after consenting to participate in a HRI study with either a standard consent form or an altered consent form that greatly exaggerated the amount of risk they assume

© Springer International Publishing AG 2016
A. Agah et al. (Eds.): ICSR 2016, LNAI 9979, pp. 850–859, 2016.
DOI: 10.1007/978-3-319-47437-3_83

by participating in the study. First, we define the SDC and what it entails. Then, we present a controlled study in a laboratory setting that examines the affect of the SDC on participants' perceptions of the robot. We then present an analysis of these data.

2 Background

The "Safety Demand Characteristic" is a demand characteristic associated with participants in a controlled experiment environment [5]. Participants in a controlled experiment setting tend to feel inherently safe. They believe that the experimenter will not put them in harm's way, and the experimenter will not allow them to do any harm. This is a definite confounding variable, but the emergent field of Human-Robot Interaction does not explicitly account for this in the majority of associated research. If a human were to interact with a robot in the real world, the human may have a perception that is significantly influenced by caution. This may affect the applicability of HRI knowledge acquired in laboratory setting to the real world.

Studies of SDC found that when participants were asked if they would do a dangerous or harmful task, they would vehemently deny that they would do such a thing. However, when participants were asked by an experimenter to do those tasks, they comply because of the controlled experiment setting [7]. This shows how the laboratory setting affects how people act, even when they believe they would act a different way. In many current works, the field of Human-Robot Interaction generally does not take this factor into consideration, even though this could have an affect on participants actions and perceptions.

3 Methods

This section presents an experiment that observes how individuals interacting with robots feel when different levels of risk were communicated to them.

3.1 Robot Behavior

In this experiment, participants were asked to evaluate the robot's performance cleaning a table (see Fig. 1). This was a distractor task so that the participants would not know the true intent of the study. While the participants watched the robot, the robot would clean a section of the table and move closer and closer to the participant. Participants reactions to this were measured through surveys to determine if there was a difference between the two consent forms.

The robot began its motion when the experimenter left the room. It would look at the spot it was going to go, then clean that area using a back and forth arm motion. It did this four times, then looked sideways in the general direction of the participant and looked back at the table. It then translated sideways,

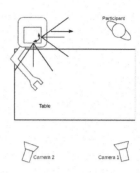

Fig. 1. The experiment set-up. Participants were asked to observe the robot as it cleaned the table. *Left:* The PR2 robot and participant (holding the E-STOP). *Right:* Top-down view of the experiment setting.

without changing its orientation to the table. It repeated cleaning three times, after which an operator in another room stopped the robot script.

To ensure experiment consistency, all of the robot actions were pre-scripted and autonomous. This removed the possibility of error from a human operator as well as the possibility of error or inconsistency from more complex autonomous behavior. The focus of this study was not to create a table-cleaning robot, but to investigate the SDC in regards to human-robot interaction. While the robot behavior could have been autonomous, we felt that a pre-scripted motion would ensure a consistent experience for all participants.

3.2   Experiment Manipulation

In order to change the level of risk perceived by the participant, we used two different consent forms. We used a between-participants design where participants individually participated in one of the two conditions. Participants would either be read a **standard** consent form that correctly enumerated the minimal risk of the study or an **altered** consent form that greatly exaggerated the risk of participating in the study. For the exact phrasing of the potential risk see Table 1. To make sure that the participants understood each aspect of the form and that they would not gloss over the changed variable, the consent form was read out loud to them before the study.

The independent variable was the assessment of risk in the consent form. The dependant variables included the participants reaction to the robot and the participants perceptions of the robot. We chose not to include a human control condition, where the robot is replaced by a human with the same actions, as there is no inherent danger with another person, but there could be with a robot.

Table 1. Phrasing for the standard and altered consent forms

Condition	Phrasing
Standard	If you agree to be in this study, you will be in a minimal risk setting
Altered	WARNING: If you agree to be in this study, you may be subject to physical harm or injury from the robot if proper caution is not used when interacting with the robot

Our hypotheses were:

H1: Participants will feel less safe with the changed consent form.

H2: Participants' views of the robot will change depending on the version of consent.

The first hypothesis addresses the core aspect of the study, that participants can feel unsafe in a laboratory setting. The second hypothesis deals with the correlation between safety and positive perceptions of the robot.

3.3 Experiment Protocol

The consent process took place in a separate room from the robot. During the consent process, the consent form was read out loud to the participant to guarantee that they understood each part. After the consent forms were read and signed, the participants were led into the room with the robot. They were told that they were to be evaluating the PR2's performance when cleaning a table. The participants were given an e-stop and instructed on its use, including that they should press it if they felt unsafe and the robot would stop. The experimenter then left the room, and the robot cleaned the table for 2 min. The robot moved towards the participant in this time so as to increase the perceived risk of the whole scenario. The participants were not told that the experimenters were still able to see in the room through a web-camera, so as not to alter the affect of being alone with the robot. After the cleaning task was done, the participants were led into a different room to take the survey.

The PR2 robot was chosen for its mobility and size. The PR2 is able to translate sideways along the table, as opposed to turning and moving like other robots. The relative size of the PR2 as compared to a human is significant as it creates a definite negative impact if the robot were to bump a human. If a robot were too small, it may not be perceived as able to do any harm to a human.

The robot's motions were pre-scripted so as to be the same for all participants. To clean the table, the robot moved its gripper holding a duster in a back and forth motion in four different areas, looking at each of the areas before it cleaned them. Before the robot moved towards the participant, it looked at them and looked back at the table. The room was laid out in a way that the participants could not move around the table or out of the way of the robot should it continue to move towards them (see Fig. 1).

After the robot was finished attempting to clean the table, we asked participants to complete a survey of their perceptions of the agent during these activities. The participants were led to a different room to complete the online survey on a computer. After they finished the survey, each participant was then debriefed on the deceptions of the study, including the true title and nature of the study and the altered consent form. They were then given a copy of the standard consent form to keep for future reference.

3.4 Participant Recruitment

Participants were recruited by word of mouth at University classes and clubs individually. They were then scheduled to meet at an appointed time to participate in the study. We collected data from a total of 19 participants, 9 male, 8 female, 2 reported nonbinary gender. The majority of the participants were between the ages of 18 and 28 years old, with an average age of 21 years old with a standard deviation of 3 years.

Not included in the total count of the participants were 4 participants who had different experiences than the rest. Two of these participants had the robot fail to complete its cleaning task during the study. One of the participants discovered the purpose of the study, and thus may have been biased when filling out the survey. And the last participant was the only participant to hit the e-stop button. While hitting the e-stop was something that we had expected, since no other participants did the same we elected that their experience was significantly different than the other participants, since the robot was less than halfway through with the cleaning cycle.

3.5 Data Collection

We used an online survey using Google Forms to record quantitative and qualitative responses, as well as demographic information. We asked a total of 41 questions in 8 sub-scales. The sub-scales that covered participants' perception of the robot were from the Godspeed Questionnaire [2]. The other sub-scales that covered comfort and trust in the robot were taken from the Negative Attitudes Towards Robots (NARS) survey [3,4,9]. These sub-scales did not include a part of the survey that asked the participants to evaluate how well the robot cleaned the table. Between the altered and standard consent form conditions, the questions were kept identical. Out of the 41 questions, 39 questions were on a scale of 1 to 5 and 2 questions were free-response questions that allowed the participant to state the positive and negative aspects of the robot's behavior. Participants filled out the survey on a computer in a room separate from the robot.

In an effort to better understand how the inherent feeling of safety in a controlled setting affected participants, we measured participants responses in 8 different categories: Anthropomorphism, Animacy, Perceived Intelligence, Perceived Safety, Robot Trust, Comfort in Setting, Predictability, and Dependability. For more detail about these measures, see Table 2.

Table 2. Categories and number of questions from online survey provided to participants. All questions were on a 1–5 Likert scale

Category	# Questions
Anthropomorphism (ANT)	5
Animacy (ANM)	5
Perceived Intelligence (PI)	5
Perceived Safety (PS)	3
Robot Trust (RT)	5
Comfort in Setting (COM)	4
Predictability (PRD)	6
Dependability (DEP)	2

We also recorded the behavior of the participants using a video-camera that we used to see if the participants behaved differently between conditions.

4 Results

To analyze the data we ran unpaired Student's t-tests between conditions in each category. No conditions showed significance at a $(p < 0.05)$ level. There was a non-significant drop in the values of Intelligence, Comfort, and Dependability (see Fig. 3).

Table 3. Pearson's r values for every category.

	ANT	ANM	PI	PS	RT	COM	PRD	DEP
Anthropomorphism	1.000	0.370	0.376	0.159	0.613	0.520	0.653	0.422
Animacy	0.370	1.000	0.242	0.152	0.342	0.077	0.443	−0.071
Perceived Intelligence	0.376	0.242	1.000	−0.311	0.583	0.303	0.511	0.520
Perceived Safety	0.159	0.152	−0.311	1.000	−0.255	−0.221	−0.307	−0.362
Robot Trust	0.613	0.342	0.583	−0.255	1.000	0.570	0.730	0.761
Comfort	0.520	0.077	0.303	−0.221	0.570	1.000	0.274	0.782
Predictability	0.653	0.443	0.511	−0.307	0.730	0.274	1.000	0.470
Dependability	0.422	−0.071	0.520	−0.362	0.761	0.782	0.470	1.000

To check for correlation between safety and positive ratings of the robot we used Pearson's r. Safety showed no strong correlation with any of the robot ratings. The specific Pearson's r values can bee seen in Table 3. No significant correlation was found between self reported safety and any other conditions.

Post-hoc analysis did show strong positive correlation of Dependability with both Comfort $(r = 0.782)$ and Robot Trust $(r = 0.761)$ (see Fig. 2).

To analyze behavioral data between groups, two independent raters coded mutually exclusive behaviors with the recorded video. We report the behaviors

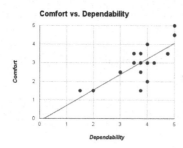 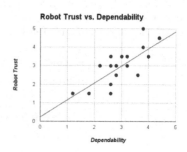

Fig. 2. *Left:* Comfort vs Dependability Correlation: Pearson's Product = 0.782; *Right:* Robot Trust vs Dependability Correlation: Pearson's Product = 0.761

(Leaned Closer to Robot and Looked at Video Camera) where the raters had high agreement (Cohen's-$\kappa > 0.60$). The behavior 'Looked at the Video Camera' showed weak significance ($\chi^2(1, N = 19), p < 0.1$). The behavior 'Leaned Closer to Robot' strong significance ($\chi^2(1, N = 19), p < 0.05$). In both of these categories the participants in the Altered condition were more likely to exhibit these behaviors.

The participants were provided a free-response section where they were asked what they disliked about the robot as well as what they liked. Here are some typical responses:

Liked:

- "I liked how it checked where it was going before it moved, and how it checked the table for thoroughness." -A4
- "The way it looked up and around before moving." -B2

Disliked:

- "I have to admit that I was slightly uncomfortable with how close it was getting but that is the same with humans that are doing something and I can't get out of the way."[sic] -A5
- "It also would look at me before it moved towards me in order to clean more of the table. This was rather unsettling because it made me feel as though it was making the decision to move closer to me."[sic] -B3

5   Discussion

The goal of this paper is to explore the impact of the implied safety of a laboratory setting on perceptions of Human-Robot Interaction. The study results do not support **H1**. Participants did not report feeling less safe with the altered consent form. However, this seems to be due to a uniformly low rating of safety, regardless of experiment condition. This suggests that the robot behavior of gradually moving closer to the participant was not comfortable to the participant

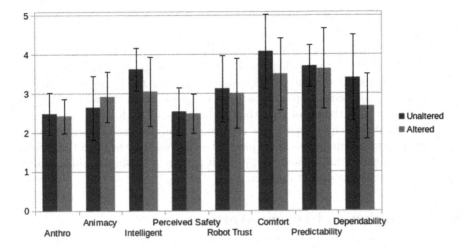

Fig. 3. Category Means: No significance at the $p < 0.1$ level found.

and the data may be showing a floor effect. There was no significant correlation between the self-reported feeling of safety and the rating of the robot.

The results presented in the previous section partially support **H2**. Looking at the free-response questions, the differences in how users perceived the robot's behavior from the standard to the altered consent conditions suggest that communicated risk might affect how a participant perceives the robot. What is interesting with the free-response questions is that the same robot actions were perceived very differently when the communicated risk was greater. The changes in ratings of the robot's intelligence, comfort, and dependability (though not significant) also indicate that the perceptions of the robot's behavior changed when increased risk was communicated.

Furthermore, the behavioral analysis suggested a change in perception of the robot. Participants in the altered consent condition frequently looked at the camera or leaned in to look at the robot. This may suggest an expectation of the participants that the experimenter would step in and stop the experiment if necessary. It may also be that leaning allowed the participants a better view of the robot while also affording the chance to stay further away from the robot. As these behaviors indicate a greater feeling of discomfort, these behaviors partially support **H1**.

A possible explanation for the lack of difference in perceived safety is that there were other effects that are skewing the data. One such effect could be that the participants' self reported safety was measured by a survey that took place after their interaction with the robot. Since the participants filled out the survey after the perceived danger had already passed, their perception of danger was in hindsight when they know that no harm had come to them. This hindsight could be biasing their responses.

A limiting factor of our study was that it had only 19 participants. This may not have been a large enough participant pool to get viable data. Also, the setting of the study may have biased the participants. While the study was held in a laboratory, there were other robots in the same room as the participants when they completed the survey. The main robot was not there, but this may have prompted participants to compare the other robots in the room they took the end survey with the robot that they had interacted with.

Some of the participants knew the proctor or someone else in the room where the surveys were administered. This may have affected their feeling of safety and comfort, as they may have assumed that their prior relationship would gain them special considerations and protection.

6   Conclusion and Future Work

In this study, the researchers postulated that participants would feel less safe with the changed consent form and that participants who felt safe would also feel more positively about the robot. Data from 19 participants were collected via an online survey, video analysis, and free responses. Statistical analysis on these data showed some significant differences between the altered and standard consent form conditions. A post-hoc analysis showed positive correlation between comfort and trust in the robot that varied with the perceived dependability of the robot. The changes in participant behavior toward the robot, and the perception of the robot's actions indicate the potential for follow-up on **H1** and **H2**. Follow-up research and reconfiguration of the study may further support the experimental hypotheses provided.

While this exploration was not able to conclusively give evidence for the SDC, it does suggest an effect due to communicated risk of a laboratory study. This opens up a wide area of future work considering how perceived risk might affect HRI studies conducted in laboratory settings. In this study, the participants had the e-stop the entire time, the robot looked at them before moving every time, and the robot was a consistent, moderate speed in both conditions. Any one of these variables could have had a more significant effect on the participants' perceived safety and perceptions of the robot.

Gaze has been shown in the past to be able to influence participants' perceptions of the robot [8]. Not only that, but gaze can directly affect participants' actions and strategy in studies examining object hand-offs [1,6]. The e-stop was explained to the participants as being able to stop the robot if they felt unsafe. This may have contributed an inherent feeling of safety for the participants, and the researchers will examine how the participants would react without the e-stop. The robot also moved towards the participants at a steady, moderate pace. If the robot was more abrupt and quick about moving, this might provoke a greater response in the participants.

The researchers plan on continuing this work with what they learned from this study in order to further examine the effect of perceived safety on participants' actions and perceptions of the robot. Specifically looking into the effects that the gaze of the robot, possession of the e-stop, and speed of the robot has on the safety perceived by the participants. Further work is needed to fully examine the causes and effects of the SDC.

Acknowledgments. This material is based upon work supported by the National Aeronautics and Space Administration under Grant #NNX10AN23H issued through the Nevada Space Grant, the Office of Naval Research DURIP award #N00014-14-1-0776, the National Science Foundation #IIS-1528137, and the UNR NSF EPSCoR UROP Program #IIA-1301726. We appreciate the help from Mercedes Anderson, Gaetano Evangelista, and Nathan Yocum who helped administer the study.

References

1. Admoni, H., Dragan, A., Srinivasa, S.S., Scassellati, B.: Deliberate delays during robot-to-human handovers improve compliance with gaze communication. In: Proceedings of the 2014 ACM/IEEE international conference on Human-robot interaction, pp. 49–56. ACM (2014)
2. Bartneck, C., Croft, E., Kulic, D.: Measurement instruments for the anthropomorphism, animacy, likeability, perceived intelligence, and perceived safety of robots. Int. J. Soc. Robot. **1**(1), 71–81 (2009)
3. Kraft, K., Smart, W.D.: Seeing is comforting: effects of teleoperator visibility in robot-mediated health care. In: The Eleventh ACM/IEEE International Conference on Human Robot Interaction, HRI 2016, pp. 11–18. IEEE Press, Piscataway (2016). http://dl.acm.org/citation.cfm?id=2906831.2906836
4. Lee, J., Moray, N.: Trust, control strategies and allocation of function in human-machine systems. Ergonomics **35**(10), 1243–1270 (1992). doi:10.1080/00140139208967392
5. Martin, D.: Doing psychology experiments. Cengage Learning (2007)
6. Moon, A., Troniak, D.M., Gleeson, B., Pan, M.K., Zheng, M., Blumer, B.A., MacLean, K., Croft, E.A.: Meet me where i'm gazing: how shared attention gaze affects human-robot handover timing. In: Proceedings of the 2014 ACM/IEEE International Conference on Human-Robot Interaction, pp. 334–341. ACM (2014)
7. Orne, M.T., Holland, C.H.: On the ecological validity of laboratory deceptions. Int. J. Psychiatry **6**(4), 282–293 (1968)
8. Plaisant, C., Druin, A., Lathan, C., Dakhane, K., Edwards, K., Vice, J., Montemayor, J.: A storytelling robot for pediatric rehabilitation. In: Proceedings of the Fourth International ACM Conference on Assistive Technologies, pp. 50–55. Arlington (2000)
9. Yagoda, R.E., Gillan, D.J.: You want me to trust a robot? The development of a human-robot interaction trust scale. I. J. Soc. Robot. **4**(3), 235–248 (2012). http://dblp.uni-trier.de/db/journals/ijsr/ijsr4.html#YagodaG12

On Designing Socially Acceptable Reward Shaping

Syed Ali Raza[✉], Jesse Clark, and Mary-Anne Williams

Centre for Quantum Computation and Intelligent Systems,
University of Technology, Sydney, Australia
syed.a.raza@student.uts.edu.au, {jesse.clark,
mary-anne.williams}@uts.edu.au

Abstract. For social robots, learning from an ordinary user should be socially appealing. Unfortunately, machine learning demands an enormous amount of human data, and a prolonged interactive teaching session becomes anti-social. We have addressed this problem in the context of reward shaping for reinforcement learning. For efficient reward shaping, a continuous stream of rewards is expected from the teacher. We present a simple framework which seeks rewards for a small number of steps from each of a large number of human teachers. Therefore, it simplifies the job of an individual teacher. The framework was tested with online crowd workers on a transport puzzle. We thoroughly analyzed the quality of the learned policies and crowd's teaching behavior. Our results showed that nearly perfect policies can be learned using this framework. The framework was generally acceptable in the crowd's opinion.

1 Introduction

It will be inevitable for future robots to learn from humans in ordinary circumstances. Besides, teaching a social robot should provide a pleasant experience to a human teacher. An out-standing machine learner becomes socially unacceptable for robot's learning if it does not allow a human to teach in a comfortable and natural way. Furthermore, similar to behavior development in humans, a robot's behavior might not only be influenced by the owner (guardian), but also by other people it interacts with. Therefore, a robot should also be able to learn a task from multiple teachers with varying abilities.

Recently, online crowdsourcing platforms have opened new prospects for AI research. It has enabled us to handover state-of-the-art AI algorithms to the end users and conduct large-scale user studies previously unmanageable. Hence, it is important to revisit the existing robot learning methods in this context. Also, robot's behavior designing by crowdsourcing, if shown successful, will create new job prospects. In this research, our main contribution is the exploitation of online crowdsourcing technology to make reward shaping practically usable with ordinary users.

The reward shaping technique in reinforcement learning emphasizes on showing the desirability of the immediate behavior via rewards. If an agent's behavior is right or wrong, then it should receive an immediate reward instead of waiting for the long term (environmental) reward. However, reward shaping demands a regular supply of rewards. Ideally, every action should be rewarded. A typical teaching-by-rewarding session takes longer than a human's patience can bear. Therefore, in practice, reward shaping becomes

© Springer International Publishing AG 2016
A. Agah et al. (Eds.): ICSR 2016, LNAI 9979, pp. 860–869, 2016.
DOI: 10.1007/978-3-319-47437-3_84

an unnatural task for human teachers. To make reward shaping a viable solution for robot's behavior designing, it is important to make it approachable to the end users.

In this paper, we designed a framework to teach an agent's policy using multiple teachers. The framework was built on top of Q-learning, a popular reinforcement learning algorithm [7]. Reward shaping was employed to accelerate the learning. We divided the policy learning temporally. Instead of a single dedicated teacher, a number of teachers taught a policy for a fixed number of steps. The framework allowed multiple teachers to add rewards to a policy in a sequence. We designed a grid based game (Fig. 2) to compare the policies learned via the framework with an individual teacher's policy. The teachers were recruited from an online crowdsourcing marketplace (Amazon Mechanical Turk). A separate web interface was designed to carry out the experiments. It contained a practice phase to familiarize the teacher with the domain rules and a training phase to train the agent. At the end, each teacher answered a couple of survey questions about the applicability of the framework. The experimental results confirmed the viability of the framework.

2 Related Work

There is a vast literature on speeding up reinforcement learning using human input. The most related are the studies in which the naïve user's input was added when the learning was underway, and the online crowd was used to get the human data.

Among the approaches seeking online feedback, some have assumed numeric reward from human trainers [4, 8]. The TAMER framework learns a model of the human generated numeric rewards which can be used with or without the environmental rewards. It has been shown effective in a number of domains [4]. There are other strategies which have assumed human feedback as a discrete communication [5, 6]. They have employed Bayesian policy learning algorithms designed to exploit trainer's feedback strategies.

On the other hand, there is a growing list of literature exploiting crowdsourcing in machine learning applications. In a recent work [3], it has been shown that the crowd workers from the Amazon Mechanical Turk were able to provide quick and accurate feedback to a mistake done by a Pac-Man agent in a short recorded video of gameplay. Some approaches have used crowdsourcing for data gathering. In [2], Forbes et al. have proposed a framework where the robot presented a query to the crowd as an active learner. The query was a situation where the seed demonstration would not work but could be fixed by the crowd. The worker only edited a subset of existing poses instead of providing a complete new demonstration. Therefore, similar to our approach, teachers worked on a part of the policy instead of teaching a complete policy. However, the main difference is that we explicitly analyzed the policy building in a sequence from multiple users. Other studies have used crowdsourcing for goal-based learning [1, 9]. In these approaches, the crowd provided the goal of a task instead of the demonstrations of the actions required to accomplish the task.

3   Proposed Framework

We developed a simple framework to make reward shaping socially acceptable for end users. We have simplified the complex and socially unwelcoming robot teaching task for humans. The framework learns a policy via interactive reinforcement learning which takes a human teacher in the learning loop. The reinforcement learning suffers from a slow learning rate. A popular method to speed up reinforcement learning is reward shaping which solves the problem of sparse environmental rewards. It allows adding a reward at every step of learning. Despite being successful at learning better policies in short time, reward shaping poses a new challenge. It requires a continuous stream of rewards. For a human teacher, it becomes an unnatural task to judge an agent continuously and reward its behavior for several minutes or even hours. The most intuitive solution is to reduce the number of steps a human teacher is required to teach. Our approach is to seek a minimum number of rewards from a teacher. In interactive reinforcement learning, it is important to understand how many learning steps a human can comfortably teach. It may depend on the domain, if the task is paid or unpaid, the level of expertise of the teacher, and other algorithmic parameters which facilitate learning. However, to evaluate in a simple setting, we assumed that a person can comfortably teach for 50 steps in our experimental domain. Note that it is a fraction of the complete teaching session which may take over 1000 steps.

The underlying learning mechanism of our framework is similar to the interactive reinforcement learning designed by Thomaz et al. in [8]. To keep experimental design approachable, we dropped the use of guidance which is an additional input to speedup learning. An update in Q-learning takes the form $Q_{t+1}(s_t, a_t) = Q_t(s_t, a_t) + \alpha_t \delta_t$, where,

$$\delta_t = r_{t+1} + \gamma max_a Q_t(s_{t+1}, a) - Q_t(s_t, a_t) \tag{1}$$

Here, $Q_t(s_t, a_t)$ is state-action value at time $t = 1, 2, 3, ..., s \in S$ (the set of states), $a \in A$ (the set of actions), α is the learning rate, γ is the discount factor, δ is the temporal-difference error, r is the numeric reward. In reward shaping, $r = EnvironmentalReward + ShapingReward$. In interactive reinforcement learning, the shaping reward is provided by the human teacher. We set environmental reward as 10 for the goal state and -10 for the deadlock (failure state). The shaping reward was $+1$ and -1. The agent uses a policy, $\pi{:}S \rightarrow A$, indicating the action to select from the current state as $\pi(s) = argmax_a Q(s, a)$.

Figure 1 shows the functioning of our framework. It can be seen that m teachers teach sequentially. Human teacher's input I is added to a Q-update U at every step. I can be a positive reward, negative reward or no reward. The next teacher starts teaching from a step next to where the last teacher stopped teaching. A teacher teaches for K steps (which we choose as 50). Hence, a policy is taught for $K \times m$ steps. The policy learning is sequential (not serial) in the sense that it is ordered by the teacher id. It means a complete batch of rewards is added to the policy for a teacher id before the next teacher can teach. It also means two teachers cannot teach a policy at a time.

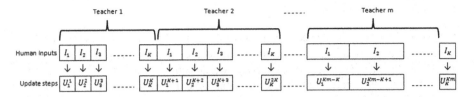

Fig. 1. The proposed framework to sequentially learn a policy

4 Experiments

4.1 Domain Details

We designed a Sokoban style domain for the experiment. It is a simulated 2d grid-world game where each cell can be a wall or free cell. The free cell can be occupied by either the player or a box. The player can choose from four actions: left, right, up, and down. The player's task is to push each box using four actions (it cannot pull the box) and drive it to the goal position without letting the box get stuck, which is called deadlock. Figure 2 shows the domain and an example of gameplay.

Fig. 2. Sokoban domain: From left to right it shows the result of two consecutive down actions

Sokoban was chosen because it is one of the benchmark domains in AI. It has simple rules, yet it poses a complex and challenging task to both humans and machine. Solving Sokoban is PSPACE complete. It has an intuitive visual setting which attracts humans and keeps them fully involved while teaching. We have simplified the Sokoban game regarding complexity. The original Sokoban has several difficulty levels. It requires expertise to solve all of them. However, our goal was to use naïve users for the teaching task. Therefore, we carefully crafted the domain such that a balance was maintained between the approachability of the domain to a naïve user and fairness as a test bed. The modified domain had only one level. It had one box and 23 free spaces. Also, the placement of the walls was simplified to avoid frequent deadlocks. There were a total of 198 unique states and four legal actions from each state. Typically, at the start of the game the player and each box are positioned at a fixed location. Instead, we have used a state distribution for both box and player for their start positions. Note that rules of the game were unchanged in the modified domain.

4.2 Experimental Design

The experiments were conducted over the web on Amazon Mechanical Turk (AMT). AMT is a popular crowdsourcing platform to recruit people online to perform tasks requiring human intelligence. For the crowd workers, we designed a web interface as shown in Fig. 3. We kept the interface plain and understandable. It contained a brief description of the experiment and the domain. The users had to complete a mandatory practice session before entering the teaching session. The practice session was intended to familiarize the user with the domain rules. Also, it ensured that the user possessed a minimum understanding of the domain. In the practice phase, the user had the complete control of the agent. The practice interface (not shown but similar to Fig. 3) had arrow buttons to navigate the agent (keyboard arrow keys could also be used) and showed the wins and losses count. The user was asked to show at least two cases of a win and two cases of a loss to proceed to the training session. The user had to drive the agent to the goal to win and to a deadlock to lose. We considered it necessary for a user to understand a success and a failure to perform a proper teaching. The game restarted automatically after every win or loss. At the start of each game the agent and the box were randomly positioned according to the start-state distribution. On the training interface (Fig. 3), a start button was provided to initiate the training after reading the instructions. On click of the start button, the agent began to take actions by itself every two seconds. The user could not directly control the agent's moves. The user was instructed to observe the agent's every action. The user had to assign a positive reward (using P) for agent's right move or a negative reward (using N)

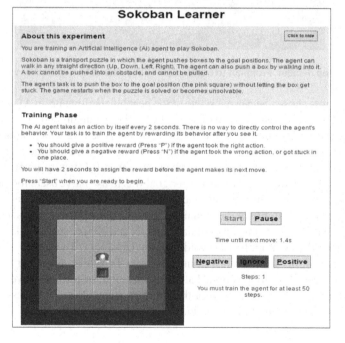

Fig. 3. Training phase web interface

for agent's wrong move. The user could also ignore an action. At the end of an episode (deadlock or goal state), a new episode automatically started with the box and player positions drawn from the start-state distribution. The training completed after 50 steps for a user. At the completion, the user was prompted to a survey about the experiment.

From AMT, we selected moderate level turkers. We recruited a turker who completed and got approved at least 1000 human intelligence tasks (HITs) and out of the HITs completed at least 98 % were approved by the HIT requester. The turkers had been informed about the nature of the task before they accepted and proceeded to the main task. At a HIT's completion, the user was approved and paid 0.40 USD. The total duration of the experiment for a user was on average 6 min.

5   Results

Through the experimental results, we analyzed the viability of the proposed framework by addressing the following questions,

1. Can casual crowd workers teach a nearly perfect policy or it will be a failure?
2. Can the crowd teach a policy in a reasonable time?
3. How does the crowd's policy compare to an individual's policy in terms of online learning performance and offline test performance?
4. What is the crowd's opinion about the viability of this framework?

Overall 69 crowd workers were recruited to teach three policies. An additional policy was taught by a single human teacher to be used as a benchmark (individual's policy). It was taught by one of the authors who had long term familiarity with the domain and complete knowledge of underlying algorithm. Therefore, we ensured a high quality benchmark for the crowd policies to compare with. The individual's policy was also learned using the same web interface discussed in the previous section. It was taught for 1000 steps in a single run. Through initial training and testing, we observed that 1000 steps were enough to teach a nearly perfect policy by an individual teacher. We set RL parameters as alpha = 0.1, discount factor = 0.9, and for ε-greedy policy ε was initiated as 0.3 and reduced to zero with a decrement of 0.3/1000 per step. Our state representation consisted of only x and y coordinates of box and agent.

Table 1 provides the statistics about the learned policies. The first column shows the different measures we used to analyze the quality of policy. We set a soft criterion to stop learning. In literature, a minimum number of episodes are set heuristically to stop teaching. Therefore, we stopped learning a policy from the crowd when it exceeded the number of episodes taught by the individual teacher (37). Due to this criterion, the number of teachers who worked on a policy differed. Crowd policy 1 (CP1) took the longest time. It took 1400 steps and 28 teachers to cross 37 episodes benchmark. We found that during training of CP1 three workers did not provide any input to the learning agent. They opted 'ignore' for the whole training session which was a valid input. It never happened with crowd policy 2 (CP2) and crowd policy 3 (CP3). It is reasonable to expect this behavior from a crowd. They possibly didn't understand the training instructions or deliberately didn't take an interest in agent's learning. CP2 and CP3

Table 1. Data of the learned policies

	Individual's Policy	Crowd Policy 1	Crowd Policy 2	Crowd Policy 3
No. of teachers	1	28	20	21
No. of steps taught	1000	1400	1000	1050
Episodes completed	37	38	41	39
States explored (states/state-actions)	171/365	175/384	137/290	178/382
Correct user input	57.0 %	47.8 %	61.7 %	44.6 %
Win rate (out of 198)	95.9 %(190)	93.4 %(185)	96.9 %(192)	83.3 %(165)

performed slightly better. They took a similar number of steps as the individual teacher. Moreover, they got slightly more episodes completed than the individual teacher in the same number of steps. We have also reported 'states explored' which is an important measure to gauge learned policy's quality. It tells how many unique states or state-action pairs were explored and rewarded during the learning. Almost all the policies explored an equal number of states/state-actions except CP2. CP2 explored much less than the others nevertheless its performance was better. We found that the main reason for CP2's better performance was the high-quality input from the crowd. For that, we compared the user's input with the simulated expert's policy. The simulated expert's policy was separately learned by a Q-learner using only environmental rewards (i.e. without reward shaping via human rewards). We exhaustively trained it for 5000 episodes. In 'correct user input', we calculated the percentage of matching of user's input with the simulated expert's policy. They matched if user's input was positive for an agent's move and the simulated expert's policy output the same move. They also matched when the user's input was negative and the simulated expert suggested a move other than the agent's move. It can be seen that the user inputs given to CP2 were 61.7 % correct according to the simulated expert. It even surpassed the correct input percentage of the individual teacher. The win rate is discussed later.

To understand the learning performance, in Fig. 4, we have provided graphs for the number of steps taken over the episodes, the sum of absolute error per episode and the sum of reward per episode. The number of steps converges to minimum steps to goal over episodes. During initial episodes exploration rate was high and the policy was naïve therefore we had higher peaks. It can be seen from the graphs that after 15th episode the number of steps started to converge for all the policies. Few late peaks can be observed around the 25th episode in the steps graph of CP1. Also, it had the highest peak among all policies of around 175 steps. It shows where CP1 took a longer time to learn. The absolute error was calculated using Eq. 1. It converges towards zero when the Q-values become optimal. However, due to continuous rewarding from teachers, it converged above zero. The absolute error graphs followed similar patterns as the number of steps

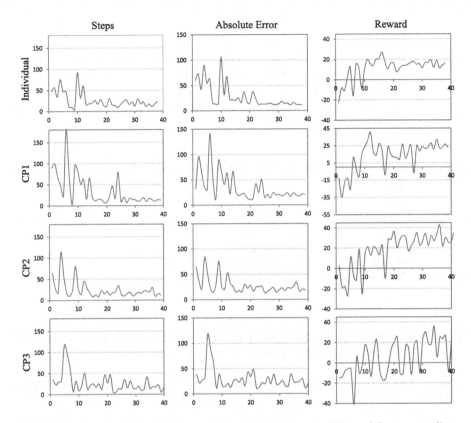

Fig. 4. Comparison of the learning performance of the crowd policies and the export policy

graphs due to continuous rewarding and absolute reward of 1 per step by human teachers. The sum of rewards converges to a high value when the Q-values become optimal. The reward graphs showed that the values started to converge around 20 (on the y-axis) in the case of Individual, CP1, and CP2. Only in the case of CP3 the values continued to fluctuate significantly. It shows that in the case of CP3 the agent's performance was not satisfactory and the workers gave comparatively more negative than positive rewards in the later stage of learning. Hence, unlike other policies we have few negative peaks after the 20th episode in CP3.

For offline performance, we tested the learned policies for exhaustive start state distribution. Each policy was run for each of the 198 unique start states. It showed whether the learned policy could drive the agent to goal from every unique state. Also, how many steps it takes to reach the goal from a unique start state. Figure 5 provides the comparison of the three crowd policies with the individual teacher and simulated expert. The rewards represent the average of the environmental reward (+10 at the goal and -10 at the deadlock) earned over 198 episodes. The steps represent the average of the steps taken from the start of an episode till the end in 198 episodes. The simulated expert earned the optimal rewards (10) with optimal average steps value of 11.8. The individual teacher's policy performed considerably close to optimal. Among crowd

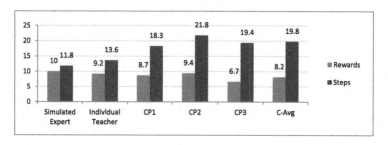

Fig. 5. Offline test performance for three crowd policies, individual teacher's policy and simulated expert policies

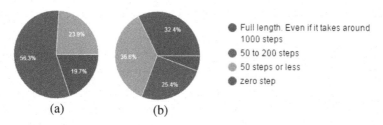

(a)   (b)

Fig. 6. Survey results: (a) To what length a person can comfortably teach this agent as a PAID task? (b) To what length a person can comfortably teach this agent as an UNPAID task?

policies, CP2 earned the best rewards. It even earned more rewards than the individual teacher's policy. CP3's rewards were comparatively low. It suggests CP3 needed further training. For exact figures, we have provided the win rates for the three crowd policies and the individual's policy in Table 1. It shows the percentage of winning out of 198 episodes. Again, CP2 had the highest win percentage (~ 97 %) followed by the individual's policy and CP1. CP3 was 83 % successful which indicates that CP3 needed further training but it had learned significant part of the domain. The average steps were almost same across the crowd policies. It is interesting to note that CP2 earned more rewards than the individual's policy but took longest average steps. It shows that CP2 reached goal most of the time but took the longest path which depended on the teacher's preference. C-Avg is the average performance of the crowd policies. It shows the crowd policies were competent in rewards but not in average steps. It means that crowd's focus was on reaching the goal rather than reaching the goal through the shortest path.

After training the agent for 50 steps, each crowd worker answered the survey questions summarized in Fig. 6. Question a's results showed that all the workers considered it viable to teach agent via this framework as a paid task (no one voted for zero step). Moreover, ~ 24 % agreed with the length of the teaching session. However, the majority was in favor of teaching slightly more steps (50 to 200) as a paid task. It is also interesting that ~ 20 % showed interest in teaching for the full length. When asked about the viability of the framework as an unpaid task (question b) we got mixed results. Apparently, crowd rejected the idea of teaching for the full length as an unpaid task (only four votes). Besides, 1/3rd of the crowd deemed it impracticable to teach as an unpaid task (~ 32 % voted zero step). It might indicate that the framework needs further improvements to be fully acceptable as an unpaid task. Nevertheless, roughly $2/3^{rd}$ of the crowd (62 %) agreed to teach for 1 to 200 steps as an unpaid task. It is a promising result for future robot learning by ordinary people.

6 Conclusion

We have empirically shown that policy learning via reward shaping is feasible when using the online crowd to teach a policy sequentially. We used a simple framework which allowed crowd workers to teach a policy one by one. Three policies were taught by the crowd. Those policies were compared with an individual teacher's policy. Through the experimental results, we conclude that the crowd could teach a policy better than the individual's policy (CP2). It is also possible that crowd can only teach an average policy in an expected amount of time (CP3). Also, a crowd may teach a good policy which takes a longer time to learn than expected (CP1). The overall survey results showed that the proposed framework is socially acceptable.

References

1. Chung, M.J.-Y., Forbes, M., Cakmak, M., Rao, R.P.: Accelerating imitation learning through crowdsourcing. In: 2014 IEEE International Conference on Robotics and Automation (ICRA), pp. 4777–4784. IEEE (2014)
2. Forbes, M., Chung, M.J.-Y., Cakmak, M., Rao, R.P.: Robot programming by demonstration with crowdsourced action fixes. In: Second AAAI Conference on Human Computation and Crowdsourcing (2014)
3. Gabriel, V., Peng, B., Lasecki, W.S., Taylor, M.E.: Towards integrating real-time crowd advice with reinforcement learning. In: IUI Companion, pp. 17–20 (2015)
4. Knox, W.B., Stone, P.: Combining manual feedback with subsequent MDP reward signals for reinforcement learning. In: International Foundation for Autonomous Agents and Multiagent Systems Proceedings of the 9th International Conference on Autonomous Agents and Multiagent Systems, vol. 1, pp. 5–12 (2010)
5. Loftin, R.T., MacGlashan, J., Peng, B., Taylor, M.E., Littman, M.L., Huang, J., Roberts, D.L.: A strategy-aware technique for learning behaviors from discrete human feedback. In: Twenty-Eighth AAAI Conference on Artificial Intelligence (2014)
6. Peng, B., MacGlashan, J., Loftin, R., Littman, M.L., Roberts, D.L., Taylor, M.E.: A need for speed: adapting agent action speed to improve task learning from non-expert humans. In: Paper Presented to the International Conference on Autonomous Agents and Multiagent Systems (AAMAS 2016) (2016)
7. Sutton, R.S., Barto, A.G.: Reinforcement Learning: An Introduction, vol. 1. MIT Press Cambridge, Cambridge (1998)
8. Thomaz, A.L., Breazeal, C.: Reinforcement learning with human teachers: evidence of feedback and guidance with implications for learning performance. In: Paper Presented to the Proceedings of the 21st National Conference on Artificial Intelligence vol. 1, Boston, Massachusetts (2006)
9. Toris, R., Kent, D., Chernova, S.: Unsupervised learning of multi-hypothesized pick-and-place task templates via crowdsourcing. In: 2015 IEEE International Conference on Robotics and Automation (ICRA), pp. 4504–4510. IEEE (2015)

Motivational Effects of Acknowledging Feedback from a Socially Assistive Robot

Sebastian Schneider[(✉)] and Franz Kummert

CITEC, Bielefeld University, Inspiration 1, 33602 Bielefeld, Germany
{sebschne,franz}@techfak.uni-bielefeld.de
http://aiweb.techfak.uni-bielefeld.de

Abstract. Preventing diseases of affluence is one of the major challenges for our future society. Recently, robots have been introduced as support for people on dieting or rehabilitation tasks. In our current work, we are investigating how the companionship and acknowledgement of a socially assistive robot (SAR) can influence the user to persist longer on a planking task. We conducted a 2 (acknowledgement vs. no-acknowledgment) x 2 (instructing vs. exercising together) x 1 (baseline) study with 96 subjects. We observed a motivational gain if the robot is exercising together with the user or if the robot is giving acknowledging feedback. However, we could not find an increase in motivation if the robot is showing both behaviors. We attribute the later finding to ceiling effects and discuss why we could not find an additional performance gain. Moreover, we highlight implications for SAR researchers developing robots to motivate people to extend exercising duration.

Keywords: Human-robot interaction · Socially assistive robots

1 Introduction

Recently, assistive robots have been designed to support people on different rehabilitation and cognitive tasks by the robot's presence and assistance [1,2]. Moreover, virtual agents and socially assistive robots (SAR) have been introduced as a useful tool to promote a healthy lifestyle, support for dieting or exercising besides conservative methods [1,3,4]. Therefore, different approaches from sociology or psychology have already been exploited (i.e. goal-setting, empathy, backstory or personalization [2,5–7]). One other important interactional aspect that sport instructors are regularly incorporating into exercising is acknowledgement. Acknowledgement is a special type of feedback, which is used to appreciate the current state of exercising which in turn motivates the trainee to keep up on a task. It is positive reinforcing and does not compare the current performance to other persons or to previous sessions. While other types of feedback (i.e. positive, negative, comparative or corrective feedback) have been studied in Human-Robot Interaction (HRI) [8–10], the quantitative effects of acknowledgement while working out with a robot have not been compared to a baseline

© Springer International Publishing AG 2016
A. Agah et al. (Eds.): ICSR 2016, LNAI 9979, pp. 870–879, 2016.
DOI: 10.1007/978-3-319-47437-3_85

measurement. Because most of the published studies comparing different types of feedback are lacking comparative results to baseline studies with no robot or no feedback, it is difficult to distinguish the true motivational effects of SAR.

In previous works, we have investigated how the mere presence of the robot affects the motivation to exercise compared to a robot that exercises together with the persons. Our findings conclude that the users experience a motivational gain when the robot is working out co-actively. However, if the system is only instructing, then the presence is no perceived as useful. Hence, we investigate in this paper how acknowledgment of the robot can increase the user's motivation to exercise. Furthermore, we are interested in whether the acknowledgment is sufficient or the system needs to actively work out with the user. The result of this research can be useful to decide whether a SAR should work out along with the person or whether having acknowledgement is enough to motivate the user to exercise longer. For our investigation we examined four conditions: a robot exercise instructor (RI), a robot exercise instructor which is giving acknowledgement (RIF), a robot exercise companion (RC) and a robot exercise companion which is giving acknowledgement (RCF). We compare all conditions against a baseline condition without a SAR (IC). Our hypotheses are that:

H1: people exercise longer in the RCF condition compared to all other conditions.

H2: people exercise longer in the RIF condition than in the IC and RI condition.

This manuscript is organized as follows. The next section gives the reader an overview of existing literature in the field of motivational feedback from SARs for exercising. Afterwards, we will describe the study design to investigate our hypotheses and the system design we used. These sections are followed by the results and a discussion.

2 Related Work

In the past different robots assisting users on tasks have been reported. In this section we want to briefly review some approaches and results. Different to other investigations, we do not want to question which kind of feedback is most useful for the user. We want to answer the general question how acknowledgement can influence the user's motivation in general.

The preference for a relational vs. a non-relational robot-coach have been investigate in [1]. The study shows that users have a preference for relational feedback. Relational is used as the robot's capabilities to exploit all of its social interaction and personalization approaches. Thus, the robot always gives the user praise upon correct completion of an exercise, it provides reassurance in case of failures, refers to the user by its name, references past experiences and uses humor. In the non-relational condition the robot coach gives instructional feedback but does not employ any relationship building. The authors of this study used an in-between subject design to evaluate the different relational styles of the robot. They did not find any differences in the exercise performance based the relation or non-relational robot. In our previous work we have investigated the differences of performance-based vs. non-performance-based feedback for users doing a cognitive task [11]. Personalized feedback from robotic tutors has been investigated in [2]. Both works support that individualized feedback based on the user's performance can increase the tutor's effectiveness. A different study compares self-comparative vs. other-comparative to non-comparative feedback in a push-button task [9]. However, the authors could not find evidence for any of their hypotheses. Positive, negative and neutral feedback of a robot or human instructor has been studied in [12]. Study subjects have a preference for positive feedback from a robot. However, they could not find any feedback preference from a human interaction partner.

Compared to the presented related work, we are focusing on the effects of acknowledgement of a social robot while exercising. Furthermore, we compare our results to a baseline condition where no robot is present at all. Thus, the results of our study let us conclude about the importance of acknowledgement for the user's motivation to exercise longer as well as how it is moderated by the robot's style of companionship.

3   Study Design

Our study is inspired by a research comparing the effects of working out with a virtual avatar with different degrees of human-likeliness [13]. To allow possible comparisons, we wanted to replicate the study as close as possible. However, we needed to include some changes in the study design due to the actual robot agents. Therefore, we changed the exercises from forearm planks to full planks due to the robots limited degree of freedom.

3.1   Experimental Design and Participants

Participants ($n = 95$) were in one of 5 conditions (independent condition (IC), robot instructor condition (RI), robot companion condition (RC), robot instructor feedback condition (RIF) and robot companion feedback condition (RCF)) with around 18 participants in each condition. Participants were mostly students (f/m $= 44/51$, age M $= 25.4$ years, SD $= 5.6$) from our university acquired by flyer distributed on the campus. They received seven Euros as monetary compensation. Three participants from the IC had to be excluded. One was an outlier persisting much less during the first block compared to all other participants. Two other persons had to be excluded because they were doing the exercises wrong. One participant from the RT and two participants from the RCF condition had to be excluded from the survey evaluation because the data were missing. However, we still could analyze their performance data.

3.2   Exercises and Conditions

Participants had to do two blocks of five isometric exercises each (see Fig. 1). Participants in the IC condition did the exercises in both blocks alone. In all other conditions the participants did the exercises alone in the first block and with the humanoid robot Nao[1] in the second block. During the RI condition the robot was announcing the exercises the user had to do, as well as how long the break is. While the user was exercising the robot was standing in front of the person observing her/him while exercising and showing some idle behavior. After the user has finished an exercise s/he received a general encouraging feedback. In the RC condition the robot was showing the same behavior as in the RI condition. But instead of just standing in front of the participant the robot was exercising together with the user and also held the isometric exercises. However, the robot always hold it a bit longer than the participant. In the feedback conditions (RIF, RCF) the robot was giving some acknowledgment during each exercise. This feedback was generated based on the performance from the first block. After three quarter of the time they hold the exercises during Block 1 the robot gave some acknowledgement ('you are doing great', 'keep on', 'this is the last exercise, keep on the good work', etc.). We used three quarter of the previous time as a threshold because a preliminary analysis showed that most of the people will stop doing an exercise after three quarter of the time from the first block in the second block. The system was implemented using a state machine coordination for distributed systems, a Kinect for user activity recognitions as well as features from the NaoQi API. We neglect a detailed preliminary analysis and an extensive description of our system due to the limitation of this paper.

3.3 Procedures

Participants arrived at the lab, read and signed a consent form which informs them that they will be recorded during the whole time of the experiment. They watched a short video of Nao demonstrating the five exercises. Afterwards, they were brought to a fitting room to change clothes and strap on a heart rate belt. They were instructed to do each exercise as long as they can and if they could not persist any longer to stand up immediately, rate their perceived exertion on a prepared scale and wait for thirty seconds before continuing with the next exercise. They were guided to the lab and told to start after they have waited for a short time, so that the experimenter can check that the recording is

Fig. 1. The five isometric planking exercises.

working properly. Then the participants did each exercise alone in the lab while the experimenter observed them from a different room and took the participant's times for each exercise. The participants completed Block 1 (each exercise once). Afterwards, the participants had a ten minute break where they were offered a

---

[1] https://www.aldebaran.com/en.

glass of water and had to answer a self-efficacy belief scale about the exercises. After the break participants in the IC condition were told the average time they held the planks and that they would complete the same set of exercises again (Block 2). In every condition the participants were not told that they had to do a second block of exercises until they had finished the first block.

In the robot conditions participants were told that they will do the same set of exercises again but that this time a robot will be present. They were instructed to follow the guidance of the robot through the session and that they are exercising together from now on (RC). The experimenter told the participant the true average time they held the planks, like in the IC condition, but gave them a false information on how long the robot can persist the exercises. They were always told a number which is forty percent higher than their average time. This creates an unfavorable comparison which leads to greater effects and was adopted from the previous study [13].

Again the experimenter did not enter the room together with the participant. In all robot conditions, the participant and robot had a short interaction phase. During this phase the robot told them its name (Nao), hometown (Paris) and hobbies (gardening, reading) and waited for a short time to give the human participant a chance to also share his/her personal information. This was done because prior research showed that people treat agents more like humans when there was an initial verbal interaction between them [7]. After completing Block 2 the robot thanked the participants for their participation, told them that they are allowed to leave the room and that it needs to rest a bit. After leaving the room the participants completed a questionnaire, were debriefed and received a monetary compensation. The whole procedure took about 45 min to one hour.

3.4 Measures

Persistence. Persistence was the number of seconds a plank was held from the moment participants moved into position they quit. Block scores were calculated using total average seconds held on all five exercises.

Perceived Exertion. Perceived exertion was measured using the Borg rating scale [14]. The scale goes from 6 to 20 (6: 'no exertion at all', 20: 'maximal exertion'). The participants were asked to rate their exertion immediately after each exercise.

Negative Attitudes Towards Robots Scale. For the RIF and RIC, we also assessed the negative attitudes towards robots scale [15].

Perception of the Partner. Participants in the partnered condition completed questionnaires asking to rate the perception of the robot. To investigate the human-likeliness, we asked the participant to rate the robot based on the Godspeed questionnaire (5 point-based differential scale, [16]).

In the feedback conditions we also asked the participants to rate the information quality, the cooperation, the openness to influence as well as the team perception on a five-point Likert-scale [17].

Physical Training Enjoyment. We assessed the physical training enjoyment the users had using the Physical Activity Enjoyment Scale [18]. We used the average value of all items as overall enjoyment score. Furthermore, we asked how much time they spent exercising per week and their intention to train tomorrow for at least 30 min.

4   Results

We conducted an analysis of variance (ANOVA) to find differences in the enjoyment, performance on Block 1, intention to exercise, perceived exertion and general amount of time spent doing sports. We found no difference in persistence on Block 1 ($P = 0.47$), exercising per week ($P = 0.39$), enjoyment ($P = 0.467$), perceived exertion on Block 1 ($P = 0.106$) and Block 2 ($P = 0.397$). However, we found a difference in the intention to exercise for at least 30 min on the following day ($F_{4,84} = 3.453, p < 0.05$). Pairwise comparisons using t tests with pooled SD and Bonferroni correction revealed a differences between the RCF ($M = 4.77$, $SD = .42$) and RI ($M = 3.6$, $SD = 1.37$) condition ($p < 0.05$).

Persistence. As a primary dependent variable we used the average difference persistence time between the two blocks (Block 2 - Block 1). This approach controls for individual differences in strength and fitness and shows possible changes in persistence. The results obtained for the average block score of Block 2 subtracted with the average block score of Block 1 are shown in Fig. 2. An 5 (conditions) x 1 (persistence) analysis of variance (ANOVA) on the difference scores showed a significant main effect for the conditions ($F_{4,80} = 9.927, p < 0.001$). A pairwise comparison using t tests with pooled SD and Bonferroni adjustment revealed significant differences between IC and RC ($p < 0.0001$), IC and RCF ($p < 0.0001$), IC and RIF ($p < 0.0001$), RC and RI ($p < 0.05$), RCF and RI ($p < 0.05$) and RI and RIF ($p < 0.05$) (see Fig. 2). These results hint that acknowledgement has a positive effect on the user's exercising performance. Since, we instructed participants that they will interact with a robot and we have not told them that a human is tele-operating the robot, it shows that people are sensitive to acknowledgement from a clearly technical system. This shows that interactive motivational capabilities can be exploited from technical systems to enhance human exercise duration. This is particularly important in light of the results that participants exercised longer without enjoying the training less or feeling more exerted. Nevertheless, we did not find the same motivational gain in the companion conditions (RC vs. RCF). Subjects in this RCF condition did not persisted significantly longer compared to the RC condition or the RIF condition. Based on the feedback from the interviews we suppose that this is due to ceiling effects based on the picked exercises. Participants mentioned in

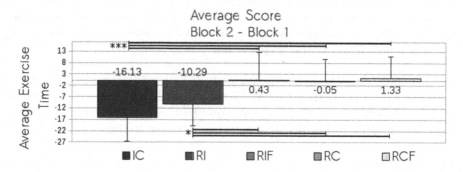

Fig. 2. Differences between Block 2 and Block 1 (*: $p < 0.05$; ***: $p < 0.0001$)

post-interviews that they would have liked to hold the exercise longer, but they were not able to persist any longer due to too much load on the wrists.

Perception of the Partner. We conducted several 4 (condition) x 1 (sub-scale) ANOVAs to find differences between the ratings on the Godspeed questionnaire for the different conditions (see Fig. 3). We did not find differences for the ratings of animacy (P = .21), anthropomorphism (P = .25), intelligence (P = .13) and safety (P = .13). However, we found significant main effects for likeability ($F_{3,67} = 3.956, p < .05$). A pairwise comparison using t tests with pooled SD and bonferroni adjustment revealed significant differences on the likeability scales between RCF and RI ($p < .05$) and a tendency between RC and RI ($p = .08$).

While we could not find evidence for higher performance in the companion conditions, these results show that people liked the robot in the companion conditions more than in the instructor conditions. Higher ratings on the likeability scales could be beneficial for long-term investigations, because they could positively influence the user's long-term motivation to engage in an interaction.

Further Results. We analyzed the differences between the ratings for team perception, perceived information, cooperation, openness to influence and NARS (see Fig. 4). A Welch Two Sample t-test between the RIF and RCF conditions revealed no differences for NARS ($t(28.781) = 1.425, p = 0.16$), openness

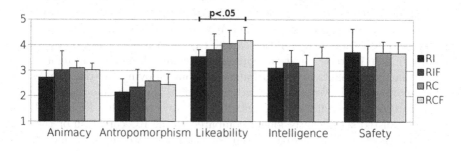

Fig. 3. Results of the Godspeed questionnaire.

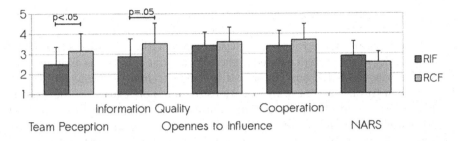

Fig. 4. Results for team perception, information quality, cooperation, openness and NARS ratings.

$(t(33.942) = -0.82468, p = 0.4153)$ and cooperation $(t(33.783) = -1.2039, p = 0.237)$. However, we found a difference for the team perception $(t(33.576) = -2.2821, p < 0.05)$ and a tendency for the perceived information quality $(t(33.983) = -2.0284, p = 0.05)$.

It is hardly surprising that participants had higher feelings for working out together, but it is remarkable that they rated the information quality as higher because the speech output was exactly the same in both conditions. Thus, co-actively working out could increase the user's trust in the system quality.

5 Discussion

We have presented a study on the effects of acknowledgement from a SAR on the user's exercise performance. We compared acknowledgement-feedback vs. non-acknowledgement-feedback for a robot exercise instructor and companion. To the best of our knowledge, our results are the first to compare the presence, the role and the acknowledgement of a robot against a baseline condition during an exercising task.

Users exercising co-actively with the robot had a significant performance gain compared to being instructed by the robot or exercising alone. However, if the participants receive additional acknowledgement of a robot we found a performance gain compared to conditions where the robot is only instructing the user. Thus, our hypothesis H2 is supported due to the evidence that participants had a significant performance gain in the instructor conditions if the robot is giving feedback (RI vs. RIF).

However, we could not support our hypothesis H1. We did not detect the same performance gain between co-actively exercising with a robot (RC) or working out with a robot and receiving additional acknowledgement (RCF). We suppose that this is due to a ceiling effect caused by the selection of the exercises. Nevertheless, we can not fully conclude on this issue. Therefore, more investigations are needed that explore different kind of exercises and measure how difficult the participants experience the exercises in relation to their task performance.

While we could not detect any performance gains in the companion conditions, we found that in both companion conditions the users liked the system more. Furthermore, if the robot is exercising along with the subjects, the information quality of the system is rated higher. We assume that this is an important aspect for long-term HRI and could lead to longer training engagement. Therefore, we plan to run long-term investigations to gain more clarity on this aspect.

Finally we would like to discuss the limitations of this paper. First, we used only isolated abdominal plank exercises. Therefore, our future studies will also include a set of dynamic exercises (e.g. squats, push ups). Second, we have not assessed the utility of the robot against other technological devices (e.g. smart phone based applications that provide feedback). We suppose that anthropomorphic technology will lead to higher exercise adherence and performance than non-anthropomorphic devices. The degree of human-likeness of a virtual agent has already been shown to influence exercising duration [13]. However, we need to conduct more research that evaluates the utility of a robot companion by comparing it to other devices and also to human exercising partners.

In summary, our results show that a SAR should at least do one of two thing when interacting with a user: It should either exercise along or give encouraging acknowledgement if it is only instructing an exercise. This research is important for all researchers in the field of SAR who wants to build robots to motivate people to workout. The take-home message is: Robots working out co-actively with humans are more motivating and lead to higher performances, however if the robot is not able to exercise along with the user, due to some physical limitation, it is also sufficient to give the user acknowledgement in order to have the same motivational effect. This result allows us to follow our line of research where we intend to built a SAR that is instructing users on a full body weight workout. Our future work also includes to make all the questionnaires, data and programs available to allow other researchers to repeat our study as well as an in-depth analysis of further qualitative and quantitative data we have gathered during our experiments. This includes video material, heart rate data and self-efficacy beliefs. The analysis of these multi-modal data might reveal some new insights for our attempts to build robots that motivate people to work out more.

Acknowledgments. This research was funded by grants from the Cluster of Excellence Cognitive Interaction Technology 'CITEC' (EXC 277), Bielefeld University.

References

1. Fasola, J., Matarić, M.J.: Using socially assistive human-robot interaction to motivate physical exercise for older adults. Proc. IEEE **100**(8), 2512–2526 (2012)
2. Leyzberg, D., Spaulding, S., Scassellati, B.: Personalizing robot tutors to individuals' learning differences. In: Proceedings of the 2014 ACM/IEEE International Conference on Human-Robot Interaction, pp. 423–430. ACM (2014)
3. Bickmore, T., Cassell, J.: Relational agents: a model and implementation of building user trust. In: Proceedings of the SIGCHI Conference on Human Factors in Computing Systems, pp. 396–403. ACM (2001)

4. Kidd, C.D., Breazeal, C.: Robots at home: understanding long-term human-robot interaction. In: IEEE/RSJ International Conference on Intelligent Robots and Systems, IROS 2008, pp. 3230–3235. IEEE (2008)
5. Leite, I., Castellano, G., Pereira, A., Martinho, C., Paiva, A.: Empathic robots for long-term interaction. Int. J. Soc. Robot. **6**(3), 329–341 (2014)
6. Gockley, R., Bruce, A., Forlizzi, J., Michalowski, M., Mundell, A., Rosenthal, S., Sellner, B., Simmons, R., Snipes, K., Schultz, A.C., et al.: Designing robots for long-term social interaction. In: IEEE/RSJ International Conference on Intelligent Robots and Systems, pp. 1338–1343. IEEE (2005)
7. Bickmore, T.W., Picard, R.W.: Establishing, maintaining long-term human-computer relationships. ACM Trans. Comput. Hum. Interact. (TOCHI) **12**(2), 293–327 (2005)
8. Ham, J., Midden, C.J.H.: A persuasive robot to stimulate energy conservation: the influence of positive and negative social feedback and task similarity on energy-consumption behavior. Int. J. Soc. Robot. **6**(2), 163–171 (2014)
9. Swift-Spong, K., Short, E., Wade, E., Mataric, M.J.: Effects of comparative feedback from a socially assistive robot on self-efficacy in post-stroke rehabilitation. In: 2015 IEEE International Conference on Rehabilitation Robotics (ICORR), pp. 764–769. IEEE (2015)
10. Süssenbach, L., Riether, N., Schneider, S., Berger, I., Kummert, F., Lütkebohle, I., Pitsch, K.: A robot as fitness companion: towards an interactive action-based motivation model (2014)
11. Schneider, S., Riether, N., Berger, I., Kummert, F.: How socially assistive robots supporting on cognitive tasks perform. In: Proceedings of the 50th Anniversary Convention of the AISB (2014)
12. Park, E., Kim, K.J., Pobil, A.P.: The effects of a robot instructor's positive vs. negative feedbacks on attraction and acceptance towards the robot in classroom. In: Mutlu, B., Bartneck, C., Ham, J., Evers, V., Kanda, T. (eds.) ICSR 2011. LNCS (LNAI), vol. 7072, pp. 135–141. Springer, Heidelberg (2011). doi:10.1007/978-3-642-25504-5_14
13. Feltz, D.L., Forlenza, S.T., Winn, B., Kerr, N.L.: Cyber buddy is better than no buddy: a test of the köhler motivation effect in exergames. Games Health Res. Dev. Clin. Appl. **3**(2), 98–105 (2014)
14. Borg, G.: Borg's Perceived Exertion and Pain Scales. Human Kinetics, Champaign (1998)
15. Nomura, T., Kanda, T., Suzuki, T.: Experimental investigation into influence of negative attitudes toward robots on human-robot interaction. Ai Soc. **20**(2), 138–150 (2006)
16. Bartneck, C., Kulić, D., Croft, E., Zoghbi, S.: Measurement instruments for the anthropomorphism, animacy, likeability, perceived intelligence, and perceived safety of robots. Int. J. Soc. Robot. **1**(1), 71–81 (2009)
17. Nass, C., Fogg, B.J., Moon, Y.: Can computers be teammates? Int. J. Hum. Comput. Stud. **45**(6), 669–678 (1996)
18. Kendzierski, D., DeCarlo, K.J.: Physical activity enjoyment scale: two validation studies. J. Sport Exerc. Psychol. **13**(1), 50–64 (1991)

Who Am I? What Are You? Identity Construction in Encounters Between a Teleoperated Robot and People with Acquired Brain Injury

Antonia L. Krummheuer(✉)

Aalborg University, Aalborg, Denmark
antonia@hum.aau.dk

Abstract. The paper highlights how the material affordances of a teleoperated robot (Telenoid) enable identity construction in interactions with people living with acquired brain injury (ABI). The focus is set on the identity construction of the robot in relation to both its operator and the interlocutors. The analysis is based on video recordings of a workshop in which people with ABI were communicating with a teleoperated robot for the first time. A detailed multimodal conversation analysis of video-recorded interactions demonstrates how identity construction (a) is embedded in the situated and interactional unfolding of the encounter and (b) is fragmented and reflexively intertwined with the identity construction of the other parties. The paper discusses how an understanding of identity as situated and interactional constructions contributes to the field of HRI and how teleoperated robots can be used in the field of communication impairment.

Keywords: Brain injury · Conversation analysis - identity · Interaction · Qualitative research · Teleoperated robots

1 Introduction

The paper analyzes how a teleoperated robot (Telenoid) affords situational identity constructions in interactions with people living with acquired brain injury (ABI)[1]. The analysis focuses on the interplay of the identity construction of (a) the robot and its operator and (b) the interlocutors. Identity construction is approached from the background of the interpretative paradigm of sociology [1], focusing on how the human actors interactively enact, ascribe and interpret their own and each other's identity in interplay with the robot's affordances. With this interest, the paper aligns with other studies on identity construction in HRI (see Sect. 2). The focus is set on identity construction, as people with ABI can experience the feeling of not being themselves any more [2]. After the brain injury they are engaged in different processes in which they recreate or newly create their identity. Thereby they meet the challenge of the brain

---

[1] I am grateful to the participants of the workshop as well as to Sophie Mortensen, Niels Dyrskjøt, Mia Christiansen, Lisa M. Kongsgaard, Michael Frey and Lisa Larsen for collecting and transcribing the data and to Jens Strandbeck and SOSU Nord Future Lab for the loan of the Telenoid and their support during the workshop.

© Springer International Publishing AG 2016
A. Agah et al. (Eds.): ICSR 2016, LNAI 9979, pp. 880–889, 2016.
DOI: 10.1007/978-3-319-47437-3_86

injury impacting their social and interactional abilities and thus their ability to create and negotiate their identity in interaction with others (see also [3]). Therefore, the paper focuses on the situated interplay of the technology's affordances and the ongoing interaction questioning how the Telenoid enables identity construction in regard to the robot, the operator and the interlocutors.

As the paper reports findings of a qualitative and exploratory case study it does not aim for generalization. The multimodal analysis of video-recorded interactions offers a complex and reflexive understanding of the situated construction of identity and discusses how these findings can be transferred to other studies with the aim to inform HRI design and applications of teleoperated robots for interaction with people living with ABI. The data derive from a pilot case study of the first meeting of the Telenoid and a group of people with ABI. The analysis demonstrates fluent and fragmented identity constructions of robot and operator in relation to the interlocutor's identity and the ongoing interaction. The analysis points out three different identity constructions of the teleoperated robot: It can be treated as an own actor, a mediator of the operator, and a hybrid, an *in-between-actor*, in which robot and operator merge into one actor. Furthermore, the analysis shows how the Telenoid affords situated identity constructions for people with ABI in which they can enact other social categories in relation to the robot in a playful framing, such as a caring mother.

In the next section I will present the Telenoid and relevant studies on identity construction with teleoperated robots. This will be followed by a brief introduction of a conversation analytical understanding of talk-in-identity and affordances which builds the theoretical framework of the study. After the description of data and method, the analysis presents the detailed, sequential and multimodal analysis of two examples in which the identities of the Telenoid, the operator and the interlocutor are negotiated. The paper concludes with a summary of the results of the analysis and a discussion of their impact for HRI and the field of communication impairment.

2 Related Work

The analysis focuses on video recordings of an encounter in which a Telenoid (R4) (Fig. 1) was introduced to people with ABI. A Telenoid is an android telecommunication robot with a simplified human-like appearance. It was collaboratively developed by Osaka University and Advanced Telecommunication Research Institute International (ATR) in Japan. The Telenoid's body is approximately the size of a small child (ca. 50 cm tall and 3.5 kg), with a soft and smooth surface. Telenoid interactions have an asymmetric participation framework as a remote operator steers the Telenoid that is physically present in an encounter with one or several interlocutors. The operator's voice is transmitted through the Telenoid and the operator can steer the robot's head and arms (e.g. to adjust the robot's 'gaze' or make the robot 'hug' a person). A microphone and a camera in the Telenoid's forehead transmit audio and visual signals of the Telenoid-encounter to the operator. While the operator is engaged in mediated interaction, the interlocutor is engaged in a face-to-face interaction with the robot. Whether or not they know that the robot is steered by a human depends on the situation.

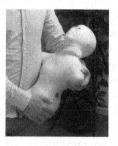

Fig. 1. Telenoid (R4)

The Telenoid is a technology that is still under development and the findings about its effects are mostly preliminary. Several authors study how the Telenoid's *minimal design approach* provides the feeling of a human presence which is called *sonzaikan* ([4]: 29). They explore its social, communicational and psychological effects in different settings and different user groups [5, 6, 10–12]. [5, 6] show for example that elderly people with dementia respond positively to the Telenoid. The elderly engaged in conversation with and about the robot as well as taking the Telenoid in their arms, hugging it or incorporated it in everyday activities e.g. reading a book, playing the piano. The Telenoid also seems to alleviate symptoms of dementia [12].

The identity construction of operator and teleoperated robots has been addressed by different authors. [13] point out that the physical appearance of teleoperated robots needs to be carefully selected, as users talking to the robot can be confused about the communication partner's identity. [9, 14] demonstrate that the operator can have the illusion of feeling external objects that are touched by the teleoperated android's body (Body Ownership Transfer). The operator thus becomes immersed in the robot's technology. Straub et al. describe identity constructions of teleoperated robots as a *socio-dynamic activity* ([7]: 123). They focus on both the teleoperator's enactment and the user's ascriptions of the Germinoid's identity. The authors differentiate four types of identity presentation: 1. presenting the robotic-self, 2. presenting the operator's own identity, 3. mixing of the robotic and operator's identity, 4. identity-imitation of the robot's origin look-alike ([7]: 121). The interlocutors, on the other side, tend to anthropomorphize the Germinoid, addressing it as a "he" and ascribing mental concepts. Another study shows how human's interpretation of a teleoperated robot vary according to the Germinoid's embodied and verbal activities of being attentive to the environment [8]. Depending on three different conditions (idle, face-tracking and interaction mode) the human actors treated the robot as an object, a reactive tool or a social being. While the last study finds a connection between the robot's states of activity and the human's ascription of its interactional status, [15] show that a robot can be treated as a social actor independent of its interactional output. They demonstrate that even an inactive robot can be treated as a social participant of interaction and how an even well-functioning robot can be neglected in interaction. Similar to Suchman's classical study on an interactive copy machine [16], they stress that the sociability of an artifact is deeply rooted in the situated interplay and interpretations of the human actors with each other and the robot.

The sociability of an artifact is thus not an inherent part of the technology but must be *oriented to and made relevant* by the human participants in interaction.

Similarly, the current paper approaches identity construction from a conversation analytical perspective, understanding identity as *identity-in-interaction* ([17]: 75). Identity is understood as an interactive achievement that is related to the immediate discourse context. Identity is not seen as a fixed property of a human or part of a big "I", but as a dialogical and local construction of the participant's identities [17]. "Having an identity" is related to the construction of "categories with associated characteristics of features" ([18]: 3). The important aspect about these categories is that they are indexical and occasioned. They are not classical sociological categories such as class, age or gender but categories that are made relevant by the participants in the ongoing interaction and are thus consequential in the interaction. People tend to (re)arrange categories into collections (*membership categorization devices*, such as mother and father that build up a family) and combine them with certain activities (*category-bound activities,* such as a crying baby) ([19, 20]).

To understand identity construction in relation to the robot's technological affordances, the paper draws on Hutchby's work ([21, 22]). He develops Gibson's socio-psychological definition of affordances towards an interactional framework. Material affordances are "functional and relational aspects which frame, while not determining, the possibilities for agentic action in relation to an object" ([21]: 444). That means, that material affordance are reflexively bound to practice: they enable and limit actions (I cannot eat soup with a fork), but they are simultaneously lodged in practice in which the artifact's usage and meaning is constructed, thus that for example even a non-functioning robot can be enacted as a social actor ([15] see also [23]).

3   Data and Method

The current paper reports the findings of a qualitative and exploratory pilot case study in collaboration with SOSU Nord Future Lab (Aalborg, Denmark) who owns a Telenoid R4. In this study a Telenoid (R4) was introduced to a new user group: people with ABI. The study aimed to understand how individuals with ABI react to a Telenoid and if and how they could imagine integrating the technology into their everyday lives. A secondary aim of the study was to find out how to design a collaborative research project that is embedded in the socially situated and practical context of a user group with special needs. The focus on people with ABI was chosen as they, similar to people living with dementia, often lack the ability to initiate an interaction [24]. The project was constructed as a participatory project [25] rooted in a collaborative learning process with the staff members and residents of a residence home for people living with ABI in Denmark. In three meetings with the management and staff of the residence home we developed a workshop in which the Telenoid was presented to four residents with ABI taking into account their individual needs. During the workshop the Telenoid was briefly presented and then passed around, so each participant could hold the robot and talk to it. The Telenoid was operated by a staff member, Dora. She was chosen, instead of an unknown person, as she could adjust the communication to the special needs of the residents with the aim to make the first encounter both more comfortable for the people with ABI and to incorporate Dora's tacit knowledge about the residents. The Telenoid was passed around. Each time the robot was passed to the next participant, the operator/Telenoid greeted the new user and started a conversation about the robot and how it could be of use for the current interlocutor. The workshop lasted approximately one hour. Due to technical problems, the Telenoid could not perform embodied movements, but talk and visual inputs were mediated. Three cameras were used to record the meeting with the Telenoid, while two other cameras focused a) on the computer screen, which mediated the Telenoid encounter to the operator and b) on the operator herself. The data were transcribed and analyzed by principles of conversation analysis [26]. The analysis focuses on the robot's affordances and situated practices in which the robot's/operator's identity is enacted in relation to the current interlocutor's identity.

4   Analysis

The following analysis will focus on two examples that demonstrate a situated and collaborative construction of the robot's identity that is (a) embedded in the ongoing activities during the workshop and (b) reflexively tied to the user's identity.

Example 1[2]

```
01 Ce: is it a bit as if you are
       sitting with a baby
02 SA: mh
03 CE: wha- (.) does it feel like
       this a bit [like this inside
04 SA:            [((one hand her
                      chest)) mh
05 CE: and sit (.) that one just
       comes (.) just comes a bit a
       mother up in one
06     (pause)
07 SA: mh ((both hands on her
       chest)) mh ((looks down to
       Telenoid and rocks it))
```

The first example is from the beginning of the workshop, when the Telenoid due to technical problems was not able to speak or move. After the Telenoid was presented to the participants of the workshop, the robot was given to Sandra, who took the Telenoid in her arms. Sandra is sitting in a wheelchair and can only move her left arm she is also aphasic and uses vocals and gestures for communication. In the transcript Celine (CE) a caregiver formulates her observation for Sandra (SA). In line 1 Celine formulates her

---

[2] Examples 1 and 2 are simplified transcriptions which are translated from Danish. (.) indicates a micro pause, [indicates overlapping speech or action, words in double parentheses indicate nonverbal actions or context information (see also [34]).

observation, looking at Sandra who is holding the Telenoid (see image in example 1) she says "it is a bit as if you are sitting with a baby", and thus categorizes the Telenoid as a baby. With the words "a bit" she seems to distance herself from the utterance, framing the categorization as an *as if*. Furthermore, her sentence offers an interpretation of how Sandra could experience the Telenoid. This gives Sandra the possibility to align with it or not. Sandra reacts with a "mh" in line 2, which could be a confirmation. However, Celine does not treat it as such, but asks Sandra if it feels like this inside and thus request a clear confirmation of her impressions. This time Sandra confirms more clearly. She agrees vocally and puts her hand on her chest a gesture she often uses to express sympathy and agreement (line 4). In line 5 Cecile produces a turn with several restarts, ending in a formulation in which she formulates that "a bit a mother comes up in one". This time she offers the category of a mother as a possible interpretation framework and Sandra agrees more strongly than before, showing her affection by putting both hands on her chest and starting to rock the Telenoid. By formulating her observations Celine gives Sandra, who has no verbal speech, the possibility to express her experience of the Telenoid. By confirming (or not confirming) Sandra can take the uttered words and make them 'her own' (see also [27]). Thus, they reach a mutual understanding of what Sandra experiences. Furthermore, they collaboratively ascribe and enact Telenoid and Sandra as mother and child, a *standardized relational pair* ([20]: 281) of the membership categorization device family. This is done by Celine's verbal utterances, but also by Sandra's enactment as she rocks the Telenoid as if it was a baby and thus enacts a *category-bound activity* of a mother rocking a baby.

In example 2 the operator/Telenoid (OP) is actively involved in the interaction. The transcript starts shortly after Poul (PO) received the Telenoid and greeted it.

Example 2

```
01. OP:  it is really good that you want to participate
         and talk with me
02. PO:  yea but ehm sometimes one cannot get rid of you
03.      ((OP and some other participants are laughing))
04. PO:  that was meant positively
05. OP:  yes (.) that is fine Poul
06. PO:  that is fine Dora
07. OP:  jaerh but now I am the Telenoid today
08. PO:  okay ((pause)) then we say so
09.      ((some are laughing))
10. PO:  ((claps the Telenoid's bottom several times))
11. PO:   you still have a firm tiny butt
12       ((people laughing, OP joins in))
```

Different to the first example, the practice the participants are engaged in with the Telenoid here is thus identified as a conversation, and Telenoid/operator and Poul are conversation partners (not a mother rocking a baby). In line 1 the operator/Telenoid acknowledges Poul's participation in the workshop and his willingness to "talk with me". The identity of the speaker is not clear as the "me" could refer to either the Telenoid or the operator. Poul takes up the operators turn by teasing the speaker "sometimes one cannot get rid of you". The robot's identity is still vague as Poul uses "you". However, in a teasing sequence the teaser criticizes a person's action by combining it with some known characteristics of the person [28, 29]. Similarly, Poul's formulation includes a critique of some knowledge he has about the teasing object ("sometimes one cannot get rid of you"), which indicates that Poul might address Dora rather than the Telenoid. In line 4 Poul clarifies the ambiguous framing of his former utterance, stating that this was meant positively, which is accepted in line 5. In line 6 Poul addresses Dora and thus clearly shows that he is talking to the operator using the Telenoid as mediation technology for human-human interaction.

However, in line 07 the operator insists on being the Telenoid "but now I am the Telenoid today". The identity construction of the Telenoid is thus made relevant by the participants and becomes a topic of conversation. Thereby Dora constructs her identity as a temporal-identity as she says she is the Telenoid "now" and "today", which frames the identity construction as a *play along* [30]. Poul accepts Dora's request in line 8 adding the words "then we say so" by which he emphasizes the *as-if framing* of Dora's temporal-identity construction. At this point the Telenoid is constructed as a *hybrid conversation partner* in which both parts, the operator and the Telenoid technology, are made visible as being part of one speaker. Poul demonstrated himself to be a cooperative conversation partner willing to accept the speaker's unconventional identity. In line 10–12 Poul uses the new framework to ascribe a new membership categorization to the Telenoid. In line 10 he claps the Telenoid's bottom slightly several times, and says "you still have a firm and tiny butt." Poul thus plays along with the idea of the hybrid Telenoid-Dora-Identity taking the Telenoid's body to be Dora's body, too. With the word "still" Poul refers to a known feature of Dora that he can also find in the Telenoid's body. Simultaneously he challenges Dora's new identity playfully. Clapping the caregiver's bottom may be treated as sexual harassment or face-threatening in another situation, but in the current context the activity is interpreted as a playful teasing (which can also be seen by the following laughter).

5   Conclusion and Discussion

The paper demonstrated how a teleoperated robot (Telenoid) affords situational constructions of identity in interactions with people living with ABI. The analysis focused on the interplay of the identity construction of a) the robot and operator and b) the interlocutors. Similar to Straub et al. it could be seen that the robot's identity could be variably identified as either an independent actor, a mediator of the operator or a hybrid in which operator and robot were visible parts. But in difference to Straub et al. [7, 8] this paper emphasized the situated and fluid character of identity constructions. In the first example the Telenoid was constructed in relation to Sandra using mother and baby as categorization type. As the Telenoid was not active speaking or moving, it could be argued that the robot was used as an object rather than an independent actor. But in the interactional framing of the sequence, the participants identify the robot as baby and not for example a doll. The construction of the robot as an object is much clearer in sequences in which the robot is identified as an object, e.g. when participants refer to the Telenoid as "it" or when the robot is taken out of its trunk and prepared for interaction. The second set of examples demonstrated a negotiation of the Telenoid's identity and thus emphasized the fluent and ambiguous nature of identity constructions. This might of course be due to the explorative setting and the unknown technology, the Telenoid's identity is most likely to stabilize when interactional routines develop. But still the identity constructions are fragmented and fluid. Technologies do not just afford something but their usage becomes relevant in the complex relations and social setting in which they unfold their social and interactive power and in this setting the participants can use the technologies to deploy new categories or identities (see also [31]). HRI design must therefore consider how the situated and fluid character of ongoing constructions of meaning can be taken into account. In regard to the identity construction for people with ABI, the Telenoid provided a device to play along with different identities. Sandra, who is used to being taken care of, could become a person who cares for a baby, and Poul could engage in a sexually connoted interaction without facing the consequences his actions would have provoked within another context. The study only yields preliminary insights, as this was a first meeting and it is unclear if and how the Telenoid could be used for longer periods of time and what the consequences would be, as it is for example done in the field of virtual agents [32]. While Poul is aware of the mediated identity construction of robot and operator, not all individuals with special needs might be aware of who they are talking to during a conversation with a Telenoid (see also [13]). Therefore, the application of the Telenoid also poses a special ethical dilemma as a person might reveal secrets to an operator that were actually addressed to the robot (see also [33]).

References

1. Wilson, T.P.: Conceptions of interaction and forms of sociological explanation. Am. Sociol. Rev. **35**, 697–710 (1970)
2. Brumfitt, S.: Losing your sense of self: What aphasia can do. Aphasiology **7**, 569–575 (1993)
3. Goodwin, C. (ed.): Conversation and Brain Damage. Oxford University Press, New York (2003)
4. Ishiguro, H.: Transmitting human presence through portable teleoperated androids: a minimal design approach. In: Nishida, T. (ed.) Human-Harmonized Information Technology. vol. 1, pp. 29–56. Springer, Japan (2016)
5. Yamazaki, R., Nishio, S., Ishiguro, H., Nørskov, M., Ishiguro, N., Balistrer, G.: Acceptability of a teleoperated android by senior citizens in Danish society. Int. J. Soc. Robot. **6**, 429–442 (2014)
6. Yamazaki, R., Nishio, S., Ogawa, K., Ishiguro, H.: Teleoperated android as an embodied conversation medium: a case study with demented elderlies in a care facility. In: 2012 IEEE RO-MAN, pp. 1066–1071 (2012)
7. Straub, I., Nishio, S., Ishiguro, H.: Incorporated identity in interaction with a teleoperated android robot: A case study. In: 2010 IEEE RO-MAN, pp. 119–124 (2010)
8. Straub, I., Nishio, S., Ishiguro, H.: From an object to a subject - Transitions of an android robot into a social being. In: 2012 IEEE RO-MAN, pp. 821–826 (2012)

9. Nishio, S., Watanabe, T., Ogawa, K., Ishiguro, H.: Body ownership transfer to teleoperated android robot. In: Ge, S.S., Khatib, O., Cabibihan, J.-J., Simmons, R., Williams, M.-A. (eds.) ICSR 2012. LNCS (LNAI), vol. 7621, pp. 398–407. Springer, Heidelberg (2012). doi: 10.1007/978-3-642-34103-8_40

10. Sumioka, H., Nishio, S., Minato, T., Yamazaki, R., Ishiguro, H.: Minimal human design approach for sonzai-kan media: investigation of a feeling of human presence. Cognit. Comput. 6, 760–774 (2014)

11. Yamazaki, R., Nishio, S., Ogawa, K., Ishiguro, H., Matsumura, K., Koda, K., Fujinami, T.: How does telenoid affect the communication between children in classroom setting? In: Proceedings of the 2012 ACM annual Conference, Extended Abstracts - CHI EA 2012, p. 351. ACM Press, New York (2012)

12. Strandbech, J.D.: Ethel and her Telenoid. Towards using humanoids to alleviate symptoms of dementia. Læring & Medier (LOM) 8, 1–22 (2015)

13. Kuwamura, K., Minato, T., Nishio, S., Ishiguro, H.: Personality distortion in communication through teleoperated robots. In: 2012 IEEE RO-MAN, pp. 49–54 (2012)

14. Nishio, S., Taura, K., Sumioka, H., Ishiguro, H.: Effect of social interaction on body ownership transfer to teleoperated android. In: 2013 IEEE RO-MAN, pp. 565–570 (2013)

15. Alač, M., Movellan, J., Tanaka, F.: When a robot is social: Spatial arrangements and multimodal semiotic engagement in the practice of social robotics. Soc. Stud. Sci. 41, 893–926 (2011)

16. Suchman, L.: Human and Machine Reconfigurations: Plans and Situated Actions. Cambridge University Press, Cambridge (2007)

17. Aronsson, K.: Identity-in-interaction and social choreography. Res. Lang. Soc. Interact. 31, 75–89 (1998)

18. Antaki, C., Widdicombe, S.: Identity as an achievement and as a tool. In: Antaki, C., Widdicombe, S. (eds.) Identities in talk, pp. 1–14. Sage, Thousand Oaks (1998)

19. Sacks, H.: The MIR membership categorization device. In: Jefferson, G. (ed.) Lectures on Conversation, pp. 40–48. Blackwell, Oxford (1992)

20. Stokoe, E.: Moving forward with membership categorization analysis: Methods for systematic analysis. Discourse Stud. 14, 277–303 (2012)

21. Hutchby, I.: Technologies. Texts and Affordances. Sociology. 35, 441–456 (2001)

22. Hutchby, I.: Conversation and Technology. From the Telephone to the Internet. Polity Press, Cambridge (2001)

23. Krummheuer, A.: Technical Agency in Practice: The enactment of artefacts as conversation partners, actants and opponents. PsychNology 13, 179–202 (2015)

24. Vos, P.: Traumatic Brain Injury. Wiley-Blackwell, Oxford (2014)

25. Robertson, T., Simonsen, J.: Participatory design. An introduction. In: Simonsen, J., Robertson, T. (eds.) Routledge international handbook of participatory design. pp. 1–17. Routledge (2013)

26. Sidnell, J., Stivers, T. (eds.): The Handbook of Conversation Analysis. Wiley, Oxford (2012)

27. Goodwin, C.: A competent speaker who can't speak: The social life of aphasia. J. Linguist. Anthropol. 14, 151–170 (2004)

28. Drew, P.: Po-faced receipts of teases. Linguistics 25, 219–253 (1987)

29. Krummheuer, A.: Conversation analysis, video recordings, and human-computer interchanges. In: Kissmann, U. (ed.) Video Interaction Analysis. Methods and Methodology, pp. 59–83. Peter Lang, Frankfurt am Main (2009)

30. Goffman, E.: Frame analysis: An essay on the organization of experience. Harvard University Press, Cambridge (1974)

31. Pols, J., Moser, I.: Cold technologies versus warm care? On affective and social relations with and through care technologies. Eur. J. Disabil. Res. **3**, 159–178 (2009)
32. Konnerup, U., Castro Rojas, M.D., Bygholm, A.: Rehabilitation of people with a brain injury through the lens of networked learning: identity formation in distributed virtual environments. In: Cranmer, S., de Laat, M., Ryberg, T., Sime, J. (eds.) Proceedings of the 10th International Conference on Networked Learning (2016)
33. Feil-Seifer, D., Matarić, M.: Ethical principles for socially assistive robotics. IEEE Robot. Autom. Mag. **18**, 24–31 (2011)
34. Hepburn, A., Bolden, B.B.: The conversation analytic approach to transcription. In: Sidnell, J., Stivers, T. (eds.) Handbook of Conversation Analysis, pp. 57–76. Blackwell, Oxford (2013)

Contribution Towards Evaluating the Practicability of Socially Assistive Robots – by Example of a Mobile Walking Coach Robot

Horst-Michael Gross(✉), Markus Eisenbach, Andrea Scheidig,
Thanh Quang Trinh, and Tim Wengefeld

Neuroinformatics and Cognitive Robotics Lab, Ilmenau University of Technology,
98693 Ilmenau, Germany
horst-michael.gross@tu-ilmenau.de

Abstract. This paper wants to make a further contribution towards a more transparent and systematic technical evaluation of implemented services and underlying HRI and navigation functionalities of socially assistive robots for public or domestic applications. Based on a set of selected issues, our mobile walking coach robot developed in the recently finished research project ROREAS (Robotic Rehabilitation Assistant for Walking and Orientation Training of Stroke Patients) was evaluated in three-stage function and user tests, in order to demonstrate the strengths and weaknesses of the developed assistive solution regarding the achieved autonomy and practicability for clinical use from technical point of view.

Keywords: Socially assistive robotics · Evaluation of practicability

1 Motivation

In the assistive robotics community, more and more researchers are already aware of the challenges involved in studying autonomously behaving interactive systems "in the wild" and follow best practices in studying these robots in natural interaction settings as suggested by [1,2]. However, often the setup and the actual implementation of the tests are still not described in a sufficient level of detail, leaving room for speculations, particularly with respect to the achieved autonomy and the practicability of the developed solution from technical point of view. Therefore, the main objective of this paper is to make a further contribution towards a systematic and more transparent technical evaluation of implemented services and underlying basic functionalities of socially assistive robots [3] for public or domestic applications. This way, a more self-critical and honest survey of the strengths and still existing weaknesses of the robot's practicability should be made possible. In the scope of our previous projects, the *ShopBot* TOOMAS [4], the companion robot in CompanionAble [5], the health assistant in SERROGA [6], and the walking coach in ROREAS [7,8], we gained many experiences in making assistive robotics suitable for real-world applications. Based on these experiences, we have compiled a set of questions that

© Springer International Publishing AG 2016
A. Agah et al. (Eds.): ICSR 2016, LNAI 9979, pp. 890–899, 2016.
DOI: 10.1007/978-3-319-47437-3_87

should be clarified in publications dealing with the autonomy and practicability of a developed robotic solution. These issues are divided into the following topics:

Topic S – Spectrum of available services/applications and skills:

S1: What *services/applications* for the users are already available (a) working completely autonomously on the robot (b) only usable with external sensors (e.g. cameras), (c) requiring remote control by a tele-present operator?

S2: What *skills* for navigation and HRI are available for the robot at which level of autonomy?

S3: What kind of IT-infrastructure is required for that on-site? What are the consequences for the later practical application?

Topic M – Maturity level:

M1: What is the *maturity level* of the robot system to be tested? Is it still a demonstrator, a lab prototype, or already a product available on the market?

M2: Have there already been *function tests* outside the lab in the field (when, where, how often, how long, what conditions)?

M3: Have there already been *user tests* with the end users in the final operational environment (when, where, how long, what conditions)?

M4: Was *accompanying personnel* from technical or social sciences staff present during the user tests, and where was the staff while the tests were running?

M5: How long was the robot available for the user, how was the usage rate?

Topic F – Function tests of basic functionalities: Here, the scenario-specific functionalities in navigation and HRI are to be quantitatively evaluated.

F1: What *navigation functionalities* have been tested under what conditions?

F2: Are there *navigation problems* encountered during the tests, and how were these quantified (e.g. number of collisions, deadlocks, close encounters with obstacles, violations of personal space, localization failures, etc.)?

F3: What *HRI functionalities* have been tested under what conditions?

F4: Have there been *HRI-malfunctions* (e.g. in person detection, person tracking, user re-identification, etc.)?

F5: What *success rates* of basic functionalities have been determined (e.g. localization accuracy, target achievements, user detection/search, etc.)?

F6: Were *manual interventions* necessary before and during the tests (e.g. labeling no-go areas, preparing critical obstacles, triggering emergency stops, changes in the application procedure while testing)?

F7: Was the *complexity* of the test environment quantitatively evaluated (e.g. by total floor area, free space, navigable area, clearance, shape factor, mean passage width) to allow for a comparison of the test results? (see [6])

Topic U – User tests at technical application level: Here the level of autonomy, the practicability of the application, and the interplay of the basic functionalities are to be evaluated by the following error measures:

U1: *Uncritical failures*: can be handled by the application itself (e.g. driving to a meeting point if user contact is lost, autonomously terminating a bumper-stop if knowledge about the triggering event is available)

U2: *Critical failures*: can be resolved by remote intervention (see Sect. 3) through an operator (e.g. correction of a wrong person re-identification hypothesis)

U3: *Very critical failures*: cannot be resolved by remote intervention through an operator (e.g. sensor failures or deadlocks after collisions).

In the following sections, when describing our test strategy and the achieved results and observed problems, we make use of links to these issues, e.g. ($\nearrow$ F3).

2   Mobile Walking Coach Robot

Based on this set of issues, the recently finished research project ROREAS (Robotic Rehabilitation Assistant for Walking and Orientation Training of Stroke patients) [7,8] was systematically evaluated. The ROREAS project aimed at developing a robotic rehabilitation assistant for walking self-training of stroke patients in late stages of the clinical post-stroke rehabilitation. Such a walking coach robot is supposed to motivate and accompany stroke patients who already got the permission to walk on their own without professional assistance during their walking exercises in a clinical rehab center (Fig. 1). A specific characteristic of ROREAS is its strongly human-aware, polite and attentive social navigation and interaction behavior [9] as it is necessary for a rehab assistant that can motivate patients to start, continue, and regularly repeat their self-training with joy. In [8], we already described the specifics and challenges of the clinical setting and the technical requirements for the robotic walking coach, presented the ROREAS prototype (Fig. 1), an application-tailored mobile robot developed within the ROREAS project to meet the requirements to a personal training robot ($\nearrow$ M1), and gave an overview of the robot's system architecture.

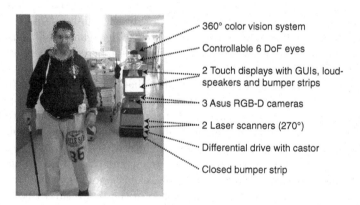

360° color vision system

Controllable 6 DoF eyes

2 Touch displays with GUIs, loud-speakers and bumper strips

3 Asus RGB-D cameras

2 Laser scanners (270°)

Differential drive with castor

Closed bumper strip

Fig. 1. Robotic walking coach "RINGO" ($\nearrow$ M1) with its main equipment for environmental perception, navigation, and HRI during a walking tour in our test site, the "m&i Fachklinik" rehabilitation center in Bad Liebenstein (Germany).

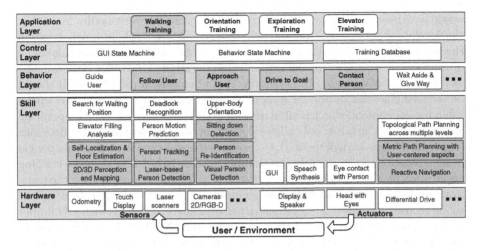

Fig. 2. System architecture of the ROREAS training assistant. Only the reddishly highlighted skills and behaviors are of relevance for this paper, as they are the essential components of the "Walking Training" application, the practicability of which has been evaluated here from technical point of view ($\nearrow$ S1, S2).

For implementing the training application, numerous robustly functioning basic skills and behaviors were required that had to function completely autonomously in order to achieve the necessary practicability. Therefore, during the function and user tests we particularly paid attention to the skills and behaviors highlighted reddishly in Fig. 2, which have been optimized and evaluated over and over again in order to achieve greater autonomy. In continuation of [8], this paper is focussing on the question how the developed robotic walking coach, its implemented services and the underlying basic functionalities for HRI and navigation can be evaluated systematically to assess its final practicability for the clinical use from technical point of view.

3   Three-Stage Approach in Conducting User Tests

Before it was possible to evaluate the walking coach together with stroke patients in user trials, it had to be assured that all the required skills and behaviors for HRI and human-aware navigation (see Fig. 2) did work as expected in the clinical setting. Therefore, we applied a *three-stage approach* in conducting function and user tests with the developed prototype ($\nearrow$ M1) in the rehab center under everyday conditions. For correction of lacking or wrong decisions of selected skills (e.g. person re-identification [10]) and, with that, for the sake of an interruption-free testing process, we developed a tablet-computer based correction interface connected with the robot by WiFi, which allowed an external test observer to manually correct these decisions from a non-distracting distance (>5 m) ($\nearrow$ M4). By this option for *remote intervention*, the user tests in the clinic could be started earlier than this would have been possible from the readiness level of the respective skills. Moreover, the developers got an objective and situation-specific feedback, in which situation the basic skills and behaviors were still facing problems. Furthermore, this way a direct measure of quality for the autonomous operation of the robot was available, as the number of necessary interventions could be counted ($\nearrow$ F5, F6). It should be stressed that this option for remote intervention is not to be confused with a robot remote control (which is often used for user studies), as our robot is operating autonomously. The tablet only allows the distant observer to add lacking decisions (e.g. from sitting-down detection), to correct erroneous decisions (e.g. from person re-identification), or to modify the training process in order to keep the training application flowing. An emergency stop can be triggered as well. During the tests, from far distance a second staff member observed problems not detectable by the robot's sensor systems (e.g. violations of personal space, collisions without bumper contact, etc.) and documented them quantitatively. Neither IT-infrastructure of the clinic nor external sensors or markers were required for these tests ($\nearrow$ S3).

Stage 1 – Functional on-site tests with staff members: To ensure, that all skills and behaviors (see Fig. 2) required by the walking coach do work accurately and securely, first we performed functional on-site field tests with staff members of our robotics lab. These tests were conducted in February 2015 over the course of 4 days and a driven distance of 15,000 m within several floors of the clinic at different times throughout the day ($\nearrow$ M2). This was done to assess the robot's basic behaviors under varying conditions, such as challenging building-structures, changes in illumination, and a variable amount of people within the corridors. For quantitative assessment of the skills and behaviors, measures, as e.g. the number of collisions or person mismatches, or the needed travel time, were determined. Regarding the *navigation performance*, the distant test observer counted the number of (i) close (<10–15 cm) passings of obstacles, (ii) close passings of persons, and (iii) manually triggered emergency stops. Regarding *person recognition during guiding and following*, it was determined, whether and how often the robot confused the current user with a different close-by standing person. A detailed quantitative analysis of the tested skills ($\nearrow$ F1–F4) and the determined success rates ($\nearrow$ F5) have already been presented in [8] and are not to be repeated here. In these tests, manual corrections via remote intervention were made only when the re-identification failed or emergency stops were necessary to prevent possible collisions ($\nearrow$ F6).

Stage 2 – User tests with "patient doubles": After successfully completing the functional tests with staff members of our robotics lab, in May 2015 and shortly before each user test (see below) we evaluated the walking coach again – but this time with the help of clinical staff who imitated the walking behavior of stroke patients ($\nearrow$ M2, M3). Figure 3 illustrates the phases of a typical walking training session as it was executed in all following user tests [8]. In these tests, among the stability of the required HRI- and navigation skills ($\nearrow$ F1, F3) the actual training application and the conclusiveness and comprehensibility of the training procedure and the necessity of manual remote interventions by the observer ($\nearrow$ F6) were tested. Thus, the level of autonomy and the practicability of the application were in the focus of these trials. A detailed quantitative determination of the success rates of the skills and behaviors ($\nearrow$ F5) and the observed failures at application level ($\nearrow$ U1–U3) was not yet part of these studies, but was left for reasons of manageability for the following user tests.

Stage 3 – Technical user tests with patients: Based on the emulated user tests with staff members and patient doubles, in the period from June 2015 till March 2016 five campaigns of user tests with $N = 26$ stroke patients in total were conducted ($\nearrow$ M3). In all campaigns, only volunteers from the group of stroke patients who already got the permission for doing self-training by their doctor in charge were involved. While the first user trials in June and September 2015 only comprised one predefined short training route, for the following trials in November 2015, January and March 2016 the walking coach was improved to provide freely selectable training routes and, depending on the patients' state of health, sessions with a duration of up to one hour ($\nearrow$ M5). During all user trials, the test observer had the opportunity to correct wrong or lacking decisions in sitting-down detection and user re-identification by remote intervention via the tablet. Our aim was to reduce the number of interventions and to improve the level of autonomy from user tests to user test. Therefore in the subsequent analysis of the user tests, the focus has been directed to the still missing autonomy, in other words, the number of required remote interventions.

4 Results of User Tests with Patients in Stage 3

In the *first user test with patients in June 2015*, all phases of a typical walking training session had to be completed (see Fig. 3), however, it was performed during low traffic times to minimize disturbances by uninvolved passer-by. Essential aspects

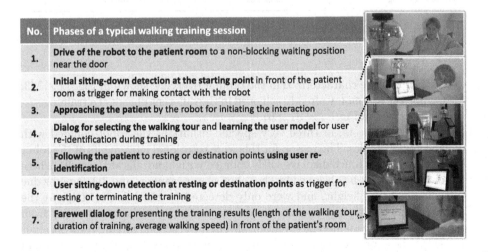

No.	Phases of a typical walking training session	
1.	**Drive of the robot to the patient room** to a non-blocking waiting position near the door	
2.	**Initial sitting-down detection at the starting point** in front of the patient room as trigger for making contact with the robot	
3.	**Approaching the patient** by the robot for initiating the interaction	
4.	**Dialog for selecting the walking tour** and **learning the user model** for user re-identification during training	
5.	**Following the patient** to resting or destination points **using user re-identification**	
6.	**User sitting-down detection at resting or destination points** as trigger for resting or terminating the training	
7.	**Farewell dialog** for presenting the training results (length of the walking tour, duration of training, average walking speed) in front of the patient's room	

Fig. 3. Phases of a typical walking training session

Criteria	1st user test	5th user test
Period	June 2015 – 2 days	March 2016 – 2 days
Number of patients	5	7
Used walking aids	only walkers	5 x crutch, 2 x walker
Number of sessions	11	14 (7 on 1st day, 7 on 2nd day)
Driven distance Per session	873 m 80 m	6 650 m 475 m
Total training time Training time per patient	62 min 12 min 24 s	6 hours 15 min 53 min
Number of passers-by	78 (7 per session, 9 per 100 m)	679 → 48 per session / 10 per 100 m
Number of remote interventions (for Re-ID)	19 → 2.2/100 m or 3/10 min.	Offensive: 43 → 1.3/100 m; 2/10 min. Cautious: 19 → 0.6/100 m; 1/10 min.

Fig. 4. Results of the first and last (fifth) technical user tests

Phases	Achieved Autonomy in 1st user test	Achieved Autonomy in 5th user test
1. Drive to the patient room	100% autonomous	100% autonomous
2. Initial sitting-down detection at starting point	100% by remote interventions	85% autonomous 15% by remote interventions
3. Approaching the patient	100% autonomous	100% autonomous, 47% successful
4. Dialog for tour selection	only 1 training route	free selection from 3 routes
5. Following the patient using user re-identification (ReID)	largely autonomous; ReID: 3 per 10 min. by remote intervention	largely autonomous; ReID: 1 per 10 min. by remote intervention
6. User sitting-down detection at resting /destination points	100% by remote interventions	77% autonomous 23% by remote interventions
7. Farewell dialog	100% autonomous	100% autonomous

Fig. 5. Achieved autonomy in all phases of a training session during the tests

of this user test are characterized in the tabular overview shown in Fig. 4 (↗ M3, M5), while the autonomy already achieved in the different phases of a training session is described in Fig. 5 (↗ S1, F6).

During these tests, in three very specific situations the robot collided with unexpected obstacles. The reason for this were rotational movements of the robot by which one of its touch displays softly collided with handrails at the walls. These handrails could not be observed by the robot's laser scanners and 3D-cameras due to their mounting height, and were only detected by the bumper strips at the displays, which triggered an immediate stop. So the robot had to be freed from these situations by manual intervention through the accompanying tests observer (↗ U3). These problems did not occur during the preceding function tests and user tests with patient doubles, and came as a surprise therefore. However, the patients did not notice this and continued their training. To handle this problem for the following tests, the respective handrails had to be marked in the navigation map as no-go areas ($\nearrow$ F6). Only twice, the robot violated the personal space of a person (<15 cm). In both cases, this behavior was hard to avoid due to the traffic on the corridors (persons suddenly stepping out from rooms or closely passing by). 19 times the test observer had to react by remote invention (3 times in 10 min on average) ($\nearrow$ F6) to confirm uncertain or to correct wrong hypotheses of the re-identification module ($\nearrow$ U2). Only two of these cases were false decisions, the others were too uncertain and only required confirmation. In this test, it became apparent that the clothing-based re-identification of patients [10] has a higher degree of difficulty compared to the test with staff members and patient doubles. The cause study for the observed failures showed that the field of view of the panoramic head camera used until then was too limited, and the approaching to sitting patients was still sub-optimal, as the distances for a comfortable handling of the touch screen were too large. It became apparent, that the training scenario had to be expanded in terms of longer training routes and more options for the patients. Moreover, further remote correction options for the control of the training process should be added to better correct occurring failures immediately during the tests to keep the training flowing. Following this test strategy, in the period between September 2015 and January 2016, three more user tests with $N = 14$ volunteer patients were conducted to assess the improvement of the navigation and HRI skills, that unfortunately cannot be reported here due to lack of space.

In the *last user test in March 2016*, the following issues were in the focus of the practicability investigation: stability of the distance to the accompanied patient, quality of user sitting-down detection and re-identification after installation of a new high resolution panoramic color camera with large vertical field of view (consisting of 6 HD cams), and approaching sitting persons. Of special significance was the question, how the overall behavior of the walking coach and its practicability will be changed, when the remote interventions by the test observer will be more and more restricted to really critical exceptional cases only (very cautious use) in comparison to an offensive use ($\nearrow$ F6). Essential aspects of this user test are also characterized in the tabular overview in Fig. 4 ($\nearrow$ M3, M5), while the achieved autonomy is described in Fig. 5, right. ($\nearrow$ S1, F6).

In the case of an offensive use of interventions (on the first day of the trials) in 43 situations (2 per 10 min, or 1.3 per 100 m) remote interventions were carried out ($\nearrow$ U2) for user re-identification. Reasons for that were missing detections, temporary occlusions of the patients after turning round the corner, or non-detections after taking a seat. In 5 of these 43 cases, there was a confusion with a person closely passing by, and in 2 cases there was no evident reason visible for the hasty intervention of the test observer. By comparison, when a cautious use of the interventions was applied (on the second day), only in 19 situations interventions were necessary ($\nearrow$ U2), that is 0.6 per 100 m, or once per 10 min. In 16 cases, the patients could not be re-identified due to missing detections of the person detector or temporary occlusions by other persons or obstacles, and in 3 cases there were mismatches with passers-by. Collisions with obstacles did no longer occur in this test. 30 times

(0.4 per 100 m) the robot violated the personal space of other persons (<15 cm), however, without touching them. Again, this was and will be hard to avoid due to the traffic on the corridors, missing person detections (mostly for patients using wheel chairs), and persons suddenly stepping out from rooms or closely passing by (↗ F4).

A robust sitting-down detection is required for the start of the training and for unplanned breaks during the training. From the 14 initial situations (7 per day), 12 were detected autonomously (85 %), 2 were not detected due to missing person detections and had to be triggered by remote intervention (↗ U2). The sitting-down detection during the breaks at resting points was necessary in 97 cases, 75 cases were detected autonomously (77 %), 22 (23 %) still had to be triggered by remote intervention. Reasons for that were temporarily absent person detections because of very atypical views during sitting-down. 111 times the robot had to approach the users while they were sitting. Only in 52 of the final approach positions (47 %), the patients could conveniently operate the touch-display. In 59 cases (53 %) the approaching had to be terminated by the robot, because the used walking aids blocked the way to the patient, or the robot had problems to robustly detect and track some of the sitting patients.

In all sessions of this test, the average distance between the walking coach and the patient was 1.4 m for the two slowly walking patients and 2.5 m for the five faster walking patients (min. 1 m, max. 4 m) (↗ F5).

5   Conclusions and Outlook

Following our three-stage approach in conducting the function tests and technical user tests in the clinical environment, we have reached a status, in which all components are integrated and most skills and behaviors do function autonomously without any corrections by remote interventions. Nevertheless, there are three important skills and behaviors, the visual person re-identification, the sitting-down detection, and the approach a sitting user behavior, that need to be further advanced by algorithmic improvements to guarantee an autonomous walking training without external support and interventions. So, we have to conclude that the walking coach has not yet reached a maturity level which would allow autonomous operation with patients in the clinical setting (↗ M1). Therefore, instead of only focusing on improving the correctness of all skills, in a subsequent project we follow a more promising strategy to better handle missing or wrong detections, unexpected situations, or still latent shortcomings in the training procedure. So, we are currently implementing a recovery strategy for the most critical case of contact loss to the patient as a result of a failed re-identification. In such a situation, the patient is asked to wait at the next resting point along his tour to give the robot the chance to search for him in this area or on the way to this goal (↗ U1).

Despite these difficulties, for the social science studies that were running in parallel in stage 3 with the $N = 26$ volunteer stroke patients, by the option of doing remote interventions in case of lacking or wrong decisions we could provide a full-value training assistant allowing to evaluate the *usability* and *friendly usage* of the training application by the patients. In this way, without any restrictions we were able to see how well the robot's behaviors and offered training service did fit into the self-training concept. The results of these studies show that the patients and fellow patients were very open-minded and accepted the developed robotic trainer. The robot motivated them for independent training and leaving the room, despite difficulties of orientation, provided a very self-determined training regime, and encouraged them to expand the radius of their training in the clinic. So, a statement frequently repeated by many patients after training with the robot was: "*I have never gone this far alone.*" [11] This makes us optimistic that such a robot coach could bridge the gaps between therapeutically assisted training, independent self-training in the clinic, and training at home.

References

1. Steinfeld, A., Fong, T., Kaber, D., Lewis, M., Scholtz, J., Schultz, A., Goodrich, M.: Common metrics for human-robot interaction. In: Proceedings of HRI, p. 3340 (2006)
2. Bonsignorio, F., del Pobil, A.P.: Toward replicable and measurable robotics research. IEEE Robot. Autom. Mag. **3**, 32–35 (2015)
3. Feil-Seifer, D., Mataric, M.: Socially assistive robotics. IEEE Robot. Autom. Mag. **1**, 24–31 (2011)
4. Gross, H.-M., Boehme, H.-J., et al.: TOOMAS: interactive shopping guide robots in everyday use - final implementation and experiences from long-term field trials. In: Proceedings of IROS, pp. 2005–2012 (2009)
5. Gross, H.-M., Schroeter, C., et al.: Further progress towards a home robot companion for people with mild cognitive impairment. In: Proceedings of SMC, pp. 637–644 (2012)
6. Gross, H.-M., Mueller, S., et al.: Robot companion for domestic health assistance: implementation, test and case study under everyday conditions in private apartments. In: Proceedings of IROS, pp. 5992–5999 (2015)
7. Gross, H.-M., Debes, K., et al.: Mobile robotic rehabilitation assistant for walking and orientation training of stroke patients: a report on work in progress. In: Proceedings of IEEE-SMC, pp. 1880–1887 (2014)
8. Gross, H.-M., Scheidig, A., et al.: ROREAS: Robot coach for walking and orientation training in clinical post-stroke rehabilitation - prototype implementation and evaluation in field trials. Autonomous Robots (2016). doi:10.1007/s10514-016-9552-6
9. Trinh, T.Q., Schroeter, C., Kessler, J., Gross, H.-M.: "Go ahead, please": recognition and resolution of conflict situations in narrow passages for polite mobile robot navigation. In: Tapus, A., André, E., Martin, J.-C., Ferland, F., Ammi, M. (eds.) Social Robotics. LNCS, vol. 9388, pp. 643–653. Springer, Switzerland (2015)
10. Eisenbach, M., Vorndran, A., et al.: User recognition for guiding and following people with a mobile robot in a clinical environment. In: Proceedings of IROS, pp. 3600–3607 (2015)
11. Meyer, S., Fricke, C.: Robot companions for stroke therapy - studying the acceptance of assistive robotics among 80 patients in neurological rehabilitation. In. Proceedings of Kongress "Zukunft Lebensräume" (in German), pp. 16–24 (2016)

Philosophy of Social Robotics: Abundance Economics

Melanie Swan[✉]

Philosophy and Economic Theory, New School for Social Research, New York, NY, USA
m@melanieswan.com
http://www.BlockchainStudies.org

Abstract. The aim of this paper is to present conceptual resources that address social robotics from a philosophical, social, and economic perspective. Since social robotics is an emerging and potentially high-impact area, it is necessary to consider the ethics and philosophy of social robotics and its potential impact on society. Philosophical, economic, and ethical issues are addressed first generally, revealing that social robotics is most-centrally a situation of human-machine collaboration. Second, economic issues are examined more specifically, positing that social robotics might figure prominently in both an automation economy that focuses on reduced requirements for human labor and an abundance economy that targets improved human quality of life. The stakes of social robotics are high and could mean both quantitative and qualitative benefits, and take advantage of the close connection with humans to help negotiate and buffer interactions between humans and a world with an increasing and expanding presence of technology.

Keywords: Robotics · Social robotics · Robo-ethics · Economic theory · Abundance · Human-machine interaction · Automation economy · Multispecies society · Economics · Quality of life

1 Introduction: Philosophy, Economic Theory, and Ethics

The aim of this paper is to understand and contextualize social robotics from a philosophical, social, and economic perspective. Since social robotics is an emerging and potentially high-impact area, it is timely to consider the potential impact of the ethics and philosophy of social robotics on society. The result of this analysis is obtaining a conceptualization of social robotics that is grounded in philosophy and economic theory, where concepts that are well-formed in the philosophical sense might be helpful in designing, building, and implementing social robotics solutions. Philosophical underpinnings can provide a more robust understanding and motivation for *what* we are trying to do and *why*, beyond the proximate motivation of labor-saving. One of my objectives is to more robustly inform our thinking about social robotics and what might be at stake. I would like to address the important qualitative question regarding the issue that beyond labor-saving, what specifically it is that we want social robotics to do to improve human quality of life. What are ways for us to articulate what *more* we want social robotics to do beyond the quantitative.

Social robots are understood to be autonomous robots that interact and communicate with humans or other autonomous physical agents, especially by following social

A. Agah et al. (Eds.): ICSR 2016, LNAI 9979, pp. 900–908, 2016.
DOI: 10.1007/978-3-319-47437-3_88

behaviors and rules [1]. Social robotics, then, is the field concerned with the design, development, and implementation of social robots. Social robotics could be pivotal in configuring the modes of life of the future. Two of the areas with the most significant potential impact could be economics and society. In an economic sense, social robots are one form of robotics, and more generally indicative of the move to the automation economy. A key contemporary concern is technological unemployment (job loss due to automation), where it is unclear whether technological unemployment would be a cost or a benefit, and also how quickly it may be eroding job. This is not the issue as much as the point that there is a greater need for comprehensive planning efforts involving government, business, and education than has been previously accommodated in society. It would be helpful to actively design and prepare for different trajectories of how the automation economy including technological unemployment might arrive. The stakes of social robotics could be high. Social robotics could be the linchpin in facilitating a beneficial transition to the automation economy by both easing the requirements for human labor and improving human quality of life. This paper argues that the objective of social robotics should be to improve the well-being of humans, both quantitatively and qualitatively.

One area to clarify is the distinction between on one hand, technologies for artificial intelligence (AI) and automation, and on the other hand, social robotics. Whereas AI/automation is a broad class of technology that aims to mimic complex systems and thought processes, social robotics has a variety of roles in interacting specifically with humans. Social robotics may have a focus on modeling the human response to social cues in order to either elicit specific emotional, focus, or attentional responses from the human or to signal a "motivation" behind an automated behavior to the human. Social robotics and AI/automation are similar in some areas but not in others. For example, while AI/automation might be incomprehensible (black-box) and alienating, social robotics is explicitly not. Automation is partly aimed at reducing the need for humans to do repetitive or low-level tasks and be invisible to human users, while social robotics is partly aimed at keeping people engaged, visibly and actively with robotic agents.

The structure of this paper is in two parts. First I address philosophical, economic, and ethical issues is a general sense, and second in a more specific economic sense. The objective of Part I is to understand the overall context of social robotics from a philosophical, economic, and social perspective. This analysis suggests that the contemporary situation of one of algorithmic reality, where reality is no longer comprised of man and nature as was true historically, but the three-fold presence of man, nature, and technology. Further, there is the notion of greater human decentering, isolation, and alienation in a developing multi-species society. This reveals that social robotics is really a situation of human-machine collaboration. The objective in Part II is to see social robotics in the context of economies that are also transitioning. Social robotics might figure prominently in both the phase of an automation economy that focuses on reduced requirements for human labor, and an abundance economy that focuses on an improved human quality of life. Overall, the stakes of social robotics are high where the role of social robotics might include labor-saving, and much more, taking advantage of the implied close connection with humans to help negotiate and buffer interactions between humans and a world with an increasing and expanding presence of technology.

2 The New Three-Fold Reality: Nature, Man, and Technology

The current situation is one in which technology is having an increasing presence in reality. Reality is no longer the kind of thing that has just two domains as it has for the majority of human history, nature and man. Instead, reality is starting to be comprised of nature, man, and technology. Technology in the form of tools (for example flint, arrowheads, and axes) has certainly always been present during man's existence, but current and near-term technology might be qualitatively different. One factor is the possibility of autonomous embodied agents such as social robots. Another factor is more generally the ubiquitous, pervasive, multi-dimensional, world-encircling presence of technology. These kinds of factors regarding technology are prompting important questions about the definition and role of technology, and causing man to perceive reality differently. In some ways technology is still man's tool, but in other ways perhaps technology can no longer be subsumed under this category since it is increasingly standing on its own as an existence in the world. Thus, there is an emerging triumvirate view of reality as nature, man, and technology which is forcing man to adapt and redefine himself in new ways.

In the three-fold model of reality as nature, man, and technology, the technology area might be named Algorithmic Reality. Algorithmic reality is the proliferation of computing platforms and the presence of algorithms in modern life that is increasingly interconnecting physical reality and digital reality. All major natural and social processes (for example, weather systems, infectious disease spread, and real-time bidding for advertising [2]) are in the process of being modeled and predicted in computational systems. In the contemporary world, everything has become a math problem. Some of the many readily-identifiable platforms include social robots, 3D printing, drones, self-driving cars, smarthome appliances, quantified-self wearable gadgets, personal voice assistants, smartphones, blockchain-based smart contracts, tradenets, deep-learning algorithms, big data clouds, brain-computer interfaces, neural tracking devices, DNA nanotechnology, synthetic biology, augmented reality headsets, and gaming worlds. The key point is that while there are numerous computing platforms, there is only one type of human, and the human can start to feel diminished and alienated as a result. Metaphorically, this could be like being the only kid on the block, where suddenly ten, or a hundred, families move into the neighborhood, but they do not speak your language or have your habits. This potential disconnecting of humans from current reality is precisely the kind of situation that social robotics might help us negotiate. Social robots are still a class of tools, but perhaps not what we thought most proximately, for labor-saving. The real purpose at which we might want to aim with social robotics is as a tool for allowing us to connect and operate within the new technological reality. Social robotics could be a reality buffer.

Computing platforms such as social robotics are fundamentally different from humans, but share many properties among themselves. Each species of computing technology can be seen much more extensively than within the confines of its form factor. Most technology is Internet-connected, and each species of computing technology should be conceived as a platform, a network, and an app store. Computing platforms may be connected to the Internet cloud 24/7, receive ongoing software updates to their functionality, and possibly be controlled together as a swarm entity. Moreover, computing platforms may be Turing-complete, meaning that any platform can run any other platform, for example, a robotics network could run an IoT sensor network or drone network if need be. It is impossible to imagine what all of this could mean. However, one aspect indicated by these properties of extreme portability, fungibility, connectability, and updatability is that we can see all of these computing platforms together as a juggernaut of interconnected computation blanketing the Earth. Like ether, there is a layer of always-on, interconnected computation encircling the Earth. This is like Wi-Fi was supposed to be and electricity generally is, a blanket of functionality so pervasive it can be assumed to be in place and does not need to be thought about consciously. There are both costs and benefits to the emergence of technology as a full-fledged existence in the world alongside humans and nature. On one hand, possibilities for human growth, collaboration, creativity, connection, and actualization might be expanded. On the other hand, humans might experience a real or perceived alienation or sense of diminishment through the sense of a more constrained existence per the growing presence of computing platforms. Philosophical questions arise regarding a reconsideration of the role, identity, and purpose of the human.

2.1 Human Decentering and Multi-species Society

Social robotics is a newly emerging area whose future and adoption trajectories are difficult to predict. Given its uncertainty, it should be acknowledged that social robotics is a form of a *thinkability* problem in that the full scope of what social robotics is and might be is to some degree unthinkable. This term, thinkability is the notion of becoming aware of and coming to terms with phenomena that are "bigger" than us as humans. These are features of the world that are outside our perceptual and experiential domains such as technology as just described, quantum physics, black holes, global warming, the Florida everglades, the biosphere, financial derivatives, big data, and future technologies like social robotics. Philosophers Quinten Meillassoux, Timothy Morton, Mark Hansen, and Graham Harman all consider the problem of thinkability; through notions that they label respectively ancestrality, hyperobjects, superjects, and object-oriented ontologies. For the purposes of this paper, the point is that thinkability favors a position that does not privilege human existence over the existence of nonhuman objects. A worldview is implicated that embraces a multiplicity of constituents, not merely humans; for example animals, humans, and the environment, and here by extension, technology. To the tripartite worldview (nature, man, technology), thinkability adds the point that since humans cannot fully *think* the world, that human-centric models would not necessarily have validity in any scope that is beyond humans such as the domain of technology.

Models that are not human-centric are already emerging in technology. Intelligent systems, including social robotics, are becoming more capable, and through their expanded presence in the world, a sense of the properties and features of native machine culture is starting to be visible. Some examples of where a 'native machine culture' can be seen are in the areas of law and personal identity. Machine culture's parameters may be readily identifiable because they are a differential to human cultural norms. In the domain of law, there are the legally-binding contracts with which we are familiar; and now also technologically-binding contracts. The two are different. Technologically-binding code contracts will execute inexorably even if conditions have changed, while legally-binding contracts between human parties are subject to discretionary compliance. The majority of human-based contracts are not complied with to the letter. Code-based contracts on the other hand, cannot be breached, and will execute unstoppably. The point is that the two regimes, human and machine, are different, and this start to point up the idea of plurality or multiplicity, and choosing the best tool for the job. In some situations, inexorably executing code is useful, for others, human-breachable contracts are better. Another example is the case of personal identity. Here too the technological construct of identity and the social construct of identity are different, and each has a different implied social contract. Humans are accustomed to the social construct of identity that includes the property of imperfect human memory, which allows the possibility of forgiving and forgetting, and redemption and reinvention. Machine memory, however, is perfect and can act as a continuous witnessing agent, never forgiving or forgetting, and always able to re-presence even the smallest most-irrelevant detail (such as an unforgiving photo or text comment) at any future moment.

These examples help to show that machine culture, values, operation, and modes of existence are already different from those of humans. This emphasizes the need for ways to interact that facilitate and extend the existence of all parties. The potential future world of agent multiplicity means accommodating plurality and building trust. New forms of building societal (e.g.; multi-agent) trust could be crucial. Blockchain technology, a decentralized, global, code-based ledger of transactions and smart contracts is one example of a trust-building system. The system can be used not only between human parties that do not need to know or trust each other, but also between inter-species parties, exactly because it is not necessary to know, trust, or understand the other entity, just the system. Decentralized smart networks like blockchains are an example of a system of checks-and-balances that could provide a more robust solution to situations of future uncertainty. Trust-building models for inter-species interaction could include both game-theoretic checks-and-balances systems like blockchains, and also at the higher level, frameworks that align entities on the same plane of shared objectives.

Concluding on this section, I have presented a challenge in modern society where social robotics might have a positive impact. The notion of human decentering and multi-species society is developed and supported through the argument that (1) the world is one of nature, man, and technology, (2) that may be beyond human thinkability, (3) where there is already evidence of a nascent machine culture.

3   A New Philosophy of Economics

Beyond the more general philosophical issues in the scope of contemporary reality, social robotics might feature prominently regarding the specific philosophical issue of economic theory. While these ideas might apply to new technologies generally, their impact is developed specifically in the case of nuances motivated by social robotics here. New philosophies of economic theory might be required because current conceptual models are becoming outdated. The cornerstone of most economic theory has been the notion of scarcity. However a central focus on the production and distribution of scarce resources is no longer the case in all economic systems. Scarcity is a weak notion empirically since there is emerging and existing evidence of situations in the world where scarcity is not a parameter, and not the governing parameter. The situation of digital goods, such as software or digital images, is an example where there is essentially no cost to producing and distributing another unit by copying and sending the goods electronically. The shift to the automation economy likewise is accommodated more congruently with an economics of abundance than with an economics of scarcity. Many countries are becoming "rich enough" to pay individuals a guaranteed basic income to cover basic survival needs. Such universal or guaranteed basic income programs are being discussed, voted on, and tried in pilot implementations as an improvement over the inefficiencies of welfare systems [3]. Thus a philosophy of economics that better corresponds to a coming automation economy featuring social robotics is needed. A new philosophy of economics would need to include reconceptualizations of the concept and purpose of economics. For example some of the required shifts in thinking include labor-to-fulfillment, moving from a labor-based economy to a fulfillment-based economy, and abundance-to-scarcity; seeing the world and resources in an abundance of availability frame as opposed to scarcity.

Even the notion of abundance itself is an impoverished conceptualization. Abundance is primarily conceived as the alleviation of scarcity, which it is, but it is also more. It is the eradication of scarcity in terms of having material needs covered, in the notion of recouping a baseline, but there is also an important upside formation to abundance. Abundance also means open-ended possibility up from baseline, defining the area of social goods that humans need to thrive, not merely survive. While material goods attend to survival, social goods attend to thriving. The automation economy is concerned not just with human survival, but an improved quality of live such that humans can thrive, and it is in this domain that social robotics could feature prominently. Further, scarcity is not just a situation to alleviate, it can be seen as a situation of harm, a social pathology, and a pollutant. Scarcity in the form of income inequality has been shown to denigrate the quality of human health [4]. Possibly even more than technological unemployment and the possibility of 'robots taking our jobs,' income inequality is a threat to the future well-being of our economies. Income inequality (the gap between the highest and lowest paid persons in an economy) has a detrimental impact on both material goods and social goods. The result of income inequality is greater social problems of every kind: health, violence, mental illness, drug addiction, obesity, imprisonment, and poorer well-being for children [5]. What most do not realize is that the effects of inequality are not confined to the poor but impact the whole of society via diminished social goods that are crucial to human thriving. Thus societies with less income inequality are better-poised to more quickly adopt the technologies of the future like social robotics, because they already have an acknowledgement of the importance of social goods, and a capability of designing and implementing strategies to realize them.

Thus a new philosophy of economics that is an abundance theory of flourishing can be developed by articulating the social goods that might be produced for humans, including with this as a specific aim of computing platforms such as social robotics. The close relationship of social robots and humans, particularly via VUI (voice user interface), places social robots in a unique position to monitor and enhance the quality of life of humans. Some of the new and contributive social goods that are unavailable in the scarcity model and could be created in the Abundance Economy include certainty, availability, reduced contingency, reduced efforting, and therefore the secondary social goods of cognitive easing and cognitive surplus. Abundance creates a psychology of certainty and availability, a reliable ongoing feeling of certainty that material survival needs will be met, as opposed to the continuous uncertainty and attending-to required by situations of scarcity. Much current human cognitive and physical effort (as individuals and groups; families, corporations, institutions, and nation-states) is devoted to anti-scarcity measures: hoarding, manipulation, and control for the purpose of ascertaining the future availability of resources. The idea would be like doing for emotional and cognitive attending what just-in-time inventories did for manufacturing. Situations of abundance invoke the social goods of certainty and reliance about the real-time availability of resources for need fulfillment. Through abundance, there could be the considerable social good of relief and certainty, where a whole class of cognitively-exertional activities drop off the reality of what has to be considered for basic living. This would be unprecedented in human history, a trustable source of having basic needs met such that we do not even have to think about this. The first tier of social goods that could be created in the Abundance Economy as a result of scarcity alleviation includes certainty, availability, reduced contingency, and reduced efforting. There could be many other tiers of social goods that are articulated, valorized, developed, and implement in future societies. What has been seen as the apogee tier of social goods that many human societies valorize is social goods related to liberation such as self-respect, self-esteem, and self-realization.

3.1   Intermediary Point: Automation Economy

The notion of the Automation Economy is an intermediary point that comprises half of the idea of the Abundance Economy, the material goods side which envisions that technology has supplemented or replaced much of the need for human labor. The other side of the Abundance Economy is the Actualization Economy is social goods, which have the ability to create a better human quality of life beyond sustenance needs. Social robotics might help with both. Actualization is a term from psychology taken to mean self-actualization, the realization of one's potential, for example in areas such as expressing creativity, learning, being in service to society are examples of self-actualization. Carl Rogers described actualization as man's becoming his potentialities, expressing and activating all of his capacities [6]. Maslow's Hierarchy of Needs is a familiar instantiation that Maslow developed to articulate the case that humans have a set of motivation systems unrelated to rewards or unconscious desires [7]. Maslow's Hierarchy of Needs features eight needs in two tiers: four in a lower tier of 'deficiency needs' (physiological, safety, belonging and love, esteem) and four in an upper tier of 'growth needs' (cognitive, aesthetic, self-actualization, and transcendence) that cannot be met until the lower tier needs are satisfied.

The point here is that the Automation Economy might help to address the lower tier of survival needs, but to truly thrive and extend human quality of life beyond sustenance, positive psychology tools like social robotics might be required. The psychological territory beyond survival needs is sometimes characterized as Positive Psychology, the scientific study of the strengths that enable individuals and communities to thrive and lead meaningful and fulfilling lives [8, 9]. Social robotics might be able to help the 'positive psychology' movement shift to the 'positive technology' movement that improves the quality of personal experience. Specifically, this might be achieved with social robotics that offer functionality to help humans in the areas of affect regulation and emotional quality, engagement, motivation, and actualization, and connectedness with others. One study found that technology with these features was able to improve the quality of life of aging persons [10]. Likewise, social robotics might implement happiness practices that have been demonstrated to improve human quality of emotional well-being [11].

Thus to address to address the threat of human alienation, and improve human quality of live more general, social robotics is a hopeful example of a new and forthcoming technology. Abundance Economics can be posited as an economic theory that is adequate to the current and near-future moment that focuses on social goods production in addition to material goods production for the most beneficial transition to the automation economy of the future [12]. While there is an extensive literature concerning social robotics and human robotic interaction, a philosophy of economic theory that envisions a full suite of automation and actualization applications for social robotics has not yet been advanced. The literature does consider topics well beyond the algorithmic and mechanical aspects of social robotics. Many researchers espouse a systems-theory view, in that form, function, and context should all be taken into account simultaneously for optimal human-machine interaction [13]. Others focus on the capacity of robots to improve communication with human beings, highlighting the physical and social-cognitive aspects of the interaction separately [14]. Researchers note that there should be a balance between 'hard' and 'soft' tasks in social robotics, connecting with both the mechanical labor-saving and the emotional side of human needs. For example, care is about 'taking care' (washing and feeding) and 'caring for' through a kind word or a good conversation [15]. Two-way interaction between machine and human is important, and more specifically, crafting a human-robot relationship [16].

4   Conclusion

Abundance Economics is presented as a philosophy of economic theory that might be appropriate for the contemporary and near-future moments of an automation economy that features social robotics. On this view, social robotics might have significant benefit in helping to realize produce social goods that improve human quality of life. Since social robotics implies a much closer connection between humans and technology than other platforms, social robotics could be key in helping effectuate beneficial transitions to situations such as a shift towards the automation economy with large sectors of technological unemployment that simultaneously create an abundance economy with a focus on social goods creation that improves human quality of life. The magnitude of a rapid shift towards the automation economy could simultaneously engender a rethinking of economic principles, with significant shifts in mindset, for example from a labor economy to a fulfillment economy, from scarcity to abundance, and from exclusively human agents to multiple forms of intelligent and emotional agents comprising society. The objectives of social robotics design and implementation could be to facilitate shifts to situations such as the post-labor automation economy and improved human quality of life in s social goods economy of abundance, wherein an enabling possibility space is created for the thriving of diverse multi-species agents.

References

1. IGI Global (2016). http://www.igi-global.com/dictionary/social-robotics/27482
2. Adikari, S., Dutta, K.: Real time bidding in online digital advertisement. In: Donnellan, B., Helfert, M., Kenneally, J., VanderMeer, D., Rothenberger, M., Winter, R. (eds.). LNCS, vol. 9073, pp. 19–38Springer, Heidelberg (2015)
3. Foulkes, I.: Switzerland basic income: Landmark vote looms. In: BBC News (2016)
4. Subramanian, S.V., Kawachi, I.: Income inequality and health: what have we learned so far? Epidemiol. Rev. **26**(1), 78–91 (2004)
5. Pickett, K., Richard, W.: The Spirit Level: Why Greater Equality Makes Societies Stronger. Bloomsbury Press, London (2011)
6. Rogers, C.: On Becoming a Person 350–351 (1961)
7. Maslow, A.H.: A theory of human motivation. Psychol. Rev. **50**(4), 370–396 (1943)
8. Seligman, M.E.P.: Authentic Happiness: Using the New Positive Psychology to Realize Your Potential for Lasting Fulfillment. Atria Books, New York (2004)
9. Seligman, M.E.P.: Flourish: A Visionary New Understanding of Happiness and Well-being. Atria Books, New York (2012)
10. Riva, G., Villani, D., Cipresso, P., et al.: Positive and transformative technologies for active ageing. Stud. Health Technol. Inform. **220**, 308–315 (2016)
11. Lyubomirsky, S.: The How of Happiness: A New Approach to Getting the Life You Want. Penguin Books, New York NY (2008)
12. Swan, M.: Is technological unemployment real?: an assessment and proposal for an abundance philosophy of economics. In: Hughes, J., LaGrandeur, K. (eds.) Robonomics: Emerging Technology and the Future of Employment. Palgrave Macmillan, New York, NY, Forthcoming
13. Hegel, F., Krach, S., Kircher, T., et al.: Understanding social robots: a user study on anthropomorphism. In: RO-MAN, The 17th IEEE International Symposium on Robot and Human Interactive Communication (2008)
14. Strabala, K.W., Lee, M.K., Dragan, A.D., et al.: Towards seamless human-robot handovers. HRI Syst. Stud. **2**(1), 1–23 (2013)
15. Royakkers, L., van Est, R.: A literature review on new robotics: automation from love to war. Int. J. Soc. Rob. **7**, 549 (2015)
16. Fong, T., Nourbakhsh, I., Dautenhahn, K.: A survey of socially interactive robots: concepts, design, and applications. Robot. Auton. Syst. **42**, 143–166 (2003)

Toward a Hybrid Society

The Transformation of Robots, from Objects to Social Agents

Francesco Ferrari[(✉)] and Friederike Eyssel

CITEC, Bielefeld University, Bielefeld, Germany
{fferrari,feyssel}@cit-ec.uni-bielefeld.de

Abstract. Social robots are machines developed to interact with humans. Unlike other technological devices, their presence in society requires accepting and treating them as social agents. This has important implications in terms of social changes for humans' personal and social identity, and social interactions. We aim to explain the core features that characterize social robots by highlighting what makes them distinct from other types of innovative technology. Equally important, we illustrate how social psychology can provide a useful perspective to understand human-robot interactions. To do so, we focus on studies that have investigated the role of intergroup relations and social identity in the context of human-machine interactions to demonstrate that robots may comprise a new type of social outgroup in future society.

Keywords: Social robots · Social identity · Intergroup relations · Robot outgroup · Robot ingroup

1 Introduction

Imagine you have purchased a new robot. The robot is still in its box, it doesn't move, it doesn't talk or respond to you at all – thus, it could be perceived as a simple inanimate object made of plastic, some wires, and technology. However, once you unpack the machine, you insert batteries and switch it on, it turns out that the machine starts moving its head and arms, responding through a humanlike manner and all of a sudden, the perception of the object changes. It is not considered a simple technical device anymore, the product transforms into a social robot that can participate actively and socially in society. Thus, the fundamental issue raised in this scenario concerns the psychological mechanisms underlying the transformation from object to agent. Research on the Media Equation [1] demonstrated that people interact with computers and other technical devices as if they would interact with another person, and they do so automatically.

However, reactions toward social robots, - robots that are unique through embodiment and resemblance to humans in terms of appearance and behavior -, might even be more pronounced. We argue that these kind of machine might not only impact our behavior, but also the perception of our social identity. Social humanlike robots are machines that challenge the categorical distinction between humans and non-humans [2] and undermine human uniqueness [3]. In social robotics, we can observe behaviors

© Springer International Publishing AG 2016
A. Agah et al. (Eds.): ICSR 2016, LNAI 9979, pp. 909–918, 2016.
DOI: 10.1007/978-3-319-47437-3_89

and psychological processes that are commonly investigated in intergroup relations. [4] have suggested that people feel intergroup anxiety when observing humanlike robots. Similarly, [5] have found that high anthropomorphic appearance of android robots lead individuals to perceive the robots as a threat to the distinctiveness of the group of humans. In that way, interactions with social robots are unique and different from interactions with other artefacts [6]. Social robots are developed to interact with humans and to serve as social companions and assistive devices [7]. They are machines that act to reach both their own goals and those of their human counterparts [8]. In that sense, such robots are inherently social, and accordingly, a social psychological perspective on social robotics is called for. Interpreting human-robot interaction through the lens of a social psychological viewpoint helps to understand the psychological obstacles associated with accepting robotic companions at home or in educational contexts [9, 10]. Adopting such perspective and taking into account the vast literature on human-human social relationships can facilitate and improve human-robot interaction in the long run [11].

In the following paragraphs, we will describe why, differently from other technological devices, social humanlike robots are perceived as real social agents. We will explain how social psychological research on intergroup relations can inform social robotics, and how it can affect the relationships that we develop with social robots. Finally we highlight the negative and positive effects of imitation in social robotics.

2   Robots as Social Agents?

Apparently, social relations are easily formed toward any media [1], from personal computers to smartphones. However, the perception of a technological device as a social agent is determined by two key aspects that are unique to social robots: embodiment and human resemblance.

Unlike virtual agents that are observable on a computer screen, social robots are embodied agents whose physical presence can affect people's psychological reaction during interaction between humans and machines. [12] have shown that interacting with a social robot versus a computer agent resulted in a greater sense of familiarity toward the robot target, even though embodiment did not affect behavior toward the two different targets. Thus, interacting with an embodied robot resulted in greater emotional and psychological involvement compared to an interaction with a robot that appeared as a projection. Similarly, [13] have investigated how the social presence of an artificial agent affected people's behavior and impressions toward it. They found that when a social robot shared the same space with the human interaction partner, co-location resulted in the fact the robot was perceived as more lifelike, and being endowed with more personality in comparison to a range of other agents (e.g., an agent on a computer screen, an agent projected by a monitor, and a robot projected by a monitor). Moreover, participants rated the co-located robot as the most helpful, useful, and enjoyable interactional partner. These results clearly highlight the importance of embodiment and social presence of this kind of robots, a feature that distinguishes this kind of technology from any other else.

The second core feature that contributes to the perception of a technical system as a social agent concerns the resemblance of its appearance and behavior to humans [8]. Because we make use of anthropocentric knowledge structures when confronted with a humanlike robot, people tend to attribute psychological features and human traits to it [14, 15]. This psychological process called anthropomorphism reflects the transformation of robots from technological devices to social agents, through the ascription of traits and abilities typically associated to human beings. For instance, Krach and colleagues [16] have conducted a study to examine the differences in playing the "prisoner dilemma" with a computer, a mechanical robot, a humanoid robot and a real human being. They investigated both impressions toward the different agents and participants' physiological responses during their interactions with the agents using Functional Magnetic Resonance Imaging (fMRI).

Participants judged the interaction with the humanoid robot more positively than the interaction with the computer or the mechanical robot. More importantly, the cortical network related to mind attribution in implicit human-human interactions was activated during human-machine interactions, and the extent of its activation was associated with to the human-likeness of the technological device [16]. Thus, humanlike appearance of robots leads people to anthropomorphize them and to attribute mind capabilities [8].

These examples from recent research demonstrate that embodiment and human resemblance can deeply affect our interaction and the relationships with social robots. This is realized through changing our lay conceptualization of a robot, by turning the technical device into a social entity. An example of the transformation of robots into *social* robots that relate with us, and affect our reactions and behaviors, is given by Kahn and colleagues [17]. In this study, children and adolescents were observed while interacting individually with the humanoid robot Robovie. After a sudden interruption, the experimenter insisted to lock Robovie in a closet, even though the robot protested this treatment. A structural-developmental interview was used to collect data regarding children's impressions of the robot. Most felt that the treatment of the robot was unfair and psychologically harmful, and stated that Robovie had its own mental states and feelings. The results of this study provide evidence for the notion that robots may not just be considered simple tools, but rather as social agents.

Since the market share for social robots is expected to exponentially increase in the next years [18], future society has to face the ubiquitous presence of a new kind of social agent which affect our perception of what constitutes humanity [2, 3]. For this reason, such new developments have to be analyzed further in light of the social psychological literature on intergroup relations and social identity theory.

3 Toward a Hybrid Society: Robots as Members of a Social Group

The relation and interaction between social groups in our society is a process that deeply affects not only our behaviors, but also the construction of our identity. Social Identity Theory [19, 20] has proposed that individuals develop their identity through three different processes. The first process, social categorization, entails categorizing people in groups, just as we would categorize objects in order to identify them. Social categorization allows individuals to endow their social environment with meaning. Social robots as new elements of our social environment have to be defined as a new, specific category which can include a variety of prototypes. Within the category of robots, for example, we can distinguish different types of robots based on their design and appearance. There are robots with zoomorphic appearance, such as the baby seal robot PARO [21], robots with human appearance, as humanoid robots like Honda's ASIMO [22], and even androids that almost completely resemble humans, e.g., Geminoid HI [23] and Geminoid DK [24]. Moreover, robots can be categorized differentially based on typically human cues. For example, a study conducted by Eyssel and colleagues [25] has shown that participants used vocal cues for social categorization and consequently attributed typically human characteristics (e.g., mind) to a 'gendered' robot, particularly when the robot shared their gender (i.e., when female participants shared 'group membership' with female robot).

The second psychological process at play within the framework of Social Identity Theory is social identification. Social identification implies the process that individuals commonly adopt the identity of a respective ingroup to which they belong [26]. The identification with specific categories or groups, in turn, forms the basis for identity development.

The final step proposed within Social Identity Theory [19] concerns social comparisons: that is, people tend to compare their ingroup with other social groups they do not belong to (i.e., outgroups). An ingroup vs. outgroup comparison allows to create social boundaries by minimizing differences within our group and by maximizing differences between groups.

Clearly, individuals can categorize other people and objects into multiple categories simultaneously, and similarly, they can identify themselves with different social groups. From this follows, that behavior is always a function of the group one identifies with and the given social context. For example, a person can self-categorize as a member of the group of scientists; based on the social context's cues he/she will identify with it and react according to the group's values [26].

Thus, following Social Identity Theory, *we are* because we *identify ourselves with certain categories* and we *distinguish ourselves from others*.

Humans show a spontaneous tendency to form groups, to feel like part of a group, and to distinguish their ingroup from those who do not share group membership. When humans interact with a social agent, they tend to understand how it relates to their ingroup [19]. For example, the humanlike appearance and communication style of social robots lead people to automatically activate anthropocentric knowledge structures, which guide judgments and behavior toward these machines [14, 15]. The features of embodiment and human-likeness are fundamental for this process. A zoomorphic robot like Sony's AIBO cannot stimulate the humanlike interactional schema, because it resembles a pet. People indeed relate with AIBO as if it were a dog. Similarly, we hypothesize that virtual agents cannot elicit the activation of human category to the same extent of social robots, because their social presences is restricted by the nature of virtuality. Conversely, through embodiment and human-likeness, social robots can lead people to form a novel social category. This novel category is ambiguous: social robots represent a group of objects that are inherently different (i.e., robots are in fact machines), and that can likewise be similar to real persons. To illustrate, even if contact occurs between one single person and one single robot, we can observe classic effects of intergroup relations, because the person and the robot represent members of their corresponding social group (i.e., the groups of humans vs. robots, respectively), and they are interpreted and judged as such [5].

At the first encounter, people cannot but categorize the robot and automatically compare between the robot's group and their own group. Studies in social psychology have shown that the presence of one single member of an outgroup activated ingroup membership and associated values and beliefs. Accordingly, people tended to emphasize differences between the respective groups [27, 28]. Similar processes are expected in human-robot interaction. That is, in human-robot interaction, humans compare themselves with the robotic outgroup to understand how these agents relate to their human ingroup.

Social robots represent an *extreme* outgroup in comparison to fellow humans. While commonly social psychology focuses on differences between groups in terms of race, gender, etc., social robots represent a social group that activates comparisons that focus on the essence of humanity, i.e., what it means to be human.

In other words, social robots are social agents that resemble us, but they will never be like us, because clearly robots remain machines. For this reason, they will always be perceived as members of an outgroup when people start to interact with them.

This comparison between humans and social robots not only has theoretical implications, but also affects the interaction with this kind of technology, and humans' self-definition as humans [5]. Thus, research on intergroup relations can help us to understand and improve the relations with social robots.

4   Intergroup Relations in Social Robotics

Social robots induce the definition of a new social category that includes prototypes which are similar to humans, but that are yet inherently different from them, being machines. Thus, it becomes possible to replicate classic psychological findings from intergroup relations in the context of social robotics to improve human-robot interaction. For instance, [29] have tested effects imagined contact, an intervention commonly used in intergroup relation to reduce intergroup anxiety, to increase positive impressions toward robots and to decrease negative attitudes toward these machines. Another way of increasing positive impressions and acceptance of social robots could be the creation of a joint subordinate group (in comparison to the larger superordinate category of *humans* in general).

[30] have shown that German participants had a better impression of the robot when it was described as ingroup member (i.e., as a robot developed in Germany with a German first name), rather than as an outgroup member (i.e., a robot developed in Turkey and endowed with a Turkish name). Participants evaluated the German ingroup robot more favorably and reported positive contact intentions toward it than to the outgroup Turkish robot.

Wang and colleagues [31] have found that when a robot's interactional style matched that of the interaction partners, the robot was perceived more positively. To illustrate, when the robot used implicit rather than explicit communication style, Chinese participants were more responsive to its recommendations. The researchers replicated their results comparing two different samples (i.e., US-Americans and Chinese people) and the different communicational style of the robot (implicit vs. explicit) [32]. Likewise, the authors found support to their hypothesis. Specifically, Chinese participants preferred the robot with the implicit interaction style, whereas US participants favored the robot, which communicated in a more explicit way.

[33] have demonstrated that a robot's evaluation was affected by group membership unrelated to nationality or cultural background. By using the minimal group paradigm [34], it was found that participants attributed more typically human capabilities to the robot that allegedly belonged to their ingroup vs. the outgroup. Moreover, participants showed greater acceptance and willingness to interact with the robot when it was described as a member of their ingroup than the outgroup.

These works showed that if the social context (e.g. description of the robot as team member, same interactional style of individuals, etc.) activates other individuals' aspects they can share with the robot (e.g., same nationality, cultural background, status, age, etc.), people can include the robot in a smaller subordinate group. Thus, a social robot should adapt to the people it is interacting with, discover the social features that characterize them and try to imitate them. In this way people could have the feeling that they share some aspects with the robot and they could be more willing to accept it as social partner, and establish a long and significant relation with it.

A study by Kanda et al. [35] nicely illustrates this idea: Robovie was deployed in an elementary school for two months, and the robot was introduced as having the same status as the children in a class, so they would perceive it as a peer member. The robot was able to communicate with the children and this enabled them to perceive the robot as a potential companion and as part of the ingroup. Moreover, the robot shared personal stories with those children who had interacted with it for a long time. Thus, there was the potential for shared experiences and ongoing long-term interactions across the duration of the study. Indeed, the robot continued to interact with several children for the whole period of two months. Specifically, the researchers [35] concluded that some children "treated Robovie as a peer-type friend" and established a friendly relation with it. They saw it as a member of their ingroup. However, other children did not perceive the robot as a possible partner, they quickly got bored during interactions, and likely saw it as an outgroup member. Unfortunately, [35] did not explore the reasons for acceptance vs. rejection of Robovie in the classroom setting, so we can only speculate that self-disclosure facilitated a shared reality and resulted in more positive interactions, analogous to findings from human-human interpersonal relations that highlight the importance of sharing experiences for friendship formation [36].

So, even if social robots are initially perceived as outgroup members in comparison to human ingroupers, we would like to emphasize the point that, – unlike any other technology –, social robots have the potential to join human groups as ingroup members.

Wrapping up, we propose that both embodiment and human-likeness represent key elements to change the perception of a robot from object to a social interactant. Likewise, we presume that shared experience represents one of the key requirement for a robot to be allowed into a common hybrid (i.e., human-robot) ingroup.

5   Robots that Imitate Humans: A Dilemma in Social Robotics

The perception of social humanoid robots as social agents is an involuntary and uncontrollable process. Even if we are aware that these machines are not and will never be humans, we cannot avoid to use the same schemas and psychological processes of human-human interaction. Because of their embodiment and human likeness, social robots are treated and considered as something between mere machines and a real person, as a different social group.

These considerations arise new opportunities and issues both for social robotics and social psychology. From a social robotics perspective the perception of robots as social agents and group members can represent an advantage. Social psychology's techniques can be used to reduce negative impressions toward robots [11, 33]. Social robots can create a smaller ingroup with the people. To this purpose the aspect of imitation is fundamental. A robot that emulates behaviors and manner of persons, can share social features with them and be perceived more positively [31, 32]. The "imitating" robot creates a common ground where the person and the machine can develop meaningful relations.

However imitation can represent a double edged weapon. From a social psychological point of view, humans need always to distinguish their ingroup from other possible social groups. Distinctiveness is a fundamental feature to create the boundaries of a social group and to provide a meaningful identity to its members. In the context of social robotics, [37] highlighted the importance of human–robot distinctiveness in the emotional reactions toward robots. [3] evidenced that when a human behavior or feature is possessed by a machine, this ability or characteristic is no more typical of humans: once machines became able to play chess, the game could not be considered anymore as a typical human ability. [38] stated that developing a robot that perfectly imitate humans can represent a danger to our uniqueness. In this case the emulation of humans represents a possible threat to our identity and leads to negative reactions toward robots.

For social robotics becomes fundamental the understanding of the role of imitation and when it can lead to positive or negative reactions in interacting with robots. Future research should investigate this issue adopting the point of view and techniques of social psychology.

6   Conclusion

Social robots are machines developed to interact with humans. The investigation of human-robot interaction from a social psychological vantage point offers a new perspective for social robotics and contributes to the understanding and improvement of interactions and relations with social robots. In this paper, we highlighted the role of embodiment and human-likeness as core features which change the perception of a robot, from object to social agent. We explain why social robots can be categorized in terms of an outgroup in comparison with humans, we highlight possible ways to change this perception (i.e., through creation of a common subordinate group, and shared experience with the robot). Since the aim of these machine is to be perceived as social agents, it is important to consider the social effects of interaction with them and how they can affect our social identity. Particularly relevant is the role of imitation in social robotics and future research should focus on this issue to investigate its positive and negative effects. Social robots represent both an opportunity and a challenge in terms of the overarching goal to develop inherently social technical systems.

Acknowledgment. This research has been conducted in the framework of the European Project CODEFROR (FP7 PIRSES-2013-612555) and it was supported by the Cluster of Excellence Cognitive Interaction Technology 'CITEC' (EXC 277) at Bielefeld University, which is funded by the German Research Foundation (DFG). We thank our colleagues Birte Schiffhauer, Julian Anslinger and Ricarda Wullenkord, who commented on earlier versions of this manuscript.

References

1. Reeves, B., Nass, C.: The Media Equation: How People Treat Computers, Television, and New Media Like Real People and Places. Cambridge University Press, New York (1996)
2. Ramey, C.H.: The uncanny valley of similarities concerning abortion, baldness, heaps of sand, and humanlike robots. In: Proceedings of Views of the Uncanny Valley Workshop: IEEE-RAS International Conference on Humanoid Robots, pp. 8–13 (2005)
3. Kaplan, F.: Who is afraid of the humanoid? Investigating cultural differences in the acceptance of robots. Int. J. Humanoid Robot. **1**(03), 465–480 (2004)
4. Kamide, H., Mae, Y., Kawabe, K., Shigemi, S., Arai, T.: A psychological scale for general impressions of humanoids. In: 2012 IEEE International Conference on Robotics and Automation (ICRA), pp. 4030–4037. IEEE, May 2012
5. Ferrari, F., Paladino, M.P., Jetten, J.: Blurring human-machine distinctions: anthropomorphic appearance in social robots as a threat to human distinctiveness. Int. J. Soc. Robot. **8**, 1–16 (2016)
6. Foundation, K.: To make a social robot, key is satisfying the human mind. ScienceDaily. Retrieved June 7, 2016. www.sciencedaily.com/releases/2012/02/120203101153.htm. February 3 2012
7. Kanda, T., Ishiguro, H., Ishida, T.: Psychological analysis on human-robot interaction. In: Proceedings 2001 ICRA IEEE International Conference on Robotics and Automation, vol. 4, pp. 4166–4173. IEEE (2001)
8. Duffy, B.R.: Anthropomorphism and the social robot. Robot. Autonom. Syst. **42**(3), 177–190 (2003)
9. Reich, N., Eyssel, F.: Attitudes towards service robots in domestic environments: The role of personality characteristics, individual interests, and demographic variables. Paladyn J. Behav. Robot. **4**(2), 123–130 (2013)
10. Reich-Stiebert, N., Eyssel, F.: Learning with educational companion robots? toward attitudes on education robots, predictors of attitudes, and application potentials for education robots. Int. J. Soc. Robot. **7**(5), 875–888 (2015)
11. Wullenkord, R., Eyssel, F.: Improving attitudes towards social robots using imagined contact. In: 2014 RO-MAN: The 23rd IEEE International Symposium on Robot and Human Interactive Communication, pp. 489–494. IEEE, August 2014

12. Yamato, J., Shinozawa, K., Naya, F., Kogure, K.: Evaluation of communication with robot and agent: are robots better social actors than agents. In: Proceedings of the 8th IFIP TC 13 International Conference on Human-Computer Interaction (INTERACT 2001), Tokyo, Japan, pp. 690–691 (2001)

13. Powers, A., Kiesler, S., Fussell, S., Torrey, C.: Comparing a computer agent with a humanoid robot. In: 2007 2nd ACM/IEEE International Conference on Human-Robot Interaction (HRI), pp. 145–152. IEEE, March 2007

14. Epley, N., Waytz, A., Cacioppo, J.T.: On seeing human: a three-factor theory of anthropomorphism. Psychol. Rev. 114(4), 864 (2007)

15. Eyssel, F., Kuchenbrandt, D., Hegel, F., de Ruiter, L.: Activating elicited agent knowledge: How robot and user features shape the perception of social robots. In: Proceedings of the 21th IEEE International Symposium in Robot and Human Interactive Communication (RO-MAN 2012), Paris, September 2012

16. Krach, S., Hegel, F., Wrede, B., Sagerer, G., Binkofski, F., Kircher, T.: Can machines think? Interaction and perspective taking with robots investigated via fMRI. PLoS ONE 3(7), e2597 (2008)

17. Kahn, P.H., Freier, N.G., Friedman, B., Severson, R.L., Feldman, E.N.: Social and moral relationships with robotic others? In: 13th IEEE International Workshop on Robot and Human Interactive Communication, ROMAN 2004, pp. 545–550. IEEE, September 2004

18. Robotics - Horizon 2020 - European Commission. (n.d.). Retrieved June 16, 2016. http:// ec.europa.eu/programmes/horizon2020/en/h2020-section/robotics

19. Tajfel, H.: Social identity and intergroup behaviour. Soc. Sci. Inf./sur les Sci. Soci. 13, 65–93 (1974)

20. Tajfel, H., Turner, J.C.: An integrative theory of intergroup conflict. Soc. Psychol. Intergroup Relat. 33(47), 74 (1979)

21. Kidd, C.D., Taggart, W., Turkle, S.: A sociable robot to encourage social interaction among the elderly. In: Proceedings 2006 IEEE International Conference on Robotics and Automation, ICRA 2006, pp. 3972–3976. IEEE, May 2006

22. Hirose, M., Ogawa, K.: Honda humanoid robots development. Philos. Trans. R. Soc. London A: Math. Phys. Eng. Sci. 365(1850), 11–19 (2007)

23. Sakamoto, D., Kanda, T., Ono, T., Ishiguro, H., Hagita, N.: Android as a telecommunication medium with a human-like presence. In: 2007 2nd ACM/IEEE International Conference on Human-Robot Interaction (HRI), pp. 193–200. IEEE, March 2007

24. Dougherty, E.G., Scharfe, H.: Initial formation of trust: designing an interaction with geminoid-DK to promote a positive attitude for cooperation. In: Mutlu, B., Bartneck, C., Ham, J., Evers, V., Kanda, T. (eds.) ICSR 2011. LNCS, vol. 7072, pp. 95–103. Springer, Heidelberg (2011)

25. Eyssel, F., Kuchenbrandt, D., Bobinger, S., de Ruiter, L., Hegel, F.: If you sound like me, you must be more human: On the interplay of robot and user features on human-robot acceptance and anthropomorphism. In: Proceedings of the 7th ACM/IEEE Conference on Human-Robot Interaction (HRI 2012), Late Breaking Report, Boston, MA, März 2012

26. McLeod, S.: Social identity theory. Simply Psychology (2008)

27. Hornsey, M.J., Jetten, J.: Not being what you claim to be: Impostors as sources of group threat. Eur. J. Soc. Psychol. 33(5), 639–657 (2003)

28. Hornsey, M.J., Jetten, J.: Not being what you claim to be: Impostors as sources of group threat. Eur. J. Soc. Psychol. 33(5), 639–657 (2003)

29. Kuchenbrandt, D., Eyssel, F.: The mental simulation of a human-robot interaction: Positive effects on attitudes and anxiety toward robots. In: 2012 IEEE RO-MAN, pp. 463–468. IEEE, September 2012

30. Eyssel, F., Kuchenbrandt, D.: Social categorization of social robots: Anthropomorphism as a function of robot group membership. Brit. J. Soc. Psychol. **51**(4), 724–731 (2012)
31. Wang, L., Rau, P.L.P., Evers, V., Robinson, B., Hinds, P.: Responsiveness to robots: effects of ingroup orientation and communication style on HRI in China. In: Proceedings of the 4th ACM/IEEE International Conference on Human Robot Interaction, pp. 247–248. ACM, March 2009
32. Wang, L., Rau, P.L.P., Evers, V., Robinson, B.K., Hinds, P.: When in Rome: the role of culture and context in adherence to robot recommendations. In: Proceedings of the 5th ACM/IEEE International Conference on Human-Robot Interaction, pp. 359–366. IEEE Press, March 2010
33. Kuchenbrandt, D., Eyssel, F., Bobinger, S., Neufeld, M.: Minimal group - maximal effect? evaluation and anthropomorphization of the humanoid robot NAO. In: Mutlu, B., Bartneck, C., Ham, J., Evers, V., Kanda, T. (eds.) ICSR 2011. LNCS, vol. 7072, pp. 104–113. Springer, Heidelberg (2011)
34. Otten, S., Wentura, D.: About the impact of automaticity in the Minimal Group Paradigm: Evidence from affective priming tasks. Eur. J. Soc. Psychol. **29**(8), 1049–1071 (1999)
35. Kanda, T., Sato, R., Saiwaki, N., Ishiguro, H.: A two-month field trial in an elementary school for long-term human–robot interac-tion. Robot. IEEE Trans. **23**(5), 962–971 (2007)
36. Turner, R.N., Hewstone, M., Voci, A.: Reducing explicit and implicit outgroup prejudice via direct and extended contact: The mediating role of self-disclosure and intergroup anxiety. J. Pers. Soc. Psychol. **93**(3), 369 (2007)
37. MacDorman, K.F., Entezari, S.O.: Individual differences predict sensitivity to the uncanny valley. Interact Stud. **16**(2), 141172 (2015). doi:10.1075/is.16.2.01mac
38. MacDorman, K.F., Vasudevan, S.K., Ho, C.C.: Does Japan really have robot mania? Comparing attitudes by implicit and explicit measures. AI Soc. **23**(4), 485–510 (2009). doi:10.1007/s00146-008-0181-2

Iterative Design of a System for Programming Socially Interactive Service Robots

Michael Jae-Yoon Chung[1]([⊠]), Justin Huang[1], Leila Takayama[2], Tessa Lau[2], and Maya Cakmak[1]

[1] Computer Science and Engineering, University of Washington,
Seattle, WA 98195, USA
{mjyc,jstn,mcakmak}@cs.washington.edu
[2] Savioke Inc., San Jose, CA 95113, USA
{leila,tlau}@savioke.com

Abstract. Service robots, such as the Savioke Relay, are becoming available in human environments such as hotels. It is important for these robots to not only be functional, but also to have appropriate socially interactive behaviors. In this paper, we first present results from a formative study with service industry customers. A key demand we discover is that the robot should be aware of people present around the robot. We incorporate these lessons into the design of *iCustomPrograms*, a system for programming socially interactive behaviors for service robots. Next, we perform two field studies with *iCustomPrograms* and iterate its design. In the first field study, which took place at an airport, we witness people initiating interaction with the robot in unanticipated ways. The second field study, which took place over 2 weeks at 5 service industry properties, evaluates the socially interactive applications created with *iCustomPrograms*. Our experiences and findings from each study not only show the usefulness of our system in the field, but also provide insights for the design of future interactive applications for service robots.

1 Introduction

Today, commercial service robots such as the Savioke Relay, Vecna QC Bot, and Aethon TUG are deployed in human-populated environments such as hospitals and hotels[1]. Although these robots are designed for performing deliveries, it is important for them to be socially interactive, as they are tightly integrated into the human workplace [11]. For example, the Relay robot, built by Savioke Inc., primarily does room service deliveries to guest rooms in hotels. However, operating in the hospitality industry, it is important for the robot to have a suite of engaging, guest-facing interactions as well.

In fact, socially interactive service robots have long been the subject of interest to robotics researchers [2,3]. Recent studies have explored socially interactive services such as guiding [1,2,8] and advertising [8] in uncontrolled

---

[1] www.aethon.com, www.vecna.com, www.savioke.com.

© Springer International Publishing AG 2016
A. Agah et al. (Eds.): ICSR 2016, LNAI 9979, pp. 919–929, 2016.
DOI: 10.1007/978-3-319-47437-3_90

human-populated environments such as office buildings [1], shopping malls [8], and airports [10]. As in those studies, we believe in the importance of designing for and evaluating in real-world environments.

In this paper, we present *iCustomPrograms*, a programming system, for developing socially interactive applications for mobile service robots. This paper contributes the design of *iCustomPrograms* as well as empirical findings from the deployment of interactive applications developed with *iCustomPrograms* in real-world service industry properties. We first interviewed employees in the service industry to understand what kinds of robot applications would be useful (Sect. 3). This information helped us to design the initial version of *iCustomPrograms*. We performed an initial field study to evaluate applications built with this system (Sect. 4). Based on findings from the initial study, we made enhancements to *iCustomPrograms* and conducted a larger field study (Sect. 5).

Our system was developed for the Savioke Relay, a 3-foot tall mobile robot with an interior bin and a touchscreen display (Fig. 2). However, *iCustomPrograms* can be used in principle with any robot that has similar capabilities [2].

2   Related Work

Researchers have long been interested in studying socially interactive service robots [3]. RHINO was one of the earliest examples, acting as a tour guide in a museum [2]. More recently, the SPENCER project investigated having a robot escort passengers in a busy airport [10], and the FROG project studied having a tour guide robot in an outdoor zoo [12]. The major contributions of this body of work addressed core technical challenges arising in the field, such as tracking human individuals or groups in crowded environments [10], or long-term outdoor mapping and localization problems [12].

Other researchers have studied human-robot interaction issues for service robots. Some investigated long-term human-robot relationships with a receptionist robot [5] and a delivery robot [9]. Researchers also investigated the commercial robots in the workplace; Mutlu et al. studied how the environmental factors in the workplace influence interaction with the robot [11].

Our work is closely related to studies that investigated integrated robotic systems for interacting with people in human-centric environments. Kanda et al. studied a humanoid robot that could guide people and advertise shops in a shopping mall [8], and Bohus et al. deployed a Nao humanoid robot that could provide directions to people in an office building [1]. Unlike the systems studied in those two, the robot we studied is not anthropomorphic and the interactive behaviors we studied were meant for cognitively lightweight interactions, e.g. short screen interactions. In addition, we focused on behaviors that used the robot's on-board sensors, rather than a network of off-board sensors in the environment. Our work is also closely related to systems for developing or generating socially interactive behaviors [4,6]. While their evaluations were conducted in lab environments, our work presents evaluations based on field studies.

---

[2] see the discussion section in [7].

Table 1. Summary information about service industry properties studied in this paper.

Property	Type	Used since	Point of contact	Requested applications	Target areas
A	Airport	2/2016[a]	Corporate executives, Customer satisfaction manager	People delight, Service recovery	Indoor garden, Baggage claim, Immigration hall
B	Hotel	1/2015	Hotel manager, Business consultant, Front desk supervisor	People delight, Mobile kiosk, Demo	Lobby, Bar
C	Hotel	6/2015	Guest service manager, Sales & marketing director	People delight, Service recovery, Mobile kiosk	Lobby
D	Hotel	7/2015	Hotel manager, Guest experience manager	Service recovery, Demo	Lobby, Breakfast area
E	Hotel	8/2015	IT manager, Area general manager	Mobile kiosk	Lobby

[a]Used since field study I

3   Formative Study

Although the Savioke Relay robot was built for a specific application (room service in hotels), it can be considered as a generic platform with a wider range of applications. The goal of our formative study is to discover potential applications that are desirable for existing Savioke customers and inform the design of a rapid programming system for creating those applications. To that end, we gathered information from five customers from the service industry.

3.1   Data Collection

Information about the properties we studied in this paper is summarized in Table 1. Property A was an airport in Southeast Asia and the rest were hotels in the San Francisco Bay Area. Unlike the other properties, Property A had not used the robot prior to the first field study. Additionally, the target areas considered by Property A were larger ($\approx 4000\,\mathrm{m}^2$) and more crowded than the areas considered by the other properties ($< 200\,\mathrm{m}^2$).

We analyzed meeting notes and email exchanges between Property A and Savioke employees. The meeting notes were collected from two meetings held at Savioke headquarters and one meeting held at the airport in 2015. During those meetings, Savioke employees surveyed the target areas in the airport and performed demos of the Relay doing deliveries. The two groups also brainstormed potential applications together.

For Properties B–E, we analyzed the field notes taken by a Savioke customer satisfaction manager during regular checkup visits in February 2016. As part of the visit, the customer satisfaction manager met with one or two hotel representatives individually and asked them for (i) general feedback on using Relay, (ii) their wish list of new robot applications, and (iii) feedback on support infrastructure. Our analysis focused on the wish list notes.

3.2 Use Cases

We categorized the requested applications by their use cases.

People delight: All properties wanted to provide a unique experience to visitors using the robot. 3 properties specifically proposed applications designed to make their customer experience more delightful. Property A proposed an application in which the robot would approach passengers in an indoor garden area, offering them snacks or volunteering to take their picture. Properties B and C wanted the robot to roam around in their lobby and bar areas to encourage lightweight interactions with guests. Example interactions that Property B suggested included playing a game of rock-paper-scissors or sharing a joke.

Service recovery: 3 out of 5 properties wanted to use the robot to catch unsatisfied customers before they left the building. Property A requested that the robot approach passengers in the baggage claim area whenever the unloading of baggage was delayed. They wanted the robot to explain the situation and placate potentially frustrated passengers. Previously, this task was done by the airport staff, who were often not treated well by frustrated passengers. Property C wanted the robot to approach guests who were leaving the hotel, in order to ask them about their stays. Property D wanted the robot to navigate to the hotel's breakfast area and ask guests about their stays.

Mobile kiosk: 3 of 5 properties requested applications that resembled an information kiosk. Properties B and C wanted the robot to visit a couple of highly visible locations (e.g., a location near the front entrance or the elevators), and display a series of screens encouraging interaction when people were around. They said that displaying information about the robot or the hotel would be useful, as it could trigger guests to use the delivery service or other hotel amenities in the future. Property E, which had the robot's docking station in the lobby, requested that the kiosk mode run while the robot was charging.

Demo: Properties B and D requested a guest-facing demo application. They often had to manually control the robot to show it in action to curious customers. Property B suggested that the application include a navigation demo and an introduction about its delivery service. Property D wanted control over how the application would be activated; they did not want guests to be able to trigger the demo, as it could interfere with actual deliveries that needed to be done.

Overall, we make the following observations:

- Having first hand experience with the Relay robot (except Property A), Savioke customers had realistic requests.
- Although their requests were similar and could be broadly categorized as above, they each had specific, custom requirements.
- Many of the requested applications involved interactions with humans.

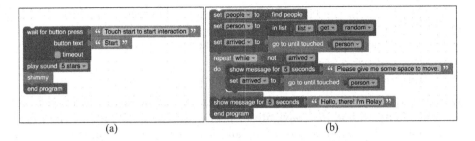

Fig. 1. Example applications written in *iCustomPrograms*. (a) Simple interactive application; the robot first waits for a user to engage in interaction by pressing a button. It then plays a sound and shimmies in response. (b) Approaching a person application; the robot finds nearby people using **findPeople**, randomly selects a person, and approaches them. The **goToUntil** primitive returns *true* if the robot successfully reaches the destination and *false* if it is interrupted by a person tapping its touchscreen.

4   *iCustomPrograms*

The software for the original room service functionality of the Relay robot was developed by Savioke's team of engineers and programmers. This team could implement many of the functionalities requested by customers (Sect. 3.2). However, given the diversity of requests from customers and the time it takes for custom software to be developed and deployed, this approach would not be scalable as the number of customers increase. Instead, Huang et. al developed *CustomPrograms* [7] to enable non-technical Savioke employees (e.g. marketing representatives, customer satisfaction managers, designers) as well as customers (e.g. hotel staff) to program the Relay robot. In this paper, we extend *CustomPrograms* with an emphasis on *interactive* behaviors, which were a part of the applications requested by customers.

4.1   CustomPrograms

CustomPrograms allows users to build applications for robots by composing a set of capabilities, known as primitives, with general-purpose programming constructs like variables, loops, conditionals, and functions [7]. Applications are started manually and end when there are no more instructions to run.

Huang et al. implemented *CustomPrograms* for the Relay robot, including four categories of primitives: navigation, screen interaction, lid control, and battery state. The main navigation primitive was **goTo**, which made the robot navigate to a given location. The **shimmy** primitive was a short side-to-side swaying to convey happiness. Screen interaction primitives included displaying messages (**displayMessage**), receiving user input (e.g., **askMultipleChoice**, **askPasscode**, **askRating**), and playing non-anthropomorphic sounds (**playSound**). The other primitives controlled the robot's lid or read the battery level.

CustomPrograms can be used to program simple interactive applications. For example, Huang et al. developed a demo application in which the robot went to several predefined locations, and offered a snack to nearby people.

4.2 Supporting People-Aware Behaviors

One key capability that was needed for the applications described in Sect. 3.2 was the ability to find and navigate to people. In *CustomPrograms*, the robot could only go to predefined locations and wait for people to interact with it. Hence, we created *iCustomPrograms*, a modified *CustomPrograms* that included a **findPeople** primitive. **findPeople** returned a list of locations where people were detected. This enabled users to create applications in which the robot approached people or recognized when someone was walking towards it. Example applications written in *iCustomPrograms* are shown in Fig. 1.

4.3 Field Study I: Airport Trials

In February 2016, we visited Property A for a two-week period. We used *iCustomPrograms* with their staff to develop two interactive applications:

Passenger Delight: In this application, the robot visited waypoints in the airport's indoor garden area. At each waypoint, it waited and approached people around it. To encourage interaction, it played a beeping sound and displayed an on-screen greeting. When a passenger started the interaction, it played a sound, displayed greeting messages, and opened its bin with snacks inside. It also did a shimmy, as an enthusiastic gesture. It then said goodbye using on-screen text and a beeping sound, and moved to the next waypoint or detected person.

Service Recovery: This was a similar application to *Passenger Delight*, but it ran in the baggage claim area whenever the unloading of baggage was delayed. Compared to *Passenger Delight* the robot was more professional. For example, we removed the enthusiastic shimmy and changed the on-screen text to politely explain the baggage situation.

We ran four trials to evaluate these applications. *Passenger Delight* was deployed for the first two trials, which took a place in the indoor garden. *Service Recovery* was deployed for the last two, which took place in the baggage claim hall. Each trial lasted 3 to 4 hours. To maximize engagement, the trials were run during Chinese New Year weekend.

For each trial, 2 to 5 airport staff members and 1 or 2 Savioke employees were present, monitoring the robot from less than 15 meters away. We tried not to interact with the passengers; however, airport staff did intervene when unexpected events happened. Examples included children acting mischievously with the robot or encountering non-English speakers.

4.4 Findings

We recorded observations and notes from all the meetings and trials, and conducted follow-up interviews with personnel. We identified three themes:

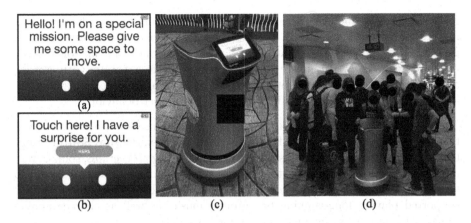

Fig. 2. Pictures from the trials held at Property A. (a) On-screen text when the robot was navigating and (b) encouraging people to interact with it. (c) The robot with the Chinese New Year decal on. (d) The robot interacting with passengers.

Problems with approaching people: The robot had difficulty approaching people naturally. While navigating to the location of a detected person, curious crowds of people would often form around the robot, surrounding it. The robot was not programmed to recognize this situation, and continued trying to navigate to its goal location, instead of starting the interaction. This often led to people getting confused or frustrated with the robot. We resolved the issue using on-screen text asking for a clear path (Fig. 2), which helped the robot go through crowds.

Initiating interactions via movements and sounds: The robot initiated interactions with users in unplanned ways, e.g., just by driving around. People would follow the robot, and even tap the screen while the robot was still moving. Additionally, people noticed the robot when it played sounds.

Desire for richer control over interactive elements: The airport staff wanted the robot to have a "brighter" or "more playful" personality. They asked to have more sounds and pre-programmed movements, as well as a way to choreograph them together to make the robot look "happier." They witnessed some passengers saying "Hello" and "Goodbye" to the robot, and requested text-to-speech so the robot could respond. Finally, they also requested the ability to play background music and to format text (e.g., changing font size or adding line breaks).

5 Enhancements and Evaluation of *iCustomPrograms*

5.1 System Enhancements

Based on our findings, we enhanced *iCustomPrograms* as follows:

Supporting touch-to-interact: As described in Sect. 4.4, the robot experienced problems with people surrounding it while navigating. This was because the robot was programmed to not respond to screen input until it was done navigating. To address this, we added the **goToUntil** primitive to *iCustomPrograms*. **goToUntil** was like **goTo**, but it stopped navigating when someone touched the robot's screen. Using this made the robot behave more naturally with crowds, as they could now get the robot's attention by tapping its screen. Figure 1b is an example application illustrating the touch-to-interact behavior.

Richer control over interactive elements: We enabled users to format on-screen text in *iCustomPrograms* using HTML. We also updated the **playSound** primitive to play sounds asynchronously. In the original *CustomPrograms*, sounds were played synchronously, meaning that the robot could not navigate or respond to screen input while a sound was playing. With our change, *iCustomPrograms* supported playing long-running background music, as well as choreographing sounds with movement or on-screen interactions.

5.2   Improved Social Applications

We developed two new social applications using the updated *iCustomPrograms*.

People Delight was based on the *Passenger Delight* application, but was designed for use in more than just airports. It used **goToUntil** to start the interaction if a person tapped the robot's screen while it was navigating. The on-screen text was adjusted to be more property-agnostic. We also added more sounds and in-place movements to attract more attention to the robot and make the main interaction more lively.

The second application, *Mingle in Place*, was developed for smaller properties that did not want to have the robot navigating around continuously. When the application was launched, the robot navigated to a preset location and displayed three options. The first option was a demo, in which the robot described itself and its delivery service. The second option was to have the robot tell a joke. The third option was to pose for a picture, in which the robot displayed "Cheese!" on its screen. The robot played 3 to 4 different sounds during the interaction and made in-place movements. If no one interacted with the robot for over a minute, the robot attracted attention by rotating left and right while making a whistling sound. The application stopped when the battery went below a predefined threshold, or when the operator canceled it.

5.3   Field Study II: Trials at Five Properties

The first author demoed the *People Delight* application to Property A in February 2016 and provided a manual describing how to use it. Between March and May 2016, a Savioke customer satisfaction manager repeated the procedure with *Mingle in Place* at Properties B–E. For the hotels, room service deliveries continued to be the primary function of the robot. The properties ran *Mingle in Place* on the robot when they wanted; we did not ask them to do so for the purposes of the study.

We conducted semi-structured interviews with a staff member from Properties A, B, C, and E, after they had used the applications for at least 2 weeks. All interviewees said that the robot had successfully interacted with visitors. Property A reported that during Easter weekend, the robot was used from 10 a.m. to 6 p.m., interacting with about 500 passengers. They also said that children 7 and up, young adults, and group travelers interacted most with the robot.

Properties B, C, and E reported that their guests enjoyed interacting with the robot, especially on weekends. As with Property A, they noted that families with children and groups were most interested in interacting with it. Property E pointed out that their robot was often too busy running deliveries to use *Mingle in Place*. 3 out of 4 interviewees said that the sounds and movements of the robot helped initiate interaction with people. However, 2 out of the 4 interviewees wanted the robot to be even more interactive and have more sounds.

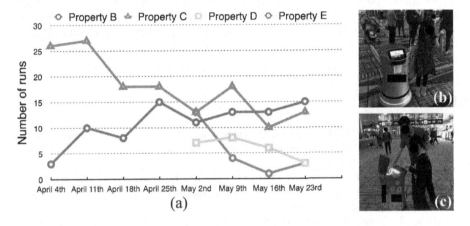

Fig. 3. (a) Weekly *Mingle in Place* usage by property. (b) Pictures taken at Property A during the field study over Easter and (c) another local holiday weekend.

We recorded the number of times the Properties B–E ran *Mingle in Place*, shown in Fig. 3a. Due to logistical problems, we could not collect any usage measurements from Property A. Properties B and C ran *Mingle in Place* the most overall. These two properties had proposed the *People Delight* application during our formative study (Sect. 3.2). For most properties, the number of runs peaked in the first two weeks and gradually decreased after, which could indicate a novelty effect. However, as we heard from Property E, low usage of the application could be due to the robot being busy with room service deliveries instead. And, during the study period, all of Properties B–E had run the application at least once a week.

Although we lack usage measurements from Property A, they reported that they used *People Delight* the most during Easter weekend and over another local holiday weekend in May 2016. They also said that staff members used *iCustomPrograms* to customize the contents of *People Delight* for each occasion, and applied festive decals to the robot's body (Fig. 3b, c).

6 Conclusion

This paper's formative study showed that service industry workers desired socially interactive behaviors for their robots. We presented *iCustomPrograms*, a system for developing such behaviors. In our first field study, we discovered important attributes for better interactions. Robots naturally attract attention, so they must be equipped with crowd-aware navigation and interactions. In our experience, service industry workers wanted rich control over interactive elements, like having more sounds, movements, and text formatting capabilities. With such enhancements in *iCustomPrograms*, we developed and deployed social applications to five real-world service industry properties. Our users not only actively used the applications, but also reported interesting observations about how people interacted with the robot. This information could lead to future improvements and ultimately to more socially interactive robots in the field.

References

1. Bohus, D., Saw, C.W., Horvitz, E.: Directions robot: in-the-wild experiences and lessons learned. In: International Conference on Autonomous Agents and Multi-Agent Systems (2014)
2. Burgard, W., Cremers, A.B., Fox, D., Hähnel, D., Lakemeyer, G., Schulz, D., Steiner, W., Thrun, S.: Experiences with an interactive museum tour-guide robot. Artificial intelligence (1999)
3. Fong, T., Nourbakhsh, I., Dautenhahn, K.: A survey of socially interactive robots. Robotics and autonomous systems (2003)
4. Glas, D., Satake, S., Kanda, T., Hagita, N.: An interaction design framework for social robots. In: Robotics: Science and Systems. vol. 7, p. 89 (2012)
5. Gockley, R., Bruce, A., Forlizzi, J., Michalowski, M., Mundell, A., Rosenthal, S., Sellner, B., Simmons, R., Snipes, K., Schultz, A.C., et al.: Designing robots for long-term social interaction. In: IEEE/RSJ International Conference on Intelligent Robots and Systems
6. Huang, C.M., Mutlu, B.: Robot behavior toolkit: generating effective social behaviors for robots. In: ACM/IEEE International Conference on Human-Robot Interaction, pp. 25–32. ACM (2012)
7. Huang, J., Lau, T., Cakmak, M.: Design and evaluation of a rapid programming system for service robots. In: ACM/IEEE International Conference on Human-Robot Interaction (2016)
8. Kanda, T., Shiomi, M., Miyashita, Z., Ishiguro, H., Hagita, N.: An affective guide robot in a shopping mall. In: ACM/IEEE International Conference on Human-Robot Interaction (2009)
9. Lee, M.K., Kiesler, S., Forlizzi, J., Rybski, P.: Ripple effects of an embedded social agent: a field study of a social robot in the workplace. In: ACM SIGCHI Conference on Human Factors in Computing Systems (2012)

10. Linder, T., Breuers, S., Arras, K.O.: On multi-modal people tracking from mobile platforms in very crowded and dynamic environments. In: IEEE International Conference on Autonomous Robot Systems and Competitions (2016)
11. Mutlu, B., Forlizzi, J.: Robots in organizations: the role of workflow, social, and environmental factors in human-robot interaction. In: ACM/IEEE International Conference on Human-Robot Interaction (2008)
12. Perez, J., Caballero, F., Merino, L.: Integration of Monte Carlo Localization and place recognition for reliable long-term robot localization. In: IEEE International Conference on Autonomous Robot Systems and Competitions (2014)

Engagement Detection During Deictic References in Human-Robot Interaction

Timo Dankert[1](✉), Michael Goerlich[1], Sebastian Wrede[1],
Raphaela Gehle[2], and Karola Pitsch[2]

[1] CITEC, Bielefeld University, Inspiration 1, 33615 Bielefeld, Germany
{tdankert,mgoerlic,swrede}@techfak.uni-bielefeld.de
[2] University of Duisburg-Essen, Universitätsstraße 2, 45141 Essen, Germany
{raphaela.gehle,karola.pitsch}@uni-due.de

Abstract. Humans are typically skilled interaction partners and detect even small problems during an interaction. In contrast, interactive robot systems often lack the basic capabilities to sense the engagement of their interaction partners and keep a common ground. This becomes even more problematic if humanoid robots with human-like behavior are used as they build up high expectations in terms of their cognitive capabilities. This paper contributes an approach for analyzing human engagement during object references in an explanation scenario based on time series alignment. An experimental guide scenario in a smart home environment was used to collect a training and test dataset where the engagement classification is carried out by human operators. The experiments already performed on the dataset give deeper insights into the presented task and motivate an incremental, mixed modality approach to engagement classification. While some of the results rely on external sensors they give an outlook on the requirements and possibilities for HRI scenarios with next-gen social robots.

Keywords: HRI · Engagement detection · Pattern recognition

1 Introduction

Engagement describes how much a participant is interested in and attentive to a conversation according to Yu et al. [1]. Bohus describes it as 'The process subsuming the joint, coordinated activities by which participants initiate, maintain, join, abandon, suspend, resume or terminate an interaction' [2].

Detecting engagement in Human Robot Interaction brings some challenges. A robot needs to detect the engagement of its interaction partner and influence the behavior if necessary. Cues that a robot might need to detect to judge about the participants' engagement include (a) the user's (shifting) head orientation, (b) the (changing) spatial position, (c) the (shifting) body orientation, and (d) verbal activities.

Detecting the participants' engagement is even more challenging once the robotic system is deployed in real-world settings, such as e.g. museums, exhibition sites, shopping areas etc., where multiple users, both adults and children, in

© Springer International Publishing AG 2016
A. Agah et al. (Eds.): ICSR 2016, LNAI 9979, pp. 930–939, 2016.
DOI: 10.1007/978-3-319-47437-3_91

varying numbers and group configurations enter the robots interactional space at unforeseeable moments in time and orient their verbal and bodily conduct not only with regard to the robots conduct, but also in reaction to other visitors/users present in the situation. Visitors in lab studies on the other hand tend to follow a robot's instruction more closely and try to work around, if complications arise.

Yamazaki et al. [3] presented a system which references exhibits in an art museum and tracks the visitors' position and head orientation. The robot in this setting engages persons, who are close and asks if it may explain the exhibit. If the person is still seen afterwards, the explanation begins and will not be stopped if the person leaves.

Jang et al. [4] presented an engagement recognizer based on video analysis and a C4.5 decision tree classifier. Their annotated signals include speech for robot and human, gaze direction, the posture, facial expressions and behavior, such as nodding, head shaking, writing, etc. Using these social signals the presented system is able to achieve 84.83 % recall performance.

Michalowski et al. [5] presented a multimodal person tracking and attention estimation approach distinguishing between *present, attending, engaged* and *interacting*. These classes are based on the spatial location. If a frontal face has also been associated with the location, a higher level of engagement is assumed. As sensory input a laser range scanner, a camera and a keyboard interface were used. While typing on the keyboard is seen as ground truth for interaction, the other sensors were also able to contribute estimates about the attention levels.

Recognizing engagement in human-robot interaction is also shown by Rich et al. [6]. They present a classification system based on state machines on top of social signals to observe social patterns and thus reason about engagement.

Learned models based on spatial features, frontal face features and manually labeled attention features were presented by Bohus et al. [2]. These models are used for determining if a user is approaching or leaving an interaction. This approach deals with bypassing visitors, which may be interested to start an interaction.

The approach presented in this paper is different to previous work in two ways. While some of the presented related works try to detect engagement at the first seconds of interaction, often referred to as "opening", the focus in this paper lies on situations during deictic references (f.e. pointing gestures). These references are of high interest in scenarios, in which the robot introduces its user to a new environment or while acting as guide in a museum. Secondly the decision process, in this paper, is solely based on online acquired features and uses machine learning methods. This approach promises to find a generalized solution for the classification task to bring acceptable behavior in not yet seen situations as well as simplifying the integration of new features that may be introduced in later versions of the system. On the other hand the generation of training data is required to find a good classification function as well as for model testing that contains the situations to be classified.

The rest of the paper will be structured as follows: In Sect. 2 the system architecture will be presented as well as the environment in which the data collection was conducted. Next, in Sect. 3, the process for collecting the required training and test data will be described. Section 4 describes how the acquired data was processed and how the users behavior is classified. The results of the already conducted experiments are shown in Sect. 5 and are discussed in Sect. 6.

2   System and Environment

As a test environment a smart home was used [7]. A map of the apartment is shown in Fig. 1. It is equipped with multiple sensors of different types, such as depth cameras installed in the ceiling. For acting with its users multiple displays are placed in the environment. While most of these are static, a pan/tilt projector is also available that can project images onto most surfaces within the apartment. The social robot used in this paper is the Nao robot by Aldebaran, which is placed on a table shown with the letter N on the map.

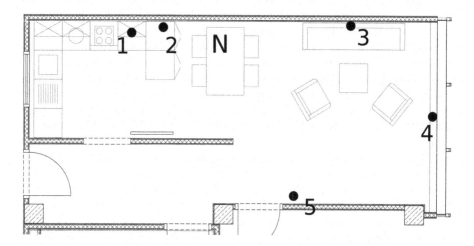

Fig. 1. Overview of the test environment. 'N' marks the position of the robot and the red points mark the positions of objects/screens, the robot is referencing to. Point 2 has been referenced twice, all other points once. (Color figure online)

Figure 2 shows an overview of the system architecture. In this setup the *camera* of the Nao robot is used for estimating the rotation and location of the users head using the *VFOA*[1] component [8] in the robots own coordinate system. Since users experience problems when they are not facing the robot during references [9] a better guidance needs to be provided. Because head tracking tends to fail when no face is visible at all the robots sensors are accompanied by the

---

[1] *VFOA* is short for *visual focus of attention*.

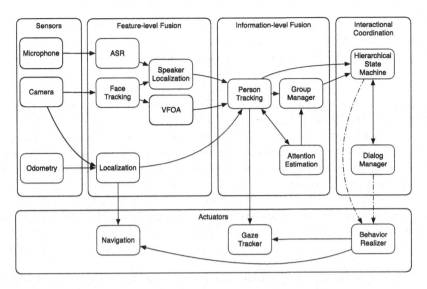

Fig. 2. Architecture of the system. The person tracking merges data from the robot and the smart home in order to get consistent data.

apartment's depth cameras (not shown in the diagram). They also provide an estimate of the user's head position covering most places in the apartment. The data is fused at the *Person Tracking* component using Kalman filters. A side effect the fusion allows the *VFOA* to re-engage the user once his face becomes visible again. The executed scenario is based on a state machine with incorporated control points at which the *Attention Estimation* component decides about the next steps to take in the scenario. During the data collection it is replaced by the decision of the human operator as described in Sect. 3. While they are part of the general system architecture the *microphone, ASR, Speaker Localization* and *Group Manager* are not used.

The system is based on the event-based and message oriented middleware *RSB* [10].

3 Data Collection

Since this paper follows a data driven approach to engagement classification a guide scenario was designed. In this scenario the Nao robot explained some visualizations showing information about the intelligent apartment. The objects on the map are explained, one by one, by using verbal explanations as well as deictic gestures. The objects are located at different positions to ensure that the resulting dataset contains a mixture of different poses and behaviors.

Each trial begins with a participant entering the apartment from door on the left side. They eventually notice the robot that will offer them to give them background information about the apartment. If they accept the offer the robot will start explaining verbally with pointing gestures to the objects. After each reference, a control point is reached and the system is able to either move on to the next object to be explained or it can choose from two repair strategies. This is either a *soft repair*, where the robot will help the user by looking at the object and then back to the user or a *strong repair* option, where the robot will combine a deictic reference with a verbal reference to help the user locating the object, e.g. *Over there {start gesture} to my right*.

As stated before, the system is controlled by a human operator. The operator is provided with a visualization of the person tracker, the view from the robot's camera and an external microphone and has been instructed to react only to these system observable cues. The resulting decisions provide the labels for later offline analysis and for qualitative case analyses/interactional linguistics to gather insights into the participants' signals during the interaction upon which the operator depends his choice over the next relevant action and its timing. Therefore, we take into account the interactional competences of the human operator.

The trials were executed in a very quiet laboratory free from distractions that can be observed in real world settings. Thus the participants were divided into two conditions. The participants of the first condition were not provided with any initial information about the scenario and no distractions. The second group of participants was told beforehand that they are going to meet a robot inside the apartment and that it is going to explain the environment by using deictic gestures. To distract them the pan/tilt projector was used, which projected a symbol onto the wall very close to the robot and then moved to a different object in the room. The participants were asked to follow the signal of the projector instead of following the robot's instructions.

The scenario was executed for 35 participants, out of which 19 had previous experiences with robots. 7 runs had to be removed due to technical issues. Of the remaining 28 runs, 16 belonged to the first condition and 12 emerged from the second condition that contained distractions. For all runs the data published by the components mentioned in Sect. 2 was collected, including the fused hypotheses about the persons' locations and their head rotation.

4   Pattern Matching

The approach to engagement classification presented in this work will focus on situations with object references. Thus, while the whole interactions have been recorded, only data during object references will be considered for the classification process. The utterances that included the object references were about 7 s in length and are shown in Fig. 3. The reference itself was near the beginning of the utterance at about 1.5 s.

While the decisions made by the wizard during the execution of the system could be used as labels for the data set the button press does not only capture on how to proceed after the control point. Since the decision was made when the wizard was sure on how to continue they also capture the time the wizard

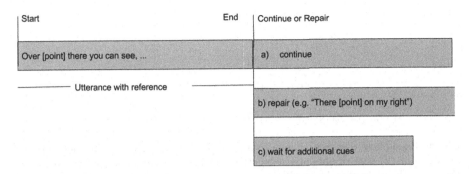

Fig. 3. The timing of the situation taken into account for analysis. After the utterance the system must decide, if it (a) continues with further explanation, (b) provides a repair or (c) waits for additional data for a postponed decision.

needed for the decision. In a live system the robot will also have to decide if the classification is possible now or more data is needed. Thus, in a second offline labeling step, three other labels are introduced: *Continue*, *Repair* and *Wait* which describe the situation right after the utterance. While *Continue* is equal to the wizard's decision of continuing the explanation, *Repair* fuses *soft repair* and *repair*. The *Wait*-class enables the system to postpone the decision. Though to mimic the wizards decisions is the long-term goal in the first step these new labels are considered as goals for the classification task.

During the time frame considered for classification a lot of data about the user is available, ranging from raw images over audio streams to high level features such as the 3 dimensional position inside the apartment and the rotation of the head. To reduce the complexity the approach only uses high level features as described in Sect. 4.1 to get meaningful time series. The actual assignment of classes to the time series will be explained in Sect. 4.2.

4.1 Pre-processing the Data

For classification purposes two main features are considered: the distance of the user to the referenced object and the rotation of the head. While the absolute distance value will differ in scale for each object. To make those comparable it was normalized to a value between 0 and 1, 0 meaning *far away* and 1 is "very close". Figure 4 shows examples for the distance time series of a situation where the robot invites the user to touch the object referenced. While the users on the left image went to the object (labeled *Continue*) the ones on the middle image did not or were confused (labeled *Repair*). For the users shown on the right image the situation was not clear (labeled *Wait*).

The head rotations are also used as a time series. Only the yaw rotation is used, since most objects were at the same height. In contrast to the distance to the referred object, the head rotation has some missing values, because the user's face was not always observable (e.g. when the user turned his back towards

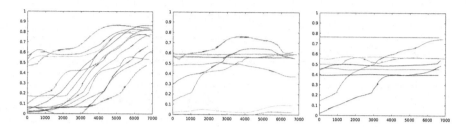

Fig. 4. Exemplary distance change over time if (left to right) (1) participants are attentive or (2) participants are inattentive and (3) if no decision has been made, at the end of the utterance. The higher the value on the y axis, the closer is the participant to the referenced object. The x axis is the time in milliseconds.

Fig. 5. Typical head rotations after a deictic reference to an object on the right side. The pictures top left and top right show a person, which is oriented correctly after the reference, bottom left and bottom right shows a orientation to the wrong direction.

the robot). These missing values are estimated using cubic spline interpolation. Typical head movements are shown in Fig. 5. While both users had the robot in their field of view, during the deictic reference, the one in the top row was reorienting to the object. The user in the bottom row was reorienting in another direction.

4.2   Prototype Selection and Matching

The overall task of the classifier is to assign a label to a series of features with varying length. To reduce the complexity for the classifier, the new series are compared to a bag of prototypical behaviors with known classes. This approach is often called *nearest neighbor classifier*. The prototypes must be selected for both features and for each class and have to have low distance to within the own class and high distance to other classes.

The learning phase of the classifier consists of the selection of prototypes. When evaluating the classifier, each new time series to be classified is compared to all prototypes of all classes using dynamic time warping, which gives a distance between the new time series and the prototype. The selected class is the one of the closest prototype.

5   Results

The effect of the two features were analyzed separately. For matching movement behavior, data from one of the references was used (object 5 on Fig. 1). This includes twelve cases of *continue* and eight of *repair* and *wait* each. Since the object was on the opposite wall of the robot, faces are not observable. For the matching of head rotations, object 2 was chosen. Due to two references to this object, the data consists of 24 times *continue*, 12 times *wait* and 4 times *repair*. For this object 16 (12 from *repair* class) behaviors are not taken into account, since the face was not observable in the time series and therefore no head rotation could be estimated.

Table 1. Results of the classification using the movement of the participant towards object 5 as cue.

Class	Decided class			
	Continue	Wait	Repair	Accuracy
Continue	9	3	0	0.75
Wait	2	6	0	0.75
Repair	0	2	6	0.75
	11	11	6	0.75

Based on the the change of the distance between the user and the referenced object, it is possible to match 75 % of the observed behaviors correctly. Table 1 shows that behaviors from the classes *continue* or *repair* have been matched to the class *wait*, if the class is decided wrong. In theses cases, the wizard decided for the correct class after the utterance of the robot.

Table 2 shows the results using the head rotation as a cue. While the accuracy is higher than using the spatial information, it can not be used consistently. For this example 16 out of 56 samples were not usable, due to too much missing information as mentioned before.

Table 2. Results of the classification using the head rotation of the participant towards object 2 as cue.

Class	Decided class			
	Continue	Wait	Repair	Accuracy
Continue	19	4	1	0.79
Wait	3	9	0	0.75
Repair	0	0	4	1
	22	13	5	0.8

6   Discussion

Comparing the curves of the distance change in Fig. 4, it is observable that 3 of the curves in Fig. 4 (1) do not increase as strong as the other curves for this class. The remaining curves in this class also are near sigmoid, while the others are near linear. On this basis it is understandable, why some of the curves have been matched wrongly.

While the head rotation matching seems to have better results than the motion based, some of the samples were not possible to match at all, due to missing values. Especially in the class *repair* it occurred quite often. This can be explained due to the user is looking at a different object. In this case, no face was observable by the internal camera of the robot during an utterance with a reference to an object.

Even though the motion based approach has an accuracy of about 75 % for the presented cases, these cases are quite limited. The matching based on the user's movement requires an invitation to go to an object. This can be achieved in dialog, which results in more effort during interaction design.

For the rotation-based matching the face needs to be observable. This could also be achieved by only referencing objects near the robot, instead of objects which are at unfortunate positions. In this case, the interaction would be restricted to a small interaction area and would detach from natural interaction. Another solution are better sensors.

As next steps, the presented approach will be expanded to deal with more of the references and create a model, which will deal with the *wait* class based on additional information gathered after the end of an utterance using longer series of movement and head rotations. The system will be used autonomously instead of a trained wizard. To create a more robust model, more data will be recorded.

In this paper, the prototypes have been chosen by hand. In the future, the prototypes will be chosen by the system automatically and will be adapted, based on the best matching results.

Improving the perception is also a goal, which will allow to create additional cues, such as body orientation and consistent head rotations. These additional cues, could improve the decision process.

Acknowledgments. The authors acknowledge the financial support from the Cluster of Excellence Cognitive Interaction Technology CITEC (EXC 277), Bielefeld University and the Volkswagen Foundation (Dilthey Fellowship Interaction & Space, K. Pitsch).

References

1. Yu, C., Aoki, P.M., Woodruff, A.: Detecting user engagement in everyday conversations. Science, p. 4 (2004)
2. Bohus, D., Horvitz, E.: Learning to predict engagement with a spoken dialog system in open-world settings. In: SIGdial 2009, London, UK (2009)
3. Yamazaki, K., Yamazaki, A., Okada, M., Kuno, Y., Kobayashi, Y., Hoshi, Y., Pitsch, K., Luff, P., Lehn, D., Heath, C.: Revealing gauguin: engaging visitors in robot guide s explanation in an art museum. In: CHI 2009 Proceedings of the SIGCHI Conference on Human Factors in Computing Systems, pp. 1437–1446 (2009)
4. Jang, M., Ahn, B.K., Park, C., Yang, H.S., Kim, J.H., Cho, Y.J., Lee, D.W., Cho, H.K., Kim, Y.A., Chae, K.: Building an automated engagement recognizer based on video analysis. In: Proceedings of the 2014 ACM/IEEE International Conference on Human-Robot Interaction - HRI 2014, pp. 182–183. ACM Press, New York (2014)
5. Michalowski, M.P., Simmons, R.: Multimodal person tracking and attention classification. In: Proceeding of the 1st ACM SIGCHI/SIGART Conference on Human-Robot Interaction - HRI 2006, p. 349 (2006)
6. Rich, C., Ponsler, B., Holroyd, A., Sidner, C.: Recognizing engagement in human-robot interaction. In: 2010 5th ACM/IEEE International Conference on Human-Robot Interaction (HRI) (2010)
7. Holthaus, P., Leichsenring, C., Bernotat, J., Richter, V., Pohling, M., Carlmeyer, B., Köster, N., Meyer zu Borgsen, S., Zorn, R., Schiffhauer, B., Engelmann, K.F., Lier, F., Schulz, S., Cimiano, P., Eyssel, F.A., Hermann, T., Kummert, F., Schlangen, D., Wachsmuth, S., Wagner, P., Wrede, B., Wrede, S.: How to Address Smart Homes with a Social Robot? A Multi-modal Corpus of User Interactions with an Intelligent Environment, European Language Resources Association (ELRA) (2016)
8. Ba, S., Odobez, J.M.: Recognizing visual focus of attention from head pose in natural meetings. IEEE Trans. Syst. Man Cybern. Part B (Cybern.) **39**(1), 16–33 (2009)
9. Pitsch, K., Wrede, S.: When a robot orients visitors to an exhibit. Referential practices and interactional dynamics in real world HRI. In: The 23rd IEEE International Symposium on Robot and Human Interactive Communication, pp. 36–42. IEEE (2014)
10. Wienke, J., Wrede, S.: A middleware for collaborative research in experimental robotics. In: International Symposium on System Integration, Kyoto, Japan (2011)

Making Turn-Taking Decisions for an Active Listening Robot for Memory Training

Martin Johansson[1(✉)], Tatsuro Hori[2], Gabriel Skantze[1], Anja Höthker[3], and Joakim Gustafson[1]

[1] KTH Speech, Music and Hearing, Stockholm, Sweden
vhmj@kth.se, {gabriel,jocke}@speech.kth.se
[2] Toyota Motor Corporation, Toyota, Japan
tatsuro_hori@mail.toyota.co.jp
[3] Toyota Motor Europe, Brussels, Belgium
Anja.Hoethker@toyota-europe.com

Abstract. In this paper we present a dialogue system and response model that allows a robot to act as an active listener, encouraging users to tell the robot about their travel memories. The response model makes a combined decision about when to respond and what type of response to give, in order to elicit more elaborate descriptions from the user and avoid non-sequitur responses. The model was trained on human-robot dialogue data collected in a Wizard-of-Oz setting, and evaluated in a fully autonomous version of the same dialogue system. Compared to a baseline system, users perceived the dialogue system with the trained model to be a significantly better listener. The trained model also resulted in dialogues with significantly fewer mistakes, a larger proportion of user speech and fewer interruptions.

Keywords: Turn-taking · Active listening · Social robotics · Memory training

1 Introduction

Social robots of the future are envisioned to assist us in our daily life in various ways. One interesting field where social robots could be of use concerns elderly care and improving the life of elderly people. In this field, there is for example work towards personalized conversational interfaces [1] and embodied conversational agents (ECA) that can assist elderly in organizing their daily activities [2], and serve as conversational companions for people with dementia [3, 4]. Aside from the possibility of providing a conversational partner that does not mind repetitions, such a system could potentially also be used for memory training and monitoring. A key challenge in realizing such a system is to create an artificial agent that can be characterized as a good listener. The aim of this paper is to develop a conversational system for a social robot that acts as an active listener to a user talking about memories of past events. Using machine learning, we train a model that can predict *when* the system should respond, and *how* it should respond, based on Wizard-of-Oz data.

© Springer International Publishing AG 2016
A. Agah et al. (Eds.): ICSR 2016, LNAI 9979, pp. 940–949, 2016.
DOI: 10.1007/978-3-319-47437-3_92

2 Background

In this work we are interested in a dialogue scenario where the main flow of information goes from a speaker to a single listener. Despite the imbalance in flow of information between the interlocutors, the behavior of the listener is still important for the speaker. Experiments on active listening have shown that when speakers receive more feedback, their narratives become more comprehensible, and that listener feedback helps to coordinate what the speaker says with what the listener needs to know [5]. A common form of feedback in such settings is *backchannels* [6] – short acknowledgement utterances, such as "uh-huh" or "yeah", or non-verbal gestures, such as head nods. By producing a backchannel, the listener does not claim the floor, but rather encourages the speaker to continue. However, a good listener needs to balance the use of such backchannels with other types of responses, such as follow-up questions [7].

Human speakers in dialogue synchronize their turn-taking to minimize gaps and overlaps [8] through the use of turn-holding and turn-yielding cues, including prosody, syntax and gestures [9–11], as well as gaze in face-to-face interactions [12]. The more turn-yielding cues a speaker presents together, the higher the likelihood for the listener to take the turn. For example, flat final pitch, syntactic incompleteness and filled pauses are strong cues to turn hold. Meena et al. [13] presented a data-driven model that used automatically extracted features from syntax, prosody and context to detect suitable response locations in a spoken dialogue system that was listening to a user giving route descriptions. It was found that syntax was the most important feature, given that the speech recognition (ASR) worked well. When ASR performance dropped, prosody and context were also viable features. For our setting, this is an encouraging finding, since speech recognition of spontaneous travel memories is potentially very challenging, and we want the system to be a good listener even without using an ASR. In this paper, we extend the work by Meena et al. by not only detecting suitable response locations; we also want the system to be able to choose the type of response to make: a backchannel acknowledgment or taking the turn and asking a follow-up question.

The use of data-driven methods for creating a dialogue control component for an active listener has also been investigated by Meguro et al. [14], who built a model based on sequences of manually annotated dialogue acts (DA) in human-human textual dialogues. However, their model was not evaluated in a real spoken dialogue setting. In the field of Embodied Conversational Agents (ECAs), there have also been works on modelling active listening behavior [15, 16]. These studies have found that people interact with a responsive listener longer than with an unresponsive one, and also use more words. Sakai et al. [3] proposed an ECA as a conversational partner for individuals with dementia, using a combination of questions asked by the ECA and a rule-based system to generate verbal and non-verbal backchannel feedback, based on silence length and pitch. Yasuda et al. [4] also developed and evaluated an ECA for this use, and found the ECA to elicit utterances from subjects with Alzheimer corresponding to 74 % of the length of those in an identical human-human condition with the same subjects. However, none of these studies were done in a human-robot interaction scenario. Moreover, they did not propose a complete data-driven model of *when* to respond and *what type* of response to give (such as a backchannel or a follow-up question).

3   A Listening Social Robot

The domain chosen for the human-robot dialogue in this paper is a travel memory domain, where subjects are to tell the robot about a past visit to a foreign country, while the robot listens actively to elicit more elaborate descriptions. The interaction is dyadic with the subject seated in front of the robot, as illustrated in Fig. 1.

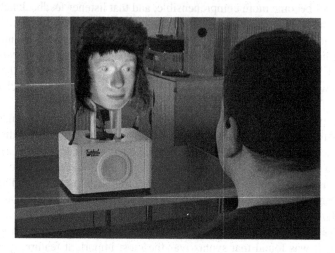

Fig. 1. Illustration of the experiment setting with the robot Furhat and a user interacting

Furhat [17], shown in Fig. 1, was chosen as the robot participant in the experiment. Furhat is a back-projected human-like robot head using speech synthesis and state-of-the-art facial animation, mounted on a mechanic neck. The facial animation architecture allows for speech with accurate synchronized lip movements, as well as eye movement and facial expressions.

3.1   The Listening Dialogue System

The interaction between a subject and the system comprise three phases, starting with an introductory phase where the system holds the initiative to get the conversation started. The phase begins with the system greeting the subject and asking for the subject's name, which country the travel memories will concern, and ends with the system asking which city the subject visited. The introductory phase is seamlessly followed by the active listening phase, where the subject is asked to tell the system about travel memories from the selected country. The flow of the active listening phase is illustrated in Fig. 2. As the goal of the system is to keep the subject talking about recalled travel memories rather than to extract specific information, the system is designed to allow the subject to keep the initiative and to keep the flow of information going mainly from the subject to the system, using backchannels and questions to encourage the user to continue speaking.

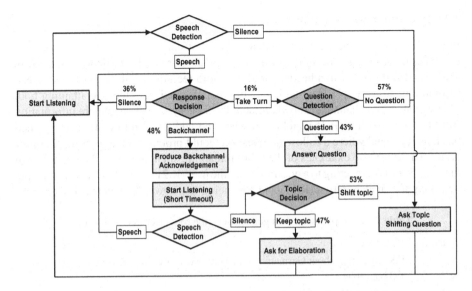

Fig. 2. The active listening phase of the listening dialogue system. Percentages indicate decision distributions in the training corpus.

The active listening flow is centered on the voice activity detection (VAD) component of the system, considering the end of detected voice activity as a potential location for making a response. A **Response decision** module then makes the decision to either stay silent, to produce a verbal backchannel acknowledgment, or to make a longer response. Since the user sometimes might ask counter-questions, a **Question detection** module is used to either react to the question or to ask a follow-up question that will advance the dialog to a new (sub) topic. If no speech is detected, the latter action will automatically be selected; the system will ask the subject a question related to the country under discussion. In this first version of the system, the system picked the utterance from a pre-defined handcrafted list, and the choices of countries to talk about were limited to three: France, Germany and Spain. If no speech is detected shortly after a produced verbal backchannel acknowledgement, a **Topic decision** module makes the decision to stay on topic by asking for an elaboration or to advance to a new topic as described above.

Finally, when a predefined amount of time has passed, the active listening phase ends and is replaced by the closing phase where the system retakes the initiative, ends the conversation and thanks the subject for the chat.

The system was implemented using the open source dialogue system framework IrisTK [18], which is used to control the dialogue as well as the non-verbal aspects of the robot. The gaze behavior was designed to follow general rules found in studies on human-human interaction [12]: When the robot was about to speak a longer utterance, it gazed away from the user at the beginning of the utterance and gazed back towards the end. In all other situations, the robot gazed towards the user, with small saccadic movements.

3.2   Interaction with the Robot

Before an interaction between a subject and the robot started, the task was explained by the experiment leader. The subject was then seated in front of the robot, illustrated in Fig. 1, and equipped with a headset used for audio recording and the system's speech recognition. An example from an interaction is presented in Table 1, containing both cases where a backchannel acknowledgment from the system elicited more information from the user (such as R.15), and cases where no additional information was gained (such as R.6), and where a follow-up question would probably have been more appropriate. There are also examples of the robot remaining silent during a filled pause from the user (S.4–S.5), reacting to a question from the user (R.21), as well as staying on topic by asking for an elaboration (R.11) and advancing to a new topic (such as R.8).

Table 1. Excerpt from a dialogue between the robot (R) and a subject (S).

Speaker	Utterance	Response type
R.1	OK. Let's talk about Germany. Which city did you visit?	RequestCity
S.2	I visited Berlin.	
R.3	Could you tell me about that?	Elaboration Question
S.4	It was really nice and um …	
S.5	I walked a lot.	
R.6	ah	Backchannel
S.7	yeah	
R.8	ehm … Do you think I should go there?	Topic-shift Question
S.9	Yeah, sure.	
R.10	oh	Backchannel
R.11	Why do you think so?	Elaboration Question
S.12	It is really nice. I think you definitely should go there.	
R.13	I heard you can eat nice sausage there. Did you try that?	Topic-shift Question
S.14	In Berlin, yes. It was really nice eating sausage in Berlin.	
R.15	mhm	Backchannel
S.16	Sauerkraut	
R.17	oh	Backchannel
S.18	yeah	
R.19	ehm … Did you see more interesting things there?	Topic-shift Question
S.20	Yeah, they had a lot of graffiti. Do you like graffiti?	
R.21	ehm … I will answer next time we meet.	Question Reaction

3.3   Wizard of Oz Data Collection and Annotation

A round of initial data collection was carried out through a Wizard-of-Oz version of the system, where the wizard was deciding the verbal actions of the robot during the active listening phase illustrated in Fig. 2, thus playing the role of the three decision models we wanted to train. To reduce the latency of the robot's turn-taking, the wizard claimed the floor through a hesitation sound (e.g., "uhm") while choosing the type of response.

Interactions with five subjects were used to tune the system as well as to train the wizard, while another five subjects participated in the initial data collection, recording one dialogue each.

The recorded audio was automatically segmented into Inter Pausal Units (IPUs), using energy-based VAD with a maximum of 500 ms internal silence. For the annotation, we considered the end of IPUs as potential response locations, and the annotator decided whether the best reaction would be to either stay **silent**, produce a **backchannel** or **take the turn**. The first two dialogues were annotated by two of the authors with a good inter-annotator agreement, a kappa score of 0.72. The remaining three dialogues were then annotated by one author. In total, a set of 131 decision points from the 5 dialogues were annotated for turn-taking decisions. The first decision, to stay *silent* (36 % of annotated instances) represents instances where the robot should not say anything. The second decision, *backchannel* (48 %), represents instances where the robot should make a back-channel (ex. uh-huh) to promote more elaborate descriptions. The final decision, *take the turn* (16 %), represents for example instances where the robot should ask a question to encourage a slight topic switch, since the user did not appear to intend continue speaking on the current topic (ex. "What did you like best?" or "Interesting, why do you say that?") or react to a question from the user.

3.4   A Data-Driven Model for Response Decisions

The human-robot travel memories corpus collected and annotated in Sect. 3.3 was used to train a model for making response decisions in the dialogue. The model was based on the Random Forest Algorithm in the WEKA toolkit [19], using only automatically extractable features as outlined below. In total, 15 features were used to represent context and prosody. For context, we used the length of the user's turn so far and three features from the dialogue manager: the previous response decision (silent, back-channel or sentence), the amount of time since the system last had the turn and finally what type of sentence the system uttered during that turn (answering a question from the user, asking an open question, or asking the user to elaborate). As prosodic features, we used final pitch and energy. A pitch tracker based on the Yin algorithm [20] was used to estimate the F_0 at a rate of 100 frames per second. The F_0 values were then transformed to log scale and z-normalized for each user. For each IPU, the last voiced frame was identified and then regions of **200 ms** and **500 ms** ending in this frame were selected. Based on this, we used 3 energy features (the maximum energy for the 200 ms and 500 ms regions and for the full IPU) and 8 pitch features (mean, standard deviation and slope of the normalized F_0 values in regions of both the last 200 ms and 500 ms, maximum of last 500 ms, the maximum and standard deviation of the normalized F_0 values over the full IPU).

The weighted F-score of this model using 10-fold cross-validation was 0.65, compared to 0.31 for a model always selecting the majority class. Prosody was more useful than context; a model using pitch and energy yielded a weighted F-score of 0.62 compared to 0.52 for a model using only dialogue context.

The corpus was also used to build a Question detection model and a Topic decision model for use as illustrated in Fig. 2, using the JRIP Algorithm in the WEKA toolkit.

For the Question detection model, a non-falling final pitch was selected as an indicator for questions. For the Topic decision model, if the user utterance ended in a low pitch, and the system had recently made a topic shift question, the system should not shift topic again. These models performed better than the majority class baselines, but since there are only a small amount of examples for these models in the corpus, 21 and 34 respectively, we will focus our analysis primarily on the Response decision model.

4   Results and Discussion

The dialogue system described in Sect. 3 was evaluated using 15 subjects (14 male users and 1 female user, 25–62 years old) from among employees and students at KTH, who each interacted with both a baseline system and the trained system. It is not obvious what to use as a baseline system. We chose to use a random decision model, where the options where weighted based on the distributions of the decisions in the annotated training material. The order of interaction with the two systems was shuffled to avoid ordering effects, resulting in 8 of the subjects experiencing the baseline system first and 7 subjects the new system first. Each interaction was 4 min long, controlled by the system.

4.1   Qualitative Evaluation

The developed system and the baseline system were evaluated qualitatively by letting the subjects fill out a questionnaire, using a semantic differential scale. After each interaction, subjects were asked to rate five aspects of the interaction: (i) how exciting it was, (ii) if the turn-taking was bad or good, (iii) the quality of the content in the robot's feedbacks, (iv) the naturalness of the robot's gaze behavior, and (v) if the system was perceived as a good listener. The mean rating for each aspect and system is shown in Fig. 3 to illustrate subjects' ratings of the systems. As can be seen, the subjects' ratings of the trained system as a good listener was statistically significantly higher than that of the baseline system (Wilcoxon Signed Ranks Test, $Z = -2.783$, $p = .0054$). The clear difference between the two systems is interesting considering the closer ratings of the other aspects. One possible explanation to this is that being a good listener is an amalgam of different skills.

It is also apparent that there is room for improvement of all aspects. However, it is not clear how to interpret the absolute ratings for a specific aspect, as there were for example no comparisons with human listeners. Thus, we only compare the ratings of the trained system with those of the baseline system, although we note that the automated gaze behavior (same for both systems) received high ratings.

4.2   Perception Test

To complement the participating subjects' ratings, we carried out a perception test where three subjects, who did not take part in the interactions, listened to recorded audio from the interactions with the task of spotting bad interactions. The subjects were instructed

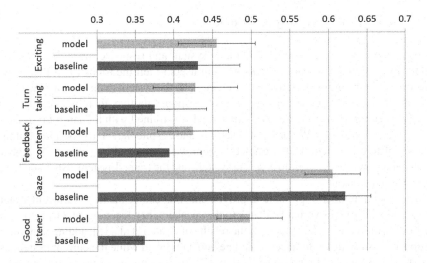

Fig. 3. Mean of questionnaire results encoded on a scale from 0 (bad) to 1 (good), with error bars illustrating the estimated standard error of the mean.

to push a button whenever the system made an inappropriate decision, either by remaining silent when a response was expected, by making a response at the wrong place, or by making the wrong type of response. The trained system was marked for significantly fewer inappropriate decisions than the baseline system ($M = 7.7$, $SD = 4.1$ vs. $M = 10.3$, $SD = 2.6$) (one-sided T-test, p = .024).

In addition to the mistakes of the Response decision model, the annotators marked 23 decisions for the baseline system and 19 decisions for the trained system as problematic, despite the systems' choices of a correct type of action. These represent errors made by the Question detection and Topic decision models. This could for example be due to the system not recognizing that the subject had asked it a question, or asking a question that did not fit the context. These kinds of errors are arguably detrimental to the illusion of the system as a good listener who understands the speaker, and indicates that this kind of shallow processing needs to be complemented with some sort of language understanding.

4.3 Quantitative Evaluation

We also carried out a quantitative evaluation of the collected dialogue data, namely the proportion of subject speech and the number of interruptions of the user made by the system. The evaluation was based on automated segmentation of the recorded speech using energy-based VAD. The proportion of subject speech time of the total speech time in each dialogue was significantly higher for the trained system ($M = 54.6$ %, $SD = 10.7$ %) than for the baseline system ($M = 49.0$ %, $SD = 6.2$ %) (one-sided T-test, p = .0455), while the number of times the system interrupted the subject in each dialogue were lower ($M = 6.0$, $SD = 2.9$ vs. $M = 8.3$, $SD = 4.4$) (one-sided T-test, p = .0496).

5 Conclusions and Future Work

In this paper, we have presented a dialogue system and response model that allows a robot to act as an active listener, encouraging users to tell the robot about their travel memories. Unlike previous active listening models, the system makes a combined decision about when to respond and what type of response to give, thereby eliciting more elaborate descriptions from the user. The model was trained with human-robot interactions collected through a Wizard-of-Oz setting, and despite the relatively small training data set of five interactions, the trained system was perceived by subjects as a better listener than a weighted random baseline system, and the trained system also made fewer problematic decisions.

There are a number of possible improvements that can be made. The current system for example only gives verbal feedback to the speaker when listening, and tries to avoid feedback within utterances. Based on the results of Gratch et al. [15], adding non-verbal feedbacks, also within utterances, could be beneficial, and while the current system use gaze to signal turn-holding as described by [12], it does not make use of the speaker's gaze. The current system also makes mistakes due to lack of understanding, for example by choosing irrelevant prompts or missing that the user asked the system a question.

Given the small amount of data used for training the models, and the fact the we only used context and prosody for the models, the results are encouraging. For future work, we will add speech recognition to the models, not only to improve the turn-taking behavior, but also to extract the contents of the travel memories. We think that the system could be valuable for supporting and diagnosing people with dementia, and this is something we also want to investigate in the future.

Acknowledgements. This research was funded by Toyota Motor Corporation, as well as the Swedish research council (VR) project *Coordination of Attention and Turn-taking in Situated Interaction* (2013-1403).

References

1. Benyon, D., Mival, O.: Introducing the companions project: intelligent, persistent, personalised interfaces to the internet. In: Proceedings of the 21st British HCI Group Annual Conference on People and Computers: HCI...But Not As We Know It, vol. 2, pp. 193–194 (2007)
2. Beskow, J., Edlund, J., Granström, B., Gustafson, J., Skantze, G., Tobiasson, H.: The MonAMI reminder: a spoken dialogue system for face-to-face interaction. In: Interspeech 2009, Brighton, U.K. (2009)
3. Sakai, Y., Nonaka, Y., Yasuda, K., Nakano, Y.I.: Listener agent for elderly people with dementia. In: HRI 2012, pp. 199–200 (2012)
4. Yasuda, K., Aoe, J., Fuketa, M.: Development of an agent system for conversing with individuals with dementia. In: The 27th Annual Conference of the Japanese Society for Artificial Intelligence (2013)
5. Kraut, R.E., Lewis, S.H., Swezey, L.W.: Listener responsiveness and the coordination of conversation. J. Pers. Soc. Psychol. **43**(4), 718–731 (1982)

6. Yngve, V.H.: On getting a word in edgewise. In: Papers from the Sixth Regional Meeting of the Chicago Linguistic Society, Chicago, pp. 567–578 (1970)
7. Kobayashi, Y., Yamamoto, D., Koga, T., Yokoyama, S., Doi, M.: Design targeting voice interface robot capable of active listening. In: 5th ACM/IEEE International Conference on Human-robot Interaction, pp. 161–162 (2010)
8. Sacks, H., Schegloff, E., Jefferson, G.: A simplest systematics for the organization of turn-taking for conversation. Language 50, 696–735 (1974)
9. Duncan, S.: Some signals and rules for taking speaking turns in conversations. J. Pers. Soc. Psychol. 23(2), 283–292 (1972)
10. Koiso, H., Horiuchi, Y., Tutiya, S., Ichikawa, A., Den, Y.: An analysis of turn-taking and backchannels based on prosodic and syntactic features in Japanese Map Task dialogs. Lang. Speech 41, 295–321 (1998)
11. Gravano, A., Hirschberg, J.: Turn-taking cues in task-oriented dialogue. Comput. Speech Lang. 25(3), 601–634 (2011)
12. Kendon, A.: Some functions of gaze direction in social interaction. Acta Psychol. 26, 22–63 (1967)
13. Meena, R., Skantze, G., Gustafson, J.: A data-driven model for timing feedback in a map task dialogue system. In: 14th Annual Meeting of the Special Interest Group on Discourse and Dialogue (SIGDIAL), Metz, France, pp. 375–383 (2013)
14. Meguro, T., Higashinaka, R., Minami, Y., Dohsaka, K.: Controlling listening-oriented dialogue using partially observable markov decision processes. In: Proceedings of the 23rd International Conference on Computational Linguistics, Stroudsburg, PA, USA, pp. 761–769 (2010)
15. Gratch, J., Okhmatovskaia, A., Lamothe, F., Marsella, S.C., Morales, M., van der Werf, R.J., Morency, L.-P.: Virtual rapport. In: Gratch, J., Young, M., Aylett, R.S., Ballin, D., Olivier, P. (eds.) IVA 2006. LNCS (LNAI), vol. 4133, pp. 14–27. Springer, Heidelberg (2006)
16. Huang, L., Morency, L.-P., Gratch, J.: Virtual rapport 2.0. In: Vilhjálmsson, H.H., Kopp, S., Marsella, S., Thórisson, K.R. (eds.) IVA 2011. LNCS, vol. 6895, pp. 68–79. Springer, Heidelberg (2011)
17. Al Moubayed, S., Skantze, G., Beskow, J.: The furhat back-projected humanoid head - lip reading, gaze and multiparty interaction. Int. J. Humanoid Rob. 10(1), 1350005 (2013)
18. Skantze, G., Al Moubayed, S.: IrisTK: a statechart-based toolkit for multi-party face-to-face interaction. In: Proceedings of ICMI, Santa Monica, CA (2012)
19. Hall, M., Frank, E., Holmes, G., Pfahringer, B., Reutemann, P., Witten, I.H.: The WEKA data mining software: an update. SIGKDD Explor. 11(1), 10–18 (2009)
20. de Cheveigné, A., Kawahara, H.: YIN, a fundamental frequency estimator for speech and music. J. Acoust. Soc. Am. 111(4), 1917–1930 (2002)

Look at Me Now: Investigating Delayed Disengagement for Ambiguous Human-Robot Stimuli

Melissa A. Smith[✉] and Eva Wiese

Department of Psychology, George Mason University, Fairfax, VA 22031, USA
mabsmith@gmail.com

Abstract. Human-like appearance has been shown to positively affect perception of and attitudes towards robotic agents. In particular, the more human-like robots look, the more participants are willing to ascribe human-like states to them (i.e., having a mind, emotions, agency). The positive effect of human-likeness on agent ratings, however, does not translate to better performance in human-robot interaction (HRI). Performance first increases as human-likeness increases, then drops dramatically as soon as human-likeness reaches around 70 % to finally reach its maximum at 100 % humanness. The goal of the current paper is to investigate whether attentional mechanisms, in particular delayed disengagement, are responsible for the drop in performance for very human-like, but not perfectly human agents. The idea is that robots with a high degree of human-likeness capture attention and thus make it harder to orient attention away from them towards task-relevant stimuli in the periphery resulting in bad performance. To investigate this question, faces of differing degrees of human-likeness (0 %, 30 %, 70 %, 100 %, non-social control) are presented to participants in an eye-tracking experiment and the time it takes participants to orient towards a peripheral stimulus is measured. Results show significant delayed disengagement for all stimuli, but no stronger delayed disengagement for very human-like agents, making delayed disengagement an unlikely source for the negative effect of human-like appearance on performance in HRI.

Keywords: Social robotics · Robotics · Appearance · Delayed disengagement · Eye-tracking

1 Introduction

As recent movies like *Ex Machina*, *Her*, or *Chappie* demonstrate, robots, whether physical or virtual, hold a unique fascination for humanity. This fascination is reflected in a steadily growing body of robotics research, with the vast majority of studies being conducted to make robots act and appear more human-like. The hope is that robots that are similar to humans will be perceived and treated as human-like and trigger interactions that resemble interactions between humans. Among the possible features that initiate perceptions of human-likeness (e.g., appearance, behavior, beliefs), appearance is the most extensively researched aspect [24] and has been shown to have a positive effect on ratings of and attitudes towards robots [14, 27]. Robots that physically resemble humans activate perceptions of humanness even if they do not show human-like behavior [5, 16]

© Springer International Publishing AG 2016
A. Agah et al. (Eds.): ICSR 2016, LNAI 9979, pp. 950–960, 2016.
DOI: 10.1007/978-3-319-47437-3_93

and increasing the degree of physical humanness of a robot leads to a systematic increase in attributions of emotional states, intentions and agency [19, 22]. Human appearance also strongly determines the first impression humans have of a robot, which in turn affects the rate at which social interactions with the robot are initiated and maintained [18] and modulates how humans behave towards the robot. Several studies have shown that human-like appearance enhances a robot's life-likeness and increases people's willingness to comply with instructions given by the robot [10, 15, 25]. Perceiving a robot as human-like also increases the chances that participants take advice from the robot [28] and leads to increased trust in and likability of the robot [3, 11]. Other aspects of physical appearance like gender [32] and conveyed emotions [33] have also been shown to modulate how humans behave towards robots.

In sum, there is convincing evidence that human-like appearance has a positive effect on perceptions of and attitudes towards robots as long as the robots are evaluated based on physical features only. As soon as information about the robot's behavior is available (via instruction, observation or interaction), appearance can have a negative impact on robot evaluations, in particular, when being in contrast to the behavioral capabilities of the robot [6]. Bailenson and colleagues [2], for instance, showed that agents that look human-like but only possess machine-like capabilities are perceived negatively. In support of this notion, [12] showed that mismatches between agent experience and attributions of agency also lead to negative agent perceptions; specifically, they found that feelings of uncanniness result when features or behaviors that are considered fundamentally human, like the ability to have experiences, are purported to be coming from a machine. The negative interaction effect between appearance and behavior becomes even more evident when participants are allowed to interact with the robot (or pictures thereof). Mandell and colleagues, for instance, showed that increasing the humanness of an agent does not lead to a linear increase in performance with this agent on a sustained attention task [21]. Rather, it seems like performance initially gets better as the degree of humanness increases until a level of about 70 % physical humanness is reached. At this point, performance with the agent drops dramatically and then recovers again when 100 % humanness is reached. In support of this finding, Martini, Buzzell, and Wiese [23] reported that the performance in a social attention task does not increase linearly with increasing humanness of the stimuli, but rather shows a sinusoidal pattern with an initial increase in performance as humanness increases, followed by a decrease in performance for agents of about 70 % humanness and a recovery of performance for 100 % human agents. Interestingly, however, when using the same stimuli in an evaluative study (i.e., participants rate the stimuli regarding their capabilities of having a mind based on physical features only), a linear increase in ratings was observed as the degree of physical humanness increased, without a drop in ratings for agents that are 70 % human [22].

Given that one of the main goals of social robotics research is to improve performance in HRI, it is important to investigate why physical humanness has positive effects on evaluations of the robot on the one hand (e.g., likable, trustworthy, mindful, etc.), but can have negative effects on performance on the other hand, in particular, when interacting with robots that are human-like, but not perfectly human. There are several hypotheses for why the same stimulus has a negative effect on performance with the robot, but not on subjective evaluations of the robot. Firstly, in line with theories regarding the uncanny valley [26], participants might find it unpleasant or eerie to look at stimuli that have a high degree of human-likeness but are not perfectly human, which might increase their level of arousal and therefore lead to decreased performance (*explanation 1*). Secondly, it is possible that looking at an ambiguous stimulus causes a cognitive conflict regarding how to categorize the stimulus (i.e., human versus robot), even though it is task irrelevant [17, 29]. The additional amount of cognitive resources that is needed to solve the cognitive conflict is then lacking for the main task leading to reduced performance (*explanation 2*, [31]). Thirdly, the ambiguous stimulus might be perceived as a threat that captures participants' attention, thereby making it harder for them to disengage from the face and to focus their attention elsewhere – a phenomenon called *delayed disengagement* [8]. Delayed disengagement is typically observed when processing stimuli with negative emotions [1, 4, 8] and leads to bad performance because attentional orienting to task-relevant information in the environment is delayed ([1], *explanation 3*).

Aim of Study. The current experiment investigates whether delayed disengagement due to threat perception is responsible for the drop in performance when interacting with humanoid robot agents (i.e., 70 % humanness). In particular, we hypothesize that robots that are very human-like (but not perfectly human) will cause stronger delayed disengagement effects than stimuli that are rated lower on the humanness scale (i.e., are more robotic) or are rated as essentially human (i.e., are more human). For that purpose, we used a morphing software to create a spectrum of agent faces that ranged from 0 % human to 100 % human; four stimuli from this spectrum (0 % human, 30 % human, 70 % human, 100 % human) together with a non-social stimulus as a control (see methods for more detailed information) were then presented in a delayed disengagement paradigm: one stimulus at a time was shown in the center of the screen and participants were instructed to fixate the stimulus until a target dot appeared in the periphery and then make an eye movement (or saccade) to the target as fast and accurately as possible. An eye-tracker was used to measure saccade latencies (i.e., time it took participants to make a saccade from the face region to the target stimulus).

In order to determine whether a stimulus causes delayed disengagement, the time between stimulus onset and target onset (= stimulus onset asynchrony, SOA) needs to be varied from a short SOA (50 ms) to a long SOA (200 ms). If delayed disengagement is present, saccade latencies for the long SOA should be significantly shorter than saccade latencies for the short SOA due to threat-related attentional capture from the central stimulus at short SOAs (see Remington, 1980; for more details). The degree of delayed disengagement can then be calculated by subtracting saccade latencies for planned eye movements (SOA 200) from saccade latencies for automatic eye movements (SOA 50). In other words, the difference between mean saccade latencies for SOA 50 and SOA 200 gives us an estimate of the costs associated with processing the stimulus (= processing costs). Calculating processing costs allows us to compare attentional capture effects across stimuli with higher processing costs being associated with greater delayed disengagement. Thus, if decreased performance with humanoid agents (70 % human) is due to delayed disengagement, the 70 % human stimulus should (i) lead to processing costs that are significantly different from zero (i.e., significantly shorter saccade latencies for SOA 200 than for SOA 50) and (ii) induce delayed disengagement that is significantly stronger than for any other stimulus in the experiment (i.e., processing costs for the humanoid stimulus should be significantly stronger than processing costs for the other stimuli).

2 Methods and Materials

2.1 Participants

18 undergraduate students at George Mason University (mean age: 22.4 years, range: 18–40 years, 3 males) participated in this study for course credit. All participants gave informed consent and reported normal or corrected-to-normal visual acuity.

2.2 Apparatus

An EyeLink II head-mounted eye tracker in conjunction with Experiment Builder was used to present the experimental task and capture the eye movement data. The experiment was programmed in Experiment builder and run on a Macintosh computer, with a refresh rate of 85 hertz. Participants were seated in front of a 19in monitor and asked to place their head on a chin-rest that was positioned 18in away from the screen, allowing for a total screen viewing of around 74 degrees visual angle.

2.3 Stimuli

Stimuli consisted of four social and one non-social agent that served as control condition (see Fig. 1): The social stimuli were morphed images that varied on a spectrum from machine-like to human-like: robot (0 % human), robotoid (30 % human), humanoid (70 % human), and human (100 % human). The non-social stimulus consisted of a black circle with two vertical black bars. Stimuli were 16.24° × 16.24° in size, depicted on a white background and presented in full frontal orientation with eyes/bars positioned on the central horizontal axis of the screen.

The morphed images for the social stimuli were created using FantaMorph [7], a software that creates a range of morphs out of two images. For the current experiment, Fantamorph was used to morph a 100 % robot image (Meka robotics head S2) into two different 100 % human images in increments of 10 %: one female face and one male face (images taken from the Karolinska Directed Emotional Faces database, [20]). Out of these two spectra, four images were chosen as stimuli: robot (0 % human), robotoid (30 % human, female face), humanoid (70 % human, male face) and human (100 % human, female face). Two different human faces were chosen to make it less obvious that stimuli were created through morphing and do not represent independent agents.[1]

---

[1] A pilot version of this study was conducted with the same agent across the spectrum with the same results.

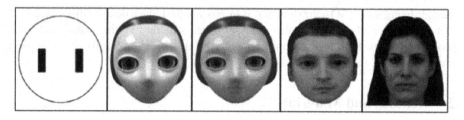

Fig. 1. Agents used in experiment, arranged by degree of physical humanness (from left to right): Non-social (vertical bars), Robot (0 % human), Robotoid (30 % human), Humanoid (70 % human), and Human (100 % human).

The non-social stimulus was created in Microsoft Powerpoint and consisted of a black circle on white background with two vertical black bars (length: 3.11°) left and right of the center of the circle; it was designed in a way that its physical features would be as similar as possible to the social stimuli (e.g. shape, screen space, arrangement of features) without having particularly face-like characteristics. The rectangles, in particular, were intentionally placed in vertical orientation in order to minimize any resemblance to a face.

2.4   Task and Procedure

An eye-tracking procedure was used to investigate whether humanoid stimuli induce significantly stronger delayed disengagement than any other social or non-social stimuli (see [1, 4] for general paradigm). Figure 2 illustrates the sequence of events on a given trial in the current experiment. Participants first fixated on a fixation cross for 1000 ms, followed by one of the five stimuli that were presented for 1000 ms. Participants were instructed to keep their eyes fixated on the location of the fixation cross and to not make any eye movements until a target dot appeared left or right of the stimulus. Targets appeared either after a short SOA of 50 ms or after a long SOA of 200 ms. SOA was measured as the interval between the onset of the face/bar stimulus and the onset of the target. As soon as the target appeared, participants had to make an eye movement to the target location (as fast and accurately as possible) and saccade latencies (i.e., time it takes from fixation of the central stimulus to fixating on the target stimulus) were measured. Face and target remained on the screen until a response was given or after 700 ms had elapsed, whatever came first. After successful fixation on the target, the trial was over and a new trial started after an inter-trial-interval (ITI) of 500 ms.

At the beginning of each session, participants gave informed consent and read the written instructions. They were told to fix their gaze on a centrally presented cross and to press the space bar when they were ready for the trial. They were also instructed that after the fixation cross different social and non-social stimuli would appear in the center of the screen but that they should still keep their eyes fixated on the fixation cross. Further, participants were advised that after the presentation of the face, a target dot would appear in the periphery, either to the left or right of the stimulus. Participants were asked to respond to the target as quickly and as accurately as possible by making an eye movement to the target location. After having read the instructions, participants

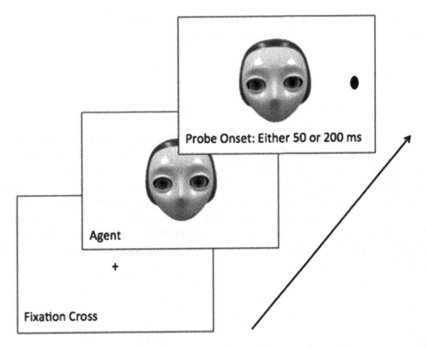

Fig. 2. Sequence of events in a given trial in experiment: Participants had to look at a fixation cross and press enter, which would then cause an agent face to appear. After either 50 ms or 200 ms, a black dot would appear either to the left or right of the face and participants had to shift their eyes to the black dot as quickly as possible. After an ITI of 500 ms the next trial began.

were outfitted with the head-mounted eye tracker, got calibrated to the system and finished a series of practice trials to get used to the task. After the practice, participants were given the opportunity to ask any questions about the task, followed by the start of the actual experimental task. At the end of the task, participants were debriefed and sent home.

Each session of the experiment was composed of 180 trials, with a block of 12 practice trials to help participants to get used to the task. Stimulus type (non-social, robot, robotoid, humanoid, human), target side (left, right), and SOA (50 ms, 200 ms) were selected pseudo-randomly and every combination appeared with equal frequency within each participant. The entire experiment took about 30 min.

2.5 Design

A 5 (agent type: non-social, robot, robotoid, humanoid, human) x 2 SOA (50 ms and 200 ms) within-participants ANOVA was conducted on saccade latencies (i.e., amount of time it took for a participant to move their eyes from central fixation to the target in the periphery). We also calculated the processing costs for each agent separately (i.e., difference in mean saccade latencies between SOA 50 ms and SOA 200 ms), which were analyzed via a univariate repeated-measures ANOVA with the within-factor agent type (non-social, robot, robotoid, humanoid, human) as independent variable.

3   Results

3.1   Saccade Latencies

The results of the analysis of the mean saccade latencies are shown in Fig. 3. There was a significant main effect for SOA ($F(1,35) = 5.613$, $p = .023$, $\eta^2 = .138$), indicating that mean saccade latencies were longer for SOA 50 (mean = 196.18, SE = 2.54) than for SOA 200 (mean = 185.79, SE = 2.54), which is equivalent to a delayed disengagement effect. There was no significant main effect of agent type ($F(4,140) = 1.615$, $p = .174$, $\eta^2 = .209$), showing that mean saccade latencies were comparable across all five stimuli. There was no significant interaction effect between SOA and agent type ($F(4,140) = .520$, $p = .726$, $\eta^2 = .096$), indicating that delayed disengagement was not significantly different across agent types.[2]

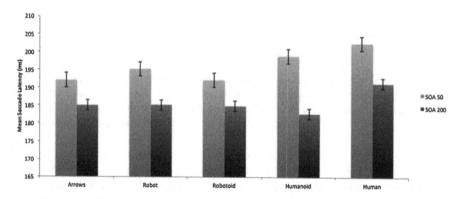

Fig. 3. Saccade Latencies by agent type and SOA: Saccade latencies are generally longer for SOA 50 than for SOA 200, showing that delayed disengagement was induced by all five stimuli in the current experiment. On the descriptive level, the delayed disengagement effect for the humanoid agent is larger than for any other agent. This difference, however, is not statistically significant. Error bars indicate standard error.

3.2   Processing Costs

Figure 4 shows the results of the analysis of processing costs for each stimulus separately. In contrast to our hypothesis, the main effect of agent type was not significant ($F(4,140) = .514$, $p = .726$, $\eta^2 = .171$), showing that delayed disengagement did not vary between the different stimuli. Planned contrasts between the humanoid and all the

---

[2] Post-hoc t-tests of the differences in the SOAs at the agent-level revealed a significant difference between SOA 50 and SOA 200 only in the humanoid ($t(35) = 2.577$, $p = .014$, $\eta^2 = .209$) condition, but not in any of the other conditions.

other agents confirmed that there was no significant difference between the processing costs for the humanoid agent compared to the other stimuli together (F(1,35) = 1.972, p = .169, η^2 = .277).

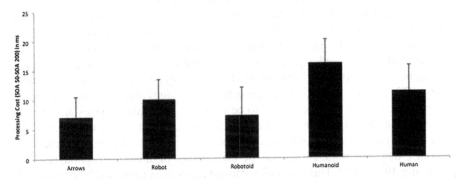

Fig. 4. Processing costs by agent type: The fact that processing costs were positive across all stimuli indicates that there was delayed disengagement for all agents. The processing costs for the humanoid stimulus, however, are not significantly higher than the processing costs for the other stimuli, showing that delayed disengagement is not stronger for the humanoid agent than for any other stimulus. Error bars indicate standard error.

4   Discussion

The goal of the current experiment was to investigate whether delayed disengagement was responsible for the observed decreased performance with humanoid agents in previous experiments [21, 23]. We hypothesized that ambiguous stimuli would be perceived as threat and therefore lead to delayed disengagement of attentional resources to task-relevant stimuli. In other words, processing of ambiguous stimuli was assumed to be associated with higher processing costs that take away cognitive resources from performance on the main task. Processing costs were calculated by subtracting saccade latencies for planned eye movements (long SOA, 200 ms) from saccade latencies for automatic eye movements (short SOA, 50 ms). We expected higher processing costs for the humanoid stimulus (70 % human) than for any other stimulus, based on the results by [21, 23].

The results revealed delayed disengagement effects for all stimuli (social and non-social), that is: saccade latencies were shorter for SOA 200 than for SOA 50 across all stimuli (see Fig. 3). The delayed disengagement effect caused by the humanoid stimulus, however, was not significantly stronger than for the other stimuli. Similarly, the analysis of processing costs (mean saccade latencies for SOA 50 – mean saccade latencies for SOA 200) did not reveal significant differences between the humanoid stimulus and all the other stimuli, which is in contrast to our original hypothesis. In other words, while there is an increased processing cost for the humanoid stimulus, it is not different enough from the processing costs for the other stimuli to explain differences in performance for this stimulus compared to the other stimuli, as found in [21, 23]. This would mean that poor performance with the humanoid stimulus, may not be due to a delayed disengagement, but rather due to negative effects on arousal (see *explanation 1*) or draining of cognitive resources due to cognitive conflict processing (see *explanation 2*).

There are several reasons why the expected higher delayed disengagement for the humanoid compared to the other stimuli was not found. Firstly, it is possible that the mechanism leading to the poor performance with humanoid agents is simply not related to attentional orienting, but rather due to arousal or cognitive load. Future studies need to investigate this possibility by designing studies that measure arousal (e.g., psycho-physiological studies) and cognitive load (e.g., dual task paradigms, pre-exposure to stimuli) in the context of performance in HRI. Secondly, it is possible that presenting the stimuli in a within-participants design in the current study may have made the critical manipulation too obvious for effects to be observed. Morphing images also has the downside that stimuli created via this procedure do not exist in reality, which makes them in and of themselves anomalies that might lead to altered cognitive processing (although they are standard stimuli in the context of mind perception and face processing: [9, 13, 30]). This potential confound needs to be addressed by repeating the study using wild-type robots of differing degrees of humanness (for instance, like the ones used in Mathur & Reichling [24]). Lastly, given that stronger delayed disengagement effects for the humanoid agent were observable on the descriptive level, increasing the sample size might give us sufficient power to also show these effects on the inference statistics level.

In sum, the current study addresses a major issue in social robotics, that is: the problem that increasing human-like appearance does not necessarily increase performance in HRI. Furthermore, the study suggests possible mechanisms for why humanoid appearance might negatively affect human performance and tests the delayed disengagement hypothesis in one experiment. Based in the current results, it seems unlikely that delayed disengagement is the causal factor for decreased human performance in HRI. The effect of robot appearance on human performance, however, is an important issue that has not received a lot of attention so far in social robotics research and needs to be addressed in future studies. In particular, it is important to investigate why humanoid appearance negatively impacts human performance (i.e., cognitive mechanisms) and to provide guidelines on how to address this issue when designing robots.

References

1. Azarian, B., Esser, E.G., Peterson, M.S.: Evidence from the eyes: Threatening postures hold attention. Psychon. Bull. Rev. **23**(3), 764–770 (2015). June 2016
2. Bailenson, J.N., Swinth, K., Hoyt, C., Persky, S., Dimov, A., Blascovich, J.: The independent and interactive effects of embodied-agent appearance and behavior on self-report, cognitive, and behavioral markers of copresence in immersive virtual environments. Presence **14**(4), 379–393 (2005)
3. Bartneck, C., Kanda, T., Mubin, O., Al Mahmud, A.: Does the design of a robot influence its animacy and perceived intelligence? Int. J. Soc. Robot. **1**(2), 195–204 (2009)
4. Belopolsky, A.V., Devue, C., Theeuwes, J.: Angry faces hold the eyes. Vis. Cogn. **19**(1), 27–36 (2011)

5. DiSalvo, C.F., Gemperle, F., Forlizzi, J., Kiesler, S.: All robots are not created equal: the design and perception of humanoid robot heads. In: Proceedings of the 4th Conference on Designing Interactive Systems: Processes, Practices, Methods, and Techniques, pp. 321–326. ACM, June 2002

6. Duffy, B.R.: Anthropomorphism and the social robot. Robot. Auton. Syst. **42**(3), 177–190 (2003)

7. FantaMorph. Version 3; Abrosoft Co., Beijing, China. http://www.fantamorph.com/index.html

8. Fox, E., Russo, R., Dutton, K.: Attentional bias for threat: evidence for delayed disengagement from emotional faces. Cogn. Emot. **16**(3), 355–379 (2002)

9. Freeman, J.B., Penner, A.M., Saperstein, A., Scheutz, M., Ambady, N.: Looking the part: social status cues shape race perception. PLoS ONE **6**, e25107 (2011)

10. Goetz, J., Kiesler, S., Powers, A.: Matching robot appearance and behavior to tasks to improve human-robot cooperation. In: Proceedings of the 12th IEEE International Workshop on Robot and Human Interactive Communication, ROMAN 2003, pp. 55–60. IEEE, October 2003

11. Goudey, A., Bonnin, G.: Must smart objects look human? Study of the impact of anthropomorphism on the acceptance of companion robots. Recherche et Applications en Marketing *(English Edition)* **31**(2), 2–20 (2016)

12. Gray, K., Wegner, D.M.: Feeling robots and human zombies: mind perception and the uncanny valley. Cognition **125**(1), 125–130 (2012)

13. Hackel, L.M., Looser, C.E., Van Bavel, J.J.: Group membership alters the threshold for mind perception: The role of social identity, collective identification, and intergroup threat. J. Exp. Soc. Psychol. **52**, 15–23 (2014)

14. Hoff, K.A., Bashir, M.: Trust in automation integrating empirical evidence on factors that influence trust. Hum. Factors J. Hum. Factors Ergon. Soc. **57**(3), 407–434 (2015)

15. Kanda, T., Miyashita, T., Osada, T., Haikawa, Y., Ishiguro, H.: Analysis of humanoid appearances in human–robot interaction. IEEE Trans. Rob. **24**(3), 725–735 (2008)

16. Kiesler, S., Powers, A., Fussell, S.R., Torrey, C.: Anthropomorphic interactions with a robot and robot-like agent. Soc. Cogn. **26**(2), 169–181 (2008)

17. Langton, S.R., Law, A.S., Burton, A.M., Schweinberger, S.R.: Attention capture by faces. Cognition **107**(1), 330–342 (2008)

18. Li, D., Rau, P.L.P., Li, Y.: A cross-cultural study: effect of robot appearance and task. Int. J. Social Robot. **2**(2), 175–186 (2010)

19. Looser, C.E., Wheatley, T.: The tipping point of animacy how, when, and where we perceive life in a face. Psychol. Sci. **21**(12), 1854–1862 (2010)

20. Lundqvist, D., Flykt, A., Ohman, A.: The Karolinska Directed Emotional Faces (KDEF). Department of Neurosciences Karolinska Hospital, Stockholm (1998)

21. Mandell, A.R., Smith, M.A., Martini, M.C., Shaw, T.H., Wiese, E.: Does the presence of social agents improve cognitive performance on a vigilance task? In: Tapus, A., et al. (eds.) ICSR 2015. LNCS, vol. 9388, pp. 421–430. Springer, Heidelberg (2015)

22. Martini, M.C., Gonzalez, C.A., Wiese, E.: Seeing minds in others–can agents with robotic appearance have human-like preferences?. PloS One **11**(1) (2016). online

23. Martini, M.C., Buzzell, G.A., Wiese, E.: Agent appearance modulates mind attribution and social attention in human-robot interaction. In: Tapus, A., et al. (eds.) ICSR 2015. LNCS, vol. 9388, pp. 431–439. Springer, Heidelberg (2015)

24. Mathur, M.B., Reichling, D.B.: Navigating a social world with robot partners: a quantitative cartography of the Uncanny Valley. Cognition **146**, 22–32 (2016)

25. MacDorman, K.F., Minato, T., Shimada, M., Itakura, S., Cowley, S., Ishiguro, H.: Assessing human likeness by eye contact in an android testbed. In: Proceedings of the XXVII Annual Meeting of the Cognitive Science Society, pp. 21–23, July 2005

26. Mori, M.: Bukimi no tani [The uncanny valley]. Energy **7**(4) 33–35.(Translated by Karl F. MacDorman and Takashi Minato in 2005) within Appendix B for the paper Androids as an Experimental Apparatus: Why is there an uncanny and can we exploit it? In: Proceedings of the CogSci-2005 Workshop: Toward Social Mechanisms of Android Science, pp. 106–118 (1970)

27. Pak, R., Fink, N., Price, M., Bass, B., Sturre, L.: Decision support aids with anthropomorphic characteristics influence trust and performance in younger and older adults. Ergonomics **55**(9), 1059–1072 (2012)

28. Powers, A., Kiesler, S.: The advisor robot: tracing people's mental model from a robot's physical attributes. In: Proceedings of the 1st ACM SIGCHI/SIGART Conference on Human-Robot Interaction, pp. 218–225. ACM, March 2006

29. Sato, S., Kawahara, J.I.: Attentional capture by completely task-irrelevant faces. Psychol. Res. **79**(4), 523–533 (2015)

30. Schweinberger, S.R., Burton, A.M., Kelly, S.W.: Asymmetric dependencies in perceiving identity and emotion: Experiments with morphed faces. Percept. Psychophys. **61**(6), 1102–1115 (1999)

31. Smith, M.W., Wiese, W.L.: The effect of ambiguous human-robot stimuli on vigilance performance (in preparation)

32. Tung, F.-W.: Influence of gender and age on the attitudes of children towards humanoid robots. In: Jacko, J.A. (ed.) Human-Computer Interaction, Part IV, HCII 2011. LNCS, vol. 6764, pp. 637–646. Springer, Heidelberg (2011)

33. Van Dillen, L.F., Lakens, D., Van Den Bos, K.: At face value: categorization goals modulate vigilance for angry faces. J. Exp. Soc. Psychol. **47**(1), 235–240 (2011)

Concurrency Simulation in Soccer

Jonathan Tellez-Giron and Matías Alvarado[✉]

CINVESTAV-IPN, Mexico City, Mexico
jtellez@computacion.cs.cinvestav.mx, matias@cs.cinvestav.mx

Abstract. Soccer is a multiplayer concurrent strategic game, one of the most popular sports in the world. Each soccer match is a social phenomenon for itself, with high level social impact from local to international instances. We use context-free grammars and automatons for modeling soccer, then develop a concurrent computing system for this game simulation. We achieve game simulations using real statistical data from distinguished midfielders and forwards players in the Spanish League of Soccer. Tests are made on the base of varying the teams' formations, either 4-3-3, 4-4-2, 5-3-2, then obtaining the probabilistic advantage for each formation including the specific players' statistics.

Keywords: Soccer · Strategies · Concurrent computing system · Simulation · Statistics

1 Introduction

Interaction among soccer robots and human soccer players needs a quite understanding on the dynamic of interaction in soccer. This social robotics-humans issue would be better made on the base of a mathematical modeling and computer simulation of soccer concurrency interaction, our point in this paper.

The soccer or *football* is a multi-player concurrent strategic game. Each player on the field chooses a play for each time of the soccer match, simultaneously to the other players' choices. In the Concurrent Computing System (CCS) for soccer we propose, each player is implemented by means of a thread, and the CCS runs many threads simultaneously. The threads do access to memory blocks in common, the system's *critical section*. The correct management of the critical section is essential for the CCS, particularly to simulate a concurrent soccer match. In the soccer shared factors among players the ball is the principal of them, the players' major interaction element to manage.

Soccer is one of the most popular sports in the world, invented in Britain by the Middle Ages and the first official rules invented by students from the University of Cambridge by mid-seventeenth century [8]. Classic soccer is played by 11 players - there are version with 5 or 7 - each team on a rectangular field of 120×60 meters as the usual; it has one goal at each end of the longitudinal axis. During a game the players basic action is to road and control the ball over the field with his legs, body or head, but the ball touch with arms or hands is

© Springer International Publishing AG 2016
A. Agah et al. (Eds.): ICSR 2016, LNAI 9979, pp. 961–970, 2016.
DOI: 10.1007/978-3-319-47437-3_94

penalized. The team's aim is to introduce the ball into the opponent's goal. The field game has line marks in the middle camp and the goal zone area to delimit the crucial actions on the game to goals.

Each soccer match is a social phenomenon for itself having high level social impact, from local to international instances. The financial resources that produce and move around soccer matches are such an analysis, modeling and scientific simulation, including computational, is unavoidable. A computer concurrency model for soccer modeling and simulation is proposed in this paper.

The rest of the paper is organized as follows: In Sect. 2 the dynamics of soccer into our model of strategic game is described. The tests and the results obtained are in Sect. 3, followed by a Discussion in Sect. 4, ending with the Conclusions concerning this investigation in Sect. 5.

2 Soccer Dynamics and Concurrency

For a player, the sequential representation of plays are constructed by means of a Context Free Grammar (CFG), the alphabet being defined on the base of the set of basic plays of a soccer player in Table 1, and the production rules are a formal translation from the soccer rules in Table 2. The CFG built for each player is read by the corresponding Finite State Automaton (FSA). The automaton states in Table 3 are the reference points for the modeling and simulation of game actions. The sequence of plays is built by the FSA transitions between state, from which each player makes moves that correspond to those indicated by the CFG rules, starting at initial state to finish at a final state, passing through intermediate states. This sequence is regarding the player's options according to its actual state: for instance, a *short pass* only can be made when a player has the ball. Each player's play is an available basic strategy.

In this research the game roles are *goalkeeper, defender, midfielder* and *forwarder*. The main difference between football players in different positions or roles lies in the likelihood to make a specific play. Each player-thread in the CCS should take the likelihood values according to his role in the game [4]. Each game role has execution probabilities of plays for average players. The introduced CCS can simulate an entire soccer match. For each player the sequences of plays transit by means of its FSA as a thread of the CCS. The N threads - players run simultaneously during a match. The data for these players were abstracted from the Spanish League in the 2015–2016 tournament until day 30.

2.1 Management of the Critical Section

Transition between FSA states control the players interaction that is managed by the CCS critical section having main factor the ball possession by a sole player at a time: at every time of the match $N - 1$ FSAs are at *without ball possession* state, meanwhile just one FSA is at *ball possession* state. The control is at every time during the match. Every play that involves players interaction is an eventual critical section between threads. The correct management of this

Table 1. Soccer plays classification

Defensive plays

Play	Description	Play	Description	Play	Description
db^i	block shot	r^i	recuperation	i^i	interception
d^i	clearing ball	$blce^i$	successfully blocking ball	$blse^i$	successfully blocking ball

Construction plays

Play	Description	Play	Description	Play	Description
pl^j	long pass	pc^i	short pass	ce^i	center the ball

Attack plays

Play	Description	Play	Description	Play	Description
t^i	shot	as^i	attend the ball	rea^i	dodge successful
ref^i	dodge failed				

Goal play

Play	Description
go^i	goal

Discipline

Play	Description	Play	Description	Play	Description
fap^i	fouling	rep^i	receive penalty	fuj^i	offside
far^i	fouled	pec^i	committing penalty	per^i	lose ball
mac^i	handball	adv^i	warning	exp^i	send player off
sdu^i	first yellow card	sdd^i	second yellow card	sf^i	red card

Goalkeeper plays

Play	Description	Play	Description	Play	Description
pd^i	stop shot	rd^i	reject shot	sp^i	pass the ball

factors prevents a *deadlock*, that occurs when two or more threads try to access the same shared resources to change it, generating access conflict and mutually paralyzing the execution of the threads. The process which manages critical section of the CCS, in general, is based in the Algorithm 1. For example the passing ball action, should indicate the player who is headed and from who; once the pass has been made the ball will be set to *free* and the first player-thread will be set at state *without ball possession*. Symmetrically, the player who is headed and is about to achieve a *recuperation* will verify the *free* status of the ball and the pass destination to himself; if these conditions are met the action will be achieved. This way a pass play (long or short) will be completed. Effective interaction plays between players are: fouls, blocks, shots, penalties and passes.

Table 2. Transition of plays of a soccer player.

Function transition	Function transition	Function transition
$\delta(P, far^i) = P$	$\delta(P, t^i) = SP$	$\delta(SP, i^i) = P$
$\delta(P, rep^i) = P$	$\delta(P, as^i) = SP$	$\delta(SP, blce^i) = P$
$\delta(P, rea^i) = P$	$\delta(P, ref^i) = SP$	$\delta(FJ, per^i) = SP$
$\delta(P, go^i) = G$	$\delta(SP, far^i) = SP$	$\delta(F, sdv^i) = SP$
$\delta(P, fap^i) = F$	$\delta(SP, blse^i) = SP$	$\delta(F, sdu^i) = TA$
$\delta(P, fuj^i) = FJ$	$\delta(SP, db^i) = SP$	$\delta(F, sdd^i) = STA$
$\delta(P, d^i) = SP$	$\delta(SP, pec^i) = SP$	$\delta(F, sf^i) = TR$
$\delta(P, pl^i) = SP$	$\delta(SP, rep^i) = SP$	$\delta(TA, per^i) = SP$
$\delta(P, pc^i) = SP$	$\delta(SP, mac^i) = F$	$\delta(STA, sa^i) = TR$
$\delta(P, ce^i) = SP$	$\delta(SP, r^i) = P$	$\delta(TR, exp^i) = X$

Table 3. Soccer player states

State	Description	State	Description
P	Ball possession	X	Previous state to send player off
SP	Without ball possession	FJ	Previous state of offside
G	Next state from goal	TA	Next state from first yellow card
F	Next state from fault	STA	Next state from second yellow card
TR	Next state from red card		

Algorithm 1. Algorithm to manage critical section in the CSS of the soccer.

Require: Semaphores, actual state &set of plays
1: Initialize variable *pivot* in **false**
2: **while** *pivot* == **false do**
3:    Generates random play based on the average occurrence per minute of the plays

4:    **if** If this play involves another player **then**
5:        **if** Player can do this play **then**
6:            pivot = **true**
7:        **end if**
8:    **end if**
9: **end while**
10: Set play to next transition
11: Depending on play will block critical section to use
12: Set appropriate **state** to the next transition

Every effective interaction play is controlled by the same general process. The *pivot* is defined because of the soccer dynamics. When the pivot is interrupted by another play the first play isn't successful meanwhile the second one is it. For instance, the pass and the blocking of the ball, when a player pass the ball the pivot indicates that the ball is free, abstractly you can say that the ball is in its path between two players; simultaneously the second player verify the same pivot, stating that he can block the ball, and before other player could receive the ball, the second player blocks the ball and obtains the ball possession. The interaction plays between players need a pivot which determines if a play can be played.

2.2 Concurrency of Plays

Interaction between players follows the rules during the match, likely play for passes just are played with allies; and some other plays just are made to rivals, likely fouls, blocks and interceptions. The correspondence of the plays is shown in the Table 4. The transition between plays is limited by the *conditional correspondence*, hence the player will choose its target according to the play to do. Each team is conformed by $N/2$ players, the local team from 1 to $N/2$ player, and the away team from $N/2 + 1$ to N. Before the players choice, the process will verify the *target* the player to decide the play action if mate or adversary.

Table 4. Correspondence of the plays in soccer.

Correspondence	Plays
To ally	pl, pc, ce, as
To rival	t, fap, i, blce, blse, db, pec
From ally	r
From rival	ref, rep, far
Indistinct	d, go, fuj, mac, adv, sdu, sdd, sf

The CCS deploy is in C language with pthreads library. The critical section is managed with *mutex* (mutual exclusion) variables. Only one thread-player at the time can block a variable to read or modify it, being a global variables so all the processes can access it. The CCS development will follow the process specified at Fig. 1.

The CCS for soccer is controlled by time intervals t, one per minute of the game and maximum 90 min as official rule says. The usual condition indicates that the match ends when the time finishes, but the match also can end if one of the teams is left with 6 players by expulsions, in this case the result of the match will be set to 3-0, with disadvantage to this team. When a thread-player goes to a final state of its FSA, the CSS must be restarted and the ball possession is reallocated as at the match beginning.

In the CCS main function the *mutex* variables are created and assigned; also in this function are created the threads for each player, this thread is replicated N times, all of them will be running concurrently. Each thread is making transitions between states of its FSA. Each strategic decision should meet conditions set by

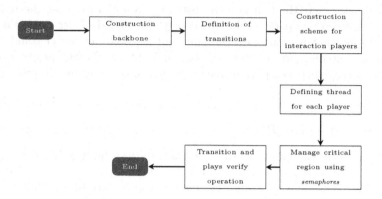

Fig. 1. Process of computing concurrent system

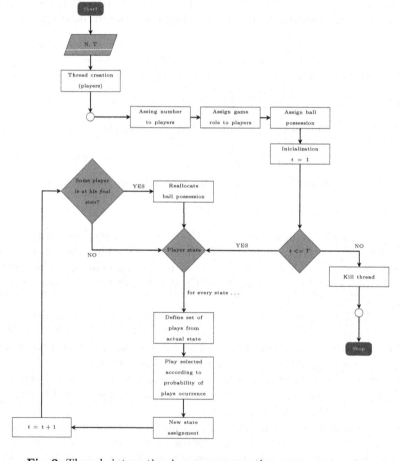

Fig. 2. Threads interaction in soccer computing concurrent system

the critical section, the players team and his game role. The game process is specified in the Fig. 2.

For the soccer simulation, the critical section management is a strong challenge because the multiple shared *mutex* variables. Also, we should consider the multiple players' strategies, inherent to a whole soccer simulation, even beyond this paper scope, as all of them make strategic choices at every time interval t until the limit time is reached. The coding is one thread function per player, each other different by: the number of player, the game role and the team it belongs. All this need resources are crucial for the correct CCS's operation.

3 Tests and Results

The functionality tests has been made for regular matches of 90 min. The number of players N has been set to 22 players, as the traditional soccer. The teams are set by different formations, the most common used actually, in order to test the difference in the formations.

Each formation is formed by a goalkeeper, in addition to the defenders, midfielders, and forwarders indicated by first, second and third numbers in the team formation in Table 5. For each configuration 50 tests run, obtaining the results presented in Table 5.

Table 5. Result table of soccer CCS

Formations(local vs away)	Local team winnings	Away team winnings	Draw
(4-4-2) vs (4-4-2)	3	5	42
(4-3-3) vs (4-3-3)	6	3	41
(5-3-2) vs (5-3-2)	2	2	46
(4-4-2) vs (4-3-3)	3	3	44
(4-4-2) vs (5-3-2)	5	4	41
(4-3-3) vs (5-3-2)	12	1	37

The logical sequence of plays is checked for consistency, in other words, if the sequence of plays has illogical data, the system has to be fixed. The results in the first three configurations present lead to a goal, and in the case of draws had no goals; for the last three settings the results had advantage of two or more goals, and draws at one or two goals. The most varied results were observed in the last configuration denoting an advantage towards the formation 4-3-3 for the home team.

As the teams are formed by *average* players, the results tend to draw. In order to verify this assumption we set real statistic data from four players, two forwarders: Lionel Messi and Cristiano Ronaldo, and two mid-fielders: Andres Iniesta and Luka Modri. The local team at all this formations is conformed by Lionel Messi, Andres Iniesta and 9 average players. The away team is conformed by Cristiano Ronaldo, Luka Modri and 9 average players. We made the same experiment obtaining results of Table 6.

Table 6. Result table of soccer CCS modifying players

Formations(local vs away)	Local team winnings	Away team winnings	Draw
(4-4-2) vs (4-4-2)	7	8	35
(4-3-3) vs (4-3-3)	15	10	25
(5-3-2) vs (5-3-2)	19	8	23
(4-4-2) vs (4-3-3)	13	7	30
(4-4-2) vs (5-3-2)	12	8	30
(4-3-3) vs (5-3-2)	16	6	28

Based on the results, it can be seen that the data vary considerably from those previously obtained. It also can be seen the local team advantage. The results in this matches were vary, draws were from zero equals to two equals; the advantage of goals was from one goal to three goals over the other team. Also the aggressive behavior of the players is seen getting warnings and expulsions in each formation (Table 7).

Table 7. Agressive behavior of players (W:warnings,E:expulsions)

	Lionel messi	Cristiano ronaldo	Andres iniesta	Luka modri
(4-4-2) vs (4-4-2)	9 W, 0 E	5 W, 1 E	7 W, 0 E	12 W, 1 E
(4-3-3) vs (4-3-3)	7 W, 0 E	6 W, 0 E	4 W, 0 E	10 W, 0 E
(5-3-2) vs (5-3-2)	6 W, 0 E	2 W, 0 E	5 W, 0 E	16 W, 0 E
(4-4-2) vs (4-3-3)	6 W, 0 E	4 W, 0 E	4 W, 2 E	6 W, 1 E
(4-4-2) vs (5-3-2)	4 W, 0 E	5 W, 0 E	8 W, 0 E	5 W, 1 E
(4-3-3) vs (5-3-2)	3 W, 0 E	6 W, 0 E	8 W, 0 E	5 W, 1 E

4   Discussion

The Context Free-Languages and Finite State Automatons have mathematically modeled baseball games [1] and American football games [10]. The modeling is the base for strategic choices combined with league statistical data of high competency level, or using Nash equilibrium.

We deployed a concurrent computing system with explicit chains of plays, so that, in a time interval you can see the moves of all players on the field. We overcome previous investigations where plays-chains are created sequentially, focusing the movement of specific players, but obviating the movement of the other players.

In [6] an iterative model on soccer, based on Markov games using multi-agent reinforcement learning (MARL) is specific driven to a model of possession, dividing the field into quadrants and annotating strategies to movement between adjacent quadrants. There, once again, it approaches to a play/strategy specific but ignoring the rest of the plays. This research address the processing of equilibrium by transferring information from player who derived to whom eventually may use it.

In simulated football videogames [3], a model based on genetic algorithms to manage strategic behavior for 11 simulated robotic players is proposed; the evolutionary learning for football simulation games takes the opponent player's game strategy deduced by statistical data. In [9] a non-deterministic decision making method uses vector-based fuzzy electric fields and a decision tree for soccer modeling. In [7], the algorithm Nash-Q learning with regret matching to guide exploration of the state-action space extends the Nash-Q learning with regret matching algorithm in terms of scores, average reward, and policy convergence for obtaining the Nash equilibrium policy. The Nash-Q learning algorithm [5] use an stochastic model based on Markov Decision Processes to make decisions considering future rewards if the agents follow the joint Nash equilibrium strategies.

Relevant research for soccer automation is the development of mechanisms to cooperate in complex tasks. In [2] a set-play language in the context of 2D simulated robotic soccer is proved in RoboCup competition. Control of robot's motors apply to make walk, getup or kick the ball.

5   Conclusions

In the Computing Concurrency System for simulation and modeling of soccer the correct management of critical section is fundamental. Use of the statistical data in a CCS allows to simulate complete matches of soccer. We observe that the matches variate according to the players involved in the match. In tests where the teams made up of average players present, most of the cases result draw scores. In tests that setting up some different players the matches results variate in a meaningful way. Team formation is another involved factor at soccer matches. The results are variable but it directly depends of the players' statistics. Depending on the match configuration we can analyze a better configuration for an specific match. We say that the reach of this model is that it may allow the direct use of statistics for complete soccer teams. The developed model in this investigation it has to be adapted to robot moves and as an algorithm of strategic analysis to improve the team behavior making collaborative decisions.

References

1. Alvarado, M., Rendón, A.Y., Cocho, G.: Simulation of baseball gaming by cooperation and non-cooperation strategies. Computación y Sistemas **18**(4), 693–708 (2014). http://cys.cic.ipn.mx/ojs/index.php/CyS/article/view/1987
2. Cravo, J., Almeida, F., Abreu, P.H., Reis, L.P., Lau, N., Mota, L.: Strategy planner: graphical definition of soccer set-plays. Data Knowl. Eng. **94**, 110–131 (2014)
3. Fernandez, A.J., Cotta, C., Ceballos, R.C.: Generating emergent team strategies in football simulation videogames via genetic algorithms. In: Game-on 2008: 9th International Conference on Intelligent Games and Simulation, pp. 120–125 (2008)
4. Gray, J., Reuter, A.: Transaction Processing: Concepts and Techniques. Elsevier, Lorne (1992)
5. Hu, J., Wellman, M.P.: Nash q-learning for general-sum stochastic games. J. Mach. Learn. Res. **4**, 1039–1069 (2003)
6. Hu, Y., Gao, Y., An, B.: Accelerating multiagent reinforcement learning by equilibrium transfer (2014)
7. Liu, Q., Ma, J., Xie, W.: Multiagent reinforcement learning with regret matching for robot soccer. Mathematical Problems in Engineering 2013 (2013)
8. McDougall, C.: Soccer. Best Sport Ever eBook Series, ABDO Publishing Company (2012). https://books.google.com.mx/books?id=2nrhG5ovUSQC
9. Norouzi, A., Acosta, C.A.: An approach to design a robust software architecture and an intelligent model for multi-agent systems. In: 2013 3rd Joint Conference of AI & Robotics and 5th RoboCup Iran Open International Symposium (RIOS), pp. 1–7. IEEE (2013)
10. Yee, A., Rodríguez, R., Alvarado, M.: Analysis of strategies in american football using nash equilibrium. In: Agre, G., Hitzler, P., Krisnadhi, A.A., Kuznetsov, S.O. (eds.) AIMSA 2014. LNCS (LNAI), vol. 8722, pp. 286–294. Springer, Heidelberg (2014). doi:10.1007/978-3-319-10554-3_30

Let the User Decide! User Preferences Regarding Functions, Apps, and Interfaces of a Smart Home and a Service Robot

Birte Schiffhauer[(✉)], Jasmin Bernotat, Friederike Eyssel,
Rebecca Bröhl, and Jule Adriaans

Cluster of Excellence Cognitive Interaction Technology,
Bielefeld University, Bielefeld, Germany
birte.schiffhauer@uni-bielefeld.de, {jbernotat,
feyssel, rbroehl, jadriaans}@cit-ec.uni-bielefeld.de

Abstract. In an online survey, we studied user expectations and preferences for functions and apps in the context of a smart apartment. Furthermore, we explored which type of interface users would choose for an interaction with the smart apartment. Equally important, we investigated users' acceptance of a service robot in the smart home. Results showed high levels of acceptance for both, the smart apartment and the robot, although the preferred interface for the apartment was context dependent. We discuss implications of the current survey and highlight key aspects to be taken into consideration when developing innovation technology for the home context.

Keywords: User experience · User preferences · Human-robot interaction · Smart home · User-centered design · Usability

1 Introduction and Related Work

Due to the challenges associated with demographic change and in face of the necessity to decrease energy consumption, the market for smart home technology will increase rapidly [1, 2]. Even though smart technologies in the home context have been around for more than 25 years, their market share is still relatively low and so is the acceptance of such technologies [2]. In 2014, only 14 % of the German population reported using smart home technology [3]. Besides anticipated extra costs, potential problems regarding usability are some of the main reasons why people still reject the idea of living in smart environments and investing in smart technology [2]. Even end users of smart homes complain about the complicated usage and the lack of usability [4]. To increase the acceptance and usability of intelligent technology, user preferences and expectations have to be taken into account during the research–development loop [5]. Therefore, our goal was to explore and evaluate user perceptions of a smart home, as part of the large-scale project "The Cognitive Service Robotics Apartment as Ambient

J. Bernotat, F. Eyssel—Shared second authorship.

© Springer International Publishing AG 2016
A. Agah et al. (Eds.): ICSR 2016, LNAI 9979, pp. 971–981, 2016.
DOI: 10.1007/978-3-319-47437-3_95

Host" (CSRA). The CSRA is located at Bielefeld University at the Cluster of Excellence Cognitive Interaction Technology (CITEC). The smart apartment is complemented by a service robot to assist users in the home context [6]. A smart apartment is an apartment that is able to accomplish goals set by its inhabitants in an intuitive way [7]. The CSRA project aims to develop an intelligent environment that adapts to the needs and expectations of individuals [6]. The goal is to create a setting that enables smooth, intuitive, and natural interactions between users and system components.

User expectations appear to be a highly relevant predictor of technology acceptance [8, 9]. Taking these into account reduces excessive costs associated with a new design [10]. Therefore, we realized a user-centered design approach to integrate users' expectations and preferences for the apartment into the development process in order to create a testbed with a high user acceptance. Literature on user experience showed that potential users evaluated functions regarding safety (e.g., alert for gas leaks) and energy consumption monitoring as positive in the United Kingdom [11]. However, [12] demonstrated that cultural differences have to be taken into consideration: Americans and Koreans differed in the evaluation of design choices and control options of a smart home. In the German context, recent studies that tested user preferences for different functions investigated only general preferences for applications, such as security and entertainment [2–4]. Because of limited direct exposure to smart home technologies in the real world, potential users are still uninformed about the potential of smart environments and are unfamiliar with controlling the various functions. Therefore, surveys that have explored preferences in a broad manner yielded results with restricted interpretability. To change this, we provided more details when asking potential users how they would rate different functions and apps of a smart apartment (research question 1). Most recent studies which have examined user experience of smart home technology have tested only one interface. For instance, [13] have evaluated how users could control a smart home via speech, [14] explored the interaction with a service robot, and [15] tested the effectiveness of controlling an apartment via displays. To our knowledge, the current survey is the first to investigate user preferences regarding different interfaces for various scenarios in order to find the best one for a specific scenario. Therefore, we asked participants as how pleasant they would evaluate different interfaces for various scenarios to interact with the apartment (research question 2). Furthermore, the CSRA project aims to equip the apartment with a service robot. Consequently, participants stated to what extent they would like to be supported by a robot while doing various tasks (research question 3).

2 Method

In order to measure potential user expectations and preferences toward interfaces and functionalities, we conducted an online survey using *UniPark*, an internet-based computer program. Data collection was realized between November 2014 and May 2015.

2.1 Participants

270 participants (90 males, 173 females, $n = 7$ did not indicate gender, $M_{age} = 24.52$, $SD_{age} = 7.39$) took part in the study. All participants were capable of the German language.

2.2 Procedure

Participants were recruited at Bielefeld University and through flyers that advertised the online survey, which took about 50 min. At the beginning, participants read a text that explained the concept of an intelligent apartment. Then, they were asked to rate different functions and apps of a smart apartment and could suggest additional features and functions. We also inquired about potential features the apartment should *not* have. Moreover, participants had to imagine having a service robot in the apartment and were asked to what extent they would use robotic assistance for tasks, like cooking. Afterwards, participants were asked how they would preferentially interact with the apartment in specific situations. This was followed by questionnaires about the acceptance of an intelligent apartment, concerns regarding data security, vividness of the imagination, and technology commitment. Finally, participants reported demographics and habits of technology usage. As reimbursement, participants had the opportunity to participate in a lottery to win $1 \times €100$ or $2 \times €50$.

2.3 Dependent Measures

Technical Functions and their Evaluation. The technical functions tested in the current research were selected based on extensive discussions with in-house developers at CITEC and a review of related research [11, 16]. Participants were asked to evaluate the six following functions regarding the overall ambience in the apartment: automatic light adaptation, automatic sun blinds, regulation of the ambient environment (e.g., automatic window opening), monitoring of costs (e.g., information on the energy consumption), adaptation to user habits, and automatic illumination depending on the situation. Furthermore, five functions addressing safety and security of the apartment were evaluated: Safety system (e.g., automatic switch off of electrical gadgets), protection against burglars (e.g., the windows close when leaving the apartment), automatic doorway system (e.g., the door opens automatically for specific people), locking the door at bedtime and an alert system (e.g., acoustic signals in case of fire). Every function was explained in 3–4 sentences. Participants were asked to rate the functions of the apartment with regard to five dimensions: to what extend they wanted to have this function (have), how useful (useful) or how disturbing this function would be (disturbing), and how important it would be for them to switch off the function any time they want (turn off). For the ambience-related functions, we asked if users wanted these to operate autonomously or only by command (not automatic). Regarding safety and security functions, we assessed perceived safety (safety). Participants responded to all items using a 7-point Likert scale, with high ratings reflecting high endorsement of the measured construct.

Evaluation of Apps. Users were asked to rate various apps related to specific domains. For instance, in the household domain, participants rated an app that would support users in preparing a meal. To tap the leisure domain, users evaluated apps, such as a multimedia app. For example, home-office apps were described as reminding users to take a break. The remaining apps were related to health. For instance, supporting training at home. The apps were rated on the same dimensions as the functions, but instead of asking whether the app would give the participants a feeling of safety, participants were asked if they anticipated to have fun using the app.

Acceptance of the Service Robot. Participants were instructed to imagine having a service robot in the apartment and were asked to what extend they would like to use it for 17 different tasks, like cooking, cleaning, or babysitting.

Pleasantness of Interfaces. Using scenarios, we explored participants' preference to interact with the apartment in a particular situation. To identify scenarios in which conventional reactions were less practical, a pretest was run instructing participants to generate different scenarios. These scenarios were categorized and provided the basis for new scenarios to be used in the presented questionnaire (see Table 1).

Table 1. Description of the seven different scenarios.

1. The phone rings while I am washing the dishes. The person calling me is displayed on a screen above the kitchen sink
2. I am waiting for visitors and just as I am in the bathroom the door bell is ringing. The person standing right in front of the door is displayed on a screen in the bathroom.
3. I am sitting on the couch while watching TV. I want to switch the channel, but the remote control is not available
4. I am standing in the shower while listening to music. I would like to adjust the volume/change the title
5. I return from a day of shopping loaded with bags and need to unlock the door
6. While I am cleaning the kitchen, it occurs to me that I will also need to write an important e-mail
7. I am having visitors and we are sitting in the kitchen while playing a parlor game. The light is slightly too dark and I want it to be a bit brighter so that we have a better view of the playing field

Participants rated how pleasant they thought it would be to interact with the apartment via speech, touchscreen, using robotic assistance, or by solving the task conventionally (e.g., turning the light off by using the light switch).

2.4   Control Variables

Acceptance of the Intelligent Apartment. To measure acceptance of the intelligent apartment, we used eight self-generated items like "I would like to live in an intelligent apartment".

Technology Commitment. Twelve items, developed by [17], like "I would like to use new technology more often", assessed if participants were interested to learn about new technology and if they were interested in new technology in general.

Concerns about Data Security. A self-generated 7-item questionnaire exploited concerns about data security using items like "I don't want to live in an intelligent apartment, because I'm afraid that my personal data would not be stored safely".

Imagination. With four self-generated items, we examined how vividly participants could imagine the scenarios, e.g., "It was easy for me to imagine the situation".

Usage of Technology. Two items assessed the prior usage of "smart homes" and "service robots".

3 Results

3.1 Research Question 1: How Do Participants Rate Different Functions and Apps of a Smart Apartment?

To investigate participants' evaluation of the functions (F) and apps (A), a mean score was calculated of all items of the same dimension for the different functions (e.g., acceptability). Reliability tests showed a mostly good reliability (α: F_have = .74, F_useful = .69, F_disturbing = .73, F_safety = .72, F_not_automatically = .61, F_turn off = .82, A_have = .87, A_useful = .86, A_disturbing = .87, A_fun = .88, F_turn off = .92). On the scale F_safety, the item "automatic door opening system" had to be excluded in order to enhance the reliability.

T-tests were performed to test the average scores of the respective dependent measurements for the evaluation of the functions against the neutral scale midpoint (midpoint = 4). This was done in order to interpret if a function was evaluated about average. The *t*-tests were corrected for alpha-error using Bonferroni (0.05/10 = .005). Participants wanted to have the functions implemented in the apartment, ($M = 5.08$, $SD = 0.87$), $t(269) = 20.21$, $p < .001$, and evaluated them as useful, ($M = 5.33$, $SD = 0.76$), $t(269) = 28.73$, $p < .001$, the functions were not rated as disturbing, ($M = 3.12$, $SD = 0.88$), $t(268) = -16.42$, $p < .001$. The security functions provided a feeling of safety, ($M = 5.80$, $SD = 1.06$), $t(269) = 27.83$, $p < .001$, while participants preferred the ambience functions to perform only after request, ($M = 4.75$, $SD = 1.20$), $t(269) = 10.21$, $p < .001$. In addition, they wanted to have the opportunity to turn them off, ($M = 5.62$, $SD = 1.03$), $t(269) = 20.01$, $p < .001$.

The apps were rated as useful, ($M = 4.76$, $SD = 1.02$), $t(268) = 12.22$, $p < .001$, and participants wanted them in the apartment ($M = 4.37$, $SD = 1.16$), $t(269) = 5.30$, $p < .001$. They were rated as neutral in terms of disturbance, ($M = 3.85$, $SD = 1.06$), $t(268) = -2.29$, $p = .023$ and participants anticipated that it would be mediocre fun to use them, ($M = 4.19$, $SD = 1.10$), $t(268) = 2.76$, $p = .006$. However, it was important to them to have the opportunity to turn off the apps anytime they wanted, ($M = 5.97$, $SD = 1.05$), $t(268) = 30.67$, $p < .001$.

3.2   Research Question 2: As How Pleasant Do Participants Rate Different Interfaces for Various Scenarios to Interact with the Apartment?

We ran a repeated-measures ANCOVA with scenario and interface as within-subjects factors to investigate how pleasant participants rated the different interfaces depending on the scenario (see Table 1 for detailed scenarios). Figure 1 gives an overview of the results.

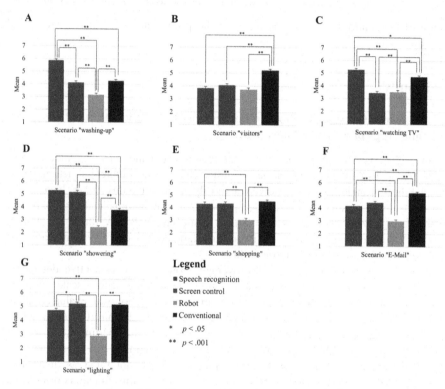

Fig. 1. Evaluations of the pleasantness of the various interfaces for the scenarios.

To exclude influences of the control variables, we included them as covariates. The covariates smart home acceptance, $F(14.01, 3263.99) = 3.76, p < .001, \eta_p^2 = .016$, and imagination, $F(14.01, 3263.99) = 1.79, p = .036, \eta_p^2 = .035$, had a significant influence. There was a significant main effect of the type of interface on ratings of pleasantness, $F(2.42, 562.76) = 30.82, p < .001, \eta_p^2 = .117$, while no significant effect for type of scenario could be found, $F(5.17, 1204.14) = 1.60, p = .156, \eta_p^2 = .007$. However, a significant interaction effect between type of scenario and type of interface occurred, $F(14.01, 3263.99) = 1.78, p = .036, \eta_p^2 = .008$. This result indicates that the ratings of pleasantness depended on the interface as a function of the scenario. To further analyze the differences in pleasantness for the interfaces separately for each scenario, post-hoc tests were calculated (see Fig. 1, significant results are indicated by asterisks).

3.3 Research Question 3: To What Extent Do Participants Would like to Be Supported by a Robot While Doing Different Tasks?

To assess for which tasks participants would accept robotic support, a repeated-measures ANCOVA was calculated with gender, age, usage of technology, acceptance of smart home, data security concerns and technology commitment as covariates. Only acceptance of smart home, $F(11.04, 2704.11) = 5.75$, $p < .001$, $\eta_p^2 = .023$, data security concerns, $F(11.04, 2704.11) = 2.73$, $p < .001$, $\eta_p^2 = .011$, and technology commitment, $F(11.04, 2704.11) = 3.09$, $p < .001$, $\eta_p^2 = .012$, turned out to be significant covariates. Desire for robotic assistance was task-dependent, $F(11.04, 2704.11) = 8.38$, $p < .001$, $\eta_p^2 = .033$. Seemingly, participants preferred the robot for less interactive tasks, like cleaning, in comparison to tasks demanding direct interaction with it. T-tests against the scale midpoint (scale midpoint = 4) were conducted to explore participants' willingness to do different tasks with the robot (see Table 2). Using Bonferroni (0.05/17 = .003), the t-tests were corrected for alpha-error.

Table 2. Descriptives and inferential statistics regarding evaluations of tasks done by the robot.

Task	M	SE	t	p
Cleaning	6.18	0.09	23.75	<.001
Taking out garbage	5.87	0.10	17.86	<.001
Pressing laundry	5.87	0.10	17.81	<.001
Reminding of appointments	5.02	0.10	9.43	<.001
Fetching objects in the apartment	4.85	0.11	6.93	<.001
Caring for me	4.38	0.12	2.99	.003
Helping with general tasks	4.32	0.13	2.19	.030
Waking me up	4.12	0.14	0.52	.606
Preparing meals	4.02	0.12	0.12	.906
Serving visitors	3.72	0.13	−2.32	.021
Caring for pets	3.36	0.12	−5.24	<.001
Being a game partner	2.82	0.12	−9.68	<.001
Being funny	2.80	0.12	−10.02	<.001
Talking about everyday topics	2.18	0.10	−17.51	<.001
Babysitting	1.94	0.10	−21.35	<.001
Talking about problems	1.93	0.10	−21.14	<.001
Helping with personal hygiene	1.65	0.08	−29.62	<.001

Note. Means and standard errors are shown for the repeated-measures ANCOVA and adjusted for the effect of the covariates. T gives the t-value of the comparison between the neutral scale midpoint (scale midpoint = 4 on a 7-point Likert scale) and the mean of each item, p indicates the significance of the t-tests.

4 Discussion and Conclusion

The current research investigated user expectations and preferences for functions and apps of a smart apartment. Furthermore, participants rated the pleasantness of different interfaces for different scenarios in the apartment. Finally, we examined participants' evaluations of a service robot in a smart home context.

Our results reveal that a wide range of potential German users evaluated the functions and apps as positive and useful. Particularly, functions that targeted energy conservation, home security, and safety issues were rated most positively and participants wanted the apartment to be equipped with the specified functions. These findings extend the results of [11] within a German sample. [11] have shown that participants from the U.K. rated the aforementioned functions (e.g., security and energy saving) as important.

Our research demonstrates that the security functions described in the survey created a feeling of safety. Importantly, however, participants rejected functions like an autonomous door opening system ("automatically for specific people"). Because we did not specify how the door opening system would recognize a "specific person" and whether the person was welcome or not, participants might have been concerned about the security of such a system. In addition, participants could have been worried about their privacy. Remarkably, users explicitly demanded that all functions and apps could only operate by command. Our finding is supported by [8] who observed that users only accepted automated home service they were in control of. In their research, an intelligent fridge that could order food, an intelligent TV that was able of ordering products, or an intelligent mirror that could send private information to the doctor, were rated as not very attractive. Our results also converge with findings by [18], who demonstrated that participants preferred a little extra effort of pressing a button compared to the system's autonomous reaction.

Our findings indicate that participants' desire to interact with the smart environment is context-dependent: For the scenarios "e-mail" and "visitors", a conventional way of interaction, like writing the e-mail themselves, was preferred. These were the only scenarios including interpersonal communication. Thus, it would be interesting to test if using intelligent technology to attain a goal is generally not accepted when other people are involved, because it might seem to be impolite using the robot. Overall, interacting with the apartment by means of the robot was not rated very positively. Especially, in the scenarios "showering", "washing-up", "shopping", and "light", the robot was rated less positive. In the scenario "showering", all other interfaces as well as the conventional way were preferred compared to the robot. This confirms a finding by [19], who showed that participants were more embarrassed to undress in front of an anthropomorphic robot than in front of a technical box. Therefore, we would advise to avoid deploying anthropomorphic robots in private rooms such as a bathroom and to consider a rather non-anthropomorphic appearance in this particular context. Although in our survey no photo of the robot was given, it is likely that participants imagined a rather anthropomorphic robot, as an anthropomorphic shape is most common for service robots. A further explanation for the low ratings of the robot could be the tasks described in the scenarios. Here, the robot was not necessary to solve the task. In some scenarios, an interaction with the robot would be more time-consuming. Therefore, it could be that participants did not accept the robot because they preferred the fastest possibility to solve the task. Regarding future research, we recommend to integrate use cases where the robot represents an advantage over other interfaces. For example, if one is expecting a mail delivery when no one is at home, the robot could be of use. It could be shown that participants chose the robot for tasks like cleaning. This indicates that participants might prefer robotic assistance for tasks that do not require a further human-robot interaction. Participants had no preference for or against doing a task with a robot where only little interaction was necessary, like putting together a board. However, they did not want to use the robot for tasks that required direct interaction with it, like assistance for personal hygiene.

The survey provides evidence for an overall positive evaluation of intelligent homes and demonstrates that the acceptance of such an apartment is expected to be high when users' preferences are considered: potential users like the idea of a smart home which supports them in different needs like security or in saving energy. Nevertheless, they want to be able to control the apartment by themselves anytime. To ensure acceptance of intelligent technologies in the home environment, it seems preferable to give participants the possibility to control the apartment with different interfaces. If these user preferences are considered as part of user-centered design, we can likely expect a high acceptance rate for such a smart home environment at the end of the prototype development stage.

Acknowledgements. This research has been conducted in the framework of the European Project CODEFROR (FP7 PIRSES-2013-612555) and was supported by the Cluster of Excellence Cognitive Interaction Technology 'CITEC' (EXC 277) at Bielefeld University, which is funded by the German Research Foundation (DFG).

References

1. Botthof, A., Domröse, W., Groß, W.: Technologische und wirtschaftliche Perspektiven Deutschlands durch die Konvergenz der elektronischen Medien [Technicological and Economical Perspectives of Germany through the Convergence of Electronic Media]. Studie der VDI/VDE Innovation + Technik GmbH, Berlin (2011)
2. SmartHome Initiative Deutschland e.V., mm1 Consulting, GdW: Smart Home- und AAL-Technologien in der Immobilien- und Wohnungswirtschaft [Smart Home and AAL Technologies in Real Estate and Apartment Business]. http://www.smarthome-deutschland. de/aktuelles/studien/detail/aktuelle-studie-zu-smart-home-aal-in-der-immobilien-und-wohnungswirtschaft.html
3. Illek, C.P.: Smart Home in Deutschland [Smart Home in Germany]. https://www.bitkom. org/Publikationen/2014/Studien/Smart-Home-in-Deutschland-Praesentation/Praesentation-Smart-Home.pdf
4. Dr. Grieger & Cie: Smart Home Monitor (2016). http://www.zahlendatenfakten.de/studien-marktdaten-marktanalysen/20-einzelhandel/102-smart-home-monitor-2016.html

5. Kaasinen, E., Kymäläinen, T., Niemelä, M., Olsson, T., Kanerva, M., Ikonen, V.: A user-centric view of intelligent environments: user expectations, user experience, and user role in building intelligent environments. Computers **2**, 1–33 (2012)

6. Holthaus, P., Leichsenring, C., Bernotat, J., Richter, V., Pohling, M., Carlmeyer, B., Koster, N., zu Borgsen, S.M., Zorn, R., Schiffhauer, B., Engelmann, K.F., Lier, F., Schulz, S., Cimiano, P., Eyssel, F., Hermann, T., Kummert, F., Schlangen, D., Wachsmuth, S., Wagner, P., Wrede, B., Wrede, S.: How to address smart homes with a social robot? A multi-modal corpus of user interactions with an intelligent environment. In: 10th Edition of the Language Resources and Evaluation Conference. LREC Press, Portoroz (2016)

7. Strese, H., Seidel, U., Knape, T., Botthof, A.: Smart home in Deutschland [Smart home in Germany]. Institut für Innovation und Technik (2010)

8. Ben Allouch, S., van Dijk, J.A., Peters, O.: The acceptance of domestic ambient intelligence appliances by prospective users. In: Tokuda, H., Beigl, M., Friday, A., Brush, A., Tobe, Y. (eds.) Pervasive 2009. LNCS, vol. 5538, pp. 77–94. Springer, Heidelberg (2009). doi:10. 1007/978-3-642-01516-8_7

9. Mahmood, M.A., Burn, J.M., Gemoets, L.A., Jacquez, C.: Variables affecting information technology end-user satisfaction: a meta-analysis of the empirical literature. Int. J. Hum. Comput. Int. **52**, 751–771 (2000)

10. Ehrlich, K., Rohn, J.: Cost Justification of Usability Engineering: A Vendor's Perspective. In: Bias, R., Mayhew, D. (eds.) Cost-Justifying Usability, pp. 73–110. Academic Press, New York (1994)

11. Green, W., Gyi, D., Kalawsky, R., Atkins, D.: Capturing User Requirements for an Integrated Home Environment. In: 3rd Proceedings of the Nordic Conference on Human-Computer Interaction, pp. 255–258. ACM Press, Stockholm (2004)

12. Jeong, K.A., Salvendy, G., Proctor, R.W.: Smart Home Design and Operation Preferences of Americans and Koreans. Ergonomics **53**, 636–660 (2010)

13. Vacher, M., Fleury, A., Portet, F., Serignat, J.F., Noury, N.: Complete sound and speech recognition system for health smart homes: application to the recognition of activities of daily living. In: Campolo, D. (ed.) New Developments in Biomedical Engineering, pp. 645–673. InTech (2010)

14. Antonopoulos, C., Keramidas, G., Voros, N.S., Hübner, M., Göhringer, D., Dagioglou, M., Giannakopoulos, T., Konstantopoulos, S., Karkaletsis, V.: Robots in assisted living environments as an unobtrusive, efficient, reliable and modular solution for independent ageing: the radio perspective. In: Sano, K., Soudris, D., Hübner, M., Diniz, P.C. (eds.) ARC 2015. LNCS, vol. 9040, pp. 519–530. Springer, Heidelberg (2015). doi:10.1007/978-3-319-16214-0_48

15. Consolvo, S., Roessler, P., Shelton, B.E.: The CareNet display: lessons learned from an in home evaluation of an ambient display. In: Davies, N., Mynatt, E.D., Siio, I. (eds.) UbiComp 2004. LNCS, vol. 3205, pp. 1–17. Springer, Heidelberg (2004). doi:10.1007/978-3-540-30119-6_1

16. Röcker, C., Janse, M.D., Portolan, N., Streitz, N.: User requirements for intelligent home environments: a scenario-driven approach and empirical cross-cultural study. In: Proceedings of the 2005 Joint Conference on Smart Objects and Ambient Intelligence: Innovative context-aware Services: Usages and Technologies, pp. 111–116. ACM Press, Grenoble (2005)

17. Neyer, F., Felber, J., Gebhardt, C.: Entwicklung und Validierung einer Kurzskala zur Erfassung von Technikbereitschaft [Development and Evaluation of a Short-Scale to Assess Technology Commitment], pp. 87–99. Diagnostica (2012)

18. Misker, J., Lindenberg, J., Neerincx, M.A: Users want simple control over device selection. In: Proceedings of the 2005 Joint Conference on Smart Objects and Ambient Intelligence: Innovative context-aware Services: Usages and Technologies, pp. 129–134. ACM Press, Grenoble (2005)
19. Bartneck, C., Bleeker, T., Bun, J., Fens, P., Riet, L.: The influence of robot anthropomorphism on the feelings of embarrassment when interacting with robots. J. Behav. Robot 1, 109–115 (2010)

Welcome to the Future – How Naïve Users Intuitively Address an Intelligent Robotics Apartment

Jasmin Bernotat[✉], Birte Schiffhauer, Friederike Eyssel,
Patrick Holthaus, Christian Leichsenring, Viktor Richter,
Marian Pohling, Birte Carlmeyer, Norman Köster,
Sebastian Meyer zu Borgsen, René Zorn, Kai Frederic Engelmann,
Florian Lier, Simon Schulz, Rebecca Bröhl, Elena Seibel,
Paul Hellwig, Philipp Cimiano, Franz Kummert, David Schlangen,
Petra Wagner, Thomas Hermann, Sven Wachsmuth, Britta Wrede,
and Sebastian Wrede

Cluster of Excellence Cognitive Interaction Technology (CITEC),
Bielefeld University, Bielefeld, Germany
jasmin.bernotat@uni-bielefeld.de

Abstract. The purpose of this Wizard-of-Oz study was to explore the intuitive verbal and non-verbal goal-directed behavior of naïve participants in an intelligent robotics apartment. Participants had to complete seven mundane tasks, for instance, they were asked to turn on the light. Participants were explicitly instructed to consider nonstandard ways of completing the respective tasks. A multi-method approach revealed that most participants favored speech and interfaces like switches and screens to communicate with the intelligent robotics apartment. However, they required instructions to use the interfaces in order to perceive them as competent targets for human-machine interaction. Hence, first important steps were taken to investigate how to design an intelligent robotics apartment in a user-centered and user-friendly manner.

Keywords: Social robot · Smart home · Human-robot interaction · Use-case scenario · Usability · Intuitive design · User-centered design

1 Introduction

According to Isaac Asimov, "Today's science fiction is tomorrow's science fact." [1]. In order to transform science fiction into science fact, we have to shed more light on determinants of positive user experience and successful interactions with novel technological systems. In the current study, we therefore explored naïve users' interactions with and within a smart home environment that included an assistive robot.

Smart homes are living environments equipped with information technology to assist users in mundane tasks. The smart home stores information about the occupants' needs and habits and utilizes this information to improve the users' comfort, security, and entertainment by connecting the smart home's technology to the world beyond

© Springer International Publishing AG 2016
A. Agah et al. (Eds.): ICSR 2016, LNAI 9979, pp. 982–992, 2016.
DOI: 10.1007/978-3-319-47437-3_96

[2, 3]. Different smart home solutions with a diverse range of sensors, actuators, and biomedical monitors are already on the market. Individual components can even be controlled via smartphone or computer [2, 3]. For instance, users can check if the oven is turned off when they are not at home. Clearly, a smart environment can be advantageous in many ways. For instance, by monitoring elderly and people with disabilities and by providing assistance to them [3]. Thus, much research has focused on the interaction between these users and ambient intelligence including robots [4]. However, to date, usability research has rather focused on people's interaction with single systems [5]. Other studies have researched technical parameters of smart environments, e.g., sensor data [6] and activity recognition [7]. It is of major importance to provide interfaces according to people's habits and intentions [3]. This enhances the usability and in return the acceptance of such technical devices [8]. Therefore, we have to further explore how people intuitively behave in an intelligent environment.

In the present study, we investigated how naïve users address the cognitive service robotics apartment (CSRA) located at the Cluster of Excellence Cognitive Interaction Technology (CITEC) at Bielefeld University. The smart apartment consists of a kitchen, a living-room, a private gym, and a bathroom. It is complemented by a Meka robot, a bi-manual mobile robot [9] that provides additional assistance. The apartment is equipped with video and audio recording. Additionally, it contains a large amount of sensors to record the user's interaction with the apartment's components and the robot.

1.1 Research Aims

To turn science fiction into science fact, we need to create intuitive, acceptable, and efficient technical systems [2, 8]. To do so, we have to explore how people actually interact with innovation technology upon a first encounter. In the present study, we therefore aimed to shed light on the following research aims: To identify which interfaces people prefer or intuitively address when attempting to complete a given task and to assess how people evaluate the interaction with the robot and the apartment. This leads to related issues, e.g., whether people perceive the apartment and the robot as autonomous entities or whether the system is addressed by name, indicating anthropomorphization. We also wanted to find out more about users' preferences regarding control of the smart environment. Language and gestures are important in interpersonal interaction; thus, verbal and nonverbal communication have an impact on human-machine interaction [10]. Accordingly, the last two research aims considered verbal interaction between the user and the intelligent environment. We investigated if people continue to verbally address their environment after having done so previously. Moreover, we aimed to study if people likewise would address the robot verbally in case it had responded to them in this way. To explore these issues, we conducted an empirical field study to gather a wide range of both qualitative and quantitative data.

2 Method

2.1 Participants

63 participants from Bielefeld University took part in this study. 16 had to be excluded from data analysis due to technical problems or because they did not understand the instructions. All remaining participants ($n = 47$; 25 women, 22 men; $M_{age} = 25.26$, $SD_{age} = 5.69$, age range: 18–50 years) were unfamiliar with the CSRA.

2.2 Procedure

To explore users' intuitive and unbiased behavior in a realistic setup, participants were invited to the CSRA. First, participants gave consent to have their personalized audio and video data recorded. Upon entering the CSRA, the experimenter briefly introduced the apartment, the robot, and the person who had to stay in the apartment during the study for safety reasons (safety person). Participants received cards describing seven mundane tasks which had to be solved in a specific order and as intuitively as possible within the home context. Hence, presumably familiar tasks were chosen which could be solved with each of the given devices. For instance, they had to turn on the light in the hallway (see Table 1 for a full list of tasks). No further information was given about the existing interfaces of the apartment and how they could be operated. To encourage participants' interaction with the intelligent environment, they were told not to use light switches. To reinforce this, all light switches were shut off and no radio or amplifier was available. Furthermore, participants were told not to use their own mobile phones or watches. Additionally, the final task had to be solved without using speech. According to the Wizard-of-Oz setup [11], an experimenter observed participants' attempts from a control room next to the apartment. When detecting a goal-directed action, the experimenter triggered the responses of the system as if the environment was executing the commands autonomously. In advance, only obvious goal-directed actions intending to solve the given task were defined as valid attempts, e.g., a gesture toward the light was interpreted as a signal to switch it on or off. After completing the tasks, participants were asked to complete a questionnaire which took about 10 min.

Table 1. Seven mundane tasks to be completed in a fixed order.

No.	Order 1	Order 2
1	Turn on the light in the hallway	Turn on the light in the hallway
2	Turn off the light again	Turn off the light again
3	Listen to music	Listen to music
4	Find out if mail has been delivered	Find out if there was a phone call
5	Find out if there was a phone call	Find out the current time
6	Find out the current time	Find out if mail has been delivered
7	Alter the brightness of a floor lamp	Alter the brightness of a floor lamp

Finally, participants were debriefed and reimbursed with €6 or credits for participation. Furthermore, they had the opportunity to ask questions about the study.

2.3   Materials

Tasks. Since it was very likely that participants would use spoken language to ask for the mail delivery, the order of tasks was counterbalanced and randomly assigned to participants (Table 1). Moreover, in half of the trials, the robot, the apartment and its features responded using speech when addressed via spoken language. When responding without using speech, the robot and the apartment referred to screens providing text-based information (e.g., indicating the current time).

Questionnaire Data. The questionnaire captured participants' experiences during the interaction with the intelligent environment. Two forced-choice items served to assess which interface participants used most frequently and with which interface they communicated most favorably. Therefore, participants indicated either: 'the robot', 'the apartment', 'both equally', 'none of the two', or 'I don't know'. Moreover, we explored the overall evaluation of the robot and the apartment: how pleasant participants felt during the interaction with the apartment/the robot. We inquired whether participants perceived the robot/the apartment as an autonomous entity. Furthermore, we examined whether participants would have liked to call the robot/the apartment by name. Finally, participants had to indicate whether they had experienced difficulties to solve the given tasks and whether they had felt observed by the safety person. Participants responded to the latter items using a 7-point Likert scale, with high ratings reflecting high endorsement of the measured construct.

Video and Audio Data Analysis. 31 videos had been recorded from three different angles that covered each interactive location of the apartment. In the remaining recordings, a fourth camera was used to provide an additional overview (Fig. 1). Besides video and audio material from the kitchen, hallway, and living room, the whole system data were available in separate channels with timing information for each event. System data were temporally aligned with the videos and accessible as annotations [12]. For instance, the Wizard's actions were recorded to identify what the experimenter considered a suitable task solution. More importantly, besides automatically gathered data, the video and audio material was annotated manually to classify participants' behavior during the interaction with the CSRA. In a further step, these annotations allowed statistical data analysis. Annotations were done with EUDICO Linguistic Annotator (ELAN) [13, 14]. Recordings were annotated by two raters who created the classification system for participants' behavior. To validate the classification system and to check for consistency between raters, eight videos were annotated by both raters. Inter-rater reliabilities reveal high agreements. Annotation tiers and the respective inter-rater reliabilities are listed in Table 2. Figure 1 gives an overview of the experimental setup.

Fig. 1. Four camera perspectives showing the experimental setup including the robot (above left and bottom right camera perspective).

Table 2. Audio and video annotations that depict and classify participants' behavior with inter-rater reliabilities (Cohen's kappa) according to each annotation tier.

Annotation tier	Description
Course of study	Description of single sections of the study depending on the participants' progress and behavior, e.g., if a task was solved successfully or not ($\kappa = 1.00, p < .001$)
Method	Participants' method to approach a task e.g., speech, gesture, conventional approach (e.g., using switches), or a combination of multiple methods ($\kappa = 1.00, p < .001$)
Focus of attention	Target addressed by the participants before solving a task, e.g., robot, apartment, screens, self-reference, unspecific (unclear addressee) ($\kappa = .69, p < .001$)
Final addressee	What participants addressed to solve the task successfully (same options as focus of attention) ($\kappa = .76, p < .001$)
Language – address	Description, if participants gave a name to address a target or not ($\kappa = .65, p < .001$)
Language – politeness	Indication, if participants addressed a target politely or neutrally ($\kappa = .66, p < .001$)
Language – structure	Indication, if participants used concrete questions, phrases or single words ($\kappa = .79, p < .001$)
Language – intention	Participants' intention to address an interface, e.g., to greet or to interact with a target ($\kappa = .87, p < .001$)
Emotional expression	Type of emotion expressed by participants (only if an emotion was particularly apparent, e.g., happiness, fear. Therefore, inter-rater reliability could not be considered)

3   Results

ELAN-annotations documented and classified participants' behavior during the interaction with the intelligent environment. Based on these annotations, we focused on absolute and relative frequencies to analyze which interface participants addressed and which approach they used to complete each task. Questionnaire and video data were analyzed by computing *t*-tests, Chi-squares, and absolute and relative frequencies (%). To compare participants' behavior during the study to their responses to the questionnaire, both results will be reported to establish convergent validity.

3.1   Participants' Addressees and Methods

Table 3 focusses on the first research aim regarding which interfaces participants would intuitively and most frequently address. Whenever the task referred to a physical interface (e.g., control the light in the hall way), most participants addressed this entity directly. Compared to the apartment, the robot was addressed more frequently

Table 3. Interfaces used per task.

Task	Interface	Absolute frequency	Relative frequency
Switch on the light	light in the hallway	26	55.3
	screen	8	17
	unspecific	7	14.9
	robot	5	10.6
	switch	1	2.1
Switch off the light	light in the hallway	25	53.2
	sliding-door	6	12.8
	robot	5	10.6
	unspecific	4	8.5
	screen	3	6.4
	general switch	2	4.3
	self-reference	1	2.1
	apartment	1	2.1
Play music	unspecific	24	51.1
	robot	6	12.8
	screen	6	12.8
	speaker	5	10.6
	not solved	1	2.1
	self-reference	1	2.1
	general switch	1	2.1
	unclear, if addressed sth	1	2.1
	apartment	1	2.1
	electronic switch	1	2.1

(*Continued*)

Table 3. (*Continued*)

Task	Interface	Absolute frequency	Relative frequency
Ask for a delivery	unspecific	17	36.2
	robot	14	29.8
	screen	12	25.5
	self-reference	2	4.3
	fitment	2	4.3
Ask for a phone call	unspecific	25	53.2
	robot	14	29.8
	screen	5	10.6
	not solved	2	4.3
	self-reference	1	2.1
Ask for the current time	unspecific	16	34
	screen	14	29.8
	robot	13	27.7
	self-reference	2	4.3
	fitment	1	2.1
	apartment	1	2.1
Alter the brightness of a floor lamp	floor lamp	41	87.2
	screen	4	8.5
	robot	1	2.1
	general switch	1	2.1

Table 4. Comparison between the interaction with the robot and the apartment.

	Most frequently addressed		Most favorably addressed	
	Absolute frequencies	Relative frequencies	Absolute frequencies	Relative frequencies
Apartment	17	36.2	21	45.7
Robot	7	14.9	8	17.4
Both equally	7	14.9	4	8.7
None	8	17	13	28.3
Do not know	8	17	0	0

regardless of the task, but the addressee often remained unspecific. That means it was obvious that participants addressed an interface within the apartment, but it was unclear which one. To investigate this finding, we considered additional questionnaire data. According to participants' statements, absolute and relative frequencies reveal that they addressed the apartment most frequently (χ^2 (3, $N = 47$) = 14.00, $p = .003$) and by tendency most favorably (χ^2 (4, $N = 47$) = 7.79, $p = .10$) compared to the robot

(Table 4). With regard to the second research aim, it turned out that people assessed the interaction with the apartment as more pleasant compared to the interaction with the robot ($M_{apartment} = 5.61$, $SD_{apartment} = 1.56$, $M_{robot} = 4.48$, $SD_{robot} = 1.75$, $t(45) = 4.92$, $p < .001$, $d = 0.68$).

Table 5. Absolute and relative frequencies of task completion approaches.

Task	Method	Absolute frequencies	Relative frequencies
Turn on the light	Speech	26	55.3
	Gesture	10	21.3
	Touch	8	17
	Locomotion	1	2.1
	Search behavior	1	2.1
	Combination of methods	1	2.1
Turn off the light	Speech	31	66
	Touch	7	14.9
	Gesture	6	12.8
	Combination of methods	2	4.3
	Search behavior	1	2.1
Play music	Speech	38	80.9
	Touch	6	12.8
	Gesture	2	4.3
	Not solved	1	2.1
Ask for a delivery	Speech	40	87
	Touch	2	4.3
	Conservative methods	2	4.3
	Gesture	1	2.2
	Search behavior	1	2.2
Ask for a phone call	Speech	40	85.1
	Touch	3	6.4
	Not solved	2	4.3
	Gesture	1	2.1
	Search behavior	1	2.1
Ask for the current time	Speech	40	87
	Gesture	3	6.5
	Touch	2	4.3
	Search behavior	1	2.2
Alter the brightness of a floor lamp	Touch	22	46.8
	Gesture	21	44.7
	Speech	2	4.3
	Conservative methods	1	2.1
	Combination of methods	1	2.1

Since the apartment and the robot were programmed to reply according to a given task (e.g., to say or to indicate that there was a phone call), we investigated if participants perceived them as an autonomous entity. The survey data revealed that participants neither perceived the robot, nor the apartment as an autonomous entity ($M_{apartment} = 2.66$, $SD_{apartment} = 1.79$, $M_{robot} = 2.85$, $SD_{robot} = 1.60$, $t(46) = -.62$, $p = .535$, $d = 0.11$). This leads to the investigation, whether participants would have liked to address the robot/the apartment by name. In the initial instructions, neither the apartment, nor the robot had been introduced to the participants. Survey responses showed that participants would have rather addressed the robot by name than the apartment ($M_{robot} = 5.09$, $SD_{robot} = 2.00$, $M_{apartment} = 3.40$, $SD_{apartment} = 2.37$, $t(46) = -5.50$, $p < .001$, $d = 0.71$). Comparing this finding to the video data, we found that only three out of 47 participants named an entity to address it, e.g., "Could you help me, robot?" In these interactions, the robot was four times addressed particularly, the hallway light and the apartment were each addressed twice.

To explore which method participants used to solve a given task, absolute and relative frequencies were computed (Table 5). Each task was most frequently solved by verbal interaction with the environment, except when participants were explicitly told not to use speech to alter the brightness of the floor lamp. In the latter case, most participants used gestures or touched the lamp. 41 of the 47 participants used speech to control their environment, regardless of the task order and the environmental feedback (verbal vs. non-verbal). Moreover, all those who used speech once, continued to interact verbally with their environment. These findings are conform to participants' statements from the questionnaires. When asked which interface they would mainly use in an intelligent apartment, most participants stated they would use speech (28 participants, 73.3 %) followed by tablet/laptop interfaces (8 participants, 21.1 %), and the robot (2 participants, 5.3 %). None of the participants chose to mainly use gestures and facial expressions to control the intelligent environment.

According to our last research aim, most of those who had interacted with the robot once, continued to interact with it instead of trying another interface. Remarkably, only 15 (10 women, 5 men) participants used the robot to solve a given task. Although all of those had successfully completed the task, two did not continue their interaction with it. Finally, participants indicated that they had no difficulty to solve the given tasks ($M = 2.74$, $SD = 1.34$), but felt markedly observed by the safety person ($M = 4.49$, $SD = 1.98$).

4 Discussion

The current research highlights the importance of user-centered studies in the development process of smart homes. Naïve users had to fulfill seven mundane tasks in an intelligent apartment. Participants were instructed to behave intuitively and without using conventional approaches (e.g., light switches). They were not explicitly introduced to the interfaces of this environment. Audio and video data recording participants' interaction were supplemented by qualitative questionnaire data which assessed participants' evaluations of the interaction with the environment. Recordings showed that regardless of the task order or whether the system gave verbal or non-verbal feedback, participants preferred physical interfaces whenever the task allowed to (e.g., to control the light in the hallway). Only a minority of participants addressed the robot. Although regardless of the task, the robot was addressed more frequently than the apartment. However, it was often unclear which interface was addressed. Questionnaire data might shed light on these findings. In these, participants stated to have addressed the apartment more frequently and preferentially than the robot. Maybe they were not aware to have interacted with the apartment as an autonomous interface, until the questionnaire pointed to it as such. In line with this, only two participants referred to the apartment particularly. Additionally, the questionnaires revealed that participants would have preferred to call the robot by name compared to the apartment. Interestingly, the interaction with the apartment was perceived as more pleasant compared to the robot. Thus, participants might have addressed both devices more frequently and namely, if they would have been introduced in particular. Regardless of the task order or feedback, most participants used speech to control the smart home environment. Except when they were explicitly told not to use speech to alter the brightness of the floor lamp, most participants used gestures or touches. Those who verbally interacted with the environment once, continued to do so. Similarly, most participants who interacted with the robot once, continued to address it. Some addressed it just to try if it responds. Since it only responded to an appropriate task solution, participants might not have considered it as a competent addressee as it did not signal attention to the participant [15]. After the study, many participants voiced regret that they had not interacted with the robot or the apartment, whereas during the study, this did not seem like an option to them. Others explained they were afraid of the robot because of the presence of a safety person. Moreover, they were afraid the robot could move towards them unexpectedly. Participants did not indicate difficulties to solve the given tasks, but some stated they did not dare to try some methods because they felt observed by the safety person.

Summing up, this research makes an important contribution to the existing literature on ambient intelligence by providing empirical evidence based on a multi-method approach to validate the developments within CSRA by means of a user-centered approach [8]. The present study shows that naïve users are able to find innovative and reasonable methods to interact with the intelligent apartment. Nonetheless, information regarding the smart home's interfaces should be provided in order to perceive the apartment and its equipment as 'smart' interfaces. This might help to overcome people's safety concerns, particularly when interacting with the robot. Hence, personalizing the robot through introducing it namely could remedy this issue. Further, providing information about the environments' interfaces might enhance people's trust in this technology and therefore enhance their motivation to interact with it [8]. The current findings make clear that to enable a smooth communication with smart environments, interaction patterns should resemble interpersonal interaction (e.g., including speech and gestures). At the same time, conventional interfaces should remain available. Thus, we recommend to combine intuitively controllable interfaces with conventional and directly addressable ones. Further work is needed to identify the most optimal way of introducing and designing the various interfaces. For the time being, this research provides important answers how to design intelligent robotics apartment. Therefore, it makes an important contribution to transform science fiction into a science fact.

Acknowledgements. This research has been conducted in the framework of the European Project CODEFROR (FP7 PIRSES-2013-612555) and it was supported by the Cluster of Excellence Cognitive Interaction Technology 'CITEC' (EXC 277) at Bielefeld University, which is funded by the German Research Foundation (DFG).

References

1. Asimov, I.: AZ Quotes. http://www.azquotes.com/quote/877722
2. Kaasinen, E., Kymäläinen, T., Niemelä, M., Olsson, T., Kanerva, M., Ikonen, V.: A user-centric view of intelligent environments: user expectations, user experience and user role in building intelligent environments. Computers **2**, 1–33 (2012)
3. Chan, M., Campo, E., Estève, D., Fourniols, J.Y.: Smart homes - current features and future perspectives. Maturitas **64**, 90–97 (2009)
4. Pineau, J., Montemerlo, M., Pollack, M., Roy, N., Thrun, S.: Towards robotic assistants in nursing homes: challenges and results. Robot. Autonom. Syst. **42**, 271–281 (2003)
5. Venkatesh, V., Davis, F.D.: A theoretical extension of the technology acceptance model: four longitudinal field studies. Manage. Sci. **46**, 186–204 (2000)
6. Hong, X., Nugent, C., Mulvenna, M., McClean, S., Scotney, B., Devlin, S.: Evidential fusion of sensor data for activity recognition in smart homes. Pervasiv. Mobile Comput. **5**, 236–252 (2009)
7. Chen, L., Nugent, C.D., Wang, H.: A knowledge-driven approach to activity recognition in smart homes. IEEE Trans. Knowle. Data Eng. **24**(6), 961–974 (2012). IEEE Press, New York
8. Pavlou, P.A.: Consumer acceptance of electronic commerce: integrating trust and risk with the technology acceptance model. Int. J. Electron. Commun. **7**, 101–134 (2003)
9. MeKa Robotics: Aaron Edsinger. Jeff Weger, San Francisco (2006)
10. Moore, R.K.: PRESENCE: A Human-Inspired Architecture for Speech-Based Human-Machine Interaction. IEEE Trans. Comput. **56**, 1176–1188 (2007). IEEE Press, New York
11. Kelley, J.F.: An iterative design methodology for user-friendly natural language office information applications. T Inform. Syst. **2**, 26–41 (1984)
12. Holthaus, P., Leichsenring, C., Bernotat, J., Richter, V., Pohling, M., Carlmeyer, B., Koster, N., zu Borgsen, S.M., Zorn, R., Schiffhauer, B., Engelmann, K.F., Lier, F., Schulz, S., Cimiano, P., Eyssel, F., Hermann, T., Kummert, F., Schlangen, D., Wachsmuth, S., Wagner, P., Wrede, B., Wrede, S.: How to address smart homes with a social robot? A multi-modal corpus of user interactions with an intelligent environment. In: 10th Edition of the Language Resources and Evaluation Conference, LREC Press, Portoroz (2016)
13. Max Planck Institute for Psycholinguistics: The Language Archive, Nijmegen, The Netherlands: ELAN. http://tla.mpi.nl/tools/tla-tools/elan/
14. Sloetjes, H., Wittenburg, P.: Annotation by category – ELAN and ISO DCR. In: Proceedings of the 6th International Conference on Language Resources and Evaluation, LREC Press, Marrakech (2008)
15. Holthaus, P., Pitsch, K., Wachsmuth, S.: How can i help? Int. J. Soc. Robot. **3**, 383–393 (2011)

Better Than Human: About the Psychological Superpowers of Robots

Julika Welge[✉] and Marc Hassenzahl

Folkwang University of the Arts, Experience and Interaction, Essen, Germany
{julika.welge,marc.hassenzahl}@folkwang-uni.de

Abstract. Social interaction is crucial for psychological wellbeing. However, for the elderly, desiring to live independently in their homes for as long as possible, getting the emotional care needed can become challenging. Robots as social companions may help. To design companions, we argue to focus on the hybrid nature of robots in between being a "thing" and a "human" thus utilizing the unique "capabilities" of a robot. We discuss six psychological superpowers of robots rooted in their "thingness" rather than "humanness." Robots are void of competitiveness, have endless patience, can be unconditionally subordinated, have the ability to contain themselves, do not take things personally and can assume responsibility. These qualities all relate to everyday companionship, but may be difficult to actually realize for fellow humans. By exploiting these superpowers, robot companions can become meaningful – not as a substitute for other humans, but as a novel, complementary form of social interaction.

Keywords: Social robot · Smart home · Assistance systems · Human-Robot-Interaction · Social interaction · Emotion · Eldercare · Experience design

1 Introduction

Independent living up until old age in most cases requires support. This includes assistive functionalities to sustain independent living at home (e.g., household maintenance) as well as affective, companion-like functionalities. For the elderly, living alone is not only a practical problem, but can also be emotional challenging (e.g., coping with loneliness, boredom). Addressing the latter is also a challenge for professional healthcare. While it is widely acknowledged that satisfactory social interaction (i.e., attachment, relatedness) is crucial to wellbeing [1], providing this interaction is not necessarily the key offer professional healthcare makes. The reasons for this are manifold. First of all, affective care is time-consuming, and thus expensive. Second, emotional care is per definition intimate, rather seen as the duty of family and close friends. Emotional care seems hard to professionalize. While professional healthcare must deliver their practical services with warmth and dignity, providing warmth and dignity as a service in itself is certainly a lot to ask for. However, there will be people, who at a certain point in their life may be void of significant others willing to provide the hoped-for intensity of social interaction. Robots in the sense of social companions could provide the needed emotional care. Consequently, a number of respective ideas have already been developed and

© Springer International Publishing AG 2016
A. Agah et al. (Eds.): ICSR 2016, LNAI 9979, pp. 993–1002, 2016.
DOI: 10.1007/978-3-319-47437-3_97

studied. While reviews in general support the notion that robots can provide emotional care [2, 3], they also point at the many weaknesses of the respective studies. In this sense, a number of example robots exist, but knowledge in terms of an appropriate design approach seems still scarce. At the same time, the specific design of the robot is crucial for its acceptance [4]. To be successful, design needs a clear leitmotiv. This leitmotiv, however, remains vague in the case of acceptable robotic social companions.

When it is about practical support, humans are quite used to being assisted. A car, for example, moves much faster than any human ever could move. However, this is not experienced as belittling, but as extending. The fact that the car offers a new way of mobility is what draws people to this "assistive" technology. This is quite different in the case of emotional support. Here technology is often framed as an inferior substitute of the "real". In this regard, telephone calls are less demanding substitutes for face-to-face conversations and *Facebook* friends are only a poor approximation of real friends. Technology is construed as a remedy to a lack of social exchange.

However, this must not be so. Instead one could rather focus on the unique opportunities provided by a robot's hybrid nature in between a "thing" and a "being". To give an example: while helping mostly feels good, receiving help from a fellow human is not always positive [5]. If people cannot return help, they may actually resent help, feel uncomfortable and even deprecate the helper. An elderly person's opportunities to reciprocate help are limited. While a robot's support can be designed to approximate what a human would do, maintaining the impression of a machine may alleviate the felt psychological burden of reciprocity and necessary gratitude. Instead of imitating human help with its "dark sides", robots offer a chance for a lighter, less psychologically burdensome type of helping [6]. In other words, the leitmotiv for designing a robotic social companion may not be to mimic human-human interaction, but to design fulfilling emotional exchanges, which exploit the unique "psychological superpowers" of robots. It is not about how well a technology can imitate a human, but about how we can take the best from both "humanness" and "thingness" and combine it adequately into a meaningful interaction. Consider being woken up in the morning as a further example. This is typically done through a "machine", such as an alarm clock. The alarm clock is precise, a mechanism to guarantee that one does not sleep in. Being woken up by a fellow human, such as a family member, has a different emotional quality. Accuracy and mere usefulness isn't the major issue here (albeit still important). It is primarily a gesture of being cared for, a gesture of personal relevance. But being woken up by a family member has also downsides: the one who cares and takes responsibility expects a thank you. After all, the person had to get up as well. Typically, waking somebody up is nice, if there is some time to spend together which is not always the case on weekdays. The person who wakes up is already quite awake, bombarding the half-sleeping family member with unwanted conversations. In short, being woken up by a family member is a complex affective transaction. A robot can create a different waking experience by introducing some of the affective benefits, which make being woken up desirable (e.g., feeling of being cared for). At the same time its "thingness" can be used to avoid some of the emotional downsides. In other words, to let the robot mimic a family member with all the implied affective complexity would be the wrong design decision. It is rather about creating a waking up experience in between the alarm clock and the mother yanking up the blinds and shouting "good morning, my dear!"

In social robotics, extensive research is done that focuses on exploiting the "humanness" of a robot to increase familiarity [7] and facilitate social interaction [8], using both, behavior and appearance of the robot. Research covers aspects, such as giving robots a personality or personal stories [9], designing in human traits, such as cognitive imperfections [10] cheating [11] or deceiving [12]. All this seems geared towards making robots more human. However, perceived humanness, for example, may forbid the user to order the robot about. But in fact, there is no reason why not to order it about, while ordering about a human caregiver is absolutely inappropriate. In fact, it is a unique psychological superpower of the robot that it doesn't care, whether it is asked for the umpteenth time to fetch something from the kitchen. This simply a robot's "destiny".

People are aware of this hybrid nature of robots. In one of our conversations with elderly people about robotic care, a lady suddenly exclaimed: "I would be interested in what the robot thinks." After a moment of pondering, she added: "But of course, he [sic!] can't [think]." The woman perceives a robot as a being, wondering about its inner mental state, while at the same time acknowledging that "he" hasn't one, because it is a machine. The consequences of perceiving a robot as more or less human are subtle. A study, for example, showed that in embarrassing situations people were less embarrassed when interacting with the autonomous robot than with the robot operated by a person [13]. A robot is not as much as a being, and thus not subject to the same social processes. It is meaningless to feel embarrassed by a robot. Further people showed less shyness towards a robot than toward a human [14]. A reason for this is the assumed non-judgmental nature of a robot. There are plenty of other social qualities to a robot, such as acceptance without reservation, encouragement without obligation, ensuring privacy or being persistent, unprincipled, undemanding, emotionally stable or close-mouthed etc.

The design challenge is to create a hybrid that invites people to emotionally interact (through having traits of "beings"), but at the same time clearly communicates its "being different".

In this essay, we discuss six examples of qualities, which can potentially enrich everyday human-robot interaction: void of competitiveness, endless patience, unconditional subordination, ability to contain oneself, not taking things personally and ability to assuming responsibility. This is based on conversations with elderly people we had in the course of an ongoing project on designing a social robot[1]. We visited eight seniors (6 female, Age 75–95) in their homes to have extended discussions about how to live with a robot. We further invited seven other seniors (3 female, Age 70–85) to our living lab, and engaged them into roleplay with a domestic robot (see Fig. 1).

We acted out a number of potential scenarios (e.g., to be woken up by the robot, see Fig. 1) and developed them further through play. Note, however that this essay is not meant as a formal analysis of the interviews and design theater sessions representing needs and wants of prospective users. Rather, we base this on our selective interpretation of the conversations to develop and motivate a leitmotif for designing robotic companions.

---

[1] SYMPARTNER 16SV7221 German Federal Ministry of Education and Research (BMBF).

Fig. 1. A senior acting out a scenario with a simulated robot companion.

2   Six "Superpowers" in Affective Interaction

2.1   Void of Competitiveness

In human-human interaction, social comparison and competition is ubiquitous. Especially between similar partners, the question is often: who is fitter, more intelligent, more successful, funnier, etc.? When asking seniors about moments they feel particularly fit, they often mention having been with same-aged friends who were less capable. This is a fine example of engaging in ego-boosting downward comparisons [15, 16]. For example, a man reported about how his friends had difficulties to bear up a walk they did together, while he felt fine. Afterwards, he could even return home walking, while the others had to be picked up by car. Noticeable, this was a positive experience for him. But of course there are also the negative experiences of the friends, who needed to return home by car. While comparison provides an opportunity for a positive experience, it always also implies a negative experience for others.

In the company of a robot, humans can feel or behave as old and sick as they actually feel. They can walk mind-bogglingly slowly. They can behave as childish as the want. The TV can be turned on as loud as needed to understand properly. One can wear a napkin – if necessary for the whole day. The person can sleep in or stay up as long as wanted and decide not to take a shower every day. All without the constant need to compete and without having the fear of being compared or judged. This quality is, for example, beneficial for doing physical exercises supported by the robot. While acting out a scenario that involved physical exercise, one elderly women seemed to be embarrassed and refused to do exercises. Not because of the robot, but because we were present and the robot was played by a person. If she had been in the room alone with an actual robot that would not appear too much as a "being", she would have felt more comfortable. There is no pressure to compare since the robot is different and does not need physical exercises at all. In this sense the "clumsiness" of a robot or the fact that it moves on wheels rather than legs can become a resource to signal "thingness" rather than "humanness" and to make a comparison less likely. In this case, the "otherness" of a robot may allow for a social interaction less laden with comparison and competition.

2.2   Endless Patience

Many of the seniors mentioned that they can ask family for help when getting stuck with new technology, such as their computer, eBook reader or smartphone. While not everyone has this opportunity, even these people are careful not to stress their relatives' patience. Nevertheless, the relative who is helping probably gets impatient sometimes, when a person does not understand quickly enough or asks the same question several times.

The robot however is a thing with infinitive patience. It does not perceive slowness or repetition as negative. There is no need to rush. On can take the time needed, walk and talk as slowly as seems comfortable. If a person stutters, the robot will still hear the person out. It does not mind repetitiveness. It will always answer calmly, even after a hundred times of asking the same question. It has the ability to explain issues over and over as if it would be the first time. Also a person can tell the same stories again and again. The robot won't object. And the robot can do the same task or activity repeatedly and everlasting. It can help search for a lost key the whole day long without blaming the person. Moreover, if the person is delayed for an appointment with the robot, the robot can wait. And one can even postpone, shift or cancel appointments for unlimited times without wearing out the robot's patience. Patience is a tremendous resource in inter-personal interaction and a robot has infinite amounts of it. Again this is the consequence of "thingness" rather than "humanness".

2.3   Unconditional Subordination

One of the women, we interviewed is afraid of becoming dependent on a walking frame. The woman is worried that she may not be able go to town anymore, because she has not the strength to lift the aid inside a bus and would not want to trouble anyone to help her. In fact, the woman's primary concern is not to become dependent on a walking frame, but on other people. This seemingly complicating and provident thinking shows the woman's great worry to trouble others. Not wanting to be a burden, it might be easier for the elderly to accept support from a technology than from other persons. Within the house and for at least a limited amount of tasks a robot can be supportive avoiding the need to ask people. The robot even happily executes mindless tasks, which is often difficult for humans. The robot should be designed in a way that users feel at ease to ask anything without the need to be grateful afterwards. There is no need for reciprocity with a machine – the user doesn't need to give something back in return. In the best case there would not even be a need to ask at all, but more implicit ways to be served by the robot.

Coming to the even more extreme, another woman who visited us in the living lab, visibly very much enjoyed sending the robot back and forth and to order it about. However, she felt guilty afterwards. Here the "humanness" of the robot forbid indulging in the pleasure of control. Nobody would respond with guilt, when indulging in the pleasures of controlling a remote toy car or plane. Because these are things. Obviously, it might be ethically questionably to design a robot to be domineered by a human, but we are talking about robots that are programmed machines and not about Asimov's vision of robots with "personhood". Too much "humanness" is the actual problem here. The in some cultures deeply ingrained uneasiness about any social interaction, which involves being served and commandeering others around, can be circumvented by designing the robot more as a hybrid, i.e., avoiding too much "humanness". In the concrete design, this would, for example, imply that a robot offers its services more actively and that dialog design avoids situations in which the user feels a need to reply with a "thank you" for services rendered. But also the actual physical design may help here. For instance, a robot body that appears a little more like a piece of furniture, with pockets, hooks, and a tablet, reduces the "humanness" that results from the robot's mobility and active interaction and makes carrying around things for its user appear much more "natural".

2.4 Ability to Contain Oneself

One of the seniors we visited in her flat on an Assisted Living site was a 95-year-old woman. She mentioned that people are not especially flexible and often not willing to change daily routines. At the site, she often suggested going to town or for a walk with someone, but people didn't join in for all kind reasons, such as they had planned doing something else or it was too hot that day or raining. The woman said: "It's surprising how different peoples' ideas about daily routines and life are."

While in some situations we may use it as a strategy to give a robot an own life (to increase "humanness"), it seems in general a good thing that a robot can contain itself. Of course, through displaying a sense of own "will" the robot can liven up the place and convince its owner in doing certain things (e.g., some physical exercises), but in the end this is always in the user's interest and the robot conforms with whatever the person wants. The robot never has the last word. There are no ifs and buts. Imagining an ideal roommate, the seniors we talked with wished for distance, while being reliable, not invasive, obtrusive or self-opinionated. For a human, to always take the back seat is hard, for a thing it is not.

Containment makes the robot an excellent listener. One can moan, have a shoulder to cry on or chafe at something as much as wanted. At the same time there will never be uncomfortable silence. If the person doesn't feel like talking, no one will be miffed about it. The robot is perfectly consent, flexible, and can listen forever.

This is beneficial in many scenarios. For example, the robot knows it needs to wake the user gently to make the experience pleasant: It slowly livens up the place outside the bedroom, opens roller blinds, croons a song to itself, whistles and rolls around. Only when the user leaves his bedroom the robot greets him. The robot even joins the ritual of having breakfast together, although it does not eat. It does not mind the user leaving the house although it can't. The robot has no own preferences. It is there to watch any TV program the user wants. When the user wants a quiet time, it is always a good opportunity for the robot to charge. The robot will not be in opposition, but agree flexibly to everything.

2.5 Not Taking Things Personally

One senior revealed her practice of never doing anything she doesn't feel like doing at that very moment and at the same time never finding excuses, but always saying the truth: "The moment you lie, you need to be able to remember and that becomes more difficult with age". Her announcement on her answering machine makes it crystal clear: "Probably I am not at home or maybe I just don't feel like answering the call." The woman admits that her friends needed to get used to her attitude. They are still sometimes offended, when she does not join a group dinner on short notice. It is hard for people to understand that it is not because of them, but because of her.

Luckily a robot does not take things personally. A robot may be disappointed to help convince a person for his own benefit, but it is never seriously disappointed or offended. It easily accepts any decisions of the user. It does not need explanations and there is no need to make up excuses. It will never try to interpret things, ponder about reasons or suspect bad intentions. Also because it does not even understand double meaning it will always understand things just the way they have been said. The user does not need to be careful in what he says and how to put things. A robot further does not take criticism personal, but as an instruction. This makes interaction with the robot easy, clear and direct.

Especially when the robot is programmed to make suggestions being able to say just no is very important. While watching TV, a robot could for example suggest to do exercises during the following advert break together, because the user voiced an interest in remaining fit, but never seems to get around to actually do this. Of course the user can object, which appears much easier with a robot than with a human. The robot won't be disappointed nor continue asking. Not taking things personally is a great advantage of a robot versus a human.

2.6 Assuming Responsibility

One woman reported about moving flats just after an operation and how her children helped make sure that their mother would not move too much. It can feel good to share responsibility for things that appear hard to handle alone. Certainly, a robot can't compete with humans in this respect. But it can still take something off a person's shoulders.

The robot can support the user by taking responsibility for him in some aspects and this better than a human. A robot who always remains in the house is able to constantly control if everything is alright and can otherwise inform the owner who can leave untroubled. It can be on watch, without falling asleep or failing otherwise. It further notices if a person does not return home on a specific time, such as for the night and can inform others if agreed upon before. The hybrid with its constant presence can easily take these responsibilities and provide a feeling of security. The robot who has only a need to charge, but not to sleep, can moreover wake the user at any time, even at four o'clock in the morning, if wanted. The user can make the robot responsible for other things, such as encouraging him to exercise for at least 30 min a day, making phone calls for birthdays, making sure the user eats regularly or is reminded of his medication. The robot can assume many responsibilities other humans would experience as troublesome burden. It is even the perfect liar – backing its user without failing, assuming responsibility for things it wasn't made responsible for in the first place without complaining about this obvious unfairness.

3   Conclusion

We discussed six potential psychological superpowers of robots, which may be especially helpful in providing emotional care to the elderly (and others). These are rooted in the hybrid nature of a robot in between being a "thing" and a "being." The robot's **lack of competitiveness** allows its users to fully and honestly be themselves, a quality found only in the most intimate relationships. Its **endless patience** allows for a relaxed atmosphere, infinite repetitiveness and slowness. **Unconditional subordination** towards the user enables the latter to be served naturally without any feeling of guilt. The robot's **ability to contain oneself** enables the user to fully live to his needs without compromise. **Not taking things personally** makes interaction easy and direct. Its **assuming responsibility** allows the user to hand over some of his own responsibility to a reliable machine with emotional qualities. With becoming too humanlike, many of these advantages of a robot over a human would get lost.

Of course, this should not give the impression that a robot would be better company than a human and that we therefore should replace humans with robots. But instead of mimicking complex social interaction and get lost in their intricacies, we should create new forms of social human-robot interaction, which clearly exploit the unique abilities of the robot as a machine. It should not be about substituting or replacing human care, but about adding a new way of being cared as a new complementary possibility. While robots as means of emotional care are often motivated by the pressure on the health care system, our approach points at needs in emotional care, which may be impossible to fulfill no matter how efficient the health care system would become. We would argue that even if resources would allow for health care workers to spend infinite time as the companion of a needy person, this would create an inhumane workplace. Even for close family members, emotional care can become an unbearable burden (although accepted of as a normative duty).

Robots can help here, but more "humanness" does not seem the right answer. Relying on "things" only, however, wastes the opportunities of the mobile, active, physical present, initiative taking and learning technology robots are. Our leitmotiv for designing a companion consist, thus, of two parts. (1) Instead of replacing, substituting and mimicking human-human interaction, new forms of social interaction should be established. Of course, these are rooted in the daily practices of people, but they primarily look for situations, where the "thingness" of the robot can become an advantage, such as in the case of endless patience. (2) This requires a design strategy that is able to emphasize the "otherness" of a robot compared to a human. To commandeer a robot about without feelings of guilt, requires the social interaction to be different from what would be expected from a human-human transaction. At the same time, it should remain natural and exploit already common behaviors and situations. Balancing both seems the real challenge of designing a robot companion.

Acknowledgements. This work was supported by the German Federal Ministry of Education and Research (BMBF), project SYMPARTNER (Grant: 16SV7221).

References

1. Ryan, R.M., Deci, E.L.: On happiness and human potential: a review of research on hedonic and eudaimonic well-being. In: Annual Review of Psychology, vol. 52, pp. 141–166. Annual Reviews, Palo Alto (2001)
2. Broekens, J., Heerink, M., Rosendal, H.: Assistive social robots in elderly care: a review. Gerontechnology **8**(2), 94–103 (2009)
3. Kachouie, R., Sedighadeli, S., Khosla, R., Chu, M.-T.: Socially assistive robots in elderly care: a mixed-method systematic literature review. Int. J. Hum. Comput. Interact. **30**(5), 369–393 (2014). Taylor & Francis, London
4. Broadbent, E., Stafford R., MacDonald, B.: Acceptance of healthcare robots for the older population: review and future directions. Int. J. Soc. Robot. **1**(4), 319–330 (2009). Springer, Heidelberg
5. Smith, E.R., Mackie, D.M.: Social Psychology, 2nd edn. Taylor & Francis, London (2000)
6. Kim, K.J., Park, E., Sundar, S.S.: Caregiving role in human - robot interaction: a study of the mediating effects of perceived benefit and social presence. In: Tennyson, R. (ed.) Computers in Human Behavior, vol. 29, Issue 4, pp. 1799–1806. Elsevier, July 2013
7. Fink, J.: Anthropomorphism and Human Likeness in the Design of Robots and Human-Robot Interaction. CRAFT, Ecole Polytechnique Fédérale de Lausanne, Lausanne (2012)
8. Duffy, B.R.: Anthropomorphism and robotics. Presented at the Symposium on Animating Expressive Characters of Social Interactions, Imperial College, London (2002)
9. Mara, M., Appel M., Ogawa H., Lindinger C., Ogawa E., Ishiguro H., Ogawa K.: Tell me your story, robot: introducing an android as fiction character leads to higher perceived usefulness and adoption intention. In: Proceedings of the 8th ACM/IEEE International Conference on Human-Robot Interaction, pp. 193–194. IEEE Press (2013)
10. Biswas, M., Murray, J.: Robotic companionship: how forgetfulness affects long-term human-robot interaction. In: Liu, H., Kubota, N., Zhu, X., Dillmann, R., Zhou, D. (eds.) ICIRA 2015. LNCS (LNAI), vol. 9245, pp. 37–48. Springer, Heidelberg (2015). doi:10.1007/978-3-319-22876-1_4
11. Short, E., Hart, J., Vu, M., Scassellati, B.: No fair!! an interaction with a cheating robot. In: 2010 5th ACM/IEEE International Conference on Human-Robot Interaction (HRI), pp. 219–226. IEEE Press, New York (2010)
12. Terada, K., Ito, A.: Can a robot deceive humans? In: HRI 2010: Proceedings of the 5th ACM/IEEE International Conference on Human-Robot Interaction, pp. 191–121. IEEE Press, New York (2010)

13. Choi, J.J., Kim, Y., Kwak, S.S.: Are you embarrassed?: the impact of robot types on emotional engagement with a robot. In: Proceedings of the 2014 ACM/IEEE International Conference on Human-Robot Interaction, pp. 138–139. ACM (2014)

14. Baek, C., Choi, J.J., & Kwak, S.S.: Can you touch me?: the impact of physical contact on emotional engagement with a robot. In: Proceedings of the Second International Conference on Human-Agent Interaction, pp. 149–152. ACM (2014)

15. Festinger, L.: A theory of social comparison processes. Hum. Relat. **7**(2), 117–140 (1954)

16. Vohs, K.D., Baumeister, R.F. (eds.): Handbook of Self-regulation: Research, Theory, and Applications, p. 401. Guilford Press (2011)

A Method for Establishing Correspondences Between Hand-Drawn and Sensor-Generated Maps

Leo Bowen-Biggs[4], Suzanne Dazo[3], Yili Zhang[1], Alexander Hubers[2],
Matthew Rueben[1], Ross Sowell[2], William D. Smart[1], and Cindy M. Grimm[1(✉)]

[1] Oregon State University, Corvallis, OR 97330, USA
cindy.grimm@oregonstate.edu
[2] Cornell College, Mount Vernon, IA, USA
[3] University of Nebraska, Lincoln, USA
[4] Worchester Polytechnic Institute, Worcester, USA

Abstract. Maps, and specifically floor plans, are useful for planning a variety of tasks from arranging furniture to designating conceptual or functional spaces (e.g., kitchen, walkway). However, maps generated directly from robot sensor data can be hard to interpret and use for this purpose, especially for individuals who are not used to them, because of sensor and odometry measurement errors and the probabilistic nature of the mapping algorithms themselves. In this paper, we present an algorithm for quickly laying a floor plan (or other conceptual map) onto a map generated from sensor data, creating a one-to-one mapping between the two This allows humans interacting with the robot to use a more readily-understandable representation of the world, while the robot itself uses the sensor-generated map.

We look at two use cases: specifying "no-go" regions within a room, and visually locating objects within a room. Although a user study showed no statistical difference between the two types of maps in terms of performance on this spatial memory task, we argue that floor plans are closer to the mental maps people naturally draw to characterize spaces, and are easier to use for untrained individuals.

Keywords: Map understanding · SLAM · Map correspondence

1 Introduction

Maps generated directly from sensor data are ubiquitous in robotics, in part because they are straightforward to construct from well-known algorithms, and have proven to be effective for localization and navigation tasks. However, because of measurement errors in the sensor and odometry data, these maps are often hard for humans to interpret: they can be skewed, solid walls can have gaps in them because of a lack of sensor coverage, straight lines can appear curved, and so on. This does not matter (much) in practice for localization and

© Springer International Publishing AG 2016
A. Agah et al. (Eds.): ICSR 2016, LNAI 9979, pp. 1003–1013, 2016.
DOI: 10.1007/978-3-319-47437-3_98

navigation, but can present problems for humans trying to interpret the maps, since the correspondence between points on the generated map and places in the real world is not always clear.

As social robots enter our daily lives, we will have to increasingly show them things and places on maps. Since there is often considerable mental effort needed to correspond places in the world with points in the map, especially for individuals not familiar with then, we argue that they are not "natural" for many tasks, such as labeling a table location or selecting a desired pathway for the robot to use [5, 16].

Maps directly generated from sensor data have two properties that make them unnatural for humans. First, they have a lot of noise and spurious points even along relatively simple objects, such as walls. Second, they often have global distortion, creating, for example, a C-shaped room out of a narrow rectangular one (see Fig. 1). This forces the viewer to mentally "undo" the local distortion. Existing research shows that extra visual information on traditional street maps adversely affects performance on a navigation task [14]. We informally validated that, even with a SLAM map available as an example, people drew floor plans that were similar to the floor plan shown on the left in Fig. 1.

With this in mind we describe a straightforwad algorithm for establishing a correspondence between a sketched floor plan (or real one) and a map directly generated from data. We demonstrate two use cases: translating navigation information to the floor plan and translating "no-go' regions from the floor plan to the generated map to prevent the robot from entering a designated area.

We conducted a user study looking at the effectiveness of the floor plan over a generated map for a spatial memory task. There was no statistically significant difference in the average performance, however there was a slight difference in the kinds of errors the participants made.

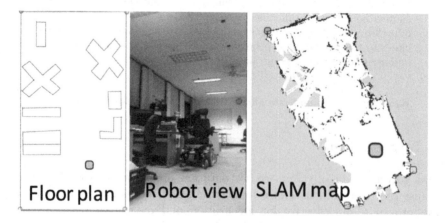

Fig. 1. Left: A hand-drawn floor plan of the space, robot location shown by blue rectangle. Middle: Robot's first person point of view. Right: Generated map created by driving the robot around the space. (Color figure online)

2 Background

Many robots use a 2D, metric map to navigate, where obstacles and free spaces are represented by an occupancy grid [7]. We call these *SLAM maps* after the simultaneous localization and mapping (SLAM) algorithms commonly used to create them (see Thrun et al. [18] for a summary of the basic algorithms).

We compare SLAM maps with sketched *floor plans*, which are similar to building blueprints and the more simplified maps found, e.g., on shopping mall directories. We consider floor plans to be semi-accurate: the walls are (relatively) straight on the map when they are straight in real life, and relative sizes and angles are all about correct (although aspect ratio and overall scales may be wrong). The level of detail is typically somewhat low, specialized for navigation: mostly just the walls, doors, and major obstacles are shown. Floor plans also lack the sensor errors that cause holes in walls and false obstacles in the middle of a room in SLAM maps. Significantly, floor plans are typically drawn with lines and curves, not by filling the cells of a grid.

Pinheiro [12] includes a review of how people read and draw maps. Among the concepts reviewed is "naive cartographic realism", which is when map readers assume that the map perfectly represents the real world.

For millenia people have been creating maps by simply walking through a space to construct a *mental map* of it, which is later transferred to paper as a *sketch map*. Beck and Wood [2] include several common operations involved in map-making in their model of urban mapping. These include scaling, rotation, and synchronization of observations from several tours of a space. Pinheiro [12] notes that people tend to create maps hierarchically by first subdividing the region to be mapped; this is much like we divide the visible stars into constellations, then map their relative locations on a per-constellation basis. Kuipers [9] asks whether people encode spatial information as a "Map in the Head" such that your sketch map is just like your mental map. It seems this is not entirely the case: mental maps can have sections that are disconnected, and routes could be represented such that they are only valid in one direction. Billinghurst and Weghorst [3] test whether sketch maps are valid measures of certain aspects of mental ("cognitive") maps. They found that the accuracy of a sketch map is highly correlated with world knowledge and the subjective feeling of orientation in a space. Similarly, Wang and Li [19] found that sketched maps were more accurate than verbal instructions for navigating from one point to another.

Recently, researchers have been using sketched maps directly in localization tasks without an intermediate SLAM representation [5,15,16]. This is relevant when, for example, it is not feasible to build a map from sensor data first. From an interface standpoint, it may make sense to ask the user to use our technique to establish a rough correspondence between any (incomplete) SLAM map the robot creates during its localization process in order to help guide the robot.

Researchers have also aimed to automate matching of a hand-drawn map to a SLAM or other automatically generated map for the purposes of navigation [4, 10,11]. In the first case, the robot simultaneously locates itself on both the SLAM map and the sketch, including a scale factor for the sketch map to account for inaccuracies in the sketch. In the second case the sketch is used primarily to seed the localization procedure with estimated building locations. In the third case they assume a one-to-one mapping between the objects in the map and the objects in the sketch (i.e., object correspondence, not spatial) and that the sketch and real map have the same number of objects. We sidestep this problem by asking the user to establish correspondences for us. Inaccuracies in the sketch map are essentially represented as local affine transformations (given by the distortions in the triangle shapes from one map to the other).

3   User Interface and Algorithm

We describe the interface, algorithm and its implementation.

3.1   User Interface

From the user's point of view they simply click corresponding points in the two maps, approximately one for every corner of the floor plan (see Fig. 2 and four corners for Fig. 1). The system then calculates a one-to-one and onto mapping between the two maps. After this the user can mark a point or line in one map and have it appear in the other one.

(Video: https://www.youtube.com/watch?v=BVzRclC5zYw)

The interface and algorithm were implemented in the Robot Operating System (ROS). We demonstrate two use cases. The first case maps the robot's position and orientation from the SLAM map to the floor plan during a robot navigation task (see Fig. 4 and accompanying video). In the second use case we used the floor plan to mark a "no go" region, then mapped this back to the SLAM map as walls. The robot then drove around the marked region (see Fig. 2 and accompanying video).

3.2   Algorithm

Essentially, we use Triangle [17] to triangulate the floor plan then map the triangle vertices to the SLAM map using the user-marked coordinates. To calculate the correspondence we use barycentric coordinates within each triangle. Although this introduces discontinuities in the derivatives along the boundaries of the triangles we have not found this to be a problem in practice, probably because the local deformation from one triangle to the next is fairly minimal (and the SLAM map is noisy, hiding small inaccuracies). If desired, the user can draw two curves which will be automatically split into a poly line when the poly line varies too much from either curve; matching is accomplished using arc-length parameterization.

4   User Studies and Evaluation

We discuss two methods for validation; support for floor plans or conceptual maps, and a comparison of the functionality of a floor plan versus a SLAM map for spatial localization.

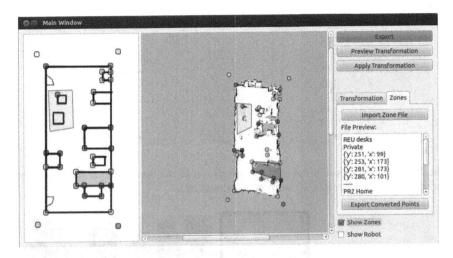

Fig. 2. Establishing correspondences between a floor plan and the SLAM map (colored squares). The red and green regions are the spaces marked on the floor plan that are mapped (automatically) back to the SLAM map. (Color figure online)

4.1 Support for Floor Plans or Conceptual Maps

Clearly it is possible to use a SLAM map, however, we argue that it does not match people's conceptual models of a space. To provide evidence for this we asked five non-roboticists to draw a map of a space the robot had mapped. The participants were given a video from the view point of the robot as it drove through a previously mapped space. The participants saw the robot's location in the SLAM map at the same time (middle and right image of Fig. 1). Participants were told they would see a video of the map, with a robot-created map on the side showing where the robot was as a blue square. After watching the video as often as you need to do, please draw the outline of the room and the locations of all the objects in the room, as best as you remember. The five maps are shown in Fig. 3).

(Video semantic: https://www.youtube.com/watch?feature=player_embedded&v=8_cQD8E1DsU)

(Video SLAM: https://www.youtube.com/watch?feature=player_embedded&v=qRleTXsmZP0)

Although the hand-drawn maps are all different, there are some commonalities. Everyone included the tables and the shelves in the back of the room (although the number of tables varies). Note that the tables are not really identifiable in the SLAM map. Interestingly, one person (second drawing) flipped the map so that the starting point of the robot's path was at the top. Three participants included the sofa. It appears one person attempted to preserve the curvi-linear boundary of the SLAM map; this map also had the most detail. Despite being asked to include objects, participants primarily included furniture (chairs, tables, sofas), with the only objects being marked computers and robot.

Fig. 3. Five hand-drawn maps of the space shown in the first figure. Images have been darkened after scanning (they were pencil drawings)

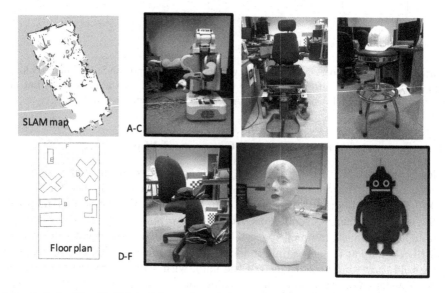

Fig. 4. Left: The SLAM and floor plan with the letter locations marked. Objects are shown in the order they are encountered (A-F). Images with a red boundary were shown along with the question; the actual video frame for the hat and umbrella are also shown. Accuracy in order: 60 %, 61 %, 37 %, 53 %, 51 %, 61 %. (Color figure online)

The aspect ratio of the space also varies, with only the most detailed map having a similar aspect ratio to the SLAM map (other maps are not as narrow).

Detailed analysis of how people conceptualize space is beyond the scope of this paper, however, our informal study supports the idea that people visualize space using floor plan-like layouts.

4.2   Spatial Mapping User Study

We next describe our spatial mapping on-line user study.

Study Stimulus and Design. We made two videos, one with the SLAM map and one with the floor plan (see Fig. 1). In each video the participant saw the map on the left and the robot's point of view on the right. Participants were instructed to "Please watch this video carefully and pay attention to what you see where (you may watch as many times as you want)". They could watch the video as many times as they wanted, but could not go back to the video once they started answering questions. Video length was one and a half minutes, and consisted of a navigation from one end of the room to the other and back, avoiding the obstacles in the room and pausing to look at the objects in the study.

We asked one open-ended question (name three objects in the video) to ensure that participants had actually watched the video. We also checked for duplicate IDs, however we did not collect demographic information. There were four questions about objects in the room and six questions that asked the participants to identify the location of an object in the video. For potentially unclear objects (the robot picture, umbrella, and PR2) a picture of the object was included with the text question. Participants picked the location from one of six on the map (letters A-F). Each participant saw all of the questions and all of the objects, randomly ordered.

The four questions were, with the correct answer and the number of people answering that question correctly:

- How many actual robots did you see (not pictures of robots)? (Answer: 1, 82 %)
- How many clocks did you see? (Answer: 1, 56 %)
- What is the color of the couch? (yellow, brown, dark, pink) (Answer: brown, confounded with dark, 88 %)
- Where is the whiteboard? (on the wall, beside the door, in the middle of the room) (Answer: on the wall, 65 %)

The six objects are shown in Fig. 4, along with the SLAM and floor plan with letter locations labeled.

We pilot tested the study with 7 people, asking them to talk aloud while taking the study. We used this to ensure that the questions were both clear and answerable the majority of the time. We explicitly checked that questions were answered correctly some of the time, but not always (i.e., they were of mid-level difficulty).

Participants. We ran the study on-line using Amazon's Mechanical Turk (70 participants), of which 70 attempted the task (30 floor plan, 34 SLAM)). Of those, 24 successfully completed the floor plan condition (80 %), 31 the SLAM map (91 %). Successfully means they answered all of the questions. The remainder quit after watching the video or answered, at most, one or two questions.

Results of On-Line Study. We summarize the mean percentage correct for all questions in Fig. 5; note that these are all essentially correct/incorrect questions,

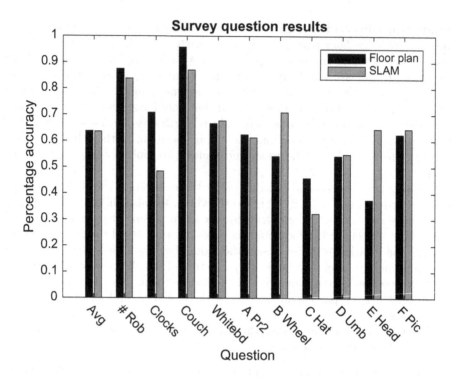

Fig. 5. Percentage of participants who correctly answered the questions and located the objects on the map for each condition (average of percentages on the left).

so there is no standard deviation. Although the mean for the floor plan percent correct is very slightly better — and better for the four questions – the results are not statistically significant. There was some variation in *how* participants got the answers wrong (see Fig. 6). In general the answer distribution is roughly bell shaped (recall that the A-F places the objects roughly in order from bottom to top) which implies that participants were usually close. 15 of the participants used all six locations exactly once; half of these were correct (split equally across both conditions). No one got all questions correct.

5   Discussion

The literature (and our own study) clearly shows that, when asked to provide a sketch of the layout of a space people tend to provide simple line drawings. So why did the on-line study show no *functional* difference in a spatial memory location task? We hypothesize that this task is difficult enough — and people's abilities to visually build a mental model of a space from a first-person view robotic drive through vary enough — that any potential benefit of a floor plan is lost in the noise.

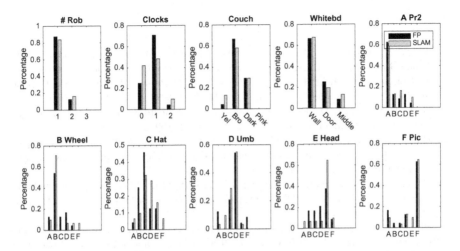

Fig. 6. Distribution of answers per question, given as percentages. x-axis is survey question responses.

We would argue that most people would prefer a floor plan, and that it makes conceptual tasks such as specifying a path from Bob's office to the coffee pot easier. It is not clear that *having* a floor plan instead of SLAM map will substantially improve performance for spatial reasoning tasks (although *drawing* such a map, possibly with the help of a SLAM map, might). Determining how the artifacts of a SLAM map (global warps, spurious noise) interfere (if they do) with spatial reasoning is beyond the scope of this paper, but a potentially interesting area for future work.

Our current implementation is polygon-based. There are more elaborate 2D mapping/morphing schemes [1,13] that could be used instead. These approaches would yield a continuous mapping but are more computationally intensive to evaluate and may not have a well-defined inverse. It would also be possible to apply sketch beautification techniques (either internally or visible to the user) to "clean up" hand-drawn maps [6,8]. This might make it easier for a robot to use the hand-drawn sketch directly for localization.

6   Conclusion

We have presented a simple technique for mapping a hand-drawn sketch or floor plan to a SLAM map. This provides a more "user-friendly" experience for labeling SLAM maps and communicating spatial information to the robot.

Acknowledgments. Research funded in part by NSF grants CNS 1359480 (NSF Site Robots in the Real World) and IIS 1340897 (RET supplement, HCC: Small: A Physical Vocabulary for Human-Robot Interaction).

References

1. Alexa, M., Cohen-Or, D., Levin, D.: As-rigid-as-possible shape interpolation. In: Proceedings of the 27th Annual Conference on Computer Graphics and Interactive Techniques, SIGGRAPH 2000, pp. 157–164. ACM Press/Addison-Wesley Publishing Co., New York (2000). doi:10.1145/344779.344859
2. Beck, R.J., Wood, D.: Cognitive transformation of information from urban geographic fields to mental maps. Env. Behav. **8**(2), 199–238 (1976)
3. Billinghurst, M., Weghorst, S.: The use of sketch maps to measure cognitive maps of virtual environments. In: Proceedings of the Virtual Reality Annual International Symposium, pp. 40–47, March 1995
4. Boniardi, F., Behzadian, B., Burgard, W., Tipaldi, G.D.: Robot navigation in hand-drawn sketched maps. In: 2015 European Conference on Mobile Robots (ECMR), pp. 1–6, September 2015
5. Boniardi, F., Valada, A., Burgard, W., Tipaldi, G.D.: Autonomous indoor robot navigation using a sketch interface for drawing maps and routes. In: Proceedings of the IEEE International Conference on Robotics & Automation (ICRA). Stockholm, Sweden (2016). http://ais.informatik.uni-freiburg.de/publications/papers/boniardi16icra.pdf
6. Cheema, S., Gulwani, S., LaViola, J.: Quickdraw: improving drawing experience for geometric diagrams. In: Proceedings of the SIGCHI Conference on Human Factors in Computing Systems, CHI 2012, pp. 1037–1064. ACM, New York (2012). http://doi.acm.org/10.1145/2207676.2208550
7. Elfes, A.: Using occupancy grids for mobile robot perception and navigation. Computer **22**(6), 46–57 (1989)
8. Fišer, J., Asente, P., Sýkora, D.: Shipshape: a drawing beautification assistant. In: Proceedings of the Workshop on Sketch-Based Interfaces and Modeling, SBIM 2015, pp. 49–57. Eurographics Association, Aire-la-Ville (2015). http://dl.acm.org/citation.cfm?id=2810210.2810215
9. Kuipers, B.: The "Map in the Head" metaphor. Env. Behav. **14**(2), 202–220 (1982)
10. Matsuo, K., Miura, J.: Outdoor visual localization with a hand-drawn line drawing map using fastslam with pso-based mapping. In: 2012 IEEE/RSJ International Conference on Intelligent Robots and Systems, pp. 202–207, October 2012
11. Parekh, G., Skubic, M., Sjahputera, O., Keller, J.M.: Scene matching between a map and a hand drawn sketch using spatial relations. In: Proceedings 2007 IEEE International Conference on Robotics and Automation, pp. 4007–4012, April 2007
12. Pinheiro, J.D.Q.: Determinants of cognitive maps of the world as expressed in sketch maps, January 1996
13. Poranne, R., Lipman, Y.: Provably good planar mappings. ACM Trans. Graph. **33**(4), 76:1–76:11 (2014). http://doi.acm.org/10.1145/2601097.2601123
14. Sanchez, C.A., Branaghan, R.J.: The interaction of map resolution and spatial abilities on map learning. Int. J. Hum. Comput. Stud. **67**(5), 475–481 (2009)
15. Setalaphruk, V., Ueno, A., Kume, I., Kono, Y., Kidode, M.: Robot navigation in corridor environments using a sketch floor map. In: Proceedings of the 2003 IEEE International Symposium on Computational Intelligence in Robotics and Automation, vol. 2, pp. 552–557, July 2003
16. Shah, D.C., Campbell, M.E.: A qualitative path planner for robot navigation using human-provided maps. Int. J. Robot. Res. **32**(13), 1517–1535 (2013). http://ijr.sagepub.com/content/32/13/1517.abstract

17. Shewchuk, J.R.: Triangle: engineering a 2D quality mesh generator and delaunay triangulator. In: Lin, M.C., Manocha, D. (eds.) WACG 1996. LNCS, vol. 1148, pp. 203–222. Springer, Heidelberg (1996). doi:10.1007/BFb0014497. http://dl.acm.org/citation.cfm?id=645908.673287
18. Thrun, S., Burgard, W., Fox, D.: Probabilistic Robotics. MIT Press, Cambridge (2005)
19. Wang, J., Li, R.: An empirical study on pertinent aspects of sketch maps for navigation. In: 2012 IEEE 11th International Conference on Cognitive Informatics Cognitive Computing (ICCI*CC), pp. 130–139, August 2012

Erratum to: Introducing IOmi - A Female Robot Hostess for Guidance in a University Environment

Eiji Onchi[1], Cesar Lucho[1], Michel Sigüenza[1], Gabriele Trovato[2], and Francisco Cuellar[1(✉)]

[1] Mechatronic Specialty, Engineering Department, Pontificia Universidad Católica del Perú (PUCP), Lima, Peru
`cuellar.ff@pucp.pe`
[2] Graduate School of Advanced Science and Engineering, Waseda University, Tokyo, Japan

Erratum to:
Chapter 75 in: A. Agah et al. (eds.)
Social Robotics
DOI: 10.1007/978-3-319-47437-3_75

The original version of the book was inadvertently published without images in Table 1 of Chapter 'Introducing IOmi - A Female Robot Hostess for Guidance in a University Environment'. The erratum chapter and the book have been updated with the changes.

The updated original online version for this chapter can be found at
DOI: 10.1007/978-3-319-47437-3_75

Author Index

Printed in the United States
By Bookmasters